# Encyclopedia of VIROLOGY

## THIRD EDITION

**EDITORS-IN-CHIEF**

Dr BRIAN W J MAHY
and
Dr MARC H V VAN REGENMORTEL

Amsterdam • Boston • Heidelberg • London • New York • Oxford
Paris • San Diego • San Francisco • Singapore • Sydney • Tokyo
Academic Press is an imprint of Elsevier

ELSEVIER

ACADEMIC PRESS

Academic Press is an imprint of Elsevier
Linacre House, Jordan Hill, Oxford, OX2 8DP, UK
525 B Street, Suite 1900, San Diego, CA 92101-4495, USA

Copyright © 2008 Elsevier Inc. All rights reserved

The following articles are US government works in the public domain and are not subject to copyright:
Bovine Viral Diarrhea Virus, Coxsackieviruses, Prions of Yeast and Fungi, Human Respiratory Syncytial Virus, Fish Rhabdoviruses, Varicella-Zoster Virus: General Features, Viruses and Bioterrorism, Bean Common Mosaic Virus and Bean Common Mosaic Necrosis Virus, Metaviruses, Crimean-Congo Hemorrhagic Fever Virus and Other Nairoviruses, AIDS: Global Epidemiology, Papaya Ringspot Virus, Transcriptional Regulation in Bacteriophage.

*Nepovirus*, Canadian Crown Copyright 2008

No part of this publication may be reproduced, stored in a retrieval system or transmitted in any form or by any means electronic, mechanical, photocopying, recording or otherwise without the prior written permission of the publisher

Permissions may be sought directly from Elsevier's Science & Technology Rights Department in Oxford, UK: phone (+44) (0) 1865 843830; fax (+44) (0) 1865 853333; email: permissions@elsevier.com. Alternatively you can submit your request online by visiting the Elsevier web site at (http://elsevier.com/locate/permission), and selecting *Obtaining permission to use Elsevier material*

Notice
No responsibility is assumed by the publisher for any injury and/or damage to persons or property as a matter of products liability, negligence or otherwise, or from any use or operation of any methods, products, instructions or ideas contained in the material herein. Because of rapid advances in the medical sciences, in particular, independent verification of diagnoses and drug dosages should be made

**British Library Cataloguing in Publication Data**
A catalogue record for this book is available from the British Library

**Library of Congress Catalog Number:** 200892260

ISBN: 978-0-12-373935-3

For information on all Elsevier publications
visit our website at books.elsevier.com

PRINTED AND BOUND IN SLOVENIA
08  09  10  11    10  9  8  7  6  5  4  3  2  1

Working together to grow
libraries in developing countries

www.elsevier.com | www.bookaid.org | www.sabre.org

ELSEVIER    BOOK AID International    Sabre Foundation

# Encyclopedia of Virology
## Third Edition

# EDITORS-IN-CHIEF

**Brian W J Mahy MA PhD ScD DSc**
*Senior Scientific Advisor,*
*Division of Emerging Infections and Surveillance Services,*
*Centers for Disease Control and Prevention,*
*Atlanta GA, USA*

**Marc H V Van Regenmortel PhD**
*Emeritus Director at the CNRS,*
*French National Center for Scientific Research,*
*Biotechnology School of the University of Strasbourg,*
*Illkirch, France*

# ASSOCIATE EDITORS

**Dennis H Bamford, Ph.D.**
Department of Biological and Environmental Sciences
and Institute of Biotechnology, Biocenter 2,
P.O. Box 56 (Viikinkaari 5),
00014 University of Helsinki,
Finland

**Charles Calisher, B.S., M.S., Ph.D.**
Arthropod-borne and Infectious Diseases Laboratory
Department of Microbiology, Immunology and Pathology
College of Veterinary Medicine and Biomedical Sciences
Colorado State University
Fort Collins
CO 80523
USA

**Andrew J Davison, M.A., Ph.D.**
MRC Virology Unit
Institute of Virology
University of Glasgow
Church Street
Glasgow G11 5JR
UK

**Claude Fauquet**
ILTAB/Donald Danforth Plant Science Center
975 North Warson Road
St. Louis, MO 63132

**Said Ghabrial, B.S., M.S., Ph.D.**
Plant Pathology Department
University of Kentucky
201F Plant Science Building
1405 Veterans Drive
Lexington
KY 4050546-0312
USA

**Eric Hunter, B.Sc., Ph.D.**
Department of Pathology and Laboratory Medicine, and
Emory Vaccine Center
Emory University
954 Gatewood Road NE
Atlanta Georgia 30329
USA

**Robert A. Lamb, Ph.D., Sc.D.**
Department of Biochemistry,
Molecular Biology and Cell Biology
Howard Hughes Medical Institute
Northwestern University
2205 Tech Dr.
Evanston
IL 60208-3500
USA

**Olivier Le Gall**
IPV, UMR GDPP, IBVM,
INRA Bordeaux-Aquitaine, BP 81,
F-33883 Villenave d'Ornon Cedex
FRANCE

**Vincent Racaniello, Ph.D.**
Department of Microbiology
Columbia University
New York, NY 10032
USA

**David A. Theilmann, Ph.D., B.Sc., M.Sc**
Pacific Agri-Food Research Centre
Agriculture and Agri-Food Canada
Box 5000, 4200 Highway 97
Summerland
BC V0H 1Z0
Canada

**H. Josef Vetten, Ph.D.**
Julius Kuehn Institute, Federal Research Centre for
Cultivated Plants (JKI)
Messeweg 11-12
38104 Braunschweig
Germany

**Peter J Walker, B.Sc., Ph.D.**
CSIRO Livestock Industries
Australian Animal Health Laboratory (AAHL)
Private Bag 24
Geelong
VIC 3220
Australia

# PREFACE

This third edition of the *Encyclopedia of Virology* is being published nine years after the second edition, a period which has seen enormous growth both in our understanding of virology and in our recognition of the viruses themselves, many of which were unknown when the second edition was prepared. Considering viruses affecting human hosts alone, the worldwide epidemic of severe acute respiratory syndrome (SARS), caused by a previously unknown coronavirus, led to the discovery of other human coronaviruses such as HKU1 and NL63. As many as seven chapters are devoted to the AIDS epidemic and to human immunodeficiency viruses. In addition, the development of new molecular technologies led to the discovery of viruses with no obvious disease associations, such as torque-teno virus (one of the most ubiquitous viruses in the human population), human bocavirus, human metapneumovirus, and three new human polyomaviruses.

Other new developments of importance to human virology have included the introduction of a virulent strain of West Nile virus from Israel to North America in 1999. Since that time the virus has become established in mosquito, bird and horse populations throughout the USA, the Caribbean and Mexico as well as the southern regions of Canada.

As in the two previous editions, we have tried to include information about all known species of virus infecting bacteria, fungi, invertebrates, plants and vertebrates, as well as descriptions of related topics in virology such as antiviral drug development, cell- and antibody-mediated immunity, vaccine development, electron microscopy and molecular methods for virus characterization and identification. Many chapters are devoted to the considerable economic importance of virus diseases of cereals, legumes, vegetable crops, fruit trees and ornamentals, and new approaches to control these diseases are reviewed.

General issues such as the origin, evolution and phylogeny of viruses are also discussed as well as the history of the different groups of viruses.

To cover all these subjects and new developments, we have had to increase the size of the Encyclopedia from three to five volumes.

Throughout this work we have relied upon the 8th Report of the International Committee on Taxonomy of Viruses published in 2005, which lists more than 6000 viruses classified into some 2000 virus species distributed among more than 390 different genera and families. In recent years the criteria for placing viruses in different taxa have shifted away from traditional serological methods and increasingly rely upon molecular techniques, particularly the nucleotide sequence of the virus genome. This has changed many of the previous groupings of viruses, and is reflected in this third edition.

Needless to say, a work of this magnitude has involved many expert scientists, who have given generously of their time to bring it to fruition. We extend our grateful thanks to all contributors and associate editors for their excellent and timely contributions.

Brian W J Mahy<br>
Marc H V van Regenmortel

# HOW TO USE THE ENCYCLOPEDIA

## Structure of the Encyclopedia

The major topics discussed in detail in the text are presented in alphabetical order (see the Alphabetical Contents list which appears in all five volumes).

## Finding Specific Information

Information on specific viruses, virus diseases and other matters can be located by consulting the General Index at the end of Volume 5.

## Taxonomic Groups of Viruses

For locating detailed information on the major taxonomic groups of viruses, namely virus genera, families and orders, the Taxonomic Index in Volume 5 (page...) should be consulted.

## Further Reading sections

The articles do not feature bibliographic citations within the body of the article text itself. The articles are intended to be a first introduction to the topic, or a 'refresher', readable from beginning to end without referring the reader outside of the encyclopedia itself. Bibliographic references to external literature are grouped at the end of each article in a Further Reading section, containing review articles, 'seminal' primary articles and book chapters. These point users to the next level of information for any given topic.

## Cross referencing between articles

The "See also" section at the end of each article directs the reader to other entries on related topics. For example. The entry *Lassa, Junin, Machupo and Guanarito Viruses* includes the following cross-references:

*See also:* Lymphocytic Choriomeningitis Virus: General Features.

# CONTRIBUTORS

**S T Abedon**
The Ohio State University, Mansfield, OH, USA

**G P Accotto**
Istituto di Virologia Vegetale CNR, Torino, Italy

**H-W Ackermann**
Laval University, Quebec, QC, Canada

**G Adam**
Universität Hamburg, Hamburg, Germany

**M J Adams**
Rothamsted Research, Harpenden, UK

**C Adams**
University of Duisburg–Essen, Essen, Germany

**E Adderson**
St. Jude Children's Research Hospital, Memphis, TN, USA

**S Adhya**
National Institutes of Health, Bethesda, MD, USA

**C L Afonso**
Southeast Poultry Research Laboratory, Athens, GA, USA

**P Ahlquist**
University of Wisconsin – Madison, Madison, WI, USA

**G M Air**
University of Oklahoma Health Sciences Center, Oklahoma City, OK, USA

**D J Alcendor**
Johns Hopkins School of Medicine, Baltimore, MD, USA

**J W Almond**
sanofi pasteur, Lyon, France

**I Amin**
National Institute for Biotechnology and Genetic Engineering, Faisalabad, Pakistan

**J Angel**
Pontificia Universidad Javeriana, Bogota, Republic of Colombia

**C Apetrei**
Tulane National Primate Research Center, Covington, LA, USA

**B M Arif**
Great Lakes Forestry Centre, Sault Ste. Marie, ON, Canada

**H Attoui**
Faculté de Médecine de Marseilles, Etablissement Français Du Sang, Marseilles, France

**H Attoui**
Université de la Méditerranée, Marseille, France

**H Attoui**
Institute for Animal Health, Pirbright, UK

**L Aurelian**
University of Maryland School of Medicine, Baltimore, MD, USA

**L A Babiuk**
University of Alberta, Edmonton, AB, Canada

**S Babiuk**
National Centre for Foreign Animal Disease, Winnipeg, MB, Canada

**A G Bader**
The Scripps Research Institute, La Jolla, CA, USA

**S C Baker**
Loyola University of Chicago, Maywood, IL, USA

**T S Baker**
University of California, San Diego, La Jolla, CA, USA

**J K H Bamford**
University of Jyväskylä, Jyväskylä, Finland

**Y Bao**
National Institutes of Health, Bethesda, MD, USA

**M Bar-Joseph**
The Volcani Center, Bet Dagan, Israel

**H Barker**
Scottish Crop Research Institute, Dundee, UK

**A D T Barrett**
University of Texas Medical Branch, Galveston, TX, USA

**J W Barrett**
The University of Western Ontario, London, ON, Canada

**T Barrett**
Institute for Animal Health, Pirbright, UK

**R Bartenschlager**
University of Heidelberg, Heidelberg, Germany

**N W Bartlett**
Imperial College London, London, UK

**S Basak**
University of California, San Diego, CA, USA

**C F Basler**
Mount Sinai School of Medicine, New York, NY, USA

**T Basta**
Institut Pasteur, Paris, France

**D Baxby**
University of Liverpool, Liverpool, UK

**P Beard**
Imperial College London, London, UK

**M N Becker**
University of Florida, Gainesville, FL, USA

**J J Becnel**
Agriculture Research Service, Gainesville, FL, USA

**K L Beemon**
Johns Hopkins University, Baltimore, MD, USA

**E D Belay**
Centers for Disease Control and Prevention, Atlanta, GA, USA

**M Benkő**
Veterinary Medical Research Institute, Hungarian Academy of Sciences, Budapest, Hungary

**M Bennett**
University of Liverpool, Liverpool, UK

**M Bergoin**
Université Montpellier II, Montpellier, France

**H U Bernard**
University of California, Irvine, Irvine, CA, USA

**K I Berns**
University of Florida College of Medicine, Gainesville, FL, USA

**P Biagini**
Etablissement Français du Sang Alpes-Méditerranée, Marseilles, France

**P D Bieniasz**
Aaron Diamond AIDS Research Center, The Rockefeller University, New York, NY, USA

**Y Bigot**
University of Tours, Tours, France

**C Billinis**
University of Thessaly, Karditsa, Greece

**R F Bishop**
Murdoch Childrens Research Institute Royal Children's Hospital, Melbourne, VIC, Australia

**B A Blacklaws**
University of Cambridge, Cambridge, UK

**C D Blair**
Colorado State University, Fort Collins, CO, USA

**S Blanc**
INRA–CIRAD–AgroM, Montpellier, France

**R Blawid**
Institute of Plant Diseases and Plant Protection, Hannover, Germany

**G W Blissard**
Boyce Thompson Institute at Cornell University, Ithaca, NY, USA

**S Blomqvist**
National Public Health Institute (KTL), Helsinki, Finland

**J F Bol**
Leiden University, Leiden, The Netherlands

**J-R Bonami**
CNRS, Montpellier, France

**L Bos**
Wageningen University and Research Centre (WUR), Wageningen, The Netherlands

**H R Bose Jr.**
University of Texas at Austin, Austin, TX, USA

**H Bourhy**
Institut Pasteur, Paris, France

**P R Bowser**
Cornell University, Ithaca, NY, USA

**D B Boyle**
CSIRO Livestock Industries, Geelong, VIC, Australia

**C Bragard**
Université Catholique de Louvain, Leuven, Belgium

**J N Bragg**
University of California, Berkeley, Berkeley, CA, USA

**R W Briddon**
National Institute for Biotechnology and Genetic Engineering, Faisalabad, Pakistan

**M A Brinton**
Georgia State University, Atlanta, GA, USA

**P Britton**
Institute for Animal Health, Compton, UK

**J K Brown**
The University of Arizona, Tucson, AZ, USA

**K S Brown**
University of Manitoba, Winnipeg, MB, Canada

**J Bruenn**
State University of New York, Buffalo, NY, USA

**C P D Brussaard**
Royal Netherlands Institute for Sea Research, Texel, The Netherlands

**J J Bugert**
Wales College of Medicine, Heath Park, Cardiff, UK

**J J Bujarski**
Northern Illinois University, DeKalb, IL, USA and Polish Academy of Sciences, Poznan, Poland

**R M Buller**
Saint Louis University School of Medicine, St. Louis, MO, USA

**J P Burand**
University of Massachusetts at Amherst, Amherst, MA, USA

**J Burgyan**
Agricultural Biotechnology Center, Godollo, Hungary

**F J Burt**
University of the Free State, Bloemfontein, South Africa

**S J Butcher**
University of Helsinki, Helsinki, Finland

**J S Butel**
Baylor College of Medicine, Houston, TX, USA

**M I Butler**
University of Otago, Dunedin, New Zealand

**S Bühler**
University of Heidelberg, Heidelberg, Germany

**P Caciagli**
Istituto di Virologia Vegetale – CNR, Turin, Italy

**C H Calisher**
Colorado State University, Fort Collins, CO, USA

**T Candresse**
UMR GDPP, Centre INRA de Bordeaux, Villenave d'Ornon, France

**A J Cann**
University of Leicester, Leicester, UK

**C Caranta**
INRA, Montfavet, France

**G Carlile**
CSIRO Livestock Industries, Geelong, VIC, Australia

**J P Carr**
University of Cambridge, Cambridge, UK

**R Carrion, Jr.**
Southwest Foundation for Biomedical Research, San Antonio, TX, USA

**J W Casey**
Cornell University, Ithaca, NY, USA

**R N Casey**
Cornell University, Ithaca, NY, USA

**S Casjens**
University of Utah School of Medicine, Salt Lake City, UT, USA

**R Cattaneo**
Mayo Clinic College of Medicine, Rochester, MN, USA

**D Cavanagh**
Institute for Animal Health, Compton, UK

**A Chahroudi**
University of Pennsylvania School of Medicine, Philadelphia, PA, USA

**S Chakraborty**
Jawaharlal Nehru University, New Delhi, India

**T J Chambers**
Saint Louis University School of Medicine, St. Louis, MO, USA

**Y Chang**
University of Pittsburgh Cancer Institute, Pittsburgh, PA, USA

**J T Chang**
Baylor College of Medicine, Houston, TX, USA

**D Chapman**
Institute for Animal Health, Pirbright, UK

**D Chattopadhyay**
University of Calcutta, Kolkata, India

**M Chen**
University of Arizona, Tucson, AZ, USA

**J E Cherwa**
University of Arizona, Tucson, AZ, USA

**V G Chinchar**
University of Mississippi Medical Center, Jackson, MS, USA

**A V Chintakuntlawar**
University of Oklahoma Health Sciences Center, Oklahoma City, OK, USA

**W Chiu**
Baylor College of Medicine, Houston, TX, USA

**J Chodosh**
University of Oklahoma Health Sciences Center, Oklahoma City, OK, USA

**I-R Choi**
International Rice Research Institute, Los Baños, The Philippines

**P D Christian**
National Institute of Biological Standards and Control, South Mimms, UK

**M G Ciufolini**
Istituto Superiore di Sanità, Rome, Italy

**P Clarke**
University of Colorado Health Sciences, Denver, CO, USA

**J-M Claverie**
Université de la Méditerranée, Marseille, France

**J R Clayton**
Johns Hopkins University Schools of Public Health and Medicine, Baltimore, MD, USA

**R J Clem**
Kansas State University, Manhattan, KS, USA

**C J Clements**
The Macfarlane Burnet Institute for Medical Research and Public Health Ltd., Melbourne, VIC, Australia

**L L Coffey,**
Institut Pasteur, Paris, France

**J I Cohen**
National Institutes of Health, Bethesda, MD, USA

**J Collinge**
University College London, London, UK

**P L Collins**
National Institute of Allergy and Infectious Diseases, Bethesda, MD, USA

**A Collins**
University of Wisconsin School of Medicine and Public Health, Madison, WI, USA

**D Contamine**
Université Versailles St-Quentin, CNRS, Versailles, France

**K M Coombs**
University of Manitoba, Winnipeg, MB, Canada

**J A Cowley**
CSIRO Livestock Industries, Brisbane, QLD, Australia

**J K Craigo**
University of Pittsburgh School of Medicine, Pittsburgh, PA, USA

**M St. J Crane**
CSIRO Livestock Industries, Geelong, VIC, Australia

**J E Crowe, Jr.**
Vanderbilt University Medical Center, Nashville, TN, USA

**H Czosnek**
The Hebrew University of Jerusalem, Rehovot, Israel

**T Dalmay**
University of East Anglia, Norwich, UK

**B H Dannevig**
National Veterinary Institute, Oslo, Norway

**C J D'Arcy**
University of Illinois at Urbana-Champaign, Urbana, IL, USA

**A J Davison**
MRC Virology Unit, Glasgow, UK

**W O Dawson**
University of Florida, Lake Alfred, FL, USA

**L A Day**
The Public Health Research Institute, Newark, NJ, USA

**J C de la Torre**
The Scripps Research Institute, La Jolla, CA, USA

**X de Lamballerie**
Faculté de Médecine de Marseille, Marseilles, France

**M de Vega**
Universidad Autónoma, Madrid, Spain

**P Delfosse**
Centre de Recherche Public-Gabriel Lippmann, Belvaux, Luxembourg

**B Delmas**
INRA, Jouy-en-Josas, France

**M Deng**
University of California, Berkeley, CA, USA

**J DeRisi**
University of California, San Francisco, San Francisco, CA, USA

**C Desbiez**
Institut National de la Recherche Agronomique (INRA), Station de Pathologie Végétale, Montfavet, France

**R C Desrosiers**
New England Primate Research Center, Southborough, MA, USA

**A K Dhar**
Advanced BioNutrition Corp, Columbia, MD, USA

**R G Dietzgen**
The University of Queensland, St. Lucia, QLD, Australia

**S P Dinesh-Kumar**
Yale University, New Haven, CT, USA

**L K Dixon**
Institute for Animal Health, Pirbright, UK

**C Dogimont**
INRA, Montfavet, France

**A Domanska**
University of Helsinki, Helsinki, Finland

**L L Domier**
USDA–ARS, Urbana, IL, USA

**L L Domier**
USDA-ARS, Urbana-Champaign, IL, USA

**A Dotzauer**
University of Bremen, Bremen, Germany

**T W Dreher**
Oregon State University, Corvallis, OR, USA

**S Dreschers**
University of Duisburg–Essen, Essen, Germany

**R L Duda**
University of Pittsburgh, Pittsburgh, PA, USA

**J P Dudley**
The University of Texas at Austin, Austin, TX, USA

**W P Duprex**
The Queen's University of Belfast, Belfast, UK

**R E Dutch**
University of Kentucky, Lexington, KY, USA

**B M Dutia**
University of Edinburgh, Edinburgh, UK

**M L Dyall-Smith**
The University of Melbourne, Parkville, VIC, Australia

**J East**
University of Texas Medical Branch – Galveston, Galveston, TX, USA

**A J Easton**
University of Warwick, Coventry, UK

**K C Eastwell**
Washington State University – IAREC, Prosser, WA, USA

**B T Eaton**
Australian Animal Health Laboratory, Geelong, VIC, Australia

**H Edskes**
National Institutes of Health, Bethesda, MD, USA

**B Ehlers**
Robert Koch-Institut, Berlin, Germany

**R M Elliott**
University of St. Andrews, St. Andrews, UK

**A Engel**
National Institutes of Health, Bethesda, MD, USA and
D Kryndushkin
National Institutes of Health, Bethesda, MD, USA

**J Engelmann**
INRES, University of Bonn, Bonn, Germany

**L Enjuanes**
CNB, CSIC, Madrid, Spain

**A Ensser**
Virologisches Institut, Universitätsklinikum, Erlangen, Germany

**M Erlandson**
Agriculture & Agri-Food Canada, Saskatoon, SK, Canada

**K J Ertel**
University of California, Irvine, CA, USA

**R Esteban**
Instituto de Microbiología Bioquímica CSIC/University de Salamanca, Salamanca, Spain

**R Esteban**
Instituto de Microbiología Bioquímica CSIC/University of Salamanca, Salamanca, Spain

**J L Van Etten**
University of Nebraska–Lincoln, Lincoln, NE, USA

**D J Evans**
University of Warwick, Coventry, UK

**Ø Evensen**
Norwegian School of Veterinary Science, Oslo, Norway

**D Falzarano**
University of Manitoba, Winnipeg, MB, Canada

**B A Fane**
University of Arizona, Tucson, AZ, USA

**R-X. Fang**
Chinese Academy of Sciences, Beijing, People's Republic of China

**D Fargette**
IRD, Montpellier, France

**A Fath-Goodin**
University of Kentucky, Lexington, KY, USA

**C M Fauquet**
Danforth Plant Science Center, St. Louis, MO, USA

**B A Federici**
University of California, Riverside, CA, USA

**H Feldmann**
National Microbiology Laboratory, Public Health Agency of Canada, Winnipeg, MB, Canada

**H Feldmann**
Public Health Agency of Canada, Winnipeg, MB, Canada

**F Fenner**
Australian National University, Canberra, ACT, Australia

**S A Ferreira**
University of Hawaii at Manoa, Honolulu, HI, USA

**H J Field**
University of Cambridge, Cambridge, UK

**K Fischer**
University of California, San Francisco, San Francisco, CA, USA

**J A Fishman**
Massachusetts General Hospital, Boston, MA, USA

**B Fleckenstein**
University of Erlangen – Nürnberg, Erlangen, Germany

**R Flores**
Instituto de Biología Molecular y Celular de Plantas (UPV-CSIC), Valencia, Spain

**T R Flotte**
University of Florida College of Medicine, Gainesville, FL, USA

**P Forterre**
Institut Pasteur, Paris, France

**M A Franco**
Pontificia Universidad Javeriana, Bogota, Republic of Colombia

**T K Frey**
Georgia State University, Atlanta, GA, USA

**M Fuchs**
Cornell University, Geneva, NY, USA

**S Fuentes**
International Potato Center (CIP), Lima, Peru

**T Fujimura**
Instituto de Microbiología Bioquímica CSIC/University of Salamanca, Salamanca, Spain

**R S Fujinami**
University of Utah School of Medicine, Salt Lake City, UT, USA

**T Fukuhara**
Tokyo University of Agriculture and Technology, Fuchu, Japan

**D Gallitelli**
Università degli Studi and Istituto di Virologia Vegetale del CNR, Bari, Italy

**F García-Arenal**
Universidad Politécnica de Madrid, Madrid, Spain

**J A García**
Centro Nacional de Biotecnología (CNB), CSIC, Madrid, Spain

**R A Garrett**
Copenhagen University, Copenhagen, Denmark

**S Gaumer**
Université Versailles St-Quentin, CNRS, Versailles, France

**R J Geijskes**
Queensland University of Technology, Brisbane, QLD, Australia

**T W Geisbert**
National Emerging Infectious Diseases Laboratories, Boston, MA, USA

**E Gellermann**
Hannover Medical School, Hannover, Germany

**A Gessain**
Pasteur Institute, CNRS URA 3015, Paris, France

**S A Ghabrial**
University of Kentucky, Lexington, KY, USA

**W Gibson**
Johns Hopkins University School of Medicine, Baltimore, MD, USA

**M Glasa**
Slovak Academy of Sciences, Bratislava, Slovakia

**Y Gleba**
Icon Genetics GmbH, Weinbergweg, Germany

**U A Gompels**
University of London, London, UK

**D Gonsalves**
USDA, Pacific Basin Agricultural Research Center, Hilo, HI, USA

**M M Goodin**
University of Kentucky, Lexington, KY, USA

**T J D Goodwin**
University of Otago, Dunedin, New Zealand

**A E Gorbalenya**
Leiden University Medical Center, Leiden, The Netherlands

**E A Gould**
University of Reading, Reading, UK

**A Grakoui**
Emory University School of Medicine, Atlanta, GA, USA

**M-A Grandbastien**
INRA, Versailles, France

**R Grassmann**
University of Erlangen – Nürnberg, Erlangen, Germany

**M Gravell**
National Institutes of Health, Bethesda, MD, USA

**M V Graves**
University of Massachusetts–Lowell, Lowell, MA, USA

**K Y Green**
National Institutes of Health, Bethesda, MD, USA

**H B Greenberg**
Stanford University School of Medicine and Veterans Affairs Palo Alto Health Care System, Palo Alto, CA, USA

**B M Greenberg**
Johns Hopkins School of Medicine, Baltimore, MD, USA

**I Greiser-Wilke**
School of Veterinary Medicine, Hanover, Germany

**D E Griffin**
Johns Hopkins Bloomberg School of Public Health, Baltimore, MD, USA

**T S Gritsun**
University of Reading, Reading, UK

**R J de Groot**
Utrecht University, Utrecht, The Netherlands

**A J Gubala**
CSIRO Livestock Industries, Geelong, VIC, Australia

**D J Gubler**
John A. Burns School of Medicine, Honolulu, HI, USA

**A-L Haenni**
Institut Jacques Monod, Paris, France

**D Haig**
Nottingham University, Nottingham, UK

**F J Haines**
Oxford Brookes University, Oxford, UK

**J Hamacher**
INRES, University of Bonn, Bonn, Germany

**J Hammond**
USDA-ARS, Beltsville, MD, USA

**R M Harding**
Queensland University of Technology, Brisbane, QLD, Australia

**J M Hardwick**
Johns Hopkins University Schools of Public Health and Medicine, Baltimore, MD, USA

**D Hariri**
INRA – Département Santé des Plantes et Environnement, Versailles, France

**B Harrach**
Veterinary Medical Research Institute, Budapest, Hungary

**P A Harries**
Samuel Roberts Noble Foundation, Inc., Ardmore, OK, USA

**L E Harrington**
University of Alabama at Birmingham, Birmingham, AL, USA

**T J Harrison**
University College London, London, UK

**T Hatziioannou**
Aaron Diamond AIDS Research Center, The Rockefeller University, New York, NY, USA

**J Hay**
The State University of New York, Buffalo, NY, USA

**G S Hayward**
Johns Hopkins School of Medicine, Baltimore, MD, USA

**E Hébrard**
IRD, Montpellier, France

**R W Hendrix**
University of Pittsburgh, Pittsburgh, PA, USA

**L E Hensley**
USAMRIID, Fort Detrick, MD, USA

**M de las Heras**
University of Glasgow Veterinary School, Glasgow, UK

**S Hertzler**
University of Illinois at Chicago, Chicago, IL, USA

**F van Heuverswyn**
University of Montpellier 1, Montpellier, France

**J Hilliard**
Georgia State University, Atlanta, GA, USA

**B I Hillman**
Rutgers University, New Brunswick, NJ, USA

**S Hilton**
University of Warwick, Warwick, UK

**D M Hinton**
National Institutes of Health, Bethesda, MD, USA

**A Hinz**
UMR 5233 UJF-EMBL-CNRS, Grenoble, France

**A E Hoet**
The Ohio State University, Columbus, OH, USA

**S A Hogenhout**
The John Innes Centre, Norwich, UK

**T Hohn**
Basel university, Institute of Botany, Basel, Switzerland

**J S Hong**
Seoul Women's University, Seoul, South Korea

**M C Horzinek**
Utrecht University, Utrecht, The Netherlands

**T Hovi**
National Public Health Institute (KTL), Helsinki, Finland

**A M Huger**
Institute for Biological Control, Darmstadt, Germany

**L E Hughes**
University of St. Andrews, St. Andrews, UK

**R Hull**
John Innes Centre, Colney, UK

**E Hunter**
Emory University Vaccine Center, Atlanta, GA, USA

**A D Hyatt**
Australian Animal Health Laboratory, Geelong, VIC, Australia

**T Hyypiä**
University of Turku, Turku, Finland

**T Iwanami**
National Institute of Fruit Tree Science, Tsukuba, Japan

**A O Jackson**
University of California, Berkeley, CA, USA

**P Jardine**
University of Minnesota, Minneapolis, MN, USA

**J A Jehle**
DLR Rheinpfalz, Neustadt, Germany

**A R Jilbert**
Institute of Medical and Veterinary Science, Adelaide, SA, Australia

**P John**
Indian Agricultural Research Institute, New Delhi, India

**J E Johnson**
The Scripps Research Institute, La Jolla, CA, USA

**R T Johnson**
Johns Hopkins School of Medicine, Baltimore, MD, USA

**W E Johnson**
New England Primate Research Center, Southborough, MA, USA

**S L Johnston**
Imperial College London, London, UK

**A T Jones**
Scottish Crop Research Institute, Dundee, UK

**R Jordan**
USDA-ARS, Beltsville, MD, USA

**Y Kapustin**
National Institutes of Health, Bethesda, MD, USA

**P Karayiannis**
Imperial College London, London, UK

**P Kazmierczak**
University of California, Davis, CA, USA

**K M Keene**
Colorado State University, Fort Collins, CO, USA

**C Kerlan**
Institut National de la Recherche Agronomique (INRA), Le Rheu, France

**K Khalili**
Temple University School of Medicine, Philadelphia, PA, USA

**P H Kilmarx**
Centers for Disease Control and Prevention, Atlanta, GA, USA

**L A King**
Oxford Brookes University, Oxford, UK

**P D Kirkland**
Elizabeth Macarthur Agricultural Institute, Menangle, NSW, Australia

**C D Kirkwood**
Murdoch Childrens Research Institute Royal Children's Hospital, Melbourne, VIC, Australia

**R P Kitching**
Canadian Food Inspection Agency, Winnipeg, MB, Canada

**P J Klasse**
Cornell University, New York, NY, USA

**N R Klatt**
University of Pennsylvania School of Medicine, Philadelphia, PA, USA

**R G Kleespies**
Institute for Biological Control, Darmstadt, Germany

**D F Klessig**
Boyce Thompson Institute for Plant Research, Ithaca, NY, USA

**W B Klimstra**
Louisiana State University Health Sciences Center at Shreveport, Shreveport, LA, USA

**V Klimyuk**
Icon Genetics GmbH, Weinbergweg, Germany

**N Knowles**
Institute for Animal Health, Pirbright, UK

**R Koenig**
Biologische Bundesanstalt für Land- und Forstwirtschaft, Brunswick, Germany

**R Koenig**
Institut für Pflanzenvirologie, Mikrobiologie und biologische Sicherheit, Brunswick, Germany

**G Konaté**
INERA, Ouagadougou, Burkina Faso

**C N Kotton**
Massachusetts General Hospital, Boston, MA, USA

**L D Kramer**
Wadsworth Center, New York State Department of Health, Albany, NY, USA

**P J Krell**
University of Guelph, Guelph, ON, Canada

**J Kreuze**
International Potato Center (CIP), Lima, Peru

**M J Kuehnert**
Centers for Disease Control and Prevention, Atlanta, GA, USA

**R J Kuhn**
Purdue University, West Lafayette, IN, USA

**G Kurath**
Western Fisheries Research Center, Seattle, WA, USA

**I Kusters**
sanofi pasteur, Lyon, France

**I V Kuzmin**
Centers for Disease Control and Prevention, Atlanta, GA, USA

**M E Laird**
New England Primate Research Center, Southborough, MA, USA

**R A Lamb**
Howard Hughes Medical Institute at Northwestern University, Evanston, IL, USA

**P F Lambert**
University of Wisconsin School of Medicine and Public Health, Madison, WI, USA

**A S Lang**
Memorial University of Newfoundland, St. John's, NL, Canada

**H D Lapierre**
INRA – Département Santé des Plantes et Environnement, Versailles, France

**G Lawrence**
The Children's Hospital at Westmead, Westmead, NSW, Australia and
University of Sydney, Westmead, NSW, Australia

**H Lecoq**
Institut National de la Recherche Agronomique (INRA), Station de Pathologie Végétale, Montfavet, France

**B Y Lee**
Seoul Women's University, Seoul, South Korea

**E J Lefkowitz**
University of Alabama at Birmingham, Birmingham, AL, USA

**J P Legg**
International Institute of Tropical Agriculture, Dar es Salaam, Tanzania,
UK and
Natural Resources Institute, Chatham Maritime, UK

**P Leinikki**
National Public Health Institute, Helsinki, Finland

**J Lenard**
University of Medicine and Dentistry of New Jersey (UMDNJ), Piscataway, NJ, USA

**J C Leong**
University of Hawaii at Manoa, Honolulu, HI, USA

**K N Leppard**
University of Warwick, Coventry, UK

**A Lescoute**
Université Louis Pasteur, Strasbourg, France

**D-E Lesemann**
Biologische Bundesanstalt für Land- und Forstwirtschaft, Brunswick, Germany

**J-H Leu**
National Taiwan University, Taipei, Republic of China

**H L Levin**
National Institutes of Health, Bethesda, MD, USA

**D J Lewandowski**
The Ohio State University, Columbus, OH, USA

**H-S Lim**
University of California, Berkeley, Berkeley, CA, USA

**M D A Lindsay**
Western Australian Department of Health, Mount Claremont, WA, Australia

**R Ling**
University of Warwick, Coventry, UK

**M L Linial**
Fred Hutchinson Cancer Research Center, Seattle, WA, USA

**D C Liotta**
Emory University, Atlanta, GA, USA

**W Ian Lipkin**
Columbia University, New York, NY, USA

**H L Lipton**
University of Illinois at Chicago, Chicago, IL, USA

**A S Liss**
University of Texas at Austin, Austin, TX, USA

**J J López-Moya**
Instituto de Biología Molecular de Barcelona (IBMB), CSIC, Barcelona, Spain

**G Loebenstein**
Agricultural Research Organization, Bet Dagan, Israel

**C-F Lo**
National Taiwan University, Taipei, Republic of China

**S A Lommel**
North Carolina State University, Raleigh, NC, USA

**G P Lomonossoff**
John Innes Centre, Norwich, UK

**M Luo**
University of Alabama at Birmingham, Birmingham, AL, USA

**S A MacFarlane**
Scottish Crop Research Institute, Dundee, UK

**J S Mackenzie**
Curtin University of Technology, Shenton Park, WA, Australia

**R Mahieux**
Pasteur Institute, CNRS URA 3015, Paris, France

**B W J Mahy**
Centers for Disease Control and Prevention, Atlanta, GA, USA

**E Maiss**
Institute of Plant Diseases and Plant Protection, Hannover, Germany

**E O Major**
National Institutes of Health, Bethesda, MD, USA

**V G Malathi**
Indian Agricultural Research Institute, New Delhi, India

**A Mankertz**
Robert Koch-Institut, Berlin, Germany

**S Mansoor**
National Institute for Biotechnology and Genetic Engineering, Faisalabad, Pakistan

**A A Marfin**
Centers for Disease Control and Prevention, Atlanta, GA, USA

**S Marillonnet**
Icon Genetics GmbH, Weinbergweg, Germany

**G P Martelli**
Università degli Studi and Istituto di Virologia vegetale CNR, Bari, Italy

**M Marthas**
University of California, Davis, Davis, CA, USA

**D P Martin**
University of Cape Town, Cape Town, South Africa

**P A Marx**
Tulane University, Covington, LA, USA

**W S Mason**
Fox Chase Cancer Center, Philadelphia, PA, USA

**T D Mastro**
Centers for Disease Control and Prevention, Atlanta, GA, USA

**A A McBride**
National Institutes of Health, Bethesda, MD, USA

**L McCann**
National Institutes of Health, Bethesda, MD, USA

**M McChesney**
University of California, Davis, Davis, CA, USA

**J B McCormick**
University of Texas, School of Public Health, Brownsville, TX, USA

**G McFadden**
University of Florida, Gainesville, FL, USA

**G McFadden**
The University of Western Ontario, London, ON, Canada

**D B McGavern**
The Scripps Research Institute, La Jolla, CA, USA

**A L McNees**
Baylor College of Medicine, Houston, TX, USA

**M Meier**
Tallinn University of Technology, Tallinn, Estonia

**P S Mellor**
Institute for Animal Health, Woking, UK

**X J Meng**
Virginia Polytechnic Institute and State University, Blacksburg, VA, USA

**A A Mercer**
University of Otago, Dunedin, New Zealand

**P P C Mertens**
Institute for Animal Health, Woking, UK

**T C Mettenleiter**
Friedrich-Loeffler-Institut, Greifswald-Insel Riems, Germany

**H Meyer**
Bundeswehr Institute of Microbiology, Munich, Germany

**R F Meyer**
Centers for Disease Control and Prevention, Atlanta, GA, USA

**P de Micco**
Etablissement Français du Sang Alpes-Méditerranée, Marseilles, France

**B R Miller**
Centers for Disease Control and Prevention (CDC), Fort Collins, CO, USA

**C J Miller**
University of California, Davis, Davis, CA, USA

**R G Milne**
Istituto di Virologia Vegetale CNR, Torino, Italy

**P D Minor**
NIBSC, Potters Bar, UK

**S Mjaaland**
Norwegian School of Veterinary Science, Oslo, Norway

**E S Mocarski**
Emory University School of Medicine, Atlanta, GA, USA

**E S Mocarski, Jr.**
Emory University School of Medicine, Emory, GA, USA

**V Moennig**
School of Veterinary Medicine, Hanover, Germany

**P Moffett**
Boyce Thompson Institute for Plant Research, Ithaca, NY, USA

**T P Monath**
Kleiner Perkins Caufield and Byers, Menlo Park, CA, USA

**R C Montelaro**
University of Pittsburgh School of Medicine, Pittsburgh, PA, USA

**P S Moore**
University of Pittsburgh Cancer Institute, Pittsburgh, PA, USA

**F J Morales**
International Center for Tropical Agriculture, Cali, Colombia

**H Moriyama**
Tokyo University of Agriculture and Technology, Fuchu, Japan

**T J Morris**
University of Nebraska, Lincoln, NE, USA

**S A Morse**
Centers for Disease Control and Prevention, Atlanta, GA, USA

**L Moser**
University of Wisconsin – Madison, Madison, WI, USA

**B Moury**
INRA – Station de Pathologie Végétale, Montfavet, France

**J W Moyer**
North Carolina State University, Raleigh, NC, USA

**R W Moyer**
University of Florida, Gainesville, FL, USA

**E Muller**
CIRAD/UMR BGPI, Montpellier, France

**F A Murphy**
University of Texas Medical Branch, Galveston, TX, USA

**A Müllbacher**
Australian National University, Canberra, ACT, Australia

**K Nagasaki**
Fisheries Research Agency, Hiroshima, Japan

**T Nakayashiki**
National Institutes of Health, Bethesda, MD, USA

**A A Nash**
University of Edinburgh, Edinburgh, UK

**N Nathanson**
University of Pennsylvania, Philadelphia, PA, USA

**C K Navaratnarajah**
Purdue University, West Lafayette, IN, USA

**M S Nawaz-ul-Rehman**
Danforth Plant Science Center, St. Louis, MO, USA

**J C Neil**
University of Glasgow, Glasgow, UK

**R S Nelson**
Samuel Roberts Noble Foundation, Inc., Ardmore, OK, USA

**P Nettleton**
Moredun Research Institute, Edinburgh, UK

**A W Neuman**
Emory University, Atlanta, GA, USA

**A R Neurath**
Virotech, New York, NY, USA

**M L Nibert**
Harvard Medical School, Boston, MA, USA

**L Nicoletti**
Istituto Superiore di Sanità, Rome, Italy

**N Noah**
London School of Hygiene and Tropical Medicine, London, UK

**D L Nuss**
University of Maryland Biotechnology Institute, Rockville, MD, USA

**M S Oberste**
Centers for Disease Control and Prevention, Atlanta, GA, USA

**W A O'Brien**
University of Texas Medical Branch – Galveston, Galveston, TX, USA

**D J O'Callaghan**
Louisiana State University Health Sciences Center, Shreveport, LA, USA

**W F Ochoa**
University of California, San Diego, La Jolla, CA, USA

**M R Odom**
University of Alabama at Birmingham, Birmingham, AL, USA

**M M van Oers**
Wageningen University, Wageningen, The Netherlands

**M B A Oldstone**
The Scripps Research Institute, La Jolla, CA, USA

**G Olinger**
USAMRIID, Fort Detrick, MD, USA

**K E Olson**
Colorado State University, Fort Collins, CO, USA

**A Olspert**
Tallinn University of Technology, Tallinn, Estonia

**G Orth**
Institut Pasteur, Paris, France

**J E Osorio**
University of Wisconsin, Madison, WI, USA

**N Osterrieder**
Cornell University, Ithaca, NY, USA

**S A Overman**
University of Missouri – Kansas City, Kansas City, MO, USA

**R A Owens**
Beltsville Agricultural Research Center, Beltsville, MD, USA

**M S Padmanabhan**
Yale University, New Haven, CT, USA

**S Paessler**
University of Texas Medical Branch, Galveston, TX, USA

**P Palese**
Mount Sinai School of Medicine, New York, NY, USA

**M A Pallansch**
Centers for Disease Control and Prevention, Atlanta, GA, USA

**M Palmarini**
University of Glasgow Veterinary School, Glasgow, UK

**P Palukaitis**
Scottish Crop Research Institute, Invergowrie, Dundee, UK

**I Pandrea**
Tulane National Primate Research Center, Covington, LA, USA

**O Papadopoulos**
Aristotle University, Thessaloniki, Greece

**H R Pappu**
Washington State University, Pullman, WA, USA

**S Parker**
Saint Louis University School of Medicine, St. Louis, MO, USA

**C R Parrish**
Cornell University, Ithaca, NY, USA

**R F Pass**
University of Alabama School of Medicine, Birmingham, AL, USA

**J L Patterson**
Southwest Foundation for Biomedical Research, San Antonio, TX, USA

**T A Paul**
Cornell University, Ithaca, NY, USA

**A E Peaston**
The Jackson Laboratory, Bar Harbor, ME, USA

**M Peeters**
University of Montpellier 1, Montpellier, France

**J S M Peiris**
The University of Hong Kong, Hong Kong, People's Republic of China

**P J Peters**
Centers for Disease Control and Prevention, Atlanta, GA, USA

**M Pfeffer**
Bundeswehr Institute of Microbiology, Munich, Germany

**H Pfister**
University of Köln, Cologne, Germany

**O Planz**
Federal Research Institute for Animal Health, Tuebingen, Gemany

**L L M Poon**
The University of Hong Kong, Hong Kong, People's Republic of China

**M M Poranen**
University of Helsinki, Helsinki, Finland

**K Porter**
The University of Melbourne, Parkville, VIC, Australia

**A Portner**
St. Jude Children's Research Hospital, Memphis, TN, USA

**R D Possee**
NERC Institute of Virology and Environmental Microbiology, Oxford, UK

**R T M Poulter**
University of Otago, Dunedin, New Zealand

**A M Powers**
Centers for Disease Control and Prevention, Fort Collins, CO, USA

**D Prangishvili**
Institut Pasteur, Paris, France

**C M Preston**
Medical Research Council Virology Unit, Glasgow, UK

**S L Quackenbush**
Colorado State University, Fort Collins, CO, USA

**F Qu**
University of Nebraska, Lincoln, NE, USA

**B C Ramirez**
CNRS, Paris, France

**A Rapose**
University of Texas Medical Branch – Galveston, Galveston, TX, USA

**D V R Reddy**
Hyderabad, India

**A J Redwood**
The University of Western Australia, Crawley, WA, Australia

**M Regner**
Australian National University, Canberra, ACT, Australia

**W K Reisen**
University of California, Davis, CA, USA

**T Renault**
IFREMER, La Tremblade, France

**P A Revill**
Victorian Infectious Diseases Reference Laboratory, Melbourne, VIC, Australia

**A Rezaian**
University of Adelaide, Adelaide, SA, Australia

**J F Ridpath**
USDA, Ames, IA, USA

**B K Rima**
The Queen's University of Belfast, Belfast, UK

**E Rimstad**
Norwegian School of Veterinary Science, Oslo, Norway

**F J Rixon**
MRC Virology Unit, Glasgow, UK

**Y-T Ro**
Konkuk University, Seoul, South Korea

**C M Robinson**
University of Oklahoma Health Sciences Center, Oklahoma City, OK, USA

**G F Rohrmann**
Oregon State University, Corvallis, OR, USA

**M Roivainen**
National Public Health Institute (KTL), Helsinki, Finland

**L Roux**
University of Geneva Medical School, Geneva, Switzerland

**J Rovnak**
Colorado State University, Fort Collins, CO, USA

**D J Rowlands**
University of Leeds, Leeds, UK

**P Roy**
London School of Hygiene and Tropical Medicine, London, UK

**L Rubino**
Istituto di Virologia Vegetale del CNR, Bari, Italy

**R W H Ruigrok**
CNRS, Grenoble, France

**C E Rupprecht**
Centers for Disease Control and Prevention, Atlanta, GA, USA

**R J Russell**
University of St. Andrews, St. Andrews, UK

**B E Russ**
The University of Melbourne, Parkville, VIC, Australia

**W T Ruyechan**
The State University of New York, Buffalo, NY, USA

**E Ryabov**
University of Warwick, Warwick, UK

**M D Ryan**
University of St. Andrews, St. Andrews, UK

**E P Rybicki**
University of Cape Town, Cape Town, South Africa

**K D Ryman**
Louisiana State University Health Sciences Center at Shreveport, Shreveport, LA, USA

**K D Ryman**
Louisiana State University Health Sciences Center, Shreveport, LA, USA

**K H Ryu**
Seoul Women's University, Seoul, South Korea

**M Safak**
Temple University School of Medicine, Philadelphia, PA, USA

**M Salas**
Universidad Autónoma, Madrid, Spain

**S K Samal**
University of Maryland, College Park, MD, USA

**J T Sample**
The Pennsylvania State University College of Medicine, Hershey, PA, USA

**C E Sample**
The Pennsylvania State University College of Medicine, Hershey, PA, USA

**R M Sandri-Goldin**
University of California, Irvine, Irvine, CA, USA

**H Sanfaçon**
Pacific Agri-Food Research Centre, Summerland, BC, Canada

**R Sanjuán**
Instituto de Biología Molecular y Cellular de Plantas, CSIC-UPV, Valencia, Spain

**N Santi**
Norwegian School of Veterinary Science, Oslo, Norway

**C Sarmiento**
Tallinn University of Technology, Tallinn, Estonia

**T Sasaya**
National Agricultural Research Center, Ibaraki, Japan

**Q J Sattentau**
University of Oxford, Oxford, UK

**C Savolainen-Kopra**
National Public Health Institute (KTL), Helsinki, Finland

**B Schaffhausen**
Tufts University School of Medicine, Boston, MA, USA

**K Scheets**
Oklahoma State University, Stillwater, OK, USA

**M J Schmitt**
University of the Saarland, Saarbrücken, Germany

**A Schneemann**
The Scripps Research Institute, La Jolla, CA, USA

**G Schoehn**
CNRS, Grenoble, France

**J E Schoelz**
University of Missouri, Columbia, MO, USA

**L B Schonberger**
Centers for Disease Control and Prevention, Atlanta, GA, USA

**U Schubert**
Klinikum der Universität Erlangen-Nürnberg, Erlangen, Germany

**D A Schultz**
Johns Hopkins University School of Medicine, Baltimore, MD, USA

**S Schultz-Cherry**
University of Wisconsin – Madison, Madison, WI, USA

**T F Schulz**
Hannover Medical School, Hannover, Germany

**P D Scotti**
Waiatarua, New Zealand

**B L Semler**
University of California, Irvine, CA, USA

**J M Sharp**
Veterinary Laboratories Agency, Penicuik, UK

**M L Shaw**
Mount Sinai School of Medicine, New York, NY, USA

**G R Shellam**
The University of Western Australia,
Crawley, WA, Australia

**D N Shepherd**
University of Cape Town, Cape Town, South Africa

**N C Sheppard**
University of Oxford, Oxford, UK

**F Shewmaker**
National Institutes of Health, Bethesda, MD, USA

**P A Signoret**
Montpellier SupAgro, Montpellier, France

**A Silaghi**
University of Manitoba, Winnipeg, MB, Canada

**G Silvestri**
University of Pennsylvania, Philadelphia, PA, USA

**T L Sit**
North Carolina State University, Raleigh, NC, USA

**N Sittidilokratna**
Centex Shrimp and Center for Genetic Engineering and Biotechnology, Bangkok, Thailand

**M A Skinner**
Imperial College London, London, UK

**D W Smith**
PathWest Laboratory Medicine WA, Nedlands, WA, Australia

**G L Smith**
Imperial College London, London, UK

**L M Smith**
The University of Western Australia,
Crawley, WA, Australia

**E J Snijder**
Leiden University Medical Center, Leiden, The Netherlands

**M Sova**
University of Texas Medical Branch – Galveston, Galveston, TX, USA

**J A Speir**
The Scripps Research Institute, La Jolla, CA, USA

**T E Spencer**
Texas A&M University, College Station, TX, USA

**P Sreenivasulu**
Sri Venkateswara University, Tirupati, India

**J Stanley**
John Innes Centre, Colney, UK

**K M Stedman**
Portland State University, Portland, OR, USA

**D Stephan**
Institute of Plant Diseases and Plant Protection, Hannover, Germany

**C C M M Stijger**
Wageningen University and Research Centre, Naaldwijk, The Netherlands

**L Stitz**
Federal Research Institute for Animal Health, Tuebingen, Gemany

**P G Stockley**
University of Leeds, Leeds, UK

**M R Strand**
University of Georgia, Athens, GA, USA

**M J Studdert**
The University of Melbourne, Parkville, VIC, Australia

**C A Suttle**
University of British Columbia, Vancouver, BC, Canada

**N Suzuki**
Okayama University, Okayama, Japan

**J Y Suzuki**
USDA, Pacific Basin Agricultural Research Center, Hilo, HI, USA

**R Swanepoel**
National Institute for Communicable Diseases, Sandringham, South Africa

**S J Symes**
The University of Melbourne, Parkville, VIC, Australia

**G Szittya**
Agricultural Biotechnology Center, Godollo, Hungary

**M Taliansky**
Scottish Crop Research Institute, Dundee, UK

**P Tattersall**
Yale University Medical School, New Haven, CT, USA

**T Tatusova**
National Institutes of Health, Bethesda, MD, USA

**S Tavantzis**
University of Maine, Orono, ME, USA

**J M Taylor**
Fox Chase Cancer Center, Philadelphia, PA, USA

**D A Theilmann**
Agriculture and Agri-Food Canada, Summerland, BC, Canada

**F C Thomas Allnutt**
National Science Foundation, Arlington, VA, USA

**G J Thomas Jr.**
University of Missouri – Kansas City, Kansas City, MO, USA

**J E Thomas**
Department of Primary Industries and Fisheries, Indooroopilly, QLD, Australia

**H C Thomas**
Imperial College London, London, UK

**A N Thorburn**
The University of Melbourne, Parkville, VIC, Australia

**P Tijssen**
Université du Québec, Laval, QC, Canada

**S A Tolin**
Virginia Polytechnic Institute and State University, Blacksburg, VA, USA

**L Torrance**
Scottish Crop Research Institute, Invergowrie, UK

**S Trapp**
Cornell University, Ithaca, NY, USA

**S Tripathi**
USDA, Pacific Basin Agricultural Research Center, Hilo, HI, USA

**E Truve**
Tallinn University of Technology, Tallinn, Estonia

**J-M Tsai**
National Taiwan University, Taipei, Republic of China

**M Tsompana**
North Carolina State University, Raleigh, NC, USA

**R Tuma**
University of Helsinki, Helsinki, Finland

**A S Turnell**
The University of Birmingham, Birmingham, UK

**K L Tyler**
University of Colorado Health Sciences, Denver, CO, USA

**A Uchiyama**
Cornell University, Ithaca, NY, USA

**C Upton**
University of Victoria, Victoria, BC, Canada

**A Urisman**
University of California, San Francisco, San Francisco, CA, USA

**J K Uyemoto**
University of California, Davis, CA, USA

**A Vaheri**
University of Helsinki, Helsinki, Finland

**R Vainionpää**
University of Turku, Turku, Finland

**A M Vaira**
Istituto di Virologia Vegetale, CNR, Turin, Italy

**N K Van Alfen**
University of California, Davis, CA, USA

**R A A Van der Vlugt**
Wageningen University and Research Centre, Wageningen, The Netherlands

**M H V Van Regenmortel**
CNRS, Illkirch, France

**P A Venter**
The Scripps Research Institute, La Jolla, CA, USA

**J Verchot-Lubicz**
Oklahoma State University, Stillwater, OK, USA

**R A Vere Hodge**
Vere Hodge Antivirals Ltd., Reigate, UK

**H J Vetten**
Federal Research Centre for Agriculture and Forestry (BBA), Brunswick, Germany

**L P Villarreal**
University of California, Irvine, Irvine, CA, USA

**J M Vlak**
Wageningen University, Wageningen, The Netherlands

**P K Vogt**
The Scripps Research Institute, La Jolla, CA, USA

**L E Volkman**
University of California, Berkeley, Berkeley, CA, USA

**J Votteler**
Klinikum der Universität Erlangen-Nürnberg, Erlangen, Germany

**D F Voytas**
Iowa State University, Ames, IA, USA

**J D F Wadsworth**
University College London, London, UK

**E K Wagner**
University of California, Irvine, Irvine, CA, USA

**P J Walker**
CSIRO Australian Animal Health Laboratory, Geelong, VIC, Australia

**A L Wang**
University of California, San Francisco, CA, USA

**X Wang**
University of Wisconsin – Madison, Madison, WI, USA

**C C Wang**
University of California, San Francisco, CA, USA

**L-F Wang**
Australian Animal Health Laboratory, Geelong, VIC, Australia

**R Warrier**
Purdue University, West Lafayette, IN, USA

**S C Weaver**
University of Texas Medical Branch, Galveston, TX, USA

**B A Webb**
University of Kentucky, Lexington, KY, USA

**F Weber**
University of Freiburg, Freiburg, Germany

**R P Weir**
Berrimah Research Farm, Darwin, NT, Australia

**R A Weisberg**
National Institutes of Health, Bethesda, MD, USA

**W Weissenhorn**
UMR 5233 UJF-EMBL-CNRS, Grenoble, France

**R M Welsh**
University of Massachusetts Medical School, Worcester, MA, USA

**J T West**
University of Oklahoma Health Sciences Center, Oklahoma City, OK, USA

**E Westhof**
Université Louis Pasteur, Strasbourg, France

**S P J Whelan**
Harvard Medical School, Boston, MA, USA

**R L White**
Texas A&M University, College Station, TX, USA

**C A Whitehouse**
United States Army Medical Research Institute of Infectious Diseases, Frederick, MD, USA

**R B Wickner**
National Institutes of Health, Bethesda, MD, USA

**R G Will**
Western General Hospital, Edinburgh, UK

**T Williams**
Instituto de Ecología A.C., Xalapa, Mexico

**K Willoughby**
Moredun Research Institute, Edinburgh, UK

**S Winter**
Deutsche Sammlung für Mikroorganismen und Zellkulturen, Brunswick, Germany

**J Winton**
Western Fisheries Research Center, Seattle, WA, USA

**J K Yamamoto**
University of Florida, Gainesville, FL, USA

**M Yoshida**
University of Tokyo, Chiba, Japan

**N Yoshikawa**
Iwate University, Ueda, Japan

**L S Young**
University of Birmingham, Birmingham, UK

**R F Young, III**
Texas A&M University, College Station, TX, USA

**T M Yuill**
University of Wisconsin, Madison, WI, USA

**A J Zajac**
University of Alabama at Birmingham, Birmingham, AL, USA

**S K Zavriev**
Shemyakin and Ovchinnikov Institute of Bioorganic Chemistry, Russian Academy of Sciences, Moscow, Russia

**J Ziebuhr**
The Queen's University of Belfast, Belfast, UK

**E I Zuniga**
The Scripps Research Institute, La Jolla, CA, USA

# CONTENTS

| | |
|---|---|
| Editors-in-Chief | v |
| Associate Editors | vii |
| Preface | ix |
| How to Use the Encyclopedia | xi |
| Contributors | xiii |

## VOLUME 1

### A

| | | |
|---|---|---|
| Adenoviruses: General Features | B Harrach | 1 |
| Adenoviruses: Malignant Transformation and Oncology | A S Turnell | 9 |
| Adenoviruses: Molecular Biology | K N Leppard | 17 |
| Adenoviruses: Pathogenesis | M Benkő | 24 |
| African Cassava Mosaic Disease | J P Legg | 30 |
| African Horse Sickness Viruses | P S Mellor and P P C Mertens | 37 |
| African Swine Fever Virus | L K Dixon and D Chapman | 43 |
| AIDS: Disease Manifestation | A Rapose, J East, M Sova and W A O'Brien | 51 |
| AIDS: Global Epidemiology | P J Peters, P H Kilmarx and T D Mastro | 58 |
| AIDS: Vaccine Development | N C Sheppard and Q J Sattentau | 69 |
| Akabane Virus | P S Mellor and P D Kirkland | 76 |
| Alfalfa Mosaic Virus | J F Bol | 81 |
| Algal Viruses | K Nagasaki and C P D Brussaard | 87 |
| *Allexivirus* | S K Zavriev | 96 |
| *Alphacryptovirus* and *Betacryptovirus* | R Blawid, D Stephan and E Maiss | 98 |
| *Anellovirus* | P Biagini and P de Micco | 104 |
| Animal Rhabdoviruses | H Bourhy, A J Gubala, R P Weir and D B Boyle | 111 |
| Antigen Presentation | E I Zuniga, D B McGavern and M B A Oldstone | 121 |
| Antigenic Variation | G M Air and J T West | 127 |
| Antigenicity and Immunogenicity of Viral Proteins | M H V Van Regenmortel | 137 |

| | |
|---|---|
| Antiviral Agents     *H J Field and R A Vere Hodge* | 142 |
| Apoptosis and Virus Infection     *J R Clayton and J M Hardwick* | 154 |
| Aquareoviruses     *M St J Crane and G Carlile* | 163 |
| Arboviruses     *B R Miller* | 170 |
| Arteriviruses     *M A Brinton and E J Snijder* | 176 |
| Ascoviruses     *B A Federici and Y Bigot* | 186 |
| Assembly of Viruses: Enveloped Particles     *C K Navaratnarajah, R Warrier and R J Kuhn* | 193 |
| Assembly of Viruses: Nonenveloped Particles     *M Luo* | 200 |
| Astroviruses     *L Moser and S Schultz-Cherry* | 204 |

# B

| | |
|---|---|
| Baculoviruses: Molecular Biology of Granuloviruses     *S Hilton* | 211 |
| Baculoviruses: Molecular Biology of Mosquito Baculoviruses     *J J Becnel and C L Afonso* | 219 |
| Baculoviruses: Molecular Biology of Sawfly Baculoviruses     *B M Arif* | 225 |
| Baculoviruses: Apoptosis Inhibitors     *R J Clem* | 231 |
| Baculoviruses: Expression Vector     *F J Haines, R D Possee and L A King* | 237 |
| Baculoviruses: General Features     *P J Krell* | 247 |
| Baculoviruses: Molecular Biology of Nucleopolyhedroviruses     *D A Theilmann and G W Blissard* | 254 |
| Baculoviruses: Pathogenesis     *L E Volkman* | 265 |
| Banana Bunchy Top Virus     *J E Thomas* | 272 |
| Barley Yellow Dwarf Viruses     *L L Domier* | 279 |
| Barnaviruses     *P A Revill* | 286 |
| Bean Common Mosaic Virus and Bean Common Mosaic Necrosis Virus     *R Jordan and J Hammond* | 288 |
| Bean Golden Mosaic Virus     *F J Morales* | 295 |
| Beet Curly Top Virus     *J Stanley* | 301 |
| *Benyvirus*     *R Koenig* | 308 |
| Beta ssDNA Satellites     *R W Briddon and S Mansoor* | 314 |
| Birnaviruses     *B Delmas* | 321 |
| Bluetongue Viruses     *P Roy* | 328 |
| Border Disease Virus     *P Nettleton and K Willoughby* | 335 |
| Bornaviruses     *L Stitz, O Planz and W Ian Lipkin* | 341 |
| Bovine and Feline Immunodeficiency Viruses     *J K Yamamoto* | 347 |
| Bovine Ephemeral Fever Virus     *P J Walker* | 354 |
| Bovine Herpesviruses     *M J Studdert* | 362 |
| Bovine Spongiform Encephalopathy     *R G Will* | 368 |
| Bovine Viral Diarrhea Virus     *J F Ridpath* | 374 |
| Brome Mosaic Virus     *X Wang and P Ahlquist* | 381 |
| Bromoviruses     *J J Bujarski* | 386 |

| | |
|---|---:|
| Bunyaviruses: General Features  *R M Elliott* | 390 |
| Bunyaviruses: Unassigned  *C H Calisher* | 399 |

# C

| | |
|---|---:|
| Cacao Swollen Shoot Virus  *E Muller* | 403 |
| Caliciviruses  *M J Studdert and S J Symes* | 410 |
| *Capillovirus, Foveavirus, Trichovirus, Vitivirus*  *N Yoshikawa* | 419 |
| Capripoxviruses  *R P Kitching* | 427 |
| Capsid Assembly: Bacterial Virus Structure and Assembly  *S Casjens* | 432 |
| Cardioviruses  *C Billinis and O Papadopoulos* | 440 |
| *Carlavirus*  *K H Ryu and B Y Lee* | 448 |
| *Carmovirus*  *F Qu and T J Morris* | 453 |
| Caulimoviruses: General Features  *J E Schoelz* | 457 |
| Caulimoviruses: Molecular Biology  *T Hohn* | 464 |
| Central Nervous System Viral Diseases  *R T Johnson and B M Greenberg* | 469 |
| Cereal Viruses: Maize/Corn  *P A Signoret* | 475 |
| Cereal Viruses: Rice  *F Morales* | 482 |
| Cereal Viruses: Wheat and Barley  *H D Lapierre and D Hariri* | 490 |
| Chandipura Virus  *S Basak and D Chattopadhyay* | 497 |
| Chrysoviruses  *S A Ghabrial* | 503 |
| Circoviruses  *A Mankertz* | 513 |
| Citrus Tristeza Virus  *M Bar-Joseph and W O Dawson* | 520 |
| Classical Swine Fever Virus  *V Moennig and I Greiser-Wilke* | 525 |
| Coltiviruses  *H Attoui and X de Lamballerie* | 533 |
| Common Cold Viruses  *S Dreschers and C Adams* | 541 |
| Coronaviruses: General Features  *D Cavanagh and P Britton* | 549 |
| Coronaviruses: Molecular Biology  *S C Baker* | 554 |
| Cotton Leaf Curl Disease  *S Mansoor, I Amin and R W Briddon* | 563 |
| Cowpea Mosaic Virus  *G P Lomonossoff* | 569 |
| Cowpox Virus  *M Bennett, G L Smith and D Baxby* | 574 |
| Coxsackieviruses  *M S Oberste and M A Pallansch* | 580 |
| Crenarchaeal Viruses: Morphotypes and Genomes  *D Prangishvili, T Basta and R A Garrett* | 587 |
| Crimean–Congo Hemorrhagic Fever Virus and Other Nairoviruses  *C A Whitehouse* | 596 |
| Cryo-Electron Microscopy  *W Chiu, J T Chang and F J Rixon* | 603 |
| Cucumber Mosaic Virus  *F García-Arenal and P Palukaitis* | 614 |
| Cytokines and Chemokines  *D E Griffin* | 620 |
| Cytomegaloviruses: Murine and Other Nonprimate Cytomegaloviruses  *A J Redwood, L M Smith and G R Shellam* | 624 |
| Cytomegaloviruses: Simian Cytomegaloviruses  *D J Alcendor and G S Hayward* | 634 |

# VOLUME 2

## D

| | | |
|---|---|---|
| Defective-Interfering Viruses | L Roux | 1 |
| Dengue Viruses | D J Gubler | 5 |
| Diagnostic Techniques: Microarrays | K Fischer, A Urisman and J DeRisi | 14 |
| Diagnostic Techniques: Plant Viruses | R Koenig, D-E Lesemann, G Adam and S Winter | 18 |
| Diagnostic Techniques: Serological and Molecular Approaches | R Vainionpää and P Leinikki | 29 |
| Dicistroviruses | P D Christian and P D Scotti | 37 |
| Disease Surveillance | N Noah | 44 |
| DNA Vaccines | S Babiuk and L A Babiuk | 51 |

## E

| | | |
|---|---|---|
| *Ebolavirus* | K S Brown, A Silaghi and H Feldmann | 57 |
| Echoviruses | T Hyypiä | 65 |
| Ecology of Viruses Infecting Bacteria | S T Abedon | 71 |
| Electron Microscopy of Viruses | G Schoehn and R W H Ruigrok | 78 |
| Emerging and Reemerging Virus Diseases of Plants | G P Martelli and D Gallitelli | 86 |
| Emerging and Reemerging Virus Diseases of Vertebrates | B W J Mahy | 93 |
| Emerging Geminiviruses | C M Fauquet and M S Nawaz-ul-Rehman | 97 |
| Endogenous Retroviruses | W E Johnson | 105 |
| *Endornavirus* | T Fukuhara and H Moriyama | 109 |
| Enteric Viruses | R F Bishop and C D Kirkwood | 116 |
| Enteroviruses of Animals | L E Hughes and M D Ryan | 123 |
| Enteroviruses: Human Enteroviruses Numbered 68 and Beyond | T Hovi, S Blomqvist, C Savolainen-Kopra and M Roivainen | 130 |
| Entomopoxviruses | M N Becker and R W Moyer | 136 |
| Epidemiology of Human and Animal Viral Diseases | F A Murphy | 140 |
| Epstein–Barr Virus: General Features | L S Young | 148 |
| Epstein–Barr Virus: Molecular Biology | J T Sample and C E Sample | 157 |
| Equine Infectious Anemia Virus | J K Craigo and R C Montelaro | 167 |
| Evolution of Viruses | L P Villarreal | 174 |

## F

| | | |
|---|---|---|
| Feline Leukemia and Sarcoma Viruses | J C Neil | 185 |
| Filamentous ssDNA Bacterial Viruses | S A Overman and G J Thomas Jr. | 190 |
| Filoviruses | G Olinger, T W Geisbert and L E Hensley | 198 |
| Fish and Amphibian Herpesviruses | A J Davison | 205 |
| Fish Retroviruses | T A Paul, R N Casey, P R Bowser, J W Casey, J Rovnak and S L Quackenbush | 212 |

| | | |
|---|---|---|
| Fish Rhabdoviruses | G Kurath and J Winton | 221 |
| Fish Viruses | J C Leong | 227 |
| Flaviviruses of Veterinary Importance | R Swanepoel and F J Burt | 234 |
| Flaviviruses: General Features | T J Chambers | 241 |
| Flexiviruses | M J Adams | 253 |
| Foamy Viruses | M L Linial | 259 |
| Foot and Mouth Disease Viruses | D J Rowlands | 265 |
| Fowlpox Virus and Other Avipoxviruses | M A Skinner | 274 |
| Fungal Viruses | S A Ghabrial and N Suzuki | 284 |
| *Furovirus* | R Koenig | 291 |
| Fuselloviruses of Archaea | K M Stedman | 296 |

# G

| | | |
|---|---|---|
| Gene Therapy: Use of Viruses as Vectors | K I Berns and T R Flotte | 301 |
| Genome Packaging in Bacterial Viruses | P Jardine | 306 |
| Giardiaviruses | A L Wang and C C Wang | 312 |

# H

| | | |
|---|---|---|
| Hantaviruses | A Vaheri | 317 |
| Henipaviruses | B T Eaton and L-F Wang | 321 |
| Hepadnaviruses of Birds | A R Jilbert and W S Mason | 327 |
| Hepadnaviruses: General Features | T J Harrison | 335 |
| Hepatitis A Virus | A Dotzauer | 343 |
| Hepatitis B Virus: General Features | P Karayiannis and H C Thomas | 350 |
| Hepatitis B Virus: Molecular Biology | T J Harrison | 360 |
| Hepatitis C Virus | R Bartenschlager and S Bühler | 367 |
| Hepatitis Delta Virus | J M Taylor | 375 |
| Hepatitis E Virus | X J Meng | 377 |
| Herpes Simplex Viruses: General Features | L Aurelian | 383 |
| Herpes Simplex Viruses: Molecular Biology | E K Wagner and R M Sandri-Goldin | 397 |
| Herpesviruses of Birds | S Trapp and N Osterrieder | 405 |
| Herpesviruses of Horses | D J O'Callaghan and N Osterrieder | 411 |
| Herpesviruses: Discovery | B Ehlers | 420 |
| Herpesviruses: General Features | A J Davison | 430 |
| Herpesviruses: Latency | C M Preston | 436 |
| History of Virology: Bacteriophages | H-W Ackermann | 442 |
| History of Virology: Plant Viruses | R Hull | 450 |
| History of Virology: Vertebrate Viruses | F J Fenner | 455 |

| | |
|---|---|
| *Hordeivirus*    J N Bragg, H-S Lim and A O Jackson | 459 |
| Host Resistance to Retroviruses    T Hatziioannou and P D Bieniasz | 467 |
| Human Cytomegalovirus: General Features    E S Mocarski Jr. and R F Pass | 474 |
| Human Cytomegalovirus: Molecular Biology    W Gibson | 485 |
| Human Eye Infections    J Chodosh, A V Chintakuntlawar and C M Robinson | 491 |
| Human Herpesviruses 6 and 7    U A Gompels | 498 |
| Human Immunodeficiency Viruses: Antiretroviral Agents    A W Neuman and D C Liotta | 505 |
| Human Immunodeficiency Viruses: Molecular Biology    J Votteler and U Schubert | 517 |
| Human Immunodeficiency Viruses: Origin    F van Heuverswyn and M Peeters | 525 |
| Human Immunodeficiency Viruses: Pathogenesis    N R Klatt, A Chahroudi and G Silvestri | 534 |
| Human Respiratory Syncytial Virus    P L Collins | 542 |
| Human Respiratory Viruses    J E Crowe Jr. | 551 |
| Human T-Cell Leukemia Viruses: General Features    M Yoshida | 558 |
| Human T-Cell Leukemia Viruses: Human Disease    R Mahieux and A Gessain | 564 |
| Hypovirulence    N K Van Alfen and P Kazmierczak | 574 |
| Hypoviruses    D L Nuss | 580 |

# VOLUME 3

## I

| | |
|---|---|
| Icosahedral dsDNA Bacterial Viruses with an Internal Membrane    J K H Bamford and S J Butcher | 1 |
| Icosahedral Enveloped dsRNA Bacterial Viruses    R Tuma | 6 |
| Icosahedral ssDNA Bacterial Viruses    B A Fane, M Chen, J E Cherwa and A Uchiyama | 13 |
| Icosahedral ssRNA Bacterial Viruses    P G Stockley | 21 |
| Icosahedral Tailed dsDNA Bacterial Viruses    R L Duda | 30 |
| *Idaeovirus*    A T Jones and H Barker | 37 |
| *Iflavirus*    M M van Oers | 42 |
| *Ilarvirus*    K C Eastwell | 46 |
| Immune Response to Viruses: Antibody-Mediated Immunity    A R Neurath | 56 |
| Immune Response to Viruses: Cell-Mediated Immunity    A J Zajac and L E Harrington | 70 |
| Immunopathology    M B A Oldstone and R S Fujinami | 78 |
| Infectious Pancreatic Necrosis Virus    Ø Evensen and N Santi | 83 |
| Infectious Salmon Anemia Virus    B H Dannevig, S Mjaaland and E Rimstad | 89 |
| Influenza    R A Lamb | 95 |
| Innate Immunity: Defeating    C F Basler | 104 |
| Innate Immunity: Introduction    F Weber | 111 |
| Inoviruses    L A Day | 117 |
| Insect Pest Control by Viruses    M Erlandson | 125 |
| Insect Reoviruses    P P C Mertens and H Attoui | 133 |
| Insect Viruses: Nonoccluded    J P Burand | 144 |

| | | |
|---|---|---|
| Interfering RNAs | K E Olson, K M Keene and C D Blair | 148 |
| Iridoviruses of Vertebrates | A D Hyatt and V G Chinchar | 155 |
| Iridoviruses of Invertebrates | T Williams and A D Hyatt | 161 |
| Iridoviruses: General Features | V G Chinchar and A D Hyatt | 167 |

# J

| | | |
|---|---|---|
| Jaagsiekte Sheep Retrovirus | J M Sharp, M de las Heras, T E Spencer and M Palmarini | 175 |
| Japanese Encephalitis Virus | A D T Barrett | 182 |

# K

| | | |
|---|---|---|
| Kaposi's Sarcoma-Associated Herpesvirus: General Features | Y Chang and P S Moore | 189 |
| Kaposi's Sarcoma-Associated Herpesvirus: Molecular Biology | E Gellermann and T F Schulz | 195 |

# L

| | | |
|---|---|---|
| Lassa, Junin, Machupo and Guanarito Viruses | J B McCormick | 203 |
| Legume Viruses | L Bos | 212 |
| Leishmaniaviruses | R Carrion Jr, Y-T Ro and J L Patterson | 220 |
| Leporipoviruses and Suipoxviruses | G McFadden | 225 |
| Luteoviruses | L L Domier and C J D'Arcy | 231 |
| Lymphocytic Choriomeningitis Virus: General Features | R M Welsh | 238 |
| Lymphocytic Choriomeningitis Virus: Molecular Biology | J C de la Torre | 243 |
| Lysis of the Host by Bacteriophage | R F Young III and R L White | 248 |

# M

| | | |
|---|---|---|
| *Machlomovirus* | K Scheets | 259 |
| Maize Streak Virus | D P Martin, D N Shepherd and E P Rybicki | 263 |
| Marburg Virus | D Falzarano and H Feldmann | 272 |
| Marnaviruses | A S Lang and C A Suttle | 280 |
| Measles Virus | R Cattaneo and M McChesney | 285 |
| Membrane Fusion | A Hinz and W Weissenhorn | 292 |
| Metaviruses | H L Levin | 301 |
| *Mimivirus* | J-M Claverie | 311 |
| Molluscum Contagiosum Virus | J J Bugert | 319 |
| *Mononegavirales* | A J Easton and R Ling | 324 |
| Mouse Mammary Tumor Virus | J P Dudley | 334 |
| Mousepox and Rabbitpox Viruses | M Regner, F Fenner and A Müllbacher | 342 |
| Movement of Viruses in Plants | P A Harries and R S Nelson | 348 |

| | | |
|---|---|---|
| Mumps Virus | B K Rima and W P Duprex | 356 |
| Mungbean Yellow Mosaic Viruses | V G Malathi and P John | 364 |
| Murine Gammaherpesvirus 68 | A A Nash and B M Dutia | 372 |
| Mycoreoviruses | B I Hillman | 378 |

# N

| | | |
|---|---|---|
| Nanoviruses | H J Vetten | 385 |
| Narnaviruses | R Esteban and T Fujimura | 392 |
| Nature of Viruses | M H V Van Regenmortel | 398 |
| *Necrovirus* | L Rubino and G P Martelli | 403 |
| *Nepovirus* | H Sanfaçon | 405 |
| Neutralization of Infectivity | P J Klasse | 413 |
| *Nidovirales* | L Enjuanes, A E Gorbalenya, R J de Groot, J A Cowley, J Ziebuhr and E J Snijder | 419 |
| Nodaviruses | P A Venter and A Schneemann | 430 |
| Noroviruses and Sapoviruses | K Y Green | 438 |

# O

| | | |
|---|---|---|
| *Ophiovirus* | A M Vaira and R G Milne | 447 |
| Orbiviruses | P P C Mertens, H Attoui and P S Mellor | 454 |
| Organ Transplantation, Risks | C N Kotton, M J Kuehnert and J A Fishman | 466 |
| Origin of Viruses | P Forterre | 472 |
| Orthobunyaviruses | C H Calisher | 479 |
| Orthomyxoviruses: Molecular Biology | M L Shaw and P Palese | 483 |
| Orthomyxoviruses: Structure of Antigens | R J Russell | 489 |
| Oryctes Rhinoceros Virus | J M Vlak, A M Huger, J A Jehle and R G Kleespies | 495 |
| *Ourmiavirus* | G P Accotto and R G Milne | 500 |

# VOLUME 4

# P

| | | |
|---|---|---|
| Papaya Ringspot Virus | D Gonsalves, J Y Suzuki, S Tripathi and S A Ferreira | 1 |
| Papillomaviruses: General Features of Human Viruses | G Orth | 8 |
| Papillomaviruses: Molecular Biology of Human Viruses | P F Lambert and A Collins | 18 |
| Papillomaviruses of Animals | A A McBride | 26 |
| Papillomaviruses: General Features | H U Bernard | 34 |
| Paramyxoviruses of Animals | S K Samal | 40 |
| Parainfluenza Viruses of Humans | E Adderson and A Portner | 47 |
| Paramyxoviruses | R E Dutch | 52 |
| Parapoxviruses | D Haig and A A Mercer | 57 |

| | |
|---|---|
| Partitiviruses of Fungi    *S Tavantzis* | 63 |
| Partitiviruses: General Features    *S A Ghabrial, W F Ochoa, T S Baker and M L Nibert* | 68 |
| Parvoviruses of Arthropods    *M Bergoin and P Tijssen* | 76 |
| Parvoviruses of Vertebrates    *C R Parrish* | 85 |
| Parvoviruses: General Features    *P Tattersall* | 90 |
| *Pecluvirus*    *D V R Reddy, C Bragard, P Sreenivasulu and P Delfosse* | 97 |
| Pepino Mosaic Virus    *R A A Van der Vlugt and C C M M Stijger* | 103 |
| Persistent and Latent Viral Infection    *E S Mocarski and A Grakoui* | 108 |
| Phycodnaviruses    *J L Van Etten and M V Graves* | 116 |
| Phylogeny of Viruses    *A E Gorbalenya* | 125 |
| Picornaviruses: Molecular Biology    *B L Semler and K J Ertel* | 129 |
| Plant Antiviral Defense: Gene Silencing Pathway    *G Szittya, T Dalmay and J Burgyan* | 141 |
| Plant Reoviruses    *R J Geijskes and R M Harding* | 149 |
| Plant Resistance to Viruses: Engineered Resistance    *M Fuchs* | 156 |
| Plant Resistance to Viruses: Geminiviruses    *J K Brown* | 164 |
| Plant Resistance to Viruses: Natural Resistance Associated with Dominant Genes      *P Moffett and D F Klessig* | 170 |
| Plant Resistance to Viruses: Natural Resistance Associated with Recessive Genes      *C Caranta and C Dogimont* | 177 |
| Plant Rhabdoviruses    *A O Jackson, R G Dietzgen, R-X Fang, M M Goodin, S A Hogenhout, M Deng and J N Bragg* | 187 |
| Plant Virus Diseases: Economic Aspects    *G Loebenstein* | 197 |
| Plant Virus Diseases: Fruit Trees and Grapevine    *G P Martelli and J K Uyemoto* | 201 |
| Plant Virus Diseases: Ornamental Plants    *J Engelmann and J Hamacher* | 207 |
| Plant Virus Vectors (Gene Expression Systems)    *Y Gleba, S Marillonnet and V Klimyuk* | 229 |
| Plum Pox Virus    *M Glasa and T Candresse* | 238 |
| Poliomyelitis    *P D Minor* | 243 |
| Polydnaviruses: Abrogation of Invertebrate Immune Systems    *M R Strand* | 250 |
| Polydnaviruses: General Features    *A Fath-Goodin and B A Webb* | 257 |
| Polyomaviruses of Humans    *M Safak and K Khalili* | 262 |
| Polyomaviruses of Mice    *B Schaffhausen* | 271 |
| Polyomaviruses    *M Gravell and E O Major* | 277 |
| *Pomovirus*    *L Torrance* | 283 |
| Potato Virus Y    *C Kerlan and B Moury* | 288 |
| Potato Viruses    *C Kerlan* | 302 |
| *Potexvirus*    *K H Ryu and J S Hong* | 310 |
| Potyviruses    *J J López-Moya and J A García* | 314 |
| Poxviruses    *G L Smith, P Beard and M A Skinner* | 325 |
| Prions of Vertebrates    *J D F Wadsworth and J Collinge* | 331 |
| Prions of Yeast and Fungi    *R B Wickner, H Edskes, T Nakayashiki, F Shewmaker, L McCann, A Engel and D Kryndushkin* | 338 |

| | | |
|---|---|---|
| Pseudorabies Virus | T C Mettenleiter | 342 |
| Pseudoviruses | D F Voytas | 352 |

# Q

| | | |
|---|---|---|
| Quasispecies | R Sanjuán | 359 |

# R

| | | |
|---|---|---|
| Rabies Virus | I V Kuzmin and C E Rupprecht | 367 |
| Recombination | J J Bujarski | 374 |
| Reoviruses: General Features | P Clarke and K L Tyler | 382 |
| Reoviruses: Molecular Biology | K M Coombs | 390 |
| Replication of Bacterial Viruses | M Salas and M de Vega | 399 |
| Replication of Viruses | A J Cann | 406 |
| Reticuloendotheliosis Viruses | A S Liss and H R Bose Jr. | 412 |
| Retrotransposons of Fungi | T J D Goodwin, M I Butler and R T M Poulter | 419 |
| Retrotransposons of Plants | M-A Grandbastien | 428 |
| Retrotransposons of Vertebrates | A E Peaston | 436 |
| Retroviral Oncogenes | P K Vogt and A G Bader | 445 |
| Retroviruses of Insects | G F Rohrmann | 451 |
| Retroviruses of Birds | K L Beemon | 455 |
| Retroviruses: General Features | E Hunter | 459 |
| Rhinoviruses | N W Bartlett and S L Johnston | 467 |
| Ribozymes | E Westhof and A Lescoute | 475 |
| Rice Tungro Disease | R Hull | 481 |
| Rice Yellow Mottle Virus | E Hébrard and D Fargette | 485 |
| Rift Valley Fever and Other Phleboviruses | L Nicoletti and M G Ciufolini | 490 |
| Rinderpest and Distemper Viruses | T Barrett | 497 |
| Rotaviruses | J Angel, M A Franco and H B Greenberg | 507 |
| Rubella Virus | T K Frey | 514 |

# S

| | | |
|---|---|---|
| *Sadwavirus* | T Iwanami | 523 |
| Satellite Nucleic Acids and Viruses | P Palukaitis, A Rezaian and F García-Arenal | 526 |
| Seadornaviruses | H Attoui and P P C Mertens | 535 |
| Sequiviruses | I-R Choi | 546 |
| Severe Acute Respiratory Syndrome (SARS) | J S M Peiris and L L M Poon | 552 |
| Shellfish Viruses | T Renault | 560 |
| Shrimp Viruses | J-R Bonami | 567 |

| | | |
|---|---|---|
| Sigma Rhabdoviruses | *D Contamine and S Gaumer* | 576 |
| Simian Alphaherpesviruses | *J Hilliard* | 581 |
| Simian Gammaherpesviruses | *A Ensser* | 585 |
| Simian Immunodeficiency Virus: Animal Models of Disease | *C J Miller and M Marthas* | 594 |
| Simian Immunodeficiency Virus: General Features | *M E Laird and R C Desrosiers* | 603 |
| Simian Immunodeficiency Virus: Natural Infection | *I Pandrea, G Silvestri and C Apetrei* | 611 |
| Simian Retrovirus D | *P A Marx* | 623 |
| Simian Virus 40 | *A L McNees and J S Butel* | 630 |
| Smallpox and Monkeypox Viruses | *S Parker, D A Schultz, H Meyer and R M Buller* | 639 |
| *Sobemovirus* | *M Meier, A Olspert, C Sarmiento and E Truve* | 644 |
| St. Louis Encephalitis | *W K Reisen* | 652 |
| Sweetpotato Viruses | *J Kreuze and S Fuentes* | 659 |

# VOLUME 5

## T

| | | |
|---|---|---|
| Taura Syndrome Virus | *A K Dhar and F C T Allnutt* | 1 |
| Taxonomy, Classification and Nomenclature of Viruses | *C M Fauquet* | 9 |
| *Tenuivirus* | *B C Ramirez* | 24 |
| Tetraviruses | *J A Speir and J E Johnson* | 27 |
| Theiler's Virus | *H L Lipton, S Hertzler and N Knowles* | 37 |
| Tick-Borne Encephalitis Viruses | *T S Gritsun and E A Gould* | 45 |
| Tobacco Mosaic Virus | *M H V Van Regenmortel* | 54 |
| Tobacco Viruses | *S A Tolin* | 60 |
| *Tobamovirus* | *D J Lewandowski* | 68 |
| *Tobravirus* | *S A MacFarlane* | 72 |
| Togaviruses Causing Encephalitis | *S Paessler and M Pfeffer* | 76 |
| Togaviruses Causing Rash and Fever | *D W Smith, J S Mackenzie and M D A Lindsay* | 83 |
| Togaviruses Not Associated with Human Disease | *L L Coffey,* | 91 |
| Togaviruses: Alphaviruses | *A M Powers* | 96 |
| Togaviruses: Equine Encephalitic Viruses | *D E Griffin* | 101 |
| Togaviruses: General Features | *S C Weaver, W B Klimstra and K D Ryman* | 107 |
| Togaviruses: Molecular Biology | *K D Ryman, W B Klimstra and S C Weaver* | 116 |
| Tomato Leaf Curl Viruses from India | *S Chakraborty* | 124 |
| Tomato Spotted Wilt Virus | *H R Pappu* | 133 |
| Tomato Yellow Leaf Curl Virus | *H Czosnek* | 138 |
| Tombusviruses | *S A Lommel and T L Sit* | 145 |
| *Torovirus* | *A E Hoet and M C Horzinek* | 151 |
| *Tospovirus* | *M Tsompana and J W Moyer* | 157 |
| Totiviruses | *S A Ghabrial* | 163 |

| Transcriptional Regulation in Bacteriophage | R A Weisberg, D M Hinton and S Adhya | 174 |
| Transmissible Spongiform Encephalopathies | E D Belay and L B Schonberger | 186 |
| Tumor Viruses: Human | R Grassmann, B Fleckenstein and H Pfister | 193 |
| Tymoviruses | A-L Haenni and T W Dreher | 199 |

## U

| *Umbravirus* | M Taliansky and E Ryabov | 209 |
| Ustilago Maydis Viruses | J Bruenn | 214 |

## V

| Vaccine Production in Plants | E P Rybicki | 221 |
| Vaccine Safety | C J Clements and G Lawrence | 226 |
| Vaccine Strategies | I Kusters and J W Almond | 235 |
| Vaccinia Virus | G L Smith | 243 |
| Varicella-Zoster Virus: General Features | J I Cohen | 250 |
| Varicella-Zoster Virus: Molecular Biology | W T Ruyechan and J Hay | 256 |
| *Varicosavirus* | T Sasaya | 263 |
| Vector Transmission of Animal Viruses | W K Reisen | 268 |
| Vector Transmission of Plant Viruses | S Blanc | 274 |
| Vegetable Viruses | P Caciagli | 282 |
| Vesicular Stomatitis Virus | S P J Whelan | 291 |
| Viral Killer Toxins | M J Schmitt | 299 |
| Viral Membranes | J Lenard | 308 |
| Viral Pathogenesis | N Nathanson | 314 |
| Viral Receptors | D J Evans | 319 |
| Viral Suppressors of Gene Silencing | J Verchot-Lubicz and J P Carr | 325 |
| Viroids | R Flores and R A Owens | 332 |
| Virus Classification by Pairwise Sequence Comparison (PASC) | Y Bao, Y Kapustin and T Tatusova | 342 |
| Virus Databases | E J Lefkowitz, M R Odom and C Upton | 348 |
| Virus Entry to Bacterial Cells | M M Poranen and A Domanska | 365 |
| Virus Evolution: Bacterial Viruses | R W Hendrix | 370 |
| Virus-Induced Gene Silencing (VIGS) | M S Padmanabhan and S P Dinesh-Kumar | 375 |
| Virus Particle Structure: Nonenveloped Viruses | J A Speir and J E Johnson | 380 |
| Virus Particle Structure: Principles | J E Johnson and J A Speir | 393 |
| Virus Species | M H V van Regenmortel | 401 |
| Viruses and Bioterrorism | R F Meyer and S A Morse | 406 |
| Viruses Infecting Euryarchaea | K Porter, B E Russ, A N Thorburn and M L Dyall-Smith | 411 |
| Visna-Maedi Viruses | B A Blacklaws | 423 |

## W

Watermelon Mosaic Virus and Zucchini Yellow Mosaic Virus    *H Lecoq and C Desbiez*    433

West Nile Virus    *L D Kramer*    440

White Spot Syndrome Virus    *J-H Leu, J-M Tsai and C-F Lo*    450

## Y

Yatapoxviruses    *J W Barrett and G McFadden*    461

Yeast L-A Virus    *R B Wickner, T Fujimura and R Esteban*    465

Yellow Fever Virus    *A A Marfin and T P Monath*    469

Yellow Head Virus    *P J Walker and N Sittidilokratna*    476

## Z

Zoonoses    *J E Osorio and T M Yuill*    485

Taxonomic Index    497

Subject Index    499

# Adenoviruses: General Features

**B Harrach**, Veterinary Medical Research Institute, Budapest, Hungary

© 2008 Elsevier Ltd. All rights reserved.

## Introduction

Adenoviruses are middle-sized, nonenveloped, icosahedral, double-stranded DNA viruses of animals. The prefix adeno comes from the Greek word ἀδήν (gland), reflecting the first isolation of a virus of this type from human adenoid tissue half a century ago. Adenoviruses have since been isolated from a large variety of hosts, including representatives of every major vertebrate class from fish to mammals. Using polymerase chain reaction (PCR) technology, a large variety of putative novel adenoviruses have been detected, but isolation of such viruses and *in vitro* propagation is hampered in most cases by the lack of appropriate permissive cell cultures. Some human and animal adenoviruses can cause diseases or even death, but most are not pathogenic in non-immuno-compromised, healthy individuals. Adenoviruses have been used as model organisms in molecular biology, and important findings of general relevance have emerged from such studies, including splicing in eukaryotes. Adenoviruses have become one of the most popular vector systems for virus-based gene therapy and vaccination and have potential as antitumor tools. Wide prevalence in diverse host species and a substantially conserved genome organization make adenoviruses an ideal model for studying virus evolution.

## Taxonomy

Adenoviruses belong to the family *Adenoviridae*. No higher taxonomical level has yet been established, despite the fact that certain bacteriophages (*Tectiviridae*), the green alga virus *Paramecium bursaria Chlorella virus 1* (*Phycodnaviridae*), and a virus of Archaea living in hot springs (sulfolobus turreted icosahedral virus) seem to have common evolutionary roots with adenoviruses.

There are four official genera in the family. Two genera (*Mastadenovirus* and *Aviadenovirus*) comprise adenoviruses that have probably co-evolved with mammals and birds, respectively. The other two genera (*Atadenovirus* and *Siadenovirus*) have a broader range of hosts. Atadenoviruses were named after a bias toward high A+T content in the genomes of the initial representatives, which infect various ruminant and avian hosts, as well as a marsupial. Every known reptilian adenovirus also belongs to the atadenoviruses, although their genomes do not show the same bias toward high A+T content. The very few known siadenoviruses were isolated from or detected by PCR in birds and a frog. This genus was named in recognition of the presence of a gene encoding a potential sialidase in the viruses concerned. The single confirmed fish adenovirus falls into a separate group that may eventually found a fifth genus; adenovirus-like particles have been described in additional fish species.

Within each genus, the viruses are grouped into species, which are named according to the host first described and supplemented with letters of the alphabet (**Table 1**). Host origin is only one of several criteria that are used to demarcate the species. Phylogenetic distance is the most significant criterion, with species defined as separated by more than 5–10% amino acid sequence divergence of hexon and DNA polymerase (pol), respectively. Further important characteristics come into play, especially if DNA sequence data are not available: DNA hybridization, restriction fragment length polymorphism, nucleotide composition, oncogenicity in rodents, growth characteristics, host range, cross-neutralization, ability to recombine, number of virus-associated (VA) RNA genes, hemagglutination properties, and organization of the genome. However, all of these ancillary data are expected to accord with the results of phylogenetic calculations. Thus, for example, chimpanzee adenoviruses are classified into human adenovirus species. Adenoviruses of humans have been studied far more intensively than those of other animals, and the six species (*Human adenovirus A* to *Human adenovirus F*; abbreviated informally to HAdV-A to HAdV-F) correspond to substantial 'groups' or 'subgenera' defined previously. Each human adenovirus serotype is abbreviated to HAdV hyphenated to a number.

**Table 1** The taxonomy of family Adenoviridae[a]

| Genus/species | Serotype | Strain | Genome |
|---|---|---|---|
| *Mastadenovirus* | | | |
| Bovine adenovirus A | BAdV-1 | | G |
| Bovine adenovirus B | BAdV-3 | | G |
| Bovine adenovirus C | BAdV-10 | | |
| Canine adenovirus | CAdV-1, 2 | | G |
| Equine adenovirus A | EAdV-1 | | |
| Equine adenovirus B | EAdV-2 | | |
| Human adenovirus A | HAdV-12, 18, 31 | | 12 |
| Human adenovirus B | HAdV-3, 7, 11, 14, 16, 21, 34, 35, 50 | | G |
| | Simian adenovirus 21 (SAdV-21) | | G |
| Human adenovirus C | BAdV-9, HAdV-1, 2, 5, 6 | | 1, 2, 5 |
| Human adenovirus D | HAdV-8–10, 13, 15, 17, 19, 20, 22–30, 32, 33, 36–39, 42–49, 51 | | 9, 17 26, 46, 48, 49 |
| Human adenovirus E | HAdV-4, SAdV-22–25 | | G |
| Human adenovirus F | HAdV-40, 41 | | G |
| Murine adenovirus A | MAdV-1 | | G |
| Ovine adenovirus A | BAdV-2, OAdV-2–5 | | BAdV-2 |
| Ovine adenovirus B | OAdV-1 | | |
| Porcine adenovirus A | PAdV-1, 2, 3 | | 3 |
| Porcine adenovirus B | PAdV-4 | | |
| Porcine adenovirus C | PAdV-5 | | G |
| Tree shrew adenovirus | TSAdV-1 | | G |
| Goat adenovirus (ts) | GAdV-2 | | |
| Guinea pig adenovirus (ts) | GPAdV-1 | | |
| Murine adenovirus B (ts) | MAdV-2 | | |
| Ovine adenovirus C (ts) | OAdV-6 | | |
| Simian adenovirus A (ts) | SAdV-3 | | G |
| Squirrel adenovirus (ts) | SqAdV-1 | | |
| ? | HAdV-52 | | G |
| ? | SAdV-1–2, 4–20 | | 1, 6, 7, 20 |
| *Aviadenovirus* | | | |
| Fowl adenovirus A | FAdV-1 | CELO | G |
| Fowl adenovirus B | FAdV-5 | 340 | |
| Fowl adenovirus C | FAdV-4, 10 | KR95, CFA20 | |
| Fowl adenovirus D | FAdV-2, 3, 9, 11 | P7-A, 75, A2-A, 380 | 9 |
| Fowl adenovirus E | FAdV-6, 7, 8a, 8b | CR119, YR36, TR59, 764 | |
| Goose adenovirus | GoAdV-1–3 | | |
| Duck adenovirus B (ts) | DAdV-2 | | |
| Pigeon adenovirus (ts) | PiAdV-1 | | |
| Turkey adenovirus B (ts) | TAdV-1, 2 | | |
| Psittacine adenovirus? | Psittacine adenovirus 1 | | |
| Falcon adenovirus? | Falcon adenovirus 1 | | |
| *Atadenovirus* | | | |
| Bovine adenovirus D | BAdV-4, 5, 8, strain Rus | | 4 |
| Duck adenovirus A | DAdV-1 | | G |
| Ovine adenovirus D | GAdV-1, OAdV-7 | | 7 |
| Possum adenovirus | PoAdV-1 | | |
| Bearded dragon adenovirus (ts) | BDAdV-1 | | |
| Bovine adenovirus E (ts) | BAdV-6 | | |
| Bovine adenovirus F (ts) | BAdV-7 | | |
| Cervine adenovirus (ts) | Odocoileus adenovirus 1 (OdAdV-1) | | |
| Chameleon adenovirus (ts) | ChAdV-1 | | |
| Gecko adenovirus (ts) | GeAdV-1 | Fat-tailed gecko | |
| Snake adenovirus (ts) | SnAdV-1 | Corn snake, python | G |
| Gekkonid adenovirus(?) | Tokay gecko adenovirus | | |
| Helodermatid adenovirus(?) | Gila monster adenovirus | | |
| Scincid adenovirus(?) | Blue-tongued skink adenovirus | | |
| Genus *Siadenovirus* | | | |
| Frog adenovirus | FrAdV-1 | | G |
| Turkey adenovirus A | TAdV-3 | | G |
| ? | Raptor adenovirus 1 | Harris hawk | |

*Continued*

**Table 1** Continued

| Genus/species | Serotype | Strain | Genome |
|---|---|---|---|
| Unassigned Viruses in the Family | | | |
| ? | White sturgeon adenovirus 1 (WSAdV-1) | | |
| ? | Crocodile adenovirus | | |

<sup>a</sup>Official genus and species names as published in the Eighth Report of the ICTV are in italics, and tentative species (ts), proposed species (marked by a query) or single isolates are not. Because of confusion in the serotype numbering in some cases (e.g. among fowl adenoviruses), certain characteristic strains are shown for easier identification. Available full genome sequences are noted by G or by the number of the sequenced serotype(s) if those listed are not all available.

To illustrate the need to proceed carefully in developing adenovirus taxonomy, the case of the newest adenovirus isolated from human samples (HAdV-52) is illuminating. This virus seems to be sufficiently different from other human adenoviruses to merit the erection of a new species. However, it is very similar to some previously characterized Old World monkey adenoviruses (simian adenoviruses 1 and 7 (SAdV-1, SAdV-7) plus others). One taxonomical proposal would be to establish a new species, *Human adenovirus G*, containing HAdV-52 and the related monkey adenoviruses. Clearly, this would depend on epidemiological data demonstrating that HAdV-52 is properly a human virus and not an occasional, opportunistic transfer from monkeys. For similar reasons and others, classification of many nonhuman adenoviruses into species is not yet resolved.

## Virion Morphology and Properties

The icosahedral capsid is 70–90 nm in diameter and consists of 240 nonvertex capsomers (called hexons), each 8–10 nm in diameter, and 12 vertex capsomers (pentons), each with a protruding fiber 9–77.5 nm in length. The members of genus *Aviadenovirus* that have been studied have two fiber proteins per vertex. Fowl adenovirus 1 (FAdV-1) even has two, tandem fiber genes of different lengths, resulting in two fibers of different sizes at each vertex. Members of species *Human adenovirus F* (and HAdV-52 and the related monkey viruses) also have two fiber genes of different lengths, but the fibers are distributed in single copies alternately on the vertices. The main capsomers (hexons) are formed by the interaction of three identical polypeptides (designated hexon, and also as polypeptide II, after a Roman numeral system based on the relative mobilities of structural proteins under reducing conditions in sodium dodecyl sulfate-polyacrylamide gel electrophoresis). Each hexon has two characteristic parts: a triangular top with three 'towers', and a pseudohexagonal base with a central cavity. Hexons, or more exactly their bases, are packed tightly to form a protein shell that protects the inner components of the virion.

In members of the genus *Mastadenovirus* 12 copies of polypeptide IX are found between the nine hexons in the center of each facet. However, polypeptide IX is not present in the members of any other genus. Two monomers of polypeptide IIIa penetrate the hexon shell at the edge of each facet, and multiple copies of polypeptide VI form a ring underneath the peripentonal hexons. Penton bases are formed at the vertices by the interaction of five copies of polypeptide III, and are tightly associated with one (or, in the aviadenoviruses, two) fibers, each consisting of three copies of polypeptide IV in the form of a shaft of characteristic length with a distal knob. Polypeptide VIII is situated at the inner surface of the hexon shell. Polypeptides VI and VIII appear to link the capsid to the virion core, which consists of a single copy of the DNA genome complexed with four polypeptides (V, VII, X, TP). Polypeptide V exists only in mastadenoviruses.

Adenoviruses are stable on storage in the frozen state. They are stable to mild acid and insensitive to lipid solvents. Heat sensitivity varies among the genera.

## Nucleic Acid, Genome Organization, and Replication

The adenovirus genome is a linear molecule of double-stranded DNA (26 163–45 063 bp) containing an inverted terminal repeat (ITR) of 36–368 bp at its termini, with the 5′ ends of the genome linked covalently to a terminal protein (TP). The nucleotide composition is 33.7–63.8% G+C. The genetic organization of the central part of the genome is conserved throughout the family, whereas the terminal parts show large variations in length and gene content (**Figure 1**). Splicing was first discovered in adenoviruses, and is a common means of expressing mRNAs in this virus family. In the conserved region, most late genes are expressed by splicing from the rightward-oriented major late promoter located in the pol gene. The early genes encoding pol, the precursor of TP (pTP), and the DNA-binding protein (DBP) are spliced from leftward-oriented promoters. Where it has been examined, splicing is also a common feature of genes in the nonconserved regions.

Replication of various human adenoviruses has been studied in detail, in particular with HAdV-2. Virus entry takes place via interactions of the fiber knob with specific receptors on the surface of a susceptible cell followed by internalization via interactions between the penton base

4    Adenoviruses: General Features

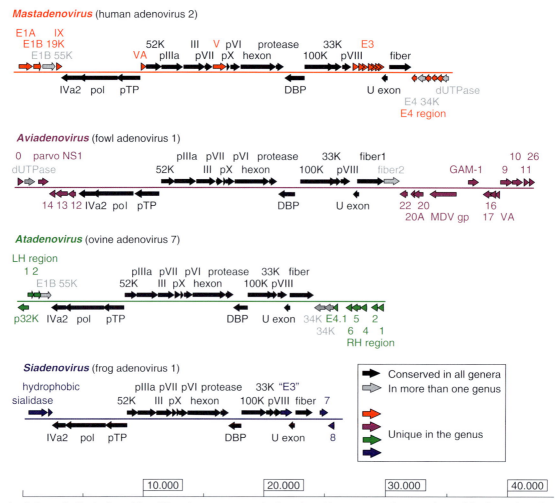

**Figure 1** Schematic illustration of the different genome organizations found in representative members of the four adenovirus genera. Black arrows depict genes conserved in every genus, gray arrows show genes present in more than one genus, and colored arrows show genus-specific genes. Reprinted from Benkö M, Harrach B, Both GW, et al. (2005) Family *Adenoviridae* In: Fauquet CM, Mayo MA, Maniloff J, Desselberger U, and Ball LA (eds.) *Virus Taxonomy: Eighth Report of the International Committee on Taxonomy of Viruses*, pp. 213–228. San Diego, CA: Elsevier Academic Press, with permission from Elsevier.

and cellular $\alpha_v$ integrins. After uncoating, the virus core is delivered to the nucleus, which is the site of virus transcription, DNA replication, and assembly. Virus infection mediates the shutdown of host DNA synthesis and later RNA and protein synthesis. Transcription of the adenovirus genome by host RNA polymerase II involves both DNA strands of the genome and initiates (in HAdV-2) from five early (E1A, E1B, E2, E3, and E4), two intermediate, and the major late (L) promoter. All primary transcripts are capped and polyadenylated, with complex splicing patterns producing families of mRNAs. In primate adenoviruses, one or two VA RNA genes are usually present upstream from the main pTP coding region. These are transcribed by cellular RNA polymerase III and facilitate translation of late mRNAs and blocking of the cellular interferon response. Corresponding VA RNA genes have not been identified in nonprimate adenoviruses, although a nonhomologous VA RNA gene has been mapped in some aviadenoviruses near the right end of the genome. More generally, the replication of aviadenoviruses has been shown to involve significantly different pathways from those characterized in human adenoviruses. This is not unexpected, given the considerable differences in gene layout between nonconserved regions of the genome.

## Proteins

About 40 different polypeptides (the largest number being in fowl adenoviruses and the smallest in siadenoviruses) are produced. Almost a third of these compose the virion, including a virus-encoded cysteine protease, which processes a number of precursor proteins (these are prefixed with p). With the exception of polypeptides V and IX, the other structural proteins are conserved in every genus. Products of the four early regions of mastadenoviruses modulate the host cell's transcriptional machinery (E1 and E4), assemble the virus DNA replication complex

(E2), and subvert host defense mechanisms (E3). The E2 region (encoding DBP, pol, and pTP) is well conserved throughout the family, while the E3 and E4 regions show great variation in length and gene content among the mastadenoviruses. E1A, E1B 19K, and the E3 and E4 regions (with the exception of 34K) occur only in mastadenoviruses. Genes encoding proteins related to 34K are also present in atadenoviruses, sometimes in duplicate. Intermediate (IX, only in mastadenoviruses, and IVa2) and late gene products (in some mastadenoviruses expressed from five transcription units, L1–L5) are concerned with assembly and maturation of the virion. Late proteins include: 52K (scaffolding protein) and pIIIa (L1); III (penton base), pVII (major core protein), V (minor core protein, only in mastadenoviruses) and pX (L2 ); pVI, hexon (II) and protease (L3); 100K, 33K (and an unspliced version, 22K) and pVIII (L4); and fiber (IV) (L5). Seemingly, there are no lipids in adenovirus particles. However, the fiber proteins and some of the nonstructural proteins are glycosylated.

## Antigenic Properties

Adenovirus serotypes are differentiated on the basis of neutralization assays. A serotype is defined as either exhibiting no cross-reaction with others or showing a homologous/heterologous titer ratio of greater than 16 (in both directions). For homologous/heterologous titer ratios of 8 or 16, a serotype assignment is made either if the viral hemagglutinins are unrelated (as shown by lack of cross-reaction in hemagglutination-inhibition tests), or if substantial biophysical, biochemical, or phylogenetic differences exist. Antigens at the surface of the virion are mainly type specific. Hexons are involved in neutralization, and fibers in neutralization and hemagglutination inhibition. Soluble antigens associated with virus infection include surplus capsid proteins that have not been assembled. The genus-specific antigen is located on the basal surface of the hexon, whereas serotype-specific antigens are located mainly on the tower region. Practical problems may arise during serotyping (and in phylogenetic calculations) from the occasional occurrence of homologous recombination in the hexon gene (e.g., HAdV-16 and SAdV-23). Under natural circumstances, recombination occurs only between members of the same species.

## Biological Properties

The natural host range of adenovirus types is usually confined to one animal species or to closely related species. This also applies for cell cultures. Some human adenoviruses (mainly in species *Human adenovirus C*) cause productive infection in various animal (rodent or ruminant) cells. Several viruses cause tumors in newborn rodents. Subclinical infections are frequent in various virus–host systems. Transmission occurs from the throat, feces, the eye, or urine, depending on the virus serotype. Certain human adenovirus types (given in parentheses below) are predominantly associated with a specific pathology, such as adenoidal–pharyngeal conjunctivitis (3, 4, 7, and 14), acute respiratory outbreaks (4, 7, 14, and 21), epidemic kerato-conjunctivitis (8, 19, and 37), or venereal disease (3). HAdV-40 and HAdV-41 can be isolated from the feces of young children with acute gastroenteritis, and only the rotaviruses are known to cause more cases of infantile viral diarrhea. HAdV-11 is associated with hemorrhagic cystitis (mostly in immuno-suppressed patients after organ transplantation). The newer types of human adenovirus (42–51) were isolated from acquired immune deficiency syndrome patients. In mammals, mastadenovirus infection is common, but disease is usually manifested only if predisposing factors (such as management problems, crowding, shipping, or concurrent bacterial infections) are also present. Canine adenovirus, which can cause hepatitis or respiratory disease in dogs, seems to be an exception and has caused epizootics in foxes, bears, wolves, coyotes, and skunks.

Certain siadenoviruses and atadenoviruses that may have undergone host switches during their evolution seem to be more pathogenic in general in their new hosts: birds, ruminants, and a marsupial. Egg drop syndrome virus (duck adenovirus 1, DAdV-1) or turkey hemorrhagic enteritis virus (turkey adenovirus 3, TAdV-3) can cause serious economical losses for the poultry industry. An atadenovirus (Odocoileus adenovirus) caused a hemorrhagic epizooty and killed thousands of mule deers in California.

Adenoviruses infecting susceptible cells cause similar cytopathology consisting of early rounding of cells, aggregation or lysis of chromatin, followed by the later appearance of characteristic eosinophilic or basophilic nuclear inclusions.

HAdV-5 has been engineered and used extensively as a gene delivery vector. Other human adenoviruses such as HAdV-35, or even nonhuman serotypes (ovine adenovirus 7 and canine adenovirus 1), are being tested as a means of overcoming the problem posed by preexisting neutralizing antibodies in the human population, and also to achieve targeting to specific organs and tissues. Bovine, porcine, canine, and fowl adenovirus types have also been tested as novel antigen delivery vectors for immunizing the cognate animal species.

## Mastadenoviruses

Mastadenoviruses occur only in mammals, and in general, mammals are host to only mastadenoviruses. However, half of the adenoviruses found in ruminant species are atadenoviruses and the only marsupial adenovirus identified so far is also an atadenovirus. Mastadenoviruses were originally distinguished from aviadenoviruses on the basis of different genus-specific complement-fixing antigens.

Virus infectivity is inactivated by heating at 56 °C for more than 10 min.

The mastadenovirus genomes that have been sequenced range in size between 30 288 bp and 37 741 bp, and in nucleotide composition from 40.8% to 63.8% G+C. The ITR is considerably longer (93–368 bp) and more complex (containing a variety of cellular factor-binding sites) than in members of the other genera. Proteins encoded only by the mastadenoviruses are polypeptides V and IX, and most of those from the E1A, E1B, E3, and E4 regions. Polypeptide IX, besides cementing the hexons on the outer surface of the capsid, has been demonstrated to act as a transcriptional activator, and also takes part in nuclear reorganization during infection. Polypeptide V is a core protein that, in association with cellular protein p32, seems to be involved in the transport of viral DNA into the nucleus of the infected cell. The E3 and E4 regions can differ markedly even among different mastadenovirus species. The E3 region is considerably shorter and simpler in nonprimate adenoviruses than in monkey, ape, and human adenoviruses. The simplest layout is found in murine adenovirus 1, where it consists of a single 12.5K gene. In the E4 region, only the 34K gene seems to be conserved in all mastadenoviruses. This gene is duplicated in bovine adenovirus 3 and porcine adenovirus 5.

The entry processes of human mastadenoviruses into the cell are well characterized. The coxsackievirus and adenovirus receptor (CAR) is the most common, but not the only, receptor for the attachment to the cell.

## Aviadenoviruses

Members of the genus *Aviadenovirus* occur only in birds and possess a common genus-specific complement-fixing antigen. However, birds can also harbor siadenoviruses and atadenoviruses. Aviadenovirus virions contain two fibers per vertex. Fowl adenovirus 1 (FAdV-1) has two fiber genes, and two projections of considerably different lengths from each penton base. Other fowl adenoviruses also have two fibers per vertex, but apparently only one fiber gene, and the projections are therefore of similar lengths. For attachment of FAdV-1 to the cell, the long, but not the short, fiber utilizes CAR. The aviadenovirus genomes are considerably larger (20–45%) than those of mastadenoviruses. FAdV-1 and FAdV-9, with genomes of 43 804 and 45 063 bp, respectively, represent the longest adenovirus DNA molecules known to date. The nucleotide composition of the partial or full genome sequences characterized so far is between 53.8% and 59% G+C. The length of the ITR is 51–72 bp.

The genes encoding polypeptides V and IX are absent from aviadenovirus genomes, as well as the genes in the mastadenovirus E1 and E3 regions (**Figure 1**). A dUTP pyrophosphatase (dUTPase) gene is situated next to the left end of the genome in the aviadenoviruses studied to date; a dUTPase gene is also present in some mastadenoviruses, but it is at the right genome end. The right end of aviadenovirus genomes contains a large number of transcription units that are confined to this genus. The majority of genes and proteins from this region have not yet been characterized in detail. The GAM-1 protein of FAdV-1 has been demonstrated to have an anti-apoptotic effect, and to activate the heat-shock response in the infected cell. GAM-1, in synergy with the protein encoded by ORF22, binds the retinoblastoma protein and can activate the E2F pathway. Some as yet uncharacterized aviadenovirus gene products are similar to proteins of other viruses, such as the nonstructural protein NS1 (Rep protein) of parvoviruses, or triacylglycerol lipase, a homolog of which occurs in an avian herpesvirus (Marek's disease virus).

Aviadenoviruses have been isolated from various poultry species including turkey, goose, and Muscovy duck, as well as a number of wild birds, such as falcon and parrots. However, chicken adenoviruses have been studied in most detail. Twelve fowl adenovirus serotypes have been classified into species *Fowl adenovirus A* to *Fowl adenovirus E* (FAdV-A to FAdV-E). Unfortunately, the type numbering of fowl adenoviruses became inconsistent when different systems were adopted in Europe and North America. The resulting confusion was partially resolved by the introduction of FAdV-8a and FAdV-8b types, though their distinctness could not be fully confirmed because of the ambiguous results of serum neutralization tests.

Avian adenoviruses have been associated with diverse disease patterns, including body hepatitis, bronchitis, pulmonary congestion, edema, and gizzard erosion in various bird species, but they generally seem to be less pathogenic than siadenoviruses and atadenoviruses in birds. An exception to this is FAdV-4, the causative agent of hydropericardium syndrome in chickens, which causes considerable losses mainly in Asia. FAdV-1 (chick embryo lethal orphan virus), FAdV-9, and FAdV-10 are studied extensively for their potential feasibility as gene delivery vectors.

## Atadenoviruses

As opposed to the mastadenoviruses and aviadenoviruses, which have clear host origins, the members of genus *Atadenovirus* represent a much broader host range spanning several vertebrate classes. This genus was originally established to cope with the classification of certain exceptional bovine and ovine adenoviruses with unusual characteristics. The name refers to the nucleotide composition of the first members of this genus, which is biased toward high A+T content. The large phylogenetic distance between the ruminant atadenoviruses and the mastadenoviruses and aviadenoviruses (**Figure 2**) inspired a hypothesis that they may have originated via tranfers of adenoviruses with lower vertebrate hosts. Indeed, recent studies

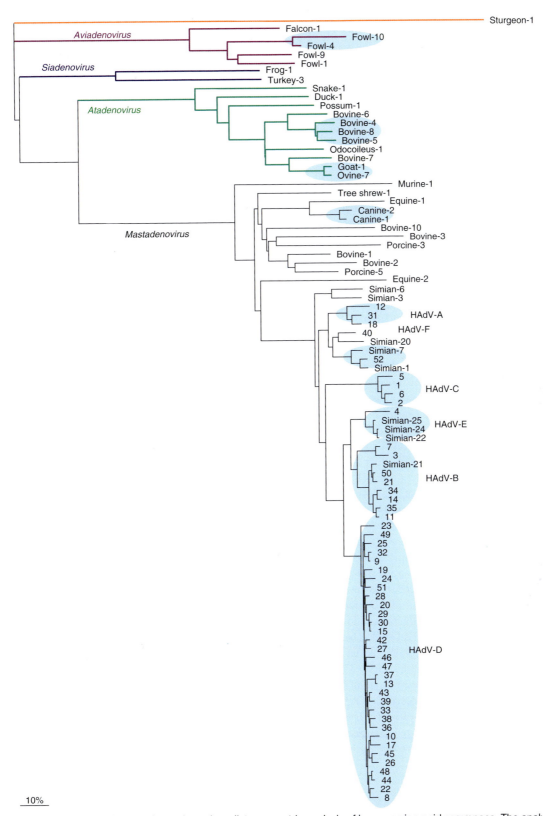

**Figure 2** Phylogenetic tree of adenoviruses based on distance matrix analysis of hexon amino acid sequences. The analysis was performed using the Protdist and Fitch programs of the PHYLIP 3.65 package. The unrooted tree was visualized with white sturgeon adenovirus as the outgroup. Adenoviruses are denoted by the name of the host and the serotype number, or only by the serotype number for human adenoviruses. Viruses that belong to the same species are grouped by light-blue ovals, and human adenovirus species are indicated by their informal abbreviations. Data for HAdV-16, HAdV-41, and SAdV-23 were excluded, as evolution of their hexon genes may have involved homologous recombination events. The bar indicates 10% difference between two neighboring sequences.

have confirmed that every reptilian adenovirus for which data are available belongs to this genus. However, atadenoviruses found in reptiles seem to have a balanced nucleotide content. Additional atadenovirus types have been detected in various species of birds and mammals, including duck, domestic and wild ruminants, as well as a marsupial, the brushtail possum. DAdV-1, which was originally isolated from flocks of laying hens that showed a sharp decrease in egg production, was moved to the genus *Atadenovirus* from the genus *Aviadenovirus*.

The virions of bovine atadenoviruses are more heat stable than those of mastadenoviruses. They retain substantial infectivity after treatment for 30 min at 56 °C.

The size of sequenced atadenovirus genomes ranges from 27 751 bp (snake adenovirus 1, SnAdV-1) to 33 213 bp (DAdV-1) with an ITR of 46 (OAdV-7) to 118 bp (SnAdV-1). For ruminant, marsupial, and avian atadenoviruses, the nucleotide composition ranges between 33.7% (OAdV-7) and 43% G+C (DAdV-1).

The common ancestry of the atadenoviruses is supported by their shared genomic organization. Putative early regions (LH and RH) on the genome ends occupy the places of the mastadenovirus E1 and E4 regions (**Figure 1**). The functions of the novel genes in these regions have been elucidated only partially. The LH region includes a gene encoding a novel structural protein (p32K), and genes LH1 (not present in DAdV-1), LH2, and LH3. LH3 seems to be weakly related to the E1B 55K gene of mastadenoviruses, and its product has also been shown to be a structural protein. In the RH region, E4.3 and E4.2 seem to be homologs of the E4 34K gene of mastadenoviruses. E4.1 is absent from DAdV-1. Closer to the right end, different numbers of RH genes are found in different viruses. Many genes seem to be the result of duplication events, with one family encoding putative F-box proteins. There is only one copy of an F-box gene in SnAdV−1, whereas five are present in ovine adenovirus 4. Surprisingly, SnAdV−1 also encodes a protein that is homologous to the gene 105R product found in tree shrew adenovirus, a mastadenovirus. The DAdV-1 genome has an extension at the right end containing seven additional genes, which are so far unique to this virus. These genes are likely to be host specific in their function. This region of the DAdV-1 genome also encodes a VA RNA gene that is seemingly homologous to that of FAdV-1. In atadenoviruses, no immuno-modulatory genes such as those in the mastadenovirus E3 region have yet been identified, though it is likely that they exist.

## Siadenoviruses

The fourth genus has been established for two adenoviruses that are of very distant host origin and yet share a similar genome organization and are phylogenetically closely related. These are frog adenovirus 1 (FrAdV-1), which was isolated from the renal tumor of a leopard frog, and TAdV-3, which was isolated from birds. Additional strains of TAdV-3, originating from turkey, pheasant and chicken, as well as avirulent (vaccine) strains, show little or no diversity in terms of nucleotide sequence or serology, respectively. The genomes of FrAdV-1 and TAdV-3 are collinear and represent the minimal gene content recognized among adenoviruses to date (**Figure 1**). In addition to the set of genes occupying the central genome region and conserved throughout the whole family, only five additional genes have been identified. The protein encoded next to the left ITR seems to be related to bacterial sialidases. Adjacent to this is a gene predicted to code for a highly hydrophobic protein. The E3 gene takes this name solely because of its position between the pVIII and fiber genes; it is not homologous to any of the mastadenovirus E3 genes, or to any other known genes. Similarly, ORF7 and ORF8, which are situated at the right end of the genome, have no recognizable homologs elsewhere. The genes encoding polypeptides V and IX are absent, as well as those in the mastadenovirus E1, E3 and E4 regions. FrAdV-1 and TAdV-3 have the shortest adenovirus genomes, 26 163 and 26 263 bp in size, respectively, with ITRs of 36 and 39 bp and nucleotide compositions of 38% and 35% G+C.

The evolutionary history of siadenoviruses awaits elucidation. In the same way that it is possible that atadenoviruses originated with reptiles (see above), the siadenoviruses may represent adenoviruses that have co-evolved with amphibians, with TAdV-3 the result of an interclass host switch.

FrAdV-1 is supposedly nonpathogenic, whereas TAdV-3 is associated with serious diseases in various hosts, including hemorrhagic enteritis in turkeys, marble spleen disease in pheasants, and splenomegaly in chickens. Recently, a putative novel siadenovirus has been detected in fatally diseased birds of prey.

## Phylogeny

The five clusters of adenoviruses defined by gene organization and sequence-based phylogenetic calculations correspond to the four recognized genera (*Mastadenovirus*, *Aviadenovirus*, *Atadenovirus*, and *Siadenovirus*) plus one proposed genus for fish adenoviruses (**Figure 2**). The evolutionary distances among the adenoviruses seem to be generally proportional to those among their hosts, supporting co-evolution of virus and host. However, there are some exceptions where very distantly related viruses infect the same host, for example, the bovine mastadenoviruses and atadenoviruses. Thus, in addition to co-evolution, several host switches might have occurred. According to this hypothesis, the mastadenoviruses and aviadenoviruses represent co-evolution, whereas reptilian adenoviruses may have switched to ruminants, birds, and marsupials. Indeed, genome organization and phylogenetic analysis indicate that SnAdV-1 most closely represents the most ancient common ancestor of the atadenoviruses. Siadenoviruses may have an amphibian origin, though the small number of viruses in this group limits analyses of this possibility. More recent host switches can also be identified within genera, for example between humans and other primates, or vice versa.

See also: Adenoviruses: Malignant Transformation and Oncology; Adenoviruses: Molecular Biology; Adenoviruses: Pathogenesis; Electron Microscopy of Viruses; Evolution of Viruses; Gene Therapy: Use of Viruses as Vectors; Phylogeny of Viruses; Recombination; Virus Databases; Virus Species.

## Further Reading

Benkö M, Harrach B, Both GW, et al. (2005) Family Adenoviridae. In: Fauquet CM, Mayo MA, Maniloff J, Desselberger U, and Ball LA (eds.) Virus Taxonomy: Eighth Report of the International Committee on Taxonomy of Viruses, pp. 213–228. San Diego, CA: Elsevier Academic Press.

Burnett RM (1997) The structure of adenovirus. In: Chiu W, Burnett RM, and Garcea RL (eds.) Structural Biology of Viruses, pp. 209–238. Oxford: Oxford University Press.

Davison AJ, Benkö M, and Harrach B (2003) Genetic content and evolution of adenoviruses. Journal of General Virology 84: 2895–2908.

Doerfler W and Böhm P (eds.) (2003) Adenoviruses: Model and vectors in virus host interactions. In: Current Topics of Microbiology and Immunology, vol. 272. Berlin: Springer.

Farkas SL, Benkö M, Élö P, et al. (2002) Genomic and phylogenetic analyses of an adenovirus isolated from a corn snake (Elaphe guttata) imply common origin with the members of the proposed new genus Atadenovirus. Journal of General Virology 83: 2403–2410.

Kovács GM, LaPatra SE, D'Halluin JC, and Benkö M (2003) Phylogenetic analysis of the hexon and protease genes of a fish adenovirus isolated from white sturgeon (Acipenser transmontanus) supports the proposal for a new adenovirus genus. Virus Research 98: 27–34.

Russell WC (2000) Update on adenovirus and its vectors. Journal of General Virology 81: 2573–2604.

Wold WSM and Tollefson A (eds.) (2006) Methods in molecular medicine. In: Adenovirus Methods and Protocols, 2nd edn., vol. 131, pp. 299–334. Totowa, NJ: Humana Press.

Zsivanovits P, Monks DJ, Forbes NA, et al. (2006) Presumptive identification of a novel adenovirus in a Harris hawk (Parabuteo unicinctus), a Bengal eagle owl (Bubo bengalensis), and a Verreaux's eagle owl (Bubo lacteus). Journal of Avian Medicine and Surgery 20: 105–112.

# Adenoviruses: Malignant Transformation and Oncology

**A S Turnell,** The University of Birmingham, Birmingham, UK

© 2008 Elsevier Ltd. All rights reserved.

## Transforming and Oncogenic Properties of Human Adenoviruses

Human adenoviruses (Ads) are small, nonenveloped DNA viruses with linear, double-stranded genomes of about 35 kbp that are associated with a broad range of infections, but are most commonly linked with acute infections of the upper respiratory and gastrointestinal tracts. The clinical manifestations of an adenovirus infection are dependent on the adenovirus and the host cell type infected. The oncogenic capacity of adenovirus was first established by John Trentin and colleagues in 1962 when they showed that human adenovirus 12 (Ad12) was tumorigenic when injected into newborn hamsters. Adenovirus tumorigenicity was shown to be dependent upon various factors, such as virus serotype, virus dose, host age at infection time, and host genetic and immune status. Indeed, the tumorigenic potential of various adenoviruses was attributed to their ability to evade the host immune system, as nontumorigenic adenoviruses were shown to induce tumors in immunocompromised hosts. Adenovirus tumorigenicity in this regard is associated with nonproductive infections of human viruses in rodent cells. It is generally believed that a productive lytic infection in human cells precludes human adenoviruses from being tumorigenic in humans (see later). Consistent with the ability of nontumorigenic adenoviruses to promote tumors in nude mice, tissue culture studies later revealed that rodent cells could be transformed by both tumorigenic and nontumorigenic adenoviruses. Adenovirus serotypes have been classified into species (*Human adenovirus A* to *Human adenovirus F*) according to various criteria and correlated broadly with oncogenic potential (**Table 1**).

Transfection studies in primarily rodent, but also human, cells have revealed that various combinations of adenovirus DNA can transform primary cells in tissue culture, and this has greatly facilitated the study of the transformation properties of specific adenovirus genes. As with other DNA tumor viruses, the oncogenic properties of the various adenovirus genes become apparent when the mechanics of the productive lytic cycle are considered. Adenoviruses normally infect quiescent epithelial cells. In order to replicate the viral genome, the virus must first create a cellular environment conducive for DNA replication. It does this by reactivating the host cell cycle to create an S-phase-like environment. Additionally, it prevents the host cell from undergoing premature apoptosis, allowing for the assembly and production of progeny virions. Thus viral genes have evolved to circumvent the normal growth restrictions placed upon the cell. It is therefore perhaps no surprise that many viruses, in addition to adenovirus, target the tumor suppressor gene products pRB and p53 and proteins involved in DNA damage response/repair pathways (see later) to overcome normal cell cycle checkpoints. The focus of the rest of this article is a consideration of the biological functions of adenovirus-transforming genes.

## Transforming and Immortalizing Properties of E1A

*E1A* expression is essential for a fully productive adenovirus infection and for adenovirus-mediated transformation. The Ad5 *E1A* gene encodes two major proteins of 289 and 243 amino acid residues (R) that differ only by an internal sequence of 46 residues present in the larger protein, namely, conserved region 3 (CR3), which is required specifically for adenovirus early gene expression during infection (**Figure 1**). As for E1A species from other Ad serotypes, the Ad5E1A 289R and 243R proteins possess transforming and immortalizing activity, although the smaller 243R protein is more efficient than the 289R protein in this regard. In isolation, however, E1A displays

**Table 1** Classification and oncogenicity of human adenoviruses

| Species | Serotype | Oncogenicity in rodents | Transformation in vitro |
|---|---|---|---|
| Human adenovirus A | 12,18,31 | High | + |
| Human adenovirus B | 3, 7, 11, 14, 16, 21, 34, 35, 50 | Moderate | + |
| Human adenovirus C | 1, 2, 5, 6 | Low or none | + |
| Human adenovirus D | 8–10, 13, 15, 17, 19, 20, 22–30, 32, 33, 36–39, 42–49, 51 | Low or none | + |
| Human adenovirus E | 4 | Low or none | + |
| Human adenovirus F | 40, 41 | Not reported | ? |

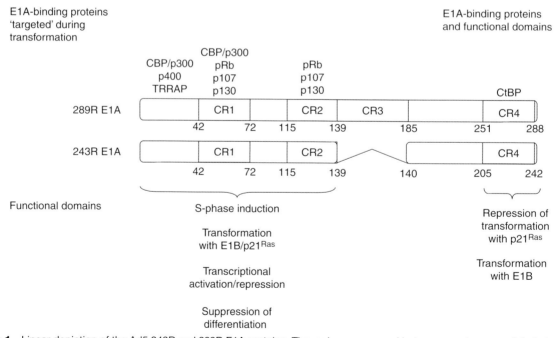

**Figure 1** Linear depiction of the Ad5 243R and 289R E1A proteins. The regions conserved between serotypes are labeled as CR1, CR2, CR3, and CR4; amino acid ordinates show CR boundaries. The E1A-binding proteins considered important for E1A-mediated transformation are listed and their relative binding sites depicted. E1A regions important in cell cycle control and cellular transformation are also shown.

only poor transforming activity. This is due specifically to E1A's potent proapoptotic activity. E1A's transforming capacity only becomes apparent upon its expression with co-operating oncogenes such as adenovirus *E1B*, adenovirus *E4*, or activated, mutant *p21$^{ras}$*. Sequence comparisons of the largest E1A proteins of several adenovirus serotypes have identified four regions of sequence conservation, designated CR1, CR2, CR3, and CR4, which are largely responsible for many of E1A's biological activities, although the less well conserved N-terminal region also participates in E1A function (**Figure 1**). The transforming properties of E1A are mainly attributable to E1A's ability, through the reorchestration of host cell transcription programmes, to force quiescent cells to enter, and, for rodent cells at least, pass through the cell cycle; E1A does not promote mitosis in human cells. E1A also blocks cellular differentiation programmes.

The N-terminal region and CR1 of E1A promote transformation by targeting the transcriptional co-activators CBP and p300 (CBP/p300); E1A mutants unable to bind CBP/p300 are transformation defective. Specifically, E1A binding to CBP/p300 allows for quiescent G$_0$ rodent cells to enter the cell cycle and progress into S-phase. CBP and p300 are highly conserved lysine acetyltransferases. In this regard, the cellular activities of sequence-specific transcription factors such as p53, c-Myc, and NF-κB are all regulated by CBP/p300-acetyltransferase activity. CBP/p300 also acetylate the core histone proteins H2A, H2B, H3, and H4 to regulate transcription. It is as yet unclear, however, whether E1A utilizes CBP/p300 acetyltransferases during transformation to reprogram cellular transcription through altered acetylation, or, alternatively, promotes transformation by inhibiting CBP/p300 acetyltransferase activities. Indeed, E1A generally inhibits CBP/p300 transcription function in transient gene reporter assays. A model has been proposed, however, whereby E1A utilizes cellular acetyltransferases to alter cellular histone acetylation status, in order to stimulate E2F-regulated transcription programmes and induce S-phase. It is suggested that E1A first facilitates the demethylation of K9 of histone H3 to promote the release of repressor E2F transcription factors from transcriptionally repressed promoters, whereupon it subsequently promotes the acetylation of K9 of histone H3 and the recruitment of activator E2F transcription factors to these promoter elements to activate transcription. The specific role played by CBP/p300 in the acetylation of K9 in this particular situation awaits clarification. Additionally, the ability of E1A to modulate CBP/p300 binding to transcription factors, independent of acetyltransferase activity, might also affect transforming activity. Theoretically, E1A could utilize other cellular acetyltransferases to promote G1–S progression. Specifically, it is possible that E1A could utilize the acetyltransferase P/CAF, to which it also binds, to regulate transcription programmes during transformation. Unfortunately, E1A point mutants have not yet been identified that unambiguously distinguish between CBP/p300 and P/CAF binding *in vivo*, such that the contribution of P/CAF in E1A-mediated transformation is unclear. It seems unlikely from the literature published thus far, however, that the interaction of other proteins with the N-terminal region of E1A spanning residues 4 and 25 is necessary to promote transformation.

Recent evidence suggests that the APC5 and APC7 components of the anaphase-promoting complex (APC/C) interact with CBP/p300 through protein interaction domains evolutionarily conserved in the N-terminal region, and CR1 (Fx$^D$/$_E$xxxL motif) of E1A. E1A specifically targets APC/C-CBP/p300 complexes in order to promote the cellular transformation of rat embryo fibroblasts. Thus, exogenous expression of wild-type (w.t.) APC5 or APC7 suppresses the ability of E1A to cooperate with either E1B or activated p21$^{Ras}$ in the transformation process, whereas the expression of APC/C mutants unable to bind CBP/p300 does not suppress E1A-induced transformation. E1A also targets endogenous CBP/p300-APC/C complexes during transformation, such that RNAi-mediated knockdown of either APC5 and/or APC7 restores transforming activity to E1A species unable to bind CBP/p300, such as R2G.

Interestingly, the N-terminal region of E1A possesses additional transforming capacity. At limiting concentrations of E1A, it has been determined that residues 26–35 also promote transformation in cooperation with activated p21$^{Ras}$. It is yet to be established whether this region also cooperates with *E1B*, or other cellular or viral oncogenes, in transformation. It has been determined, however, that this region of E1A binds specifically to two cellular proteins, p400 and TRRAP, in order to promote transformation. p400 is a member of the SWI2/SNF2 chromatin-remodeling family, which associates specifically with TRRAP, the DNA helicases TAP54α/β, actin-like proteins, and the human homolog of the *Drosophila* enhancer of polycomb protein, EPc, while TRRAP is a Myc-binding protein and also a component of distinct GCN5/SAGA and Tip60/NuA4 acetyltransferase complexes. *In vitro* binding studies have indicated that E1A binds to p400 through a central region (residues 951–2048) of the protein, which also encompasses the SWI2/SNF2 homology domain and TAP54-binding site. Interestingly, E1A can also associate with p400 at a distinct C-terminal site *in vivo* (residues 2033–2484), through its interaction with TRRAP. Similar binding studies indicate that E1A binds TRRAP through residues 1360–2260. It has been determined that E1A utilizes p400 complexes to promote transformation, as co-expression of p400 fragments with E1A enhances E1A-dependent transformation. However, co-expression of a TRRAP fragment that binds E1A suppresses

E1A-dependent transformation. Collectively, these data suggest that multiple and distinct E1A/p400/TRRAP complexes exist within cells which contribute differently to the transformation process. It has been proposed that during transformation E1A perturbs normal p400 and TRRAP function by forming E1A/p400 and E1A/TRRAP complexes with distinct subunit compositions to regulate chromatin structure and function.

The ability of CR1 and CR2 of E1A to promote cellular transformation resides in its capacity to interact with the protein product of the pRB tumor suppressor gene. Similar to its ability to bind CBP/p300, it has been established that E1A binding to pRB allows for quiescent $G_0$ rodent cells to enter the cell cycle and progress into S-phase; however, E1A must bind both CBP/p300 and pRB in order to promote mitosis in primary rodent cells. The integrity of the core DLxCxE motif in CR2 as well as CR1 is essential for E1A interaction with pRB and pRB family members p107 and p130. E1A binds specifically to the A and B pocket domains of these proteins. Like E1A's interaction with CBP and p300, any specific requirement for E1A binding to particular pRB family members during transformation is currently unknown. It has been proposed that E1A interaction with pRB and either p107 and/or p130 is important in driving quiescent cells into S-phase. In a temporally coordinated manner, E1A disrupts p130–E2F4 transcriptional repression complexes, allowing for chromatin remodeling, histone demethylation, and histone acetylation. E1A's ability to mimic cyclin-dependent kinase (CDK)-dependent phosphorylation of pRB through binding to pRB disrupts pRB–E2F1 complexes and allows the recruitment of activator E2Fs to promoters. This recruitment induces E2F-regulated genes important in S-phase induction, such as CDK2, CDC6, and cyclins E and A. The ability of E1A to disrupt pRB–E2F1 complexes also allows for E1A to overcome pRB-induced senescence programs initiated by CDK inhibitors, $p21^{CIP1/WAF1}$ and $p16^{INK4}$.

Interestingly, E1A has been found to reorganize, and associate with, PML-containing nuclear oncogenic bodies (PODs), nuclear structures implicated in regulating apoptosis, tumor suppression, antiviral responses, DNA repair, and transcriptional regulation. E1A recruitment to PODs is dependent on the pRB-interaction domain (pRB can be found associated with PODs). Whether E1A similarly associates with CBP in PODs is unclear. The precise requirement for E1A–POD interaction in the transformation process is unclear, though it is suspected that E1A might be associated with multiple POD functions.

E1A has been shown to target the CtBP family of transcriptional co-repressors through a conserved PxDLS motif located toward the C-terminal region of the protein. E1A's interaction with CtBP is believed to be important in the transformation process. The role of the region encoded by exon 2 in cellular immortalization and cellular transformation is, however, context dependent. In cooperation with exon 1, exon 2 enhances the E1A-dependent immortalization of rodent cells and E1A-dependent transformation with E1B, such that certain exon 2 deletion mutants have reduced immortalization/transformation capacities relative to w.t. 12S (243R) E1A. In contrast, exon 2 suppresses 12S E1A/$p21^{Ras}$-mediated immortalization and transformation, such that exon 2 deletion mutants enhance the E1A-dependent immortalization and transformation of primary cells in culture, and, moreover, their tumorigenicity and metastatic potential in nude mice. These differences in exon 2 functions presumably reside in the differential capacities of E1B and $p21^{Ras}$ to activate/suppress cellular pathways involved in promoting immortalization and/or transformation.

It has been suggested that exon 2 promotes immortalization and transformation with E1B through its ability to complement exon 1 immortalization functions and evade Mortality stage 2 (crisis). It has also been proposed that the ability of exon 2, in the context of w.t. 12S E1A, to function as a tumor suppressor and suppress $p21^{Ras}$-mediated transformation resides in its ability to promote mesenchymal–epithelial cell transitions through the regulation of specific epithelial-promoting transcription programs. The ability of exon 2 mutants to promote hypertransformation, enhanced tumorigenicity and metastatic potential, relative to w.t. 12S E1A, of both epithelial and fibroblastic cells is believed to be mediated, in part, by the activation of the small Ras-related proteins, Rac and Cdc42, suggesting that w.t. 12S E1A may normally serve to suppress these pathways. The current literature suggests that the ability of exon 2 to regulate oncogenesis is mediated at least in part by its ability to bind to CtBP. A more thorough examination however of exon 2-binding proteins and exon 2 deletion mutants is required to determine whether E1A binding to CtBP accounts for all of these cellular phenotypes.

## Oncogenicity of E1A

E1A from oncogenic serotypes specifically regulates pathways involved in immune regulation and contributes toward immune evasion – a major determinant of adenovirus oncogenicity. It is well established that E1A from the tumorigenic serotype, Ad12, downregulates the presentation of major histocompatibility complex class I (MHC-I) antigens in rodent cells, whereas E1A from the nontumorigenic serotype, Ad5, does not. MHC-I downregulation allows for adenovirus-infected cells to avoid clearance by cytotoxic T-lymphocytes (CTLs). A major difference between oncogenic and nononcogenic serotypes in this regard is in their ability to affect transcription mediated through the MHC-I enhancer sites R1 and R2. Oncogenic serotypes repress transcription mediated through the MHC-I enhancer by upregulating the transcriptional repressor COUP-TFII, which binds to the R2 site, while concomitantly reducing the level of NF-κB binding to the R1 site by reducing the level of phosphorylation of the p50 subunit of NF-κB. Nononcogenic serotypes do not upregulate COUP-TFII levels, and NF-κB binds to the R1 site with high affinity. Oncogenic serotypes have also been shown to interfere with MHC-I presentation by downregulating components of the immune proteasome, such as LMP2 and LMP7, and downregulating the transporter proteins associated with antigen presentation, TAP1 and TAP2. A further feature of Ad12 E1A that facilitates immune evasion is that it does not produce immunological epitopes recognized by CTLs, whereas Ad5 E1A does. Ad12-infected cells are also resistant to natural killer cell lysis. These functions of E1A are largely attributed to an alanine-rich 20-residue spacer region present in Ad12 E1A between CR2 and CR3, which is not present in the nontumorigenic adenovirus serotypes. Other regions of Ad12 E1A do possess tumorigenic properties, but how these manifest mechanistically awaits further investigation.

## Transforming Properties of Adenovirus E1B and E4 Regions

In the absence of cooperating oncogenes, such as $p21^{ras}$, *E1B*, or *E4*, *E1A* has little or no capacity to transform and immortalize embryonic human cells in tissue culture systems and only limited capacity to transform and immortalize embryonic rodent cells; E1A-immortalized clones tend to harbor mutations in the p53 tumor suppressor gene. Although the *E1B* and *E4* genes have no transforming capacity alone, they can cooperate with *E1A* in the transformation process through their ability to neutralize E1A-induced, p53-dependent and p53-independent, apoptosis. The biological properties of E1B and E4 proteins, in this regard, will be discussed in turn.

## E1B-19K

The *E1B* gene expresses two principle proteins: E1B-19K and E1B-55K. E1B-19K was initially identified as a homolog of the cellular antiapoptotic protein, BCL-2. Despite sequence conservation between the two proteins, it is of interest to note that E1B-19K and BCL-2 differ in the methods they employ to neutralize the proapoptotic functions of BCL-2 family members BAX and BAK. Specifically, BCL-2 inhibits the activation of BAX/BAK by binding to BID and preventing its caspase-8 dependent cleavage to active tBID, which, under normal circumstances, promotes BAX/BAK heterooligomerization at mitochondrial membranes and the subsequent induction of apoptosis. E1B-19K, on the other hand, inhibits apoptosis by binding directly to tBID-activated BAX and/or BAK, preventing BAX/BAK heterooligomerization and the subsequent downstream events that lead to apoptosis (**Figure 2**). Significantly, the ability of E1A to induce p53-independent apoptosis relies, in part, upon its ability to specifically activate BAK/BAX. In this context, and akin to ultraviolet (UV)-induced DNA damage-induced apoptotic pathways, E1A can promote apoptosis by specifically downregulating the BCL-2 antiapoptotic

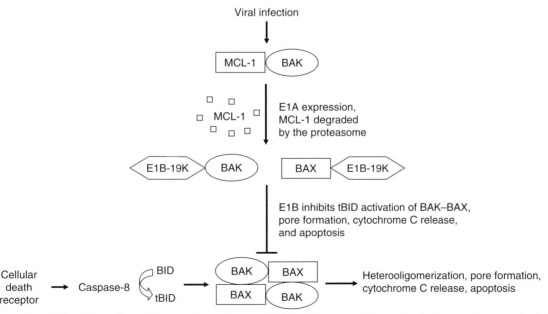

**Figure 2** Role of E1B-19K in antiapoptotic signaling pathways. E1B-19K inhibits E1A-induced, p53-independent apoptosis by binding tBID-bound BAK and BAX, and specifically inhibiting the mitochondrial-release apoptosis pathway. See text for details.

family member MCL-1 through proteasomal-mediated degradation. MCL-1 normally serves to functionally inhibit BAK, but upon E1A-mediated degradation of MCL-1, BAK can activate apoptosis. In this regard, E1B-19K cooperates with E1A in the transformation process by inhibiting E1A-activated BAK (**Figure 2**).

## E1B-55K and E4orf6

The Ad5 E1B-55K and 34K E4orf6 proteins functionally cooperate during viral infection to regulate p53, the MRE11-RAD50-NBS1 (MRN) complex (**Figure 3**), and potentially late-phase viral mRNA nuclear export and translation. For this reason, it is perhaps best to consider their roles in cellular transformation together. The current perception is that the independent, and combined, abilities of both E1B-55K and E4orf6 to repress p53-transactivation of proapoptotic genes underlies their respective, and synergistic, capacities to co-operate with E1A in the transformation process. The requirement for E1B-55K and E4orf6 binding and regulation of the MRN complex in the transformation process awaits clarification.

Specifically, the central region of E1B-55K encompassing residues 250–310, and a small region around residue 180, bind to the N-terminal transactivation domain of p53, blocking co-activator recruitment and further inhibiting p53 transcription function through specific C-terminal transcriptional repression domains. In Ad5 E1A/E1B-55K transformed cells, E1B-55K and p53 are sequestered in cytoplasmic 'phase-dense' aggresome structures that also contain microfilaments, centrosomal proteins, hsp70, and WT1. The cytoplasmic location of E1B-55K is mediated in part by a leucine-rich nuclear export signal (NES) located between residues 83 and 93; the nuclear export of E1B-55K is dependent upon the CRM1 cellular export receptor. Significantly, mutation of the critical leucine residues in the E1B-55K NES potentiates E1A/E1B-mediated transformation through enhanced inhibition of p53 transactivation function and the accumulation of mutant E1B-55K and p53 in PODs. Consistent with a role for E1B-55K localized to PML-containing nuclear bodies in p53 repression and cellular transformation, it has been demonstrated through mutational studies that SUMOylation of K104 in E1B-55K not only promotes its recruitment to PML bodies, but augments p53 transcriptional repression and enhances cellular transformation. Other functions of E1B-55K might also contribute toward its transforming potential. Indeed, the Ad12 large E1B protein (equivalent to Ad5 E1B-55K) can extend the lifespan of human embryo fibroblasts, bypassing Mortality stage 1 through ALT maintenance of telomere length.

The C-terminal region of E4orf6 can, independently of E1B-55K, bind to the C-terminal oligomerization domain (residues 318–360) of p53 to repress specifically the p53 N-terminal transactivation function; E4orf6 disrupts p53 interaction with TFIID component TAFII31. In contrast to E1B-55K, however, E4orf6 also possesses an independent and distinct ability to relieve C-terminal p53 transcriptional repression through binding. Akin to E1B-55K, E4orf6 cooperates with E1A in the transformation of primary baby rat kidney (BRK) cells. E1A/E4orf6 transformants grow more slowly than E1A/E1B-55K transformants, but are essentially morphologically indistinguishable. As already indicated, E4orf6 will also synergize with E1B-55K,

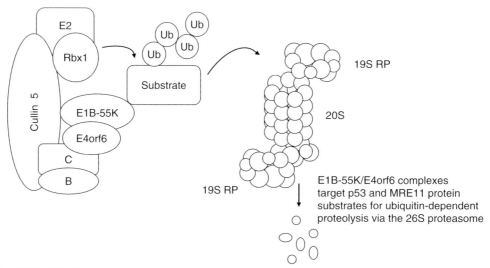

**Figure 3** Role of E1B-55K/E4orf6 complexes in the regulation of p53 and MRE11. E1B-55K/E4orf6 form complexes with the Cullin 5-based E3 ubiquitin ligase to promote the poly-ubiquitylation, and hence 26S proteasomal-mediated destruction, of p53 and MRE11. See text for more detail. B, C, Elongin B, C; E2, ubiquitin conjugating enzyme; Ub, ubiquitin.

or E1B-55K and E1B-19K, to enhance the frequency of E1A-dependent transformation; E1A/E1B/E4orf6 mutants express lower levels of p53 than E1A/E1B and E1A/E4orf6 transformants. Interestingly, E1A/E1B/E4orf6 transformants arise more rapidly than E1A/E1B and E1A/E4orf6 transformants, and when injected subcutaneously into the nude mouse promote more rapid tumor formation. In support of the tumorigenic properties of the E4orf6 protein, it also converts human 293 cells from nontumorigenic to tumorigenic in nude mice. Interestingly, and in stark contrast to E1B-55K, E4orf6 does not cooperate with E1A and E1B-19K in the transformation process; in fact, E4orf6 suppresses E1A/E1B-19K mediated transformation. It has been suggested that the additional ability of E4orf6 to relieve p53 transcriptional repression contributes toward its ability to suppress transformation in this instance.

As indicated earlier, E1B-55K and E4orf6 also regulate the activity of the MRN complex. Functionally, the MRN complex is integral to both the ATM- and ATR-DNA damage response/repair signaling pathways initiated in response to ionizing and UV irradiation, respectively. It has been suggested that adenovirus circumvents these checkpoint controls through the ability of E1B-55K and E4orf6 to target the MRE11 component of the MRN complex for ubiquitin-mediated, proteasome-dependent degradation. Inactivation of MRE11 by E1B-55K and E4orf6 ensures, in the context of viral infection, that linear viral DNA is not concatemerized by double-strand break–repair pathways initiated by the MRN complex.

The functional cooperativity displayed by E1B-55K and E4orf6 is mediated through protein–protein interaction. It has been suggested that the N-terminal 55 residues of E4orf6 are required for its binding to E1B-55K, although others suggest that the amphipathic α-helical region toward the C-terminus governs the interaction; the central region of E1B-55K encompassing residues 262–326, and a region around residue 143, are required for interaction with E4orf6. Although the E4orf6-binding site on E1B-55K overlaps with the p53-binding site, E1B-55K and E4orf6 bind independently of p53. Crucial to establishing the biochemical basis of this functional cooperativity between the two proteins, mass spectrometric identification of cellular proteins bound to E4orf6 in p53-null H1299 cells has revealed that E4orf6 can bind to the Cullin-containing, Cul5-Elongin B/C-Rbx1 E3 ubiquitin ligase complex (**Figure 3**). Significantly, E1B-55K is only found in complexes with E4orf6 bound to the Cul5-Elongin B/C-Rbx1 complex, suggesting that E1B-55K and E4orf6 might not bind directly. E4orf6 contains within its primary sequence two functional Elongin B/C-interaction (BC) boxes located between residues 46–55 and 121–130, which define its direct interaction with Elongin C. E1B-55K also contains a putative BC-box located between residues 179 and 188, though this region does not bind Elongin B or C directly. It has been postulated that E4orf6 might also bind Cul5 directly, though this awaits confirmation.

Functional studies have indicated that E4orf6–E1B-55K utilizes the Cul5-Elongin B/C-Rbx1 ubiquitin ligase to promote the poly-ubiquitylation and hence 26S proteasomal-mediated degradation of p53 and MRE11 during viral infection and cellular transformation. E4orf6 is postulated to recruit the functional E3 ligase complex, whereas E1B-55K is proposed to interact with the substrate protein targeted for degradation. The absolute depletion of functional p53, MRE11, and potentially other cellular proteins by the E4orf6–E1B-55K cellular ubiquitin ligase complexes might explain why E4orf6 and E1B-55K are functionally more adept at enhancing the frequency of E1A-dependent transformation than either protein in isolation.

E1B-55K binds the hnRNP, RNA-binding protein E1B-AP5 independently of E4orf6. The central region of E1B-55K encompassing residues 262–326, and regions around residues 180, 380, and 484 all participate in E1B-AP5 binding; E4orf6 can modulate E1B-AP5 binding to E1B-55K. However, E1B-AP5 is not targeted for degradation during viral infection by E1B-55K and E4orf6, suggesting that E1B-55K is more than just a substrate adaptor for the E4orf6–Cul5-containing E3 ubiquitin ligase. The E1B-55K/E1B-AP5 complex is proposed to regulate the shutdown of host cell mRNA export during infection. Interestingly, overexpression of E1B-AP5 suppresses E1A/E1B-55K-mediated transformation, suggesting that E1B-55K might target E1B-AP5 directly to facilitate transformation.

## E4orf3

The 11K E4orf3 gene product operates in redundant pathways with E4orf6 and E1B-55K during viral infection to regulate RNA processing and viral DNA replication. Mechanistically, E4orf3 cooperates with E4orf6 in the inhibition of DNA damage repair, presumably through binding DNA-PK and inactivation of the MRN complex. Indeed, Ad5 E4orf3 redistributes the MRN complex to nuclear PODs away from viral replication centers. Ad4 and Ad12 E4orf3 do not cause relocalization of the MRN complex, suggesting serotype-specific modulation of MRN function. E4orf3 similarly binds E1B-55K and targets it to PODs. Akin to E4orf6, E4orf3 will co-operate with E1A and E1A/E1B-55K to transform primary BRK cells in tissue culture and increase the tumorigenicity of E1A/E1B-55K transformed BRKs in nude mice. E4orf3 will also co-operate with E1A/E4orf6 and E1A/E1B-55K/E4orf6 in transformation. Like E4orf6, E4orf3 increases the frequency of transformation, the rate of cell growth, and the saturation densities to which transformants grow.

## Ad9 E4orf1

Ad9 is unique among the adenovirus family in that it promotes estrogen-dependent mammary tumors in rodents. The oncogenic capacity of Ad9 has been mapped to a single protein, E4orf1. Although conserved among different adenoviruses (45–51% identity), only the Ad9 protein is oncogenic, although other adenovirus E4orf1 proteins do display some transforming characteristics (e.g., anchorage-independent growth). It has been suggested that the oncogenic capacity of the Ad9 protein resides in its differential ability to bind the first PDZ domain of the PDZ-containing, MAGUK homology protein, and tumor suppressor protein, ZO-2. The C-terminal PDZ-binding domain of Ad9 E4orf1 selectively targets ZO-2 and redistributes it to the cytoplasm. Consistent with a role in oncogenicity, ZO-2 inhibits Ad9 E4orf1 transformation of the rat embryo fibroblast cell line, CREF. Activation of phosphoinosite 3-kinase (PI3K) by the PDZ-binding domain of E4orf1 at the plasma membrane is also crucial in the transformation process. PI3K activation presumably allows for the specific initiation of downstream survival signaling pathways, and the activation of other potential oncoproteins (e.g., $p21^{Ras}$).

## 'Hit and Run' Transformation

As outlined earlier, the prevailing, generally held view is that adenovirus infection is not a causative agent of human malignancy. Intriguingly, however, and consistent with early studies looking at human simplex virus- and Abelson leukemia virus-mediated transformation, studies have suggested that certain adenovirus genes may promote transformation by a 'hit and run' mechanism whereby adenovirus DNA, mRNA, or protein are only rarely retained in the adenovirus-transformed cells. Incidentally, this makes it very difficult to tell whether adenovirus infection actually promotes human tumorigenesis, or whether 'hit and run' transformation is merely an *in vitro* phenomenon of no clinical relevance. It has been suggested that adenovirus genes are required for the initiation of transformation, but are not required for the maintenance of the transformed phenotype. Although mechanistically ill-defined, it is suggested that adenovirus genes could promote gene mutation and/or direct genetic instability through interaction with cellular proteins. Specifically, studies have shown that when E4orf3 and/or E4orf6 cooperate with E1A in transformation, in the majority of instances the transformed cells do not retain E1A, E4orf6, and/or E4orf3, though there are clear exceptions where these proteins can be detected. Interestingly, co-expression of these proteins with E1B-55K subsequently facilitates their detection in transformed cells, suggesting that E1B-55K antagonizes the genetic instability promoted by E1A, E4orf6, and E4orf3. Significantly, a role for these proteins in promoting genetic instability is already well established. The N-terminal region of E1A and CR3 have been previously implicated in causing both random and nonrandom host cell chromosome aberrations, as well as abnormal mitoses generating both aneuploid and polyploid cells in human cells and rodent cells, whereas E4orf3 and E4orf6 have been implicated in promoting genetic instability through the interaction with MRN and DNA-PK and inhibiting the repair of damaged DNA.

## Conclusions

Adenovirus has long served as a faithful model system for investigating the molecular mechanisms of cellular immortalization, transformation, and tumorigenicity. During this time, studies on adenovirus E1A and E1B have identified the functions of key tumor suppressor proteins, not to mention the functions of proteins not directly involved in oncogenesis. In more recent years, the roles of E4 proteins in the transformation process have become increasingly apparent. It is reasonable to anticipate that future studies investigating adenovirus-mediated transformation will enhance further our understanding of the molecular mechanisms underlying oncogenesis.

*See also:* Adenoviruses: General Features; Adenoviruses: Molecular Biology; Adenoviruses: Pathogenesis.

## Further Reading

Berk AJ (2005) Recent lessons in gene expression, cell cycle control, and cell biology from adenovirus. *Oncogene* 24: 7673–7685.

Chinnadurai G (2004) Modulation of oncogenic transformation by the human adenovirus E1A C-terminal region. *Current Topics in Microbiology and Immunology* 273: 139–161.

Endter C and Dobner T (2004) Cell transformation by human adenoviruses. *Current Topics in Microbiology and Immunology* 273: 163–214.

Frisch SM and Mymryk JS (2002) Adenovirus-5 E1A: Paradox and paradigm. *Nature Reviews. Molecular Cell Biology* 3: 441–452.

Gallimore PH and Turnell AS (2001) Adenovirus E1A: Remodelling the host cell, a life or death experience. *Oncogene* 20: 7824–7835.

White E (2006) Mechanisms of apoptosis regulation by viral oncogenes in infection and tumorigenesis. *Cell Death and Differentiation* 13: 1371–1377.

Williams JF, Zhang Y, Williams MA, Hou S, Kushner D, and Ricciardi RP (2004) E1A-based determinants of oncogenicity in human adenovirus groups A and C. *Current Topics in Microbiology and Immunology* 273: 245–288.

# Adenoviruses: Molecular Biology

**K N Leppard,** University of Warwick, Coventry, UK

© 2008 Elsevier Ltd. All rights reserved.

## Introduction and Classification

Adenoviruses were first discovered during the 1950s, in studies of cultures from human adenoids and virus isolation from respiratory secretions. Their study accelerated dramatically in the 1960s after it was demonstrated that a human adenovirus could cause tumors in experimental animals, leading to the hope that such work would lead to a better understanding of human cancer. What has been learned subsequently, as well as making a significant contribution in this area, has also had a major influence on our understanding of fundamental eukaryotic cell processes such as RNA splicing and apoptosis.

The adenoviruses are classified in the family *Adenoviridae*. Historically, this family was divided into two genera, *Aviadenovirus* and *Mastadenovirus*, containing viruses that infected avian and mammalian hosts, respectively, but the advent of large-scale DNA sequencing has led to two further genera being established, based on comparisons of genome sequences. Thus, for example, the genus *Atadenovirus* contains two of five species of bovine adenovirus and one of four species of sheep adenovirus, the others being members of the genus *Mastadenovirus*. The fourth genus, *Siadenovirus*, so far contains a frog adenovirus and a turkey adenovirus.

There are six human adenovirus species, *Human adenovirus A* through *F*, all of which are classified in the genus *Mastadenovirus*. These species correspond to the subgroups of human adenovirus established earlier on the basis of the hemagglutination characteristics of their virions, their genome nucleotide composition, and their oncogenicity in rodents. There are 51 known human adenovirus serotypes that are distributed between these species. Each causes characteristic disease symptoms in man. Although there has been an increasing level of interest in other adenoviruses in recent years, the majority of what is known about adenovirus molecular biology comes from the study of human adenoviruses 2 and 5, two very closely related members of species *Human adenovirus C*. The following account is therefore based on these two viruses; important differences from this picture that pertain to other adenoviruses are pointed out where appropriate.

## Particle Structure

Adenovirus particles have icosahedral capsids containing a core that is a complex of proteins and DNA (**Figure 1**). There is no lipid envelope and no host proteins are found in the virion; thus all components of the particle are encoded by the virus (see the section of titled 'Gene expression'). The structural proteins of the particle are known by roman numerals in order of decreasing apparent size on polyacrylamide gels.

The capsid shell is built of 252 capsomers. Twelve of these are known as pentons. They occupy the vertices of the shell and are formed of two components, penton base, which is a pentamer of the penton polypeptide

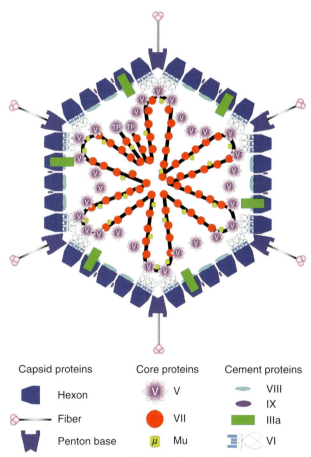

**Figure 1** A cut-through diagram of a mature adenovirus particle. The positions of proteins within the shell of the particle are defined based on cryoelectron microscopy analyses. However, the indicated structure of the DNA core is largely hypothetical; the core appears disordered in structural analyses. Several of these proteins (IIIa, VI, VII, μ, and TP) are synthesized in precursor form (pIIIa, pVI, pVII, pX, and pTP, respectively) and assembled as such into immature particles before processing by the adenovirus protease. Reproduced from Russell WC (2000) Update on adenovirus and its vectors. *Journal of General Virology* 81: 2573–2604, with permission from Society for General Microbiology.

(protein III), and fiber, which is a trimer of the fiber polypeptide (protein IV). This fiber unit projects outward from each vertex of the particle; it has a long shaft, formed of a repeating β-sheet module, and a distal knob domain. Both the penton base and fiber components play specific roles in attachment and internalization (see the section of titled 'Attachment and entry'). Each of the remaining 240 capsomers, which are known as hexons, are trimers of the hexon polypeptide (protein II), which is thus the major capsid protein of the virus.

Inside the shell, the viral core comprises the genome in complex with three proteins: VII, V and μ, the latter being a processed form of polypeptide X. The capsid shell is stabilized by proteins IIIa, VI, VIII, and IX, some of which bridge between the shell and the core. Protein IX trimers occupy crevices on the external faces of the shell; when IX is absent, the particle is less stable and hence its capacity for DNA is reduced. Finally, the particle carries a virus-coded, sequence-specific protease of 23 kDa that is essential both for the maturation of newly formed particles to an infectious state and for particle uncoating after infection. Thus, adenovirus particles are considerably more complex than those of the smaller icosahedral DNA viruses of the families *Polyomaviridae* and *Papillomaviridae*.

## Genome Organization

All adenovirus genomes are linear, double-stranded DNA molecules. They vary in length from genus *Siadenovirus* members at ~26 kbp to genus *Aviadenovirus* members at ~45 kbp; genomes of members of the genera *Mastadenovirus* and *Atadenovirus* are 30–36 kbp in length. The genomes are all thought to have a virus-coded protein, known as the terminal protein (TP), covalently attached to each genome 5′-end. This linkage is a consequence of the mechanism of DNA replication (see below). Also at each end of the linear genome is an inverted terminal repeat sequence varying in length from ~40 to 200 bp; the terminal repeats of human adenovirus 5 are 103 bp.

Among the human adenoviruses, there is a high level of conservation of genome organization. Both genome strands are used to encode proteins from genes that are designated E1A, E1B, E2, E3, E4, IX, IVa2, and the major late transcription unit (MLTU), while two small noncoding RNAs are expressed from the VA genes (**Figure 2**). The E2 and E3 genes are divided into regions A and B while the MLTU is divided into five regions, L1–L5, each with a distinct polyadenylation site. Within each gene/region there are typically multiple proteins encoded. This genome organization applies to all human adenoviruses with only subtle variations, such as the presence of one rather than two VA genes, or the presence of a second fiber coding region in L5, in a few types. Considering adenoviruses in general, the organization of the center

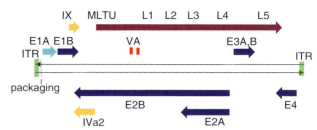

**Figure 2** A schematic representation of the genes and major genome features of human adenoviruses (not to scale). The duplex DNA is represented by black lines, with arrowheads indicating 3′-ends. Inverted terminal repeats (ITRs) and the packaging sequence are indicated as green and gray boxes, respectively. Protein-coding genes are indicated as colored arrows, the color indicating the phase of infection when expression commences (see also **Figure 3(a)**), and RNA pol III transcripts as red boxes.

of the genome (where the structural proteins and replication proteins are coded) is conserved but the regions closer to the genome termini do not necessarily contain recognizable E1A, E1B, and E4 sequences. However, it is presumed that the products encoded here perform the same types of function as these genes supply in human adenoviruses. Unlike the other adenovirus genes, E3 is highly divergent in both sequence and the number and nature of its protein products, even between different human adenoviruses. This may be responsible for aspects of viral pathogenicity and tropism that differ between adenovirus types.

## Attachment and Entry

Primary attachment of particles to the target cell surface is thought to be mediated, for all adenoviruses, by the knob domains at the distal ends of the fibers that project from the virion vertices. For human viruses in species *Human adenovirus A, C, E,* and *F,* the cell receptor for attachment appears to be a molecule designated CAR (coxsackie adenovirus receptor), a cell-surface protein of the immunoglobulin superfamily that is present on a variety of cell types. CAR mediates cell-to-cell interactions, being present at tight junctions and other types of intercellular contact, and is also used as a receptor by members of the unrelated human coxsackie viruses (members of the genus *Enterovirus*, family *Picornaviridae*). The location of CAR has consequences for virus entry and exit (see the section titled 'Assembly and release'). Other adenoviruses, such as members of species *Human adenovirus D* and certain animal adenoviruses, may also use this receptor. However, members of species *Human adenovirus B* clearly use a distinct receptor, the CD46 molecule. CD46 is involved in regulating complement activation and is widely distributed on human cells. Other cell surface molecules may also serve as receptors for some or all of these viruses under particular circumstances.

Following primary attachment, there is a secondary interaction between the penton base component of the particle and cell-surface integrins. These are a family of heterodimeric cell adhesion molecules, several of which can bind the penton base polypeptide via a conserved RGD peptide sequence motif in the viral protein. The interaction between fiber and CAR or CD46 and between the penton base and integrin together lead to internalization of the particle in an endocytic vesicle. Acidification of the endosome then causes conformational changes in the virion that lead to the loss of several proteins from the particle vertices, including fiber and VI. It is this latter protein that is now thought to cause lysis of the endosomal membrane, permitting the residual particle to enter the cytoplasm. The ability of the particle to undergo these uncoating steps depends crucially on the prior action of the encapsidated viral protease, which cleaves several particle proteins during final maturation (see the section titled 'Assembly and release'). The function of protein VI in particular is controlled by sequential specific cleavage events. Cleavage of particle proteins is thought to render the particle metastable, meaning that it is primed for uncoating during the subsequent round of infection.

Once the residual virion has escaped from the vesicle, it is rapidly transported along microtubules toward the nucleus. During this process, there is further loss of virion proteins. Ultimately, the core of the particle is delivered into the nucleus where all the remaining steps of the replication cycle, excepting actual protein synthesis, take place.

## Gene Expression

Once the incoming genome has reached the nucleus, it is used as template for transcription. The principal enzyme involved is host RNA polymerase II, although the VA genes are transcribed by RNA polymerase III. A program of viral gene expression is established, during which essentially the entire length of each genome strand is transcribed. However, not all the regions of the genome are expressed at the same time during infection. Instead, gene expression is tightly regulated, resulting in a temporal cascade of gene activation and hence protein production (**Figure 3(a)**). Key to this is the action of a combination of viral and host proteins as specific activators of viral genes. Thus, initially upon infection, the E1A gene is the only one to be transcribed. If protein production from these mRNAs is prevented with metabolic inhibitors, little further gene expression is seen, indicating that E1A proteins are important activators of the remainder of viral gene expression. One of the E1A proteins, known as the 13 S RNA product or the 289 residue protein, has a potent transactivation function although it lacks specific DNA-binding activity. Instead, its recruitment to promoters is thought to be through protein–protein interactions. The action of this protein, together with host transcription factors that E1A activates indirectly, leads to activation of the remaining early gene promoters. Among the proteins produced from these genes are the ones required for DNA replication. Other early proteins modulate the host environment (see the section titled 'Effects on the host cell'). The production of large numbers of replicated DNA templates, together with the action of the intermediate gene product IVa2 and L4 33 kDa/22 kDa proteins working in a feed-forward activation mechanism, dramatically activates the MLTU, ultimately resulting in high levels of transcription of the genes that encode the structural proteins needed to form progeny particles.

**Figure 3(b)** shows a detailed transcript map of human adenovirus 5. There are only nine promoters for RNA polymerase II in the genome, one for each of the eight genes excepting E2, which has distinct promoters for the early and late phases of infection. Of these eight protein-coding genes, only the one encoding protein IX has a single product. Extensive use of alternative splice sites and polyadenylation sites within the primary transcripts of the other genes leads to multiple mRNAs, each with distinct coding potential; in total, at least 50 different proteins are made. The MLTU, for example, encodes around 18 proteins, each with a unique and essential role as either a virion component, assembly factor, or late gene expression regulator.

As well as allowing the expression of a wide range of proteins, the use by adenovirus of differential RNA processing also opens up the possibility of further controls on the temporal pattern of gene expression during the course of infection. For all the genes that show alternative splicing, there is a trend toward the removal of larger or greater numbers of introns as infection proceeds. A well-documented example is the MLTU L1 region. This limited segment of the MLTU is actually expressed at low level even in advance of replication beginning, but only the 5′-proximal reading frame (encoding the 52/55 kDa proteins) is accessed by splicing. Later in infection, the production of L1 mRNA switches to access the distal reading frame encoding protein IIIa. This is achieved through virus-directed modification of host-splicing components and the direct involvement of L4 33 kDa protein; this protein is also needed to generate the full pattern of protein expression from other regions of the MLTU.

## DNA Replication

Adenoviral DNA replication follows a mechanism that is completely distinct from that of other animal DNA viruses,

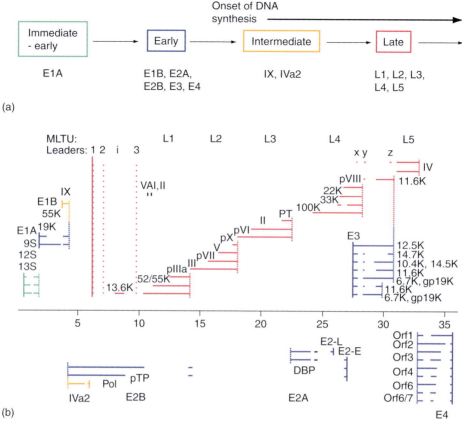

**Figure 3** Gene expression by human adenovirus 5. (a) The phases of gene expression. The numbers E1A, L1, etc. refer to regions of the viral genome from which transcription takes place. (b) A transcription map of the genome. The genome is represented at the center of the diagram as a line scale, numbered in kbp from the conventional left end, with rightward transcription shown above and leftward transcription below. Genes or gene regions are named in boldface. Promoters of RNA polymerase II transcription are shown as solid vertical lines and polyadenylation sites as broken vertical lines. VAI and VAII are short RNA polymerase III transcripts. Individual mRNA species are shown as solid lines, color-coded according to the temporal phase of their expression in (a), with introns indicated as gaps. The protein(s) translated from each mRNA is indicated above or adjacent to the RNA sequence encoding it. Structural proteins are shown by roman numerals; PT, 23 kDa virion proteinase; DBP, 72K DNA-binding protein; pTP, terminal protein precursor; Pol, DNA polymerase. (a) Reproduced from Dimmock NJ, Easton AJ, and Leppard KN (2006) *Introduction to Modern Virology*, 6th edn., figure 9.6. Oxford: Blackwell Scientific, with permission from Blackwell Publishing. (b) Adapted from Leppard KN (1998) Regulated RNA processing and RNA transport during adenovirus infection. *Seminars in Virology* 8: 301–307. Academic Press, with permission from Elsevier.

and from that of its host organism. To achieve this, it encodes three proteins that function directly in replication: a DNA polymerase (DNApol), the terminal protein precursor (pTP), and a single-stranded DNA-binding protein (DBP), all from the E2 gene. Three host proteins also contribute to the replication process: two transcription factors and topoisomerase I. Aside from this, none of the host enzymes that function in DNA replication are needed.

The requirements for adenovirus replication were established using a cell-free replication system in which accurate initiation and elongation were obtained from the viral origin of replication. This is a sequence of 51 bp that is present at the two ends of the genome as part of the inverted terminal repeat. Within this sequence, only the terminal 18 bp (the core origin) is essential for initiation of replication; these bind a complex of DNApol and pTP. The remainder of the 51 bp origin is accessory sequences that bind the two host transcription factors. This binding alters the conformation of the DNA so as to facilitate binding of the DNApol/pTP complex to the adjacent core origin. Once bound at the origin, DNApol uses a serine residue hydroxyl side chain in pTP as a primer for DNA synthesis. The 5′-residue is almost invariably a cytosine, templated by the guanidine residue at the fourth position from the 3′-end of the template strand. After initiation has occurred, the complex then slips back to pair with the residue at the 3′-end of the template strand (also a guanidine), and synthesis then proceeds 5′ to 3′ away from the origin. DBP and topoisomerase I are not needed for initiation of replication, but are essential for elongation.

A scheme for the complete replication of adenoviral DNA is shown in **Figure 4**. A key feature of the process is that there is no synchronous replication of the leading and lagging strands of the template duplex at a single

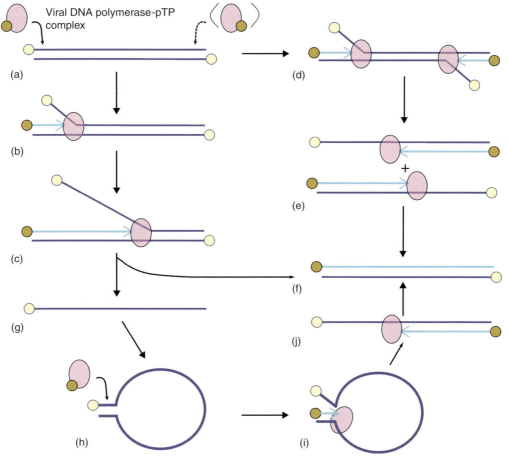

**Figure 4** A general scheme for adenovirus DNA replication. Dark blue, parental DNA; light blue, new DNA. Arrowheads on new DNA strands represent 3′-ends. Yellow circles represent the terminal protein (TP) attached to parental DNA 5′-ends and brown circles the terminal protein precursor molecules (pTP) which prime new DNA synthesis. The viral DNA polymerase is represented in pink. Reproduced from Dimmock NJ, Easton AJ, and Leppard KN (2006) *Introduction to Modern Virology*, 6th edn., figure 6.7. Oxford: Blackwell Scientific, with permission from Blackwell Publishing.

replication fork, as is seen in eukaryotic DNA replication. Instead, the complementary strand of the template DNA duplex is displaced until it can be replicated later. This requires large amounts of DBP to stabilize it. Replication of this displaced single strand is achieved in one of two ways. Either its replication will already have been initiated, via the second origin at the other genome end, prior to its displacement from the duplex by the advancing replication fork, or else it can form a pan-handle structure in which its two ends come together to form a functional origin on which its replication can be initiated subsequently.

## Assembly and Release

Electron microscopy performed on nuclei of cells late in the infectious cycle reveals the formation of paracrystalline arrays of adenovirus particles; this is the site of adenovirus assembly. The initial step in particle formation is the bringing together of the necessary components. Progeny genomes are generated within prominent replication centers in the nucleus while the proteins necessary to package this DNA are produced in the cytoplasm and then transported back to the nucleus. Hexon polypeptides require the assistance of a large 100 kDa protein, encoded in the MLTU L4 region, and protein pVI, to form trimers and reach the nucleus efficiently. It is generally believed that immature adenovirus particles assemble with the aid of scaffolding proteins (proteins that appear in intermediate virion forms but not the mature virion), prior to the insertion of the DNA genome. DNA packaging into these particles requires specific sequences in the genome that lie between the left-end copy of the inverted terminal repeat and the E1A gene (**Figure 2**). Packaging also involves the viral IVa2, L1 52/55 kDa, and L4 22 kDa proteins, and probably other host factors, to achieve efficient encapsidation of the DNA, which is accompanied by the ejection of scaffolding proteins. Sequence-specific binding of IVa2 to elements in the packaging sequence is important in this process. The final step in particle maturation is the cleavage of a number of the component proteins, including the genome-attached pTP, by the virion 23 kDa protease, an L3 gene product. The protease is stimulated by a peptide cleaved from pVI. Maturation cleavage is essential for the particles to acquire infectivity (see the section titled 'Attachment and entry').

In immortalized cell lines such as HeLa, infected cells show severe cytopathic effect, with loss of cytoskeleton leading to rounding and detachment from the monolayer within 1–2 days at high multiplicity of infection. However, the release of virus particles is rather inefficient, with considerable amounts of virus remaining cell associated. One of the E3 gene products, an 11.6 kDa protein (ADP, an abbreviation for adenovirus death protein) that is produced late in infection, facilitates loss of cell viability and particle release. In its absence, although cytopathic effect occurs normally, cell death is delayed and virus release is even less effective. However, this protein is not widely conserved, even among human adenoviruses, so the general significance of this function is uncertain.

The time taken from the initial addition of virus to the point of maximum progeny yield in cultures of HeLa cells is 20–24 h. However, in cultures of normal fibroblasts, this period is considerably longer, at around 72 h. The precise factors that determine the rate of adenovirus replication in different cell types are not known, but the availability of host functions needed for adenovirus replication may be higher in established cell lines than in normal cells.

The tissues that are targeted by adenovirus *in vivo* are epithelial cell sheets. Individual cells are linked by tight junctions that define distinct apical and basolateral membrane surfaces (externally facing and internally facing, respectively; **Figure 5**). In a culture model of such epithelia, virus infection leads to initial virus release via the basolateral surface. However, fiber–CAR interactions (see the section titled 'Attachment and entry') break down the CAR–CAR adhesion between adjacent cells, leading to increased permeability across the epithelium and escape of virus to the apical surface through gaps between the cells. The location of CAR on the basolateral surface of the cell means that when virus first enters the body and encounters the apical surface of an epithelium, it cannot access the receptors necessary to initiate infection. The initial round of infection must depend either on a break in epithelial integrity exposing the basolateral surface or on an alternative receptor that the virus can use to enter the first target cell in the epithelium.

## Effects on the Host Cell

The interaction between adenovirus and its host is complex and subtle. Infection leads to a variety of effects on the molecular and cell biology of the host. These effects can usually be rationalized as favoring the production of progeny virus. A very large number of such effects have been studied; this section mentions some of those that are best understood.

One area of intervention by viral proteins is in reacting to aspects of the host response to infection. Several viral proteins act to block the pro-apoptotic signaling that is activated early in the infectious process. These include the E1B 19 kDa and 55 kDa proteins, the E3 14.7 kDa protein, and the E3 RID protein complex. The adaptive immune response is also impaired by the E3 gp19 kDa protein, which prevents export to the cell membrane of mature major histocompatibility complex class I antigens. A third intervention is by the VA RNA, which blocks activation of protein kinase R, an aspect of the interferon response that is triggered by double-stranded RNA produced during the infection by hybridization between

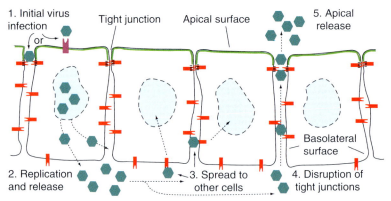

**Figure 5** A schematic diagram of a polarized epithelial cell sheet, showing the sequence of events (1–5) by which adenovirus is thought to infect the epithelium and then spread both locally and to distant sites. CAR receptor molecules are shown in red, and a hypothetical alternative receptor in pink. See text for further details.

transcripts produced from the two genome strands; this would otherwise stop protein synthesis. Susceptibility of adenovirus infections to interferon is also apparently blocked via the action of E4 Orf3 protein on PML nuclear bodies (promyelocytic leukemia protein-containing structures also known as ND10). Finally, the host cell stress and DNA damage sensor, p53, is inactivated and targeted for degradation, through the action of E1B 55 kDa and E4 Orf6 proteins, while these same viral proteins (as well as E4 Orf3 in species *Human adenovirus C*) also block the double-stranded DNA break repair pathway, which otherwise leads to end-to-end joining of progeny genomes and a substantial failure of replication.

Adenovirus proteins also have effects on the host gene expression apparatus that impact on viral gene expression. First, through the action of E1A and E4 Orf6/7 proteins, the host transcription factor E2F is released from inhibitory complexes with Rb (retinoblastoma protein) family members so that it can activate the E2 promoter; this also has the effect of activating cell-cycle progression. Second, the activity of protein phosphatase 2A is altered through the action of E4 Orf4 protein, leading to a variety of effects including changes in the activity of SR splicing proteins and consequent changes in viral mRNA splicing pattern during the course of infection. Third, the E1B 55 kDa and E4 Orf6 proteins cause a block to the export of host cell mRNA from the nucleus during the late phase of infection and instead favor the export of viral mRNA. Finally, the L4 100 kDa protein is involved in allowing efficient translation of viral mRNA while the normal process of cap-dependent recognition of mRNA by ribosomes is inhibited.

## Adenoviruses as Vectors

Several features of human adenoviruses have led to interest in them as potential gene delivery vectors for human cells. Applications are under development in which such vectors are designed to elicit a specific immune response to a delivered gene product (i.e., a recombinant vaccine), to give tumor-specific cell killing for cancer therapy, or to give long-term persistence and expression of a delivered gene (a gene therapy approach). Initial attention focused on human adenovirus 5 because of its experimental tractability (ease of making recombinants, ability to grow virus to very high titers in cell culture, and high level of genetic characterization), its ability to deliver genes into a variety of cell types whether or not they were actively dividing, and its mild disease course in a natural context. Difficulties experienced with these vectors, including the relative nonaccessibility of the CAR receptor on target tissues (see the section titled 'Assembly and release'), have since led to the development of vectors based on several other adenoviruses.

To convert human adenovirus 5 into a passive gene delivery vector, genome modifications are required to render it replication-defective and to create sufficient spare capacity for the insertion of foreign DNA; otherwise, only an additional 1–1.5 kbp of DNA can be accommodated. The E3 region is usually deleted, as its functions are dispensable for growth in cell culture, and the E1A and E1B genes are also removed. This latter deletion renders the virus largely replication defective, but the deficiency can be readily complemented in widely available cell lines to allow growth of the vector in the laboratory. Adding further deletions in other essential genes, such as E2 and E4, further impairs replication, giving even lower levels of late gene expression. Ultimately, all the viral genes can be stripped from the vector genome to leave only the replication origins and packaging sequence from the termini. These vectors have given the best performance among adenovirus vectors for long-term gene delivery but are more difficult to complement for growth in culture and hence much lower yields of vector particles are achieved.

*See also:* Adenoviruses: General Features; Adenoviruses: Malignant Transformation and Oncology; Adenoviruses: Pathogenesis; Coxsackieviruses; Gene Therapy: Use of Viruses as Vectors.

## Further Reading

Akusjärvi G and Stévenin J (2003) Remodelling the host cell RNA splicing machinery during an adenovirus infection. *Current Topics in Microbiology and Immunology* 272: 253–286.

Berk AJ (2005) Recent lessons in gene expression, cell cycle control, and cell biology from adenovirus. *Oncogene* 24: 7673–7685.

Berk AJ (2006) *Adenoviridae*: The viruses and their replication. In: Knipe DM and Howley PM (eds.) *Fields Virology*, 5th edn., ch. 63. Philadelphia: Lippincott Williams and Wilkins.

Davison AJ, Benkö M, and Harrach B (2003) Genetic content and evolution of adenoviruses. *Journal of General Virology* 84: 2895–2908.

Dimmock NJ, Easton AJ, and Leppard KN (2006) *Introduction to Modern Virology*, 6th edn. Oxford: Blackwell Scientific.

Fabry CMS, Rosa-Calatrava M, Conway JF, et al. (2005) A quasi-atomic model of human adenovirus type 5 capsid. *EMBO Journal* 24: 1645–1654.

Leppard KN (1998) Regulated RNA processing and RNA transport during adenovirus infections. *Seminars in Virology* 8: 301–307.

Liu H, Naismith JH, and Hay RT (2003) Adenovirus DNA replication. *Current Topics in Microbiology and Immunology* 272: 131–164.

Meier O and Greber UF (2003) Adenovirus endocytosis. *Journal of Gene Medicine* 5: 451–462.

Ostapchuk P and Hearing P (2005) Control of adenovirus packaging. *Journal of Cellular Biochemistry* 96: 25–35.

Russell WC (2000) Update on adenovirus and its vectors. *Journal of General Virology* 81: 2573–2604.

Zhang Y and Bergelson JM (2005) Adenovirus receptors. *Journal of Virology* 79: 12125–12131.

# Adenoviruses: Pathogenesis

**M Benkő**, Veterinary Medical Research Institute, Hungarian Academy of Sciences, Budapest, Hungary

© 2008 Elsevier Ltd. All rights reserved.

## History

Adenoviruses were first recognized as distinct agents in 1953, although certain disease conditions that were later clarified as being caused by adenoviruses had been described previously in humans as well as in animals. These include epidemic keratoconjunctivitis, the viral etiology of which had been suggested as early as 1930. It had also been recognized before the discovery of adenoviruses that the so-called fox encephalitis and dog hepatitis were caused by the same virus. Also, the first avian adenovirus (chicken embryo lethal orphan (CELO) virus) isolate had erroneously been obtained when bovine samples of lumpy skin disease were examined for an infectious etiological agent by inoculation into embryonated eggs.

The actual discovery of human adenoviruses was the result of targeted investigations into the etiology of an acute, respiratory illness frequently affecting young military recruits in the USA. The first isolates (those assigned low type numbers) were obtained either from diseased soldiers or from adenoid tissues removed by tonsillectomy (hence the name adenovirus). Morphologically indistinguishable viruses were classified as new serotypes on the basis of the results of cross-neutralization tests. The number of human adenovirus serotypes grew rapidly during the 1960s and 1970s. There are 51 human adenovirus serotypes known to date, and an additional candidate has been discovered recently. Nonhuman animal adenoviruses were initially described largely in domesticated species of economic importance. Although a considerable number of new human or animal adenoviruses were isolated from diseased individuals, many were recovered from primary tissue cultures, such as simian adenoviruses from Vero cells and bovine adenoviruses from calf kidney or testicle cell cultures.

The striking diversity of human adenoviruses was recognized early on, and various biological properties, most of which have obvious influences on pathogenicity, were used to establish a basis for subclassification (**Table 1**). The criteria used for grouping human adenoviruses included differences in hemagglutination properties, nucleotide composition of the genomic DNA (i.e., G+C content), and oncogenicity *in vivo* and *in vitro* (the latter indicated by malignant transformation in tissue culture). Four hemagglutination groups (I–IV) and six human adenovirus subgroups (A–F) were established. In the most recent taxonomy, the latter groups, which were often referred to as subgroups or subgenera, are now correctly termed species *Human adenovirus A* (abbreviated informally to HAdV-A) through *Human adenovirus F* (HAdV-F). Phylogenetic analyses of nucleotide and protein sequences and characteristic features of genome organization fully support this classification. Each species consists of a collection of serotypes; for example, human adenovirus 12 (HAdV-12) is a member of species HAdV-A. Species are also being erected for animal adenoviruses in a similar manner.

**Table 1** Features of human adenoviruses

| Species Serotype | Hemagglutination group | Tumors in animals | Transformation in tissue culture | G+C content in DNA[a] | Associated disease and affected organs |
|---|---|---|---|---|---|
| Human adenovirus A HAdV-12, 18, 31 | IV (no or weak hemagglutination) | High | Positive | 45–47 (48–49) | Cryptic enteric infection |
| Human adenovirus B HAdV-3, 7, 11, 14, 16, 21, 34, 35, 50 | I (complete agglutination of monkey erythrocytes) | Moderate | Positive | 48–52 (50–52) | Conjunctivitis Acute respiratory disease Hemorrhagic cystitis Central nervous system |
| Human adenovirus C HAdV-1, 2, 5, 6 | III (partial agglutination of rat erythrocytes) | Low or negative | Positive | 55–56 (57–59) | Endemic infection Respiratory symptoms |
| Human adenovirus D HAdV-8–10, 13, 15, 17, 19, 20, 22–30, 32, 33, 36–39, 42–49, 51 | II (complete agglutination of rat erythrocytes) | Low or negative (mammary tumors) | Positive | 48–57 (57–61) | Keratoconjunctivitis In immunocompromised and AIDS patients |
| Human adenovirus E HAdV-4 | III | Low or negative | Positive | 57 (57–59) | Conjunctivitis Acute respiratory disease |
| Human adenovirus F HAdV-40, 41 | IV | Negative | Negative | 51 | Infantile diarrhea |

[a]Values are based on complete genome sequences. Estimates made previously using other methods are shown in parentheses.

## Prevalence and Spread

Adenoviruses are abundant and widespread, and in principle every vertebrate species could host at least one species. Endemic or enzootic infections that are frequently inapparent seem to be very common. In contrast, epidemic or epizootic occurrence is relatively rare and normally has much more significant consequences. The nonenveloped virions shed by infected individuals are rather stable in the environment and can survive for long periods. Virions are not only able to tolerate adverse environmental effects, such as drought and moderate temperature or pH changes, but are also resistant to lipid solvents and simple disinfectants. There is evidence that human adenoviruses can persist and perhaps even retain infectivity in natural or communal water sources. Moreover, freshwater and marine bivalves have also been found to accumulate adenoviruses. Because of their stability, adenoviruses have become a major indicator of viral pollution in the environment. Vertical transmission has also been documented in several host species, including birds and cattle.

Although different species or types of adenovirus have different affinities for various organs and tissues, primary virus shedding is likely through the intestines. Adenoviruses are often transmitted by person-to-person contact, particularly among young children where fecal–oral spread is common. Aerosol transmission is also possible and is probably not rare in crowded populations. Adenovirus DNA may be found in tonsillar tissue, peripheral blood lymphocytes, and lung epithelial cells long after clinical disease has abated. Swimming pool-related outbreaks, particularly of strains causing keratoconjunctivitis or pharyngitis, are not uncommon. In poultry and other farm animals, transportation, crowding, and mixing of different populations can lead to mass infections.

## Epidemiology

The tropism and pathogenicity of human adenoviruses are largely type- and species-dependent. Respiratory pathogens are common, although infections of a large variety of different organs have been described. The fiber protein is likely to mediate primary tissue tropism, but the specific determinants, epitopes, and receptors have been mapped only preliminarily. The most common clinical manifestations connected to each human adenovirus species are listed below (also see **Table 1**). HAdV-B contains two phylogenetic lineages, which are informally termed subspecies HAdV-B1 and HAdV-B2.

### HAdV-A

HAdV-12, HAdV-18, and HAdV-31 can replicate efficiently in the intestines, and, based on serological surveys, are common in the population, especially in children with gastrointestinal disease. However, their role in the etiology of infant diarrhea is yet to be determined. Serotypes belonging to this species are generally difficult to isolate and culture.

### HAdV-B1

Respiratory pathogens HAdV-3, HAdV-7, HAdV-16, and HAdV-21 belong here. Seasonal outbreaks of febrile respiratory diseases, mainly during winter, can be caused in infants by HAdV-7 in most parts of the world. Similar diseases, usually with a less severe outcome, can be seen among school children. In some countries, HAdV-3 is the dominant serotype, whereas HAdV-7 seems to have a number of different genotypes that shift occasionally in certain geographic areas. HAdV-4 (from species HAdV-E) and HAdV-14 (from subspecies HAdV-B2) are most often implicated in the etiology of acute respiratory outbreaks among freshly enlisted military recruits in the USA, in addition to HAdV-7 and HAdV-21.

### HAdV-B2

HAdV-11, HAdV-34, and HAdV-35 cause persistent interstitial infection in the kidney and hemorrhagic cystitis. These, along with one of the most recently described human adenoviruses (HAdV-50), are most often shed in the feces or urine of acquired immune deficiency syndrome (AIDS) patients and organ or tissue transplantation recipients. Severe respiratory infections caused by a novel variant of HAdV-14 with seemingly elevated pathogenicity have emerged in the USA during 2006 and 2007.

### HAdV-C

The low serotype designations (HAdV-1, HAdV-2, HAdV-5, and HAdV-6) of viruses in this group reflect the relative ubiquity of the species. They can easily be isolated and, indeed, comprise approximately half of all human adenovirus serotypes reported to the World Health Organization. HAdV-1, HAdV-2, and HAdV-5 are known to maintain endemic infections, and most teenagers will have had infections with more than one serotype. The site of persistence is lymphoid tissue, and shedding can last for a couple of years after primary infection.

### HAdV-D

This species contains 32 serotypes and therefore encompasses well over one half of all known human adenoviruses. HAdV-8, HAdV-19, and HAdV-37 cause epidemic keratoconjunctivitis, especially in dry climates or densely populated areas. Other serotypes are rarely isolated except from immunocompromised patients.

### HAdV-E

This is the only species that has a single serotype, namely HAdV-4, although different genotypes of this virus have been described. HAdV-4 strains have been most often implicated in respiratory diseases among military recruits in the USA. In Japan, HAdV-4 is the second most important cause of adenovirus-associated eye disease after HAdV-8.

### HAdV-F

The so-called enteric or fastidious HAdV-40 and HAdV-41 are classified here. The genetic distance between these two serotypes comes close to meriting separation into different species, but, for practical reasons including indistinguishable pathology, they remain in a single species.

These viruses are a major cause of infantile diarrhea all over the world. In Europe, a shift in dominance from HAdV-40 in the 1970s to HAdV-41 after 1992 has been observed. Because of a deletion in the E1 region near the left end of the genome, HAdV-40 and HAdV-41 can be isolated and cultured only in complementing, transformed cell lines such as 293 or A547. The most recently described human adenovirus serotype, namely HAdV-52, was also recovered from patients with diarrhea. HAdV-52 has not yet been classified into a species.

## Epizootiology

### Simian Adenoviruses

A series of simian viruses (termed SVs) were isolated from various tissue cultures prepared from apes and monkeys. These were numbered serially, irrespective of the virus family to which they belong, since identification and allocation to a family were performed later. A considerable number of the SV isolates have been identified as adenoviruses. In many cases, unfortunately, the original SV numbers have been retained as the numbers of the adenovirus type, and thus a somewhat confusing system of SAdV numbering can still be encountered in publications as well as in the records of the American Type Culture Collection. To date, there are 25 simian adenoviruses (SAdV-1 through SAdV-25). SAdV-1 to SAdV-20 originate from Old World monkeys, whereas SAdV-21 to SAdV-25 are from chimpanzees. There are no known examples of New World monkey adenoviruses, though it is likely that they exist.

There are few data concerning prevalence or pathogenicity of simian adenoviruses in their natural hosts. An interesting mixing and grouping can be observed among the primate adenoviruses, suggesting that they can sometimes be less host specific, apparently readily infecting closely related hosts.

### Canine Adenoviruses

Only two serotypes of adenovirus have been isolated from carnivores, both termed canine adenoviruses. CAdV-1 is the causative agent of infectious canine hepatitis, a life-threatening disease of puppies. Regular vaccination worldwide has decreased the number of clinical cases. Serologically indistinguishable viruses have been shown to cause encephalitis in numerous other carnivore species such as foxes, raccoons, bears, and skunks. The disease caused by a genetically closely related but serologically distinct virus, CAdV-2, is called kennel cough and is common among breeder stocks.

Interestingly, in spite of a limited number of inclusion-body-hepatitis-like conditions in felids, focused attempts to find a distinct feline adenovirus remain unsuccessful. It is likely, however, that cats living in close proximity with humans can harbor HAdV-1 or HAdV-5 from species HAdV-C.

### Porcine Adenoviruses

At least five porcine adenovirus types (PAdV-1 to PAdV-5) are recognized. These represent three different species. Species PAdV-A contains three serotypes (PAdV-1 to PAdV-3) that are fairly similar to each other and cause no specific diseases. PAdV-4 (in species PAdV-B) has been described as associated with neurological disease, and PAdV-5 (in species PAdV-C) is most distantly related to PAdV-A. The pathogenic roles of porcine adenoviruses need further investigation.

### Equine Adenoviruses

Two equine adenovirus serotypes (EAdV-1 and EAdV-2) were described several decades ago, but their genetic characterization has lagged behind that of other adenoviruses. Only very short, partial sequences from the hexon gene are available, and apparently these contain poorly resolved areas that will need thorough revisiting. Certain Arabian horse lineages carry a genetic defect that, when present in the homozygous state, causes severe combined immunodeficiency disease. Affected animals are incapable of mounting an immune response, and foals usually die of pneumonia due to equine adenovirus, which is harmless in immunocompetent animals.

### Bovine, Ovine, Caprine, and Other Ruminant Adenoviruses

There is an amazing diversity of adenoviruses that can infect ruminant hosts. The adenoviruses described above all belong to various species in genus *Mastadenovirus*, as do a number of ruminant adenoviruses. Unusual bovine adenoviruses that are substantially different from mastadenoviruses, as judged for example by the absence of the genus-specific antigen, were described almost four decades ago. Out of the roughly one dozen bovine adenovirus serotypes (in the BAdV series) recognized to date, six are now classified in a separate genus, *Atadenovirus*. One ovine (OAdV-7) and one goat (GAdV-1) adenovirus are also classified as atadenoviruses.

Bovine mastadenoviruses were recognized early on, and were found occasionally to cause enzootic bronchopneumonia or calf pneumo-enteritis. Oddly enough, experimental infection of young calves with the isolated adenoviruses seldom, if at all, resulted in convincing reproduction of disease. BAdV-10 is one of the most interesting ruminant mastadenoviruses in that all five isolates originated from diseased or dead animals. Genomic analysis of these isolates, one from New Zealand and four from Northern Ireland, revealed genetic variations in and around the fiber gene, suggesting that the virus might be undergoing some kind of adaptation process. Propagation of BAdV-10 strains has been difficult, with most success achieved on either primary testicle cells or a pulmonary alveolar cell line. It is noteworthy that BAdV-10 not only clusters together with mouse adenovirus (MAdV-1) on phylogenetic trees but also that it has a very simple E3 region. BAdV-10 has been proposed as an example of an adenovirus that may be in the process of switching from one host to another. Crossing the host barrier has also been suggested for BAdV-2, with two subtypes, one of which is capable of infecting calves and the other sheep. Based on evolutionary relationships, BAdV-2 is presently classified into species OAdV-A together with OAdV-2 to OAdV-5.

Bovine atadenoviruses have been found more often in the intestines, and disease reproduction was more frequently successful. Serological evidence is available for the occurrence of mastadenoviruses and atadenoviruses in a large number of free living and wild ruminant hosts. A novel atadenovirus was recently recovered from mule deer during an epizootic in California causing high mortality.

## Avian Adenoviruses

Adenoviruses isolated from poultry and waterfowl were initially classified into genus *Aviadenovirus*. In addition to the criterion of host origin, aviadenoviruses can be distinguished from mastadenoviruses on the basis of a lack of the genus-common complement-fixing antigen. A large number of serotypes have been described from chicken. Some of these viruses have been isolated from other species as well. For example, strains serologically identical to CELO virus have also been found to cause bronchitis in quails. The pathologic roles of aviadenoviruses are not fully understood. The results of experimental infections have been ambiguous in the majority of cases. A specific condition referred to as hydro-pericardium syndrome has been clearly associated with fowl adenovirus 4 (FAdV-4) from species FAdV-C. Another specific disease called gizzard erosion, however, could not be linked to a definite aviadenovirus type, and experimental infections gave contradictory results.

Duck adenovirus 1 (DAdV-1) is the causative agent of egg drop syndrome (EDS), which results in depressed egg production accompanied by the production of abnormal (soft-shelled or deformed) eggs. The disease was first experienced in Europe in 1976, and soon became known worldwide. Retrospective serological studies confirmed the presence of hemagglutination inhibitory antibodies to EDS virus in archive sera originating from a large number of wild and domestic bird species, with a predominance in waterfowl, which are now considered as the main reservoir. The virus was initially assigned to genus *Aviadenovirus*, but was recorded as an exception because of its lack of common complement-fixing antigens with other fowl adenoviruses. Genome analysis of the EDS virus revealed its relatedness to atadenoviruses, and DAdV-1 has consequently been moved into this genus, which is hypothesized to be the adenovirus lineage of reptilian hosts.

Turkey adenovirus 3 (TAdV-3), better known as turkey hemorrhagic enteritis virus, shows cross-reactivity with neither aviadenoviruses nor DAdV-1, and used to be viewed as another exception among the aviadenoviruses. Serologically indistinguishable strains have been associated with various pathological entities in turkey, pheasant, and chicken. Based on genome analyses and phylogenetic calculations, TAdV-3 is a member of genus *Siadenovirus*, along with the single known adenovirus from an amphibian host, frog adenovirus 1 (FrAdV-1). This lineage has been tentatively proposed to represent adenoviruses that have coevolved with amphibians. Interestingly, a recently described sensitive polymerase chain reaction (PCR) method led to the detection of a novel type (and likely new species) of siadenovirus in diseased birds of prey. A novel avian adenovirus has also been isolated from falcons.

## Reptilian Adenoviruses

Every adenovirus found in reptiles thus far is an atadenovirus. Adenovirus infections in various snake and lizard species have been described repeatedly, but the number of virus strains isolated is very limited. In Germany, captive boid snakes from multiple collections and breeders seem to be infected with the same type of adenovirus, and also frequently by a parvovirus. Neutralizing antibodies to the adenovirus have been found in 13% of more than 100 serum samples originating from free-living and captive snakes of miscellaneous species. Bearded dragons, a reptile pet of growing popularity in North America and Europe, seem frequently to have an adenovirus infection. Partial sequence analyses have revealed that identical viruses are present on both continents. At least six additional atadenovirus types have been detected by PCR in various lizard species.

### Frog and Fish Adenoviruses

Data concerning adenovirus infection of aquatic vertebrates are scarce. In fact, only a single adenovirus has been isolated from each of the two host classes, amphibians and fish. Interestingly, FrAdV-1 was isolated from a renal tumor of a leopard frog on a reptilian cell line (TH1) prepared from turtle heart tissue. FrAdV-1 has a genome organization similar to that of TAdV-3, and the common evolutionary origin of these two viruses in genus *Siadenovirus* is supported by phylogenetic calculations. The only adenovirus isolate available from fish was obtained from an ancient chondrostei species, the white sturgeon. Partial genomic characterization implies a likely new genus. Nuclear inclusion bodies typical of adenoviruses and adenovirus-like particles have been observed by light and electron microscopy, respectively, in damaged tissues of a couple of other fish species, but molecular confirmation of adenovirus infection has not yet been successful.

### Pathogenesis

Adenovirus infections of man generally occur in childhood, and the outcome varies in severity from asymptomatic to explosive outbreaks of upper or lower respiratory tract manifestations. Less commonly, adenoviruses cause gastrointestinal, ophthalmic, urinary, and neurological diseases. The vast majority of adenovirus-caused diseases are self-limiting. However, immunocompromised patients, above all organ transplant recipients, individuals infected with human immunodeficiency virus (HIV) developing AIDS, and those receiving radiation and chemotherapy against tumors, represent special populations that are prone to experience grave, frequently fatal consequences of adenovirus infection. In numerous cases, the organ to be transplanted itself proves to be the source of invasive adenoviruses. Sporadic fatal infections may occasionally occur in healthy, immunocompetent individuals. In such cases, the presence of certain predisposing immunogenetic factors cannot be excluded. Cellular immune responses are also important for the recovery from acute adenovirus infection. Peripheral blood mononuclear cells have been found to exhibit proliferative responses to HAdV-2 antigen. This function is mediated by CD4+ T cells, which seem to recognize conserved antigens among different human adenovirus serotypes.

The incubation period from infection to clinical symptoms is estimated to be 1–7 days and may be dose dependent. Clinical symptoms during the initial viremia are dominated by fever and general malaise. Recovery from infection is associated with the development of serotype-specific neutralizing antibodies that protect against disease or reinfection with the same serotype of the virus but do not cross-react with other serotypes.

Pathology by adenoviruses is partially the consequence of viral replication and cell lysis. Correspondingly, in various tissues and organs, such as bronchial epithelium, liver, kidney, and spleen, disseminated necrotic foci can be observed upon necropsy or histopathological examination. Characteristic intranuclear inclusion bodies and so-called smudge cells contain large amounts of adenovirus capsid proteins. Besides the lesions caused by virus replication, direct toxic effects of high doses of structural proteins, as well as the host's inflammatory respond, may contribute to the aggravation of pathology. Experimental infection of animals with various adenovirus types seldom results in a pathology similar to that experienced with the same virus under natural circumstances. One of the few exceptions is turkey hemorrhagic enteritis, the pathogenesis of which has been studied in detail and is due to TAdV-3, which causes intestinal hemorrhages and immunosuppression. By *in situ* DNA hybridization and PCR, the presence of virus-specific DNA as evidence for virus replication has been demonstrated in the immunoglobulin M-bearing B lymphocytes and macrophage-like cells, but not in CD4+ or CD8+ T lymphocytes. Interestingly, fewer virions were present in the intestines, which are the principal site of pathology, than in the neighboring lymphoid organs including spleen and cecal tonsils. This finding strongly suggests that the intestinal lesions induced by TAdV-3 are mediated by the immune system. Systemic or intestinal hemorrhagic disease of ruminants seems to be related to virus replication in endothelial cells.

### Detection and Identification

With the development of modern techniques, especially PCR and direct DNA sequencing, the number of adenoviruses detected in various organ samples of human or animal origin has increased rapidly. However, there is no official agreement or convention on the criteria that would be prerequisites for the approval of new serotypes. The conventional methods, including virus isolation in tissue culture, raising antisera, and performing a large set of cross-neutralization tests, are cumbersome, and the majority of medical and veterinary diagnostic laboratories do not possess appropriate prototype strain and serum collections. Full genomic sequences should validate new types even in the absence of isolated virus strains. However, the value of short sequences from PCR fragments is still a topic of debate. In principle, PCR and sequencing should be able to replace serotyping if appropriate targets are identified.

Human medical laboratories use commercially available tests, such as complement fixation and enzyme immunoassay, to detect adenovirus-specific antibodies that cross-react with all serotypes. Nearly all adults have serologic evidence of past infection with one or more adenoviruses. For the detection of human or animal adenovirus-specific DNA, the most common target in PCR methods was initially the gene encoding the major capsid protein, the hexon. With subsequent restriction enzyme digestion, typing systems for human adenoviruses have also been elaborated. Recently, a novel nested PCR method targeting the most conserved region of the adenovirus DNA polymerase gene has been published. The highly degenerate consensus primers seem to be capable of facilitating amplification of DNA from every adenovirus known, irrespective of genus affiliation. A major drawback of this exceptionally sensitive method is the relaxed specificity required. Although its application cannot be recommended for routine diagnostic purposes, it may come in handy for finding novel adenoviruses, especially in cases where adenovirus involvement is strongly suspected from other evidence, such as electron microscopy.

## Prevention and Therapy

In the USA, orally administrable, live, enteric-coated vaccines against HAdV-4, HAdV-7, and HAdV-21 were used in military units for a couple of decades. After the cessation of vaccine production in 1996, the impact of adenovirus infection among military recruits increased, and re-emergence of HAdV-7 and especially HAdV-4 has been verified. Since 1999, 12% of all recruits were affected by adenovirus disease. Efforts to resume vaccination are in progress.

In the veterinary practice, dog vaccination schedules all over the world invariably include a live or killed CAdV-1 component against dog hepatitis. Inactivated vaccine for horses against equine adenoviruses has been prepared in Australia. In farm animals, inactivated bivalent vaccines (containing one mastadenovirus and one atadenovirus) have been in use in several countries for controlling enzootic calf pneumonia or pneumo-enteritis. In poultry practice, commercially available or experimental vaccines for the prevention of EDS or turkey hemorrhagic enteritis are applied occasionally. There are several attempts ongoing for the production of recombinant subunit vaccines, which should be safer than vaccines derived from infected birds or tissue culture.

No specific anti-adenovirus therapy has yet been established. Recent advances in understanding the pathophysiology of fulminant adenovirus diseases in immunocompromised patients have prompted the consideration of applying donor lymphocyte infusions after transplantation.

Cidofovir is a monophosphate nucleotide analog that, after undergoing cellular phosphorylation, competitively inhibits incorporation of dCTP into virus DNA by the virus DNA polymerase. Incorporation of the compound disrupts further chain elongation. Cidofovir demonstrates activity *in vitro* against a number of DNA viruses, including adenoviruses. There are a limited number of experiences with using this drug against adenovirus infections, and its clinical utility remains to be determined.

*See also:* Adenoviruses: General Features; Adenoviruses: Malignant Transformation and Oncology; Adenoviruses: Molecular Biology; Gene Therapy: Use of Viruses as Vectors.

## Further Reading

Barker JH, Luby JP, Sean Dalley A, et al. (2003) Fatal type 3 adenoviral pneumonia in immunocompetent adult identical twins. Clinical Infectious Diseases 37: 142–146.

Jones MS, Harrach B, Ganac RD, et al. (2007) New adenovirus species found in a patient presenting with gastroenteritis. Journal of Virology 81: 5978–5984.

Kojaoghlanian T, Flomenberg P, and Horwitz MS (2003) The impact of adenovirus infection on the immunocompromised host. Reviews in Medical Virology 13: 155–171.

Leen AM, Bollard CM, Myers GD, and Rooney CM (2006) Adenoviral infections in hematopoietic stem cell transplantation. Biology of Blood and Marrow Transplantation 12: 243–251.

Leen AM, Myers GD, Bollard CM, et al. (2005) T-cell immunotherapy for adenoviral infections of stem-cell transplant recipients. Annals of the New York Academy of Sciences 1062: 104–115.

Neofytos D, Ojha A, Mookerjee B, et al. (2007) Treatment of adenovirus disease in stem cell transplant recipients with cidofovir. Biology of Blood and Marrow Transplantation 13: 74–81.

Schrenzel M, Oaks JL, Rotstein D, et al. (2005) Characterization of a new species of adenovirus in falcons. Journal of Clinical Microbiology 43: 3402–3413.

Tang J, Olive M, Pulmanausahakul R, et al. (2006) Human CD8+ cytotoxic T cell responses to adenovirus capsid proteins. Virology 350: 312–322.

Wellehan JFX, Johnson AJ, Harrach B, et al. (2004) Detection and analysis of six lizard adenoviruses by consensus primer PCR provides further evidence of a reptilian origin for the atadenoviruses. Journal of Virology 78: 13366–13369.

# African Cassava Mosaic Disease

**J P Legg,** International Institute of Tropical Agriculture, Dar es Salaam, Tanzania, UK and Natural Resources Institute, Chatham Maritime, UK

© 2008 Elsevier Ltd. All rights reserved.

## Glossary

**Pandemic** An outbreak of a disease over a whole country or large part of the world.
**Pseudo-recombinant** A viral infection involving complementary genome components from different virus species.

## History

Cassava (*Manihot esculenta* Crantz) is a root crop that is grown widely throughout the tropics, primarily for its value as a starchy staple food. From its origins in Latin America, cultivated cassava was introduced to Africa in the sixteenth century by Portuguese seafarers, and subsequently spread through much of Africa south of the Sahara. Fortuitously perhaps, none of the viruses that affect cassava in the Americas seems to have been co-introduced with the crop. Over time, however, the crop became infected by indigenous viruses. The first report of a virus-like disease in African cassava was made in 1894 from what is now northeastern Tanzania. The original German descriptor, 'Krauselkrankheit', made reference to the characteristic mosaic symptoms elicited in affected plants. It was not until just over a decade later, however, that the first firm indication was given that the disease had a viral etiology. In spite of these early reports, there seems to have been little concern about the impact of cassava mosaic disease (CMD) until the 1920s. Between 1929 and 1937, however, numerous reports were made of the spread and damaging effect on cassava crops of CMD from diverse locations throughout the continent, from the island of Madagascar off the southeastern shores of the African mainland, to Sierra Leone in West Africa. These developments provided the stimulus for the earliest concerted efforts to develop approaches to controlling the viruses that caused this damaging crop disease. Although substantial progress was made in the development of cassava varieties that were resistant to cassava mosaic-causing viruses, in the 1930s and 1940s, the viruses themselves remained poorly understood, and it was not until 1983 that the first definitive study confirming the viral etiology of CMD was published. Geminate virions were shown to encapsidate a bipartite genome of single-stranded circular DNA, leading to the designation of these viruses as geminiviruses in the genus *Begomovirus*.

The group is now commonly referred to as the cassava mosaic geminiviruses (CMGs). Although the earliest studies of diversity indicated the occurrence of two species in Africa, based on serological characterization, more detailed genetic studies conducted from 1993 up to the present day have provided evidence for the occurrence of seven species, albeit with varying levels of geographic coverage. In common with all begomoviruses, CMGs are transmitted by the whitefly vector, *Bemisia tabaci*, a co-evolutionary partnership that has been a key factor in the success of this group of viruses.

## Properties of the Virion, Its Genome, and Replication

### Structural Features

CMGs comprise two single-stranded circular DNA molecules (DNA-A and DNA-B) each encapsidated within a twinned or geminate icosahedral coat, approximately 22 nm × 35 nm in size. The three-dimensional structure has been resolved for one of the CMGs, African cassava mosaic virus (ACMV) (genus *Begomovirus*, family *Geminiviridae*), based on comparisons with related viruses and through electron cryomicroscopy and image reconstruction. Through these analyses, the structure has been shown to resemble that of virions of maize streak virus (MSV) (genus *Mastrevirus*, family *Geminiviridae*) in having two $T=1$ symmetry icosahedra joined at a position at which one subunit is missing from each. This gives a total of 22 capsomers made up of 110 30.2 kDa coat protein molecules. Capsomers of the two halves of CMG particles are twisted to each other by 20° so that capsomers of one half are apposed to gaps between two capsomers of the other half. A six-residue insertion in the βD/βE loop protrudes from the coat protein and has been shown to be required for whitefly transmission. This appears to be an important structural feature of the *B. tabaci* transmitted begomoviruses. By contrast, the leafhopper-transmitted MSV and viruses of the species *Beet curly top virus* (genus *Curtovirus*, family *Geminiviridae*) have their own characteristic 14-residue insertion in the βF/βG loop.

### The Genome

Analyses of the two single-stranded DNA (ssDNA) molecules that make up the CMG genome have revealed the presence of six open reading frames (ORFs) on DNA-A and two on DNA-B. Of the two virion sense ORFs on

DNA-A, AV1 (*CP*) codes for the coat protein and plays an important role in vector specificity, while AV2 is involved in virus movement. Complementary sense DNA-A ORFs include AC1 (*Rep*) which plays a central role in virus replication, AC2 (*TrAP*) which transactivates virion-sense genes and acts as a suppressor of post-transcriptional gene silencing (PTGS), AC3 (*REn*) which enhances virus replication and AC4, whose function is yet to be fully elucidated, but which, like AC2, seems to function in suppressing host plant PTGS. On DNA-B, BV1 (*NSP*) is involved in shuttling virus between the nucleus and cell cytoplasm, while BC1 (*MP*) is implicated in longer distance intercellular movements of CMGs. All CMGs have a conserved intergenic 'common' region of *c.* 200 bp which is also shared between the two genome components.

## Replication

Most of the processes of CMG replication take place within the nuclei of infected cells. In common with other geminiviruses, CMGs multiply using a combination of rolling circle and recombination mediated replication. The mechanisms behind the initiation of negative-strand synthesis have yet to be elucidated, although it is thought that wounding responses at the point of primary infection may lead to the induction of host genes required for cell cycle reentry and DNA replication. Positive-strand nicking followed by synthesis has been mapped to the conserved nonanucleotide TAATATT/AC within the intergenic region (IR). The nonanucleotide sequence sits within the loop of a hairpin-loop structure in the 3′ IR, a feature common to all geminiviruses. Strand nicking and positive-strand synthesis are catalyzed by oligomerized *Rep* in concert with host factors. *Rep* binds specifically to repeated 'iteron' sequences within the IR. Importantly, *Rep* also binds to retinoblastoma related protein (*pRBR*), a process which catalyzes a series of intracellular reactions that move the cell from G to S phase and in so doing mobilizes host factors essential for DNA replication. Double-stranded DNA (dsDNA) produced through the replication process is then thought to assemble into transcriptionally active mini-chromosomes, under the control of *TrAP* and *REn*. Nuclear export is mediated by *NSP*, movement out of the cell through plasmodesmata, and longer-distance transport through *MP*. These processes ultimately lead to the multiplication of viral DNA within initially infected cells, and subsequent systemic movement through phloem tissue. The life cycle is then sustained both through the multiplication of CMG virions in the young newly emerged cassava leaves most favored by adult *B. tabaci*, as well as by the distribution of CMG particles throughout the length of mature stems (in susceptible cultivars), ensuring that when cuttings are taken, the presence of CMG components will ensure the continued infection of the newly sprouting plant.

## Diversity and Distribution

### Distribution of the African CMGs

The earliest studies of CMG diversity and distribution used serological techniques to characterize variability, and then utilized that variability to develop diagnostic tests based on the use of monoclonal antibodies in the enzyme-linked immunosorbent assay (ELISA). Using these methods, two principal groups of CMGs were recognized that were subsequently confirmed as distinct species following the sequencing of DNA-A molecules. These were ACMV and East African cassava mosaic virus (EACMV) (genus *Begomovirus*; family *Geminiviridae*). Two other viruses have been shown to cause CMD in south Asia, namely: Indian cassava mosaic virus (ICMV) and Sri Lankan cassava mosaic virus (SLCMV). The earliest distribution maps of the African CMGs showed EACMV to occur in the coastal east African areas of Kenya, Tanzania, and Madagascar, as well as Malawi and Zimbabwe. ACMV, by contrast, occurred throughout the remainder of the cassava-growing areas of Africa, from South Africa and Mozambique in the southeast, to Senegal in the northwest. Significantly, at this time there was no reported zone of co-occurrence of the two virus species. With the increased use of polymerase chain reaction (PCR)-based diagnostics from 1990s onward, it became possible to identify differences not solely associated with the coat protein. This led to two important developments in the understanding of CMGs in Africa. First, several new species were identified, all of which were more closely related to EACMV than to ACMV. Second, it was shown that virus mixtures belonging to different species occurred frequently. A notable consequence of this finding was the concomitant evidence for the more widespread distribution of the EACMV-like viruses than had hitherto been recognized. New virus identifications included: South African cassava mosaic virus (SACMV) (1998), East African cassava mosaic Malawi virus (EACMMV) (1998), East African cassava mosaic Cameroon virus (EACMCV) (2000), East African cassava mosaic Zanzibar virus (EACMZV) (2004), and East African cassava mosaic Kenya virus (EACMKV) (2006). Significantly, all of the CMGs occur in different parts of East and Southern Africa, while ACMV predominates in West Africa (**Figure 1**). EACMCV, the only EACMV-like virus occurring in West Africa, is nevertheless infrequent and generally occurs in mixed infections together with ACMV. ACMV is absent from coastal areas of Kenya and Tanzania. Although there has been very little CMG characterization in many of the cassava-growing countries of Africa, and it seems clear that much of the variability of this group of viruses still remains to be revealed, an assessment of the data currently available has led to the conclusion that East Africa is the center of diversity for the EACMV-like CMGs and is probably the home for yet-to-be-identified wild hosts of the proto-CMGs that

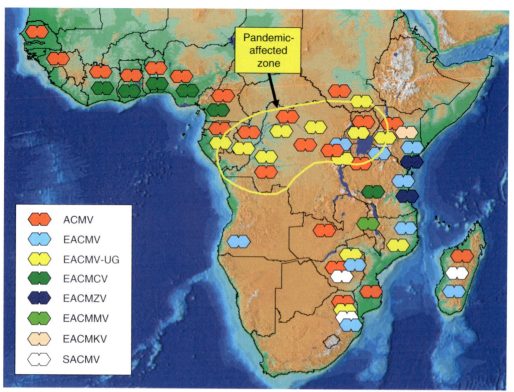

**Figure 1** Diagrammatic representation of the distribution of CMGs in Africa, 2006.

were first introduced to cassava by *B. tabaci* sometime between the earliest introductions of cassava to this part of Africa in the eighteenth century, and the first report of CMD in 1894.

## Recombination and the CMD Pandemic

The CMGs represent a very dynamic group of viruses, and evidence has been presented for the occurrence of both pseudo-recombinants, in which the DNA-A of one virus species co-replicates with the DNA-B of another species, as well as true recombinants, in which portions of the DNA-A or -B of one species have been spliced into the DNA-A or -B of another. In fact, all of the EACMV-like viruses other than EACMV show evidence for recombination events either with known or as yet unknown begomoviruses. One of the most important developments in the study of CMGs in Africa in recent years was the recognition that an unusually damaging strain, referred to as EACMV-UG, had arisen through a recombination event between EACMV and ACMV in which a 340 nt portion of ACMV AV1, had replaced the equivalent portion of the DNA-A of EACMV. The obvious consequence of this was that ELISA-based diagnostic tests erroneously identified this strain as ACMV. This is one of the reasons why PCR-based diagnostic methods are now used almost exclusively for CMG monitoring work. EACMV-UG has been associated with the rapid epidemic-like spread of CMD in East and Central Africa through the latter part of the twentieth and into the current century. Its continually expanding distribution currently includes: Burundi, Democratic Republic of Congo (DRC), Gabon, Kenya, Republic of Congo, Rwanda, southern Sudan, Tanzania, and Uganda (**Figure 1**). Additional countries in which it is likely to be widespread, but has yet to be formally identified, include Central African Republic and Angola. Localized identifications have also been made from South Africa, Swaziland, and Zimbabwe.

The dynamic character of CMG diversity and distribution cannot be overemphasized, and the propensity that this group of viruses shows to produce both naturally occurring pseudo-recombinants, as well as novel true recombinants, ensures that the patterns of diversity and distribution will continue to evolve rapidly.

## Transmission

CMGs, in common with all other members of the genus *Begomovirus*, are transmitted by the whitefly vector *B. tabaci*. Transmission can also be achieved through grafting, and relatively inefficiently through mechanical inoculation of indicator plants, but there is no seed-borne transmission. Cassava is normally propagated through the use of vegetative cuttings, and this is perhaps the most frequent source of infection in new crops under field conditions. *Bemisia tabaci* adults feed, mate, and reproduce preferentially on the upper newly emerged leaves of cassava plants, and almost all transmission occurs here. Transmission is persistent, and ACMV has been shown to be retained by B. tabaci adults for up to 9 days. Transtadial, but not transovarial transmission, has been demonstrated, although the larval instars are unimportant in the epidemiology of CMGs in view of their sessile nature. Minimum periods for each of the stages of transmission for ACMV by B. tabaci adults are: acquisition (3 h), latent period (3 h), and inoculation (10 min). Inoculated plants begin to show symptoms of infection after 3–5 weeks. Varying levels of transmission efficiency have been reported, ranging from 0.3% to 10%, depending primarily on the nature of the CMG infection in the plants from which virus is acquired. There is evidence for a limited degree of co-evolutionary adaptation between the whitefly vector and CMGs, as Indian whiteflies are significantly better at vectoring Indian than African CMGs and vice versa. However, within Africa, there is currently no indication that whiteflies from a particular location are more efficient at vectoring locally occurring CMGs than they are at vectoring CMGs from another part of the continent. There is, however, an important balance between the pattern of transmission and the nature of the virus infection. More severe infections, caused either by more virulent virus species/strains or by mixed virus infections, lead to a greater frequency of whitefly-borne infection but diminished propagation through cuttings by farmers. Conversely, more moderate or mild infections, caused by less virulent virus species/strains occurring in single infections, lead to less frequent whitefly-borne infection and increased propagation through cuttings by farmers. This balance dictates the epidemiological characteristics of CMG infection in any given location, area, or region.

## Virus–Plant Interactions

### Viral Infection Mechanisms and Host Plant Responses

Following initial infection of a previously uninfected cassava plant by a CMG, viral DNA moves through the phloem to the newly developing tissues immediately behind the meristem where rapid multiplication of virus particles takes place. Plants have developed defense responses to virus multiplication through the process of PTGS, but in response, CMG species/strains have developed various effective mechanisms to overcome this, and the degree of this effectiveness seems to be the main factor in determining the severity of disease resulting from infection. Both AC2 and AC4 have been shown to act as suppressors of PTGS. Significantly, in mixed ACMV + EACMV-like virus infections, the responses of the two viruses to host plant PTGS seem to be complementary, leading to a synergistic interaction between the two viruses, greatly increased titers of both, and a concomitant increase in severity of the disease symptoms expressed in the plant. Molecular studies of these interactions have shown that the abundance of short interfering RNA (siRNA) molecules associated with the host plant PTGS response increases over time in pure ACMV infections, yet remains low over similar time periods for ACMV + EACMV-UG co-infections. Synergism has been described for ACMV + EACMCV mixtures in Cameroon, but its significance is greatest in the widely occurring ACMV + EACMV-UG mixed infections that cause severe CMD in much of East and Central Africa.

## Field-Level Epidemiology

Epidemiological studies of the CMGs can be broadly categorized into two groups: those that describe patterns of infection at the field level and those that relate to area or region-wide spread. In the case of the former, external sources of infection have been shown to be most important for determining the rate and quantity on infection in initially CMD-free plantings in the normal field environment. Gradients of infection occur in which new infections are most frequent on the windward sides of cassava crops and these gradients are matched by similar patterns of vector distribution. Multiple regression relationships have been used to describe the association between measures of inoculum pressure and final CMD incidence in test plots. Gompertz curves model patterns of infection increase in plots of CMG-susceptible cassava cultivars, and incidence increases rapidly to 100% over the first 3–6 months of growth. For resistant varieties, however, under similar conditions of inoculum pressure, rates of infection increase are much lower and final incidences typically range from 0% to 50%.

## Regional Epidemiology

Following the earliest 'first colonization' descriptions of CMD epidemics in the 1920s and 1930s, a few reports were made of rapid area-wide spread of severe CMD at other times during the twentieth century, some of the most notable of which were epidemics in Cape Verde and southeastern Nigeria in the 1990s. Of much greater importance, however, has been the African CMD pandemic that was first reported from the northern-central part of Uganda in the late 1980s. CMD associated with the epidemic was unusually severe and was being rapidly spread by superabundant populations of B. tabaci. During the 1990s it became apparent that the zone affected by this severe CMD was expanding southwards at a rate of 20–30 km per year, and in 1997 molecular studies revealed that the severe disease phenotype was associated with the occurrence and spread of an 'invasive' recombinant CMG, EACMV-UG, commonly in mixed infection with the locally occurring, but now synergized ACMV. Regular monitoring surveys conducted throughout the East and Central African region through the 1990s and early part of the new century have used PCR-based diagnostics to map the spread of the EAMCV-UG associated with this 'pandemic' of severe CMD. This work has given rise to first reports of EACMV-UG and resulting severe CMD in eight additional countries in East and Central Africa: Kenya (1996), Sudan (1997), Tanzania (1999), DRC (1999), Republic of Congo (1999), Rwanda (2001), Burundi (2003), and Gabon (2003). Significantly, the pandemic 'front' advanced through northwestern Tanzania from the Uganda border a distance of c. 400 km between 1999 and 2004. If these rates of virus spread are sustained, EACMV-UG might be expected to arrive in Zambia by 2012 and in Malawi by 2015. Similarly, westwards spread of the EACMV-UG-associated pandemic threatens both Cameroon and Nigeria in the near future.

### Economic Importance

In common with many viruses causing plant disease, the main features of the damage caused by CMGs are a mosaic-like chlorosis on the leaf lamina, leaf puckering, twisting and distortion, and an overall reduction in plant size. These effects lead to a reduction in the quantity of photosynthetic assimilates channeled into the tuberous roots, and through this, a reduction in yield. The degree of yield loss varies greatly, depending primarily on the susceptibility of the cassava cultivar, the virulence of the CMG species/strain, and the stage of growth at which infection occurred (being most severe for cutting-infected plants). Typically, however, individual plants sustain yield losses ranging from 20% to 100%. The only study of the effects of specific viruses on yield showed average yield losses to be 42% for ACMV alone, 12% for EACMV-UG mild, 68% for EACMV-UG severe, and 82% for mixed ACMV + EACMV-UG.

The relatively moderate losses attributable to ACMV infection, coupled with moderate to low incidences, have been the reasons for the apparent lack of concern through large parts of cassava-growing Africa about the impact of CMGs and CMD. This outlook has changed markedly, however, following the emergence and regional spread of the CMD pandemic. Extensive surveys of CMD incidence, CMD severity, and the occurrence of the causal viruses have made it possible to make continent-wide yield loss estimates for CMD. The most recent of these (2006) provides estimates for losses in pandemic-recovered (16% loss), pandemic-affected (47%), and as yet unaffected (18%) countries in sub-Saharan Africa, leading to an overall loss figure of 34 million tons per year, roughly a third of total African production of fresh cassava roots. Significantly, as the pandemic-affected area of the continent increases beyond the current estimate of 2.6 million sq. km, losses are likely to continue to rise through the first quarter of the twenty-first century.

## Control

The most widely practiced approaches to controlling the CMGs that cause CMD include the deployment of host plant resistance and the use of cultural methods, particularly phytosanitation. More recently, considerable attention has been directed toward the use of genetic engineering techniques to produce transgenic virus-resistant cassava plants.

### Host Plant Resistance

The potential value of introgressing virus resistance genes from wild relatives of cassava was recognized from the earliest days of CMD research in the 1920s/30s, and interspecific crosses combining cultivated cassava with Ceara rubber (*Manihot glaziovii*) were developed independently through breeding programs in modern-day Tanzania and Madagascar. F1 progeny were triple backcrossed with cultivated cassava to produce plants that combined acceptable food quality with significantly enhanced resistance to CMGs. Germplasm developed in this way formed the basis for the later continental breeding program run from the Nigeria-based International Institute of Tropical Agriculture (IITA), which from its establishment in 1967, to the present day, has developed thousands of CMD-resistant cassava clones. Many of these have been sent to cassava-producing countries in Africa for use either specifically for CMD management programs, or more generally for cassava development. Germplasm derived from the initial interspecific crosses uses the name prefix 'TMS' for 'tropical Manihot species'. This resistance source is multigenic and has provided high levels of resistance which have been very durable when used against all CMGs and CMG combinations. Four distinct mechanisms of resistance are recognized: resistance to infection, resistance to virus multiplication, resistance to virus movement (leading to incomplete systemicity), and resistance of normal plant function to the effects of virus infection. During the 1990s, resistant landraces from West Africa, given the name prefix 'TME' for 'Tropical Manihot Esculenta', were incorporated into the breeding program. These have been shown to possess alternative sources of resistance, one of the most important of which has been characterized through genetic analyses as a single dominant gene designated *CMD2*. Molecular marker approaches have been used subsequently to combine the multigenic *M. glaziovii*-derived resistance with *CMD2*. Looking further into the future, there is hope that genomics approaches will lead to the elucidation of additional CMG-resistance genes that can be jointly incorporated into cassava germplasm through pyramiding.

CMG-resistant varieties have been widely disseminated throughout the cassava-growing areas of sub-Saharan Africa. However, their adoption by farming communities has been most widespread in CMD pandemic-affected countries, where they have provided the only effective means of restoring cassava production to pre-pandemic levels.

## Cultural Methods

A range of cultural methods has been proposed for the control of CMGs. The methods most widely recommended have been the removal of infected plants (roguing) or the selection of disease-free planting material for the establishment of a new crop (selection). Crop isolation, adjusting crop disposition in relation to the prevailing wind, varying planting date, varietal mixtures, and intercropping cassava with other 'putative' protective crops have all been suggested at various times as potentially useful control options for CMGs. No convincing experimental evidence has yet been presented to confirm the value of any of these methods, however, and current field practice is restricted to selection and occasional roguing. Roguing is considered to be of value within the framework of institutional programs for the multiplication of CMD-resistant germplasm, in view of the requirement for the production of high-quality planting material. Experiments conducted in 'post-epidemic' areas of East Africa, first affected by the CMD pandemic 5 or more years previously, have provided clear evidence for the value of selection of CMD-free stems when choosing planting material. Local cultivars treated in this way provided equivalent yields to those of CMD-resistant varieties after two cropping cycles. A key drawback to the wider adoption of this approach, however, is the variability in effectiveness of the approach in relation both to the virus inoculum pressure of the location, as well as the relative susceptibility of the cultivar.

## Transgenic Resistance

There are four strategies that are currently being explored for transgenic resistance to geminiviruses. These include: the expression of viral proteins, the expression of nonviral proteins, DNA interference, and RNA interference.

### *Expression of viral proteins*

Coat protein transformation has been widely used as a means of interfering with coat protein assembly and also inter- and intracellular transport, but this approach seems to be less effective for the bipartite geminiviruses, such as the CMGs, in which movement functions are controlled by BC1 and BV1. The most thoroughly tested approach has been transformation with AC1 (*Rep*). AC1 transformed plants were shown to have high levels of siRNAs homologous to ACMV AC1 prior to virus challenging and AC1 mRNA levels were reduced by up to 98% postinfection compared to controls. This PTGS-associated mechanism also seemed to be broadly active against other CMG species and strains. Current (2006) information, however, suggests that this resistance has been lost over time, apparently through methylation of the transgene. Further studies are currently underway to determine how to prevent this acquired interference.

### *Expression of nonviral proteins*

A number of nonviral proteins have been investigated for their potential use as transgenes conferring resistance through enhancing the hypersensitive response at the initial site of infection. These include: barnase and barstar genes under the control of the ACMV A bidirectional promoter, and the gene for the ribosome-inactivating protein *dianthin*, sited downstream of an ACMV transactivatable promoter. 'Antibodies' that interfere with the protein and DNA-binding epitopes of viral proteins also offer promise as potential transgenes, and one such example is the development of artificial zinc-finger proteins (AZPs) which have a high affinity and selectivity for the *Rep* dsDNA binding site.

### *DNA interference*

Naturally occurring defective interfering (DI) subgenomic (*c.* 1500 bp) DNA molecules of CMGs have been shown to ameliorate symptoms in plants infected by wild-type CMGs. Although virus symptom reduction has also been achieved in plants transformed with tandem repeats of DI ACMV DNA-A, potential problems with this approach include virus specificity and the possibility of deleterious effects following the integration of DI virus DNA into the host genome.

### *RNA interference*

RNA interference (RNAi) results from the expression of artificial dsRNAs homologous to viral sequences which when processed into siRNAs direct silencing complexes to target RNA or DNA sequences. For ACMV, transgenic plants expressing dsRNAs homologous to the DNA A promoter have reduced levels of virus replication. An alternative RNAi approach uses antisense RNA (asRNA) constructs. Transgenic cassava plants have been produced expressing AC1, AC2, and AC3 in the antisense direction, and both *in vitro* assays and infection experiments have shown greatly reduced ACMV replication and symptom expression.

Although virus-resistant transgenic cassava plants have yet to be tested under field conditions in Africa, this approach to control, based as it is on a detailed and fundamental understanding of CMG function, offers much promise for the future.

## Other Control Methods

Although ACMV and EACMV-like CMGs interact synergistically, there is evidence that EACMV-like CMGs may hinder infection by other EACMV-like viruses and interfere with their replication. Studies in Uganda showed that plants initially infected by mild strains of EACMV-UG were much less likely to become severely diseased when exposed in the field than plants initially CMG-free. This cross-protective effect seems to be an important cause of symptom amelioration in post-CMD pandemic areas of East and Central Africa. A thorough understanding of the molecular mechanisms underlying this phenomenon will be required, however, before an assessment can be made of the potential utility of cross protection for CMG management.

Biological control and whitefly resistance are being investigated for their potential to reduce the impact of *B. tabaci* that comes from its transmission of CMGs as well as the physical damage it causes to cassava plants. The latter is a particularly important feature of the super-abundant *B. tabaci* populations found in many of the pandemic-affected areas. Biological control efforts are hindered by the fact that *B. tabaci* is considered to be indigenous to Africa, and already has a well-developed (albeit ineffective), natural enemy fauna. Cassava cultivars are variably attractive and susceptible to *B. tabaci*, however, there is a poor correlation between these characters and patterns of CMG infection. Alternative potential resistance sources are currently being sought, both from Latin American cassava germplasm (some of which is highly resistant to non-*Bemisia* whitefly pests) and from wild relatives. Much remains to be done, however, before effective vector control tactics are ready for incorporation into CMG management strategies.

## Future Perspectives

Rapid advances within the field of molecular biology, from the latter part of the twentieth century onwards, have enabled researchers to make significant progress in furthering understanding of how geminiviruses interact with both their plant host and insect vector in causing CMD. This is particularly important in view of the great and increasing economic and social impact that this disease has on the more than 300 million people in sub-Saharan Africa who depend on cassava for their subsistence. Much progress has been made in the development and deployment of conventional sources of resistance to CMGs, and this work is playing a central role in managing the effects of the EACMV-UG-associated CMD pandemic that continues to spread in East and Central Africa. Although diverse approaches have been used to develop transgenic resistance, and CMG-resistant transgenic cassava plants have been produced, none has yet been evaluated under field conditions. Although transgenics offer great promise for the future, regulatory concerns in many African countries are likely to slow the progress of this work in the near term. The relative effectiveness of conventionally bred resistance also means that CMG-resistant transgenic cassava plants will almost certainly also have to carry with them additional transgene-derived traits such as improved nutritional characteristics and resistance genes to other key pests and diseases. Set against this current and possible future progress in developing control approaches to CMGs is the remarkable ability of this group of viruses to adapt to a changing agro-ecological environment through the processes of virus–virus cooperation, pseudo-recombination, and true recombination. There can be few more dramatic examples of the impact of virus evolution on agricultural systems than that of the recombinant virus-driven African CMD pandemic. Clearly therefore, a sustained commitment to all stages of the research to development continuum will be essential if the 'balance of power' between host plant and pathogenic virus is to be tipped back in favor of the cassava host. Only when this happens will cassava be able to fulfill its true potential in sub-Saharan Africa as a key component in the continent's agricultural and broader economic development.

*See also:* Plant Resistance to Viruses: Geminiviruses; Plant Virus Diseases: Economic Aspects; Satellite Nucleic Acids and Viruses.

## Further Reading

Bock KR and Woods RD (1983) The etiology of African cassava mosaic disease. *Plant Disease* 67: 994–995.

Fauquet CM and Stanley J (2003) Geminivirus classification and nomenclature: Progress and problems. *Annals of Applied Biology* 142: 165–189.

Hanley-Bowdoin L, Settlage SB, Orozco BM, Nagar S, and Robertson D (1999) Geminiviruses: Models for plant DNA replication, transcription and cell cycle regulation. *Critical Reviews in Plant Sciences* 18: 71–106.

Legg JP and Fauquet CM (2004) Cassava mosaic geminiviruses in Africa. *Plant Molecular Biology* 56: 585–599.

Legg JP, Owor B, Ndunguru J, and Sseruwagi P (2006) Cassava mosaic virus disease in East and Central Africa: Epidemiology and management of a regional pandemic. *Advances in Virus Research* 67: 356–419.

Ndunguru J, Legg JP, Aveling TAS, Thompson G, and Fauquet CM (2005) Molecular biodiversity of cassava begomoviruses in Tanzania: Evolution of cassava geminiviruses in Africa and evidence for East Africa being a center of diversity of cassava geminiviruses. *Virology Journal* 2: 21.

Storey HH and Nichols RFW (1938) Studies on the mosaic of cassava. *Annals of Applied Biology* 25: 790–806.

Thresh JM and Cooter RJ (2005) Strategies for controlling cassava mosaic virus disease in Africa. *Plant Pathology* 54: 587–614.

Zhou X, Liu Y, Calvert L, *et al.* (1997) Evidence that DNA-A of a geminivirus associated with severe cassava mosaic disease in Uganda has arisen by interspecific recombination. *Journal of General Virology* 78: 2101–2111.

# African Horse Sickness Viruses

**P S Mellor and P P C Mertens,** Institute for Animal Health, Woking, UK

© 2008 Elsevier Ltd. All rights reserved.

## Glossary

**Ascites** Abnormal collection of fluid in the abdominal cavity.
**Cecum** The first part of the large intestine.
**Choroid plexus** A highly vascular membrane and part of the roof of the brain that produces cerebrospinal fluid.
*Culicoides* Genus of blood-feeding dipterous insects also known as biting midges.
**Cyanosis** Bluish discoloration of the skin or mucus membranes caused by lack of oxygen in the blood.
**Ecchymotic** Diffuse type of hemorrhage larger than a petechia.
**Fascial** A band of fibrous tissue that covers the muscles and other organs.
**Hydropericardium** Excessive collection of serous fluid in the pericardial sac.
**Hydrothorax** Excessive collection of serous fluid in the thoracic cavity.
**Petechiae** Pinpoint to pinhead-sized red spots under the skin that are the result of small bleeds.
**Purpura hemorrhagica** Hemorrhages in skin, mucous membranes and other tissues. First shows red then darkening into purple, then brownish-yellow.
**Supraorbital fossae** Holes in the skull situated above the eye socket.
**TCID$_{50}$** 50% tissue culture infective dose.

## Introduction

African horse sickness virus (AHSV) causes a noncontagious, infectious, insect-borne disease of equids (African horse sickness – AHS) that was first recognized in Africa in the sixteenth century. The effects of the disease, particularly in susceptible populations of horses, can be devastating with mortality rates often in excess of 90%. Although AHS is normally restricted to Africa (and possibly north Yemen), the disease has a much wider significance as a result of the ability of AHSV to spread, without apparent warning, beyond the borders of that continent. For these reasons the virus has been allocated OIE 'serious notifiable disease' status (i.e., communicable diseases which have the potential for very rapid spread, irrespective of national borders, which are of serious socioeconomic or public health consequence and which are of major importance in the international trade of livestock or livestock products).

## Taxonomy, Properties of the Virion and Genome

*African horsesickness virus* is a species of the genus *Orbivirus* within the family *Reoviridae*. The virus is nonenveloped, approximately 90 nm in diameter and has an icosahedral capsid that is made up of three distinct concentric protein layers (**Figure 1**), and which is very similar to the structure of bluetongue virus (the prototype orbivirus). Nine distinct serotypes of AHSV have been identified by the specificity of interactions between the more variable viral proteins that make up the outermost layer of the virus capsid (VP2 and VP5), and neutralizing antibodies that are normally generated during infection of a mammalian host. The outer capsid layer surrounds the AHSV core particle ($\sim$70 nm diameter), which has a surface layer composed of 260 trimers of VP7 attached to the virus subcore. These VP7 trimers form a closed icosahedral lattice, which is made up of five- and six-membered rings that are visible by electron microscopy, giving rise to the genus *Orbivirus* (from the Latin '*orbis*' meaning ring or cycle – **Figure 2**). The VP7 trimers synthesized in infected cells sometimes form into large hexagonal crystals, composed entirely of six-membered rings, which can be observed by both electron and light microscopy. The VP7 lattice on the core surface helps to stabilize the thinner and more fragile subcore layer, which is composed of 120 copies of VP3 arranged as 12 dish-shaped decamers that interact, edge to edge, to form the complete innermost capsid layer. This subcore shell also contains the three minor viral proteins (VP1, VP4, and VP6) that form approximately 10 transcriptase complexes, associated with the 10 linear segments of dsRNA that make up the virus genome.

The five viral proteins present in the AHSV core particle and two of the nonstructural proteins (NS1 and NS2) that are also synthesized within the cytoplasm of infected cells are relatively more conserved than the outer capsid proteins. NS1 forms long tubules within the infected cell cytoplasm that are characteristic of orbivirus infections. NS2 is a major component of the granular matrices (viral inclusion bodies or VIBs) that represent the major site of viral RNA synthesis and particle assembly during the replication of AHSV and other orbiviruses (**Figure 3**). These more conserved AHSV proteins contain serogroup-specific epitopes, which cross-react between different AHSV serotypes and can therefore be used as a basis for serological assays to distinguish AHSV from the members of other *Orbivirus* species (e.g., *Equine encephalosis virus* (EEV)).

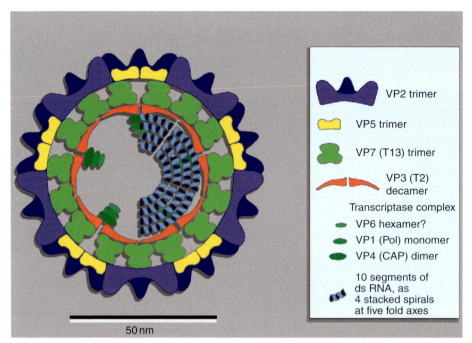

**Figure 1** Diagram of the African horse sickness virus particle structure, constructed using data from biochemical analyses, electron microscopy, cryo-electron microscopy, and X-ray crystallography. Courtesy of P.P.C. Mertens and S. Archibald – reproduced from Mertens PPC, Maan S, Samuel A, and Attoui H (2005) Orbivirus, Reoviridae. In: Fauquet CM, Mayo MA, Maniloff J, Desselberger U, and Ball LA (eds.) *Virus Taxonomy: Eighth Report of the International Committee on Taxonomy of Viruses*, pp. 466–483. San Diego, CA: Elsevier Academic Press, with permission from Elsevier.

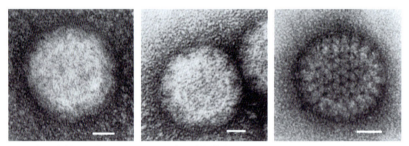

**Figure 2** Electron micrographs of African horse sickness virus (AHSV) serotype 9 particles stained with 2% aqueous uranyl acetate (left) virus particles, showing the relatively featureless surface structure; (center) infectious subviral particles (ISVP), containing chymotrypsin cleaved outer capsid protein VP2 and showing some discontinuities in the outer capsid layer; (right) core particles, from which the entire outer capsid has been removed to reveal the structure of the VP7(T13) core-surface layer and showing the ring-shaped capsomeres (line represents 20 nm). Reproduced from Mertens PPC, Maan S, Samuel A, and Attoui H (2005) Orbivirus, Reoviridae. In: Fauquet CM, Mayo MA, Maniloff J, Desselberger U, and Ball LA (eds.) *Virus Taxonomy: Eighth Report of the International Committee on Taxonomy of Viruses*, pp. 466–483. San Diego, CA: Elsevier Academic Press, with permission from Elsevier.

AHSV genome segment 10 encodes two small but largely similar proteins, NS3 and NS3a, that are translated from two in-frame start codons near the upstream end of the genome segment (see **Figure 4**). These proteins, which (by analogy with bluetongue virus) are thought to be involved in the release of virus particles from infected cells, are also highly variable in their amino acid sequence, forming into three distinct major clades. The biological significance of sequence variation in NS3/3a is uncertain, although it is clearly independent of virus serotype.

AHSV serotypes 1–8 are typically found only in restricted areas of sub-Saharan Africa while serotype 9 is more widespread and has been responsible for virtually all epizootics of AHS outside Africa. The only exception is the 1987–90 Spanish–Portuguese outbreak that was due to AHSV serotype 4.

AHSV is relatively heat resistant; it is stable at 4 and −70 °C but is labile between −20 and −30 °C. It is partially resistant to lipid solvents. At pH levels below 6.0 the virus loses its outer capsid proteins, reducing its infectivity for mammalian cell systems, although the

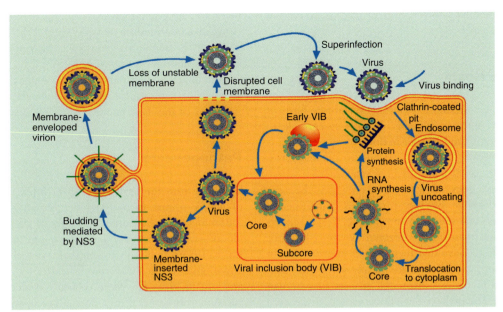

**Figure 3** Diagram of the AHSV replication cycle, based primarily on that of BTV and other members of the family *Reoviridae*. Virus adsorption involves components of the outer capsid, although cell entry may also involve VP7(T13). VP2 (possibly also VP5) is involved in cell attachment. VP5 may be involved in penetration of the cell membrane (release from endosomes into the cytoplasm) and the expressed protein can induce cell fusion. The outer capsid layer is lost during the early stages of replication, which activates the core-associated transcriptase complexes. These synthesize mRNA copies of the 10 genome segments, which are then translated into the viral proteins. These mRNAs are also thought to combine with newly synthesized viral proteins, during the formation and maturation of progeny virus particles. The viral inclusion bodies (VIBs) are considered to be the sites of viral morphogenesis and viral RNA synthesis. Negative RNA strands are synthesized on the mRNA templates, within nascent progeny particles, reforming the dsRNA genome segments. The smallest particles containing RNA that are observed within VIBs are thought to represent progeny subcore particles. The outer core protein (VP7(T13)) is added within the VIB and the outer CP at the periphery of the VIB. Reproduced from Mertens PPC, Maan S, Samuel A, and Attoui H (2005) Orbivirus, Reoviridae. In: Fauquet CM, Mayo MA, Maniloff J, Desselberger U, and Ball LA (eds.) *Virus Taxonomy: Eighth Report of the International Committee on Taxonomy of Viruses*, pp. 466–483. San Diego, CA: Elsevier Academic Press, with permission from Elsevier.

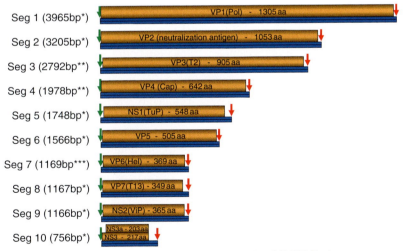

**Figure 4** The genome organization of the 10 linear dsRNA genome segments of AHSV. Each genome segment encodes a single viral protein, with the exception of genome segment 10 which has two in-frame and functional initiation codons near the upstream end of the segment. *Data derived from AHSV-9. **Data derived from AHSV-4. ***Data derived from AHSV-6 (see: www.iah.bbsrc.ac.uk/dsRNA_virus_proteins/AHSV.htm). Like other members of the family *Reoviridae*, each AHSV genome segment contains conserved terminal sequences immediately adjacent to the upstream and downstream termini (+ve strand (green arrow) 5′-GUU$^A$/$_U$A$^A$/$_U$.........AC$^A$/$_U$UAC-3′ (red arrow)) (www.iah.bbsrc.ac.uk/dsRNA_virus_proteins/CPV-RNA-Termin.htm).

core particle retains a lower level of infectivity until it is disrupted at ~pH 3.0.

## Vertebrate Hosts

Equids are by far the most important vertebrate hosts of AHSV and the horse is the species most susceptible to disease, with mules and European donkeys somewhat less so. African donkeys are fairly resistant to clinical AHS, while zebra are usually only affected subclinically.

Occasionally, dogs or wild carnivores may become infected with AHSV by ingesting virus-contaminated equid meat and can die from the disease. Some reports also suggest that they can be infected by insect bite but most authorities believe that they play little or no part in the epidemiology of AHS and are merely dead-end hosts.

AHS is not a zoonosis. Although at least four human cases of severe disease have been documented, these were all infections acquired in an AHSV vaccine plant under conditions unlikely to be duplicated elsewhere.

## Clinical Signs

AHSV can cause four forms of disease in equids and these are discussed in ascending order of severity.

Horse sickness fever is the mildest form of disease involving only a rise in temperature and possibly, edema of the supraorbital fossae; there is no mortality. It occurs following the infection of horses with less virulent strains of virus, or when some degree of immunity exists. It is usually the only form of disease exhibited by the African donkey and zebra.

The cardiac or subacute form of disease has an incubation period of about 7–14 days and then the first clinical sign is fever. This is followed by edema, first of the supraorbital fossae and surrounding ocular tissues (which may also exhibit hemorrhage), then extending to other areas of the head, neck, and chest. Petechial hemorrhages may appear in the conjunctivae and ecchymotic hemorrhages on the ventral surface of the tongue. Colic is also a feature of the disease. The mortality rate in horses from this form of disease may be as high as 50% and death usually occurs within 4–8 days of the onset of fever.

The next most severe is the mixed form of AHS which is a combination of the cardiac and pulmonary forms with mortality rates in horses as high as 80%.

The pulmonary form is peracute and may develop so rapidly that an animal can die without prior indication of disease. Usually, there will be marked depression and fever (39–41 °C) followed by onset of respiratory distress. Coughing spasms may also occur, the head and neck tend to be extended, and severe sweating develops. There may be periods of recumbence and terminally, frothy fluid or foam may be discharged from the nostrils. Death is from congestive heart failure or asphyxia and the mortality rate in horses is frequently over 90%. During epizootics in naive populations of horses all forms of disease can occur but the mixed and pulmonary forms usually predominate, so mortality rates well in excess of 80% are likely, making AHS one of the most lethal of all horse diseases.

## Pathogenesis

On entry into the vertebrate host, initial multiplication of AHSV occurs in the regional lymph nodes. This is followed by dissemination throughout the body via the blood (primary viremia) and subsequent infection of the lungs, spleen, and other lymphoid tissues, and certain endothelial cells. Virus multiplication in these tissues and organs gives rise to secondary viremia, which is of variable duration and titer dependent upon a number of factors including host species. Under natural conditions, the incubation period to the commencement of secondary viremia is less than 9 days, although experimentally it has been shown to vary between 2 and 21 days. In horses, a virus titer of up to $10^{5.0}$ $TCID_{50}$ $ml^{-1}$ may be recorded but viremia usually lasts for only 4–8 days and has not been detected beyond 21 days. In zebra, viremia occasionally extends for as long as 40 days but peaks at a titer of only $10^{2.5}$ $TCID_{50}$ $ml^{-1}$. Viremia in donkeys is intermediate between that in horses and zebra in titer and duration, while in dogs it is considered to be very low level and transitory.

In experimentally infected horses, high concentrations of AHSV accumulate in the spleen, lungs, caecum, pharynx, choroid plexus, and most lymph nodes. Subsequently, virus is found in most organs, probably due to their blood content. In the blood, virus is associated with the cellular fraction (both red blood cells and the buffy coat) and very little is present in the plasma. This may be similar to the situation that occurs with bluetongue virus, in infected ruminants where virus is sequestered in the cell membrane of infected red blood cells and is thereby protected from the effects of humoral antibody. This leads to both virus and antibody circulating in the system together. In ruminants, this leads to extended viremia. This seems not to occur with AHSV in horses although viremia in the presence of circulating antibody has been reported in zebra. For AHSV, the onset of viremia usually corresponds with the appearance of fever and persists until it disappears.

In experimentally infected horses, exhibiting the peracute form of disease, antigen is found primarily in the cardiovascular and lymphatic systems and to a lesser extent throughout the body. In animals with horse sickness fever, antigen is concentrated in the spleen, with lesser amounts elsewhere. The main locations of antigen are endothelial cells (suggesting that they are a primary target for the virus) and large cells of the red pulp of the spleen. The presence of antigen in large mononuclear cells and surrounding lymphoid follicles suggests that these cells might also be involved in virus replication and in the transport of viral protein to the lymphoid follicles.

## Pathology

### Macrolesions

These vary in accordance with the type of disease. In the pulmonary form, the most conspicuous lesions are interlobular edema of the lungs and hydrothorax. The subpleural and interlobular tissues are infiltrated with a yellowish gelatinous exudate and the entire bronchial tree may be filled with a surfactant, stabilized froth. Ascites can occur in the abdominal and thoracic cavities and the stomach mucosa may be hyperemic and edematous.

In the cardiac form, the most prominent lesions are gelatinous exudate in the subcutaneous, subfascial and intramuscular tissues, and lymph nodes. Hydropericardium is seen and hemorrhages are found on the epicardial and/or endocardial surfaces. Petechial hemorrhages and/or cyanosis may also occur on the serosal surfaces of the cecum and colon. In these instances, a distinct demarcation can often be seen between affected and unaffected parts. This may be due to a selective involvement of endothelial cells. As in the pulmonary form, ascites may occur but edema of the lungs is usually absent.

In the mixed form of AHS, lesions common to both the pulmonary and cardiac forms of the disease occur.

### Microlesions

The histopathological changes are a result of increased permeability of the capillary walls and consequent impairment in circulation. The lungs exhibit serous infiltration of the interlobular tissues with distension of the alveoli and capillary congestion. The central veins of the liver may be distended, with interstitial tissue containing erythrocytes and blood pigments while the parenchymous cells show fatty degeneration. Cellular infiltration can be seen in the cortex of the kidneys while the spleen is heavily congested. Congestion may also be seen in the intestinal and gastric mucosae, and cloudy swelling in the myocardial and skeletal muscles.

## Epidemiology and Transmission

AHSV is widely distributed across sub-Saharan Africa. It is enzootic in a band stretching from Senegal and Gambia in the west to Ethiopia and Somalia in the east, and reaching as far south as northern parts of South Africa. The virus is probably also enzootic in northern Yemen, the only such area outside the African continent. From these zones, the virus makes seasonal extensions both northward and southward in Africa. The degree of extension is dependent mainly upon the climatic conditions and how these affect the abundance, prevalence, and seasonal incidence of the vector insects. More rarely, the virus has spread much more widely and has extended as far as Pakistan and India in the east and Spain and Portugal in the west. However, prior to the 1987–91 Spanish, Portuguese, and Moroccan outbreaks, AHSV had been unable to persist for more than 2–3 consecutive years in any area outside sub-Saharan Africa or Yemen.

AHSV is transmitted between its vertebrate hosts almost exclusively via the bites of hematophagous arthropods. Various groups have been implicated over the years, ranging from mosquitoes to ticks, but certain species of *Culicoides* biting midge are considered to be by far the most significant vectors. Biting midges act as true biological vectors and support virus replication by up to 10 000-fold. Subsequent to feeding upon a viremic equid, susceptible species of *Culicoides* become capable of transmission after an incubation period of 8–10 days at 25 °C. This period lengthens as the temperature falls, and becomes infinite below 15–18 °C. The incubation or prepatent period in the vector is the time interval necessary for ingested virus to escape from the gut lumen by entering and replicating in the mid-gut cells, and then for progeny virus particles released into the hemocoel to reach and replicate in the salivary glands. Transovarial or vertical transmission of AHSV by biting midge vectors does not occur.

*Culicoides imicola*, a widely distributed species found across Africa, southern Europe, and much of Asia, is the major vector of AHSV and has long been considered to be the only important field vector. However, a closely related species, *C. bolitinos*, has recently been identified as a second vector in Africa, and the North American *C. sonorensis* (= *variipennis*) is a highly efficient vector in the laboratory. The identification of additional vectors is likely.

In general, *Culicoides* species have a flight range of less than a few kilometers. However, in common with many other groups of flying insects, they have the capacity to be transported as 'aerial plankton' over much greater distances. In this context, a considerable body of evidence suggests that the emergence of AHSV from its enzootic zones may sometimes be due to long-range dispersal flights by infected vectors carried on the prevailing winds.

## Diagnosis

In enzootic areas, the typical clinical features of AHS (described earlier) can be used to form a presumptive diagnosis. Laboratory confirmation should then be sought. The specimens likely to be required are:

1. *Blood for virus isolation.*
2. *Tissues for virus isolation (or for antigen detection by ELISA or RT-PCR-based assays)*: Spleen is best, followed by lung, liver, heart, and lymph nodes.
3. *Serum for serological tests*: Preferably, paired samples should be taken 14–28 days apart.

Confirmation of AHS is by one or more of the following:

1. Identification of the virus in submitted samples by the group specific, antigen detection ELISA or RT-PCR–based assays. AHSV RNA can be identified by RT-PCR assays using virus-species-specific oligonucleotide primers. This identification can be confirmed by sequence analyses of the resulting cDNA products and comparison to sequences previously determined for reference strains of AHSV and other orbiviruses.
2. Isolation of infectious virus in suckling mice or embryonating hens' eggs identification first by the group-specific antigen-detection ELISA, and then by the serotype-specific, virus neutralization or RT-PCR tests.
3. Identification of AHSV-specific antibodies by the group-specific antibody detection ELISA, CF, or the serotype-specific virus neutralization tests.

## Differential Diagnosis

The clinical signs and lesions reported for AHS can be confused with those caused by the closely related EEV. Many aspects of the epidemiology of the diseases caused by these two viruses are also similar. They have a similar geographical distribution and vertebrate host range and the same vector species of *Culicoides*. As a result, both can occur simultaneously in the same locations and even in the same animal. Fortunately, rapid, sensitive, and specific ELISAs are available to enable the detection of the antigen and antibody of both the AHSV and EEV, and if used in conjunction can provide a rapid and efficient differential diagnosis.

Several other diseases may also be confused with one or other of the forms of AHS. The hemorrhages and edema reported in cases of purpura hemorrhagica and equine viral arteritis may be similar to those seen in the pulmonary form of AHS, although with AHS the edema tends to be less extensive and the hemorrhages are less numerous and widespread. The early stages of babesiosis (*Babesia equi* and *B. caballi*) can be confused with AHS, particularly when the parasites are difficult to demonstrate in blood smears.

## Treatment

Apart from supportive treatment, there is no specific therapy for AHS. Affected animals should be nursed carefully, fed well, and given rest as even the slightest exertion may result in death. During convalescence, animals should be rested for at least 4 weeks before being returned to light work.

## Control

Importation of equids from known infected areas to virus-free zones should be restricted. If importation is permitted, animals should be quarantined for 60 days in insect-proof accommodation prior to movement

Following an outbreak of AHS in a country or zone that has previously been free of the disease, attempts should be made to limit further transmission of the virus and to achieve eradication as quickly as possible. It is important that control measures are implemented as soon as a suspected diagnosis of AHS has been made and without waiting for confirmatory diagnosis. The control measures appropriate for outbreaks of AHS in enzootic and epizootic situations are described in Mellor and Hamblin.

*See also:* Orbiviruses; Reoviruses: General Features; Reoviruses: Molecular Biology.

## Further Reading

Coetzer JAW and Guthrie AJ (2004) African horsesickness. In: Coetzer JAW and Tustin RC (eds.) *Infectious Diseases of Livestock*, 2nd edn, pp. 1231–1246. Cape Town: Oxford University Press.

Hess WR (1988) African horse sickness. In: Monath TP (ed.) *The Arboviruses: Epidemiology and Ecology*, vol. 2, pp. 1–18. Boca Raton, FL: CRC Press.

Howell PG (1963) African horsesickness. In: *Emerging Diseases of Animals*, pp. 71–108. Rome: FAO Agricultural Studies.

Lagreid WW (1996) African horsesickness. In: Studdert MJ (ed.) *Virus Infections of Equines*, pp. 101–123. Amsterdam: Elsevier.

Meiswinkel Venter GJ and Nevill EM (2004) Vectors:*Culicoides* spp. In: Coetzer JAW and Tustin RC (eds.) *Infectious Diseases of Livestock*, 2nd edn, pp. 93–136. Cape Town: Oxford University Press.

Mellor PS (1993) African horse sickness: Transmission and epidemiology. *Veterinary Research* 24: 199–212.

Mellor PS (1994) Epizootiology and vectors of African horse sickness virus. *Comparative Immunology, Microbiology and Infectious Diseases* 17: 287–296.

Mellor PS, Baylis M, Hamblin C, Calisher CH, and Mertens PPC (eds.) (1998) *African Horse Sickness*. Vienna: Springer.

Mellor PS and Hamblin C (2004) African horse sickness. *Veterinary Research* 35: 445–466.

Mertens PPC and Attoui H (eds.) (2006) Phylogenetic sequence analysis and improved diagnostic assay systems for viruses of the family *Reoviridae*. http://www.iah.bbsrc.ac.uk/dsRNA_virus_proteins/ReoID/AHSV-isolates.htm (accessed July 2007).

Mertens PPC and Attoui H (eds.) (2006) The dsRNA genome segments and proteins of African horse sickness virus (AHSV).

http://www.iah.bbsrc.ac.uk/dsRNA_virus_proteins/AHSV.htm (accessed July 2007).

Mertens PPC, Attoui H, and Bamford DH (eds.) (2007) The RNAs and proteins of dsRNA viruses. http://www.iah.bbsrc.ac.uk/dsRNA_virus_proteins/orbivirus-accession-numbers.htm (accessed July 2007).

Mertens PPC, Duncan R, Attoui H, and Dermody TS (2005) *Reoviridae*. In: Fauquet CM, Mayo MA, Maniloff J, Desselberger U, and Ball LA (eds.) *Virus Taxonomy: Eighth Report of the International Committee on Taxonomy of Viruses*, pp. 447–454. San Diego, CA: Elsevier Academic Press.

Mertens PPC, Maan S, Samuel A, and Attoui H (2005) *Orbivirus, Reoviridae*. In: Fauquet CM, Mayo MA, Maniloff J, Desselberger U, and Ball LA (eds.) *Virus Taxonomy: Eighth Report of the International Committee on Taxonomy of Viruses*, pp. 466–483. San Diego, CA: Elsevier Academic Press.

Sellers RF (1980) Weather, host and vectors: Their interplay in the spread of insect-borne animal virus diseases. *Journal of Hygiene Cambridge* 85: 65–102.

Walton TE and Osburn BI (eds.) (1992) *Bluetongue, African Horse Sickness and Related Orbiviruses*. Boca Raton, FL: CRC Press.

## Relevant Website

http://www.oie.int – OIE data on AHSV outbreaks.

# African Swine Fever Virus

**L K Dixon and D Chapman,** Institute for Animal Health, Pirbright, UK

© 2008 Elsevier Ltd. All rights reserved.

### Glossary

**Multigene family** Genes that are derived by duplication and therefore related to each other.
**Hemadsorption** Binding of red blood cells around infected cells.

## History and Geographical Distribution

African swine fever virus (ASFV) infection has been established over very long periods in areas of eastern and southern Africa, specifically in its wildlife hosts the warthog (*Phacochoerus aethiopicus*), the bushpig (*Potamochoerus porcus*), and the soft tick vector (*Ornithodoros moubata*). The virus is well adapted to these hosts, in which it causes inapparent persistent infections.

The disease caused by the virus, African swine fever (ASF), was first reported in the 1920s when domestic pigs came into contact with infected warthogs. Since then, ASF has spread to most sub-Saharan African countries. The first trans-continental spread of the virus occurred in 1957 to Portugal, via infected pig meat. Following a reintroduction of virus in 1960, ASF remained endemic in Spain and Portugal until the 1990s. During the 1970s and 1980s, ASF spread to other European countries, as well as Brazil and the Caribbean. Outside Africa, ASF is now endemic only in Sardinia, but within Africa ASF continues to cause major economic losses and has spread to countries such as Madagascar which were previously free from infections.

Analysis of the genomes of different virus isolates showed that those from wildlife sources in eastern and southern Africa are very diverse, reflecting long-term evolution in geographically separated host populations. Isolates from domestic pigs in western and central Africa, Europe, the Caribbean, and Brazil obtained over a 40 year period were all very closely related, suggesting that they were derived from a few introductions from wildlife reservoirs that have spread through pig populations. It is possible that these virus strains have been introduced into previously uninfected wildlife reservoirs in western and central Africa. Domestic pig isolates are more diverse in eastern and southern Africa. This suggests that several introductions of virus from wildlife hosts into domestic pigs have occurred in these regions.

## Transmission

In its sylvatic cycle, ASFV is maintained by a cycle of infection involving warthogs and the soft tick vector *O. moubata*. Ticks are thought to become infected by feeding on young warthogs, which develop transient viremia. Virus replicates to high titers in ticks and can be transmitted between different developmental stages, sexually between males and females, and transovarially. Warthogs can become infected by bites from infected ticks. Although virus is present in adult warthog tissues, high viremia is not detected, and direct transmission between adult warthogs may therefore be limited. For this reason the tick vector is thought to play an important role in the transmission cycle involving these hosts (**Figure 1**).

In many African countries, ASF has become established as an enzootic disease in domestic pigs and is maintained in the absence of contact with warthogs. Within pig populations, virus can spread by direct contact between

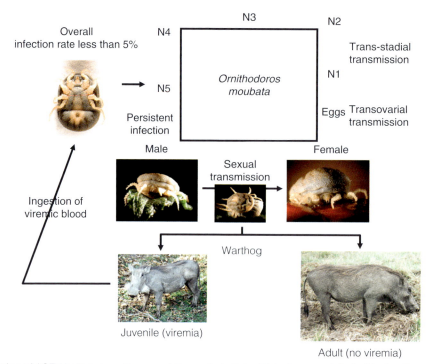

**Figure 1** Transmission of ASFV between warthogs and *O. moubata* ticks. Ticks become infected by feeding on young warthogs, which develop a transient viremia. Virus replicates in ticks and is transmitted between different nymphal stages (N1–N5), transovarially, and transexually between males and females. Ticks remain infected over long time periods and can transmit virus to the warthogs on which they feed. Direct transmission between warthogs is thought to be infrequent since viremia is low in adult warthogs. In contrast, direct transmission between infected pigs occurs readily.

pigs, which develop a high viremia, or by ingestion of infected meat or other material. However, tick vectors may also play an important role in maintaining infection in areas where they are present. The virus can replicate in other species of *Ornithodoros* including *O. erraticus*, which played a role in maintaining infection in southern Spain and Portugal. There is no vaccine available, and disease control relies on rapid diagnosis and implementation of quarantine.

## Pathogenesis

Most ASFV isolates cause an acute hemorrhagic fever with mortality approaching 100% in domestic pigs and wild swine (*Sus scrofa domesticus* and *S. s. ferus*) within 8–14 days post infection. Some moderately virulent isolates have been described that have a reduced mortality of around 30–50%. Low virulence isolates, which cause few disease signs and very low mortality, were also identified in the Iberian peninsula.

In pigs infected with virulent or moderately virulent ASFV isolates, viremia can peak at over $10^8$ hemadsorption units $50 \, \text{ml}^{-1}$. However, pigs that recover generally have lower levels of viremia and reduced replication in tissues. Only sporadic low viremia is observed in pigs infected with low virulence isolates, although moderate levels of virus replication are detected in lymphoid tissues. Persistence for long periods in recovered pigs demonstrates that the virus has effective mechanisms to evade host defense systems.

Virus replication in macrophages is observed at early times post infection, and only at later stages of disease has infection been reported in a variety of cell types, including endothelial cells, megakaryocytes, platelets, neutrophils, and hepatocytes. Thus, virus infection of macrophages is probably the primary event leading to the hemorrhagic pathology.

In common with other viral hemorrhagic fevers, ASF is characterized by damage involving induction of apoptosis in vascular endothelial cells, and this contributes to vascular permeability. This is thought to be caused by factors released from virus-infected macrophages in the early stages of disease. In the later stages of disease, the appearance of fibrin degradation products and the presence of numerous fibrin thrombi in blood indicate the development of disseminated intravascular coagulation.

Massive apoptosis of lymphocytes is observed in lymphoid tissues in pigs infected with virulent ASFV isolates. Since neither T nor B lymphocytes are infected directly by the virus, apoptosis is presumed to be caused by factors released from or on the surface of virus-infected macrophages. Observations that ASFV can act as a B-cell mitogen have suggested a model to explain lymphocyte apoptosis. This model proposes that dramatic depletion of T cells occurs by apoptosis induced by factors from infected macrophages early in infection. B cells are activated by virus infection but, because of T-cell depletion, do not receive survival signals from T cells (such as CD154 interaction), and this results in B-cell apoptosis. This dramatic depletion of T and B lymphocytes impairs the immune response to infection.

## Virus Structure

Virus particles are approximately 200 nm in diameter and have a complex multilayered structure (**Figure 2**). The nucleoprotein core contains the virus genome and enzymes and other proteins that are packaged into virus particles and used in the early stages of infection following virus entry. These include a DNA-dependent RNA polymerase, mRNA capping and polyadenylation enzymes, and other factors required for early gene transcription. The core is surrounded by a core shell and an internal envelope onto which the icosahedral capsid is assembled. Although earlier reports suggested that this internal membrane consists of a collapsed double membrane layer, it has also been suggested that only one membrane layer is present. Extracellular virions contain a loosely fitting external envelope which is derived by budding through the plasma membrane.

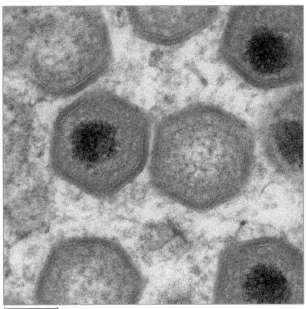

**Figure 2** ASFV structure. An electron micrograph of mature and immature virions is shown. Mature particles contain a dense nucleoprotein core containing the DNA genome and proteins that include enzymes required for early gene transcription. This is surrounded by an internal envelope derived from the endoplasmic reticulum on which the icosahedral capsid is assembled. Immature particles differ in that the nucleoprotein core has not condensed to a dense structure and the icosahedral structure is not complete. Extracellular virus particles contain an additional envelope layer derived from the plasma membrane. Electron micrographs were kindly provided by Dr. Paul Monagahan and Pippa Hawes, Institute for Animal Health, Pirbright.

The capsid consists of a hexagonal arrangement of capsomers that appear as 13 nm long hexagonal prisms each with a central hole. The intercapsomer distance is about 8 nm, and the triangulation number has been estimated at between 189 and 217, corresponding to a capsomer number between 1892 and 2172. The number of virion proteins has been estimated at over 50 by two-dimensional gel electrophoresis. In addition to the virus-encoded enzymes, which are packaged into virions, 16 virus genes that encode virion structural proteins have been identified. Of these, two encode polyproteins (pp220 and pp60) which are processed by proteolytic cleavage by a virus-encoded SUMO-like protease into virion proteins p150, p37, p14, p34, or p35, and p15, respectively. Some of these proteins have been localized in virus particles by immunogold electron microscopy. The products of the pp220 polyprotein are localized in the core shell. Proteins p12, p24, and the CD2-like protein EP402R are present in the external region of virions. Seven of the virion proteins identified contain transmembrane domains.

## Genome Structure

The virus genome consists of a single molecule of linear, covalently close-ended double-stranded DNA that is 170–192 kbp in size. The end sequences are present in two flip-flop forms that are inverted and complementary with respect to each other. Adjacent to both termini are inverted repeats, which consist of tandem repeat arrays and vary in length between 2.1 and 2.5 kbp. The complete genome sequences of one tissue-culture adapted ASFV isolate and nine field isolates have been determined. The number of open reading frames (ORFs) encoding proteins ranges between 160 and 175 depending on the isolate, and these are closely spaced and distributed on both DNA strands. The genome is A+T rich (~61%).

## Replication

### Virus Entry

The main target cells for ASFV replication are macrophages. Those macrophages that express cell surface markers characteristic of intermediate and late stages of differentiation are permissive for infection. Virus enters cells by receptor-mediated endocytosis (**Figure 3**). Recombinant scavenger receptor CD163 has been shown to bind to virus particles and to inhibit virus infection, suggesting that it may act as a virus receptor. Antibodies against recombinant virus proteins p12, p72, and p54 inhibit virus binding to cells, and those against p30 inhibit virus internalization. This suggests that these proteins have a role in these processes. Virus entry requires a fusion event between the viral envelope and the limiting membrane of the endosome at low pH. Following entry, virus cores are transported to perinuclear assembly sites via the microtubule network.

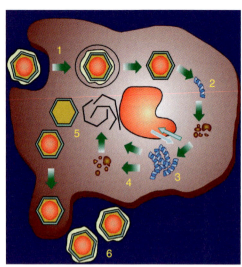

**Figure 3** ASFV replication cycle. Virus particles enter cells by receptor-mediated endocytosis (1), and early gene expression begins in the cytoplasm using enzymes and factors packaged in the nucleoprotein core (2). Replication of full-length genomes takes place in cytoplasmic perinuclear factory regions, although an early stage of replication of subgenomic fragments occurs in the nucleus (3). Following the start of DNA replication, late gene expression occurs and virus structural proteins are synthesized (4). Virion assembly takes place in the virus factories (5) and virus particles are released by budding through the plasma membrane, during which process they gain an extra membrane (6).

## DNA Replication and mRNA Transcription

Early virus gene transcription begins in the cytoplasm immediately following virus entry, using virus-encoded RNA polymerase, mRNA capping and polyadenylation enzymes, and factors packaged in the nucleoprotein core. Proteins encoded by early genes include enzymes and factors needed for later stages of the replication cycle.

A shift in the pattern of gene transcription occurs after the onset of DNA replication in the cytoplasm. Intermediate and late gene classes have been identified, and transcription of the latter is dependent on the onset of DNA replication. The similarity in the temporal pattern of gene expression suggests that ASFV transcription follows a similar regulation cascade to that of poxviruses. In general, transcription factors required for the next phase of transcription are synthesized in the previous phase. Late genes encode virion structural proteins and enzymes, and include factors such as RNA polymerase and early transcription factors that are packaged into virus particles for use during the next round of infection. Transcription of some early genes continues throughout the replication cycle.

Those virus promoters that have been mapped are short A+T-rich sequences that are located close upstream from the translation initiation codons. Transcription of all gene classes terminates at a sequence consisting of at least seven consecutive T-residues, which are often located within downstream coding regions. Increasing the number of T-residues causes transcription to terminate more efficiently. Virus mRNAs are capped at the 5' end and polyadenylated at the 3' end.

Although ASFV transcription is independent of host RNA polymerase II, productive infection requires the presence of the cell nucleus. The reasons for this are unknown, although a role in the early stages of DNA replication has been indicated by data showing replication of subgenomic length DNA fragments in the nucleus. One suggestion is that a nuclear primase may be necessary to initiate virus DNA replication.

Replication of full-length genomes occurs in the cytoplasmic factory regions via head-to-head concatamers, which are resolved to unit-length genomes and packaged into virus particles in the factories. The mechanism of DNA replication and transcription is similar to that of poxviruses, although an early phase of poxvirus DNA replication in the nucleus has not been detected.

## Assembly

Virus morphogenesis takes place in perinuclear factory regions that are adjacent to the microtubule organizing center. Virus factories resemble aggresomes since they are surrounded by a vimentin cage and increased numbers of mitochondria. Aggresomes are formed in response to cell stress and function to remove misfolded proteins, and the virus may take advantage of this cellular stress response to form its assembly sites. Progeny virus particles are assembled from precursor membranes that are thought to be derived from the endoplasmic reticulum. These membranes are incorporated as an inner envelope into virus particles and become an icosahedral structure by the progressive assembly of viral capsid protein p72. Assembly of the p72 protein into virions requires a chaperone encoded by the B602L protein. Expression of virus protein p54 (encoded by the E183L gene) is required for recruitment of envelope precursors to factory regions. This transmembrane protein is inserted into the endoplasmic reticulum when expressed in cells. It also binds to the LC8 chain of the microtubule motor dynein via a motif in its cytoplasmic domain, and this may provide a mechanism for the recruitment of membranes to virus assembly sites. The core shell is formed beneath the inner envelope by the consecutive assembly of the core shell and the DNA-containing nucleoid. Envelopment and capsid formation require calcium gradients and ATP. Expression and processing of the p220 polyprotein is required for packaging of the nucleoprotein core, and when its expression is suppressed empty virus particles accumulate in factories and can be observed budding through the plasma membrane. Both ASFV and poxviruses assemble in the reducing environment of the cell cytosol and encode proteins involved in a redox pathway that is involved in formation of disulfide bonds in some proteins in the factory. The ASFV B119L protein has been demonstrated to be a flavine adenine dinucleotide-linked sulfhydrl oxidase that is not incorporated into virus particles but is required for efficient virion maturation. The B119L protein interacts with the A151R protein, which contains a CXXC motif similar to that found in thioredoxins, and this binds to virion structural protein E248R. Possibly, all these proteins are components of a system for formation of disulfide bonds in virions.

The virus protein E120R binds to DNA and to capsid protein p72, suggesting a possible role in packaging of the genome. This protein is also required for transport from assembly sites to the plasma membrane, which occurs on microtubules using the conventional kinesin motor. Extracellular virus has an additional loose fitting external lipid envelope that is probably derived by budding through the plasma membrane.

## Virus-Encoded Proteins

Comparison of the complete genomes of ten virus isolates shows that 109 ORFs are present as single copies.

The known functions of the proteins encoded include enzymes involved in replication and transcription of the virus genome, virion structural proteins, and proteins involved in evading host defenses. Many virus encoded-proteins are not essential for replication in cells but have roles in host interactions, which are important for virus survival and transmission (**Figure 4** and **Table 1**).

## Proteins Involved in DNA Replication and Repair and mRNA Transcription and Processing

The virus encodes enzymes involved in nucleotide metabolism, such as thymidine and thymidylate kinases, ribonucleotide reductase and deoxyuridine triphosphatase. Although several of these are nonessential for replication in dividing tissue-culture cells, they are required for efficient replication in macrophages, which are nondividing and have small pools of precursor deoxynucleotide triphosphates required for incorporation into DNA. The virus encodes a DNA polymerase type B and a PCNA-like DNA clamp involved in replication of the genome. ASFV also encodes a putative DNA primase and helicase that is related to enzymes involved in binding to origins of replication and are probably involved in initiating DNA replication. An ERCC4-like nuclease is related to the principle Holliday junction resolvase, Mus81, of eukaryotes, and the virus also encodes a lambda-type exonuclease. These enzymes may be involved in resolution of the concatamers formed during virus replication.

A DNA polymerase type X, which is the smallest known, together with an ATP-dependent-DNA ligase and AP endonuclease comprise the components of a minimalist DNA base excision repair mechanism. The requirement of the AP endonuclease for efficient replication in macrophages, but not tissue-culture cells, supports the hypothesis that this repair system is an adaptation to virus replication in the highly oxidizing environment of the macrophage cytoplasm, which is likely to cause high levels of DNA damage.

Transcription of virus genes does not require the host RNA polymerase II, indicating that the virus encodes all of the enzymes and factors required. Genes with similarity to five subunits of RNA polymerase have been identified. The gene encoding the mRNA capping enzyme contains all three domains required for this function, namely a triphosphatase, a guanyl transferase, and a methyltransferase. Since transcription takes place in the cytoplasm, no introns are present in genes and the virus does not encode enzymes involved in splicing. An FTS-J-like RNA methyltransferase is encoded, which could play a role in stabilizing rRNA in infected cells.

The virus specifies several enzymes that might be involved either in regulating the virus replication cycle or in modulating the function of host proteins or cellular compartments. These enzymes include a serine/threonine protein kinase, a ubiquitin conjugating enzyme, and a prenyl transferase. The latter two enzymes are not encoded by other viruses.

## Proteins Involved in Evading Host Defenses

A number of conserved ORFs encode proteins involved in evading host defense systems. The virus replicates in macrophages, which have important roles in activating and orchestrating the innate and adaptive immune responses. By interfering with macrophage function the virus can thus disrupt both of these types of host response. One protein, A238L, inhibits activation of the host transcription factor nuclear factor kappa B (NF-κB) and also inhibits calcineurin phosphatase activity. Calcineurin-dependent pathways, such as activation of NFAT transcription factor, are therefore inhibited. This single protein may prevent transcriptional activation of the wide spectrum of immunomodulatory genes whose expression depends on these transcription factors. So far, A238L has been shown to inhibit transcription of cyclooxygenase 2 (COX2) mRNA, thus inhibiting production of prostaglandins, which have a pro-inflammatory role. A238L has also been shown to inhibit transcription from the tumor necrosis factor (TNF)-α promoter.

In addition, the virus encodes a transmembrane protein (EP402R or CD2v), which has an extracellular domain that resembles the host CD2 protein. The host CD2 protein is involved in stabilizing the interaction between T cells and antigen-presenting cells. CD2v causes binding of red blood cells to infected cells and extracellular virions, and this may help to hide virus particles and infected cells from components of the host immune system. Deletion of the CD2v gene reduces virus dissemination in infected pigs and *in vitro* abrogates the ability of ASFV to inhibit

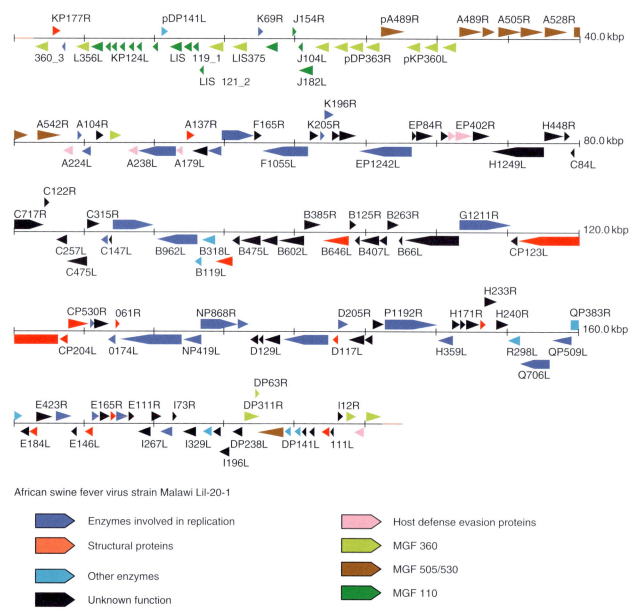

**Figure 4** Genome map of the Malawi LIL20/1 isolate. Arrows show the size and direction in which protein-coding regions are transcribed. The colors indicate protein functional class or membership of MGFs.

proliferation of bystander lymphocytes in response to mitogens. The cytoplasmic tail of this protein differs from that of the host protein and binds to a cellular adaptor protein (SH3P7/mabp1), which has roles in endocytosis, transport through the Golgi, and signaling pathways. This interaction may modulate functions of SH3P7/mabp1.

One protein (designated NL-S, l14L, or DP71L) is related in its C-terminal domain to a herpes simplex virus-encoded neurovirulence factor (ICP34.5) and host GADD34 proteins. These proteins act as regulatory subunits of protein phosphatase 1 (PP1). The ASFV protein is required for virus-induced activation of PP1, and its predominantly nuclear location indicates a possible function in regulating host gene transcription.

Three virus proteins are known to inhibit apoptosis and therefore are predicted to prolong survival of infected cells and facilitate virus replication. These include proteins which are similar to the cellular apoptosis inhibitors Bcl-2 and IAP. The Bcl-2 homolog is expressed early and thought to be essential for virus infection, whereas the IAP homolog is nonessential, is expressed late, and packaged into virus particles. The IAP homolog may play a more critical role in the tick vector.

### Multigene Families

A large proportion of the ASFV genome encodes multigene families (MGFs) consisting of related protein-coding

**Table 1** Functions of ASFV genes

| | Gene name in BA71V isolate | Alternative names | Predicted protein size (kDa) |
|---|---|---|---|
| *Nucleotide metabolism, transcription, replication, and repair* | | | |
| Thymidylate kinase | A240L | | 27.8 |
| Thymidine kinase | K196R | | 22.4 |
| dUTPase* | E165R | k1R | 18.3 |
| Ribonucleotide reductase (small subunit) | F334L | | 39.8 |
| Ribonucleotide reductase (large subunit) | F778R | | 87.5 |
| DNA polymerase β | G1211R | | 139.8 |
| DNA topoisomerase type II* | P1192R | i7R | 135.5 |
| Proliferating cell nuclear antigen (PCNA) like | E301R | j15R | 35.3 |
| DNA polymerase family X* | O174L | | 20.3 |
| DNA ligase* | NP419L | g3L | 48.2 |
| Putative DNA primase | C962R | | 111.2 |
| AP endonuclease class II* | E296R | k4R | 33.5 |
| RNA polymerase subunit 2 | EP1242L | g2L | 139.9 |
| RNA polymerase subunit 6 | C147L | g6L | 16.7 |
| RNA polymerase subunit 1 | NP1450L | | 163.7 |
| RNA polymerase subunit 3 | H359L | j1L | 41.3 |
| RNA polymerase subunit 5 | D205R | i2R | 23.7 |
| Helicase superfamily II | A859L | | 27.8 |
| Helicase superfamily II similar to origin binding protein | F1055L | | 123.9 |
| Helicase superfamily II | B962L | | 109.6 |
| Helicase superfamily II | D1133L | g10L | 129.3 |
| Helicase superfamily II | Q706L | j10L | 80.4 |
| Helicase superfamily II | QP509L | | 58.1 |
| Transcription factor SII | I243L | k9L | 28.6 |
| Guanyl transferase* | NP868R | g4R | 29.9 |
| Poly A polymerase large subunit | C475L | | 54.7 |
| FTS-J-like methyl transferase domain | EP424R | | 49.3 |
| ERCC4 nuclease domain | EP364R | | 40.9 |
| Lambda-like exonuclease | D345L | | 39.4 |
| *Other enzymes* | | | |
| Prenyl transferase* | B318L | | 35.9 |
| Serine protein kinase* | R298L | j8L | 35.1 |
| Ubiquitin conjugating enzyme* | I215L | k13L | 24.7 |
| Nudix hydrolase* | D250R | g5R | 29.9 |
| *Host cell interactions* | | | |
| IAP apoptosis inhibitor* | A224L | 4CL | 26.6 |
| Bcl-2 apoptosis inhibitor* | A179L | | 21.1 |
| IkB homolog and inhibitor of calcineurin phosphatase* | A238L | 5EL | 28.2 |
| C-type lectin-like* | EP153R | | 18.0 |
| CD2-like. Causes hemadsorbtion to infected cells* | EP402R | CD2v, Mw8R | 45.3 |
| Similar to HSV ICP34.5 neurovirulence factor | DP71L | I14L, NL | 8.5 |
| Nif S-like | QP383R | j11R | 42.5 |
| *Structural proteins and proteins involved in morphogenesis* | | | |
| P22 | KP177R | P22 | 20.2 |
| Histone-like | A104R | | 11.5 |
| P11.5 | A137R | | 21.1 |
| P10 | A78R | P10 | 8.4 |
| P72 major capsid protein. Involved in virus entry | B646L | P72, P73 | 73.2 |
| P49 | B438L | P49 | 49.3 |
| Chaperone. Involved in folding of capsid | B602L | 9RL | 45.3 |
| ERV1-like. Involved in redox metabolism* | B119L | 9GL | 14.4 |
| SUMO 1-like protease. Involved in polyprotein cleavage | S273R | | 31.6 |
| P220 polyprotein precursor of p150, p37, p14, p34. Required for packaging of nucleoprotein core | CP2475L | | 281.5 |
| P32 phosphoprotein. Involved in virus entry | CP204L | P30, P32 | 23.6 |
| P60 polyprotein precursor of p35 and p15 | CP530R | | 60.5 |
| P12 attachment protein | O61R | P12 | 6.7 |

Continued

**Table 1** Continued

| | Gene name in BA71V isolate | Alternative names | Predicted protein size (kDa) |
|---|---|---|---|
| P17 | D117L | i11L | 13.1 |
| J5R | H108R | j5R | 12.5 |
| P54 (j13L) Binds to LC8 chain of dynein, involved in virus entry | E183L | j13L, p54 | 19.9 |
| J18L | E199L | j18L | 22.0 |
| pE248R | E248R | k2R | |
| P14.5 DNA binding. Required for movement of virions to plasma membrane | E120R | k3R | 13.6 |

The functions of the virus encoded-genes are shown. Those which have been experimentally confirmed are marked with an asterisk. The designated gene names in the BA71V isolate sequence are shown and alternative names used in the literature. The predicted sizes of the encoded proteins are shown.

regions that are present in multiple copies and vary in number between different isolates. MGF 360 is the largest and contains between 11 and 19 copies in different genomes; MGF 505/530 contains between 8 and 10 copies; MGF 110 between 5 and 13; MGF 300 between 3 and 4; and MGF 100 between 2 and 3. In addition, the virus genomes contain between 1 and 3 copies of one of the virus structural proteins, p22. Comparison of the genomes of high and low virulence isolates has identified a fragment encoding 6 copies of MGF 360 and 2 copies of MGF 530 which are absent from the genome of the nonpathogenic isolate. These genes were also implicated in virulence and control of interferon-alpha production by deletion from the genome of a virulent isolate. The tissue-culture adapted isolate also has a deletion from a region close to the right end of the genome encoding five ORFs. These ORFs may facilitate virus replication in macrophages since the tissue-culture adapted isolate replicates poorly in primary macrophages.

The large investment in MGFs implies that they offer a selective advantage to the virus. However, their roles are largely unknown.

## Classification and Relationship with Other Virus Families

ASFV was first classified as a member of the family *Iridoviridae* because of its large size, cytoplasmic location, and double-stranded DNA genome. However, studies of replication strategy and the genome revealed similarities with the *Poxviridae*, although the viruses differ structurally. ASFV was therefore placed as the species *African swine fever virus* into a separate virus family, the *Asfarviridae*, of which it is the sole member of the single genus, the *Asfivirus* genus. ASFV has also been considered to be part of a larger grouping of nucleo-cytoplasmic large DNA viruses (NCLDV) which, apart from the families mentioned above, includes the family *Phycodnaviridae* (large DNA viruses that infect blue-green algae) and the genus *Mimivirus* (which infect amebae). Replication of all families in the NCLDV grouping involves at least some stage in the cytoplasm, although each family has varying requirements for host nuclear functions. For example, both ASFV and poxviruses encode their own RNA polymerase which is packaged into virus particles so that transcription of early genes begins immediately following virus entry. Members of the *Iridoviridae* have a greater requirement for the nucleus since virus particles do not contain an RNA polymerase and early virus gene transcription and replication take place in the nucleus and are initiated by host enzymes. At later stages, virus DNA replication, transcription, and virus assembly take place in the cytoplasm. Less is known about the replication strategies of the *Phycodnaviridae* and *Mimivirus*, although they exhibit a greater involvement of nuclear functions in replication compared to the *Poxviridae* and *Asfarviridae*.

Analysis of the gene complements of different families in the grouping has indicated that the ancestral NCLDV may have encoded at least 40 genes involved in replication, transcription, packaging, and assembly. Each family has evolved to encode genes that represent adaptations to its particular ecological niche. Genome analysis suggests there are two major lineages, one consisting of the *Poxviridae* and *Asfarviridae* and other of the *Iridoviridae*, *Phycodnaviridae*, and *Mimivirus*.

Comparison of the NCLDV families suggests that genes have been acquired by horizontal transfer from eukaryotic and prokaryotic hosts as well as possibly from other viruses. However, there are few genes which show evidence of recent acquisition. Another feature of the NCLDV is the presence of MGFs which have evolved by processes of gene duplication and sequence divergence. The remarkable adaptation of ASFV to replicate in its tick vector suggests that its ancestor may have replicated only in arthropods and later acquired the ability to replicate also in mammalian hosts. The independence of the virus from host transcriptional machinery facilitates virus replication in both mammalian and arthropod hosts since the gene promoters do not have to be recognised by both the mammalian and arthropod host transcriptional machinery. This could have facilitated the jump from arthropod to mammalian hosts. However, ASFV also encodes proteins (such as CD2v) that are clearly derived from a higher eukaryotic host, suggesting that growth in such hosts has substantially influenced ASFV evolution. Replication in macrophages provides the virus with opportunities to manipulate the host response to infection. This advantage may offset difficulties encountered by replicating in the harsh, microbiocidal environment of the macrophage cytoplasm.

## Future Prospects

Over the last decade our knowledge of ASFV-encoded proteins involved in virus entry and assembly, as well as those with roles in evading host defenses and causing virulence, has increased dramatically. Likewise, our knowledge of host protective immune responses and of some of the virus proteins important in their induction has increased. These advances will help the development of effective vaccines to control this economically important disease. As with other large DNA viruses, the ASFV genome may be viewed as a repository of genes that have co-evolved with its hosts and serve to manipulate host defense responses. These genes will continue to provide tools for understanding host antiviral pathways and potential leads for discovery of new immunomodulatory drugs.

Investigations of the unique replication strategy and evolutionary niche of ASFV will aid our understanding of many aspects of virus host interactions and pathogenesis, as well as mechanisms of virus replication and evolution.

*See also:* Crenarchaeal Viruses: Morphotypes and Genomes; Emerging and Reemerging Virus Diseases of Vertebrates; Enteric Viruses; Epidemiology of Human and Animal Viral Diseases; Poxviruses; Viruses Infecting Euryarchaea.

## Further Reading

Afonso CL, Piccone ME, Zaffuto KM, *et al.* (2004) African swine fever virus multigene family 360 and 530 genes affect host interferon response. *Journal of Virology* 78: 1858–1864.

Andres G, Alejo A, Salas J, and Salas ML (2002) African swine fever virus polyproteins pp220 and pp62 assemble into the core shell. *Journal of Virology* 76: 12473–12482.

Borca MV, Carrillo C, Zsak L, *et al.* (1998) Deletion of a CD2-like gene, 8-DR, from African swine fever virus affects viral infection in domestic swine. *Journal of Virology* 72: 2881–2889.

Brun A, Rivas C, Esteban M, Escribano JM, and Alonso C (1996) African swine fever virus gene A179L, a viral homologue of bcl-2, protects cells from programmed cell death. *Virology* 225: 227–230.

Carrascosa JL, Carazo JM, Carrascosa AL, Garcia N, Santisteban A, and Vinuela E (1984) General morphology and capsid fine-structure of African swine fever virus-particles. *Virology* 132: 160–172.

Cobbold C, Whittle JT, and Wileman T (1996) Involvement of the endoplasmic reticulum in the assembly and envelopment of African swine fever virus. *Journal of Virology* 70: 8382–8390.

Dixon LK, Escribano JM, Martins C, Rock DL, Salas ML, and Wilkinson PJ (2005) Asfarviridae. In: Fauquet CM, Mayo MA, Maniloff J, Desselberger U, and Ball LA (eds.) *Virus Taxonomy: Eighth Report of the International Committee on Taxonomy of Viruses*, pp. 135–143. San Diego, CA: Elsevier Academic Press.

Gomez-Puertas P, Rodriguez F, Oviedo JM, Brun A, Alonso C, and Escribano JM (1998) The African swine fever virus proteins p54 and p30 are involved in two distinct steps of virus attachment and both contribute to the antibody-mediated protective immune response. *Virology* 243: 461–471.

Iyer LM, Balaji S, Koonin EV, and Aravind L (2006) Evolutionary genomics of nucleo-cytoplasmic large DNA viruses. *Virus Research* 117: 156–184.

Miskin JE, Abrams CC, and Dixon LK (2000) African swine fever virus protein A238L interacts with the cellular phosphatase calcineurin via a binding domain similar to that of NFAT. *Journal of Virology* 74: 9412–9420.

Powell PP, Dixon LK, and Parkhouse RME (1996) An IκB homolog encoded by African swine fever virus provides a novel mechanism for downregulation of proinflammatory cytokine responses in host macrophages. *Journal of Virology* 70: 8527–8533.

Revilla Y, Callejo M, Rodriguez JM, *et al.* (1998) Inhibition of nuclear factor κB activation by a virus-encoded IκB-like protein. *Journal of Biological Chemistry* 273: 5405–5411.

Takamatsu H, Denyer MS, Oura C, *et al.* (1999) African swine fever virus: A B cell-mitogenic virus *in vivo* and *in vitro*. *Journal of General Virology* 80: 1453–1461.

Tulman ER and Rock DL (2001) Novel virulence and host range genes of African swine fever virus. *Current Opinion in Microbiology* 4: 456–461.

Yanez RJ, Rodriguez JM, Nogal ML, *et al.* (1995) Analysis of complete nucleotide sequence of African swine fever virus. *Virology* 208: 249–278.

# AIDS: Disease Manifestation

**A Rapose, J East, M Sova, and W A O'Brien,** University of Texas Medical Branch – Galveston, Galveston, TX, USA

© 2008 Elsevier Ltd. All rights reserved.

## Introduction

Human immunodeficiency virus (HIV) infection has had a staggering global impact despite its emergence only 25 years ago. It is estimated that there were around 40 million people in the world living with HIV infection in the year 2006; there were approximately 4.5 million new cases, and 3 million individuals died of acquired immunodeficiency syndrome (AIDS). Since the introduction of highly active antiretroviral therapy (HAART) in the USA and in western Europe, around 1996, there has been a significant reduction in mortality and morbidity among patients with HIV infection. The incidence of the three major opportunistic infections (OIs) associated with AIDS, namely *Pneumocystis carinii* (now called *Pneumocystis jiroveci*) pneumonia, *Mycobacterium avium complex* infection, and *Cytomegalovirus* infection is markedly reduced. Life expectancy for individuals with AIDS has increased from an estimated 4 years in 1997, to more than 24 years in 2004. Mortality rates for patients who are initiated on HAART appropriately are now comparable to populations successfully treated for other chronic conditions like diabetes. This paradigm shift has made OIs less common in populations able to access HAART. Cardiovascular disease, renal diseases, and malignancy are now the more common causes of death.

Since OIs commonly seen before the HAART era are rarely seen in patients with viral suppression, these are discussed only in passing, with citations referring to reviews or treatment guideline documents. This article focuses on clinical manifestations that pose problems in the management of HIV disease in the HAART era, including manifestations and consequences of HIV infection as well as toxic effects of treatment. There is overlap in these manifestations in that some HIV-related syndromes can be exacerbated by some antiretroviral medications. In addition, with prolonged survival, diseases associated with advancing age are becoming more prominent in HIV infection, most notably, the metabolic syndrome, which is associated with increased risk of morbidity and mortality from cardiovascular disease.

**Table 1** Common opportunistic diseases in AIDS

*Fungal infections*
Candidiasis
Pneumocystis jirovecii pneumonia
Cryptococcal meningitis
Disseminated histoplasmosis
Disseminated coccidioidomycosis
*Mycobacterial infections*
Mycobacterium tuberculosis
Mycobacterium avium complex
Mycobacterium kansasii
*Viral infections*
Herpes simplex virus (HSV)
Varicella zoster virus (VZV)
Cytomegalovirus (CMV)
Epstein–Barr virus (EBV-associated with oral hairy leukoplakia and lymphoma)
Human papilloma virus (HPV – associated with cervical dysplasia and ano-genital squamous cancers)
Human herpes virus-8 (HHV-8 – associated with Kaposi's sarcoma)
JC polyomavirus (JCPyV – associated with progressive multifocal leukoencephalopathy)
*Bacterial infections*
Salmonella septicemia
Listeriosis
Bartonella henselae (associated with bacillary angiomatosis)
*Parasitic infections*
Cerebral toxoplasmosis
Cryptosporidiosis
Isosporiasis
*Malignancies*
Kaposi's sarcoma (associated with HHV-8)
Primary CNS lymphoma
Other non-Hodgkin's lymphomas
Cervical cancer (associated with HPV)

## Opportunistic Infections and Malignancies Associated with HIV Infection

OIs and malignancies in patients with HIV infection emerge as a consequence of immune deficiency related to CD4+ T-lymphocyte depletion. A CD4+ T-lymphocyte count below 200 cells per mm$^3$ is defined as AIDS even in the absence of other diseases or symptoms since this represents such an increase in risk for OI and malignancy.

OIs and malignancies seen most commonly in HIV patients are listed in **Table 1**, and are discussed in detail in the 'AIDS surveillance case definition'. Although many of these diseases can occur in immunocompetent individuals, they are more common and often more severe in patients with AIDS. Detailed information regarding these diseases can be obtained from the website of the National Institutes of Health and the US National Library of Medicine.

Updated guidelines on the prevention of OIs in patients with HIV are available at 'Relevant website' section. While OIs are rarely seen in patients with viral suppression secondary to HAART, they may still be manifest on initial presentation. *Pneumocystis jirovecii* infection remains an important and serious clinical manifestation as initial presentation of many HIV patients. Since HIV can be a sexually transmitted disease (STD), other STDs also occur and can pose problems in patients with AIDS. These have been recently reviewed by Jeanne Marrazzo. The differential diagnosis of oral lesions in HIV patients is vast; this has been reviewed by Baccaglini L *et al.* Eye involvement in patients with HIV infection in the era of HAART has been studied in the longitudinal study of ocular complications of AIDS (LSOCA) supported by the National Eye Institute.

Malignancies classically associated with HIV infection include Kaposi's sarcoma, primary brain lymphoma, and other non-Hodgkin's lymphomas. These have been reviewed by Mathew Cheung. While some of the AIDS-defining cancers are seen less frequently in patients on HAART, it should be noted that the incidence of anal squamous intraepithelial lesions and squamous cell carcinomas is increasing in HIV-positive individuals receiving HAART.

An important phenomenon to recognize is 'immune reconstitution inflammatory syndrome' (IRIS). Following initiation of HAART, recovery of the immune system can be associated with an apparent worsening of an

HIV-associated OI or malignancy, or less commonly 'uncovering' of previously unrecognized and untreated diseases. Shingles (re-emergence of the chicken pox virus, *Varicella zoster virus*), other *Herpes* viruses, and mycobacterial infections are especially common in IRIS. With better global access to HAART, including Africa and Asia, an increasing number of cases of IRIS are being reported especially in areas with high prevalence of diseases like tuberculosis.

## The Metabolic Syndrome in HIV-Infected Individuals

A constellation of laboratory and physical abnormalities, termed the metabolic syndrome, is associated with increased risk of cardiovascular morbidity and mortality (**Table 2**). This syndrome is seen in 24% of the US population overall; its prevalence is increasing in the US. The metabolic syndrome encompasses disturbances in glucose, insulin, and lipid metabolism, associated with abdominal obesity. The presence of the metabolic syndrome roughly doubles cardiovascular disease mortality. In other studies, the risk for cardiovascular disease with metabolic syndrome is even higher.

The metabolic syndrome may be even more common in HIV-infected individuals; there are many possible reasons. HIV infection on its own may exacerbate many of the manifestations of the metabolic syndrome, particularly elevation in serum triglycerides; this was seen in HIV-infected individuals prior to the advent of antiretroviral therapy. In addition, many HIV-infected individuals smoke (50% vs. 25% in the US population overall), and many HIV-infected individuals have hypertension. The metabolic syndrome may be exacerbated by some of the drugs used to treat HIV infection, including thymidine analog reverse transcriptase inhibitors and some protease inhibitors; both tend to increase triglycerides and cholesterol, and may be associated with glucose intolerance.

There are several studies, however, that show a lower incidence of metabolic syndrome in HIV-infected individuals. In a cross-sectional study examining a cohort of 788 HIV-infected adults, metabolic syndrome prevalence was

**Table 2** Universal classification of the metabolic syndrome, as defined by the International Diabetes Foundation (IDF)[41]

| Characteristic | Measurement | |
|---|---|---|
| Waist circumference | | |
| in women | >80 cm | (31.5 in) |
| in men | >94 cm | (37 in) |
| Triglycerides | >1.7 mmol l$^{-1}$ | (>150 mg dl$^{-1}$) |
| HDL | <1.29 mmol l$^{-1}$ | (<50 mg dl$^{-1}$) |
| Glucose | >5.6 mmol l$^{-1}$ | (>100 mg dl$^{-1}$) |
| Systolic blood pressure or | >130 mm Hg | |
| Diastolic blood pressure | >85 mm Hg | |

14% by IDF criteria. Despite this low overall number, many patients in this study (49%) had at least two features of the metabolic syndrome but were not classified as having the metabolic syndrome, typically because waist circumference was not in the metabolic syndrome range. The metabolic syndrome was more common in those individuals currently receiving protease inhibitors. Although the formally defined metabolic syndrome may occur less commonly in HIV-infected individuals, components of this syndrome associated with increased risk of cardiovascular disease certainly are increased in HIV-infected individuals.

In contrast to the decreased risk of OIs seen in HIV-infected individuals who achieved virologic suppression, the metabolic syndrome and lipodystrophy (described below) appear to be more common in patients receiving HAART. The higher presence of cardiovascular disease risk factors and administration of drugs that may induce it, together with the increased survival from improved outcomes from antiretroviral therapy, make this syndrome an important one for primary care of HIV-infected individuals.

Management of the metabolic syndrome initially involves improvement in diet, increase in exercise, and avoidance of drugs that are more likely to cause the perturbation in lipid and glucose metabolism and girth. Unfortunately, switch from medications associated with higher risk of the metabolic syndrome to those that appear to be less toxic, typically, only result in minor, partial reversal of both laboratory abnormalities and abdominal fat accumulation.

## HIV-Specific Diseases

### Lipodystrophy

Lipodystrophy was identified and characterized by Carr and colleagues in 1998. In addition to laboratory abnormalities associated with the metabolic syndrome, there can also be subcutaneous lipodystrophy, which can involve either fat accumulation in the abdomen, neck and upper back, (**Figure 1**) or lipoatrophy, involving the extremities and the face; (**Figure 2**), and they can both be seen together. The lipoatrophy in the face is highly recognizable and can be stigmatizing. There is deepening of the nasolabial folds and loss of subcutaneous tissue in the temples and cheeks. Lipodystrophy has been associated with treatment with thymidine nucleoside analog-based (stavudine or zidovudine) regimens, and co-administration of a thymidine analog with some protease inhibitors may further accelerate fat loss. Studies investigating thymidine-sparing regimens typically show normal limb fat mass and lower incidence of clinical lipoatrophy, even over prolonged follow-up.

Lipodystrophy is extremely difficult to manage, as specific treatments including rosiglitazone do not appear to be effective. Switching therapy from a thymidine analog to abacavir has shown modest improvement but not resolution of lipodystrophy. Recombinant growth hormone may

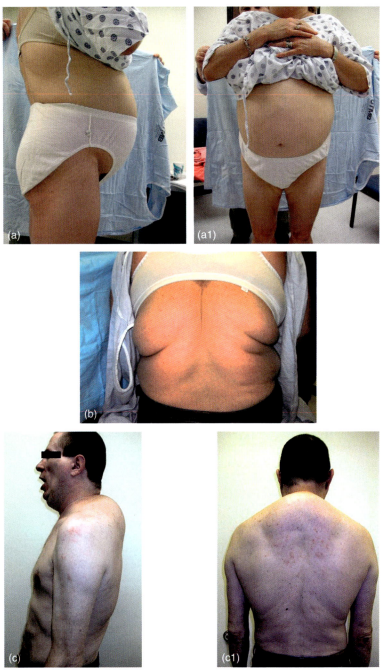

**Figure 1** Lipodystrophy – fat accumulation. (a) Visceral fat accumulation, (b) flank fat accumulation, and (c) dorsal cervical fat-pad – 'Buffalo hump'.

increase subcutaneous fat but effects may not persist. Abacavir, lamivudine, and newer nucleotide reverse transcriptase inhibitors (NRTIs) do not appear to be associated with development of lipodystrophy, and to prevent development of lipodystrophy, the older NRTIs are now commonly avoided. An exception is zidovudine, which may still retain activity in the presence of drug resistance mutations that render other drugs in this class ineffective. Nonetheless, toxicities of HIV medications are now of paramount importance since there appear to be many ways to effectively reduce viral load.

## HIV and Kidney Disease

Kidney disease related to AIDS was described as early as 1984 in reports from New York and Florida. Since then, a wide spectrum of acute and chronic renal syndromes has been reported. HIV-associated kidney disease was initially thought to occur late in the course of the infection, but it is now known that the kidneys may be involved in all the stages of HIV disease including acute infection. Renal glomerular and tubular epithelial cells may be directly infected by HIV. Effective therapies for HIV infection and

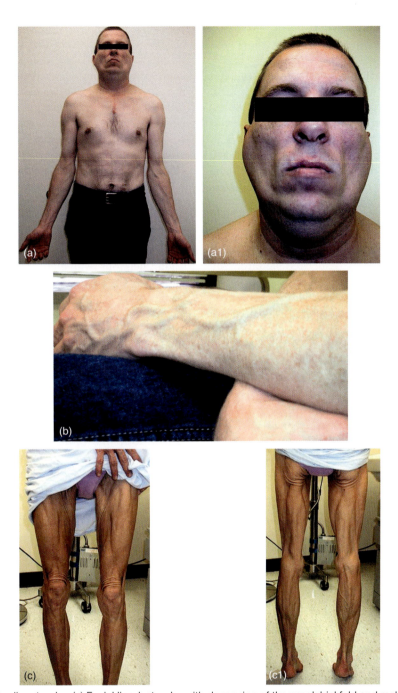

**Figure 2** Lipodystrophy–lipoatrophy. (a) Facial lipodystrophy with deepening of the nasolabial fold and malar lipoatrophy, (b) partial fat atrophy with prominent veins, and (c) buttocks and leg wasting.

the associated OIs have led to improved patient survival, which in turn has resulted in an increased number of HIV-infected individuals who require renal replacement therapy. Antiretroviral treatment has also resulted in increased reports of drug-related nephrotoxicities. IRIS may also involve the kidneys. Mortality rates for kidney diseases in HIV-infected individuals are increasing, and there is evidence that HAART may slow or prevent progression. Appropriate screening for renal dysfunction and early intervention may reduce the incidence and progression of renal disease in patients with HIV infection.

### Acute renal failure

Many of the causes of acute renal failure in HIV-infected patients are the same as for HIV-negative individuals, with a similar incidence of 5.9 cases per 100 patient-years. It is associated with a nearly sixfold increased risk of in-hospital mortality in HIV-infected patients. Factors associated with increased incidence in HIV-infected patients include advanced stage of HIV disease, exposure to antiretroviral therapy, and co-infection with hepatitis C virus. Pre-renal causes include hypovolemia, hypotension, or hypoalbuminemia. Intrinsic kidney diseases including acute tubular necrosis may occur secondary to hypotension, sepsis, or nephrotoxic drugs. Some agents used for treatment of OIs in HIV-infected patients, such as amphotericin B, aminoglycosides, foscarnet, and trimethoprim-sulphamethoxazole require careful renal monitoring. Post-renal etiologies in the setting of HIV may include outflow obstruction secondary to tumor, lymphadenopathy or fungus balls, and medication related nephrolithiasis as seen with the antiretroviral medications indinavir and atazanavir, and with sulfadiazine and acyclovir. Acute renal failure secondary to interstitial nephritis as a manifestation of IRIS has also been reported.

### Chronic kidney disease
Three syndromes of chronic renal disease are associated with HIV infection: (1) HIV-associated nephropathy (HIV-AN), (2) HIV-associated immune complex disease (HIV-ICD), and (3) HIV-associated thrombotic microangiopathies. A kidney biopsy is required to make these diagnoses.

### HIV-associated nephropathy
HIV-associated nephropathy (HIV-AN) is directly caused by HIV infection and is the most common form of chronic renal disease in HIV-infected patients. Although this syndrome has been reported in all stages of HIV disease including acute infection, advanced immunosuppression is strongly associated with HIV-AN risk. In one study, 83% of HIV-infected patients with microscopic albuminuria who were biopsied, had HIV-AN. The prevalence of HIV-AN is variable among different ethnic and racial groups suggesting that there may be genetic determinants of the disease. Casanova *et al.* did not find HIV-AN on biopsies in 26 Italian HIV-positive patients with renal disease. Also, none of 26 HIV-infected positive individuals with proteinuria were found to have HIV-AN in a study from Thailand by Praditpornsilpa *et al.* The majority of patients in the U.S. who have HIV-AN are African-American males (more than 85%). However, in one study on Ethiopian HIV patients, none of the patients fulfilled criteria for HIV-AN. This suggests that even among individuals with African heritage there may be genetic differences. Alternatively, factors other than race may play an important role in the epidemiology of this condition. Patients typically present with nephrotic syndrome in which large amounts of protein pass abnormally in the urine. However, peripheral edema, hypertension, and hematuria are often absent. Microalbuminuria is an early marker for HIV-AN and screening for microalbumniuria is recommended for early diagnosis of HIV-AN. Renal biopsy should be considered for HIV-seropositive African-American patients who present with microalbuminuria even if they have normal creatinine clearance. Classic HIV-AN is associated with focal and segmental glomerulosclerosis on histopathology. These patients can exhibit rapid progression to end-stage renal disease and prognosis is poor if left untreated. HAART may reduce the risk of development of HIV-AN and may also reduce progression of HIV-AN to end-stage renal disease, but this is controversial and antiretroviral therapy for patients with HIV-AN who do not otherwise have indications for treatment is not recommended. The best evidence for benefit of HAART in HIV-nephropathy is the reduction in incidence in the HAART era compared with earlier periods.

### HIV-associated immune complex disease
HIV-associated immune complex disease (HIV-ICD) occurs less frequently than HIV-AN. There is a higher incidence in Caucasians. Four different categories have been described: (1) immune-complex mediated glomerulonephritis (with diffuse proliferative and crescentic forms), (2) IgA nephritis (with diffuse or segmental mesangial proliferation), (3) mixed sclerotic/inflammatory disease, and (4) lupus-like syndrome. The precise role of HIV infection in the pathogenesis of these entities has not been established, and glomerular inflammation may be due to the abnormal immune responses associated with HIV infection, or secondary to superinfections. The clinical presentation is often very different from HIV-AN. Patients present with hematuria and mild proteinuria. The course is more indolent, with low rates of progression to end-stage renal disease.

### Thrombotic microangiopathy
Thrombotic microangiopathy in the setting of HIV infection is being increasingly recognized and has even been proposed as an AIDS-defining illness. It is seen more often in Caucasians as compared with African-Americans or Hispanics. Features include fever, diarrhea, hemolytic anemia, thrombocytopenia, renal failure, and neurological symptoms. Mortality rates are high even in the setting of aggressive treatment like plasma exchange and relapse is often seen in survivors. The pathogenesis of HIV-associated thrombotic microangiopathies is unknown.

### Renal disease associated with HAART
In some cases, HAART can reverse or at least control nephropathy associated with HIV infection. However, many antiretroviral medications have been associated with renal toxicity including acute and chronic renal disease. The newer antiretroviral agents commonly in use are associated with few side effects. Adefovir was the first NRTI shown to have variable antiretroviral efficacy. However, it was highly nephrotoxic in doses (60–120 mg per day) used for treatment of HIV infection and for the first time an FDA advisory committee voted against the approval of an antiretroviral drug. It was subsequently used to treat hepatitis B infection, and appears to be safer at the lower dose. The follow-up NRTI tenofovir is associated with a modest decline in renal function, but this did not lead to greater rates of discontinuation of therapy. Also, in several large randomized trials, tenofovir did not show adverse effects on overall renal function.

However as clinical use of tenofovir has widened, there have been reports of tenofovir-induced, acute renal failure, Fanconi's syndrome, renal tubular damage, and diabetes insipidus. The majority of cases have occurred in patients with underlying systemic or renal disease, or in patients taking other nephrotoxic agents including other antiretroviral agents like didanosine, lopinavir-ritonavir, or atazanavir. Patients receiving tenofovir should have creatinin clearance monitored closely. The protease inhibitor indinavir is associated with nephrolithiasis (the drug can crystallize in the urine) but cases of kidney stones associated with saquinavir, nelfinavir, and atazanavir often associated with dehydration have also been reported. The majority of medication-related, adverse events are reversible on discontinuation of the offending drug.

Although HAART can decrease the incidence of some HIV-related kidney diseases, the drugs used may cause renal problems on their own. Hence, patients should be carefully monitored clinically and by laboratory testing for microalbuminuria. A diagnosis of HIV-AN could be an indication for early initiation of HAART in an attempt to prevent further progression and potentially reverse renal disease. Other pathologies such as microangiopathy-associated renal disease carry a poor prognosis in spite of aggressive interventions. Long-term survival in patients on HAART will be associated with increased prevalence of metabolic alterations, diabetes, hypertension, and cardiovascular disease which in turn may be associated with increased secondary renal disease in these patients.

## Neurologic Manifestations of HIV

Neurological complications of HIV infection are common, with more than 50% HIV patients ultimately developing some clinical manifestations. The spectrum of the disease is broad. Classical neurologic complications of HIV infection recognized in the 1980s include: (1) aseptic meningitis, (2) HIV-associated dementia (HAD), (3) vacuolar myelopathy, and (4) distal symmetric sensory polyneuropathy. More recently, neurologic manifestations are most commonly associated with the therapeutic agents used in HAART. The most dramatic neurologic manifestations occur in HIV patients not on HAART. These include manifestations secondary to opportunistic diseases like progressive multifocal leukoencephalopathy (PML), cerebral toxoplasmosis, cryptococcal meningitis, tuberculosis, and malignancies like CNS lymphomas.

### Aseptic meningitis or encephalitis

Aseptic meningitis or encephalitis may be seen in up to 50–70% of patients who develop acute antiretroviral syndrome. Manifestations may recur during the course of the disease or as part of IRIS. As seen with other acute viral infections, symptoms are nonspecific and may consist of fever, headache, malaise, lymphadenopathy, and skin rash. Sometimes they may be severe enough to require hospitalization. Less common manifestations of early HIV infection include cranial nerve involvement (most commonly facial nerve), brachial plexus, and cauda equina syndromes, Guillian-Barré-like demyelinating polyneuropathy, mononeuritis multiplex, and radiculopathy.

### HIV-associated dementia

HIV-associated dementia (HAD) is the most common neurological manifestation of chronic HIV infection and developed in 20–60% HIV patients before effective antiretroviral treatments became available; recently, the incidence of HAD has declined. However, the prolonged life span of individuals with HIV may ultimately lead to an increased prevalence of HAD. Even in the HAART era, HIV-related neuropsychologic deficits have significant influence on the lives of the HIV-infected patients, with rates of unemployment and dependence for activities of daily living being higher among these individuals. Minor forms of cognitive and motor abnormalities that do not progress to severe dementia are also seen. HAD manifests itself in the form of progressive impairment of attention, learning, memory, and motor skills, often accompanied by a variety of behavioral changes. In the early stages, symptoms include poor concentration, mental slowing, and apathy-mimicking depression. As the disease progresses, there is worsening memory loss, personality changes (either reduced emotion or increased irritability, or disinhibition), loss of fine motor control, tremors, slowing and unsteadiness of gait, urinary incontinence, generalized hyperreflexia, and cerebellar and frontal release signs. HAD is characterized by a waxing and waning course over months to years. Neuropsychologic testing is required to make the diagnosis and follow its progression. The effects of HAD on daily activities can be measured using standardized functional evaluations. Cerebrospinal fluid (CSF) levels of HIV RNA are strongly predictive of HIV-related cognitive disorders. Low peripheral CD4+ lymphocyte counts and high levels of HIV viral load in plasma may also predict future dementia. HAD should be a diagnosis of exclusion, meaning that other causes of cognitive impairment like depression, metabolic disorders, thyroid disorders, OIs and malignancies, and drugs should be excluded. Efavirenz, a nonnucleoside reverse transcriptase inhibitor, is associated with neuropsychiatric side effects including dizziness, confusion, impaired concentration, amnesia, hallucinations, and insomnia. Most of these symptoms occur early after initiation of efavirenz and tend to resolve within a few weeks of continued treatment. Delayed onset psychiatric symptoms have also been described. This is readily distinguished from classical HAD. Proximal muscle weakness without sensory changes suggests myopathy. Polymyositis characterized by myalgias and proximal muscle weakness can result from HIV infection itself, or secondary to zidovudine therapy.

## Distal symmetric sensory polyneuropathy

Distal symmetric sensory polyneuropathy (DSPN) was recognized as a complication of HIV infection, but now is more commonly seen as a toxicity of some antiretroviral medications. In the pre-HAART era the prevalence rate was about 35% and at autopsy, 95% patients had sural nerve involvement. Low CD4+ lymphocyte counts and high viral load are significant risk factors for the development of DSPN. It is characterized by burning or aching pain, paresthesias, along with numbness or hyperpathia. It is seen more often in the lower extremities, starting in the toes and progressing proximally. On examination, there may be reduced sensation to pain, temperature, and vibration, along with reduced ankle reflexes. Strength is usually preserved. It is important to rule out other causes like diabetes, alcohol, vitamin deficiencies, and drug related injury including HIV medications such as stavudine, didanosine, and the discontinued drug zalcitabine. Medications used to treat OIs such as the TB drug isoniazid can also cause neuropathy. Clinical presentation of neuropathy secondary to antiretroviral agents is similar to that of DSPN, though it is often more painful and may develop rapidly.

*See also:* Human Immunodeficiency Viruses: Antiretroviral agents; Human Immunodeficiency Viruses: Origin; Human Immunodeficiency Viruses: Pathogenesis.

## Further Reading

Baccaglini L, Atkinson JC, Patton LL, Glick M, Ficarra G, and Peterson DE (2007) Management of oral lesions in HIV-positive patients. *Oral Surgery Oral Medicine Oral Pathology Oral Radiology and Endodontics* 103(supplement): S50 e1–e23.

Carr A, Samaras K, Chisholm DJ, and Cooper DA (1998) Pathogenesis of HIV-1-protease inhibitor-associated peripheral lipodystrophy, hyperlipidaemia, and insulin resistance. *Lancet* 351(9119): 1881–1883.

Cheung MC, Pantanowitz L, and Dezube BJ (2005) AIDS-related malignancies: Emerging challenges in the era of highly active antiretroviral therapy. *Oncologist* 10(6): 412–426.

Cho ME and Kopp JB (2004) HIV and the kidney: A status report after 20 years. *Current HIV/AIDS Reports* 1(3): 109–115.

Daugas E, Rougier J-P, and Hill G (2005) HAART-related nephropathies in HIV-infected patients. *Kidney International* 67(2): 393–403.

Friis-Moller N, Reiss P, Sabin CA, et al. (2007) Class of antiretroviral drugs and the risk of myocardial infarction. *New England Journal of Medicine* 356(17): 1723–1735.

Kaplan JE, Hanson D, Dworkin MS, et al. (2000) Epidemiology of human immunodeficiency virus-associated opportunistic infections in the United States in the era of highly active antiretroviral therapy. *Clinical Infectious Diseases* 30(supplement 1): S5–S14.

Manji H and Miller R (2004) The neurology of HIV infection. *Journal of Neurology, Neurosurgery and Psychiatry* 75(supplement 1): i29–i35.

Marrazzo J (2007) Syphilis and other sexually transmitted diseases in HIV infection. *Topics in HIV Medicine* 15(1): 11–16.

Moyle GJ, Sabin CA, Cartledge J, et al. (2006) A randomized comparative trial of tenofovir DF or abacavir as replacement for a thymidine analogue in persons with lipoatrophy. *AIDS* 20(16): 2043–2050.

Palella FJ, Jr., Baker RK, Moorman AC, et al. (2006) Mortality in the highly active antiretroviral therapy era: Changing causes of death and disease in the HIV outpatient study. *Journal of Acquired Immune Deficiency Syndromes* 43(1): 27–34.

Roling J, Schmid H, Fischereder M, Draenert R, and Goebel FD (2006) HIV-associated renal diseases and highly active antiretroviral therapy-induced nephropathy. *Clinical Infectious Diseases* 42(10): 1488–1495.

Triant VA, Lee H, Hadigan C, and Grinspoon SK (2007) Increased acute myocardial infarction rates and cardiovascular risk factors among patients with HIV disease. *Journal of Clinical Endocrinology and Metabolism* 92: 2506–2512.

Weisberg LA (2001) Neurologic abnormalities in human immunodeficiency virus infection. *Southern Medical Journal* 94(3): 266–275.

## Relevant Website

http://aidsinfo.nih.gov – Guidelines for the Use of Antiretroviral Agents in HIV-infected Adults and Adolescents.

# AIDS: Global Epidemiology

**P J Peters, P H Kilmarx, and T D Mastro,** Centers for Disease Control and Prevention, Atlanta, GA, USA

Published by Elsevier Ltd.

## Glossary

**Adult prevalence** Prevalence among the proportion of the population 15–49 years old (adults of reproductive age).

**Antenatal** Occurring before birth.

**Concentrated epidemic** HIV prevalence consistently over 5% in at least one defined subpopulation but below 1% in pregnant women. Implies that the epidemic is not established in the general population.

**Generalized epidemic** HIV prevalence consistently over 1% in pregnant women, a sentinel population used to assess trends in HIV prevalence and to estimate the adult HIV prevalence. Implies that sexual networking in the general population is sufficient to sustain the epidemic independent of high-risk subpopulations.

**Incidence** Number of new cases arising in a given time period (usually 1 year).
**Nosocomial** Relating to a hospital.
**Pandemic** An epidemic over a wide geographic area that affects a large population.
**Prevalence** Number of cases in the population at a given time point divided by the total population (a proportion).

## Brief History of HIV/AIDS

Although the first cases of acquired immune deficiency syndrome (AIDS) were recognized in the United States in 1981, phylogenetic analysis of human immunodeficiency virus (HIV) sequences suggest that HIV may have been initially transmitted to humans around 1930. By 1985, HIV had been identified in every region of the world and an estimated 1.5 million people were infected globally. Since then, unprecedented scientific advances have been made in the epidemiology, basic science, and treatment of this newly identified virus. Despite these advances, the global HIV pandemic has expanded rapidly. By 2007, an estimated 33.2 million people were living with HIV and greater than 20 million people had died of AIDS. AIDS is now the leading cause of death among people 15–59 years old and the world's most urgent public health challenge. The implementation of effective prevention strategies has proven challenging but there have been notable successes. In the US and Western Europe, extensive prevention programs in the 1980s reduced rates of infection among men who had sex with men, and systematic screening of blood donations since 1985 has virtually eliminated the risk of HIV transmission from blood transfusion. Several middle- and low-income countries have also had successful prevention initiatives. Prevention campaigns in Thailand and Uganda, for example, have resulted in substantial reductions in HIV prevalence since the 1990s. Unfortunately, in many parts of the world, stigma, discrimination, and denial about issues such as sexuality and drug use have hampered attempts to contain this epidemic.

## Molecular Epidemiology

HIV is an extremely genetically diverse virus. There are three phylogenetically distinct groups of HIV-1 based on genomic sequencing, groups M (main), O (outlier), and N (non-M, non-O). Each group has likely evolved from independent cross-species transmission events of chimpanzee simian immunodeficiency virus (SIVcpz) to humans. HIV-1 group M has spread to every region of the world and caused the global AIDS pandemic. Group O infections are uncommon and limited to people living in or epidemiologically linked to Central Africa (especially Cameroon). Group N infections have only rarely been described in Cameroon.

HIV-2 is a distinct primate lentivirus related to HIV-1 that is both less pathogenic and less transmissible. HIV-2 evolved from the cross-species transmission of sooty mangabey SIV (SIVsm). HIV-2 is highly concentrated in the West African country of Guinea-Bissau where the adult prevalance has been estimated to be as high as 8–10%. A lower prevalence (<2%) is found in surrounding West African countries. HIV-2 has remained geographically isolated to West Africa and countries with strong links to the region (Portugal, India, Angola, Mozambique, and France). Dual infection with HIV-1 and HIV-2 has been described but the viruses do not appear to recombine with each other.

High rates of viral replication coupled with continuous mutation and recombination events have resulted in the rapid genetic diversification of HIV-1 group M viruses. M group strains have diversified into nine distinct subtypes (or clades) and over 34 circulating recombinant forms (CRFs). The number of described CRFs has grown rapidly and is cataloged at the Los Alamos HIV sequence database. There are also a variety of unique recombinant forms (URFs) that have only been identified in a single person or an epidemiologically linked pair. The precise implications of variation between HIV-1 subtypes on pathogenesis, transmission, drug resistance, and immune control are not well understood. However, HIV diagnostic tests (enzyme-linked immunosorbent assay (ELISA), polymerase chain reaction (PCR), Western blot) have been able to accurately detect the vast majority of emerging subtypes and CRFs.

The initial genetic diversification of HIV-1 group M viruses likely occurred in Central Africa where the greatest diversity and earliest cases of HIV-1 have been identified. Subsequently, HIV-1 subtypes have spread with a geographically heterogeneous distribution (**Figure 1**). Subtype C, the dominant subtype in Southern Africa, Ethiopia, and India, causes 50% of the HIV infections worldwide. The predominance of subtype C, especially in countries with high-prevalence epidemics driven by heterosexual sex, has led to speculation that it may have an increased fitness for transmission. Subtype A accounts for 12% of infections worldwide and has a broad geographic distribution. CRF01_AE and CRF02_AG are two additional recombinant viruses involving subtype A that are epidemiologically important in Southeast Asia and West Africa, respectively. The emergence of these CRFs has raised concern that recombination may contribute to the selection of viruses with increased fitness, immune escape, or transmissibility. Subtype B predominates in the Americas and Western Europe. Finally, URFs are important components of the epidemics in East Africa, Central Africa, and South America. Undoubtedly, some of these URFs will emerge as important CRFs in the future.

**Figure 1** The global distribution of HIV-1 group M subtypes, 2004, by region. The world is subdivided into regions. Countries forming a region are shaded in the same color. Pie charts representing the distribution of HIV-1 subtypes and recombinants are superimposed on or connected by a line to the relevant region. The colors representing the different HIV-1 subtypes are indicated in the legend. The relative surface areas of the pie charts correspond to the relative number of individuals living with HIV in the region. Adapted from Hemelaar J, Gouws E, Ghys PD, and Osmanov S (2006) Global and regional distributions of HIV-1 genetic subtypes and recombinants in 2004. *AIDS* 20(16): W13–W23, with permission from Lippincott Williams and Wilkins.

## Modes of Transmission

HIV can be transmitted by sexual contact, exposure to blood, and from mother to child with variable efficiency (**Table 1**). Although HIV has been isolated from a variety of body fluids, only blood, semen, genital fluids, and breast milk have been proven as sources of infection.

Unprotected sexual contact is the predominant mode of HIV transmission throughout the world. Despite a relatively low efficiency of transmission per sexual act, numerous factors can enhance transmission. Receptive anal intercourse often results in microtrauma to the rectal mucosa and therefore facilitates HIV transmission by exposing damaged mucosa to HIV-infected semen. Likewise, receptive vaginal intercourse probably transmits HIV (male-to-female) more efficiently than insertive vaginal intercourse. In general, concurrent sexual partners (and not simply the absolute number of partners) augment HIV's spread in a community. The probability of sexual transmission is also augmented by factors that affect the infectiousness of the source partner and the susceptibility of the recipient partner. A high HIV viral load, genital ulcerative disease (and other sexually transmitted diseases), and blood contact during sex (due to trauma or menstruation) can all increase the probability of transmission. Male circumcision reduces female-to-male transmission of HIV and may reduce male-to-female transmission to a lesser extent. Certain genetic factors can also decrease the probability of HIV transmission.

Among injection-drug users, HIV is transmitted by exposure to HIV-infected blood through shared contaminated needles and other injection equipment. Nosocomial transmission of HIV in hospitals from reuse of syringes and needles has also been documented and the risk of acquiring HIV from a transfusion with HIV-contaminated blood products approaches 100%.

Mother-to-child transmission can take place during pregnancy, labor and delivery, and during breast-feeding. The majority of transmissions (excluding breast-feeding) occur in

**Table 1** Estimated per-act risk for acquisition of HIV, by exposure route

| Exposure route | Risk per 100 exposures to an infected source[a] |
|---|---|
| Blood transfusion | 90 |
| Needle-sharing injection-drug use | 0.67 |
| Percutaneous needle stick | 0.3 |
| Receptive anal intercourse | 0.5 |
| Receptive penile-vaginal intercourse | 0.1 |
| Insertive anal intercourse | 0.065 |
| Insertive penile-vaginal intercourse | 0.05 |
| Receptive oral intercourse | 0.01 |
| Insertive oral intercourse | 0.005 |
| Mother-to-child transmission (without breast-feeding) | 30 |
| Breast-feeding for 18 months | 15 |

[a]Estimates of risk for transmission from sexual exposure assumes no condom use.
Adapted from Smith DK, Grohskopf LA, Black RJ, *et al.* (2005) Antiretroviral postexposure prophylaxis after sexual, injection-drug use, or other nonoccupational exposure to HIV in the United States: Recommendations from the US Department of Health and Human Services. *MMWR Recommendations and Reports.* 54: 1–20 and Kourtis AP, Lee FK, Abrams EJ, Jamieson DJ, and Bulterys M (2007) Mother-to-child transmission of HIV-1: Timing and implications for prevention. *The Lancet Infectious Diseases* 6(11): 726–732.

the short interval during which the placenta detaches, labor occurs, and the infant passes through the birth canal. Overall rates of transmission are 15–40% without preventative interventions.

## Epidemiology

An estimated 33.2 million people (2.5 million children) were living with HIV in 2007 (**Figure 2**), which is an increase of 4.2 million people since 2001. Every region of the world has had an increase in the number of people living with HIV from 2001 to 2007. The prevalence of HIV varies dramatically worldwide with a disproportionate number of infections in sub-Saharan Africa (**Figure 3**). Despite these statistics, there are some promising recent developments. The incidence of new HIV infections has peaked in many countries. There has also been a decline in the HIV prevalence among young women attending antenatal clinics (a sentinel population in generalized HIV epidemics) in several high-prevalence countries. These declines have correlated with reductions in high-risk sexual behavior and increased condom usage. General trends should be interpreted cautiously, however, as even within countries there can be tremendous variability in the HIV epidemic. Improvements in surveillance techniques over time, which include expanding surveillance sites to antenatal clinics in rural areas and conducting population-based surveys, can also make trends difficult to interpret. Furthermore, the epidemics in countries with large populations, such as Nigeria, Ethiopia, Russia, India, and China (which together comprise 44% of the world's population), are still evolving and expanding in some populations.

## Sub-Saharan Africa

HIV has caused a generalized epidemic in many parts of sub-Saharan Africa. In 2007, almost 22.5 million people were living with HIV in sub-Saharan Africa (68% of the global infections but only 11–12% of the world's population), and although considerable efforts have been made to improve access to anti-retrovirals in recent years,

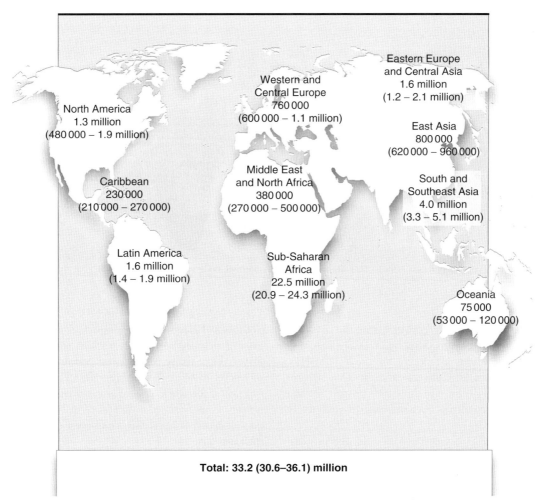

**Figure 2** Adults and children estimated to be living with HIV, 2007. Reproduced from UNAIDS/WHO (2007) AIDS Epidemic update: December 2007. http://www.unaids.org/en/HIV_data/2007Epiupdate/default.asp (accessed November 2007).

## 62 AIDS: Global Epidemiology

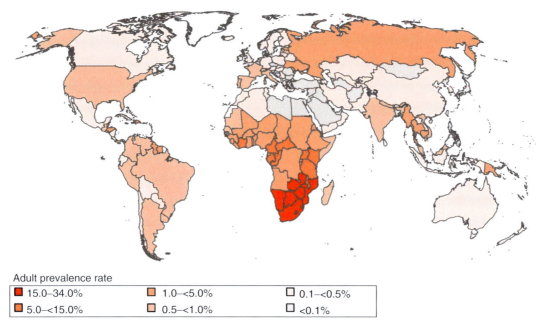

**Figure 3** Adult prevalence of HIV, 2005. Reproduced with permission from UNAIDS/WHO (2006) 2006 report on the global AIDS epidemic. Geneva: UNAIDS.

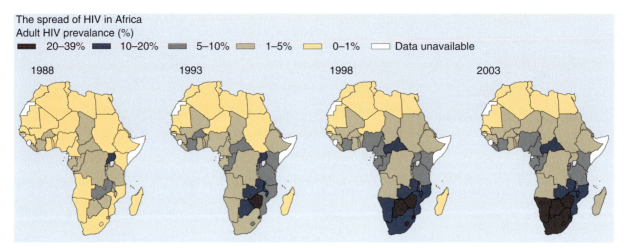

**Figure 4** The spread of HIV in Africa, 1988–2003. Reproduced with permission from UNAIDS (2005) *AIDS in Africa: Three Scenarios to 2025*. Geneva: UNAIDS.

1.6 million Africans still died of AIDS in 2007 (76% of the AIDS deaths worldwide). Within Africa, the distribution of HIV is heterogeneous. Although HIV originated in Central Africa, Southern Africa now has the highest HIV prevalence in the world. Conversely, many East African countries have seen declines in their HIV prevalence and most West African countries have maintained a relatively low HIV prevalence.

HIV arrived late in Southern Africa. In 1988, South Africa had an HIV prevalence of less than 1% and the epidemic was centered in East and Central Africa (**Figure 4**). Unfortunately in the ensuing years HIV spread to unprecedented levels. Swaziland now has the most intense HIV epidemic in the world with an estimated 1 in 4 (26%) adults living with HIV in 2007. Several other Southern African countries also had an HIV prevalence greater than 20% in 2005. Various social and biological factors have likely predisposed certain African countries to these massive HIV epidemics. High rates of men migrating for work, concurrent sex partners, and genital herpes, in addition to low rates of male circumcision and gender-based inequalities, combine to fuel the epidemic in this region.

From 2003 to 2005, the adult HIV prevalence has remained stable but high (19–24%) in Botswana, Lesotho, and Namibia and has continued to increase in South Africa from 18.6% (2003) to 18.8% (2005). This stability, however, masks extremely high rates of new infections

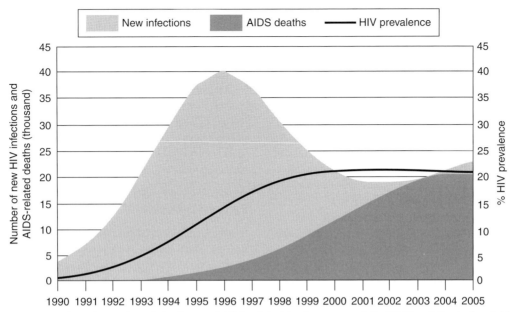

**Figure 5** Estimated number of annual new infections and AIDS-related deaths among adults (15+) in relation to the stabilizing trend of estimated prevalence among adults (15–49), Lesotho, 1990–2005. Reproduced with permission from UNAIDS/WHO (2006) AIDS Epidemic Update: December 2006. Geneva: UNAIDS.

that are being balanced by high rates of AIDS deaths (**Figure 5**). The impact of HIV in Southern Africa cannot be overstated. The average life expectancy in Botswana has dropped from 65 years (1985) to 34 years (2006) as a result of HIV. Denial continues to be a factor driving high rates of new infections as many people in the region believe that they are at low risk of infection. In a survey by Shisana *et al.* in 2005 in South Africa, 50% of people who tested HIV-positive stated that they had no risk of acquiring HIV.

East Africa was among the regions most severely affected by HIV in the early 1990s (**Figure 4**), but subsequently the HIV prevalence has dropped. Uganda's HIV epidemic was the first to stabilize and then fall sharply during the 1990s. Strong political leadership and major prevention campaigns have been credited with this success. Uganda's national adult HIV prevalence was still 6.7% in 2005, however, and new evidence of increasing infections in rural areas reinforces the need to adapt prevention campaigns to changes in the epidemic. Kenya provides another encouraging example of a serious HIV epidemic that has declined from a national prevalence of 10% in the late 1990s to 7% in 2003 and 6% in 2006. In Ethiopia, the prevalence of HIV has declined slowly in the capital city of Addis Ababa (4.7% in 2005) since the mid-1990s and has remained low in rural areas. A concerning trend has been a recent increase in the number of new infections in rural parts of the country where 80% of the population lives. Halting the spread of HIV in rural communities will be one of the most important prevention priorities in East Africa.

Although the HIV prevalence in most of West Africa is comparatively low, HIV remains a serious problem in certain communities. The HIV epidemic in Côte d'Ivoire rapidly expanded in the 1980s and the adult prevalence has remained greater than 4% as of 2005. Fortunately, data from 2005 suggest a declining prevalence in the capital city of Abidjan as well as several other West African cities. As of 2005, an estimated 2.9 million people were living with HIV in Nigeria where the national prevalence is 4%. Within the country, there is considerable variation in rates of HIV for reasons not clear. In many low-prevalence West African countries, such as Senegal with a national prevalence of 1%, commercial sex work appears to be driving the epidemic. In Central Africa, incomplete data have hidden the exact nature of the HIV epidemic. Cameroon (5% national prevalence), the Central African Republic (11%), and the Democratic Republic of the Congo (3%), however, all have significant epidemics based on estimates of prevalence from 2005.

## Asia

The adult HIV prevalence is lower in Asian countries (**Figure 3**) than sub-Saharan Africa and concentrated in high-risk groups. The epidemic continues to spread considerably, however, with over 400 000 new infections and over 300 000 deaths in 2007.

The highest regional HIV prevalence has occurred in Southeast Asia where commercial sex work, sex between men, and injection-drug use combine to fuel the epidemic. In Thailand and Cambodia, HIV spread rapidly in the late

1980s and early 1990s. Initial cases were seen in men who have sex with men and injection-drug users before spreading to female sex workers. HIV then spread to male clients who transmitted the virus to their wives and girlfriends. Extensive public education and prevention campaigns, however, have had some success in the region. Among young men in Thailand, increased knowledge of HIV and changes in sexual behavior, such as an increase in condom usage and a decrease in visits to sex workers, have correlated with reductions of the HIV prevalence in new military conscripts. Thailand's '100% Condom' campaign, which educated sex workers and promoted condom use, has also decreased HIV transmission from female commercial sex workers to their clients. Encouragingly, the epidemics in Thailand and Cambodia have been declining since the 1990s.

India is experiencing a complex HIV epidemic. Although the precise number of HIV infections in India has been debated, the Indian National AIDS Control Organization estimates that 2.5 million people were living with HIV in 2006 (an adult HIV prevalence of 0.36%). The majority of infections in India result from unprotected heterosexual sex and a large proportion of women are getting infected by their regular partners who became infected during commercial sex. Fortunately, there is evidence of a reduction in the prevalence of HIV among women attending antenatal clinics in south India from 2000 to 2004 which has correlated with increased prevention efforts in several southern Indian states. HIV prevention targeting sex workers could have a dramatic effect on the epidemic but law enforcement barriers and stigma complicate these prevention efforts. Injection-drug use is the major risk factor for HIV in the northeast and has become increasingly important in several major cities (Chennai, Mumbai, New Delhi).

In China, the HIV epidemic has followed an unusual pattern by beginning in certain rural areas, then spreading to cities. HIV was first identified among injection-drug users in the Yunnan Province (near China's southwestern border) in 1989. In the mid-1990s, a major HIV outbreak occurred in paid plasma donors in China's rural east–central provinces (Henan). During plasma donation, blood is taken from the donor and using a technique called apheresis the plasma is separated from the red blood cells. The red blood cells are then reinfused into the donor to prevent anemia. Reuse of tubing and mixing of red blood cells from multiple donors prior to reinfusion led to thousands of HIV infections in these paid donors. Since the mid-1990s, HIV has spread to all 31 of China's provinces and an estimated 650 000 people are living with HIV (0.05% prevalence in 2005). Although injection-drug use has been the predominant mode of transmission, sexual transmission has become increasingly important and accounted for an estimated 50% of new infections in 2005. China's response to its HIV epidemic was initially slow but has accelerated recently to incorporate evidence-based interventions from other countries (such as Australia's needle-exchange program and Thailand's condom campaign among sex workers).

Recently, the HIV epidemic has expanded among injection-drug users in several Asian countries, such as Pakistan, Afghanistan, Indonesia, and Malaysia. Given the overlap between injection-drug users and commercial sex workers, there is concern that the HIV epidemic could spread rapidly in these countries.

## Eastern Europe and Central Asia

Until the late 1990s, the prevalence of HIV in Eastern Europe and Central Asia was extremely low. In the last decade, however, this region has experienced the fastest growing HIV epidemic and an estimated 1.7 million people are now living with HIV (over 90% of cases are in the Russian Federation and Ukraine). Injection drug use is the predominant method of HIV transmission but high rates of HIV are also emerging in the sex partners of injection drug users, female sex workers, and prisoners. Although the epidemic is still concentrated in high-risk groups in the Russian Federation and Ukraine, there is evidence that HIV is starting to spread into the general population. In the Ukraine, the proportion of people infected through heterosexual sex increased from an average of 14% from 1999–2003 to 35% in 2006. The Ukraine's HIV epidemic is also yet to peak despite an estimated adult HIV prevalence of 1.5% in 2005. The HIV epidemic is also starting to grow rapidly in the injection-drug user populations of several other Central Asian countries (Uzbekistan, Kazakhstan, and Tajikistan).

## Caribbean

The HIV prevalence is high in many Caribbean countries where an estimated 1–4% of adults are living with HIV. The epidemic is fueled by concurrent heterosexual partners and commercial sex work. Sex between men is a hidden phenomenon but may account for 10% of cases. Since 2002, the adult HIV prevalence in urban areas of Haiti and the HIV prevalence among pregnant women in the Bahamas have declined, and these declines have correlated with high rates of condom use by sex workers.

## Latin America

Brazil, the most populous country in Latin America, has 620 000 people living with HIV in 2005. In the 1990s, many experts predicted that Brazil's epidemic would rapidly accelerate. Brazil's sustained campaign, however, to promote sex education, condom use, harm reduction, and HIV testing, and to provide anti-retroviral therapy halted HIV's expansion. Overall, the adult prevalence has remained stable at 0.5%. The majority of HIV infections in South America is occurring among men who have sex with men and injection-drug users. Incarcerated men, in particular, are at high risk for infection. Recently, rates of HIV have also increased among poor women from unprotected heterosexual sex with injection-drug users and men who also have sex with men.

The countries of Central America have the highest estimated HIV prevalences (1–2.5% in 2005) in Latin America and the epidemics appear to be growing. Sex between men and commercial sex are the major risk factors, but there is evidence that the epidemics are generalizing. Mexico has a relatively low (0.3% in 2005) HIV prevalence, but increasing rates of infection in cities along the United States border are being driven by injection-drug use and commercial sex work.

### North America

The total number of people living with HIV continues to grow in North America due to the combination of the life-prolonging anti-retroviral treatment (with a corresponding decline in AIDS deaths) and a steady rate of new infections. The main risk factors for HIV are estimated to be unsafe sex between men (44%), unprotected heterosexual sex (34%), and injection-drug use (17%). In the US, women are estimated to account for over 25% of the AIDS cases and minority populations have been disproportionately affected by the epidemic. The rate of new HIV infections in the US was seven times higher in African–American men than white men and 21 times higher among African–American women compared to white women. In Canada, aboriginal people are also disportionately affected by the epidemic. An estimated 250 000 people living with HIV in the United States are not aware of their diagnosis. This large group presents a major obstacle toward controlling the HIV epidemic since people who are unaware of their status may continue to engage in high-risk behaviors.

### Western Europe

In the 1980s, HIV spread widely among men who had sex with men and injection-drug users in Western Europe. Subsequent prevention programs have had success in reducing the incidence of HIV in these risk groups although the HIV prevalence in injection-drug users in Southern Europe remains high. The major modes of HIV transmission are now unprotected heterosexual sex (56%), unsafe sex between men (35%), and injection drug use (8%). Although heterosexual transmission has emerged as an important mode of HIV transmission in Western Europe, there is no evidence of a generalized epidemic. In many Western European countries, including Belgium, Sweden, and the United Kingdom, immigrants from sub-Saharan Africa account for the majority of heterosexually acquired infections. Unprotected sex with injection-drug users or men who have sex with men represents the other major risk for heterosexual transmission.

### Oceania

There is a serious HIV epidemic in Papua New Guinea with an estimated adult prevalence of 1.8% in 2005. High rates of concurrent sexual partners, transactional sex, and violence against women have allowed the epidemic to grow rapidly. Sex between men is the predominant mode of HIV transmission in Australia and New Zealand.

### Middle East and North Africa

Uneven HIV surveillance makes it difficult to gauge the epidemics in this region. Sudan has a generalized epidemic with an HIV prevalence of 1.6% in 2005. High rates of HIV have also been observed in injection-drug users in Iran, Libya, Algeria, Egypt, Morocco, and Lebanon.

### Global Response

The world was initially slow to recognize the severity of the HIV pandemic but global efforts have increased dramatically in recent years. Between 1996 and 2005, annual funding for AIDS in low- and middle-income countries increased from $300 million to $8.3 billion. This acceleration of the world's response was prompted by increased human rights advocacy from people living with HIV and a concern that AIDS could destabilize global economic systems and threaten global security. Although these increases in funding are impressive, an effective global response will depend on sustained growth in annual funding of effective prevention and treatment programs until the epidemic can be stopped.

### HIV Prevention

Comprehensive and sustained prevention programs have been proven to reduce HIV transmission. Unfortunately, HIV prevention strategies have not reached the majority of people at high risk. The major challenge for prevention has been generating the political will and economic resources to effectively implement proven strategies that address issues such as sex, sexuality, and drug use.

Behavioral interventions to promote safer sexual behavior are essential to reduce HIV transmission.

In Zimbabwe, sexual behavior change (increased condom use, delayed onset of sexual activity, and a reduction in sexual relations with nonregular partners) has contributed to declines in HIV prevalence. Successful behavioral interventions incorporate educational messages with prevention skills such as how to negotiate and use condoms and how to refuse sex. These interventions have been successfully implemented in school-based sex education classes, peer-led small group discussion groups, in the context of HIV counseling and testing, and as national structured interventions, such as Thailand's '100% Condom' program. For successful implementation, prevention programs must identify and engage high-risk populations.

Globally, an estimated 80% of people living with HIV are unaware of their status. Since people who are unaware that they are HIV-infected are more likely to continue transmitting HIV, an urgent priority is to increase access to HIV testing. Botswana has implemented a national, opt-out HIV-testing program in medical settings. Opt-out HIV testing has also been implemented in certain clinical settings in Kenya, Malawi, Uganda, and Zambia. In African countries with a high HIV prevalence a major risk factor for acquiring HIV is having a stable heterosexual partner who is HIV-infected (and often unaware of their status). Couples testing and counseling where both the male and female partners are counseled and HIV tested together has gained increased acceptance in certain African countries (Rwanda, Zambia) as a prevention tool in this high risk population. Universal HIV testing could both reduce transmission of HIV among newly identified infections and reduce the stigma of HIV testing.

An effective HIV vaccine would be a major advance in HIV prevention and is the subject of intense research efforts. Several other biomedical strategies also have the potential to dramatically improve HIV prevention. Male circumcision reduces a man's risk of acquiring HIV heterosexually by 50–60%. Since the highest prevalence of HIV is in countries where men are uncircumcised, large-scale implementation of male circumcision has the potential to prevent millions of new infections if provided in a safe and culturally appropriate way. Herpes simplex virus type 2 (HSV-2) infection is the most common cause of genital ulcers and is associated with an increased risk of HIV transmission. Clinical trials are now evaluating whether antiviral prophylaxis (to suppress HSV-2) will reduce the rate of HIV transmission. Interventions under a woman's control such as barrier diaphragms and vaginal microbicides which incorporate anti-retroviral drugs to block HIV transmission are also being evaluated in clinical trials. Daily oral administration of anti-retroviral drugs to high-risk, HIV-negative individuals, a strategy called pre-exposure prophylaxis (PrEP), is also being assessed. Successful biomedical interventions must still be packaged with effective behavioral prevention strategies to avoid behavioral disinhibition (an increase in high-risk behavior driven by a perceived decrease in risk for HIV infection).

Prevention of mother-to-child transmission (PMTCT) strategies have almost eliminated pediatric HIV in high-income countries. Treating the HIV-infected mother with anti-retrovirals and replacing breast milk with formula reduces the rate of mother-to-child HIV transmission from as high as 41% to less than 2%. Even a single dose of nevirapine given to the mother in labor and to the infant at birth significantly decreases the rate of HIV transmission. Despite these effective prevention strategies, there are still an estimated 1800 new mother-to-child HIV infections per day. These pediatric infections are occurring predominantly in sub-Saharan Africa due to a lack of access to PMTCT services. Breast-feeding also presents a difficult problem in sub-Saharan Africa since substituting formula for breast milk results in higher mortality among both HIV-infected and HIV-negative infants born to HIV-infected mothers. Treating HIV-infected breast-feeding women and/or their children with anti-retrovirals to prevent breast milk transmission is an innovative strategy being investigated.

Treatment of drug addiction, education about safe injection practices, and access to clean needles are effective prevention strategies in injection-drug users. In hospitals, infection-control programs that prohibit the reuse of injection equipment are necessary to prevent nosocomial HIV transmission. Finally, maintaining the safety of the blood supply by screening voluntary donors for HIV risk factors, testing each unit of blood for HIV, and reducing unnecessary transfusions, is another critical intervention to prevent HIV transmission.

## HIV Care and Treatment

HIV mortality dropped precipitously in high-income countries after the introduction of highly active anti-retroviral therapy (HAART) in 1996, but HAART remained unaffordable for many low- and middle-income countries. Brazil, however, made the pioneering decision to initiate a highly successful universal HIV treatment program which has inspired similar efforts in low-income countries. Since 2001, global financing for HIV care and treatment in low-income countries has increased markedly driven by funding from the Global Fund to Fight AIDS, Tuberculosis, and Malaria, the US President's Emergency Plan for AIDS Relief (PEPFAR), and the World Bank. By December 2006, over 2 million people living with HIV/AIDS were receiving treatment in low- and middle-income countries which represents 28% of the estimated 7.1 million people in need of treatment (**Table 2**). Treatment coverage remains strongest in Latin America (72%) driven by Brazil's leadership. The most encouraging improvements have been in sub-Saharan

**Table 2** Estimated number of people receiving anti-retroviral therapy, people needing anti-retroviral therapy, and percentage coverage in low- and middle-income countries according to region, Dec. 2003–Dec. 2006

| Geographical region | Estimated number of people receiving antiretroviral therapy, Dec. 2006 (range)[a] | Estimated number of people needing antiretroviral therapy, 2006 (range)[a] | Antiretroviral therapy coverage, Dec. 2006 (range)[b] | Estimated number of people receiving antiretroviral therapy, Dec. 2005 (range)[a] | Estimated number of people receiving antiretroviral therapy, Dec. 2003 (range)[a] |
|---|---|---|---|---|---|
| Sub-Saharan Africa | 1 340 000 (1 220 000–1 460 000) | 4 800 000 (4 100 000–5 600 000) | 28% (24–33%) | 810 000 (730 000–890 000) | 100 000 (75 000–125 000) |
| Latin America and the Caribbean | 355 000 (315 000–395 000) | 490 000 (370 000–640 000) | 72% (55–96%) | 315 000 (295 000–335 000) | 210 000 (160 000–260 000) |
| East, South, and South-east Asia | 280 000 (225 000–335 000) | 1 500 000 (1 000 000–2 100 000) | 19% (13–28%) | 180 000 (150 000–210 000) | 70 000 (52 000–88 000) |
| Europe and Central Asia | 35 000 (33 000–37 000) | 230 000 (160 000–320 000) | 15% (11–22%) | 21 000 (20 000–22 000) | 15 000 (11 000–19 000) |
| North Africa and the Middle East | 5000 (4000–6000) | 77 000 (43 000–130 000) | 6% (4–12%) | 4000 (3 000–5000) | 1000 (750–1250) |
| Total | 2 015 000 (1 795 000–2 235 000) | 7 100 000 (6 000 000–8 400 000) | 28% (24–34%) | 1 330 000 (1 200 000–1 460 000) | 400 000 (300 000–500 000) |

Note: Some numbers do not add up due to rounding.
[a]Data on children (when available) are included.
[b]The coverage estimate is based on the estimated numbers of people receiving and needing antiretroviral therapy.
Reproduced with permission from the World Health Organization. Data from World Health Organization (2007) Towards Universal Access. Scaling up Priority HIV/AIDS Interventions in the Health Sector. Progress Report, April 2007. Geneva: WHO.

Africa where the number of people being treated has increased from 100 000 in 2003 to 1.3 million by December 2006. Sustained international funding from high-income countries and political will in low-income countries will be necessary to maintain high medication adherence, to monitor for drug toxicities and drug resistance, and to improve treatment regimens.

The HIV epidemic intersects with other important diseases. By weakening the immune system, HIV predisposes to opportunistic infections such as tuberculosis (TB). Globally, more than 21 million people are co-infected with TB and HIV. Although TB infection remains latent (neither contagious nor symptomatic) in most people without HIV, with TB/HIV co-infection the risk of progression to active, contagious disease increases 100-fold. TB and HIV synergistically worsen the burden of both diseases. The global TB incidence has increased, fueled by a rapid increase in HIV-associated TB in Africa. TB is also a leading cause of death among people living with HIV. Increased collaboration between TB-elimination and HIV-treatment programs is necessary to control both of these diseases.

Although there has been debate regarding the relative cost-effectiveness of prevention strategies compared to treatment programs, combined prevention and treatment strategies have synergistic benefits. Increased treatment access, for example, enhances awareness, reduces stigma, and increases the use of HIV testing and counseling services. Increased HIV testing, in turn, identifies HIV-infected people who can be counseled to prevent further transmission and treated with HAART when indicated, which also decreases infectiousness. Only effective prevention programs can reduce the incidence of new infections and thereby reduce the number of people who will need treatment in the future.

## Future Prospects

In the past 25 years, HIV has emerged as the world's most serious public health problem. Every region of the world has been affected by AIDS, but sub-Saharan Africa has been disproportionately affected. Recent successful efforts to provide HIV care and treatment in low-income countries must now be expanded and sustained. In addition, prevention programs need to expand in a similar manner into low-income countries. These prevention strategies must include evidence-based interventions such as safer sex education programs with access to condoms and HIV-testing and counseling services. These programs must also be able to incorporate new biomedical prevention techniques (such as male circumcision) without compromising existing effective prevention services. Most importantly, sustained financial and political commitments will be required from both high- and low-income countries to control the HIV pandemic.

*See also:* AIDS: Disease Manifestation; AIDS: Vaccine Development; Human Immunodeficiency Viruses: Antiretroviral agents; Human Immunodeficiency Viruses: Molecular Biology; Human Immunodeficiency Viruses: Origin.

## Further Reading

Centers for Disease Control, Prevention (CDC) (2006) The Global HIV/AIDS Pandemic, 2006. *Morbidity Mortality Weekly Report* 55(31): 841–844.

Galvin SR and Cohen MS (2004) The role of sexually transmitted diseases in HIV transmission. *Nature Reviews Microbiology* 2(1): 33–42.

Global HIV Prevention Working Group (2006) New Approaches to HIV Prevention: Accelerating Research and Ensuring Future Access, Aug. 2006. http://www.kff.org/hivaids/hiv081506pkg.cfm (accessed August 2007).

Gupta K and Klasse PJ (2006) How do viral and host factors modulate the sexual transmission of HIV? Can transmission be blocked? *PLoS Medicine* 3(2): e79.

Hemelaar J, Gouws E, Ghys PD, and Osmanov S (2006) Global and regional distributions of HIV-1 genetic subtypes and recombinants in 2004. *AIDS* 20(16): W13–W23.

Kourtis AP, Lee FK, Abrams EJ, Jamieson DJ, and Bulterys M (2007) Mother-to-child transmission of HIV-1: Timing and implications for prevention. *Lancet Infectious Diseases* 6(11): 726–732.

Sawires SR, Dworkin SL, Fiamma A, Peacock D, Szekeres G, and Coates TJ (2007) Male circumcision and HIV/AIDS: Challenges and opportunities. *Lancet* 369(9562): 708–713.

Smith DK, Grohskopf LA, Black RJ, et al. (2005) Antiretroviral postexposure prophylaxis after sexual, injection-drug use, or other nonoccupational exposure to HIV in the United States: Recommendations from the US Department of Health and Human Services. *MMWR Recommendations and Reports* 54: 1–20.

UNAIDS (1998) Partners in Prevention: International Case Studies of Effective Health Promotion Practice in HIV/AIDS. http://data.unaids.org/Publications/IRC-pub01/JC093-PartnersInPrevention_en.pdf (accessed August 2007).

UNAIDS (2005) *AIDS in Africa: Three Scenarios to 2025.* Geneva: UNAIDS.

UNAIDS/WHO (2006) 2006 report on the global AIDS epidemic. http://www.unaids.org/en/HIV_data/2006GlobalReport/default.asp (accessed August 2007).

UNAIDS/WHO (2006) AIDS Epidemic Update: December 2006. http://www.unaids.org/en/HIV_data/epi2006/default.asp (accessed August 2007).

UNAIDS/WHO (2007) AIDS Epidemic update: December 2007. http://www.unaids.org/en/HIV_data/2007Epiupdate/default.asp (accessed November 2007).

Valdiserri RO, Ogden LL, and McCray E (2006) Accomplishments in HIV prevention science: Implications for stemming the epidemic. *Nature Medicine* 9(7): 881–886.

World Health Organization (2007) Towards Universal Access. Scaling up Priority HIV/AIDS Interventions in the Health Sector. Progress report, April 2007. http://www.who.int/hiv/mediacentre/universal_access_progress_report_en.pdf (accessed August 2007).

## Relevant Website

http://www.hiv.lanl.gov – The Circulating Recombinant Forms (CRFs), HIV Sequence Database, Los Alamos National Laboratory.

# AIDS: Vaccine Development

**N C Sheppard and Q J Sattentau,** University of Oxford, Oxford, UK

© 2008 Elsevier Ltd. All rights reserved.

## Glossary

**Adjuvants** Immune modulators designed to increase immune responses to vaccine antigens. These are of particular use for nonreplicating vaccine antigens such as killed microorganisms or antigenic subunits of microorganisms.

**AIDS (acquired immunodeficiency syndrome)** A pathological state that can be defined by the emergence of opportunistic infections or neoplasms in HIV seropositive patients, which coincides with a decline in the peripheral blood CD4 T-cell count below 200 cells $\mu l^{-1}$.

**Antigen presenting cells** Dendritic cells, macrophages, and B-cells are specialized cells capable of internalizing antigen and presenting it to lymphocytes together with the co-stimulatory molecules necessary to induce immune responses.

**Breakthrough infection/vaccine failure** Denotes that recipients of candidate vaccines have become infected and/or that viremia has escaped control by vaccine-induced immune responses.

**CD4+ T cells** A subset of T lymphocytes that coordinate the adaptive immune response, including antibody production and cytotoxic T-cell activation. They are also the principal target cell for HIV-1 replication *in vivo*.

**Cytotoxic T cells (CTL)** A subset of T lymphocytes that have the ability to kill virally infected cells, thus controlling viral replication and spread. They may also have additional functions that reduce the replication of viruses, such as production of soluble inhibitory factors, for example, cytokines and chemokines.

**Envelope glycoproteins** The HIV-1 envelope glycoproteins are used by the virus to gain entry into receptor-bearing target cells. The surface glycoprotein, gp120, binds the primary receptor CD4 and the co-receptor, usually CCR5. The transmembrane glycoprotein gp41 mediates fusion of the viral membrane with the cell membrane, allowing infection. The envelope glycoproteins are the principal targets of neutralizing antibodies (nAbs).

**Exposed-uninfected (EU) individuals** Those that are highly exposed to HIV-1 and therefore at high risk of infection, but fail to seroconvert and have undetectable viral RNA load.

**Long-term nonprogressors (LTNP)** Individuals that have been infected, normally for at least 10–12 years without ever experiencing HIV-related disease or requiring antiretroviral therapy. The CD4 T-cell count remains stable and $>600 \mu l^{-1}$.

**Neutralizing antibodies** Strictly defined as antibodies (Abs) that reduce viral infectivity without additional factors such as complement. These Abs usually function by preventing virus entry into permissive cells. The activity of nAbs may be augmented by factors such as complement, and non-neutralizing antibodies may become antimicrobial in the presence of complement.

**Primary isolates** HIV-1 strains that have undergone minimal *in vitro* culture in primary leukocytes and are thus not adapted to cell culture. Adaptation to continued cell culture often increases the growth rate of a virus while dramatically reducing its resistance to neutralizing antibodies. An exaggerated phenotype of this kind is seen with viruses adapted to growth in immortalized lymphocytic cell lines.

**SIV/SHIV challenge models** Rhesus macaques can be used to model HIV/AIDS by experimental infection with simian Immunodeficiency Virus (SIV) or chimeric SIV recombinant for the envelope glycoprotein from HIV-1 (SHIV). A rapid loss of CD4 T cells often results in comparatively early onset of an AIDS-like syndrome in these models.

## Introduction

Since the discovery of HIV-1, the causative agent of AIDS in 1983, researchers have attempted to develop a prophylactic vaccine to control the pandemic. Early attempts to achieve protective immunity made use of recombinant viral envelope glycoproteins (Env) to attempt to induce neutralizing antibody (nAb) responses. To date such approaches have not been successful, probably due to the inability of such subunit proteins to induce antibody (Ab) responses capable of cross-neutralizing the innumerable number of viral strains encountered in the field. Nevertheless, some seropositive cohorts have Ab responses capable of neutralizing diverse isolates and broadly neutralizing monoclonal antibodies (NmAbs) have been isolated from such patients. When these NmAbs are infused into animal models such as rhesus macaques prior to virus challenge, sterilizing immunity is often afforded. These observations reveal that the induction of nAbs could form a component of a protective vaccine strategy. A second component that has received much attention is the induction of CD8 cytotoxic T-lymphocyte (CTL) responses. Such responses play a major role in the control of HIV-1 replication *in vivo* and, when sufficiently induced by vaccination, can suppress viremia and delay clinical progression in macaques challenged with virus. If such reductions in viremia are achieved in humans, viral transmission between infected and uninfected individuals might be prevented or reduced. Results from a plethora of Ab and CTL strategy-based vaccine trials conducted to date suggest that poor immunogenicity is often an obstacle to achieving robust responses; thus, novel adjuvants and immunogenic recombinant vectors are likely to be required for a successful vaccine.

## Lessons from Successful Vaccines against Other Pathogens

Studies of the immune correlates of protection associated with existing vaccines reveal that Ab responses play a dominant role. Only BCG vaccination against *Mycobacterium tuberculosis* induces immunity in which Abs appear to have no role. Since CTL responses are likely to be required for an effective HIV-1 vaccine, traditional vaccine technologies are likely to be suboptimal, or counter-indicated due to safety concerns. Many existing vaccines are based on live-attenuated microorganisms. Such vaccines have potential hazards such as transmission to unvaccinated persons and the risk of reversion to pathogenicity, both of which have been observed with the Sabin oral polio vaccine. Killed microorganism vaccines do not have these drawbacks. However, they are occasionally associated with disease when the inactivation process is incomplete, and are inefficient at inducing CTL responses. Due to these factors, neither the live-attenuated nor killed vaccine approaches appear to be suitable for the development of a prophylactic vaccine against HIV-1. Therefore, the majority of HIV-1 vaccine candidates are based on variations of the subunit approach whereby one or more of the viral proteins is produced for delivery either as a protein, or alternatively encoded in a DNA plasmid or a live-attenuated vector.

Although there are now successful vaccines against persistent viral infections (HBV and HPV) and even one against a virus capable of latency (VZV), retroviruses such as HIV-1 present new challenges. Probably the most challenging is the unparalleled mutation rate of HIV-1. The enormous genetic diversity of this virus means that if the first rounds of replication are not aborted by vaccine-induced immune responses, the virus will escape the immune response and establish long-term reservoirs of latent infection refractory to immune clearance. If HIV-1 infection cannot be prevented by vaccination, then reducing the rapid diversification of HIV-1 in a new host by controlling replication to the lowest possible levels is a second major challenge.

## Clinical Trials of HIV-1 Vaccine Candidates

In 2003, VaxGen announced that phase III efficacy trials of its gp120-based AIDSVAX candidate, aimed at inducing nAb responses, had failed to protect volunteers. The inadequacy of AIDSVAX and emergence of data suggesting CTLs play the major role in the control of HIV-1 heralded a paradigm shift away from the induction of nAb responses toward inducing CTL responses. Reflecting this, most of the vaccines currently on trial are based on strategies that have, as their primary or sole aim, the induction of CTL and T-helper responses against HIV-1. Thus, the main hypothesis currently being tested in clinical trials is that CTL responses can be induced safely in humans and will impact upon viral load and disease progression as they have done in primate models of simian-human immunodeficiency virus (SHIV) infection. In 2007, the phase IIb trials of the MRKAd5 trivalent vaccine developed by Merck & Co, which aimed to give 'proof of concept' for the CTL-based approach in man were discontinued. An interim investigation revealed that the vaccine did not reduce the likelihood of infection or impact on viral load after infection. Therefore, the CTL-based approach remains to be validated in man. The vaccine pipeline will likely respond to these disappointing findings by testing vaccine candidates that induce CTL responses to additional viral targets with a further paradigm shift putting equal emphasis on nAb and CTL-based approaches.

## Defining the Correlates of Protection

A major difficulty in HIV-1 vaccine research has been in determining which host immunological factors, if any, correlate with protection from infection or protection from disease progression. The influence of human leukocyte antigen (HLA) genotype on HIV progression is a clear indicator that the adaptive immune response is implicated in the control of infection. However, only rare cohorts of individuals seem to resist HIV-1 infection despite frequent exposure to the virus, or control an established infection robustly. Groups of highly exposed seronegative (ES) individuals display great heterogeneity. They may have increased natural killer (NK) cell activity, HIV-1-specific CTL responses, HIV-specific CD4 T-cell production of MIP-1$\beta$, IFN-$\gamma$, and IL-2, neutralizing IgA responses at the mucosae or no detectable responses. So far, the most frequent finding is the presence of CTL responses to HIV-1 antigens in ES cohorts. Where CTL activity exists, it shows marked differences in epitope specificity from responses seen in infected individuals and is often restricted only by HLA alleles associated with reduced risk of infection. Findings of neutralizing IgA responses are controversial due to the difficulty in ruling out contamination of the mucosal samples of ES study subjects with Env-specific IgA from an infected partner's seminal fluid.

Although ES cohorts are defined as lacking serum IgG responses and not having detectable virus by PCR, it remains to be determined whether these individuals have never been infected, have had abortive infections, or are persistently infected at extremely low levels. Studies have detected low levels of HIV provirus in peripheral blood mononuclear cells (PBMCs) taken from seronegative infants born to seropositive mothers, high-risk seronegative homosexual men, and individuals several years prior to their seroconversion. These findings favor a model of early control of infection prior to seroconversion. In the cohorts of ES homosexual men, lower T-cell responsiveness and immune activation were seen when compared with homosexual men who later seroconverted, suggesting T-cell hypo-responsiveness might be a correlate of protection, potentially by depriving HIV-1 of susceptible target cells (activated CD4 T cells upregulate CCR5, the principal co-receptor for HIV-1 and become more permissive for viral replication).

A second group of individuals who have attracted intense scrutiny are the long-term nonprogressors (LTNPs), who maintain high CD4 T-cell counts and low-to-undetectable viral loads for many years after HIV-1 infection. Again, these people are not a homogeneous group but consist of: those of certain HLA types; groups co-infected with hepatitis G (GB virus C); cohorts who have poly-functional CTL responses, capable of lytic function as well as releasing antiviral cytokines, β-chemokines, and/or CD8 antiviral factor (CAF), which suppresses HIV-1 replication at the level of transcription; those possessing TCR clones capable of tolerating mutation in the target epitope; a few individuals infected with naturally attenuated viruses that are mutated in regions such as *nef* and *vpr*; and rare individuals who have broad nAb responses.

CTLs probably have the central role in controlling viremia in acute and chronic infection and have been most frequently correlated with natural immunity among ES individuals and LTNPs. The case for nAbs is weaker as their role in controlling established virus infection *in vivo* is not clearly defined. The rapid diversification of Env is known to be driven by nAb responses with CTLs making little contribution. NAbs exert 'soft' selection, with the rate of mounting an autologous nAb response strongly correlated with the rate of phenotypic escape. Multiple mechanisms allow the virus to rapidly escape at minimal fitness cost, meaning that the nAb response is rarely active against contemporaneous isolates. Nevertheless, the presence of nAbs in some ES and LTNP cohorts and the ability of NmAb infusions to provide sterilizing immunity in macaque challenge models supports them as a correlate of protective immunity. In addition, studies in which acutely or chronically infected individuals were transfused with a cocktail of NmAbs showed that viremia was controlled in some cases. These data suggest that NmAbs alone can control established infection, but that the dose required to do so is high and the virus can still escape, especially in patients who harbor a greater diversity of viral strains. Therefore, nAb responses are likely to be more important at preventing infection or controlling early infection, than at reducing later viral replication.

The induction of nAbs to Env and robust CTL responses to multiple components of the virus in resistant cohorts implies that CD4 T-cell responses must also have been induced. It is known that the ability of HIV-1-specific CD4 T-cells to respond to Gag epitopes by secreting IL-2 alone, or together with IFN-γ, is negatively correlated with virus load in HIV-1-infected patients. This implies that vaccines should induce good helper CD4 T-cell responses in order to optimize processes such as affinity maturation and class switching of Abs, proliferation of CTLs, and induction of appropriate memory T- and B-cell responses.

Studies of the 'correlates of disease' rather than 'protection' in nonhuman primates infected with SIV in the wild reveal that unlike their counterparts experimentally infected with SIV viruses isolated from other monkey species, natural SIV infection in the host monkey species rarely leads to an AIDS-like syndrome. Most animals remain healthy with normal CD4 T-cell levels despite a persistently high viral load. Comparative analysis of naturally and experimentally infected animals has demonstrated a lack of generalized activation of the immune response in the naturally infected animals, which consequently avoid the immune-mediated damage observed in pathogenic SIV infection. Gut-associated lymphoid tissue (GALT)-resident memory CD4 T cells are normally lost early in pathogenic SIV/HIV infection, compromising immune integrity of the GALT and resulting in the translocation of bacteria and associated inflammatory mediators from the gut lumen into the circulation. These might contribute to chronic immune activation, which eventually exhausts the homeostatic mechanisms that replenish CD4 T cells. Microbial translocation is not detected in nonpathogenic SIV infection, concurring with the lack of chronic immune activation. Interestingly, similar observations have been made with respect to reduced immune activation in the less-pathogenic HIV-2 infection of humans and T-cell hypo-responsiveness correlates with ES status in several cohorts. As a corollary to the 'immune-activation' hypothesis, activation of SIV-specific CD4 T cells by vaccination leads to exacerbated SIV-mediated disease upon challenge in the natural host. These findings shed light not only on potential mechanisms of immune pathology associated with HIV-1 infection of humans, but also send a warning that suboptimal vaccination protocols might be detrimental by encouraging immune pathology.

## Abs and HIV-1

### The Role of Abs *In Vivo*

Env is the principal target of Ab-mediated immunity since it is the sole virus antigen expressed on virions and productively infected cells. Neutralization is achieved when Ab binds any exposed epitope on the functional Env spike, thus sterically blocking the gp120–CD4 interaction, or in a few cases, preventing the necessary conformational changes required to induce fusion. Theoretically, all spikes may need to be inactivated since the binding of an nAb to one spike does not affect adjacent spikes and a single fusogenic spike may be sufficient to mediate entry. The entropy and kinetics of nAb–Env association affect neutralization potency as they determine which of many conformations that Env samples can be recognized, and the stability of the interaction. Neutralization is the most important, but not the only, antiviral mechanism mediated by Env-specific Abs. Both Ab-dependent complement-mediated inactivation (ADCMI) and Ab-dependent cellular cytotoxicity (ADCC) correlate with positive clinical indicators in some studies. In addition, Ab-mediated opsono-phagocytosis of virions might also play a significant role.

Tat is not present at the cell/virion surface, but is secreted by infected cells and is believed to transactivate bystander cells increasing HIV-1 replication. Abs that neutralize the ability of Tat to transactivate cells might reduce viral load and prevent disease progression. Clinical trials are in progress to assess whether Tat-based vaccines are capable of inducing Tat-neutralizing Ab responses and whether these correlate with beneficial outcomes.

### Env Evolution in Response to NAbs

Over the course of HIV-1 evolution, the unique mutational plasticity of Env has allowed it to accrue multifaceted immunoevasive properties to escape humoral immunity (**Table 1**).

In light of such elegant strategies, it is not surprising that HIV-1 fails to induce high nAb titers in most patients and rapidly escapes narrow-specificity nAbs as they arise. The preferential loss of HIV-1-specific CD4 T cells due to their unique ability to be both activated by and infected with the virus no doubt plays an important role in viral escape by reducing T-cell help for B cells (and CTLs). Nevertheless, in some patients, chronic exposure to Env variants leads to the production of broadly reactive nAbs that target the portions of Env that are highly conserved in order to allow virus–receptor interactions. Thus far, only a handful of such NmAbs have been isolated from patients, namely b12, 2G1, 2F5, and 4E10 (**Table 2**).

These NmAbs have been subject to intense scrutiny and the X-ray crystal structures for b12, 2G12, 2F5, and 4E10 in complex with Env or peptides have been solved. Extraordinary characteristics have been observed, including: extended complementarity-determining region (CDR) H3 loops (b12, 2F5, and 4E10); dimerization of the heavy-chain variable domain $V_H$ (2G12); the use of only the heavy chain in Ab–antigen interactions (b12); and the ability to make contacts with the epitope-utilizing interactions between main-chain atoms (b12). Whether such unusual Abs can be elicited by vaccination is a central question to the field. The different NmAbs act at one or more phases to achieve neutralization; b12 is most

**Table 1** Immunoevasion strategies utilized by HIV-1 Env to escape NAb responses

| Strategy | Mechanism |
| --- | --- |
| Decoys | Irrelevant epitope exposure on monomeric gp120, gp4l that has shed gp120, gp120-gp4l monomers, dimers, and tetramers, uncleaved gp160, and alternate trimeric forms. |
| Entropic masking | Inherent conformational flexibility of Env imposes a significant entropic penalty on Ab binding. Most Abs appear to bind a limited number of conformations and thus are at a kinetic disadvantage. |
| Glycan shield | Shifting glycans can sterically inhibit Abs from reaching their epitopes. |
| Protection of the CD4-binding site (CD4bs) | Part of the CD4bs is located in a hydrophobic cavity that limits access to CD4bsAbs. The remainder is induced by CD4 binding and in the unliganded form lies near or within a long cavity formed by portions of the outer and inner domains as well as part of the bridging sheet. The V1N2 loop on one gp120 molecule appears to obscure part of the CD4bs on the inner domain in a neighboring gpl20 in the trimer. Similarly, the V3-loop partly obscures the bridging sheet portion of the CD4bs. Contact with CD4 involves some main-chain atoms and thus mutation is tolerated. |
| Protection of the co-receptor binding site | The co-receptor binding site is composed by elements of the V3-loop stem and bridging sheet. The necessary conformation is rarely sampled by Env prior to CD4-ligation and afterwards it lies in close apposition to target cell membrane and thus bulky IgG molecules are sterically excluded. The V1N2 loop on one gp120 molecule obstruct elements of co-receptor-binding site on neighboring gp120 molecules in the trimer. |
| Sequence variability | Five V-loops that are highly tolerant of amino acid substitutions, additions, and deletions induce strain-specific Abs. (NB: V3-loop length is conserved). |
| Silent face | A hyperglycosylated region of the outer domain of gp120 which imparts low immunogenicity to gp120. |

**Table 2** The properites of the broadly specific NmAbs

| Antibody | Target | Epitope |
|---|---|---|
| b12 | gp120 | A recessed and conformationally invariant region overlapping the CD4-binding site, which is involved in the metastable interaction with CD4. All three complementarity determining region (CDR) loops of the heavy chain are involved but the light chain is not. Many contacts involve main-chain atoms. |
| 2G12 | gp120 | Mannose c(1→2) mannose-linked glycans attached to N295 and N332 and stabilized by other proximal glycans. The Fab domains form a unique dimer, which extends the surface for interaction with the dense glycan cluster on gp120. |
| 2F5 | gp41 | Centered on the core epitope ELDKWAS, which is potentially helical in nature and is situated in the membrane proximal extracellular region (MPER). An extended CDR H3 loop (22 aa) may play a role in destabilizing some proximal structure but interaction with the membrane is not necessary. Polyspecific autoreactivity shown with cardiolipin. |
| 4E10 | gp41 | Binds the helical epitope WFX(I/L)(T/S)XX(L/I)W, which lies C-terminal side of the 2F5 epitope in the MPER and potentially associated with the membrane. The apex of the long CDR H3 loop extends beyond the peptide and is likely to contact the membrane. Polyspecific autoreactivity shown with cardiolipin. |

effective before CD4 binding, whereas 2G12 still neutralizes after CD4 binding and 2F5 and 4E10 are active even after co-receptor engagement. By comparison, sera from LTNPs with high nAb titers have been shown to act mostly before CD4 engagement, 2G12-like Abs correlate with the broadest nAb activity in LTNP sera, whereas 2F5 and 4E10-like Abs appear to be absent. This is in keeping with the finding that 2F5 and 4E10 appear to be examples of polyspecific autoreactive Abs, which may be difficult to induce in most people due to self-tolerance mechanisms.

In a new host, the homogenization of Env sequences is observed during acute infection with evolution toward an ancestral sequence occurring for upto 2–3 years before diversification begins. In this process, Env may adopt shorter variable loops and reduced glycosylation, possibly increasing its fusogenicity. Such virus is unusually sensitive to neutralization. During acute infection, the patient reaches a peak in its infectivity, potentially transmitting neutralization-sensitive strains to new recipients. However, neutralization-resistant virus can also be transmitted since virus is often passed between mother and baby despite the transfer of maternal nAbs. In the context of vaccination, this suggests that acutely infected patients might pass on neutralization-sensitive virus more vulnerable to vaccine-induced nAbs, whereas neutralization-resistant virus is likely to be passed on later in infection. If sterilizing immunity is not achieved, pre-existing humoral immunity might thwart the outgrowth of neutralization-sensitive strains, perhaps limiting the rate of dissemination and magnitude of peak viremia.

## Optimizing Antigens to Induce Abs with Antiviral Activity

The problem of inducing Abs of the correct specificities that will have protective roles *in vivo* has proved difficult to solve. It is clear from clinical trials conducted thus far that vaccination with native gp120, gp160, gp41, or Env peptides fails to induce nAbs against more than a handful of isolates. Therefore, improved strategies that aim to circumvent the immuno-evasive properties of Env are being developed. The use of consensus sequence Env reduces the antigenic distance between the vaccine strain and isolates that will be encountered in the field. So far, consensus sequence Env has induced nAb responses of equal or slightly better breadth compared to naturally occurring strains, suggesting at best, only an incremental improvement. Attempts to recapitulate the native structure of the fusogenic virion-associated Env trimeric spike have also produced incremental improvements when compared with monomeric gp120. Since Gag drives the self-assembly of the virion, virus-like particles (VLPs) resembling the native virion structure but devoid of nucleic acid are generated when Gag is expressed along with other structural proteins. The presentation of stabilized trimers in the context of VLPs appears a useful approach since the presence of the cytoplasmic tail of gp160 is known to alter the antigenicity of the external portions of Env and the membrane proximal extracellular region (MPER) epitopes might be appropriately presented only when anchored proximal to the membrane. Soluble gp140 immunogens, containing the gp120 subunit and the extracellular portion of gp41, lack the intracellular portion of gp41 and thus have subtly different antigenic profiles. Studies examining variable loop deletion or truncation and the removal of glycosylation sites have revealed that partial truncation of the V2 loop gives an incremental improvement in the breadth of nAbs, whereas the removal of glycans does not. Multivalent approaches in which several strains are used simultaneously show moderate improvement in the breadth of the nAb response in animal models. However, this increase appears to be due to the additive effect of nAbs specific to each individual component and not due to the induction of Abs with broader neutralizing properties. Although there are undoubtedly fewer neutralization serotypes than genotypes of HIV-1, it will not be practical to cover all neutralization serotypes in a multivalent cocktail if each component only induces nAbs of narrow specificity. An alternative approach is the development of mimics of the epitopes of NmAbs that might induce similar broadly neutralizing specificities after immunization. Combining approaches that give incremental improvements when used alone might provide substantial additive improvements in the breadth of response.

## Inducing Cell-Mediated Immunity by Vaccination

In macaque models, vaccination with DNA plasmids or recombinant viruses bearing SIV/SHIV antigenic inserts leads to significant attenuation of viremia after subsequent infection with the highly pathogenic X4-tropic hybrid isolate SHIV89.6P. Although they do not afford sterilizing immunity under any regimen tested thus far, such vaccines enable long-term suppression of viremia and prevent disease progression. Nonetheless, eventual escape of the challenge virus from CTL control after acquisition of single amino acid mutations in critical immunodominant epitopes is sometimes observed. This results in rapid rebound of viremia and normal disease progression, and underlies the necessity of targeting the most conserved epitopes critical for viral fitness, as well as targeting multiple epitopes if 'breakthrough' replication is to be avoided. A critical caveat of the work in macaques with the frequently used SHIV-89.6P is that this virus appears to be unusually sensitive to vaccine-induced immunity. Experience using R5-tropic SIVmac239 or SIVmac251 challenge models demonstrates that viremia is often controlled only transiently and no control of HIV-1 infection was observed in the clinical trials of MRKAd5 vaccine despite earlier trials showing promising immunogenicity of this candidate.

HIV-1 evades T-cell responses in a number of ways (Table 3). Once an epitope has mutated to a form not recognized by a particular CTL clone, a new population of antigenic variants can rapidly replace those sensitive to that clone. This occurs if the mutation is less detrimental to viral fitness than remaining sensitive to the CTL clone. Even in vaccinated and challenged macaques, a degree of viral replication remains despite strong CTL responses. From this small pool of viral variants, escape mutants can arise. The current challenge is for vaccination to induce optimal cellular immunity to HIV-1, including sufficient breadth of responses to decrease the chances of viral escape. Since an individual's CTL responses are entirely governed by their HLA class I alleles, vaccines must include enough material from the virus to allow the majority of the population to make multiple responses. This might be achieved by expressing strings of epitopes, or whole or truncated HIV-1 proteins, or even fusions of several proteins from DNA plasmids or recombinant vectors. However, the process is not straightforward and, to date, the immune responses induced by experimental vaccines have been disappointingly narrow and often dominated by a response to a single epitope.

## Selecting the Best Target Antigens

The obvious advantage of CTL-based vaccines over those that aim to induce nAbs is that all HIV-1 proteins are targets for cellular immunity and some of these are considerably less tolerant of mutation than Env. CTLs escape mutations with a high cost to viral fitness and revert after transmission to a new host where the epitope is not targeted. Conversely, CTL epitopes restricted by common HLA class I alleles that are not critical to viral fitness will drive the accumulation of escape mutants in a population, since the chance of mutant virus being transmitted to individuals with the same restricting HLA

**Table 3** Immunoevasive strategies used by HIV-1 to escape T-cell response

| Strategy | Mechanism |
| --- | --- |
| Bystander apoptosis | HIV can induce bystander apoptosis of CD4 and CD8 T-cells. Activated CD4 T-cells are especially vulnerable to apoptosis induced by highly pathogenic X4-tropic isolates in a CXCR4-Env dependent manner, whereas the CD8 T-cells are killed by soluble factors released by CD4 T-cells. Fas-mediated mechanisms and virion-associated MHC class II and CD86 have been implicated. |
| Dysregulation of T-cell migration | The V2- and V3-loops of X4 using gp120 have been shown to repel T-cells, including CTL via a CD4-independent, receptor-mediated mechanism. CTLs also accumulate in the blood rather than lymphatics in acute infection, suggesting that non-X4-tropic virus mechanisms are also active. |
| Mutation | Amino acid substitutions can prevent anchoring to MHC molecules, can prevent recognition by or antagonize responding TCRs, or can alter proteolytic processing and peptide transport preventing surface expression of peptide-MHC complexes. |
| Nef-mediated downregulation of MHC class I | One of the functions of Nef is the downregulation of MHC class I molecules and subsequent reduction in the ability of infected cells to display any MHC class I epitopes from HIV-1. This results in increased resistance to CTL-mediated killing in an epitope-independent fashion. |

class I allele is high. This necessitates the rational design of vaccine immunogens to provide epitopes for the less common HLA alleles that may prove more effective at combating HIV-1 and delineating which epitopes of a mutant virus may still be presented by the common HLA alleles. Targeting early antigens such as Tat, Rev, and Nef allows CTLs the greatest amount of time to act on an infected cell. By contrast, targeting highly expressed structural proteins such as Gag would increase the chances of the CTLs recognizing infected cells given that HLA class I expression is downregulated by Nef. The question of whether CTL responses to highly conserved epitopes exist in HIV-1-infected patients is one of some controversy since such responses are only detected in some studies. It is possible that vaccination will induce such responses, but it is more likely that the most highly conserved regions of HIV-1 protein have evolved to be weakly immunogenic and poorly processed in infected cells. At present, a strategy that induces multiple responses, perhaps to consensus or mosaic sequences constructed to optimize CTL epitope coverage across clades, seems most likely to invoke prolonged control of viremia independent of virus clade.

## Achieving Strong Immune Responses to Candidate HIV-1 Vaccines

### Class of Immunogen

One of the major obstacles facing the design of HIV-1 vaccine candidates is poor immunogenicity. The methods of delivery, vectors, and adjuvants used require considerable optimization in order to induce robust and persistent immune responses. Env is considered poorly immunogenic in absolute terms, requiring high doses and repeat-boosting simply to induce substantial titers of binding Abs. Factors such as glycosylation, restricted numbers of CD4 T-cell epitopes, and tertiary structure limit the immunogenicity of Env. The low innate immunogenicity of Env and other HIV-1 antigens can be overcome by optimizing delivery and providing adjuvants. Each delivery method comes with distinct advantages and disadvantages. DNA vaccines are conceptually simple vaccine candidates that can induce both T-cell and Ab responses, but immune responses in humans and other primates have proved disappointing when compared to small animal models. Despite their promise at inducing CTL responses, DNA vaccines are expensive, requiring doses of 2–5 mg to induce weak CD4 T-cell responses in macaques. However, DNA immunization has shown promise at priming immune responses that can later be boosted by recombinant vectors and benefit from expressing only the target antigen and not multiple vector antigens. Due to the expense and poor results achieved thus far with DNA vaccines, some researchers are looking to replace DNA priming immunizations with recombinant viral or bacterial vectors, or protein. However, since the immunogenicity of DNA vaccines can be substantially improved *in vivo* by electroporation or transfection reagents, more efficient use of DNA may be eminently feasible.

Recombinant vectors, mostly based on attenuated viruses, are among the lead HIV-1 vaccine candidates currently on trial. They show particular promise at inducing T-cell responses and are especially immunogenic when used in heterologous prime/boost regimens. Vectors based on poxviruses and adenoviruses show good immunogenicity but may be sensitive to pre-existing antivector immunity, which is common in many populations. The development of vectors with low seroprevalence should counter this problem. Other vectors with potential for greater immunogenicity are under development, including: measles virus, vesicular stomatitis virus, Venezuelan equine encephalitis virus, and poliovirus.

Protein immunogens are generally excellent at inducing Ab responses, but poor at inducing CTL responses unless adjuvants that favor cross-presentation are included. However, with the correct adjuvant they can be used to induce Abs, CD4, and CD8 T-cell responses. Protein immunogens have the disadvantage of being relatively expensive to produce and, like recombinant vectors, require cold-chain storage.

### Adjuvants

Adjuvants fall into two broad classes: (1) Immune potentiators, including host factors such as cytokines, chemokines, heat-shock proteins, and complement components or pathogen associated molecular patterns (PAMPs) such as modified lipopolysaccharide (LPS), peptidoglycans, nucleic acids, or bacterial toxin subunits. (2) Delivery systems, including liposomes, synthetic microparticles, mineral salts, and oil-in-water emulsions. Adjuvants can be co-administered with, or incorporated into all the vaccine types discussed. For example, genes for pro-inflammatory cytokines can be cloned into DNA or recombinant vectors or they can be given in protein form together with the vaccine. Importantly, the direct conjugation of adjuvant to antigen appears to optimize its effects by ensuring co-delivery to antigen-presenting cells. Novel delivery systems can be combined with the novel immune potentiators to optimize their effects. Defining which combinations of these immune modulators, PAMPs, and novel delivery systems will synergize to impart the greatest magnitude and duration of immune responses is a problem that needs considerable attention.

*See also:* AIDS: Disease Manifestation; AIDS: Global Epidemiology; Antigenic Variation.

## Further Reading

Brenchley JM, Price DA, and Douek DC (2006) HIV disease: Fallout from a mucosal catastrophe? *Nature Immunology* 7(3): 235–239.

Burton DR, Stanfield RL, and Wilson IA (2005) Antibody vs. HIV in a clash of evolutionary titans. *Proceedings of the National Academy of Sciences, USA* 102(42): 14943–14948.

Douek DC, Kwong PD, and Nabel GJ (2006) The rational design of an AIDS vaccine. *Cell* 124(4): 677–681.

Pantaleo G and Koup RA (2004) Correlates of immune protection in HIV-1: What we know, what we don't know, what we should know. *Nature Medicine* 10: 806–810.

## Relevant Website

http://www.iavireportonline.org – New! Vax Special Issue – 2004: Year in Review.

# Akabane Virus

**P S Mellor,** Institute for Animal Health, Woking, UK
**P D Kirkland,** Elizabeth Macarthur Agricultural Institute, Menangle, NSW, Australia

© 2008 Elsevier Ltd. All rights reserved.

## Glossary

**Arthrogryposis (AG)** Rigid fixation of the joints, usually in flexion but occasionally in extension.
**Culicoides** Blood-feeding dipterous insects also known as biting midges.
**Dystocia** Abnormal or difficult birth.
**Encephalomyelitis** Inflammation both of the brain and the spinal cord.
**Porencephaly** Disease of the brain with the formation of small cavities in the brain substance.
**Hydranencephaly (HE)** A condition in which the brain's cerebral hemispheres are almost or completely absent and replaced by sacs filled with cerebrospinal fluid.
**Kyphosis** Outward curvature of the spine causing a humped back.
**Nystagmus** Rapid involuntary oscillation of the eyes.
**Scoliosis** Lateral deviation of the normal vertical line of the spine.
**Torticollis** Twisting of the neck that causes the head to rotate or tilt.

## Introduction

Akabane virus was originally isolated from mosquitoes in Japan, in the summer of 1959, 'Akabane' being the name of the village where the virus was first isolated. Subsequently, it has been shown to occur widely in Africa, Australia, throughout Asia and the Middle East, its distribution being determined by the occurrence of its insect vectors which are predominantly biting midges from the genus *Culicoides*.

The virus is important in veterinary pathology because it is able to cross the placenta of cattle, sheep, and goats causing a range of congenital defects, principally arthrogryposis (AG) and hydranencephaly (HE), and abortion. The virus can be widespread without evidence of disease because it usually produces asymptomatic infections in adult animals. It is only when pregnant, serologically naive adults are infected early in pregnancy that the virus may cross the placenta to cause damage to the fetus. Evidence of the damage only becomes apparent some months later with the birth of the affected young – by which time the viruses have been eliminated from both mother and offspring and so cannot be isolated from them. This meant that for many years Akabane virus was not connected to these clinical manifestations, until careful epidemiological observations and the development of appropriate serological tests confirmed the association.

## Taxonomy, Properties of the Virion and Genome

Akabane virus is a member of the genus *Orthobunyavirus* in the family *Bunyaviridae*. Several related viruses (Aino, Cache Valley, Peaton, and Tinaroo) have been shown under experimental conditions to have the potential to cross the ruminant placenta but only Aino in Australia and Japan and Cache Valley in the USA have been recognized as pathogens in the field.

The Akabane virion is spherical in shape and is *c.* 90–100 nm in diameter. A lipid envelope with projecting glycoprotein peplomers surrounds the viral genome which consists of three separate segments of single-stranded RNA. Each segment is of a different length. The large segment encodes the viral transcriptase, the medium segment two glycoproteins, and the small segment the internal nucleoprotein. Each of the segments also encodes a nonstructural protein.

The virus is acid labile and is readily inactivated by chloroform, ether, and trypsin. It is also very heat labile and is inactivated at 56 °C in a few minutes and loses about 0.3 log of infectivity per hour at 37 °C. However, it remains viable in blood samples kept at 4 °C for several months and can be stored indefinitely at −80 °C or lower.

## Vertebrate Hosts

The virus has been isolated from cattle, sheep, and goats. Antibodies have also been detected in horses, pigs, camel, red deer, and a wide selection of African wildlife ranging from various species of antelope to hippopotamus, elephant, and giraffe.

There are no known cases of human disease caused by Akabane virus even though some of the mosquito species from which the virus has been isolated (*Aedes vexans, Culex tritaeniorhynchus*, and *Anopheles funestus*) regularly bite humans. However, the role of mosquitoes as vectors remains unclear and limited serological surveys have not found evidence of human infection. Further, the *Culicoides* species that are known vectors of the virus rarely bite humans.

## Clinical Signs

When Akabane virus infects pregnant cattle, sheep, or goats, the virus is able to replicate in and cross the ruminant placenta causing a variety of congenital abnormalities in the fetus. The range and severity of these abnormalities are dependent upon the stage of gestation at infection. In adult animals, however, infection is usually subclinical and in endemic areas most breeding age animals will have acquired an active immunity during early life sufficient to prevent the virus from reaching the developing fetus. In these situations, the virus exists as a 'silent' infection and no evidence of disease is seen. The pathogenic effects of Akabane infection are usually observed only when the vector spreads beyond the limits of its endemic areas under favorable conditions, to enter regions where many adult animals are still susceptible and therefore able to be infected during pregnancy. In such situations, an epidemic in cattle, sheep, or goats may be noticed by an increased incidence of abortions and premature births in late autumn or early winter. Calves infected close to term may be born with encephalitis, often apparent clinically as a flaccid paralysis. Some strains of Akabane virus may also cause encephalitis in newborn calves and, infrequently, in older animals. This is followed by the birth of calves, lambs, or kids with a range of congenital defects, principally AG and HE (see **Figures 1** and **2**). Young with

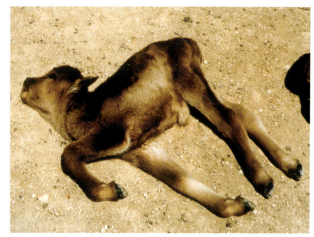

**Figure 1** Calf with AG and torticollis due to Akabane virus infection.

**Figure 2** Akabane virus-induced HE in a calf.

these defects may be stillborn or delivered alive at term. AG is characterized by fixation of the joints in flexion (most frequently) or extension. The birth of some arthrogrypotic animals is associated with dystocia, necessitating embryotomy or cesarean section to save the dam. Young born with HE may show blindness, nystagmus, deafness, dullness, slow suckling, paralysis, and lack of coordination. They may survive for several months if hand-reared but often die from misadventure. In cattle herds with animals at different stages of pregnancy, there will be a succession of defects, initially with cases of AG and later calves with HE. In sheep and goats, due to the shorter period of gestation and narrower range at which the fetus is susceptible, affected progeny born may show one or more of the defects of AG and HE concurrently.

Epidemics of AG/HE disease due to Akabane virus have been recorded in Japan, Australia, Israel, and Turkey, the most severe involving 30 000 calves in Japan and approximately 8000 calves in Australia. The incidence of affected progeny varies depending on the strain of virus. In naturally infected cattle, the incidence of AG and HE varies from 25% to 50% but in sheep that have been experimentally infected with different virus strains, the incidence of abnormal lambs can vary from 15% to 80%.

## Pathogenesis

In naturally infected cattle, there is an incubation period of a few days followed by a viremia of 3–6 days duration. This is not usually associated with clinical signs in the adult animal. From about day 12 post infection, virus-neutralizing antibodies begin to be detected in the circulation.

The sequence of events leading to fetal infection involves virus replication in the endothelial cells of the placenta, then the trophoblastic cells, and finally in the fetus itself. The ability of the virus to produce congenital damage depends upon the stage of gestation at which fetal infection occurs. Fetal Akabane virus infection is characterized by a predictable chronological sequence of congenital defects directly referable to the fetal age at which infection has occurred. The bovine fetus is susceptible from about 90 days through to term while the ovine and caprine are most susceptible between 28 and 56 days of gestation. In sheep and goats, defects are less common than in cattle because of both the limited gestational range and the fact that small ruminants are less likely to be pregnant during the virus transmission period.

In cattle, the defects occur over a longer period and are seen in consistent sequence. The first abnormalities to appear in live neonates are nonsuppurative encephalitis and encephalomyelitis, the outcome of infection late in gestation. Arthrogrypotic neonates follow as a result of infection in the fifth and sixth months of gestation, then more severe malformations including porencephaly and HE reflecting infection in the third and fourth months. When AG is first observed, only one joint on a single limb might be affected, but in cases that are born later (infected earlier) the abnormalities are more severe with changes in several joints on all limbs. Calves affected with severe AG after infection early in the fifth month of gestation may show early changes in the brain with small cystic lesions of porencephaly. The lesions of HE then develop as a result of infection in the third and fourth months of gestation. As calves are born later in an outbreak, the lesions of HE will progress in severity, initially being seen as small focal cystic cavities but soon progressing to large fluid-filled cavities. In calves with severe lesions, the cerebral hemispheres are completely absent and only fluid-filled meninges remain but the brainstem and cerebellum remain unaffected.

The replication of Akabane virus and its tissue tropisms in the fetus are determined by fetal age, the presence of rapidly dividing cells, and the development of immunocompetence. The most severe lesions occur in fetuses infected during the early stages of organogenesis. In cattle the most severe lesions occur after infection between 70 and 90 days of gestation and in sheep and goats between 28 and 36 days. Immunocompetence develops in calves from about 90 days and in lambs from 65–70 days of gestation, and subsequent to these times virus is rapidly eliminated from the fetal tissues. As a result, virus cannot be isolated from calves and lambs delivered at term but may be isolated from a fetus that is aborted soon after infection.

## Pathology

AG and HE are the main gross lesions of Akabane disease but cervical scoliosis, torticollis, and kyphosis also occur. In sheep and goats, there may also be pulmonary hypoplasia. Arthrogrypotic calves exhibit restrictions to movements of the limb joints exclusively as a result of changes to the soft tissues. The joint surfaces are normal and the bones are unaffected. In affected limbs, the muscles are usually reduced in size and paler in color than normal. This is a result of both neurogenic muscular atrophy and also from primary infection of the muscles.

Microscopically, there is a severe loss of myelinated fibres in the lateral and ventral funiculi of the spinal cord, and of ventral horn neurons and nerve fibers in the ventral spinal nerves. However, white matter is unaffected.

In HE calves, the cerebral cortex is represented only by a thin shell of brain tissue, perhaps only the meninges, filled with fluid (**Figure 2**). The meninges may be thickened. In most cases, the brainstem is intact and the cerebellum appears normal.

During the early stages of an outbreak of Akabane disease, there may be calves which are incoordinate or unable to stand at birth. No gross pathological lesions are found in such animals but microscopically there is a mild to moderate nonsuppurative acute encephalomyelitis, most evident in the gray matter of the mid- and posterior brain.

## Epidemiology and Transmission

Most of Africa, the Middle East, southern Asia, Japan, Korea, and Australia may be regarded as being endemic for Akabane virus. Papua New Guinea and the island countries of the Pacific, however, are free from infection. The geographical distribution of this virus is controlled completely by the distribution, seasonal activity, and abundance of its insect vectors.

Akabane virus was first isolated in Japan in 1959 from mosquitoes of the species *Ae. vexans* and *Cx. tritaeniorhynchus*. Then, in 1968, it was isolated in Australia from the biting midge *Culicoides brevitarsis*. More recently, isolations have also been made from *An. funestus* in Kenya, and from *Culicoides* species such as *C. oxystoma* in Japan,

**Figure 3** *Culicoides imicola*, a vector of Akabane virus.

*C. imicola* (**Figure 3**) and *C. milnei* in Zimbabwe, *C. imicola* in Oman, *C. brevitarsis* and *C. wadai* in Australia, and a mixed pool consisting mainly of *C. imicola* in South Africa. The virus has been shown to replicate in *C. sonorensis* (=*C. variipennis*) by up to 1000-fold, and transmission occurs after 7–10 days incubation at 25 °C. As *C. sonorensis* is widely distributed and abundant in North America, this suggests that should Akabane virus gain entry to that continent, transmission would be likely. The virus has also been shown to replicate in *C. brevitarsis* and reaches the salivary glands of infected individuals after 10 days incubation. In Australia, *C. brevitarsis* is the only biting insect whose distribution correlates closely with the distribution of Akabane antibodies in cattle. The development of antibodies also coincides with the seasons when *C. brevitarsis* is active. Furthermore, *C. brevitarsis* feeds upon sheep and horses in Australia, in which species antibodies to Akabane virus have been found. These findings suggest that in Australia *C. brevitarsis* is the principal vector of Akabane virus. No close correlation has yet been reported between any mosquito species and Akabane virus distribution, and as replication and transmission of the virus have not been demonstrated in any mosquito species, this suggests that *Culicoides* species are likely to be the major vectors of Akabane virus and mosquitoes are of lesser importance. Akabane virus persists in nature by alternate cycling of the virus between the midges and its mammalian hosts after biting of the mammalian host by the insects. Transovarial (vertical) transmission of Akabane virus has not demonstrated.

In Australia, the principal vector *C. brevitarsis* spreads Akabane virus throughout an endemic area that extends across the north of the country and southward along the east coast. Within this area, virus transmission occurs annually, so breeding animals are usually infected early in life before becoming pregnant and few Akabane cases occur. Under harsh climatic conditions, usually drought, the distribution of the midge may retreat northward and toward the coast, reducing the extent of infection for one or perhaps two seasons. On resumption of the usual pattern of spread, a susceptible population of pregnant females will be infected at the margins of the range of the midge, resulting in small disease outbreaks. Under the influence of very favorable conditions of higher rainfall and mild temperatures, the insect may spread further south and inland than is usual and result in virus transmission in highly susceptible livestock populations. Large outbreaks will then occur with abortions noticed in late autumn and calves with AG/HE begin to appear in midwinter and spring, from July through to September.

Outbreaks of Akabane disease in Japan also show seasonal and geographic clustering, with most cases occurring from September to March. In Israel, disease outbreaks have occurred from November to June and in western Turkey during early spring (March). The differences in timing are presumably related to the abundance of the vector insects as influenced by climate.

## Diagnosis

Sporadic cases of AG/HE due to Akabane virus often remain undiagnosed. However, when there is a cluster of cases, a teratogenic virus should be considered as a possible cause.

Because Akabane virus does not persistent in the fetus, attempts to isolate virus from affected newborn calves, lambs, and kids are uniformly unsuccessful. Nevertheless, virus may be isolated from the tissues (e.g., brain, spinal cord, skeletal muscle) of fetuses that are aborted early in pregnancy. Virus isolation is usually conducted by inoculation of a continuous cell line and the polymerase chain reaction (PCR) assay has been used to reliably detect viral RNA in fetal tissues.

Akabane infection is most frequently confirmed serologically. Serum from affected calves, lambs, and kids should be collected prior to suckling to enable detection of specific antibodies to Akabane virus. As virus-specific antibodies are produced in fetuses after they become immunocompetent, the detection of antibodies to this virus in serum and body fluids prior to suckling is considered to confirm an *in utero* infection. An examination of the abomasum for milk curd should be made to confirm that the calf has not suckled. As a further check, a serum sample from the dam should also be taken and the two samples tested in parallel for antibodies to Akabane and other common viruses. If the serum of the neonate contains a similar range of antibodies to the dam, it would suggest that suckling has taken place. If only antibodies to Akabane virus are detected in the serum of the neonate but there are antibodies to several in its dam's serum then the evidence is strong that it has been infected *in utero*.

The virus neutralization test is the most specific serological test available but enzyme-linked immunosorbent assays (ELISAs) have also been developed that are either Akabane specific or Simbu serogroup specific.

## Differential Diagnosis

Other viral teratogens may be associated with *in utero* infections and the delivery of neonates with congenital defects. Of the bunyaviruses, Aino virus has been linked with congenital AG/HE in cattle, though the lesions of the cerebrum are less symmetrical than those due to Akabane virus and the cerebellum may also be affected. In sheep in the USA (where Akabane virus is exotic), Cache Valley virus is a possible cause. Other viral agents that may cause defects include bluetongue, Rift Valley fever, Wesselsbron, border disease, and bovine viral diarrhea viruses. An important differential feature of Akabane infection in cattle is that the cerebellum is unaffected. Individual agents may be confirmed by using virus-specific serology to test precolostral serum or fluid from body cavities (e.g., pericardial or pleural fluid).

A range of noninfectious causes of congenital AG and HE in calves, lambs, and kids should also be considered and include maternal intoxication (e.g., by ingestion of toxic plants) or an inherited defect. The detection of elevated immunoglobulin G (IgG) levels in precolostral serum or fluids will differentiate between infectious and noninfectious causes of congenital defects.

## Prophylaxis and Control

There are two main approaches to the prophylaxis of Akabane virus infections: vaccination and management strategies to control or avoid vectors.

## Vaccination

Vaccination is the main method of prophylaxis and Akabane vaccines have been produced in Japan and Australia.

In Japan and Australia, inactivated vaccines have been developed that induce high antibody titers after two doses given with a 4 week interval. In trials, these vaccines prevented development of a viremia and fetal infection on challenge or after natural exposure and were safe when used in pregnant animals. As antibody titers decline fairly rapidly, annual revaccination is recommended. In Japan, a live attenuated vaccine has also been produced by serial passage of Akabane virus in cell culture. When used in cattle this vaccine induces high titers of neutralizing antibodies without pyrexia, leucopoenia, or viremia and virus was not recoverable from the organs or the fetuses of vaccinated animals. It also prevented infection of the bovine fetus on challenge with virulent Akabane virus. However, in pregnant ewes, the vaccine induced a viremia and frequently caused intrauterine infection of the fetus, so this vaccine is not recommended for use in sheep.

## Vector Control and Management Strategies

At present, vector control is considered to be impractical partly because of inadequate knowledge of vector biology over much of the global range of Akabane virus and partly because vector control can be prohibitively expensive. Further, absolute control of vectors is impossible and animals only need to be bitten by a single virus-infected vector for infection to occur. Possible control measures involve elimination or reduction in the vector-breeding sites (damp organically enriched soil or large herbivore dung) and direct insecticide attack on adult vectors. Susceptible hosts may also be protected from the bites of vectors by the use of insect repellents or by screening animal housing with insect-proof nets or mesh.

An important measure to minimize losses due to Akabane virus is to avoid the movement of susceptible pregnant animals into virus-endemic areas during the vector season. Such inadvertent movements have resulted in significant losses.

*See also:* Bunyaviruses: General Features; Vector Transmission of Animal Viruses.

## Further Reading

Al-Busaidy SM, Mellor PS, and Taylor WP (1988) Prevalence of neutralising antibodies to Akabane virus in the Arabian Peninsula. *Veterinary Microbiology* 17: 11–149.

Charles JA (1994) Akabane virus. *Veterinary Clinics of North America: Food Animal Practice* 10: 525–546.

Inaba Y and Matumoto M (1990) Akabane virus. In: Dinter Z and Morein B (eds.) *Virus Infections of Ruminants*, pp. 467–480. Oxford: Elsevier.

Jagoe S, Kirkland PD, and Harper PAW (1993) An outbreak of Akabane virus induced abnormalities in calves following agistment in an endemic region. *Australian Veterinary Journal* 70: 56–58.

Jennings DM and Mellor PS (1988) *Culicoides*: Biological vectors of Akabane virus. *Veterinary Microbiology* 21: 125–131.

Kirkland PD and Barry RD (1986) The economic impact of Akabane virus and cost effectiveness of Akabane vaccine in New South Wales. In: St. George TD, Kay BH, and Blok J (eds.) *Arbovirus Research in Australia*, pp. 229–232. Brisbane: CSIRO.

Kirkland PD, Barry RD, Harper PAW, and Zelski RZ (1988) The development of Akabane virus induced congenital abnormalities in cattle. *Veterinary Record* 122: 582–586.

Matumoto M and Inaba Y (1980) Akabane disease and Akabane virus. *Kitasato Archives of Experimental Medicine* 53: 1–21.

St. George TD and Kirkland PD (2004) Diseases caused by Akabane and related Simbu-group viruses. In: Coetzer JAW and Tustin RC (eds.) *Infectious Diseases of Livestock*, 2nd edn., vol. 2, pp. 1029–1036. Cape Town: Oxford University Press.

St. George TD and Standfast HA (1989) Simbu group viruses with teratogenic potential. In: Monath TP (ed.) *The Arboviruses: Epidemiology and Ecology*, vol. 4, pp. 145–166. Boca Raton, FL: CRC Press.

Taylor WP and Mellor PS (1994) The distribution of Akabane virus in the Middle East. *Epidemiology and Infection* 113: 175–185.

# Alfalfa Mosaic Virus

**J F Bol,** Leiden University, Leiden, The Netherlands

© 2008 Elsevier Ltd. All rights reserved.

## Glossary

**Agroinfiltration** Infiltration of plant leaves with a suspension of Agrobacterium tumefaciens for transient expression of genes from a T-DNA vector.

**T-DNA vector** Plasmid with the T-DNA (transferred DNA) sequence of the Ti plasmid of Agrobacterium tumefaciens, which is transferred to plant cells where it can be transiently expressed or become integrated in the plant genome.

## History

Alfalfa mosaic virus (AMV) was identified in 1931 by Weimer as the causal agent of an economically important disease in alfalfa. Purification of AMV around 1960 showed that virus preparations contained bacilliform particles of different length. Fractionation of these particles by sucrose-gradient centrifugation revealed that the four major components each contained a specific type of RNA, termed RNAs 1, 2, 3, and 4. Initially, an analysis of the biological activity of the viral nucleoproteins and RNAs resulted in a puzzle. A mixture of the three largest viral particles was fully infectious, but a mixture of RNAs 1, 2, and 3, purified from these particles, was not. At the RNA level, RNA 4 was required in the inoculum to initiate infection. In 1971 it became clear that the AMV genome consisted of RNAs 1, 2, and 3, and that a mixture of these genomic RNAs became infectious only after addition of coat protein (CP) or its subgenomic messenger, RNA 4. High-affinity binding sites for CP were identified in the AMV RNAs in 1972 and could be localized to the 3′ termini of the RNAs in 1978. From 1975 onwards it became clear that, similar to AMV, ilarviruses required CP to initiate infection. Moreover, CPs of AMV and ilarviruses could bind to the 3′ termini of each other's RNAs, and could be freely exchanged in the initiation of infection. Currently, AMV is the type species of the genus *Alfamovirus* and is classified together with the genus *Ilarvirus* in the family *Bromoviridae*.

## Taxonomy and Classification

In addition to the genera *Alfamovirus* and *Ilarvirus*, the family *Bromoviridae* contains the genera *Bromovirus*, *Cucumovirus*, and *Oleavirus*. The tripartite RNA genomes of bromo- and cucumoviruses are infectious as such and do not require CP to initiate infection. Although a mixture of the genomic RNAs of AMV and ilarviruses has a low intrinsic infectivity, initiation of infection is stimulated approximately 1000-fold by addition of CP or RNA 4. This phenomenon has been termed 'genome activation'. CP of bromo- and cucumoviruses is unable to initiate infection by AMV or ilarvirus genomic RNAs. Oleaviruses have been studied in less detail but are believed to resemble bromo- and cucumoviruses in their replication strategy. The 3′ termini of the RNAs of bromo- and cucumoviruses contain a tRNA-like structure (TLS) that can be charged with tyrosine, whereas the RNAs of AMV and ilarviruses cannot be charged with an amino acid. Many isolates of AMV are known which are closely related by serology and nucleotide sequence similarity. The complete nucleotide sequence of AMV isolate 425 (Leiden [L] and Madison [M] isolates) has been determined, and partial sequences of the Strasbourg (S) and yellow spot mosaic virus (Y) isolates have been published. The available data indicate that sequence similarity between isolates is over 95%.

## Host Range and Economic Significance

AMV occurs worldwide. Strains of this virus have been found in natural infections of about 150 plant species representing 22 families. The experimental and natural host ranges include over 600 species in 70 families. Recently, 68 *Arabidopsis thaliana* ecotypes were analyzed for their susceptibility to AMV infection. Thirty-nine ecotypes supported both local and systemic infection, 26 ecotypes supported only local infection, and three ecotypes could not be infected. Although AMV infects mostly herbaceous plants, several woody hosts are included in its natural host range. AMV is an economically important pathogen in alfalfa and sweet clover and may affect pepper, pea, tobacco, tomato, soybean, and celery. Worldwide, AMV causes calico mosaic in potato but its economic importance in this crop is limited.

## Particle Structure and Composition

**Figure 1** shows an electron micrograph of AMV. The four major classes of particles in AMV preparations are called bottom component (B), middle component (M), top component *b* (Tb), and top component *a* (Ta). B, M, and Tb are bacilliform and contain the genomic RNAs 1, 2, and 3, respectively. Ta contains two molecules of the subgenomic

RNA 4 and can be subdivided into bacilliform Ta-b and spheroidal Ta-t particles. The bacilliform particles are all 19 nm wide and have lengths of 56 nm (B), 43 nm (M), 35 nm (Tb), and 30 nm (Ta-b). The RNAs are encapsidated by a single type of CP which in the case of strain AMV-L has a length of 220 amino acids (mol. wt 24 280). In solution, AMV CP occurs as dimers which under appropriate conditions of pH and ionic strength form a $T=1$ icosahedral structure built from 30 dimers. This structure can be crystallized and has been studied by X-ray diffraction, cryoelectron microscopy, and image reconstruction methods. The CP was found to have the canonical eight-stranded β-barrel fold with the N- and C-terminal arms as extended chains. Dimer formation in the $T=1$ particle is based on the clamping of the C-terminal arms of the subunits.

From particle weight measurements and analysis of electron microscopic images, it has been concluded that the number of CP monomers in the major viral components is $60 + (n \times 18)$, $n$ being 10 (B, 240 subunits), 7 (M, 186 subunits), 5 (Tb, 150 subunits), or 4 (Ta, 132 subunits). By gel electrophoresis at least 13 minor components have been resolved which probably represent other $n$ values and contain monomers of genomic RNAs, multimers of genomic RNAs or RNA 4, or specific degradation products of RNA 3. Although details of the arrangement of the protein monomers have not been established, electron microscopical studies indicate that the cylindrical parts of the bacilliform particles have a hexagonal surface lattice with dimers of CP associated with the twofold symmetry axes. The cylindrical part is believed to be capped by two halves of an icosahedron by changing the axes from sixfold symmetry in the cylinder into axes of fivefold symmetry. Neutron scattering data suggested that the capsid structure of spheroidal Ta-t particles is represented by a deltahedron with 52-point group symmetry built from 120 subunits. The percentage of RNA in the virions decreases from 16.3 in B to 15.2 in Ta-b. The buoyant density in CsCl of the major components fixed with formaldehyde varies from 1.366 (Ta) to 1.372 (B) g cm$^{-3}$. The protein shell has an inner radius of 6.5 nm and an outside radius of 9.4 nm. The RNA is uniformly packed within the 6.5 nm radial limit, occupies about 20% of the interior volume available, and slightly penetrates the protein shell. The particles are mainly stabilized by protein–RNA interactions. The RNA is easily accessible to ribonucleases A and T1 through holes in the protein shell. At slightly alkaline pH the particles unfold reversibly.

## Genome Structure

The genome structure of the L isolate of AMV is shown in **Figure 2**. RNAs 1 and 2 encode the replicase proteins P1 and P2, respectively. P1 contains an N-terminal methyltransferase-like domain and C-terminal helicase-like domain, whereas P2 contains a polymerase-like domain. RNA 3 is dicistronic and encodes the 5′-proximal movement protein (MP) gene and the 3′-proximal CP gene. MP and CP are both required for cell-to-cell

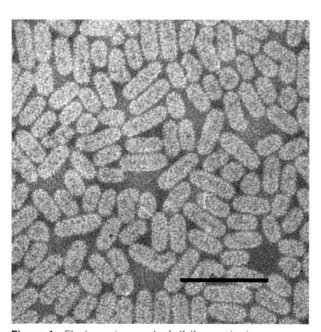

**Figure 1** Electron micrograph of alfalfa mosaic virus. Scale = 100 nm.

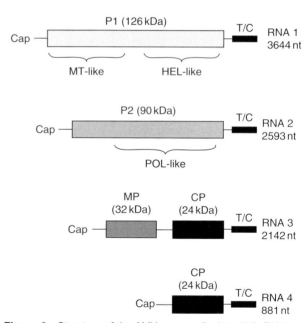

**Figure 2** Structure of the AMV genome (isolate 425). RNAs 1 and 2 encode the replicase proteins P1 and P2; RNA 3 encodes the movement protein (MP) and coat protein (CP). CP is translated from subgenomic RNA 4. The bar labeled T/C represents the 3′-terminal 112 nt of the RNAs, which can adopt either a tRNA-like structure (TLS) or a structure with a high affinity for CP (CPB).

transport of the virus. CP is translated from RNA 4, which is identical in sequence to the 3′-terminal 881 nucleotides (nt) of RNA 3. The length of the intercistronic region in RNA 3 is 52 nt including the leader sequence of RNA 4 of 36 nt. At the 5′ end, all four AMV RNAs are capped. The organization of the leader sequence of RNA 3 varies between strains. For strains M, S, L, and Y the length of this leader sequence is 240, 313, 345, and 391 nt, respectively. The increased length of the last three strains is due to the presence of direct repeats of 56 (S), 75 (L), or 149 (Y) nt. At their 3′ termini, the viral RNAs contain a homologous sequence of 145 nt. The 3′ 112 nt of this sequence can adopt two mutually exclusive conformations, one representing a strong CP binding site (CPB) and the other representing a TLS resembling the TLS of bromo- and cucumoviruses. In the CPB structure, the 112 nt sequence consists of four hairpins (designated A, B, C, and D from 3′ to 5′ end) flanked by tetranucleotide sequences AUGC (or a UUGC derivative). Base-pairing between a four nucleotide sequence in the loop of the 5′-proximal hairpin D and a four nucleotide sequence in the stem of the 3′-proximal hairpin A results in a pseudoknot interaction that generates the TLS structure. Structures similar to the AMV CPB and TLS conformations have been identified at the 3′ termini of the RNAs of prunus necrotic ring-spot ilarvirus. Although the TLS of AMV and ilarviruses cannot be charged with an amino acid, the 3′ end of AMV RNAs is specifically recognized by the host enzyme that adds CCA to the 3′ termini of cellular tRNAs (CTP/ATP:tRNA nucleotidyl transferase). The CCA-adding enzyme was able to adenylate the AMV TLS structure *in vitro*, but not the CPB structure. The AMV CPB and TLS structures have been shown to be required for translation of viral RNA and minus-strand promoter activity, respectively.

## Interaction between Viral RNA and Coat Protein

*In vitro*, the 3′-untranslated region (UTR) of AMV RNAs can bind several dimers of CP. A minimal CP binding site consists of the 3′-terminal 39 nt of the RNAs with the structure 5′-AUGC-[hairpin B]-AUGC-[hairpin A]-AUGC-3′. This structure contains two overlapping binding sites for the N-terminal peptides of the two subunits of a CP dimer. The consensus binding site in the RNA is UGC-[hairpin]-RAUGC (in which R is a purine). In addition to dimers of native CP, N-terminal peptides of CP can bind the 39 nt RNA fragment in a 2:1 ratio. The N terminus of CP contains basic residues at positions 5, 6, 10, 13, 16, 17, 25, and 26, but only arginine-17 appeared to be critical for binding of CP to RNA. This Arg residue is part of a Pro-Thr-x-Arg-Ser-x-x-Tyr (PTxRSxxY) RNA binding domain conserved among AMV and ilarvirus CPs.

The complex of the 3′-terminal 39 nt RNA fragment and a peptide corresponding to the N-terminal 26 amino acids of CP has been crystallized and its structure was solved to 3 Å resolution. Co-folding altered the structure of both the peptide and the RNA. In the co-crystal, hairpins A and B are oriented at approximately right angles and each hairpin is extended by 2 bp formed between nucleotides from adjacent AUGC sequences. If the AUGC motifs in the 39 nt fragment are numbered 1–3, starting from the 3′ end of the RNA, hairpin B is extended by a duplex formed by base-pairing between the U and C residue of motif 3 and the A and G residue of motif 2. Similarly, hairpin A is extended by a duplex formed by base-pairing between the U and C residue of motif 2 and the A and G residue of motif 1. The two peptides in the co-crystal each form an α helix with residues 12–26 ordered in peptide 1 and residues 9 to 26 ordered in peptide 2. The data provided insight into the role of the PTxRSxxY-motif in RNA binding. *In vitro* selection of RNA fragments with a high affinity for full-length AMV CP from a pool of randomized RNAs yielded fragments that maintained the unusual inter-AUGC base pairs observed in the crystal structure, although the primary sequences diverged from the wild-type RNA.

## Translation of Viral RNA

Extension of the 3′ end of AMV genomic RNAs with an artificial poly(A)-tail increased the basal level of infectivity of these RNAs 50-fold, compared to a 1000-fold increase caused by binding of CP to the RNAs. A role of CP in translation of viral RNAs was investigated by extension of the 3′ end of a luciferase reporter RNA with the AMV 3′-UTR [Luc-AMV], a poly(A)-tail [Luc-poly(A)], or a plasmid-derived sequence [Luc-control]. Transfection of plant protoplasts with these transcripts in the absence or presence of AMV CP showed that CP did not affect translation of Luc-control or Luc-poly(A), but stimulated translational efficiency of Luc-AMV 40-fold. Moreover, CP had only a minor effect on the half-life of Luc-AMV. GST-pull-down assays and Far Western blotting revealed that AMV CP specifically interacted with the eIF4G-subunit from the eIF4F initiation factor complex from wheat germ (and with the eIFiso4G-subunit from the eIFiso4F complex that is present in plants). eIF4F consists of the helicase eIF4A, the cap-binding protein eIF4E, and the multifunctional scaffold protein eIF4G. **Figure 3** illustrates that translational efficiency of cellular messengers is strongly enhanced by the formation of a closed-loop structure, due to interactions of the poly(A) binding protein (PABP) with the poly(A)-tail and with the eIF4G subunit of the eIF4F complex. Due to its binding to the CPB structure at the 3′ end of viral

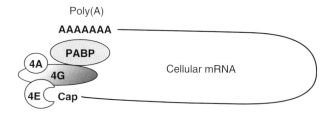

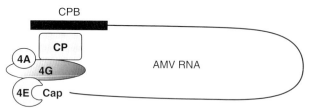

**Figure 3** Model for the role of coat protein in translation of AMV RNAs. Translation of cellular mRNAs is strongly enhanced by the formation of a closed-loop structure by interactions of the poly(A) binding protein (PABP) with the 3' poly(A) tail and with the eIF4G subunit of the eIF4F complex of initiation factors bound to the 5' cap structure (upper panel). AMV coat protein (CP) enhances translational efficiency of viral RNAs 40-fold by binding to the CP binding site (CPB) at the 3' end of the viral RNAs. The finding that CP also interacts with eIF4G indicates that CP mimics the function of PABP in formation of the closed-loop structure (lower panel).

RNAs and its interaction with eIF4G, AMV CP could stimulate translation of viral RNAs by a similar mechanism (**Figure 3**).

Protoplasts transfected with wild-type AMV RNA 4 accumulated CP at a detectable level, although accumulation was 100-fold lower than in productively infected protoplasts. However, translation of the RNA 4 mutant R17A was below the detection level. This mutant encodes CP with arginine-17 replaced by alanine, and this CP is unable to bind to the 3' end of AMV RNAs. Translation of mutant R17A could be rescued to wild-type levels by expression in *trans* of CP that was functional in RNA binding. Also, translation of mutant RNA 4 was rescued to wild-type levels by replacing the 3'-UTR of this mutant by the 3'-UTR of brome mosaic virus (BMV) RNA 4. BMV is the type species of the genus *Bromovirus* in the family *Bromoviridae*. Apparently, the 3'-UTR of AMV requires binding of CP to stimulate translation, whereas the BMV 3'-UTR stimulates translation independently of CP, possibly by the binding of host factors.

## Replication of Viral RNA

In tobacco, AMV replication complexes are associated with vacuolar membranes. Template-dependent replicase preparations have been purified from infected plants, which specifically accept exogenous AMV RNA as template. These preparations have been used to study viral minus-strand and plus-strand RNA synthesis *in vitro*.

The identification in 1978 of high-affinity binding sites for CP at the 3'-termini of AMV RNAs led to the hypothesis that binding of CP was required to permit initiation of minus-strand RNA synthesis. However, all experiments done to test this hypothesis showed that CP is not required for the synthesis of AMV minus-strand RNA *in vivo* or *in vitro*: (1) expression of AMV RNAs 1 and 2 from a T-DNA vector in agroinfiltrated leaves results in wild-type levels of minus-strand RNA synthesis in the complete absence of CP; (2) a CP-free replicase, purified from these agroinfiltrated leaves, transcribes AMV RNAs *in vitro* into minus-strand RNAs as efficiently as does the replicase purified from infected leaves; (3) in an *in vitro* replicase assay, addition of CP inhibits minus-strand RNA synthesis; (4) mutation of the 3'-terminal AUGC-motifs in AMV RNAs interferes with binding of CP *in vitro* but does not affect minus-strand RNA synthesis by the purified viral replicase. However, transcription of such mutant RNA by purified AMV replicase is no longer inhibited by CP.

An analysis of sequences in AMV RNAs which direct minus-strand RNA synthesis by the purified viral replicase *in vitro* showed that the entire 3'-terminal homologous sequence of 145 nt is required for minus-strand promoter activity. This promoter consists of the TLS structure formed by the 3'-terminal 112 nt and a hairpin structure, termed hairpin E, between nt 112 and 145 from the 3' end of the RNAs. Hairpin E is probably the primary element recognized by the viral replicase, and the TLS serves to direct the bound replicase to the very 3' end of the template. If the TLS structure is disrupted by mutations affecting the pseudoknot interaction or if the TLS is completely deleted, hairpin E directs initiation of minus-strand synthesis to a position located 5' from the hairpin. In the absence of the TLS, the mechanism of RNA synthesis directed by hairpin E is very similar to that directed by the subgenomic promoter hairpin. This subgenomic promoter is located in minus-strand RNA 3 and directs plus-strand RNA 4 synthesis. The finding that the subgenomic promoter hairpin could be replaced by hairpin E without loss of infectivity illustrates the functional equivalence between the two hairpins.

Although a knockout mutation of the CP gene does not affect AMV minus-strand RNA synthesis in infected protoplasts or in agroinfiltrated leaves, such a mutation results in a 100-fold drop in the accumulation of viral plus-strand RNAs. Initially, it was proposed that this reflected a role of CP in *de novo* synthesis of plus-strand RNA. The observation that CP stimulated RNA 4 synthesis on a minus-strand RNA 3 template by the purified replicase *in vitro* supported this hypothesis. However, this observation could not be reproduced in later experiments. Recently, it was shown that expression of CP in agroinfiltrated leaves not only enhances the accumulation of replication-competent RNAs but also the accumulation of replication-defective viral RNAs. Thus, it is possible that the stimulation of the accumulation of plus-strand RNAs by CP in infected cells reflects a role of CP in protection of the RNAs from degradation.

RNA 3 can be replicated in *trans* by P1 and P2 proteins expressed from replication-defective RNAs 1 and 2 in agroinfiltrated tobacco leaves or by P1 and P2 expressed from nuclear transgenes in transgenic P12 tobacco. However, RNAs 1 and 2 are unable to make use of these transiently or transgenically expressed replicase proteins and are dependent for replication on their encoded proteins in *cis*. This requirement in *cis* explains why RNA 3 can initiate replication in protoplasts of transgenic P12 plants without a requirement for CP in the inoculum, whereas initiation of replication of RNA 1 or RNA 2 in this system requires CP to permit efficient translation of these RNAs into their encoded proteins. Moreover, replication of RNAs 1 and 2 is strictly coordinated. If RNAs 1, 2, and 3 replicate in plant cells in the presence of a replicase transiently expressed from a T-DNA vector, a mutation in RNA 1 that is lethal to the function of the encoded P1 protein blocks the replication of both mutant RNA 1 and wild-type RNA 2, but does not affect the replication of RNA 3 by the transiently expressed replicase. Similarly, a mutation in RNA 2 that affects the conserved GDD polymerase sequence in the P2 protein blocks replication of RNAs 1 and 2 in this system, but not replication of RNA 3. Apparently, replication of RNA 1 controls replication of RNA 2 and vice versa.

## Virus Encapsidation and Movement

Expression of mutant RNAs 1 and 2 with their 3'-UTRs deleted and wild-type RNA 3 in agroinfiltrated leaves results in replication of RNA 3 by the transiently expressed replicase proteins and encapsidation of all three viral RNAs into virions. This indicates that the high-affinity binding sites for CP in the 3'-UTRs of RNAs 1 and 2 are dispensable for encapsidation of these RNAs. Possibly, assembly of virions initiates on internal CP binding sites that have been identified in AMV RNAs.

The RNA 3 encoded MP and CP are both required for cell-to-cell movement of AMV in infected plants. The MP is able to form tubular structures on the surface of infected protoplasts, suggesting that virus movement in plants involves transport of virus particles through tubules, which traverse the cell wall through modified plasmodesmata. Such a mechanism has been found for viruses of the genera *Comovirus*, *Caulimovirus*, and *Badnavirus*. However, tubular structures filled with virus particles have not been observed in AMV infected leaf tissue. Moreover, a CP mutant has been reported that is defective in the formation of virions but is able to move cell-to-cell at a reduced level. Possibly, AMV moves cell-to-cell as viral ribonucleoprotein complexes that structurally differ from virions.

## Role of Coat Protein in the AMV Replication Cycle

A natural infection with AMV will be initiated by viral particles B, M, and Tb, containing the genomic RNAs 1, 2, and 3, respectively. Inoculation of plants in the laboratory with a mixture of these three genomic RNAs results in very low, barely detectable levels of infection and infectivity of this mixture is increased 1000-fold by addition of four to ten molecules of CP per RNA molecule. Alternatively, infectivity can be increased by addition of RNA 4, the subgenomic messenger for CP. It is assumed that AMV infection starts with the three genomic RNAs, each associated with one or more dimers of the viral CP. These RNA/CP complexes can be generated by partial disassembly of particles B, M, and Tb in inoculated cells, by mixing purified viral RNAs and CP *in vitro*, or by translation of RNA 4 in the inoculated cells and subsequent binding of *de novo* synthesized CP to the viral RNAs. Infection can be initiated independently of CP when the three genomic RNAs are extended with an artificial 3'-terminal poly(A) tail or are transcribed in the plant nucleus from viral cDNA by polymerase II. It has been shown that binding of CP to the 3' end of AMV RNAs strongly stimulates translation of the RNAs in plant cells, and that CP specifically binds to the eIF4G subunit of the eIF4F complex of plant initiation factors. Based on these observations, it has been proposed that the role of CP in the inoculum is to stimulate translation of AMV (and ilarvirus) RNAs by mimicking the function of the poly(A) binding protein in translation of cellular mRNAs (**Figure 3**). A similar function in translation of nonpolyadenylated viral RNA has been reported for the NSP3 protein of rotaviruses (family *Reoviridae*). Efficient translation of AMV RNA 4 in protoplasts requires the ability of its encoded CP to bind to the 3' end of its own messenger. Available data support the notion that after inoculation of plants with AMV RNAs 1, 2, 3, and 4, RNA 4 is initially translated with low efficiency until translation of this messenger is stimulated by its own translation product.

*In vitro*, binding of CP to the 3' end of AMV RNAs blocks minus-strand RNA synthesis by the purified viral replicase. It is possible that early in infection the binding of CP to the 3' end of the RNAs not only promotes translation but also blocks premature initiation of replication to prevent a collision between ribosomes and replicase, traveling along the RNA in opposite directions. However, this possibility is not supported by experimental evidence. The switch from translation to replication requires dissociation of parental CP from the 3' end of the RNAs to allow the formation of the TLS structure, as the TLS conformation has been shown to be required for minus-strand promoter activity. Replicase proteins translated from inoculum RNAs 1 and 2 may target the viral RNAs to vacuolar membranes where replication complexes are formed and may bind to minus-strand promoter hairpin E (between nt 112 and 145 from the 3′ end of these RNAs) to promote dissociation of CP. Another possibility is that embedding of viral ribonucleoprotein complexes in the vacuolar membrane promotes dissociation of CP.

The mechanism of the switch from minus-strand to plus-strand RNA synthesis is not yet clear. Although the promoter for plus-strand subgenomic RNA 4 synthesis is structurally similar to hairpin E in the minus-strand promoter, the promoters for plus-strand genomic RNA synthesis have not yet been characterized. The requirement of CP for efficient accumulation of plus-strand RNA in infected cells may reflect a role of CP in protection of the RNAs from degradation, but a role in plus-strand RNA synthesis has not yet been ruled out. There is growing evidence for AMV and other viruses in the family *Bromoviridae* that many steps in the viral replication cycle are tightly linked.

## Virus–Host Relationships

The AMV group is a large conglomerate of strains infecting a high number of susceptible hosts. This accounts for the tremendous range of symptoms displayed by AMV-infected plants. Mutations in the coat protein gene and 5′-UTR of RNA 3 have been shown to affect symptom formation in tobacco. Cytological modifications in AMV-infected plants occur only in cells of organs showing symptoms. In these cells fragmentation of the ground cytoplasm and an increased accumulation of membrane-bound vesicles has been observed. Sometimes the lamellar system of chloroplasts is affected and invaginations of the nuclear membrane have been reported. The MP protein has been localized in the middle lamella of walls of those cells that had just been reached by the infection front and in which viral multiplication had just begun. The P1 protein is exclusively localized at the tonoplast, whereas P2 was found both at the tonoplast and other locations in the infected cell. Virus particles are mainly found in the cytoplasm with a few records of particles in the nucleus. Depending on the strain, different types of intracellular aggregates of virus particles may occur.

## Transmission

AMV is easily transmissible manually. Field spread occurs predominantly by aphid transmission. At least 15 aphid species are known to transmit the virus in the stylet-borne or nonpersistent manner. Acquisition of the virus occurs within 10–30 s and is followed by immediate transmission without a latent period. The ability to continue transmission is lost by the aphid within 1 h. The variability of individual aphid species in their capacity to transmit different AMV strains suggests a specific virus–vector relationship which is probably governed by the structural properties of the CP. Seed transmission of AMV has been reported for alfalfa and seven other plant species with rates of transmission varying from 0.1% to 50%. Transmission of the virus between plants by parasitic dodder has been observed with five *Cuscuta* species.

## Epidemiology and Control

Although there have been reports on resistance and tolerance of alfalfa to AMV, control of the virus in this crop can be done mainly by using virus-free seed and avoiding reservoir hosts of the virus. Because the virus occurs naturally in many different plant species, this is practically impossible.

Tobacco plants transformed with the CP gene of AMV were found to be highly resistant to the virus when infection was done by mechanical inoculation. Resistance to transmission of virus by aphids has not yet been tested. The resistance was clearly protein mediated as plants with the highest level of CP accumulation were the most resistant. Plants with the highest level of CP were resistant to infection with inocula consisting of either viral particles or RNAs, whereas plants with lower levels of CP were resistant to infection with particles only. A mutation in the transgene that affected the N-terminal sequence of the encoded CP destroyed resistance to the wild-type virus but the mutant transgene conferred resistance to virus expressing the mutant CP.

*See also*: Ilarvirus.

## Further Reading

Balasubramaniam M, Ibrahim A, Kim B-S, and Loesch-Fries S (2006) *Arabidopsis thaliana* is an asymptomatic host of alfalfa mosaic virus. *Virus Research* 121: 215–219.

Bol JF (2005) Replication of alfamo- and ilarviruses: Role of the coat protein. *Annual Review of Phytopathology* 43: 39–62.

Boyce M, Scott F, Guogas LM, and Gehrke L (2006) Base-pairing potential identified by *in vitro* selection predicts the kinked RNA backbone observed in the crystal structure of the alfalfa mosaic virus RNA–coat protein complex. *Journal of Molecular Recognition* 19: 68–78.

Guogas LM, Filman DJ, Hogle JM, and Gehrke L (2004) Cofolding organizes alfalfa mosaic virus RNA and coat protein for replication. *Science* 306: 2108–2111.

Hull R (1969) Alfalfa mosaic virus. *Advances in Virus Research* 15: 365–433.

Jaspars EMJ (1985) Interaction of alfalfa mosaic virus nucleic acid and protein. In: Davies JW (ed.) *Molecular Plant Virology,* vol.1, pp. 155–230. Boca Raton, FL: CRC Press.

Krab IM, Caldwell C, Gallie DR, and Bol JF (2005) Coat protein enhances translational efficiency of alfalfa mosaic virus RNAs and interacts with the eIF4G component of initiation factor eIF4F. *Journal of General Virology* 86: 1841–1849.

Kumar A, Reddy VS, Yusibov V, *et al.* (1997) The structure of alfalfa mosaic virus capsid protein assembled as a $T = 1$ icosahedral particle at 4.0-Å resolution. *Journal of Virology* 71: 7911–7916.

Olsthoorn RCL, Haasnoot PC, and Bol JF (2004) Similarities and differences between the subgenomic and minus-strand promoters of an RNA plant virus. *Journal of Virology* 78: 4048–4053.

Olsthoorn RCL, Mertens S, Brederode FT, and Bol JF (1999) A conformational switch at the 3′ end of a plant virus RNA regulates viral replication. *EMBO Journal* 18: 4856–4864.

# Algal Viruses

**K Nagasaki,** Fisheries Research Agency, Hiroshima, Japan
**C P D Brussaard,** Royal Netherlands Institute for Sea Research, Texel, The Netherlands

© 2008 Elsevier Ltd. All rights reserved.

## Glossary

**Algal viruses** Viruses infecting eukaryotic algae.
**HaRNAV** A positive-sense single-stranded RNA virus infecting the bloom-forming microalga *Heterosigma akashiwo* (Raphidophyceae).
**VLPs** Virus-like particles that are identified by transmission electron microscopy.

## Introduction

With the realization during the last two decades that viruses are highly abundant in various aquatic environments (both marine and freshwater), interest in aquatic viruses significantly increased. Viruses are now recognized as important biological agents not only regulating population dynamics, succession, and diversity of the host organisms in marine systems, but also influencing the functioning of aquatic food webs and biogeochemical cycling (energy and matter fluxes). Organisms within the microbial food web can become infected by viruses, including eukaryotic algae, cyanobacteria, and heterotrophic protists. Indeed, viruses or virus-like particles (VLPs) have been found in more than 60 algal species for 12 recognized classes of eukaryotic algae; to date, about 40 algal viruses have been isolated and characterized (**Table 1**). Although their classification and nomenclature has been improved on, only two algal virus families are officially established to date: *Phycodnaviridae* and *Marnaviridae*. The family *Phycodnaviridae* comprises large genome-sized double-stranded DNA (dsDNA) viruses infecting eukaryotic algae, and the family *Marnaviridae* was just recently established based on the analysis of a positive-sense single-stranded RNA (ssRNA) virus infecting a raphidophyte *Heterosigma akashiwo*.

Most recently, four algal viruses were discovered that do not belong to the above two virus families – RsRNAV, HcRNAV, MpRV and CsNIV; they are infectious to the single-celled marine phytoplankton species *Rhizosolenia setigera*, *Heterocapsa circularisquama*, *Micromonas pusilla*, and *Chaetoceros salsugineum*, respectively. In this article, we summarize the characteristics of these algal viruses, and discuss their putative classification.

## ssRNA Viruses Infecting Eukaryotic Algae

### RsRNAV

Rhizosolenia setigera RNA virus (RsRNAV) is the first diatom-infecting virus that was isolated and characterized. It is a positive-sense ssRNA virus infecting the bloom-forming diatom *R. setigera*. Viral replication occurs within the host cytoplasm, the virus particle is icosahedral, lacking a tail, and 32 nm in diameter (**Figure 1**).

The RsRNAV genome is a 3′ polyadenylated linear RNA lacking a cap structure at the 5′ terminus. Although the genome size was initially estimated to be 11.2 kbp using denaturing gel electrophoresis, full genome sequencing later revealed that the genome is 8877 nt in length, excluding the polyA tail (DDBJ accession number: AB243297). This may be due to the additional length of the 3′ poly(A) tail or the addition of a viral genome-linked protein (VPg). Further analysis is required to explain this disagreement in genome length estimates. RsRNAV genome has two open reading frames (ORFs): ORF-1 coding for a putative polyprotein containing the RNA helicase domain and the RNA-dependent RNA polymera (RdRp) domain, and ORF-2 encoding at least three major

**Table 1** Viruses infecting eukaryotic algae

| Virus | Host | Size (nm) | Genome type | Genome size (kbp) |
|---|---|---|---|---|
| *Viruses infecting unicellular algae* | | | | |
| BtV | *Aureococcus anophagefferens* | 140 | dsDNA | |
| CbV | *Chrysochromlina brevifilum* | 145–170 | dsDNA | |
| CdDNAV[a] | *Chaetoceros debilis* | 30 | ssDNA (fragmented?) | |
| CeV | *Chrysochromlina ericina* | 160 | dsDNA | 510 |
| CgNIV[a] | *Chaetoceros* cf. *gracilis* | 30 | ssDNA (?) | 3–4 |
| *Chlorella* virus (e.g., ATCV-1, -2) | *Chlorella* SAG 3.83 (symbiont of *Acanthocystis turfacea*) | 140–190 | dsDNA | 288 |
| *Chlorella* virus (e.g., PBCV-1, NY-2A, AR158) | *Chlorella* NC64A (symbiont of *Paramecium bursaria*) | 190 | dsDNA | 331–369 |
| *Chlorella* virus (e.g., MT325, FR483) | *Chlorella* Pbi (symbiont of *Paramecium bursaria*) | 140–150 | dsDNA | 314–321 |
| *Chlorella* virus (e.g., HVCV) | *Chlorella*-like alga (symbiont of *Hidra viridis*) | 170–180 | dsDNA | 200 |
| CnRNAV[a] | *Chaetoceros neogracilis* | 31 | ssRNA | |
| CsNIV | *Chaetoceros salsugineum* | 38 | (ss + ds)DNA | 6 |
| CspNIV | *Chaetoceros* cf. *gracilis* | 25 | | |
| CsRNAV[a] | *Chaetoceros socialis* | ~30 | RNA (?) | |
| EhV | *Emiliania huxleyi* | 170–200 | dsDNA | 410–415 |
| HaNIV[a] | *Heterosigma akashiwo* | 30 | ssDNA | |
| HaV | *Heterosigma akashiwo* | 202 | dsDNA | 294 |
| HaRNAV | *Heterosigma akashiwo* | 25 | ssRNA | 9.1 |
| HcRNAV | *Heterocapsa circularisquama* | 30 | ssRNA | 4.4 |
| HcV | *Heterocapsa circularisquama* | 197 | dsDNA | 356 |
| MpRV (originally MpRNAV-01B) | *Micromonas pusilla* | 50–60 | dsRNA | 24.6 |
| MpV (e.g., MpV-PL1, PB7) | *Micromonas pusilla* | 115 | dsDNA | 200 |
| MpV[a] (e.g., MpV-02T,-03T) | *Micromonas pusilla* | ~110–120 | dsDNA | 191–217 |
| MpV[a] (e.g., MpV-01T, 15T) | *Micromonas pusilla* | ~70–90 | | |
| PgV Group I | *Phaeocystis globosa* | 150 | dsDNA | 466 ± 4 |
| PgV Group II | *Phaeocystis globosa* | 100 | dsDNA | 177 ± 3 |
| PgV-102P | *Phaeocystis globosa* | 98 | dsDNA | 176 |
| PoV | *Pyramimonas orientalis* | 180–220 | dsDNA | 560 |
| PpV | *Phaeocystis pouchetii* | 130–160 | dsDNA | 485 |
| RsRNAV | *Rhizosolenia setigera* | 32 | ssRNA | 11.2 |
| *Viruses infecting multicellular algae* | | | | |
| EsV | *Ectocarpus siliculosus* | 130–150 | dsDNA | 336 |
| EfasV | *Ectocarpus fasciculatus* | 135–140 | dsDNA | 340 |
| FlexV | *Feldmannia simplex* | 120–150 | dsDNA | 170 |
| FirrV | *Feldmannia irregularis* | 140–170 | dsDNA | 180 |
| FsV | *Feldmannia* sp. | 150 | dsDNA | 170 |
| HincV | *Hincksia hinckiae* | 140–170 | dsDNA | 220 |
| MclaV | *Myriotrichia clavaeformis* | 170–180 | dsDNA | 340 |
| PlitV | *Pilayella littoralis* | 161 | dsDNA | 280 |

[a]Y Tomaru, Y Eissler, Y Shirai, CPD Brussaard, JE Lawrence, personal communication.

capsid proteins (MCPs) with molecular mass of 41.5, 41.0, and 29.5 kDa (**Figure 2**). Although a significant similarity in amino acid sequences of the nonstructural and structural proteins between RsRNAV and HaRNAV was found by BLAST search, these viruses differ in the number of ORFs (two and one, respectively) and AU content (63.7% and 53.1%, respectively); hence, RsRNAV was concluded not to be a member of the family *Marnaviridae*. Phylogenetic analysis of the amino acid sequences of the RNA helicase and RdRp domains suggests that RsRNAV belongs to a new previously unrecognized virus group (**Figure 3**). Smaller unidentified RNA molecules ranging in size of 0.6, 1.2, and 1.5 kbp were occasionally included in RsRNAV virions; one possibility is that they are subgenomic RNAs.

RsRNAV has a high degree of strain specificity. One-step growth experiment using an exponentially growing host culture showed $3.1 \times 10^3$ infectious units are released from an infected host cell within 48 h post infection. When a stationary phase culture was used as host for the one-step growth experiment, the burst size decreased to $1.0 \times 10^3$ infections units per cell. This shows that viral propagation

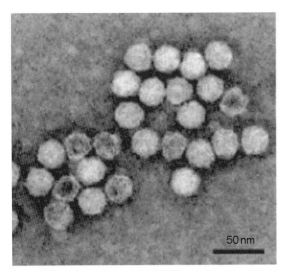

**Figure 1** Negatively stained RsRNAV particles. Reprinted from Nagasaki K, Tomaru Y, Katanozaka N, et al. (2004) Isolation and characterization of a novel single-stranded RNA virus infecting the bloom-forming diatom *Rhizosolenia setigera*. *Applied and Environmental Microbiology* 70: 704–711, with permission from American Society for Microbiology.

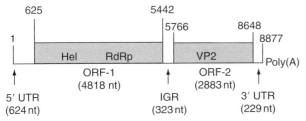

**Figure 2** Schematic genome structure of RsRNAV. Numbers indicate base positions from the 5′ terminus in the nucleotide sequence. Hel, RNA helicase domain; RdRp, RNA-dependent RNA polymerase domain; UTR, untranslated region; IGR, intergenic region. Reprinted from Shirai Y, Takao Y, Mizumoto H, et al. (2006) Genomic and phylogenetic analysis of a single-stranded RNA virus infecting Rhizosolenia setigera (Stramenopiles: Bacillariophyceae). *Journal of the Marine Biological Association of the United Kingdom* 86: 475–483, with permission from The Marine Biological Association of the UK.

is considerably affected by the physiological condition of the host cells. Although the entry mechanism of RsRNAV into the host cell is unknown, the host's frustule pores (ellipses; 91 nm × 73 nm) are considered to be a possible route of infection.

## HcRNAV

Heterocapsa circularisquama RNA virus (HcRNAV) is a positive-sense ssRNA virus infecting the bivalve-killing bloom-forming dinoflagellate *H. circularisquama*. This virus is icosahedral, ~30 nm in diameter, and propagates in the host cytoplasm often forming a crystalline array (**Figure 4**). Its genome is linear positive-sense ssRNA ~4.4 kbp long, lacking 5′ cap structure and 3′ polyA tail, and has a strong stem–loop structure at the 3′ end (**Figure 5**).

HcRNAV clones can be divided into two ecotypes (types UA and CY) based on their intraspecies host specificity patterns. Each type shows its own strain-specific infectivity that is complementary to each other; that is, the *H. circularisquama* strains sensitive to HcRNAV type UA were resistant to HcRNAV type CY, and vice versa, showing that HcRNAV is not species specific, but strain specific. These two HcRNAV ecotypes can coexist in natural water. Typical HcRNAV clones of type UA and CY (HcRNAV 34 and 109, respectively) were fully sequenced (DDBJ accession numbers: AB218608 and AB218609, respectively); they are ~97% identical at the nucleotide sequence level. Each genome has two ORFs (**Figure 5**). ORF-1 encodes a putative polyprotein having at least the serine protease domain and the RdRp domain, but no RNA helicase domain was identified. A specific 15 nt deletion was found in about half of the tested virus clones. This, however, is not involved in determining the intraspecies host specificity. ORF-2 coding for the single MCP is unlikely a polyprotein gene because of the molecular mass directly estimated by sodium dodecyl sulfate (SDS) polyacrylamide gel electrophoresis is in good agreement with the value predicted by the deduced amino acid sequence of ORF-2. Between the two virus clones tested, the stem–loop structures at the 3′ end are different in stem length and loop size; this may affect the replication efficiency.

Similarity analysis for the deduced amino acid sequences of ORF-1 revealed that HcRNAV is evolutionarily quite distant from any of the land and aquatic viruses that have been genetically studied. This is also supported by a phylogenetic analysis of the RdRp amino acid sequence (**Figure 3**(b)). Although HcRNAV shares some characteristics with the typical marnavirus HaRNAV in infectivity (lytic to marine eukaryotic microalgae) and genome structure (a linear positive-sense ssRNA), they still differ in the number of ORFs (two and one, respectively) and 3′ polyadenylation of the genome RNA (polyadenylated and nonpolyadenylated, respectively). Hence, HcRNAV is not a member of the family *Marnaviridae*.

Genomic comparison revealed complementary host ranges of the two HcRNAV ecotypes (UA and CY: mentioned above) which may be related to the amino acid substitution patterns in ORF-2, the MCP-encoding gene. In addition, the tertiary structure of the MCPs predicted by using computer modeling indicated many of the amino acid substitutions were located in regions on the outside of the viral capsid proteins exposed to the ambient water environments. This suggests that the intraspecies host specificity of HcRNAV is determined by small structural differences of the viral surface that may affect its binding affinity to the host cell and the uncoating process. The results of the transfection experiment also supported this idea: (1) the intraspecies host specificity of HcRNAV

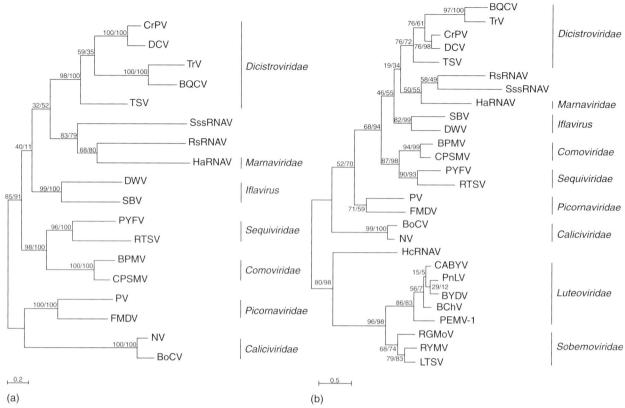

**Figure 3** Maximum likelihood (ML) trees calculated from confidently aligned regions of amino acid sequences of concatenated amino acid sequences of RNA helicase domain and RNA-dependent RNA polymerase (RdRp) domain (a), and of RdRp whole domain (b). ML bootstrap values (%) from 100 samples are shown at the nodes followed by bootstrap values based on neighbor-joining analysis (%) from 100 samples. The ML distance scale bars are shown. Amino acid sequences used for comparison in the analyses are as follows with the organism's scientific names, with abbreviations in parentheses if necessary, and the database accession numbers (referring to the US National Center for Biotechnology Information (NCBI) unless otherwise stated): beet chlorosis virus (BChV), AAK49964; bovine enteric calicivirus (BoCV), AJ011099; bean pod mottle virus (BPMV), NC__003496; black queen cell virus (BQCV), NC__003784; barley yellow dwarf virus (BYDV), BAA01054; cucurbit aphid-borne yellows virus (CABYV), CAA54251; cowpea severe mosaic virus (CPSMV), M83830; cricket paralysis virus (CrPV), NC__003924; drosophila C virus (DCV), NC__001834; deformed wing virus (DWV), NC__004830; foot-and-mouth disease virus (FMDV), P03306; heterosigma akashiwo RNA virus (HaRNAV), NC__005281; heterocapsa circularisquama RNA virus 109 (HcRNAV), DDBJ accession number: AB218609; lucerne transient streak virus (LTSV), NP__736596; Norwalk virus (NV), M87661; pea enation mosaic virus 1 (PEMV-1), AAA72297; poinsettia latent virus (PnLV), CAI34771; human poliovirus 1 Mahoney (PV), V01149; parsnip yellow fleck virus (PYFV), D14066; ryegrass mottle virus (RGMoV), NP__736587; rhizosolenia setigera RNA virus (RsRNAV), DDBJ accession number: AB243297; rice turgo spherical virus (RTSV), AAA66056; rice yellow mottle virus (RYMV), CAE81345; sacbrood virus (SBV), NC__002066; schizochytrium single-stranded RNA virus (SssRNAV), BAE47143; triatoma virus (TrV), NC__003783; taura syndrome virus (TSV), NC__003005. Reprinted with minor modification from Shirai Y, Takao Y, Mizumoto H, et al. (2006) Genomic and phylogenetic analysis of a single-stranded RNA virus infecting Rhizosolenia setigera (Stramenopiles: Bacillariophyceae). *Journal of the Marine Biological Association of the United Kingdom* 86: 475–483, with permission from The Marine Biological Association of the UK.

is determined by the upstream events of virus infection; (2) the host intracellular condition is permissive for HcRNAV replication even in incompatible host–virus combinations.

By the cross-reactivity test, *H. circularisquama* clones are also divided into two ecotypes according to the sensitivity spectra to the two HcRNAV ecotypes; however, the two host ecotypes are indistinguishable when comparing their morphology or the sequences of the internal spacer regions of the ribosomal RNA genes. These two host ecotypes coexist in natural water. Thus, there are at least two distinct (and independent) host/virus systems between *H. circularisquama* and HcRNAV; that is, multiple ecotypes of host and virus coexist within natural blooms of *H. circularisquama* and their combinations are regulated with exquisite molecular mechanisms.

## dsRNA Viruses Infecting Eukaryotic Algae

### MpRV

The micromonas pusilla reovirus (MpRV, originally abbreviated as MpRNAV) is a dsRNA virus infecting the cosmopolitan picoprasinophyte *M. pusilla* (**Figure 6**).

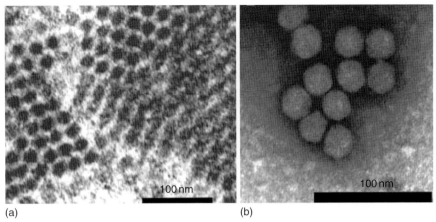

**Figure 4** Transmission electron micrographs of intracellular crystalline array formation of HcRNAV (a) and negatively stained HcRNAV particles (b). Reprinted from Tomaru Y, Katanozaka N, Nishida K, et al. (2004) Isolation and characterization of two distinct types of HcRNAV, a single-stranded RNA virus infecting the bivalve-killing microalga Heterocapsa circularisquama. *Aquatic Microbial Ecology* 34: 207–218, with permission from Inter-Research Publication.

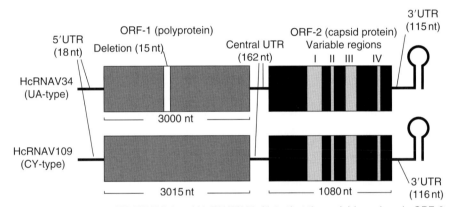

**Figure 5** Schematic genome structure of HcRNAV34 and HcRNAV109. Note that the variable regions in ORF-2 are remarkably different between these two virus clones. Reprinted from Nagasaki K, Shirai Y, Takao Y, et al. (2005) Comparison of genome sequences of single-stranded RNA viruses infecting the bivalve-killing dinoflagellate *Heterocapsa circularisquama*. *Applied and Environmental Microbiology* 71: 8888–8894, with permission from American Society for Microbiology.

This is the first algal virus having a dsRNA genome that was isolated and characterized. The virus can coexist with a large genome-sized dsDNA virus infecting the same *M. pusilla* strain. It has a narrow host range, a latent period of 36 h (based on a one-step lytic virus growth cycle and transmission electron microscopy (TEM) enumeration of the virus particles), and sensitivity to temperatures >35 °C. Its genome is composed of 11 segments ranging between 741 and 5792 bp in length, with a total size of 25 563 bp (**Figure 7** and **Table 2**). The polysegmented dsRNA genome of MpRV identified it as a member of the family *Reoviridae*, which was confirmed by sequence analysis, morphological and physiochemical properties. The size of intact virus particles (90–95 nm) is, however, large. The subcore particles size of 50 nm and its smooth surface indicate that MpRV belongs to non-turreted *Reoviridae*.

Comparison of the genome sequence of MpRV to those of characterized members of the family *Reoviridae* indicated that MpRV could not be classified within any of the existing genera of the family (including *Rotavirus* and *Aquareovirus* that both contain viruses with 11-segmented dsRNA genomes). The maximum amino acid identity with other *Reoviridae* proteins was 21%, which is compatible with differences existing between distinct genera.

Within the phylogenetic tree built with reovirus RdRp sequences, the branch of MpRV dissects the tree, separating the group of turreted and nonturreted viruses (**Figure 8**). As *M. pusilla* is evolutionarily older than the hosts of other members of the family *Reoviridae*, the topology of the tree suggests that the branch of MpRV could be ancestral. An interesting feature is the unusual length of segment 1, encoding a protein of 200 kDa (VP1). The many repeats within this sequence suggest that VP1 may have arisen from amino acid fragment duplication, followed by diversification of the sequence. The mechanism and the constraints, which have driven such an evolution, are not clear, although

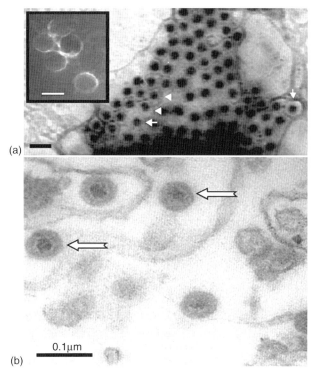

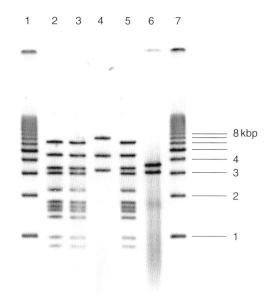

**Figure 6** (a) Electron micrographs of MpRV. Particles pelleted from the clarified lysate of infected *M. pusilla*. Some particles (indicated by arrowheads) have a larger diameter. At the upper left corner (inset), core particles treated with 1.5 M CaCl$_2$ are shown to have a smooth outline (turrets are absent). Scale = 100 nm (main image); 50 nm (inset). (b) A TEM image of thin-sectioned *M. pusilla* cells infected with MpRV. The arrows point at intracellular virus particles consisting of a thick outer layer and a smaller electron-dense inner core. (a) Reprinted from Attoui H, Jaafar FM, Belhouchet M, et al. (2006) *Micromonas pusilla* reovirus: A new member of the family *Reoviridae* assigned to a novel proposed genus (*Mimoreovirus*). *Journal of General Virology* 87: 1375–1383, from the Society for General Microbiology. (b) Reprinted from Brussaard CP, Noordeloos AA, Sandaa RA, et al. (2004) Discovery of a dsRNA virus infecting the marine photosynthetic protist *Micromonas pusilla*. *Virology* 319: 280–291, with permission from Elsevier.

**Figure 7** Total nucleic acid patterns of MpRV (lanes 2, 3, and 5). Molecular size standards: 1 kbp dsDNA Molecular Ruler (lanes 1 and 7), dsRNA bacteriophage Phi-6 (lane 4; segments of 6374, 4074, 2948 bp), and dsRNA Infectious Bursal Disease Virus IBDV-V4 (lane 6; segments of 3260 and 2827 bp). Precise lengths of the 11 dsRNA segments are shown in **Table 2**. Reprinted from Brussaard CP, Noordeloos AA, Sandaa RA, et al. (2004) Discovery of a dsRNA virus infecting the marine photosynthetic protist *Micromonas pusilla*. *Virology* 319: 280–291, with permission from Elsevier.

**Table 2** Lengths of the dsRNA segments 1–11 of MpRV

| Segment | Segment length (bp) |
| --- | --- |
| 1 | 5792 |
| 2 | 4175 |
| 3 | 3129 |
| 4 | 2833 |
| 5 | 2027 |
| 6 | 1687 |
| 7 | 1556 |
| 8 | 1449 |
| 9 | 1296 |
| 10 | 878 |
| 11 | 741 |
| Total | 25 563 |

Adapted from Attoui H, Jaafar FM, Belhouchet M, et al. (2006) *Micromonas pusilla* reovirus: A new member of the family *Reoviridae* assigned to a novel proposed genus (*Mimoreovirus*). *Journal of General Virology* 87: 1375–1383, from the Society for General Microbiology.

very recently a model of stem–loop formation (based on the crystal structure of the polymerase of a mammalian orthoreovirus) was proposed for explaining such duplications.

The structural proteins in the purified particles were analyzed and a protein having a molecular weight of approximately 200 kDa was identified in relatively intact particles. This size is compatible with the VP1 which should represent an additional coat protein layer. It is noteworthy that the sequence of VP1 bears many glycosylation sites and matched the envelope proteins of many bacteria and viruses.

Based on particle morphology and sequence analysis, MpRV was classified as the representative of a new genus within the family *Reoviridae*, for which the name *Mimoreovirus* has been proposed.

## ssDNA Viruses Infecting Eukaryotic Algae

### CsNIV

*Chaetoceros salsugineum* nuclear inclusion virus (CsNIV) infects the small-sized (2.0–9.5 μm wide) diatom *C. salsugineum*, which forms short or long straight chains.

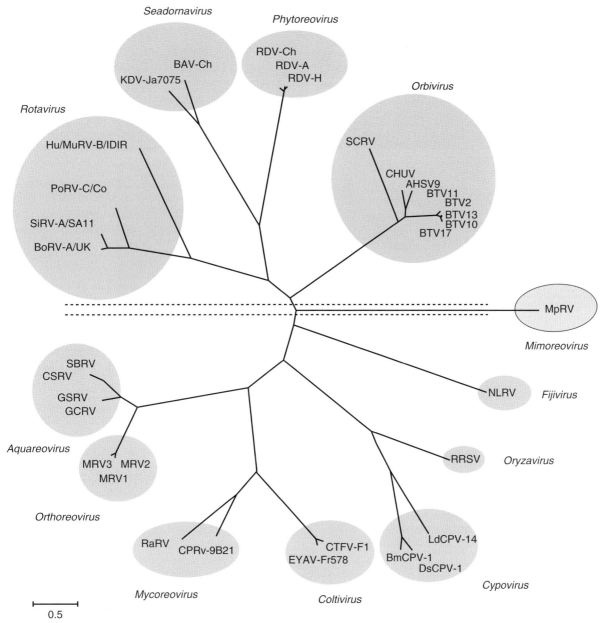

Figure 8 Phylogenetic relationships of MpRV to other members of the family *Reoviridae* based on the sequence of the putative RdRp. Neighbor-joining phylogenetic tree built with available polymerase sequences (using the Poisson-correction or gamma-distribution algorithms) for representative members of 11 genera of the family *Reoviridae*. Scale bar represents the number of substitutions per site. Reprinted from Attoui H, Jaafar FM, Belhouchet M, et al. (2006) Micromonas pusilla reovirus: A new member of the family Reoviridae assigned to a novel proposed genus (Mimoreovirus). *Journal of General Virology* 87: 1375–1383, with permission from the Society for General Microbiology.

Its bloom occurs in brackish lakes and estuarine waters. CsNIV is a ~38 nm icosahedral virus that replicates within the host nucleus (**Figure 9**). The genome structure is unlike that of other viruses that have been described to date; that is, it consists of a single molecule of covalently closed circular ssDNA (6000 nt (previously reported as 6005 nt)) as well as a segment of linear ssDNA (997 nt); the linear segment is complementary to a portion of the closed circle creating a partially double-stranded genome (**Figure 10**) (DDBJ accession number: AB193315). Sequence analysis revealed, within one of the six ORFs identified in the genome, only a low similarity to the replicase of circoviruses that have a covalently closed circular ssDNA genome. One-step growth experiment showed that ~330 infectious units are released from an infected cell within 24 h; however, considering that

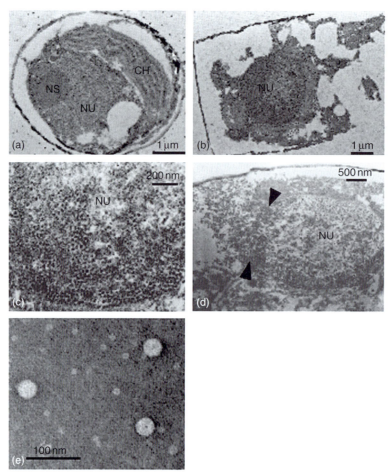

**Figure 9** Transmission electron micrographs of *Chaetoceros salsugineum*. (a) Thin section of a healthy cell; (b) thin section of a cell 24 h after inoculation with CsNIV; (c) close-up view of intranuclear CsNIV particles in (b); (d) thin section of a CsNIV-infected cell in which the nuclear envelope was partially ruptured (arrowheads); (e) negatively stained CsNIV particles in the culture lysate. CH, a chloroplast; NU, a nucleus; NS, a nucleolus. Reprinted from Nagasaki K, Tomaru Y, Takao Y, et al. (2005) Previously unknown virus infects marine diatom. *Applied and Environmental Microbiology* 71: 3528–3535, with permission from American Society for Microbiology.

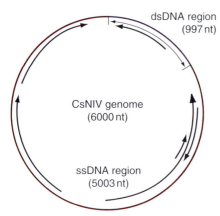

**Figure 10** Schematic genome structure of CsNIV. Bold arrows indicate putative ORFs. Adapted from Nagasaki K, Tomaru Y, Takao Y, et al. (2005) Previously unknown virus infects marine diatom. *Applied and Environmental Microbiology* 71: 3528–3535, with permission from American Society for Microbiology.

several hundreds of virus particles were included even in a thin section for TEM observation, the resulting burst size is assumed to be underestimated. This may reflect aggregation of virions or dominance of defective particles.

## Implications

Phytoplankton is important in maintaining oxygen levels in the atmosphere and sustaining the primary nutritional production of the aquatic environment. Its ecological dynamics is affected at different levels by various factors: physical factors (e.g., temperature, salinity, and irradiation), chemical factors (e.g., nutrients and metals), and biological factors (e.g., grazing, virus infection, and algicidal effects of bacteria). Among them, viruses are considered significant mortality agents, resulting in cell loss comparable to grazing and/or sedimentation. Whether algae sink out, are grazed upon, or die due to viral-induced cell lysis has different implication for the flow of matter and energy in the aquatic ecosystems. Viral-mediated mortality will force the food web toward a more regenerative system as a result of enhanced production of dissolved organic matter.

Viral control of the algal host's population occurs for *M. pusilla*, a eukaryotic marine phytoplankter that is globally abundant but almost never forms massive blooms. *M. pusilla* belongs to the Prasinophyceae, a class of phytoplankton that is considered to have given rise to green algae and land plants. The dsRNA virus MpRV described earlier in this chapter has been found to coexist with large genome-sized dsDNA viruses, of which some infect the same clone of *M. pusilla*. The various different virus model systems infecting *M. pusilla* that are available in culture permit to investigate virus competition, intraspecies diversity, co-evolution, and genetic flux in more detail.

A good example of viral control on a finer scale, that is, the clonal composition of the algal host population, is the relationship between HcRNAV and its host *H. circularisquama*. Occurrence of *H. circularisquama* blooms in natural waters is accompanied by increased abundance of HcRNAV. Its abundance in the water column rapidly decreases following the termination of host blooms, but the viruses that are supplied to the sediment show only a very gradual decrease. Dynamics of the type UA viruses and the type CY viruses having complementary host ranges are different, supporting the idea that there are two independent host/virus systems between *H. circularisquama* and HcRNAV. Changes in abundance of each HcRNAV ecotype are considered to reflect the fluctuation of its suitable host ecotype *in situ*. By comparing the dynamics of HcRNAV and *H. circularisquama*, the amount of HcRNAV accumulated in the sediment just prior to the host's blooming season is suggested to be a significant factor in determining the size or term length of *H. circularisquama* blooms; that is, when concentration of HcRNAV is high in sediments, occurrence of dense *H. circularisquama* blooms appears suppressed. HcRNAV infection seems, thus, to affect the population dynamics of *H. circularisquama* not only in quantity (biomass) but also quality (clonal composition).

In contrast, ecological implication of viruses infecting diatoms has not been sufficiently understood. Considering that diatoms are one of the most widespread plant groups on earth, the impact viruses have on the diatom's population dynamics should be more intensively studied.

In this article, algal viruses having ssRNA, dsRNA, or ssDNA genomes are described. Given the large variety of algae and the diversity of viruses isolated and characterized to date, marine and freshwaters provide a treasury of undiscovered viruses.

*See also:* Marnaviruses; Phycodnaviruses.

## Further Reading

Attoui H, Jaafar FM, Belhouchet M, *et al.* (2006) *Micromonas pusilla* reovirus: A new member of the family *Reoviridae* assigned to a novel proposed genus (*Mimoreovirus*). *Journal of General Virology* 87: 1375–1383.

Bettarel Y, Kan J, Wang K, *et al.* (2005) Isolation and characterisation of a small nuclear inclusion virus infecting the diatom *Chaetoceros* cf. *gracilis*. *Aquatic Microbial Ecology* 40: 103–114.

Brussaard CP (2004) Viral control of phytoplankton populations – a review. *Journal of Eukaryotic Microbiology* 51: 125–138.

Brussaard CP, Noordeloos AA, Sandaa RA, *et al.* (2004) Discovery of a dsRNA virus infecting the marine photosynthetic protist *Micromonas pusilla*. *Virology* 319: 280–291.

Lang AS, Culley AI, and Suttle CA (2004) Genome sequence and characterization of a virus (HaRNAV) related to picorna-like viruses that infects the marine toxic bloom-forming alga *Heterosigma akashiwo*. *Virology* 320: 206–217.

Lawrence JE, Chan AM, and Suttle CA (2001) A novel virus (HaNIV) causes lysis of the toxic bloom-forming alga *Heterosigma akashiwo* (Raphidophyceae). *Journal of Phycology* 37: 216–222.

Mizumoto H, Tomaru Y, Takao Y, *et al.* (2007) Intraspecies host specificity of a single-stranded RNA virus infecting a marine photosynthetic protist is determined at the early steps of infection. *Journal of Virology* 81: 1372–1378.

Nagasaki K, Shirai Y, Takao Y, *et al.* (2005) Comparison of genome sequences of single-stranded RNA viruses infecting the bivalve-killing dinoflagellate *Heterocapsa circularisquama*. *Applied and Environmental Microbiology* 71: 8888–8894.

Nagasaki K, Tomaru Y, Katanozaka N, *et al.* (2004) Isolation and characterization of a novel single-stranded RNA virus infecting the bloom-forming diatom *Rhizosolenia setigera*. *Applied and Environmental Microbiology* 70: 704–711.

Nagasaki K, Tomaru Y, Nakanishi K, *et al.* (2004) Dynamics of *Heterocapsa circularisquama* (Dinophyceae) and its viruses in Ago Bay, Japan. *Aquatic Microbial Ecology* 34: 219–226.

Nagasaki K, Tomaru Y, Shirai Y, *et al.* (2006) Dinoflagellate-infecting viruses. *Journal of the Marine Biological Association of the United Kingdom* 86: 469–474.

Nagasaki K, Tomaru Y, Takao Y, *et al.* (2005) Previously unknown virus infects marine diatom. *Applied and Environmental Microbiology* 71: 3528–3535.

Shirai Y, Takao Y, Mizumoto H, *et al.* (2006) Genomic and phylogenetic analysis of a single-stranded RNA virus infecting *Rhizosolenia setigera* (Stramenopiles: Bacillariophyceae). *Journal of the Marine Biological Association of the United Kingdom* 86: 475–483.

Tomaru Y, Hata N, Masuda T, *et al.* (2007) Ecological dynamics of the bivalve-killing dinoflagellate *Heterocapsa circularisquama* and its infectious viruses in different locations of western Japan. *Environmental Microbiology* 9: 1376–1383.

Tomaru Y, Katanozaka N, Nishida K, *et al.* (2004) Isolation and characterization of two distinct types of HcRNAV, a single-stranded RNA virus infecting the bivalve-killing microalga *Heterocapsa circularisquama*. *Aquatic Microbial Ecology* 34: 207–218.

# Allexivirus

**S K Zavriev,** Shemyakin and Ovchinnikov Institute of Bioorganic Chemistry, Russian Academy of Sciences, Moscow, Russia

© 2008 Elsevier Ltd. All rights reserved.

## History

The genus *Allexivirus* of plant viruses belongs to a new family of plant viruses, *Flexiviridae*, which also includes the genera *Capillovirus, Carlavirus, Citrivirus, Foveavirus, Potexvirus, Trichovirus, Vitivirus,* and *Mandarivirus*. The first publication on an allexivirus as a member of a new plant virus group appeared in 1992 after analysis of the complete genome structure of a filamentous virus isolated from shallot plants in the Institute of Agricultural Biotechnology (Moscow, Russia). The virus thus discovered and described was named shallot virus X (ShVX). Allexiviruses acquired their name by fusion of the host (family Alliaceae) and the type member (ShVX) names. The allexivirus-associated diseases are usually mild and in many cases symptomless. Very often, allexiviruses persist in the infected plants as a part of multiple infection induced by viruses with similarly restricted host ranges. Since allexiviruses do not cause serious plant diseases (as well as the majority of related carlaviruses), they do not attract special attention of phytopathologists, but they are quite interesting for taxonomy studies, genome structure, and relationships between genera, species, and subspecies of distinct viruses.

## Taxonomy and Classification

Table 1 lists definite and possible or insufficiently characterized members of the genus *Allexivirus*. Practically, all viruses have been included in the group according to the characteristics of their genome structure (primarily, a unique virus-specific protein encoded by ORF4) as well as the serological relatedness to ShVX and the common host range.

Poorly characterized viruses serologically related to ShVX have been found in several *Allium* species, as well as in tulip and narcissus plants, but it is unclear whether these should be regarded as strains of ShVX or as distinct viruses. The serological relationships of ShVX to other well-characterized allexiviruses or to the viruses in other genera have not yet been studied.

## Viral Structure and Composition

### Virions

Virions of allexiviruses are highly flexible filamentous particles, about 800 nm in length and 12 nm in diameter. They resemble potyviruses in size and closteroviruses in flexibility and cross-banded substructure (**Figure 1**).

Virus preparations obtained from individual *Allium ascalonicum* plants infected only with ShVX were found to contain virions of two morphological types: (1) typical allexivirus particles with characteristic cross-striation; and (2) thinner (6 nm), more flexible, and aggregation-prone particles. The latter were minor in most preparations, but constituted the bulk of the viral population in about 5% of plants. Particles of both types were formed of homologous genomic RNAs and serologically close coat proteins (CPs). Sequence analysis of the 2500 3′-terminal nucleotides of the genomic RNA of type 1 and 2 particles showed 87% homology; the amino acid sequences of the coat and the 15 kDa (OPF6) proteins had a single difference each, whereas the 42 kDa (ORF4) proteins had about 15% changes.

**Table 1** Definite and tentative species of the genus *Allexivirus*

| Species in the genus | GenBank accession no | |
|---|---|---|
| Garlic mite-borne filamentous virus | [X98991, AY390254] | GarMbFV |
| Garlic virus A | [AB010300, F478197] | GarVA |
| Garlic virus B | [AB010301, F543829] | GarVB |
| Garlic virus C | [AB010302, D49443] | GarVC |
| Garlic virus D | [AB010303, AF519572, L38892] | GarVD |
| Garlic virus E | [AJ292230] | GarVE |
| Garlic virus X | [AJ292229, U89243] | GarVX |
| Shallot virus X (397[a]) | [M97264, L76292] | ShVX |
| Tentative species in the genus | | |
| Garlic mite-borne latent virus | | GarMbLV |
| Onion mite-borne latent virus | | OMbLV |
| Shallot mite-borne latent virus | | ShMbLV |

[a]Number of the CMI/AAB Plant Virus Description (see http://www.dpvweb.net).

## Nucleic Acid

Virions contain a single molecule of linear single-stranded RNA (ssRNA), about 9.0 kb in length, with a 3′ poly(A) tract. ShVX RNA preparations, besides genomic ssRNA, contain molecules of 1.5 kb double-stranded RNA (dsRNA), whose genesis and function(s) are unknown. The complete nucleotide sequences of the genomic RNA of ShVX, garlic viruses A (GarVA), C (GarVC), E (GarVE), and X (GarVX), and the partial sequences of the RNA of garlic miteborne filamentous virus (GarMBFV) and garlic virus B (GarVB) and D (GarVD) have been determined.

## Proteins

Virions are composed of a single polypeptide of 28–37 kDa. It was reported that the ORF6 protein is a minor component in ShVX virions.

## Physicochemical and Physical Properties

The sedimentation constant of the ShVX virion is about 170S (0.1 M Tris-HCl, pH 7.5 at 20 °C). The buoyant density in CsCl is 1.33 g cm$^{-3}$.

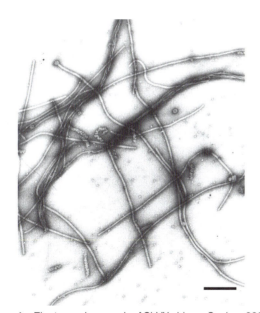

**Figure 1** Electron micrograph of ShVX virions. Scale = 200 nm. Courtesy of V. Vishnichenko.

## Transmission

Allexiviruses are supposed to be mite-borne; the only vector known is the eriophyd mite *Aceria tulipae*, which was proved to transmit GarVC and GarVD. All allexiviruses are manually transmissible by sap inoculation of healthy host plants. None could be transmitted by aphids or any other insects.

## Genome Organization and Replication

The genome organization of allexiviruses resembles that of carlaviruses, with the major exception of an 'additional' ORF between the triple gene block (TGB) and the CP ORF. The genomic RNA of allexiviruses contains six large ORFs and noncoding sequences of about 100 nucleotides at the 5′ terminus and about 100 nucleotides followed by a poly(A) tail at the 3′ terminus (**Figure 2**). Determination of the complete genomic RNA sequence of several allexiviruses proves their genome organization to be almost identical. The type member, ShVX, codes for polypeptides of 195, 26, 11, 42, 28, and 15 kDa, respectively, from the 5′ end to the 3′ end. Gene arrangement of other incompletely sequenced allexiviruses is similar. The 195 kDa polypeptide is a viral RNA polymerase, and most probably it is the only virus-encoded protein required for replication. In comparison of the amino acid sequences of methyltransferase, helicase, or RNA-dependent RNA polymerase (RdRp) motifs, these conserved domains of allexiviruses are most similar to those of potexviruses. The 26 and 11 kDa proteins are similar to the first two proteins encoded by the TGB of potexviruses and carlaviruses and are probably involved in cell-to-cell movement of the virus. There is a coding sequence for a small (7–8 kDa) TGB protein but it lacks the initiation AUG codon. The 42 kDa polypeptide has no significant homology with any proteins known, though it is expressed in plants infected with ShVX in relatively large amounts. Immunoelectron microscopy using polyclonal antisera against the recombinant ShVX 42 kDa protein showed reaction with certain regions of the virions, with the immune complexes nonuniformly distributed along the particle. The 42 kDa protein was supposed to act as a cofactor to provide proper interaction of the CP with the genomic RNA in the virion assembly process. The 28 kDa polypeptide is the CP. In polyacrylamide gel electrophoresis (PAGE)

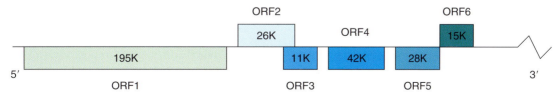

**Figure 2** Genome organization of shallot virus X (genus *Allexivirus*).

it migrates as a 32–36 kDa protein, which can be due to its high hydrophilicity evident from amino acid sequence. The 15 kDa protein is similar to the 11–14 kDa proteins encoded by the 3′-proximal ORFs of carlaviruses, and has a zinc-binding-finger motif with affinity for nucleic acids. The exact function of this polypeptide is unknown, but it may be involved in virion assembly.

At least two subgenomic mRNAs (sgRNA) are used for translating the 5′-distal ORFs: one for the movement proteins (ORF2 and ORF3) and the other for the CP (ORF5) and the 3′-distal nucleic acid-binding protein (ORF6). Although the ORF4 protein may be expressed in the infected plants in relatively large amounts, no evidence is available for an ORF4-specific sgRNA. It is suggested that the ORF4-encoded protein is translated from the sgRNA for the ORF2 and ORF3 proteins.

## Geographical Distribution

Allexiviruses have been identified in Russia, Japan, France, Germany, the UK, The Netherlands, Korea, China, Taiwan, Thailand, and Argentina. Most probably, they are all distributed across the world, especially in the regions where bulbous plants are widespread.

## Phylogenetic Information

Phylogenetic trees based on the nucleotide sequence depend on the genome region used for analysis. In the family *Flexiviridae*, the allexiviruses in general occupy an intermediate position in putative phylogeny between carlaviruses, potexviruses, and mandariviruses.

See also: Capillovirus, Foveavirus, Trichovirus, Vitivirus; Carlavirus; Cytomegaloviruses: Murine and Other Nonprimate Cytomegaloviruses; Flexiviruses; Vegetable Viruses.

## Further Reading

Adams MJ, Antoniw JF, Bar-Joseph M, et al. (2004) The new plant virus family Flexiviridae and assessment of molecular criteria for species demarcation. *Archives of Virology* 149: 1045–1060.

Arshava NV, Konareva TN, Ryabov EV, and Zavriev SK (1995) The 42K protein of shallot virus X is expressed in the infected *Allium* plants. *Molecular Biology* (Russia) 29: 192–198.

Barg E, Lesemann DE, Vetten HJ, and Green SK (1994) Identification, partial characterization and distribution of viruses infecting crops in south and south-east Asian countries. *Acta Horticulturae* 358: 251–258.

Chen J, Chen J, and Adams MJ (2001) Molecular characterization of a complex mixture of viruses in garlic with mosaic symptoms in China. *Archives of Virology* 146: 1841–1853.

Kanyuka KV, Vishnichenko VK, Levay KE, Kondrikov DYu, Ryabov EV, and Zavriev SK (1992) Nucleotide sequence of shallot virus X RNA reveals a 5′-proximal cistron closely related to those of potexviruses and a unique arrangement of the 3′-proximal cistrons. *Journal of General Virology* 73: 2553–2560.

Song SI, Song JT, Kim CH, Lee JS, and Choi YD (1998) Molecular characterization of the garlic virus X genome. *Journal of General Virology* 79: 155–159.

Sumi S, Tsuneyoshi T, and Furutani H (1993) Novel rod-shaped viruses isolated from garlic, possessing a unique genome organization. *Journal of General Virology* 74: 1879–1885.

Van Dijk P and van der Vlugt RA (1994) New mite-borne virus isolates from Rakkyo, shallot and wild leek species. *European Journal of Plant Pathology* 100: 269–277.

Vishnichenko VK, Konareva TN, and Zavriev SK (1993) A new filamentous virus in shallot. *Plant Pathology* 42: 121–126.

Vishnichenko VK, Stelmashchuk VY, and Zavriev SK (2002) The 42K protein of the Shallot virus X participates in formation of virus particles. *Molecular Biology* (Russia) 36: 1080–1084.

Yamashita K, Sakai J, and Hanada K (1996) Characterization of a new virus from garlic (*Allium sativum* L.), garlic mite-borne virus. *Annals of Phytopathology Society of Japan* 62: 483–489.

# *Alphacryptovirus* and *Betacryptovirus*

**R Blawid, D Stephan, and E Maiss**, Institute of Plant Diseases and Plant Protection, Hannover, Germany

© 2008 Elsevier Ltd. All rights reserved.

## History

The common symptoms induced by plant virus infections are most easily recognized as mosaics, mottles, and ringspots. However, in 1968 and 1969, Pullen showed that all apparently healthy plants of seven beet species (*Beta* sp.) contained isometric virus-like particles (VLPs) of c. 29–30 nm in diameter with unusual properties. The particles could not be transmitted to other herbaceous plants and were not eliminated by heat therapy. At that time VLPs were supposed to be of plant origin until a few beet plants were found not containing VLPs. In 1977, Kassanis *et al.* rediscovered these virus particles and showed that they were present in 90% of commercially grown sugar beet plants. They referred to the VLPs as beet cryptic virus (BCV) and were the first to develop a purification method of BCV particles. Nevertheless, evidence from serology, analytical ultracentrifugation, and CsCl isopycnic centrifugation suggested that more than one type of particles might have been present. This was the first time that the name 'cryptic' virus was suggested. The name 'cryptic' comes from Greek and means 'hidden, secret' as no symptom can be visualized from plants carrying cryptic viruses.

During the 1980s, two other VLPs were discovered, with properties similar to those of BCV. These were named vicia cryptic virus (VCV) and carnation cryptic virus (CarCV). In 1978, Kenten *et al.* showed that VCV was up to 88% seed borne in some *Vicia faba* cultivars but was not transmitted to VLP-free plants by aphids, dodder, sap inoculation, or grafting. Furthermore, VCV was not eliminated from seedlings by heat treatment. In field trials, VCV infection had little or no effect on plant growth and yield. Abou-Elnasr *et al.* purified isometric particles of VCV from *Vicia faba* seedlings. In 1981, Lisa *et al.* identified the nucleic acids of CarCV as double-stranded RNA (dsRNA) using serological methods, electron microscopy, and DNase and RNase treatments. In 1988, Marzachi *et al.* purified CarCV particles, which subsequently were shown to possess RNA-dependent RNA polymerase (RdRp) activity. Analysis of the RNAs synthesized *in vitro* showed that the CarCV putative RdRp enzyme acts as a replicase, catalyzing the synthesis of dsRNA. In the mid-1980s, Natsuaki and colleagues found also VLPs in alfalfa, beet, and white clover that had common characteristics to those referred to as cryptic viruses. These 'temperate' viruses cause little or no symptoms; the virus concentration in plants was often very low and they were not graft-transmitted or mechanically transmissible but highly seed-transmitted, showing 80–100% transmission rates through ovules and pollen to the seed. It became clear that 'temperate' and 'cryptovirus' were belonging to the same group of viruses.

In 1985, Boccardo *et al.* presented striking evidence that cryptic viruses are in fact viruses of plants. They purified from apparently healthy plants white clover cryptic virus (WCCV) 1, WCCV 2, and WCCV 3 with yields of about 200 µg of virus per kg of tissue. After isopycnic centrifugation in CsCl, WCCV 1 and 2 formed homogeneous bands at densities of 1.392 and 1.375 g ml$^{-1}$, respectively, whereas WCCV 3 was not recovered. Particles of WCCV 1 and 3 were about 34 nm in diameter, whereas WCCV 2 particles were 38 nm in diameter, with prominent morphological subunits. Each of the three viruses contained two segments of dsRNA (molecular weight from 1.03 to 1.49 × 10$^6$ Da). Again, particles of WCCV 1 and 2 were neither transmitted by grafting nor by mechanical inoculation. Boccardo *et al.* showed also that the incidence and concentration of WCCV 1 and 2 in white clover plants were unaffected by high levels of systemic and surface fungicides and also that the viruses were already present in minute amounts in plantlets grown from surface-sterilized seeds, indicating that the virus does not reside in fungal hyphae, which are able to colonize the interior and surface of a plant.

Until recently, only little work on the molecular characterization of cryptic plant viruses was done. In 1993, Xie *et al.* determined the nucleotide sequence of the BCV 3 dsRNA 2 that encodes a putative RdRp. In spring 2006, only the complete nucleotide sequences of two assigned members of the genus *Alphacryptovirus*, namely VCV and WCCV 1, were available in GenBank.

## Taxonomy and Classification

Cryptic viruses belong to the family *Partitiviridae*, which includes three genera: *Partitivirus*, *Alphacryptovirus*, and *Betacryptovirus*. Viruses of the last two genera in this family infect plants. The *VIII Report of the International Committee on Taxonomy of Viruses* (ICTV) listed 16 species of plant cryptic viruses and ten species that are unassigned members in the genus *Alphacryptovirus*. The genus *Betacryptovirus* contains only four species and one tentative, whereas no nucleotide sequence is available up to now. Cryptoviruses have been found in different plants like, sugar beet, ryegrass, carnation, alfafa, radish, spinach, fire trees, green and brown algae, red pepper, carrot, hop trefoil, and broad beans. **Table 1** shows a list of species in the genera *Alphacryptovirus* and *Betacryptovirus* and tentative species, virus abbreviations, and accession numbers. In addition, **Table 1** shows a list of unassigned plant partitiviruses found in GenBank.

## Virion Properties

Viruses with dsRNA genomes are classified into seven families, namely *Birnaviridae*, *Cystoviridae*, *Hypoviridae*, *Partitiviridae*, *Reoviridae*, *Totiviridae*, and *Chrysoviridae*. Only the families *Partitiviridae* and *Reoviridae* include plant-infecting viruses. In addition, only recently the ICTV accepted the *Endornavirus* as a new genus of plant dsRNA virus.

The family *Partitiviridae* is characterized by morphological, physical, and physicochemical properties. All candidate viruses consist of small isometric particles, 37–38 nm in diameter. The virion buyant density in CsCl is in the range of 1.34–1.39 g cm$^{-3}$. Virions are stable in butanol and chloroform. *In vitro* transcription/replication occurs by a semiconservative mechanism and virions accumulate mainly in the cytoplasm and rarely in nuclei and nucleoli of parenchyma cells.

All cryptic viruses that have been analyzed to date possess at least two classes of dsRNAs with molecular weights between 0.8 and 1.6 × 10$^6$ Da. In 1990, Accotto *et al.* purified and characterized from *Medicago sativa* the alfalfa cryptic virus (ACV). The genome of ACV-1 consists of two dsRNAs, one with an estimated molecular weight of 1.27 × 10$^6$ Da (RNA 1) and the other of 1.17 × 10$^6$ Da (RNA 2). In a different paper, ACV was designated as ACV_M, showing besides the two earlier described dsRNA segments a third larger dsRNA of an estimated molecular weight of 2.70 × 10$^6$ Da. The RdRp, able to replicate the genomic dsRNAs *in vitro*, was associated with purified ACV particles. In addition, further examples

**Table 1** Definite members (italics) and tentative members (Roman) in the genera *Alphacryptovirus* and *Betacryptovirus*

| Genus | Abbreviation | Accession number |
|---|---|---|
| *Alphacryptovirus* | | |
| Official virus species names | | |
|   *Alfalfa cryptic virus 1* | (ACV-1) | |
|   *Beet cryptic virus 1* | (BCV-1) | |
|   *Beet cryptic virus 2* | (BCV-2) | |
|   *Beet cryptic virus 3* | (BCV-3) | (S63913) |
|   *Carnation cryptic virus 1* | (CCV-1) | |
|   *Carrot temperate virus 1* | (CTeV-1) | |
|   *Carrot temperate virus 3* | (CTeV-3) | |
|   *Carrot temperate virus 4* | (CTeV-4) | |
|   *Hop trefoil cryptic virus 1* | (HTCV-1) | |
|   *Hop trefoil cryptic virus 3* | (HTCV-3) | |
|   *Radish yellow edge virus* | (RYEV) | |
|   *Ryegrass cryptic virus* | (RGCV) | |
|   *Spinach temperate virus* | (SpTV) | |
|   *Vicia cryptic virus* | (VCV) | (NC_007241) |
|   *White clover cryptic virus 1* | (WCCV-1) | (NC_006275) |
|   *White clover cryptic virus 3* | (WCCV-3) | |
| Tentative virus species names | | |
|   Carnation cryptic virus 2 | (CCV-2) | |
|   Cucumber cryptic virus | (CuCV) | |
|   Fescue cryptic virus | (FCV) | |
|   Garland chrysanthemum temperate virus | (GCTV) | |
|   Mibuna temperate virus | (MTV) | |
|   Red pepper cryptic virus 1 | (RPCV-1) | |
|   Red pepper cryptic virus 2 | (RPCV-2) | |
|   Rhubarb temperate virus | (RTV) | |
|   Santosai temperate virus | (STV) | |
| *Betacryptovirus* | | |
| Official virus species names | | |
|   *Carrot temperate virus 2* | (CTeV-2) | |
|   *Hop trefoil cryptic virus 2* | (HTCV-2) | |
|   *Red clover cryptic virus 2* | (RCCV-2) | |
|   *White clover cryptic virus 2* | (WCCV-2) | |
| Tentative virus species names | | |
|   Alfafa cryptic virus 2 | (ACV-2) | |
| Unassigned virus in the family | | |
|   Pyrus pyrifolia cryptic virus | (PyrR-1) | (AB012616) |
|   Fragaria chiloensis cryptic virus | | (DQ093961) Partial sequence |
|   Pepper cryptic virus 1 | | (DQ361008) Partial sequence |
|   Pinus sylvestris partitivirus NL-2005 | | (AY973825) Partial sequence |
|   Raphanus sativus cryptic virus 1 | | (DQ181926) |
| | | (AY949985) |
|   Raphanus sativus cryptic virus 2 | | (DQ218036) |
| | | (DQ218037) |
| | | (DQ218038) |

exist, where three or more classes of dsRNAs have been found, namely from *Beta vulgaris*, *Festuca pratensis*, *Vicia faba*, *Medicago lupulina*, and *Trifolium repens*. These plant species may contain mixtures of two or more viruses.

## Genome Organization, Molecular Biology, and Replication

The partitivirus genome consists of two monocistronic dsRNA segment. One open reading frame (ORF) of dsRNA 1 codes for the RdRp, whereas the other of dsRNA 2 codes for the coat protein (CP).

**Figure 1** shows a schematic representation of the genome organization of VCV. Sequence analysis of dsRNA 1 revealed two ORFs. The larger ORF initiation codon AUG at nucleotide position 93–95 is probably the start codon of the VCV polymerase. The termination codon UAA is located at position 1941–1943. Therefore, the larger ORF consists of 1848 nucleotides and translates into a protein of 616 amino acids with a calculated molecular mass of 72.9 kDa. In addition, an internal ORF was found which translates into a putative protein with a calculated molecular mass of 18 kDa. No significant hits were obtained using Blastp to any other protein, therefore its synthesis and function in the plant remains unclear.

Sequence analysis of dsRNA 2 of VCV revealed also two ORFs. The larger ORF consists of 1461 nucleotides and translates into a protein of 487 amino acids with a calculated molecular mass of 53.9 kDa. The smaller ORF located on the complementary strand translates into a putative protein with a calculated molecular mass of 16 kDa. Like the small ORF of dsRNA 1, no information is available on its synthesis and function in plants.

**Figure 2** shows the 5′ nontranslated region (NTR) of the dsRNA 1 and dsRNA 2 of VCV and WCCV 1. The 5′ NTR consists of 92 nucleotides up to the initiation codon of the RdRp ORF, which is longer in comparison with the 5′ NTR of WCCV 1. The 5′ NTRs of VCV and WCCV 1 genomes share 52 nucleotides that are conserved between both genomes (**Figure 2**).

One interesting feature concerning the NTR of several cryptic viruses is the presence of 'cytosine-adenine-adenine (CAA)' repeats, which are thought to be associated with initiation and enhancement of translation. It appears that the 5′ end conserved regions are involved in replication and packaging of the respective dsRNAs as well as in translation of the RdRp and CP genes. These 'CAA' repeats are highly conserved and can be found in RdRp and/or CP 5′ NTRs of partitiviruses like gremmeniella abietina virus, heterobasidion annosum virus, discula destructiva virus 2, and fusarium poae virus 1. Many partitiviruses contain interrupted poly(A) tails at their 3′ ends. The VCV dsRNA 1 reveals 33 A residues out of 40 bases at its 3′ end and the dsRNA 2 34 A residues. The polyadenylation signal AAUAAA was identified in the 3′ NTRs of both VCV dsRNAs.

The calculated molecular mass of RdRps of viruses belonging to the genus *Alphacryptovirus* varies from 55 to 73 kDa. Amino acid comparisons of RdRp's from plant cryptic viruses showed 12 different conserved motifs. Five slightly different motifs out of the eight standard motifs found in RdRp sequences of single-stranded and double-stranded RNA viruses are present in the RdRp's of cryptic viruses: motif I 'L', motif III 'KXR3XG', motif IV 'DW2XFD', motif V 'SG3XT4XS2XN', and motif VI 'GDD'.

Currently, there are not many nucleotide sequences of plant cryptic virus CPs available at the GenBank. A Blast search using the CP of VCV identified no well-known conserved domains. However, multiple alignments using CPs of different partitiviruses reveled two major conserved domains (SQLY and PGPL3XF), but it is still unknown if these motifs have any essential function for the virus. Even if there is little information on how replication and encapsidation of plant cryptic viruses occur,

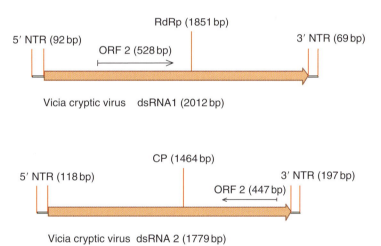

**Figure 1** Genome organization of vicia cryptic virus (VCV). The RdRp ORF and the CP ORF are represented by orange arrows. Thin black arrows indicate putative ORFs.

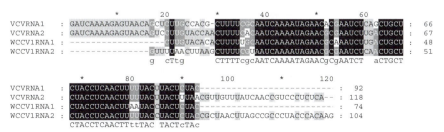

**Figure 2** Alignment of the 5′ NTRs of the dsRNA 1 and dsRNA 2 of VCV and WCCV 1. Adapted from Blawid R, Stephan W, and Maiss E (2007) Molecular characterization and detection of Vicia cryptic virus in different *Vicia faba* cultivars. *Archives of Virology* 152(8): 1477–1488.

CsCl buoyant density values may indicate that only one dsRNA molecule can be encapsidated in each particle. The genome replication is catalyzed by the virus-encoded RdRp. It is very likely that plant cryptic viruses replicate inside the capsid, as known for others dsRNA viruses, to protect themselves from dsRNA-induced host defense mechanisms. RdRp activity has been associated with purified VLPs of CarCV, WCCV, and ACV. But it is still not clear if the same RdRp that acts in viral replication and assembly is also responsible for transcription. It is likely that plant factors are also involved in the process of transcription of plant cryptic viruses. The model of replication and transcription of plant cryptic viruses might be very similar to those described for members of the genus *Partitivirus* that infect fungi. In this model, a newly synthesized positive strand displaces the parental strand (semiconservative replication). The RdRp catalyzes the synthesis of newly negative strands that later will serve as template for synthesis of the positive strand.

## Transmission

In the past, many unsuccessful attempts have been made to transmit cryptic viruses either by mechanical inoculation or by grafting to virus-free plants. Attempts to transmit BCV by mechanical inoculation of sugar beet sap to healthy sugar beet or other plant species like *Nicotiana tabacum* cv. Xanthi-nc, *Nicotiana clevelandii*, *Cucumis sativus*, *Petunia hybrida*, and *Chenopodium* ssp. failed. In addition, different cultivars of *Vicia faba* (Beryl, Maris Bead, and Minica) that were not containing VCV dsRNA were inoculated with sap from plants containing VCV. No VCV dsRNA could be extracted from these inoculated plants. Experiments involving transmission of WCCV to different plant species (*Chenopodium, Glycine max, Gomphrena globosa, N. clevelandii, N. tabacum,* and *Vigna unguiculata*) by grafting or mechanical inoculation also failed.

However, experiments involving BCV transmission through seed were successful. It was shown that cryptic viruses are transmissible through pollen as well as through the female plant. Cross-pollination experiments performed by Kassanis *et al.*, Kenten *et al.*, and Natsuaki *et al.* demonstrated that the percentage of progeny infection differed if the cryptic virus was obtained from the female parent or from the pollen donor. The progenies derive from female carriers were presenting higher infection percentages (80% carrying BCV) in comparison to those originated from pollen carriers (40%).

## Virus–Host Relationships

It is known that cryptic viruses do not induce symptoms in their hosts. Plants carrying cryptic viruses are symptomless. However, in 2005, Nakatsukasa-Akune *et al.* reported that TrEnodDR1 encoding the CP gene of WCCV 1 expressed in *Lotus japonicus* after *Agrobacterium* spp. transformation increased the concentration of endogenous abscisic acid and thereby suppresses root nodulation. Plant cryptic viruses can affect the growth and development of the host plant in mixed infections. In 1989, Kühne and Stanarius measured the interaction between mixed infections of BCV 2 and beet mild yellowing virus (BMYV) as well as BCV 1 and beet yellows virus (BYV). The amount of BCV 2 increased significantly 23 weeks after infection with BMYV. The same was observed if BCV 1 carrying plants were co-infected with BYV. Similar results are reported for ryegrass mosaic virus (RgMV) and ryegrass cryptic virus (RGCV) mixed infection, meaning that RGCV particle concentration increased after infection with RgMV.

As cryptic viruses are present in very low concentrations in plants, localization in different plant tissues is difficult. Kühne and Stanarius observed that BCV 1 was present mainly in the mesophyll, whereas BVC 2 was found in the vascular bundle. In addition, BCV 1 and BCV 2 might occur separately or together in different cultivars of *Beta vulgaris* L. Many different plant cultivars of carnation, white clover, radish, and *Lolium* species were found to be free or containing only sparse amounts of cryptic virus particles. Abou-Elnasr group detected dsRNAs from VCV in extracts from leaves, stems, roots, flowers, seeds, and mesophyll protoplasts from different *Vicia* cultivars. Although they could extract dsRNA from stems and flowers, we were not able to extract dsRNA from stems of *Vicia faba* 'Hangdown'. It is still uncertain, if cryptic viruses are present in all organs and tissues of their hosts in equal amounts. However, it is clear that different serological unrelated cryptic viruses might be present in the same plant species.

No serological relationship has been detected between cryptic viruses isolated from plants in different families, but relationships have been found between cryptic viruses in related host species. Thus, an antiserum to RYEV did not react with ACV, BCV 1, CTeV, GCTV, MTV, RTV, STV, and SpTV (for abbreviations, see **Table 1**). Other reports show that cryptic viruses from plant species belonging to different families are not serologically related, for example, WCCV 1 and WCCV 2 did not react when tested with antisera to CarCV and RCV. Even CarCV was not detected with antisera obtained with BCV and RCV as well as many others. In contrast, serological relationships have been demonstrated between HTCV1 and ACV1, between HTCV2 and RCCV2, and finally between HTCV 3 and VCV. In addition, cryptic viruses from red clover appeared to be serologically related to WCCV 1 and WCCV 2.

## Evolutionary Relationships

During cloning and analysis of the VCV genome, seven different cultivars ('Hangdown', 'Dreifach Weisse', 'Frühe Weisskeimige', 'Osnabrück', 'The Sutton', 'Divine', and 'Major') were investigated for the presence of VCV. dsRNA molecules were extracted from all cultivars and the presence of RdRp and CP of VCV was confirmed by reverse transcription-polymerase chain reaction (RT-PCR) in all cultivars, with exception of the cultivar 'Major'. For a better understanding of the virus–host relationship between VCV and *Vicia faba* plants, we sequenced the complete ORFs of VCV RdRp and CP from four different cultivars of *Vicia faba*. Interestingly, comparisons based on nucleic acid as well as amino acid level showed a high degree of identity (98–99%), indicating that these cultivars are infected probably by one virus, which may have a common cryptic virus ancestral, thereby demonstrating the high evolutionary relationship between VCV and *Vicia faba* varieties.

VCV RdRp shows high amino acid identity (84%) to WCCV 1, also a member of the genus *Alphacryptovirus*. In 2005, Chen *et al.* isolated dsRNA from radish plants. Although radish yellow edge virus (RYEV) and raphanus sativus cryptic virus 1 and 2 were isolated from the same plant species (*Raphanus sativus*), no significant similarity was found between the deduced amino acid sequence of the putative RdRp's from raphanus sativus cryptic virus 1 (RasR 1) or raphanus sativus cryptic virus 2 to VCV or WCCV 1 (**Figure 3**). A Blastp search using RasR 1 delivered an identity of 51% and 32% to the RdRp of helicobasidium mompa virus V1–2 and WCCV 1, respectively. However, this might change when more plant partitiviruses will be characterized at the molecular level. Osaki *et al.* detected dsRNA from Japanese pear (*Pyrus pyrifolia*) that presented conserved RdRp motifs found in genes that encode putative RdRp of RNA viruses. Amino acid analysis revealed an identity of 53% with BCV 3 dsRNA 2 and 38% with raphanus sativus cryptic virus 2. In 2006, Veliceasa *et al.* isolated dsRNA from *Pinus sylvestris*. Analysis of partial amino acid sequences showed 41% and 40% of identity to the unassigned pyrus pyrifolia cryptic virus and BCV 3, respectively.

RdRp's and CPs of VCV, WCCV1, and raphanus cryptic virus 1 form a separate branch within the *Partitiviridae* (**Figures 3** and **4**). However, the question of the origin of cryptic viruses and their relationship to fungal infecting viruses still remains open. Further molecular cloning and characterization of cryptic viruses might help in obtaining a better view on their relatedness. Therefore, the current phylogenetic trees based on the amino acid sequences of RdRp's and CPs may change in future when additional sequence information becomes available, resulting probably in a different relationship among these viruses.

*See also:* Partitiviruses: General Features; Partitiviruses of Fungi.

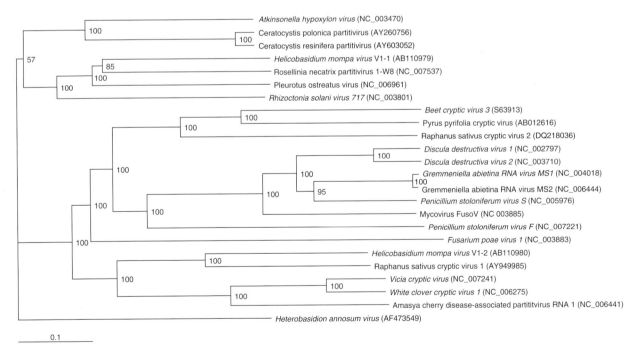

**Figure 3** Phylogenetic tree based on RdRp amino acid sequences of definitive members (italics) and tentative members (Roman) of the family *Partitiviridae*. Numbers at nodes indicate bootstrap values for 100 replicates. Accession numbers are given in brackets.

## Alphacryptovirus and Betacryptovirus

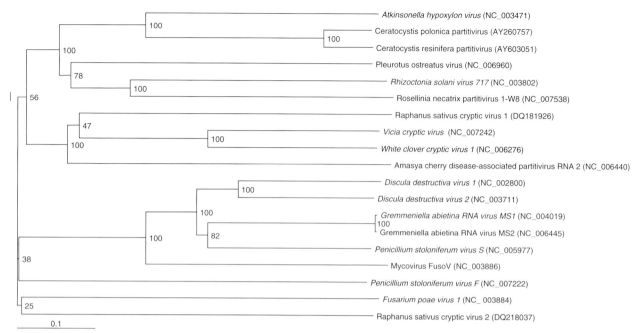

**Figure 4** Phylogenetic tree based on CP amino acid sequences of definitive members (italics) and tentative members (roman) of the family Partitiviridae. Numbers at nodes indicate bootstrap values for 100 replicates. Accession numbers are given in brackets. Adapted from Blawid R, Stephan W, and Maiss E (2007) Molecular characterization and detection of Vicia cryptic virus in different *Vicia faba* cultivars. *Archives of Virology* 152(8): 1477–1488.

### Further Reading

Antoniw JF, White RF, and Xie WS (1990) Cryptic viruses of beet and other plants. In: Frasser RSS (ed.) *Recognition and Response in Plant–Virus Interactions*, pp. 273–285. Heidelberg: Springer.

Boccardo G and Candresse T (2005) Complete sequence of the RNA 1 of an isolate of White clover cryptic virus 1, type species of the genus *Alphacryptovirus*. *Archives of Virology* 150: 399–402.

Boccardo G and Candresse T (2005) Complete sequence of the RNA 2 of an isolate of White clover cryptic virus 1, type species of the genus *Alphacryptovirus*. *Archives of Virology* 150: 403–405.

Boccardo G, Lisa V, Luisoni E, and Milne RG (1987) Cryptic plant viruses. *Advances in Virus Research* 32: 171–214.

Chen L, Chen JS, Liu L, *et al.* (2006) Complete nucleotide sequences and genome characterization of double-stranded RNA 1 and RNA 2 in the *Raphanus sativus*-root cv. *Yipinghong*. *Archives of Virology* 151: 849–859.

---

Alphaviruses *See:* Togaviruses: Alphaviruses.

---

# Anellovirus

**P Biagini and P de Micco,** Etablissement Français du Sang Alpes-Méditerranée, Marseilles, France

© 2008 Elsevier Ltd. All rights reserved.

### Glossary

**Apoptosis** Mechanism that allows cells to self-destruct when stimulated by the appropriate trigger.

**Hepatitis** Disease or condition marked by inflammation of the liver.

**Rhinitis** Inflammation of the mucous membrane of the nose.

**Tamarin** American monkey that is related to the marmoset.

**Tupaia** Small mammal native to the tropical forests of Southeast Asia, also known as the tree shrew.

**Unassigned genus** Genus that is not assigned taxonomically to any existing virus family.

## Introduction

In 2004, the International Committee on Taxonomy of Viruses (ICTV) officially created the genus *Anellovirus* (from latin 'Anello', the ring) to accommodate circular single-stranded DNA viruses isolated from humans and some other animal species. The genus *Anellovirus* is an unassigned genus and its members are thereby distinguished from viruses belonging to other families with circular single-stranded DNA genomes that infect bacteria, plants, and vertebrates, such as circoviruses, nanoviruses, and geminiviruses.

The genus *Anellovirus* officially includes one species *Torque teno virus* (TTV), one tentative species Torque teno mini virus (TTMV) and several yet unclassified animal viruses. Recently, a third group of anelloviruses infecting humans was further identified and called 'small anellovirus' in anticipation of its official designation.

This group of viruses is characterized by a very high genomic variability, a high prevalence in human populations, a still unknown significance for host health and the absence of well-defined mode(s) of transmission.

## Historical Overview of Anelloviruses

TTV was the first virus with circular single-stranded DNA genome identified in humans. It was initially discovered in 1997 by means of a subtractive technique (representational difference analysis, RDA) in the serum of a Japanese patient (initials T.T.) with post-transfusion non-A–G hepatitis. The short nucleotide sequence obtained (~500 nt) was initially extended to ~3700 nt; subsequently, the resolution of an additional GC-rich region of about 120 nt permitted to complete the TTV genome sequence and highlighted the circular nature of the viral genome.

TTMV was discovered at the end of 1999 during a study of TTV prevalence in blood donors. Some unexpected amplification products were sequenced and identified as highly divergent when compared to known TTV sequences. A circular genome of about 2900 nt was further characterized and initially designated TTV-like mini virus (TLMV) by analogy with TTV. Following a taxonomic re-evaluation of the status of this virus, it has been officially named TTMV by ICTV.

A putative third member of the genus *Anellovirus*, the 'small anellovirus', has been identified in 2005 using a sequence-independent polymerase chain reaction (PCR) amplification method. Two highly divergent circular sequences (~2200 and ~2600 nt) were initially characterized in human plasma samples by this approach.

Using a primer system located on a relatively well-conserved region of the genome, extremely divergent anellovirus sequences were also characterized in several animal species: examples were found in cats (~2100 nt), in dogs (~2800 nt), in pigs (~2900 nt), and in tupaia (~2200 nt). Circular genomes, highly divergent or similar to those identified in humans, were also identified in nonhuman primates.

## Virion Properties, Genome, and Replication

Members of the genus *Anellovirus* are nonenveloped viruses, with an estimated diameter of about 30–32 nm for TTV and slightly less than 30 nm for TTMV. The buoyant density of virions in CsCl is 1.31–1.33 g cm$^{-3}$ for TTV and 1.27–1.28 g cm$^{-3}$ for TTMV. The genomes of anelloviruses are negative stranded. This has been demonstrated by hybridization studies using sense or reverse nucleic acid probes. They are hydrolyzed by DNase I and Mung Bean nuclease as well.

The genome organization shows: (1) a coding region with at least two main open reading frames (ORF1 (long) and ORF2 (short)) deduced directly from the nucleotide sequence, generally overlapping, and (2) a noncoding region containing a GC-rich zone that forms a stem–loop structure (**Figure 1**). Respective sizes differ widely depending on the isolates studied. On the basis of the TTV prototype 1a, the coding region is about 2600 nt long, with the ORF1 and ORF2 composed of 770 and 202 amino acids respectively, while the noncoding region is composed of about 1200 nt with a short zone (~110 nt) of high GC content (~90%). Sequences analysis reveals a variable G+C content for anelloviruses: calculated values from full-length sequences are ~52%, ~38%, and ~39% for TTV, TTMV, and the 'small anellovirus', respectively.

The replication mechanism of anelloviruses is not well known. However, some studies highlighted the presence of TTV mRNA forms and double-stranded DNA in bone marrow and in various human tissues and organs (including the liver), suggesting an active replication in these locations. The presence of double-stranded DNA forms would suggest, as for other circular single-stranded DNA viruses, a rolling-circle mechanism for replication. Three types of TTV mRNAs (2.9, 1.2, and 1.0 kbp), generated by an alternative splicing, have been detected in bone marrow cells and were also obtained following transfection of a permuted whole-genome into African green monkey COS cells. Identification of these mRNAs permitted to establish the functionality of both ORF1 and ORF2 and confirmed the implication of additional ORFs. Transfection into the 293 cell line (human embryonic kidney cells) of a full-length TTV clone not only confirmed the existence of the three mRNA classes but also further demonstrated the expression of six different proteins following an alternative translation strategy. Importance of the untranslated region (UTR) of TTV was highlighted by the identification of a basal promoter ~110 nt upstream the transcription initiation site,

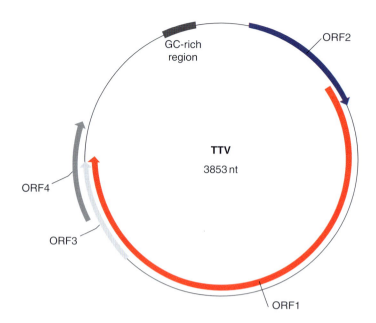

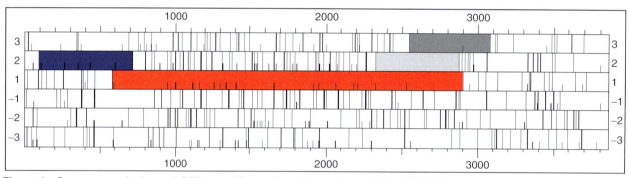

**Figure 1** Genome organization and ORF map of *Torque teno virus* prototype (isolate TTV1a).

and by the presence of enhancer elements in a ~490 nt region upstream this promoter.

The absence of a tissue culture system that would support efficient replication of anelloviruses hindered for a long time the studies of virus–host cell interactions. Interestingly, a human liver cell line (the Chang Liver cell line) does support TTV replication following infection with contaminated sera, and releases significant and persistent levels of infectious viral particles into culture fluid supernatants.

Information concerning the functions of the various proteins expressed during the anellovirus cell-cycle infection is fragmentary. The ORF1 might encode a single polypeptide combining structural and functional roles, that is, the capsid and replication functions of the virus. The presence of conserved motifs related to the Rep protein in the ORF1 would confirm the involvement of a rolling-circle mechanism for the anellovirus replication. Expression studies identified TTMV ORF2 as a dual-specificity protein phosphatase which could be involved in mechanisms of immune evasion and virus persistence.

The TTV ORF3 gene would also generate two variants of a protein with a different serine-phosphorylation state which could be involved in the virus-replication cycle. Finally, a putative TTV protein located upstream the ORF1 was shown to induce apoptosis in hepatocellular carcinoma-derived cells.

## Phylogenetic and Taxonomic Aspects

The genetic diversity among anelloviruses is far larger than within any other defined group of ssDNA viruses. The considerable genetic heterogeneity is exemplified by the large number of highly divergent full-length sequences progressively identified as TTV, TTMV, and 'small anellovirus' genomes.

Historically, primer extension of the initial sequence (~500 nt, N22 clone) to about 3700 nt (TA278 clone) primarily suggested a distant resemblance of TTV to parvoviruses, based on the apparent linear nature of the characterized genome. The circular nature of the genome was subsequently elucidated (TTV-1a clone), leading to the possible assignment of TTV members to the families of viruses possessing a circular single-stranded DNA genome. This initiated studies on circular single-stranded DNA viruses infecting humans.

Concomitant comparisons of short nucleotide sequences obtained by PCR in the N22 region allowed to identify three distinct genotypes (differing by 27–30% nucleotide divergence) describing TTV genetic diversity in early 1999. The progressive characterization of partial and complete nucleotide sequences had not only demonstrated the existence of a large number of genotypes but has also allowed to classify these genotypes into five distinct clusters (~50% nucleotide divergence) representing the TTV major phylogenetic groups, as defined in 2002 (**Figure 2**). The creation of the genus *Anellovirus* by ICTV in 2004 has officially presented such classification, but it is possible that the next ICTV report will bring significant changes to the taxonomic status of anelloviruses, such as the creation of a specific family hosting several genera accommodating many species, and modify phylogenetic clusters due to the description of new genomic sequences.

Despite the fact that genomic sequences from TTMV and 'small anellovirus' are not as well described as those of TTV, they revealed a high genetic heterogeneity, at least of the same magnitude of that identified for TTV or greater. In 2005, the available full-length TTMV sequences clustered in four major phylogenetic groups (~40% nucleotide divergence), whereas the only two 'small anellovirus' complete sequences described exhibited a nucleotide divergence reaching 46%. Extremely divergent isolates (as compared to human TTV, TTMV, and 'small anellovirus') have been also identified in nonhuman primates and low-order mammals.

A low degree of sequence homology exists between TTV, TTMV, the 'small anellovirus', and animal isolates. However, there is a ~130 nt long sequence that is relatively well conserved between the viral groups, which is located within the UTR downstream of the GC-rich region (**Figure 3**). Moreover, the genome organization of anelloviruses globally appears similar. It includes a coding region containing at least two main ORFs, and a UTR generally having a GC-rich stretch. The respective sizes of each of these components is however variable between isolates.

Accurate phylogenetic analyses inside each viral group are feasible by comparing full-length ORF1, and to a lesser extent ORF2 nucleotide sequences. By contrast, phylogenetic analysis of short nucleotide sequences located in the N22 region or on the UTR proved to be unreliable. The latter approach is also biased by the occurrence of recombination events which are statistically more frequent in this location than in the rest of the genome.

The amino acid sequence comparisons of translated ORF1 or ORF2 proved to be reliable for phylogenetic analyses, and is the only approach for taxonomic studies combining all *Anellovirus* members which markedly differ in sequence and size. Such comparisons at the amino acids level have also highlighted that most, if not all, of the anelloviruses possess an ORF1 with an arginine-rich N-terminal part and motifs related to the Rep protein, while the ORF2 exhibits the well-conserved motif $WX_7HX_3CX_1CX_5H$. Interestingly, the same features are found in the chicken anemia virus (CAV), the type species of the genus *Gyrovirus* of the family *Circoviridae*. CAV possesses a negative-sense genome, a similar genome organization to anelloviruses, with overlapping ORFs, and several viral proteins with functions supposedly similar to their counterparts in members of the genus *Anellovirus*.

## Distribution, Epidemiology, and Transmission

In 1997, the identification of TTV DNA in the blood of Japanese patients with hepatitis of unknown etiology was the starting point of the research on this new group of viruses. Despite its initial identification in populations with liver disorders, further epidemiological studies not only identified the virus in populations with parenteral risk exposure, including intravenous drug users, hemophiliacs, or HIV infected patients, but also in populations without proven pathology like blood donors. It was also clearly demonstrated that TTV is distributed worldwide, as the virus was detected in rural or urban populations in Africa, Americas, Asia, Europe, and Oceania.

Due to the enormous genetic variability characterizing anelloviruses, the choice of the viral DNA target for PCR amplification proved to be highly important for the sensitivity of PCR assays. It was in direct correlation with the progressive determination of full-length genomic sequences and the identification of highly conserved regions between viral isolates. So, early estimated prevalence values for TTV DNA ranged from 1% to 5% in the general population but increased dramatically within 1 year to ~80% in blood donor cohorts, revealing a wide and intriguing dispersion of anelloviruses in human populations without any apparent disease. Higher prevalence values (~90%) are generally identifiable in cohort of patients with health disorders, such as cancer, diabetes mellitus, HIV infection, or under hemodialysis treatment. It was also shown that the prevalence of viremia increased slightly with age. Data relating to TTMV and the 'small anellovirus' tend to reveal similar features concerning their prevalence in human populations.

Subsequent information concerning anellovirus infection in humans has been obtained by the analysis of the distribution of the five major TTV phylogenetic groups in blood donors: interestingly, it revealed a nonrandom pattern of group distribution and a predominant prevalence

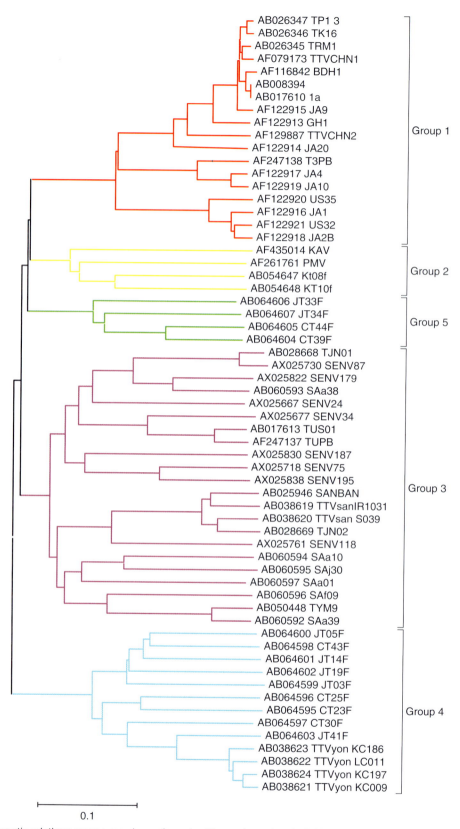

**Figure 2** Phylogenetic relations among members of species *Torque teno virus* (neighbor-joining tree built with full-length nucleotide sequences).

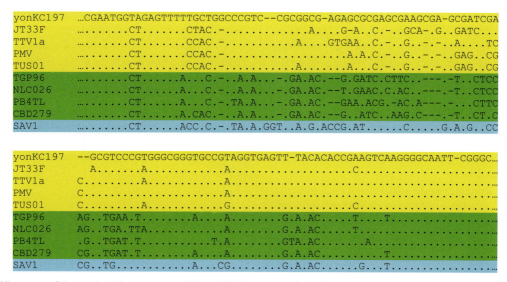

**Figure 3** Alignment of the nucleotide sequence of TTV, TTMV, and 'small anellovirus' representative isolates identified in humans (conserved zone, UTR region).

of phylogenetic groups 1, 3, and 5 in Brazilian and French cohorts. For the latter, the first study of TTMV distribution revealed a predominance of phylogenetic group 1 over the three other groups considered.

Mixed infections appeared as an important characteristic of the *Anellovirus*–host relationship. The possible co-infection of an individual by multiple viral strains of TTV, TTMV, or 'small anellovirus', even highly divergent, is frequent. As demonstrated for TTMV with the analysis of multiples clones of a single PCR, intra-individual viral genetic diversity could reach values above 40%.

Initially described in plasma and serum samples, TTV, TTMV, and 'small anellovirus' DNA have been subsequently isolated in various biological specimens such as peripheral blood mononuclear cells (PBMCs), saliva and nasal secretions, feces, and in other body fluids including semen and tears. Identification of anelloviruses in breast milk, in cord blood samples, and in the blood of newborns, was also described.

These data would suggest that the high rates of prevalence of anelloviruses in various body fluids of healthy subjects might be due to a possible transfusion, fecal–oral, saliva droplet, sexual, or mother-to-child route(s) of transmission.

The presence of anelloviruses members is not limited to human hosts since they were also detected in nonhuman primates such as chimpanzees and tamarins, in tupaias, in popular pets like cats and dogs, and in domesticated farm animals such as pigs and cows. Viral genomic sequences identified in these hosts were either highly divergent or very similar when compared to those obtained from humans. The fact that TTV partial sequences already identified in human blood were found in the blood of farm animals suggests that interspecies transmission could occur. In the same way, intravenous inoculation of TTV-positive human serum or infected fecal supernatant into chimpanzees demonstrated that TTV can be transmitted to primates.

## Detection Approaches/Viral Load

Genomic DNA detection has been the mostly developed approach for the identification of anelloviruses in biologic samples. Historically, due to the limited knowledge of TTV genomic sequences, initial PCR systems were restricted to a short portion of the genome in the center of the ORF1. This region was extensively targeted in nested or hemi-nested PCR protocols, but despite improvements in the design of degenerate primers or optimization of the PCR protocols, detection of TTV DNA appeared restricted to certain genotypes and revealed reduced sensitivity when the virus content in samples was low. Extension of the TTV genomic sequence allowed the design of alternative PCR systems located in different parts of the viral genome and their use in combined detection approaches; such strategy permitted to significantly improve the estimated TTV prevalence in the general population. The growing knowledge of virus genome diversity, as revealed by the characterization of highly divergent full-length genomic sequences, permitted to identify a short and relatively well-conserved zone suitable for optimum primer design. The current 'gold' method for viral DNA detection is based on the use of highly conserved primers, genotype-independent, located in the UTR of the genome upstream the ORF1. This PCR system increased the positive detection rates of samples for TTV DNA and also proved to be applicable to the efficient detection of TTMV, 'small anellovirus' DNA, and highly divergent genomes identified in animals as well.

Real-time PCR assays were developed using such conserved PCR systems in order to gain insights on quantitative aspects of viral infection; such an approach demonstrated that plasma viral loads vary widely, ranging generally between $\sim 10^3$ and $\sim 10^8$ genomes ml$^{-1}$ in individuals.

Distinct amplification systems, designed using a representative dataset of nucleotide sequences specific to the major phylogenetic groups of TTV and TTMV, have also been used for the analysis of the genetic distribution of members of the genus *Anellovirus* in some human cohorts, including blood donors.

Alternative strategies in the diagnosis of viral infection were proposed in a few studies describing serologic (IgG and IgM detection tests, Western blot test using partial ORF1 recombinant protein) and *in situ* hybridization approaches.

## Clinical Significance

Since their discovery, the question of a possible implication of anelloviruses in a particular disease is still a matter to debate.

Historical presentation of TTV as associated with elevated transaminase levels in post-transfusion hepatitis of unknown etiology suggested that the virus was able to induce non-A–G hepatitis. Therefore, TTV was suspected as a possible cause of some forms of acute and chronic hepatitis and fulminant liver failure, and could be involved in liver diseases. The identification of TTV in the general population seems to refute this interpretation and led to the suggestion that the virus may cause only occasional liver injury, either by the implication of hepatotropic variants or by the presence of host determinants enhancing the pathogenicity of TTV.

The further identification of TTMV and 'small anellovirus' DNA in humans has increased the number of hypotheses concerning the impact of anelloviruses in human health. Hence, based on studies involving patient cohorts with well-defined health disorders, it has been suggested that this class of viruses may also be implicated in various diseases such as pancreatic cancer, systemic lupus eythematosus, idiopathic inflammatory myopathies, or chromosomal translocations. The implication of anelloviruses in respiratory diseases has also been proposed following studies involving cohorts of children suffering from asthma or rhinitis. It has also been suggested that the respiratory tract could be a site of primary infection in young children and a site of continual replication of TTV and related viruses.

Other studies suggested a possible link between the viral load and the immune status of the host because of high TTV or TTMV titers in plasma samples from immunocompromised patients, but this remains hypothetical since the loads of TTV and TTMV can differ extensively among individuals in the general population.

To overcome the limitations of a diagnosis of anellovirus infection based only on highly conserved amplification systems, it has been suggested that it would be useful to compare individuals with specific diseases with a reference population such as blood donors. Such an approach may lead to the recognition of a possible pathogenic role of these viruses.

Alternatively, anelloviruses may be considered as part of the 'normal' human flora.

## Further Reading

Bendinelli M, Pistello M, Maggi F, et al. (2001) Molecular properties, biology, and clinical implications of TT virus, a recently identified widespread infectious agent of humans. *Clinical Microbiology Reviews* 14: 98–113.

Biagini P, Charrel RN, de Micco P, and de Lamballerie X (2003) Association of TT virus primary infection with rhinitis in a newborn. *Clinical Infectious Diseases* 36: 128–129.

Biagini P, Gallian P, Attoui H, et al. (2001) Genetic analysis of full-length genomes and subgenomic sequences of TT virus-like mini virus human isolates. *Journal of General Virology* 82: 379–383.

Biagini P, Gallian P, Cantaloube JF, et al. (2006) Distribution and genetic analysis of TTV and TTMV major phylogenetic groups in French blood donors. *Journal of Medical Virology* 78: 298–304.

Biagini P, Todd D, Bendinelli M, et al. (2005) Anellovirus. In: Fauquet CM, Mayo MA, Maniloff J, Desselberger U, and Ball LA (eds.) *Virus Taxonomy: Eighth Report of the International Committee on Taxonomy of Viruses*. San Diego, CA: Elsevier Academic Press.

Hino S and Miyata H (2007) Torque teno virus (TTV): Current status. *Reviews in Medical Virology* 17: 45–57.

Jones MS, Kapoor A, Lukashov VV, et al. (2005) New DNA viruses identified in patients with acute viral infection syndrome. *Journal of Virology* 79: 8230–8236.

Maggi F, Pifferi M, Fornai C, et al. (2003) TT virus in the nasal secretions of children with acute respiratory diseases: Relations to viremia and disease severity. *Journal of Virology* 77: 2418–2425.

Nishizawa T, Okamoto H, Konishi K, et al. (1997) A novel DNA virus (TTV) associated with elevated transaminase levels in posttransfusion hepatitis of unknown etiology. *Biochemical and Biophysical Research Communications* 241: 92–97.

Okamoto H, Nishizawa T, Kato N, et al. (1998) Molecular cloning and characterization of a novel DNA virus (TTV) associated with posttransfusion hepatitis of unknown etiology. *Hepatology Research* 10: 1–16.

Okamoto H, Nishizawa T, Takahashi M, et al. (2001) Heterogeneous distribution of TT virus of distinct genotypes in multiple tissues from infected humans. *Virology* 288: 358–368.

Okamoto H, Takahashi M, Nishizawa T, et al. (2002) Genomic characterization of TT viruses (TTVs) in pigs, cats and dogs and their relatedness with species-specific TTVs in primates and tupaias. *Journal of General Virology* 83: 1291–1297.

Pifferi M, Maggi F, Caramella D, et al. (2006) High torquetenovirus loads are correlated with bronchiectasis and peripheral airflow limitation in children. *Pediatric Infectious Disease Journal* 25: 804–808.

Takahashi K, Hoshino H, Ohta Y, Yoshida N, and Mishiro S (1998) Very high prevalence of TT virus (TTV) infection in general population of Japan revealed by a new set of PCR primers. *Hepatology Research* 12: 233–239.

Takahashi K, Iwasa Y, Hijikata M, and Mishiro S (2000) Identification of a new human DNA virus (TTV-like mini virus, TLMV) intermediately related to TT virus and chicken anemia virus. *Archives of Virology* 145: 979–993.

# Animal Rhabdoviruses

**H Bourhy,** Institut Pasteur, Paris, France
**A J Gubala,** CSIRO Livestock Industries, Geelong, VIC, Australia
**R P Weir,** Berrimah Research Farm, Darwin, NT, Australia
**D B Boyle,** CSIRO Livestock Industries, Geelong, VIC, Australia

© 2008 Elsevier Ltd. All rights reserved.

## Glossary

**Dipterans** Insects having usually a single pair of functional wings (anterior pair) with the posterior pair reduced to small knobbed structures and mouth parts adapted for sucking or lapping or piercing.
**Hematophagous** Feeding on blood.
**Homopterans** Insects having membranous forewings and hind wings.
**Orthopterans** Insects having leathery forewings and membranous hind wings and chewing mouthparts.

## Introduction

RNA viruses of the family *Rhabdoviridae* comprise arthropod-borne agents that infect plants, fish, and mammals, as well as a variety of non-vector-borne mammalian viruses. The *Rhabdoviridae* presently comprises six genera, and members of three of these genera – *Vesiculovirus*, *Lyssavirus*, and *Ephemerovirus* – have been obtained from a variety of animal hosts and vectors, including mammals, fish, and invertebrates. The remaining three rhabdovirus genera are more taxon-specific in their host preference. Novirhabdoviruses infect numerous species of fish, while cytorhabdoviruses and nucleorhabdoviruses are arthropod-borne and infect plants. Rhabdoviruses are the etiological agents of human diseases that cause serious public health problems. Some of them can also cause important economic loss in plants and livestock. Other than the well-characterized rhabdoviruses that are known to be important for agriculture and public health, there is also a constantly growing list of rhabdoviruses (presently 85), isolated from a variety of vertebrate and invertebrate hosts, that are partially characterized and are still awaiting definitive genus or species assignment.

## History and Classification

Commencing in the 1950s, monitoring programs have been established in tropical regions of Africa, America, Southeast Asia, and northern Australia to detect and identify arboviruses of medical or veterinary importance. In Australia, the use of sentinel cattle herds monitoring associated with insect and vertebrate trapping has provided a means of monitoring the ecology of endemic arboviruses that infect livestock and humans, and an early warning system for incursions of exotic arboviruses and insect vectors. In most other regions, these programs have been based on the collection of viruses from terrestrial vertebrates and hematophagous arthropods. As a result, many new and uncharacterized rhabdoviruses have been isolated. The considerable range of likely vectors and hosts and the wide geographical distribution seen among these isolates highlights the diversity of animal rhabdovirus evolution and ecological adaptation.

The isolation of viruses in the early years was primarily performed in suckling mice by intracranial inoculation with clarified supernatant fluid obtained after grinding of pools of collected arthropods. The mice were then observed daily for at least 14 days and those that showed signs of illness were euthanized and the brains were removed for extraction and subsequent passage in mice to amplify the virus. In more recent times, virus isolations have been made by routine propagation through embryonated eggs or mosquito (*Aedes albopictus* C6/36 or *Aedes pseudoscutellaris* AP 61), baby hamster kidney (BHK-21), African green monkey kidney epithelial (Vero), hamster kidney (CER), or swine kidney (PS) cell lines. Monolayers are observed daily until cytopathic effect (CPE) is observed, at which point the cell culture medium is clarified at low-speed centrifugation and aliquots of virus are stored frozen at $-70\,°C$. Generally, the presence of virus has been verified by indirect immunofluorescent assays (IFAs) and electron microscopy.

Unknown viruses have been classified as members of the *Rhabdoviridae* by electron microscopy, based on their bullet-shaped morphology – a characteristic trait of members of this family (**Figure 1**). Subsequently, the assessment of antigenic relationships between these unknown viruses and other viruses worldwide has been performed using serological tests. Immune reagents have been developed for tentative assignment of the unclassified viruses by IFA, complement fixation (CF), virus neutralization (VNT) assays, hemagglutination inhibition (HI), and more recently enzyme-linked immunosorbent assays (ELISAs). However, many of these isolates have revealed no relationship with any known rhabdovirus. Up to seven different antigenic groups have been defined. Gene sequencing and phylogenetic relationships have been progressively applied to complete this initial virus taxonomy.

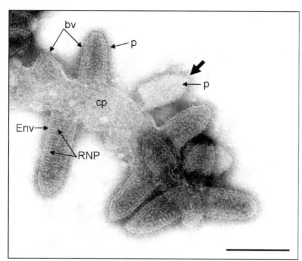

**Figure 1** Electron micrograph of a negative-stained cell infected with Wongabel virus, representative of the nonassigned rhabdoviruses. Viruses are budding (bv) from a cell process (cp). The viruses are bullet-shaped, possess an envelope (Env), surface projections (peplomers) (p), and an internal helical ribonucleocapsid (RNP). The peplomers (p) consisting of trimers of viral glycoprotein can also be observed on the surface of the virus not penetrated by stain (short thick arrow). Scale = 200 nm. Electron micrograph courtesy 'Electron Microscopy and Iridoviruses Group', CSIRO-AAHL.

Other than the approved species, tentative species have been proposed in each of the four currently recognized genera of animal rhabdoviruses: 19 in the genus *Vesiculovirus*, 5 in the genus *Lyssavirus*, 3 in the genus *Ephemerovirus*, and 2 in the genus *Novirhabdovirus*. There remain six serogroups of animal rhabdoviruses (altogether comprising 20 different viruses) that have not yet been assigned to an existing genus: the Bahia Grande group, Hart Park group, Kern Canyon group, Le Dantec group, Sawgrass group, and Timbo group. A further 43 unassigned animal rhabdoviruses are listed in the *Eighth Report of the International Committee on Taxonomy of Viruses* (Eighth ICTV Report). Serological surveys conducted using sera collected from livestock, various wildlife and humans from Australia, Asia, and the Pacific region have revealed potential animal hosts and the geographical distribution for some of these viruses. However, for most of the uncharacterized rhabdoviruses information is limited. Although links between disease and the uncharacterized rhabdoviruses have not been made, recent studies have provided insights into their genetic composition and have revealed that wide genetic diversity exists among them, beckoning more intensive study of this family of viruses.

## Virion Properties

### Morphology

Rhabdovirus virions are 100–400 nm long and 50–100 nm in diameter (**Figure 1**). Viruses appear bullet-shaped. From the outer to the inner side of the virion, one can distinguish the envelope covered with viral glycoprotein spikes and, internally, the nucleocapsid with helical symmetry consisting of the nucleoprotein tightly bound to genomic RNA.

## Genome Organization and Genetics

All rhabdoviruses contain a genome consisting of a nonsegmented single-stranded negative-sense RNA molecule with a size in the range of approximately 8.9–15 kbp. This RNA molecule contains at least five open reading frames (ORFs) encoding five virion proteins in the order ($3'$–$5'$): nucleoprotein (N); phosphoprotein (P); matrix protein (M); glycoprotein (G); and polymerase (L). Viruses in the genus *Ephemerovirus* contain several additional ORFs between the G and L genes, which encode a second glycoprotein ($G_{NS}$) and several other nonstructural proteins. Similarly, in the genus *Novirhabdovirus*, a sixth functional cistron between the G and L genes encodes a nonstructural protein (NV) of unknown function. The unclassified rhabdoviruses, sigma virus infecting flies (*Drosophila* spp.) and plant rhabdoviruses in the genera *Cytorhabdovirus* and *Nucleorhabdovirus* also contain an additional ORF which is located between the P and M genes. Flanders virus from mosquitoes (*Culista melanura*) has a complex arrangement of genes and pseudogenes in the same genome region. Nucleotide sequence analysis of Tupaia virus from the tree shrew (*Tipiai belangeri*) has identified an additional gene encoding a small hydrophobic protein between the M and G genes, and genome sequence analysis of Wongabel virus, an unassigned rhabdovirus isolated from biting midges (*Culicoides austropalpalis*), has revealed that it contains five additional genes that appear to be novel. The function of these other proteins (including additional glycoproteins) is not yet known. Therefore, despite preservation of a characteristic particle morphology, the *Rhabdoviridae* includes viruses that display a wide genetic diversity (**Figure 2**).

Relatively low sequence identities across the *Rhabdoviridae* prevent the construction of a family phylogeny. One approach to determining the phylogenetic relationships among the rhabdoviruses, as well as the identification of new viral species, is to utilize the conserved regions that have been identified in alignments of the N and L genes.

## Evolutionary Relationships

A molecular phylogenetic analysis of 56 rhabdoviruses, including 20 viruses which are currently unassigned or assigned as tentative species within the *Rhabdoviridae*, has been reported by using the sequences from a region of block III of the L polymerase (**Table 1** and **Figure 3**). Block III is predicted to be essential for RNA polymerase function because it is conserved among all L proteins, and mutations

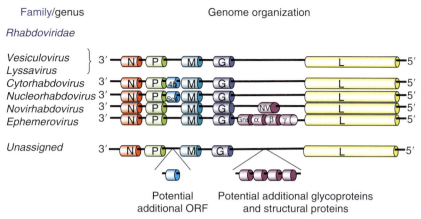

**Figure 2** Genome organization of rhabdoviruses.

**Table 1** Unassigned rhabdoviruses from Australia and Papua New Guinea

| Virus name | Isolated from | Neutralizing antibody detected | Year of isolation | Location |
|---|---|---|---|---|
| Almpiwar | *Ablepharus boutonii virgatus* (skink) | Skink, cattle, horse, sheep, kangaroo, bandicoot, various birds, human | 1966 | Mitchell River, Queensland |
| Charleville | *Phlebotomus* spp. (sand fly) *Gehyra australis* (gecko) *Lasiohelea* spp. (biting midge) | Human | 1969 | Charleville, Queensland |
| Coastal Plains | Cattle | Buffalo, cattle (Australia and Papua New Guinea) | 1981 | Beatrice Hill[a], Northern Territory |
| Humpty Doo | *C. marksi, Lasiohelea* spp. (biting midge) | Unknown | 1975 | Beatrice Hill, Northern Territory |
| Joinjakaka | Mixed Culicines (mosquito) | Cattle | 1966 | Sepik District, Papua New Guinea |
| Koolpinyah | Cattle (*Bos indicus* x *Bos taurus*) | Cattle | 1985 | Berrimah Farm, Northern Territory |
| Kununurra | *Ad. catasticta* (mosquito) | Unknown | 1973 | Kununurra, Western Australia |
| Ngaingan | *C. brevitarsis* (biting midge) | Wallaby, kangaroo, cattle | 1970 | Mitchell River, Queensland |
| Oak Vale | *Cx. edwardsi* (mosquito) | Ferral pigs | 1981 | Peachester, Queensland |
| Parry Creek | *Cx. annulirostris* (mosquito) | Unknown | 1973 | Kununurra, Western Australia |
| Tibrogargan | *C. brevitarsis* (biting midge) | Buffalo, cattle | 1976 | Peachester, Queensland |
| Wongabel | *C. austropalpis* (biting midge) | Sea birds | 1979 | Wongabel, Queensland |
| CSIRO75 (Harrison Dam virus) | *Cx. annulirostris* (mosquito) | Unknown | 1975 | Beatrice Hill, Northern Territory |
| CSIRO1056 | *C. austropalpis* (biting midge) | Unknown | 1981 | Samford, Queensland |
| DPP1163 (Holmes Jungle virus) | *Cx. annulirostris* (mosquito) | Cattle, buffalo, humans | 1987 | Darwin, Northern Territory |
| OR559 (Little Lilly Creek virus) | *Cx. annulirostris* (mosquito) | Unknown | 1974 | Kununurra, Western Australia |
| OR1023 (Ord River virus) | *Cx. annulirostris* (mosquito) | Unknown | 1976 | Kununurra, Western Australia |

[a]Beatrice Hill, NT has in the past been known as Coastal Plains.

in this region abolish polymerase activity. This phylogenetic analysis produced an evolutionary tree that generally, although not entirely, conforms to accepted serological groupings and taxa within the *Rhabdoviridae*. In particular, members of four genera – *Lyssavirus, Novirhabdovirus, Cytorhabdovirus*, and *Nucleorhabdovirus* – obtained from a variety of host species, including mammals, fish, arthropods, and plants, can be easily distinguished and fall into

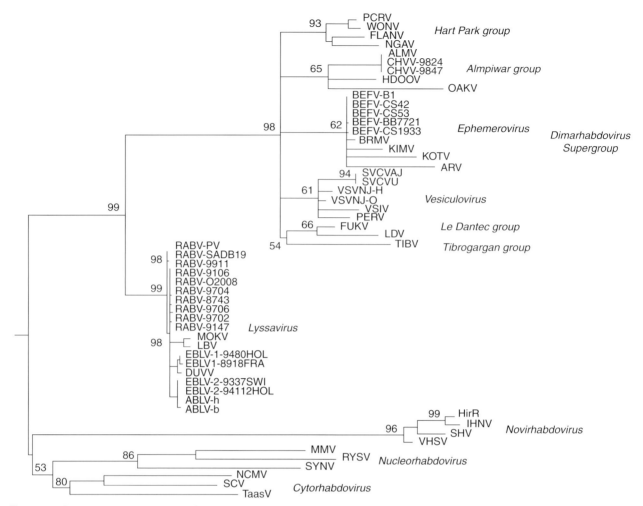

**Figure 3** Phylogenetic relationships of members of the *Rhabdoviridae* based on a maximum likelihood analysis of a 158-amino-acid residue alignment of the L polymerase region. The established rhabdovirus genera as well as the newly proposed groups are indicated. Horizontal branches are drawn to scale and quartet puzzling frequencies are shown for key nodes (values in italics are for genera, groups, and supergroups, while all other quartet puzzling frequencies are shown in normal font). The tree is midpoint rooted for purposes of clarity only and all potential outgroup sequences were deemed too divergent to include in the analysis. Adapted from Bourhy H, Cowley JA, Larrous F, Holmes EC, and Walker PJ (2005) Phylogenetic relationships among rhabdoviruses inferred using the L polymerase gene. *Journal of General Virology* 86: 2849–2858.

relatively well-supported clades. Although the vesiculoviruses and ephemeroviruses also fall into clear monophyletic groups, they are less well supported, and each genus contains some unclassified viruses. Furthermore, kotonkan virus, which causes clinical ephemeral fever in cattle but has previously been classified as a lyssavirus, is very clearly clustered with members of the genus *Ephemerovirus*. Lastly, there is some evidence that the two groups of plant rhabdoviruses – the cytorhabdoviruses and nucleorhabdoviruses – form a distinct clade. Taastrup virus which remains unassigned is related to cytorhabdoviruses.

Strikingly, this phylogenetic analysis has also identified four more monophyletic groups of currently unclassified rhabdoviruses, which have variable support values. First, Wongabel, Parry Creek, Flanders, and Ngaingan viruses formed a distinct cluster, with high levels of support. We refer to this as the Hart Park group, based on the serologic grouping of Flanders virus in the Hart Park serological group. Second, a tentatively named Almpiwar group containing Almpiwar, Humpty Doo, Charleville, and Oak Vale viruses was also identified. Although this grouping had a lower support, the Almpiwar and Charleville viruses possess almost indistinguishable sequences in the L gene region and each has been associated with infection in lizards. Another group, consisting of the Le Dantec and Fukuoka viruses, and herein referred to as the Le Dantec group, was also seen to form a distinct cluster. Finally, the phylogenetic position of Tibrogargan virus was ambiguous.

Common rhabdovirus sequence motifs have also been found in the central region of the N gene. Phylogenetic analysis of partial N gene sequences indicated that two viruses isolated from bats from various regions of the world, Oita virus and Mount Elgon bat virus, were grouped in a monophyletic cluster. Kolongo virus (Africa) and Sandjimba virus (Asia) also formed a distinct clade together with Tupaia virus (Asia). In this analysis, Flanders virus, sea trout virus, sigma virus, Kern Canyon virus, and Rochambeau virus remained isolated on the phylogenetic tree.

Phylogenetic analysis of both partial N and L gene sequences of Obodhiang virus and kotonkan virus has indicated that they should be classified as members of the genus *Ephemerovirus*.

## Host Range and Virus Propagation
### Transmission

Many of the rhabdoviruses replicate in and are transmitted by insect vectors. Plant-infecting rhabdoviruses are transmitted by vectors including aphids, planthoppers, and leafhoppers. Several animal rhabdoviruses (e.g., ephemeroviruses and vesiculoviruses) are known to be transmitted by hematophagous insects such as mosquitoes and biting midges. The widespread ability of rhabdoviruses to infect insects has led to the hypothesis that this virus family has evolved from an ancestral insect virus and that the host range is largely determined by the insect host. Sigma virus (SIGMAV) does not appear to have any vertebrate host but can be transmitted congenitally in flies (*Drosophila* spp.).

A number of important biological conclusions can be drawn from the rhabdovirus phylogeny based on the L gene. First, assuming a midpoint rooting of the tree, there is major split between those viruses that infect fish (novirhabdoviruses) or infect plants and employ arthropods as vectors (cytorhabdoviruses and nucleorhabdoviruses) – and those viruses that infect both mammals or lizards and dipterans (dimarhabdoviruses, sigla for dipteran-mammal-associated rhabdovirus). Such a division illuminates the biology of a number of key rhabdoviruses. For example, although vesicular stomatitis virus (VSV) is responsible for a disease of horses, cattle, and pigs, and can be transmitted directly by transcutaneous or transmucosal routes, the virus replicates in a wide range of hosts, including insects, and there is good evidence that VSV may employ insect vectors as at least one of its mechanisms of transmission. Similarly, bovine ephemeral fever virus, which is frequently found in Australasia and Africa, is also dipteran-transmitted, using vectors such as biting midges and culicine and anopheline mosquitoes. Finally, viruses assigned by the L-based phylogenetic analysis to the four new groups (Le Dantec, Tibrogargan, Hart Park, and Almpiwar groups) have all been found to infect dipterans and in some cases also mammals (Tibrogargan, Le Dantec, and Ngaingan viruses) or lizards (Charleville virus).

Importantly, there is, as yet, no evidence for a virus making the link between plant rhabdoviruses, novirhabdoviruses of fish, or dimarhabdoviruses. Furthermore, the uncertainty over branching order at the root of the tree makes it difficult to determine whether the ancestral mode of transmission in rhabdoviruses was vector or nonvector transmission. However, the major phylogenetic division between these groups indicates that the biology of the rhabdoviruses could be strongly influenced by mode of transmission and by the host (plant, fish, or mammals) and vector species (orthopterans, homopterans, or dipterans).

### Clinical Features and Pathology

For most of the uncharacterized rhabdoviruses, links with disease have not yet been made. The only virus isolated from a natural infection in humans is Le Dantec virus. As indicated above, kotonkan virus causes clinical ephemeral fever in cattle and deer. Fukuoka virus has also been isolated from the blood of febrile calves.

## Serogroups of Nonassigned Rhabdoviruses (As Recognized in the Eighth ICTV Report)
### Hart Park Group

Phylogenetic analyses of the L gene have indicated that Flanders virus (FLANV), Ngaingan virus (NGAV), Parry Creek virus (PCV), and Wongabel virus (WONV) cluster with Hart Park virus (HPV) and are possible members of the Hart Park group. FLANV was isolated in 1961 from a pool of engorged female *Culista melanura* mosquitoes that were collected in Long Island, New York, USA. Similar viruses have been collected from different mosquito species in various parts of the United States. FLANV has also been isolated from the blood of house sparrows, red-winged blackbirds, and from the spleen of an oven bird. HPV was isolated in 1955 from a pool of female *Culex tarsalis* collected at Hart Park, California. NGAV (strain MRM14556) was isolated in 1970 from a pool of biting midges (*Culicoides brevitarsis*) that were collected at Mitchell River Aboriginal community in northern Queensland. This arbovirus has also been found to multiply in experimentally infected mosquitoes (*Aedes aegypti*). Serological surveys have indicated that NGAV infects wallabies, and possibly kangaroos and cattle, although its role in disease in these animals is unknown. IFA, CF, and VNT results place NGAV in the Tibrogargan antigenic group.

PCV (strain OR189) was isolated in 1973 from mosquitoes (*Culex annulirostris*) that were collected at Parry Creek near Kununurra, Western Australia. WONV (strain CSIRO264) was isolated in 1979 from biting midges (*Culicoides austropalpalis*) collected at Wongabel on the Atherton Tableland of northern Queensland. Morphological examination has revealed bullet-shaped particles $(80-90) \times (160-180)$ nm in dimension (**Figure 1**). This species of biting midge has been observed to have a feeding preference for birds. Although no link has been established between WONV and disease, neutralizing antibodies were detected in sea birds collected off the Great Barrier Reef. No neutralizing antibodies were detected in human sera from island residents in this region. Two other viruses, Mossuril virus (MOSV) and Kamese virus (KAMV), have been classified with the Hart Park group based on antigenic relationships only, as they were not included in any phylogenetic analysis. MOSV was first identified in 1959 in Mozambique and later in Central African Republic. It was isolated from mosquitoes (*Culex sitiens, Culex decens, Culex perfusus, Culex pruina, Culex telesilla, Culex Weschei*, and *Culex tigripes*) and birds (*Andropadus virens* and *Cliuspasser maccrourus*). KAMV was first identified in Uganda in 1967 and then in Central African Republic. It has been isolated from culicine mosquitoes (*Culex annulioris, Aedes africanus, Culex decens, Culex perfuscus, Culex pruina*, and *Culex tigripes*).

## Le Dantec Group

At the L gene amino acid level, Fukuoka virus (FUKAV) and Le Dantec virus (LDV) appear to be related, although FUKAV was previously classified in the Kern Canyon group on the basis of its antigenic properties (see below). LDV was originally recovered in 1965 from a patient with a febrile illness, headaches, and spleen and liver hypertrophy in Senegal. In CF tests with other known rhabdoviruses, Le Dantec virus was found to be antigenically related to Keuraliba virus, a previously ungrouped agent isolated from rodents (*Tatera kempi*) in Senegal in 1968. FUKAV was first isolated from biting midges, *Culicoides punctalis* and *Culex tritaeniorhynchus* in 1986, and subsequently isolated from blood of calves with fever and leucopoenia.

## Bahia Grande Group

Bahia Grande (BGV) (prototype strain TB4–1054), Reed Ranch (RRV)(TB4–222), and Muir Springs (MSV) viruses (76V-23524) were first obtained from salt-marsh mosquitoes (*Culex, Aedes, Anopheles,* and *Psorophora* spp.) collected between 1972 and 1979 in west Texas, New Mexico, Louisiana, Colorado, North Dakota, and south Texas. Structural analysis of the prototype strain of BGV from Texas has revealed five proteins. Comparative oligonucleotide fingerprint maps has shown 51–86% sharing of the large oligonucleotides between BGV (strain TB4–1054) and 11 other antigenically related isolates but not with MSV (strain 76V-23524), an antigenically distinct isolate from mosquitoes collected in Colorado. A serological survey for antibody to BGV has shown that humans, cattle, sheep, reptiles, and wild mammals from south Texas have neutralizing antibodies to this virus.

## Timbo Group

Chaco virus (CHOV) was isolated from *Ameiva ameiva ameiva* and *Kentropyx calcaratus* lizards in Brazil in 1962. Timbo virus (TIMV) was isolated from *Ameiva ameiva ameiva* lizards in Brazil. The optimal growth temperature of these viruses is approximately 30 °C.

## Sawgrass Group

New Minto virus (NMV) was isolated from *Haemaphysalis leporis-palustris* (Packard) ticks removed from snowshoe hares (*Lepus americanus* Erxleben) in east central Alaska in 1971. This virus is serologically (complement fixation and neutralization tests) related to Sawgrass virus. Sawgrass virus (SAWV) was isolated for the first time in Florida in 1964 from *Dermacentor variabilis* ticks removed from a raccoon, and later from *Haemophysalis leporis-palustris* ticks. Connecticut virus (CNTV) was first isolated in 1978 from a pool of nymphal *Ixodes dendatus* ticks removed from eastern cottontail rabbits (*Sylvilagus floridanus*) trapped in Connecticut, USA. Neutralizing antibodies have been detected in the eastern cottontail population in Connecticut, suggesting tick–rabbit maintenance cycle.

## Kern Canyon Group

Kern Canyon virus (KCV) was first isolated in 1956 from a pool of spleen and heart tissues from *Myotis yumanensis* bats in Kern County, California. KCV is not related to other groups or viruses classified in established genera according to the N gene phylogeny. Barur virus (BARV) was isolated from rodents (*Mus booduga*) and from ixodid ticks (*Haemaphysolis intermedia*) in 1961 in Mysore State, India and later, in 1971, from a pool of *Mansonia uniformis* mosquitoes in Kono Plain, Kenya. FUKAV was first grouped with KCV based on its antigenic properties. According to recent phylogenetic studies, it seems more related to the Le Dantec group. Nkolbisson virus (NKOV) was first isolated from *Aedes* sp., *Culex* sp., and *Eretmapodites* sp. mosquitoes in Cameroon in 1965. It was later isolated from Culicidea in the Ivory Coast, and from humans in the Republic of Central Africa.

## Other Groups as Proposed by Phylogenetic Analyses but Not Yet Recognized in the Eighth ITCV Report

### Almpiwar Group

Recent analysis of the L gene of different rhabdoviruses has suggested that Almpiwar virus (ALMV), Charleville virus (CHVV), Oak Vale virus (OVRV), and Humpty Doo virus (HDOOV) share a high genetic similarity, and they have been proposed to constitute the Almpiwar group.

ALMV (strain MRM4059) was isolated in 1966 from the skink *Ablepharus boutonii virgatus* at the low-lying plains of the Mitchell River Aboriginal community on the Gulf of Carpentaria in northern Queensland (**Figure 4**). Antibody to this virus has been detected in this species of skink. The virus optimally replicates at 30 °C in cell culture, which further supports the presumption that it is a reptilian virus. Although ALMV has never been isolated from arthropods, evidence of multiplication and passage of the virus in experimentally infected mosquitoes (*Culex fatigans*) supports the assumption that it is an arbovirus. No antigenic relationship has been found by complement fixation or neutralization test to any other known or suspected arboviruses. Although evidence indicates the presence of neutralizing antibodies to ALMV in sera from several different vertebrates, including humans, the significance of these results is considered uncertain. While ALMV and Charleville virus (CHVV) (see below) were each isolated from lizards (of different species) captured at Mitchell River, the viruses appear to share no serological relationship.

CHVV (strain Ch9824) was initially isolated in 1969 from sandflies (*Phlebotomus* spp.) collected near Charleville in southern Queensland. The following year, CHVV was isolated from the heart, liver, and lung of a lizard (*Gehyra australis*), and from a pool of biting midges (*Lasiohelea* spp.), each collected at Mitchell River in northern Queensland. Multiplication of this virus was demonstrated in experimentally infected mosquitoes (*A. aegypti*), supporting the possibility that it may be an arbovirus. CHVV was analyzed by complement-fixation test against all known Australian arboviruses (including ALMV, see above) but no relationships were found. Limited evidence suggests the presence of neutralizing antibody to CHVV in humans (1 of 30 sera tested).

OVRV (strain CSIRO1342) was isolated on nine occasions in 1984 from mosquitoes (*Culex edwardsi*) that were collected near a sentinel cattle herd located at Peachester near Brisbane in Queensland. OVRV was also isolated once from mosquitoes (*Aedes vigilax*) in Darwin, Northern Territory in January 1984. OVRV is not neutralized by any antiserum prepared against known Australian arboviruses. However, antiserum to this virus has been shown to cross-react with Kimberley virus (genus *Ephemerovirus*) by IFA. No antibodies to OVRV were detected in cattle sera collected at the same time from the same area, but neutralizing antibodies were found in feral pigs. Although it has been suggested that OVRV may be cycling between mosquitoes and an undefined avian host, this virus did not successfully replicate in cattle egrets, unlike Kununurra virus (KNAV) which is also suspected to have an avian host.

Humpty Doo virus (strains CSIRO79 and 80) was isolated between the years 1974 and 1976 from biting midges (*Culicoides marksi* and *Lasiohelea* spp.) that were collected at Beatrice Hill, near Darwin in the Northern Territory.

### Tibrogargan Group

At the L amino acid level, Tibrogargan virus (TIBV) is likely to be a member of the proposed 'dimarhabdovirus supergroup'. However, it appears to be considerably different from the other members of this supergroup. Complement fixation tests have indicated that Coastal Plains virus (CPV) is serologically related to TIBV, but cross-neutralization tests using rabbit antisera prepared to each of these viruses indicated that they are distinct.

TIBV (prototype strain CSIRO132) was isolated from a pool of biting midges (*Culicoides brevitarsis*) that were collected during a 2-week summer period in 1976 at a farm near Peachester, near Brisbane, Queensland. Shortly after isolation, analysis of the virus by complement fixation and hemagglutination inhibition tests did not reveal any relationships with other known or suspected arboviruses from Australia, Papua New Guinea, or elsewhere in the world. However, further studies have indicated that TIBV and CPV are antigenically related, but distinguishable by VNT tests. Subsequent tests performed using sera collected during the mid-1970s indicated the presence of neutralizing antibodies to TIBV in cattle from New Guinea, and in a region from northern Australia spanning as far south as the central coast of New South Wales. This distribution mirrors the geographical distribution of the biting midge *C. brevitarsis*. Some sentinel cattle herds have been found to be up to 100% seropositive for neutralizing antibodies. Neutralizing antibodies were also found in water buffalo in far north Australia, but no evidence of neutralization has been found in sera from a range of other animal species including humans. Despite the high prevalence of neutralizing antibodies in the tested cattle and water buffalo, there are no records of the direct isolation of

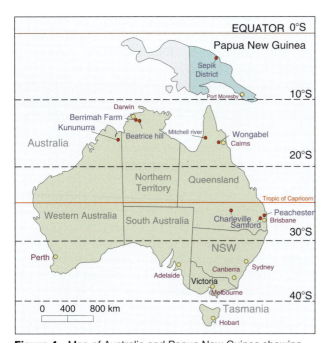

**Figure 4** Map of Australia and Papua New Guinea showing locations from which unassigned animal rhabdoviruses were isolated.

this virus from any vertebrate, and it has not been directly linked with disease.

CPV (strain DPP53) was isolated from the blood of a healthy steer (*Bos taurus*) in 1981 at Coastal Plains Research Station near Darwin in the Northern Territory. Although TIBV was isolated from biting midges, CPV was isolated directly from cattle and it is therefore not conclusively an arbovirus. However, the geographic distribution of neutralizing antibodies to CPV in cattle sera in Australia and Papua New Guinea appears to correspond to the distribution of the biting midge *C. brevitarsis*, which is the known vector of TIBV. Neutralizing antibody to CPV has also been detected in the sera of water buffalo, dogs, and one horse, but not in sera from humans, deer, pigs, or wallabies collected in the same areas.

## Mount Elgon Bat Virus Group

Nucleoprotein gene sequence analysis has indicated that Mount Elgon Bat virus (MEBV) and Oita virus (OITAV) form a distinct clade. Although they do not group by this analysis with the other vesiculoviruses, MEBV is currently classified as a tentative species of the genus *Vesiculovirus*. OITAV (296/1972) was isolated in 1972 from the blood of a wild horseshoe bat *Rhinolophus cornutus* (*Temminck*) in Japan. This virus causes lethal encephalitis in mice through the intracerebral route. MEBV was first isolated in 1964 in Kenya from bats (*Rhinolophus hildebrandtii*).

## Kolongo and Sandjimba Group

Kolongo virus (KOLV) and Sandjimba virus (SJAV) form a monophyletic clade based on analysis of the central region of the N gene. KOLV and SJAV were isolated in 1970 in Central African Republic from birds (*Euplected afra* and *Acrocephalus schoenbaeus*, respectively). Tupaia virus (TUPV) (a tentative species of the genus *Vesiculovirus* according to the Eighth ITCV Report) has been shown to join the same clade, although quite distantly.

## Other Nonassigned Rhabdoviruses

'Sigma virus' (SIGMAV) is the agent responsible for $CO_2$ sensitivity in *Drosophila melanogaster*. It is a noncontagious rhabdovirus which is transmitted through gametes.

Rochambeau virus (RBUV) was first isolated from mosquitoes (*Coquillettidia albicosta*) in French Guiana in 1973. It is classified as a tentative species of the genus *Lyssavirus* in the Eighth ICTV Report and is not related to any other dimarhadovirus, according to the phylogenetic analysis of the central region of the nucleoprotein.

Joinjakaka virus (strain MK7837) (JOIV) was isolated in 1966 from a pool of mixed mosquitoes (*Culicines*) that were aspirated from human bait at Joinjakaka, Sepik River District of Papua New Guinea. No relationship has been found between JOIV and any other known or putative arboviruses by complement fixation tests. Evidence suggests that cattle in Queensland have tested positive for antibody to JOIV by serum neutralization tests, but this study is not well documented.

Koolpinyah virus (KOOLV) was isolated from heparinized blood collected from two bulls (*Bos indicus-taurus* cross) at Berrimah Farm near Darwin in the Northern Territory in 1985 (strain DPP819) and 1986 (strain DPP883). In 1985, three additional bulls in the herd located at Berrimah developed neutralizing antibody to KOOLV, as did a sheep experimentally inoculated with blood from one of the bulls. KOOLV has been reported to be related serologically to Parry Creek virus (PCRV) and kotonkan virus (KOTV) by cross-neutralization tests, and to KOTV, PCRV, Obhodiang virus, and SJAV by indirect immunofluorescent antibody tests.

Kununurra virus (strain OR194) (KNAV) was isolated in 1973 from a pool of mosquitoes (*Aedeomyia catasticta*) that were collected in the Ord River Valley near Kununurra, Western Australia. Although KNAV is accepted as a new virus, no other serologically related rhabdovirus has since been identified. Under experimental conditions, KNAV has been shown to multiply in cattle egrets. Additionally, the mosquito host from which KNAV was isolated has a bird-feeding preference. Collectively, these results suggest that this virus may circulate in birds.

Several other apparently novel viruses with morphological characteristics of rhabdoviruses have been isolated from hematophagous insects in Australia but little or no information is available on their serological relationships to other viruses (**Table 2** and **Figure 1**). Isolate CS1056 was obtained in 1981 from a pool of biting midges (*C. austropalpalis*) collected at Samford near Brisbane in southeast Queensland. Isolate DPP1163 (tentatively named Holmes Jungle virus) was obtained in 1987 from mosquitoes (*C. annulirostris*) collected at Palm Creek near Darwin in the Northern Territory. Neutralizing antibody to this virus has been detected in cattle, buffalo, and humans, but there is no known association with disease. Isolate CSIRO75 (tentatively named Harrison Dam virus) was isolated in 1975 from mosquitoes (*C. annulirostris*) at Beatrice Hill, near Darwin in the Northern Territory. No information is available on the prevalence of antibodies in domestic or native animals. Isolate OR559 (tentatively named Little Lily Creek virus) was isolated from mosquitoes (*C. annulirostris*) collected at Kununurra in the Ord River region of Western Australia in 1974. Isolate OR1023 (tentatively named Ord River virus) was obtained in 1976, also from mosquitoes (*C. annulirostris*) collected at Kununurra. No other information is yet available about these viruses. Recent evidence suggests that KNAV might not be a rhabdovirus, but further studies are required to confirm this.

Two unassigned rhabdoviruses were isolated from birds during surveillance for arboviral encephalitis in the

Table 2  Representative isolates of the genera and groups of rhabdoviruses used for the phylogenetic analysis of the L gene

| Genus | Name | UA /TS[a] | Abbreviation | Species from which it was isolated and from which neutralizing antibodies were identified | Origin[b] | Year of first isolation |
|---|---|---|---|---|---|---|
| Nucleorhabdovirus | Rice yellow stunt virus | | RYSV | Leafhopper | | |
| | Sonchus yellow net virus | | SYNV | Aphid | | |
| | Maize mosaic virus | | MMV | Leafhopper (*Peregrinus maidis*) | | |
| Cytorhabdovirus | Northern cereal mosaic virus | | NCMV | Leafhopper (*Laodelphax striatellus*) | Japan | |
| | Strawberry crinkle virus | | SCV | Aphid (*Fragaria* spp.) | Chile | |
| | Taastrup virus | UA | TaasV | Leafhopper (*Psammottetix alienus*) | Denmark | |
| Novirhabdovirus | Infectious hematopoietic necrosis virus | | IHNV | Rainbow trout (*Onchorynchus mykiss*)/ Invertebrate reservoirs? | | |
| | Viral hemorragic septicemia virus | | VHSV | Rainbow trout (*Onchorynchus mykiss*)/ Invertebrate reservoirs? | | |
| | Snakehead rhabdovirus | TS | SHV | Sneakhead fish (*Ophicephalus striatus*) | | |
| | Hirame rhabdovirus | UA | HirR | | | |
| Ephemerovirus | Adelaide River virus | | ARV | *Bos taurus* | Australia | 1981 |
| | Berrimah virus | | BRMV | *Bos taurus* | Australia | 1981 |
| | Kimberley virus | TS | KIMV | *Bos taurus* | Australia | 1980 |
| | Kotonkan virus | UA | KOTV | *Culicoides* species | Nigeria | 1967 |
| | Bovine ephemeral fever virus | | BEFV | *Bos taurus, An. Bancrofti* | Australia, China | 1968 |
| | Obodhiang virus | UA | OBOV | *Mansonia uniformis* | Sudan | 1963 |
| Almpiwar Group | Humpty Doo virus | UA | HDOOV | *Lasiohelea* species, *Culicoides marski*. cattle | Australia | 1975 |
| | Charleville virus | UA | CHVV | *Phlebotomus* species (sand fly), *Lasiohelea* species (biting midge) lizard (*Gehyra australis*), human | Australia | 1969 |
| | Almpiwar virus | UA | ALMV | *Ablepharus boutonii virgatus*, skink, cattle, horse, sheep, kangaroo, bandicoot, birds, human | Australia | 1966 |
| | Oak-Vale virus | UA | OVRV | *Culex* species, *Culex edwardsi, Aedes vigilax*, ferral pigs | Australia | 1981 |
| Tibrogargan Group | Tibrogargan virus | UA | TIBV | *Culicoides brevitarsis*, water buffaloes, cattle | Australia | 1976 |
| | Coastal Plains | UA | CPV | Cattle | Australia, Papua New Guinea | 1981 |
| Hart Park Group | Parry Creek virus | UA | PCRV | *Culex annulirostris* | Australia | 1972 |
| | Hart Park virus | UA | HPV | *Culex tarsalis*, birds | USA | 1955 |
| | Wongabel virus | UA | WONV | *Culicoides austropalpalis*, sea birds | Australia | 1979 |
| | Flanders virus | UA | FLANV | *Culiseta melanura, Culex pipiens quinquefasciatus, Cx. salinarus, Cx. territans, Cx. restuans, Cx. tarsalis, Seiurus aurocapillus*, birds | New York, USA | 1961 |

Continued

**Table 2** Continued

| Genus | Name | UA /TS[a] | Abbreviation | Species from which it was isolated and from which neutralizing antibodies were identified | Origin[b] | Year of first isolation |
|---|---|---|---|---|---|---|
| | Ngaingan virus | UA | NGAV | Culicoides brevitarsis, wallaby, kangaroo, cattle | Australia | 1970 |
| | Mossuril virus | UA | MOSV | Culex sitiens, Culex decens, Culex perfusus, Culex pruina, Culex telesilla, Culex weschei, Culex tigripes, birds (Andropadus virens and Cliuspassrer maccrourus) | Mozambique, Central African Republic | 1959 |
| | Kamese virus | UA | KAMV | Culex annulioris, Aedes africanus, Culex decens, Culex perfuscus, Culex pruina and Culex tigripes | Uganda, Central African Republic | 1967 |
| Le Dantec and Kern Canyon Group | Le Dantec virus | UA | LDV | Human | Senegal | 1965 |
| | Fukuoka virus | UA | FUKV | Culicoides punctatus, calves | Japan | 1982 |
| | Keuraliba virus | UA | KEUV | Rodents (Tatera kempi, taterillus sp.) | Senegal | 1968 |
| Vesiculovirus | Perinet virus | TS | PERV | Mosquitoes: Anopheles coustani, Culex antennatus, Culex gr pipiens, Mansonnia uniformis Others: Phlebotomus berentensis | Madagascar | 1978 |
| | Vesicular stomatitis virus New Jersey | | VSNJV | Sus scrofa; Bos taurus, Culex nigripalpus, Culicoides species, Mansonia indubitans | USA | 1949 |
| | Vesicular stomatitis virus Indiana | | VSIV | Bos taurus | USA | 1925 |
| | Spring viremia of carp virus | TS | SVCV | Cyprinus carpio | Yougoslavia | 1971 |
| Lyssavirus | Mokola virus | | MOKV | Cat | Zimbabwe | 1981 |
| | Lagos bat virus | | LBV | Bat (Eidolon helvum) | Nigeria | 1956 |
| | European Bat Lyssavirus subtype 1 | | EBLV-1 | Bat (Eptesicus serotinus) | | |
| | European Bat Lyssavirus subtype 2 | | EBLV-2 | Bat (Myotis daubentonii, Myotis dasycneme) | | |
| | Duvenhage virus | | DUVV | Human | Rep. South Africa | 1986 |
| | Australian bat lyssavirus | | ABLV | Human, bat (Pteropus species) | Australia | 1996 |
| | Rabies virus | | RABV | | | |

[a]UA, unassigned species and unclassified viruses; TS, tentative species.
[b]Precise location of isolates from Australia and Papua New Guinea is given on **Figure 4**.

northeastern United States. Rhode Island virus strains RI-166 and RI-175 were each isolated from brain tissue of dead pigeons (*Columba livia*) collected at two localities in Rhode Island in summer 2000. Farmington virus designated CT-114 was originally isolated from an unknown wild bird captured in central Connecticut in 1969. Both viruses infect birds and mice, as well as monkey kidney cells in culture.

## Future Perspectives

The list of viruses described here is not complete and more viruses will certainly be characterized in the near future. Although there is strong phylogenetic support for the dimarhabdovirus supergroup, the precise branching order within this group cannot be resolved on the L or

N gene data. Indeed, there is a clear need for further phylogenetic studies within the dimarhabdovirus supergroup, particularly with respect to the demarcation of genera, which currently is influenced more by genome structure than host/vector relationships. There is some evidence that some of these viruses contain additional genes that are not present in lyssaviruses and vesiculoviruses. Although the functions of these additional proteins are not understood, revealing the evolution of genome complexity may be an important factor in resolving the taxonomy of this supergroup.

*See also:* Chandipura Virus; Rabies Virus.

## Further Reading

Bourhy H, Cowley JA, Larrous F, Holmes EC, and Walker PJ (2005) Phylogenetic relationships among rhabdoviruses inferred using the L polymerase gene. *Journal of General Virology* 86: 2849–2858.

Calisher CH, Karabatsos N, Zeller H, *et al.* (1989) Antigenic relationships among rhabdoviruses from vertebrates and hematophagous arthropods. *Intervirology* 30: 241–257.

Fu ZF (2005) Genetic comparison of the rhabdoviruses from animals and plants. In: Fu ZF (ed.) *Current Topics in Microbiology and Immunology, Vol. 292: The World of Rhabdoviruses*, p. 1. Berlin: Springer.

Hogenhout SA, Redinbaugh MG, and Ammar ED (2003) Plant and animal rhabdovirus host range: A bug's view. *Trends in Microbiology* 11: 264–271.

Karabatsos N (1985) *International Catalogue of Arboviruses Including Certain Other Viruses of Vertebrates*. San Antonio, TX: American Society of Tropical Medicine and Hygiene.

Kuzmin IV, Hughes GJ, and Rupprecht CE (2006) Phylogenetic relationships of seven previously unclassified viruses within the family *Rhabdoviridae* using partial nucleoprotein gene sequences. *Journal of General Virology* 87: 2323–2331.

## Relevant Websites

http://www.pasteur.fr – CRORA Report. This report involves all the data collected by Pasteur Institute and ORSTOM, since 1962, more than 6000 isolated strains of 188 arboviruses or mixed arboviruses. For each virus identified by the CRORA, all the observed hosts or vectors are given, with the number of collected strains in each country, viral properties of collected strains, and bibliographical references concerning them.

http://www.ncbi.nlm.nih.gov – ICTV database on *Rhabdoviridae*, International Committee on Taxonomy of Viruses, NCBI.

# Antigen Presentation

**E I Zuniga, D B McGavern, and M B A Oldstone,** The Scripps Research Institute, La Jolla, CA, USA

© 2008 Elsevier Ltd. All rights reserved.

## Introduction

The immune system is responsible for the tremendous task of fighting a wide range of pathogens to which we are constantly exposed. This system can be broadly subdivided in innate and adaptive components. The innate immune system exists in both vertebrate and invertebrate organisms and represents the first barrier against microbial invasion. This arm of the immune system rapidly eliminates the vast majority of microorganisms that we daily encounter and is responsible for limiting early pathogen replication. The adaptive response is a more sophisticated feature of vertebrate animals involving a broad repertoire of genetically rearranged receptors that specifically recognize microbial antigens (antigen is a generic term for any substance that can be recognized by the adaptive immune system). The hallmark of the adaptive response is the generation of a potent and long-lasting defense specifically directed against the invading pathogen.

B and T lymphocytes represent the effector players of adaptive immunity and carry on their surface antigen-specific receptors, B-cell receptors (BCRs) and T-cell receptors (TCRs), respectively. There are two major classes of T lymphocytes: CD8 cytotoxic and CD4 helper T cells. Upon antigen encounter, lymphocytes undergo clonal expansion and differentiation of their unique functional features. B cells differentiate into plasma cells and secrete antibodies that specifically bind the corresponding antigen. CD8 T cells directly kill infected cells or release cytokines that interfere with viral replication, while CD4 T cells activate other cells such as B cells and macrophages. Unlike B cells, which can directly bind native free antigen, T cells only recognize antigen-derived peptides displayed on cell surfaces in the context of major histocompatibility complex (MHC) class I (MHC-I, CD8 T cells) or class II (MHC-II, CD4 T cells) molecules.

Different pathogens preferentially replicate in distinct cellular compartments. While viruses and intracellular bacteria replicate in the cytosol, microbes such as mycobacterium and protozoan parasites are intravesicular and colonize the endosomal and/or lysosomal compartments. In addition, extracellular bacteria release antigens, such as toxins, that are engulfed by antigen-presenting cells (APCs) to also reach the endosomal pathway. Antigenic peptides derived from these sources are exhibited on cell surfaces by MHC molecules. This process, which represents the major focus of this article, is named 'antigen presentation' and is a fundamental pillar of antimicrobial host defense.

## Antigen-Presenting Cells

For initiation of an immune response, naive T cells need to be activated or 'primed'. For that, they require both the recognition of the specific MHC–peptide complex (signal 1) and simultaneous co-stimulation (signal 2). Although all nucleated cells express MHC-I and can potentially display MHC-I –microbial peptide complexes after infection, only a specialized group of leukocytes, named APCs, expresses both MHC-I and MHC-II as well as co-stimulatory molecules. The best-characterized co-stimulatory molecules are B7-1 and B7-2, which bind to the CD28 molecule on the T-cell surface. In addition, T cells express CD40 ligand, which interacts with CD40 on APC further enhancing co-stimulation and enabling T-cell response. Finally, there is another group of adhesion molecules such as lymphocyte function-associated antigen-1 (LFA-1) on APCs which binds to ICAM-1 on T cells that seal the APC–T-cell interface. During APC–T-cell interactions, all these molecules cluster together forming a highly organized supramolecular adhesion complex (SMAC), enabling the intimate contact between the two cells that is referred to as the immunological synapse (**Figure 1**).

APCs are composed of macrophages, B cells, and dendritic cells (DCs). They differ in location, antigen uptake, and expression of antigen-presenting and co-stimulatory molecules. Macrophages are localized in connective tissues, body cavities, and lymphoid tissues. Within the secondary lymphoid tissues, macrophages are mainly distributed in the marginal sinus and medullary cords. They specialize in phagocytosis and engulf particulate antigens through scavenger germline receptors such as the mannose receptor. On the other hand, B cells form follicular structures within secondary lymphoid organs and recirculate through the blood stream and lymph seeking their specific antigen. B cells recognize antigens specifically through a rearranged BCR. DCs are the most professional and robust of the APCs. They are widely distributed through the body at an 'immature' stage of development, acting as sentinels in peripheral tissues. They continuously sample the antigenic environment by both phagocytosis and macropinocytosis,

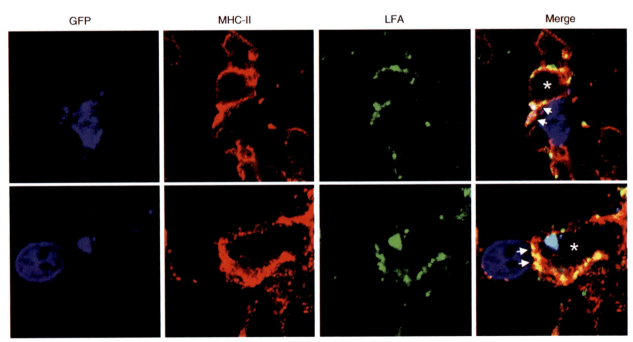

**Figure 1** Interactions between virus-specific T cells and APCs. Three-color confocal microscopy was used to demonstrate immunological synapse formation between lymphocytic choriomeningitis virus-specific T cells (blue) and MHC-II$^+$ APCs (red) in the central nervous system. Immunological synapses were indicated by the polarization of the adhesion molecule LFA-1 (green) between the CTL and APC. Asterisks denote the engaged APC, and arrows denote the contact point between the two cells. LFA-1 is expressed on both CTLs and APCs, but note that all of the CTL-associated LFA-1 is focused toward a contact point at the CTL–APC interface. Reproduced from Lauterbach H, Zúñiga EI, Truong P, Oldstone MBA, and McGavern DB (2006) Adoptive immunotherapy induces CNS dentritic cell recruitment and antigen presentation during clearance of a persistent viral infection. *Journal of Experimental Medicine* 203 (8): 1963–1975, with permission from Rockefeller University Press.

which is the engulfment of large volume of surrounding liquid. Within the secondary lymphoid organs, some DCs strategically localize within T-cell areas where they can optimally encounter circulating naive T lymphocytes that actively scan the DC network.

APCs are able to detect components of invading pathogens which trigger their activation/maturation. Specifically, pathogen-associated molecular patterns (PAMPs), as these components are termed, range from lipoproteins to proteins to nucleic acids carried by potential invaders. These PAMPs are recognized by evolutionary conserved 'pattern recognition receptors' (PRRs) on APCs. Among PRRs, the Toll-like receptors (TLRs) have emerged as critical players in determining APC imprinting on the ensuing immune response. TLR triggering has pleiotropic effects on APCs, promoting survival, chemokine secretion, expression of chemokine receptors, migration, cytoskeletal and shape changes, and/or endocytic remodeling. After interacting with these pathogen signatures, the microbial antigens are processed and presented as peptides associated with MHC molecules and activated APCs upregulate both antigen-presenting and co-stimulatory molecules initiating a 'maturation' process. As part of this process, APCs in peripheral tissues change their chemokine receptors and initiate migration to secondary lymphoid organs where the adaptive immune response is initiated.

The strategic migration and location of DCs into T-cell areas of the secondary lymphoid organs coupled to their superior antigen-presenting capacity make them the most powerful APCs. Indeed, DCs are about 1000 more efficient than B cells or macrophages in stimulating naive T cell. This has been shown by several experiments in which elimination of DCs prevented the initiation of antigen-specific T-cell responses. Interestingly, DCs are a heterogeneous cell population composed of different subtypes which present unique and overlapping functions. As many as six different subsets of DCs occupy the lymph nodes. Three major defined populations of DCs have been recognized in mouse spleen and humans: CD8+ conventional DCs (cDCs), CD11b+ cDCs, and plasmacytoid DCs. These subpopulations differ not only in surface phenotype but also in functional potential and localization. In this regard, cDCs are potent activators of naive T lymphocytes, as CD8+ DCs are believed to be specialized in cross-presentation of exogenous antigens. A recent study suggests that CD8+ and CD11b+ cDCs differ from each other in their intrinsic antigen-processing capacity being specialized in MHC-I and MHC-II antigen presentation, respectively. In contrast, plasmacytoid DCs are poorer activators of T cells, even after stimulation *in vitro*. They likely play a more protagonist role during innate immunity by secreting specific cytokines and chemokines, such as type 1 interferons (IFNs an important antiviral mediator), and activation of a broad range of effector cells, such as natural killer (NK) cells. Thus, the heterogeneity inherent to DC populations significantly influences the varieties of immune responses to different pathogens, which are subsequently amplified by cross talk between the various subsets.

## Major Pathways of Antigen Presentation

Although in all healthy individuals MHC molecules play the same crucial role of antigen presentation, these molecules are highly polymorphic. There are hundreds of different alleles encoding the MHC molecules in the whole population and each individual exhibits only few of them. The major allelic variants of MHC are found in key amino acids forming the peptide-binding cleft. Thus, although a given MHC-I molecule can bind several different peptides; particular amino acids are preferred in certain positions of the peptide resulting in differential peptide sets for particular MHC variants. Importantly, T-cell specificity involves co-recognition of a particular antigenic peptide together with a particular MHC variant, a feature known as T-cell MHC restriction.

Like other polypeptide chains of proteins destined to arrive at the cell surface, MHC molecules are translocated to the lumen of the endoplasmic reticulum (ER) during synthesis. In this compartment, the subunits of MHC molecules are assembled together and the peptide-binding groove or cleft is formed. However, MHC molecules are unstable in the absence of bound peptide. In the following sections, we will consider how MHC molecules are folded and generated peptides are bound to MHC-I or MHC-II molecules. After binding, MHC–peptide complexes travel to the cell surface where they are recognized by antigen-specific T cells. Although not discussed in this article, it should be noted that other MHC-like molecules (i.e., CD1) also display peptide and lipid antigens contributing to antigen presentation, especially during mycobacteria infections.

### MHC-I Antigen Presentation

MHC-I molecules are expressed in most if not all nucleated cells. MHC-I molecules are heterodimers of a highly polymorphic α-chain (43 kDa) that binds noncovalently to β2-microglobulin (12 kDa), which is nonpolymorphic. The α-chain contains three domains. The α3 domain crosses the plasma membrane while the α1 and α2 domains constitute the antigen-binding site. The peptides that bind the MHC-I molecule are usually 8–10 amino acids long and contain key amino acids at two or three positions that anchor the peptide to the MHC pocket and are called anchor residues.

As mentioned above, the peptide-binding site of MHC molecules is formed in the ER. However, all proteins, including viral-derived antigens, are synthesized in the cytosol. Numerous studies in the last years outlined the molecular events connecting the antigen generation in the cytosol with the peptide binding to the MHC-I molecule in the ER. A highly conserved multicatalytic proteasome complex is in part responsible for cytosolic protein degradation into small peptides. The proteasome contains 28 subunits forming a cylindrical structure composed of four rings, each of seven units. Under normal conditions, the proteasome complex exists in a constitutive form. During viral infections, IFNs released by cells of the innate immune system induce the synthesis of three different proteasome subunits, which replace their constitutive counterparts to form the immunoproteasome. This inflammatory form of the proteasome favors the production of peptides with a higher chance of MHC binding. Moreover, IFNs can also enhance the rate of proteasome peptide degradation increasing the availability of peptides and reducing their excessive cleavage. It is important to highlight that other cytosolic proteases also contribute to MHC-I peptide generation and further cleavage can occur within the ER before MHC binding. The source of peptides for MHC-I complexes still holds its secrets. Proteasome substrates may encompass *de novo* synthesized, mature stable, and/or defective proteins. It is believed that defective ribosomal products (DRiPs), which are proteins targeted for degradation due to premature termination or misfolding, constitute an important source of MHC-I peptides.

Peptides available in the cytosol are transported into the lumen of the ER by ATP-dependent transporters-associated antigen-processing 1 and 2 (TAP-1 and TAP-2) proteins. TAP proteins are localized in the ER membrane forming a channel through which peptides can pass. Within the ER, the newly synthesized MHC-I α-chain binds to a chaperone molecule called calnexin, which retains the incomplete MHC molecule in the ER. After binding to the β2-microglobulin, calnexin is displaced and the emerging MHC molecule binds to a loading complex composed by the chaperone protein calreticulin, TAP, the thiol oxidoreductase Erp57, and tapasin, which bridges MHC-I molecule and TAP. After peptide binding, the fully folded MHC-I molecule and its bound peptide are released from the complex and transported to the cell membrane. Importantly, under steady-state conditions, the MHC-I molecules in ER are in excess with respect to peptides allowing the rapid appearance of microbial peptides onto the cell surface during infection. However, since MHC-I molecules are unstable without bound peptide, they also present self antigens under normal conditions. Because of the absence of microbial signatures, antigen presentation of self peptides by inactivated/immature APCs leads to T-cell tolerance rather than activation. This is one of the important ways anti-self or autoimmune responses are controlled.

For several years, intracellular peptides were believed to be the only source of MHC-I molecules. However, it is now clear that exogenous proteins also have access to the cytosolic compartment and bind MHC-I in the ER. This mechanism is known as cross-presentation and is believed to be particularly important for enabling MHC-I presentation by cells that are not directly infected by the virus but instead are engulfing viral particles by phagocytosis or micropinocytosis. The molecular mechanism by which MHC-I molecules access exogenous peptides is of considerable interest. Different nonexclusive possibilities have been proposed, including sampling of phagosome-generated peptides by MHC-I molecules, transference of ER molecules (including MHC-I and its loading complex) into phagosomes, re-entry of plasma membrane MHC-I into recycling endosomes with the subsequent peptide exchange, and finally the acquisition of peptides from other cells through GAP junctions.

## MHC-II Antigen Presentation

The MHC-II molecule is composed by two noncovalently bound transmembrane glycoprotein chains, α (34 kDa) and β (29 kDa). Each chain has two domains and altogether form a four-domain heterodimer similar to the MHC-I molecule. α1 and β1 domains form the peptide-binding cleft resulting in a groove which is open at the ends, which is different from the MHC-I groove in which the extremes of the peptide are buried at the ends. Peptides that bind to MHC-II are larger than those that bind to MHC-I molecules, being 13–17 amino acids long or even much longer.

Since MHC-I is a surface protein, its biosynthesis is initiated in the ER. To prevent newly synthesized MHC-II molecules from binding cytosolic peptides that are abundant in the ER, its peptide-binding cleft is covered by a protein known as MHC-II-associated invariant chain (Ii). Through a targeting sequence in its cytoplasmic domain, the Ii also directs MHC-II molecules to acidified late endosomal compartments, where Ii is cleaved leaving only the Ii pseudopeptide (CLIP) covering the peptide-binding groove. MHC-II molecules bound to CLIP cannot bind other peptides, indicating that CLIP must be dissociated or displaced by the antigenic peptide.

Proteins that enter the cell through endocytosis or are derived from pathogens that replicate in vesicles are degraded by endosome proteases. These proteases become activated as the endosome pH progressively decreases. The final set of peptides available in the endosomal compartment is a result of antigen processing by several acid proteases that exist in endosomes and lysosomes. For instance, the cathepsin S is a very predominant acid protease and mice deficient in this enzyme have a compromised antigen-processing capacity. Vesicles carrying peptides fuse with the vesicles carrying MHC-II molecules, achieving CLIP dissociation and the incorporation of antigenic peptides to MHC-II molecules. An MHC-II-like molecule that is predominant in the endosome facilitates this process. This molecule contributes to 'peptide editing', removing weakly bound peptides and assuring that the emerging MHC-II–peptide complexes are stable enough to be scanned by CD4 T cells.

MHC-II molecules seem to be in excess and are rapidly degraded unless microbial peptides become available to fill the groove. This excess is important to permit MHC-II availability upon infection. However, during infection, APCs are exposed to both self and microbial peptides. How APCs discriminate between self and nonself represents a fundamental question in immune biology. Recent evidence suggests that the efficiency of presenting antigens from phagocytosed cargo is dependent on the presence of a TLR ligand within the cargo. Thus, TLR signaling would mark a particular phagosome for an inducible mode of maturation dictating the fate of the cargo-derived peptides and favoring their presentation by MHC-II molecules in a phagosome autonomous fashion.

Because they travel through the endocytic pathway, which can be considered a topological continuation of the extracellular space, MHC-II molecules were believed to be specialized in the presentation of exogenous antigens. However, the analysis of MHC-II peptidome revealed many peptides of cytosolic or even nuclear origin. Autophagy or 'self-eating' explains MHC-II access to cytosolic peptides. This highly conserved pathway could be accomplished by several mechanisms including microphagy (when lysosomal invagination sequesters cytosolic componets), macrophagy (when a double membrane structure that encloses and isolates cytoplasmic components and eventually fuses with lysosome), and chaperone-mediated autophagy (when cytosolic proteases generate peptides that are transported into lysosomes).

## Viral Subversion of Antigen Presentation

Considering the crucial role of antigen presentation for host defense, it is not surprising that many viruses have evolved maneuvers to evade or divert this process. Particularly, the essential role played by APCs in host defense to pathogens makes them an ideal target for viruses to suppress the immune response, thereby maximizing their chances of survival, replication, and transmission. Indeed, many viruses that cause major health problems are able to interfere with the ability of APCs to prime an efficient and effective antiviral immune response. In fact, many viruses have developed different mechanisms to subvert each stage of APC biology. Furthermore, with the greater understanding of antigen presentation pathways comes the discovery of novel viral immune-evasion strategies. In this section, we illustrate selective viral strategies to subvert antigen presentation by describing particular cases.

An interesting example of virus blockade of antigen presentation from very initial steps is the ability of the prototypic arenavirus lymphocytic choriomeningitis virus (LCMV), to dramatically block DC development from early hematopoietic progenitors. Fms-like tyrosine kinase 3 ligand (Flt3L) is known to induce the expansion of undifferentiated progenitors into DCs within the spleen and bone marrow (approximate 20-fold increase), both in mice and humans. In contrast, LCMV-clone (CL)-13 that suppresses the immune response and causes a persistent infection in mice is associated with DC early progenitors that become refractory to the stimulatory effects of Flt3L.

TLRs function in APCs as an early sensor against pathogens; therefore, impairment in TLR signaling confers another selective advantage to certain infectious agents. As an example, vaccinia virus (VV) blocks TLR signaling and the subsequent maturation of APCs. Specifically, two proteins of VV suppress intracellular signaling of interleukin-1 (a potent pro-inflammatory host factor) and TLR-4.

Migration of DCs is a crucial step in initiating the adaptive immune response. Examples of viruses that have developed mechanisms to prevent migration of infected DCs to lymphoid organs are herpes simplex virus (HSV) 1 and human cytomegalovirus (HCMV). In both cases, there is an inhibition of complete DC maturation and subsequent expression of chemokine homing receptors. In addition, HCMV inhibits DC migration one step further by preventing APCs from arriving at a site of infection by producing homologs of chemokines that interfere with host pro-inflammatory chemokine gradients.

Another effective immune-evasion strategy used by viruses to disrupt APCs is the prevention of or interference with antigen-specific T-cell activation. The ability to disrupt MHC–peptide binding has evolved in many different virus species including adenovirus and human immunodeficiency virus (HIV). Herpesviruses have also evolved to block host cell antigen presentation. Some mechanisms utilized by herpesviruses to disrupt the antigen-presentation pathway include blocking peptide transport to the ER through interference with TAP proteins (HSV ICP47, HCMV US6), transport of particular MHC-I heavy chains from the ER to the cytosol where they are destroyed (HCMV US11, HCMV US2), retention of specific MHC-I heavy chains in the ER (HCMV US3, murine CMV–MCMV-m152), and disruption of T-cell recognition of MHC-I on the cell surface (MCMV m04). That viruses have independently evolved numerous mechanisms to disrupt MHC–peptide presentation indicates the effectiveness and importance of this strategy to the survival of viruses with different infectious life cycles.

Maturation of APCs results in upregulation of co-stimulatory molecules and expression of cytokines that enable them to stimulate naive T cells. Viruses that can impair T-cell stimulation by preventing the upregulation of co-stimulatory molecules include Ebola virus, Lassa fever virus (LFV), HSV-1, and HIV. Additionally, a number of viruses (hepatitis C virus (HCV), HIV, measles virus (MV), and dengue virus (DV)) are also able to inhibit interleukin (IL)-12 production, which is often required for effective T-cell response. HCV does this through the action of its core and nonstructural protein 3 (NS3), which induces production of IL-10. DV, on the other hand, is able to inhibit IL-12 production through an IL-10 independent mechanism. In addition, compelling evidence showed that *in vivo* persistent infection of mice with LCMV, as well as persistent HCV infection in humans, induces IL-10 production by APCs resulting in the blunting of the CD8 T-cell response and chronic viral infection. Remarkably, antibodies blocking IL-10/IL-10R interactions correct T-cell exhaustion by restoring T-cell function, which results in purging of virus from mice persistently infected with LCMV.

Finally, a novel immunosuppressive molecule, programmed cell death-1 (PD-1), is upregulated in nonfunctional CD8 T cells during chronic infections (LCMV, HIV, HCV). Interaction of PD-1 with PD-ligands on APCs (or parenchymal cells) inhibits lymphocyte activation. As for IL-10 blockade, antibodies interfering with PD-1/PDL interactions also promote viral clearance from a persistently infected host.

The fact that not all viruses are able to block APC maturation does not necessarily represent a failure of the pathogen or a success for the host. A good example of this is observed following MV infection that exploits the ability of DCs to mature and migrate to lymphoid organs in response to infection. MV benefits greatly by having infected DCs home to lymphoid compartments where the infected cells are able to actively suppress T-cell proliferation (mediated through T-cell contact with surface viral glycoproteins) and also facilitate virus spread to more lymphoid cells. Therefore, the full understanding of the virus–host relationship requires not only studying the active mechanisms that viruses use to disable the immune system, but by also asking how a virus benefits by not altering a particular immune function.

## Concluding Remarks

Co-evolution of certain hosts and pathogens for millions of years has resulted in a fine-tuned equilibrium that enables survival of both. Antigen presentation is one of the critical elements in this balance. While antigen presentation is an essential process for long-term effective host defense, targeting APCs represents a common maneuver of many viruses to avoid host surveillance and establish a chronic or persistent infection. A major challenge in biomedical research is to thwart microbial APC subversion to promote eradication of the pathogen. A better understanding of the mechanisms used by APC to display microbial antigens as well as the virus strategies to subvert APC functions during immune responses will provide new tools for designing novel vaccination approaches and immunotherapeutic treatments for human infectious diseases.

## Acknowledgments

This is publication no. 18909 from Molecular and Neuroscience Integrative Department, The Scripps Research Institute (TSRI). This work was supported by NIH grants AI 45927, AI 05540, and AI 09484.

E. I. Zuniga is a Pew Latin American Fellow.

*See also:* Cytokines and Chemokines; Immune Response to viruses: Antibody-Mediated Immunity; Immune Response to viruses: Cell-Mediated Immunity; Immunopathology; Persistent and Latent Viral Infection; Vaccine Strategies.

## Further Reading

Bevan MJ (2006) Cross-priming. *Nature Immunology* 7: 363–365.

Blander JM and Medzhitov R (2006) On regulation of phagosome maturation and antigen presentation. *Nature Immunology* 7: 1029–1035.

Dudziak D, Kamphorst AO, Heidkamp GF, *et al.* (2007) Differential antigen processing by dendritic cell subsets *in vivo*. *Science* 315: 107–111.

Itano AA and Jenkins MK (2003) Antigen presentation to naive CD4 T cells in the lymph node. *Nature Immunology* 4: 733–739.

Janeway CA, Travers P, Walport M, and Shlomchik MJ (2005) *Immunobiology: The Immune System in Health and Disease*. New York: Garland Science Publishing.

Lauterbach H, Zúñiga EI, Truong P, Oldstone MB, and McGavern DB (2006) Adoptive immunotherapy induces CNS dendritic cell recruitment and antigen presentation during clearance of a persistent viral infection. *Journal of Experimental Medicine* 203(8): 1963–1975.

Menendez-Benito V and Neefjes J (2007) Autophagy in MHC class II presentation: Sampling from within. *Immunity* 26: 1–3.

Norbury CC and Tewalt EF (2006) Upstream toward the 'DRiP'-ing source of the MHC class I pathway. *Immunity* 24: 503–506.

Oldstone MB (2007) A suspenseful game of 'hide and seek' between virus and host. *Nature Immunology* 8: 325–327.

Reis e Sousa C (2006) Dendritic cells in a mature age. *Nature Reviews Immunology* 6: 476–483.

Shen L and Rock KL (2006) Priming of T cells by exogenous antigen cross-presented on MHC class I molecules. *Current Opinion in Immunology* 18: 85–91.

Shortman K and Liu YJ (2002) Mouse and human dendritic cell subtypes. *Nature Reviews Immunology* 2: 151–161.

Strawbridge AB and Blum JS (2007) Autophagy in MHC class II antigen processing. *Current Opinion in Immunology* 19: 87–92.

Yewdell JW and Nicchitta CV (2006) The DRiP hypothesis decennial: Support, controversy, refinement and extension. *Trends in Immunology* 27: 368–373.

# Antigenic Variation

**G M Air and J T West**, University of Oklahoma Health Sciences Center, Oklahoma City, OK, USA

© 2008 Elsevier Ltd. All rights reserved.

## Glossary

**Envelope protein (Env)** The surface spike of HIV-1.
**Gp120** The subunit of HIV-1 Env that binds to receptors on cells to initiate infection.
**Gp41** The subunit of HIV-1 Env that carries fusion activity.
**Hemagglutinin** Influenza surface glycoprotein that binds cell surface receptors to initiate infection.
**Neuraminidase** Influenza surface glycoprotein that cleaves sialic acid receptors to spread virus to new cells.
**Neutralizing antibody** An antibody that blocks virus infection.

## Manifestations of Antigenic Variation

Antigenic variation refers to the observation that different isolates of a single virus species may show variable cross-reactivity when tested with a standard serum. The homologous virus (the isolate that was used to raise the antiserum) usually shows the highest reactivity. Cross-reactivities with other viruses of the same species may vary from high to zero. While zero cross-reaction is due to a different type or subtype of the species, intermediate reactivities define antigenic groups or serotypes.

Immune response to viruses is generally thought of as negative selection – that is, elimination of the pathogen. There are a few examples where immune recognition has a positive influence on the continued proliferation of the recognized pathogen, due to viral mechanisms to exploit immune elimination. Some viruses can use Fc receptors or mannose receptors to internalize antibody-bound virus into a replication mode instead of a destructive one. Alternatively, the virus may sabotage the immune response by changing antigenicity. A primary basis of antigenic variation is selection of virus mutants by antibodies. These are known as escape mutants, and since the escaped virus is resistant to antibody neutralization it possesses a fitness advantage in the presence of antibody. Escape mutants may also be selected by CD4+ or CD8+ T cells. There is certainly variation in the T-cell epitopes, but a lack of definitive examples that these have been selected by T cells as an immune evasion mechanism in the same way as occurs with antibody selection.

There are several mechanisms that allow mutations to occur with sufficiently high frequency to be selected by antibodies. Viruses with an RNA genome show the highest degrees of antigenic variation. The mutations are not induced by antibody but are present in the population and can be selected out by antibodies when wild-type virus particles are neutralized. The origin of this rather high rate of mutation is the viral replicase. RNA polymerases have no $3'$ editing function and so insertion of an occasional mismatched base is not corrected and quasispecies with random mutations are present in any virus population. The mystery of RNA viruses is not why they show so much antigenic variation, but why some RNA viruses are antigenically almost invariant. Rhinoviruses exist in over 100 serotypes while poliovirus, another picornavirus, is stable enough that the vaccine did not need updating during the WHO Global Eradication campaign. Dengue virus exists in four serotypes and shows large variations in antigenic cross-reactivity within each serotype, but another flavivirus, yellow fever virus, is antigenically stable. Respiratory syncytial virus exists as two distinct antigenic groups with high antigenic diversity within each group, but other paramyxoviruses, such as measles, mumps, and rubella, are stable and therefore amenable to easier vaccination protocols.

It is clear from the above observations that vaccine success is inversely related to the degree of antigenic variation in the targeted virus. The measles-mumps-rubella vaccine has reduced those diseases almost to zero in the developed world and where vaccine compliance is high, but there is still no vaccine against respiratory syncytial virus. It does not follow that an antigenically stable virus necessarily leads to a successful vaccine, since viruses have evolved many different mechanisms to evade the immune system. Some viruses remain hidden from antibodies – for example, in neural tissues. Some make cytokine mimics that block immune signaling, or code for proteins that inhibit signaling pathways. But it is antigenic variation that causes as yet insurmountable barriers to making effective vaccines for many pathogens where the diversity is high. If all antigenic groups are represented in the vaccine, the dose becomes extremely large with consequent danger of adverse side effects. The recently licensed human papillomavirus vaccine contains the three genotypes considered oncogenic, but difficulties in culturing human papillomaviruses has precluded a serological classification of the more than 100 genotypes that exist and it remains unclear how broadly effective the current vaccine will be in areas where different or unique subtypes co-circulate.

DNA viruses do not generally incorporate mutations during replication. DNA polymerases can only extend from a base that is correctly hydrogen bonded to its partner on the template strand. A mismatched nucleotide causes synthesis to stop until the wrong base is excised by the 3′ nuclease activity. Nevertheless, DNA viruses have accumulated mutations during evolution, and some have a high rate of recombination that leads to antigenic diversity. Some small DNA viruses are quite diverse in sequence (papillomaviruses) and canine parvovirus has evolved different antigenic properties since it first appeared in the 1970s. There is evidence that the Epstein–Barr and Kaposi's sarcoma herpesviruses undergo some degree of antigenic variation.

There are distinct patterns of antigenic diversity among viruses. Most exist in several serotypes that 'rotate' in the human population, resulting in waves of disease. As antibodies accumulate in the population against one serotype, another can move in because it is not neutralized by those antibodies. Presumably, there is little antigenic memory against these viruses, because even with >100 serotypes of rhinoviruses, a person would eventually become exposed to all of them. Alternatively, the lack of immunity against all serotypes could be because the immune response focuses on single or a few immunodominant but noncross-reactive epitopes (a phenomenon known as original antigenic sin) or rapid clearance of the antigen may result in failure to achieve affinity maturation of the antibody response. Influenza viruses, on the other hand, are constantly evolving into new antigenic variants and the old ones never seem to return. It is not clear if other viruses, even very variable RNA viruses, undergo the same style of antigenic drift as influenza: a progressive, unidirectional evolution. The most variable virus of all is human immunodeficiency virus (HIV), the cause of AIDS. During HIV infection each infected individual generates a unique swarm of virus variants known as a quasispecies that diversify in a radial fashion. For many virus species, the pattern of variation is not well understood, due to infrequent epidemics, geographic isolation, or lack of sufficiently extensive molecular analysis. Here we compare and contrast the antigenic variability of the most-studied viruses, influenza virus and HIV.

## Antigenic Variation in Influenza Viruses

Influenza viruses are classified as types A, B, or C based on cross-reactivity of internal antigens. Type A influenza viruses, those most commonly associated with human infection, show two distinct mechanisms of antigenic variation: antigenic shift and antigenic drift. Antigenic shift is a replacement of one subtype of surface antigen with another. There are two surface glycoprotein antigens, hemagglutinin (HA) and neuraminidase (NA) that exist in multiple serotypes. To date there are 16 subtypes of HA (H1–H16) and nine subtypes of NA (N1–N9), based on lack of antigenic cross-reactivity. H1, H2, H3, N1, and N2 have been found in human epidemic viruses. All subtypes are found in avian influenza viruses and thus birds are considered the natural viral reservoir, providing new antigens that occasionally are transferred into human viruses. The segmented influenza genome facilitates this replacement, since a new antigen gene can be readily exchanged into an existing human virus during a mixed infection without the need for RNA recombination. However, avian viruses do not easily infect mammals, so one hypothesis is that such mixing can only occur in a species that can host both of the parental viruses, such as the pig. Pigs have not been implicated in the transmission of avian influenza H5N1 to humans, which continues to be a rare event, and so far no reassortant viruses with H5 HA have been found. Antigenic shifts are shown in **Figure 1**. In recent years influenza viruses circulating in humans have been H1N1, H3N2, and type B.

Antigenic drift is a less dramatic form of antigenic variation but is responsible for annual epidemics of influenza. The term refers to a gradual change in serological cross-reactivity when compared to the original pandemic virus. Drift has been continuous in H3 HA since 1968, in N2 NA since 1957, and in H1 and N1 since 1977. Drift is detected using ferret antiserum, which discriminates better than antisera of other species, raised against a particular H3N2 strain. There is some degree of cross-reactivity to viruses isolated before and after the appearance of that strain, typically over a period of about 10 years, but for better protection, the vaccine strain is changed as soon as cross-reactivity is decreased.

The progressive, unidirectional changes in antigenic cross-reactivity correlate with progressive accumulation of changes in amino acid sequence. **Figure 2** shows the sequence changes that have occurred in the H3 HA since it appeared in the human population in 1968. Comparison of isolates in any given year shows the same antigenic properties and virtually the same sequence worldwide. This is commonly attributed to air travelers carrying the virus across and between continents, but it is also possible that the same mutations have been independently selected in different places.

## Mechanism of Antigenic Drift

The mechanism of antigenic drift in influenza viruses is relatively well understood, although it does not allow one to predict the next epidemic strains. The development of monoclonal antibodies, and the demonstration that these can be used to select antigenic variants that escape from antibody neutralization by a single amino acid change, led to the mapping of neutralizing epitopes on both HA and

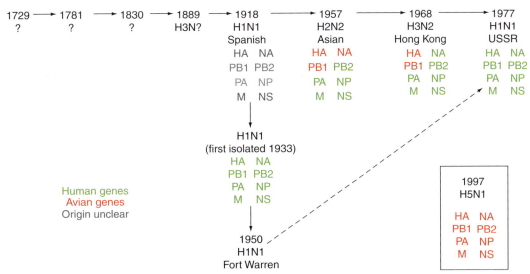

**Figure 1** Influenza pandemics are very distinctive in their epidemiology and descriptions can be found back to ancient times. The figure shows some well-described pandemics (antigenic shifts) of more recent times. Retrospective serology of samples taken before the H3N2 pandemic of 1968 from people who had lived through the 1889 pandemic showed the presence of H3 antibodies. The virus was first isolated in 1933 and antibodies raised against it showed no break back to the pandemic of 1918, classified as H1N1. The H2 HA and N2 NA introduced in 1957 were both of avian origin, as was the H3 HA in 1968. The return of H1N1 in 1977 is not regarded as a pandemic since people over the age of 20 still had antibodies. The genes have been sequenced from the 1918 virus and many authors say they were of avian origin, but in the absence of an essentially identical avian influenza sequence from the same time period, this conclusion is not substantiated. The H5N1 virus that has occasionally infected people since 1997 is wholly avian in its genes. Potentially it could provide new antigens to human influenza viruses, or mutate to spread more widely into the human population.

NA. Crystal structures of HA and NA bound to neutralizing antibodies show that the amino acids that change in escape mutants are within the epitope as defined by amino acids that contact the antibody. Potentially, escape mutants could have amino acid sequence changes outside the epitope causing a conformational change that is transmitted to the epitope, but this has not been observed. The reason is that a widespread conformational change would almost certainly affect the function of the HA or NA, debilitating the virus, and the selection of escape mutants can only occur if the mutants are close to wild type in fitness.

The neutralizing epitopes, determined crystallographically, consist of 11–18 amino acids that are separated in the linear sequence but come together in the three-dimensional structure to form a binding surface that exists only when the antigen is folded into its native structure (**Figure 2**). The footprint of an antibody on HA and NA is quite large and so it is not surprising that antibodies contact more than one linear segment of the protein. Many years of effort have not yielded a peptide-based vaccine for influenza and it seems unlikely that this will be possible. In contrast, for HIV there is still very active research ongoing to develop peptide-based vaccines or therapeutic antibodies that bind to linear peptides, because the surface antigen gp160 may be more dynamic than influenza HA or NA and linear segments may become transiently accessible to antibodies (see below).

## Mechanisms of Neutralization

The basis of antibody-mediated neutralization of influenza virus (and HIV, see below) is generally considered to be blocking of biological function rather than Fc-mediated mechanisms such as phagocytosis or complement activation. The most commonly used assay to assess effectiveness of vaccination is the hemagglutination inhibition (HAI) test, which measures blocking of the receptor-binding site by antibodies. This might seem to be an oversimplification of many potential mechanisms of neutralization, but the correlation between HAI and neutralization or protection has held up for several decades. Similarly, although NA antibodies are not as much studied, monoclonal antibodies do not neutralize or protect from infection unless they inhibit the NA activity. Thus, neutralizing antibodies against influenza bind close enough to the sialic acid binding site (HA) or the enzyme active site (NA) to inhibit binding to receptors or substrates. The binding sites themselves have constraints on tolerating changes because they must retain function, but the polypeptide loops surrounding them are able to mutate so that the neutralizing antibody can no longer bind, or binds with such low affinity that it no longer inhibits function. Site-directed mutagenesis of all side chain contacts of the epitope has shown that a subset contributes the major energy of the interaction. These two or three critical contacts are the ones that are found to change in escape mutants, and are the only changes in site-directed mutants that abolish antibody inhibition. In a few cases, the

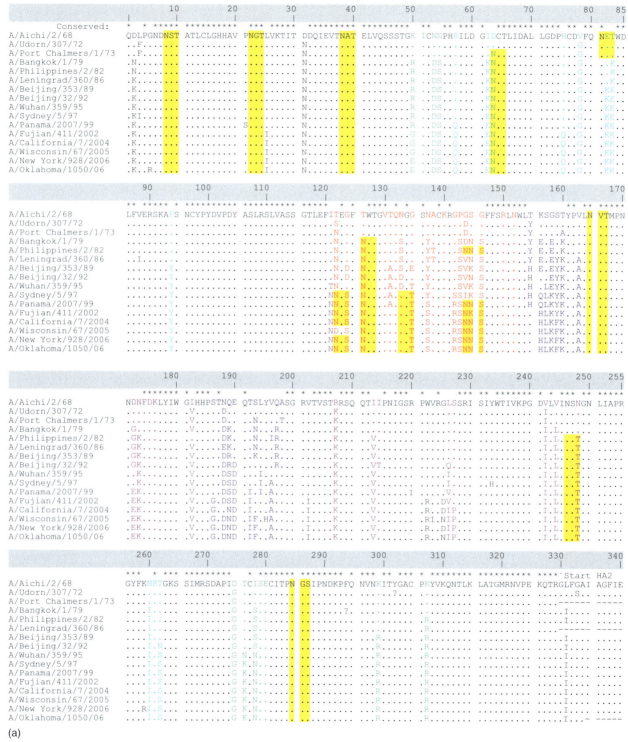

**Figure 2** Continued

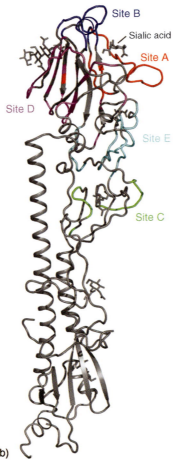

**Figure 2** (a) The amino acid sequence of A/Aichi/2/68, representative of the first human H3N2 isolates. The HA precursor peptide is cleaved into two polypeptides called HA1 and HA2. Only HA1 is shown here since there are no known antigenic sites on HA2 and it shows very little sequence drift. Changes in HA1 of later viruses are shown, with dots where there is no change. Mutations that map in the five antigenic regions are colored: red, site A; blue, site B; green site C, magenta, site D; cyan, site E. The assignment of residues to antigenic sites is accessible at the Influenza Sequence Database (ISD). Sequons that predict N-linked glycosylation are shown as yellow bands. The progressive nature of antigenic (and genetic) drift is clearly seen; previous mutations are generally retained in later isolates. The sequences chosen were in most cases used as vaccine components, and essentially the same sequence (and antigenic character) was found worldwide. Two 2006 isolates are included for comparison. (b) Alpha carbon trace of a monomer of HA showing one or more sugars at each site of N-linked glycosylation and a bound sialic acid in the receptor binding site. Each peptide segment of the antigenic sites in (a) are seen to come together in the three-dimensional structure. Antigenic sites A through E are colored as in (a). Site D is partly buried in the trimer interface when the whole molecule is present. Site C is some distance from the receptor binding site but the crystal structure of a complex with antibody shows that the constant domain of the antibody reaches to and obscures the sialic acid binding site.

antibody still binds but does not inhibit function, but most commonly there is no detectable binding of antibody to a mutant that has a change in a critical contact. This explains how a change in only one out of the 11–19 amino acids that make up the epitope can eliminate the antibody inhibition, and greatly reduces the number of changes needed for a new antigenic variant.

## How Does Antigenic Drift Occur?

The surface area of the HA accommodates a multitude of epitopes that have been grouped into antigenic sites A through E (**Figure 2**) by determining if binding of one antibody blocks the binding with another. If it does, the two antibodies are considered to belong to the same site. In the face of a human polyclonal antibody response, it would seem necessary to mutate each of the five distinct sites to generate an escape mutant. The early H3 variants show changes in all sites, but more recently the changes have been confined to fewer sites. The difference is likely to be in immunogenicity, and suggests that the human antibody response to recent viruses is not very polyclonal. The possible reasons include previous substitutions rendering a site nonimmunogenic, or blocking of sites by the addition of N-linked carbohydrates. A/Aichi/68 HA1 contained six carbohydrate chains but by 1999 there were 11 sites for carbohydrate addition. Site A may now be completely shielded from antibodies. The mobilization and addition of carbohydrate is a primary mechanism of immune avoidance in HIV (see below). The antigenically significant transition from California/04 to Wisconsin/05 was due to changes only in site B.

### Is There Significant Antigenic Drift in NA?

Antibodies against NA have been shown to contribute to protection. They do not inhibit entry of human viruses into cells but they inhibit spread of progeny viruses because sialic acid is not removed from the N-linked carbohydrates of the surface glycoproteins, causing aggregation of virus particles by HA–HA interaction. The rate of amino acid substitution in NA is about half that in HA, suggesting a lesser antigenic selection. Ferret antisera that show clear differences in HA antigenicity from 1 year to the next may not discriminate between drifted NAs. However, the crystal structure of the NA of A/Memphis/31/98 complexed with a neutralizing antibody suggests that NA may be contributing significantly to antigenic drift in H3N2 viruses. Of the 11 amino acids that make contact through side chains to the antibody, five have changed in isolates up to 2006 (**Figure 3**).

## Antigenic Variation in HIV

### Antibody and Antigenic Variation in HIV-1 Infection

Nearly all HIV-1-infected individuals develop antibodies capable of mediating some level of neutralization. The presence of strong, broadly neutralizing responses in some long-term nonprogressors suggests that neutralizing antibodies may be a correlate of protection, but their absence during acute infection, when the escalating HIV-1 viral load is reduced, suggests they play a minor role in initial immune control and that cellular immune responses are more effective. HIV-1 neutralizing responses take months or even years to develop in patients but during a long and chronic infection, antibody may be critical in limiting both virus-to-cell and cell-to-cell spread of virus. It is possible that non-neutralizing responses such as antibody-dependent cell-mediated cytotoxicity (ADCC) are important in acute infection.

In contrast to influenza, there seems to be no basis for a serological definition of subtypes in HIV-1. Instead, subtypes are defined based on genetic relatedness. HIV-1 exhibits intra-subtype sequence differences of up to 20% in the envelope glycoprotein gene (*env*) and differences of up to 35% between subtypes. Only antibodies directed against the viral envelope glycoprotein that limit or prevent virus entry into CD4-bearing target cells are effective in reducing viral replication. HIV-1 is also selected by cellular immune mechanisms, and recent studies suggest that CTLs target Env more often than previously appreciated; however, here we will focus on the antigenic variation in Env and its interplay with humoral immune selection.

### HIV-1 Evolves to Create Global and Intra-Patient Antigenic Variation

HIV-1, a member of the genus *Lentivirus* of the subfamily *Orthoretrovirinae* of the family *Retroviridae*, is one of the most rapidly evolving pathogens ever studied. Through a combination of rapid production of highly genetically diverse progeny and the extreme plasticity of many of its viral proteins, HIV-1 has managed to elude efforts aimed at control through drug treatment or vaccination. The most recent common ancestor of the HIV-1 pandemic subtypes came into the human population less than a century ago. Over the course of that relatively short interval, the virus has become not only one of the most significant human pathogens, but also one of the most diverse. The pandemic HIV-1 can be segregated into subtypes A–K (minus E, I, and K) (**Figure 4**). Most subtypes co-circulate in West-Central Africa pointing to this region, where the geographical habitat of chimpanzees and humans overlap, as a nexus for the phylogenetically inferred initial transmission events into humans. Circulating recombinant forms (CRFs) are rapidly emerging in areas where more than one subtype co-circulates. These recombinants pose a serious concern, as they can be thought of as analogous to reassortants in influenza and could rapidly obviate the effectiveness of vaccination unless sufficiently broad cross-reactivity can be generated. Antigenic

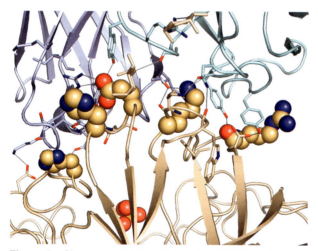

**Figure 3** Changes in the epitope of a 1998 N2 NA up to 2006. The NA is brown, Fab H chain is dark blue, and L chain is light blue. Amino acid side chains that are in contact in the interface are shown as stick models, except for those on NA that have changed in later isolates, which are shown in space-filling form. Reproduced from Venkatramani L, Bochkareva E, Lee JT, *et al.* (2006) An epidemiologically significant epitope of a 1998 human influenza virus neuraminidase forms a highly hydrated interface in the NA-antibody complex. *Journal of Molecular Biology* 356(3): 651–663 (cover picture), with permission from Elsevier.

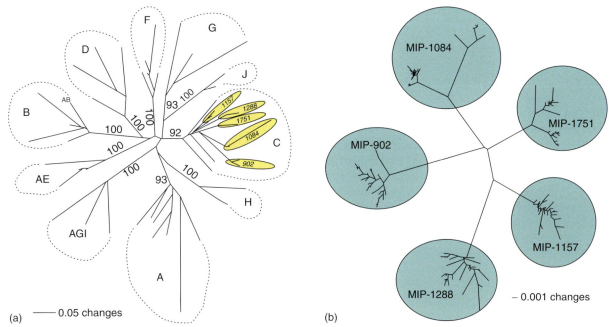

**Figure 4** (a) Shows that the various HIV-1 subtypes segregate one from another when compared using envelope glycoprotein sequences and that individual viruses cluster within the subtype into distinct branches. (b) An unrooted phylogenetic tree shows that each of five mother–infant pair's glycoprotein sequences form an independent lineage despite all being subtype C viruses isolated from the same geographic region. Over time, infant and maternal sequences will also segregate as they undergo independent viral antigenic variation and selection. HIV-1 diversification appears perpetual.

variation within the global HIV-1 population, as well as within each infected individual, is a primary mechanism underlying the success of HIV-1 as a human pathogen.

## The HIV-1 Replicase and the Generation of Diversity/Variation

In order to understand the mechanisms that produce variation, it is necessary to become familiar with some of the basic processes of retroviral replication. HIV-1 contains two copies of message (+)-sense RNA. Replication of the HIV-1 genome proceeds through a dsDNA intermediate that becomes permanently integrated into the target cell genome. The HIV-1 replicase is an RNA-dependent DNA polymerase or reverse transcriptase (RT). The HIV-1 RT lacks any proofreading capacity. The rate of misincorporation of nucleotides by HIV-1 RT is between $10^{-3}$ and $10^{-4}$/position/round of replication. Given a genome size of approximately 10 000 base pairs this suggests that HIV-1 introduces at least one mutation with each genome copy.

## HIV-1 Diversity Influences Its Population Genetics

The viral loads in plasma for individuals undergoing acute infection are as high as $10^8$ copies ml$^{-1}$. This means that the virus has the potential to produce all possible single mutations each day, as well as a population containing multiple mutations. The replication rate is accompanied by rapid viral and infected cell turnover, pressure to avoid immune recognition and elimination, all coupled with fitness maximization. The population of closely related but distinct viruses generated by replication is referred to as a quasispecies. For HIV, this term refers to the dynamic group of viral genotypes found in an infected individual at a given sampling. It is the creation of this swarm of subtly different progeny viruses and the resultant selection of those progeny that drives HIV natural selection and antigenic variation.

In addition to single-base mutations, HIV readily undergoes recombination but this does not occur through reassorting genomic segments as in influenza. HIV-1 does not possess a segmented genome, but instead packages two copies of viral genomic RNA. If co-infection of target cells and co-packaging of distinct RNA genomes occurs, reverse transcriptase can switch viral RNA templates during synthesis of DNA. The result is the creation of large insertions and deletions that introduce antigenic changes in a single replication cycle. While co-infection has proved difficult to monitor *in vivo*, the globally increasing number of HIV-1 CRFs suggests that co-infection occurs with relatively high frequency. In the absence of a vaccine the abundance of such recombinants will only increase given the continuation of human cultural, medical, and sexual practices that promote virus transmission, and the occurrence of infection rates in the developing world as high as 35%. Both vaccine development and drug treatment efforts will be further complicated as these recombinants become more widely intermixed and distributed.

## The HIV-1 Envelope Glycoprotein: Variable Loops and Decoys

The target of HIV-neutralizing responses is the envelope glycoprotein. The glycoprotein complex exists in the virion membrane as a trimer of noncovalently associated heterodimers of surface protein (SU) and transmembrane protein (TM). The gp120 component is the most exposed component of the complex and contains five hypervariable loops (V1–V5) and five relatively invariant or constant regions (C1–C5). Vaccination with gp120 peptides, glycoprotein, or glycoprotein-expression constructs has met with only limited success, as defined by elicitation of broadly neutralizing antibody responses. Neutralization of HIV-1 correlates with blocking virus entry into cells. The diversity of HIV sequences means that the vast majority of epitopes are not shared and thus the only hope for vaccination is to target conserved epitopes. Much of what we know about Env epitopes that are cross-reactive comes from studies of a handful of human monoclonal antibodies that have broad neutralizing activity for multiple viral strains and subtypes. Structural studies of some of these antibodies in complex with gp120 have revealed their binding sites and it is now clear that they do not bind to linear epitopes but to conserved structural or conformational epitopes that the virus has evolved to rarely reveal to the immune system (**Figures 5** and **6**).

HIV-1 entry is a series of steps mediated through interactions between Env and receptors on the target cell

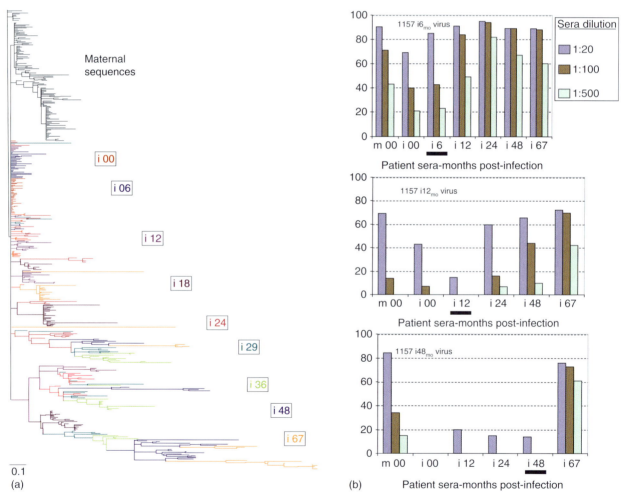

**Figure 5** (a) A maximum-likelihood phylogenetic tree displaying the diversifying evolution of HIV-1 Env in a single individual. The numbers in color indicate the months post infection when samples were obtained for sequence analysis. Note that divergence from a maternal root sequence continues with time and that at later time points the virus appears to resample earlier genotypes. (b) Neutralization of viruses derived from some of the time points in (a) by autologous contemporaneous and noncontemporaneous sera suggests that recognition of early (6 month) viruses is maintained even though the genotype is clearly replaced, and that the neutralizing response generated early is incapable of neutralizing virus generated after selection (12 month serum is ineffective on 12 month virus or subsequent time points). In total, these data suggest that diversity generates an antibody response that is never 100% effective, not broadly reactive and perhaps mistargeted by 'original antigenic sin' as discussed in the text.

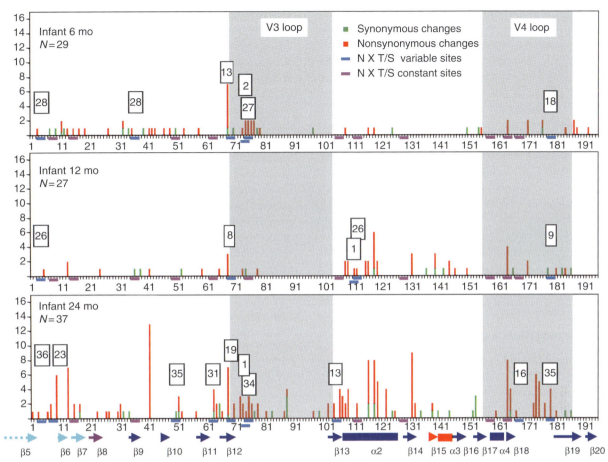

**Figure 6** Amino acid sequence changes found in the HIV-1 subtype C Envelope glycoprotein in a newly infected child. C2-V4 diversity (amino acids 228–427 in HXB2 numbering) for sequences in a single patient ($N$ = number of sequences per time point) in relation to potential glycosylation sites that are either constant in the population at that time point (blue) or show variability (pink). The proportion of the glycans maintained, for variable sites, is shown in the boxes above the sites. Red bars indicate the presence of nonsynonymous changes (codon level) and their height indicates the abundance in the population. Green bars indicate synonymous polymorphism. Note that variability increases with time, sites undergoing positive selection (red bars) change with time in some instances but not in others, and that glycan addition sites change both in abundance and position over the course of infection. Designations at the bottom of the figure shows relationships to structural elements identified in the unliganded SIV envelope glycoprotein compared with HIV-1 structures PDB id 2BF1.

plasma membrane. The first of these steps is a high-affinity interaction between gp120 and the primary receptor CD4. Thus, antibodies that interfere with CD4 binding would be anticipated to be effective at inhibiting infection. The interaction with CD4 induces substantial conformational changes in Env that lead to exposure of the chemokine co-receptor (usually CCR5 or CxCR4) binding site in the V3 loop and other induced structures. The CD4-induced conformation is another step at which it is thought that neutralizing antibodies might exert a selective and protective influence. The interaction with CD4 and chemokine co-receptor induces further changes in the gp120 structure that promote exposure of gp41 and insertion of its N-terminal fusion domain into the target cell membrane. This final step is also potentially a target for recognition of epitopes and the development of fusion-inhibiting antibody responses. However, despite considerable antigenic diversity, the HIV-1 glycoprotein has evolved to be an exceedingly poor neutralization immunogen; at least it is a very poor inducer of broadly neutralizing responses particularly at conserved sites most pertinent to the functional processes of Env defined above.

Some of the hypervariable loops of gp120 tolerate extensive sequence variation as well as insertion and deletion, and are excellent inducers of antibody, but rarely neutralizing antibody. In the event a loop-induced neutralizing response is generated, it is isolate specific and not cross-reactive; the virus utilizes the diversity it generates through replication to escape the neutralizing response and repopulate. Thus such loop-targeted antibodies are unlikely to have durable neutralizing capacity even in the individual. It is therefore likely that at least one function of the loops is to decoy the immune response, misdirecting it to portions of the protein that can continually support variation to no discernable functional detriment for the virus. In support of this concept, HIV-1 with deletion of V1–V2 is still replication competent but is more susceptible to neutralization.

The CD4 binding site is conserved in all HIV subtypes and is contained within a deep and inaccessible pocket of the protein that is inaccessible to antibodies (the canyon hypothesis). Ab that could interact with Env to prevent CD4 association would be neutralizing. However, it appears that the loops also serve to mask important targets for neutralization such as the CD4 binding site and CD4-induced epitopes such as the co-receptor binding regions by sterically inhibiting the access of antibody to potential epitopes or by limiting the affinity with which such Ab can associate. The loops are also extensively glycosylated and this also impacts recognition as discussed below.

### Env Glycosylation and Antibody Recognition

HIV-1 Env gp120 and gp41 have been shown to be extensively modified by an average of ~30 carbohydrate additions for both proteins, and glycosylation of the HIV-1 Env is essential for folding and oligomerization. Moreover, glycosylation participates in protecting HIV-1 from recognition by antibodies. This concept, termed glycan shielding, suggests that the concentration of carbohydrate additions on gp120 block antibody access to the protein domains beneath that are critical for CD4 recognition, co-receptor binding, and exposure of the fusion peptide. Highly conserved glycans often play important structural and functional roles, but others are highly polymorphic between subtypes and within infected individuals. Thus, variability in the protein sequence in or around glycosylation sites affects immune recognition by subtly altering the primary sequence, the fold of the structure, and/or accessibility to epitopes. The continual diversification leads to a dynamic of appearing and disappearing glycosylation sites followed by humoral selection and outgrowth of the escape mutant.

It is important to realize that the nature of the shield itself is also a component of protection. Unlike the highly immunogenic cell wall components in prokaryotes (e.g., lipopolysaccharide), the sugars in HIV-1 are synthesized and modified as host components and, therefore, indistinguishable from carbohydrate additions to a multitude of other membrane proteins. Vaccine strategies that stimulate recognition of glycans will need to take into consideration targeting what is essentially 'self' carbohydrate. Some broadly neutralizing anti-HIV-1 envelope monoclonal antibodies have polyspecific or self-reactivity to host antigens but are not routinely made due to elimination by host B-cell tolerance mechanisms.

### Concluding Remarks

Though influenza virus and HIV both show significant antigenic variation in response to neutralizing antibody, the mechanisms they use to generate variation are both similar and distinct. Influenza virus diversification by drift is unidirectional whereas HIV tends to generate radial diversity. Both viruses undergo shift to induce completely novel antigens; influenza virus by reassorting genomic segments and HIV by RT-mediated recombination. A new influenza variant spreads through the human population whereas a new HIV variant merely repopulates the infected individual. Both viruses utilize carbohydrate additions to avoid the antibody response, but in HIV the number of modifications is more substantial. In influenza, variation is utilized to avoid antibody binding whereas HIV diversity promotes recognition of epitopes lacking functional significance, and therefore non-neutralizing responses.

*See also:* AIDS: Vaccine Development; Antigenicity and Immunogenicity of Viral Proteins; Immune Response to viruses: Antibody-Mediated Immunity; Influenza; Neutralization of Infectivity; Orthomyxoviruses: Structure of antigens; Quasispecies; Vaccine Strategies.

### Further Reading

Air GM, Laver WG, and Webster RG (1987) Antigenic variation in influenza viruses. *Contribution Microbiology and Immunology* 8: 20–59.
Hammond AL, Lewis J, May J, Albert J, Balfe P, and McKeating JA (2001) Antigenic variation within the CD4 binding site of human immunodeficiency virus type 1 gp120: Effects on chemokine receptor utilization. *Journal of Virology* 75: 5593–5603.
Knipe DM, Howley PM, Griffin DE, *et al.* (2001) *Field's Virology*, 4th edn. Philadelphia: Lippincott Williams and Wilkins.
Korber B, Gaschen B, Yusim K, Thakallapally R, Kesmir C, and Detours V (2001) Evolutionary and immunological implications of contemporary HIV-1 variation. *British Medical Bulletin* 58: 19–42.
Skehel JJ and Wiley DC (2000) Receptor binding and membrane fusion in virus entry: The influenza hemagglutinin. *Annual Review of Biochemistry* 69: 531–569.
Venkatramani L, Bochkareva E, Lee JT, *et al.* (2006) An epidemiologically significant epitope of a 1998 human influenza virus neuraminidase forms a highly hydrated interface in the NA-antibody complex. *Journal of Molecular Biology* 356(3): 651–663.
Wilson IA and Cox NJ (1990) Structural basis of immune recognition of influenza virus hemagglutinin. *Annual Review of Immunology* 8: 737–771.
Yusim K, Kesmir C, Gaschen B, *et al.* (2002) Clustering patterns of cytotoxic T-lymphocyte epitopes in human immunodeficiency virus type 1 (HIV-1) proteins reveal imprints of immune evasion on HIV-1 global variation. *Journal of Virology* 76: 8757–8768.

### Relevant Websites

http://www.hiv.lanl.gov – A series of freely accessible reviews on HIV-1 diversity, drug resistance and immunology.
http://www.flu.lanl.gov – The Influenza Sequence Database (ISD). To see the sequence diversity in influenza viruses.

# Antigenicity and Immunogenicity of Viral Proteins

**M H V Van Regenmortel**, CNRS, Illkirch, France

© 2008 Elsevier Ltd. All rights reserved.

## Glossary

**B-cell epitope** Surface region of a native protein antigen that is specifically recognized by the binding sites of free and membrane-bound antibody molecules. The membrane-bound antibodies are the B-cell receptors that recognize the antigen during the immunization process.

**Cryptotope** Antigenic site or epitope hidden in polymerized proteins and virions because it is present on the surfaces of subunits that become buried. Cryptotopes are antigenically active only after dissociation of protein aggregates and virions.

**Mimotope** A peptide possessing similar binding activity as that of a peptide epitope but showing little or no sequence similarity with it. Originally, mimotopes were defined as peptides able to bind an antibody but showing no sequence similarity with the protein antigen used to induce the antibody, usually because the antibody was directed to a discontinuous epitope.

**Neotope** Antigenic site or epitope specific for the quaternary structure of polymerized proteins or virions. Neotopes arise from the juxtaposition of residues from neighboring subunits or through conformational changes in the monomers resulting from intersubunit interactions.

**Paratope** Binding site of an antibody molecule that binds specifically to an epitope of the antigen. Paratopes are constituted of residues from six complementarity-determining regions (CDRs) located on the heavy and light chains of immunoglobulins. The CDRs vary greatly in sequence and in length in individual antibodies.

## Virus Antigenicity

The antigenic reactivity or antigenicity of viruses corresponds to their capacity to bind specifically to the functional binding sites of certain immunoglobulin molecules. Once such binding has been observed experimentally, the particular immunoglobulin becomes known as an antibody specific for the virus.

The antigenicity of nonenveloped viruses resides in the viral proteins that form the viral capsid, whereas the antigenicity of enveloped viruses resides mostly in the exposed proteins and glycoproteins that are anchored in the viral, lipid membrane. The oligosaccharide side chains of viral glycoproteins contribute significantly to the antigenic properties of enveloped viruses.

The antigenic sites or B-cell epitopes of viral proteins correspond to those parts of viral capsids and envelope proteins that are specifically recognized by the binding sites or paratopes of free and membrane-bound antibody molecules. Antibody molecules anchored in the outer membrane of B cells correspond to the receptors of B cells that recognize the antigen when it is administered to a vertebrate host during immunization. The B-cell receptors recognize the native tertiary and quaternary structure of viral proteins and the antibody molecules that are released when the B cells have matured into plasmocytes, also recognize the native conformation of the proteins.

During a natural viral infection or after experimental immunization, the immune system of the host may also encounter dissociated viral protein subunits and it will elicit antibodies specific for these components. In addition, during a viral infection, nonstructural viral proteins that are not incorporated into the virions will also induce the production of specific antibodies in the infected host. Diagnostic tests that detect antibodies to nonstructural viral proteins are useful for differentiating animals infected with, for instance, foot-and-mouth disease virus from vaccinated animals that possess only antibodies directed to the capsid proteins of the virus. In this case, the ability to differentiate vaccinated animals from infected animals by a suitable immunoassay is an important prerequisite for convincing trading partners that a cattle-exporting country is free of foot-and-mouth disease.

## Viral Antigenic Sites

In the absence of further qualification, the term epitope used in the present article refers to an antigenic site of a protein recognized by B cells. Immune responses are also mediated by T cells, that is, by lymphocytes that recognize protein antigens through T-cell receptors, after the antigen has been processed into peptide fragments. T-cell responses to viral antigens which involve the so-called T-cell epitopes will not be discussed here.

Because of the additional antigenic complexity that arises in viral proteins as a result of the quaternary structure of virions, it is useful to distinguish two categories of viral epitopes known as cryptotopes and neotopes. Cryptotopes are epitopes that are hidden in the intact, assembled capsid. They are located on the surfaces of viral subunits that become buried when the subunits associate into capsids, and they are antigenically active only after dissociation or denaturation of virions. Neotopes are epitopes that are specific for the quaternary structure of virions and which are absent in the dissociated protein subunits. Neotopes arise either through conformational changes in the monomers induced by intersubunit interactions or result from the juxtaposition of residues from neighboring subunits. These terms are useful when it is important to distinguish between the epitopes carried by different aggregation states of viral proteins. The availability of monoclonal antibodies has made it easy to identify neotopes and cryptotopes in many viruses whose antigenic structure has been analyzed in detail. There is evidence, for instance, that the trimeric form of the gp160 protein of human immunodeficiency virus (HIV) possesses neotopes that are not present in the monomeric form but are important for the induction of neutralizing antibodies.

Another important category of viral epitope is the neutralization epitopes that are recognized by antibodies able to neutralize viral infectivity. Since it is the antibodies that bring about neutralization, it is appropriate to talk of neutralizing antibodies and not of neutralizing epitopes. These epitopes are best referred to as neutralization epitopes.

Infectivity neutralization depends on properties of the virus, the antibody, and the host cell. The ability of an antibody to interfere with the process of infection always takes place in a specific biological context and it cannot be described adequately only in terms of a binding reaction between virus and antibody. In many cases loss of infectivity occurs when the bound antibody molecules inhibit the ability of the virus to attach to certain receptors of the host cell. It can happen that the antibody prevents the virus from infecting one type of host cell but not another type.

Some picornaviruses such as the rhinoviruses occur in the form of as many as 100 different variants known as serotypes, each serotype being neutralized by its own antibodies but not by antibodies specific for other serotypes. In contrast, another picornavirus, that is, poliovirus, exists only as three serotypes. The structural basis for this difference in the number of neutralization serotypes between different picornaviruses has not yet been clarified.

## Antibodies to Viral Antigens

The capacity of antibodies to recognize viral epitopes resides in their functional binding sites or paratopes. The most common type of antibody is an immunoglobulin known as IgG which possesses two identical binding sites. These paratopes are constituted of three complementarity-determining regions (CDRs) located in a heavy chain and three CDRs in a light chain. The CDRs comprise a total of about 50–70 amino acid residues and form six loops that vary greatly not only in sequence but also in length from one antibody to another. Each individual paratope is made up of atoms from not more than 15–20 CDR residues which make contact with a specific epitope. This means that as many as two-thirds of the CDR residues of an antibody molecule do not participate directly in the interaction with an individual epitope. These residues remain potentially capable of binding to other epitopes that may have little structural resemblance with the first epitope, a situation which gives rise to antibody multi-specificity. The relation between an antibody and its antigen is thus never of an exclusive nature and antigenic cross-reactivity will be observed whenever it is looked for. The so-called antibody footprint corresponds to an area on the surface of the protein antigen of about 800 $\text{Å}^2$.

## Continuous and Discontinuous Epitopes

Protein epitopes are usually classified as either continuous or discontinuous depending on whether the amino acids included in the epitope are contiguous in the polypeptide chain or not (**Figure 1**). This terminology may give the impression that the elements of recognition operative in epitope–paratope interactions are individual amino acids, whereas it is in fact at the level of individual atoms that the recognition occurs. The distinction between continuous and discontinuous epitopes is not clear-cut since discontinuous epitopes often contain stretches of a few contiguous residues that may be able, on their own, to bind to antibodies directed to the cognate protein. As a result, such short stretches of residues may sometimes be given the status of continuous epitopes. On the other hand, continuous epitopes often contain a number of indifferent residues that are not implicated in the binding interaction and can be replaced by any of the other 19 amino acids without impairing antigenic activity. Such continuous epitopes could then be considered to be discontinuous.

Since discontinuous epitopes consist of surface residues brought together by the folding of the peptide chain, their antigenic reactivity obviously depends on the native conformation of the protein. When the protein is denatured, the residues from distant parts of the sequence that collectively made up the epitope are scattered and they will usually no longer be individually recognized by antibodies raised against the discontinuous epitope.

Discontinuous epitopes are often called conformational epitopes because of their dependence on the intact conformation of the native protein. However, this terminology may be confusing since it seems to imply that continuous epitopes, also called sequential epitopes, are conformation independent. In reality the linear peptides that constitute continuous epitopes necessarily also have a

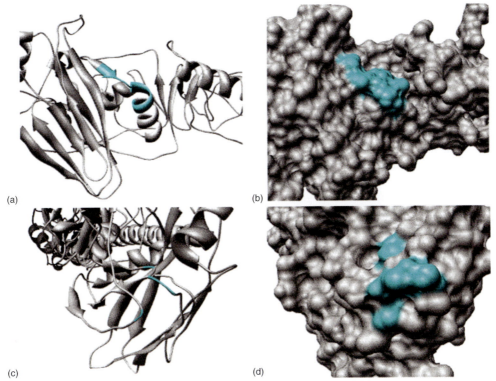

**Figure 1** Illustration of the difference between a continuous (a, b) and discontinuous epitope (c, d) in the hemagglutinin protein of influenza virus. The residues that are part of epitopes are colored in blue; the remaining residues are gray. (a) Ribbon representation of a continuous epitope. (b) Surface representation of the epitope shown in (a). (c) Ribbon representation of a discontinuous epitope. Residues distant in the sequence are brought close together by the folding. (d) Surface representation of the epitope in (c). Reproduced from Greenbaum JA, Andersen PH, Blythe M, et al. (2007) Towards a consensus on datasets and evaluation metrics for developing B cell epitope prediction tools. Journal of Molecular Recognition 20: 75–82, with permission from Wiley.

conformation which, however, is mostly different from that of the corresponding region in the intact protein.

Most of our knowledge of the structure of discontinuous epitopes has been obtained from a small number of X-ray crystallographic studies of antibody–antigen complexes. It is important to realize that the structure of epitopes and paratopes seen in the complex may be different from the structure of the respective binding sites in the free antigen and antibody molecules, that is, before they have been altered by the mutual adaptation or induced fit that occurs during the binding interaction. For this reason, the structure of epitopes after complexation tends to be an unreliable guide for identifying the exact epitope structure that was recognized by the B-cell receptors during the immunization process. Crystallographic studies of antigen–antibody complexes have shown that the vast majority of protein epitopes are discontinuous and consist of residues from between two and five segments of the polypeptide chain of the antigen.

Most of our knowledge of protein antigenicity has not been obtained by X-ray crystallography but was derived from studies of the antigenic cross-reactivity between the intact protein and peptide fragments. In such studies, antibodies raised against the virus or isolated viral proteins are tested for their ability to react in various immunoassays with natural or synthetic 6–20-residue peptides derived from the protein sequence. Any linear peptide that is found to react in such an assay is labeled a continuous epitope of the protein. It is customary to test peptides of decreasing size and to give the status of epitope to the smallest peptide that retains a measurable level of antigenic reactivity. Usually this leads to the identification of continuous epitopes with a length of 5–8 residues, although the lower size limit tends to remain ill-defined. It is not unusual for certain di- or tripeptides to retain a significant binding capacity in particular types of solid-phase immunoassays.

On the other hand, increasing the length of peptides does not always lead to a higher level of cross-reactivity with antiprotein antibodies since longer peptides may adopt a conformation that is different from the one present in the intact protein. There is also no reason to assume that short peptides will have a unique conformation mimicking that of the corresponding region in the protein. Cross-reactivity of peptides with antiprotein antibodies is commonly observed because of antibody multispecificity and of the induced fit and mutual adaptation capacity of the two partners.

Many investigators believe that the majority of so-called continuous epitopes described in the literature and listed in immunological databases actually correspond to unfolded regions of denatured protein molecules, that is, that they are not genuine epitopes of native proteins. They argue that there is usually little experimental evidence to show that short peptides are actually able to bind to antibodies specific for the native state of the cognate protein. It is indeed very difficult to demonstrate that the protein sample used in an immunoassay does not contain at least some denatured molecules that could be responsible for the observed binding reaction. Furthermore, claims made in the early 1980s that immunization with peptides always leads to a high frequency of induction of antibodies that recognize the native cognate protein are no longer considered valid. It is now accepted that such claims arose because the ability of antipeptide antibodies to recognize proteins was tested in solid-phase immunoassays in which the proteins had become denatured by adsorption to plastic.

## Mimotopes

The multispecificity of antibody molecules is illustrated by the existence of so-called mimotopes. The term mimotope was coined in 1986 by Mario Geysen and was originally defined as a peptide that is able to bind to a particular antibody, but is unrelated in sequence to the protein antigen used to induce that antibody, usually because the antibody is directed to a discontinuous epitope. Currently, the term mimotope is applied to any epitope mimic, irrespective of whether the epitope being mimicked is continuous or discontinuous. It is indeed possible to mimic a continuous epitope with a cross-reactive mimotope peptide that shows little sequence similarity with the original epitope. This cross-reactivity of mimotopes is due to the fact that dissimilar amino acid residues may actually contain a sufficient number of identical atomic groups to allow the peptide to cross-react with atoms of the antibody-combining site.

To qualify as a mimotope, a peptide should not only be able to bind a particular antibody but it should also be capable of eliciting antibodies that recognize the epitope being mimicked. This requirement stems from the fact that a single immunoglobulin always harbors a number of partly overlapping or nonoverlapping paratopes, each one capable of binding related or unrelated epitopes. When different subsites in the immunoglobulin-binding pocket partly overlap, binding to one epitope may prevent a second unrelated epitope from being accommodated at a nearby location. Therefore, when a peptide is labeled a mimotope of epitope A because of its capacity either to bind to an anti-A antibody or to inhibit the binding of epitope A to the antibody, it cannot be excluded that the putative mimotope actually binds to a different subsite from the one that interacts with epitope A. This is why it is necessary to show that a putative peptide mimotope is also able to induce antibodies that cross-react with epitope A, thereby demonstrating that it really is a mimotope of epitope A.

Nowadays, mimotopes are often identified by testing combinatorial peptide libraries, obtained by chemical synthesis or phage display, for their capacity to bind monoclonal antibodies specific for viral proteins. It is also possible to screen a phage library by means of sera collected from individuals that recovered from a viral infection and have seroconverted. Mimotopes identified in this manner may find applications in peptide-based diagnostic assays. It is also believed that mimotopes could be used for developing peptide-based vaccines.

The antigenic reactivity of viral proteins discussed so far is based on chemical reactions between epitopes and paratopes and it can be described using parameters such as the structural and chemical complementarity of the two partners, electrostatic and hydrogen bond interactions between them, the kinetics and equilibrium affinity constants of the interaction, the discrimination potential of individual antibody molecules, etc. Such immunochemical descriptions take the existence of antibodies for granted and do not ask questions about the biological origin, synthesis, or maturation of antibody molecules. The situation is different when it comes to investigations of the immunogenicity of viral proteins since this property cannot be analyzed outside of the biological context of an immunized host.

## Immunogenicity

Whereas the antigenicity of proteins is a purely chemical property, their immunogenicity is a biological property that has a meaning only in the context of a particular host. Immunogenicity is the ability of a protein to give rise to an immune response in a competent host and it depends on extrinsic factors such as the host immunoglobulin repertoire, self-tolerance, the production of cytokines, and various cellular and regulatory mechanisms definable only in a given biological context.

When a number of continuous epitopes of a protein have been identified, this knowledge does not provide information on which particular immunogenic structure, present in the antigen used for immunization, was recognized by B-cell receptors and initiated the production of antibodies. Our ignorance of the exact immunogenic stimulus is often referred to as the black box conundrum and this makes the study of immunogenicity very much an empirical endeavor.

Although any peptide identified as a continuous epitope will readily elicit antipeptide antibodies, it is only rarely able to induce antibodies that also recognize the cognate, native protein antigen. In an immunoassay, an antibody raised against the native protein may be able to select one conformation in a peptide or it may induce a reactive conformation by an induced fit or mutual adaptation mechanism, the result in both cases being the occurrence of a cross-reaction between the peptide and the antiprotein antibody. On the other hand, when the same peptide meets a variety of B-cell receptors during the immunization process, it may not be able to bind preferentially to those rare paratopes in the receptors that, in addition to recognizing the peptide, also cross-react with the native protein. As a result most elicited antipeptide antibodies will only recognize the peptide and will not cross-react with the cognate protein.

As far as the immunogenicity of discontinuous epitopes is concerned, the situation is even more uncertain since the residues belonging to the epitope were identified by crystallographic analysis at the end of a process of mutual adaptation and conformational change in the two binding partners. The discontinuous epitope with its unique conformational features cannot be dissected out of the three-dimensional assembly of residues in the native protein and its immunogenic potential cannot therefore be studied independently of the rest of the protein antigen. As a result, the exact three-dimensional structure of the immunogen which was recognized by B-cell receptors and initiated the immune response cannot be known with certainty. Furthermore, reconstituting a discontinuous epitope in the form of a linear peptide that would include all the epitope residues from distant parts of the antigen sequence and would assemble them in the correct conformation appears to be an extremely difficult task.

## Implications for Vaccine Development

Our increasing knowledge of the structure of viral epitopes has given rise in some quarters to the expectation that it should be possible to develop peptide-based viral vaccines. However, the results so far have been disappointing. In spite of the hundreds of viral epitopes that have been identified by studying the antigenicity of peptide fragments of viral proteins, no commercial peptide vaccine has yet reached the marketplace.

Although most peptide fragments are immunogenic in the sense that they readily induce antibodies that react with the peptide immunogen, this type of immunogenicity is irrelevant for vaccination purposes. What is required is the induction of antibodies which, on the one hand, recognize the cognate, native viral antigen (so-called cross-reactive immunogenicity) and which in addition also neutralize the infectivity of the virus (so-called cross-protective immunogenicity). Unfortunately, very few of the continuous epitopes of viruses that have been described possess the required cross-reactive and cross-protective immunogenicity. Many attempts have been made to increase the conformational similarity between peptide and intact protein, for instance, by constraining the peptide conformation by cyclization but this has not resulted in peptides possessing adequate cross-protective immunogenicity. There is also some evidence that the intrinsic disorder in certain loop regions of viral proteins may be responsible for the finding that peptides corresponding to such disordered regions are sometimes better vaccine immunogens than peptides with a constrained conformation.

The difficulties that must be overcome to transform a continuous epitope into an effective vaccine immunogen are illustrated by the many studies of the peptide ELDKWAS corresponding to residues 662–668 of the gp41 protein of HIV-1. This peptide which is recognized by the anti-HIV-1 broadly cross-reactive neutralizing monoclonal antibody (Mab) 2F5 has, for a long time, been regarded as a promising vaccine candidate because it is located in a conserved region of gp41 necessary for envelope-mediated fusion of the virus.

The ELDKWAS peptide has been incorporated into a variety of immunogenic constructs in an attempt to have it elicit antibodies with the same neutralizing capacity as Mab 2F5. When additional gp41-derived flanking residues were added to the peptide or when its conformation was constrained, peptides were obtained which had a tenfold higher affinity for the 2F5 antibody than the unconstrained peptide. However, in spite of their improved antigenicity, the peptide constructs, when used as immunogens, were still unable to induce antibodies with detectable neutralizing capacity.

In an attempt to ascertain which structural elements close to the ELDKWAS residues in the gp41 immunogen may have influenced the induction of the neutralizing 2F5 antibody, the crystal structure of the 2F5 antibody in complex with various gp41 peptides was determined. The conformation of the bound peptides was found to differ significantly from the corresponding region in the gp41 protein, indicating that the putative ELDKWAS epitope was able to assume various conformations depending on the fusogenic state of gp41. However, it was not clear which conformation should be stabilized in the peptide constructs intended for vaccination.

It seems that the viral epitopes involved in the immunogenic stimulus may in fact be dynamic structures with variable conformations and it has been suggested that such epitopes should be referred to as transitional epitopes. Such a label is an appropriate reminder that it is necessary to include the fourth dimension of time in the description of antigenic specificity.

The findings obtained with the ELDKWAS peptide suggest that these residues are part of a more complex discontinuous epitope that elicited the neutralizing Mab 2F5. There is also evidence that the hydrophobic membrane environment close to the ELDKWAS sequence played a role in the induction of the neutralizing antibody 2F5.

Unfortunately, only some of the antibodies induced by viral antigens possess neutralizing activity and it is not known which structural features in the immunogen are responsible for the appearance of neutralizing rather than non-neutralizing antibodies. In recent years, attempts to find out have relied mainly on the crystallographic analysis of viral proteins complexed with neutralizing Mabs. Although such studies reveal the structure of the epitopes present in the complexes, they do not provide information on which transitional immunogenic epitopes were able to induce neutralizing antibodies. It seems likely that this type of information will only be obtained by systematic empirical trials in which numerous candidate immunogens are tested for their ability to induce protective immune responses.

*See also:* AIDS: Vaccine Development; Antigenic Variation; Immune Response to viruses: Antibody-Mediated Immunity; Neutralization of Infectivity; Vaccine Strategies.

## Further Reading

Clavijo A, Wright P, and Kitching P (2004) Developments in diagnostic techniques for differentiating infection from vaccination in foot-and-mouth disease. *Veterinary Journal* 167: 9–22.

Geysen HM, Rodda SJ, and Mason TJ (1986) *A priori* delineation of a peptide which mimics a discontinuous antigenic determinant. *Molecular Immunology* 23: 709–715.

Greenbaum JA, Andersen PH, Blythe M, *et al.* (2007) Towards a consensus on datasets and evaluation metrics for developing B cell epitope prediction tools. *Journal of Molecular Recognition* 20: 75–82.

Klenk H-D (1990) Influence of glycosylation on antigenicity of viral proteins. In: Van Regenmortel MHV and Neurath AR (eds.) *Immunochemistry of Viruses II*, pp. 25–37. Amsterdam: Elsevier.

Sheppard N and Sattentau Q (2005) The prospects for vaccines against HIV-1: More than a field of long-term nonprogression? *Expert Review Molecular Medicine* 7(2): 1–21.

Uversky VN, Oldfield CJ, and Dunker AK (2005) Showing your ID: Intrinsic disorder as an ID for recognition, regulation and cell signaling. *Journal of Molecular Recognition* 18: 343–384.

Van Regenmortel MHV (1996) Mapping epitope structure and activity: From one dimensional prediction to four-dimensional description of antigenic specificity. *Methods: Companion to Methods in Enzymology* 9: 465–472.

Van Regenmortel MHV (1998) From absolute to exquisite specificity. Reflections on the fuzzy nature of species, specificity and antigenic sites. *Journal of Immunological Methods* 216: 37–48.

Van Regenmortel MHV (1999) Molecular design versus empirical discovery in peptide-based vaccines: Coming to terms with fuzzy recognition sites and ill-defined structure-function relationships in immunology. *Vaccine* 18: 216–221.

Van Regenmortel MHV (2006) Immunoinformatics may lead to a reappraisal of the nature of B cell epitopes and of the feasibility of synthetic peptide vaccines. *Journal of Molecular Recognition* 19: 183–187.

Van Regenmortel MHV and Muller S (1999) In: *Synthetic Peptides as Antigens*, pp. 1–381. Amsterdam: Elsevier.

Wyatt R and Sodroski J (1998) The HIV-1 envelope glycoproteins: Fusogens, antigens and immunogens. *Science* 280: 1884–1888.

Yuan W, Bazick J, and Sodroski J (2006) Characterization of the multiple conformational states of free monomeric and trimeric human immunodeficiency virus envelope glycoproteins after fixation by cross-linker. *Journal of Virology* 80: 6725–6737.

Zolla-Pazner S (2004) Identifying epitopes of HIV-1 that induce protective antibodies. *Nature Reviews Immunology* 4: 199–210.

Zwick MB, Jensen R, Church S, *et al.* (2005) Anti-human immunodeficiency virus type 1 (HIV-1) antibodies 2F5 and 4E10 require surprisingly few crucial residues in the membrane-proximal external region of glycoprotein gp41 to neutralize HIV-1. *Journal of Virology* 79: 1252–1261.

# Antiviral Agents

**H J Field,** University of Cambridge, Cambridge, UK
**R A Vere Hodge,** Vere Hodge Antivirals Ltd., Reigate, UK

© 2008 Published by Elsevier Ltd.

## Introduction

There have been two previous articles in *Encyclopedia of Virology*, the latter being in 1999. That review, by A. K. Field and C. A. Laughlin, gave a good update, especially in the treatment of HIV. In order to avoid repetition, we have focused on antiviral targets. We have not attempted to cover vaccines nor immunomodulating agents, such as interferons, except when these are mentioned briefly in those cases in which they are the therapy of choice. These approaches which utilize a cellular target are outside the scope of this article but the reader is referred to a recent review by L. Schang.

The term 'virucidal' has long been used to describe an agent that destroys virions while outside the host cell. A few compounds inactivate viruses by highly specific mechanisms but many, such as alcohol wipes, have a broad spectrum of activity against many infectious entities. Virucidal agents have an important role, especially in hospitals, in preventing the transmission of viruses and may be useful in reducing the risk of virus transmission between individuals, for example, during sexual contact. These types of agent will not be considered further in this article.

The focus of this article is those antiviral compounds which target virus replication selectively. After a brief historical overview, we discuss the concepts and challenges which have guided and stimulated progress with antiviral therapy – selectivity, spectrum, viral targets, prodrugs, resistance and drug combinations, viral fitness, latency. A summary of the most important antiviral agents is given in **Table 1**. This article outlines the principles

**Table 1** List of important current antiviral agents

*(a) Primarily active against Herpes viruses*

| Generic name (abbreviation) Trade name (company) | Structural information | Mechanism viral target | Fig 1 stage | Target viruses |
|---|---|---|---|---|
| Valaciclovir (VACV) Valtrex® (GSK) | Prodrug of acyclovir (ACV) Zovirax® | Activated by viral TK, inhibits viral polymerase | VI | HSV-1 and -2 VZV (CMV) |

Notes: ACV is extremely safe and, apart from some problems related to high doses and crystallization in the kidney, no serious side effects have been encountered. Drug-resistance in immunocompetent patients has remained rare ($<1\%$) in over two decades of clinical experience. In immunocompromised patients, resistant variants (generally in TK and occasionally in DNA polymerase) occur in up to 5% patients. Fortunately, TK-deficient HSV appears to be less pathogenic and less able to reactivate from latency than wt HSV.
VACV is a valine ester of ACV. In addition to the main target viruses, VACV is sometimes used to prevent CMV infections in transplant patients. During suppressive therapy for genital herpes, VACV reduced the HSV-transmission rate by about 50%.

| | | | | |
|---|---|---|---|---|
| Famciclovir (FCV) Famvir® (Novartis) | Prodrug of penciclovir (PCV) Denavir®/ Vectavir® | Activated by viral TK, inhibits viral polymerase | VI | HSV-1 and -2 VZV |

Notes: This was the first antiviral agent to be developed as the prodrug and stimulated the search for an ACV prodrug. FCV was shown to be highly effective, vs. placebo, against HSV-1 and -2 and VZV. In the only clinical trial comparing FCV and VACV for the treatment of shingles, they appeared to be equally effective. Interestingly, PCV has a higher affinity than ACV for HSV TK but is a less potent inhibitor of herpes DNA polymerase. However, the intracellular half-life of PCV triphosphate is significantly longer than ACV triphosphate and this may help to explain its potent antiviral activity. The dynamics of resistance selection may differ subtly from ACV but clinical resistance in immunocompetent patients remains extremely rare.

| | | | | |
|---|---|---|---|---|
| Foscarnet (PFA) Foscavir® (Astra Zeneca) | Pyrophosphate analog | Polymerase inhibitor | VI | HSV-1 and -2 VZV |

Notes: PFA is an effective inhibitor of HSV replication. However, the compound must be administered iv and suffers from toxicity (e.g., tendency to accumulate in bone). Therefore it is used in immunocompromised patients with herpes infections resistant to ACV and PCV.

| | | | | |
|---|---|---|---|---|
| Valganciclovir (VGCV) Valcyte® (Roche) | Prodrug of ganciclovir (GCV) Cymmevene® | Activated by kinase encoded by UL 97, polymerase inhibitor | VI | CMV |

Notes: VGCV has become the drug of choice for the prevention and treatment of CMV in immunocompromised patients.

*(b) Primarily active against RNA viruses*

| Generic name Trade name (company) | Structural information | Mechanism viral target | Fig 1 stage | Target viruses |
|---|---|---|---|---|
| Zanamivir Relenza® (GSK) | Sialic acid analog | Neuraminidase inhibitor | X | Influenza A and B |

Notes: Zanamivir prevents the influenza virions leaving the infected cell. It can be used either for the treatment or prophylaxis of influenza infections. It has a good safety record but its use is limited by the need for inhalation. It is currently being stockpiled, as the second line agent, in preparation for a possible outbreak with H5N1 virus.

| | | | | |
|---|---|---|---|---|
| Oseltamivir Tamiflu® (Roche) | Sialic acid analog | Neuraminidase inhibitor | X | Influenza A and B |

Notes: Like zanamivir, oseltamivir has good, selective activity against influenza viruses. In contrast to zanamivir, it has good bioavailability which has led to it becoming the first line agent for treatment of influenza. In adult patients, resistance is rare and it appears that those resistant strains are partially disabled. It is the main agent for stockpiling for an H5N1 outbreak.
An inhibitor of influenza RNA polymerase, T-705, has demonstrated excellent activity in cell culture assays and animal models. It is expected to be entering clinical trials soon. With the threat of an H5N1 outbreak, it will be useful to have another treatment option, possibly used in combination with oseltamivir.

*Continued*

**Table 1** Continued

*(b) Primarily active against RNA viruses*

| Generic name<br>Trade name<br>(company) | Structural information | Mechanism viral target | Fig 1 stage | Target viruses |
|---|---|---|---|---|
| Ribavirin<br>Virazole®<br>(Schering-Plough) | Nucleoside analog | Possibly, no direct viral target | ? | HCV<br>RSV |

*Notes*: HCV – Current therapy is with one of several interferon forms, usually with ribavirin which seems to enhance the chance for long-term response. Being immune modulators, these are outside the scope of this review. However, antiviral agents, targeting either HCV protease or polymerase, are being developed.

RSV – Although ribavirin has been used to treat serious cases of RSV, its efficacy has been questioned. Palivizumab and RSV immune globulin (RSV-IGIV) are FDA approved.

*(c) Primarily active against Hepadnaviruses (HBV)*

| Generic name<br>(abbreviation)<br>Trade name | Structural information | Mechanism viral target | Fig 1 stage | Company |
|---|---|---|---|---|
| Lamivudine (3TC)<br>Zeffix®,<br>Heptovir® | NA inhibitor | Polymerase | VI | GSK |

*Notes*: Lamivudine has become the first line choice for therapy of chronic HBV although viral resistance appears in about 50% of patients after two years monotherapy. Lamivudine is unusual as it is in routine use for both HBV and HIV.

| Adefovir dipivoxil<br>(ADV)<br>Hepsera® | Prodrug of adefovir<br>NA inhibitor | Polymerase | VI | Gilead |
|---|---|---|---|---|

*Notes*: In comparison with lamivudine, the rate of appearance of resistance is much slower, in about 2.5% patients after 2 years monotherapy.

| Entecavir<br>Baraclude® | NA inhibitor | Polymerase | VI | BMS |
|---|---|---|---|---|

*Notes*: Entecavir was approved by FDA for HBV in 2006. It is very active with tablet sizes of 0.5 and 1 mg.

| Telbivudine<br>Tyzeka® | NA inhibitor | Polymerase | VI | Idenix/Roche |
|---|---|---|---|---|

*Notes*: Telbivudine is the most recent (2007) of FDA approved agents for HBV. Telbivudine has an excellent safety record in animal toxicity tests. Although the tablet size (600 mg) is larger than with the other compounds, it is perhaps the most effective in reducing HBV DNA levels but it should not be used to treat patients failing on lamivudine therapy. It may replace lamivudine as the first line therapy.

*(d) Primarily active against Retroviruses (HIV)*

| Generic name<br>(abbreviation)<br>Trade name | Structural information | Mechanism viral target | Fig 1 stage | Company |
|---|---|---|---|---|
| Zidovudine/Lamivudine<br>(AZT/3TC)<br>Combivir® | Two NRTIs | Polymerase | VI | GSK |

*Notes*: As single agents, the trade names are Retrovir®/Epivir® respectively.

| AZT/3TC/abacavir<br>Trizivir® | Three NRTIs | Polymerase | VI | GSK |
|---|---|---|---|---|

*Notes*: The trade name for abacavir as a single agent is Ziagen® and, when combined with lamivudine, Epzicom® or Kivexa®.

| Emtricitabine/tenofovir/<br>efavirenz<br>Atripla® | Two NNRTIs and one NNRTI | Polymerase | VI/VI/VII | Gilead & BMS<br>(Jointly) |
|---|---|---|---|---|

*Notes*: The trade name for the combination emtricitabine/tenofovir is Truvada®. As single agents, the trade names are Emtrive®/Viread®/Sustiva® respectively.

Continued

**Table 1** Continued

**(d) Primarily active against Retroviruses (HIV)**

| Generic name (abbreviation) Trade name | Structural information | Mechanism viral target | Fig 1 stage | Company |
|---|---|---|---|---|
| Nevirapine Viramune® | NNRTI | Polymerase | VII | Boehringer |
| *Notes*: Should be used with at least two other agents, one of which should be an NRTI. | | | | |
| Fosamprenavir Lexiva® | PI | Protease | V | GSK |
| *Notes*: Fosamprenavir is the oral prodrug of amprenavir (Agenerase®) and is replacing it in clinical use. | | | | |
| Saquinavir mesylate Invirase® Fortovase® (Roche) | PI | Protease | V | Roche |
| *Notes*: Invirase capsules and tablets should be used only with ritonavir which significantly inhibits the metabolism of saquinavir to provide plasma saquinavir levels at least equal to those achieved with Fortavase capsules which should be use, if at all, as the sole PI in a regimen. | | | | |
| Lopinavir/ritonavir Kaletra® | 2 PIs | Protease | V | Abbott Lab |
| *Notes*: Lopinavir is not marketed separately but combined with ritonavir (ratio 4:1) which enhances the plasma levels of lopinavir. The trade name for ritonavir is Norvir®. | | | | |
| Indinavir Crixivan® | PI | Protease | V | Merck |
| *Notes*: Indinavir should be used only with other agents. | | | | |
| Darunavir Prezista® | PI | Protease | V | Tibotec |
| *Notes*: Darunavir was given fast track status and was approved in 2006. Darunavir has less than tenfold decreased susceptibility in cell culture against 90% of 3309 clinical isolates resistant to other PIs. | | | | |
| Efuviritide (T-20) Fuzeon® (Roche) | Fusion inhibitor | gp41 envelope protein | II | Roche |
| *Notes*: Administered sc bid. Mainly used in salvage therapy. It is the only FDA approved fusion inhibitor. | | | | |

Further information, structures, and references may be obtained from AVCC FactFile and company web sites.

underlying the successful development of antiviral compounds for the many clinical applications available today. Some prospects for the future are discussed.

## Historical Perspective

### The First Uncertain Steps

Following the discovery of antibiotics for treating bacterial infections, for several decades it was thought that 'safe' antiviral chemotherapy would be difficult if not impossible. The earliest antiviral compounds only emphasized the problems. Marboran was introduced in the early 1960s to treat smallpox and vaccinia; its effectiveness was equivocal and its use short-lived. Amantadine and rimantadine were used to treat influenza with relatively little side effects but the influenza virus became resistant within a few days of treatment and the resistant virus spread readily to contacts. The first nucleoside analogs, with antiviral activity, emerged during the search for drugs to treat cancer. The first of these, idoxuridine (IDU), was discovered to be active against herpes viruses by Dr. William (Bill) Prusoff in 1959. However, its toxicity limited its use to topical treatments only (e.g., infections of the eye). Vidarabine (Ara A) was slightly more selective; its main systemic use was to treat herpes encephalitis. No convincing selective antiviral compound had yet been discovered.

### The Major Breakthrough

The discovery of the antiherpes drug, acyclovir (ACV) in 1978, was the major milestone in antiviral therapy. For the first time, it was demonstrated that an effective, non-toxic, antiviral drug is an achievable aim. Moreover, chronic suppressive acyclovir therapy for several years, to prevent the misery of recurrent herpes, is possible without adverse effects. More than two decades of worldwide use has proved that acyclovir is one of the safest drugs in clinical therapy. Acyclovir proved that not all nucleoside analogs had to be mutagenic and/or carcinogenic. The early doubts, about the potential for discovering safe antiviral agents, were dispelled for ever.

The selectivity of acyclovir for the herpes viruses is dependent on it being activated only in herpes virus-infected cells. The critical initial step in that activation is its phosphorylation to ACV-monophosphate by the viral thymidine kinase. Cellular enzymes convert the monophosphate to the triphosphate which inhibits selectively herpes viral DNA polymerase. Following the discovery of acyclovir, other selective antiherpes virus compounds were discovered, including BVDU, BvaraU, and penciclovir (PCV). Currently, ACV and PCV, and their prodrugs valaciclovir (VACV) and famciclovir (FCV), respectively, are the most commonly used drugs to treat infections with HSV-1 and -2 and VZV. Another compound, ganciclovir (GCV), was discovered to have good activity against CMV. Like ACV and PCV, GCV is activated specifically in virus-infected cells but by the kinase encoded by the CMV UL97 gene. The prodrug of GCV, valganciclovir (VGCV), is the preferred therapy for CMV prevention and treatment in immunocompromised patients.

### Impetus from the Challenge of HIV

Following the emergence of AIDS, an enormous research effort worldwide was directed at the search for inhibitors of HIV, resulting in a large number of new drugs. Initially, these inhibitors were nucleoside analogs. These had to be phosphorylated (activated) to their triphosphate entirely by cellular enzymes; HIV does not encode a thymidine kinase. Therefore, the selectivity was due entirely to the greater inhibition of HIV reverse transcriptase than the cellular polymerases. Because the medical need was so great, several inhibitors were brought quickly into clinical use even though they were associated with long-term toxicity problems, for example, mitochondrial toxicity with didanosine (ddI) and zalcitibine (ddC) and nephrotoxicity with cidofovir (HPMPC), used to treat CMV in AIDS patients. More recently, great progress has been made in selecting nucleoside and nucleotide analogs (**Table 1**) with much less potential for toxicity and so long-term therapies became normal practice. Once HIV protease had been shown to be essential for HIV replication, protease inhibitors (PIs) were discovered. Although these inhibitors have a high degree of selectivity for the HIV protease, they are not without some side effects, including the development or redistribution of fatty masses. However, one of the PIs is commonly included in the combination therapy now known as HAART (highly active anti-retroviral therapy).

### Expansion into Therapies for Other Viruses

A spin-off from this work with HIV has led to inhibitors of other families of viruses, particularly hepatitis B virus (HBV). Lamivudine and adefovir have become the treatments of choice for HBV. Meanwhile, there seemed to be progress discovering compounds active against picornaviruses (which include rhinoviruses causing the common cold). These compounds bind into a pocket within the viral capsid. The best example is pleconaril although unacceptable side effects stopped its development. However, the viruses quickly became resistant and some strains even became dependent on the 'antiviral' compound. Therefore, this approach has not resulted in any clinically useful drugs.

Once the structure of influenza virus neuraminidase was known, new inhibitors of influenza viruses were discovered. Recent advances with replication systems for hepatitis C virus (HCV) have allowed the discovery of anti-HCV compounds. Currently, the threat of bio-terrorism has prompted the successful search for drugs active against poxviruses and, with the prospect of the next influenza pandemic looming, there is renewed research effort directed to anti-influenza drugs. There are at least 25 (**Table 1**) widely used licensed antiviral agents and that number looks set to increase rapidly in the coming years.

## Key Concepts

### Definition of Selective Activity Index

Viruses are obligate intracellular parasites. Thus, inhibitors of virus replication must do so without toxic effects on the cells, tissues, and organs of the host. This is the concept of 'selective activity' against viruses.

For those viruses which form plaques in a cell monolayer, reduction in plaque formation has long been regarded as the 'gold standard' assay. Antiviral activity is usually defined as the concentration of the inhibitor which reduces viral plaques by 50% (50% effective concentration; $EC_{50}$). Alternatively, the reduction in the yield of infectious virus can give useful information by using different multiplicities of infection (MOI); in such assays, the concentration of inhibitor to reduce replication by 99% ($EC_{99}$) is usually reported. For some viruses, for example, HBV, reduced production of a measurable virus product (e.g., nucleic acid or protein) is the only way to assess viral replication. For inhibition of a viral enzyme, the 50% inhibitory concentration ($IC_{50}$) is commonly reported.

Irrespective of the antiviral assay, it is necessary to assess cellular toxicity of the test compound. This is essential to eliminate the possibility that the lack of virus replication is due to the destruction of the cells. However, the ratio of the concentration to inhibit the replication of the virus to the concentration to destroy nondividing cells in a monolayer is *not* an indication of selective activity.

To determine selective activity, it is essential to compare like with like, replicating virus with replicating cells. To test for an effect on replicating cells, it is necessary to aim for a tenfold increase (just over three doublings) so that at the 50% cytotoxic concentration ($CC_{50}$), at least one doubling has occurred and the second round of replication has started. The ratio of 50% viral inhibitory concentration to 50% cytotoxic concentration is defined as the selective index (SI).

In the literature, authors may not abide by the above definition of selective index. Therefore, published values for selectivity should be considered with some skepticism unless the methodology is fully described. We urge editors of antiviral articles to be aware of this difficulty.

The determination of $CC_{50}$ values in a single assay may be sufficient to select compounds in a primary screen. For assessing compounds worthy of progression into development, it is prudent to assess the compounds in a variety of cytotoxicity tests using different cells. For example, inhibition of mitochondrial DNA can be indicated by using granulocyte-macrophage (CFU-GM) and erythroid (BFU-E) cells. Human bone marrow stem cells in primary culture are a good predictor for potential hematoxicity. A particular class of compounds may be known to have a potential for a particular toxicity, such as phosphonates for nephrotoxicity. In this case, the established pattern of nephrotoxicity in the clinic was reflected in a cell culture assay using primary human renal proximal tubule cells (RPTCs).

When antiviral agents are evaluated against virus infections *in vivo*, the ratio of the effective dose ($ED_{50}$) to the minimum toxic dose is sometimes referred to as the 'therapeutic ratio'. These measurements are widely used during the preclinical development of new antiviral agents and are often helpful in making comparisons between alternative compounds.

## Toxicity of Antiviral Compounds

As mentioned above, SIs provide a numerical estimate of the degree of selectivity for antiviral compounds and may be useful in proceeding with structure–activity relationships (SARs) from a lead compound. However, these results must be treated with great caution when extrapolating to man. Although a high ($>1000$) SI value may be encouraging, a low SI should not be ignored. During preclinical evaluation, many potential toxic effects can be assessed. For example, pharmacokinetic studies can identify drug–drug interactions. For nucleoside analogs, it is not sufficient to study the possible interactions in plasma; it is also necessary to study drug–drug interactions on phosphorylation within cells. There are well-established tests for indicating potential genetic toxicity of nucleosides and nucleotides and toxicity for mitochondrial enzymes. However, there always remains the possibility for unexpected effects such as lipodystrophy which may not become apparent until the drug is used in man – possibly for several years. In some cases, these complications once identified can be avoided or managed. In other cases, for example, ddI and ddC, the development of safer options eclipses the older compounds. For those infections requiring long-term therapies, over a year for HBV, several years for suppression of recurrent genital herpes, over 10 years for HIV, the test for clinical safety is very exacting. It is a credit to research workers in antiviral chemotherapy that compounds, which have stood the test of time, have been brought into clinical practice.

## Spectrum of Antiviral Action

There are few, if any, truly broad spectrum antivirals. Given the highly specific nature of the mechanism of action of most antivirals, this is unsurprising. Most antiviral compounds affect a narrow range of viruses – often restricted to a single virus family or particular subfamily. Telbivudine is highly specific for the HBV (and the related duck and woodchuck viruses) and is inactive against all other viruses tested. Acyclovir is typical in being active against some viruses within one family, the herpes viruses. Lamivudine is unusual in becoming a common treatment for both HIV and HBV.

Broad-spectrum activity against all enveloped viruses does seem to be a possibility. Rep 9 is a phosphorothioate oligonucleotide with a random order of bases (A, C, G, and T), the antiviral activity being dependent not on the base sequence but on the size, the optimal being about 40–50 bases. This oligonucleotide has both hydrophobic (along the backbone) and hydrophilic (bases) surfaces which seem to be essential for activity. Various enveloped viruses have a surface protein (e.g., HIV gp41, influenza hemagglutin) with an alpha-helix which matches the length of the 40 nucleotides in REP 9. The viruses that have been shown to be sensitive to REP 9 include members of the *Herpes-*, *Orthomyxo-*, *Paramyxo-*, and *Retroviridae*.

## Virus Replication Cycle Presents Antiviral Targets

### The Virus Replication Cycle

All viruses are dependent upon a host cell for protein synthesis. Thus, all viruses replicate via a broadly similar sequence of events (**Figure 1**). The virus must first attach ('adsorb') to the cell. The virus or virus nucleic acid genome then enters ('penetrates') the cytoplasm. The genome is liberated from the protective capsid ('uncoats'). In some cases the viral genome enters the cell nucleus; in all cases the genome is transcribed and thus the viral mRNA directs protein synthesis. Virus products include proteins that regulate further transcription; enzymes

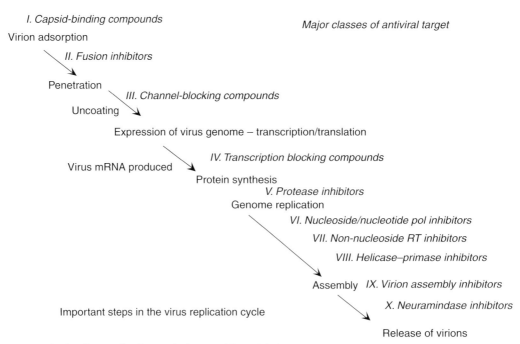

**Figure 1** Key steps in the virus replication cycle that provide antiviral targets.

including those involved in genome replication; proteases that process other virus-induced proteins; and many proteins that subvert host-defense mechanisms within and without the infected cell. The virus undergoes 'genome replication' and together with virus structural proteins form new virions ('assembly') which are 'released' from the cells. In some cases release is associated with budding through cellular membranes that have been decorated with virus glycoproteins enabling the virus to acquire its envelope. All stages are highly regulated both temporarily and quantitatively, and vary for each particular virus.

## Important Classes of Antiviral Targets and Compounds

The major targets and compounds are illustrated in **Figure 1**.

### Virus attachment

Currently, there are no major drugs which prevent virus attaching to the host cell. The problem has been illustrated by the picornavirus capsid-binding compounds such as pleconaril. Other compounds are being tried for HIV but with little success so far. It seems that many sites on the surface of a virus can undergo changes without great loss of viral viability; the compounds present too low a genetic barrier (see below).

### HIV fusion inhibitors

To date, only one entry inhibitor has been approved by the US Food and Drug Administration (FDA). This is efuvirtide (T-20), a synthetic peptide that targets the HIV gp41 envelope protein to prevent fusion. The broad-spectrum antiviral agent, Rep 9, is progressing through development.

### Influenza channel-blocking compounds

Amantadine and rimantadine act by binding to M2 and thus interfering with the penetration of hydrogen ions into the virion, a process essential for uncoating the single-strand RNA genome. Key resistant mutations map to the *M2* gene (and to a lesser extent, HA gene). Such resistant variants are selected quickly and are fully viable. This has severely limited the clinical usefulness of these drugs. During the 2005/06 season in the USA, 109/120 (91%) of H3N2 isolates were resistant to amantadine and rimantadine. Combination with other drugs now seems unlikely to be a viable option and so their future role may be minimal.

### Antisense binding to viral mRNA

The antisense approach has the potential for great selectivity due to the sequence of bases in the oligonucleotide being complementary to a particular viral sequence. Conceptually, it may be possible to interfere with latent virus, even when the HIV genome has integrated into the DNA of a long-lived human cell. In practice, there have been many difficulties in progressing antisense compounds to the clinic. So far, only formivirsen has been licensed by the FDA. Its use is limited to intravitreal inoculation in patients with CMV retinitis.

### Protease inhibitors

Many viruses encode one or more enzymes involved in the cleavage of virus or cellular proteins. HIV induces an enzyme that specifically cleaves protein precursors in the maturation of gag and pol polyproteins. Peptide mimetics were designed specifically to target HIV protease. The resulting compounds, several of which are now standard HIV drugs, are among the most active antiviral compounds known and inhibit HIV in the nanomolar range. Like other HIV antivirals, drug-resistant mutants arise and so the compounds are used in combination with reverse transcriptase inhibitors. Other families of viruses encode proteases; except for HCV, the search for effective inhibitors has yielded few useful compounds.

### Viral DNA/RNA polymerase inhibitors

Nucleoside and nucleotide analogs play a pivotal role in antiviral chemotherapy. The ultimate target is the viral DNA/RNA polymerase which is well conserved within different strains of a particular virus and sometimes within a virus family. For those nucleosides which are activated initially to their monophosphates by a viral enzyme, there is an additional potential for selectivity. In general, the viral enzymes are less stringent in their structural requirements than the corresponding host (e.g., human) equivalents. Because the viral polymerase is crucial to the replication of that virus, there are few potential mutations which allow the polymerase to retain full function but reject the antiviral compound. This results in a relatively high genetic barrier (see below).

Nucleoside/nucleotide analogs are the treatment of choice for herpes viruses and HBV. They are included as two or three components of HAART for HIV. For influenza, T-705 is a good candidate progressing through development. For HCV, viral polymerase inhibitors, such as R1626, are being developed.

### Non-nucleoside RT inhibitors

In the search for new inhibitors of retrovirus reverse transcriptase (RT), several chemically unrelated compounds emerged from screens that bind to RT at sites other than the nucleotide binding site. Some are highly selective and inhibit HIV-1 RT at nanomolar concentrations. Enzyme studies showed that these compounds inhibit RT non-competitively. The compounds are generally thought to be allosteric inhibitors. An early problem associated with these compounds was the rapid emergence of drug-resistant strains. Because the drug-resistant mutations do not directly involve the active sites on the enzyme, they are less likely to compromise enzyme function. Therefore, the genetic barrier is lower than that corresponding to nucleosides targeting the catalytic site. However, adding non-nucleoside RT inhibitor (NNRTI) to NRTI(s) raises the genetic barrier and so the NNRTIs play an important role in HIV combination therapy (see below).

### Helicase inhibitors

A new target for effective inhibition of HSV has been discovered. These compounds are being developed by Bayer AG Pharma (e.g., BAY 57–1293) Boehringer-Ingelheim (e.g., BILS 22 BS) and Tularik Inc. (e.g., T157602), and others. To date, they are all aminothiazole derivatives and interact with the helicase–primase complex. This comprises a group of three proteins concerned with unwinding the dsDNA and priming the daughter strand during DNA replication. The specific target is thought to be the product of the HSV UL5 gene, the helicase proteins. The compounds appear to be highly potent in tissue culture and in animal models where they appear to be superior to the nucleosides, ACV, and PCV, especially their efficacy upon delayed therapy. However, it remains to be seen whether toxicity will be an issue; furthermore, preliminary evidence suggests that the development of drug resistance may impede their development.

### Virus assembly

As yet, there are no FDA-approved drugs targeting virus assembly but ST-246 is progressing through development. It was discovered as part of the program to prepare for bioterrorism, pox viruses being one of the perceived threats. Investigations with ST-246 resistant virus indicated that the mechanism involves the F13L gene which codes for a major virus envelope protein, p37. Intracellular enveloped viruses (IEVs) are formed from intracellular mature viruses (IMVs) and p37 participates in this wrapping process. ST-246 inhibits this stage and so IEVs are not available to be transported to the cell surface to produce extracellular virus. Exactly how ST-246 acts is being elucidated. Deleting F13L from vaccinia resulted in a virus (delta F13L-Vac) which replicates in cell culture although producing smaller plaques than wild-type (wt) virus. In mice infected with delta F13L-Vac, no lesions were produced but there was a good immune response which protected mice from subsequent challenge with wt virus.

### Neuraminidase inhibitors

Mature virions of influenza virus comprise enveloped particles. The envelope membrane is decorated with two glycoproteins which form morphologically distinct 'spikes'. The most prominent of these are hemagglutinin (HA) molecules. Much less numerous but with a distinctive long-stalked 'mushroom' shape are the neuraminidase (NA) molecules. The function of the latter is to cleave sialic acid from sialylated proteins. Sialic acid is the receptor for the influenza virus on mammalian cells. Thus, NA is thought to help the release of virus from the host cell, preventing re-adsorption to the same cell; the enzyme may also aid the passage of virus through mucus to facilitate the colonization of respiratory or intestinal tissues.

During the early 1990s, the crystal structures of NA and its binding to sialic acid were solved. The NA active site was shown to contain well-formed large and relatively rigid pockets. Molecules were designed to interact with this pocket leading to a series of potent inhibitors. Two compounds with this mechanism of action are zanamivir and oseltamivir. Both are potent inhibitors of influenza

A virus replication attesting to the important functional role of NA in virus replication. Other compounds targeting this function are currently in development.

## Challenges for Antiviral Therapies

### Antiviral Prodrugs

The efficacies of several important antiviral drugs (including the nucleoside analogs acyclovir and penciclovir) are severely limited by poor oral bioavailability. One approach to this has been the synthesis of chemically modified derivatives that are rapidly converted to the nucleosides by the host metabolic enzymes. Valaciclovir, the prodrug for acyclovir, and famciclovir, the prodrug for penciclovir, are readily absorbed yielding high blood levels of the parent compound. Both are used widely for treating herpes simplex and varicella zoster infections. Similarly, other oral prodrugs have been developed: the influenza drug oseltamivir, the HIV drugs tenofovir disoproxil fumarate and fosamprenavir, the HBV drug adefovir dipivoxil and the HCV drug R1626. When new compounds are identified with antiviral activity, this kind of approach may be considered at an early stage of development. For example, famciclovir, rather than penciclovir, was used in the clinical trials and currently R1626, the oral prodrug of R1479, is being developed for HCV therapy.

Besides considering prodrugs for improving oral bioavailability, there are many other potential aims. In order to improve cell permeability of GS9148, a prodrug (GS9131) was selected. GS9131 showed potent anti-HIV activity with mean $EC_{50}$ of 50 nM against multiple HIV-1 isolates. After oral administration to dogs at $3\,mg\,kg^{-1}$, GS9131 effectively delivered the active metabolite, GS9148-DP, at high levels for a prolonged period into PBMCs ($C_{max} = 9.2\,\mu M$, $T_{1/2} > 24\,h$).

A new approach was reported in the *Prusoff Award Lecture* by Dr. Tomas Cihlar (Gilead Sciences, Foster City, CA, USA) at ICAR 2006. The concept is well established that nucleoside analogs, as their triphosphates, are trapped within the cell and therefore give prolonged activity against virus replication. Could the same strategy be applied to HIV protease inhibitors (PI)? Is it possible to make a phosphonate prodrug of a known PI so that, after entering the cell, it is converted to a charged active phosphonate metabolite with enhanced intracellular retention? This approach is currently under investigation.

### Potential for Mutations Leading to Drug Resistance

Most antiviral compounds are highly selective and target a single virus protein. A natural consequence is that point mutations in the virus genome can result in drug-resistant variants. Double-stranded DNA viruses have a relatively low intrinsic error rate ($1 \times 10^{-7}$ to $1 \times 10^{-8}$ errors per nucleotide) and a $3'-5'$ exonuclease proofreading function can edit out errors. In contrast, some RNA viruses have very high intrinsic mutation rates with no proofreading. This high rate of mutation ($\geq 1 \times 10^{-4}$), coupled with a large population of virions, can quickly lead to enormous genetic diversity within a single infected host. For example, HIV has a single-strand RNA genome of approximately 9000 nucleotides. The replication rate in an infected individual has been estimated to be approximately $10^9$ daily; thus, $10^{-4} \times 9000 \times 10^9 = 9 \times 10^8$ mutants occur each day. This means that, in theory, every point mutation occurs $10^5$ times per day in an HIV-infected individual and every double mutant 10 times per day! As a result, HIV actually exists as a 'quasispecies' or 'swarm' around a particular consensus sequence. Similarly, HCV exists as 'quasispecies'; it has the fastest known daily replication rate of $10^{12}$ virions daily. HBV, which has a reverse transcriptase step within its replication cycle, mutates readily but it has overlapping reading frames so that a change in the gene encoding the surface protein may also change the gene for the polymerase. This may limit the number of viable mutations. Influenza virus has an additional mechanism for the spread of resistance. The influenza virion contains eight separate segments of single-stranded RNA genome. When two influenza strains co-infect a single cell, the segments can re-assort and pass a resistance mutation from one strain to another.

Based upon virus mutation rates alone, viruses would evolve to resist any specific inhibitor if it were not highly effective in reducing virus replication, ideally to zero. In practice, this is, almost, achieved with drug combinations but patients have a responsibility for their own therapy – any missed dose gives the virus a chance. Also, resistance mutations usually have a biological price; virus fitness may be compromised.

Mutations, that give rise to amino acid substitutions at the interaction site and cause reduced binding of the inhibitor, are termed 'primary mutations'. Such mutations usually arise early upon exposure to the inhibitor. Further mutations, termed 'secondary mutations', may accumulate and contribute to the level of resistance. Yet further mutations may appear which are apparently unrelated to the interaction site and may have no effect on the resistance level. However, these can increase enzyme efficiency so as to compensate for the deleterious effects of the primary and secondary mutations. Many compensating mutations are suspected but often their precise role has yet to be elucidated.

### Drug Resistance

#### Genetic barrier

When a virus is being inhibited by an antiviral compound, resistance mutations are selected but the ease with which this is done depends upon how many potential mutations can give resistance without compromising viral fitness.

This has become known as the genetic barrier. With monotherapies, resistance appears quickly with NNRTIs but more slowly with NRTIs which target the same reverse transcriptase (RT) but at the catalytic site. For PIs, which target the protease catalytic site, the rate of appearance of resistance is about comparable to that for NRTIs. It was thought that combining one NRTI and one PI would delay the appearance of resistance greatly, but clinical practice showed that the delay was modest. It is more effective to combine two or three compounds which target the same HIV enzyme but have differing mutation patterns. One high genetic barrier is more effective in delaying resistance than two low genetic barriers.

The anti-influenza M2 channel blockers and antipicornarvirus capsid binding compounds are examples of agents which appear to create too low a genetic barrier to become useful clinical therapies.

## Mechanism of Antiviral Action

The most important classes of virus inhibitors are shown in **Figure 1**. This defines the mechanism of action which may be determined by various approaches. The time-of-addition of the compound in relation to its activity may give important clues. Patterns of cross-resistance to previously characterized agents give useful information. Drug-resistant mutants may be selected by passage of virus in suboptimal concentrations of the inhibitor. The location of drug resistance, found by virus genome sequencing followed by site-directed mutagenesis or marker transfer, provides direct evidence for the mechanism and may give information about the precise interaction between compound and the target protein; this may be at or near the active site of an enzyme. In the search for new antiviral compounds, inhibition of a viral enzyme may be the test screen. Computer modeling of compounds binding to their target proteins may aid further optimization and define the mechanism more precisely.

## Virus Fitness

Drug-resistant mutant viruses are often attenuated or 'less fit'. The efficiency of virus replication can be reduced leading to smaller plaque size in cell cultures or reduced pathogenicity when tested in suitable animal infection models. The $TK^-$ strains of herpes viruses resistant to ACV and PCV are usually less pathogenic and less able to reactivate from latency than wt HSV. The His274Tyr influenza virus, resistant to oseltamavir, is less pathogenic in ferrets than wt influenza. In chronic HBV patients, continued therapy with lamivudine, even after the appearance of the YMDD resistant strain, used to be better than no treatment although now patients would be switched to alternative therapy. In HIV patients being treated with PIs, the initial resistant mutations give HIV protease with reduced catalytic activity. Later, a constellation of mutations gives rise to more successful drug-resistant viruses that will then dominate the population.

## Antiviral Combination Therapy

The problem of antiviral resistance led directly to the introduction of antiviral combination therapy as a crucial feature to control chronic virus infections, notably those caused by HIV and probably will be with HBV and HCV. Initially, there was much opposition to the introduction of antiviral combinations, probably best explained by an underlying fear of enhanced toxicity: this was the problem that dominated the early phase of HIV therapy but eased as newer, better tolerated drugs became available. HIV presented a new urgency; very high levels of HIV genome RNA turnover in the patient and the high mutation rate in the HIV genome meant that antiviral resistance inevitably led to the failure of monotherapies. The concept of genetic barrier was developed at this time. The aim is to create the highest possible genetic barrier and stop all virus replication so that the virus has no chance to overcome that barrier by mutation. This ultimately led to HAART involving the use of triple and quadruple combinations of active compounds. Experience has shown that sequential use of monotherapies is undesirable. Switching from one drug to another enables the sequential development of multiple resistant strains. Current guidelines recommend that at least two RT inhibitors are included in any HAART regimen. This has been very successful in controlling the disease but put a heavy burden on patients who had to take many pills at varying times through the day. So companies developed combinations of their drugs, the first being Combivir (**Table 1(d)**). Recently, two major pharmaceutical companies have been cooperating in the development of a joint formulation in which three antivirals are combined within a single pill (Atripla) leading to a greatly improved convenience and, hopefully, compliance. This first example has started a trend. A similar approach is considered applicable to HBV. For example, telbivudine and valtorcitabine together are being evaluated in clinical trials.

Favorable drug/drug interaction is another rationale for combining drugs. Nearly all current selective antiviral compounds are considered to be virustatic agents. This concept has had major implications in the treatment of HIV. Continual antiviral blood levels of drugs, with intervening troughs, are not sufficient for control of HIV replication. There must be continuous high levels which well exceed the antiviral concentration for that drug. Ritonavir, by blocking the elimination of other PIs, enhances both the concentration and duration of the PI in patients. Lopinavir is co-formulated with ritonavir in a 4:1 ratio primarily to enhance the pharmacokinetics of lopinavir (**Table 1(d)**).

When the symptoms of an infection require treatment, then combining an antiviral agent with an immunomodulator, anesthetic, or anti-inflammatory substance may give added benefit. Thus, acyclovir has been combined with cortisone to simultaneously reduce pain and irritation of the herpes lesion at the same time reducing the replication of infectious virus which would otherwise be prolonged by the anti-inflammatory component. This concept is likely to be widened to other cases in which the inflammatory response plays an important role in the disease, for example, influenza.

## Virus Latency

Virus latency remains an obstacle to successful antiviral chemotherapy for which no solution appears to be in sight. Members of the *Herpesviridae* are notable for being able to establish latent infections; with HSV and VZV, latency is established in neuronal cells. During latent HSV-1 infection, the latency-associated transcript (LAT) of HSV-1 is the only known viral gene expressed. A micro-RNA (miR-LAT), which is generated from the exon 1 region of the LAT gene, reduces the induction of apoptosis and so contributes to the survival of that latently infected neuron. Few, if any, HSV proteins are expressed, so affording little chance for either the immune system or antiviral drugs to control the infection. Latency can be established despite ongoing antiviral chemotherapy although there is some evidence from animal models that famciclovir may be more effective than valaciclovir in reducing the load of latent virus when given very early during experimental infections. It still remains to be demonstrated whether or not this has any practical value in patients where therapy shortly after primary exposure is unlikely to be possible. While long-term therapy using nucleoside analogs such as valaciclovir or famciclovir are very successful in suppressing recrudescent lesions in a latently infected individual, on termination of suppressive therapy reactivation leading to recurrent disease is prone to occur. Other viruses, notably HIV, also establish latency in a few long-lived cells, thus making the complete eradication of these infections especially difficult to achieve.

## Future Prospects

### Herpes

Some bicyclic nucleoside analogs (BCNAs) had been found to be exceptionally active and selective for VZV. A lead compound is Cf 1743. Like ACV and PCV, Cf 1743 is activated by VZV TK but, unlike ACV and PCV, it seems not to act via its triphosphate. Therefore, it will be particularly interesting to evaluate this highly active compound for its ability to reduce postherpetic neuralgia (PHN) in patients with shingles. Cf 1743 as its valyl ester prodrug (FV 100) has recently entered development and progress has been reported on large-scale synthesis and initial toxicity testing.

Helicase-primase inhibitors (see above) represent a new class of compounds active against the herpes viruses. During the last decade, there has been a notable lack of new classes of antiherpes compounds. It will be interesting to see how they progress.

### HBV

Since 1999, four drugs have received FDA approval, lamivudine, adefovir dipivoxil, entecavir, and telbivudine (**Table 1(c)**). Although entecavir is the most active compound on a weight basis, the greatest reductions in HBV DNA levels have often been achieved with the most recently (2007) approved compound, telbivudine. In addition, some interesting compounds are progressing through development.

Telbivudine, in phases IIb and III, has been consistently more active than lamivudine against HBV. Because HBV strains (with rtM204I and rtL180M mutations in the polymerase) are cross-resistant to both telbivudine and lamivudine, telbivudine should not be used to treat patients with HBV resistant to lamivudine. However, it may be expected that the greater control of HBV replication with telbivudine than with lamivudine would lead to a lower proportion of patients with resistant virus. Telbivudine, in both pre-clinical toxicological tests and clinical trials, has shown a remarkably good safety record. Therefore, telbivudine may become the first line drug of choice, replacing lamivudine. Since telbivudine and valtorcitabine showed synergy in both cell culture assays and in the woodchuck hepatitis model, these two compounds are being evaluated together in phase IIb trials. Only time will tell if this combination is also able to reduce the rate of appearance of resistance. Telbivudine preferentially inhibits the second strand of HBV DNA synthesis and valtorcitabine inhibits both the first and second strand of HBV DNA synthesis; HBV would have to overcome two stages in its replication cycle and the genetic barrier (especially for the second strand synthesis) would be increased.

With most HBV therapies, although there may be a short lag period after treatment, HBV DNA levels return to baseline. This is typical for drugs which are virustatic. In contrast, clevudine showed a prolonged effect. In a phase I/II trial, at a daily dose of 100 mg, HBV DNA levels were reduced by $3.0 \log_{10}$ at the end of 4 weeks of therapy and were still $2.7 \log_{10}$ below baseline at 6 months. However, until an antiviral mechanism has been elucidated for this prolonged effect, one is not sure if this is a great benefit or a warning sign. If there is an antiviral mechanism, then clevudine clearly differs from other anti-HBV drugs and cannot be regarded as a virustatic drug. We suggest that an antiviral compound, which acts via inhibition of the replication cycle and destroys the ability of the virus to continue replication, should be described by a new term, a virureplicidal drug.

## HCV

Progress has been severely restricted by the lack of suitable test systems. It seems ironic that, in patients, the production rate of HCV is $10^{12}$ virions/day, the highest known rate of any virus to date and yet it has taken so long to discover a replication system in cell culture. In 1999, an HCV replicon system became available. No infectious particles are formed but viral RNA synthesis can be followed. Alternatively, HCV pseudoparticles can be obtained by adding the HCV genes for E1(GP35) and E2(gp66) to a partial HIV. The resulting hybrid virus can generate pseudoparticles which incorporate E1 and E2 as heterodimers. These pseudoparticles can be used to raise neutralizing antibody against HCV.

Only in 2005 was full HCV virus replication achieved by Wakita *et al*. JHF1 cDNA was cloned from an individual with fulminant HCV, strain 2a. After transfection into human hepatoma cell line (Huh7), the JFHI genome replicates and leads to the secretion of viral particles. These virions were then used to infect Huh7 cells and chimpanzees. Essentially, only this combination of JHF1 and Huh7 cells has been successful although some chimeric viruses, with the E1 and E2 genes of JFH1 replaced by those from another strain, can also replicate in Huh7 cells.

The availability of the HCV replicon system allowed screening for HCV RNA polymerase inhibitors. One example is R1479 which has progressed through to phase I/II trials. Its activity in HCV-infected patients was sufficient to encourage the development of the oral prodrug, R1626. An alternative approach is the inhibition of the HCV NS3/NS4A protease. Two compounds in phase II studies are VX-950 and SCH 503034. Both gave reductions in HCV RNA levels when used as monotherapies and both appear to have additive or synergistic activity with interferons. A potential limitation of these compounds is the requirement to be dosed 3 times daily but the addition of ritonavir improves the pharmacokinetics such that less frequent dosing may be possible, just as for the HIV protease inhibitors.

## Influenza

Since 1999, two neuraminidase inhibitors, zanamivir and oseltamivir, have received FDA approval. Further development of other compounds became slow due to lack of interest but the threat of avian influenza A (H5N1) has stimulated renewed interest. Zanamivir dimer has a long duration of activity, about a week. Peramivir failed to lessen the duration of symptoms in a phase III trial but an oral prodrug may be worth trying.

T-705 (6-fluoro-3-hydroxy-3-pyrazine carboxamide), as its ribo-triphosphate (T-705RTP), inhibits the influenza virus polymerase in a GTP-competitive manner. In cell culture assays, it is active against many strains of A, B, and C influenza viruses. In mice infected with an H5N1 strain, T-705 was highly effective and, on rechallenge with 100-fold more virus on day 21, all the T-705 treated mice survived. T-705 seems to be a likely future therapy.

## HIV

Over the last decade, there have been some big gains (reductions in morbidity and mortality) and harsh realities (replication rate $10^{10}$ virions/day, long-lived latent pool, side effects and viral resistance). Prospects for a vaccine remain poor but new compounds are being developed, including entry and integrase inhibitors. Great efforts have been made to make HIV therapies available to resource-limited areas. Because so many factors differ between areas, outcomes are most likely to differ. Most importantly, research, studying clinical outcomes in these resource-limited settings, should be continued. Otherwise resistant virus may be unknowingly spread around. Several firms have agreed to support the creation of centers of excellence. A great challenge is to investigate the role that therapeutics might play in preventing transmission.

A novel therapeutic concept was described by Jan Balzarini (Rega Institute for Medical Research). HIV gp120 is heavily glycosylated with about 11 'high-mannose' type glycans/gp120 molecule. Pradimicin A (PRM-A) can be considered as a prototype. A high genetic barrier is created by several PRM-A molecules binding to each HIV gp120 protein; resistance can occur in cell culture but only when there have been a large number (>5) of mutations leading to the loss of glycosylation sites. In HIV-infected patients, every loss of a glycosylation site presumably exposes part of the HIV gp120 protein surface to the immune system. Thus, the more resistant the HIV becomes toward PRM-A, the more immunogenic it may become. This is a 'hard choice' which has not yet been presented to HIV with any current therapy. Could the same approach be successful with HBV and HCV?

## Conclusions

### Niche Targets for Antivirals

Many viruses are well controlled by vaccines; the most outstanding instances are smallpox and polio. Cases of rubella, munps, and measles have been greatly reduced by routine vaccination. Vaccines, mainly used to protect travellers, are available for hepatitis A, rabies, yellow fever, and others. New vaccines can prevent those cervical cancers caused by papillomavirus types 16 and 18. Although these two types account for the majority of cervical cancers, activity against many other types would be required to give full protection. An antiviral could possibly be active against many strains and be effective in an established infection. However, the papillomavirus genome is so small, not even encoding a DNA polymerase, there are few targets. New infections with VZV and HBV are preventable with vaccines but there remains a large pool of infected patients for whom antiviral therapy is required. The threat of avian influenza has led to the stockpiling of anti-influenza drugs. The lack of effective vaccines for HSV, HIV, and HCV has encouraged antiviral therapies. West Nile virus is an example of a newly emerging virus for which there is neither vaccine nor specific antiviral.

## Broad Spectrum Antivirals?

Antiviral chemotherapy has come a long way since the first faltering steps with marboran, amantadine and rimantadine, idoxuridine (IDU), and vidarabine (Ara A). Acyclovir set the standard for effectiveness together with clinical safety. For over two decades, clinical safety has been associated with high selectivity for a single virus or a few closely related viruses. So far, there are no FDA-approved antiviral drugs, targeting viral replication, which are active against a broad spectrum of viruses. However, Rep 9 may be the first example.

*See also:* Human Immunodeficiency Viruses: Antiretroviral agents.

## Further Reading

Anderson RM and May RM (1992) *Infectious Diseasese of Humans Dynamics and Control.* New York: Oxford University Press.
De Clercq E (2005) Recent highlights in the development of new antiviral drugs. *Current Opinion in Microbiology* 5: 552–560.
De Clercq E (2005) Antiviral drug discovery and development where chemistry meets with biomedicine. *Antiviral Research* 67: 56–75.
De Clercq E, Branchale A, Vere Hodge A, and Field HJ (2006) Antiviral chemistry and chemotherapy's current antiviral agents FactFile 2006. *Antiviral Chemistry and Chemotherapy* (1st edition) 17: 113–166.
De Clercq E and Field HJ (2006) Antiviral prodrugs – the development of successful prodrug strategies for antiviral chemotherapy. *British Journal Pharmacology* 147: 1–11.
Field AK and Loughlin CA (1999) Antivirals. In: Granoff A and Webster RG (eds.) *Encyclopedia of Virology*, 2nd edn., pp. 54–68. New York: Academic Press.
Field HJ and De Clercq E (2004) Antiviral drugs: A short history of their discovery and development. *Microbiology Today* 31: 58–61.
Field HJ and Whitley RJ (2005) Antiviral chemotherapy. In: Mahy BWJ and ter Meulen V (eds.) *Topley and Wilson's Microbiology and Microbial Infections,* 10th edn., vol. 2, pp. 1605–1645. London: Hodder Arnold.
Schang LM, St. Vincent MR, and Lacasse JJ (2006) Five years of progress on cyclin-dependent kinases and other cellular proteins as potential targets for antiviral drugs. *Antiviral Chemistry and Chemotherapy* 17(6): 293–320.
Wakita T, Pietschmann T, Kato T, *et al.* (2005) Production of infectious hepatitis C virus in tissue culture from a cloned viral genome. *Nature Medicine* 11(7): 791–796.
Wutzler P and Thust R (2001) Genetic risks of antiviral nucleoside analogues – a survey. *Antiviral Research* 49: 55–74.

# Apoptosis and Virus Infection

**J R Clayton and J M Hardwick,** Johns Hopkins University Schools of Public Health and Medicine, Baltimore, MD, USA

© 2008 Elsevier Ltd. All rights reserved.

## Glossary

**Bcl-2** Founding member of anti-apoptotic Bcl-2 protein family; homologs found in cells and viruses.
**Caspase** Family of cysteine proteases that cleave after aspartate residues.
**DD/DED/CARD** Structurally related protein–protein interaction domains found in a variety of different proteins (e.g., death receptors, adaptor molecules, and caspases).
**FLICE** Caspase-8 (original name: Fas-associated death domain protein interleukin-1beta-converting enzyme (ICE)-like).
**FLIP** Family of caspase-8 (FLICE)-like inhibitory proteins encoded by viruses (vFLIPs) and cells (cFLIPs).
**IAP (inhibitor of apoptosis)** Family of proteins with BIR (baculovirus IAP repeats) and a RING finger; homologs found in cells and viruses.
**Micro-RNA** Small hairpin RNAs that regulate mRNAs through RNAi or translational inhibition.
**NF-κB** Transcription factor that when activated translocates to the nucleus and regulates a variety of cellular and viral functions.
**RNAi (RNA interference)** Mechanism by which RNA is targeted for degradation in a sequence-specific manner.
**TRAILR** TRAIL (tumor necrosis factor-related apoptosis inducing ligand) receptor.

## Introduction

Programmed cell death is any process by which cells participate in their own death. Cell suicide programs are facilitated by the actions of gene products encoded by the cell destined to die. These death-promoting genes evolved, at least in part, for the purpose of orchestrating cell autonomous death. The term apoptosis describes the morphology of naturally occurring programmed cell death during development and following certain pathological stimuli. In contrast to this deliberate cell death, the term necrosis describes cell morphology resulting from accidental cell death caused by acute injury. However, necrosis is less well defined, and is also used to describe the appearance of cells undergoing nonapoptotic programmed cell death. In addition, the term apoptosis can be used as a mechanistic term that refers to all types of programmed cell death. In multicellular organisms, programmed cell death is essential for development, tissue homeostasis, and regulating immune responses. In humans, it is estimated that billions of cells normally die per day by programmed death. Insufficient or excessive cell death characterizes most human disease states, including virus infections.

Classical apoptosis is characterized by a number of morphological and biochemical changes, including condensation of chromatin, cleavage of DNA between nucleosomes (DNA laddering), plasma membrane blebbing, exposure of phosphatidyl-serine on the outer leaflet of the plasma membrane (detected by annexin V binding), and fragmentation/division of mitochondria. These classic apoptotic events are caused by the activity of a subset of intracellular proteases known as caspases. Caspases are a family of cysteine proteases that cleave after specific aspartate residues in selected cellular proteins. A second subset of caspases includes the proteases that activate inflammatory mechanisms and may promote nonapoptotic programmed cell death. Other nonapoptotic programmed cell death pathways involve other types of proteases such as lysosomal enzymes, but less is known about these pathways, and little is known about their role in virus infections.

Any perturbation of the cell can potentially lead to activation of programmed cell death as a host defense mechanism to eliminate aberrant or damaged cells. Similarly, virus infection of a cell usually triggers the activation of programmed cell death. The ability of cells to recognize intruding viruses and activate cell suicide provides an important host defense mechanism for eliminating viruses by eliminating virus-infected cells prior to mounting host immune responses. However, viruses have developed myriad mechanisms to adapt to and regulate cellular death processes. As a consequence, many viruses cause disease primarily by inducing host cell death.

## Cellular Death Pathways

Of the multiple pathways leading to cell death, caspase-dependent apoptosis is the best characterized. There are two general pathways for activating caspases, the receptor-activated (extrinsic) pathway, and the mitochondrial (intrinsic) pathway. Different viruses regulate multiple steps in both of these pathways.

Receptor-mediated pathways for the activation of caspases can be further subdivided into two groups: cell surface death receptors of the tumor necrosis factor receptor (TNFR) superfamily, and the intracellular pattern recognition receptors that activate primarily inflammatory caspases. The cytoplasmic portions of death receptors recruit and activate long pro-domain caspases (e.g., caspase-8). In turn, caspase-8 cleaves and activates the short pro-domain caspases (e.g., caspase-3). Caspase-3 is responsible for cleaving most of the known cellular substrates to facilitate cell death. Death receptor pathways are regulated by both viral and cellular factors that bind the receptor or other associated components in a manner that prevents caspase activation (**Figure 1**). The second receptor subgroup includes the pattern recognition receptors (PRRs), which can be further subdivided into three groups: the plasma membrane toll-like receptor family and two intracellular receptor families, the NOD-like receptors (NLRs), and the RIG-like helicases (RLHs), which are capable of detecting intracellular invaders. NLRs are activators of inflammatory caspases. Of particular relevance to virus infections, the intracellular RLH receptors recognize viral nucleic acids and trigger host defense mechanisms.

The mitochondrial pathway for activating caspases is triggered when cytochrome $c$ is released from the intermembrane space of mitochondria into the cytoplasm. Released cytochrome $c$ and ATP bind to Apaf-1 resulting in oligomerization of Apaf-1 and its binding partner caspase-9. Like the oligomerization-induced activation of caspase-8 by death receptors, oligomerization of the DED-like caspase-recruitment domain (CARD) motifs of caspase-9 also serves to activate the protease. Thus, initiator caspases (e.g., caspases-8 and -9) are activated by oligomerization, while effector caspases (e.g., caspases-3 and -7) are activated by proteolytic processing. Like caspase-8, activated caspase-9 also cleaves and activates caspase-3 to mediate apoptotic cell death. The mitochondrial pathway is regulated by pro- and anti-apoptotic cellular Bcl-2 family proteins that promote or inhibit cytochrome $c$ release. Pro-apoptotic Bcl-2 family members Bax and Bak form homo-oligomers in the mitochondrial outer membrane that directly or indirectly release cytochrome $c$ and other mitochondrial factors. This function of Bax and Bak is inhibited directly or indirectly by anti-apoptotic Bcl-2 family members Bcl-2, Bcl-$x_L$, and others. A more diverse subclass of Bcl-2-related proteins, the BH3-only proteins, inhibits antideath family members and/or promotes the functions of Bax and Bak. Cross-talk between the mitochondrial pathway and the receptor-activated pathway occurs when receptor-activated caspase-8 cleaves and activates the BH3-only protein Bid, leading to Bax- or Bak-mediated cytochrome $c$ release. Cross-talk serves to amplify the apoptotic death pathway. Viruses also modulate multiple steps in the mitochondrial cell death pathway (**Figure 1**). For example, many viruses encode homologs of cellular anti-apoptotic Bcl-2 family proteins or other antagonists of

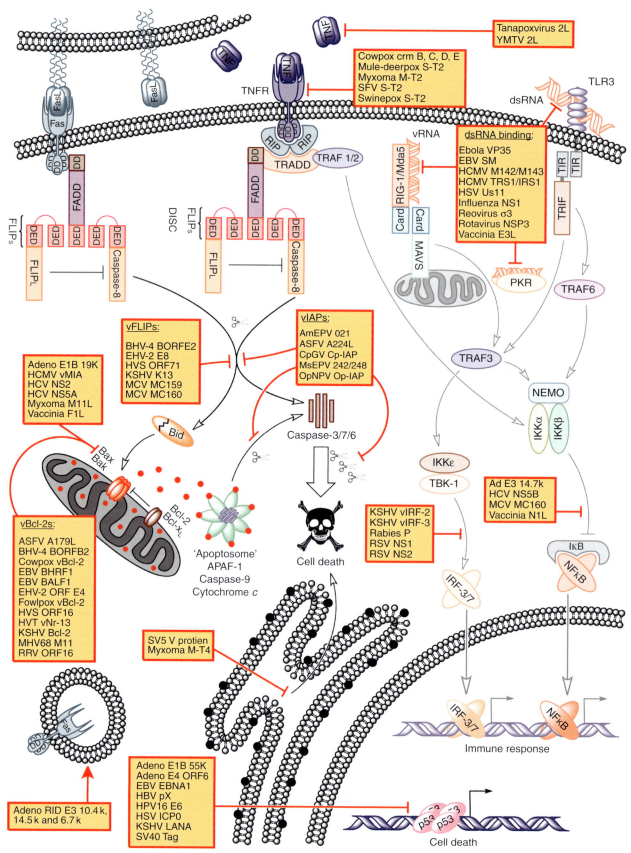

**Figure 1** Cell death signalling during virus infection. Abbreviations are as in **Table 1**. Grayed areas indicate innate immunity signalling. Viral cell death regulators (yellow boxes, red arrows); cellular death regulators (black arrows); proteolytic cleavage events (scissors); ribosomes (black circles); cytochrome c (red circles).

Bax/Bak-mediated cell death. Still other viruses encode direct inhibitors of caspases, both pro-apoptotic and pro-inflammatory caspases.

Viruses also cause nonapoptotic cell death, but the mechanisms are less well characterized. Nonprogrammed necrotic cell death may simply be a consequence of extensive virus-induced cell injury after the virus has successfully inhibited apoptosis. Nonapoptotic cytopathologies induced by viruses include the formation of membranous vacuoles (often the sites of virus replication), activation of autophagy (lysosome-mediated degradation), and plasma membrane rupture/lysis. Viruses also trigger inflammation, which is mediated in part by caspases that cleave and activate pro-inflammatory cytokines, such as caspase-1 cleavage and activation of IL-1β. However, even if nonapoptotic death mechanisms are facilitated by cellular gene products, they may or may not cause programmed cell death. For example, cell death caused by lysosomal proteases that are released accidentally in damaged cells is not programmed if the pro-death function of lysosomal proteases did not evolve, at least in part, for the purpose of promoting cell suicide (to benefit the whole organism). However, this pathway is programmed if cathepsin cleavage sites in the pro-apoptotic BH3-only protein Bid, or other substrates, were selected during evolution to facilitate purposeful cell death during infection. Both programmed and non-programmed cell death mechanisms induced by viruses can be useful targets for therapeutic intervention.

## Death Receptor Signaling

Death-inducing receptors on the cell surface can be placed into two subgroups, one containing Fas/CD95 and the other containing tumor necrosis factor receptor 1 (TNFR1). Fas recruits caspase-8 to directly and potently activate cell death. TNFR1 recruits different factors to form a distinct complex to inefficiently induce apoptosis or to activate other cellular signaling pathways.

Binding of Fas ligand (FasL) to Fas, or binding of TRAIL to death receptors 4 and 5 (DR4 and DR5), triggers the cytoplasmic tails of these receptors to form a complex known as the death-inducing signaling complex (DISC). The death domain (DD) in the cytoplasmic tail of the receptor binds directly to the DD domain of the small adaptor protein Fas-associated death domain (FADD). The DD-related death-effector domain (DED) of FADD binds directly to the DED domains in the N-terminus of caspase-8 to oligomerize caspase-8 in the DISC. In the other subgroup of death-inducing receptors, tumor necrosis factor-alpha (TNF-α) binds to TNFR1, allowing TNFR1 to recruit intracellular adaptor molecules receptor-interacting protein (RIP) kinase, TNFR-associated factors (e.g., TRAF2), and TNFR1-associated death domain (TRADD) protein to assemble a complex that dissociates from the membrane, recruits FADD and caspase-8 that in turn cleaves and activates the effector caspases (e.g., caspase-3,-7) to cause cell death. Both cellular and viral factors regulate cell death and other functions of TNFR1 and related receptors. TNFR1 can recruit additional factors, including the cellular inhibitors of apoptosis (e.g., cIAP1 and cIAP2), and can cause pleiotropic effects on cells, including the activation of the pro-inflammatory transcription factor NF-κB.

## Viral Regulators of Receptor Signaling

Viruses encode a plethora of molecules that interfere with signaling by extracellular ligands and their respective receptors. In some cases, these viral factors mimic TNFR1, such as the cytokine response modifier (CRM) proteins of cowpox virus, M-T2 protein of myxoma virus, and S-T2 orthologs encoded by several poxviruses. These soluble or membrane-bound receptor mimics act as TNFα-sinks, thereby diminishing the frequency of the interaction between TNFα and the host receptor. Similarly, the viral TNF-binding protein (vTNF-BP) 2L from tanapox (and several other poxviruses) sequesters TNF directly, but bears amino acid similarity to MHC class I rather than TNFR. Another TNFR-superfamily receptor mimic found in cowpox and mousepox viruses is vCD30, which diminishes signaling to the host cell and the proliferation of B and T cells via CD30L, dampening the immune response to infection (**Table 1**).

Other viruses alter intracellular mechanisms of death receptor signaling. The latency membrane protein 1 (LMP1) of Epstein–Barr virus (EBV) acts like a constitutively active (ligand-independent) TNFR-superfamily receptor capable of interacting with multiple TRAFs to inhibit apoptosis. Herpes simplex virus (HSV) glycoprotein D binds a TNFR-related protein as a receptor (HVEM) for entry into the cell, resulting in the activation of NF-κB. Wild strains of human cytomegalovirus (HCMV) encode a HVEM-like molecule UL144 that is missing in laboratory strains. Adenovirus encodes proteins E3–10.4k, E3–14.5k, and E3–6.7k that can form the 'receptor internalization and degradation' or RID complex that binds to the cytoplasmic portion of several different death receptor-superfamily members (including TNFR1, Fas, and TRAILR). Several different viruses target the adaptor TRAF2 to redirect cell death signals toward activation of NF-κB. Herpesvirus saimiri (HVS) StpC, KSHV K15, and rotavirus VP4 also activate NF-κB. Similarly, hepatitis C virus (HCV) nonstructural protein 5A (NS5A) can redirect the death signal by interacting with TRADD to inhibit its association with FADD, thus inhibiting death signaling, while HCV core protein selectively recruits FADD to TNFR1 to promote cell death. NS5A is also known to inhibit TRAF2, further directing the death signal away from NF-κB activation and survival signaling. These are only a few of the many examples

**Table 1** Virus-encoded cell death modulators, listed alphabetically by virus family

| Family | Name | Viral gene | Function |
|---|---|---|---|
| Adenoviridae | Adenovirus | E1A | Downregulates cFLIP$_S$, induces TNF-mediated caspase-8 activation, inhibits NF-κB |
| | | E1B-19k | Binds and inhibits Bax/Bak |
| | | E1B-55k | Ubiquitination of p53 |
| | | E3 (10.4k, 14.5k, 6.7k) | Internalization of TNF-related death receptors and their subsequent degradation |
| | | E3-14.7k | Binds IKK and p50, suppresses NF-κB; inactivates caspase-8 |
| | | E4-ORF6 | Ubiquitination of p53 |
| | | VAI | Inhibits Dicer as pseudo-substrate |
| | | VAI | Inhibits PKR as pseudo-substrate |
| Asfarviridae | ASFV[a] | A179L | Bcl-2 homolog, inhibits MMP and cell death |
| | | A224L | vIAP, activator of IKK |
| | | A238L | Non-phosphorylatable IκB ortholog |
| Baculoviridae | AcMNPV[b] | P35 | Broad spectrum caspase inhibitor |
| | BmNPV[c] | P35 | Broad spectrum caspase inhibitor |
| | CpGV[d] | Cp-IAP | Binds caspases; Ub-ligase |
| | OpNPV[e] | Op-IAP | Binds caspases, Ub-ligase |
| | SlNPV[f] | P49 | P35-related caspase inhibitor |
| Bunyaviridae | LAC[g] | NSs | Suppresses RNAi |
| | SAV[h] | NSs | Inhibits transcription/translation; Drosophila reaper-like |
| Filoviridae | Ebola virus | VP35 | Binds dsRNA |
| Flaviviridae | HCV[i] | Core | Enhances Bid cleavage and TRAIL signaling, binds TNFR1 |
| | | E2 | PKR pseudo-substrate, interacts with PKR |
| | | IRES | vRNA; competitive inhibitor of PKR |
| | | NS2 | Interacts with CIDE-B and prevents MMP |
| | | NS3/4A | Cleaves MAVS, inhibits interferon response; cleaves TRIF, inhibits recognition by TLR3 |
| | | NS5A | Direct interaction with PKR; inhibits TRADD/FADD association; interacts with TRAF2, inhibits NF-κB; binds Bax |
| | | NS5B | IKKα-specific inhibitor of IKK complex |
| Hepadnaviridae | HBV[j] | pX | Direct binding to and inhibition of p53, activation of Src, MAPK, and NF-κB |
| Herpesviridae | BHV-4[k] | BORFB2 | Bcl-2 homolog, inhibits MMP and cell death |
| | | BORFE2 | vFLIP, modulates caspase-8 activation |
| | EBV[l] | BALF1 | Suppresses BHRF1 indirectly |
| | | BHRF1 | Bcl-2 homolog; inhibits cell death |
| | | BZLF1 | Downregulates TNFR1 expression |
| | | EBER1,2 | PKR pseudoactivator |
| | | EBNA1 | Transcriptionally downregulates HAUSP, leads to p53 poly-Ub |
| | | LMP1 | Constitutively active CD40, important for transformation |
| | | miR-BHRF1 | Latency-related |
| | | SM | Binds dsRNA, interacts with PKR |
| | EHV-2[m] | E8 | vFLIP, modulates caspase-8 activation |
| | | ORF E4 | Bcl-2 homolog, inhibits MMP |
| | HCMV[n] | IE1 | Transcriptional regulator of NF-κB effectors |
| | | IE1/IE2 | Affects TNFR trafficking |
| | | M142/M143 | Binds dsRNA |
| | | TRS1, IRS1 | Binds dsRNA |
| | | UL144 | TNFR Superfamily; NF-κB activation |
| | | UL36/vICA | Non-FLIP caspase-8 inhibitor |
| | | UL37/vMIA | Binds ANT, GADD45; promotes mitochondria fragmentation but inhibits cell death |
| | HSV[o] | γ(1)34.5 | eIF2α phosphatase |
| | | ICP0 | Sequesters the deubiquitinating enzyme HAUSP, inactivates p53 |
| | | LAT | Micro-RNA of TGF-β and SMAD3, suppresses stress response |
| | | Us11 | Binds dsRNA, interacts with PKR |
| | | gD | Binds TNFR-related HVEM, activates NF-κB |
| | | US3 | vSer/Thr Kinase; modifies Bad, blocks Bad induced death |
| | HVS[p] | ORF 16 | Bcl-2 homolog, inhibits MMP |
| | | ORF 71 | vFLIP, modulates caspase-8 activation |
| | | StpC | Interacts with TRAF2, activates NF-κB |

Continued

**Table 1** Continued

| Family | Name | Viral gene | Function |
|---|---|---|---|
| | | Tip | Lck Adaptor; NF-κB activation |
| | HVT[q] | vNr-13 | Bcl-2 homolog, inhibits MMP |
| | KSHV[r] | K1 | vTyrosine kinase, constitutive Lyn activation |
| | | K13 | vFLIP, modulates caspase-8 activation |
| | | K15 | Interacts with TRAF2, activates NF-κB |
| | | K7 | Complexes with Bcl-2/caspase-3; regulates $Ca^{2+}$ levels |
| | | KSBcl-2 | Bcl-2 homolog, inhibits MMP |
| | | LANA | Suppresses p53 transcription |
| | | vIRF-2 | Direct interaction with PKR |
| | | vIRF-3 | Inhibits IKKβ and NF-κB activation |
| | Marek's disease virus | MEQ | Transcriptional inhibitor of apoptosis through TNFR, ceramide, UV and serum withdrawal |
| | MCMV[s] | Unknown | Disregulates TNFR levels |
| | MHV68[t] | M11 | Bcl-2 homolog, inhibits MMP |
| Iridoviridae | RRV[u] | ORF16 | Bcl-2 homolog, inhibits MMP |
| | ATV[v] | ReIF2H | v-eIF2α; not phosphorylatable |
| Nodaviridae | FHV[w] | B2 | Binds dsRNA |
| | | | Orthomyxoviridae |
| Influenza virus | NS1 | Binds dsRNA | |
| | | NS2 | Triggers RIG-I |
| | | PB1-F2 | Mitochondrial fragmentation, MMP loss and cytochrome c release |
| Papillomaviridae | HPV16[x] | E6 | eIF2α phosphatase; targets p53 for ubiquitination and degradation |
| | | E6/E7 | Interferes with DISC formation |
| Paramyxoviridae | Hendra virus | V | Inhibits Mda5 |
| | Mumps virus | V | Inhibits Mda5 |
| | RSV[y] | M2-1 | Associates with RelA, activates NF-κB |
| | | NS1, NS2 | Interferes with IRF3 activation |
| | Sendai virus | C | Abolishes IFN/STAT signaling |
| | | V | Inhibits Mda5 |
| | SV5[z] | V | Inhibits Mda5; suppressor of ER-stress-mediated cell death |
| Picornaviridae | Coxsackie B3 | Unknown | Degradation of Cyclin D1, p53, and β-catenin |
| | Poliovirus | 3A | Affects TNFR trafficking |
| | | 3C | Cleaves RelA |
| Polydnaviridae | MdBV[aa] | H4, N5 | vIκB, inhibit NF-κB translocation |
| Polyomaviridae | SV40[bb] | miR-S1 | Suppresses TAg and early transcripts in late stages; immune evasion |
| | | TAg | eIF2α phosphatase; binds and inactivates p53 |
| Poxviridae | AmEPV[cc] | 021 | vIAP; IAP-binding motif antagonist |
| | Cowpox virus | CrmA/B12R | Serpin and caspase-1, 8, 10 inhibitor |
| | | CrmB,C,D,E | TNFR/death receptor mimic |
| | | vBcl-2 | Bcl-2 homolog, inhibits MMP |
| | | vCD30 | CD30 homolog that dampens proliferative signal to B and T Cells |
| | DPV[dd] | S-T2-like | TNF-receptor decoy; sequesters TNF |
| | Fowlpox virus | vBcl-2 | Bcl-2 homolog, inhibits MMP |
| | MCV[ee] | MC066L | vGlutathione-peroxidase; protects from UV and peroxide stress |
| | | MC159 | vFLIP, interacts with FADD, TRAF3 and caspase-8; promotes NF-κB activation |
| | | MC160 | IKKα-specific inhibitor of IKK complex and NF-κB activation; vFLIP; inhibits Fas signaling; binds FADD and caspase-8 |
| | Mousepox virus | P28/012 | Anti-apoptotic ubiquitin ligase |
| | | vCD30 | CD30 homolog that dampens proliferative signal to B and T Cells |
| | MsEPV[ff] | 242 | vIAP; IAP-binding motif antagonist |
| | | 248 | vIAP; IAP-binding motif antagonist |
| | Myxoma virus | E13L | Interacts with ASC; inhibits caspase-1 activation |
| | | M11L | Inhibits MMP and cell death |
| | | MNF | vIκB, inhibits NF-κB translocation |
| | | M-T2 | TNF-receptor decoy; sequesters TNF |
| | | M-T4 | Inhibits apoptosis, localizes to ER |
| | | Serp2/M151R | Caspase and granzyme B inhibitor |
| | SFV[gg] | P28/N1R | Anti-apoptotic ubiquitin ligase |
| | | S-T2 | TNF-receptor decoy; sequesters TNF |

Continued

## Table 1 Continued

| Family | Name | Viral gene | Function |
|---|---|---|---|
| | Swinepox virus | S-T2-like | TNF-receptor decoy; sequesters TNF |
| | Tanapox virus | 2L | vTNF-BP; sequesters TNF |
| | Vaccinia virus | A46R | TLR3 decoy receptor, TIR domains only |
| | | CrmA/B13R | Serpin and caspase-1, 8, 10 inhibitor |
| | | E3L | Binds dsRNA, independently inhibits PKR |
| | | F1L | Interacts with Bim and Bak to inhibit MMP |
| | | K1L | Prevents IκB degradation, inhibits NF-κB |
| | | K3L | v-eIF2α, inhibits PKR function |
| | | N1L | Inhibits TRAF2/6 and IKKα/β |
| | YMTV[hh] | 2L-like | vTNF-BP; sequesters TNF |
| Reoviridae | Reovirus | σ3 | Binds dsRNA |
| | | Unknown | Prevents IκB degradation |
| | Rotavirus | NSP3 | Binds dsRNA |
| | | VP4 | TRAF2 binding, IKK activating impairs caspase-8 activation |
| Retroviridae | BLV[ii] | Unknown | Disregulation of TNFR cell surface expression |
| | HIV[jj] | gp120 | Interacts with CXCR4, causes apoptosis in CTLs; cross-links CD4, induces Fas expression in uninfected cells |
| | | miR-N367 | Suppresses Nef, suppresses its own replication |
| | | Nef | Enhances NF-κB signaling and TNF-mediated apoptosis; reduces MHC-I, CD4 expression; phosphorylation and inhibition of Bad |
| | | Protease | Cleaves caspase-8, degradation of Bcl-2 |
| | | Tar | vRNA binds TRBP; suppresses RISC |
| | | Tat | PKR pseudo-substrate; enhances NF-κB signaling, TNF-apoptosis |
| | | Vpr | G2/M arrest; glucocorticoid-like; Bax-dependent apoptosis |
| | | Vpu | Suppresses NF-κB activation |
| | HTLV-1[kk] | Tax | IKKγ adaptor ; enhancer of Bcl-x$_L$, repressor of Bax, |
| | REV-T[ll] | v-Rel | Activated c-Rel |
| Rhabdoviridae | Rabies virus | P | Interferes with IRF3 activation |

[a]African swine fever virus; [b]*Autographa californica* multiple nucleopolyhedrovirus; [c]*Bombyx mori* nucleopolyhedrovirus; [d]*Cydiapomonella* granulovirus; [e]*Orgyia pseudotsugata* multiple nucleopolyhedrovirus; [f]*Spodoptera litura* nucleopolyhedrovirus; [g]La Crosse virus; [h]San Angelo virus; [i]Hepatitis C virus; [j]Hepatitis B virus; [k]Bovine herpesvirus-4; [l]Epstein Barr virus; [m]Equid herpesvirus-2; [n]Human cytomegalovirus; [o]Herpes simplex virus; [p]Herpesvirussaimiri; [q]Herpesvirus of turkeys; [r]Kaposi's sarcoma-associated herpesvirus; [s]Mouse cytomegalovirus; [t]Mouse herpesvirus strain 68; [u]Regina ranavirus; [v]*Ambystoma tigrinum stebbensi* virus; [w]Flock house virus; [x]Human papillomavirus 16; [y]Respiratory syncytial virus; [z]Simian virus 5; [aa]*Microplitis demolitor* bracovirus; [bb]Simian virus 40; [cc]*Amsacta moorei* entomopoxvirus; [dd]Mule-deer poxvirus; [ee]*Molluscum contagiosum* virus; [ff]*Melanoplus sanguinipes* entomopoxvirus 'O'; [gg]Shope fibroma virus; [hh]Yaba monkey tumorvirus; [ii]Bovine leukemia virus; [jj]Human immunodeficiency virus; [kk]Human T-lymphotropic virus 1; [ll]Reticuloendotheliosis virus strain T.

where viral factors alter death receptor signaling apparently to alter programmed cell death and other signaling pathways (**Figure 1**).

## FLIP Proteins Interfere with Receptor-Induced Death

Viral FLIP proteins (vFLIPs), identified in herpesviruses, modulate the activation of caspase-8 by competitively binding with the DED domain of FADD and preventing the oligomerization of preformed caspase-8-containing complexes that otherwise activate cell death (**Figure 1**). Cellular counterparts of cFLIPs were later identified and take the form of catalytically inactive caspase-8 or shorter forms that, like viral FLIPs, consist of only two DED domains, thereby mimicking the pro-domain of caspase-8. Thus, vFLIPs function more as suppressors than as modulators of caspase activation. Many different viruses encode FLIP proteins, including several herpesviruses and the molluscum contagiosum virus (MCV) that encodes two distinct vFLIPs, MC159 and MC160. The HCMV gene product vICA (inhibitor of caspase activation) encodes an unrelated caspase-8 inhibitor (**Table 1**). In contrast, adenovirus E1A induces tumor formation and cell death, perhaps in part by downregulating cFLIP$_S$. However, just as cytochrome *c* has both cell survival (mitochondrial electron carrier) and cell death (apoptosome formation) functions, cellular caspases also have alternative functions, including cell proliferation, at least at low activity levels. Therefore, caspase activation by E1A may have dual functions.

## Viral Caspase Regulators

The P35 protein of baculoviruses is a caspase pseudo-substrate that effectively inhibits caspases by binding into the caspase active site. In addition to being the first molecule of its kind to be discovered, P35 has proved to be exceptional in its ability to inhibit a wide variety of caspases. The related viruses BmNPV and SlNPV encode P35 homologs. To date, there are no cellular homologs of these viral proteins that directly inhibit caspases. Viral IAP proteins related to cellular cIAPs are another class of caspase regulators discovered in baculoviruses. IAPs are characterized by their baculovirus inhibitory repeat (BIR) domains and by a C-terminal RING finger domain with ubiquitin ligase activity that targets its substrates for proteasomal degradation. Exemplars from this class of caspase inhibitor include Cydiapomonella granulovirus (CpGV) Cp-IAP and Orgyia pseudotsugata nuclear polyhedrosis virus (OpNPV) Op-IAP. Viral IAPs are also encoded by poxviruses; these include Amsacta moorei entomopoxvirus (AmEPV) gene 021 and Melanoplus sanguinipes entomopoxvirus genes 242 and 248. Viral and cellular IAP proteins were suggested to directly inhibit caspases, but this model is currently challenged by evidence suggesting more indirect mechanisms to explain their inhibitory effects. Other poxviruses encode RING finger proteins that are suggested to have E3 ubiquitin ligase activity, though they lack overall amino sequence similarity to the IAP family.

Another important poxvirus caspase inhibitor produced by vaccinia and the closely related cowpox virus is CrmA, a member of the serine protease inhibitor (serpin) superfamily. CrmA inhibits both serine and cysteine proteases, including caspases-1,-8, and -10. Myxoma virus (Serp-2) can inhibit caspases as well as the related cysteine protease granzyme B. Myxoma virus E13L also acts at the level of caspase activation by binding and inhibiting the apoptosis-associated speck-like protein containing a CARD (ASC) protein, a component of the caspase-1-activating inflammasome complex (**Table 1**).

## Regulatory Viral RNAs and RNA-Binding Proteins

### Virus Recognition by Cellular PRRs

Early recognition of several viruses is mediated by PRRs and results in the activation of interferon and NF-κB. Recent advances have been made in this area through the discovery of RIG-I and Mda5 as well as their downstream signaling partner, the CARD-containing mitochondrial antiviral signaling protein (MAVS). The protease NS3/4A from the flavivirus hepatitis C virus (HCV) cleaves and inactivates MAVS, thereby short-circuiting virus recognition and signaling through the CARD. Meanwhile, the same protease also cleaves the TIR domain-containing adaptor-inducing IFN-β (TRIF) and short-circuits TLR signaling as well. Vaccinia virus uses a clever trick to dampen TLR signaling; A46R contains a TIR domain to sequester TRIF away from TLR3, thus preventing the recognition of viral antigens. Other virus factors modulate RIG-I and Mda5 by alternative mechanisms, including NS2 from Influenza virus and V proteins of paramyxoviruses.

B2 protein of the betanodavirus flock house virus (FHV) is an important determinant of virulence that acts by sequestering double-strand RNA and preventing its recognition by the host. RNA-binding proteins encoded by both RNA viruses (influenza, Ebola virus, and reovirus) and DNA viruses (vaccinia, CMV, and HSV) have also been implicated in modulating cellular strategies for controlling RNA levels.

The nonstructural protein (NSs) encoded on the small genome segment of a subset of mosquito-borne bunyaviruses (e.g., La Crosse virus and San Angelo virus) shares limited sequence similarity with the *Drosophila* pro-death molecule Reaper, and suppresses host cell gene expression, perhaps through an RNA interference (RNAi) mechanism, in both mammalian and cell extracts. These related proteins are also known to inhibit translation, suggesting that they act at multiple levels to direct cell death. The virus-associated interfering RNA (VAI) encoded by adenovirus acts at least in part by forming secondary structures that competitively inhibit the dsRNA binding by the endonuclease Dicer. Similarly, the RNA element TAR of HIV can bind TRBP (TAR RNA-binding protein), a cellular component of the Dicer-containing dsRNA-induced silencing complex (RISC), to inhibit RNA degradation activity. Therefore, post-transcriptional gene silencing through RNAi is involved in regulating host cell responses to RNA virus infections.

### Virus-Encoded Micro-RNAs

Micro-RNAs (miRNAs) are small hairpin-forming RNAs that are processed to mature forms both in the nucleus by the nuclease Drosha and in the cytoplasm by Dicer. Just as endogenous cellular miRNAs regulate cellular protein expression, the number of known virus-encoded miRNAs is growing rapidly. HIV miR-N 367 targets and suppresses the translation of Nef and negatively regulates virus replication. The polyomavirus simian virus 40 (SV40) produces an miRNA that suppresses the T-antigen (TAg), effectively evading immune surveillance. The Bcl-2 homolog BHRF-1of EBV is regulated by three distinct miRNAs that are encoded in the 3′-UTR of BHRF1 on the opposite strand. The HSV latency-associated transcript (LAT) also encodes an miRNA, whose targets were recently identified as components of the TGF-β signaling pathway, including TGF-β itself and the downstream adapter SMAD3, the regulation of which results in the suppression of cellular stress responses.

## Viral Regulators of the Mitochondrial Pathway

The mitochondrial cell death pathway is regulated by viral and cellular Bcl-2 proteins. Cellular Bcl-2 proteins are generally divided into three functional subgroups, anti-apoptotic (Bcl-2, Bcl-$x_L$, Mcl-1, Bcl-w), pro-apoptotic Bcl-2 family members (Bax and Bak), and the more distantly related pro-apoptotic BH3-only proteins (Bad, Bid, and others). While these classifications generally apply, there are definitive examples where the endogenous functions of these proteins are opposite to expected. For example, Bax, Bak, and Bad protect neurons from Sindbis virus-induced apoptosis, while Bcl-2 and Bcl-$x_L$ can promote cell death when they are cleaved by caspases.

### Viral Bcl-2 Family Proteins

Viral homologs of the cellular anti-apoptotic protein Bcl-2 are encoded by the γ-herpesviruses, including Epstein–Barr virus (EBV) and Kaposi's sarcoma-associated herpesvirus (KSHV/HHV8), and several different poxviruses including the myxoma virus M11L protein and fowlpox virus-Bcl-2. Viral Bcl-2 proteins have the same three-dimensional structure as their cellular counterparts, but differ from the cellular factors in that vBcl-2 proteins are resistant to normal cellular regulatory mechanisms, such as cleavage by caspases, which can convert cellular Bcl-2 into a pro-death factor. Like cellular Bcl-2, vBcl-2 prevents permeabilization of the outer mitochondrial membrane by Bax and Bak, thereby preventing cytochrome *c* release and caspase activation (**Figure 1**).

### Other Viral Proteins That Target Pro-Death Cellular Bcl-2 Proteins

Adenovirus E1B-19K, vaccinia virus F1L, cytomegalovirus vMIA (viral mitochondrial inhibitor of apoptosis) and vICA (viral inhibitor of caspase-8 activation) proteins also function at the same step as Bcl-2 as they bind and regulate cellular Bax and/or Bak. vMIA is encoded by exon 1 of immediate early gene UL37. vMIA localizes to mitochondria and causes mitochondrial fragmentation/division that is generally associated with apoptotic cell death, yet vMIA potently suppresses apoptosis triggered by diverse stimuli (**Table 1**).

### Yeast Viruses Induce Yeast Programmed Cell Death

Programmed cell death was assumed to arise during evolution with the origin of multicellular organisms. However, unicellular species such as yeast and bacteria also undergo programmed cell death. In mammals, virus-induced apoptosis can be responsible for the pathogenesis caused by viruses such as HIV, mosquito-borne encephalitis viruses, and many others. Most wild yeast strains are persistently infected with viruses, such as the dsRNA L-A virus and associated satellite dsRNA M viruses. These viruses are transmitted to a new host cell by cell–cell fusion. Yeast viruses are more analogous to sexually transmitted human viruses such as retroviruses and herpesviruses that remain in latently infected cells throughout the life of the host organism.

The M1 and M2 viruses of *Saccharomyces cerevisiae* encode a preprotoxin that is cleaved by the Kex proteases to produce the α- and β-chains of active M1- and M2-encoded toxins. Interestingly, yeast M1 virus preprotoxin is organized analogously to mammalian preproinsulin. The M1-encoded K1 toxin induces programmed cell death in neighboring susceptible cells by activating the cell death function of mitochondrial factors encoded by the cells destined to die. Therefore, mitochondrial factors may regulate an evolutionarily conserved cell death pathway that provides single cell organisms with a type of host defense or innate immunity. This virus–host relationship in yeast is analogous to pathogenic human viruses that persist throughout life and can occasionally reactivate to cause disease.

### Concluding Remarks

There are nearly as many strategies by which viruses interface with the cell death/survival machinery of host cells as there are viruses, from initial recognition of a virus to the ultimate proteolytic dismantling of the cell by proteases. A large number of virus-encoded cell death modulators have been reported in recent years (partial list in **Table 1**), and many more are likely to be discovered. The importance of viral regulators of cell death is clear from the number of these regulators encoded by a single virus, particularly DNA viruses with large genomes, while small RNA viruses must develop additional strategies. Common themes uncovered through the study of virus-modulated cell death pathways have proved to be broadly applicable to biology, as well as to the study of virus infection, pathogenesis, and evolution.

*See also:* Capillovirus, Foveavirus, Trichovirus, Vitivirus.

### Further Reading

Benedict CA, Norris PS, and Ware CF (2002) To kill or be killed: Viral evasion of apoptosis. *Nature Immunology* 3(11): 1013–1018.

Chen YB, Seo SY, Kirsch DG, Sheu T-T, Cheng W-C, and Hardwick JM (2006) Alternate functions of viral regulators of cell death. *Cell Death and Differentiation* 13: 1318–1324.

Flint SJ, Enquist LW, Racaniello VR, and Skalka AM (eds.) (2004) *Principles of Virology: Molecular Biology, Pathogenesis, and Control of Animal Viruses*, 2nd edn. Washington, DC: ASM Press.

Hiscott J, Nguyen T-LA, Arguello M, Nakhaei P, and Paz S (2006) Manipulation of the nuclear factor-κB pathway and the innate immune response to viruses. *Oncogene* 25: 6844–6867.

Rahman MM and McFadden G (2006) Modulation of tumor necrosis factor by microbial pathogens. *PLoS Pathogens* 2(2): e4 0066–0077.

# Aquareoviruses

**M St. J Crane and G Carlile,** CSIRO Livestock Industries, Geelong, VIC, Australia

© 2008 Elsevier Ltd. All rights reserved.

## Introduction

*Aquareovirus*, as the name implies, is a genus (one of 12 genera) of the family *Reoviridae*. Species in this genus infect aquatic animals – finfish, crustaceans, and mollusks – from both marine and freshwater environments. The name *Reoviridae* is derived from *r*espiratory *e*nteric *o*rphan viruses, and 'orphan' viruses are those viruses that are not associated with any known disease. Although, originally, reoviruses commonly may have been associated with subclinical or asymptomatic infections, it is now known that many are associated with disease. Reo-like viruses have been reported from aquatic animals since the 1970s and, as for other reoviruses, the majority of aquareoviruses are of low pathogenicity and have not attracted much attention. However, others have been isolated from populations with mortality rates higher than normal and some with mortality rates as high as 80%. There are at least six (A–F) recognized aquareovirus species based on RNA–RNA blot hybridization, RNA electrophoresis, antigenic properties, and nucleic acid sequence analysis (**Table 1**). More than 40 aquareoviruses have been isolated to date, with the majority classified within species groups A and B. However, some aquatic reoviruses such as the marine crab reoviruses, P and W2, do not fit this classification and these and several other aquareoviruses are yet to be classified.

## Taxonomy and Classification

The genus *Aquareovirus* forms one of 12 genera (*Orthoreovirus*; *Orbivirus*; *Rotavirus*; *Coltivirus*; *Seadornavirus*; *Aquareovirus*; *Idnoreovirus*; *Cypovirus*; *Fijivirus*; *Phytoreovirus*; *Oryzavirus*; *Mycoreovirus*) that make up the family *Reoviridae*. The *Reoviridae* is one of the six families of viruses possessing a double-stranded RNA (dsRNA) genome. Of these, the *Reoviridae* is one of only two families that infect vertebrates and the only family that infects mammals. Within the family, each genus can be placed in a subset based on whether the virion is turreted (*Orthoreovirus*; *Aquareovirus*; *Idnoreovirus*; *Cypovirus*; *Fijivirus*; *Oryzavirus*; *Mycoreovirus*) or unturreted (*Orbivirus*; *Rotavirus*; *Coltivirus*; *Seadornavirus*; *Phytoreovirus*). The reovirus genome is made up of 10–12 segments of dsRNA. Each segment encodes one to three (usually one) proteins. Aquareovirus genomes are made up of 11 segments of dsRNA. Except for segment 11, which usually encodes two proteins, each segment encodes a single protein. Interestingly, segment 11 of chum salmon reovirus is tricistronic. Aquareoviruses have been reported from finfish, crustaceans, and mollusks and from both freshwater and marine host species.

For aquareoviruses, the primary species demarcation criteria are RNA cross-hybridization and antibody-based cross-neutralization. Currently, six genogroups or species (A–F) have been identified with several aquareoviruses remaining unclassified (**Table 1**). Ultrastructural similarity between orthoreoviruses and aquareoviruses has been noted and the relatively high level of sequence homology between some isolates of these two genera indicates a common evolutionary lineage. Despite this relatedness, other properties, such as host range, different numbers of genome segments, and absence of antigenic similarity, support classification into two genera.

## Virion Structure and Morphology

The structure and morphology of aquareovirus virions are similar to other members of the family *Reoviridae*, particularly members of the genus *Orthoreovirus*. Virions are spherical in appearance, have icosahedral symmetry, and are *c.* 80 nm in diameter (**Figure 1**). The capsid consists of two concentric shells made up of three layers of protein. The outer capsid surrounds the inner core which is approximately 60 nm in diameter. For members of the genus *Aquareovirus*, the core is turreted – turrets or spikes project from the surface of the inner core and interconnect with the outer capsid layers. The inner protein layer of the core surrounds the 11 segments of the dsRNA.

A study of the turbot aquareovirus has indicated that its replication and morphogenesis generally follow that typical of other reoviruses. Virions were observed to be internalized by direct penetration of the host cell plasma membrane. In infected cells, two sizes of virus particles were observed in the cytoplasm: particles, 30 nm in diameter, which were probably inner cores; and single-shelled particles, 45 nm in diameter, located in the endoplasmic reticulum. Viral replication was observed to occur in the cytoplasm and immature virions (45 nm in diameter) formed complete, mature, double-shelled virions (75–80 nm in diameter) by budding through the plasma membrane so that complete viral particles were only observed outside host cells.

**Table 1** Aquareovirus species, host species, and geographical distribution

| Virus | Host species | Pathogenicity | Mortality rate | Country |
|---|---|---|---|---|
| *Aquareovirus A* | | | | |
| Geoduck clam aquareovirus (CLV) | *Panope abrupta* | NS | 0% | USA |
| Herring aquareovirus (HRV) | *Clupea harengus* | NS | 0% | USA |
| Striped bass aquareovirus (SBRV) | *Morone saxatilis* | L | NR | USA |
| American oyster reovirus (13$_{p2}$RV) | *Crassostrea virginica* | NS | 0% | USA |
| | *Lepomis macrochirus* | H | 44% | Exp |
| | *Oncorhynchus mykiss* | H | 60% | Exp |
| Angelfish aquareovirus (AFRV) | *Pomacanthus semicirculatus* | L | L | USA |
| Chum salmon aquareovirus (CSRV) | *Oncorhynchus keta* | L | 0% to L | Japan |
| Smelt aquareovirus (SRV) | *Osmerus mordax* | H | H | Canada |
| Atlantic salmon reovirus (ASRV) | *Salmo salar* | NS | NR | Canada |
| Atlantic salmon reovirus (TSRV) | *S. salar* | L | L | Australia |
| Atlantic salmon reovirus (HBRV) | *S. salar* | NS | NR | USA |
| Chinook salmon reovirus (DRCRV) | *Oncorhynchus tshawytscha* | NS | NR | USA |
| Guppy aquareovirus (GRV) | *Poecilia reticulata* | L | L | Singapore |
| *Aquareovirus B* | | | | |
| Coho salmon aquareovirus (CSRV) | *Oncorhynchus kisutch* | L | L | USA |
| Coho salmon reovirus (LBS) | *O. kisutch* | NS | NR | USA |
| Chinook salmon reovirus (YRCV) | *O. tshawytscha* | NS | NR | USA |
| Chinook salmon reovirus (ICRV) | *O. tshawytscha* | NS | NR | USA |
| Coho salmon reovirus (GRCV) | *O. kisutch* | NS | NR | USA |
| Coho salmon reovirus (CSRV) | *O. kisutch* | NS | NR | USA |
| Coho salmon reovirus (ELCV) | *O. kisutch* | NS | NR | USA |
| Coho salmon reovirus (SSRV) | *O. kisutch* | NS | NR | USA |
| *Aquareovirus C* | | | | |
| Golden shiner aquareovirus (GSRV) | *Notemignous crysoleucas* | L | L | USA |
| | *Semotilus atromaculatus* | L | NR | USA |
| | *Pimephales promelas* | M | M | USA |
| Grass carp reovirus (GCRV) (GCHV) | *Ctenapharyngodon idellus* | H | 70–95% | China |
| *Aquareovirus D* | *Mylopharyngodon piceus* | H | 70–80% | China |
| Channel catfish reovirus (CCRV) | *Ictalurus punctatus* | L | <5% | USA |
| *Aquareovirus E* | | | | |
| Turbot aquareovirus (TRV) | *Scophthalmus maximus* | L | <5% | Spain |
| *Aquareovirus F* | | | | |
| Chum salmon reovirus (PSRV) | *O. keta* | NR | NR | NR |
| Chinook salmon reovirus (SCRV) | *O. tshawytscha* | NR | NR | NR |
| *Unclassified aquareoviruses* | | | | |
| Haddock reovirus | *Melanogrammus aeglefinus* | NR | 0% | UK |
| Golden ide reovirus (GIRV) | *Leuciscus idus* | NR | 0% | Germany |
| Gilthead seabream reovirus | *Sparus aurata* | M | M | Spain |
| Red grouper reovirus | *Plectropomus maculatus* | H | 30–90% | Singapore |
| Landlocked salmon virus | *Oncorhynchus masou* (Brereoort) | L | L | Taiwan |
| Tench reovirus (TNRV) | *Tinca tinca* | NS | 0% | Germany |
| Japanese eel reovirus | *Anguilla japonica* | NR | NR | Japan |
| Fancy carp reovirus | *Cyprinus carpio* | NR | NR | Japan |
| Halibut reovirus | *Hippoglossus hippoglossus* | H | H | Canada, UK |
| Chub reovirus (CHRV) | *Leuciscus cephalus* | NS | 0% | Germany |
| Marine threadfin fish reovirus | *Eleutheronema tetradactylus* | H | Up to 100% | Singapore |
| | *Lates calcarifer* | H | 60% | Exp |
| Brown trout reovirus | *Salmo trutta* | L | 0% | UK |
| | *O. mykiss* | L | 0% | Exp |
| Snakehead reovirus (SKRV) | *Channa striata* | NS | 0% | Thailand |
| Mediterranean shore crab reovirus (RC84) | *Carcinus mediterraneus* | L | 0% | France |
| Mediterranean shore crab reovirus (W2) | *C. mediterraneus* | M | NR | France |
| Mediterranean swimming crab reovirus (P) | *Macropipus depurator* | L | NR | France |
| Chinese mitten crab reovirus | *Eriocheir sinensis* | NR | NR | China |
| Tiger shrimp reovirus | *Penaeus japonicus* | M | M | France |
| Tiger prawn reovirus | *Penaeus monodon* | H | 95% | Malaysia |

Exp, Experimental infections; L, low; M, moderate; H, high; NS, no significant pathology; NR, not reported.

# Aquareoviruses 165

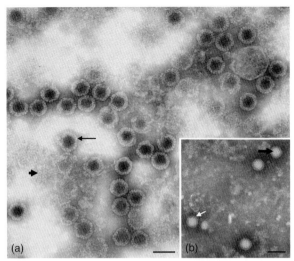

**Figure 1** Transmission electron micrographs of negative-stained preparations of Tasmanian Atlantic salmon aquareovirus (TSRV). (a) Image of a purified preparation where the stain has penetrated the inner capsid. (b) Image of similar aquareovirus particles derived from the supernatant of infected cells. Thin arrows (black and white): surface projections associated with the outer capsid. Thick arrows indicate locations of hexameric (or tetrameric) clusters of surface projections (thin arrows). Scale = 100 nm (a, b). Courtesy of Electron Microscopy Group, Australian Animal Health Laboratory, CSIRO Livestock Industries, Geelong, VIC, Australia.

## Genome Organization and Transcription Strategy

The genus *Aquareovirus* is relatively diverse, being made up of at least six species. The genomes of viruses in each species consist of 11 segments of dsRNA which migrate as three size classes in 1% agarose gels. Segments 1–3 are the large-size class, segments 4–6 medium, and segments 7–11 small. The complete genomic sequence of golden shiner reovirus (GSRV), comprising 23 696 bp, and grass carp reovirus (GCRV), comprising 23 695 bp, have been determined. The sequences of some segments of other aquareovirus isolates have also been determined. Apart from segments 7 and 11, each segment is monocistronic, containing a single open reading frame (ORF). For most aquareoviruses, segment 11 encodes two proteins (**Table 2**); segment 11 from chum salmon reovirus (CSRV) encodes three proteins. Thus, there are seven structural proteins (VP1–VP7) and five nonstructural proteins (NS1, NS2, NS4, NS15, and NS29). There is variation in the estimated molecular mass of each polypeptide for different aquareoviruses. For example, when compared by electrophoresis in a single polyacrylamide gel, the estimated molecular masses of the five major structural viral proteins for four aquareoviruses ($13_{p2}$, GSRV, CSRV, channel catfish reovirus (CCRV)) varied in the ranges 132–137 kDa (VP1), 126–130 kDa (VP2), 63–72 kDa (VP5), 43–45 kDa (VP6), and 32–36 kDa (VP7). **Table 2** shows the data reported for

**Table 2** Aquareovirus A genome segments and encoded polypeptides with the estimated mass and location

| Genome segment | Protein | Estimated Mass (kDa) | Protein location |
|---|---|---|---|
| 1 | VP1 | 130 | Core turret |
| 2 | VP2 | 127 | Inner capsid |
| 3 | VP3 | 126 | Inner capsid |
| 4 | NS1 | 97 | NS |
| 5 | VP5 | 71 | Inner capsid |
| 6 | VP4 | 73 | Outer capsid |
| 7[a] | NS4 | 28 | NS |
| 8 | VP6 | 46 | Inner capsid |
| 9 | NS2 | 39 | NS |
| 10 | VP7 | 34 | Major outer capsid protein |
| 11[b] | NS29 | 29 | NS |
|  | NS15 | 15 |  |

[a]Segment 7 of GCRV and GSRV has been reported as bicistronic encoding two nonstructural proteins with estimated masses of 16 (NS5) and 31 kDa (NS4).
[b]Segment 11 of CSRV is tricistronic encoding three nonstructural proteins with estimated masses of 13, 15, and 16 kDa.

*Aquareovirus A*. As stated above, segment 11 of CSRV has been reported to be tricistronic encoding three polypeptides with estimated masses of 16.9, 15.1, and 13.0 kDa, and segment 7 of GCRV/GSRV appears to be bicistronic encoding two nonstructural proteins.

Putative functions of the aquareovirus proteins have been predicted based on the functions established for the equivalent mammalian orthoreovirus proteins. Accordingly, VP1 is the homologue of λ2 protein, pentamers of which form the turrets. This protein has important functions in particle assembly, as well as guanylyltransferase and methyltransferase activities in mRNA capping. VP2, equivalent to λ3, is located in the inner capsid or core and has RNA-dependent RNA polymerase activity associated with transcription and replication. VP3 (λ1) is the main component of the inner capsid and possesses NTPase, RTPase, and helicase activities. VP4 (μ1) is located in the outer capsid and is believed to function in penetration of the cellular membrane to allow virus entry into the cell. The function of VP5 (μ2) is not fully understood but it is known to bind RNA and possess NTPase activity, indicating that VP5 may be an RNA polymerase cofactor. VP6 (σ2), together with VP3, forms the core shell. VP6 functions are poorly understood; it binds to RNA and is thought to be involved in replicase-particle assembly. VP7 (σ3) is the major outer capsid protein and is thought to play important roles in outer capsid assembly and in stabilizing virions in extracellular environments. VP7 also binds dsRNA and functions in translational control in infected cells. The nonstructural proteins NS1 and NS2 are likely to be involved in RNA replication and packaging during virus particle assembly and the function of NS4 is unclear thought to be the cell attachment protein. The functions of the nonstructural proteins encoded by RNA segment 11 are also unknown.

## Geographic Distribution and Host Range

Most aquareoviruses have been isolated from healthy aquatic animals and are likely to be of low pathogenicity. Thus, the presence of aquareoviruses is likely to be under-reported. Moreover, it is unlikely that routine surveillance for aquareoviruses has been undertaken in any systematic manner. Many isolations of aquareoviruses have occurred incidentally during surveillance activities for other, more significant, pathogens. It is during these surveillance activities that several aquareoviruses have been isolated from apparently healthy aquatic animals. **Table 1** lists most of the known aquareoviruses, their hosts, and country of isolation, cited in the readily accessible scientific literature. Some of the reoviruses listed have not been fully characterized and it may eventuate that some of the viruses will be classified outside of the currently recognized taxa.

Clearly, both the host range and the geographical distribution of aquareoviruses are broad and it is probable that, with ongoing surveillance, the known host range and geographical distribution will be extended. As aquaculture expands and the interaction between wild and farmed aquatic animal species increases, it can be expected that the number of known viruses, including reoviruses, will increase. A good example is virus $13_{p2}$, originally isolated from healthy American oysters has been shown to infect and cause disease in more than one finfish species, indicating that the primary host and wild reservoir of this virus are unclear.

While *Aquareovirus A* appears to be the most diverse species with virus isolates obtained from several host species from very different parts of the world, it appears that *Aquareovirus B* has very limited host and geographical ranges, based on the several available isolates. The isolates of *Aquareovirus B* listed in **Table 1** may be the same virus merely isolated on different occasions. However, such a conclusion may be premature since many of the aquareoviruses have not been fully characterized and a large number of aquareoviruses remain unclassified with respect to species assignment.

## Pathology

As stated above, the majority of known aquareoviruses are of low pathogenicity and have been isolated from apparently healthy animals. Nevertheless, some of the known aquareoviruses are highly pathogenic. For example, GCRV and others that have been isolated from normal animals may, under certain environmental conditions, be pathogenic. There are several examples in which stocking density and water temperature influence the pathogenicity of the virus. A good example is GSRV. Mortality rates associated with GSRV are normally around 5% but, under crowded conditions and high temperatures, acute epizootics with mortality rates of 50–75% have been reported.

It is also of note that isolation of aquareoviruses often occurs during investigations of conditions with apparently mixed etiology, involving either other viruses or a concomitant bacterial infection. In mixed infections with bacteria, treatment with antibiotics does not always resolve the condition, suggesting that the reovirus may play some role in the disease, if not as the primary pathogen. In other cases of mixed infections, the respective roles of each agent are not at all clear and it is difficult to ascribe disease signs to either agent.

Several reoviruses isolated from apparently healthy finfish with no gross external signs have been shown to cause low-level pathology in experimental infections. While these viruses have been shown to replicate in experimental fish, mortality is rare. The pathology is characterized by a diffuse multifocal necrosis of the liver which, in some cases, subsequently resolves.

For those aquareoviruses that have been isolated from disease outbreaks in finfish hosts, external signs are those typically found for systemic infections and include lethargy, inappetance, anorexia, abnormal swimming behavior, petechial hemorrhages on the body surface, lateral recumbency, distended abdomen, and high mortality rates. Internal signs included discoloration of the liver. Histological examination may reveal hepatic lesions with varying degrees of severity. Syncytial giant cell formations of hepatocytes have been reported.

GCRV is one of two highly pathogenic aquareoviruses isolated to date. The virus, first reported in 1984, is responsible for an acute hemorrhagic disease affecting grass carp (*Ctenopharyngodon idella*) and black carp (*Mylopharyngodon piceus*) in China. Disease outbreaks occur in the summer with mortality rates in the range of 70–95%. As with most viral diseases of finfish, younger age classes are more susceptible than older fish. External signs include exophthalmia and hemorrhages at the fin bases and gill covers. Internally, hemorrhages have been reported to occur in all the major organs – intestinal tract, liver, spleen, kidneys, and throughout the musculature. Recent studies indicate that GCRV and GSRV are variants of the same virus.

The other highly pathogenic reovirus was isolated in 1998 from cultured threadfin (*Eleutheronema tetradactylus*) fingerlings undergoing a mass mortality at a farm in Singapore. Following isolation, the virus was used in experimental infections of threadfin fingerlings and sea bass (*Lates calcarifer*). Clinical signs (dark pigmentation, lethargy, recumbency with sudden bursts of swimming) developed within 1 day post infection (d.p.i.), resulting in 100% cumulative mortality in the infected threadfin within 4 d.p.i. and 60% cumulative mortality in the sea bass within 7 d.p.i.

Several reoviruses or reo-like viruses have been isolated from crustaceans such as shrimp and crabs, some of which harbored mixed infections with other agents. As commonly reported for other viral diseases, external signs have included discoloration and abnormal behavior such as lethargy and inappetance. While virus particles have been observed in different tissues/organs (hepatopancreas, gills, digestive tract, lymphoid organ), a common feature appears to be involvement of the hepatopancreas where the most obvious lesions occur.

## Host Response to Infection

There are no detailed studies of the immune response of finfish to infection by aquareoviruses but, from observations that have been reported, it is clear that cellular and humoral immune reactions are stimulated. While not consistent, local infiltration of host inflammatory cells, particularly in the liver of infected fish, has been observed during aquareovirus infections. Moreover, in studies in which sera have been collected from infected fish, specific neutralization titers of >1:4000 have been obtained in neutralization tests. Further evidence of an immune response comes from studies with a vaccine produced in China for use against GCRV. A crude inactivated GCRV vaccine was produced in the late 1970s and was reported to provide good protection (70% survival of fingerlings).

Of further interest is a study reporting aquareovirus interference-mediated resistance of rainbow trout to infectious hematopoietic necrosis. Pre-exposure of rainbow trout to coho salmon reovirus (CSRV) induced a protective response to subsequent challenge with the rhabdovirus infectious hematopoietic necrosis virus. Protection appeared to be mediated by innate immune factors.

## Transmission

The natural transmission cycle has not been studied in detail for any of the aquareoviruses. Experimentally, infection and disease can be transmitted horizontally by injection and by immersion in diluted tissue culture supernatants from reovirus-infected fish cell lines. It is not known whether vertical transmission occurs.

## Genetic Diversity

Within the family *Reoviridae*, viruses in only three genera, *Mycoreovirus*, *Rotavirus*, and *Aquareovirus*, possess an 11-segmented dsRNA genome. Other properties such as host range, RNA sequence, and serological differences justify the classification of the aquatic reoviruses into a separate genus. Interestingly, there are some relatively high sequence identities in some proteins of aquareoviruses and orthoreoviruses, indicating these two genera share a common ancestral lineage. It is also noteworthy that some reoviruses, or reo-like viruses, that infect aquatic animals, such as viruses P and W2, isolated from marine crabs, possess 12 segments of dsRNA and may represent a new reovirus genus.

Currently, there are six genogroups/species (A–F) within the genus *Aquareovirus*. This classification is based on RNA–RNA blot hybridization, RNA electrophoresis, antigenic properties, and nucleotide sequence analysis (**Table 1**). While reciprocal RNA–RNA hybridization and cross-neutralization assays provide useful data for the classification of aquareoviruses, genomic sequence provides more precise data. For example, based on RNA–RNA hybridization, it had been suggested that GCRV may represent a seventh species group (G) but subsequent nucleotide sequence analysis has clearly indicated that it should be placed within genogroup C.

In a study that analyzed the relative electrophoretic mobility of the 11 RNA segments of 19 aquareoviruses, distinct electropherotypes were observed. Whilst there was no correlation with the host species from which the viruses were isolated, there was a correlation with geographical location. Further studies are required to determine whether, as for other members of the *Reoviridae*, electrophoretic mobility will be useful for strain identification within the aquareoviruses.

A comparison of deduced amino acid sequences of segment 10 (encoding major outer capsid protein VP7) of viral isolates representing the species *Aquareovirus A*, *Aquareovirus B*, and *Aquareovirus C*, has indicated sequence identities in the range 19–100%. Within a species, sequence identities were >80% and, for some, >99%. Between viruses in different species, identities were in the range 19–22%. Similar results have been obtained in other studies.

With the current level of knowledge, it is difficult to determine any correlation between *Aquareovirus* species, host and geographical ranges, and pathogenicity.

## Diagnosis and Disease Management

External signs of disease associated with infection by highly pathogenic aquareoviruses, including lethargy, inappetance, anorexia, abnormal swimming behavior, petechial hemorrhages, lateral recumbency, distended abdomen, high mortality rates, are not pathognomonic and laboratory investigation is required for definitive diagnosis. Several of the known aquareoviruses have been detected by virus isolation on any of a number of fish cell lines in common use in diagnostic laboratories. Depending on the aquareovirus isolate and local conditions, cell lines used for viral isolation and replication have included bluegill fry cell line (BF-2), chinook salmon embryo cell line (CHSE-214), fathead minnow cell line (FHM), *Epithelioma papillosum cyprini* cell line (EPC), channel catfish ovary cell line (CCO), brown bullhead cell line (BB), grass carp kidney cell line (CIK), Asian seabass cell line (SB), rainbow trout mesothelioma cell line (RTM), striped snakehead cell line (SSN-1), and rainbow trout gonad cell line (RTG-2). The appearance of the cytopathic effect (CPE) caused by aquareoviruses in fish cell lines is quite variable, depending on the aquareovirus isolate and the cell line used for isolation. Examples of CPE caused by Tasmanian Atlantic salmon reovirus (TSRV) in two cell lines are shown in **Figure 2**. Where pathogenic aquareoviruses are endemic, more rapid, sensitive, and specific diagnostic tests are highly desirable. Species-specific antisera are currently not available but would provide useful diagnostic reagents in a variety of immunoassays. A rapid test, based on a reverse-transcription polymerase chain reaction (RT-PCR), has recently been developed for local pathogenic strains of threadfin aquareovirus and grass carp reovirus that are endemic in China and Singapore. Tentative assignment of species/genogroup can be achieved by comparing isolates with known aquareovirus species by reciprocal RNA–RNA hybridizations and cross-neutralization assays. Precise genogroup assignment can only be achieved by RT-PCR, subsequent sequence analysis, and comparison with sequences of other known genogroup isolates.

Apart from the GCRV vaccine, no vaccines or therapeutics are currently available for the control of diseases

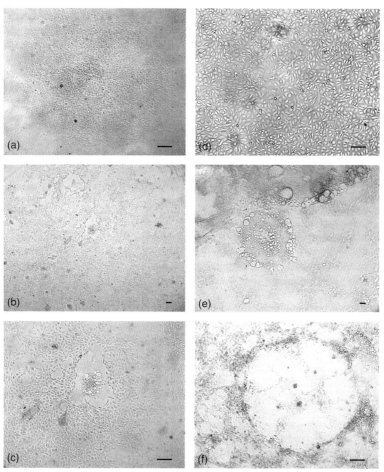

**Figure 2** Photomicrographs of cytopathic effect (CPE) produced by infection by TSRV. (a–c) CHSE-214 and (d–f) EPC cell cultures. (a, d) Uninfected cultures; (b, e) low magnification image; (c, f) high magnification image. Scale = 100 μm (a–f). Courtesy of Nette Williams, Australian Animal Health Laboratory, CSIRO Livestock Industries, Geelong, VIC, Australia.

associated with aquareovirus infections. As for other viral diseases of aquatic animals, avoidance is the preferred strategy. Thus, when available, pathogen exclusion can be attempted using good biosecurity and sanitary protocols in hatcheries and on farms. These can include the selection of specific pathogen-free (SPF) broodstock using virus isolation on cell lines and RT-PCR as screening tools, disinfection of fertilized eggs, and rearing water with, for example, ozone. In addition, stress reduction by maintaining high water quality and low stocking density is important. As aquareoviruses appear to have broad host and geographical ranges, the farming of aquatic animals where they are exposed to wild host species could present risks. Vaccines, when available, are likely to provide an additional level of protection.

## Current Status

The classification of aquareoviruses is not fully resolved, with many viral isolates remaining unclassified, and this is an important area for further research. For many of the species/genogroups, there are insufficient numbers of available isolates to allow determination of the key biological properties within each species. Thus, there does not appear to be correlation between host species, geographical range, pathogenicity, and species group. Many of the aquareoviruses, including the pathogenic threadfin aquareovirus, remain unclassified. As sequence data are available for only a few of the aquareoviruses, phylogenetic analysis has been limited. Further research is required not only to clarify the phylogenetic relationships within the genus *Aquareovirus* but also to assist in a better understanding of the role each segment plays in viral pathogenesis. Furthermore, the classification of the crustacean reo-like viruses and their relationship to other reoviruses, including aquareoviruses, requires clarification.

The isolation of different *Aquareovirus* species from both healthy and diseased fish indicates that, as for many other viruses of fish, pathogenicity is influenced by host (e.g., species, concomitant infections, immunocompetency) and environmental factors (e.g., water temperature, water quality, stocking density). In addition to elucidation of virulence factors (virus factors), further investigation on the respective roles of host and environmental factors on the pathogenicity of aquareoviruses is required. Based on knowledge to date, it seems likely that, for some *Aquareovirus* species at least, disease outbreaks will occur under adverse conditions of water quality/stocking density/temperature, and possibly in new host species. Thus, as the aquaculture industry expands globally, the interactions between farmed species and wild aquatic animal species will become more numerous, increasing the chance of virus transfer between wild and farmed animals. With this situation, taken together with their broad geographic range, it appears inevitable that aquareoviruses will produce disease in an increasing number of aquatic animal species in the future.

*See also:* Fish Viruses; Reoviruses: General Features; Reoviruses: Molecular Biology.

## Further Reading

Attoui H, Billoir F, Cantaloube JF, Biagini P, de Micco P, and de Lamballerie X (2000) Strategies for the sequence determination of viral dsRNA genomes. *Journal of Virological Methods* 89: 147–158.

Attoui H, Fang Q, Mohd Jaafar F, et al. (2002) Common evolutionary origin of aquareoviruses and orthoreoviruses revealed by genomic characterization of golden shiner reovirus, grass carp reovirus, striped bass reovirus and golden ide reovirus (genus *Aquareovirus*, family *Reoviridae*). *Journal of General Virology* 83: 1941–1951.

Dopazo CP, Bandin I, Rivas C, Cepeda C, and Barja JL (1996) Antigenic differences among aquareoviruses correlate with previously established genogroups. *Diseases of Aquatic Organisms* 26: 159–162.

Goodwin AE, Nayak DK, and Bakal RS (2006) Natural infections of wild creek chubs and cultured fathead minnow by Chinese grass carp reovirus (golden shiner virus). *Journal of Aquatic Animal Health* 18: 35–38.

Kim J, Tao Y, Reinisch KM, Harrison SC, and Nibert ML (2004) Orthoreovirus and aquareovirus core proteins: Conserved enzymatic surfaces, but not protein–protein interfaces. *Virus Research* 101: 15–28.

Lupiani B, Hetrick FM, and Samal SK (1993) Genetic analysis of aquareoviruses using RNA–RNA blot hybridization. *Virology* 197: 475–479.

Lupiani B, Subramanian K, and Samal SK (1995) Aquareoviruses. *Annual Review of Fish Diseases* 5: 175–208.

McEntire ME, Iwanowicz LR, and Goodwin AE (2003) Molecular, physical and clinical evidence that golden shiner virus and grass carp reovirus are variants of the same virus. *Journal of Aquatic Animal Health* 15: 257–263.

Meyers TR (1980) Experimental pathogenicity of reovirus 13$_{p2}$ for juvenile American oysters *Crassostrea virginica* (Gmelin) and bluegill fingerlings *Lepomis macrochirus* (Rafinesque). *Journal of Fish Diseases* 3: 187–201.

Rivas C, Noya M, Cepeda C, Bandin I, Barja JL, and Dopazo CP (1998) Replication and morphogenesis of the turbot aquareovirus (TRV) in cell culture. *Aquaculture* 160: 47–62.

Samal SK, Attoui H, Mohd Jaafar F, and Mertens PPC (2005) *Reoviridae – Aquareovirus*. In: Fauquet CM, Mayo MA, Maniloff J, Desselberger U, and Ball LA (eds.) *Virus Taxonomy: Eighth Report of the International Committee on Taxonomy of Viruses*, pp. 511–516. San Diego, CA: Elsevier Academic Press.

Seng EK, Fang Q, Chang SF, et al. (2002) Characterisation of a pathogenic virus isolated from marine threadfin fish (*Eleutheronema tetradactylus*) during a disease outbreak. *Aquaculture* 214: 1–18.

Seng EK, Fang Q, Lam TJ, and Sin YM (2004) Development of a rapid, sensitive and specific diagnostic assay for fish *Aquareovirus* based on RT-PCR. *Journal of Virological Methods* 118: 111–122.

Subramanian K, Hetrick FM, and Samal SK (1997) Identification of a new genogroup of aquareovirus by RNA–RNA hybridization. *Journal of General Virology* 78: 1385–1388.

Subramanian K, McPhillips TH, and Samal SK (1994) Characterization of the polypeptides and determination of genome coding assignments of an aquareovirus. *Virology* 205: 75–81.

# Arboviruses

**B R Miller,** Centers for Disease Control and Prevention (CDC), Fort Collins, CO, USA

© 2008 Elsevier Ltd. All rights reserved.

## Glossary

**Arbovirus** A virus that is biologically transmitted to vertebrates by infected hematophagous arthropods.
**Hematophagous arthropods** Arthropods, including mosquitoes, ticks, sandflies, blackflies, biting-midges, etc., that gain nourishment by feeding on vertebrate blood.
**$M_r$** The relative molar mass of a virion.
**Transovarial transmission** Transmission of an arbovirus from an infected arthropod female to her progeny via viral infected ovaries.
**Vertical transmission** Transmission of an arbovirus from an infected female vector to her progeny.
**Zoonoses** An infectious disease transmissible from vertebrate animals to humans under natural conditions.

## Introduction

The term 'arbovirus' is used to define viruses transmitted by blood-feeding arthropods. The word was coined by virologists investigating taxonomically diverse viruses that shared this biological feature. These agents are maintained in nature in various transmission cycles, from simple to complex, by replicating in blood-feeding arthropods that transmit the virus in their saliva to vertebrates. The virus then replicates or is 'amplified' in the vertebrate where it becomes accessible in peripheral blood to other blood-feeding arthropods, completing the cycle. These viruses thus are, 'arthropod-borne viruses', or arboviruses.

The *International Catalog of Arboviruses Including Certain Viruses of Vertebrates* lists 545 viruses that are known or suspected to be arthropod-borne as well as viruses with 'no known vector'. Many of the viruses in the catalog infect and replicate in vertebrates other than humans and livestock. Arboviruses circulate in nature, going largely unnoticed until they produce disease in humans and/or domestic animals or wildlife. Of the 545 viruses listed in the catalog, 134 are known to cause human disease. Arboviruses including *Yellow fever virus* (genus *Flavivirus*, family *Flaviviridae*), *Crimean-Congo hemorrhagic fever virus* (genus *Nairovirus*, family *Bunyaviridae*), and *Japanese encephalitis virus* (genus *Flavivirus*, family *Flaviviridae*) are capable of producing severe disease in humans; while other arboviruses including *Rift Valley fever virus* (genus *Phlebovirus*, family *Bunyaviridae*), *Venezuelan equine encephalitis virus* (genus *Alphavirus*, family *Togaviridae*), and *West Nile virus* (genus *Flavivirus*, family *Flaviviridae*) can cause clinical disease in humans and domestic animals.

Although arboviruses are found throughout the world, the majority is found in the tropics, where arbovirus transmission can occur without interruption, thus increasing opportunities for these viruses to evolve. The biological characteristic of arthropod transmission has evolved independently in diverse virus lineages; for instance, in the virus family, *Flaviviridae*, some members are transmitted by ticks, others by mosquitoes, others have no known vector and may be exclusively viruses of vertebrates, and still others appear to have been transmitted by vectors but secondarily have lost this ability over evolutionary time. Viruses in another primitive lineage in the family only replicate in mosquitoes.

## Taxonomy and Classification

The most important human and veterinary arboviral pathogens listed in the catalog are found primarily in select genera of three virus families: the *Bunyaviridae* (three of five genera), the *Flaviviridae* (one of three genera), and the *Togaviridae* (one of two genera). Arboviruses also belong to other virus families: the *Rhabdoviridae* (two of six genera), the *Reoviridae* (two of 12 genera), and the *Orthomyxoviridae* (one of five genera) (**Table 1**). The viruses in these families all have genomes consisting of RNA molecules in various configurations; there is only a single arbovirus, *African swine fever virus* (genus *Asfivirus*), in the family *Asfarviridae* that has a DNA genome. Viruses have been classified by using serological and molecular methodologies.

## Virion: Physical Properties, Composition, and Genome Organization

African swine fever virus has a sedimentation coefficient $S_{20,W}$ of about 3500 S and a buoyant density of $1.095 \text{ g cm}^{-3}$ in Percoll and $1.19–1.24 \text{ g cm}^{-3}$ in CsCl. Virions are susceptible to irradiation, ether, deoxycholate, and chloroform, and they are inactivated at 60 °C for 30 min. Virions can survive for years at 4 °C and they are stable over a wide range of pH. The asfivirus particle consists of a nucleoprotein core that is 70–100 nm in diameter, surrounded by two internal lipid layers, and a 170–190 nm icosahedral capsid ($T = 189–217$) with an external, lipid-containing envelope. The mature virion has 1892–2172 capsomers and a diameter of 175–215 nm. The genome of African swine

**Arboviruses** 171

**Table 1** Representative arboviruses associated with human and livestock illnesses

| Virus | Genus | Family | Host/associated illness | Arthropod vector | Primary host | Distribution |
|---|---|---|---|---|---|---|
| African swine fever | *Asfivirus* | *Asfarviridae* | Swine, bush pigs/hemorrhagic fever | Soft ticks, *Ornithodoros* | Wild boar | Africa, Spain, Portugal |
| Crimean-Congo hemorrhagic fever | *Nairovirus* | *Bunyaviridae* | Human/hemorrhagic fever | Hard ticks, *Hyalomma* | Wild and domestic animals | Africa, Asia, Europe |
| La Crosse | *Orthobunyavirus* | *Bunyaviridae* | Human/encephalitis | Mosquitoes, *Aedes triseriatus* | Chipmunks, squirrels | North America |
| Rift Valley fever | *Phlebovirus* | *Bunyaviridae* | Human/hemorrhagic fever, hepatitis, retinitis, encephalitis Livestock/abortion, fever, hepatitis | Mosquitoes, *Culex* and *Aedes* | Cattle and sheep | Africa, Arabian Peninsula |
| Dengue 1–4 | *Flavivirus* | *Flaviviridae* | Human/fever, hemorrhagic fever | Mosquitoes, *Aedes aegypti*, *Aedes albopictus* | Humans, primates | Worldwide in tropics |
| Japanese encephalitis | *Flavivirus* | *Flaviviridae* | Human/fever, encephalitis | Mosquitoes, *Culex* | Wading birds, pigs | Asia, Pacific Islands |
| Kyasanur Forest disease | *Flavivirus* | *Flaviviridae* | Humans, primates/fever, encephalitis, hemorrhagic fever | Hard ticks, *Hemaphysalis* | Camels, rodents | India, Saudi Arabia |
| Murray Valley encephalitis | *Flavivirus* | *Flaviviridae* | Humans/fever, encephalitis | Mosquitoes, *Culex* | Birds | Australia |
| Rocio | *Flavivirus* | *Flaviviridae* | Humans/fever, encephalitis | Mosquitoes, *Culex* | Birds | South America |
| West Nile | *Flavivirus* | *Flaviviridae* | Humans/fever, encephalitis | Mosquitoes, *Culex* | Birds | Africa, Asia, Europe, Americas |
| Yellow fever | *Flavivirus* | *Flaviviridae* | Humans, New World primates/fever, hemorrhagic fever, hepatitis | Mosquitoes, *Aedes*, *Hemagogus*, *Sabethes* | Primates | Africa, South America |
| Thogoto | *Thogotovirus* | *Orthomyxoviridae* | Humans/fever, optic neuritis, encephalitis | Hard ticks | Wild and domestic vertebrates | Africa, Europe |
| Colorado tick fever | *Coltivirus* | *Reoviridae* | Humans/fever, myalgia, encephalitis | Ticks, *Dermacentor* | Ground squirrels | Western North America |
| Banna | *Seadornavirus* | *Reoviridae* | Humans/fever, myalgia, encephalitis | Mosquitoes, *Culex*, *Aedes*, *Anopheles* | Pigs, cattle | Asia |
| Vesicular stomatitis Indiana | *Vesiculovirus* | *Rhabdoviridae* | Humans/fever Domestic animals/ fever, vesicles in the mucosa | Sandflies, blackflies, mosquitoes | Wild and domestic vertebrates | Americas |
| Chikungunya | *Alphavirus* | *Togaviridae* | Humans/fever, arthralgias, rarely encephalitis | Mosquitoes, *Aedes* | Humans, primates | Africa, Asia |
| O'nyong-nyong | *Alphavirus* | *Togaviridae* | Humans/fever, arthralgias, lymphadenopathy | Mosquitoes, *Anopheles* | Unknown | Africa |
| Ross River | *Alphavirus* | *Togaviridae* | Humans/fever, arthralgia | Mosquitoes, *Aedes*, *Culex* | Macropods | Australia, Pacific Islands |
| Sindbis | *Alphavirus* | *Togaviridae* | Humans/fever, arthralgia | Mosquitoes, *Culex*, *Culiseta*, *Aedes* | Passerine birds | Europe, Africa, Asia, Australia |
| Mayaro | *Alphavirus* | *Togaviridae* | Humans/fever, arthralgia | Mosquitoes, *Haemagogus* | Primates | Americas |
| Eastern equine encephalitis | *Alphavirus* | *Togaviridae* | Humans, equines, pheasants /fever, encephalitis | Mosquitoes, *Culiseta*, *Culex* | Birds | Americas |
| Western equine encephalitis | *Alphavirus* | *Togaviridae* | Humans, equines/fever, encephalitis | Mosquitoes, *Culex* | Birds | Americas |
| Venezuelan equine encephalitis | *Alphavirus* | *Togaviridae* | Humans, equines/fever, encephalitis | Mosquitoes, *Culex* (*Melanoconion*) | Rodents | Americas |

fever virus is a single molecule of double-stranded DNA that is covalently close-ended and ranges from 170 to 190 kbp in size. Mature virions contain more than 50 proteins, including an inhibitor of apoptosis homolog protein, guanyltransferase, poly A polymerase, protein kinase, and RNA polymerase. There are 150 open reading frames (ORFs) that are read from both strands. Early mRNA synthesis begins in the cytoplasm of the infected cell using enzymes in the virus core; viral DNA replication and assembly take place in perinuclear areas. Viral transcripts are capped at the $5'$-end, and the $3'$-end is polyadenylated. Formation of the capsid takes place on two layers of membranes derived from the endoplasmic reticulum (ER); extracellular virus acquires a membrane by budding through the cellular plasma membrane.

In the *Bunyaviridae*, virion $M_r$ is $(3-4) \times 10^8$ with sedimentation coefficients ($S_{20W}$) in the range of 350–500 S. Buoyant densities in sucrose and CsCl are 1.16–1.18 and 1.20–1.21 g cm$^{-3}$, respectively. Virions contain 50% protein, 20–30% lipid, and they are sensitive to heat, irradiation, formaldehyde, and lipid solvents. Bunyavirus virions are spherical with a diameter of 80–120 nm and have a lipid bilayer that contains glycoprotein projections (5–10 nm). Virion envelopes are obtained from Golgi membranes and the ribonucleocapsids exhibit helical symmetry. The three arboviral genera in the family *Bunyaviridae* (*Nairovirus*, *Orthobunyavirus*, and *Phlebovirus*) have slightly different genome organizations; however, all have segmented, single-stranded, negative-sense RNA genomes with an L-segment that codes for the viral polymerase, an M-segment that codes for two envelope glycoproteins, and an S-segment that codes for the viral nucleocapsid protein. Synthesis of viral mRNA takes place in the cytoplasm using host cellular capped primers ('cap-stealing' or 'cap-snatching'). Mature virions are produced by budding into the Golgi cisternae, although *Rift Valley fever virus* has been observed to bud at the cell surface.

Flavivirus virion $M_r$ is about $6 \times 10^7$; virions have sedimentation coefficients of approximately 200 S and have buoyant densities of about 1.19 g cm$^{-3}$ in sucrose. Virions are stable at weakly alkaline pH but they are inactivated at acidic pH values, at heat above 40 °C, in organic solvents, and on irradiation. Virions are spherical with icosahedral symmetry and about 50 nm in diameter. There are two glycoproteins in the lipid bilayer: M and E in mature, extracellular virions and prM and E in immature virions; prM is cleaved to M by host-cell enzymes during maturation. The E-glycoprotein is a rod-shaped, dimeric structure that does not form spike-like projections; rather, it lies parallel to the viral membrane in a neutral pH environment. The flavivirus genome is capped and consists of single-stranded, positive-sense RNA that codes ($5' \rightarrow 3'$) for three structural proteins and seven nonstructural proteins contained in a single ORF; the $5'$- and $3'$-ends are noncoding regions (NCRs) necessary for viral replication and translation. A nascent polyprotein is transcribed and cleaved co- and post-translationally by cellular and viral proteases. Viral RNA synthesis, virion assembly, and acquisition of a lipid envelope take place in the ER; mature virions are released by exocytosis.

Orthomyxovirus virion $M_r$ is $250 \times 10^6$ and buoyant density is 1.19 g cm$^{-3}$ in sucrose. $S_{20,W}$ is 700–800 S and virions are sensitive to irradiation, solvents, heat, and detergents. Virions are spherical (80–120 nm) and sometimes pleomorphic with surface glycoproteins projecting 10–14 nm from the surface. The nucleocapsid is segmented and has helical symmetry. Within the genus, *Thogotovirus*, the negative-sense, single-stranded, segmented RNA genome is comprised of six segments for *Thogoto virus* (family *Orthomyxoviridae*) and seven segments for *Dhori virus* (family *Orthomyxoviridae*); the genome is about 10 kb in size. Interestingly, the glycoprotein, coded by the fourth segment, shares sequence similarity with a surface glycoprotein of insect baculoviruses. Thogoto virus nulceocapsids are transported to the nucleus where viral enzymes synthesize $5'$-capped mRNA that are polyadenylated. Protein synthesis occurs in the cytoplasm, although early on in infection certain proteins are found in the nucleus and later transported to the cytoplasm. Viral membrane proteins are transported through the Golgi cisternae to the cellular plasma membrane; nulceocapsids bud through the plasma membrane in regions populated by viral membrane proteins.

In the *Reoviridae*, coltiviruses and orbiviruses have a buoyant density of 1.38 and 1.36 g cm$^{-3}$, respectively, in CsCl. Viruses are sensitive to low pH, heat, and detergent but are stable for long periods at 4 °C in the presence 50% fetal bovine serum. Coltivirus virions are 60–80 nm in diameter consisting of two concentric capsid shells with a core structure that is about 50 nm in diameter while the orbivirus, bluetongue virus is 90 nm in diameter with core particles 73 nm in diameter. Virus particles have icosahedral symmetry and are closely associated with granular matrices and filamentous formations in the cytoplasm. The mosquito-borne seadornaviruses and tick-borne coltiviruses have genomes composed of 12 double-stranded (dsRNA) segments that are, respectively, 21 000 and 29 000 bp in size; each genome segment is flanked $5'$ and $3'$ by similar noncoding sequences. Coltivirus VP1 (the largest segment) encodes the viral RNA-dependent RNA polymerase. There is no indication of virus release prior to cell death; more than 95% of coltivirus progeny virions and 60% of seadornavirus progeny virions remain cell associated. Orbivirus core particles are arranged as hexameric rings composed of capsomeres. The viral genome is composed of ten segments of dsRNA surrounded by the inner capsid shell. Inclusion bodies and tubules are produced during viral replication. Flat hexagonal crystals formed from the major outer core protein may also be produced during orbivirus replication. Unpurified virus is associated with cellular membranes although mature virus lacks a lipid envelope. Virions can leave the host cell by budding through the plasma membrane and transiently acquiring an envelope that is unstable and is rapidly lost.

Rhabdovirus virion $M_r$ is $3–10 \times 10^8$ and $S_{20,W}$ is 550–1045 S. Buoyant density in sucrose is $1.7–1.9 \text{ g cm}^{-3}$ and $1.19–1.20 \text{ g cm}^{-3}$ in CsCl. Heat above 56 °C, lipid solvents, and irradiation will inactivate virions. Rhabdoviruses are 'bullet-shaped' or 'cone-shaped', 100–430 nm in length, and 45–100 nm in diameter. Trimers of the viral glycoprotein (G), 10 nm long and 3 nm in diameter, known as 'peplomers' cover the outer surface of the virus. The nucleocapsid is 30–70 nm in diameter and shows helical symmetry. It is composed of RNA and N (nucleocapsid proteins) collectively with the L-(transcription and replication factors) and P-(polymerase cofactor) proteins that interact with the envelope G-protein via the M-(binds to N and G) protein. The genomes are negative-sense, single-stranded RNA molecules that encode five major polypeptides. Viral genes are transcribed as monocistronic mRNAs that are capped and polyadenylated. After the nucleocapsid is released into the cytoplasm, the viral genomic RNA is transcribed by the virion transcriptase. Genome replication occurs in the cytoplasm by full-length positive-strand RNA synthesis followed by full-length negative-strand RNA synthesis. Nucleocapsids are synthesized in the cytoplasm and viral membranes are acquired as viruses bud from the host-cell plasma membranes.

Togavirus virion $M_r$ is approximately $52 \times 10^6$. Virions have a buoyant density of $1.22 \text{ g cm}^{-3}$ in sucrose and an $S_{20,W}$ is of 280 S. Virions are sensitive to heat, acidic pH, solvents, and irradiation. Togavirus viruses are spherical, 70 nm in diameter, and they have heterodimer spikes in the envelope consisting of two virus glycoproteins, E1 and E2. The envelope surrounds a 40 nm nucleocapsid that has icosahedral symmetry ($T = 4$). Togaviruses have a genome that is positive-sense, single-stranded RNA; the mRNA is capped and polyadenylated. There are four genes coding for nonstructural proteins situated at the 5′-end of the genome and five genes coding for structural proteins located at the 3′-end of the genome. Negative-sense RNA, synthesized during RNA replication, is utilized as the template for production of genome-length, positive-sense RNA as well as for creation of a 'subgenomic' 26S mRNA that is capped and polyadenylated and is translated to make the viral structural proteins. The polyprotein is cleaved by a combination of cellular and viral proteases. Noncoding sequences located at the 5′- and 3′-termini of the genome are required for both negative- and positive RNA synthesis. The nucleocapsids are assembled in the cytoplasm; viral surface glycoproteins are processed in the ER and translocated through the Golgi apparatus to the cellular plasma membrane where they are acquired as the nucleocapsids bud through the membrane.

## Evolutionary Relationships Among Arboviruses

The biological characteristic of viral replication in arthropods followed by transmission to vertebrates in the course of blood feeding has arisen independently in seven families of viruses. Taxonomic relationships within the arboviruses were originally based on analyses of antigenic cross-reactivity data obtained from neutralization, complement fixation, and hemagglutination tests. Virion morphology, determined by electron microscopy, was also important in taxonomic classification.

The availability of viral, genomic, nucleic acid sequences and sophisticated nucleic acid analyses and phylogenetic software has enabled hypothesis testing of arboviral evolution. Defining the deeper nodes in evolutionary trees has been difficult because of the limited number of sequences from different virus species in each virus family. Predictably, the arboviruses that are most well-characterized are those associated with disease. It will be difficult to formulate an accurate and robust theory of arboviral evolution until this sampling bias is resolved by obtaining data from other viruses, those not necessarily associated with disease. Brief summaries of the best-studied arbovirus groups are presented below.

## Evolutionary Relationships of the Flaviviruses

The genus, *Flavivirus* is currently composed of about 70 different viruses that infect a wide variety of vertebrate hosts from birds to bats, transmitted by ticks and mosquitoes; some have no known vector associations and others are found only in mosquitoes.

Flavivirus-related sequences have been detected in the genomes of *Aedes* mosquitoes suggesting flavivirus sequence integration into an eukaryotic genome. Another curious finding is the discovery of defective Dengue virus 1 (genus *Flavivirus*, family *Flaviviridae*) virions that are transmitted by *Aedes aegypti* to humans over long periods of time. The defective genomes have acquired a stop codon in the envelope glycoprotein gene resulting in a truncated E-protein. This defective lineage persists through complementation by co-infection of host cells with functional viruses. The relevance of these findings on virulence and pathogen transmission are unknown.

The detection of diverse quasispecies populations of Dengue virus 3 in *Aedes aegypti* mosquitoes and in humans suggests that flaviviral mutation frequencies are similar to those of other RNA viruses. Estimates of flavivirus evolutionary rates suggest they are generally less than those for single-host RNA viruses, reflecting the evolutionary constraints of obligatory replication in vertebrate and invertebrate hosts. Approximation of the degree of amino acid divergence between mosquito-borne and tick-borne viruses indicates that flaviviruses in the tick-borne group evolve two to three times more slowly than do the mosquito-borne viruses. This is probably the result of limited virus replication and the lengthy tick life cycle, which can be measured in years. Ixodid (hard) ticks feed only three times during their life cycle (larva→nymph→adult) with months passing between ecdysis (molting) and feeding on a subsequent vertebrate host, thus allowing virus lineages to persist for long periods of time in a quiescent arthropod. Also, an infected tick can directly infect another tick co-feeding on the same uninfected animal host, eliminating the need for a viremic vertebrate as had been demonstrated for tick-borne encephalitis virus and *Ixodes ricinus* ticks. This 'nonviremic' transmission may result in viruses persisting for long periods of time in the tick, further constraining rates of evolutionary change. This is in marked contrast to mosquitoes where viral replication occurs rapidly and to high titers over a life span measured in days. The increased amount of RNA replication in mosquitoes generates more virus variation, thereby increasing opportunities for viral evolution.

Data sets from alignments of sequences from the nonstructural genes NS3 and NS5 and whole-genome alignments have been examined for phylogenetic signal in determining evolutionary relationships among most viruses in the family. In general, data alignments from the NS3 genes and from the full-length genomes have proven to be the most useful in elucidating the basal divergence of derived lineages. Hypotheses based on recent studies suggest that the absence of a vector is the ancestral condition for the family as a whole (genera *Hepacivirus*, *Pestivirus*, and *Flavivirus*); within this family, arboviruses occur only in the *Flavivirus* genus.

Within the genus *Flavivirus*, determination of the divergence of the three main groups (no known vector, mosquito-borne, and tick-borne) is equivocal. Phylogenetic analysis of the NS5 data set suggests that the no-known vector group diverged before the arthropod-borne viruses while analysis of the NS3 and full-genome data sets imply that the mosquito-borne flaviviruses diverged first and form a sister clade to the no-known vector and tick-borne viruses. If the mosquito-borne viruses were the first divergent group, we would expect the existence of a lineage of unknown flaviviruses occurring in mosquitoes. The recent discovery of the flavivirus, *Kamiti River virus*, which replicates in mosquitoes exclusively, adds support to the speculation that there may be a large number of unidentified flaviviruses existing in nature that are the descendants of a primitive, mosquito–host lineage.

Attempts have been made to associate epidemiology, disease pattern, and biogeography with the major virus clades using either partial sequences of the NS5 gene or of the E-gene. The mosquito-borne flaviviruses partition into two groups: viruses that cause neuroinvasive disease (Japanese encephalitis virus and West Nile virus) and are transmitted by *Culex* vectors to bird amplification reservoirs, and viruses that cause hemorrhagic disease (Yellow fever virus and dengue viruses) and are transmitted by *Aedes* vectors to primate amplification reservoirs. The tick-borne viruses also formed two groups, one associated with seabird hosts and another associated with rodents. The no-known vector viruses split into three groups, two associated with bats and one with rodents. The above associations undoubtedly reflect the complex interactions and selection pressures between arboviruses, arthropod vectors, and vertebrate hosts over evolutionary time.

Regardless of which group diverged first, the flaviviruses probably originated in the Old World. Within the *Aedes* mosquito-borne clade, only Yellow fever virus, the four dengue viruses, and West Nile virus are established in the Americas; the presence of these viruses represents relatively recent introductions from Africa or Asia. With the exception of the *Powassan virus* (family *Flaviviridae*), all the tick-borne flaviviruses occur in the Old World. Evidence for the Old World origin of the no-known vector group is less convincing. The rodent-borne viruses are only found in the New World with the exception of *Apoi virus* (family *Flaviviridae*); the bat-borne group has representatives that occur worldwide. These observations could be explained by a single dispersion event of an ancestral virus from the Old World into the New World that resulted in infection of resident rodents. Rodents generally occupy a small geographic range and are less likely to widely disperse viruses, increasing the likelihood of founder effects, genetic drift, and viral allopatric speciation. Migratory bats on the other hand are capable of transporting viruses over long distances.

## Evolutionary Relationships of the Alphaviruses

The genus *Alphavirus* in the family *Togaviridae* is comprised of 29 described species. Within the genus, viruses have been grouped into eight antigenic complexes by classic serological techniques. Phylogenetic relationships in the genus have been estimated using nucleic acid and amino acid alignments from the E1, nsP1, and nsP4 codons. Phylogenetic trees produced from analyses of alignments of nucleic acids and amino acids generally support the antigenic groups as monophyletic clades. The antigenic complexes are also associated with disease syndromes, as described for the flaviviruses with the exception of hemorrhagic fevers. For example, members of the Eastern equine encephalitis and Venezuelan equine encephalitis complexes produce neuroinvasive disease in humans and equines, while members of the *Semliki Forest virus* complex commonly cause mild to severe arthralgias. Interestingly, the Western encephalitis virus complex contains viruses capable of causing both disease syndromes; arthralgias in the *Sindbis virus* clade and neuroinvasive disease caused by Highlands J and Western equine encephalitis viruses. This dichotomy in disease associations is the probable result of an ancient, genomic recombination event. Two of the viruses in this complex, Highlands J and Western equine encephalitis, are descendents of a recombinant ancestral alphavirus; presumably, the genetic material responsible for the potential to cause encephalitis came from an Eastern equine encephalitis virus-like ancestor and a Sindbis-like virus.

Molecular studies using phylogenetic analyses of alphaviruses have shown that the majority of genome sequence divergence results in synonymous substitutions, indicating that rates of evolutionary change are constrained by purifying selection. This conservation is revealed in the 26S subgenomic promoter sequences; they are functionally interchangeable between Sindbis virus and other alphaviruses. Again, the hypothesis to explain the lower rates of evolution in these viruses in contrast to nonarboviral RNA viruses is the restraint imposed by required replication in hosts with and hosts without a backbone.

An examination of alphavirus phylogenetic divergence demonstrates the importance of host switching and host mobility. An intriguing example is the relationship between the closely related Chikungunya and O'nyong-nyong viruses. *Chikungunya virus* is thought to have originated in Africa in a sylvan, *Aedes*–primate transmission cycle. *O'nyong-nyong virus* (family *Togaviridae*) is genetically very similar to Chikungunya virus yet it has become adapted to *Anopheles funestus* and *Anopheles gambiae* mosquitoes. Molecular studies with chimeric Chikungunya–O'nyong-nyong viruses indicate that an ancestor of Chikungunya virus became adapted to replication in anopheline mosquitoes. This is significant because *Anopheles* species as arboviral vectors are the exception, rather than the rule.

In general, alphavirus divergence is highest when virus transmission occurs focally, usually in small vertebrate hosts, such as rodents which have restricted geographic ranges. Virus transmission in these isolated foci can produce new genotypes as has been elegantly demonstrated for viruses in the Venezuelan equine encephalitis complex. Conversely, genomic divergence is less for viruses that are amplified in wide-ranging vertebrates like birds where genotypic mixing and selection maintains a dominant genotype.

## Disease Associations

Although arboviruses are represented in diverse viral taxa, the disease syndromes they produce in humans can be arbitrarily classified into three major groups: systemic febrile illness, meningitis and/or encephalitis, and hemorrhagic fever. It is important to note that these syndromes represent a spectrum of disease severity: from sub-clinical infections to fatal outcomes. The factors responsible for differential disease severity are unknown but age and the immune status of the host are important as well as the amount of virus inoculated (dose) and the relative pathogenicity of the infecting arbovirus strain. Disease is most often correlated with infection of dead-end hosts including humans and their domestic animals; there is little evidence of these viruses causing harm in their natural hosts.

Febrile illness caused by infection with dengue viruses (serotypes 1–4), West Nile, Chikungunya, and other viruses consists of fever, headache, muscle and joint pain, and sometimes with rash. In a certain subset of cases, infection with West Nile virus progresses to encephalitis, resulting in fatalities or neurological sequelae in survivors, whereas infection with the dengue viruses rarely results in encephalitis or a hemorrhagic syndrome. Age-associated disease syndromes have been noted with West Nile virus infection: inapparent infection in children, typical fever syndrome in adults, and neuroinvasive disease in the elderly. Other arboviruses that can produce a febrile illness associated with polyarthritis are the alphaviruses, Chikungunya, O'nyong-nyong, Sindbis, Ross River, Barmah, and Mayaro. These febrile illnesses are impossible to differentiate on a clinical basis and are often confused with malaria and other common viral infections, including influenza. No fatalities are associated with uncomplicated febrile illnesses although recovery time can be prolonged and sequelae are commonly observed.

More serious illnesses are caused by arboviruses that have a tropism for the central nervous system, including Japanese encephalitis, Saint Louis encephalitis, West Nile, Eastern equine encephalitis, Western equine encephalitis, Venezuelan equine encephalitis, and *LaCrosse* (genus *Orthobunyavirus*, family *Bunyaviridae*) viruses. Illness begins with fever, chills, headache, malaise, drowsiness, anorexia, myalgia, and nausea. The syndrome can progress to confusion, stupor, and coma. These infections can be life-threatening and survivors may suffer from severe neurological sequelae.

Another serious manifestation of infection by some arboviral pathogens is hemorrhagic fever. Infection with the flavivirus Yellow fever virus can produce a clinical spectrum ranging from a mild febrile illness to fulminating fatal disease. Flaviviruses in the mammalian tick-borne virus group that can cause hemorrhagic fevers include Kyasanur Forest disease and Omsk hemorrhagic fever viruses. Infection with Rift Valley fever and Crimean-Congo hemorrhagic fever viruses can also result in hemorrhagic fever.

The above-mentioned examples are all for human infections. As might be expected, domestic animals may exhibit similar disease syndromes following arboviral infection. Infections that can produce febrile illness in animals include the alphavirus *Getah* in horses, *Bovine ephemeral fever virus* (genus *Ephemerovirus*, family *Rhabdoviridae*) in cattle, and Crimean-Congo hemorrhagic fever virus in cattle, sheep, and goats. More severe disease with encephalitis results from infection with West Nile, Eastern equine encephalitis, Western equine encephalitis, and Venezuelan equine encephalitis viruses in horses. Infection with the flavivirus *Wesselsbron virus*, *Bluetongue virus* (genus *Orbivirus*, family *Reoviridae*) and Nairobi sheep disease virus (genus *Nairovirus*, family *Bunyaviridae*) in sheep, and African swine fever virus in pigs can result in hemorrhages. Infection of sheep, goats, and cattle with Rift Valley fever virus can result in abortions while infection with the orthobunyavirus *Akabane virus* (family *Bunyaviridae*) can result in congenital malformations.

## Transmission Cycles

Human and livestock encroachment into natural transmission cycles is the most common means of exposure to arboviral infections. This virus transmission scenario is sometimes termed 'spillover' because the principal enzootic vector(s) has a wide host feeding range not limited to the natural, amplifying, vertebrate hosts. Arboviruses in this category of transmission include, but are not limited to, West Nile, sylvan Yellow fever, Western equine encephalitis and La Crosse viruses. African swine fever virus is maintained in sub-Saharan Africa in a cycle involving *Ornithodoros* ticks and warthogs. Infected ticks infesting warthog burrows can also transmit the virus to their progeny by transovarial transmission. Infected warthogs do not show signs of the disease; however, domestic swine are severely affected with mortality rates approaching 100%. In addition to tick bite, transmission between domestic pigs can occur through direct contact and the ingestion of infected offal.

Domestication of animals for human use has provided arboviruses with new and abundant hosts. For example, Rift Valley fever virus circulates in extensive areas of Africa and is transmitted by certain species of *Aedes* and *Culex* mosquitoes to various native vertebrates. The virus is thus maintained by horizontal transmission in the vertebrates and by transovarial and/or vertical transmission of the virus in mosquito eggs during periods of drought. El Nino events can lead to sustained rainfall and flooding, resulting in the production of massive numbers of other species of *Aedes* and *Culex* mosquitoes that enter into the existing, endemic transmission cycle. Infected vectors disperse and bloodfeed on livestock, initiating a secondary amplification cycle. Viremic livestock become the source for a dramatic increase in the numbers of infected vectors, which in turn engenders human epidemics and livestock epizootics.

The third pattern of arboviral transmission is one in which humans serve as both the reservoir and amplification hosts. Arboviruses maintained in human→vector→human cycles are generally transmitted by domestic mosquito vectors, most importantly, *Aedes aegypti*. The dengue viruses, the most important arboviruses in terms of human disease, are the classic example of an arbovirus that has adapted to the human host. Other examples include Chikungunya and Yellow fever viruses which can cause spectacular outbreaks in areas infested with *Aedes aegypti* vectors. These latter viruses generally disappear after a sufficient numbers of humans become immune, but persist in sylvan transmission cycles until nonimmune human hosts become available or when human herd immunity wanes.

*See also:* Akabane Virus; Crimean-Congo Hemorrhagic Fever Virus and Other Nairoviruses.

## Further Reading

Calisher C (2005) A very brief history of arbovirology focusing on contributions by workers of the Rockefeller Foundation. *Vector Borne and Zoonotic Diseases* 5: 202.

Cook S and Gould E (2006) A multigene analysis of the phylogenetic relationships among the flaviviruses (Family: *Flaviviridae*) and the evolution of vector transmission. *Archives of Virology* 151: 309.

Karabatsos N (1985) *International Catalog of Arboviruses Including Certain Other Viruses of Vertebrates,* 3rd edn. San Antonio: American Society of Tropical Medicine and Hygiene.

Knipe DM and Howley PM (eds.) (2001) *Fields Virology,* 4th edn. 2 vols. Philadelphia: Lippincott.

Monath TP (ed.) (1988) *The Arboviruses: Epidemiology and Ecology.* 5 vols. Boca Raton: CRC.

Theiler M and Downs WG (1973) *The Arthropod-Borne Viruses of Vertebrates.* New Haven, CT: Yale University Press.

Weaver SC and Barrett ADT (2004) Transmission cycles, host range, evolution and emergence of arboviral disease. *Nature Reviews Microbiology* 2: 789.

# Arteriviruses

**M A Brinton,** Georgia State University, Atlanta, GA, USA
**E J Snijder,** Leiden University Medical Center, Leiden, The Netherlands

© 2008 Elsevier Ltd. All rights reserved.

## History

Equine arteritis virus (EAV) was first isolated in 1953 in Bucyrus, Ohio from the lung tissues of an aborted fetus during an epidemic of abortions and arteritis in pregnant mares. However, an equine disease similar to that caused by EAV was first observed in the late 1800s. At the time of its discovery, EAV was distinguished from equine (abortion) influenza virus.

Lactate dehydrogenase-elevating virus (LDV) was discovered by accident in 1960 during a study to find methods for early detection of tumors in mice. A five- to tenfold increase in lactate dehydrogenase (LDH) levels in serum 4 days after inoculation of mice with either Ehrlich carcinoma cells or cell-free extracts suggested that an infectious agent was responsible for the observed LDH elevation.

Porcine respiratory and reproductive syndrome was first observed in North America in 1987 and in Europe in 1990. This disease has also been referred to as porcine epidemic abortion and respiratory syndrome (PEARS), swine infertility and respiratory syndrome (SIRS), and mystery swine disease (MSD). The causative agent of the disease is now referred to as porcine respiratory and reproductive syndrome virus (PRRSV).

Simian hemorrhagic fever virus (SHFV) was isolated in 1964 during outbreaks of a fatal hemorrhagic fever disease in macaque colonies in the US, Russia, and Europe. A number of additional SHFV outbreaks in macaque colonies have occurred since the 1960s. The most 'famous' of these was the one in Reston, Virginia, which occurred in conjunction with an Ebola virus outbreak in the same facility.

## Taxonomy and Classification

On the basis of virion size and morphology as well as the positive polarity of the RNA genome, LDV and EAV were originally classified within the family *Togaviridae*. In 1996, following the sequence analysis of their genomes, EAV, LDV, SHFV, and PRRSV were classified as species within a new family, *Arteriviridae*, genus *Arterivirus*. EAV was designated the prototype of this family. At the same time, the family *Arteriviridae* was classified together with the family *Coronaviridae* in the order *Nidovirales*. This order also includes two additional virus groups, the toroviruses (a genus in the *Coronaviridae* family) and the family *Roniviridae*. The arterivirus genome shares similar general organizational features and conserved replicase motifs with corona-, toro-, and ronivirus genomes, but is only about half their size. In addition, arterivirus particles are smaller than those of other nidoviruses, differ from them morphologically, and are the only ones to have an isometric nucleocapsid structure.

## Geographic Distribution

Viruses with biological properties identical to those of LDV have been isolated from small groups of wild mice (*Mus musculus*) in Australia, Germany, the US, and UK. Natural infections with EAV and EAV-induced disease in horses and donkeys have been documented in North America, Europe, and Japan and anti-EAV antibodies have been detected in horse sera from Africa and South America, indicating that EAV infection is geographically widespread. Natural PRRSV infections in pigs have been reported in North America, Europe, and Asia. SHFV infection in captive patas monkeys has been documented and this virus has also been detected in the blood of wild-caught patas and African green monkey as well as baboons, suggesting that these African primates are the natural hosts for SHFV.

## Host Range and Virus Propagation

Natural infections with EAV occur only in horses and donkeys. Field isolates of EAV can be readily obtained from field samples (semen, fetal tissues, and buffy coats) using RK-13 cells. Laboratory strains of EAV have been successfully grown in primary cultures of horse macrophages and kidney cells, rabbit kidney cells, and hamster kidney cells and also in cell lines, such as BHK-21, RK-13, MA-104, and Vero.

LDV replicates efficiently in all strains of laboratory and wild *Mus musculus* and somewhat less efficiently in the Asian mouse *Mus caroli*. Numerous attempts to infect other rodents such as rats, hamsters, guinea pigs, rabbits, deer mice (*Peromyscus maniculatus*), and dwarf hamsters (*Phodopus sungorus*) with LDV have not been successful. LDV replicates only in primary murine cell cultures that contain macrophages, such as spleen, bone marrow, embryo fibroblast, and peritoneal exudate cell cultures. Although peritoneal cultures prepared from starch-stimulated adult mice contain 95% phagocytic cells, only 6–20% of these cells support LDV replication as demonstrated by autoradiographic, *in situ* hybridization, immunofluorescence and electron microscopic techniques, suggesting that LDV infects an as yet uncharacterized subpopulation of macrophages. A much higher percentage of cells in peritoneal exudate cells obtained from infant mice are susceptible to virus infection.

Natural infections with PRRSV were thought to be restricted to pigs. However, one report suggested that chickens and mallard ducks may be susceptible to the virus. PRRSV can replicate in primary cultures of porcine alveolar macrophages and macrophages from other tissues. Some, but not all, isolates of PRRSV can be adapted to replicate in a subclone of the MA-104 cell line.

Natural infections with SHFV occur in several species of African primates, namely *Erythrocebus patas*, *Cercopithecus aethiops*, *Papio anuibus* and *Papio cynocephalus*. SHFV infection of members of the genus *Macaca* has occurred in primate facilities in a number of countries and was associated with a fatal hemorrhagic fever. Isolates of SHFV can replicate in primary cultures of rhesus aveolar lung macrophages or peripheral macrophages and some isolates can replicate efficiently in the MA-104 cell line.

Maximum arterivirus yields after infection of susceptible cell cultures are observed by 10–15 h after infection. The maximum titers obtained for LDV and PRRSV are $10^6$–$10^7$ ID$_{50}$ ml$^{-1}$ and can exceed $10^8$ PFU ml$^{-1}$ for EAV and SHFV.

## Properties of the Virion

Arterivirus particles are spherical, enveloped, and 40–60 nm in diameter (**Figures 1(a)** and **1(b)**). Unfixed

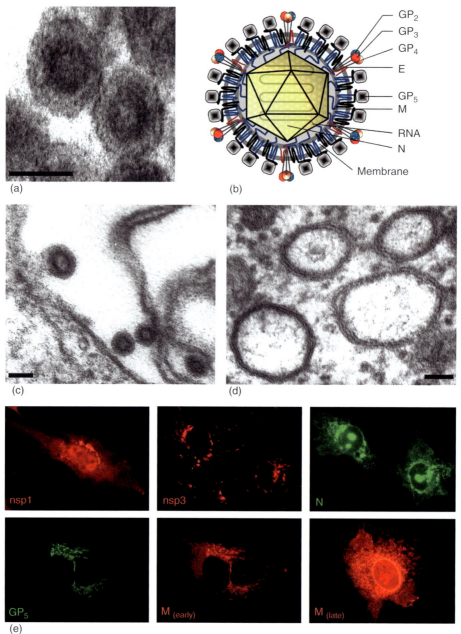

**Figure 1** (a) An electron micrograph of extracellular PRRSV particles. (b) Schematic representation of an arterivirus particle and its seven structural proteins. (c) An electron micrograph of EAV particles budding from smooth membranes in (or close to) the Golgi region of infected baby hamster kidney cells (BHK-21) cells. (d) An electron micrograph of typical double-membrane vesicles found in the cytoplasm of arterivirus-infected cells that have been implicated in replication complex formation and viral RNA synthesis. (e) Localization of selected EAV nonstructural and structural proteins in infected BHK-21 cells by immunofluorescence microscopy. In contrast to all other nsps (e.g., nsp3), the N-terminal replicase subunit nsp1 only partially localizes to the perinuclear region and is partially targeted to the nucleus. The double-labeling for nsp3 and the N protein shows that a considerable part of the latter co-localizes with the viral replication complex, whereas another fraction of the N protein is targeted to the nucleus. Early in infection, double-labeling for the major glycoprotein $GP_5$ and the M protein showed almost complete co-localization of the two proteins in the Golgi complex, in the form of the heterodimer that has been found to be critical for virus assembly. Later in infection, the M protein accumulates in the endoplasmic reticulum. Scale = 50 nm (a, c–d). (a) Reprinted from Snijder EJ and Meulenberg JM (1998) The molecular biology of arteriviruses. *Journal of General Virology* 79: 961–979. (b) Reprinted from Snijder EJ, Siddell SG, and Gorbalenya AE (2005) The order Nidovirales. In: Mahy BWJ and ter Meulen V (eds.) *Topley and Wilson's Microbiology and Microbial Infections, Vol. 1: Virology*, 10th edn., pp. 390–404. London: Hodder Arnold. (c, d) Reprinted from Snijder EJ and Meulenberg JM (1998) The molecular biology of arteriviruses. *Journal of General Virology* 79: 961–979. (e) Images courtesy of Yvonne van der Meer and Jessika Zevenhoven, Leiden University Medical Center, The Netherlands.

virions undergo distortion and disintegration during standard negative staining procedures. The virion surface appears rather smooth. The virion capsid is icosahedral and about 25–35 nm in diameter. Buoyant densities of 1.13–1.17 g cm$^{-3}$ and sedimentation coefficients of 214S to 230S have been reported for arteriviruses. Virions can be stored indefinitely at $-70\,°C$ but are heat labile. For instance, the infectivity of LDV samples in plasma decreased by half after 4 weeks at $-20\,°C$ and by about 3.5 logs after storage for 32 days at $4\,°C$. Virus in media supplemented with 10% serum is stable for 24 h at room temperature, but completely inactivated by heating at $58\,°C$ for 1 h. Virions are fairly stable between pH 6 and pH 7.5, but are rapidly inactivated by high or low pH. Virus is efficiently disrupted by low concentrations of nonionic detergent.

The locations of the seven structural proteins in an EAV virion are indicated schematically in **Figure 1(b)**. The icosahedral capsid is composed of the nucleocapsid (N) protein. The major envelope glycoproteins, $GP_5$ and M, form a disulfide-linked heterodimer. The minor glycoproteins $GP_2$, $GP_3$, and $GP_4$ form a disulfide-linked heterotrimer. $GP_2$–$GP_4$ heterodimers have also been detected in EAV virions. Although all six of these proteins were shown to be required for EAV and PRRSV infectivity, not all of the minor structural proteins have been identified so far in the other arteriviruses and the nomenclature for the SHFV structural proteins differs due to an insertion in the 3′ region of the SHFV genome. Virions bud into the lumen of cytoplasmic vesicles (**Figure 1(c)**).

## Properties of the Genome

Arterivirus genomes are single-stranded RNAs of positive polarity that contain a 3′ poly(A) tract of approximately 50 nucleotides in length and a type I cap at the 5′ end. The genome lengths are 12.7 kb for EAV, 14.1 kb for LDV, 15.1 kb for PRRSV, and 15.7 kb for SHFV.

The large nonstructural or 'replicase' polyproteins are encoded by open reading frames (ORFs) 1a and 1b and occupy the 5′ three-fourths of the genome. ORF 1b is translated only when a −1 ribosomal frameshift occurs in the short ORF 1a/ORF 1b overlap region. A 'slippery sequence' upstream of a pseudoknot directs frameshifting and for EAV, a frameshifting efficiency of 15–20% has been reported. ORF 1a encodes three or four proteases that post-translationally cleave the pp1a and pp1ab polyproteins at multiple sites into the mature viral nonstructural proteins (**Figure 2**). The lengths of the ORF 1a regions of the different arteriviruses vary. ORF 1b encodes major conserved domains, in particular an RNA-dependent RNA polymerase, a putative zinc-binding domain, an RNA helicase, and a nidovirus uridylate-specific endoribonuclease. The multiple 3′-proximal ORFs (**Figure 2**) encode the structural proteins. There are six such ORFs in the genomes of EAV, PRRSV, and LDV, while SHFV contains nine ORFs downstream of ORF 1b. Limited sequence homology suggests that the SHFV ORFs 2a, 2b, and 3 may be duplications of ORFs 4, 5, and 6, respectively. In most cases, adjacent structural protein genes of arteriviruses are in different reading frames and overlap. Conserved

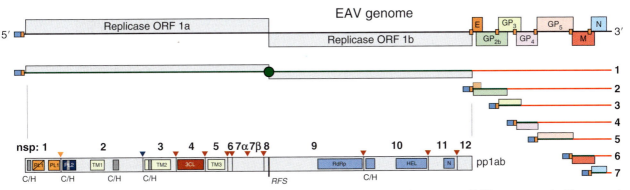

**Figure 2** Arterivirus genome organization and expression are illustrated using the family prototype EAV as an example. The genomic open reading frames are indicated and the names of the corresponding proteins are given. Below the genome, the nested set of mRNAs found in infected cells is depicted, with RNA1 being identical to the viral genome and subgenomic RNAs 2–7 being used to express the structural protein genes located in the 3′-proximal quarter of the genome. With the exception of the bicistronic mRNA2, the subgenomic mRNAs are functionally monocistronic. The EAV replicase gene organization is depicted in the polyprotein pp1ab from the replicase (pp1a is identical to the nsp1–8 region of pp1ab). Ribosomal frameshift (RFS) delineates the boundary between amino acids encoded in ORF 1a and ORF 1b and arrows represent sites in pp1ab that are cleaved by papain-like proteases (yellow and blue) or the main (3CL) protease (red). The proteolytic cleavage products (nsps) are numbered and the locations of various conserved domains are highlighted. These include domains with conserved Cys and His residues (C/H), putative transmembrane domains (TM), protease domains (PL1, PL2, and 3CL), the RNA-dependent RNA polymerase domain (RdRp), helicase (HEL), and uridylate-specific endoribonuclease (N). Adapted from Siddell SG, Ziebuhr J, and Snijder EJ (2005) Coronaviruses, toroviruses, and arteriviruses. In: Mahy BWJ and ter Meulen V (eds.) *Topley and Wilson's Microbiology and Microbial Infections, Vol. 1: Virology*, 10th edn., pp. 823–856. London: Hodder Arnold.

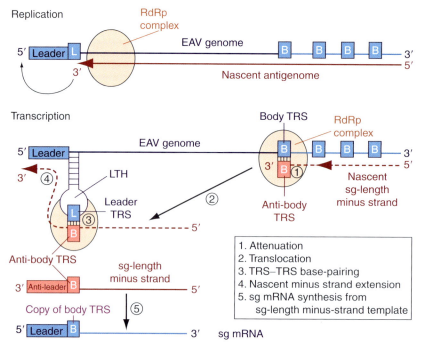

**Figure 3** Arterivirus RNA synthesis (using a hypothetical virus that produces four subgenomic (sg) mRNAs). Both replication (top panel) and transcription (bottom panel) are depicted. In the 'replication mode', the RdRp produced a full-length minus-strand RNA (antigenome) that serves as the template for synthesis of new genomic RNA. In the 'transcription mode', minus-strand RNA synthesis is thought to be discontinuous and regulated by transcription-regulating sequences (TRSs). Body TRSs (B) in the genome would act as attenuators of minus-strand RNA synthesis. Subsequently, the nascent minus strand, with an anti-body TRS at its 3' end, would be redirected to the 5'-proximal region of the genomic template by a base-pairing interaction with the leader TRS (L) that (for EAV) has been shown to reside in a leader TRS hairpin (LTH) structure. Following the addition of the anti-leader to the nascent minus strands, the sg-length minus strands would then serve as templates for sg mRNA synthesis.

transcription-regulating sequences (TRSs; **Figure 3**) are located at the 3' end of the genomic leader sequence (leader TRS) and upstream of each of the 3'-proximal ORFs (body TRSs). RNA hairpin structures are located near the 5' end of the genome (including a leader TRS-presenting hairpin) and also in the 3' NTR.

## Properties of the Viral Proteins

Arterivirus proteins that are encoded at the 5' end of the genome are translated directly from the genomic RNA as polyproteins (pp1a and pp1ab; **Figure 2**). The proteins generated from these ORFs contain all functions required for viral RNA synthesis. Both pp1a/pp1ab contain multiple papain-like cysteine proteases and a chymotrypsin-like (or '3C-like') serine protease (**Figure 2**). The EAV papain-like cysteine proteases each cleave at a single site. The nsp4 serine protease, or main protease, cleaves at six sites in the pp1a region and at three additional sites in the ORF 1b-encoded part of pp1ab. Due to the existence of two alternative processing cascades (minor and major pathways) a variety of processing intermediates and mature proteins are generated from the C-terminal half of pp1a. In total, 13 (EAV) or 14 (PRRSV/LDV) mature proteins are (predicted to be) generated from the arterivirus replicase polyproteins. Three hydrophobic regions in pp1a are thought to be important for membrane association of the viral replication–transcription complexes. With the exception of nsp1, which is partly found in the nucleus, the rest of the nonstructural proteins localize to endoplasmic reticulum-derived double-membrane structures (**Figure 1(d)**) in the perinuclear region (**Figure 1(e)**). The mature ORF 1b-encoded proteins (nsp9–nsp12) are thought to be the primary enzymes of the viral replication–transcription complexes that direct viral RNA synthesis.

The proteins encoded in the 3'-proximal quarter of the genome are expressed from six (nine for SHFV) overlapping subgenomic mRNAs (a 3' co-terminal nested set; **Figure 2**). Although the subgenomic mRNAs, with the exception of the smallest one, are structurally polycistronic, in general only the 5' terminal ORF is translated. An exception is mRNA 2, which is bicistronic encoding $GP_{2b}$ and E. The nucleocapsid (N) protein is encoded by ORF 7 (ORF 9 in SHFV). Analysis of the crystal structure of the C-terminal domain of the PRRSV N protein suggests that arteriviruses have a unique capsid-forming domain. The M protein which is encoded by ORF 6 (ORF 8 in SHFV), is a triple-membrane-spanning protein and the major nonglycosylated envelope protein. The major envelope glycoprotein is encoded by ORF 5 (ORF 7 in SHFV). $GP_{2b}$, $GP_3$, and $GP_4$ are each about

20 kDa in size and are minor envelope glycoproteins (EAV, **Figure 2**). $GP_{2b}$ and $GP_4$ ($GP_{4b}$ and $GP_6$ in SHFV) are class I integral membrane proteins. A soluble, non-virion-associated form of PRRSV $GP_3$ is released from infected cells. E is an unglycosylated small hydrophobic minor envelope protein. The PRRSV E protein has been shown to possess ion channel protein-like properties.

The nonstructural protein nsp1 and the N protein have been detected in the nucleus as well as the cytoplasm (**Figure 1(e)**). However, the biological significance of the nuclear localization of these two viral proteins is currently not known.

## Replication

Cell tropism is determined in part at the level of a receptor on the cell surface, since in some cases cells that are nonpermissive for an arterivirus have been reported to be productively infected after transfection of viral genomic RNA. Evidence for a specific saturable, but as yet unidentified receptor for LDV on a subpopulation of murine macrophages has been reported. LDV-immune complexes are also infectious and can infect macrophages via Fc receptors. Sialoadhesin (sialic acid-dependent lectin-like receptor 1), a macrophage-restricted, cell surface protein, has been shown to mediate the internalization of PRRSV by alveolar macrophages. Heparin sulphate on the cell surface and sialic acid on the virion are also thought to play a role in entry. PRRSV has been reported to enter cells via a low pH-dependent endocytic pathway. Soon after infection, both EAV and LDV particles have been observed in small vesicles that appear to be clathrin-coated. The existence of an additional level of host restriction at the endosomal membrane fusion or uncoating steps was suggested by the observation that nonsusceptible cells expressing recombinant sialoadhesin could internalize virus but rarely became productively infected.

The arterivirus replication cycle occurs in the cytoplasm of infected cells. After the incoming genomic RNA is uncoated, it is translated to produce polyproteins pp1a (1727–2502 amino acids) and pp1ab (3175–3959 amino acids) and then becomes the template for minus-strand synthesis (**Figure 3**). Either a full-length minus-strand RNA, that then serves as a template for genomic RNA synthesis, or subgenomic minus strands, that then serve as the templates for subgenomic mRNA synthesis, can be produced. The subgenomic RNAs are thought to be produced by a mechanism of discontinuous minus-strand RNA synthesis that utilizes conserved primary (TRSs) and higher-order RNA structures as signals for producing a subgenomic minus-strand template for each subgenomic mRNA. Subgenomic mRNAs are 3′-coterminal and contain a common 5′ leader sequence that is identical to the 5′ terminus of the genomic RNA. *cis*-Acting regulatory signals required for arterivirus replication have been mapped to the ~300 nt at each end of the genome. Host proteins also appear to be involved in the regulation of arterivirus RNA synthesis.

The co-localization of N with replicase complexes (**Figure 1(e)**) suggests that genome encapsidation may be coordinated with genome synthesis. New virions form via budding of preformed capsids into the lumens of smooth endoplasmic reticulum and/or Golgi complex membranes (**Figure 1(c)**). Arterivirus envelope proteins localize to intracellular membranes and recent data suggest that the formation of the $GP_5$–M heterodimer is required for budding (**Figure 1(e)**). Mature virions in the lumens of these vesicles are then transported to the exterior of the cell and released.

The formation of cytoplasmic double-membrane vesicles (**Figure 1(d)**), which have been implicated in viral RNA synthesis, is characteristic of arterivirus-infected cells. Infection of primary macrophages by EAV, SHFV, PRRSV, and probably also LDV is cytocidal. Laboratory strains of EAV, SHFV, and PRRSV cause obvious cytopathology in the continuous cell cultures that they infect, such as MA-104 cells. Infected cells become rounded by 24–36 h after infection and release from the tissue culture flask. Apoptosis has been reported in PRRSV-infected primary porcine alveolar macrophages and MA-104 cell cultures as well as in testicular germ cells in pigs. However, other studies with PRRSV showed that necrosis, not apoptosis, was the main cause of death of infected cells.

## Genetics

Evidence for virulence variants of all arteriviruses has been obtained. One strain of LDV isolated from C58 tumor-bearing mice and designated LDV-C was shown to efficiently induce neurologic disease in a few susceptible inbred mouse strains, such as AKR and C58, both of which are homozygous for the $Fv-1^n$ allele. Subsequent studies showed that neuropathogenic and non-neuropathogenic isolates coexist in most LDV pools. The number of glycosylation sites in the ectodomain of $GP_5$ varies in different LDV strains and it has been postulated that antibodies bind less efficiently to virions with extensive glycosylation in this region. A neurovirulent strain of PRRSV has also been reported. Virulent and avirulent mutants of EAV have been identified on the basis of the severity of the diseases they cause. Attenuated vaccine strains of EAV and PRRSV and a number of temperature-sensitive mutants of EAV have been selected. SHFV isolates that produce acute asymptomatic infections and ones that cause persistent, asymptomatic infections in patas monkeys have been reported. EAV and PRRSV infectious cDNA clones have been constructed and provide a means for analyzing the virulence determinants via reverse genetics.

## Evolution

Evidence of RNA recombination has been obtained by genome sequencing for both LDV and PRRSV after co-infections with different strains of the same virus type and it is thought that RNA recombination is the mechanism responsible for the observed gene duplication in the SHFV genome. Sequence comparisons of various field isolates of either PRRSV or EAV indicate that the sequences of the M and N proteins are more conserved than those of the virion glycoproteins. The extent of the divergence of the sequences of European and North American PRRSV isolates indicates that these two virus populations represent subspecies and also suggests that the ability to cause porcine disease arose independently in geographically separated virus populations. Phylogenetic analysis of the arteriviruses indicated that PRRSV is most closely related to LDV and that SHFV is more closely related to both of these viruses than to EAV. Although the host specificity of LDV has experimentally been shown to be restricted to mouse species, it has been postulated that PRRSV arose when wild boars became infected with LDV after they ate infected wild mice and that wild boars then introduced a divergent 'LDV' virus into domestic pigs independently in North America and Europe.

The nidoviruses represent a distinct evolutionary lineage among positive-strand RNA viruses. Although the organization of the conserved replicase motifs in the arterivirus genome is very similar to that of the other nidoviruses (coronaviruses, toroviruses, and roniviruses), the structural protein genes of these viruses are apparently unrelated (**Figure 1**). This level of divergence may be related to a high frequency of RNA recombination that appears to be a characteristic of nidovirus replicases and may be a byproduct of the mechanism of discontinuous RNA synthesis utilized by these viruses for subgenomic RNA production. The ancestral nidovirus has been postulated to have had an icosahedral capsid. If via a recombination event with another type of virus, the progenitor of the coronavirus/torovirus lineage acquired an N protein that could form a helical nucleocapsid, then packaging restrictions on genome size would have been lost, allowing genome size expansion via further recombination events and divergence from the arterivirus branch.

## Serologic Relationships and Variability

Attempts to demonstrate antigenic cross-reactivity between EAV, LDV, PRRSV, and SHFV have not been successful with one exception. Antibodies produced to a single linear LDV neutralization site located in $GP_5$ neutralized both LDV and PRRSV. Monoclonal antibodies elicited by one strain of LDV did not bind to the proteins of most other LDV isolates in Western blots, suggesting that serologic variants of LDV exist. Variation in PRRSV N epitopes has been observed between North American and European virus isolates and a high degree of heterogeneity has been observed between strains of PRRSV within the ectodomain of $GP_4$, which contains a secondary neutralization epitope.

## Transmission

There is no evidence for transmission of any of the arteriviruses via insect vectors. Horizontal transmission of both EAV and PRRSV occur via the respiratory route as well as via the venereal route by virus in the semen of persistently infected 'carrier' males. Vertical transmission of PRRSV *in utero* has been reported.

Nothing is currently known about the incidence of transmission of LDV in wild mouse populations. In the laboratory, unless the cage mates are fighting males, LDV is rarely transmitted between mice housed in the same cage, even though infected mice excrete virus in their feces, urine, milk, and saliva. However, transmission of LDV from mother to the fetus via the placenta and to pups via breast milk/saliva has been documented within the first week after infection of the mother. Since LDV in mice and SHFV in patas monkeys is produced throughout the lifetime of persistently infected animals, the transfer of fluids or tissues from an infected animal to an uninfected one results in the inadvertent transfer of infection. Historically the most frequent mode of transmission of LDV among laboratory mice and SHFV from patas monkeys to macaques has been through experimental procedures such as the use of the same needle for sequential inoculation of several animals. Currently, the most frequent sources of LDV contamination are pools of other infectious agents or tumor cell lines that have been repeatedly passaged in mice, especially those first isolated in the 1950s. Such materials should be checked for the presence of LDV. Infectious agent stocks can be 'cured' of LDV by passage in a continuous cell line or a different animal species. Tumor cell stocks can be 'cured' by *in vitro* culture for several passages. It has been suggested but not proven that SHFV can be transmitted between macaques via the respiratory route.

## Tissue Tropism

The primary target cells for all four arteriviruses are macrophages. Measurement of the amount of virus in various tissues during natural EAV infections indicated that lung macrophages and endothelial cells were the first host cells to be infected. Bronchial lymph nodes subsequently became infected and then the virus spread throughout the body via the circulatory system. In fatally infected horses, lesions are found in subcutaneous tissues, lymph nodes, and viscera. The progression of PRRSV infection in pigs is likely to be similar to that observed with EAV. However, although PRRSV is thought to be naturally transmitted by aerosols, experimental transmission by this route has been difficult to achieve. PRRSV has been reported to replicate in testicular germ cells which could result in excretion of virus into the semen. LDV replicates in an uncharacterized subpopulation of murine macrophages. Virus target cells are located in tissues as well as in the blood. Cells containing LDV-specific antigen have been identified in sections of liver and spleen by indirect immunofluorescence. In spleen, the virus-infected cells are nucleated and located in the red pulp. In liver, only Kupffer cells contained LDV-specific antigen. In C58 and AKR mice infected with a neurotropic strain of LDV, virus replication was demonstrated in ventral motor neurons by *in situ* hybridization.

## Pathogenicity and Clinical Features of Infection

Serological evidence indicates that even though EAV is widespread in the horse population, it rarely causes clinical disease. Both EAV and PRRSV can cause either persistent asymptomatic infections or induce various disease symptoms such as respiratory disease, fever, necrosis of small muscular arteries, and abortion. The severity of disease caused by EAV and PRRSV depends on the strain of virus as well as the condition and age of the animal. The most common symptoms of natural EAV infections in horses are anorexia, depression, fever, conjunctivitis, edema of the limbs and genitals, rhinitis, enteritis, colitis, and necrosis of small arteries. If clinical symptoms occur, they are most severe in young animals and pregnant mares. Infections in pregnant mares are often inapparent, but result in a high percentage (50%) of abortions. Young animals occasionally develop a fatal bronchopneumonia after infection, but natural infections are not usually life-threatening. In contrast, about 40% of pregnant mares and foals experimentally inoculated with EAV die as a result of the infection. Horses infected with virulent EAV isolates develop a high fever, lymphopenia, and severe disease symptoms. Symptoms observed in PRRSV-infected pigs include fever, anorexia, labored breathing, and lymphadenopathy. Lesions are observed in the lungs and infected pregnant sows produce weak or stillborn piglets.

Mice infected with LDV usually display no overt symptoms of disease. A distinguishing feature of LDV infections is the chronically elevated levels of seven serum enzymes, LDH (eight- to tenfold), isocitrate dehydrogenase (five- to eightfold), malate dehydrogenase (two- to threefold), phosphoglucose isomerase (two- to threefold), glutathione reductase (two- to threefold), aspartate transaminase (two- to threefold), and glutamate-oxaloacetate transaminase (two- to threefold). A decrease in the humoral and cellular immune response to non-LDV antigens is observed during the first 2 weeks following LDV infection. Thereafter, the immune response to other antigens is normal. In immunosuppressed C58 and AKR mice, neurovirulent isolates of LDV can induce a sometimes fatal poliomyelitis. In these mice, immunosuppression is required to delay antibody production until after virus has reached the central nervous system (CNS) and infected susceptible ventral motor neurons. LDV-infected neurons become the targets of an inflammatory response. In mice 6 months of age or older, paralysis of one or both hindlimbs and sometimes a forelimb is observed. In younger C58 mice, poliomyelitis is usually subclinical.

Isolates of SHFV that induce persistent, asymptomatic infections and ones that cause acute, asymptomatic infections of patas monkeys have been reported. All SHFV isolates cause fatal hemorrhagic fever in macaque monkeys. Infected macaques develop fever and mild edema followed by anorexia, dehydration, adipsia, proteinuria, cynosis, skin petechia, bloody diarrhea, nose bleeds, and occasional hemorrhages in the skin. The pathological lesions consist of capillary-venous hemorrhages in the intestine, lung, nasal mucosa, dermis, spleen, perirenal and lumbar subperitoneum, adrenal glands, liver, and periocular connective tissues. These signs and symptoms are not unique to SHFV-infected animals, since they are also observed after infection of macaques with other types of hemorrhagic fever viruses such as Ebola virus. Although the SHFV-induced lesions are widespread in infected animals, the level of tissue damage is not severe. Even so, mortality in macaques infected with SHFV approaches 100% and occurs within 1 or 2 weeks after infection.

## Pathology and Histopathology

In horses experimentally or fatally infected with EAV, the most common gross lesions are edema, congestion, and hemorrhage of subcutaneous tissues, lymph nodes, and viscera. Microscopic investigation of tissues from chronically infected horses, which had mildly swollen lymph nodes and slightly increased volumes of pleural and peritoneal fluids, revealed extensive lesions consisting of generalized endothelial damage to blood vessels of all sizes as well as severe glomerulonephritis. Both types of lesions are thought to be caused by the deposition of viral immune complexes. Extensive capillary necrosis leads to a progressive increase in vascular permeability and volume, hemoconcentration, and hypotension. During the terminal stages of the disease, lesions are also found in the adrenal cortex, and degenerative changes are observed in the bone marrow and liver. Focal myometritis is observed in infected pregnant mares and is thought to be the cause of deficiencies in the fetal and placental blood supply. The resulting anoxia is probably the cause of abortion.

Although most LDV infections are inapparent in mice, some histopathogenic changes are observed in infected animals. As described above, the serum levels of seven enzymes are chronically elevated in LDV-infected mice. Normally, an increase in serum levels of tissue enzymes is the result of tissue damage, but in LDV-infected animals little tissue damage is observed. Although there are five naturally occurring LDH isozymes in mouse plasma, only the level of isozyme LDVV is elevated in LDV-infected mice. Studies have indicated that the increase in enzyme levels is primarily the result of a decreased rate of enzyme clearance. A subpopulation of Kupffer cells involved in receptor-mediated endocytosis of LDH is severely diminished in mice by 24 h after LDV infection. It has been postulated that LDV replication in these cells causes their death and results in increased LDH serum levels. Splenomegaly, characterized by a greater than 30% increase in spleen weight, occurs in about 40% of the mice infected with LDV. The increase in spleen weight is observed by 24 h after infection and persists for up to a month. A marked necrosis of lymphocytes in thymic-dependent areas occurs during the first 4 days after LDV infection together with a transient decrease in the number of circulating T lymphocytes between 24 and 72 h after infection. A transient decrease in peritoneal macrophages is also observed between the first and tenth day of infection. Despite the lifelong presence of circulating viral immune complexes and the demonstration of some LDV antibody deposits in the kidneys of LDV-infected mice as early as 7 days after infection, these animals do not develop kidney disease. It has been suggested that nephritis does not develop in these chronically infected mice because of the inability of the majority of the LDV–antibody complexes to bind C1q. Low levels of C1q-binding activity can only be detected between days 10 and 18 after LDV infection. LDV infection can alter the outcome of concomitant autoimmune disease, probably through modulation of the host-immune responses. LDV infection can also trigger the spontaneous production of different types of autoantibodies, possibly as a result of polyclonal B-lymphocyte activation.

The CNS lesions in neurovirulent LDV-infected C58 and AKR mice are located in the gray matter of the spinal cord and, occasionally, in the brainstem and consist of focal areas of inflammatory mononuclear cell infiltrates in the ventral horn. Virus-specific protein and nucleic acid have been detected in ventral motor neurons, and maturing virions in these neurons have been observed by electron microscopy.

## Immune Response

Antibodies in sera obtained from animals infected with EAV or PRRSV recognize virion proteins N, M, $GP_5$, and $GP_2$. Neutralizing antibodies are primarily directed to $GP_5$ and the neutralization epitopes of EAV, LDV, and PRRSV have been mapped to the ectodomain of this protein. For EAV, there are four major $GP_5$ conformational neutralization sites that are interactive. Also, for EAV, interaction between $GP_5$ and M is required for neutralization. For LDV and PRRSV, the major neutralization site is located in the N-terminus of $GP_5$. A secondary neutralization site for PRRSV has been mapped to the ectodomain of $GP_4$. The neutralization epitopes of SHFV have not yet been studied.

Anti-EAV antibodies can be detected in horses 1–2 weeks after infection with virulent or avirulent strains of the virus. Complement-fixing, antiviral antibodies peak 2–3 weeks after the initiation of infection and then decline. Neutralizing antibody peaks between 2 and 4 months after infection. An increase in neutralizing antibodies usually leads to virus clearance. Often after 8 months, anti-EAV antibody can no longer be detected by complement-fixation or neutralization assays. However, in some animals the virus persists and viral immune complexes continue to circulate.

The first month of infection of pigs with PRRSV is characterized by high viremia and disease symptoms. A vigorous antiviral antibody response can be detected by ELISA beginning 7–9 days after infection but these antibodies have little neutralizing activity. Beginning at about a month after infection, neutralizing antibody can be detected and peaks between 1 and 2 months after infection. Viremia is reduced to very low levels but virus continues to be produced from infected cells in tissues for at least 5 months. Usually, the infection eventually is completely cleared, but in some cases it continues to persist.

In LDV-infected mice, which always become persistently infected, antiviral $GP_5$ and N antibody that is primarily of the IgG2a subclass is produced as early as 6–10 days after infection. The production of this antibody is dependent on functional T helper cells. Plasma from LDV-infected mice has a much higher nonspecific binding activity than plasma from uninfected mice; virus-specific binding measured by enzyme-linked immunosorbent assay (ELISA) usually cannot be detected until the plasma has been diluted at least 1:400. Some virus neutralization by this early antibody has been demonstrated but is incomplete due to the presence of virus quasi-species that are resistant to neutralization. Anti-LDV antibody that is not complexed to virus can be detected by 15 days after infection, indicating that antibody is present in excess of virus in chronically infected mice. Although the presence of anti-LDV antibodies does not prevent infection of macrophages, it does effectively neutralize neurovirulent LDV strains and so protects motor neurons from becoming infected. LDV-infected mice display a polyclonal humoral response and anti-LDV antibody apparently accounts for only a small portion of this polyclonal response. The mechanism by which LDV infection activates B cells polyclonally is currently not known, but mice immunized with inactivated virus do not develop a polyclonal response. Autoantibodies to a variety of cellular components (autoimmune antibodies) have been detected in mice chronically infected with LDV.

SHFV isolates that produce acute infections in patas monkeys induce high levels of neutralizing antibody, whereas SHFV isolates that induce persistent infections induce low titers of non-neutralizing antibody. Antibodies to virus that causes acute infection do not cross-neutralize virus that causes persistent infection. In macaques, death from an SHFV infection occurs before an effective adaptive immune response can be elicited.

LDV-infected animals develop cytotoxic T cells that can specifically recognize and lyse virus-infected macrophages. However, this cytotoxic response is not able to clear the infection. Whether anti-LDV cytotoxic T cells persist indefinitely in chronically infected mice or eventually disappear due to clonal exhaustion is disputed. A cytotoxic T-cell response as indicated by IFNγ-producing T cells can be detected in PRRSV-infected pigs starting about a month (the same time that neutralizing antibody appears) after infection and lasts at least a year.

Although the production of neutralizing antibody and cytotoxic T cells is delayed in EAV- and PRRSV-infected animals, these responses are usually effective in clearing the infection. However, in some EAV and PRRSV infections and in all LDV infections, persistent infections develop even though good levels of neutralizing antiviral antibodies and helper and cytotoxic T-cell responses are elicited. The mechanisms by which these viruses evade clearance by the adaptive immune system include extensive glycosylation of the major glycoprotein $GP_5$ that masks the major neutralization epitope of some strains of the virus, and enhanced infection of macrophages by infectious viral immune complexes via cell surface Fc receptors. Neutralization escape virus variants have been reported to arise during persistent LDV infections and may also arise during persistent infections with other arteriviruses. An immunodominant decoy epitope located just upstream of the neutralization epitope in PRRSV $GP_5$ induces a strong non-neutralizing antibody response and may be responsible for the weak/absent induction of neutralizing antibodies during the first month of infection.

## Prevention and Control

Avirulent and virulent strains of EAV and PRRSV have been isolated. A number of live attenuated vaccines and killed vaccines are commercially available for both EAV and PRRSV. The live vaccines are more efficacious in providing protection and induce a longer-lasting immunity than the killed vaccines. Although these vaccines induce immunity against disease, immunized animals are not completely protected from reinfection. Animals immunized with live vaccines can spread virus and can become persistently infected. Outbreaks of disease due to reversion of live PRRSV vaccines have been reported. To allow discrimination between natural and vaccine infections, markers have been engineered into one EAV live vaccine. Recent vaccine development efforts have focused on the utilization of recombinant virus vectors, such as Venezuelan equine encephalitis virus and pseudorabies, or DNA vectors to express $GP_5$ or both $GP_5$ and M.

The current lack of rapid diagnostic assays for the detection of LDV and SHFV in persistently infected animals means that it is still a time-consuming task to identify animals with inapparent infections. Care should be taken not to inadvertently transfer arteriviruses from a persistently infected animal to other susceptible animals. Cells and infectious agent pools obtained from animals that might be persistently infected with an arterivirus should be checked for viral contamination before they are injected into a susceptible animal.

## Future Perspectives

Arteriviruses have so far been isolated from mice (LDV), horses (EAV), pigs (LV), and monkeys (SHFV). It seems likely that other host species, including humans, harbor additional members of this virus family. However, such viruses will be difficult to find if the natural hosts develop asymptomatic infections. Little is yet known about the functional roles of the arterivirus proteins in the virus life cycle. Recent studies suggest that the arterivirus nucleocapsid and envelope proteins have unique properties. The intense current interest in dissecting the structure and function of the SARS-coronavirus replicase may provide new insights for similar analyses of the arterivirus replicase. The availability of reverse genetic systems for several of the arteriviruses will not only aid the further molecular characterization of these viruses but will also facilitate the study of viral pathogenesis and antiviral immunity as well as the development of improved vaccines.

## Further Reading

Balasuriya UB and Maclachlan NJ (2004) The immune response to equine arteritis virus: Potential lesions for other arteriviruses. *Veterinary Immunology and Immunopathology* 102: 107–129.

Coutelier J-P and Brinton MA (2006) Lactate dehydrogenase-elevating virus. In: Fox JG, Newcomer C, Smith A, Barthold S, Quimby F, and Davidson M (eds.) The Mouse in Biomedical Research, 2nd edn., pp. 215–234. San Diego, CA: Elsevier.

Delputte PL, Costers S, and Nauwynck HJ (2005) Analysis of porcine reproductive and respiratory syndrome virus attachment and internalization: Distinctive roles for heparan sulphate and sialoadhesin. *Journal of General Virology* 86: 1441–1445.

Godeny EK, de Vries AAF, Wang XC, Smith SL, and de Groot RJ (1998) Identification of the leader–body junctions for the viral subgenomic mRNAs and organization of the simian hemorrhagic fever virus genome: Evidence for gene duplication during arterivirus evolution. *Journal of Virology* 72: 862–867.

Gorbalenya AE, Enjuanes L, Ziebuhr J, and Snijder EJ (2006) Nidovirales: Evolving the largest RNA virus genome. *Virus Research* 117: 17–37.

Lopez OJ and Osorio FA (2004) Role of neutralizing antibodies in PRRSV protective immunity. *Veterinary Immunology and Immunopathology* 102: 155–163.

Murtaugh MP, Xiao ZG, and Zuckermann F (2002) Immunological responses of swine to porcine reproductive and respiratory syndrome virus infection. *Viral Immunology* 15: 533–547.

Pasternak AO, Spaan WJM, and Snijder EJ (2006) Nidovirus transcription: How to make sense...? *Journal of General Virology* 87: 1403–1421.

Plagemann PGW and Moennig V (1992) Lactate dehydrogenase-elevating virus, equine arteritis virus, and simian hemorrhagic fever virus: A new group of positive-strand RNA viruses. *Advances in Virus Research* 41: 99–192.

Siddell SG, Ziebuhr J, and Snijder EJ (2005) Coronaviruses, toroviruses, and arteriviruses. In: Mahy BWJ and ter Meulen V (eds.) *Topley and Wilson's Microbiology and Microbial Infections, Vol. 1: Virology*, 10th edn., London: Hodder Arnold.

Snijder EJ and Meulenberg JM (1998) The molecular biology of arteriviruses. *Journal of General Virology* 79: 961–979.

Snijder EJ, Siddell SG, and Gorbalenya AE (2005) The order Nidovirales. In: Mahy BWJ and ter Meulen V (eds.) *Topley and Wilson's Microbiology and Microbial Infections, Vol. 1: Virology*, 10th edn., London: Hodder Arnold.

Snijder EJ and Spaan WJM (2007) Arteriviruses. In: Knipe DM, Howley PM, Griffin DE, *et al.* (eds.) *Fields Virology*, 5th edn., pp. 1337–1356. Philadelphia, PA: Lippincott Williams and Wilkins.

Wieringa R, de Vries AAF, van der Meulen J, *et al.* (2004) Structural protein requirements in equine arteritis virus assembly. *Journal of Virology* 78: 13019–13027.

Ziebuhr J, Snijder EJ, and Gorbalenya AE (2000) Virus-encoded proteinases and proteolytic processing in the *Nidovirales*. *Journal of General Virology* 81: 853–879.

# Ascoviruses

**B A Federici,** University of California, Riverside, CA, USA
**Y Bigot,** University of Tours, Tours, France

© 2008 Elsevier Ltd. All rights reserved.

## Glossary

**Apoptosis** Genetically programmed cell death.
**Apoptotic bodies** Cell vesicles resulting from apoptosis.
**Caspase** Protease that activates a major portion of programmed cell death.
**Endoparasitic wasps** Species of insect parasites belonging to the order Hymenoptera, which lay their eggs in insects where the wasp larvae develop.
***Per os* infection** Infection by feeding.
**Programmed cell death** Genetically programmed cascade proteases and nucleases that cleave DNA and proteins within a cell leading to its death.
**Reniform** Shaped like a kidney.
**Transovarial transmission** Transmission of virus inside the egg.
**Virion-containing vesicles** Vesicles containing virions formed by ascoviruses by rescue of apoptotic bodies induced by ascovirus infection.

## Introduction

The family *Ascoviridae* is one of the newest families of viruses, established in 2000 to accommodate several species of a newly recognized type of DNA virus that attacks larvae of insects of the order Lepidoptera. Viruses of this family produce large, enveloped virions, measuring 130 nm in diameter by 300–400 nm in length, and when viewed by electron microscopy have a reticulated appearance. They are typically bacilliform or reniform in shape, and contain a circular double-stranded DNA genome that, depending on the species, ranges from ∼120 to 185 kbp. Whereas the virions of ascoviruses are structurally complex like those of other large DNA viruses that attack insects, such as those of iridoviruses (family *Iridoviridae*) and entomopoxviruses (family *Poxviridae*), they differ from these in two significant aspects. First, ascoviruses are transmitted from diseased to healthy lepidopteran larvae or pupae by female endoparastic wasps when these lay eggs in their hosts. Second, ascoviruses have a unique cell biology and cytopathology in which shortly after infecting a cell, they induce apoptosis and then rescue the developing apoptotic bodies and convert these into virion-containing vesicles. This aspect of viral reproduction apparently evolved to disseminate virions to the larval blood where they could contaminate the ovipositors of female wasps so that the virus could be transmitted to new hosts. Ascoviruses appear to occur worldwide, wherever there are endoparasitic wasps and larvae of species belonging to the lepidopteran family Noctuidae. However, as these viruses have been discovered relatively recently and their signs of disease are not commonly known in the scientific community, relatively few ascovirus species have been described.

# History

The first ascoviruses were discovered during the late 1970s in southern California where they were found causing disease in larvae of moths belonging to the lepidopteran family Noctuidae. Diseased larvae were recognized by the presence of blood that was very white and opaque, in marked contrast to the blood of healthy larvae which is translucent and slightly green (**Figure 1**). The color and opacity of the blood in diseased larvae was shown to be due to the presence of high concentrations of vesicles that contained virions (**Figure 2**). The white blood and virion-containing vesicles are diagnostic for the disease, and the name for this group, ascoviruses (derived from the Greek *asco* meaning 'sac'), was chosen on the basis of the latter characteristic. Since the discovery of the first ascovirus, ascoviruses have been isolated as the cause of disease in many species of noctuid larvae. In addition, an ascovirus that attacks the pupal stage of a species belonging to the family Yponomeutidae was discovered in the 1990s in France.

# Distribution and Taxonomy

With respect to distribution, ascoviruses have been reported from the United States, Europe, Australia, and Indonesia, and it is highly probable that they occur worldwide. This is because their most common hosts, larvae of lepidopteran species belonging to the family Noctuidae, the largest family of the order Lepidoptera, as well as their most common vectors, endoparasitic wasps of the families Braconidae and Ichnuemonidae, are distributed throughout the world. Although only a few ascovirus species have been described to date, there are probably many, including variants, that occur worldwide. Thus, given the common occurrence of their hosts and vectors, it is possible that ascoviruses are very common insect viruses. That they have not been discovered in large numbers, for example, like baculoviruses (family *Baculoviridae*), is probably because they cause a chronic disease with few easily detectable signs, making it difficult for individuals not familiar with the disease to recognize diseased larvae in field populations.

At present, five species of ascoviruses are officially recognized based on a combination of properties including the relatedness of key genes coding for the DNA polymerase and major capsid protein, the degree to which their genomic DNA cross-hybridizes under conditions of low stringency, their lepidopteran host range, and their tissue tropism (**Table 1**). The type species is the *Spodoptera frugiperda ascovirus* (SfAV-1a), with the other species being *Tricoplusia ni ascovirus* (TnAV-2a), *Heliothis virescens ascovirus* (HvAV-3a), and *Diadromus pulchellus ascovirus* (DpAV-4a). The Arabic numeral reflects the order in which each species was formally recognized, whereas the lower case letter indicates the type species of the variants. Variants from the type species are recognized by different consecutive lower case numbers; for example, TnAV-2b and 2c would represent two different variants of TnAV-2a recognized subsequently. Herein, these viruses are referred to by their acronyms without the numerical and lower case suffix.

Ascoviruses have been isolated from many more insect species than those listed above, but these isolates have turned out to be variants of known ascoviruses, and therefore they have not been named after the host from which they have been isolated. For example, ascoviruses related to TnAV and HvAV have been isolated from noctuid species such as *Autographa precationis*, *Helicoverpa zea*, *Helicoverpa armigera*, and *Helicoverpa punctigera*; however, they do not bear the name of their host of isolation. What this implies is that ascoviruses belonging to the TnAV and HvAV species have a broad and overlapping host range among different noctuid species, although this has only been tested experimentally to a limited extent.

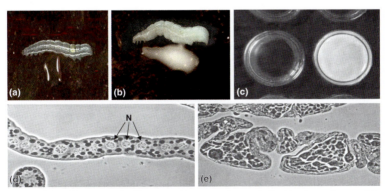

**Figure 1** Major characteristics of the disease typically caused by ascoviruses in lepidopteran larvae. (a) and (b) Healthy and ascovirus-infected larvae, respectively, of the cabbage looper, *Trichoplusia ni*, infected with TnAV. Note the opaque white blood in the infected larva. (c) Spot plate containing blood from healthy (left) and infected larvae (right). (d, e) Sections through lobes of fat body from a healthy and infected larva, respectively, of the fall armyworm, *Spodoptera frugiperda*, infected with SfAV ascovirus. Note the greatly hypertrophied cells in the fat body of the infected larva. The cells in most of this tissue have already cleaved into viral vesicles. N, nuclei.

## 188 Ascoviruses

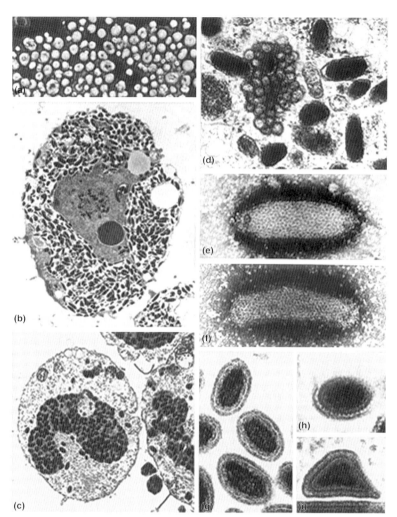

**Figure 2** Structural and morphological characteristics of ascovirus virions and virion-containing vesicles. (a) Wet mount preparation viewed with phase microscopy of blood from a *Spodoptera frugiperda* larva infected with SfAV. The spherical refractile bodies are virion-containing vesicles. (b, c) Electron micrographs of ultrathin sections through viral vesicles produced by the trichoplusia (TnAV) and spodoptera (SfAV) ascoviruses, respectively. (d) Matrix of the occlusion body produced by the spodoptera (SfAV) ascovirus. The occlusion consists of virions, protein, and small spherical vesicles. (e, f) Negatively stained virions of the spodoptera (SfAV) and trichoplusia (TnAV) ascoviruses, respectively. Note the reticulate appearance of the virions. (g, h) Electron micrographs of ultrathin cross sections through inner particles of the spodoptera (SfAV) ascovirus after formation (g) and during envelopment (h). (i) Ultrathin cross section through a fully developed virion of the trichoplusia (TnAV) ascovirus.

**Table 1** Members in the genus *Ascovirus* belonging to the family *Ascoviridae*

| Species name | Virus abbreviation | Accession number |
| --- | --- | --- |
| Recognized species | | |
| *Spodoptera frugiperda ascovirus 1a* | SfAV-1a | [AM3988432] |
| *Trichoplusia ni ascovirus 2a* | TnAV-2a | [AJ312707] |
| *Heliothis virescens ascovirus 3a* | HvAV-3a | |
| *Diadromus pulchellus ascovirus 4a* | DpAV-4a | [AJ279812] |
| Tentative species | | |
| Spodoptera exigua ascovirus 5a | SeAV-5a | |
| Spodoptera exigua ascovirus 6a | SeAV-6a | |
| Helicoverpa armigera ascovirus 7a | HaAV-7a | |
| Helicoverpa punctigera ascovirus 8a | HpAV-8a | |

## Virion Structure and Composition

Depending on the species, the virions of ascoviruses are either bacilliform or reniform in shape, with complex symmetry, and very large, measuring about 130 nm in diameter by 300–400 nm in length. The virion consists of an inner particle surrounded by an outer envelope (**Figure 2**). The inner particle is complex containing a DNA/protein core as well as an apparent internal lipid bilayer surrounded by a distinctive layer of protein subunits. Thus, the virion appears to contain two lipid membranes: one associated with the inner particle and the other forming the lipid component of the envelope. In negatively stained preparations, virions have a distinctive reticulate appearance, which is thought to be due to superimposition of subunits on the surface of the internal particle with those in the envelope.

As indicated by the size and complexity of the virions, the genome of ascoviruses is large, and consists of a single molecule of double-stranded circular DNA. The genomes of three species have been sequenced, the type species SfAV, TnAV, and HvAV. The SfAV genome is 157 kbp and codes for at least 120 proteins (**Figure 3**), whereas the TnAV 2a genome is slightly larger, 174 kbp, and codes for at least 134 proteins. The genome of HvAV is 186 kbp and codes for approximately 180 potential proteins. Based on gel analyses, ascovirus virions contain at least 12 structural polypeptides ranging in size from 12 to 200 kDa. In addition to proteins and the DNA genome, the presence of an envelope as detected by electron microscopy, as well as experiments with detergents and organic solvents, indicate that virions contain a substantial lipid component. And, as in other enveloped viruses of eukaryotes, it is likely that the virion also contains carbohydrate in the form of glycoproteins, though none have been identified.

## Transmission and Ecology

One of the most interesting features of ascoviruses is that their transmission from host to host appears to be dependent on their being vectored by female endoparasitic wasps belonging to the families Braconidae and Ichneumonidae (order Hymenoptera). Ascoviruses are extremely difficult to transmit *per os*, with typical infection rates averaging less than 15% even when larvae are fed as many as $10^5$ virion-containing vesicles in a single dose. In contrast to this, infection rates for caterpillars injected with as few as 10 virion-containing vesicles are typically greater than 90%. Moreover, experiments with parasitic wasps show that they can effectively transmit ascoviruses to their noctuid hosts. For example, when females are allowed to lay eggs in ascovirus-infected noctuid caterpillars, thereby contaminating their ovipositor, and then allowed to lay eggs in healthy larvae, the majority of the latter contract ascovirus disease. Interestingly, though the parasite eggs hatch in their infected noctuid hosts, the parasite larvae die as the ascovirus disease develops in the caterpillar. Under field conditions, the prevalence of ascovirus disease in caterpillars is correlated with high rates of parasitization by endoparasitic wasps. When wasps from these populations are collected in the field and allowed to oviposit in healthy caterpillars reared in the laboratory, the latter often exhibit ascovirus disease within a few days. Thus, laboratory and field studies provide sound evidence that the primary mechanism for the transmission of ascoviruses attacking noctuid larvae is through being vectored mechanically by parasitic wasps. No evidence has been found in the lepidopteran hosts for transovum or transovarial transmission.

In the case of DpAV, the association of the virus with its wasp and caterpillar hosts is much more intimate. DpAV DNA is carried in wasp nuclei as a circular molecule, and small numbers of virions are produced in the oviducts of females. However, the virus does not cause noticeable pathology in the wasp host. The females lay eggs in the pupal stage of the lepidopteran host, *Acrolepiopsis assectella*, introducing small numbers of ascovirus virions along with the wasp eggs. These virions invade lepidopteran host cells, replicate, and initiate destruction of major host tissues. The wasp larva then emerges from the egg and feeds on the host tissues and ascovirus virions. The DpAV genome is carried by both male and female wasps, where it is apparently transmitted from generation to generation transovarially. These observations make ascoviruses the only known group of viruses pathogenic to insects primarily dependent on vectors for their transmission.

Now that the characteristics of the disease are known, field studies in the southeastern United States and California are beginning to show that ascoviruses are probably the most common type of virus to occur during most of the year in populations of several important noctuid pests, including the cabbage looper, *T. ni.*, fall armyworm, *S. frugiperda*, and the corn earworm, *H. zea*. Prevalence rates range from 10% to 25%, depending on the species and time of the year, with the highest rates of infection, as noted above, being correlated with high levels of parasitization. In South Carolina, ascovirus infection rates as high as 60% have been reported in populations of noctuid larvae at the end of summer.

## Host Range

The experimental host range of ascoviruses varies with the viral species. TnAV, HvAV, and SeAV have a broad host range and are capable of replication in a variety of noctuid

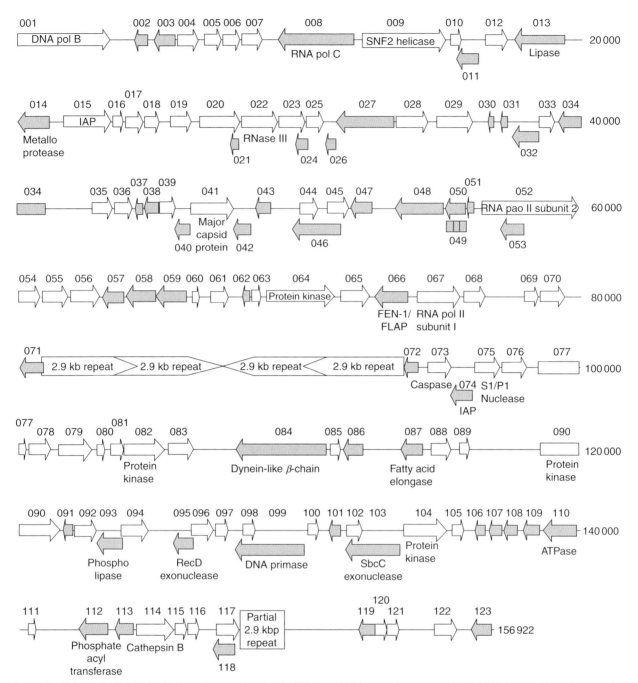

**Figure 3** Schematic illustration in linear form of the circular DNA 156 922 base pair genome of the SfAV 1a ascovirus, the ascovirus type species, isolated from a larva of the fall armyworm, *Spodoptera frugiperda*. The illustration identifies the relative positions of key genes including those coding for the DNA polymerase, major capsid protein, executioner caspase, as well as several genes coding for proteins involved in lipid metabolism. White and dark arrows represent, respectively, ORFs in forward and reverse orientations along the genome. From Bideshi DB, Demattei MV, Rouleux-Bonnin F, *et al*. (2006) Genomic sequence of spodoptera frugiperda ascovirus 1a, an enveloped, double-stranded DNA insect virus that manipulates apoptosis for viral reproduction. *Journal of Virology* 80: 11791.

species, as well as in selected species belonging to other families of the order Lepidoptera. Alternatively, the experimental host range of SfAV is limited to other species of the genus *Spodoptera*. DpAV can replicate in hymenopteran and lepidopteran hosts closely related to its natural host species, *A. assectella*. To propagate virus in the laboratory, all ascoviruses can be grown in their larval or pupal hosts. To infect caterpillars, they are injected with virus in the fourth or early fifth instar, and virion-containing vesicles are harvested from the blood 5–7 days later.

## Pathology and Pathogenesis

### Signs of Disease

The signs of ascovirus disease are very subtle, and this probably accounts for why ascoviruses were discovered only recently. The most obvious sign of disease within 24 h of infection is a decrease in the normal rate of feeding. The feeding rate continues to slow as the disease progresses, and as a result larvae fail to gain weight or advance in development. Healthy larvae, particularly in the early stages of development, will easily quadruple their weight and size in a period of 3 or 4 days, whereas ascovirus-infected larvae cease to grow and may actually lose weight. This feature of ascovirus disease is almost impossible to detect in infected larvae in the field. However, it is easily noticed under laboratory conditions when infected and healthy larvae are reared side by side over a period of a few days. A second feature easily noted in the laboratory is that ascovirus diseases are chronic, though usually fatal. When infected during early stages of development, ascovirus-diseased larvae often survive for 2 or 3 weeks beyond the time at which most healthy larvae have completed their development and pupated. Signs of disease other than these are minor, but include the inability to completely cast the molted cuticle, a bloated thoracic region, and a white or creamy discoloration and hypertrophied appearance of the larval body at advanced stages of disease development.

### Cytopathology and Cell Biology

In comparison to all other known viruses, the most unique property of ascoviruses is the unusual cytopathology that leads to the formation of the virion-containing vesicles. This process resembles apoptosis, and recent studies of the SfAV genome have shown that it encodes an executioner caspase, synthesized 9 h after infection, which by itself is capable of inducing apoptosis (**Figure 4**).

At the cellular level *in vivo*, the disease begins with extraordinary hypertrophy of the nucleus accompanied by invagination of sections of the nuclear envelope, followed by a corresponding enlargement of the cell. Cells typically grow from 5 to 10 times the diameter of uninfected cells. As the nucleus enlarges, the nuclear envelope ruptures and disintegrates into fragments. At about this stage, the cell plasmalemma begins to invaginate along 'planes' toward the now anucleate cell center. Concomitantly, sheets of membrane form closely adjacent to mitochondria that accumulate along the planes. As this process continues, the membrane sheets coalesce and join the invaginating plasmalemma, thereby cleaving the cell into a cluster of 20 to more than 30 vesicles, ranging in size from 5 to 10 µm in diameter. This aspect of ascovirus cellular pathology resembles the formation of apoptotic

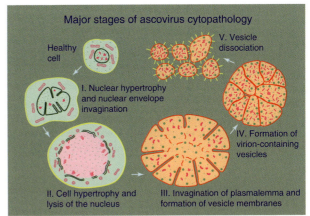

**Figure 4** Major stages of cellular pathogenesis caused by a typical ascovirus, a process that resembles apoptosis. After infection, the nucleus enlarges and the nuclear membrane invaginates, and then lyses. Subsequently, the plasma lemma of the cell invaginates and coalesces with cytoplasmic membranes, apparently formed *de novo*, thereby dividing the cell into a cluster of virion-containing vesicles. These vesicles dissociate and are liberated into the blood as the basement membrane of infected tissues degenerates. Virion assembly becomes apparent as the nuclear membrane lyses, and continues throughout all subsequent stages of vesicle formation.

bodies during apoptosis. However, rather dissipate as the cell dies, the developing apoptotic bodies are rescued by the virus and progress to form vesicles in which virions continue to assemble. These virion-containing vesicles, also referred to as viral vesicles, typically remain in the tissue until the basement membrane ruptures, though on occasion cell hypertrophy can be so great that the enlarging cell erupts out through the basement membrane of the infected tissue, releasing large fragments of the infected cell directly into the blood. Analysis of both the SfAV and TnAV genomes shows that, unlike many other large DNA viruses, ascoviruses encode several lipid-metabolizing enzymes that are likely involved in the process of converting developing apoptotic bodies into virion-containing vesicles.

Although the process by which viral vesicles are cleaved from cells varies among different ascoviruses, the histopathology is similar among virtually all viruses. Vesicles accumulate in the tissues where they are formed, but as these tissues degenerate during disease progression, the basement membrane of infected tissues deteriorates and ruptures, allowing the vesicles to spill out into the blood. There they accumulate reaching concentrations as high as $10^7$–$10^8$ vesicles ml$^-$ within 3–4 days of infection. There is some evidence that viral replication proceeds within the vesicles as they circulate in the blood, and thus this tissue must also be considered one of the tissues attacked by ascoviruses. If fact, because such high concentrations of viral vesicles are found in the blood, this tissue could be considered a major site of infection, particularly if it is eventually shown that these viruses continue to replicate in the vesicles as they circulate in the blood.

Despite the chronic nature of the disease caused by ascoviruses, virion-containing vesicles are present in the blood within 2 or 3 days of infection. When the virus replicates in cells *in vitro*, the vesicles are formed within 12–16 h of infection. The rapid development and circulation of the viral vesicles in the blood probably evolved to enhance transmission of the virus by parasitic wasps.

### Tissue Tropism

The cytopathology of ascoviruses is consistent among different viral species; however, considerable variation occurs with respect to the tissues attacked, that is, in which replication occurs. TnAV, HvAV, and SeAV exhibit a relatively broad tissue tropism infecting the tracheal matrix, epidermis, fat body, and connective tissue. Differences exist between these species in that some HvAV variants infect the epidermis much more extensively than TnAV variants, whereas some of the latter can also replicate more extensively in fat body cells, but appear only to do this when larvae are infected early in their development. Alternatively, the type species, SfAV, and its variants have a very narrow tissue tropism, with the fat body being the primary site of infection. DpAV occurs in the nuclei of all tissues of its wasp host, but appears to only produce progeny in ovarial tissues. In its lepidopteran pupal host, it attacks and replicates in a wide variety of tissues.

### Replication and Virion Assembly

Although there have been few biochemical studies of viral DNA replication or protein synthesis, studies carried out with ascoviruses *in vivo* and *in vitro* show that progeny virions first appear about 12 h after infection. Virion assembly is initiated after the nucleus ruptures, and occurs prior to and during the cleavage of the cell into viral vesicles. The first recognizable structural component of the virion to form is the multilaminar layer of the inner particle. Based on its ultrastructure, this layer consists of a unit membrane and an exterior layer of protein subunits. As the multilaminar layer assembles, a dense nucleoprotein core aggregates on the interior surface. This process continues until the inner particle is complete. After formation, the inner particle is enveloped by membranes within the cell or vesicle. These membranes are apparently synthesized *de novo*. Thus, the assembly of the virions is reminiscent of that in other viruses with complex virions, such as the ridoviruses, herpesviruses, and poxviruses, where the virions differentiate after association of the precursors of virion structural components.

After formation, the virions of the TnAV ascovirus accumulate toward the periphery of the vesicle where they often form inclusion bodies, that is, aggregations of virions (**Figure 2**). In SfAV, occlusion bodies are formed in which the virions are actually occluded in a 'foamy' vesicular matrix that consists of a mixture of protein and minute spherical vesicles. When viewed with phase microscopy, these viral inclusion and occlusion bodies are phase bright, and are largely responsible for the highly refractile appearance of the vesicles. Ascoviruses do not typically form the types of occlusion bodies characteristic of other types of DNA insect viruses, such as baculoviruses and entomopoxviruses.

### Origin and Evolution

The subject of viral evolution over millions of years has received relatively little study due to the lack of a fossil record. Moreover, viruses are considered polyphyletic, and thus most of the more than 70 families of viruses are thought to have originated independently. In this regard, ascoviruses may provide a unique opportunity to obtain insights into virus evolution over long periods. Phylogenetic comparisons of ascovirus genes sequenced to date including those coding DNA polymerase and major capsid protein as well as several enzymes indicate that these viruses evolved from a lepidopteran iridovirus (family *Iridoviridae*). Iridoviruses, in turn, appear to have originated from phycodnavirsues (family *Phycodnaviridae*), which attack certain ciliates and algae. On the other end of the evolutionary scale, ascovirus virions are structurally and morphologically similar to the particles formed by ichnoviruses of the family *Polydnaviridae*. Ichnovirus particles are produced in the reproductive tracts of endoparasitic wasps of the family Ichneumonidae, and the wasp vector and host of DpAV is a member of this family. Thus, there is a reasonable possibility that ascoviruses and ichnoviruses are related phylogenetically, and share a common ancestor. This possibility is currently under investigation, and should be resolved over the next several years through a comparative analysis of the molecular evolution of genes of ascoviruses and ichnoviruses, after more structural genes from the latter viruses have been cloned and sequenced. A major question to be addressed is whether the DpAV represents an early ascovirus branch that evolved from an iridovirus or is representative of an ascovirus branch that eventually led to the origin of ichnovirus particles. With respect to the ichnoviruses and bracoviruses, recent data on the DNA contained by the particles of these putative viruses suggest that these are not viruses after all, but rather are an unusual highly evolved type of organelle that evolved from DNA viruses, which are used by endoparasitic wasps to suppress the internal defense responses of their insect hosts.

## Future Perspectives

As present, too little is known about ascoviruses to assess whether they are or will turn out to be of economic importance. Their poor infectivity *per os* makes it highly unlikely they will ever be developed as viral insecticides, especially given the successful advent of insect-resistant transgenic crops. However, as more entomologists become familiar with the disease caused by ascoviruses, it may be shown that in habitats rarely treated with chemical insecticides, such as transgenic crops, these viruses are responsible for significant levels of natural pest suppression, particularly where parasitic wasps are abundant. Such findings would encourage even greater emphasis on the development of biological control and other more environmentally sound methods of pest control. With respect to the cell biology of viral vesicle formation, ascoviruses provide an interesting model for how apoptosis can be manipulated at the molecular level. Additionally, study of the unusual process by which ascoviruses rescue the developing apoptotic bodies to form viral vesicles could lead to insights into how cells manipulate the cytoskeleton and mitochondria. Finally, it is possible that viral vesicles will provide a unique anucleate cellular system for studying the replication of a complex type of enveloped DNA virus *in vitro*.

*See also:* Baculoviruses: Apoptosis Inhibitors; Baculoviruses: General Features; Entomopoxviruses; Phycodnaviruses; Polydnaviruses: General Features.

## Further Reading

Asgari S (2006) Replication of Heliothis virescens ascovirus in insect cell lines. *Archives of Virology* 151: 1689.

Asgari S, Davis J, Wood D, Wilson P, and McGrath A (2007) Sequence and Organization of the Heliothis virescens ascovirus genome. *Journal of General Virology* 88: 1120–1132.

Bideshi DB, Demattei MV, Rouleux-Bonnin F, *et al.* (2006) Genomic sequence of spodoptera frugiperda ascovirus 1a, an enveloped, double-stranded DNA insect virus that manipulates apoptosis for viral reproduction. *Journal of Virology* 80: 11791.

Bideshi DB, Tan Y, Bigot Y, and Federici BA (2005) A viral caspase contributes to modified apoptosis for virus transmission. *Genes and Development* 19: 1416.

Bigot Y, Rabouille A, Doury G, *et al.* (1997) Biological and molecular features of the relationships between Diadromus pulchellus ascovirus, a parasitoid hymenopterna wasp (*Diadromus pulchellus*) and its lepidopteran host, Acrolepiopsis assectella. *Journal of General Virology* 78: 1149.

Cheng XW, Wang L, Carner GR, and Arif BM (2005) Characterization of three ascovirus isolates form cotton insects. *Journal of Invertebrate Pathology* 89: 193.

Federici BA (1983) Enveloped double-stranded DNA insect virus with novel structure and cytopathology. *Proceedings of the National Academy of Sciences, USA* 80: 7664.

Federici BA and Bigot Y (2003) Origin and evolution of polydnaviruses by symbiogenesis of insect DNA viruses in endoparasitic wasps. *Journal of Insect Physiology* 49: 419.

Federici BA, Bigot Y, Granados RR, *et al.* (2005) Family *Ascoviridae*. In: Fauquest CM, Mayo MA, Maniloff J, Desselberger U, and Ball LA (eds.) *Virus Taxonomy. Eighth Report of the International Committee on Taxonomy of Viruses*, pp. 269–274. San Diego, CA: Elsevier Academic Press.

Federici BA and Govindarajan R (1990) Comparative histopathology of three ascovirus isolates in larval noctuids. *Journal of Invertebrate Pathology* 56: 300.

Federici BA, Vlak JM, and Hamm JJ (1990) Comparison of virion structure, protein composition, and genomic DNA of three ascovirus isolates. *Journal of General Virology* 71: 1661.

Govindarajan R and Federici BA (1990) Ascovirus infectivity and the effects of infection on the growth and development of Noctuid larvae. *Journal of Invertebrate Pathology* 56: 291.

Pellock BJ, Lu A, Meagher RB, Weise MJ, and Miller LK (1996) Sequence, function, and phylogenetic analysis of an ascovirus DNA polymerase gene. *Virology* 216: 146.

Stasiak K, Renault S, Demattei MV, Bigot Y, and Federici BA (2003) Evidence for the evolution of ascoviruses from iridoviruses. *Journal of General Virology* 84: 2999.

Wang L, Xue J, Seaborn CP, Arif BM, and Cheng XW (2006) Sequence and organization of the Trichoplusia ni ascovirus 2c (*Ascoviridae*) genome. *Virology* 354: 167.

# Assembly of Viruses: Enveloped Particles

**C K Navaratnarajah, R Warrier, and R J Kuhn**, Purdue University, West Lafayette, IN, USA

© 2008 Elsevier Ltd. All rights reserved.

## Glossary

**Core-like particle (CLP)** Subviral particles assembled from recombinantly expressed capsid protein are referred to as core-like particles. They are morphologically similar to authentic cores isolated from viruses or infected cells.

**Cryo-EM** Cryo-electron microscopy and image reconstruction techniques are used to elucidate the structures of viruses and other macromolecular structures.

**Immature virion** Viruses usually produce noninfectious particles which require a maturation step in order to form the infectious, mature virus. The maturation step often involves the proteolytic cleavage of a precursor protein.

**Nucleocapsid core (NC)** Capsid protein packages the nucleic acid genome to form a stable protein–nucleic acid complex which is then enveloped with a lipid bilayer.

## Introduction

Viruses have long been distinguished by their physical features, usually visualized by electron microscopy or analyzed biochemically. One feature that has been frequently used to categorize viruses is the presence or absence of a lipid bilayer. Many animal viruses are surrounded by a lipid bilayer that is acquired when the nucleocapsid buds through cell membranes, usually at a late stage of virus assembly. While the protein coat of nonenveloped viruses plays a crucial role in protecting the genome from the environment, for enveloped viruses the lipid membrane partially fulfills this role. The lipid membranes are decorated with virus-encoded envelope proteins that are important for the subsequent infectivity of the virus, although some viruses also incorporate cellular proteins in their membrane. Virus envelopment can take place after the assembly of an intact nucleocapsid structure (betaretroviruses) or capsid assembly and envelopment can occur concomitantly (orthomyxovirus). Specific or nonspecific interactions between the viral envelope glycoproteins and the proteins that make up the nucleocapsid mediate the envelopment of the core or nucleoprotein–nucleic acid complex. Enveloped viruses acquire their lipid bilayer from a variety of locations within the cell, but a given virus will usually bud from one specific cellular membrane (**Table 1**). Enveloped viruses can take advantage of the cellular secretory pathway in order to assemble and bud out of the cell. In contrast, nonenveloped viruses usually exit infected cells by disrupting the plasma membrane. Thus, budding provides enveloped viruses with a nonlytic method of exiting infected cells, and they must do so while the cell is still alive.

## Viral Envelope

The main component of the viral envelope is the host-derived lipid bilayer. The precise composition of this lipid membrane varies, as different viruses acquire their envelopes from different cellular membranes. The choice of membrane from which the virus buds is often determined by the specific targeting and accumulation of the envelope proteins at a particular site in the secretory pathway (**Table 1**). There are examples of viruses that bud from the plasma membrane (togaviruses, rhabdoviruses,

**Table 1** List of enveloped virus families and the origin of the envelope

| Virus family[a] | Membrane |
|---|---|
| *Arenaviridae* | Plasma membrane |
| *Arterivirus* | Endoplasmic reticulum |
| *Asfarviridae* | Endoplasmic reticulum and plasma membrane |
| *Baculoviridae* | Plasma membrane |
| *Bunyaviridae*[b] | Golgi complex |
| *Coronaviridae* | ER Golgi intermediate compartment |
| *Cystoviridae*[c] | Plasma membrane |
| *Deltavirus* | Endoplasmic reticulum |
| *Filoviridae* | Plasma membrane |
| *Flaviviridae* | Endoplasmic reticulum |
| *Fuselloviridae*[d] | Plasma membrane |
| *Hepadnaviridae* | Endoplasmic reticulum |
| *Herpesviridae* | Nuclear envelope |
| *Hypoviridae* | Plasma membrane |
| *Iridoviridae* | Plasma membrane |
| *Lipothrixviridae*[d] | Plasma membrane |
| *Orthomyxoviridae* | Plasma membrane |
| *Paramyxoviridae* | Plasma membrane |
| *Plasmaviridae*[c] | Plasma membrane |
| *Polydnaviridae* | Plasma membrane |
| *Poxviridae* | ER Golgi intermediate compartment |
| *Retroviridae* | Plasma membrane |
| *Rhabdoviridae*[b] | Plasma membrane |
| *Togaviridae* | Plasma membrane |

[a]Animal viruses unless otherwise noted.
[b]This group also includes plant viruses.
[c]Bacteriophage.
[d]Archaea.

paramyxoviruses, orthmyxoviruses, and retroviruses), endoplasmic reticulum (ER) (coronaviruses and flaviviruses) and the Golgi complex (bunyaviruses). There are also examples of viruses that undergo transient envelopment and reenvelopment (herpesvirus).

Viral proteins are found embedded in the lipid membrane. The majority of these proteins are transmembrane glycoproteins. The viral envelope glycoproteins mediate the interaction of the virus with cell receptors and promote the fusion of the viral and cellular membranes during infection of susceptible cells. Viral glycoproteins are also crucial for the assembly of the virion. They can make important lateral contacts with each other, thus driving oligomerization and also capturing other viral components such as the capsid or matrix protein. The majority of enveloped viruses contain one or more glycoproteins that are usually found as oligomers embedded within the lipid membrane. High-resolution structural information is available for many glycoproteins such as the hemagglutinin and neuraminidase proteins of influenza A virus, the gp120 of HIV, and the E protein of dengue virus. Based on these structural and biochemical studies, it has been shown that most glycoproteins are primed for the conformational changes that are required in order to gain entry to the host cell during an infection.

Some enveloped viruses contain integral membrane proteins that have multiple membrane-spanning regions that oligomerize to form channels in the membrane. The influenza A virus M2 protein forms an ion channel and plays an important role in the assembly and entry of the virus particle. Some viruses such as the retroviruses also incorporate cellular membrane proteins into the viral envelope. In a majority of the cases the host proteins that are present at the sites of assembly or budding are incorporated in a passive, nonselective manner. However, there are examples where the virus actively recruits specific host proteins that may help in evading the defenses of the immune system or enhance infectivity.

## Icosahedral Enveloped Viruses: Alphaviruses and Flaviviruses

Alphaviruses, and more recently, flaviviruses have served as model systems to study the assembly and budding of simple enveloped viruses. These positive-strand RNA viruses consist of a single RNA genome that is encapsidated by multiple copies of a capsid protein to form the nucleocapsid core (NC). The envelopment of the NC is mediated by the interaction between the envelope glycoproteins and this core. The assembly and budding of these two simple enveloped viruses will be described in detail in order to present common themes in the assembly of icosahedral enveloped viruses.

## Alphavirus Assembly

### Alphavirus life cycle
Alphaviruses are members of the family *Togaviridae*, which also includes the genus *Rubivirus*. Alphaviruses enter the host cell by receptor-mediated endocytosis via the clathrin-coated endocytic pathway. Following fusion at low pH with the endosomal membrane, the NC is released into the cytoplasm. The NC has been proposed to uncoat by transfer of capsid proteins (CPs) to ribosomes. This releases the genome RNA into the cytoplasm which is translated to produce the nonstructural proteins. The nonstructural proteins transcribe a negative-sense copy of the genome RNA. This RNA serves as template for genomic and subgenomic RNA. The subgenomic RNA, which is synthesized in greater amounts than the genomic RNA, codes for the structural proteins of the virus. CP is found at the N-terminus of the structural polyprotein, followed by proteins PE2 (E3+E2), 6K, and E1. Two hundred and forty copies of CP, E2, and E1 assemble to form the alphavirus virion (**Figure 1(a)**). The transmembrane E1 glycoprotein functions during entry to mediate the fusion of the viral membrane with that of the endosomal membrane, while the transmembrane E2 glycoprotein is responsible for cell receptor binding. E1 and PE2, a precursor of E2 and E3, are processed together as a heterodimer in the ER and Golgi, and are transported to the cell surface in the form of spikes that are each composed of three heterodimers of E1/E2. E3 serves as a chaperone to promote the correct folding of E2, as well as to prevent the premature fusion of E1 in the acidic environment of the late Golgi. The maturation cleavage of PE2 to generate E3 and E2 by a furin-like protease in a late Golgi compartment primes the glycoprotein spike complex for subsequent fusion during virus entry. The function of 6K is unclear but it does promote infectivity of the particle. A single copy of the genome RNA is packaged by 240 copies of the CP to form an icosahedral NC in the cytoplasm of infected cells. The NC interaction with the E1/E2 trimeric spikes at the plasma membrane results in the budding of the mature virus from the cell membrane.

### Alphavirus virion structure
Cryo-electron microscopy (cryo-EM) and image reconstruction techniques have revolutionized the understanding of the molecular architecture of alphaviruses (**Figure 1(a)**). Studies with Ross River, Semliki Forest, Venezuelan equine encephalitis, Aura, and Sindbis have shown that these viruses consist of an outer protein layer made up of the glycoproteins E1 and E2 (**Figures 1(a) and 1(c)**). The membrane spanning regions of these glycoproteins traverse a host-derived lipid bilayer that surrounds the NC of the virus. The CP and glycoprotein layers interact with one another and are arranged symmetrically in a $T=4$ icosahedral configuration. Fitting of the atomic coordinates of the crystal structures of the ectodomain of E1 and amino acids 106–264 of the CP into the cryo-EM density of Sindbis virus allowed a pseudo-atomic model of the virus to be generated. The fitting of the E1 structure into the cryo-EM density reveals that E1 forms an icosahedral scaffold on the surface of the viral membrane. E1 is positioned almost tangential to the lipid bilayer, whereas E2 has a more radial arrangement. The bulk of E2 lies on top of E1 and caps the fusion peptide, thereby preventing premature fusion with cell membranes. This arrangement of the glycoproteins is in agreement with the function of each protein, where the surface-exposed E2 interacts with cellular receptors and protects E1 until it is required for fusion. The fusion peptide is only exposed when the E1–E2 heterodimer dissociates in the presence of low pH in the endosome. Fitting of amino acids 106–264 of the CP into the cryo-EM density of Sindbis virus showed that each subunit of the projecting pentamers and hexamers (known as capsomeres) observed in the NC layer is made up of the CP protease domain consisting of amino acids 114–264. There is very little interaction between amino acids 114–264 of the CP either within the capsomere or in between capsomeres. Thus, the major contributors to the stability of the NC in the absence of glycoproteins are

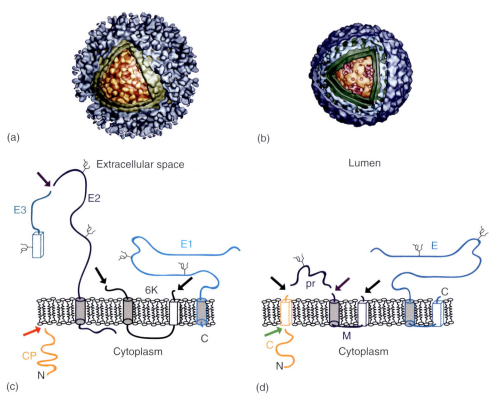

**Figure 1** Cryo-EM reconstructions and glycoprotein topology of alphaviruses and flaviviruses. (a) Surface shaded view of a Cryo-EM reconstruction of Ross River virus (alphavirus). The glycoprotein envelope shown in blue has been cut away to reveal the lipid bilayer (green) and the nucleocapsid core (orange). (b) Cryo-EM reconstruction of dengue virus (flavivirus). The glycoprotein envelope shown in blue has been cut away to reveal the lipid bilayer (green) and the nucleocapsid core (orange/purple). (c) The topology of the alphavirus glycoproteins is shown. The rectangular cubes represent signal sequences while the cylinders represent stop transfer sequences. The glycosylation sites are represented by branched structures. The arrows indicate cleavage sites. The black arrows indicate signalase cleavage sites, the purple arrow indicates the furin cleavage site, and the red arrow indicates the CP cleavage site. (d) The topology of the flavivirus structural proteins is shown. The rectangular cubes represent signal sequences while the cylinders represent stop transfer sequences. The glycosylation sites are represented by branched structures. The arrows indicate cleavage sites. The black arrows indicate signalase cleavage sites, the purple arrow indicates the furin cleavage site, and the green arrow indicates the NS2B/NS3 (viral protease) cleavage site.

CP–RNA and RNA–RNA interactions that take place in the RNA–protein layer below the projecting capsomeres.

### Alphavirus assembly and budding

Alphavirus virions always contain an NC and it is likely that this promotes and is required for budding through direct interactions with the glycoproteins. Thus, the first step in assembly is for the alphavirus CP to specifically recognize and encapsidate the genome RNA to form NCs in the cytoplasm of infected cells (**Figure 2**). The N-terminus of the CP (amino acids 1–80, SINV numbering) is largely basic and thought to be involved in charge neutralization of the genome RNA. Amino acids 38–55 are conserved uncharged residues that form a coiled-coil alpha-helix (helix I) important in dimerization of the CP during the assembly process. While the process of virus assembly is difficult to study in the complex cellular environment, the development of an *in vitro* assembly system based on bacterial expression of CP has led to advances in understanding NC assembly. These studies suggest that the initial event of NC assembly is the binding of CP amino acids 81–112 to the encapsidation signal on the genome RNA corresponding to nucleotides 945–1076 (SINV numbering) (**Figure 2**). This interaction exposes a second site on the encapsidation signal where another molecule of CP binds and forms a dimer with the first CP molecule. Amino acids 114–264 constitute the previously mentioned chymotrypsin-like serine protease that autoproteolytically cleaves CP from the nascent structural polyprotein. This region is involved in binding residues from the cytoplasmic domain of E2, thus linking the outer icosahedral glycoprotein shell with the NC across the lipid bilayer (**Figure 1(a)**).

Other lines of evidence support the dimer model of NC assembly. Helix I of CP, which is required for core accumulation in infected cells, may be functionally substituted by a GCN4 helix that forms dimeric coiled–coil interactions but not by a GCN4 helix that has a propensity to form trimeric coiled–coil interactions. In addition, helix I acts as a checkpoint in NC assembly whereby incompatible helices

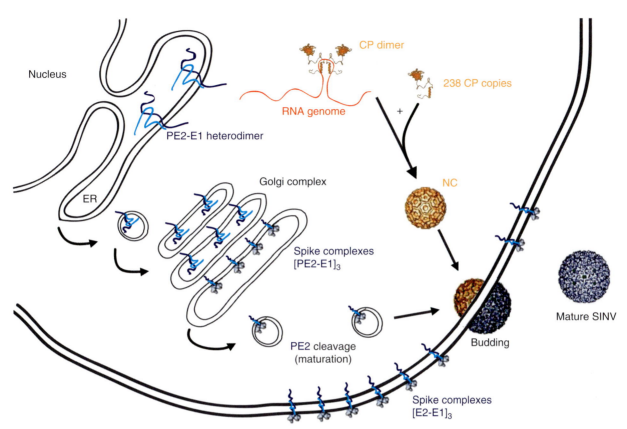

**Figure 2** Alphavirus assembly. Two capsid proteins (CPs) bind the encapsidation signal of the genome RNA to form a CP dimer-RNA complex in the cytoplasm. The CP dimer is stabilized by coiled–coil helix I interactions. The subsequent steps of nucleocapsid core assembly have not been elucidated but cores form and accumulate in the cytoplasm. The glycoproteins PE2 and E1 form heterodimers in the endoplasmic reticulum (ER). The glycoproteins are folded, glycosylated, and palmitoylated as they are transported through the ER and Golgi. The PE2–E1 heterodimers form spike complexes [PE2-E1]$_3$ in the Golgi. E3 is cleaved from the spikes by a furin-like protease before they are transported to the plasma membrane. The NC interacts with the cytoplasmic domain of E2 in the spike complexes, driving the budding of the mature virus from the plasma membrane.

prevent the formation of core-like particles (CLPs) *in vitro*. Furthermore, a portion of CPs in either NCs or CLPs may be cross-linked into dimers by DMS, a lysine specific cross-linker with a 12 Å cross-linking distance. Cross-linking enabled an assembly deficient helix mutation of CP to assemble into NCs, suggesting that the cross-link can functionally replace the helix interaction.

While assembly of the CP into NCs proceeds in the cytoplasm, the processing and assembly of the glycoproteins occur in the ER and Golgi (**Figures 1(c)** and **2**). The autocatalytic cleavage of the CP reveals a signal sequence on the N-terminus of the newly cleaved structural polyprotein that directs it to the ER (**Figure 1(c)**). PE2 is translocated into the ER until it reaches a 26-amino-acid stop transfer signal which anchors PE2 in the membrane. The C-terminal 33 residues of E2 then act as a second signal sequence to direct the next protein, 6K (55 amino acids), into the ER. 6K possesses a stop transfer sequence which anchors it in the membrane, and the C-terminus of 6K acts as signal sequence for the translocation of E1. E1 is anchored in the ER membrane by a final stop transfer sequence close to its C-terminus. The release of PE2 and E1 by cellular signalase cleavage allows the formation of PE2-E1 heterodimers in the ER (**Figure 2**). PE2 and E1 are each glycosylated in all alphaviruses, but the number and location of the modifications vary. In addition to glycosylation, the glycoproteins are palmitoylated in the Golgi apparatus. As the heterodimers are processed and transported through the ER and Golgi, they undergo a series of folding intermediates that are mediated by disulfide exchange and chaperones. Ultimately, they associate to form spikes which are composed of trimers of PE2-E1 dimers (**Figure 2**).

The final maturation event is the cleavage of PE2 into E3 and E2 by a furin-like protease (**Figure 2**). This cleavage occurs in a late Golgi or post-Golgi compartment and results in the destabilization of the heterodimer enabling the mature virus to fuse more readily with the target membrane. In most alphaviruses including Sindbis, E3 is released and not found in the mature virion. The final destination for the spike complexes is the plasma membrane, where the cytoplasmic domain of E2 (cdE2) recruits NCs assembled in the cytoplasm (**Figure 2**). Structural studies show that cdE2 residues Tyr400 and Leu402 bind into a hydrophobic pocket of the CP. Mutation of Tyr400 negatively impacted virus budding, while protein translation and core accumulation were at wild-type levels. This interaction of cdE2 with the hydrophobic pocket in the CP is thought to drive the budding of the mature virus at the cell membrane.

## Flavivirus Assembly

### Flavivirus life cycle

Flaviviruses belong to the family *Flaviridae* of positive-strand RNA viruses which also consist of the pestiviruses and the hepaciviruses. The flaviviruses comprise more than 70 members including important human pathogens such as yellow fever virus, dengue virus, and West Nile virus. Flaviviruses enter cells via receptor-mediated endocytosis. The low pH environment of the endosomal membrane triggers the conformational change of the envelope glycoprotein which results in the fusion of the viral and endosomal membranes releasing the genome into the cytoplasm. The viral proteins are translated from the RNA genome as a single polyprotein. Signal sequences and stop transfer sequences result in the translocation of the nascent polyprotein to the ER membrane. The polyprotein is processed by a combination of cellular and viral proteases to produce the mature structural and nonstructural proteins. Genome replication and virion assembly occur in ER membrane-bound vesicles. The structural proteins and the genome bud into the lumen of the ER to form the immature virion which is transported through the secretory pathway. Prior to secretion of the virion, a furin cleavage converts the immature virus into the mature, infectious form of the virus.

### Flavivirus virion structure

The flavivirus virion is made up of three structural proteins: capsid (C), pre-membrane (prM), and envelope (E) that are translated from the 5′ one-third of the RNA genome (**Figures 1(b)** and **1(d)**). Signal sequences at the C-terminus of the C protein and prM serve to translocate prM and E respectively into the ER (**Figure 1(d)**). The role of the highly basic C protein (12 kDa) is to encapsidate the viral genome during virion assembly. In contrast to the alphavirus CP which exhibits no membrane association, the flavivirus C protein is anchored to the membrane, at least transiently (**Figure 1(d)**). However, Rubella virus, the sole member of the genus *Rubivirus* within the family *Togaviridae*, has a membrane anchored CP, perhaps indicating a common origin for the *Togaviridae* and *Flaviviridae*. prM is a glycoprotein that associates with the E protein and serves as a chaperone to facilitate the proper folding of E. The immature virions that bud into the ER consist of prM-E heterodimers. prM prevents premature fusion from occurring in the acidic environment of the ER and Golgi. Thus, prM has several functions analogous to the E3 glycoprotein of the alphaviruses. Cleavage of prM into pr and M by a furin-like protease triggers the rearrangement of the prM-E heterodimers into E–E homodimers, resulting in a radical change in size and shape required for the formation of the mature virus particle. The E glycoprotein is responsible for host cell receptor binding and for fusion of the viral and cellular membranes. The E glycoprotein is also critical for the assembly of the virion. High-resolution structures of the ectodomains of several flavivirus E proteins are available. The ectodomain is divided into three domains. Domain II constitutes the dimerization domain as it contains most of the intradimeric contacts between E–E homodimers. Domain II also contains the fusion peptide, a glycine-rich hydrophobic sequence that initiates fusion by insertion into the target cell membrane. Domain III comprises the immunoglobulin-like domain responsible for receptor binding. In addition to the dramatic conformational and translational changes that the E protein undergoes during the virion maturation process, it also changes conformation during membrane fusion. The low pH of the endosome during infection triggers a conformational change which results in the formation of E homotrimers. In this arrangement, the fusion peptides are exposed and available to insert into cellular membranes. Interestingly, the structure of the E protein was found to be very similar to the structure of the Semliki Forest virus E1 protein, the fusion protein of the alphaviruses.

The structures of two flaviviruses, dengue and West Nile virus, have been solved by cryo-EM and image reconstruction techniques and have been shown to be similar (**Figure 1(b)**). The mature virion is ∼50 nm in diameter and exhibits a smooth outer surface in contrast to the alphaviruses which have distinctive spike structures (cf. **Figures 1(a)** and **1(b)**). The E proteins are arranged parallel to the surface of the virus, with 90 E dimers arranged in groups of three to form a 'herringbone' pattern on the viral surface. This arrangement of the E proteins completely covers the surface of the virus, thus rendering the lipid bilayer inaccessible. Domain III of E protrudes slightly from the viral surface, allowing interaction with cell receptors. The membrane-spanning regions of E and M proteins form antiparallel helices while the stem regions are arranged parallel to the membrane.

The immature virus particle exhibits a dramatically different glycoprotein organization compared to the mature virion. Cryo-EM and image reconstruction of dengue and yellow fever virus immature virions have revealed that these particles are larger (∼60 nm) and have spikes that protrude from the surface of the virus. These spikes are composed of trimers of prM-E heterodimers. The pr peptide covers the fusion peptide of E in this arrangement, similar to E2 covering E1 in alphaviruses, thus protecting it from premature fusion as the immature particle is transported through the acidic environment of the secretory pathway.

The NC is found below the viral envelope and is composed of a single copy of the genome RNA and multiple copies of the C protein. Cryo-EM reconstructions of the virion have shown that in contrast to the alphaviruses, there is no apparent organization to the flavivirus NC. This may be because there is no direct interaction between the C proteins in the core and the glycoproteins in the viral envelope since the E and M proteins do not penetrate below the inner leaflet of the membrane. Furthermore, no NCs have been observed in the cytoplasm of infected cells and attempts to establish an *in vitro* assembly system analogous to the alphavirus *in vitro* assembly system have failed. The lack of coordination between the C protein and the viral envelope proteins suggests that the assembly of virions is driven by the lateral interactions of the E and M proteins in the viral envelope and not by the C protein. This is supported by the observation that flavivirus infections result in the production of noninfectious subviral particles which are composed of just the viral envelope (E and M) and the lipid bilayer. Thus, the flavivirus glycoproteins are sufficient to induce particle budding.

### Flavivirus assembly and budding

Virus-induced membrane structures called vesicle packets, which are continuous with the ER membrane, are the sites of flavivirus replication and assembly (**Figure 3**). Within these structures the structural proteins are in intimate contact with the genome RNA. The C protein associates with the genome RNA via interactions between the positive charges distributed throughout the protein and the negatively charged phosphate backbone of the RNA. It is not yet clear how the C protein specifically recognizes the genome RNA; unlike for alphaviruses, a packaging signal has not been conclusively identified for flaviviruses. Coupling between genome replication and assembly within the vesicle packets has been proposed as a mechanism to ensure the specific encapsidation of the genome RNA. It has been shown that one or more nonstructural proteins (NS2A and NS3) are involved in genome packaging and NC assembly. The NC lacks a defined icosahedral structure as described above. Therefore, core formation is probably concomitant with the association of the C protein and RNA genome with the viral glycoproteins and budding into the ER lumen, thus giving rise to the immature particle (**Figure 3**). The immature virion is transported from the ER to the Golgi where the viral glycoproteins are post-translationally modified. The cleavage of the prM protein in the trans-Golgi network triggers the dramatic reorganization of the viral glycoproteins that results in the formation of the mature virion (**Figure 3**). The mature virion is then released from the host cell by exocytosis.

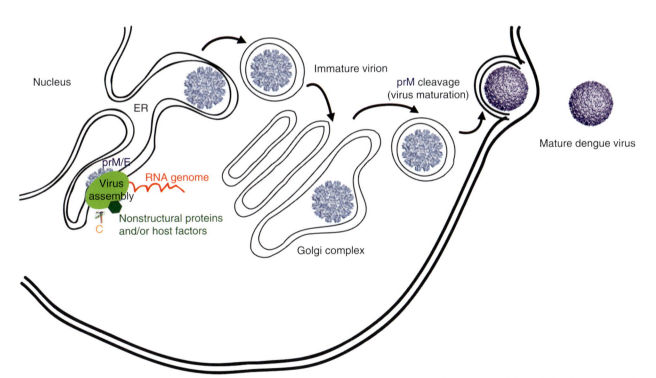

**Figure 3** Flavivirus assembly. Flavivirus assembly occurs on ER-associated membranes known as vesicle packets. Assembly and genome replication are coupled and the sites of assembly consist of the capsid proteins (C), the glycoproteins prM and E, the RNA genome, and one or more nonstructural proteins and/or host factors. The immature virion buds into the ER and is transported to the Golgi and trans-Golgi network. The glycoproteins are post-translationally modified as the immature virus is transported through the secretory pathway. Furin cleavage of prM results in the formation of the mature virus which then exits the cell by exocytosis.

## Conclusion

Following from the discussion of alphavirus and flavivirus assembly, it is apparent that the assembly of even the simplest enveloped viruses requires the complex interaction of viral and host factors in order to produce a virus particle which is at once stable and at the same time primed for disassembly. The whole range of the cell's machinery including the translation apparatus, polymerases, chaperones, and post-translational modification enzymes are co-opted by viruses in order to replicate the viral components necessary for assembly. Enveloped viruses have evolved to utilize different cellular membranes and cellular compartments for assembly and they take advantage of the secretory pathway to produce their viral glycoproteins. A majority of viruses bud from the plasma membrane (**Table 1**). This is the case with alphaviruses, where NC assembly occurs in the cytoplasm and the final assembly of the mature virion occurs at the plasma membrane. The high concentration of viral proteins, often concentrated at specific sites allows for the efficient interaction and assembly of virions. In contrast to alphaviruses, NC assembly and glycoprotein assembly is coupled in the flaviviruses and occurs in vesicle packets associated with the ER. Thus, the whole flavivirus virion is transported through the ER and Golgi while in the case of the alphaviruses only the glycoproteins are transported through the secretory pathway. These exit strategies are not unique and thus serve as model systems to study enveloped virus assembly and release.

Proteolytic cleavage of glycoproteins in order to convert them from stable oligomeric structures to metastable structures primed for fusion are common themes in enveloped virus structure and assembly. Cleavage of PE2 into E3 and E2 by a furin-like protease primes the alphavirus spike complex for fusion. A similar cleavage of prM triggers a dramatic conformational change of the flavivirus glycoproteins resulting in the formation of the mature virion which is now infectious.

Alphavirus budding requires the specific interaction of the NC with the E1–E2 spike complexes at the plasma membrane, thus ensuring that all virions have a genome packaged into them. However, the flaviviruses only require the interaction of the envelope proteins for budding, giving rise to subviral particles devoid of the C protein and genome RNA. Thus, the flavivirus envelope proteins alone are sufficient to drive budding of virus particles and the close coupling of genome replication and the C protein (perhaps mediated by replication proteins and host factors) is required to package the genome into virus particles. A third strategy for budding is exhibited by the retroviruses where capsid assembly has been shown to be sufficient to drive budding of the virus. In this case, targeting of the envelope proteins to these sites of CP assembly is essential to ensure the incorporation of the glycoproteins into the virion.

Although much has already been discovered about enveloped virus assembly, there are still many processes yet to be described. There is an increasing interest in the assembly pathway of viruses partly fueled by the potential to develop successful therapeutic agents targeting virus specific assembly processes. Advances in the field of structural biology will further help attempts to understand the assembly pathway of this important class of viruses.

*See also:* Assembly of Viruses: Nonenveloped Particles.

## Further Reading

Garoff H, Hewson R, and Opstelten DJ (1998) Virus maturation by budding. *Microbiology and Molecular Biology Reviews* 62(4): 1171–1190.

Harrison SC (2006) Principles of virus structure. In: Knipe DM, Howley PM, Griffin DE, et al. (eds.) *Fields Virology*, 5th edn., pp. 53–98. Philadelphia, PA: Lippincott Williams and Wilkins.

Hunter E (2006) Virus assembly. In: Knipe DM, Howley PM, Griffin DE, et al. (eds.) *Fields Virology*, 5th edn., pp. 141–168. Philadelphia, PA: Lippincott Williams and Wilkins.

Mukhopadhyay S, Kuhn RJ, and Rossmann MG (2005) A structural perspective of the flavivirus life cycle. *Nature Reviews Microbiology* 3(1): 13–22.

# Assembly of Viruses: Nonenveloped Particles

**M Luo,** University of Alabama at Birmingham, Birmingham, AL, USA

© 2008 Elsevier Ltd. All rights reserved.

## Architecture of Viruses

Nonenveloped viruses have two essential components: protein and nucleic acid. The protein forms a coat called 'capsid' that packages the nucleic acid, which may be DNA or RNA. This complex constitutes a virion or a virus particle. The nucleic acid is the viral genome that encodes all the virus-specific genes required for viral replication. The protein capsid packages the viral genome during replication, and transmits it for the next round of infection. When the virion reaches the host cell, the capsid usually recognizes a specific receptor that helps the virion to enter. Once inside the host cell, the capsid has to release the viral genome so replication can begin. The size of a viral genome is usually limited, so only a few genes can be encoded. It is more efficient if only one or a small number of genes encode for capsid proteins that can self-assemble into a complete shell by use of many copies of the same proteins. The assembly of the capsid proteins follows a specific type of symmetry that allows a small protein unit to assemble into a large particle. The protein capsid can have a helical (filamentous virus) or icosahedral (spherical virus) symmetry. Helical symmetry is described by the diameter, $d$, the pitch, $P$, and the number of subunits per turn. There are as many capsid proteins as necessary for completely covering the nucleic acid genome. Icosahedral symmetry is defined by 12 fivefold axes, 20 threefold axes, and 30 twofold axes. A number, $T$, called the 'triangulation number', indicates how many quasi-symmetrical subunit interactions are within one asymmetrical region of the icosahedron. There are a total of 60 T copies of the capsid proteins in one icosahedral capsid.

For example, a picornavirus (pico-(small)-RNA virus) has a single positive-stranded RNA genome of about 8000 nucleotides. The RNA genome encodes a long polyprotein that is processed into individual viral proteins after translation directly from the positive RNA genome. Three of the 10 proteins are capsid proteins: viral protein 1 (VP1), viral protein 3 (VP3), and viral protein 0 (VP0). VP0 is processed to VP2 and VP4 after virus assembly. VP1, VP2, and VP3 are the three capsid proteins that form the body of the coat while VP4 is entirely inside the coat. There are 60 copies of each capsid protein in the coat. A picornavirus is, therefore, a $T=1$ particle. However, the structure of VP1, VP2, and VP3 is highly homologous. If the small differences in each major capsid protein are ignored, the three proteins can be considered as the same building block. A picornavirus is thus called a pseudo-$T=3$ particle (**Figure 1**). The interactions between the capsid proteins in the coat are similar despite their different symmetry locations. For instance, VP1 (blue) interacts around the fivefold symmetry axis, whereas VP2 (green) and VP3 (red) interact with each other around the threefold symmetry axis. Since VP2 and VP3 are similar, the symmetry around the threefold axis is like a sixfold symmetry axis. This symmetry axis is therefore called a quasi-sixfold axis. The interactions between the neighboring VP1 proteins are considered quasi-equivalent (means more or less similar) to the interactions between VP2 and VP3 even though VP1 subunits are around a fivefold axis, and VP2/VP3 subunits are around a quasi-sixfold axis. This quasi-equivalence allows the closed capsid to be assembled using the same building block, the capsid proteins. It should be emphasized, however, that the true differences that exist in the capsid proteins are critically correlated with functions such as virus entry and release or packaging the viral genome.

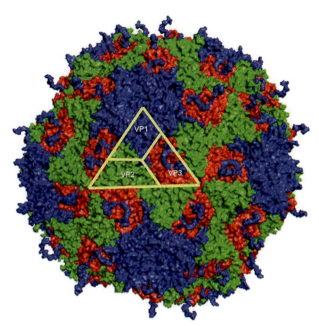

**Figure 1** The capsid of Theiler's murine encephalomyelitis virus (TMEV), a picornavirus. The triangle outlines one set of 60 copies of each capsid protein.

## Methods of Structure Determination

X-ray diffraction is the common technique used for studying the atomic structure of proteins and nucleic acids. When X-rays strike on the electrons of the atoms in a stationary specimen, a diffraction pattern of spots with different intensities is generated and recorded. By analysis of the diffraction pattern and the spot intensities, a three-dimensional electron density map (EDM) can be calculated by Fourier transformation. A three-dimensional chemical structure could be built based on the interpretation of the EDM. Two types of X-ray diffraction experiments are useful for virus structure studies: fiber diffraction (for filamentous viruses) and crystallography (for spherical viruses and globular viral proteins).

Another common technique is electron microscopy. Recent advances in electron microscopy allow researchers to determine relatively high-resolution three-dimensional structures of viral particles by use of image reconstruction of cryo-electron micrographs and electron tomography. This technique is particularly suitable for large viral particles that are difficult to crystallize. Electron microscopy and X-ray crystallography are therefore complementary to each other.

## Atomic Structure of Helical Viruses

The disk of the tobacco mosaic virus (TMV) coat protein has been crystallized and its atomic structure resolved by

X-ray crystallography. The intact TMV structure containing its nucleic acid could only be determined by X-ray fiber diffraction experiments, as also was that of Pf2 phage. The coat proteins of TMV and Pf2 contain mainly α-helices and the nucleic acid interacts with the coat protein by one base (Pf2) or three bases (TMV) per protein unit. The protein subunit is arranged in a super-helical structure coincident with a super-helical structure of the nucleic acid. The protein subunits form the outer layer of the helix to protect the nucleic acid from the environment. The nucleic acid is embedded between the protein layers that are stacked up in the spiral structure. The axis of the coat protein helix coincides with that of the nucleic acid. Since the symmetry of the super-helix puts no limit on the length of the helix, the virus particle can grow as long as the length of the nucleic acid genome; in this way all bases in the genome are covered by the coat protein. The coat proteins of TMV have many aggregation forms, depending on pH or ionic strength. In natural conditions, the TMV coat protein forms a disk with two layers of 17 protein subunits in each. The subunit of the TMV capsid protein contains four antiparallel α-helices with connecting long loops. Each protein subunit takes the shape of a shoe. When the disks of the capsid proteins are stacked, the RNA fits into the grooves formed by the protein subunits. The backbone of the RNA has charged interactions with the side chains of the amino acids while the bases of the RNA are accommodated by the hydrophobic space. The TMV RNA genome is inserted into the center of the coat protein disk to begin virus assembly. More and more disks are added to the top of the growing virus particle to pull the RNA through the center of the super-helix. The assembly is completed when the complete RNA is pulled into the virus particle. The coat protein of Pf2 has a different shape that contains an extended α-helix. The coat protein is added one by one to the DNA helix emerging from the bacterial membrane. The length of the α-helix is in the direction of the virus particle.

## Atomic Structure of Spherical Viruses

Nonenveloped spherical viruses form large single crystals under proper conditions. Their atomic structure can be determined by X-ray crystallography with aid of fast computers and synchrotron X-ray sources. Since 1978, there are numerous atomic structures of viruses reported. Today, atomic structures have been determined for every major family of nonenveloped viruses, especially human pathogens.

Most capsid proteins of these viruses contain an antiparallel, eight-stranded β-barrel folding motif (**Figure 2**). The motif has a wedge-shaped block with four β-strands

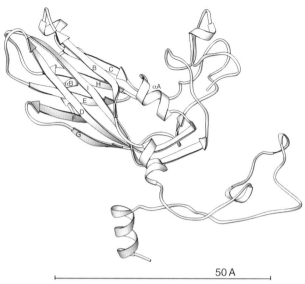

**Figure 2** The β-barrel fold found in spherical viruses. The fold contains eight essential β-strands: BIDG on one side and CHEF on the other side. The figure was generated with rhinovirus 16 VP1 (PDB code 1AVM).

(BIDG) on one side and another four (CHEF) on the other. There are also two conserved α-helices (A and B): one is between βC and βD, and the other between βE and βF. In animal viruses, there are large loops inserted between the β-strands. These loops form the surface features of individual viruses. The common presence of the β-barrel motif in viral capsid proteins is the result of structural requirements for capsid assembly. It also points to a common ancestor of different virus families.

A virus capsid may contain multiple copies of the β-barrel fold with the same amino acid sequence (such as $T=3$ calicivirus or $T=1$ canine parvovirus) or with different amino acid sequences (such as pseudo $T=3$ picornavirus). In some cases, there are two β-barrel folds in a single polypeptide (such as adenovirus hexon). Capsid proteins of spherical viruses can have other motifs such as α-helices in reovirus and hepadnavirus.

## Nucleic Acid–Protein Interaction

The viral nucleic acid genome is always packaged inside the protein capsid. Positively charged patches formed by the side chains of arginines and lysines are found on the interior surface of the protein capsid. These positively charged areas are the preferred binding sites for the nucleotides. Usually the structure of the nucleic acid cannot be observed in a single-crystal X-ray diffraction experiment because of the random orientation of the icosahedral particles in the crystal. However, in rare cases, the nucleic acid might assume icosahedral symmetry by

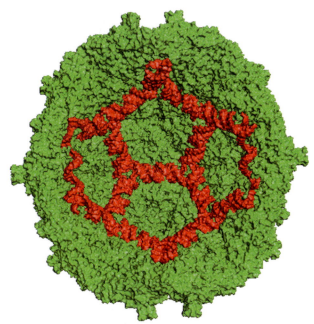

**Figure 3** The interior of the pariacoto virus capsid (green). The structured RNA genome is shown in red. The figure was generated with coordinates of pariacoto virus (PDB code 1F8V).

interacting with the protein capsid. Fragments of the complete genome assume the same conformation, although with different nucleotide sequences, at locations related by icosahedral symmetry. Such structures have been seen in bean pod mottle virus, tobacco mosaic virus, pariacoto virus and flock house virus (RNA viruses), and canine parvovirus (DNA virus). The bases are stacked either as an A-type RNA helix or form a coiled conformation to fit the interactions with the protein capsid. In the case of pariacoto virus, the RNA genome forms a cage that reassembles the icosahedral symmetry of the capsid (**Figure 3**). These viruses readily form empty virus particles and have a hydrophobic pocket on the interior surface of the capsid. The nucleic acid generally interacts nonspecifically with the capsid protein.

## Evolution

The highly conserved β barrel motif of the viral capsid protein indicates that many viruses must have evolved from a single origin. The unique three-dimensional structure of this motif is required for capsid assembly and it is generally conserved over a longer period of time than the amino acid sequence. The superposition of the capsid proteins from different viruses can be used to estimate the branch point in the evolutionary tree for each virus group. The structure alignment not only relates plant viruses to animal viruses, RNA viruses to DNA viruses, but also viruses to other proteins such as concanavalin A that has a similar fold and competes with poliovirus for its cellular receptor. The evolutionary relationship of these viruses is supported by amino acid sequence alignment of more conserved viral proteins such as the viral RNA polymerase. The structural similarities of the capsid proteins support the notion of a common evolutionary origin among nonenveloped viruses.

## Assembly

The icosahedral capsid is assembled from smaller units made of several protein subunits. For picornaviruses, a protomeric unit is first formed with one copy of each polypeptide after translation. The termini of the subunits are intertwined with each other to hold the subunits together in the protomer. The protomers are then associated as pentamers which in turn form the complete icosahedral virion while encapsidating the viral RNA. In $T=3$ or $T=1$ plant RNA viruses, the pentamers are formed by dimers of the capsid proteins. In adenovirus and SV40, the capsid proteins form hexon units (three polypeptides, each has two β-barrels) or pentamers before they assemble into an icosahedral shell. The nucleic acid could be packaged at different stages of assembly.

## Host Receptor Recognition Site

Animal viruses have to recognize a specific host cellular receptor for entry during infection. Host receptor binding is the initial step of virus life cycle and could be an effective target for preventing virus infection. Based on the atomic structure of animal viruses, it was found that the receptor recognition site is located in an area surrounded by hyper-variable regions of the antigenic sites. Usually, the area is in a depression (called the 'canyon') on the viral surface that may be protected from recognition by host antibodies. This structural feature is, for instance, present in human rhinovirus (also known as the common cold virus), and the active site of influenza virus hemagglutinin (HA). The receptor-binding site on influenza virus HA does not have a deep depression, but it is surrounded by antigenic sites. The receptor-binding area on the surface of the viral capsid is conserved for recognition by the receptor, whereas the sites recognized by antibodies are distinct from the receptor-binding site and keep changing from strain to strain. By this mechanism, the virus can escape the host immune system by mutating the antibody epitopes, and at the same time maintain a constant receptor-binding site to continue its infection of the host cells. Evidence supports that this is a general mechanism that viruses use to evade the host immune defense.

## Antigenic Sites

Antibodies are the first line of defense by the immune system against a viral infection. The epitopes combined with the neutralizing antibodies are mapped on a few isolated locations on the surface of viral proteins. The structure of human rhinovirus complexed with Fab fragments showed that the antibody makes contact with an area about 6 nm$^2$ and that the epitope spans different discontinuous polypeptides. Therefore, an effective vaccine usually needs to include a complete viral protein or a large fragment. The binding of the antibodies does not significantly change the structure of the antigen. The exact mechanism by which antibodies neutralize antigens is dependent upon the binding site and processes of the virus replication.

## Antiviral Agents

Viral infectious diseases can be cured if an agent can be administered to stop viral infection. Such agents have been synthesized and shown to bind to the capsid of rhinovirus in the crystal structure. The compounds were inserted into the hydrophobic pocket within the β-barrel of the major capsid protein VP1. Binding of the compounds stops uncoating of the virion and the receptor binding, which resulted in the failure of releasing the viral RNA into the cytoplasma. These compounds inhibit infections of several other RNA viruses and may be effective against other viruses after modification since the β-barrel structure exists in many viruses.

The most successful antiviral drugs are the HIV protease inhibitors which are developed based on the atomic structure of the protease. Through iterative cycles of computer modeling, chemical synthesis and structural studies of the protein-inhibitor complexes, a panel of clinical effective drugs has been brought to the market and has shown great benefits to patients. Inhibitors of influenza virus neuraminidase have also been developed by the same method and marketed as antiviral drugs.

*See also:* Assembly of Viruses: Enveloped Particles; Theiler's Virus.

## Further Reading

Knipe DM, Howley PM, Griffin DE, *et al.* (eds.) (2002) *Fields Virology*, 4th edn. Philadephia, PA: Lippincott Williams and Wilkins.

Rossmann MG and Johnson JE (1989) Icosahedral RNA virus structure. *Annual Review of Biochemistry* 58: 533–573.

Schneemann A (2006) The structural and functional role of RNA in icosahedral virus assembly. *Annual Review of Microbiology* 60: 51–67.

# Astroviruses

**L Moser and S Schultz-Cherry,** University of Wisconsin – Madison, Madison, WI, USA

© 2008 Elsevier Ltd. All rights reserved.

## Glossary

**Enterocytes** Epithelial cells lining the intestines.
**Intussusception** Obstruction of the intestine.
**Interstitial nephritis** Inflammation of the kidney.
**Poult** Young turkey.
**Villi** Finger-like intestinal projections lined with enterocytes.

## Introduction

Astroviruses are enteric viruses first identified in the feces of children with diarrhea. Detection was originally based on a five- to six-pointed star morphology of virions by electron microscopy (EM). However, only about 10% of viral particles display these structures; the remaining 90% of particles have a smooth surface and a size similar to other small, round-structured viruses like picornaviruses and caliciviruses. Thus, accurate diagnostics were difficult to obtain and the true prevalence of astrovirus within a population was difficult to assess. Development of much more sensitive detection techniques like real time reverse transcription-polymerase chain reaction (RT-PCR), cell culture RT-PCR, and astrovirus-specific enzyme-linked immunosorbent assays (ELISAs) have made detection more accurate and specific, even allowing diagnosis of specific serotypes. Utilizing these techniques, astroviruses have been found in approximately 3–8% of children with diarrhea. Astroviruses can also be isolated in a subset of asymptomatic individuals, suggesting that a proportion of infected individuals shed the virus asymptomatically or for some time after the resolution of other symptoms of infection. Asymptomatic carriers may be a major reservoir for astroviruses in the environment and could contribute to dissemination of the virus.

The release of astroviruses into the environment is a concern due to the extreme stability of the virus. Astroviruses are resistant to inactivation by alcohols (propanol, butane, and ethanol), bleach, a variety of detergents, heat treatment including 50 °C for an hour or 60 °C for 5 min, and UV treatment up to 100 mJ cm$^{-2}$. Human astroviruses are known to survive up to 90 days in both marine and tap water, with survival potential increasing in colder temperatures. Studies have described the isolation of infectious virus from water treatment facilities. Furthermore, astroviruses can be concentrated by filter-feeding shellfish like oysters and mussels in marine environments. Astroviruses are transmitted fecal–orally, and contaminated food and water have been linked to astrovirus outbreaks.

## History and Classification

Astroviruses were originally observed by Appleton and Higgins in 1975 as a small round virus in stools. Later that year, Madeley and Cosgrove identified the virus in association with diarrhea in children and bestowed the name astrovirus (from the Greek *astron*, meaning star) for the star-like morphology of a proportion of viral particles seen by EM (**Figure 1(a)**). Because of genomic similarities, astroviruses were originally thought to belong to either the families *Picornaviridae* or *Caliciviridae*. However, the lack of a helicase and use of a frameshifting event during replication (discussed below) distinguish astroviruses so completely that, in 1993, the International Committee on Taxonomy of Viruses (ICTV) classified astroviruses as a unique family, *Astroviridae*, composed of a single genus *Astrovirus*. Continued investigation into newly discovered astroviruses led to the division of the family into two genera, *Mamastrovirus* and *Avastrovirus*, by the ICTV in 2005. ICTV nomenclature abbreviates astrovirus AstV, with a single-letter abbreviation for the species type (i.e., human astrovirus; HAstV; turkey astrovirus; TAstV, etc.). Successive, serologically distinct isolates of astroviruses are named sequentially within that species (i.e., HAstV-1 through HAstV-8).

## Epidemiology

### Humans

Astroviruses have been detected throughout the world. While the exact incidences of infection vary from study to study, community-acquired astroviruses are found in 3–6% of children with infectious gastroenteritis. In some developing countries, infection rates as high as 20% have

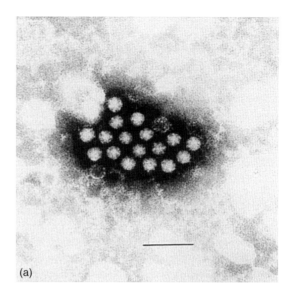

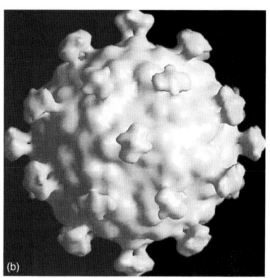

**Figure 1** (a) Astroviruses have historically been identified by the five- to six-pointed star morphology visible by electron microscopy. Scale = 100 nm. (b) A reconstruction of the astrovirus virion, based on cytoelectron microscopy, along the twofold axis of symmetry. Reprinted from Matsui SM and Greenberg HB (1996) Astroviruses. In: Fields BN, Knipe DM, and Howley PM (eds.) *Fields Virology*, 3rd edn., pp. 875–893. Philadelphia: Lippencott-Raven.

been observed. In many cases, astroviruses are the second most commonly detected viral pathogen in young children after rotavirus. Astrovirus infections are identified in up to 2% of asymptomatic individuals. These data may underrepresent actual astrovirus infections, as studies generally survey individuals visiting medical care centers. Because astrovirus disease is generally mild in humans (see the section discussing pathogenesis), hospital cases may represent only a slight proportion of actual infections in the community. In support of this, serological studies have demonstrated that up to 90% of children have been exposed to at least one strain of astrovirus by age 9.

Eight serotypes of human astrovirus have been identified to date, with all eight circulating within the global population to various levels. HAstV-1 is by far the most prevalent serotype, comprising 25–100% of astroviruses in a region, and the most prevalent reactivity of antibodies detected, although serological surveys of all serotypes have not been undertaken. HAstV-6, -7, and -8 are the least frequently detected, although three to four serotypes of HAstV are often detected in a region at any given time. The differing prevalence of serotypes could be a reflection of severity; perhaps HAstV-1 infection results in a higher frequency of hospital visits than other serotypes and is therefore overrepresented in hospital-based epidemiological studies. Alternatively, serotypes may be restricted by region. For example, one Mexican study identified HAstV-1 as the predominant serotype throughout the country, but HAstV-3 and -8 were prominent in select regions.

Viral infection occurs with equal frequency in boys and girls and predominantly in children under the age of 2. Infection is not restricted to young children, however, and has been noted in individuals of all ages, including immunocompetent adults and the elderly. Immunodeficient individuals, particularly those that are HIV-positive, appear to be at an increased risk of astrovirus infection.

Astrovirus infection occurs year-round, but with the highest frequency during the autumn and early winter months. In tropical climates, infection correlates with the rainy season. These seasonal correlations likely reflect the indoor confinement of the population as well as the increased stability of astroviruses in cold, damp conditions. Astrovirus outbreaks have also been associated with high-density environments, including childcare centers, primary and junior high schools, military recruiting centers, elderly care centers, and swimming pools. Astrovirus as a cause of hospital-acquired viral diarrhea in young children is second only to rotavirus, occurring at rates of 4.5–6%, and, in some studies, surpasses rotavirus in rates of nosocomial infections.

Interestingly, astrovirus infection occurs quite frequently (up to 50%) as a co-infection with other enteric pathogens. The most frequent co-pathogens are noroviruses and rotaviruses, but infections with adenoviruses, parasites, and enteric bacteria are often detected as well. The importance of this in humans is not entirely clear. In a study specifically examining co-infections, astrovirus co-infection with rotavirus increased the duration of diarrhea and vomiting over either virus alone, although whether this difference was statistically significant is unknown.

### Animals

Most animals are not routinely screened for astrovirus infection, so our knowledge of the prevalence of infection is limited to surveillance studies. Astroviruses have been found in association with most animals examined, although the effect of infection varies with species (see below). While astroviruses were originally identified in humans, they have since been identified in both mammalian and avian species, including rabbits, mice, calves, sheep, piglets, dogs, red-tailed deer, kittens, mink, turkeys, ducklings, chicken, and guinea fowl. At least three serotypes of bovine astroviruses are postulated to exist based on distinct neutralizing antibodies (one in the United States and two in the United Kingdom). In addition, astroviruses have been isolated from mink across Scandinavia, and serological studies have demonstrated that astroviruses were prevalent in chicken flocks in the 1980s as well as in 2001. Interestingly, two very different manifestations of chicken astrovirus infection have been described (see the section on pathogenesis), suggesting that distinct chicken astroviruses may circulate; however, this is yet to be proven. The best epidemiologically characterized animal astroviruses are the turkey astroviruses. Surveillance of turkey flocks in the 1980s isolated astrovirus from 78% of diseased flocks, but only 29% of normal flocks. Astroviruses were the first pathogen detected in many flocks and were most commonly detected in birds less than 4 weeks of age. Similar to human infections, turkey astrovirus was frequently isolated with other pathogens, most commonly rotavirus-like viruses. The early age of infection and the prevalence of co-infections led one group to postulate that astrovirus infection may predispose birds to infection by other viruses.

## Virus Propagation

Attempts at *in vitro* propagation of astroviruses have been met with varying degrees of success. The most successful techniques utilize cultured cells from the host species and provide exogenous trypsin in the culture. Successful propagation of human astroviruses was originally achieved by repeated passage through primary human embryonic kidney cells; it was later discovered that direct passage through the human intestinal cell line Caco-2 would also yield infectious virus. Propagation of porcine, bovine, and chicken astroviruses has been successful in their respective host cells *in vitro*. However, many astroviruses still have not been adapted to propagation *in vitro* for unknown reasons, while others lose infectivity with subsequent passages and therefore cannot be maintained continuously. This problem has been circumvented in some systems by passing the virus through an animal system, as is the case for the turkey astrovirus, in which highly concentrated virus can be obtained from infected turkey embryos *in ovo*.

## Molecular Virology and Protein Expression

Astroviruses contain one copy of positive-sense, single-stranded RNA. The genome is approximately 6.8 kb long and contains three open reading frames (ORFs), ORF1a, -1b, and -2, as well as 5′ and 3′ untranslated regions (UTRs) (**Figure 2**). The RNA is polyadenylated, but lacks a 5′ cap structure. The 5′ and 3′ UTRs are highly conserved and are believed to contain signals important for genome replication.

Astroviruses initiate infection by binding to an unknown receptor and entering the cell via receptor-mediated endocytosis. The plus-strand genome is released into the cytoplasm by unknown mechanisms and ORF1a and -1b are immediately translated by the host machinery. ORF1a is 2.8 kb and encodes a polypeptide of approximately 110 kDa. This polypeptide contains a variety of conserved motifs, including several putative transmembrane domains, a bipartite nuclear localization sequence (NLS), and a serine protease motif. The translated polypeptide is cleaved by both cellular protease(s) and the viral protease into at least five peptides. The actual function of each protein remains largely unknown. The transmembrane domains may localize to the endoplasmic reticulum (ER) membrane to facilitate replication, as all plus-strand RNA viruses have been shown to replicate in association with a membrane. One peptide, NSP1a/4, colocalizes with the viral RNA at the ER membrane; mutations in NSP1a correlate with increased viral titers *in vitro* and *in vivo*, suggesting a role for this protein in viral replication. The role for the NLS remains unclear; some reports suggest viral antigen is observed in the nucleus, while others find that it is excluded.

The second reading frame, ORF1b, overlaps ORF1a by 70 nucleotides and has no detectable start codon. Intensive research has determined that ORF1b is translated by a frameshift into the −1 frame. This frameshifting event is unique among plus-strand animal RNA viruses and requires a highly conserved shifty heptameric sequence ($A_6C$) as well as a downstream hairpin structure. This event, which occurs with frequencies up to 25% in cells, results in an ORF1a/1b fusion peptide. Cleavage near the 1a/1b border releases the ORF1b gene product: the viral RNA-dependent RNA polymerase (RdRp). Astrovirus polymerase is a supergroup 1 RdRp, a group which generally utilizes a VPg to initiate transcription. Although a VPg is postulated to exist and a putative VPg genomic linkage site has been identified, its existence is yet to be empirically proven.

Expression of the RdRp results in production of a minus-strand viral template. This generates multiple copies of the plus-strand genome as well as a polyadenylated subgenomic RNA (sgRNA) containing short 5′ and 3′ UTRs and ORF2. ORF2 is in the 0 frame and overlaps ORF1b slightly (four nucleotides) in human astroviruses. Production of the capsid protein from a sgRNA not only temporally restricts capsid production to later in the viral replication cycle, but also allows for massive capsid protein expression; it is estimated that sgRNA is produced in tenfold excess of the viral genome by 12 h post infection (hpi). The sgRNA is about 2.4 kb and encodes the single structural protein of approximately 87 kDa. This peptide is cleaved by an intracellular protease to approximately 79 kDa; mutational analyses suggest that this 8 kDa stretch is required for efficient expression of the capsid protein. Individual capsid proteins multimerize spontaneously to form icosahedral structures of about 32 nm (**Figure 1(b)**).

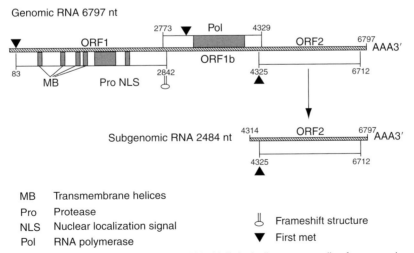

**Figure 2** The genomic organization of astroviruses (based on HAstV-1), including open reading frames and encoded protein features, is shown. Reprinted from *Virus Taxonomy: Sixth Report of the International Committee on Taxonomy of Viruses*, 1995, p. 365, *Astroviridae*, Murphy FA, Fauquet CM, Bishop DHL, *et al.* (eds.), copyright 1995, with kind permission of Springer Science and Business Media.

Positive-sense genomes are packaged into these viral-like particles (VLPs), possibly through interactions with the first 70 amino acids of the capsid protein. The virions are released by an unknown mechanism, which may involve cellular caspases, after which the capsid undergoes an extracellular trypsin-mediated maturational cleavage. This increases infectivity up to $10^5$ fold, condenses the virion to approximately 28 nm, and transforms the 79 kDa capsid protein into at least three smaller peptides of approximately 34, 29, and 26 kDa. Computational predictions suggest that VP34 may comprise the core of the virion while VP29 and VP26 form spike-like projections that may be important for viral tropism and receptor binding. This is corroborated by studies suggesting that VP26 is only loosely associated with the virion. These spikes are also thought to be responsible for the star morphology visible by EM (**Figure 1(a)**).

## Evolution

Examination of nucleotide changes and nonsynonymous amino acid changes from the whole genome and across species suggests that an ancient divergence between avian and mammalian astroviruses occurred approximately 310 million years ago. Mammalian astroviruses split more recently into two distinct clades: human astroviruses and feline/mink-associated astroviruses. Phylogenetic clustering of the human astroviruses together argues against continual human–animal interspecies transmission. It is hypothesized that at least two interspecies transmission events (avian to porcine, porcine to feline) led to the current division of viruses. Further comparison of synonymous mutations by codon usage generates an astrovirus evolutionary pattern which mirrors the evolution of respective hosts, suggesting that recent evolution of the virus may have been in adaptation to the host. As RNA viruses, astroviruses are expected to undergo frequent genetic changes. However, nucleotide changes occur at rates of approximately 5% in human viruses over time, despite the co-circulation of muptiple serotypes within a region. Nucleotide and amino acid comparisons of ORF1a of human astroviruses demonstrate two distinct lineages, known as genogroup I (HAstV-1 to -5) and genogroup II (HAstV-6 and -7). Comparisons of ORF1b or ORF2 lack these distinct groups, leading investigators to postulate that a recombination event at the ORF1a/1b junction occurred before HAstV-6 or -7 diverged.

## Clinical Features, Pathology, and Pathogenesis

### Mammalian Astroviruses

Astrovirus infection in mammals presents clinically as gastroenteritis. Disease has been most closely studied in humans and, in volunteer studies, astrovirus-infected individuals develop diarrhea, the most prominent symptom, as well as vomiting, nausea, anxiety, headache, malaise, abdominal discomfort, and fever. Onset of symptoms at 2–3 days post infection (dpi) correlates with shedding of the virus in feces, although shedding can continue after resolution of other symptoms. Astrovirus infection has also been associated with intussusception, although a causative role has not been established.

The earliest studies of astrovirus pathogenesis utilized gnotobiotic sheep and calves as models. In calves, astrovirus infection was localized to the dome epithelial cells overlying Peyer's patches. These cells appeared flat or rounded and released cells were identified in the intestinal lumen. Astrovirus infection in calves was shown to be specifically targeted to M cells and led to the sloughing of necrotic M cells into the intestinal lumen. Enterocytes were never observed to be infected. Specific tropism of the virus for immune cells suggests that astrovirus may have an immunomodulatory role in calves. While the virus replicated in these animals and could be detected in their feces, the calves displayed no clinical signs. In most bovine studies, viral infection is asymptomatic, although changes in the feces from solid and brown to soft and yellow were noted in one study. Mild villus atrophy and slight changes in villus-to-crypt ratios have been noted but no changes in xylose absorption were observed. Despite the lack of symptoms, viral shedding continued until the termination of the experiment.

Studies in sheep have shed more light on histological changes associated with infection. Astrovirus-infected sheep developed a transient diarrhea as early as 2 dpi, but virus was detected at early as 14 hpi and initially confined to the lumenal tips of the intestinal villi. By 23 hpi, virus was observed coating the microvilli and infection had spread to the apical two-thirds of the villi. This correlated with sloughing of degenerate cells from the apical portion of the villi, which continued through 38 hpi. At this time, villus blunting was apparent in the ileum and midgut. Furthermore, normal epithelial cells lining the villi were replaced with immature, cuboidal cells reminiscent of crypt cells. Neither these immature cells nor crypt cells were ever observed to be infected, suggesting that only mature enterocytes are susceptible to infection. By 5 dpi, viral infection had cleared and intestinal histology had returned to normal.

Volunteer studies in humans have not explored the underlying causes of astrovirus pathogenesis; our knowledge is therefore limited to intestinal biopsies taken for other reasons, but generally support the observations described above. In a biopsy from a child shedding large quantities of astrovirus, slight histological changes, including mild villous blunting and irregular epithelial cells, were observed. Infection increased distally through the small intestine. Similarly to animal models, astrovirus infection was restricted to the apical two-thirds of intestinal villi and could be identified in infected cells.

## Avian Astroviruses

In avian species, astrovirus infection has a much broader range of disease than in mammals. While astrovirus does cause gastroenteritis in turkeys and chickens, it can also cause nephritis in chickens and a severe, often fatal, hepatitis in young ducklings.

Turkey astrovirus was the first discovered avian astrovirus and remains the best characterized in terms of pathogenesis, due in part to the development of the turkey as a small animal model. In these animals, virus could be detected from 1 to 12 dpi in the intestines. Viral replication was limited to the enterocytes on the apical portion of the villi, but the virus could be detected throughout the body, including the blood. The development of viremia is rare among enteric viruses and its function remains unclear. Infected turkeys developed a yellow, frothy, gas-filled diarrhea from 1 to 12 dpi. Diarrhea occasionally contained undigested food, but never blood. The intestines of infected birds became thin walled, flaccid, and distended. Despite these changes, histological examination suggested that only mild changes occur during infection. A mild crypt hyperplasia and shortening of the villi were noted from 4 or 5 to 9 dpi, and single degrading enterocytes could be identified. However, TUNEL staining suggested that the amount of cell death in infected intestines is similar to control birds. D-xylose absorption, a measure of intestinal absorption, was significantly decreased from 2 to 5 dpi in one study and up to 13 dpi in another. This effect was exacerbated in the presence of another enteric pathogen, turkey coronavirus. Astrovirus infection also caused a significant growth depression in turkey poults by 5 dpi; infected birds never recovered from this, leading to flock unevenness. Infected birds also demonstrated a transient (3–9 dpi) reduction of the thymus, which returned to normal by 12 dpi.

Avian infection by astroviruses can present with nonenteric symptoms as well. Infection of ducklings with duck astrovirus causes a severe hepatitis. Infected birds develop liver hemorrhage, swollen kidneys, and hepatocyte necrosis. On farms, infection leads to mortality rates of 10–25% in adult (4–6-week-old) ducks, but can reach 50% in ducklings under 14 days of age. In chickens, infection with the astrovirus avian nephritis virus (ANV) results in discoloration of the kidney, development of renal lesions, and interstitial nephritis by 3 dpi. Pathogenesis is age dependent, with 1-day-old chicks the most susceptible and adult birds the least. ANV infection can result in mortality rates of up to 33%, although rates appear to be strain specific.

## Immune Response

The immunological response to astrovirus infection is poorly defined; however, observations in humans and animal models suggest that both the adaptive and innate responses play important roles in controlling and eliminating the virus.

The humoral immune response likely plays a major role in astrovirus immunity. The biphasic infection pattern of young children and the elderly suggests that antibodies are protective during the middle of life. Indeed, serological studies have indicated that approximately 50% of neonates have maternally acquired antibody to HAstV, which wane by 4–6 months of age. Children then acquire anti-HAstV antibodies rapidly due to astrovirus exposure. By the age of 9, up to 90% of the population has been exposed to HAstV-1. Furthermore, volunteer experiments demonstrate that astrovirus exposure generally leads to an increase in anti-astrovirus antibody titer. While astrovirus antibodies protected individuals from symptoms associated with infection, virus was identified in the feces, suggesting that such antibodies do not necessarily prevent viral replication. Additionally, immunoglobulin treatment has been attempted as a treatment for severe or chronic astrovirus infection. The results have been mixed and difficult to interpret, as the presence of astrovirus-specific antibodies in the immunoglobulin treatment was not always confirmed.

Cellular immunity may also play a role in controlling and/or preventing astrovirus infection. Studies have demonstrated that most individuals possess HLA-restricted, astrovirus-specific T cells. When stimulated with astrovirus *in vitro*, these cells produce tumor necrosis factor, interferon gamma, and occasionally interleukin (IL)-5 but not IL-2 or IL-4. These cytokines are typical of the T-helper-type response thought to be important in controlling viral infections. Individuals deficient in T and B-cell functions are unable to control infection, shedding virus to very high titers ($\geq 10^{14}$ particles ml$^{-1}$) and for extended periods of time (up to 18 months), further supporting the importance of cellular immunity.

Experiments in a turkey model demonstrate that the adaptive response is not the only important immunological response. In this model, no increase in T cells (CD4$^+$ or CD8$^+$) could be demonstrated after TAstV-2 infection. Moreover, while infected turkeys produced a slight increase in antibody production, these antibodies were not neutralizing and did not prevent against future infection. However, it was noted that macrophages from TAstV-2 infected turkeys produced significantly higher levels of nitric oxide (NO) both *in vivo* and upon stimulation *ex vivo*. Inhibition of NO *in vivo* led to a significant increase in viral production, while addition of exogenous

NO decreased viral production to below the detection limit, suggesting that NO is an important factor in controlling astrovirus infection. The importance of macrophages and their role in astrovirus infection has been corroborated by observations in astrovirus-infected lambs, where EM showed virions within macrophages. Furthermore, it is possible that astroviruses have a mechanism to combat this response, as macrophages in astrovirus-infected turkeys demonstrate a reduced ability to phagocytose.

## Treatment, Prevention, and Control

Because astrovirus infection is generally mild and self-limiting in humans, treatment is generally restricted to fluid rehydration therapy. This can often be accomplished at home; thus, hospital admittance is rare. No vaccine is yet available for humans, and as noted above, immunoglobulin treatment for immunocompromised individuals has been met with varying degrees of success. Additionally, no treatment for astrovirus-infected animals exists. The best solution, therefore, is prevention of transmission, which is best done in humans by conscientious hand and food washing. The stability of astroviruses and their resistance to inactivation make them difficult to eliminate after introduction. This is a significant problem in hospitals, where individuals are generally immunocompromised and therefore more susceptible to infection. One outbreak in a bone marrow transplant ward prompted the hospital to scrub the entire ward with warm, soapy water. However, surveillance of the subsequent inhabitants demonstrated fecal shedding of astroviruses, underscoring the difficulty in removing the virus. This is also a significant problem in commercial farming, where astrovirus infection of animals significantly decreases productivity. Its introduction and maintenance in this environment can mean drastic financial losses. In each of these environments, early detection and thorough disinfection are keys to limiting transmission and controlling infection.

*See also:* Caliciviruses; Enteric Viruses; History of Virology: Vertebrate Viruses; Picornaviruses: Molecular Biology; Replication of Viruses; Rotaviruses; Viral Pathogenesis; Virus Particle Structure: Nonenveloped Viruses.

## Further Reading

Koci MD (2005) Immunity and resistance to astrovirus infection. *Viral Immunology* 18: 11–16.

Koci MD and Schultz-Cherry S (2002) Avian astroviruses. *Avian Pathology* 31: 213–227.

Matsui SM and Greenberg HB (1996) Astroviruses. In: Fields BN, Knipe DM, and Howley PM (eds.) *Fields Virology,* 3rd edn., pp. 875–893. Philadelphia: Lippincott-Raven.

Monroe SS, Carter MJ, Herrmann JE, Kurtz JB, and Matsui SM (1995) *Astroviridae*. In: Murphy FA, Fauquet CM, Bishop DHL, *et al.* (eds.) *Virus Taxonomy: Sixth Report of the International Committee on Taxonomy of Viruses*, pp. 364–367. Vienna: Springer.

Monroe SS, Jiang B, Stine SE, Koopmans M, and Glass RI (1993) Subgenomic RNA sequence of human astrovirus supports classification of *Astroviridae* as a new family of RNA viruses. *Journal of Virology* 67: 3611–3614.

# Baculoviruses: Molecular Biology of Granuloviruses

**S Hilton,** University of Warwick, Warwick, UK

© 2008 Elsevier Ltd. All rights reserved.

## Glossary

**Defective interfering particle** Virus particles that are missing part of their genome. These deletions in their genome mean that they cannot sustain an infection by themselves and depend on coinfection with a suitable helper virus.

**Occlusion body** A crystalline protein matrix which surrounds the virions of some insect viruses.

**Origin of replication** A unique DNA sequence at which DNA replication is initiated.

**Palindrome** A sequence of DNA equal to its complementary sequence read backwards.

## Introduction

The *Baculoviridae* are a family of invertebrate viruses with large circular, double-stranded DNA genomes (80–180 kbp). The genomes are packaged into nucleocapsids, which are enveloped and embedded in proteinaceous occlusion bodies. Baculoviruses are divided taxonomically into two genera, *Nucleopolyhedrovirus* (NPV) and *Granulovirus* (GV). GVs and NPVs show major differences, not only in the morphology of their occlusion bodies but also in their tissue tropism and cytopathology. NPVs have large polyhedral-shaped occlusion bodies measuring 0.15–15 μm, comprised predominantly of a single protein called polyhedrin and typically have many virions embedded. GVs have small ovicylindrical occlusion bodies, which average about 150 nm × 500 nm. These are comprised predominantly of a single protein called granulin, which is related in amino acid sequence to polyhedrin. GVs normally have a single virion consisting of one nucleocapsid within a single envelope (**Figure 1(a)**). They infect both agricultural and forest pests, making them important as potential biological insecticides. To date, eight complete GV genomes have been sequenced (**Table 1**).

## Infection Cycle

Baculoviruses produce two distinct virion phenotypes, occlusion-derived virus and budded virus, which are responsible for infection of insects (*per os* infection) and insect cells (cell-to-cell spread), respectively. Infection for both GVs and NPVs begins with the ingestion and solubilization of the occlusion bodies in the host larval midgut. The released occlusion-derived virus then fuses with the microvilli of midgut epithelial cells, and nucleocapsids pass through the cytoplasm to the nucleus. At the nuclear pore, the GV DNA is thought to be injected into the nucleus leaving the capsid in the cytoplasm. As replication proceeds, the nucleus enlarges and the nucleoli and chromatin move to its periphery (**Figure 1(b)**). The nuclear membrane ruptures early in infection in GVs, following the production of limited numbers of nucleocapsids. This is followed by extensive virogenesis, nucleocapsid envelopment and the formation of occlusion bodies in the mixed nuclear-cytoplasmic contents. Some nucleocapsids continue to form and pass out of the cell, forming budded virus which initiates secondary infection (**Figure 1(c)**). Baculoviruses encode two different major budded virus envelope glycoproteins, GP64 for group I NPVs and F protein for group II NPVs and GVs, which mediate membrane fusion during viral entry. Other nucleocapsids are enveloped and occluded within granulin-rich sites throughout the cell. As the number of occlusion bodies increase, the cell greatly enlarges and eventually lyses (**Figure 1(d)**). When the larvae die, the remaining cells lyse and the occlusion bodies are released back into the environment.

## Taxonomy and Classification

GVs have been isolated only from the insect order Lepidoptera (butterflies and moths). They have been isolated from over 100 species belonging to at least 10 different host families within Lepidoptera, mainly Noctuidae and

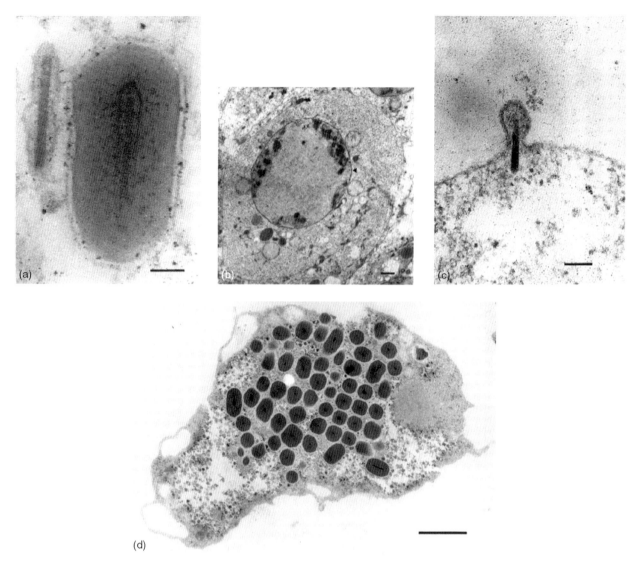

**Figure 1** GV morphology. (a) Enveloped GV virion and adjacent occluded virion. (b) Early stages of GV replication. Nucleocapsids are present throughout the nucleus prior to disintegration of the nuclear membrane. The nuclear membrane is indicated by an arrowhead. (c) virus budding from plasma membrane. (d) *In vitro* replication of CpGV. *C. pomonella* cell late in infection containing occluded virus particles. Scale = 100 nm (a, c); 1 μm (b, d). (a–c) Reprinted from Winstanley D and Crook NE (1993) Replication of *Cydia pomonella* granulosis virus in cell cultures. *Journal of General Virology* 74: 1599–1609, with permission from Society for General Microbiology.

Tortricidae. There has been no formal classification of subgroups within the GVs. However, it is recognized that there are three types of GVs based on their tissue tropism, although these groups are not supported by phylogenetic analyses.

The first type of GV is slow-killing and predominantly infects larvae belonging to the family Noctuidae, but also includes the Torticidae-specific virus adoxophyes orana GV (AdorGV). They infect only the fat body of the larvae after the virus has passed through the midgut epithelium. Slow-killing GVs tend to kill in the final instar regardless of the instar in which the larvae are infected. These infected larvae tend to take longer to die than other GV- or NPV-infected larvae, typically from 10 to 35 days. A possible explanation for this is that important tissues such as the tracheal matrix and the epidermis are not infected.

The second type of GV includes the GV type species, *Cydia pomonella* GV (CpGV). They are relatively fast-killing, taking only 5–10 days to kill. The fat body is infected along with the epidermis, malpighian tubules, tracheal matrix, hemocytes, and many other tissues to a lesser extent. The faster speed of kill could be due to the infection of a wider range of body tissues. Fast-killing GVs tend to kill in the instar in which the larvae are infected or within the following instar. These viruses infect larvae from a variety of families of Lepidoptera including Tortricidae, Pieridae, Yponomeutidae, Pyralidae, and Gelechiidae.

The third type of GV contains only one species to date, *Harrisina brillians* GV (HabrGV). HabrGV replicates only in the midgut epithelium cells of *H. brillians* larvae, from the family Zygaenidae. The larvae develop diarrhoea which consists of a discharge containing occlusion bodies. This can lead to a rapid spread of infection. The younger the larvae are infected, the longer they take to die, which is also typical of the slow-killing GVs.

## Genome Organization

Phylogenetic analysis of the completely sequenced baculoviruses (**Figure 2**) shows that there is one clade which contains all of the Tortricidae-specific GVs. Although there are no other clearly defined clades, it is clear that the GVs tend to group based on the family of their host insect and not on their tissue tropism type. The only known type 3 GV, HabrGV, which infects insects from the family Zygaenidae, has not been completely sequenced and so was not included in these analyses. Previous phylogenetic analysis using the granulin gene placed HabrGV in a clade with the Tortricidae-specific viruses. However, baculovirus relationships based solely on occlusion body protein sequence can disagree with other

**Table 1** Features of the eight completely sequenced granuloviruses

| Granulovirus | Abbreviation | Length (nt) | AT content (%) | Number of ORFs | Host family |
|---|---|---|---|---|---|
| Adoxophyes orana granulovirus | AdorGV | 99 657 | 65.5 | 119 | Tortricidae |
| Agrotis segetum granulovirus | AgseGV | 131 680 | 62.7 | 132 | Noctuidae |
| Choristoneura occidentalis granulovirus | ChocGV | 104 710 | 67.3 | 116 | Tortricidae |
| Cryptophlebia leucotreta granulovirus | CrleGV | 110 907 | 67.6 | 128 | Tortricidae |
| Cydia pomonella granulovirus | CpGV | 123 500 | 54.8 | 143 | Tortricidae |
| Phthorimaea operculella granulovirus | PhopGV | 119 217 | 64.3 | 130 | Gelechiidae |
| Plutella xylostella granulovirus | PlxyGV | 100 999 | 59.3 | 120 | Plutellidae |
| Xestia c-nigrum granulovirus | XecnGV | 178 733 | 59.3 | 181 | Noctuidae |

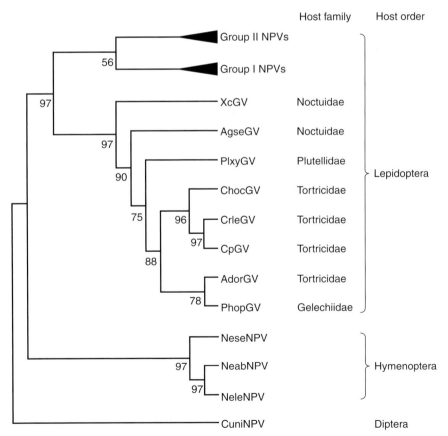

**Figure 2** Baculovirus phylogeny (maximum pasimony) of completely sequenced baculoviruses based on the LEF-8 and LEF-9 concantenated sequences. Bootstrap percentage support values are indicated.

gene phylogenies, and further HabrGV genes need to be analyzed before evolutionary conclusions can be made.

The genomic organization of the GVs has been compared using gene parity plot analysis (**Figure 3**). Gene parity plots compare the relative positions of homologous genes in different genomes and illustrate conservation between baculovirus genomes. Gene arrangement in baculoviruses reflects their evolutionary history, with the more closely related viruses sharing a higher degree of gene collinearity. The gene order among the sequenced GVs is virtually identical, with between only one and five common genes in different positions along the genome. One of these, the *iap-3* gene, is in three different positions in the five genomes which contain it. This suggests that this gene may be a more recent acquisition. The GVs appear to be far more collinear than the NPVs. There are many inversions and rearrangements among the NPV genomes relative to each other. The only GV genome to show any inversion of a block of genes is phthorimaea operculella GV (PhopGV) which has a block of six genes (Phop113–118) inverted relative to the other GVs. This block of genes flanks a putative homologous repeat region, which in NPV genomes, are sites where most major rearrangements, insertions, and deletions are found. The gene order among GVs is quite different from that of NPVs (**Figure 3**). GVs share only one main region of collinearity with NPVs, although this is inverted with respect to the granulin/polyhedrin gene. This region is from autographa californica NPV (AcMNPV) open reading frames (ORFs) Ac103-61.

The gene content of CpGV is shown in **Figure 4** and **Table 2**. There are 62 genes common to all Lepidopteran baculoviruses sequenced to date. There are a further 28 genes common to all GVs, 22 of which are absent from NPVs. A further 22 GV genes are found in more than one GV, but not in NPVs. The GV-specific genes could be responsible for the biological differences between these two baculovirus groups. The relationship between biological differences and gene content of GVs and NPVs awaits the characterization of further genus specific genes. CpGV also contains 11 CpGV-specific genes, which may encode host-specific factors. There are also 20 genes which are present in some GVs and some NPVs, these are likely to be auxiliary genes, which are not essential for viral replication, but provide the virus with some selective advantage. These include *chitinase, cathepsin, iap-3, pe38, bro, ptp-2, rr1* and *rr2a, lef-10*, and *egt*. The GV auxiliary genes which have been assigned putative functions are shown in **Table 3**.

## GV Replication *In Vitro*

Several NPVs replicate well in cell culture and some cell lines are commercially available. In contrast, there are only a few laboratories worldwide that have managed to maintain GV-permissive cell lines, which yield low virus

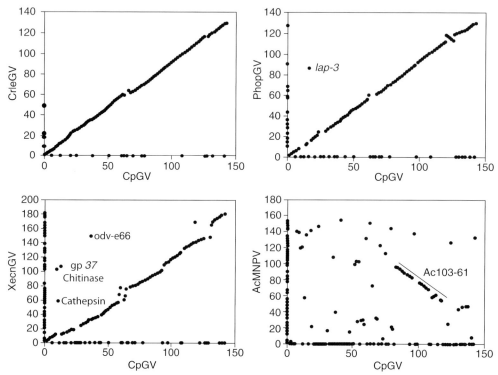

**Figure 3** Gene parity plots of CpGV gene organization versus CrleGV, PhopGV, XecnGV, and AcMNPV.

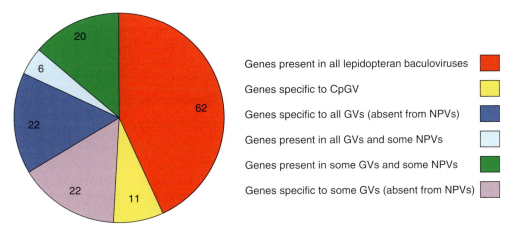

**Figure 4** CpGV gene content. The number of CpGV genes (from a total of 143) are shown in each section of the pie chart.

**Table 2** Genes present in all lepidopteran baculoviruses and those present in all GVs

| | Genes present in all lepidopteran baculoviruses | Additional genes present in all GVs (genes in bold italic are absent from NPVs) |
| --- | --- | --- |
| Transcription | 39K, p47, lef-5, lef-4, lef-6, lef-8, lef-9, lef-11, vlf-1 | |
| Replication | lef-1, lef-2, dnapol, helicase, dbp1, lef-3, ie-1, me53, | Cp120 (dnaligase), Cp126 (helicase-2) |
| Structural | F protein, gp41, odv-ec27, odv-e56, p6.9, p74, vp91, vp39, vp1054, pif, fp25K, odv-e18, odv-e25, odv-e66, pk1, polh | ***pep/p10 (Cp22)***, Cp71(p24capsid) |
| Auxiliary | alk-exo, fgf, ubiquitin | ***mp-nase (Cp46), fgf-1 (Cp76), iap-5 (Cp116), fgf-3 (Cp140)***, Cp59(sod) |
| Unknown | 38K, ac22, ac29, ac38, ac53, ac66, ac68, ac75, ac76, ac78, ac81, ac82, ac92, ac93, ac96, ac106, ac109, ac110, ac115, ac142, 38.7K, ac145, ac146, p40, p12, p45 | ***Cp2, Cp4, Cp5, Cp20, Cp23, Cp29, Cp33, Cp39, Cp45***, Cp47(p13), ***Cp50, Cp56***, Cp79(Ac150), ***Cp99, Cp100, Cp115, Cp122, Cp135, Cp136*** |
| Total | 62 | 28 (**22**) |

**Table 3** Auxiliary gene content of GVs

| | egt | enhancin | rr1 | rr2a | pe38 | ptp-2 | chitinase | cathepsin | Cp94 (iap) | iap-3 | bro | lef-10 |
| --- | --- | --- | --- | --- | --- | --- | --- | --- | --- | --- | --- | --- |
| AdorGV | ✓ | ✗ | ✗ | ✗ | ✗ | ✗ | ✗ | ✗ | ✗ | ✓ | ✗ | ✗ |
| AgseGV | ✓[a] | ✓ | ✗ | ✗ | ✗ | ✗ | ✓ | ✓ | ✗ | ✓ | ✗ | ✓ |
| ChocGV | ✓ | ✗ | ✗ | ✗ | ✗ | ✗ | ✗ | ✗ | ✗ | ✓ | ✗ | ✓ |
| CrleGV | ✓ | ✗ | ✗ | ✗ | ✓ | ✓ | ✓[b] | ✓ | ✓ | ✓ | ✗ | ✓ |
| CpGV | ✓ | ✗ | ✓ | ✓ | ✓ | ✓(2) | ✓ | ✓ | ✓ | ✓ | ✓[a] | ✓ |
| PhopGV | ✓ | ✗ | ✗ | ✗ | ✗ | ✗ | ✗ | ✗ | ✓ | ✗ | ✓ | ✓ |
| PlxyGV | ✓ | ✗ | ✗ | ✗ | ✗ | ✗ | ✗ | ✗ | ✗ | ✗ | ✗ | ✗ |
| XecnGV | ✗ | ✓(4) | ✗ | ✗ | ✗ | ✗ | ✓ | ✓ | ✗ | ✗ | ✓(7) | ✓ |

Numbers in parenthesis indicate gene copy number if greater than one.
[a]Likely not functional due to truncation.
[b]Likely not functional due to chitinase active site absent.

titers. As a result, there are very few gene expression studies and little biochemical characterization of the GV genes. Currently, little is known regarding the molecular basis for phenotypic differences between GVs and NPVs. Consequently, some GV genes have been studied using NPV systems. These include the CpGV *cathepsin* and *iap-3* genes, trichoplusia ni GV (TnGV) *helicase* and *enhancin* genes and the xestia c-nigrum GV (XecnGV) *mmp* gene. However, the study of GV genes in their native virus and host cells is preferable. Only three GVs have replicated in cell culture. TnGV was reported to infect *Trichoplusia ni* cells but their permissive character was not maintained. Complete viral replication of PhopGV has been reported in a *P. operculella* cell line but has not led to any molecular studies. A *C. pomonella* embryonic cell line permissive for CpGV has enabled molecular studies on GVs to progress and recombinant CpGV viruses to be produced.

## Gene Expression

The gene expression of NPVs has been found to be sequential and coordinated. It is essentially split into two stages. Early gene expression precedes DNA replication and late/very late gene expression occurs after the onset of DNA replication. The successive stages of viral gene expression are dependent on the previous gene products. Early genes are thought to be transcribed using the host RNA polymerase II. Therefore, promoters of NPV early genes tend to mimic promoters of eukaryotic genes. Late and very late genes are transcribed by virus-induced RNA polymerase that recognizes the late promoter motif (A/T/G)TAAG. Temporal studies on GV gene expression are limited due to a lack of highly permissive cell lines. However, GV genes possess promoters similar to the early and late promoters described above, and transcription of a small number of genes has been shown to initiate from them. It is therefore likely that GVs follow the same cascade of gene expression as NPVs.

## AT Content

The sequenced GV genomes are AT-rich, ranging from 54.8% in CpGV, to 67.6% in cryptophlebia leucotreta GV (CrleGV), with an average of 62.6% (**Table 1**). CpGV and CrleGV are the most closely related sequenced GVs, although their AT content is highly divergent. The difference in AT content is mainly due to the base composition of the third nucleotide position within the codon of the coding regions. The host-cell machinery has to provide the virus with sufficient suitable t-RNA to replicate. The viruses may therefore have adapted to the base composition of the host. CrleGV is only infective to *C. leucotreta*, whereas CpGV is infective to both hosts, *C. pomonella* and (to a lesser degree) *C. leucotreta*. However, the AT content of these hosts is not yet known. CpGV is the only sequenced GV to contain the genes encoding the large and small subunits of ribonucleotide reductase (*rr1* and *rr2a*), which are also found in several NPVs and other DNA virus genomes. These enzymes are involved in nucleotide metabolism and catalyze the reduction of host cell rNTPs to dNTPs. It may be advantageous for CpGV to maintain these genes in order to alter the composition of dNTPs, allowing for replication in hosts with a divergent AT composition.

## GV Specific Genes

### Metalloproteinase

Baculovirus genomes have been shown to contain a number of proteases. Those studied so far are classified as nonessential auxiliary genes. Enhancin is a metalloproteinase which digests the peritrophic membrane in the insect midgut to facilitate virus infection. It has so far been found in all Noctuidae-specific GVs and one Tortricidae-specific GV (choristoneura fumiferana GV) and also in several group II NPVs. A cysteine proteinase (cathepsin) is involved in postmortem degradation of the infected host, in conjunction with a viral chitinase homolog. Cathepsin is found in most lepidopteran NPVs and four of the sequenced GVs (**Table 3**). A third group of proteases has been identified exclusively in the GVs, with homology to the matrix metalloproteinases (MMP). MMPs are active against components of the extracellular matrix and are usually secreted. All eight GV genomes possess an MMP homolog, suggesting it plays an important role in GV infection. Very little is known about MMP activity in infected larvae. A role for MMP in the breakdown of infected host tissue, potentially facilitating virus dispersal, and stimulation of postmortem melanization, has been proposed for XecnGV MMP. This study was performed using XecnGV MMP, overexpressed in occlusion-negative bombyx mori NPV (BmNPV), and may not be representative of its function in GV infection. However, results suggested that XecnGV MMP did encode a functional MMP. Putative signal peptide sequences are found in all of the GV MMP sequences except XecnGV. This suggests that these proteins, in keeping with MMPs, may enter the secretory pathway. Analysis of the predicted GV MMP amino acid sequences shows that residues predicted to be essential for metalloproteinase activity are conserved between all of the GV MMP sequences. The greatest conservation is in the region of the predicted zinc-binding active site of the enzyme. The conservation of MMP in all sequenced GVs implies that MMP plays an important role during GV infection, possibly in the breakdown of the basement membrane. As MMP is found exclusively in the GVs, its activity may lead to a difference in the infection process between GVs and NPVs.

### Inhibitors of Apoptosis

Baculoviruses possess two families of genes that suppress apoptosis, the P35/P49 family and the inhibitor of apoptosis (IAP) family. The IAP-3 protein of CpGV was the first member of the baculovirus IAP family of proteins to be identified. It has been shown to block apoptosis in diverse systems and substitute for the *p35* gene in blocking AcMNPV-induced apoptosis in SF21 cells. Since then all of the GVs (and lepidopteran NPVs) have been found to contain IAP homologs. The GVs contain between one and three IAP homologs. Phylogenetic analyses have demonstrated that there are three clades of GV IAPs suggesting that they are not gene duplications. The first group contains the previously characterized IAP-3, which is found in CpGV, CrleGV, choristoneura occidentalis GV (ChocGV), AdorGV, and agrotis segetum GV (AgseGV). A second group includes Cp94 which has homologs, in the same genome position, in CrleGV and PhopGV. This IAP has so far been shown not to have any anti-apoptotic activity. A third group contains IAP-5 which is specific to all of the GVs. All GV genomes sequenced to date contain IAP-5, which is absent from NPV genomes. The IAP-5 genes are located in the same positions on the GV genomes suggesting an ancient acquisition before the GVs diverged. This is in contrast to IAP-3 which is one of the only GV genes found in different positions within the GV genomes, suggesting a more recent acquisition. The CpGV IAP-5 has not been shown to have anti-apoptotic activity by itself, but it can stimulate the anti-apoptotic activity of CpGV IAP-3. However, its function in genomes lacking IAP-3 is unknown. The P35/P49 family was thought to be NPV specific but recently a P35/P49 homolog has been identified in ChocGV, although functionality has yet to be ascertained.

## Fibroblast Growth Factors

Fibroblast growth factors, or FGFs, are a family of growth factors involved in tissue repair and cell proliferation and differentiation. It has been suggested that baculovirus FGFs may be required for efficient spread of infection beyond the midgut. FGFs are conserved among vertebrate and invertebrate organisms, but within the viruses they have only been identified in baculoviruses. All of the GV genomes contain three *fgf* genes, whereas there is only one *fgf* gene present in the NPV genomes (homologs of AcMNPV Ac32). Ac32 is transcribed both early and late in infection and is secreted when expressed in insect cells, which suggests that it acts as an extracellular ligand. An Ac32-null mutant was constructed and compared to wild-type virus in cell culture. No obvious differences were observed in protein or DNA synthesis, although its pathogenesis *in vivo* has not yet been evaluated.

All of the GV FGF proteins show homology to the FGF domain in Ac32. Phylogenetic analyses have demonstrated that there are three clades of GV FGFs suggesting that they are not gene duplications. The NPV FGFs did not group clearly with any GV FGF clade, but FGF-2 is usually annotated as the Ac32 homolog. It has not been determined whether all copies are functional and the advantage of three GV *fgf* genes with respect to pathogenesis is not yet known.

## DNA Ligase and Helicase-2

All GV genomes contain the genes, *DNA ligase* and *helicase-2*. The *helicase-2* gene is part of the helicase superfamily I which includes PIF1 from eukaryotes and RecD from *Escherichia coli*. These enzymes are typically involved in DNA metabolism, such as replication, recombination, and repair. The only NPVs to contain either of these genes are lymantria dispar (LdMNPV) which contains a *DNA ligase* and a *helicase-2*, and mamestra configurata NPV (MacoNPV) and spodoptera litura NPV (SpltNPV) which both contain a *helicase-2*. The LdMNPV DNA ligase displays catalytic properties of a type III DNA ligase. However, in LdMNPV, neither the *helicase-2* nor the *DNA ligase* gene stimulates DNA replication in transient assays. The roles of *DNA ligase* and *helicase-2* in viral replication are unknown but based on their homology to genes from other organisms, it is likely that they are involved in DNA recombination or repair systems.

## PEP/P10

Three GV genes form part of a highly conserved GV gene family. These are the homologs of the CpGV genes, Cp20, Cp22, and Cp23. Two of the genes, Cp20 and Cp23, are 35% identical to each other. Phylogenetic analyses of Cp20 and Cp23 and their homologs in other GVs clearly suggest that they are likely to be paralogous genes which were duplicated in a common ancestor before the GVs differentiated. The three genes show a significant similarity to domains of the baculovirus polyhedron envelope/calyx protein (PEP). PEP is thought to be an integral part of the polyhedron envelope. It is concentrated at the surface of polyhedra, and is considered important for the proper formation of the periphery of polyhedra. It is thought that PEP may stabilize polyhedra and protect them from fusion or aggregation. Electron microscopic evidence exists for a GV calyx and PEP may play a similar role in GVs.

The GV homologs of Cp22 share a number of motifs with P10, including a proline-rich domain and a heptad repeat sequence. In NPV-infected cells, P10 forms fibrillar structures in the nucleus and cytoplasm. These structures have yet to be identified in electron micrographs of GVs. It is possible that GV fibrous bodies are smaller, more granular structures, which may have been overlooked. The P10 protein is implicated in occlusion body morphogenesis and disintegration of the nuclear matrix, resulting in the dissemination of occlusion bodies. P10 is also crucial for the proper formation of the polyhedron envelope. The Cp22 homologs are over twice the size of most NPV P10 proteins and much of the sequence identity is between sequences of low complexity. The large size of the Cp22 homologs may be due to the presence of domains similar to the calyx/pep protein of NPVs. It has been suggested that in GVs the PEP and P10 may be conserved in a single protein (Cp22) and that the other members of the gene family (Cp20 and Cp23) may also be involved in the formation of the occlusion body envelope.

## Origins of Replication

Within the genomes of NPVs, sequences have been identified that function as origins of replication (*ori*s) when cloned into plasmids and transfected into infected cells. In some NPVs, these also function as enhancers of early gene transcription. These regions, called homologous regions (*hr*s), typically contain one or more copies of an imperfect palindrome sequence and are located in several positions on the genome. They have been identified in most NPVs sequenced to date and have been found to act as *ori*s in several NPVs using infection-dependent DNA replication assays. In addition, complex structures containing multiple direct and inverted repeats of up to 4000 bp have also been identified that act as *ori*s. These have similar structural characteristics to eukaryotic *ori*s and are found only once per genome; they are therefore called non-*hr ori*s. The discovery that serial passage of some NPVs results in the appearance of defective interfering particles (DIs) containing reiterations of a non-*hr ori*, is further evidence that non-*hr ori*s may be involved in genome replication. It has also been suggested that the expansion of non-*hr ori*-like regions is not restricted to DI particles. A 900 bp region within the non-*hr ori*-like region of CrleGV was amplified in submolar populations in CrleGV *in vivo*, resulting in the expansion of the hypervariable region.

The CpGV genome contains 13 imperfect palindromes of 74–76 bp (**Figure 5**), which are found in 11 regions of the genome. Six of these palindromes are within putative genes. The 13 palindromes have been characterized using an infection-dependent replication assay in *C. pomonella* cells and all have been found to replicate. The palindromes are most conserved at their ends and have a highly AT-rich center. The entire 76 bp palindrome is required for replication, with no replication occurring when 10 bp from each end of the palindrome are removed. The specific flanking DNA of each palindrome is required for optimal replication, even though there was no homology between the flanking sequences. A region reminiscent of the non-*hr* type *ori* in NPVs has also been identified in the CpGV genome. However, this does not replicate in the infection-dependent replication assay and its function in GVs is not yet known.

Singleton imperfect palindromes are also found in CrleGV, ChocGV, PhopGV, and AdorGV. These GVs group in a clade with CpGV and it is likely that these palindromes descended from a common viral ancestor. The palindromes range in size from 63 bp in CrleGV to 320 bp in PhopGV and are often found in putative *hr*s which also consist of short direct repeats. The number of palindromes in each putative *hr* ranges from typically one up to four, which is less than the number of repeated palindromes in most NPV *hr*s. The ends of the palindromes are conserved in all of these GVs having a consensus of 13 bp (A(C/T)GAGTCCGANTT). The centers of the palindromes differ in sequence but all are AT-rich. These palindromes may have the potential to form secondary structures, such as cruciform (stem-loop or hairpin) structures through intrastrand base pairing. These structures may allow the initiation of DNA replication via the binding of specific protein factors. They may also remain in the linear state, acting as binding sites for protein dimers. This appears to be the case for the AcMNPV *hr*5 palindrome and the regulatory protein IE-1. NPV *hr*s have been demonstrated to be *cis*-acting enhancers of transcription of baculovirus early genes including *39K*, *ie-2*, *p35*, and *helicase*. It is not yet known if CpGV *hr*s act as enhancers of early gene transcription.

The other GVs which do not group in the CpGV clade (PlxyGV, XecnGV, and AgseGV) have quite different putative *hr* regions. PlxyGV has large *hr* regions of up to 2383 bp consisting of direct repeats of 101–105 bp, repeated up to 23 times with a 15 bp palindrome near the center. XecnGV contains direct repeats of about 120 bp, repeated 3–6 times without any palindromes. So far been no putative *hr*s have been identified in the AgseGV genome. The only shared

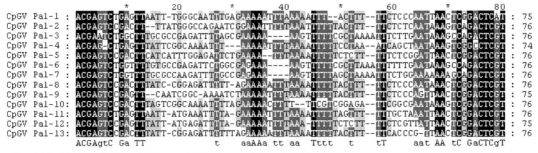

**Figure 5** The 13 CpGV imperfect palindromes that acts as origins of replication.

feature between all of the GV *hr*s is the relative positions of some of the *hr*s within the genome. For example, there is often a putative *hr* between the *desmoplakin* and *lef-3* genes, between *sod* and *p74* and flanking or within the *vp91* gene. Within the NPVs, *hr* regions are also often found flanking *sod* and *vp91*, which suggests that some of the *hr*s may have originated before the GVs and NPVs diverged.

*See also:* Baculoviruses: Apoptosis Inhibitors; Baculoviruses: Expression Vector; Baculoviruses: General Features; Baculoviruses: Molecular Biology of Mosquito Baculoviruses; Baculoviruses: Molecular Biology of Nucleopolyhedroviruses; Baculoviruses: Molecular Biology of Sawfly Baculoviruses; Baculoviruses: Pathogenesis.

## Further Reading

Escasa SR, Lauzon HAM, Mathur AC, Krell PJ, and Arif BM (2006) Sequence analysis of the *Choristoneura occidentalis* granulovirus genome. *Journal of General Virology* 87: 1917–1933.

Hashimoto Y, Hayakawa T, Ueno Y, Fujita T, Sano Y, and Matsumoto T (2000) Sequence analysis of the *Plutella xylostella* granulovirus genome. *Virology* 275: 358–372.

Hayakawa T, Ko R, Okano K, Seong S-I, Goto C, and Maeda S (1999) Sequence analysis of the *Xestia c-nigrum* granulovirus genome. *Virology* 262: 277–297.

Herniou EA, Olszewski JA, Cory JS, and O'Reilly DR (2003) The genome sequence and evolution of baculoviruses. *Annual Review of Entomology* 48: 211–234.

Hess RT and Falcon LA (1987) Temporal events in the invasion of the codling moth *Cydia pomonella* by granulosis virus: An electron microscope study. *Journal of Invertebrate Pathology* 50: 85.

Hilton SL and Winstanley D (2007) Identification and functional analysis of the origins of DNA replication in the *Cydia pomonella* granulovirus genome. *Journal of General Virology* 88: 1496–1504.

Jehle JA (2002) The expansion of a hypervariable, non-*hr* ori-like region in the genome of *Cryptophlebia leucotreta* granulovirus provides *in vivo* evidence for the utilization of baculovirus non-*hr* oris during replication. *Journal of General Virology* 83: 2025–2034.

Jehle JA (2006) Molecular identification and phylogenetic analysis of baculoviruses from lepidoptera. *Virology* 346: 180–193.

Lange M and Jehle JA (2003) The genome of the *Cryptophlebia leucotreta* granulovirus. *Virology* 317: 220–236.

Luque T, Finch R, Crook N, O'Reilly DR, and Winstanley D (2001) The complete sequence of *Cydia pomonella* granulovirus genome. *Journal of General Virology* 82: 2531–2547.

Miller LK (ed.) (1997) *The Baculoviruses*. New York: Plenum.

O'Reilly DR and Vilaplana L (2003) Functional interaction between *Cydia pomonella* granulovirus IAP proteins. *Virus Research* 92: 107–111.

Winstanley D and Crook NE (1993) Replication of *Cydia pomonella* granulosis virus in cell cultures. *Journal of General Virology* 74: 1599–1609.

Wormleaton S, Kuzio J, and Winstanley D (2003) The complete sequence of the *Adoxophyes orana* granulovirus genome. *Virology* 311: 350–365.

# Baculoviruses: Molecular Biology of Mosquito Baculoviruses

**J J Becnel,** Agriculture Research Service, Gainesville, FL, USA
**C L Afonso,** Southeast Poultry Research Laboratory, Athens, GA, USA

© 2008 Elsevier Ltd. All rights reserved.

## Glossary

**Epizootiology** The study of the causes and forms of diseases at all levels of intensity in an animal population.

**Hypertrophy** Denotes greater bulk through an increase in the size of a cell but not in the number of individual tissue elements.

**Occlusion body** Proteinaceous bodies that occlude the virions of certain insect viruses expecially baculoviruses (DNA) and cypoviruses (RNA).

**Patent infection** External signs of infection with a pathogen that are obvious at the macro level of diagnostics.

## Introduction

Perhaps the first scientific description of a baculovirus was made in 1808 from silkworms in France but it was more than 150 years later before the first baculovirus was isolated from a mosquito. Mosquito baculoviruses are of growing interest as they may represent a separate branch within the family *Baculoviridae* that existed prior to the split of lepidopteran nucleopolyhedroviruses (NPVs) and granuloviruses (GVs). They may also be ancestral to the baculoviruses from hymenoptera which form a branch distinct from the lepidopteran baculoviruses. Mosquitoes are also important vectors of numerous human and veterinary diseases and baculoviruses offer the opportunity to investigate specific virus–host interactions at the molecular level. This possibility has been facilitated by the availability of the complete genomes for one mosquito baculovirus and several important mosquito vectors, especially the malaria vector *Anopheles gambiae*.

## History and Classification

The family *Baculoviridae* currently contains the genera *Nucleopolyhedrovirus* and *Granulovirus*. Members of the genus *Nucleopolyhedrovirus* have occlusion bodies (OBs) that contain many virions while members of the genus *Granulovirus* have OBs that contain one or rarely two virions. The baculoviruses found in mosquitoes thus far have been mainly assigned to the *Nucleopolyhedrovirus* group although it has been proposed that they represent a new genus in the family *Baculoviridae*. (Jehle and collaborators have recently proposed that a new genus, *Deltabaculovirus*, should be created in the family *Baculoviridae* for dipteran-specific baculoviruses.) The first baculovirus (OcsoNPV) to be isolated from mosquitoes was from *Ochlerotatus sollicitans* (formerly *Aedes sollicitans*) in Louisiana, USA. Since then, naturally occurring NPVs have been isolated from 11 additional mosquito species (**Table 1**) in the genera *Aedes, Anopheles, Culex, Ochlerotatus, Psorophora, Uranotaenia*, and *Wyeomyia*. More than 20 mosquito species have been found to be susceptible to NPV infections.

Before 2001, the only detailed studies on development and transmission of mosquito NPVs were conducted with an isolate from *O. sollicitans* (OcsoNPV) found in several different mosquito hosts in Louisiana. Recently, a great deal of new information has been compiled for a mosquito NPV from *Culex nigripalpus* (CuniNPV) and to a lesser extent for an NPV from *Uranotaenia sapphirina* (UrsaNPV). Therefore, information presented here will draw heavily on CuniNPV to describe current knowledge of the biology, morphology, and genomics of mosquito baculoviruses.

## Virion and OB Properties

Mature OBs are relatively large (5–20 μm) and polyhedral in shape or small (0.5 μm) and globular in shape (**Figure 1**). The OBs lack an envelope (the polyhedron envelope of NPVs), a characteristic feature of most other baculoviruses. OBs can contain a variable number of virions depending on the species, ranging from very few (1–2) in *Anopheles crucians* to many (50+) in some *Ochlerotatus* spp. (**Table 1**). Nucleocapsids are singly enveloped and the resulting virions are approximately 200 nm × 40 nm. Each virion consists of the nucleocapsid, intermediate layer, and an outer envelope. In some species (*Ochlerotatus sollicitans nucleopolyhedrovirus, Uranotaenia sapphirina nucleopolyhedrovirus*) early OBs are irregularly shaped and seem to subsequently coalesce to form large polyhedra with many virions while in other species (*Culex nigripalpus nucleopolyhedrovirus*) the OBs remain as relative small globular particles with 4–8 virions. Mature OBs of UrsaNPV have a unique dumbbell shape, lack the polyhedron envelope, and measure up to 10–15 μm in length and 2–3 μm in diameter.

Occlusion bodies of CuniNPV have a density of 1.14–1.18 g ml$^{-1}$ on a Ludox continous gradient. Dissolution of the OBs requires a pH > 12.0. Temperatures below 50 °C have a minimal effect on infectivity. Temperatures above 55 °C result in a total loss of activity.

## Pathology and Histopathology

Mosquito NPVs are highly specific for midgut tissues (primarily midgut epithelium) of mainly larval mosquitoes but infections have been found in midguts of adult mosquitoes. Patent infections are detected by the hypertrophied nuclei of midgut cells that appear white due to the accumulation of occlusion bodies (**Figure 2**).

CuniNPV affects the development, behavior, and appearance of infected *C. quinquefasciatus* and *C. nigripalpus*

**Table 1** Mosquito baculovirus names and corresponding abbreviations

| Virus | Abbreviation | Accession number |
|---|---|---|
| Anopheles crucians nucleopolyhedrovirus | AncrNPV | |
| Culex nigripalpus nucleopolyhedrovirus | CuniNPV | AF403738 |
| Culex pipiens nucleopolyhedrovirus | CupiNPV | |
| Culex salinarius nucleopolyhedrovirus | CusaNPV | |
| Ochlerotatus sollicitans nucleopolyhedrovirus | OcsoNPV | |
| Ochlerotatus taeniorhynchus nucleopolyhedrovirus | OctaNPV | |
| Ochlerotatus triseriatus nucleopolyhedrovirus | OctrNPV | |
| Psorophora confinnis nucleopolyhedrovirus | PscoNPV | |
| Psorophora ferox nucleopolyhedrovirus | PsfeNPV | |
| Psorophora varipes nucleopolyhedrovirus | PsvaNPV | |
| Uranotaenia sappharina nucleopolyhedrovirus | UrsaNPV | |
| Wyeomyia smithii nucleopolyhedrovirus | WysmNPV | |

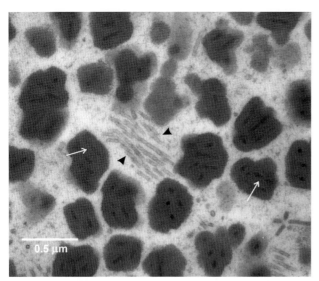

**Figure 1** Ultrastructural features of CuniNPV OBs and virions. Arrow indicates rod-shaped virons within an OB. Arrowhead indicates empty capsids.

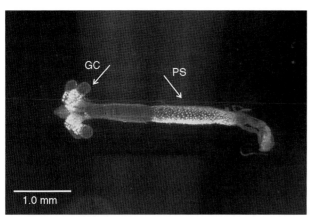

**Figure 2** Gross pathology of CuniNPV in a dissected midgut of a *Culex quinquefasciatus* larvae. Arrows indicate infected nuclei in the gastric caecae and the posterior stomach.

larvae. Within 24 h postinfection (p.i.), infected larvae are typically stunted in size when compared to unexposed individuals, indicating a failure to molt or problems with nutrient uptake. Larvae actively feed through about 48 h p.i., but by 72 h p.i. larvae are lethargic and often remain suspended at the water surface. By 48 h p.i., nuclei of most cells in the gastric ceca and posterior stomach (**Figure 2**) are opaque to white in color due to the proliferation of OBs. Cells in the anterior stomach rarely support viral development of CuniNPV but the entire midgut may support infection in some species as occurs in OcsoNPV. Death of the larvae usually occurs within 72–96 h p.i., at which time most susceptible nuclei in midgut and gastric ceca are infected.

## Life Cycle

A cascade of events is required for the invasion, replication, and spread of CuniNPV in the mosquito host. Infections are initiated when larval mosquitoes ingest OBs together with the appropriate divalent cation, usually magnesium (see next section). Occlusion-derived virions (ODVs) are released from OBs due to the alkaline conditions of the midgut together with other possible factors. The released ODVs attach and pass through the peritrophic matrix (PM) followed by attachment to the membrane of the microvilli and entry into the cytoplasm of midgut epithelial cells. Invasion of the nucleus occurs when nucleocapsids attach to the nuclear envelope where DNA is released into the nucleoplasm to initiate viral replication. This process of ingestion of infectious particles to the invasion of the nucleus occurs within 2–4 h post exposure to OBs. The next phase of the process involves the rapid spread of CuniNPV to other midgut cells. The initial viral replication produces nucleocapsids that are released from the nucleus that form extracellular virions (the budded virion, BV), which are released out of the midgut cells into the ectoperitrophic space. The BVs can then attach to microvilli of other midgut cells and make their way to the nucleus to initiate another round of replication. Newly synthesized nucleocapsids acquire *de novo* envelopes in the nucleus and become occluded by the OB protein. The nuclei become packed with OBs which are released upon death of the host and cause infection when ingested by a susceptible mosquito larva. The spread within the mosquito midgut and production of OBs occurs within 14–48 h post exposure.

## Transmission and Host Range

Initial attempts to transmit CuniNPV to larval mosquitoes in the laboratory were unsuccessful. Studies using deletion analysis of the most abundant cations present in the field water were used to determine the water-borne factors critical for transmission of CuniNPV. Assays of *C. quinquefasciatus* larvae exposed to CuniNPV were conducted in an artificial salt mixture based on the analysis of the ion composition of field water. The addition of salts to deionized water significantly improved the infection levels in larvae. Salt mixtures without magnesium resulted in less than 1.0% infections. Salts with magnesium or salts without calcium increased infection levels to 80–100%. Further investigations revealed conclusively that transmission is mediated by divalent cations: magnesium is essential, whereas the presence of calcium inhibits the magnesium's activity to mediate transmission. In addition, other divalent cations can function as activators and inhibitors of CuniNPV transmission. Activators include barium, cobalt, nickel, and strontium, while additional inhibitors of transmission are copper, iron, and zinc. Transmission of CuniNPV has only been successful to mosquitoes within the genus *Culex*, subgenus *Culex*. These include *C. nigripalpus, C. quinquefasciatus, C. salinarius, C. pipiens, C. pipiens* f. *molestus*, and *C. restuans*. *Culex territans* (subgenus *Neoculex*) is the only *Culex* species tested that is not susceptible to CuniNPV. No infections have been found with any species of *Aedes, Anopheles, Ochlerotatus, Culiseta*, or *Toxorhynchites* exposed in laboratory assays. Transmission of UrsaNPV is also driven by cations similar to that found for CuniNPV with a host range restricted to species of *Uranotaenia*. The very restricted host range of CuniNPV and UrsaNPV differs from transmission studies conducted with OcsoNPV, showing that *Aedes, Ochlerotatus*, and *Psorophora* spp. are susceptible but not *Culex* and *Anopheles* spp.

## Field Epizootiology

Mosquito baculoviruses have been less commonly reported from field populations than other baculoviruses and generally at very low prevalence rates (<1.0%). Recently, natural epizootics of CuniNPV have been studied at two field sites in Florida: one, a swine wastewater site with *C. nigripalpus* dominant during the warmer months and *C. quinquefasciatus* dominating during the cooler months. Regular and extended epizootics of CuniNPV in *C. nigripalpus* were documented at this site with an average infection rate of 20.1% and a maximum rate of 60%. *C. quinquefasciatus* larvae have an average CuniNPV infection rate of 8.6% and a maximum infection rate of 20%. The second site was a dairy wastewater facility where *C. quinquefasciatus* was the dominant species. Although high larval populations of *C. quinquefasciatus* were present, CuniNPV infected larvae were collected rarely and never at epizootic levels (0.08% infection). The $Mg^{2+}/Ca^{2+}$ ratios at the swine wastewater site are 1.9/0.8 mM while these ratios at the dairy wastewater site were 3.7/3.0 mM. It appears that the favorable $Mg^{2+}/Ca^{2+}$ ratios at the swine wastewater site mediated transmission of CuniNPV while the high calcium levels at the dairy wastewater site were unfavorable for transmission.

## Genome Structure

CuniNPV contains a circular double-stranded DNA genome packaged into rod-shaped singly enveloped nucleocapsids. Complete nucleotide sequencing (GenBank accession no. AF403738) has determined a size of 108 252 base pairs encoding densely arranged bidirectional nonoverlapping clusters of two to ten open reading frames (ORFs). CuniNPV encodes at least 109 putative proteins, some of which are homologous to those from other baculoviruses involved in early and late gene expression, DNA replication, structural and auxiliary functions. Only 36 of the 109 putative CuniNPV predicted proteins demonstrate clear homology to proteins from other baculoviruses, and 72 of the CuniNPV ORFs show no homology to any other known baculovirus ORFs. Amino acid conservation between CuniNPV and other baculovirus proteins is low (18–44% to hymenopteran, and 18–54% to lepidopteran viruses), with an average of 28% identity to autographa californica multiple NPV (AcMNPV). CuniNPV genes display clusters of only two to four genes in similar transcriptional orientation or genomic position in comparison to other known baculoviruses, but 46 putative genes contain typical early and late baculovirus promoters upstream from the initiation codons. The CuniNPV genome contains distinctive baculovirus features such as homologous regions (*hr*s) that have been implicated as enhancers of viral transcription and also as origins of DNA replication. Four *hr*s located in intergenic regions are composed of 64–85-bp repeats but lack sequence similarity to *hr*s from other baculoviruses.

## RNA Transcription

Baculovirus infections are initiated by the host RNA polymerases that transcribe early promoters in conjunction with the baculovirus transcriptional transactivators. Following DNA replication, baculovirus RNA polymerases transcribe genes containing late promoters. The large number of CuniNPV homologs of baculovirus genes involved in RNA transcription and DNA replication suggests that CuniNPV may utilize similar mechanisms, but with species-specific differences. CUN096 (*lef-4*), CUN026 (*lef-8*), CUN059 (*lef-9*), and CUN073 (*p47*) are homologs of genes that encode the multi-subunit RNA polymerases involved in late transcription. In addition, CUN088 (*lef-5*) has been implicated in late gene transcription. CUN018 is similar to very late transcription factor-1 (*vlf-1*), the major transactivator of very late gene expression. Although CuniNPV contains the minimal complex necessary for late polymerase activity (*lef-4, lef-8, lef-9*, and *p47*), it lacks homologs of other lepidopteran baculovirus expression factors which are required for optimum levels of late transcription (*lef-6, lef-10, lef-11, lef-12*, and *pp31* (*39k*)) in other baculovirus–host systems. Some of the transcription factors absent in CuniNPV are species specific (*lef-7, lef-10, lef-12*, and *host cell specific factor-1*), but others are conserved in many completely sequenced lepidopteran baculoviruses (*lef-6, lef-11, pk-1*, and *pp31* (*39k*)). Since promoters of early baculovirus genes are transcribed by a combination of host RNA polymerase II and virally encoded transcription factors, the absence of *ie-0, ie-1, ie-2*, and *pe38* homologs suggests that CuniNPV is either completely dependent on host factors for early transcription, or that unique CuniNPV genes are involved in these functions. Alternatively, sequence conservation of certain CuniNPV genes may be extremely low, thus making identification of homologs from other baculoviruses impossible.

## DNA Replication

CuniNPV replicates in the nuclei of the infected host. Of the six virally encoded AcMNPV genes implicated as being essential for replication, four homologs are present in CuniNPV (CUN045 (*lef-1*), CUNO25 (*lef-2*), CUN089 (*helicase-1*) and CUN091 (*dna-pol*)). In addition, CUN054 (*alk-exo*) and CUN018 (*vlf-1*) are also likely involved in DNA replication. The CuniNPV genome contains six baculovirus-repeated ORFs (*bro* ORFs) (CUN001, CUN004, CUN005, CUN095, CUN108, and CUN109). All six CuniNPV *bro* ORFs contain the sequence for the 41-amino-acid motif common to all *bro* proteins and two additional amino acid motifs conserved in group I and III *bro* proteins. Some *bro* proteins have been shown to bind DNA and may be involved in viral DNA replication. CuniNPV is distinguished from other hymenopteran and lepidopteran baculoviruses by the existence of significant amino acid differences in proteins involved in viral replication and by the absence of a number of genes considered essential and nonessential for DNA replication. CuniNPV lacks recognizable homologs for *lef-3*, which has the properties of a single-stranded DNA-binding protein (*dbp*), for *lef-7*, proliferating cell nuclear antigen (*pcna*), for *DNA ligase, rnr1*, and *rnr2* (large and small subunits of ribonucleotide reductase), and for *dUTPase*.

## Virion and Other Structural Functions

CuniNPV has two virion phenotypes: an occluded form (ODV) which initiates infection in the midgut, and a budded form (BV) which spreads the infection within the midgut. CuniNPV contains only 8 of the 17 virion structural protein genes conserved among lepidopteran baculoviruses (*vp39, vp91, vp1054, odv-ec27, odv-e56, p6.9, gp41,* and *p74*). CUN024 is similar to the *vp39* capsid protein gene, CUN035 resembles the *vp91* gene encoding a protein found in both the capsid and envelope of ODVs, and CUN008 resembles the *vp1054* gene for a virion-associated protein that functions in nucleocapsid formation. Homologs of the genes encoding capsid-associated proteins p87, p24, and orf1629 are absent or not identifiable in CuniNPV. Interestingly, orf1629 has been shown to be an AcMNPV essential protein that is associated with the basal structure of the capsid. While CuniNPV lacks a homolog of AcMNPV *gp64*, it does encode for a homolog of the gene for LdMNPV envelope protein Ld130 (CUN104), which is also a fusion protein and can functionally replace *gp64*. The CUN104 also demonstrates relatedness to gene for the envelope fusion protein Se8, of spodoptera exigua multiple NPV (SeMNPV), and CG4715, a *Drosophila* gene product related to Ld130. CUN032 is the homolog of *odv-ec27*, which encodes a protein present in ODV nucleocapsids and envelopes and may function as a cyclin. CUN074 is similar to the gene for p74, an ODV protein required for oral infectivity. CUN102 is the homolog of *odv-e56*, which encodes a protein associated with both viral-induced intranuclear vesicles and envelopes. CUN033 is the homolog of the gene for the ODV tegument protein gp41. The protein gp41 is required for the production of polyhedra and the egress of nucleocapsids from the nucleus is blocked. The absence of multiple lepidopteran baculovirus gene homologs suggests involvement of another set of viral or perhaps cellular proteins in CuniNPV ODV assembly and cellular entry functions. Structural genes absent or not identifiable in CuniNPV include *orf1629, p87, gp64, p24, odv-e18, odv-e25, odv-e66, p10, pp34, polyhedrin/granulin,* and *fp25k*.

## Virion Occlusion

CuniNPV OBs are globular, lack a polyhedron envelope and calyx structure, and typically contain about four individually enveloped virions. In addition, although they appear to be composed of a peptide similar in size to lepidopteran polyhedrin or granulin (about 30 kDa), N-terminal amino acid sequence analysis did not reveal homology to other baculovirus OB proteins. However, the sequence did match positions 693–709 of CUN085, an ORF with no homology to any other known baculovirus gene. The large size of the CUN085 product (882 amino acids) suggests that it is cleaved to produce components of CuniNPV OBs. CuniNPV lacks homologs of the genes encoding polyhedrin, granulin, or the polyhedron-associated proteins p10, FP25 (a conserved protein involved in polyhedra formation), and pp34, the polyhedron envelope and calyx protein. The lack of a *pp34* homolog is reflected in the absence of a polyhedron envelope and calyx surrounding the OB. The small size of the CuniNPV OBs, the limited number of virions occluded, and the lack of an envelope and calyx structure are reminiscent of the structure of the OBs of GVs. The absence of a P10 gene in the CuniNPV genome is supported by observed morphological differences from cells infected with lepidopteran NPVs. In infected lepidopteran cells, p10 is expressed as an abundant protein that is associated with nuclear and cytoplasmic fibrillar bodies during the terminal stages of infection. Fibrillar material produced during CuniNPV infection, however, does not resemble the fibrillar bodies associated with other lepidopteran baculovirus infections. Most notably, the cytoplasmic features of CuniNPV-infected cells are microtubule bundles and irregular cisternae of smooth endoplasmic reticulum.

## Auxiliary Functions

Auxiliary genes are nonessential for viral replication in cell culture, although they likely provide selective advantages in insects. CuniNPV contains one gene (CUN075) with homology to the baculovirus anti-apoptosis *p35* gene, but it lacks homologs of the inhibitor of apoptosis (*iap*) family of genes. CUN075 lacks sequence identity to *p35* in the 110-amino-acid C-terminal region, which in AcMNPV p35 mediates anti-apoptotic activity, suggesting a different mode of action or function for this gene. CUN039 shows relatedness to genes for insulin-binding proteins of *Spodoptera frugiperda*. It is also related to a *Caenorhabditis elegans* gene of unknown function and the *Drosophila* gene IMP-L2. Insulin and related peptides are very important hormones for the regulation of host growth and metabolism. The insulin-related peptide binding protein secreted from *S. frugiperda* cells is composed of two immunoglobulin-like C2 domains, is capable of binding human insulin, and inhibits insulin signaling through the insulin receptor. The closest homolog of the *S. frugiperda* binding protein is the essential protein IMP-L2 found in *Drosophila melanogaster*. IMP-L2 also binds insulin and related peptides and is implicated in neural and ectodermal development. CUN039 may have a similar role in modifying host metabolism and/or development. CuniNPV lacks homologs of at least 14 other baculovirus auxiliary genes, including those encoding for superoxide dismutase, ubiquitin, inhibitor of apoptosis, protein kinase 1, and viral enhancing factors, but the existence of a large number of genes of unknown functions in the CuniNPV genome suggest that other auxiliary genes of unknown function may exist.

## Phylogenetic Relationships to Other Baculoviruses

Striking differences in genome organization, the lack of conservation in gene order, the low level of amino acid conservation for homologous genes, and the absence of many genes that are conserved in other baculoviruses suggest that CuniNPV is a distant relative to lepidopteran and hymenopteran baculoviruses (**Figure 3**). Phylogenetic analyses using DNA polymerase, p74, and concatenated gene sequences place CuniNPV in a baculovirus lineage distinct from the three other lineages (lepidopteran NPVs and GVs and hymenopteran NPVs). The most striking differences with other baculoviruses are the absence of homologs of genes that have been found

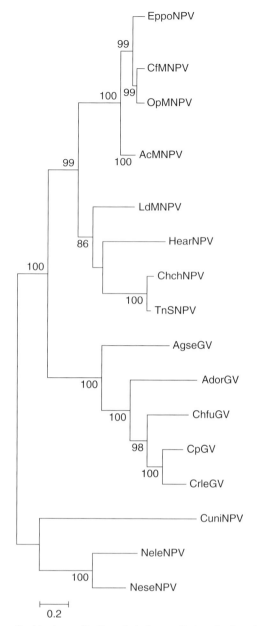

**Figure 3** Maximum likelihood phylogenetic tree for the p74 gene of CuniNPV and baculoviruses representative of different lineages (obtained using PhyMl discrete gamma model and 100 bootstraps). Abreviations are as follows: EppoNPV, epiphyas postvittana NPV; CfMNPV, choristoneura fumiferana MNPV ; OpMNPV, orgyia pseudotsugata MNPV; AcMNPV, autographa californica MNPV; LdMNPV, lymantria dispar MNPV; HearNPV, helicoverpa armigera NPV. ChchNPV, chrysodeixis chalcites NPV; TnSNPV, trichoplusia ni SNPV; AgseGV, agrotis segetum GV; AdorGV, adoxophyes orana GV; ChfuGV, choristoneura fumiferana granulovirus; CpGV, cydia pomonella GV; CrleGV, cryptophlebia leucotreta GV; CuniNPV, culex, nigripalpus NPV; NeleNPV, neodiprion lecontei NPV; NeseNPV, neodiprion sertifer NPV.

to be essential for or stimulatory to both DNA replication and gene transcription in lepidopteran baculoviruses and the lack of homologs of genes involved in the formation of virogenic stroma, nucleocapsid, envelope of occluded virions, and polyhedral formation. It has been suggested that some viruses, including baculoviruses, may co-evolve with their hosts in a process known as host-dependent evolution. Molecular evidence indicates that the ancestral Lepidoptera and Diptera separated about 280 million years ago and the extent of the genomic differences between the lepidopteran baculoviruses and CuniNPVs may reflect this ancient separation.

*See also:* Baculoviruses: General Features; Baculoviruses: Molecular Biology of Granuloviruses; Baculoviruses: Molecular Biology of Nucleopolyhedroviruses; Baculoviruses: Pathogenesis.

## Further Reading

Afonso CL, Tulman ER, Lu Z, *et al.* (2001) Genome sequence of a baculovirus pathogenic for *Culex nigripalpus*. *Journal of Virology* 75: 11157–11165.

Andreadis TG, Becnel JJ, and White SE (2003) Infectivity and pathogenicity of a novel baculovirus, CuniNPV from *Culex nigripalpus* (Diptera : Culicidae) for thirteen species and four genera of mosquitoes. *Journal of Medical Entomology* 40: 512–517.

Anthony DW, Hazard EI, and Crosby SW (1973) A virus disease in *Anopheles quadrimaculatus*. *Journal of Invertebrate Pathology* 22: 1–5.

Becnel JJ, White SE, Moser BA, *et al.* (2001) Epizootiology and transmission of a newly discovered baculovirus from the mosquitoes *Culex nigripalpus* and *C. quinquefasciatus*. *Journal of General Virology* 82: 275–282.

Becnel JJ, White SE, and Shapiro AM (2003) Culex nigripalpus nucleopolyhedrovirus (CuniNPV) infections in adult mosquitoes and possible mechanisms for dispersal. *Journal of Invertebrate Pathology* 83: 181–183.

Federici BA (1980) Mosquito baculovirus: Sequence of morphogenesis and ultrastructure of the virion. *Virology* 100: 1–9.

Federici BA (1985) Viral pathogens of mosquito larvae. *Bulletin of the American Mosquito Control Association* 6: 62–74.

Federici BA and Anthony DW (1972) Formation of virion-occluding proteinic spindles in a baculovirus disease of *Aedes triseriatus*. *Journal of Invertebrate Pathology* 20: 129–138.

Federici BA and Lowe RE (1972) Studies on the pathology of a baculovirus in *Aedes triseriatus*. *Journal of Invertebrate Pathology* 20: 14–21.

Herniou EA, Olszewski JA, Cory JS, and O'Reilly DR (2003) The genome sequence and evolution of baculoviruses. *Annual Review of Entomology* 48: 211–234.

Jehle JA, Blissard GW, Bonning BC, *et al.* (2006) On the classification and nomenclature of baculoviruses: A proposal for revision. *Archives of Virology* 151: 1257–1266.

Moser BA, Becnel JJ, White SE, *et al.* (2001) Morphological and molecular evidence that Culex nigripalpus baculovirus is an unusual member of the family *Baculoviridae*. *Journal of General Virology* 82: 283–297.

Perera OP, Valles S, Green TB, *et al.* (2006) Molecular analysis of an occlusion body protein from Culex nigripalpus nucleopolyhedrovirus (CuniNPV). *Journal of Invertebrate Pathology* 91: 35–42.

Shapiro AM, Becnel JJ, and White SE (2004) A nucleopolyhedrovirus from *Uranotaenia sapphirina* (Diptera: Culicidae). *Journal of Invertebrate Pathology* 86: 96–103.

Stiles B, Dunn PE, and Paschke JD (1983) Histopathology of a nuclear polyhedrosis infection in *Aedes epactius* with observations in four additional mosquito species. *Journal of Invertebrate Pathology* 41: 191–202.

Stiles B and Paschke JD (1980) Midgut pH in different instars of three *Aedes* mosquito species and the relation between pH and susceptibility of larvae to a nuclear polyhedrosis virus. *Journal of Invertebrate Pathology* 35: 58–64.

# Baculoviruses: Molecular Biology of Sawfly Baculoviruses

**B M Arif**, Great Lakes Forestry Centre, Sault Ste. Marie, ON, Canada

© 2008 Published by Elsevier Ltd.

## Glossary

***Anopheles gambiae*** Mosquito from the genus Anopheles.

***Apis mellifera*** Western or European honeybee.

***Gene parity plot*** A plot of the sequential organization and collinearity of homologous genes within two genomes.

***Gilpinia hercyniae*** European spruce sawfly.

***Hymenoptera*** Large order of insects comprising of sawflies, ants, bees, and wasps.

***Neodiprion abietis*** Balsam fir sawfly.

***Neodiprion lecontei*** Red-headed pine sawfly.

***Neodiprion sertifer*** European pine sawfly.

***Synteny*** Preserved order of genes between related species.

## Introduction

Hymenoptera having existed since the Triassic period ($205–248 \times 10^6$ years ago) represent a more ancient order of insect than Lepidoptera, which appeared on earth much later during the Cretaceous period of approximately $65–144 \times 10^6$ years ago. Sawflies are considered primitive

Hymenoptera and have existed since the Mesozoic period. Nucleopolyhedroviruses (NPVs) that infect sawflies have always been known to be excellent biological control agents. For example, the European spruce sawfly, *Gilpinia hercyniae*, was introduced to North America early in the twentieth century and became a serious forest pest of spruce in northeastern Canada and the United States. A singly embedded NPV infecting *Gilpinia hercyniae* was imported from Finland and upon introduction to infested stands, the virus spread rapidly in the population and caused the collapse of the infestation. Viruses of sawflies are considered highly host specific and efficacious pathogens. One of the main reasons for its efficacy is the insect habitat. Sawflies are gregarious in nature and the virus causes what has been described as infectious diarrhea which is the sloughing off of infected cells causing rapid spread of infection in the population. The use of sawfly viruses in the control of infestations in Canada and Europe has been well documented. Recently, it has been shown that a decline in infestation of the balsam fir sawfly, *Neodiprion abietis*, was consistently associated with the presence of an NPV of this insect (NeabNPV). It was later proved that an epizootic by NeabNPV initiated population decline of the balsam fir sawfly.

In the laboratory of Dr. James Maruniak at the University of Florida, we became interested in the genomics of sawfly viruses for three main reasons. First, since sawflies represent a more ancient order of insects, we wondered if their viruses actually represent ancestors to the more recent baculoviruses of Lepidoptera. Second, restriction endonuclease analyses have shown that these viruses contain a much smaller genome than baculoviruses of Lepidoptera. Third, while lepidopteran baculoviruses generally infect a wide variety of larval tissues, tropism of sawfly viruses is limited to the larval midgut and we wondered whether there is a relationship between tissue tropism and the small number of genes in the viral genome.

In order to reveal the gene content and organization in the genomes of sawfly viruses, the DNAs of two viruses were fully sequenced; namely those of the *Neodiprion lecontei* and the *Neodiprion sertifer* NPVs (NeleNPV and NeseNPV, respectively). In this brief review we highlight features that distinguish them from baculoviruses of Lepidoptera and show how these differences might play a role in the biology of the virus.

## Gene Content and Organization

The open reading frame (ORF) organization in the genome of NeleNPV is shown in **Figure 1**. The genome of this virus contains 89 predicted ORFs while that of NeseNPV contains 90 ORFs. These are the smallest baculovirus genomes sequenced so far. The G+C content is quite low of the order of 33%. Both viruses have 44 ORFs that are homologous to known baculovirus genes and additional 43 that are only shared with each other. Alignment of the two genomes reveals that they are mostly collinear but a noticeable difference is seen between the genes encoding polyhedrin (ORF1) and DNA binding protein (*dbp*, ORF14) **Figure 2**. This region has been termed nonsyntenic region (NSR) due to the lack of conserved synteny. Within the NSR, NeseNPV contains 15 ORFs that are absent from the genome of NeleNPV. The latter has eight ORFs within this region that are absent from the genome of NeseNPV. Also within the NSR, *nese18* and *nese19* encode proteins that are homologs and share 71.2% amino acid identity. They lack an identifiable promoter, are in the opposite orientation, and are flanked either by homologous repeated sequences (*hrs*) or direct repeated sequences (*drs*). The region also contains the NeseNPV methyltransferase homolog (NeseNPV ORF 5), which is absent from the genome of NeleNPV, and the inhibitor of apoptosis (*iap*) homologs (*nese17* and *Nele11*). It has also been noted that the closest BLAST matches of these ORFs were to insect proteins rather than to baculovirus proteins. The NeleNPV NELE11 had its highest match to an IAP from *Sprodoptera frugiperda* and the NeseNPV NESE5 had its highest match to a honeybee protein. The NeseNPV NESE17 was closest to a *Bombyx mori* IAP. The NSR also has homologs to trypsin-like proteases from insects (*nese7* and *nele6*). It has been hypothesized that the NSR probably arose by horizontal transfer of a cluster of ORFs from an insect host or hosts and only those genes that survived selection processes remained in the genomes.

## Specific Genes

### Inhibitors of Apoptosis

NeseNPV and NeleNPV contain iap genes that have their closest identity to insect-specific *iap* genes indicating that the viral genes were acquired by a transfer from an insect host. This hypothesis has been previously suggested for other baculoviruses. Both *iap* homologs (*nese17* and *nele11*) were closest to insect IAPs than to each other or to IAPs from other baculoviruses. The NESE17 contained one IAP repeat (BIRs) and a zinc finger while the NELE11 contained two BIRs and lacked a ring-finger motif.

### Trypsin-Like Serine Protease

Until the sequencing of the NeleNPV and the NeseNPV genomes, no trypsin-like serine protease has been reported in the genomes of baculoviruses. The predicted protein of *nele6* has the trypsin catalytic triad of histidine, aspartic, and serine and the six conserved cysteines in NESE7. Not surprisingly these ORFs share maximum identity with insect trypsin-like serine proteases. Phylogenetic analysis

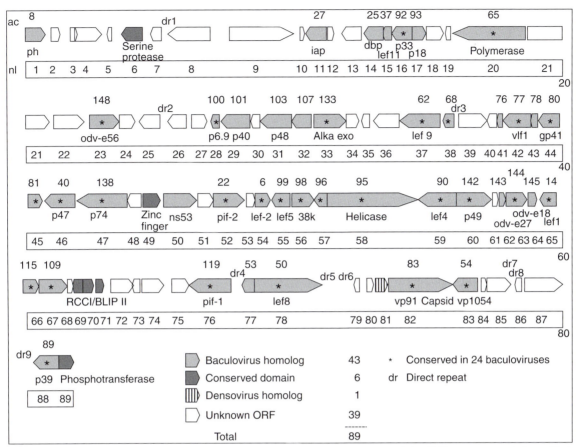

**Figure 1** A linearized map of the NeleNPV genome. Arrows indicate ORF orientation and the ORF numbers are indicated below the arrows. Reprinted from Lauzon HAM, Lucarotti CJ, Krell PJ, Feng Q, Retnakaran A, Arif BM (2004) Sequence and organization of the *Neodiprion lecontei* nucleopolyhedrovirus genome. *Journal of Virology* 78: 7023–7035.

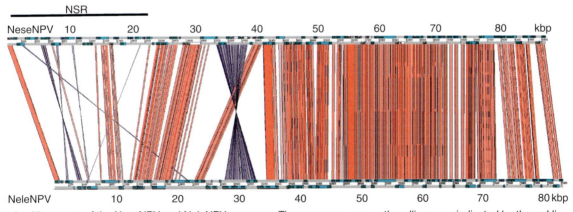

**Figure 2** Alignments of the NeseNPV and NeleNPV genomes. The genomes are mostly collinear, as indicated by the red lines. Blue lines show inversions. The nonsyntenic region (NSR) is indicated. Reprinted from Lauzon HAM, Garcia-Maruniak A, Zanotto PMA, *et al*. (2006) Genomic comparison of neodiprion sertifer and *Neodiprion lecontei* nucleopolyhedroviruses and identification of potential hymenopteran baculovirus-specific open reading frames. *Journal of General Virology* 87: 1477–1489.

showed that NELE6 and NESE7 grouped with the homolog from *Apis mellifera* and exhibited relatedness to other insect trypsin-like serine proteases. It was suggested that NeleNPV and NeseNPV acquired the gene from an ancestral baculovirus by a horizontal transfer from a host.

## Phosphotransferase

Phosphotransferases have been reported in eukaryotes and in limited numbers of eubacteria and archaeal organisms. To date, they have not been reported in viruses until the sequencing of the two sawfly viruses. This enzyme is involved in tRNA splicing. Phylogenetic analysis showed that the hymenopteran baculovirus homologs grouped with the eukaryotic enzymes. The closest identity of the nele89 was to a phosphotransferase from *D. melanogaster*.

### Regulator of Chromosome Condensation Proteins (RCC1)

These proteins play a role in chromosome condensation and are associated with chromatin or bind directly to DNA. Both NeleNPV and NeseNPV genomes contain three RCC1 homologs that are not found in other baculoviruses. In fact, these are the only viruses that have RCC1/BLIP domains mentioned in the InterPro taxonomic treatment for RCC1 proteins. Predicted proteins from *nese74* and *nele71* have strong matches to insect RCC1 proteins but they share very low amino acid identity with each other.

### Zinc-Finger Protein

Zinc fingers are domains known to bind nucleic acids and comprise a varied superfamily. The proteins encoded by *nese52* and *nele49* contain four C2H2 zinc-finger domains and share an amino acid identity of 35.6%. The C2H2 zinc-finger domains contain 25–30 amino acids including two cysteines and two histidines that are conserved in a motif of C2C12H3H. Top matches for the proteins from *nese52* and *nele49* were to *Anopheles gambiae* (36% aa identity) to *D. melanogaster* zinc-finger protein (36% aa identity), respectively.

### Membrane Fusion Proteins (MFPs)

Baculoviruses encode either a GP64 (group I NPVs) or an F protein which is a homolog of the *Lymantria dispar* NPV LD130 (group II and in some group I NPVs). A homolog to LD130 was also reported in the mosquito, *Culex nigripalpus*, NPV (CuniNPV), which is a baculovirus of mosquitoes. These proteins are a diagnostic feature of the budded virus phenotype and mediate the fusion of the virus to cell membrane and facilitate the release of nucleocapsids into the cell. The budded virus phenotype is responsible for the dissemination of infection within larval tissues and in tissue culture cells. These proteins have a distinct structure of a signal peptide, transmembrane domain, conserved cysteines, and a furin cleavage site. Thorough searches of all the potential ORFs in the genomes of NeseNPV and NeleNPV revealed the absence of an MFP. Several ORFs were found to contain transmembrane domains but lacked the other features of MFPs. It, therefore, appears that the budded virus phenotype may not play a role in the biology of sawfly baculoviruses and, indeed, this phenotype may not be needed by these viruses).

### Immediate Early Proteins (IE1)

IE1 is a main transcriptional transactivator that activates early and late baculovirus gene expression. The *ie-1* gene itself is transactivated by host cell factors. It has been known that this is an absolutely essential gene in baculoviruses that basically initiates the infection process within the cell. Thorough searches of ORF sequence and structure in the genomes of NeleNPV and NeseNPV did not expose the presence of this gene. It appears that these two viruses may not need an IE-1 or utilize a hitherto unknown protein to substitute its function.

### Auxiliary Genes

Normally, baculoviruses contain a number of auxiliary genes that are not needed for infectivity but appear to give the virus a selective advantage in nature. Practically none of the baculovirus genes classified as auxiliary have been found in the genome of NeleNPV and NeseNPV. Only the gene encoding an alkaline exonuclease (*alk-exo*) was located in these genomes.

### Densovirus Capsid Protein

Densoviruses contain a linear single-stranded 4–6 kbp DNA genome that can be either positive or negative sense. Interestingly, ORFs *nele81* and *nese83* had strong BLAST matches to densovirus structural proteins. Highest matches were to proteins 1–4 from *Casphalia extranea* densovirus to a capsid protein from *Bombyx mori* densovirus. It is interesting to note that densoviruses infect a variety of larval tissues; however, the *Bombyx mori* densovirus is restricted to the midgut. It is not possible at the present time to ascertain if these proteins are functional in the hymenopteran baculoviruses.

### Gene Parities and Phylogeny

Gene parity plots, have shown that genes are basically organized in a collinear manner in closely related baculovirus genomes. Many viruses have distinct rearrangements such as deletions, inversions, insertions, etc. that are revealed when comparing two genomes. Data have shown that in comparisons of the NeleNPV genome with those from representatives group I NPVs (AcMNPV) and group II NPVs (HaSNPV), a GV (PxGV) and CuniNPV, the gene order was not generally conserved. Except for a central portion of four genes (*lef-5, 38k, ac96, helicase*), there was no conservation of gene order (**Figure 3**).

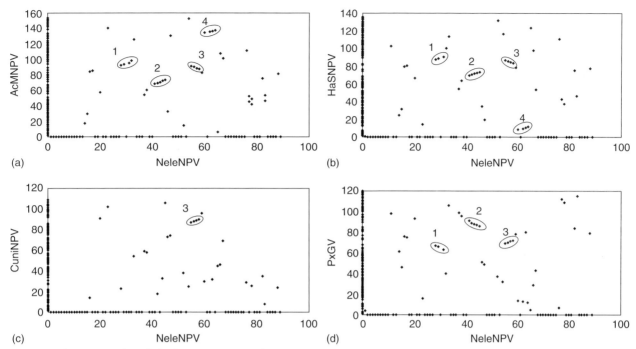

Figure 3  Gene parity plots. Pairwise comparisons of gene order of (a) NeleNPV/AcMNPV; (b) NeleNPV/HaSNPV; (c) NeleNPV/CuniNPV, and (d) NeleNPV/PxGV. The most conserved clusters have been circled. 1. *p6.9, p40, p48*; 2. *ac76, vlf-1, ac78, gp41*; 3. *lef-5,38k, ac96, helicase*; 4. *p49, odv-ec27, ac145*. Reprinted from Lauzon HAM, Lucarotti CJ, Krell PJ, Feng Q, Retnakaran A, Arif BM (2004) Sequence and organization of the *Neodiprion lecontei* nucleopolyhedrovirus genome. *Journal of Virology* 78: 7023–7035.

Phylogenies based on single genes have generally been found to be less robust than those based on combined sequences of several shared genes. With the complete sequencing of two hymenopteran baculoviruses, the total number of conserved genes among all baculoviruses sequenced so far is at least 29. A maximum parsimony tree was generated by using a combined data set of the 29 conserved proteins in 24 sequenced baculovirus genomes (**Figure 4**). The tree demonstrates that the hymenopteran and dipteran baculoviruses do not fit into the NPV groups I or II and, in fact, do not fit into the present classification genera of baculoviruses. This led us to suggest a new classification scheme to accommodate the dipteran and hymenopteran baculoviruses and any other future baculovirus. The authors suggested adding two new genera to the classification system and changing the names of the *Baculoviridae* genera to *Alphabaculovirus, Betabaculovirus, Gammabaculovirus,* and *Deltabaculovirus*. This system will accommodate the hymenopteran and dipteran baculoviruses and has the flexibility to accept more genera and is consistent with other groups of viruses such as the *Herpesviridae* and *Entomopoxvirinae*.

## DNA Repeat Regions

Most baculovirus genomes contain what has been described as homologous regions (*hrs*) that consist of multiple tandem repeats of perfect or imperfect palindromes within direct repeated sequences. They have been implicated as origins of DNA replication and as enhancers of gene expression. A major difference between NeleNPV and NeseNPV is in the structure of the repeated sequences. The NeseNPV genome contains baculoviral *hrs* interdispersed in the genome as well as direct repeats (*drs*). The latter repeats do not bear resemblance to baculoviral *hrs*. NeleNPV genome contains only *drs*. Most of the repeated regions are located in the NSR of both genomes. Within the NSR, the NeseNPV had five *hrs* and two *drs* and NeleNPV had one *dr* and three ORFs that contained repeats (*nele2, 7,* and *8*).

## Biology of the Virus

The tissue tropism of sawfly viruses is quite restricted and they appear to replicate only in midgut cells. As a defense mechanism against invaders, the insect sloughs off infected cells resulting in infectious diarrhea. Since these insects are gregarious in nature (**Figure 5**), the infected sloughed-off cells usually result in rapid dissemination of the virus within the larval population causing collapse within a few days to a week. The infection process within midgut cells is still not totally clear. For example, it is not known if progeny virus spreads from one cell to another. Since the virus does not contain a MFP, it is unlikely that a

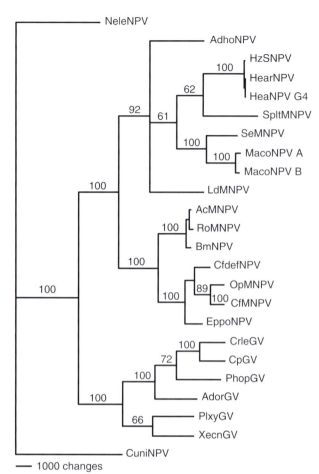

**Figure 4** A phylogenetic map based on a data set of 29 conserved baculovirus genes. The dipteran CuniNPV and the hymenopteran NeleNPV do not branch with NPV of Lepidoptera.

**Figure 5** The gregarious nature of the red-headed pine sawfly, *Neodiprion lecontei*.

budded virus phenotype is produced and spreads infection from one cell to another. It is interesting to note that these viruses encode a GP41 protein that has been termed an egress protein, which allows the virus to exit from an infected cell. It is conceivable that these viruses spread by crossing over to adjacent cells but this is yet to be demonstrated. The need for a complement of auxiliary genes may also not be necessary in this case. For example, expression of chitinase and v-cathepsin at the end of baculovirus replication cycle causes dissolution of the integument and melting of the insect. Restricted replication to midgut cells precludes the need for these two enzymes. Baculoviruses also encode an ecdysteroid UDP-glucosyltransferase that conjugates the hormone ecdysone with sugars and renders it inactive. In this way the virus prevents larval molting and prolongs the life within a certain instar, resulting in a more prolific replication of the virus in various larval tissues. Again, because of their midgut restriction, hymenopteran viruses do not need to inactivate ecdysone in the larval hemolymph.

*See also:* Baculoviruses: Molecular Biology of Granuloviruses; Baculoviruses: Molecular Biology of Nucleopolyhedroviruses; Insect Pest Control by Viruses.

## Further Reading

Afonso CL, Tulman ER, Lu Z, *et al.* (2001) Genome sequence of a baculovirus pathogenic for *Culex nigripalpus*. *Journal of Virology* 75: 11157–11165.

Büchen-Osmond C (ed.) (2003) *Densovirus*. In: *ICTVdB – The Universal Virus Database*, version 3, chapter 00.050.2.01. ICTVdB Management, Columbia University, New York, USA. http://www.ncbi.nlm.nih.gov/ICTVdb/ICTVdB/index.htm (accessed July 2007).

Cunningham JC and Entwistle PF (1981) Control of sawflies by baculoviruses. In: Burges HD (ed.) *Microbial Control of Pests and Plant Diseases 1970–1981*, pp. 379–408. Academic Press: New York.

Garcia-Maruniak A, Maruniak JE, Zanotto PMA, *et al.* (2004) Sequence analysis of the genome of the *Neodiprion sertifer* nucleopolyhedrovirus. *Journal of Virology* 78: 7036–7051.

Guarino LA, Gonzales MA, and Summers MD (1986) Complete sequence and enhancer function of the homologous DNA regions of *Autographa californica* nuclear polyhedrosis virus. *Journal of Virology* 60: 224–229.

Herniou E, Olszewski JA, O'Reilly DR, and Cory JS (2004) Ancient coevolution of baculoviruses and their insect hosts. *Journal of Virology* 78: 3244–3251.

Herniou EA, Olszewski JA, Cory JS, and O'Reilly DR (2003) The genome sequence and evolution of baculoviruses. *Annual Review of Entomology* 48: 211–234.

Hughes AL and Friedman R (2003) Genome-wide survey for genes horizontally transferred from cellular organisms to baculoviruses. *Molecular Biology and Evolution* 20: 979–987.

Kool M, Ahrens CH, Vlak JM, and Rhormann GF (1995) Replication of baculovirus DNA. *Journal of General Virology* 76: 2103–2118.

Kuzio J, Pearson MN, Harwood SH, *et al.* (1999) Sequence and analysis of the genome of a baculovirus. *Virology* 253: 17–34.

Lauzon HAM, Garcia-Maruniak A, Zanotto PMA, *et al.* (2006) Genomic comparison of neodiprion sertifer and *Neodiprion lecontei* nucleopolyhedroviruses and identification of potential hymenopteran baculovirus-specific open reading frames. *Journal of General Virology* 87: 1477–1489.

Lauzon HAM, Lucarotti CJ, Krell PJ, Feng Q, Retnakaran A, and Arif BM (2004) Sequence and organization of the *Neodiprion lecontei* nucleopolyhedrovirus genome. *Journal of Virology* 78: 7023–7035.

Moreau G, Lucarotti CJ, Kettela EG, *et al.* (2005) Aerial application of nucleopolyhedrovirus induces decline in increasing and peaking populations of *Neodiprion abietis*. *Biological Control* 33: 65–73.

Ohtsubo M, Okazaki H, and Nishimoto T (1989) The RCC1 protein, a regulator for the onset of chromosome condensation locates in the nucleus and binds to DNA. *Journal of Cell Biology* 109: 1389–1397.

Pearson MN and Rohrmann GF (2002) Transfer, incorporation, and substitution of envelope fusion proteins among members of the *Baculoviridae*, *Orthomyxoviridae*, and *Metaviridae* (insect retrovirus) families. *Journal of Virology* 76: 5301–5304.

Theilmann D and Stewart S (1991) Identification and characterization of the IE-1 gene of *Orgyia pseudotsugata* multicapsid nuclear polyhedrosis virus. *Virology* 180: 492–508.

# Baculoviruses: Apoptosis Inhibitors

**R J Clem,** Kansas State University, Manhattan, KS, USA

© 2008 Elsevier Ltd. All rights reserved.

## Glossary

**AcMNPV** Autographa californica multiple nucleopolyhedrovirus, the most studied baculovirus.

**Annihilator mutant** The first *p35* mutant of AcMNPV described.

**Apoptosis** A type of programmed cell death commonly observed in multicellular animals that involves cell shrinkage, plasma membrane blebbing, nuclear fragmentation, and chromatin degradation. Important in many different physiological and pathological situations.

**Apoptosome** A large protein complex consisting of multiple copies of an initiator caspase (DRONC in *Drosophila*, caspase-9 in mammals) bound to multiple copies of adaptor protein (DARK in *Drosophila*, Apaf-1 in mammals). In mammals, cytochrome *c* is also bound to Apaf-1. Apoptosome formation allows for activation of the initiator caspase.

**BIR domain** Baculovirus IAP repeat, the signature domain of IAP proteins.

**Blebbing** A term used to describe the outward bulging of plasma membrane often observed during apoptosis. Blebbing results from changes in the cytoskeleton that occur during apoptosis, and eventually leads to the cell breaking up into membrane-bound fragments called apoptotic bodies.

**Caspase** A family of cysteine proteases that are centrally involved in the execution of apoptosis.

**Effector caspase** One of several caspases that are activated by initiator caspases. Responsible for cleavage of numerous substrates that lead directly to apoptosis.

***Iap*** inhibitor of apoptosis gene.

**Initiator caspase** One of the several caspases that are the first to be activated following an apoptotic signal. Activate effector caspases.

**RING domain** A type of zinc finger domain found in many different proteins, including some IAP proteins. Important in E3 ubiquitin ligase activity.

## Introduction

In order to successfully replicate and survive, viruses must evade the immune responses of their hosts. One aspect of the innate immune response that was largely ignored by virologists until relatively recently is apoptosis. Apoptosis is a type of programmed cell death that is important in the clearance of unwanted or potentially harmful cells, including pathogen-infected cells. We now appreciate that many viruses carry genes that can inhibit apoptosis, indicating that host cell apoptosis is an important evolutionary pressure that viruses have had to overcome. Members of the virus family *Baculoviridae* were some of the first viruses that were shown to carry apoptosis-inhibiting genes, and these insect viruses continue to provide an excellent model for understanding the role of apoptosis in antiviral defense in both invertebrates and higher animals.

## Apoptosis Pathways in Insects

The ability to commit suicide appears to be a universal characteristic of all cells, whether they be prokaryotic, single-celled eukaryotic, plant, or animal cells. As it is strictly defined morphologically, the specific type of cell death known as apoptosis is confined to metazoans (multicellular animals), and is just one type of programmed cell death found in these organisms. The biochemical pathways that lead to apoptosis are more or less conserved in metazoans, and can be triggered by a wide variety of different stimuli. While a myriad of interconnected pathways are involved in transmitting the death signal to the core apoptotic machinery, the core machinery itself is well conserved. What follows is a highly simplified description of the basic apoptotic pathway in the only insect where it has been examined in detail, *Drosophila melanogaster*. For more in-depth information, the reader is encouraged to consult recent reviews on the subject.

The engine that runs the apoptotic core machinery consists of a family of proteases called caspases (**Figure 1**). Caspases are cysteine proteases that cleave their target substrates at highly specific recognition sites. Cleavage involves recognition and proteolysis following specific aspartate residues. The amino acids immediately surrounding the aspartate cleavage site are crucial in determining caspase substrate specificity. Caspases are present in cells as pro-enzymes which need to undergo proteolytic cleavage to become active. Caspases can be roughly divided into two groups, the initiator caspases and the effector caspases. Initiator caspases are the first to be activated following an apoptotic stimulus. In most cases, caspase activation involves caspase-mediated cleavage, whether it is autocatalytic, in the case of initiator caspases, or mediated by other caspases, as with effector caspases. An apoptotic stimulus causes the aggregation of adaptor proteins, which bind to pro-initiator caspases and cause them to self-associate and autoactivate. Once activated, initiator caspases then recognize and cleave pro-effector caspases. Activated effector caspases then cleave a number of cellular proteins that directly cause the morphological changes associated with apoptosis, including plasma membrane blebbing, nuclear condensation and fragmentation, chromatin degradation, and cellular disintegration.

The existence of proteases that are capable of autoactivation and whose activity leads to death is obviously a dangerous situation for the cell and must be tightly regulated. There are many checks and balances in the pathways that lead to caspase activation, but the main proteins that directly oppose caspase activity are the Inhibitor of apoptosis (IAP) proteins, some of which are able to directly bind and inhibit the activity of initiator and effector caspases. In *Drosophila*, one IAP protein in particular, DIAP1, is crucial for cell survival, as downregulation of DIAP1 causes rapid, spontaneous apoptosis (**Figure 1**). The main target of DIAP1 appears to be an initiator caspase called DRONC. DRONC undergoes continuous autoactivation in *Drosophila* cells. DRONC activation requires association with a protein called DARK, which is homologous to the Apaf-1 protein in mammals. Together, DRONC and DARK form a structure called the apoptosome. In mammals, apoptosome formation requires the release of cytochrome *c* from mitochondria, and binding of cytochrome *c* to DARK. In *Drosophila*, however, cytochrome *c* does not appear to play a role in formation of the apoptosome. DIAP1 can directly bind to activated DRONC and cause DRONC to be ubiquitinated and degraded by proteasomes. Thus, as long as DIAP1 levels remain constant, the continuous DRONC activation that occurs in normal cells is kept in check.

In order for *Drosophila* cells to die, DIAP1 inhibition of DRONC must be removed. This is often accomplished by upregulation of proteins which antagonize the DIAP1-DRONC interaction, called Reaper, Hid, Grim, and Sickle (RGH proteins). RGH proteins bind to DIAP1 at the same site as DRONC, allowing active DRONC to accumulate, which leads to apoptosis through the activation of effector caspases. However, signals that disrupt

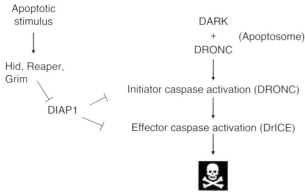

**Figure 1** A simplified apoptosis pathway in *Drosophila*. Unlike mammalian cells, where it is thought that initiator caspases are not activated until the cell receives a death signal, in *Drosophila* cells the initiator caspase DRONC appears to spontaneously autoactivate all of the time, even in unstimulated cells. In order to become activated, DRONC must bind to the adapter protein DARK, forming a structure known as the apoptosome. DRONC activity is normally kept at very low levels because of the ability of the DIAP1 protein to bind and ubiquitinate activated DRONC, resulting in DRONC degradation by the proteasome. Interruption of the DIAP1/DRONC interaction by IAP antagonists such as Hid, Reaper, or Grim results in accumulation of activated DRONC, which leads to cleavage and activation of effector caspases such as DrICE. Effector caspase activity then leads to death of the cell.

DIAP1 expression, such as metabolic inhibitors or stress signals, also trigger spontaneous apoptosis in *Drosophila* cells by allowing DRONC activation.

## Induction of Apoptosis by Baculoviruses

The first realization that a baculovirus could stimulate apoptosis arose from characterization of a spontaneous mutant of the baculovirus autographa californica multiple nucleopolyhedrovirus (AcMNPV). This mutant, which was called the annihilator mutant, causes rapid and widespread apoptosis in Sf-21 cells, a cell line derived from the lepidopteran insect *Spodoptera frugiperda* (**Figure 2**). Mapping of the mutation that caused this apoptotic phenotype revealed that the mutation was a 754 bp deletion, which resulted in premature truncation of an open reading frame (ORF) in the AcMNPV genome called the 35K ORF. This gene, which was renamed *p35*, turned out to encode a highly effective inhibitor of effector caspases from not only insects but also other organisms including nematodes and mammals.

Infection of Sf-21 cells with wild-type AcMNPV does not result in apoptosis, but rather in passive cell lysis. Permissive baculovirus infections are most often lytic, but whether cell lysis occurs by apoptosis or by a passive mechanism is a complicated matter that is determined by several factors. In a typical permissive infection, such as Sf-21 cells infected with wild-type AcMNPV, the infected cells produce large amounts of budded virus over the first day or so, and then switch to producing occluded virus particles. The cells usually survive for several days before lysing. By this time, the cell has been largely subverted by the virus, and many cellular processes such as host mRNA and protein synthesis have long since declined or ceased. The mechanism of death does not appear to be apoptotic, but rather passive lysis, presumably due to impaired cellular functions.

However, things are not always so simple. Even in a normal infection, some of the initial stages of apoptosis can occur. For example, in Sf-21 cells infected with AcMNPV, a transient blebbing of the plasma membrane similar to what is observed in early apoptosis is observed at approximately 9–12 h post infection, and effector

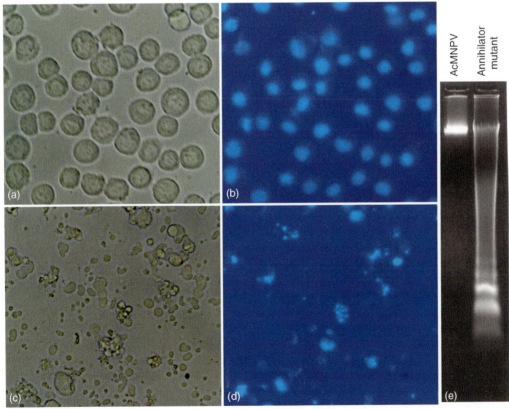

**Figure 2** The apoptotic morphology of Sf-21 cells infected with the annihilator mutant of AcMNPV, a mutant virus lacking the *p35* gene. In panels (a)–(d), cells were stained with a fluorescent dye that binds to DNA and viewed by phase contrast ((a) and (c)) or by fluorescence microscopy ((b) and (d)). Panels (a, b) are uninfected Sf-21 cells, while panels (c) and (d) are Sf-21 cells infected with the annihilator mutant. Note the extensive cell blebbing, chromatin condensation, and nuclear fragmentation in the annihilator-infected cells. Panel (e) shows the oligonucleosomal DNA fragmentation associated with apoptosis of Sf-21 cells infected with the annihilator mutant. Cellular DNA was isolated from infected cells and analyzed by agarose gel electrophoresis.

caspases become activated. Effector caspase activity is kept low, however, by the P35 protein.

In some cases, full-blown apoptosis can occur even though the virus encodes one or more functional apoptosis inhibitors. Whether or not apoptosis occurs depends on the particular baculovirus and cell line, but is usually observed in nonpermissive or semipermissive situations. One example is seen with wild-type AcMNPV, which causes apoptosis in nonpermissive *Spodoptera littoralis* SL2 cells and *Choristoneura fumiferana* CF-203 cells, despite the presence of *p35*. In the case of SL-2 cells infected with AcMNPV, apoptosis has been shown to be due to low expression of the *p35* gene, and apoptosis can be prevented by expressing *p35* using a stronger promoter. The complicated nature of the interactions between virus and host is further illustrated by the fact that infection of another *S. littoralis* cell line results in extensive cytopathic effect, but little apoptosis. Also, nonpermissive infection does not always result in apoptosis. Infection of *Drosophila* cell lines with wild-type AcMNPV does not result in virus replication, and apoptosis is not observed. However, apoptosis does occur in *Drosophila* cells if the infecting strain of AcMNPV lacks the *p35* gene. Thus, the response of cells to baculovirus infection varies widely, depending on the interaction between the gene products expressed by the virus and the particular cellular environment.

The mechanisms by which baculoviruses induce apoptosis are incompletely understood, and our knowledge is entirely based on studies with AcMNPV in Sf-21 cells. What is clear from these studies is that mere attachment of the virus to a cellular receptor is not sufficient to induce apoptosis. Rather, the virus must enter the cell and express at least some of its early genes. The viral transcriptional activator *ie-1* has been reported to be capable of inducing apoptosis when expressed in Sf-21 cells, and another transactivator, *pe-38*, can enhance the ability of *ie-1* to stimulate apoptosis, although the ability of these genes to cause apoptosis is somewhat controversial. Even so, expression of these transactivators may affect the transcription of cellular genes that lead to triggering of apoptosis, or at least making the cells more sensitive to apoptotic stimuli.

While these early viral gene products may play a role in apoptosis, full induction of apoptosis appears to also require late events in the virus replication cycle. The transition from the early to late phases of gene expression involves several inter-related processes including the onset of viral DNA replication, a block in the cell cycle at the G2 phase, the shut-off of host mRNA and protein synthesis, and the initiation of transcription of viral late genes by a viral RNA polymerase. Evidence exists implicating each of these events in triggering apoptosis, so it appears that there are multiple stimuli that contribute to the onset of the apoptotic cascade during baculovirus infection. This probably explains why most DNA viruses need to carry apoptosis inhibitors in their genomes:

apoptosis may be an unavoidable consequence of DNA virus replication.

## Baculovirus Anti-Apoptotic Genes: P35 and Its Relatives

The P35 protein is a remarkable molecule. This baculovirus protein is the most widely acting apoptosis inhibitor known; it can inhibit apoptosis in nematodes, insects, and mammals. As such, P35 has proven to be a highly useful tool to investigators studying apoptosis in a wide variety of systems ranging from developmental events in *Drosophila* to immune system function in mice. Thanks to structural biology, we know a great deal about how P35 functions as a stoichiometric caspase inhibitor.

The crystal structure of P35 was published in 1999 and revealed a teapot-like structure with a body, neck, and handle (**Figure 3**). The most interesting aspect of the structure was the handle, or reactive site loop, which contains a caspase cleavage site at its apex. Further studies demonstrated that cleavage at this site by caspases results in a conformational shift in the P35 protein, causing the N-terminus of P35 to interact with the caspase active site, and part of the reactive site loop to swing down and interact with a β-sheet in the body of the protein. A thioester bond is formed between P35 and the caspase, resulting in the P35 cleavage products becoming covalently

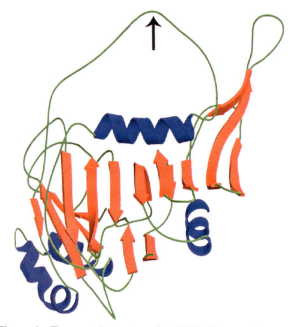

**Figure 3** The crystal structure of AcMNPV P35 reveals a teapot-like structure, with a solvent-exposed loop that contains the caspase cleavage site (arrow). Reproduced from Clarke TE and Clem RJ (2003) Insect defenses against virus infection: The role of apoptosis. *International Reviews of Immunology* 22: 401–424, figure 3, with permission from Taylor & Francis.

bound to the caspase, permanently inactivating the caspase. While somewhat reminiscent of the mechanism used by serpins, this represents a unique mechanism of protease inhibition, and one that may have application in designing novel therapeutic drugs.

Although it is often referred to as a pan-caspase inhibitor, AcMNPV P35 actually prefers to inhibit effector caspases, and has little inhibitory activity against most initiator caspases. However, a homolog of P35 found in spodoptera littoralis M nucleopolyhedrovirus called P49 is able to inhibit both effector and initiator caspases. The mechanistic basis for this difference in specificity between P35 and P49 is not known at this time. While other baculovirus P35 homologs have not been characterized biochemically, recently the first homolog of p35 outside of the baculoviruses was characterized in the entomopoxvirus of *Amsacta moorei*. This gene, called p33, encodes a protein with low but significant homology to P35 and P49. P33 appears to function more similarly to P35, as it also preferentially inhibits effector caspases.

## P35 Proteins: More Than Just Caspase Inhibitors?

While the ability of P35 to inhibit caspases is well documented, evidence also exists that hints at additional functions for this protein. P35 has been reported to bind to RNA polymerase II and have effects on gene transcription in both insect and human cells. P35 has also been reported to promote transformation of mouse embryo fibroblasts by a mechanism that does not appear to be due to inhibiting apoptosis. Similarly, P35 expression is able to stimulate translation arrest in an AcMNPV-infected cell line, but chemical caspase inhibitors do not. Finally, recent unpublished data indicate that *p35* mutant virus is less efficient than wild-type virus at initiating infection in TN-368 cells from *Trichoplusia ni*, which do not undergo apoptosis when infected with either virus. Together, these observations suggest that P35 may do more than just inhibit caspases, although exactly what else it does remains unclear.

## The IAPs: Apoptosis Inhibitors Found in Both Viral and Cellular Genomes

Shortly after the discovery of *p35*, a genetic complementation assay was used in an attempt to identify *p35* homologs in two other baculoviruses, orgyia pseudotsugata M nucleopolyhedrovirus (OpMNPV) and cydia pomonella granulovirus (CpGV). While this assay did identify genes in both OpMNPV and CpGV that could complement the *p35* mutation in AcMNPV (i.e., inhibit AcMNPV-induced apoptosis), the complementing genes turned out not to be homologs of *p35*. They were, however, related to each other. These genes were named *inhibitor of apoptosis* (*iap*) genes, and the genes from OpMNPV and CpGV were named Op-*iap* and Cp-*iap*, respectively.

The sequencing of baculovirus genomes (more than 30 to date) has revealed that *iap* genes are widespread among the baculoviruses. Almost all baculoviruses carry at least one, and usually more, *iap* homologs. This includes viruses that also carry *p35* homologs, such as AcMNPV. Phylogenetic analysis has classified the baculovirus *iap* genes into five groups, *iap*-1 through *iap*-5. The *iap*-5 group has so far been found only in granuloviruses, while the other four groups are present in both NPVs and GVs. None of the viruses sequenced to date have more than one member of each family.

Many of the baculovirus *iap* genes do not seem to have anti-apoptotic function. While less than half of known baculovirus *iap* homologs have been tested, most of the ones that inhibit apoptosis are found in the *iap*-3 group, including the original Op-*iap* and Cp-*iap*. For example, the Ac-*iap* gene does not have the ability to inhibit apoptosis, at least not in any of the situations where it has been tested, and a study of four *iap* homologs from epiphyas postvittana nucleopolyhedrovirus demonstrated that only two of the four genes could inhibit apoptosis, and another one was only able to delay apoptosis. The question of what role these 'nonfunctional' *iap* homologs play during a viral infection is an interesting one. As viruses do not normally retain genes that fail to confer a selective advantage, presumably they have some type of function. It is possible that they may inhibit apoptosis in specialized tissues or insect hosts that have not been tested yet, or they may have other, as yet unknown functions.

The distinguishing feature of *iap* genes is that the proteins they encode contain between one and three copies of a type of domain known as a BIR (baculovirus IAP repeat) (**Figure 4**). BIR domains are approximately 70 amino acids in length and are always found at the N-terminus of IAP proteins. They contain several highly conserved residues, including cysteines and histidines. Based on this, it was predicted that BIR domains were a type of zinc finger domain, and structural studies have borne this out, with each BIR coordinating one atom of zinc. Many, but not all, IAP proteins also contain a second type of zinc finger domain at the C-terminus known as a RING domain. RING domains, which are also found in many other types of proteins besides IAPs, have been shown to be involved in ubiquitin ligase activity. In fact, in every case to date where a RING-containing protein has been tested, including two baculovirus IAPs, they have been shown to possess this activity.

Expression of Op-*iap* or Cp-*iap* inhibits apoptosis triggered by various stimuli, in both lepidopteran and mammalian cells. For some unknown reason, Op-IAP does not inhibit apoptosis in *Drosophila* cells, and actually

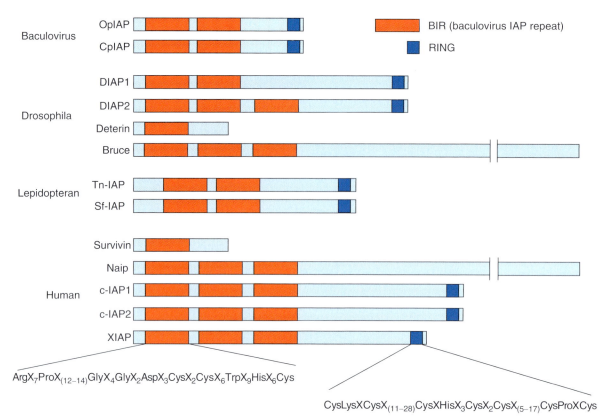

**Figure 4** Schematic representation of IAP proteins from select baculoviruses, insects, and humans. The consensus sequences of BIR and RING domains are shown below the schema.

has been shown to sensitize *Drosophila* S2 cells to AcMNPV-induced apoptosis. However, in most cells, Op-IAP is an efficient apoptosis inhibitor. Exactly how Op-IAP functions is not known, and may be different in insect versus mammalian cells. The available evidence indicates that Op-IAP does not act by directly binding to caspases and inhibiting their activity, but may simply be a sink for IAP antagonists such as Reaper, Hid, and Grim in insects or Smac/DIABLO in mammals. However, mutants of Op-IAP exist that still appear to bind to Hid normally, but fail to inhibit Hid-induced apoptosis, indicating that binding to Hid is necessary but not sufficient to inhibit apoptosis induced by Hid. Thus, questions remain about how baculovirus IAP proteins inhibit apoptosis.

Genes with homology to the baculovirus *iap* genes are found in the genomes of a few other invertebrate viruses such as entomopoxviruses, iridoviruses, and African swine fever virus (which has an obligate stage in ticks). Homologs of *iap* genes are also found in cellular genomes including those of yeast, nematodes, insects, and mammals. The *iap* genes in yeast and nematodes appear to be involved in regulation of cytokinesis, while those in higher animals are primarily involved in regulating apoptosis. However, the mammalian *survivin* gene, which is most similar in structure to the yeast and nematode *iap* genes, also regulates cytokinesis. Other mammalian IAP proteins also appear to act in other types of signal transduction pathways, suggesting that IAPs are more pleiotropic in function than their name implies.

## Apoptosis as an Antiviral Immune Response

Like all other living organisms, insects are constantly challenged by bacteria, fungi, and viruses. In vertebrate animals, immunity can be broken down into innate and adaptive immunity. Insects do not possess the adaptive immune response that is characteristic of vertebrates including antibodies and T cells. That does not mean, however, that insects are defenseless against invading microorganisms. Insects have robust innate immune systems that are in many ways similar to the innate immune mechanisms found in vertebrates. Over the past 10–15 years, enormous strides have been made in understanding the molecular basis for insect innate immunity against bacteria and fungi. Despite these advances, the understanding of how insects defend themselves against viruses has lagged behind. Work with baculoviruses has demonstrated that apoptosis can serve as an innate immune mechanism in insects.

As soon as the annihilator mutant was discovered, it was suspected that apoptosis would have negative consequences for baculovirus replication. This was because the mutant virus replicated poorly in Sf-21 cells, which became apoptotic when infected with the mutant, but the mutant replicated normally in TN-368 cells, which did not die by apoptosis. Growth curve experiments revealed that the replication of AcMNPV mutants that lack the *p35* gene is reduced by 100-fold or more (depending on the multiplicity of infection) in Sf-21 cells, but not in TN-368 cells. In addition, viral late and very late gene expression was substantially reduced in Sf-21 cells infected with *p35* mutants, but was normal in TN-368 cells.

These cell culture results predicted that *p35* mutant viruses might also be less able to infect insect larvae. Indeed, successful infection of *S. frugiperda* or *Spodoptera exigua* larvae by feeding of occluded virus (the natural route of infection) or by intrahemocoelic injection of budded virus requires much higher doses of *p35* mutant virus than control viruses containing *p35*, and analysis of infected tissues reveals widespread apoptosis in infected tissues of the *p35* mutant-infected insects. However, *T. ni* larvae are equally susceptible to *p35* mutant and wild-type infection, and apoptosis is not observed in *p35* mutant-infected tissues. Additional evidence that the defect in infectivity is due to apoptosis comes from the demonstration that insertion of a different anti-apoptotic gene, Cp-*iap*, into a *p35* mutant background rescues the infectivity defect in *S. frugiperda* larvae.

The importance of apoptosis in antiviral immunity in insects is only beginning to be understood, and many questions remain. For example, apoptosis has thus far been examined in only a small number of virus–host combinations. Also, there is evidence suggesting that apoptosis can be more effective when combined with other types of immune responses, but almost nothing is known about these other responses. Whether certain tissues within the insect are more important than others is also not known. What is apparent, however, is that the apoptosis inhibitors carried by baculoviruses are important virulence factors, and that apoptosis probably plays a role in determining the host range of these viruses.

*See also:* Apoptosis and Virus Infection; Baculoviruses: General Features; Baculoviruses: Molecular Biology of Granuloviruses; Baculoviruses: Molecular Biology of Nucleopolyhedroviruses; Baculoviruses: Pathogenesis; Innate Immunity: Defeating; Innate Immunity: Introduction; Polydnaviruses: Abrogation of Invertebrate Immune Systems.

## Further Reading

Clarke TE and Clem RJ (2003) Insect defenses against virus infection: The role of apoptosis. *International Reviews of Immunology* 22: 401–424.

Clem RJ (2001) Baculoviruses and apoptosis: The good, the bad, and the ugly. *Cell Death Differentiation* 8: 137–143.

Clem RJ (2005) The role of apoptosis in defense against baculovirus infection in insects. *Current Topics in Microbiology and Immunology* 289: 113–130.

Clem RJ, Fechheimer M, and Miller LK (1991) Prevention of apoptosis by a baculovirus gene during infection of insect cells. *Science* 254: 1388–1390.

Fisher AJ, Cruz W, Zoog SJ, Schneider CL, and Friesen PD (1999) Crystal structure of baculovirus P35: Role of a novel reactive site loop in apoptotic caspase inhibition. *EMBO Journal* 18: 2031–2039.

Hay BA and Guo M (2006) Caspase-dependent cell death in *Drosophila*. *Annual Review of Cell and Developmental Biology* 22: 623–650.

Kornbluth S and White K (2005) Apoptosis in *Drosophila*: Neither fish nor fowl (nor man, nor worm). *Journal of Cell Science* 118: 1779–1787.

Okano K, Vanarsdall AL, Mikhailov VS, and Rohrmann GF (2006) Conserved molecular systems of the *Baculoviridae*. *Virology* 344: 77–87.

Salvesen GS and Duckett CS (2002) IAP proteins: Blocking the road to death's door. *Nature Reviews. Molecular Cell Biology* 3: 401–410.

Wang L and Ligoxygakis P (2006) Pathogen recognition and signaling in the *Drosophila* innate immune response. *Immunobiology* 211: 251–261.

Xu G, Cirilli M, Huang Y, *et al.* (2001) Covalent inhibition revealed by the crystal structure of the caspase-8/p35 complex. *Nature* 410: 494–497.

# Baculoviruses: Expression Vector

**F J Haines,** Oxford Brookes University, Oxford, UK
**R D Possee,** NERC Institute of Virology and Environmental Microbiology, Oxford, UK
**L A King,** Oxford Brookes University, Oxford, UK

© 2008 Elsevier Ltd. All rights reserved.

## Glossary

**Bacterial artifical chromosome (BAC)** An artificial chromosome based on the bacterial fertility (F) plasmid which permits replication of large DNA molecules within bacterial cells.

**Co-transfection** Introduction of both transfer vector and viral DNA into eukaryotic cells to mediate

homologous recombination and ultimately recombinant virus production.
**Homologous recombination** The process by which the exchange or replacement of a DNA region occurs between homologous flanking regions of genetic material.

## Introduction

Baculoviruses are a diverse group of insect-specific viruses, predominantly infecting insect larvae of the order Lepidoptera. By far the most widely studied member of this family is autographa californica multiple nucleopolyhedrovirus (AcMNPV) for which the complete genome sequence has been determined. AcMNPV has a circular, double-stranded, super-coiled DNA genome of *c.* 130 kbp packaged in a rod-shaped nucleocapsid. These nucleocapsids can be extended lengthways and thus the virus genome can effectively accommodate large insertions of foreign DNA. Such insertions of foreign genes into the AcMNPV genome have resulted in production of baculovirus expression vectors, that is, recombinant viruses genetically modified to contain a foreign gene of interest, which can be expressed in insect cells under the control of a baculovirus gene promoter. The baculovirus expression vector system is now one of the most popular methods of protein production, largely due to the high yields of good-quality protein that are obtained with a wide variety of different protein types.

AcMNPV has a biphasic replication cycle resulting in the production of two virus phenotypes: budded virus (BV) and occlusion-derived virus (ODV) (**Figure 1**). BVs consist of a single, rod-shaped nucleocapsid enveloped in a host-derived membrane, enriched in a virally encoded membrane fusion protein, GP64, which is incorporated into the BV particle during virus budding and release. The budded form of the virus is responsible for the cell-to-cell transmission of the virus both *in vivo* and *in vitro*. During the later stages of infection, large numbers of occlusion bodies (OBs) or polyhedra are formed. The major component of the OB matrix is polyhedrin, a virus-encoded protein produced by the powerful transcriptional activity of the polyhedrin (*polh*) gene promoter. This protein protects the virus in the environment and allows virus particles to survive outside of their natural host.

## Insect Cell Culture

The traditional cell lines for baculovirus-mediated expression studies are Sf-21 and Sf-9 cells. Both cell lines were originally derived from the pupal ovarian cells of *Spodoptera frugiperda* (fall army worm) termed IPLB-Sf-21 cells, with Sf-9 cells (IPLB-Sf-21-AE) being a clonal isolate of Sf-21 cells. High-Five cells (BTI-TN-5B14), a clonal isolate from *Trichoplusia ni* (cabbage looper) cell lines, are also used for expression of recombinant proteins using the baculovirus expression system and have been found to produce greater levels of recombinant protein compared to Sf-21 and Sf-9 cells in some circumstances. In general, Sf-21 or Sf-9 cell lines are used for co-transfections, virus amplification, and plaque assays while *T. ni* cell lines are employed for protein expression.

Insect cell lines are generally considered simple to sustain compared to mammalian cell lines and can be maintained as either suspension cultures, in shake or stirred flasks, or in monolayer cultures in T-flasks or culture dishes. Most insect cell culture media utilize a phosphate buffering system rather than the carbonate-based buffers commonly used for mammalian cell lines, negating the requirement for $CO_2$ incubators. The range for growth and infection of most cultured insect cells is between 25 and 30 °C, although 28 °C is generally considered the optimal growth temperature.

Certain insect cells, for example, Sf-21 cell lines, require serum supplement. Fetal bovine serum (FBS) is most frequently used as the primary growth supplement in insect cell culture and although it promotes cell growth and provides shear force protection important in shake and stirrer cultures, it can also cause excessive foaming, interfere with transfection reagents and is costly. For this reason, a number of commercially available cell lines have now been adapted and optimized for growth in serum-free media, such as Sf-900 II SFM and EXPRES-FIVE SFM for Sf-9 cell growth. Such media contain amino acids, carbohydrates, vitamins, and lipids essential for insect cell growth that reduce the effect of rate-limiting nutritional restrictions or deficiencies present within serum-supplemented media. Serum-free media also tend to support faster cell-doubling times and permit cell growth to higher densities than serum-supplemented media, which, when used for recombinant virus and/or protein production, ultimately results in higher virus titers or protein yields.

## Baculovirus Expression Vector Systems

Since the application of baculovirus expression systems for the safe and abundant expression of foreign proteins in insect cells in the early 1980s, this expression system has become one of the most popular methods for the production of large quantities of recombinant proteins within eukaryotic cells. The majority of baculovirus expression systems exploit the *polh* promoter to drive high-level production of foreign proteins. The baculovirus *polh* gene is nonessential for virus replication in insect cell cultures and therefore can be removed from the virus genome with no detrimental effect to BV production.

# Baculoviruses: Expression Vector 239

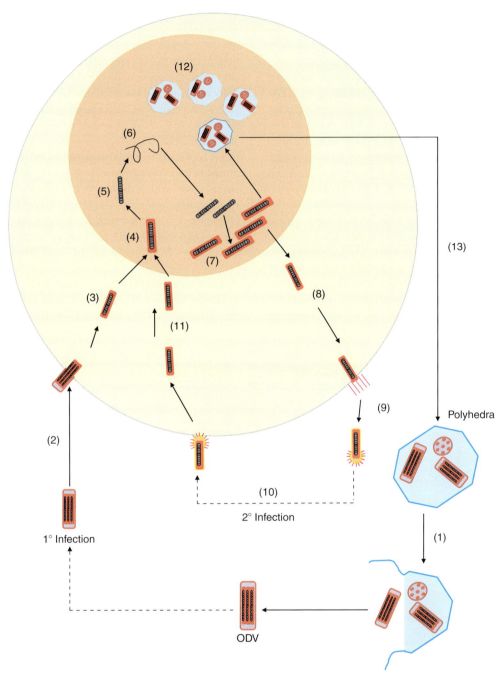

**Figure 1** Schematic of the biphasic AcMNPV replication cycle. Polyhedra persist in the environment until ingested by a susceptible larval host. The alkaline environment of the insect midgut dissolves the polyhedra (1) to release occlusion-derived virions (ODVs). ODVs then fuse with midgut epithelial cells (2) and enter the cell. Nucleocapsids (NCs) then travel to the nucleus (3), enter the nucleus most likely via the nuclear pore (4), where the NC is uncoated (5) and viral transcription initiated (6). *De novo* production of NCs ensues (7) and NCs leave the host cell nucleus (8). NCs are then transported to the plasma membrane where they bud from the host cell acquiring an envelope derived from the host cell membrane, to produce budded virus (BV) (9). BVs then attach to further susceptible cells (10), are uncoated within the cell cytoplasm, and NCs then traffic to the nucleus (11), following which translation, production of *de novo* NCs and additional BV occurs (4–10). During the later stages of infection, NCs remaining within the nucleus are occluded within polyhedra (13) which are released from the cell on cell death and host liquefaction.

As the baculovirus genome is generally considered too large for direct insertion of foreign genes, the gene of interest is first cloned into a transfer vector containing sequences that flank the *polh* gene in the viral genome. Virus DNA and transfer vector are co-transfected into the host insect cell and homologous recombination between the flanking sequences common to both DNA molecules occurs. This causes the insertion of the gene of interest into the viral genome at the *polh* locus, resulting in the production of a recombinant virus genome. The genome then undergoes replication within the host nucleus, generating recombinant BV containing the foreign gene under the control of the strong, late viral polyhedrin promoter. Polyhedra are not produced as the *polh* gene is no longer functional, having been replaced by the gene of interest.

Recombinant baculovirus expression vectors were originally produced using the highly inefficient homologous recombination process between a transfer vector containing the gene of interest and parental virus DNA. Insect cells were co-transfected with the baculovirus and transfer plasmid DNA, producing mixture of both recombinant and parental viruses, with recombinant viruses comprising only *c.* 0.1% of total virus produced. Isolation of recombinant virus then required plaque purification where recombinant clones were identified by their characteristic occlusion-negative plaque phenotypes.

To improve the efficacy of recombinant virus production, a unique *Bsu*361 restriction enzyme site was engineered into the *polh* locus of baculovirus DNA to permit linearization of the viral genome prior to co-transfection, giving rise to a higher frequency of recombinant virus production. Further improvements employed multiple *Bsu*361 sites, with digestion of the viral DNA resulting in a partial deletion within an essential gene, ORF1629. The deletion within this gene prevents replication of parental virus, increasing the yield of recombinant virus to more than 90%. Insertion of the *Escherichia coli lacZ* gene into the *polh* locus, replacing the *polh* gene, produced the commercially available BacPAK6 (**Figure 2**). A *Bsu*361 restriction enzyme site is located within the *lacZ* gene along with two additional sites situated in the two flanking genes either side of *lacZ*. Digestion of BacPAK6 with *Bsu*361 removes the *lacZ* gene and a fragment of ORF1629 (**Figure 2(b)**), resulting in linear virus DNA incapable of replicating within insect cells.

Co-transfection of insect cells with BacPAK6 DNA and a transfer vector containing the gene of interest restores the deletion in ORF1629 and re-circularizes the virus

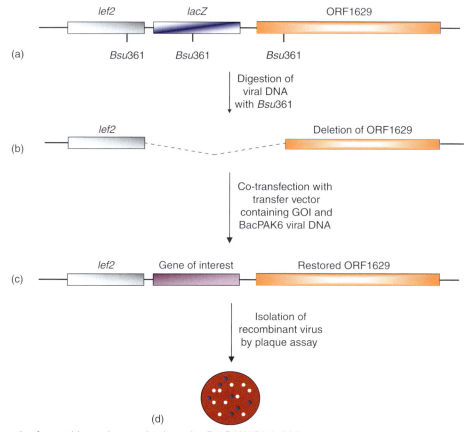

**Figure 2** Schematic of recombinant virus production using BacPAK6 DNA. (a) The commercially available BacPAK6 baculovirus DNA contains the *E. coli lacZ* gene inserted at the polyhedrin (*polh*) locus. A *Bsu*361 restriction enzyme site is located within the *lacZ* gene in addition to 2 BSU36q sites found in the flanking *lef2* and ORF1629 coding regions. (b) Restriction enzyme digestion of the viral DNA results in the removal of the *lacZ* gene and partial deletion of the ORF1629 coding region, a gene essential in viral replication, resulting in linear DNA incapable of replication within insect cells. (c) Co-transfection of insect cells with linear BacPAK6 DNA and a transfer vector containing a gene of interest results in insertion of the foreign gene into the virus DNA and restoration of the ORF1629 deletion, circularizing the DNA and permitting virus replication within insect cells and recombinant virus production. (d) Isolation of recombinant virus is achieved via plaque assay. GOI, gene of interest.

DNA by allelic replacement (**Figure 2(c)**). This restoration of the essential virus gene permits replication within insect cells, followed by assembly of nucleocapsids within the nucleus and ultimately production of recombinant viruses. Although the additional *Bsu*361 sites in BacPAK6 lead to an increased proportion of recombinants compared to previous baculovirus expression systems by reducing parental background, restriction enzyme digestion of virus DNA is never 100% efficient and co-transfection still results in the mixture of both parental and recombinant virus. Isolation of solely recombinant virus also requires plaque purification. The presence of *lacZ*, however, allows the selection of colorless, recombinant virus plaques against a background of parental, nonrecombinant, blue plaques in the presence of X-gal (5-bromo-4-chloro-3-indolyl-β-D-galactopranoside).

Efforts to remove the requirement for plaque purification and increase both the speed and ease of recombinant virus production have resulted in the development of a number of unique baculovirus expression systems that involve various mechanisms to transfer foreign genes into the viral genome and shall be reviewed here.

## Bac-to-Bac

The Bac-to-Bac baculovirus expression system (invitrogen) provides a rapid and efficient method to generate recombinant baculoviruses (**Figure 3**). The system is based on *in vivo* bacterial site-specific transposition of an expression cassette into a baculovirus shuttle vector, also known as a bacmid, propagated in *E. coli*. The bacmid contains a low-copy number mini-F replicon, a kanamycin-resistance gene, and the *lacZ*α gene. Located at the N-terminus of the *lacZ*α gene is an attachment site for the bacterial transposon Tn7 that does not interrupt the reading frame of the *lacZ*α peptide. The bacmid is propagated in *E. coli* cells as a large plasmid resistant to kanamycin that demonstrates blue selection in the presence of X-gal and isopropyl-beta-D-thiogalactopyranoside (IPTG).

Prior to the production of a recombinant baculovirus using the Bac-to-Bac system, the gene of interest is cloned into a donor plasmid containing either a *polh* or *p10* promoter region (**Figure 3(a)**). The expression cassette is flanked by the left and right arms of the Tn7 transposon and contains a gentamicin-resistance gene and an SV40 polyadenylation signal to form a mini-Tn7. The recombinant transfer vector is then transformed into *E. coli* cells containing the bacmid, and a helper plasmid encoding a transposase (**Figure 3(b)**). The mini-Tn7 element from the recombinant transfer vector is transposed to the mini-*att*Tn7 attachment site on the bacmid DNA aided *in trans* by helper plasmid-encoded transposase. Insertion of the gene of interest into the bacmid disrupts the *lacZ*α peptide resulting in white recombinant colonies when exposed to X-gal and IPTG compared to blue parental colonies containing the unaltered bacmid DNA.

Having created the recombinant bacmid, the viral DNA must then be extracted from bacterial cells (**Figure 3(d)**) followed by transfection into insect cells (**Figure 3(e)**). The recombinant virus DNA undergoes transcription and translation, resulting in the production of recombinant BV particles which are then harvested and amplified through repeated rounds of infection to generate a high-titer baculovirus working stock in large volume.

Using such site-specific transposition of genes into bacmid DNA has a number of advantages over using homologous recombination within insect cells for the production of recombinant baculoviruses. Primarily, the viral DNA isolated from selected bacterial colonies is not mixed with parental DNA, eliminating the need to isolate recombinant from nonrecombinant virus and multiple rounds of plaque purification. This substantially reduces the time taken to isolate and purify recombinant virus from 4–6 weeks to 7–19 days. Second, the Bac-to-Bac system easily permits simultaneous production and isolation of several recombinant viruses by one person. In addition to this, the Bac-to-Bac expression system is compatible with the Invitrogen Gateway cloning vectors, a system comprising a multitude of plasmids permitting both single and multiple insertions of genes into a single baculovirus genome.

## BaculoDirect

The BaculoDirect baculovirus expression system (Invitrogen) uses the Gateway technology to permit the direct transfer of a gene of interest into the baculovirus genome, without the need for production of recombinant bacmid DNA. Gateway technology is a universal cloning method which provides a rapid cloning step, allowing the efficient transfer of DNA sequences into multiple vector systems. The technology is based on the site-specific recombination properties of the bacteriophage lambda, which facilitates the integration of lambda into the *E. coli* chromosome (**Figure 4(a)**). This integration is mediated by a mixture of enzymes, the lambda recombinase, termed integrase, and the *E. coli*-encoded integration host factor (IHF) and occurs via intermolecular DNA recombination at specific attachment (*att*) sites.

During lambda integration into the *E. coli* chromosome, recombination occurs between *att*B and *att*P sites, present in the *E. coli* chromosomal and lambda phage DNA, respectively, to give rise to *att*L and *att*R sites which flank the integrated phage DNA. This reaction can also be reversed utilizing another lambda-encoded enzyme, excisionase, and recombination of *att*L and *att*R sites, resulting in the excision of the lambda DNA from the *E. coli* chromosome, recreating the *att*B site in *E. coli* and the *att*P site present in lambda phage.

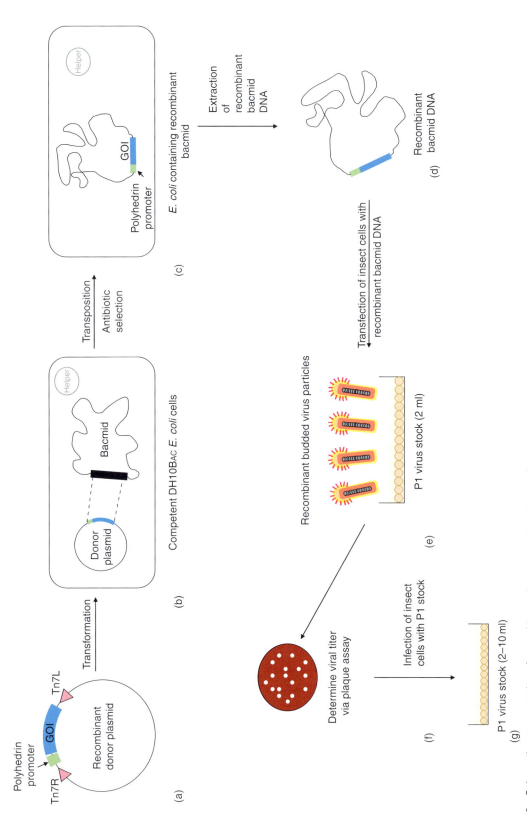

**Figure 3** Schematic representation of recombinant virus generation using the Bac-to-Bac system. (a) The gene of interest is first cloned into a suitable donor plasmid containing a polyhedrin (*polh*) promoter and a mini-Tn7 element. (b) The recombinant donor plasmid is then transformed into *E. coli* cells which contain a bacmid with a mini-attTn7 target site and a helper plasmid. The mini-Tn7 element on the recombinant donor plasmid facilitates transposition of the gene of interest (GOI) into the target site on the bacmid along with transposition proteins provided by the helper plasmid, resulting in insertion of the GOI into the bacmid DNA (c). Recombinant virus DNA is then extracted and purified (d) following which insect cells are transfected with the recombinant bacmid DNA (e) to produce a P1 virus stock. (f) The virus titer is determined via plaque assay and further insect cell cultures infected with the P1 stock to create a P2 stock of 2–10 ml (g). GOI, gene of interest.

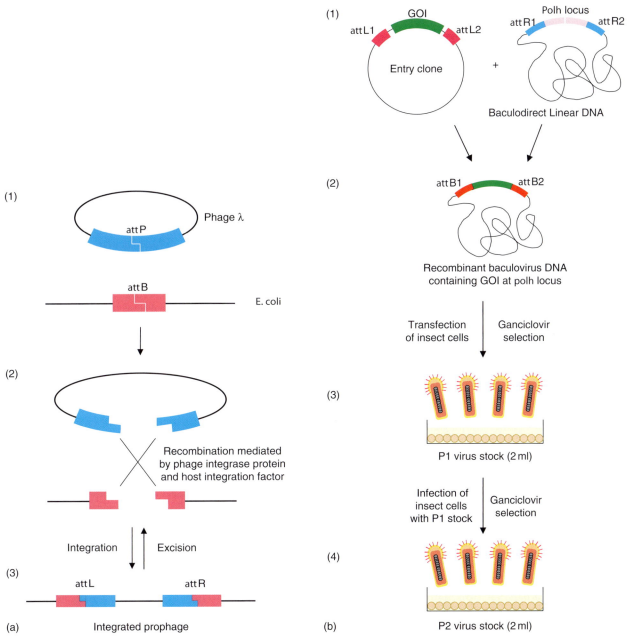

**Figure 4** (a) Schematic representation of lambda bacteriophage site-specific transposition in *E. coli*. Following infection of *E. coli* cells, the lambda phage DNA integrates into the host genome between integration sites *att*P and *att*B found in the phage and *E. coli* DNA, respectively (1). The integration reaction is mediated by both phage (intergase) and host (host integration factor) proteins (2) and results in the production of two new sites, termed *att*L and *att*R and the integration of phage DNA into the host genome (3). The integration reaction is reversible mediated by the protein integrase, host integration factor, and excisionase, resulting in the removal of the phage DNA from the viral genome, recreating the *att*B and *att*P sites in *E. coli* and phage DNA, respectively. (b) Schematic representation of recombinant baculovirus production using the BaculoDirect system. Following successful production of an entry clone containing a gene of interest (GOI), an LR reaction between the entry clone and the BaculoDirect linear DNA (1) generates recombinant baculovirus DNA (2) containing the GOI at the polyhedrin (*polh*) locus. Insect cells are then transfected with the recombinant baculovirus DNA and recombinant BV selected via the addition of ganciclovir, resulting in the production of a 2 ml P1 recombinant virus stock (3). This P1 virus stock is then used to infect further insect cell cultures in the presence of the selective compound ganciclovir to produce a high titer, 2 ml P2 virus stock (4).

The Gateway system mimics the lambda integration system in an *in vitro* environment and combined with the BaculoDirect system, provides a suitable system for high-throughput production of multiple recombinant viruses (**Figure 4(b)**). First, the gene of interest must be cloned into a suitable vector, termed an entry clone, with the foreign DNA flanked by two recombination sites, *att*L1 and *att*L2. BaculoDirect linear DNA has been modified to contain *att*R1 and *att*R2 recombination sites located at the *polh* locus, along with a herpes simplex virus thymidine kinase gene (HSV1 tk) and a *lacZ* gene located between *att*R sites. To create a recombinant baculovirus, it is necessary to perform the recombination reaction utilizing the Gateway LR clonase solution, an enzyme mix comprised of integrase, IHF, and excisionase, to transfer the gene of interest into the baculovirus DNA. The *att*R sites present within the virus DNA undergo recombination with the *att*L sites within the entry clone, resulting in the integration of the foreign gene into the virus DNA to create an *att*B-containing expression virus.

Having performed the LR reaction, baculovirus DNA is then used to directly transfect insect cells. Purification of recombinant viruses ensues following addition of medium containing the selective nucleoside analog, ganciclovir. Ganciclovir is enzymatically phosphorylated by HSV1 tk present in non-recombinant genomes, and results in the incorporation of this nucleoside into DNA, thereby inhibiting their DNA replication. Recombinant viruses that have lost the counterselectable marker via homologous recombination are capable of replication within insect cells, promoting the amplification of recombinant viruses and production of a low-titer, P1 viral stock. Screening of P1 stocks can then be undertaken to ensure recombinant protein expression and the absence of parental background. Alternatively, the P1 stock can be used to directly infect insect cells in the presence of ganciclovir to produce a larger volume P2 working stock.

The high-throughput application of BaculoDirect for the production of multiple recombinant viruses appears to offer significant time savings compared to other systems such as Bac-to-Bac. However, this system is currently restricted to the production of recombinant single gene baculoviruses preventing co-expression of multiple proteins. In addition, the necessity to selectively isolate recombinant from parental virus via the addition of ganciclovir is detrimental to total cell numbers and therefore consequently negative to final viral titers, ultimately requiring multiple virus amplification steps before a high-titer working virus stock is produced.

### *flash*BAC

A new platform technology for the production of recombinant baculoviruses has recently emerged that has significant advantages over the other vector systems described above. The *flash*BAC expression system (Oxford Expression Technologies, Oxford, UK) combines the advantages of the traditional homologous recombination system within insect cells with those of the more recent bacterial-based systems. *Flash*BAC has been developed to remove the requirement for isolation of recombinant from parental virus via plaque purification, resulting in a one-step procedure for recombinant virus production and ultimately significantly reducing the time taken to produce a working high-titer recombinant virus stock.

The *flash*BAC expression system is based around a modified AcMNPV genome that contains a bacterial artificial chromosome (BAC) at the *polh* locus replacing the *polh* gene. The presence of the BAC within the virus genome allows the virus genome to be maintained and propagated within bacterial cells from which the circular DNA is then isolated and purified. Additionally, the modified genome contains a deletion in the essential gene ORF1629 which renders the virus inactive and unable to replicate within insect cells (**Figure 5(a)**).

Homologous recombination between *flash*BAC DNA and a suitable transfer vector containing the gene of interest, following transfection into insect cells, restores the deletion within, and thus the function of, the essential ORF1629 gene (**Figure 5(b)**). Simultaneously, the gene of interest is inserted into the viral genome at the *polh* locus under the control of the *polh* promoter, concomitantly removing the BAC replicon. The recombinant virus genome, with the restored essential gene, is then capable of replication within insect cells, producing recombinant BV particles that can be harvested from the culture medium of the transfected cells (**Figure 5(c)**). The deletion present within ORF1629 of *flash*BAC DNA prevents replication of any nonrecombinant, parental virus; therefore, there is no requirement for recombinant virus isolation. Following transfection, virus can be added directly to insect cells to produce a high-titer working recombinant virus stock.

The *flash*BAC system is back-compatible with all baculovirus transfer vectors based on homologous recombination in insect cells at the *polh* locus, including vectors using the *polh* promoter, dual triple and quadruple expression vectors, and those that use other gene promoters such as *p10*, *ie1*, or *gp64*. However, vectors such as pFastBac which are designed for site-specific transposition in *E. coli* systems such as Bac-to-Bac are not compatible with the *flash*BAC system.

The *flash*BAC system is unique among baculovirus expression systems as it maximizes protein secretion and membrane protein targeting. Within the baculovirus genome are a number of auxiliary genes which are nonessential in the replication and production of BV particles *in vitro*. One such gene codes for chitinase (*chiA*), an enzyme with exo- and endochitinase activity, essential in the host-to-host transmission of the virus. Following infection of a host insect, chitinase works in synergy with another virally encoded protein, termed cathepsin, to break down the host cuticle, ultimately resulting in tissue liquefaction and release of polyhedra to infect more hosts. Confocal and electron microscopy analysis has shown the localization of chitinase within the endoplasmic reticulum during baculovirus infection. This severely compromises

# Baculoviruses: Expression Vector 245

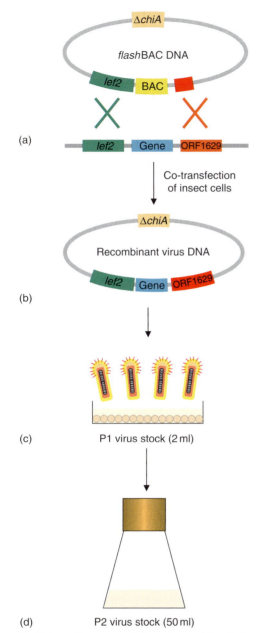

**Figure 5** Schematic representation of recombinant virus production using the *flash*BAC system. The *flash*BAC expression system relies on a modified baculovirus genome containing a BAC at the polyhedrin (*polh*) locus and a partial deletion of the essential ORF1629 viral gene (a). Homologous recombination within insect cells between the *flash*BAC DNA and a transfer vector containing a gene of interest (GOI) flanked by *lef2* and the complete ORF1629 gene coding regions results in the insertion of the GOI at the *polh* locus and restoration of ORF1629 gene (b). The deletion within ORF1629 results in a P1 stock containing only recombinant baculovirus production with no parental background (c). This P1 stock is then used to infect a larger insect cell culture resulting in the production of a high-titer, high-volume (50 ml) P2 working recombinant virus stock (d).

the function efficiency of the secretory pathway, consequently affecting the levels of recombinant protein production. Deletion of *chiA* from *flash*BAC has substantially improved the efficacy of the secretory pathway for producing recombinant proteins in insect cells, enhancing the yield of secreted of membrane-bound proteins in comparison to proteins produced in recombinant viruses that synthesize chitinase.

The one-step *flash*BAC technology facilitates the use of robotic systems for the generation of multiple recombinant viruses without the requirement for plaque purification. It has been demonstrated that co-transfection mixtures from multiple recombinant viruses can be used to infect insect cells for automated high-throughput protein screening in 96-well format. Additionally, this process can be performed solely in insect cells, removing the risk of cross-contamination between insect and bacterial cell cultures.

## Baculovirus-Mediated Gene Delivery in Mammalian Cells

The baculovirus expression system has been extensively used for the expression of recombinant proteins within insect cells for a number of years. More recently, recombinant baculovirus vectors have been developed to permit transient and stable gene delivery into a number of mammalian cell lines. Such vectors contain baculovirus promoters alongside mammalian cell-active expression cassettes to permit amplification of recombinant viruses in insect cells and expression of recombinant proteins in mammalian cells, respectively.

Baculovirus-mediated gene delivery into mammalian cells, known as BacMam technology, was first demonstrated in hepatic cell lines and subsequent studies have reported transduction of human cell lines including HeLa, pancreatic β-cells and primary neural cells, Chinese hamster ovary and porcine kidney cell lines. The transduction efficiencies of different cell lines vary considerably, with baby hamster kidney cells having a transduction efficiency of 95% while mouse NIH 3T3 cells demonstrate only 10% efficiency. Transduction efficiency can be enhanced by the addition of various compounds such as trichostain A or sodium butyrate that act as histone deacetylase inhibitors; however, these drugs have cytotoxic effects on cell cultures. The majority of recombinant viruses used to transduce mammalian cell lines contain hybrid promoters consisting of a chicken β-actin gene promoter and a cytomegalovirus immediate early gene enhancer element along with a *p10* baculovirus promoter for expression of the gene in insect cell lines.

One significant advantage of BacMam technology is that successful gene delivery of foreign DNA into mammalian cells is possible by simply adding recombinant baculovirus inoculum to a culture of mammalian cells. The entry mechanism of baculovirus into mammalian cell is poorly understood; however, it is thought that the viral surface glycoprotein, GP64, plays a role in viral entry and endosomal release. Following viral entry, nucleocapsids induce actin filament formation within mammalian cells, a process also known to occur within insect cells involved in the propulsion of virus particles to the nucleus. Reports have demonstrated an increase in nuclear localization of nucleocapsids in human hepatocytes following the disruption of the microtubule network. This suggests that microtubules constitute a barrier to baculovirus transport toward the nucleus and thus disruption of these filaments may provide a simple method with which to increase nuclear localization of recombinant nucleocapsids and ultimately increase foreign gene expression within mammalian cell lines. Transduction of mammalian cells with recombinant baculovirus is generally considered nontoxic and has no apparent effect on cell growth, even at high multiplicities of infection.

The inability of baculoviruses to replicate within mammalian cells provides an attractive vector for *in vivo* applications, including gene function studies and gene therapy. However, baculovirus is rapidly inactivated by human serum complement, destroying the ability of the recombinant virus to transfer genes *in vivo*. A number of methods have been developed to alleviate this problem, including production of baculovirus particles pseudotyped with the vesicular stomatitis virus (VSV)-G protein, which provides a virus more resistant to complement than unmodified virus. Additionally, the use of baculoviruses to express antigens under the control of mammalian promoters has been shown to elicit an immune response *in vivo*. Recombinant baculoviruses expressing the gB protein from pseudorabies produced an immune response in mice intramuscularly inoculated with the modified baculovirus. Additionally, recombinant baculoviruses expressing the influenza hemagglutinin protein elicited immune responses in mice when delivered by intramuscular injection and provided immunity to further infection following subjection to a lethal dose of influenza virus. The possibility of utilizing baculoviruses for gene therapy and vaccine applications is still a very new, albeit extremely promising, technology and much research continues in this area to optimize baculoviruses as potential gene therapy vectors.

## Conclusions and Future Perspectives

Manipulation of the baculovirus genome provides a powerful tool for the expression of recombinant proteins in both insect and mammalian cell lines. Advances in vector design and the various commercially available baculovirus expression systems permit simple, recombinant virus production, with some systems negating the requirement of virus isolation and purification. This, coupled with the automation and high-throughput possibilities of virus production using robotic systems for simultaneous multiple-virus production, has resulted in more laboratories employing baculoviruses as their expression system of choice.

One of the main disadvantages of the baculovirus expression system is the dissimilarity of the insect cell protein-processing pathways compared to those of higher eukaryotes, for example, the N-glycosylation pathway. The production of transgenic insect cell lines expressing humanized protein glycosylation pathways offers a way to overcome this potential problem, enabling the production of recombinant polypeptides which demonstrate greater similarity to their native mammalian proteins.

Recombinant baculoviruses have become a widely used system for the production of recombinant proteins within insect cells. The availability of the entire baculovirus sequence will continue to enable further manipulation of the virus genome to increase and further optimize recombinant protein expression in both insect and mammalian cell lines. It is hoped that the use of recombinant baculoviruses as gene delivery vectors for higher eukaryotic cell lines will become as routine as the use of such viruses for recombinant protein expression within insect cells and that advances in knowledge and technology will continue to expand the possibilities and applications of the baculovirus expression system.

*See also:* Baculoviruses: General Features; Baculoviruses: Molecular Biology of Mosquito Baculoviruses; Baculoviruses: Pathogenesis.

## Further Reading

Hu Y (2005) Baculovirus as a highly efficient expression vector in insect and mammalian cells. *Acta Pharmacologica Sinica* 26: 405–416.

Hunt I (2005) From gene to protein: A review of new and enabling technologies for multi-parallel protein expression. *Protein Expression and Purification* 40: 1–22.

King LA and Possee RD (1992) *Baculovirus Expression System – A Laboratory Guide*. London: Chapman and Hall.

Kitts PA and Possee RD (1993) A method for producing recombinant baculovirus expression vectors at high frequency. *Biotechniques* 14: 810–817.

Kost TA and Condreay JP (2002) Recombinant baculovirus as mammalian cell gene-delivery vectors. *Trends in Biotechnology* 20: 173–180.

Kost TA, Condreay JP, and Jarvis DL (2005) Baculovirus as versatile vectors for protein expression in insect and mammalian cells. *Nature Biotechnology* 23: 567–575.

Pennock GD, Shoemaker C, and Miller LK (1984) Strong and regulated expression of *Escherichia coli* beta-galactosidase in insect cells with a baculovirus vector. *Molecular and Cellular Biology* 4: 399–406.

Smith GE, Summers MD, and Fraser MJ (1983) Production of human beta interferon in insect cells infected with a baculovirus expression vector. *Molecular and Cellular Biology* 3: 2156–2165.

# Baculoviruses: General Features

**P J Krell,** University of Guelph, Guelph, ON, Canada

© 2008 Elsevier Ltd. All rights reserved.

## Glossary

**Apoptosis** Programmed cell death.
**Caspase** Protease involved in the apoptotic response.
**Enhancin** A baculovirus protein which enhances the infectivity of another virus.
**Granulin** The major protein in GV granules.
**IAP** A baculovirus inhibitor of apoptosis.
**P35** A baculovirus inhibitor of apoptosis.
**Polyhedrin** The major protein in NPV polyhedra.
**VLF** Very late expression factor.

## Historical Perspectives

Diseases reminiscent of baculovirus infections appear in the historical record going back to at least 1527 when Bishop Marco Girolamo Vida in Italy wrote a verse which referred to a jaundice disease of the silkworm *Bombyx mori*. In 1856, Maestri and Cornalia independently described refractile bodies, reminiscent of the polyhedra of nucleopolyhedroviruses (NPVs), within cells of infected silkworm larvae demonstrating a 'polyhedrosis' disease. Glaser, Paillot, and colleagues, between 1913 and 1928, demonstrated that the polyhedral disease of caterpillars (a polyhedrosis) was due to a filterable agent and therefore must have a viral etiology. With the perfection of insect tissue culture methodology, William Trager in 1935 reported on the ability of the Grasserie virus (now known as bombyx mori NPV (BmNPV)) to replicate in silkworm tissue culture cells. This led to the opportunity of studying the replication of these viruses under more controlled conditions and free of other microbial agents.

## Nomenclature, Taxonomy, and Classification

The family *Baculoviridae* consists of insect viruses with a large, 80–180 kbp circular dsDNA genome and a distinct rod-shaped virion morphology. It is composed of two genera, *Nucleopolyhedrovirus* for NPVs with *Autographa californica nucleopolyhedrovirus* as the type species and *Granulovirus* for the granuloviruses (GVs), with *Cydia pomonella granulovirus* (CpGV) as the type species (refer to **Table 1** for virus abbreviations). Though not an accepted taxonomic designation, but based on comparison of genomic sequences, the NPVs can further be subdivided into group I, which encode *gp64* but not *Fusion protein* (F) or *p13*, and group II, which encode *F* and *p13* but not *gp64*.

The family *Baculoviridae*, infecting invertebrate, mostly insect, hosts, owes its name to the stick-shaped morphology of the virions (*baculum*, Latin for stick). The rod-shaped double-stranded DNA (dsDNA)-containing nucleocapsids measuring around 30 nm × 100 nm are surrounded by a membrane envelope. The *Baculoviridae* are unique in having two morphologically distinct forms and a biphasic replication cycle. Early in infection, budded viruses are produced by nucleocapsids budding through the cell membrane. This budded form is responsible for cell-to-cell transmission within the infected insect or in tissue culture. Later in infection, nucleocapsids remain in the nucleus where they are enveloped either singly or in multiples in a nuclear-derived envelope and are subsequently embedded in a polyhedral (*Nucleopolyhedrovirus*) or granular (*Granulovirus*)-shaped proteinaceous, alkaline-sensitive, crystalline matrix that is highly resistant to environmental influences. The occluded form is responsible for the horizontal, insect-to-insect spread.

In the past, the baculovirus family was a repository for many arthropod viruses with a similar rod-shaped viral morphology including the polydnaviruses of parasitic wasps and the whispoviruses of shrimp that now have their own taxa, families *Polydnaviridae* and *Nimaviridae*, respectively. Oryctes virus (OrV; formerly named oryctes baculovirus) and Hz-1 baculovirus are also baculovirus-like but lack an occlusion body whereby the genus *Nudivirus* has been suggested for them. As the characteristics needed to define the *Baculoviridae* have crystallized, the number of species recognized has become more manageable. For example, in the VIth Report of the International Taxonomy of Viruses, a total of 15 species and an amazing 372 tentative species were assigned to the genus *Nucleopolyhedrovirus*, while the genus *Granulovirus* numbered four species and a tentative species list of 65. As many of these were too poorly characterized, original stocks were not available, and some may have simply been the same virus but isolated from a different host, most of the tentative species have been dropped. In the VIIIth report, there are now 24 *Nucleopolyhedrovirus* species recognized with only five tentative species and 15 *Granulovirus* species but no tentative species.

Analysis of complete baculovirus genome sequences provides a powerful basis for refining this classification from two to four genera to reflect the differences in biology and genome phylogeny of members of the *Baculoviridae*.

**Table 1** Taxonomy, names, abbreviations, and genome sizes of of baculoviruses used in the text

| Subgroup/host | Species name[a] | Abbreviation for virus name | Genome size (bp) |
|---|---|---|---|
| Genus *Nucleopolyhedrovirus* | | | |
| Lepidopteran | | | |
|   Group I | *Anticarsia gemmatalis M nucleopolyhedrovirus* | AgMNPV | 132 239 |
| | *Autographa californica M nucleopolyhedrovirus* | AcMNPV | 133 894 |
| | *Bombyx mori nucleopolyhedrovirus* | BmNPV | 128 413 |
| | *Choristoneura fumiferana M nucleopolyhedrovirus* | CfMNPV | 129 593 |
| | *Epiphyas postvittana nucleopolyhedrovirus* | EppoNPV | 118 584 |
| | *Hyphantria cunea nucleopolyhedrovirus* | HycuNPV | 132 959 |
| | *Orgyia pseudotsugata M nucleopolyhedrovirus* | OpMNPV | 131 995 |
|   Group II | *Adoxophyes honmai nucleopolyhedrovirus* | AdhoNPV | 113 220 |
| | *Helicoverpa armigera nucleopolyhedrovirus* | HearNPV | 130 759 |
| | *Leucania separata nucleopolyhedrovirus* | LeseNPV | 168 041 |
| | *Lymantria dispar M nucleopolyhedrovirus* | LdMNPV | 161 046 |
| | *Spodoptera exigua M nucleopolyhedrovirus* | SeMNPV | 135 611 |
| | *Spodoptera littoralis nucleopolyhedrovirus* | SpliNPV | |
| | *Spodoptera litura nucleopolyhedrovirus* | SpltNPV | 139 342 |
| Dipteran | *Culex nigripalpus nucleopolyhedrovirus* | CuniNPV | 108 252 |
| Hymenopteran | *Neodiprion abietis nucleopolyhedrovirus* | NeabNPV | 84 264 |
| | *Neodiprion lecontei nucleopolyhedrovirus* | NeleNPV | 81 755 |
| | *Neodiprion sertifer nucleopolyhedrovirus* | NeseNPV | 86 462 |
| Genus *Granulovirus* | | | |
| Lepidopteran | *Adoxophyes orana granulovirus* | AdorGV | 99 657 |
| | *Choristoneura occidentalis granulovirus* | ChocGV | 104 710 |
| | *Cydia pomonella granulovirus* | CpGV | 123 500 |
| | *Phthorimaea operculella granulovirus* | PhopGV | 119 217 |
| | *Xestia c-nigrum granulovirus* | XecnGV | 178 733 |

[a]Species names in italics are approved species, names in Roman script are tentative or as yet an unassigned species according to Eighth Report of the International Committee on Taxonomy of Viruses.

The four genera suggested are lepidopteran-specific NPVs (*Alphabaculovirus*), dipteran-specific NPVs (*Deltabaculovirus*), hymenopteran-specific NPVs (*Gammabaculovirus*), and lepidopteran-specific GVs (*Betabaculovirus*).

Ever since their earliest discovery, baculoviruses have been named according to the insect from which the virus was derived and the type of occlusion body associated with it. While this might cause some confusion, since the same virus might occur in different insect species, this approach is still applied today. Originally baculovirus abbreviations used the first letter of the genus and species name of the host insect followed by NPV or GV depending on the type of occlusion body. Hence the multiple-embedded NPV (MNPV) from the alfalfa looper *Autographa californica* is abbreviated as AcMNPV and the MNPV from *Choristoneura fumiferana* is abbreviated CfMNPV. Those baculoviruses which have historically had a two-letter species code continue to maintain those abbreviations. However, as more baculoviruses were being described, a two-letter species designation system was untenable and now a four-letter species identification based on the first two letters of each of the genus and species names of the host insect from which the virus was first isolated is used. Hence the more recently described NPV from *Helicoverpa armigera* is abbreviated as HearNPV and that from *Neodiprion lecontei* is abbreviated as NeleNPV. If identical letterings might result, the third or later letters are used instead of the second as for the NPV from *Spodoptera litura* which is abbreviated SpltNPV since an NPV from *Spodoptera littoralis* is abbreviated SpliNPV. Although the S and M are useful morphological descriptors in the names indicating if nucleocapsids exist either singly or in multiples within virions, they do not appear to have any taxonomic relevance and so in recently described baculoviruses these designations have been dropped but have been maintained for those baculoviruses in which there is a strong historical precedence.

## Morphological Characteristics

The *Baculoviridae* are currently comprised of two genera, *Nucleopolyhedrovirus* and *Granulovirus*, based initially on different morphological characteristics. Viruses in both genera have dual phenotypes and biphasic replication cycles, though with similar virion and genomic features. Nucleocapsids are rod shaped with diameters of 30–60 nm and lengths, dependent on the size of the dsDNA genome, of 250–300 nm. Nucleocapsids are structurally polar with a flat base plate on one end and a nipple at the other. The nucleocapsids are enveloped in either *de novo*-derived membrane in the nucleus for occlusion-derived viruses or viral-modified plasma membrane of the infected cell for budded viruses. The budded virions of NPVs also have polarity with the nipple end showing a distinct peplomeric structure consisting of viral glycoproteins such as GP64 or F proteins.

For the genus *Nucleopolyhedrovirus*, several virions containing either single (SNPVs) or multiple nucleocapsids within a single envelope (MNPVs) are embedded in a proteinaceous polyhedral-shaped occlusion body measuring 0.5–15 μm in diameter but most measure around 1 μm in diameter. The major structural protein in the polyhedra is aptly named polyhedrin.

The *Granulovirus* virions contain only a single nucleocapsid per virion and only one virion is embedded in a much smaller proteinaceous ovocylindrical occlusion body referred to as a granule which measures $c.$ 120–300 nm × 300–500 nm. The major structural protein in the granules is aptly named granulin but is, however, evolutionarily related to polyhedrin. GVs tend to have a more restricted host range and limited tissue tropism compared to the NPVs.

## Genomes, Gene Content, Organization, and Evolution

Of 41 full-length baculovirus genome sequences deposited in the NCBI Virus Genome database, the genome sizes of the GVs range from 99 657 bp for adoxophyes orana granulovirus to 178 733 for xestia c-nigrum granulovirus, which is also the largest genome of any baculovirus; and for the NPVs range from 81 755 bp for neodiprion lecontei nucleopolyhedrovirus to 168 041 bp for leucania separata nucleopolyhedrovirus. The total number of computational open reading frames (ORFs, usually defined as those encoding proteins of at least 50 amino acids) identified among baculoviruses ranges from 89 (neodiprion lecontei nucleopolyhedrovirus) to 181 (xestia c-nigrum granulovirus) per genome. The guanine–cytosine (GC) content of baculovirus genomes is very broad ranging for NPVs from 33% for neodiprion lecontei nucleopolyhedrovirus to 58% for lymantria dispar M nucleopolyhedrovirus, and for GVs from 32% for choristoneura occidentalis granulovirus to 45% for cydia pomonella granulovirus.

Over 500 unique genes have been identified in baculoviruses but all genomes sequenced to date encode a common set of 29 core genes including those involved in DNA replication like *dnapol* and *hel*; transcription, such as for the four proteins making up the late viral RNA polymerase and late expression factor-4 (LEF-4), a protein which might be involved in capping; structural proteins such as the small basic P6.9 protein associated with the genomic DNA, the major capsid protein VP39, and proteins VP1054 and VP91 associated with both budded and occluded virus; auxiliary proteins like alkaline exonuclease and the two *per os* infectivity factors; and some unknown proteins identified only by an ORF sequence such as the AcMNPV ac98 ORF.

Transcription is fairly equally distributed to both strands. There is no particular clustering of genes on the genome on the basis of, for example, function or temporal transcription class although there is some conservation of gene order among closely related viruses.

Baculovirus genomes can contain multiple homologous repeat regions (*hr*'s) containing differing numbers of imperfect palindromic sequences. The analysis of *Eco*R1 digestion patterns in AcMNPV DNA initially led to the discovery of *hr*'s. AcMNPV has 8 *hr*'s, with each *hr* having two to eight highly conserved regions of *c.* 72 bp. AcMNPV *hr*5 has six, 60–70 bp direct repeats, most with an *Eco*R1 restriction site within a 28 bp imperfect palindrome. *Hr*'s act as enhancers of transcription and as replication origins for transient plasmid DNA replication assays.

The availability of baculovirus complete genome sequences, currently 41, along with a better understanding of the biology of the baculoviruses allowed for the modeling of potential evolutionary pathways. Throughout evolution, individual genes doubtless underwent mutations, but there also appeared to be extensive horizontal gene transfer between virus and host and between virus and virus in shared environments. Moreover, most viruses in nature exist as a mixture of genotypes. It is this genotype plasticity in concert with horizontal gene transfer that complicates efforts at defining a clear evolutionary pathway for the baculoviruses.

Evolutionary models were based initially on phylogeny of individual core genes such as polyhedrin and granulin, and then, of 'fused' core protein sequences as well as conservation of gene order. This phylogeny results in four evolutionary groups which is in concert with the proposal for creating four genera. The most ancient baculoviruses are thought to be the dipteran ones, with culex nigripalpus nucleopolyhedrovirus as the sole member. The next most ancient are hymenopteran baculoviruses, like neodiprion sertifer NPV (NeseNPV) and NeleNPV which infect sawflies and have the smallest genomes of all baculoviruses. The lepidopteran NPVs represent a more recent lineage with initial branches separating the GVs such as CpGV and xestia c-nigrum granulovirus (XecnGV) from the NPVs and the NPVs separating easily into two groups, group II including lymantria dispar MNPV (LdMNPV) and spodoptera exigua MNPV (SeMNPV) and the most recent group, group I NPVs, including AcMNPV and CfMNPV.

## Virus Replication Cycle

### Attachment, Entry, and Uncoating

The initial steps leading to virus entry in insects differ somewhat from that in cell culture. In insects, following *per os* ingestion of polyhedra, the polyhedra dissolve in the alkaline environment of the insect gut releasing the occlusion-derived virions. In order to access the target columnar midgut epithelial cells, the released virus must first penetrate the peritrophic membrane lining the gut. Some viruses, like some GVs, encode enhancin or virus-enhancing factor (VEF), which increases the infectivity in larvae of other baculoviruses. As a metalloproteinase which degrades mucin, VEF is thought to disrupt the peritrophic membrane barrier to the midgut epithelium, allowing easier penetration by the virus. VEF is also thought to increase the fusion efficiency between the viral and cellular membranes during infection. The membrane of the virus then fuses with the membranes of the microvilli of the epithelial cell, releasing the nucleocapsid into the cytoplasm. The host cell receptor is not known and may be a generalized one. The attachment and fusion require the highly conserved baculovirus envelope protein p74 and proteins such as *per os* infectivity factors 1 or 2 (PIF-1, PIF-2).

In insect cell cultures, virus attachment is facilitated by a viral attachment protein, for example, either the group I NPV glycoprotein GP64 or its homologs, or the group II NPV F (fusion) protein or its homologs. The host cell receptor is not known, though it is likely a glycoprotein. Following virus entry by receptor-mediated endocytosis, virus is released from the acidified endosome, presumably as a consequence of acid-mediated fusion between the viral and the endosomal membranes.

The released nucleocapsids, either in insects or in tissue culture cells, migrate to the nuclear pore, perhaps via actin cables induced by the nucleocapsids. Some nucleocapsids which enter midgut epithelial cells bypass the nucleus and bud out into the hemocoel through the basal cell membrane. The nucleocapsid is uncoated either at the nuclear pore or as it enters the nuclear pore to release the DNA core into the nucleus.

### Temporal Regulation of Transcription

Baculovirus transcription is temporally regulated into at least three classes, early, late, and very late, in which expression of genes of one temporal phase is dependent on the expression of genes in the preceding phase. The expression of some genes is not clearly delineated and extend through several phases. Early genes are expressed prior to DNA replication and late genes are defined as those starting at or after the onset of viral DNA replication.

#### Early gene expression

Early genes can be divided into immediate early and delayed early genes. The early genes can have promoters containing TATA-like or CAGT and initiator motifs and some have enhancers allowing them to be transcribed by the host transcription. Immediate early genes, like *ie1*, *ie2*, and *ie0*, the only baculovirus gene yet shown to produce spliced RNAs that result in different protein products, are transcribed within minutes of entry by host RNA pol II. IE0, IE1, and IE2 act as transregulators of downstream genes. For example, IE1 transregulates early genes such as *dnapol* and *hel* for viral DNA replication and genes important for suppression of apoptosis, such as *p35* or inhibitor of apoptosis (*iap*). Some baculoviruses, such as those from sawflies, do not have identifiable early transregulatory genes and may have alternate mechanisms for early gene regulation. Another set of early genes are those for auxiliary proteins important for modifying the cell or host insect environment to the benefit of the virus including the production of ecdysteroid UDP glucosyl transferase (EGT), which prevents larval molting and pupation.

Transcription of some of these early genes is increased by *hr* enhancer sequences, allowing for enhanced transcription of genes under conditions early in infection when only a small amount of input viral DNA is available.

#### DNA replication

Once viral proteins involved in viral DNA replication, like DNApol, helicase, and some of the late expression factors (LEFs), are produced, viral DNA replication can initiate. The timing varies with the virus and with the host insect or host cell. In AcMNPV-infected *Spodoptera frugiperda* cells, for example, DNA replication begins at around 6 h post infection and increases until about 18 h post infection. In comparison in CfMNPV-infected *C. fumiferana* cells, DNA replication does not initiate until about 24 h post infection. Little is known about the *cis*-sequences which might act as origins of DNA replication. The baculovirus *hr*'s act as DNA *ori*'s in transient, plasmid DNA replication assays. Other *non hr ori*'s have also been implicated as origins of DNA replication, for example, the *Hin*dIII K fragment of AcMNPV and the *Xba*I-F2 fragment of SeMNPV. Several viral genes, such as AcMNPV *ie-1*, *ie-2*, *hel*, *dnapol*, *lef-1*, *lef-2*, *lef-3*, and *p35*, are essential for NPV viral DNA replication and replication is augmented by *pe38*. LEF-1 is a DNA primase, LEF-2 might help stabilize binding of LEF-1 to the DNA, LEF-3 is a single-stranded DNA (ssDNA) binding protein and acts to transport the viral helicase into the nucleus, and P35 is an apoptotic suppressor which might act both by inhibiting apoptosis and by activating expression of the *hel*, *lef-3*, and *DNA pol* genes among others. It is not currently known if all baculoviruses require similar virally encoded genes for viral DNA replication.

In NPV-infected cells, viral DNA and at least four viral proteins (IE1, LEF3, DNA helicase and another viral ssDNA binding protein, DBP) in association with *hr*'s accumulate in discrete areas within the nucleus, forming a virogenic stroma in which viral DNA replication, occluded virus morphogenesis, and occlusion body formation are thought to occur.

### Late gene expression

The late phase of baculovirus transcription begins after the initiation of viral DNA replication and appears to be at least in part dependent on DNA replication. The late genes are mostly for the virion structural proteins including the AcMNPV major capsid protein VP39, a basic protein P6.9, and the virion envelope glycoprotein GP64. Chitinase and viral cathepsin, both involved in the liquefaction of the infected insects, are also synthesized late in infection.

All baculovirus late and very late genes are distinguished from the early genes by their promoters, which have a canonical DTAAG as an essential element, and their expression by a viral RNA polymerase. These promoters are differentially recognized by a viral RNA polymerase comprised of four late proteins, LEF-4, LEF-8, LEF-9, and P47. LEF-4 has an essential guanyltransferase activity and so is presumably responsible for 5′ capping of viral transcripts. LEF-8 and LEF-9 have motifs suggestive of an active site for RNA polymerase. The role of P47 is still unknown. Coincident with late gene expression is an almost global decrease in host gene expression. This depression appears to be due to a downregulation of host mRNA synthesis such as transcripts for initiation factors eIF2, eIF2α, eIF4E, and eIF5A, ribosomal protein L15 (rpL15), histone, and actin. However, a few host genes, such as the heat shock cognate Sf Hsc70 gene, appear to be upregulated, at least transiently, following infection with AcMNPV.

*p10* and *polyhedrin* (or *granulin*) are expressed very late in infection. In addition to the DTAAG element typical of late gene promoters, very late promoters have a 'burst element' downstream of the DTAAG site. In addition, a very late expression factor 1 (*vlf-1*) gene product, VLF1, preferentially stimulates very late gene expression. P10 is involved in cellular lysis and is associated with a nuclear/ cytoplasmic fibrillar structure in the area of polyhedral morphogenesis in the intranuclear virogenic stroma and is thought to stabilize them. Though expendable, P10 might mediate production of tubular structures found in infected cells. As both polyhedrin for NPVs and granulin for GVs are needed in high amounts and only very late in infection after budded virus is produced, their genes are expressed both at high levels and very late in infection.

## Morphogenesis and Release

During viral morphogenesis, nucleocapsids start forming in the nuclear virogenic stroma, first by the production of capsid sheaths later filled in by core material in a polar manner with a cap structure at one end. The virus then undergoes two different morphogenetic pathways, one leading to budded virus, the other leading to enveloped virions, which become occluded to form polyhedra (NPVs) or granules (GVs). The budded viruses are responsible for cell-to-cell transmission *in vitro* or *in vivo*, while the occluded form is responsible for insect-to-insect spread.

Early in infection, some of these intranuclear nucleocapsids follow the budding route. The NPV nucleocapsids first bud through the nuclear membrane by synhymenosis, acquiring a double membrane. By an unknown mechanism (e.g., disintegration of the membrane, or fusion of inner and outer membranes), the nucleocapsids shed that membrane to release the nucleocapsids into the cytoplasm. In GV-infected cells, the nuclear envelope disintegrates and the nucleocapsids are released directly into the cytoplasm. For NPV and possibly GV, nucleocapsids align in a polar fashion with the nipple end of the nucleocapsids aligned with the plasma membrane. They then bud through the plasma membrane such that the peplomers, that may contain GP64 or F, cover only the apical end, providing polarity to the virions.

For some baculovirus species, such as the sawfly NPVs and some GVs, the infection remains within the midgut. However, for most baculovirus species, virus disseminates from the midgut epithelial cells, throughout the insect to infect all tissues. For these, the virus buds through the basal layer of the midgut epithelial cells and then must pass through the basal lamina to the hemolymph in the open circulatory system of the larvae where it is circulated to other tissues. Baculoviruses are also thought to be disseminated via the tracheal system.

Later in infection, the intranuclear nucleocapsids become enveloped in a *de novo*-derived membrane. Some baculovirus nucleocapsids become enveloped singly (SNPVs and GVs) and for some, multiple nucleocapsids are enveloped within a common envelope (MNPVs). The SNPVs and the GVs generally have only a single nucleocapsid per virion. For the MNPV, several nucleocapsids are found within each virion. The virions then become occluded within polyhedra (NPVs) or granules (GVs). The systemic infection is so effective that following death, up to 25% of the larval mass is due to polyhedra.

Also, late in infection, chitinase and cathepsin, which have been produced in an inactive form, are activated. They are interdependent for activation, release, and mediating liquefaction, which is critical to release, spread, and transmission of the infection in nature. If either of these proteins is missing, as in a knockout virus or viruses that lack one or both of these genes, there is no liquefaction. These enzymes are not encoded by certain GVs such as adoxophyes orana granulovirus (AdorGV), phthorimaea operculella granulovirus (PhopGV), and XecnGV, or the dipteran, mosquito baculovirus culex nigripalpus nucleopolyhedrovirus (CuniNPV), and the hymenopteran sawfly viruses NeleNPV, NeseNPV, and neodiprion abietis nucleopolyhedrovirus (NeabNPV). For these, much of the virus is released by sloughing of infected gut epithelial cells and excretion of viruses, so there is no need for liquefaction to spread the virus.

## Modulation of Host Insect Defenses and Physiology

Many viruses are now known to modulate host defenses and host physiology through specific viral proteins which disrupt the normal signaling pathways. One of these is the inhibition of apoptosis (cellular suicide) and another is manipulation of the ecdysone-mediated insect molting program.

### Apoptosis and Baculovirus Defense against It

Apoptosis, also known as programmed cell death, can be activated in response to detection of cell damage and infection with viruses. Apoptosis is a regulated process in response to an apoptotic signal and effected through the activation of initiator caspases which in turn activate effector caspases. The protease activities of these activated caspases, in concert with that of certain enzymes, like caspase-activated DNase (CAD), activated by them, leads to DNA fragmentation, membrane blebbing, and disruption of the cellular architecture resulting in the formation of apoptotic bodies and cell death. Though apoptosis is a normal part of development and tissue homeostasis, during a virus infection, a cell often initiates an apoptotic response, preventing the cell from supporting virus replication and halting further virus spread. However, many viruses are now known to have an anti-apoptotic defense to prevent this. As shown in 1991, one of the first viral proteins demonstrated to have this anti-apoptotic ability is the P35 protein of AcMNPV.

The AcMNPV P35 is a pancaspase inhibitor which targets one of the *S. fugiperda* host cell effector caspases, Sf-caspase-1, thereby preventing completion of the apoptotic cascade. AcMNPVs lacking P35, as in the annihilator mutant, do not replicate efficiently due to premature apoptosis of infected cells. Interestingly, the AcMNPV P35 also protects *Ceanorhabditis elegans*, *Drosophila melanogaster*, and mammalian cells against apoptotic stimuli, thereby increasing the stature of this baculovirus P35 in the wider apoptosis community. Although BmNPV has a *p35* homolog, it is not functional. However, *p49*, the SpliNPV *p35* homolog, can inhibit apoptosis and moreover can rescue *p35* deletion mutants of AcMNPV and inhibits the *Drosophila* initiator caspase DRONC which P35 is unable to.

The inhibitor of apoptosis proteins (IAPs) are another class of baculovirus apoptosis inhibitors. Some act by binding to and inhibiting the function of various pro-apoptotic factors as well as more directly inhibiting caspase activity. First discovered in CpGV and orgyia pseudotsugata nucleopolyhedrovirus (OpMNPV), as many as 20 other *iap*'s have been described among both GVs and NPVs. IAPs were shown to be functional in CpGV, OpMNPV, and epiphyas postvittana nucleopolyhedrovirus (EppoNPV). IAPs are metalloproteins with an essential amino-terminal zinc-binding motif, BIR (baculovirus IAP repeat), and a carboxy-terminal zinc-binding RING motif. As for P35, some IAPs, like Op-IAP from OpMNPV, also target Sf caspase-1. Op-IAP binds HID and GRIM, two *Drosophila* cell death activators, through the BIR domains. The RING domain appears to be involved with ubiquination, but whether the RING is functional in inhibition of apoptosis is not yet known.

### Baculovirus EGT and Inhibition of Molting

While many of the virus–host interactions occur at the cellular level, baculoviruses can also influence the overall physiology of infected insects to their advantage. One example is the disruption of the hormonally regulated metamorphosis through larval molts and pupation by modification of the hormone, ecdysone, which regulates this process. Many baculoviruses produce EGT, a UDP ecdysteroid glucosyltransferase resulting in glycosylation of ecdysone, thereby disrupting its normal regulatory function. Though overly simplistic, essentially the insect remains in the larval stage and continues to grow during infection, resulting in a higher yield of virus than would otherwise be possible and resulting in even more extensive feeding damage. *Egt* is nonessential for growth in cell culture and so is a common site for foreign gene insertion. Depending on the virus–host system, deletion of *egt* enhances virus infectivity. For example, deleting the *egt* of HearNPV decreased the $LT_{50}$ for this virus by 27% compared to nonmodified HearNPV.

## Baculovirus Host Range Determinants

Depending on the virus species, different baculoviruses have host ranges from one or only a few host insect species, like many of the GVs and the CfMNPV attacking the spruce budworm, to a very broad host range such as for AcMNPV. Although the host range and tissue tropism of many viruses is dependent on interaction with specific host receptors, this does not appear to be the case for baculoviruses and no specific cell receptors have yet been identified. There are several other factors which influence the host range of different baculoviruses. In some cases, the exchange of as little as a single gene (or even only a small part of the gene) is sufficient to alter the host range. For example, AcMNPV can replicate in

*S. frugiperda* cells but very poorly in *B. mori* cells. However, if the AcMNPV helicase gene is exchanged for the helicase from BmNPV, the host range of AcMNPV is expanded to include *B. mori* larvae. In order for AcMNPV to replicate in TN368 cells, it requires a host-cell-specific factor gene 1 (*hcf-1*); however, *hcf-1* is dispensable for replication in Sf21 cells. HCF-1 is thought to act at the level of gene repression requiring a RING finger domain. Similarly, a host range factor gene, *hrf-1*, was identified in LdMNPV, which allowed replication of LdMNPV in Gypsy moth *Lymantria dispar* cells and insects. When *hrf-1* was introduced into three NPVs – AcMNPV, hyphantria cunea nucleopolyhedrovirus (HycuNPV), and BmNPV – all of which normally cannot replicate in an *L. dispar*-derived cell line (Ld652Y), all three expanded their host range to include cell type. The mechanism of action of HRF-1 to expand the host range is not yet known.

The relative ability of anti-apoptotic proteins, like P35 and IAPs or their homologs to inhibit apoptosis in different species of insects, might also affect the host range. For example, P35 deletion mutants of AcMNPV impairs the ability of that virus to grow in Sf21 cells but not in TN368 cells. Similarly, the ability of viruses to replicate in cells of certain insect species depends in part on the types of caspases encountered. For example, AcMNPV can replicate in Sf9 cells due to its ability to inactivate the effector caspase, Sf caspase-1; however, in Sl2, a cell line from *S. littoralis*, AcMNPV actually induces apoptosis presumably as a result of the upregulation of a different effector caspase, Sl caspase-1. Consequently, the AcMNPV host range does not extend to Sl2 cells.

## Baculovirus-Revolutionized Expression Vector Technology

Early work on the molecular biology and the sequence of the polyhedrin gene, the first baculovirus gene ever sequenced, led to the discovery, in Max Summers' group in 1981, that the polyhedrin promoter is a powerful promoter. The first heterologous gene tested by a recombinant baculovirus based on the *polh* promoter was the human beta interferon which was expressed to very high levels. Since this pioneering work on the baculovirus expression vector system (BEVS), it has become the basis for the expression of many thousands of proteins for use as research reagents, as antigens, in diagnostics, as vaccines, and as therapeutic proteins like monoclonal antibodies. As a baculovirus is a eukaryotic expression vector, many of the common post-translational modifications that occur to produce mammalian proteins also occur in the BEVS system. However, whereas the glycosylation in insect cells involves simpler N-glycans with terminal mannose residues, in mammalian cells, more complex N-glycans with terminal sialic acids are made. Nevertheless, biological activity is retained in most BEVS-expressed proteins, even if the glycosylation is not that of the native protein. Attempts are now underway to 'mammalianize' the glycosylation pathway in insect cells to improve the authenticity of baculovirus-expressed mammalian proteins even more. Since baculoviruses are also capable of transduction into mammalian cells, they are also being studied for their potential in gene therapy.

## Baculoviral Insecticides

The major impetus for the study of baculoviruses has been and continues to be their real and theoretical value as potent, biologically based insect pesticides. As these viruses already occur in nature, their use should leave less of an environmental imprint compared to the use of synthetic chemicals in the management of agricultural and forest pest insects. There are many examples of insect baculoviruses being effective in controlling pest populations. In Canada, for example, during the late 1930s, the 12 000 mi$^2$ infestation of the European sawfly *Gilpinia hercyniae* was brought completely under control by an NPV specific to this pest, brought in from Finland, and by 1945, the pest was totally eradicated. In Brazil, and elsewhere, the velvetbean caterpillar anticarsia gemmatalis is a major pest of soybeans. However, use of the native baculovirus anticarsia gemmatalis MNPV (AgMNPV) reduced the larval populations by 80%, the same level as for insecticide treatment and reduced the need for and costs of chemical insecticides. Up to 1.2 million ha has been treated on an annual basis in Brazil, with excellent results using virus produced by farm cooperatives and commercial sources. In the case of Brazil, the extension services informing farmers of this opportunity to reduce the use of chemical pesticides and the clear success shown by the use of AgMNPV was critical to the overall implementation of this form of biological control. Some of the major drawbacks on the commercialization of baculoviruses for pest control remain the relatively high cost of production, the extensive safety documentation needed for registration, the slow speed of kill, and the reluctance to implement new insect pest control strategies. Nevertheless, the development of pesticide resistant insects combined with the decreased public acceptance of chemical control methods, the biologically based baculovirus alternatives should become more attractive. Much work is also ongoing to improve the efficacy of a variety of baculoviruses in controlling pest insects. Some of these strategies include introduction of entomotoxic protein genes into the baculovirus genome, and other genetic manipulations, such as deletions of genes, like *egt* or exchange of genes among different baculoviruses and reorganization of the genome.

*See also:* Baculoviruses: Apoptosis Inhibitors; Baculoviruses: Expression Vector; Baculoviruses: Molecular Biology of Granuloviruses; Baculoviruses: Molecular Biology of Mosquito Baculoviruses; Baculoviruses: Molecular Biology of Nucleopolyhedroviruses; Baculoviruses: Molecular Biology of Sawfly Baculoviruses; Baculoviruses: Pathogenesis.

## Further Reading

Arif BM (2005) A brief journey with insect viruses with emphasis on baculoviruses. *Journal of Invertebrate Pathology* 89: 39–45.

Clem RJ (2005) The role of apoptosis in defense against baculovirus infection in insects. *Current Topics in Microbiology and Immunology* 289: 113–129.

Cory JS and Myers JH (2003) The ecology and evolution of insect baculoviruses. *Annual Review of Ecology, Evolution, and Systematics* 34: 239–272.

Harrison RL and Jarvis DL (2006) Protein N-glycosylation in the baculovirus-insect cell expression system and engineering of insect cells to produce 'mammalianized' recombinant glycoproteins. *Advances in Virus Research* 68: 159–191.

Hu YC (2006) Baculovirus vectors for gene therapy. *Advances in Virus Research* 68: 287–320.

Inceoglu AB, Kamita SG, and Hammock BD (2006) Genetically modified baculoviruses: A historical overview and future outlook. *Advances in Virus Research* 68: 323–360.

Jehle JA, Blissard GW, Bonning BC, et al. (2006) On the classification and nomenclature of baculoviruses: A proposal for revision. *Archives of Virology* 151: 1257–1266.

Miller LK (ed.) (1997) *The Baculoviruses.* New York: Plenum Press.

Moscardi F (1999) Assessment of the application of baculoviruses for control of Lepidoptera. *Annual Review of Entomology* 44: 257–289.

Summers MD (2006) Milestones leading to the genetic engineering of baculoviruses as expression vector systems and viral pesticides. *Advances in Virus Research* 68: 3–73.

Theilmann DA, Blissard GW, Bonning B, et al. (2005) Baculoviridae. In: Fauquet CM, May MA, Maniloff J, Desselberger U and Ball LA (eds.) *Virus Taxonomy: Eighth Report of the International Committee on Taxonomy of Viruses,* San Diego, CA: Elsevier Academic Press.

Van Oers MM (2006) Vaccines for viral and parasitic diseases produced with baculovirus vectors. *Advances in Virus Research* 68: 193–253.

# Baculoviruses: Molecular Biology of Nucleopolyhedroviruses

**D A Theilmann**, Agriculture and Agri-Food Canada, Summerland, BC, Canada
**G W Blissard**, Boyce Thompson Institute at Cornell University, Ithaca, NY, USA

© 2008 Elsevier Ltd. All rights reserved.

## Glossary

**Budded virus** The type of baculovirus virion that is formed by budding from the cell plasma membrane and mediates the systemic spread of infection in the infected insect.

**Homologous repeat element** A DNA sequence that is typically comprised of a series of imperfect palindromes and repeated at various locations around a baculovirus genome. Homologous repeat elements function as origins of DNA replication and as transcriptional enhancers.

**Occlusion derived virus** The type of baculovirus virion that is assembled in the nucleus and becomes occluded with the paracrystalline matrix of the occlusion body.

**Occlusion body** A crystalline protein matrix which surrounds the virions of some insect viruses.

**Polyhedrin** The major occlusion body protein of NPV occlusion bodies.

## Introduction

The family *Baculoviridae* currently consists of two genera, *Nucleopolyhedrovirus* and *Granulovirus*. The nucleopolyhedroviruses (NPVs) currently include all baculoviruses that are not classified as a granulovirus (GV), including those isolated from hymenopterans (sawflies) and dipterans (mosquitoes). In recent years, it has been recognized that the hymenopteran and dipteran baculoviruses are distinct from all other baculoviruses and it has been proposed that the *Baculoviridae* be subdivided into four new genera: the alphabaculoviruses (which include the current lepidopteran NPVs), betabaculoviruses (which include the current GVs), gammabaculoviruses (which include the current hymenopteran NPVs), and deltabaculoviruses (which include the current dipteran NPVs). This article focuses primarily on the lepidopteran NPVs (the proposed alphabaculoviruses).

Lepidopteran NPVs are the most widely studied of the *Baculoviridae*, primarily due to the availability of insect cell culture systems that are permissive for infection.

A number of lepidopteran NPVs have been developed as bioinsecticides, and in addition lepidopteran NPVs are used widely as protein expression vectors and more recently as vectors for mammalian cell transduction. Also, there is a great deal of recent interest in developing certain lepidopteran NPVs as mammalian gene therapy vectors.

## Life Cycle

Like all baculoviruses the lepidopteran NPVs contain genomes of double-stranded circular DNA, with genomes ranging in size from ~111 to 168 kbp. Two distinct types of virions are produced in the life cycle: the budded virus (BV) and the occlusion-derived virus (ODV) (**Figure 1**). BVs generally contain a single nucleocapsid and obtain an envelope by budding from the plasma membrane of the infected cell. In contrast, ODVs are formed in the nucleus, acquire a membrane that is derived from the inner nuclear membrane, and virions consist of either single (S) or multiple (M) nucleocapsids per envelope. The S and M designation (SNPV and MNPV) does not appear to hold any taxonomic significance above the species level but this characteristic is clearly associated with certain viral species. As the infection progresses, ODVs are assembled in the nucleus and become embedded in a paracrystalline proteinaceous matrix to form occlusion bodies (OBs) (also known as polyhedra) that can range in size from 0.15 to 15 μm. A single matrix protein called polyhedrin makes up the majority of the mass of the polyhedra or OB. The OB as a whole is surrounded by a carbohydrate-protein layer known as the polyhedral envelope.

Because the ODVs and BVs have lipid bilayer envelopes that are derived from different sources (nuclear vs. plasma membranes, respectively), the envelopes of ODV and BV have a significantly different protein composition. Proteomic analyses of ODV from the archetype NPV, autographa californica MNPV (AcMNPV), and from helicoverpa armigera NPV (HearNPV) have resulted in the identification of up to 46 viral proteins of which eight or more may be found in the ODV envelope (summarized in **Figure 1**). ODV-specific proteins include P74, PIF-1 and -2, ODV-E18, ODV-EC27, ODV-E56 (ODVP-6E), ODV-E66, and P96. Such detailed proteomic analyses have not been performed on BV but the BV virion

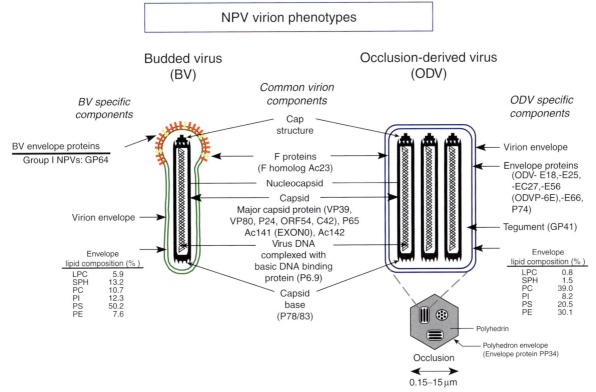

**Figure 1** Schematic diagram of NPV virion phenotypes showing BV and ODV structures. Components common to both virion phenotypes are shown in the center and components unique to each phenotype are indicated on the left and right. The lipid compositions of the BV and ODV envelopes are indicated. LPC, lysophosphatidylcholine; SPH, sphingomyelin; PC, phosphatidylcholine; PI, phosphatidylinositol; PS, phosphatidylserine; PE, phosphatidylethanolamine. Adapted from Cytotechnology, vol. 20(1), 1996, pp. 73–93, Baculovirus–insect cell interactions, Blissard GW, Copyright 1996. With kind permission from Springer Science and Business Media.

phenotype is known to contain BV-specific genes such as GP64. The NPV BV envelope appears to contain fewer viral proteins than ODV envelopes.

## Infection Cycle

The infection cycle of NPVs is initiated when insects ingest polyhedral OBs (**Figure 3(a)**). The OBs pass through the foregut, enter the midgut, and dissolve or disassemble in response to the alkaline environment of the lepidopteran insect midgut. This process releases the ODVs which then traverse the peritrophic membrane (a protein–chitin structure that lines the gut) and infect the columnar epithelial cells of the lepidopteran midgut. This infection process is aided by virally encoded metalloproteases called enhancins that degrade the peritrophic membrane, thus helping virions cross this barrier and interact with the microvilli of the midgut epithelial cells. At the microvillar surface ODVs bind to the cell surface and are believed to enter the cell by direct fusion of the ODV envelope with the plasma membrane at the cell surface. ODV entry has not been examined in great detail although ODV-specific protein P74 appears to be necessary for ODV entry.

Upon entry into the cell, nucleocapsids are transported to the nucleus where they appear to interact with the nuclear pore complex enter the nucleus, and deliver the DNA genome to the interior of the nucleus. The interaction of baculovirus nucleocapsids with the nuclear pore complex and genome delivery to the nucleus is not clearly understood and requires further study. After the initial uncoating of the viral genome, transcription of viral early genes is initiated by host RNA polymerase II, resulting in early transcription and the subsequent production of proteins associated with viral DNA replication and late gene expression. A variety of additional regulatory and even structural protein genes are also transcribed in the early phase. Production of early gene products results in the initiation of viral DNA replication, and the assembly and activity of a virus-encoded RNA polymerase. This viral 'late' RNA polymerase recognizes unique viral late promoters resulting in the transcription of viral structural and other genes. With only rare exceptions, NPV late promoters contain the sequence 'DTAAG' at the late transcription start site. The DTAAG motif and the sequences immediately flanking it comprise the late promoter. Viral DNA replication and nucleocapsid assembly occurs within the nucleus in a viral structure known as the virogenic stroma (**Figure 2(a)**). Nucleocapsids are transported from the cell nucleus to the plasma membrane, where they bud from the cell surface to form the BVs, which are responsible for the systemic spread of the viral infection to other tissues within the insect (**Figure 3(b)**). Nucleocapsids are also retained in the nucleus and become singly or multiply enveloped to form the ODV, which become embedded in polyhedrin to form the OBs (**Figure 2**). Many or most cells and tissues of the host insect are permissive for viral infection and become infected and filled with OBs, which are released when the cells lyse. At least two viral proteins, chitinase and cathepsin, aid in the disruption of infected cells and tissues and the release of OBs into the environment. OBs released into the environment can remain stable for months or even many years if protected from ultraviolet (UV) light. Over $1 \times 10^8$ OBs can be produced from an individual infected larva depending on the host species, the virus, and the larval instar. The dissemination of OBs is believed to be aided by viral modification of host behavior. Many NPV-infected insects will exhibit a behavior of climbing to the upper regions of plants during the late stages of infection. Release of the OBs from such elevated locations is believed to enhance the spread of the virus by making the released OBs more accessible to other larvae feeding on lower regions of the plant.

## Genome Organization (Genomes, Gene Content, Organization, Evolution)

To date the complete genomic sequences have been determined for 30 lepidopteran NPVs. The smallest genome is that from the maruca vitrata NPV (MaviNPV) at 111 953 bp and the largest from the leucania separata NPV (LeseNPV) at 168 041 bp. The archetype and best-studied lepidopteran NPV, AcMNPV, has a genome size of 133 894 bp (**Figure 4**). The numbers of predicted genes (open reading frames (ORFs) of 50 amino acids or greater) range from 126 (MaviNPV) to 169 (LeseNPV). Taxonomic analysis of lepidopteran NPV genomes has led to the subdivision of the NPVs into two distinct clades: group I and group II NPVs (**Figure 5**). The group I clade appears to represent a more cohesive and well-defined phylogeny, whereas the group II clade represents a less-homogenous group of viruses (**Table 1**).

A distinct feature of nearly all lepidopteran NPVs is regions of homologous repeats (*hr*'s) distributed throughout the genome. The *hr* sequences can vary significantly in size. Each *hr* region contains a series of repeated palindromic sequences. In the AcMNPV genome, for example, seven *hr*'s are present and they all contain repeats of an imperfect palindrome that contains an *Eco*RI restriction site at its core (**Figure 4**). Transient transcription and plasmid replication assays have shown that the *hr* sequences function as transcription enhancers and as replication origins. Single copy non-*hr* replication origins have also been identified in the genomes of AcMNPV, orgyia pseudotsugata MNPV (OpMNPV), and spodoptera exigue MNPV (SeMNPV). The non-*hr* origins appear more similar to eukaryotic cellular replication

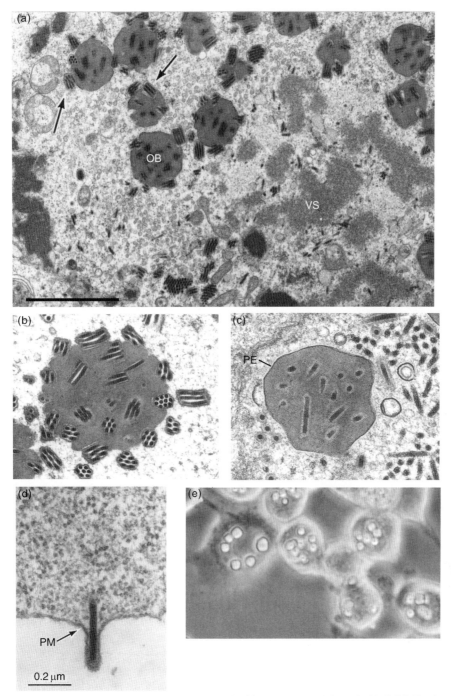

**Figure 2** Electron and light micrographs of NPV-infected insect cells. (a) An insect cell infected with AcMNPV showing the nuclear virogenic stroma (VS) that contains developing nucleocapsids and the OBs showing ODV in the occlusion process (arrows). Scale = 1 μm. (b) An OB of the MNPV type, showing the process of occlusion of ODV-containing multiple nucleocapsids in each virion. Many virions are found in an OB. (c) A mature OB of the SNPV type, showing a single nucleocapsid in each virion (ODV) and many ODVs per occlusion body. Also note the formation of the polyhedral envelope (PE). (d) An NPV nucleocapsid budding from the plasma membrane (PM) of an infected cell. (e) Tn5b1–4 cells infected with AcMNPV producing large relatively uniform OBs. (a–d) Courtesy of R. Granados.

origins and contain multiple short repeated elements. The AT content of the lepidopteran NPVs can be quite variable, ranging from 42.4% (lymantria dispar MNPV; LdMNPV) to 64.4% (adoxophyes hommai NPV; AdhoNPV) and the significance of this variation is not currently known.

## DNA Replication

Baculovirus DNA replication has been best characterized in the lepidopteran NPVs. Transient assays or the use of gene knockout viruses have shown that a number of genes are

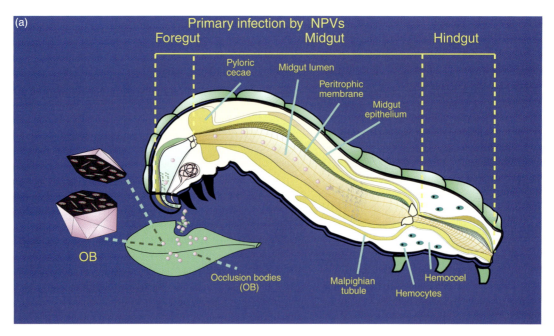

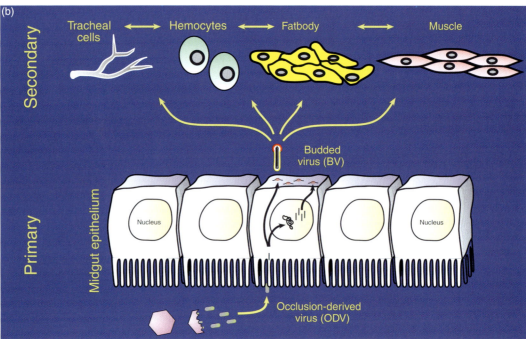

**Figure 3** Schematic diagrams illustrating the infection process in a lepidopteran larva after ingestion of NPV OBs from a contaminated food source. (a) Within the midgut, OBs dissolve or 'disassemble' in the alkaline environment and the embedded ODVs are released into the lumen of the midgut. ODVs traverse the lining of the midgut, the peritrophic membrane, aided by virally encoded proteins and establish primary foci of infection in the midgut epithelial cells. (b) Upon infection of midgut epithelial cells by ODV interactions with microvilli at the apical surfaces of midgut epithelial cells, nucleocapsids are transported to the nucleus and initiate a primary round of viral replication, producing BVs that bud from the basal surface and subsequently infect cells of secondary tissues including tracheal cells, hemocytes, fatbody, and muscle. Alternatively, there is evidence that nucleocapsids can traverse the midgut cell and bud directly from the basal surface without uncoating or replicating in the nucleus, potentially avoiding cellular defense mechanisms associated with the midgut and accelerating the infection of secondary tissues. (a) Reproduced from Slack J and Arif BM (2006) The baculoviruses occlusion-derived virus: Virion structure and function. *Advances in Virus Research* 69: 99–165, with permission from Elsevier.

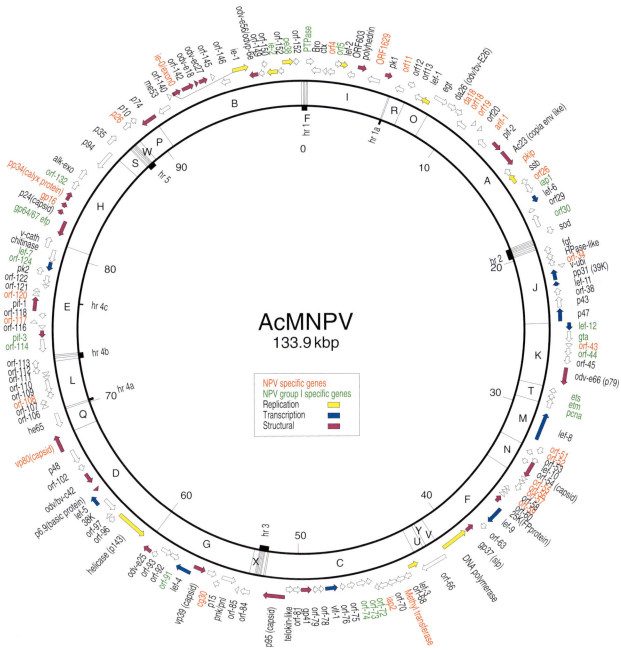

**Figure 4** Genomic map of the archetype lepidopteran NPV, AcMNPV, which contains 154 open reading frames that encode predicted proteins of 50 or more amino acids. Functional groups of genes are highlighted (colored arrows) as well as genes specific to lepidopteran NPVs and group I lepidopteran NPVs (text colors). The inner circle shows the EcoRI restriction map of the C6 strain of AcMNPV. Locations of homologous repeat or hr regions (which contain repeats of EcoRI restriction sites) are indicated.

involved in viral DNA replication and viral genome processing. This includes *lef-1*, *lef-2*, *lef-3*, *lef-11*, *DNA polymerase*, *helicase*, *vlf-1*, *dna binding protein* (*dbp*), *alkaline exonuclease* (*ae*), *me53*, and *ie0* or *ie1*. *lef-1* has been shown to have primase activity and potentially to interact with *lef-2* which has an unknown function. The LEF-3 protein has a single-stranded DNA-binding (SSB) activity, forms homotrimers, and is required for transport of helicase to the nucleus. Helicase is believed to be involved in unwinding of viral DNA. DNA polymerase has homology to other known polymerases and is essential when assayed within the context of a viral infection. In addition to LEF-3, DBP has also been shown to have SSB activity and localizes at sites of viral DNA replication. In the absence of DBP, viral DNA is not processed into full-length genomes and is not packaged correctly into nucleocapsids. In transient replication assays, LEF-11 is not required for viral DNA replication; however, a *lef-11* knockout virus is unable to replicate DNA. LEF-11 has been shown to be a nuclear

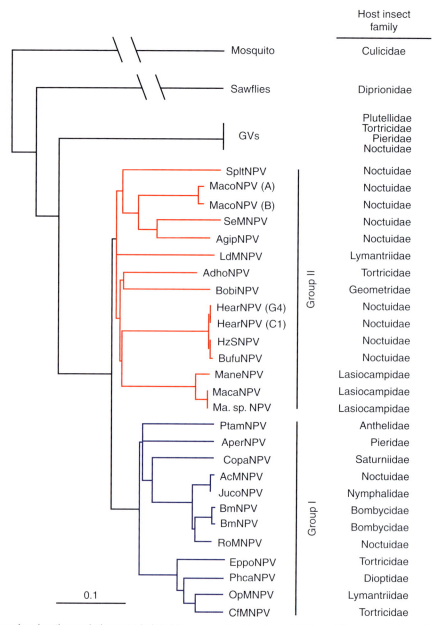

**Figure 5** Phylogram showing the evolutionary relationships among lepidopteran NPVs and their lepidopteran hosts. A concatenated partial amino acid sequence of the common proteins polyhedrin/granulin, lef-8, and lef-9 was used to generate the neighbor-joining (NJ) distance tree and CuniNPV was used as outgroup. Blue branches identify group I NPVs, red identify group II NPVs, and green identify GVs. The lepidopteran family which is the known host of each respective baculovirus is shown on the right. Reprinted from Lange M, Wang H, Zhihong H, and Jehle JA (2004) Towards a molecular identification and classification system of lepidopteran-specific baculoviruses. *Virology* 325: 36–47, with permission from Elsevier.

protein but its role in NPV DNA replication is unknown. Similar to *lef-11*, the deletion of *me53* also results in a virus that is unable to replicate viral DNA. VLF-1 was originally identified due to its impact on baculovirus very late gene transcription. However, in the absence of VLF-1, viral DNA is not packaged into nucleocapsids correctly. In transient assays or recombinant viruses, either of the viral transactivators IE0 or IE1 can support viral DNA replication but both are required to achieve full levels of viral replication. IE-1 and presumably IE-0 bind *hr* sequences and therefore may function as origin binding proteins.

A number of viral proteins have been shown to augment viral DNA replication but are not essential. They include *iap-1* and the group I specific NPV genes *pe38* and *ie2*. All of these genes are RING finger proteins and PE38 and IE2 have ubiquitin ligase activities. However, it is not known whether this enzymatic activity plays a role in DNA replication.

**Table 1** Summary of genes specific to lepidopteran NPVs, and the group I and group II NPV clades

| Gene | Gene function |
|---|---|
| Lepidopteran NPV specific | |
| arif-1 (ac21) | Rearrangement of cellular actin |
| iap-2 (ac71) iap3 | RING domain protein-possible apoptosis inhibition |
| orf1629 (ac9) | Essential structural gene of the nucleocapsid |
| pkip (ac24) | Stimulates the activity of the viral protein kinase-1 *in vitro* |
| pp34 (ac131) | Polyhedral envelope-associated phosphoprotein |
| vp80 (ac104) | Capsid protein |
| cg30 (ac88) | RING domain protein |
| exon0 (ac141) | BV nuclear egress |
| ie0 (ac141 and ac147) | Transcriptional transactivator, DNA replication |
| ac4, ac11, ac17, ac18, ac19, ac26, ac34, ac43, ac51, ac52, ac55, ac56, ac57, ac59, ac69, ac108, ac117, ac120, gp16(ac130), p26 (ac136) | Unknown function |
| Group I NPV specific | |
| ptp1 (ac1) | Protein tyrosine/serine phosphatase, affects OB formation |
| odve26 (ac16) | ODV envelope protein |
| iap1 (ac27) | RING domain protein-possible apoptosis inhibition |
| ie2 (ac151) | RING domain protein, transcriptional activation |
| lef7(ac125) | Late gene expression |
| vp80a (ac87) | Capsids protein |
| lef12 (ac41) | Late gene expression |
| gp64 (ac128) | BV specific glycoprotein required for virion entry |
| pe38 (ac153) | RING domain protein, transcriptional activation |
| ac5, ac30, ac72, ac73, ac114, ac124, ac132, gta (ac42), etm (ac48), ets (ac47), ac44, ac74, ac91 | Unknown function |
| Group II NPV specific | |
| orf4PE | Polyhedral envelope protein |
| rr2b (ld120) | Ribonucleotide Reductase, R2/beta subunit |
| rr1 (ld148) | Ribonucleotide Reductase small subunit |
| parg (ld141) | Possible poly (ADP-ribose) glycohydrolase |
| ld55, ld129, ld111, ld124, ld127, ld142, ld144, ld138 | Unknown function |

The mechanism by which the viral genome replicates is not known but high molecular weight DNAs that are suggestive of genomic concatemers have been reported. It has therefore been suggested that NPV DNA may replicate as a rolling circle. More recently however, due to presence of high molecular weight non-unit lengths of viral DNA in infected cells, it has been proposed that baculoviruses may use a recombination-dependent mechanism of viral DNA replication.

## Temporal Regulation of Transcription

Transfected NPV DNA is infectious indicating that no viral proteins are required to initiate or mediate early transcription from the viral genome. Baculovirus genes are expressed in a temporal cascade beginning with early gene expression and followed by late gene expression. Late gene transcription requires prior or concomitantly with the onset of DNA replication. Very late gene expression occurs at the terminal part of the replication cycle and includes hyperexpressed viral genes such as polyhedrin.

## Early Gene Expression

Early NPV expression is dependent upon the host cell RNA polymerase II complex and is sensitive to α-amanitin. NPV early genes have been divided into two categories, immediate early (IE) and delayed early (DE) genes. IE genes require only cellular factors for expression whereas DE genes are either dependent on, or substantially upregulated by prior viral gene expression. Early genes are primarily involved in gene regulation, host modification, DNA replication, and factors required for late gene expression.

Many IE genes have a common motif, TATA-$N_{24-26}$-CAGT, at their transcription start site similar to many motifs found in insect genomes. The promoters of IE and DE genes resemble typical eukaryotic RNA Pol II promoters and contain host cell transcription factor binding sites.

The primary viral regulatory protein for early transcription is the IE-1/IE-0 complex. The *ie0* gene is the only known spliced baculovirus gene that produces two viral protein products. IE0 contains the entire IE1 coding sequence but in addition has an N-terminal extension that is variable in length depending on the NPV species. IE0 has peak expression prior to viral DNA replication whereas IE1 continues to increase in steady-state levels until the final stages of the infection cycle. IE0 appears to be specific to the lepidopteran NPV as bioinformatics analysis has been unable to identify homologs in non-lepidopteran NPVs or GVs.

Viral gene knockouts have shown that either IE1 or IE0 is essential for viral replication and in the absence of this gene no viral infection is initiated. IE1/0 is an acidic domain transcriptional transactivator, similar to the herpesvirus VP16 protein; and activates viral gene transcription by both enhancer-dependent and independent mechanisms. IE1/0 forms dimers that bind to the *hr* sequences which serve as the transcription enhancers. IE1/0 is also essential for the viral replication complex where it is believed to play the role of an origin binding protein. In support of this, it has been shown that a transcriptionally inactive OpMNPV IE1 is able to support transient viral DNA replication.

Additional IE genes include *ie2* and *pe38*, which have been shown to augment viral transcription and, as indicated previously, viral DNA replication. Unlike IE1/0, both *ie2* and *pe38* can be deleted from the genome and the virus remains capable of replicating. In both cases however, dependent on the cellular environment, BV production and viral DNA replication are reduced. Some of the most highly expressed viral IE genes are *he65*, *me53*, and *gp64*. The localization of G-actin within the nucleus has been shown to be in part due to HE65 function, whereas ME53 is essential for DNA replication. Overall the IE genes play crucial regulatory roles that coordinate the NPV infection cycle.

DE expression requires additional viral factors to achieve nonbasal levels of expression. An example of a DE promoter is the AcMNPV *pp31* promoter which contains a TATA-CAGT sequence at the transcription start site and is dependent on IE1 for activated transcription (above basal-level) expression. The promoters of DE genes have not been clearly defined and it is not clear whether IE1/0 or other regulatory proteins may bind directly to the DE gene promoter.

## Late Gene Expression

Baculovirus late genes are transcribed by a viral-encoded DNA-dependent α-amanitin-insensitive RNA polymerase that recognizes baculovirus late promoters. With only very rare exceptions, the baculovirus late promoters contain the sequence DTAAG and late transcription initiates within this seqeunce. The DTAAG sequence is considered to be at the core of the late promoter. In model late promoters that have been examined, only very short sequences immediately up- and downstream of the DTAAG sequence were identified as components of the late promoter. Regions farther downstream have been shown to affect levels of very late gene expression. Using transient assays 20 NPV genes or late expression factors (*lef*'s) were identified as important or necessary for late gene expression (*lef-1 to -12, pp31, p47, DNApol, vlf-1, helicase, p35, ie1/0, ie-2*). Many of the *lef* genes have been shown to be required for DNA replication, a prerequisite for late gene expression. Three of these genes are only found in the group I lepidopteran NPVs (*lef-7, lef-12,* and *ie-2*). Studies of purified NPV RNA polymerase have shown that it is comprised of four major proteins, LEF-4, LEF-8, LEF-9, and P47, and homologs of all these genes have been identified in all baculovirus genomes sequenced to date. LEF-4 has been shown to have enzymatic activity for RNA capping and LEF-8 and LEF-9 contain conserved motifs found in known RNA polymerases.

At very late times post infection, transcription of the hyperexpressed late genes (*polyhedrin* and *p10*) is upregulated, resulting in expression of those gene products at extremely high levels. The burst of very late gene transcription is mediated by the viral protein VLF-1, which binds to an A/T-rich region called a burst sequence located downstream of the DTAAG motif. This hyperexpression during the very late phase forms the basis of the baculovirus expression system. The *vlf-1* gene is a core baculovirus gene found in all sequenced baculovirus genomes.

## Morphogenesis

### Viral Structures – Nucleocapsids

The NPV life cycle is biphasic and produces two virion phenotypes, the BV and the ODV. Nucleocapsids of both virion phenotypes are believed to be structurally identical. NPV nucleocapsids are rod-shaped and approximately 30–60 nm by 250–300 nm (**Figures 1** and **2**) and are assembled in the ring zone of the nuclear virogenic stroma. DNA is densely packed within the nucleocapsid and small highly basic protein P6.9 is associated with the viral DNA. P6.9 is a core baculovirus protein that contains *c.* 40% arginine and 30% serine/threonine. The highly basic nature of this protein functions in neutralizing the positive charges of the nucleic acid and aids in the condensation and packaging of the viral DNA. Surrounding the viral DNA is the major capsid protein, VP39 (AcMNPV *orf89*). In the virus AcMNPV, the following proteins are associated with the nucleocapsid: P24, VP80, ORF54, BV/ODV-C42, ORF142, P95, ORF1629, and EXON0. ORF1629 is a phosphoprotein that is essential and is found associated with the basal end of the nucleocapsid. EXON0 is required for efficient egress of nucleocapsids from the nucleus. Interestingly, the very late gene transcription factor VLF-1 has also been shown to associate with the ends of the nucleocapsids and the data suggest that it facilitates the packaging of viral DNA and the formation of correctly sized nucleocapsids. Deletion of BV/ODV-C42 showed that this gene is essential and BV is not produced though viral DNA replication is not affected.

## Viral Structures – BV and ODV Envelopes

The virion envelopes of BV and ODV are obtained by different mechanisms and contain arrays of proteins that are generally specific to each virion type. Nucleocapsids are believed to acquire an initial envelope when they egress from the nucleus to the cytoplasm. By an unknown mechanism, the envelope is lost and nucleocapsids are transported to the cell surface, potentially utilizing cytoskeletal structures such as microtubules. At the cell surface, the nucleocapsids bud from the plasma membrane acquiring the modified plasma membrane as the BV envelope. Nucleocapsids destined to be incorporated into ODV are retained in the nucleus and are enveloped by a membrane that is derived from the inner nuclear membrane. Analysis of the lipid composition of AcMNPV ODV and BV envelopes show that they differ significantly, reflecting the differences of their origins.

The BV envelope of group I NPVs contains the major envelope glycoprotein, GP64, which is essential for viral attachment and membrane fusion during entry (see below). GP64 is not present in ODV and is therefore a BV-specific protein. GP64 proteins are not found in group II NPVs but a protein that is functionally equivalent, the F (fusion) protein, is present in group II NPVs and GVs. The best-studied GP64 proteins are those of AcMNPV and OpMNPV. F proteins have been studied most extensively from SeMNPV (ORF8), HaNPV (ORF133), and LdMNPV (ORF133). All viruses that encode a GP64 protein (group I NPVs) also encode a homolog of the F protein, although the F homolog in group I NPVs is not essential for virion entry or viral replication (see below). Unlike GP64, F proteins (and homologs in the group I NPVs) may not be specific to BVs. Proteomic analyses have identified F proteins in the ODVs of AcMNPV, HearNPV, and CuniNPV. More detailed studies will be necessary to understand any potential role of F proteins in the ODV.

Viral entry by budded virions of the NPVs is mediated primarily by the major envelope protein, GP64 or F. The GP64 proteins are highly conserved, with approximately 80% amino acid sequence identity among all GP64 protein ectodomains examined. Structurally, GP64 is a type I integral membrane glycoprotein that is phosphorylated, palmytoylated, and heavily glycosylated. It is found on the cell surface and on the virion as a disulfide-linked trimer of GP64 monomers. GP64 functions in both viral entry and exit. In viral entry, GP64 is important for host cell receptor binding and membrane fusion. Little is known of the details of GP64 interactions with host cell receptor(s) and the specific molecule(s) that serves as the host cell receptor for virion binding is unknown. However, GP64 binding appears to be highly promiscuous and this feature has been exploited in the use of baculoviruses as mammalian transduction vectors, as potential gene therapy vectors, and in the use of GP64 for pseudotyping other viruses such as retroviruses and paramyxoviruses. F proteins from group II NPVs presumably serve a similar role in host receptor binding although they may recognize a different host cell receptor. After host cell binding, BVs of viruses such as AcMNPV are internalized via endocytosis and the low pH of the endosome triggers a conformational change in the GP64 protein, resulting in activation of membrane fusion. Unlike many other membrane fusion proteins, the GP64 protein does not require a prior internal cleavage (within the ectodomain) for maturation or activation of the functional fusion protein. Studies of GP64-mediated membrane fusion indicate that large short-lived complexes of approximately 10 or more GP64 trimers form immediately prior to membrane merger and are likely the unit structure of the membrane fusion mechanism. In addition, the opening of the fusion pore occurs rapidly after triggering.

F proteins from group II NPVs such as SeMNPV, HaNPV, and LdMNPV are also low-pH-activated membrane fusion proteins. These F proteins require an internal cleavage by a cellular proprotein convertase, for fusion activity. F proteins from group II NPVs share general structural features with paramyxovirus F proteins.

Studies of a *gp64* gene knockout AcMNPV virus showed that GP64 is also necessary for efficient budding of progeny BV. Because GP64 is found on the surface of infected cells, concentrated in discrete areas, it is thought that these concentrations of GP64 represent the sites of virion budding. GP64 is found on the virion in a polarized manner with the GP64 spikes found concentrated at the end of the virion that corresponds to the end where budding initiated (**Figure 2**). An important question in the biology of this virus is whether GP64 accumulation determines the sites of BV budding, or whether it is simply targeted to the same sites. Like the VSV G protein, GP64 is required for efficient budding. However, the precise domains required for virion assembly or budding are not yet known.

Group I NPVs encode both GP64 and an F homolog (Ac23 in AcMNPV, Op21 in OpMNPV). In AcMNPV, GP64 is essential whereas the F homolog (Ac23) is not and Ac23 can be deleted with no substantial effect in cell culture infections. However, the conservation of F homologs in the genomes of the group II NPVs suggests an important function. Deletion of the Ac23 gene from the AcMNPV genome results in delayed mortality of infected larvae. Thus, while F proteins found in group II NPVs are essential entry proteins, the F homologs in group I NPVs are not essential but appear to serve an accessory role that may be important in the pathogenicity or virulence of the virus.

The GP64 protein has been used for several biotechnological applications, including peptide display on baculovirus particles, and pseudotyping by replacing the envelope protein of another virus with GP64. Gene therapy vectors derived from retroviruses can be effectively pseudotyped with AcMNPV GP64 or other GP64 proteins. Further study of these viral envelope proteins should yield important and useful new tools for applications in biotechnology, agriculture, and medicine.

ODV envelopes have been shown to contain a number of specific proteins that are not present in BV. They include the following AcMNPV proteins (or their homologs): ODV-E18, ODV-E25, ODV-EC27, ODV-E56 (ODVP-6E), ODV-E66, P74, and ORF142. Viruses containing gene knockouts have shown that AcMNPV ORF142 and ODV-EC27 are essential for successful virus assembly and production of infectious virus.

Four genes encode ODV envelope proteins that are required for oral infectivity. *Per os infectivity factor-1, -2* and *-3* (*pif-1, pif-2, pif-3*) and *p74* have been shown to be required for oral infectivity of SeMNPV and AcMNPV. ODVs produced from viruses that do not express P74 are not orally infectious and evidence suggests that P74 is required for receptor binding on the midgut cell surface. Similar results have been obtained with PIF-1, -2, and -3. None of these proteins is required for infection of cells by BV in tissue culture.

## Viral Structures – Tegument Proteins and Polyhedra

GP41 is an ODV-specific glycoprotein that does not fractionate with either the nucleocapsid fraction or the envelope fraction and is therefore believed to be a so-called tegument protein. GP41 is a core baculovirus gene, conserved in all baculovirus genomes examined. A temperature-sensitive mutant of the AcMNPV GP41 protein shows that it plays a critical role in viral development. When grown at the nonpermissive temperature, *ts* mutants of *gp41* fail to produce ODVs and polyhedra, and in addition nucleocapsids fail to egress from the nucleus to form BV. GP41 therefore appears to play a key role in the assembly of both virion phenotypes even though it is a component of only the ODV. Tegument proteins appear to be acquired within the nucleus when ODV nucleocapsids are enveloped. BVs do not contain GP41 as it is believed that all nuclear proteins surrounding nucleocapsids are lost when the nucleocapsids migrate from the nucleus to the plasma membrane.

OBs represent a feature that is common to all the currently classified baculoviruses (**Figure 2**) but the shape and size of the OB can vary substantially. Lepidopteran NPVs produce some of the largest OBs (polyhedra) and, unlike the GVs or the dipteran or hymenopteran NPVs, the embedded ODV can contain single or multiple nucleocapsids per envelope (**Figure 2**). The major component of the OBs of NPVs is the OB matrix protein, polyhedrin, which forms the bulk of the paracrystalline array (**Figure 2**). Surrounding the OB is a structure known as the calyx or envelope which is thought to be comprised of carbohydrate and protein. A lepidopteran NPV-specific phosphoprotein, PP34, is the major protein associated with this structure. To date no specific function has been attributed to the polyhedra calyx in the infection cycle of NPV. However, the OBs of an AcMNPV virus with a *pp34* deletion were found to have increased sensitivity to alkali disruption and enhanced virulence in fourth instar *Spodoptera exiguae* larvae, suggesting that the polyhedra calyx may stabilize the OB.

## NPV-Specific Genes

The lepidopteran NPVs have 28 genes that are specific to all members of this genus and a further 13 genes specific to group II NPVs and 21 genes specific to group I NPVs (**Figure 4**). A number of these genes have been characterized and are known to impart specific functionality on these viruses but many remain to be investigated to determine their role in NPV biology.

*See also:* Baculoviruses: Apoptosis Inhibitors; Baculoviruses: Expression Vector; Baculoviruses: General Features; Baculoviruses: Molecular Biology of Granuloviruses; Baculoviruses: Molecular Biology of Mosquito Baculoviruses; Baculoviruses: Molecular Biology of Sawfly Baculoviruses; Baculoviruses: Pathogenesis; Insect Pest Control by Viruses.

## Further Reading

Blissard GW (1996) Baculovirus–insect cell interactions. *Cytotechnology* 20: 73–93.

Clem RJ (2005) The role of apoptosis in defense against baculovirus infection in insects. *Current Topics in Microbiology and Immunology* 289: 113–129.

Guarino LA, Xu B, Jin J, and Dong W (1998) A virus-encoded RNA polymerase purified from baculovirus-infected cells. *Journal of Virology* 72: 7985–7991.

Herniou EA, Olszewski JA, Cory JS, and O'Reilly DR (2003) The genome sequence and evolution of baculoviruses. *Annual Review of Entomology* 48: 211–234.

Jehle JA, Blissard GW, Bonning BC, *et al.* (2006) On the classification and nomenclature of baculoviruses: A proposal for revision. *Archives of Virology* 151: 1257–1266.

Lange M, Wang H, Zhihong H, and Jehle JA (2004) Towards a molecular identification and classification system of lepidopteran-specific baculoviruses. *Virology* 325: 36–47.

Miller LK (ed.) (1997) *The Baculoviruses*. New York: Plenum.

Okano K, Vanarsdall AL, Mikhailov VS, and Rohrmann GF (2006) Conserved molecular systems of the *Baculoviridae*. *Virology* 344: 77–87.

Oomens AGP and Blissard GW (1999) Requirement for GP64 to drive efficient budding of autographa californica multicapsid nucleopolyhedrovirus. *Virology* 254: 297–314.

Slack J and Arif BM (2006) The baculoviruses occlusion-derived virus: Virion structure and function. *Advances in Virus Research* 69: 99–165.

Smith GE, Fraser MJ, and Summers MD (1983) Molecular engineering of the *Autographa californica* nuclear polyhedrosis virus genome: Deletion mutations within the polyhedrin gene. *Journal of Virology* 46: 584–593.

Stewart TM, Huijskens I, Willis LG, and Theilmann DA (2005) The *Autographa californica* multiple nucleopolyhedrovirus *ie0–ie1* gene complex is essential for wild-type virus replication, but either IE0 or IE1 can support virus growth. *Journal of Virology* 79: 4619–4629.

Theilmann DA, Blissard GW, Bonning B, et al. (2005) The *Baculoviridae*. In: Fauquet CM, Mayo MA, Maniloff J, Desselberger U, and Ball LA (eds.) *Virus Taxonomy: Eighth Report of the International Committee on Taxonomy of Viruses*, pp. 177–185. San Diego, CA: Elsevier Academic Press.

# Baculoviruses: Pathogenesis

**L E Volkman,** University of California, Berkeley, Berkeley, CA, USA

© 2008 Elsevier Ltd. All rights reserved.

## Viral Architecture, Phenotypes, and Host Distribution

The family *Baculoviridae* is comprised of large, enveloped, bacilliform, double-stranded DNA viruses with juvenile insect hosts. The family currently contains the genera *Granulovirus* and *Nucleopolyhedrovirus* though others are proposed. Granuloviruses (GVs), with exclusively lepidopteran hosts, and nucleopolyhedroviruses (NPVs) with hymenopteran, dipteran, and lepidopteran hosts, fall into distinct clusters phylogenetically. The lepidopteran NPVs are the most abundant of the baculoviruses, and fall into two clades, groups I and II. GVs are packaged as single nucleocapsids per envelope, and the NPVs are packaged either as single nucleocapsids per envelope (the single nucleopolyhedroviruses, SNPVs), or one through multiple nucleocapsids per envelope (the multiple nucleopolyhedroviruses, MNPVs) (**Figure 1**). The latter feature, referred to as the M trait in this article, is unique to lepidopteran NPVs, and occurs in both groups I and II. The M and S traits have not been linked to any viral genes, and nothing is known about factors that regulate nucleocapsid distribution, but the M and S traits reflect differences in strategies for establishing systemic infection.

Baculovirus-enveloped virions are occluded within protein matrices that protect them, enabling them to remain viable in soil for years. The *granulin* gene encodes the major occluding protein for the GVs, and the *polyhedrin* gene does the same for the NPVs. These homologous genes, along with the NPVs' *p10*, are hyperexpressed during the very late phase of infection as late gene expression subsides. Occlusion formation is not unique to baculoviruses. It evolved independently at least three times among insect viruses as a viability-maintenance mechanism. Entomopox viruses in the family *Poxviridae* and cytoplasmic polyhedrosis viruses in the family *Reoviridae* both form viral occlusions that persist in the environment.

GVs pack a single enveloped virion per occlusion while the NPVs can pack hundreds of virions per occlusion. NPV occlusions range from 1 to 15 µm in diameter and can have over 1000-fold the volume of the much smaller GV occlusions. All of the GVs and most of the NPVs infect lepidopteran larvae. Close to 300 lepidopteran species are known to support NPV infections, and 150 species support GV infections. The GVs are known to have very narrow host ranges, and are limited to one or two species within a family. The SNPVs tend to have narrow host ranges as well. In contrast, some MNPVs have impressive arrays of alternative hosts. *Autographa californica* multiple nucleopolyhedrovirus (AcMNPV) has an extremely wide host range and mortally infects at least 43 species within 11 families of lepidopteran insects.

Less than 10% of the NPVs have nonlepidopteran hosts, and all that do, have the SNPV phenotype. The hosts of sequenced NPVs in this category are sawflies and mosquitoes representing orders Hymenoptera and Diptera, respectively, all of which evolutionarily predate the Lepidoptera. Most baculoviruses appear to have radiated along with the Lepidoptera, the most recent of the insect orders. Infection of nonlepidopteran hosts is limited to midgut cells, and midgut cells produce occlusions containing infectious virus. These cells are shed as new midgut cells are infected, and the cycle continues until the death of the insect.

Infections are initiated when viral occlusions, consumed by a susceptible insect host, dissolve and release occlusion-derived virus (ODV) into the midgut lumen. In lepidopteran hosts, occlusion dissolution is triggered by the high pH (8–11) of the digestive fluids. Primary infection of lepidopteran larvae occurs in cells of the columnar epithelial lineage. Fully mature cells typically are targeted but differentiating and regenerative cells also can be infected. Secondary (systemic) infection is established by budded virus (BV), a second phenotype consisting of a single nucleocapsid (even for the MNPVs) with an envelope derived from the basal plasma membrane of the infected columnar cell (**Figure 2**). The BV envelope contains a newly synthesized, viral-encoded fusion protein not present in ODV. BV infects a variety of tissues within the host and uses the fusion protein for productive entry. GVs and most NPVs are thought to use F proteins for BV fusion, but AcMNPV and other group I NPVs use GP64 instead. GP64 is thought to have entered the group I lineage subsequent to F, and to have taken over its fusion function.

## NPV Pathogenesis in Lepidopteran Larvae

### Primary Infection

Occlusion dissolution in the gut lumen is a host range factor that begins a deadly race for the body of the caterpillar. Both virus and insect want to use the larval mass for the same ultimate purpose: reproduction. However, virus and insect reproduction are mutually exclusive events; hence, it is a life-or-death struggle.

The larval protection strategy of covering most of its epithelium with cuticle gives a virus little opportunity for achieving contact with a living cell. The midgut is accessible via ingestion, however, and it is free of cuticle. Moreover, the load maximizing feeding behavior of larval lepidopteran hosts means ingestion is likely to occur if occlusions contaminate their food supply, plant tissue. Even though the midgut is free of cuticle, it secretes a gel-like structure called the peritrophic membrane (PM) that acts as a protective barrier. Some GVs and a few NPVs contain enhancins, viral-encoded metalloproteases, thought to improve access to target cells by compromising the PM. Most baculoviruses also contain products of the '11K gene family' that improve primary infection as well, but by an unknown mechanism.

ODV initiates infection by binding at or near the tips of apical columnar cell microvilli where they fuse with the microvillar membranes. Evidence suggests that receptor binding on midgut cells is mediated by three ODV proteins. For AcMNPV ODV, the viral attachment proteins are *per os* infectivity factors PIF1, PIF2, and P74. The proteins are encoded by genes conserved among all

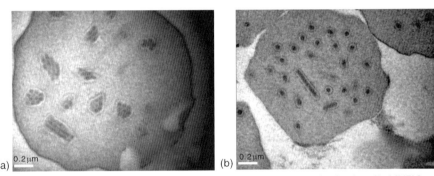

**Figure 1** Comparison of M and S traits. (a) Autographa californica multiple nucleopolyhedrovirus (AcMNPV) and (b) Helicoverpa zea single nucleopolyhedrovirus (HzSNPV) occlusions. Note that HzSNPV occlusions are smaller and more densely packed with virions, but the virions have only a single nucleocapsid compared to multiple nucleocapsids for AcMNPV.

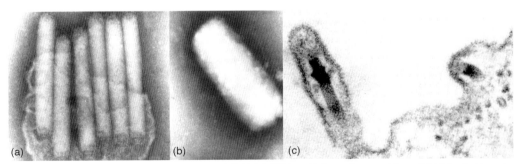

**Figure 2** Comparison of ODV and BV phenotypes of AcMNPV. (a) ODV multiple nucleocapsids and (b) intact virion showing typical smooth envelope. (c) BV in the process of budding showing envelope spikes composed of GP64.

baculoviruses (regardless of the order of their host), suggesting that a conserved mechanism exists among baculoviruses for establishing primary infection. The virus attachment proteins and PIF3, another highly conserved protein, are essential for ODV infectivity but dispensable for BV infectivity. The role of PIF3 is not known, but it probably does not include ODV primary binding or fusion. Curiously, fusion is unimpaired by the absence of any one of these four proteins, suggesting that a novel fusion mechanism may exist for primary infection, or that the ODV fusion protein has not yet been identified.

Columnar cell microvilli of larval lepidopterans may be well over 10 µm long, but typically they are only 100 nm across. Both GV and NPV nucleocapsids, by comparison, are large, measuring 30–60 nm × 250–300 nm. A single nucleocapsid, therefore, occupies about one-third the cross-sectional volume of an infected columnar cell microvillus. Moreover, the MNPVs pack several nucleocapsids within the same virion, a trait with a demonstrated advantage for achieving mortal infection. A virion containing five nucleocapsids (an average number) fused with a single microvillus would greatly exceed the normal cross-sectional volume of the microvillus, and, indeed, this uncomfortable situation has been observed by electron microscopy. Transport of the nucleocapsids down the microvillus would seem improbable as well because of the bulky microvillar F-actin core and cross filaments that seemingly would prevent or severely impair nucleocapsid transport. Because microvilli have no microtubules, nucleocapsid transport must be actin-based and directed toward the minus end of the microvillar F-actin filaments. Only one class of minus-end-directed actin-based motors has been described to date – myosin VI – but its involvment remains to be demonstrated. For the moment, the existing data present an interesting conundrum and suggest that modulation of microvillar architecture to accommodate nucleocapsid transport could be a likely activity of the highly conserved *p74* and *pif* gene products.

### Systemic Infection

Tracheal cells are the primary targets of AcMNPV BV produced in midgut cells in six hosts with widely varying susceptibilities to mortal infection, including *Trichoplusia ni*, *Heliothis virescens*, *Helicoverpa zea*, *Manduca sexta*, *Spodoptera frugiperda*, and *Spodoptera exigua*. Tracheal cells produce cytoplasmic extensions that breach basal laminal barriers to achieve close proximity to the cells they service for efficient gas exchange (**Figure 3**). BV contact with these extensions is key for systemic infection because basal laminae are effective barriers to direct BV passage. In fully permissive hosts, BVs produced in midgut cells infect tracheal cells, and tracheal cell-produced BVs, in turn, bud into the hemolymph where they have direct access to hemocytes, the only insect tissue without a protective basal lamina. Infected hemocytes produce most of the BVs that accumulate within the hemolymph. High titers are needed to infect tracheal cells at distant sites so that they, in turn, can transmit infection to the tissues they are servicing. In this manner, tracheal cell infection is key for viral passage through basal laminal barriers. Sequential infection of midgut cells, tracheal cells, and hemocytes is required for hosts to be fully permissive. Some insects have permissive hemocytes but not permissive midgut cells, while others have permissive midgut and tracheal cells but not permissive hemocytes. Such insects are semipermissive hosts under laboratory conditions, but of doubtful significance to the baculovirus in question in the field.

BVs are much more efficient units of infection than ODVs and have just one nucleocapsid per particle. Intrahemocoelically, in four different species, an $LD_{50}$ dose of AcMNPV BV is less than 0.1 PFU. AcMNPV BV has a physical to infectious particle ratio of 128:1 in cell culture; therefore, very few intrahemocoelic particles (less than 12 and possibly only 1) are needed to achieve fatal infection in fully permissive hosts. Such hosts have no resistance to infection beyond the midgut (see below), so the payoff for infecting even 1 tracheal cell is enormous. Other hosts, such as *H. zea* and *M. sexta*, can mount effective immune responses to systemic infection by AcMNPV. Such responses increase the BV concentration required for an $LD_{50}$ dose considerably.

## Primary Host Defense and Virus Counter Response

The first line of defense of lepidopteran larvae against baculovirus infection is shedding ODV-infected midgut cells. As larvae age, the rate of midgut cell sloughing increases, and, at the end of the instar in which infection is initiated, any remaining infected midgut cells are shed into the gut lumen and voided during the molt. As a consequence of this behavior, larvae become increasingly resistant to fatal infection by NPVs as they age. This phenomenon is known as developmental resistance, and it is a widespread characteristic among lepidopteran larvae. This behavior also leads to a healthy midgut in the instar following the molt that can process food needed for maximizing host size and virus yield (**Figure 3(e)**).

The easily distinguishable M trait observed in some lepidopteran NPVs and nowhere else among extant viruses represents a counter response to developmental resistance. The M trait is one component of a two-component strategy used to minimize the time required to establish systemic infection via midgut cells. AcMNPV cheats the clock by bringing in extra parental nucleocapsids that can be repackaged within midgut cells as BV. Some of the entering nucleocapsids are targeted directly

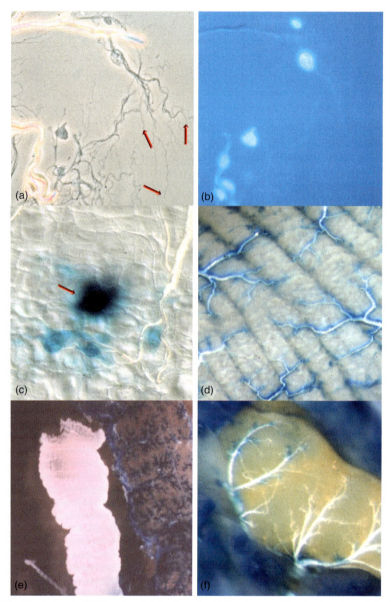

**Figure 3** Tracheal cell infection is key to NPV systemic infection. (a, b) Trachea and tracheal cells, uninfected. Note the long cytoplasmic branches (arrows) relative to the 4′,6-diamidino-2-phenylindole (DAPI)-stained nuclei (b) of the cells. (c) Establishing secondary (systemic) infection in *M. sexta* by AcMNPV expressing the reporter gene β-galactosidase. Focus of infection involves central midgut cell (arrow) and surrounding tracheal cells. (d, e) *Heliothis virescens*-infected tracheal cells against a backdrop of uninfected midgut post clearance. (f) Blue foci of infection of *H. virescens* testis are associated with infected trachael cells, demonstrating their importance in overcoming basal lamina barriers.

through the cell to the basal plasma membrane where they bud as soon as other nucleocapsids targeted to the nucleus begin early gene expression. The genes encoding GP64 and a surprising number of F proteins have early and late promoter elements. Importantly, early expression enables ODV nucleocapsids to bud as BV before late gene expression even begins. In five of the six previously listed insect species, AcMNPV establishes infection of tracheal cells in 4 h or less after having established primary infection in midgut cells. Because both multiple ODV nucleocapsids and early expression of the BV fusion protein are required for this pass-through strategy to work optimally, this mechanism cannot be used by GVs and SNPVs, a factor that may contribute to the relatively wider host ranges enjoyed by MNPVs compared to those of GVs and SNPVs.

## GV Pathogenesis

GVs can be divided into three types, according to their tissue tropisms. The first type, Trichoplusia ni granulovirus (TnGV), can replicate in both the midgut and the fat body. However, nothing is known about how infection is transmitted from the midgut to the fat body, or whether tracheal cells may be involved. TnGV-infected larvae take 10–14 days to die, giving the feeding host time to increase its mass so more virus can be produced. At death, the cadaver does not liquefy. Type II GVs, such as Cydia pomonella granulovirus (CpGV), have the widest tissue tropism of the GVs, and can infect midgut, hemocytes, tracheal cells, fat body, and epidermis similar to the NPVs. Midgut infection for these GVs is transient, and the gross pathology associated with type II GV mortal infections is similar to that seen with lepidopteran NPV mortal infections (see below).

Infection by harrisinia brillians (Western grapeleaf skeletonizer) granulovirus (HabrGV), as typical of type III GVs, is confined to the midgut and is not cleared by molting. Interestingly, HabrGV is more pathogenic in older instars than younger instars, the exact opposite pattern of pathogenesis induced by NPVs and type I and II GVs. First instars, for example, may die after 17 days of infection compared to 8 days for third instars. This pathology is consistent with increased midgut cell sloughing and increased organismal growth observed in later instars. The need to generate many centers of differentiating columnar cells could be problematic if stem and/or early differentiating midgut cells are widely infected. *Harrisinia brillians* are gregarious feeders but the later instars develop a watery diarrhea when infected, and change from gregarious to solitary 'wandering' behavior. As they wander, they leave trails of highly infectious watery excrement. In a plexiglass cage, the effect is very striking. In healthy colonies, the caterpillars stay on the plants almost exclusively, whereas in infected colonies, the cage walls become completely coated with excrement.

## Wandering Behavior and Terminal Disease

Wandering behavior, or enhanced locomotory activity, is normal behavior for lepidopteran larvae anticipating pupation. Wandering behavior induced by baculovirus infection is symptomatic of the terminal stages of infection and serves to spread the virus in the environment. The induction of wandering behavior by bombyx mori nucleopolyhedrovirus (BmNPV), a group I NPV, is caused by the expression of a virus-encoded protein tyrosine phosphatase (*ptp*). *Bombyx mori* larvae infected with BmNPV *ptp* null mutants, do not wander, making *ptp* an unusual and exciting example of a virus gene that controls host behavior.

Infected larvae eventually cease feeding, but not before infected tissues, especially the fat body, are replete with occlusions. Infected larvae in this stage of disease respond more slowly to touch, appear to be swollen, and are more fragile. Within a day of taking on this appearance, the larvae liquefy, an astonishing sight to witness (**Figure 4**). The entire infection process can take 5–12 days. Under laboratory conditions, AcMNPV-infected larvae begin liquefaction 5–7 days post infection. This process is dependent on expression of V-CATH, a cathepsin-like cysteine protease that is activated at cell death. V-CATH cannot be activated, however, if expressed in the absence of viral-encoded chitinase, CHI-A, encoded by a gene found in the same baculovirus genomes that harbor *v-cath*.

There is only one way a fully susceptible larval lepidopteran host can possibly escape death after having been systemically infected, and that is to pupate. Pupae and adults are much less susceptible to infection than larvae, and occlusion yields correspondingly drop precipitously. It is in the best interest of the virus, therefore, to control host development, and baculoviruses do just that. Baculoviruses counter the escape route pupation offers by secreting ecdysteroid UDP glucosyltransferase (EGT) into the hemolymph, an enzyme that catalyzes the transfer of galactose to ecdysome, thereby inactivates it. Ecdysone is a hormone needed to signal the beginning of the molting process, and in lower concentrations, the beginning of pupation. Secretion of EGT, therefore, masks the hormone signal and thereby prevents both pupation and molting. Interestingly, viral mutants lacking functional EGT kill hosts faster than wild-type viruses do. The earlier death is thought to be related to the degeneration of malpighian tubules, but increased anoxic stress might also be a factor. Infected tracheal cells are no doubt incapable of replacing chitinous tracheae lost during the molting process; thus, hosts that molt likely experience increased anoxic stress compared to hosts that do not molt and retain the use of existing trachae.

## Pathogenesis *In Vitro*

Columnar epithelial cells are dead-end cells that experience two extremes of pH at once. Apically, they are bathed in highly alkaline gut fluids and basally they are bathed in slightly acidic hemolymph. These conditions are difficult to mimic in cell culture, a fact that may account for the paucity of information available on primary infection.

**Figure 4** The cadaver of an *H. virescens* larva killed by AcMNPV.

Cell lines that support GV replication are rare and currently only exist for CpGV. Consequently, we know little more about mechanisms of GV infection than we do about ODV infection. Even so, we know that during GV nucleocapsid morphogenesis, the nuclear membrane of the host cell disintegrates, an event that does not disrupt nucleocapsid processing. The disappearance of the nuclear membrane is distinctive and does not occur in NPV-infected host cells.

A number of insect cell lines have been established that support NPV infections. Most of the cell lines are derived from caterpillar ovaries or embryos. AcMNPV can replicate in cell lines derived from several insect species, a fact that contributes to its being the best-studied baculovirus. In a standard one-step growth curve, AcMNPV BVs typically are produced 12–20 h post infection (hpi), and occlusions, 20–48 hpi. Replication of most other NPVs in cell culture takes hours to days longer.

AcMNPV BV is 1000-fold more infectious than ODV, primarily due to the presence of its fusion protein, GP64. BVs enter their target cells by clathrin-mediated endocytosis. BV nucleocapsids penetrate deep into the cytoplasm of target cells by being released from endosomes. Using this strategy, nucleocapsids bypass the possibly hazardous environment that lies just beneath the plasma membrane and enter the cell closer to the nucleus than fusion at the plasma membrane would allow. The pH-sensitive conformational shifts experienced by the BV fusion proteins serve to trigger a fusion event between the viral envelope and endosomal membrane.

AcMNPV nucleocapsid's escape from the endosome is immediately followed by its association with F-actin cables. Cable formation is transient, and occurs during the time viral nucleocapsids move to and enter into the nucleus. During transit, viral nucleocapsids co-localize with one end of an actin cable, possibly via P78/83, an F-actin-binding protein thought to be located at the base of nucleocapsids. The association of nucleocapsids with F-actin cables and the delay of reporter gene expression in the presence of drugs that disrupt actin/myosin function suggest that the cables may facilitate the transport of nucleocapsids to the nucleus, or mediate their passage through nuclear pores.

NPV and some GV nucleocapsids enter nuclei via nuclear pores. Some GVs uncoat at the nuclear pores but most baculovirus uncoating occurs within the nucleus. Early gene transcription begins immediately, mediated by host RNA polymerase II (Pol II). Among the early genes expressed is a subset whose expression leads to efficient G-actin transport to and accumulation within the nucleus (**Figure 5**). This spectacular manipulation of actin localization is critical for NPV progeny production and has not been described for any other pathogen. The genes involved in the nuclear localization of actin, identified in transient transfection experiments, include *ie1*, *pe38*, *Ac004*, *Ac152*, *he65*, and *Ac102*. Two of the genes, *ie1* and *Ac102*, are conserved among all lepidopteran NPVs and GVs, and both are essential.

Transition to the late stage of infection is the beginning of the BV progeny production period; cellular activities are geared toward synthesizing viral components at maximal rates, assembling them into nucleocapsid products and then exporting them. Host macromolecular synthesis is shut down during this period but host chromatin structure remains intact until cell death. Late and very late viral genes are expressed by an RNA polymerase encoded by the virus.

Microtubules, reorganized during the early phase of infection, are depolymerized by factors produced during the late phase leading to cell rounding (**Figure 6**). Similarly, G-actin, accumulated within the nucleus during the early phase, polymerizes during the late phase, concurrent with nuclear swelling and disappearance of visible host nuclear structures (**Figure 7**). The virogenic stroma, the site of viral DNA synthesis, forms at the center of the nucleus and is interspersed and surrounded by an electron-lucent zone called the 'ring zone', a place where capsids are assembled and tethered during genome loading. Nuclear F-actin co-localizes with the major AcMNPV capsid protein in the ring zone (**Figure 7**).

Nuclear F-actin is required for BV production. In the presence of F-actin-disrupting drugs, viral capsids are

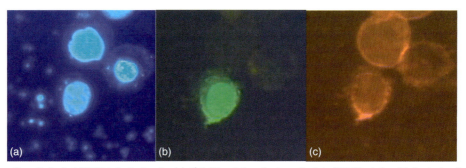

**Figure 5** Nuclear G-actin in AcMNPV-infected cells. (a) 4′,6-Diamidino-2-phenylindole (DAPI)-stained infected cells; (b) infected cell expressing transfected green fluorescent protein (GFP)-actin. (c) Lack of nuclear rhodamine phalloidin staining indicates that the nuclear actin is monomeric.

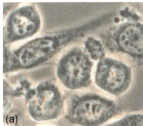
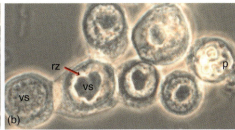

**Figure 6** AcMNPV-infected Sf21 cells demonstrating different phases of cytopathic effect (CPE). (a) Uninfected cells. Note spindle shape and presence of nucleochromatin structures. (b) AcMNPV-infected cells showing early and late CPE. Cells are rounded due to depolymerizing microtubules, nuclei are larger and cleared of host structures, and virogenic stroma (vs) is surrounded by ring zone (rz). Virogenic stromas become denser as the late stage of infection progresses. Polyhedra fill the nucleus during the very late stage of infection (p).

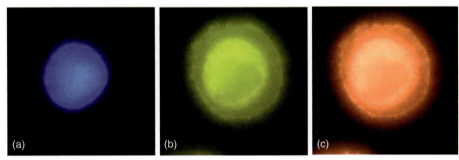

**Figure 7** Nuclear F-actin in AcMNPV-infected cell during late gene expression. (a) DAPI stain indicates location of the nucleus. (b) Green fluorescence indicates capsid protein is primarily localized in the ring zone of the nucleus. (c) Rhodamine phalloidin staining shows that F-actin co-localizes with capsid in the ring zone.

malformed and appear as long tubular structures juxtaposed to the inner nuclear membrane with infrequent patches of electron-dense material. Base plates and cap structures normally present are not evident, and an excess of membranous profiles are produced. Viral DNA synthesis occurs at the normal rate but genomes are not packaged, and the virogenic stroma has a 'relaxed' appearance compared to normal stroma. Interestingly, the phenotype for the absence of AcMNPV very late factor-1 (VLF-1) is similar, suggesting that VLF-1 may be involved in tethering capsids to the virogenic stroma, and that F-actin is a component of the stroma to which capsids are attached, directly or indirectly.

P78/83, a minor capsid protein expressed late during infection, is essential for AcMNPV viability. This feature was noted over 30 years ago and exploited in the first commercial baculovirus expression kits. P78/83 (78 kDa when unphosphorylated and 83 kDa when phosphorylated) is thought to be part of the base plates of both BV and ODV capsids. P78/83 is an F-actin-binding protein, and this activity may help tether capsids to the nuclear matrix during assembly. Interestingly, P78/83 contains another activity that explains why it is essential; it promotes actin polymerization within the nucleus. P78/83 contains domains conserved in Wiskott–Aldrich syndrome protein (WASP) family members, factors that promote nucleation of actin filaments. WASP family members positively regulate the actin-nucleating activity of the Actin Related Protein (Arp)-2/3 complex. The seven-subunit complex, conserved across eukaryotes, is translocated to the nucleus in AcMNPV-infected cells and is activated by P78/83. Mutations in P78/83 that lead to decreased ability in promoting actin nucleation correspondingly lead to decreased ability in producing infectious BV.

Immediately upon entry into the nucleus, AcMNPV DNA adopts a nucleosomal-like structure and uses nucleosomes and nucleosome-related processes in genome replication. A component of the viral replication strategy, therefore, appears to be hijacking components of, if not the entire, host chromatin remodeling capacity. Recent evidence suggests the BRO (baculovirus repeated orf) family of proteins are likely participants in this process. The BRO proteins are expressed early during infection, bind single-stranded DNA (ssDNA) and core histones, and partition with histones in fractionation experiments. BmNPV, orgyia pseudotsugata multiple nucleopolyhedrovirus, and lymantria dispar multiple nucleopolyhedrovirus all have multiple *bro* genes. AcMNPV carries only one *bro* gene, related to BmNPV *bro-d*, which is essential.

AcMNPV-encoded P6.9, the highly basic genome packaging protein, begins accumulating during the late stage of infection and an alternate chromatin structure emerges.

The P6.9–DNA interactions are controlled by the phosphorylation state of P6.9. During genome packaging, the genomic DNA from the virogenic stroma binds with P6.9 as P6.9 is dephosphorylated, condensing into a preformed capsid sheath through an opening in a conical end structure. The conical structures lie proximal to the virogenic stroma with capsid sheaths capped by base plates extending distally into a less-electron-dense space, the ring zone. F-actin and capsid protein co-localize in the ring zone, along with P78/83 and Arp2/3 complex. It is this stage of replication that is affected both by drugs that disrupt F-actin and the absence of VLF-1.

With the onset of the very late phase of infection there is a reduction in BV production and there is an onset of ODV production. Newly assembled nucleocapsids stay within the nucleus where they are wrapped in envelopes thought to be derived from the inner nuclear membrane. Enveloped virions, but not unenveloped nucleocapsids, become occluded in a protein matrix to form capsules or polyhedra. The cells eventually lyse, releasing occlusions into the medium.

*See also:* Baculoviruses: Apoptosis Inhibitors; Baculoviruses: General Features; Baculoviruses: Molecular Biology of Granuloviruses; Baculoviruses: Molecular Biology of Nucleopolyhedroviruses; Baculoviruses: Molecular Biology of Sawfly Baculoviruses.

## Further Reading

Adams JR and Bonami JR (eds.) (1991) *Atlas of Invertebrate Viruses*. Boca Raton, FL: CRC Press.

Bonning BC (2005) Baculoviruses: Biology, biochemistry and molecular biology. In: Gilbert LI, Iatrou K, and Gill SS (eds.) *Comprehensive Molecular Insect Science*, pp. 233–270. Oxford: Pergamon.

Goley ED, Ohkawa T, Woodruff JB, D'Alessio JA, Volkman LE, and Welch M (2006) Nuclear actin assembly by Arp2/3 complex and a viral WASP-like protein is essential for nucleopolyhedrovirus replication. *Science* 314: 464–467.

Kamita SG, Nagasaka K, Chuya JW, et al. (2005) A baculovirus-encoded ptotein tyrosine phosphatase gene induces enhanced locomotory activity in a lepidopteran host. *Proceedings of the National Academy of Sciences, USA* 102: 2584–2589.

Miller LK (ed.) (1997) *The Baculoviruses*. New York: Plenum.

Ohkawa T and Volkman LE (1999) Nuclear F-actin is essential for AcMNPV nucleocapsid morphogenesis. *Virology* 264: 1–4.

Ohkawa T, Washburn JO, Sitapara R, Sid E, and Volkman LE (2005) Specific binding of AcMNPV ODV to midgut cells of *Heliothis virescens* larvae is mediated by products of *pif* genes Ac119 and Ac022, but not Ac115. *Journal of Virology* 79: 15258–15264.

Okano K, Vanarsdall AL, Mikhailov VS, and Rohrmann GF (2006) Conserved molecular systems of the *Baculoviridae*. *Virology* 334: 77–87.

Trudeau D, Washburn JO, and Volkman LE (2001) Central role of hemocytes in Autographa californica M nucleopolyhedorivrus pathogenesis in *Heliothis virescens* and *Helicoverpa zea*. *Journal of Virology* 75: 996–1003.

Washburn JO, Chan EY, Volkman LE, Aumiller JJ, and Jarvis DL (2003) Early synthesis of budded virus envelope fusion protein gp64 enhances Autographa californica multicapsid nucleopolyhedrovirus virulence in orally infected *Heliothis virescens*. *Journal of Virology* 77: 280–290.

# Banana Bunchy Top Virus

**J E Thomas**, Department of Primary Industries and Fisheries, Indooroopilly, QLD, Australia

© 2008 Elsevier Ltd. All rights reserved.

## Glossary

**Cell-cycle link protein** A plant virus protein which most probably subverts the cell-cycle control of the host, forcing cells into DNA synthesis or S phase favorable to viral replication.

**Circulative transmission** Mode of transmission whereby the virus is acquired from plant sieve tubes, via the insect's stylet, and traverses a number of specific barriers as it passes from the hindgut, into the hemocoel and then to the salivary glands, from where it can be reinjected into a plant during feeding. It is characterized by a latent period before which re-transmission cannot occur and by retention of infectivity by the vector for periods ranging from several days to the entire life span of the insect. The virus does not replicate in the insect.

**Nuclear shuttle protein** A virus-encoded protein that transports the viral single-stranded DNA, as a complex with the viral movement protein, to and from its site of replication in the nucleus and into adjacent cells.

## Introduction

Banana bunchy top disease is the most economically important virus disease of banana and plantain (*Musa* spp.) worldwide, due to its devastating effect on crop yield, and the importance of banana and plantain as both a staple food and a major export commodity in much of the developing world. The causal agent is banana bunchy top virus (BBTV). Edible bananas are derived from wild progenitors including *M. acuminata*, *M. balbisiana*, and

*M. schizocarpa*, which have a center of origin in the South and Southeast Asian–Australasian region, and it is likely that BBTV also originated within this area. BBTV has a multipartite circular single-stranded (ssDNA) genome, encapsidated in small isometric virions, and is transmitted by the banana aphid, *Pentalonia nigronervosa*.

## Disease Symptoms

The symptoms of bunchy top disease in banana are characteristic, especially in the Cavendish subgroup of cultivars, and easily distinguished from all other virus diseases of banana. Plants can become infected at any stage of growth, and the initial appearance of symptoms can depend on the manner of infection. BBTV is systemic within the banana plant, and following aphid inoculation, symptoms do not appear until at least two more new leaves have been produced (bananas produce single new leaves sequentially from a basal meristem). The first symptoms comprise a few dark green streaks and dots on the lower part of the lamina and on the petiole, becoming more general on subsequent leaves. These streaks form hooks as they enter the midrib, and are best viewed from the underside of the leaf, with transmitted light (**Figure 1**). However, these dark streaks can be rare or absent in some cultivars. Successive leaves become shorter and narrower, and have a brittle lamina with upturned, chlorotic, ragged margins. Leaves fail to emerge fully, giving the plant a bunched appearance (**Figure 2**). Plants derived from infected planting material (suckers, bits) develop severe symptoms from the first leaf to emerge.

Infected plants seldom produce a bunch, though if infected late in the current cropping cycle, a small, distorted bunch may result. With very late infections, the only symptoms to appear in the current season may be a few dark streaks on the tips of the flower bracts. No fruit is produced in subsequent years, and plants generally die within a couple of years.

Histological examination suggests that BBTV is restricted to the phloem tissue, which shows hypertrophy and hyperplasia and a reduction in the development of the fibrous sclerenchyma sheaths surrounding the vascular bundles. The cells surrounding the phloem contain abnormally large numbers of chloroplasts, resulting in the macroscopic dark green streak symptom.

Using RNA probes and polymerase chain reaction (PCR), it has been demonstrated that BBTV replicates briefly at the site of aphid inoculation, then moves down the pseudostem to the basal meristem, subsequently infecting the newly formed leaves, the corm, and the roots. The virus apparently does not replicate in leaves formed prior to infection, consistent with the lack of symptoms on these leaves and an inability to recover the virus from them via the aphid vector.

**Figure 1** Dark green dot-dash, hooking and vein clearing fleck symptoms on a Cavendish banana leaf

**Figure 2** Young BBTV-infected banana plants, showing stunting, and successively shorter narrow leaves with upturned, chlorotic margins. Leaves have failed to emerge fully, giving the plant a bunched appearance.

From Taiwan, symptomless strains of BBTV, and mild strains that produce only limited vein clearing and dark green streaks have been reported. Also, some plants of the Cavendish subgroup cultivar Veimama from Fiji have been observed to initially show severe symptoms, then to recover and display few if any symptoms.

## Host Range

Confirmed hosts of BBTV are confined to the family Musaceae. Known susceptible hosts include *Musa* species, cultivars in the Eumusa and Australimusa series of edible banana and *Ensete ventricosum*. Susceptible *Musa* species include *M. balbisiana*, *M. acuminata* ssp. *banksii*, *M. textilis*, *M. velutina*, *M. coccinea*, *M. jackeyi*, *M. ornata*, and *M. acuminata* ssp. *zebrina*. There are some reports of hosts outside the Musaceae, though in independent tests none has been confirmed.

## Causal Agent

The viral nature of bunchy top disease was established by C. J. P. Magee, in Australia, in the 1920s. However, it was not until 1990 that the virus particles were first isolated, in part due to lack of a suitable herbaceous experimental host, its low titer in infected plants, and its restriction to the phloem in the fibrous vascular tissue of *Musa*. BBTV has icosahedral particles, 18–20 nm in diameter (**Figure 3**), a buoyant density of 1.29–1.30 in cesium sulfate, and a sedimentation coefficient of 46S.

BBTV has a multicomponent genome comprising six transcriptionally active components (**Figure 4**; **Table 1**), each *c.* 1 kbp in size. DNA-R encodes the master replication initiation protein (Rep), which can initiate and terminate replication, and a potential second internal ORF of unknown function, for which a transcript only has been identified. All other components are monocistronic; DNA-S encodes the coat protein, DNA-C the cell-cycle link protein, DNA-M the cell-to-cell movement protein, DNA-N the nuclear shuttle protein, and the function of the protein potentially encoded by DNA-U3 is unknown. The untranslated regions of all six components share two areas of high sequence identity, both concerned with the rolling-circle mode by which BBTV replicates. A stem–loop common region (CR-SL) of 69 nt includes a nonanucleotide sequence (TAT/GTATTAC) which is shared between plant-infecting circular ssDNA nanovirids and geminivirids, and which is the site of the origin of viral replication. It also contains iterated sequences, thought to act as recognition sites for the replication initiation protein. The major common region (CR-M) varies between 66 and 92 nt in length among the various components. Virions contain a heterogeneous population of DNA primers, *c.* 80 nt in length, which bind to the CR-M and prime complementary strand synthesis.

The coat protein gene of BBTV is highly conserved, with a maximum difference of 3% at the amino acid level between isolates. No serological differences have been detected between any isolates using polyclonal or monoclonal antibodies.

The intergenic regions of all six integral components of BBTV have been shown to have promoter activity. The highest activity in banana embryonic cells was shown by promoters from DNA-C and DNA-M, components thought to be intrinsic to the infection process. Studies on the DNA-N promoter demonstrated that expression in banana embryonic cells is limited to phloem tissue, consistent with the circulative mode of transmission by the aphid vector and observations by Magee in the 1920s that histological effects were confined to the phloem and phloem parenchyma cells. Circulative transmission by aphid vectors usually involves specific feeding on the phloem tissue for virus acquisition. The CR-M and CR-SL are not essential for promoter activity. All essential elements of the promoter are located 3′ of the stem–loop and within 239 bp of the translation start codon and within this region are an ASF-1 motif (TGACG), a hexamer motif (ACGTCA), *rbc*S-I-box (GATAAG), G-boxcore, and the TATA box, all associated with promoter activity in other genomes.

Some isolates of BBTV from Taiwan and Vietnam contain additional Rep-encoding DNAs that are thought to be capable only of self-replication and to behave like satellite molecules. These molecules have a CR-SL, though the stem sequence is not conserved with the six integral DNA components. Unlike DNA-R, the putative satellites lack the internal ORF, their TATA boxes are 5′ of the stem loop, and they generally lack the CR-M. Interestingly, the amino acid sequences of satellite Reps BBTV-S1, BBTV-S2, and BBTV-Y1 are actually more closely related phylogenetically to the Reps encoded by nanovirids outside the genus *Babuvirus* than they are to DNA-R of BBTV. Similar molecules have been detected by Southern hybridization in isolates from the Philippines, Tonga, and Western Samoa, but not from Australia, Egypt, Fiji, and India.

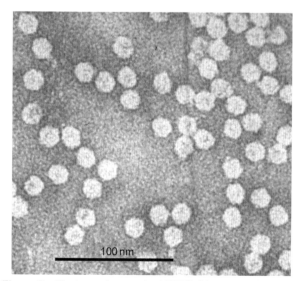

**Figure 3** Electron micrograph of BBTV virions, negatively contrasted with 1% ammonium molybdate.

## Taxonomy and Phylogenetic Relationships

BBTV is a member of the genus *Babuvirus*, in the family *Nanoviridae*. Other members of this family are classified in the genus *Nanovirus*, and include faba bean necrotic yellows virus, milk vetch dwarf virus, and subterranean clover stunt virus.

BBTV isolates worldwide fall into two broad phylogenetic groups (**Figure 5**), called the South Pacific group

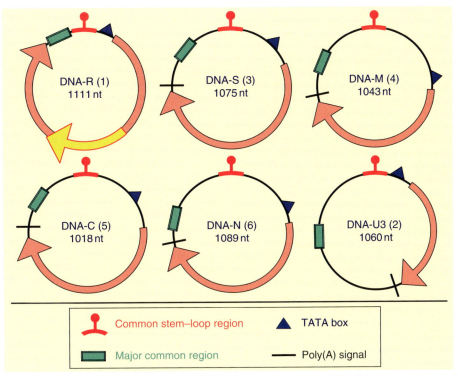

**Figure 4** Diagram of the integral genome components of BBTV, showing transcribed ORFs (black arrows) and other main genome features. Reproduced from Vetten HJ, Chu PWG, Dale JL, et al. (2005) *Nanoviridae*. In: Fauquet CM, Mayo MA, Maniloff J, Desselberger U, and Ball LA (eds.) *Virus Taxonomy: Eighth Report of the International Committee on Taxonomy of Viruses*, p. 349, figure 3. San Diego, CA: Elsevier Academic Press, with permission from Elsevier.

**Table 1** Size and function of the six integral components of BBTV

| Genome component | Size (nt) | Size of encoded protein(s) (kDa) | Function |
|---|---|---|---|
| DNA-R | 1111 | (1) 33.6 | (1) Replication initiation protein |
|  |  | (2) 5.2[a] | (2) Unknown |
| DNA-S | 1075 | 20.1 | Capsid protein |
| DNA-M | 1043 | 13.7 | Movement protein |
| DNA-C | 1018 | 19.0 | Cell-cycle link protein |
| DNA-N | 1089 | 17.4 | Nuclear shuttle protein |
| DNA-U3 | 1060 | 10.4[b] | Unknown |

[a]Expression has not been demonstrated for the transcript of this internal ORF.
[b]Most isolates of BBTV do not encode a functional ORF on DNA-U3, and where present there is no evidence for its expression.

(isolates from Australia, Fiji, Hawaii, Western Samoa, Tonga, India, Pakistan, Burundi, Egypt, Malawi) and the Asian group (China, Philippines, Japan, Vietnam, Taiwan). Sequence differences between the two groups have been demonstrated across all genome components. Most striking is the variation in the CR-M of DNA-R, where the mean sequence difference between the groups was 30%, and up to a maximum of 55% for individual pairs of isolates. Within group sequence diversity is much greater for the Asian group, suggesting a longer evolutionary period and possible evolutionary origin for the virus. The presence of the virus in many countries from the South Pacific group can be traced to introductions within the last century. A Taiwanese isolate (TW4) causing only mild symptoms was recently identified. Interestingly, it apparently lacks DNA-N and is clearly a chimeric isolate, having DNA-M of the Asian group and DNA-S, -C, and -U3 of the South Pacific group. It also contains two DNA-R molecules, one from each group.

Recently, two new viruses have been identified which are clearly distinct members of the genus *Babuvirus* (**Figure 5**). Abaca bunchy top virus has homologous components for

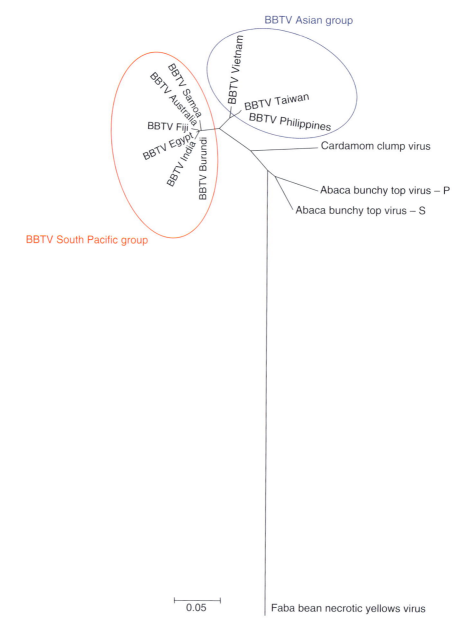

**Figure 5** Phylogenetic tree based on the amino acid sequences of babuvirus Rep proteins, using the nanovirus faba bean necrotic yellows virus as an outgroup. The Asian and South Pacific groups of BBTV are circled in blue and red, respectively.

all six genome components of BBTV and also infects *Musa* in Southeast Asia. The sequence and organization of DNA-R of cardamom clump virus (syn. cardamom bushy dwarf virus), associated with Foorkey disease of cardamom in India, also suggests that it is also a genome component of a novel babuvirus.

## Geographical Distribution

### Historical Records and Possible Origins

The origin of BBTV is unclear. The first records were from Fiji in 1889, and though reports and photographs clearly indicate that it was present at least 10 years prior, soon after the establishment of an export industry based on Cavendish cultivars, evidence suggests that the disease did not originate there. Other early records are from Egypt (1901 – source unknown) and Australia and Sri Lanka (both in 1913, and probably from planting material imported from Fiji).

The wild progenitors of modern edible bananas originated in the South and Southeast Asian–Australasian region, and the Cavendish cultivars of international trade, associated with early outbreaks of bunchy top disease, are thought to have originated in Vietnam. These factors lead to speculation that BBTV also originated and evolved in this region. The recent discovery of a chimeric BBTV isolate from Taiwan containing genome components of both geographic groups could support this hypothesis.

## Current Distribution

BBTV has a widespread, but scattered distribution in many of the banana-growing countries of the Asia-Pacific regions and Africa, but at present is not found in the Americas. In some countries, the occurrence is localized, probably due to geographic isolation. For example, in Australia, it is present in southern Queensland and northern New South Wales, but not in the major production area of north Queensland. Banana bunchy top disease has been recorded in the following countries, and for those marked with asterisk (*), the presence of the virus has been confirmed by serological or molecular assays:

*Asia.* China*, Japan (Okinawa*, Bonin Is.), India*, Indonesia*, Iran*, Myanmar, Malaysia (Sarawak*), Pakistan*, Philippines*, Sri Lanka*, Taiwan*, Vietnam*.

*Pacific.* Australia*, Fiji*, Guam, Kiribati (formerly Gilbert Is.), New Caledonia*, Samoa (American, Western*), Tonga*, Tuvalu (formerly Ellice Is.), USA (Hawaii*), Wallis Is.

*Africa.* Angola, Burundi*, Central African Republic, Congo* Democratic Republic of the Congo (formerly Zaire), Egypt*, Gabon*, Malawi*, Rwanda, Zambia*.

## Virus Transmission

The banana aphid (*Pentalonia nigronervosa*) has a worldwide distribution, and in 1925 was shown to be a vector of BBTV. It remains the only known insect vector of this virus. Hosts of the aphid include species in the family Musaceae and several related families, including Araceae, Heliconiaceae, Strelitzeaceae, and Zingiberaceae, though the aphid shows a degree of host preference and can be difficult to transfer between host species. On banana, they commonly colonize the base of the pseudostem at soil level and for several centimeters below the soil surface, beneath the outer leaf sheaths and newly emerging suckers.

Transmission by *P. nigronervosa* is of the persistent, circulative, nonpropagative type, and individuals from areas where the virus is endemic and from where it is absent, both transmit the virus with equal efficiency. There is a minimum acquisition access period of 4 h, a minimum latent period of a few hours, and a minimum inoculation access period of 15 min. Aphids retain infectivity, after removal from a virus source, for at least 20 days and probably for life. Both nymphs and adults can acquire the virus, though more efficiently by the former, and reported transmission rates for individual aphids are in the range 46–67%. There is no evidence for transmission of BBTV to parthenogenetic offspring or for replication of the virus in the aphid.

BBTV is also efficiently transmitted in vegetative planting material, both conventional corms, corm pieces (bits) and suckers, and through micropropagation. All meristems from an infected corm will eventually become infected.

## Epidemiology and Control of BBTV

### Epidemiology

Outbreaks of bunchy top disease can have a devastating effect on banana production, especially industries based on Cavendish cultivars. Production in Fiji fell by more than 80% from 1892 to 1895, primarily due to bunchy top disease. By 1925, nearly 10 years since the introduction of BBTV to Australia, the banana industry in northern New South Wales and southern Queensland had collapsed, with most plantations affected and production decreased by 90–95%. Magee noted at the time "It would be difficult for anyone who has not visited these devastated areas to visualize the completeness of the destruction wrought in such a short time by a plant disease."

More recently, a severe outbreak of banana bunchy top disease occurred in Pakistan. From 1991 to 1992, production area fell by 55% and total production by 90%, as a direct result of the disease. The disease has also recently appeared in Hawaii, New Caledonia, Angola, Zambia, and Malawi.

The epidemiology of banana bunchy top disease is simplified by the occurrence of a single insect vector species and a limited host range for the virus, usually cultivated or feral edible bananas. Long-distance spread is usually via infected planting material, and local spread via aphids and planting material.

Analysis of actual outbreaks of bunchy top disease in commercial banana plantations in Australia showed that the average distance of secondary spread by aphids was only 15.5–17.2 m, with nearly two-thirds of new infections less than 20 m from the nearest source of infection and 99% less than 86 m. Isolation of new plantations has a marked effect on reducing the risk of infection. New plantations situated adjacent to affected plantations had an 88% chance of recording infections in the first year. This was reduced to 27% if the plantations were separated by 50–1000 m, and to 5% if separated by more than 1000 m. The disease latent period (i.e., period from inoculation of a plant until an aphid can transmit the virus from this plant to another) is equivalent to the time taken for 3.7 new leaves to emerge from the plant. The actual time varies seasonally.

### Control

Control strategies were devised by C. J. P. Magee, in the 1920s, and these measures still form the basis of the very successful control program in Australia today. The two major elements of the strategy are (1) exclusion of the disease from unaffected and lightly affected areas and (2) eradication of infected plants from both lightly and heavily affected areas.

These measures require the participation of all growers and are unlikely to succeed if left to the goodwill of growers alone and are thus enforced by legislation in Australia. The measures include:

- registration of all banana plantations;
- establishment of quarantine zones;
- restrictions on the movement and use of planting material;
- regular inspections of all banana plantations for bunchy top;
- prompt destruction of all infected plants; and
- ongoing education and extension programs for growers.

When adopted, these measures allowed the complete rehabilitation of the Australian banana industry. Occasionally, total eradication of BBTV from a district has been achieved, but in most cases, incidence has been reduced to very low, manageable levels. Such successful control of bunchy top is rarely achieved in other countries, in most cases due to an inability to enforce an organized control program across whole districts.

## Detection Assays

Polyclonal and monoclonal antibodies to BBTV are used in enzyme-linked immunosorbent assay (ELISA) to detect the virus in field and tissue culture plants and can detect the virus in single viruliferous aphids. PCR was shown to be about a thousand times more sensitive than ELISA or dot blots with DNA probes. Substances in banana sap inhibitory to PCR can be circumvented by simple extraction procedures or by immunocapture PCR.

BBTV has been detected in most parts of the banana plant, including the leaf lamina and midrib, pseudostem, corm, meristems, roots, fruit stalk, and fruit rind.

## Resistance

There are no confirmed reports of immunity to BBTV in *Musa*. However, it has frequently been observed that there are differences in susceptibility between cultivars to both field and experimental infection. Edible bananas have diploid, triploid, or tetraploid genomes containing, predominantly, elements of the *M. acuminata* (A) or *M. balbisiana* (B) genomes. Cultivars in the Cavendish subgroup (AAA genome), which dominates the international export trade, and many other A genome cultivars are highly susceptible and show severe disease symptoms. By contrast, Gros Michel (AAA) displays resistance to the disease under both experimental inoculation and field conditions. Compared with highly sensitive cultivars such as Cavendish, the cultivar is less susceptible to aphid inoculation, contains a lower level of virions in infected plants, and symptoms are less severe and develop more slowly. These factors may contribute to a reduced rate of aphid transmission and field spread in plantations of Gros Michel, and introduction of this cultivar may explain the partial recovery of the Fijian banana industry after devastation of the Cavendish-based industry in the early 1900s. Field observations and glasshouse inoculations suggest that some B genome-containing cultivars are less susceptible to infection and/or display more limited symptoms, but this needs to be further investigated.

Despite concerted efforts to generate transgenic resistance to BBTV, no successful glasshouse or field results have yet been reported. However, some promising strategies are being developed at Queensland University of Technology involving the following steps.

1. Transdominant negative strategies to interfere with replication, by constitutive overexpression of mutated Rep proteins, are employed. Single mutations in either of two motifs involved with rolling-circle replication render the Rep inactive, and in transient assays with constitutive overexpression, virus replication is significantly reduced, but not abolished.
2. Rep-activated cell death is carried out using a so-called suicide gene. DNA-R intergenic sequence is cloned within an intron and flanked by a split barnase gene construct. The suicide gene is only activated in the presence of the Rep protein, resulting in cell death and containment of the virus.

*See also:* Nanoviruses.

## Further Reading

Allen RN (1987) Further studies on epidemiological factors influencing control of banana bunchy top disease, and evaluation of control measures by computer simulation. *Australian Journal of Agricultural Research* 38: 373–382.

Bell KE, Dale JL, Ha CV, Vu MT, and Revill PA (2002) Characterisation of Rep-encoding components associated with banana bunchy top nanovirus in Vietnam. *Archives of Virology* 147: 695–707.

Burns TM, Harding RM, and Dale JL (1995) The genome organization of banana bunchy top virus: Analysis of six ssDNA components. *Journal of General Virology* 76: 1471–1482.

Dugdale B, Becker DK, Beetham PR, Harding RM, and Dale JL (2000) Promoters derived from banana bunchy top virus DNA-1 to -5 direct vascular-associated expression in transgenic banana (*Musa* spp.). *Plant Cell Reports* 19: 810–814.

Geering ADW and Thomas JE (1997) Search for alternative hosts of banana bunchy top virus in Australia. *Australasian Plant Pathology* 26: 250–254.

Hafner GJ, Harding RM, and Dale JL (1995) Movement and transmission of banana bunchy top virus DNA component one in bananas. *Journal of General Virology* 76: 2279–2285.

Hu J-M, Fu H-C, Lin C-H, Su H-J, and Yeh H-H (2007) Reassortment and concerted evolution in banana bunchy top virus genomes. *Journal of Virology* 81: 1746–1761.

Hu JS, Wang M, Sether D, Xie W, and Leonhardt KW (1996) Use of polymerase chain reaction (PCR) to study transmission of banana bunchy top virus by the banana aphid (*Pentalonia nigronervosa*). *Annals of Applied Biology* 128: 55–64.

Karan M, Harding RM, and Dale JL (1994) Evidence for two groups of banana bunchy top virus isolates. *Journal of General Virology* 75: 3541–3546.

Magee CJ (1953) Some aspects of the bunchy top disease of banana and other *Musa* spp. *Journal and Proceedings of the Royal Society of New South Wales* 87: 3–18.

Magee CJP (1927) *Investigation on the Bunchy Top Disease of the Banana. Bulletin No. 30*. Melbourne: Council for Scientific and Industrial Research.

Thomas JE and Dietzgen RG (1991) Purification, characterization and serological detection of virus-like particles associated with banana bunchy top disease in Australia. *Journal of General Virology* 72: 217–224.

Thomas JE and Iskra-Caruana ML (2000) Bunchy top. In: Jones DR (ed.) *Diseases of Banana, Abaca and Enset*, pp. 241–253. Wallingford: CABI Publishing.

Thomas JE, Smith MK, Kessling AF, and Hamill SD (1995) Inconsistent transmission of banana bunchy top virus in micropropagated bananas and its implication for germplasm screening. *Australian Journal of Agricultural Research* 46: 663–671.

Vetten HJ, Chu PWG, Dale JL, *et al.* (2005) *Nanoviridae*. In: Fauquet CM, Mayo MA, Maniloff J, Desselberger U, and Ball LA (eds.) *Virus Taxonomy, Eighth Report of the International Committee on Taxonomy of Viruses*, pp. 343–352. San Diego, CA: Elsevier Academic Press.

Wu R-Y and Su H-J (1990) Purification and characterization of banana bunchy top virus. *Journal of Phytopathology* 128: 153–160.

# Barley Yellow Dwarf Viruses

**L L Domier,** USDA-ARS, Urbana-Champaign, IL, USA

© 2008 Elsevier Ltd. All rights reserved.

## Glossary

**Hemocoel** The primary body cavity of most arthropods that contains most of the major organs and through which the hemolymph circulates.

**Hemolymph** A circulatory fluid in the body cavities (hemocoels) and tissues of arthropods that is analogous to blood and/or lymph of vertebrates.

## Introduction

Barley yellow dwarf (BYD) is the most economically important virus disease of cereals, and is found in almost every grain growing region in the world. Widespread BYD outbreaks in cereals were noted in the United States in 1907 and 1949. However, it was not until 1951 that a virus was proposed as the cause of the disease. The causal agents of BYD are obligately transmitted by aphids, which probably delayed the initial classification of BYD as a virus disease. Subsequently, BYD was shown to be caused by multiple viruses belonging to the species barley yellow dwarf virus (BYDV) and cereal yellow dwarf virus (CYDV). Depending on the virulence of the virus strain, infection may contribute to winter kill in regions with harsh winters, induce plant stunting, inhibit root growth, reduce or prevent heading, or increase plant susceptibility to opportunistic pathogens and other stresses. Yield losses to wheat in the United States alone are estimated at 1–3% annually, exceeding 30% in certain regions in epidemic years. The effects of BYD in barley and oats typically are more severe than in wheat; sometimes resulting in complete crop losses. The existence of multiple strains of viruses that are transmitted in strain-specific manner has made BYDV and CYDV model systems to study interactions between viruses and aphid vectors in the circulative transmission of plant viruses. In addition, the compact genomes of the viruses have provided useful insights into the manipulation of host translation machinery by RNA viruses.

## Taxonomy and Classification

The viruses that cause BYD are members of the family *Luteoviridae*, and were first grouped because of their common biological properties. These properties included persistent transmission by aphid vectors, the induction of yellowing symptoms in grasses, and serological relatedness. Different viruses are transmitted more efficiently by different species of aphids, a fact that was originally used to distinguish the viruses. Around 1960, the viruses were separated into five 'strains' (now recognized as distinct species) based on their primary aphid vector(s). BYDVs transmitted most efficiently by *Sitobion* (formerly *Macrosiphum*) *avenae* were assigned the acronym MAV, for *Macrosiphum avenae* virus. Similarly, viruses transmitted most efficiently by *Rhopalosiphum maidis* and *Rhopalosiphum padi* were assigned the acronyms RMV and RPV, respectively. Viruses transmitted most efficiently by *Schizaphis graminum* were assigned the acronym SGV. Finally vector-nonspecific viruses, that is, viruses transmitted efficiently by both *R. padi* and *S. avenae* were assigned the acronym PAV.

Based on genome organization and predicted amino acid sequence similarities, BYDV-MAV, -PAS, and -PAV have been assigned to the genus *Luteovirus*, and CYDV-RPS and -RPV to the genus *Polerovirus*. The RNA- dependent RNA polymerases (RdRps) encoded by open reading frames (ORFs) 1 and 2 of BYDVs resemble those of members of the family *Tombusviridae* (**Figure 1**). In contrast, the predicted amino acid sequence of the RdRps encoded by ORFs 1 and 2 of CYDVs resemble those of viruses in the genus *Sobemovirus*. The two polymerase types are distantly related in evolutionary terms. For this reason, viruses for which RdRp sequences have not been determined (BYDVs GPV, RMV, and SGV) have not been assigned to a genus. These observations suggest that the genomic RNAs of BYDVs and CYDVs resulted from recombination between RNAs expressing a common set of structural and movement proteins and RNAs expressing two different sets of replication proteins. Because of these differences, it has been suggested that BYDVs should be placed in the family *Tombusviridae* and CYDVs in the genus *Sobemovirus*.

## Virion Properties and Composition

All BYDVs and CYDVs have nonenveloped icosahedral particles with diameters of 25–28 nm (**Figure 2**). Capsids are composed of major (22 kDa) and minor (65–72 kDa) coat proteins (CPs), which is formed by a carboxy-terminal extension to the major CP called the readthrough domain (RTD). According to X-ray diffraction and molecular mass analysis, virions consist of 180 protein subunits, arranged in $T = 3$ icosahedra. Virus particles do not contain lipids or carbohydrates, and have sedimentation coefficients $s_{20,w}$ (in Svedberg units) that range from 115–118S. Buoyant densities in CsCl are approximately $1.4 \text{ g cm}^{-3}$. Virions are moderately stable, insensitive to freezing, and are insensitive to treatment with chloroform or nonionic detergents, but are disrupted by prolonged treatment with high concentrations of salts.

The single encapsidated genomic RNA molecule is single-stranded, positive-sense, and lacks a 3′-terminal poly(A) tract. A small protein (VPg) is covalently linked to the 5′-terminus of CYDV RNAs. CYDV-RPV also encapsidates a 322-nucleotide satellite RNA that accumulates to high levels in the presence of the helper virus. Complete genome sequences have been determined for BYDV-MAV, -PAS, and -PAV and CYDV-RPS and -RPV (**Table 1**). For several viruses, notably BYDV-PAV, genome sequences have been determined from multiple isolates.

## Genome Organization and Expression

Genomic RNAs of BYDVs and CYDVs for which complete nucleotide sequences are available contain five to six ORFs (**Figure 3**). ORFs 1, 2, 3, and 5 are shared among all BYDVs and CYDVs. BYDVs lack ORF0. Genomic sequences of

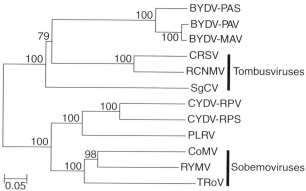

**Figure 1** Phylogenetic relationships of the predicted amino acid sequences of RNA-dependent RNA polymerases (RdRps; ORFs 1 and 2) of barley yellow dwarf viruses (BYDVs) and cereal yellow dwarf viruses (CYDVs) and members of the genus *Sobemovirus* and family *Tombusviridae*. The RdRps of BYDV-MAV, -PAS, and -PAV are more similar to those of members of the family *Tombusviridae* (carnation ringspot virus (CRSV), red clover necrotic mosaic virus (RCNMV), and saguaro cactus virus (SgCV)) than to those of CYDVs. The RdRps of CYDV-RPS and -RPV and potato leaf roll virus (PLRV, type member of the genus *Polerovirus*) are more similar to those of members of the genus *Sobemovirus* (cocksfoot mottle virus (CoMV), rice yellow mottle virus (RYMV), and turnip rosette virus (TRoV)) than to those of the BYDVs. The resulting consensus tree from 1000 bootstrap replications is shown. The numbers above each node indicate the percentage of bootstrap replicates in which that node was recovered.

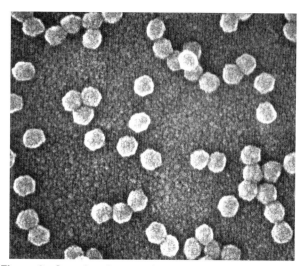

**Figure 2** Scanning electron micrograph of barley yellow dwarf virus-PAV particles, magnified 200 000×. Virions are c. 25 nm in diameter, hexagonal in appearance, and have no envelope.

some BYDVs contain one or two small ORFs, ORFs 6 and 7, downstream of ORF5. In CYDVs, ORFs 0 and 1 and ORFs 1 and 2 overlap by more than 600 nucleotides. In BYDVs, ORF1 overlaps ORF2 by less than 50 nucleotides. In BYDV and CYDV genome sequences, ORF4 is contained within ORF3. An amber (UAG) termination codon separates ORFs 3 and 5.

**Table 1** Viruses causing barley yellow dwarf in cereals

| Genus | Virus (alternative name) | Abbreviation |
| --- | --- | --- |
| *Luteovirus* | Barley yellow dwarf virus-MAV | BYDV-MAV |
| | Barley yellow dwarf virus-PAS | BYDV-PAS |
| | Barley yellow dwarf virus-PAV | BYDV-PAV |
| | (Barley yellow dwarf virus-RGV) (rice giallume) | |
| *Polerovirus* | Cereal yellow dwarf virus-RPS | CYDV-RPS |
| | Cereal yellow dwarf virus-RPV | CYDV-RPV |
| Unassigned | Barley yellow dwarf virus-GAV | BYDV-GAV |
| | Barley yellow dwarf virus-GPV | BYDV-GPV |
| | Barley yellow dwarf virus-RMV | BYDV-RMV |
| | Barley yellow dwarf virus-SGV | BYDV-SGV |

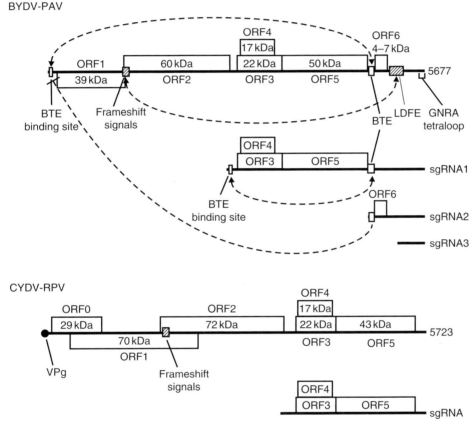

**Figure 3** Genome organizations of barley yellow dwarf virus-PAV (BYDV-PAV) and cereal yellow dwarf virus-RPV (CYDV-RPV). Individual open reading frames (ORFs) are shown as staggered open boxes. The predicted sizes of the protein products are indicated. The genome-linked protein (VPg) attached to the 5′ terminus of CYDV RNA is indicated by a solid circle. Based on homology to other viruses ORF0 encodes a silencing suppressor and ORFs 1 and 2 encode replication-related proteins. ORFs 3 and 5 encode the major coat protein and readthrough domain, respectively. ORF4 encodes a protein required for virus cell-to-cell movement. The BYDV translation enhancer (BTE) facilitates translation initiation of BYDV-PAV genomic RNA and subgenomic RNA1 (sgRNA1). In both BYDV-PAV and CYDV-RPV, ORF2 is expressed as a translational fusion with the product of ORF1 via a −1 frameshift. In BYDV-PAV, frameshifting requires interaction between the 5′ frameshift signals and the long-distance frameshift element (LDFE). Dashed lines indicate long-distance RNA–RNA interactions.

BYDVs and CYDVs have relatively short 5′ and intergenic noncoding regions. ORFs 2 and 3 are separated by about 200 nucleotides in BYDVs and CYDVs. The lengths of noncoding sequences downstream of ORF5 are very different between BYDVs and CYDVs. BYDV-PAV contains over 860 nucleotides downstream of ORF5 compared to just 170 nucleotides for CYDV-RPV.

The expression of BYDV-PAV RNA has been studied in detail and has revealed a complex set of RNA–RNA and RNA–protein interactions that are employed to express and replicate the virus genome. Less experimental data are available for CYDVs. However, expression and replication strategies and gene functions can be inferred from those of closely related poleroviruses, particularly beet western yellows virus (BWYV) and potato leaf roll virus (PLRV). ORFs 0, 1, and 2 are expressed directly from genomic RNAs. Downstream ORFs are expressed from subgenomic RNAs (sgRNAs) that are transcribed from internal initiation sites by virus-encoded RdRps from negative strand RNAs and are 3′-coterminal with the genomic RNA. Since the initiation codon for ORF0 of CYDVs is upstream of that of ORF1, translation of ORF1 is initiated by 'leaky scanning' in which ribosomes bypass the AUG initiation codon of ORF0 and continue to scan the genomic RNA until they reach the initiation codon of ORF1. The protein products of ORF2 are expressed only as a translational fusion with the product of ORF1. At a low frequency during the expression of ORF1, translation continues into ORF2 through a −1 frameshift that produces a large protein containing sequences encoded by both ORFs 1 and 2 in a single polypeptide. In BYDV-PAV, frameshifting between ORFs 1 and 2 is dependent upon the interaction of RNA sequences close to the site of frameshifting and a long-distance frameshift element (LDFE) located 4000 nucleotides downstream in the 3′ noncoding region of genomic RNAs. Mutations that disrupt the interactions between these two distal regions suppress frameshifting and abolish RNA replication.

ORFs 3, 4, and 5 are expressed from sgRNA1, the 5′ terminus of which is located about 200 nucleotides upstream of ORF3, and extends to the 3′ terminus of the genome. BYDVs produce a second sgRNA that contains ORF6. BYDV-PAV also produces a third sgRNA, which does not appear to encode a protein. ORF3 is translated from the 5′ terminus of sgRNA1. ORF4 of BYDVs and CYDVs, which encodes a 17 kDa protein, is contained within ORF3 and is expressed from the same sgRNA as ORF3 through a leaky scanning mechanism much like that used to express ORF1 of CYDVs. In BYDVs and CYDVs, ORF5 is expressed only as a translational fusion with the products of ORF3 by readthrough of the UAG termination codon at the end of ORF3. This produces a protein with the product of ORF3 at its amino terminus and the product of ORF5 at its carboxyl terminus.

While genomic RNAs of CYDVs contain 5′ VPgs that interact with translation initiation factors, BYDV-PAV RNA contains only a 5′ phosphate. Unmodified 5′ termini usually are recognized poorly for translation initiation. To circumvent this problem, a short sequence located in the noncoding region just downstream of ORF5 in the BYDV-PAV genome, called the BYDV translation enhancer (BTE), interacts with sequences near the 5′ termini of the genomic and subgenomic RNAs to promote efficient cap-independent translation initiation.

Functions BYDV and CYDV proteins have been ascribed based on homology to virus proteins with known functions and mutational characterization of protein coding regions. Similarity to proteins encoded by BWYV and PLRV suggests that the 28–29 kDa proteins encoded by ORF0 of CYDVs are inhibitors of post-transcriptional gene silencing (PTGS). PTGS is an innate and highly adaptive antiviral defense found in all eukaryotes that is activated by double-stranded RNAs (dsRNAs), which are produced during virus replication. The ORF1-encoded proteins of CYDVs contain the VPg and a chymotrypsin-like serine protease that is responsible for the proteolytic processing of ORF1-encoded polyproteins. The protease cleaves the ORF1 protein in *trans* to liberate the VPg, which is covalently attached to genomic RNAs. ORF2s of BYDVs and CYDVs, which are expressed as translational fusions with the product of ORF1, have coding capacities of 59–72 kDa and predicted amino acid sequences that are very similar to known RdRps and hence likely represent the catalytic portion of the viral replicase.

ORF3 encodes the major 22 kDa CP for both BYDVs and CYDVs. ORF5 has a coding capacity of 43–50 kDa, which is expressed only as a translational fusion with the product of ORF3 when translation reads through the termination codon at the end of ORF3 and continues through to the end of ORF5. The ORF5 portion of this readthrough protein has been implicated in aphid transmission and virus stability. Recombinant viruses that do not express ORF5 produce virions assembled from the major CP alone, which are not transmitted by aphid vectors and are less efficient in systemic infection of host plants than wild-type viruses. The amino-terminal portions of ORF5 proteins are highly conserved among BYDVs and CYDVs while the carboxyl termini are much more variable.

ORF4 of both BYDVs and CYDVs is contained within ORF3 and encodes a 17 kDa protein. Viruses that contain mutations in ORF4 are able to replicate in isolated plant protoplasts, but are deficient or delayed in systemic movement in whole plants. Hence, proteins encoded by ORF4 are thought to facilitate intra- and intercellular virus movement.

Some BYDV genomic sequences contain small ORFs (ORF6) downstream of ORF5. The predicted sizes of the proteins expressed by ORF6 range from 4 to 7 kDa. The predicted amino acid sequences of the proteins encoded by ORF6 are poorly conserved among BYDV-PAV isolates. Repeated attempts to detect protein products of ORF6 have been unsuccessful. In addition, BYDV-PAV genomes into which mutations have been introduced that disrupt ORF6 translation are still able to replicate in protoplasts. Based on these observations, it has been concluded that ORF6 is not translated *in vivo*.

## Host Range and Transmission

BYD-causing viruses infect over 150 species of annual and perennial grasses in five of the six subfamilies of the Poaceae. The feeding habits of vector aphids have a major impact on the host ranges of virus species. Hence the number of species naturally infected by the viruses is much lower than the experimental host range. As techniques for infecting plants with recombinant viruses have improved, the experimental host ranges of BYDVs and CYDVs have been expanded to include plants on which aphid vectors would not normally feed. For example, BYDV-PAV and CYDV-RPV have been shown to infect *Nicotiana* species when inoculated using *Agrobacterium tumefaciens* harboring binary plasmids containing infectious copies of the viruses, which had not been described previously as experimental hosts for the viruses.

Viruses that cause BYD are transmitted in a circulative strain-specific manner by at least 25 aphid species. Circulative transmission of the viruses is initiated when the piercing–sucking mouthparts of aphids acquire viruses from sieve tubes of infected plants during feeding. Aphids that do not probe into and feed from the vascular tissues of infected plants do not transmit the viruses. The virions of BYDVs and CYDVs travel up the stylet, through the food canal, and into the foregut (**Figure 4**). After 12–16 h, virions then are actively transported across the cells of the hindgut into the hemocoel in a process that involves receptor-mediated endocytosis of the viruses and the formation of tubular vesicles that transport viruses through epithelial cells and into the hemocoel. Virions then passively diffuse through the hemolymph to the accessory salivary gland where virions must pass through the membranes of accessory salivary gland cells in a similar type of receptor-mediated transport process to reach the lumen of the gland. The accessory salivary gland produces a watery saliva, containing few or no enzymes, that is thought to prevent phloem proteins from clogging the food canal. Once in the salivary gland lumen, virions are expelled with the watery saliva into vascular tissues of host plants. Typically hindgut membranes are much less selective than those of the accessory salivary glands. Consequently, viruses that are not transmitted by a particular species of aphid often are transported across gut membranes and accumulate in the hemocoel, but do not traverse the membranes of the accessory salivary gland. The specificity of aphid transmission and gut tropism has been linked to the RTD of the minor capsid protein. Even though large amounts of virions can accumulate in the hemocoel, there is no evidence for virus replication in their aphid vectors. Aphids may retain the ability to transmit virus for several weeks.

Genetic and biochemical studies have been conducted to identify aphid determinants of strain-specific transmission of BYDV-MAV and BYDV-PAV. Protein–protein and protein–virus interaction experiments were used to isolate two proteins from heads of vector aphids that bind BYDV-MAV that were not detected in nonvector aphids. These two proteins are good candidates for the cell-surface receptors that are thought to be involved in strain-specific transport of viruses into accessory salivary gland lumens. In addition, endosymbiotic bacteria that reproduce in specialized cells called mycetocytes in abdomens of aphids express chaperonin-like proteins that bind BYDV particles and the amino-terminal region of recombinant BYDV-PAV RTD proteins. However, the role of these proteins in aphid transmission is unclear since they are found in both vector and nonvector aphid species. Interactions of virus particles with these proteins seem to be essential for persistence of the viruses in aphids. The proteins may protect virus particles from degradation by aphid immune systems.

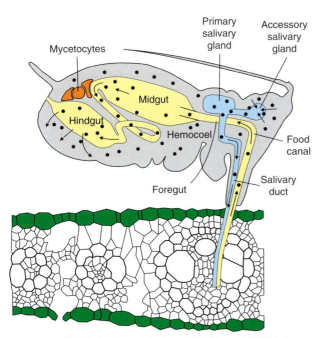

**Figure 4** Circulative transmission of BYDVs and CYDVs by vector aphids. While feeding from sieve tubes of an infected plant, an aphid (shown in cross section) acquires virions, which travel up the stylet, through the food canal, and into the foregut. Virions are actively transported across cells of the hindgut into the hemocoel. Virions then passively diffuse through the hemolymph to the accessory salivary gland where they are again actively transported into the lumen of the gland. Once in the salivary gland lumen, the virions are expelled with the saliva into the vascular tissue of host plants. Viruses that are not transmitted by a particular species of aphid often accumulate in the hemocoel, but do not traverse the membranes of the accessory salivary gland.

## Replication

Like other viruses of the family *Luteoviridae*, BYDVs and CYDVs infect and replicate in sieve elements and companion cells of the phloem and occasionally are found in phloem parenchyma cells (**Figure 5**). BYDVs and CYDVs induce characteristic ultrastructural changes in infected cells. BYDV-MAV, -PAV, and -SGV induce

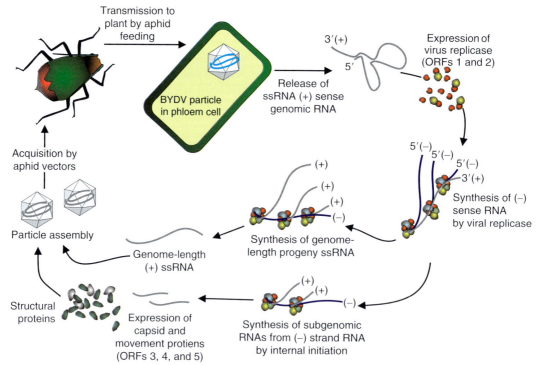

**Figure 5** Barley yellow dwarf virus-PAV life cycle. Virus particles are deposited in sieve elements by aphid vectors. By a yet unknown process, single-stranded messenger-sense genomic RNA is released from virus particles and translated by host translation machinery, which is facilitated by long-distance RNA–RNA interactions. Open reading frames (ORFs) 1 and 2, which encode the viral replicase, are expressed first because of their proximity to the 5′ termini of genomic RNAs. Virus encoded replicase then synthesizes negative-sense RNAs that are used as templates for the production of new full-length positive-sense genomic RNAs and subgenomic RNAs. Production of subgenomic RNAs results in synthesis of structural and cell-to-cell movement proteins. Subgenomic RNA2 suppresses translation from genomic RNAs, furthering the switch from early to late gene expression. Full-length positive-sense genomic RNAs and structural proteins then assemble into virions in cells of phloem tissues where they can be ingested by aphid vectors to start the process again.

single-membrane-bound vesicles in the cytoplasm near plasmodesmata early in infection. Subsequently, filaments are observed in nuclei, and virus particles are first observed in the cytoplasm. In contrast, BYDV-RMV and CYDV-RPV induce double-membrane-bound vesicles in the cytoplasm that are continuous with the endoplasmic reticulum. Later, filaments and tubules form in the cytoplasm, and BYDV-RMV and CYDV-RPV particles are first observed in nuclei.

The subcellular location of viral RNA replication has not been determined unequivocally. However, early in infection, negative-strand RNAs of BYDV-PAV are first detected in nuclei and later in the cytoplasm, which suggests that at least a portion of the BYDV-PAV replication occurs in the nucleus. A nuclear location for replication is supported by the observation that the movement protein encoded by ORF4, which also binds single-stranded RNA, localizes to the nuclear envelope and is associated with virus RNA in nuclei of infected cells. Synthesis of negative-strand RNA, which requires tetraloop structures at the 3′ end of BYDV-PAV genomic RNAs, is detected in infected cells before the formation of virus particles. Because tetraloops have been implicated in RNA–protein interactions, these structures could be binding and/or recognition sites for BYDV replication proteins. BYDV-PAV sgRNAs are synthesized by internal initiation of RNA synthesis on negative-strand RNAs from three dissimilar subgenomic promoters. Late in infection, the BTE near the 5′ terminus of BYDV-PAV sgRNA2 inhibits translation from genomic RNA, which may promote a switch from translation to replication and packaging of genomic RNAs. In addition to genomic RNAs, CYDV-RPV replicates a satellite RNA by a rolling-circle mechanism that generates multimeric satellite RNAs that self-cleave to unit length.

## Virus–Host Relationships

Visible symptoms induced by BYDVs and CYDVs vary greatly depending on the host and strain of the virus. The most common symptoms are stunting and chlorosis. While some infected plants display no obvious symptoms, most BYDVs and CYDVs induce characteristic symptoms that include stunting, leaves that become thickened, curled or serrated, and yellow, orange or red leaf discoloration, particularly of older leaves of infected plants. These symptoms result from phloem necrosis that spreads from inoculated sieve elements and causes symptoms by inhibiting translocation, slowing plant growth, and inducing the loss of chlorophyll. Symptoms may persist, may vary seasonally, or may disappear soon after infection. Temperature and light intensity often affect symptom severity and development. In addition, symptoms can vary greatly with different virus isolates or strains and with different host cultivars. Yield losses caused by BYD are difficult to estimate because the viruses are so pervasive and symptoms often are overlooked or attributed to other agents. In Australia alone, losses in barley production have been valued at over 100 million US dollars annually. Plants infected with BYD at early developmental stages suffer the most significant yield losses, which often are linearly correlated with the incidence of virus infection.

## Epidemiology

BYD infections have been reported from temperate, subtropical, and tropical regions of the world. Even though the incidence of infections of individual viruses varies from year to year and can differ among annual and perennial hosts, BYDV-PAV usually is the most prevalent of the viruses causing BYD in small grains worldwide followed by CYDV-RPV or BYDV-MAV. The remaining BYD-causing viruses are typically much less prevalent. BYDVs and CYDVs must be reintroduced into annual crops each year by their aphid vectors. Alate, that is, winged aphids may transmit viruses from local cultivated, volunteer, or weed hosts. Alternatively, alate aphids may be transported into crops from distant locations by wind currents. These vectors may bring the virus with them, or they may first have to acquire virus from locally infected hosts. In temperate regions of Europe and North America moderate and long-distance migration of viruliferous aphids is important to development of BYD epidemics. In Australasia, and other regions with Mediterranean climates, alate aphids usually transmit viruses from relatively close infected plants. Secondary spread of the viruses is often primarily by apterous, that is, wingless aphids. The relative importance of primary introduction of viruses by alate aphids and of secondary spread of viruses by apterous aphids in disease severity varies with the virus, aphid species, crop, and environmental conditions.

## Diagnosis

Accurate diagnosis of infections has been important in understanding the transmission and epidemiology of BYDVs and CYDVs and developing control strategies for BYD. Because BYD symptoms resemble those caused by other biotic and abiotic factors, visual diagnosis is unreliable and other methods have been developed. Initially, infectivity, or biological, assays were used to diagnose infections. In bioassays, aphids are allowed to feed on infected plants and then are transferred to indicator plants. These techniques have also been used to determine vector specificities of viruses causing BYD and to identify viruliferous vector aphids in epidemiological studies. These techniques are very sensitive, but they can require several weeks for symptoms to develop on indicator plants. The viruses causing BYD are strongly immunogenic, which has facilitated development of genus- and even strain-specific antibodies that have been used extensively in BYD diagnosis. Because the viruses causing BYD are present in infected tissues at very low levels, mice have been used to produce monoclonal antibodies against the viruses. Mice typically require much less viral antigen per immunization than rabbits, and hybridoma cell lines that produce monoclonal antibodies can be stored for extended periods and used for many years, which further reduces the amount of antigen needed to produce diagnostic antibodies. Techniques have also been developed to detect viral RNAs from infected plant tissues by reverse transcription polymerase chain reaction, which can be more sensitive and discriminatory than serological diagnostic techniques. Even so, serological tests are the most commonly used techniques for the detection of infections because of their simplicity, speed, and relatively low cost.

## Control

Planting of insecticide-treated seeds that protect emerging seedlings from aphid infestation has been shown to reduce losses caused by BYD in North America, Australasia and Africa. Foliar applications of insecticides on older plants typically have been less effective. Alternatively, planting of tolerant or resistant cereals has proved to be a much more cost-effective and sustainable management strategy for BYD. Breeding programs have successfully integrated genes conferring high levels of tolerance into barley and oat and to a lesser extent in wheat. Even though a limited number of single genes for BYD resistance/tolerance have been identified in cultivated barley and rice, in most instances, tolerance to BYD is conditioned by multiple genes in a quantitative fashion, which has made moving BYD tolerance into new plant lines challenging. Particularly in barley, molecular markers have begun to facilitate the process of breeding for BYD tolerance. Because of a lack of effective single-gene resistance in cultivated wheat, some researchers have moved BYD resistance genes from wheat grasses (*Thinopyrum intermedium* and *Thinopyrum ponticum*) into wheat, which have provided high levels of resistance. The lack of naturally occurring resistance in cereals to BYD has made transgene-mediated resistance very attractive. Even though the expression of CP sequences in transgenic plants has conferred resistance in several other plant-virus systems, it has not provided significant resistance to BYD in barley, oat, or wheat. In contrast, transgenic barley and oat plants have been produced that express either intact or inverted copies of BYDV-PAV replicase genes, which conferred high levels of resistance to BYDV-PAV and closely related viruses.

In many small grain growing regions, viruliferous aphids arrive at similar times each spring and fall even though sizes of the aphid populations can vary significantly from year to year. In these areas, it is sometimes possible to plant crops so that young, highly susceptible plants are not in the field when the seasonal aphid migrations occur. However, crops planted later typically do not yield as well as those planted early in the growing season. Consequently, growers must weigh the probability of obtaining higher yields against possible yield losses caused by BYD. In some instances, biological control agents such as predatory insects and parasites have reduced aphid populations significantly.

*See also:* Cereal Viruses: Wheat and Barley; Luteoviruses.

## Further Reading

Crasta OR, Francki MG, Bucholtz DB, *et al.* (2000) Identification and characterization of wheat-wheatgrass translocation lines and localization of barley yellow dwarf virus resistance. *Genome* 43: 698–706.

D'Arcy CJ and Burnett PA (1994) *Barley Yellow Dwarf Virus, Forty Years of Progress*. St. Paul, MN: American Phytopathological Society Press.

Falk BW, Tian T, and Yeh HH (1999) Luteovirus-associated viruses and subviral RNAs. *Current Topics in Microbiology and Immunology* 239: 159–175.

Figueira AR, Domier LL, and Darcy CJ (1997) Comparison of techniques for detection of barley yellow dwarf virus PAV-IL. *Plant Disease* 81: 1236–1240.

Gray S and Gildow FE (2003) Luteovirus–aphid interactions. *Annual Review of Phytopathoogy* 41: 539–566.

Koev G, Mohan BR, Dinesh-Kumar SP, *et al.* (1998) Extreme reduction of disease in oats transformed with the 5′ half of the barley yellow dwarf virus PAV genome. *Phytopathology* 88: 1013–1019.

Miller WA, Liu SJ, and Beckett R (2002) Barley yellow dwarf virus: *Luteoviridae* or *Tombusviridae*? *Molecular Plant Pathology* 3: 177–183.

Miller WA and Rasochova L (1997) Barley yellow dwarf viruses. *Annual Review of Phytopathology* 35: 167–190.

Miller WA and White KA (2006) Long-distance RNA–RNA interactions in plant virus gene expression and replication. *Annual Review of Phytopathology* 44: 447–467.

Nass PH, Domier LL, Jakstys BP, and D'Arcy CJ (1998) In situ localization of barley yellow dwarf virus PAV 17-kDa protein and nucleic acids in oats. *Phytopathology* 88: 1031–1039.

Ordon F, Friedt W, Scheurer K, *et al.* (2004) Molecular markers in breeding for virus resistance in barley. *Journal of Applied Genetics* 45: 145–159.

Zhu S, Kolb FL, and Kaeppler HF (2003) Molecular mapping of genomic regions underlying barley yellow dwarf tolerance in cultivated oat (*Avena sativa* L.). *Theoretical and Applied Genetics* 106: 1300–1306.

# Barnaviruses

**P A Revill**, Victorian Infectious Diseases Reference Laboratory, Melbourne, VIC, Australia

© 2008 Elsevier Ltd. All rights reserved.

## Glossary

**Casing** A layer of peat moss placed on top of the growing beds to encourage sporophore formation.
**Sporophore** Mushroom fruiting body.

## Introduction

In the USA in 1948, a disease of the cultivated mushroom *Agaricus bisporus* was discovered on a property in Pennsylvania that had a major impact on the mushroom industry and fungal pathology in general. It was characterized by poor colonization of compost by mycelium and misshapen sporophores with long thin stipes and small globular caps producing a drumstick-like appearance, or they were thickened with a barrel-like appearance. The poor colonization of the compost and casing by infected mycelium often produced characteristic bare patches on the growing beds and reduced yields. The disease was named La France disease and a virus was implicated as a possible cause in 1960, after it was shown that the disease could be transmitted to healthy cultures by hyphal anastamosis. In 1962, three different virus-like particles were identified, two of which were spherical (25 and 29 nm), and the third was a 19 nm × 50 nm elongated or bacilliform particle with rounded ends. Subsequently a 34–36 nm particle with a double-stranded (ds)RNA genome (La France infectious virus (LFIV)) has been identified as the causal agent of La France disease.

The bacilliform virus particle was of particular interest as almost all mycoviruses identified to that point had a spherical or isometric morphology. Bacilliform 19 nm × 48 nm virus-like particles were subsequently identified in the ascomycete *Microsphaera mougeotii*, and

17 nm × 35 nm bacilliform particles were also observed in the deuteromycete *Verticilium fungicola*, itself a pathogen of *A. bisporus*. However no relationship with the mushroom bacilliform virus (see below) and these bacilliform particles has been established.

Originally named mushroom virus 3 (MV3), the virus was subsequently renamed mushroom bacilliform virus (MBV). The viral genome was identified as positive-sense single-stranded (ss)RNA, and the virus was classified as the type member of the genus *Barnavirus* in the family *Barnaviridae*. The family name derives its roots from Bacilliform RNA virus. MBV remains the only barnavirus identified to date.

## Virion Properties

The MBV virion $M_r$ is $7.1 \times 10^6$, with a buoyant density in $Cs_2SO_4$ of $1.32$ g cm$^{-3}$. Virions are stable between pH 6 and 8 and ionic strength of 0.01 to 0.1 M phosphate.

## Virion Structure and Composition

MBV has a bacilliform or bullet-shaped morphology, with particles generally 19 nm × 50 nm in size (**Figure 1**). Virions contain a single major capsid protein (CP) of 21.9 kDa and there are approximately 240 molecules in each capsid.

Virions encapsidate a single linear molecule of a positive-sense ssRNA, 4.0 kb in size. The complete 4009 nt sequence is available (GenBank accession No. U07551). The RNA has a 5′-linked VPg and lacks a poly(A) tail. RNA constitutes about 20% of virion weight.

## Genome Organization and Expression

The RNA genome (4009 nt) contains four major and three minor open reading frames (ORFs) and has 5′- and 3′-untranslated regions (UTRs) of 60 and 250 nt, respectively (**Figure 2**). ORFs 1–4 encode polypeptides of 20, 73, 47, and 22 kDa, respectively. The deduced amino acid sequence of ORF2 contains three conserved chymotrypsin-related serine protease sequence motifs. Blast searches of the deduced ORF2 amino acid sequence show similarity to serine proteases encoded by plant sobemoviruses. ORF2 also encodes the VPg. ORF3 contains the **GX$_3$TX$_3$NX$_n$GDD** amino acid sequence shared by the putative RNA-dependent RNA polymerases (RdRps) of positive-sense ssRNA viruses and has similarity to the RdRps of sobemoviruses, enamoviruses, and poleroviruses. ORF4 encodes the CP. ORFs 5–7 encode 8, 6.5, and 6 kDa polypeptides, respectively. The polypeptides potentially encoded by ORFs 1, 5, 6, and 7 show no significant similarity to known polypeptides. The negative strand of MBV contains seven small ORFs of unknown significance. These potentially encode polypeptides ranging from $M_r$ 6.5K to $M_r$ 10.5K.

The genome arrangement and transcription/translation strategies of MBV are strikingly similar to those of a number of plant viruses, particularly poleroviruses and sobemoviruses. MBV probably also uses similar strategies to express its gene products, including leaky ribosomal scanning for expression of ORF2, ribosomal frameshifting for expression of the RdRp, and subgenomic RNA for expression of the CP. Of these, only subgenomic RNA has been confirmed *in vivo*. In a cell-free system, genomic-length RNA directs the synthesis of major 21 and 77 kDa polypeptides and several minor polypeptides of 18–60 kDa. The full-length genomic RNA and a sgRNA (0.9 kb) encoding ORF4 (CP) are found in infected cells. Virions accumulate singly or as aggregates in the cytoplasm. However the MBV life cycle has yet to be determined.

## Evolutionary Relationships

The MBV genome sequence has no similarity with any other mycovirus genome characterized to date. However the deduced ORF2 and ORF3 amino acid sequences share striking similarity with those of plant viruses, particularly sobemoviruses, poleroviruses, and enamoviruses (**Figure 3**). This, together with the similarity of the MBV and sobemovirus/polerovirus genome arrangements, suggests that MBV may have shared a common ancestor with these plant virus groups.

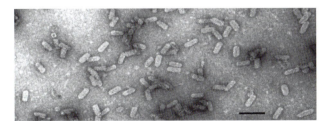

**Figure 1** Electron micrograph *Mushroom bacilliform virus* particles. Scale = 100 nm. Reprinted from *Virus Taxonomy: Seventh Report of the International Committee on Taxonomy of Viruses*, Wright PJ and Revill PA, The *Barnaviridae*, copyright 2004, with permission from Elsevier.

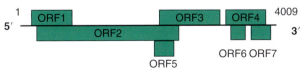

**Figure 2** The MBV genome arrangement.

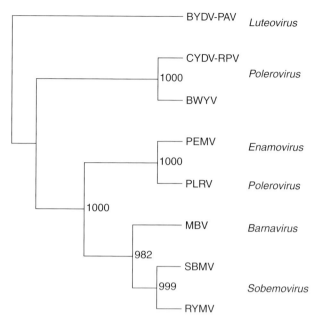

**Figure 3** Neighbor-joining tree of the MBV RdRp compared to RdRps of a number of plant viruses. Sequences were aligned using Clustal X (1000 bootstrap replicates) and the tree was constructed with Treeview. The tree was rooted using the outgroup (BYDV-PAV, barley yellow dwarf virus). CYDV-RPV, cereal yellow dwarf virus-RPV; SBMV, southern bean mosaic virus; RYMV, rice yellow mottle virus; PEMV, pea enation mosaic virus; BWYV, beet western yellows virus; PLRV, potato leafroll virus. The GenBank accession numbers of the sequences used in the analysis were PLRV (D00530), CYDV-RPV (NC004751), BWYV (NC004756, PEMV (NC003629), RYMV (NC001575), SBMV (DQ875594), MBV (U07557), BYDV-PAV (EF043235).

## Transmission and Host Range

MBV is transmitted horizontally through infected mycelium and it is yet to be determined if the virus can be transmitted in spores. There is no known insect vector.

Although morphologically similar viruses to MBV have been identified in the field agaric, *A. campestris*, it is unknown whether these particles are related to MBV. Consequently MBV remains the only barnavirus identified to date.

*See also:* Fungal Viruses; Luteoviruses; *Sobemovirus*.

## Further Reading

Goodin MM, Schlagnhaufer B, and Romaine CP (1992) Encapsidation of the La France disease specific double-stranded RNAs in 36 nm isometric virus-like particles. *Phytopathology* 82: 285–290.

Moyer JW and Smith SH (1976) Partial purification and antiserum production to the $19 \times 50$ nm mushroom virus particle. *Phytopathology* 66: 1260–1261.

Moyer JW and Smith SH (1977) Purification and serological detection of mushroom virus-like particles. *Phytopathology* 67: 1207–1210.

Revill PA, Davidson AD, and Wright PJ (1994) The nucleotide sequence and genome organization of mushroom bacilliform virus: A single-stranded RNA virus of *Agaricus bisporus* (Lange) Imbach. *Virology* 202: 904–911.

Revill PA, Davidson AD, and Wright PJ (1998) Mushroom bacilliform virus RNA: The initiation of translation at the 5′-end of the genome and identification of the VPg. *Virology* 249: 231–237.

Revill PA, Davidson AD, and Wright PJ (1999) Identification of a subgenomic mRNA encoding the capsid protein of mushroom bacilliform virus, a single-stranded RNA mycovirus. *Virology* 260: 273–276.

Romaine CP and Schlagnhaufer B (1991) Hybridization analysis of the single-stranded RNA bacilliform virus associated with La France disease of *Agaricus bisporus*. *Phytopathology* 81: 1336–1340.

Tavantzis SM, Romaine CP, and Smith SH (1980) Purification and partial characterization of a bacilliform virus from *Agaricus bisporus*: A single-stranded RNA mycovirus. *Virology* 105: 94–102.

Tavantzis SM, Romaine CP, and Smith SH (1983) Mechanism of genome expression in a single-stranded RNA virus from the cultivated mushroom *Agaricus bisporus*. *Phytopathology* 106: 45–50.

Wright PJ and Revill PA (2004) The *Barnaviridae*. In: Fauquet CM (ed.) *Virus Taxonomy: Seventh Report of the International Committee on Taxonomy of Viruses*, London: Elsevier Academic Press.

# Bean Common Mosaic Virus and Bean Common Mosaic Necrosis Virus

**R Jordan and J Hammond**, USDA-ARS, Beltsville, MD, USA

Published by Elsevier Ltd.

## Glossary

**Pathogroup** A series of isolates of the same strain or pathotype.

**Pathotype** An isolate or strain of a virus that is biologically distinct from other isolates of the same virus by virtue of differential host reactions.

**Serogroup** A series of isolates or strains of the same serotype.

**Serotype** An isolate or group of isolates that are distinguished from biologically related isolates by reaction (or lack of reaction) with key serological reagents such as defined polyclonal antisera or monoclonal antibodies.

## History

Bean common mosaic virus (BCMV) was first reported from the US in 1917, and the associated disease initially known as bean mosaic; it was renamed bean common mosaic in 1934 to differentiate it from bean yellow mosaic, caused by bean yellow mosaic virus (BYMV). Several pathotypes or strains of BCMV were distinguished in the 1970s by differential reactions of a number of bean cultivars, and in the early 1980s strains were further divided by serology into serotype A and serotype B. Serogroup A isolates were also biologically differentiated by temperature-insensitive induction of necrosis in bean cultivars carrying the dominant I gene. Peptide profiling of the coat protein (CP) and sequence analysis of the CP gene and 3′-untranslated region (UTR) demonstrated that the serotypes represented distinct viruses. Serotype A became known as bean common mosaic necrosis virus (BCMNV), while serotype B retains the name BCMV. 'Bean necrosis mosaic virus' was used for serogroup A isolates prior to acceptance of the species name *Bean common mosaic necrosis virus*. BCMV and BCMNV also differ in particle length, CP size, and biological properties (see below). Complete genome sequences support the distinction between BCMV and BCMNV proposed using CP sequences and peptide profiles.

## Taxonomy and Classification

The current criteria for differentiation of potyvirus species are CP amino acid identity of less than *c.* 80%, and nucleotide identity of less than 85% over the complete genome; different species also often have distinct polyprotein cleavage sites. Using these criteria, *Bean common mosaic virus* and *Bean common mosaic necrosis virus* are distinct potyviral species; BCMV also now includes strains previously described by other names, including azuki bean mosaic virus, blackeye cowpea mosaic virus, dendrobium mosaic virus, guar green sterile virus, peanut chlorotic ring mottle virus, peanut mild mottle virus, and peanut stripe virus. At least 19 different strains of BCMV have been biologically authenticated as distinct on the basis of host range and host response, whereas only four biologically distinct strains of BCMNV have been recognized.

BCMV and BCMNV, together with CerMV, CABMV, DsMV, FrVY, IFBV, PWV, SaVY, SMV, TrVY, WMV, WVMV, ZaMMV, and ZYMV, form the BCMV subgroup of the genus *Potyvirus* within the family *Potyviridae* (for full virus names, see **Table 1**). Although 'subgroup' has no formal taxonomic meaning, it is a useful concept to describe a subset of virus species that cluster together in a phylogenetic tree (see **Figure 1**) and also has common serological properties. Some confusion exists in the literature over isolates described as CABMV; true CABMV isolates are distinct from BCMV, but some isolates initially described as CABMV were later shown to be synonymous with BlCMV, which is now recognized as a strain of BCMV.

**Table 1** The viruses included in the bean common mosaic virus subgroup

| Virus species recognized as the bean common mosaic virus subgroup | Acronym |
|---|---|
| ***Bean common mosaic virus**[a]* | **BCMV** |
| Azuki bean mosaic virus[b] | (AzMV) |
| Bean common mosaic virus serotype B | BCMV |
| Blackeye cowpea mosaic virus | (BlCMV) |
| Dendrobium mosaic virus | (DeMV) |
| Guar green sterile virus | (GGSV) |
| Peanut chlorotic ring mottle virus | |
| Peanut mild mottle virus | (PMMV) |
| Peanut stripe virus | (PStV) |
| ***Bean common mosaic necrosis virus*** | **BCMNV** |
| Bean common mosaic necrosis virus | BCMNV |
| Bean common mosaic virus serotype A | |
| (Bean necrosis mosaic virus)[c] | (BNMV) |
| ***Ceratobium mosaic virus*** | **CerMV** |
| ***Cowpea aphid-borne mosaic virus*** | **CABMV** |
| Cowpea aphid-borne mosaic virus | CABMV |
| Sesame mosaic virus | |
| South African passiflora virus | (SAPV) |
| ***Dasheen mosaic virus*** | **DsMV** |
| (Vanilla mosaic virus)[d] | (VanMV) |
| ***Fritillary virus Y**[e]* | ***FrVY*** |
| ***Impatiens flower break virus*** | ***IFBV*** |
| ***Passionfruit woodiness virus*** | **PWV** |
| ***Sarcochilus virus Y*** | **SaVY** |
| ***Soybean mosaic virus*** | **SMV** |
| ***Tricyrtis virus Y*** | ***TrVY*** |
| ***Watermelon mosaic virus*** | **WMV** |
| Watermelon mosaic virus | WMV |
| Watermelon mosaic virus 2 | (WMV-2) |
| Vanilla necrosis virus | (VNV) |
| ***Wisteria vein mosaic virus*** | ***WVMV*** |
| ***Zantedeschia mild mosaic virus*** | ***ZaMMV*** |
| ***Zucchini yellow mosaic virus*** | **ZYMV** |

[a]Names in bold italic font are species currently recognized by the ICTV.
[b]Names in regular font are no longer recognized as potyvirus species, but are indented to indicate their position as strains of the species listed above. Acronyms in parentheses are no longer valid as species, but refer to strains of the virus in bold font.
[c]'Bean necrosis mosaic virus' was a name used for strains of the A serotype prior to ICTV designation of these isolates as strains of BCMNV.
[d]Vanilla mosaic virus is recognized in the ICTV Eighth Report as a tentative species in the genus *Potyvirus*; recent reports suggest that it is an isolate of DsMV.
[e]Viruses with names in bold font appear to meet the criteria for distinct virus species, but have not yet been recognized by the ICTV; their acronyms are indicated in bold italics.

## Geographic Distribution

BCMV probably originated in Latin America, the center of diversity of *Phaseolus vulgaris*, and is distributed

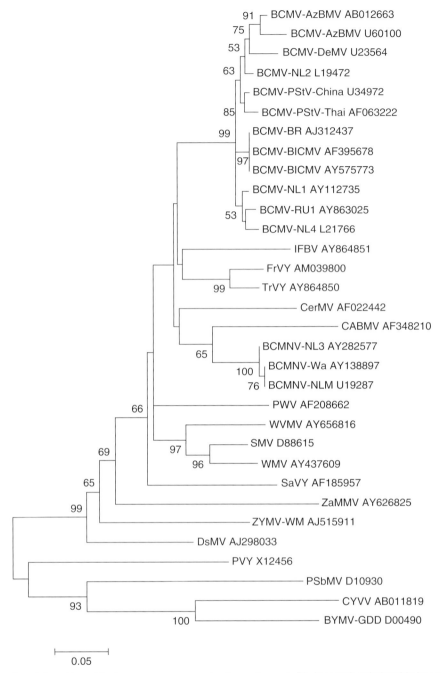

**Figure 1** Phylogenetic relationships of the bean common mosaic virus subgroup. Phylogenetic relationship based on analysis of coat protein amino acid sequences of bean common mosaic virus and bean common mosaic necrosis virus with members of the BCMV subgroup, other legume-infecting potyviruses, and the genus *Potyvirus* type member potato virus Y (PVY). The tree was constructed by the neighbor-joining algorithm based on calculations from Clustal W pairwise amino acid sequence distances. The horizontal branch lengths are proportional to the genetic distance. The data set was subjected to 1000 bootstrap replicates. All nodes supported by >50% confidence values are shown. GenBank sequence sources are listed. Virus acronyms are as listed in **Table 1**. BYMV, bean yellow mosaic virus; CYVV, clover yellow vein virus; PSbMV, pea seed-borne mosaic virus.

worldwide wherever beans are grown, and in association with other bean types, as well as soybean, pea, and cowpea. Blackeye cowpea mosaic isolates are also found worldwide. Peanut stripe isolates are found in peanuts in Asia and the United States, and also infect lupin, soybean, and sesame. In many areas of the world BCMV is the most important potyvirus affecting beans, although in some areas BCMNV or BYMV may predominate. Before major efforts were made to eradicate BCMV from US germplasm collections, more than 60% of germplasm accessions were found to be infected.

BCMNV is presumed to originate in Africa and is more important than BCMV in much of eastern, central, and southern Africa, where BCMNV is common in a wide variety of wild and forage legumes as well as beans grown for human consumption. BCMNV has also been distributed worldwide, largely through germplasm introductions, and has increased in importance in the United States partly as a consequence of deployment of resistance genes against BCMV (see below).

The increase in diversity and severity of bean-infecting geminiviruses in many regions has displaced BCMV and BCMNV as the most important viruses infecting beans in some areas.

## Host Range and Transmission

BCMV naturally infects *P. vulgaris* (kidney bean), *P. acutifolius* (tepary bean), *P. atropurpureus, P. coccineus* (runner bean), *P. mungo* (black gram), *Glycine max* (soybean), *Macroptilium lathyroides* (horse gram), *Pisum sativum* (pea), *Rhynchosia minima, Vicia faba* (broad bean), *Vigna radiata* (mung bean, green gram), *V. angularis* (azuki bean), and *V. unguiculata* (cowpea). Peanut stripe isolates naturally infect *Arachis hypogea* (peanut), *Dolichos lablab* (hyacinth bean), *Indigofera amoena, G. max, Lupinus albus* (lupin), *Pueraria phaseoloides, Sesamum* spp. (sesame), *Stylosanthes capitata*, and *S. craba*. Soybean isolates are adapted to soybean, and rarely infect peanuts. Dendrobium mosaic isolates infect *Dendrobium* orchids, while guar green sterile isolates infect guar (*Cyamopsis tetragonoloba*). In addition to its natural hosts, BCMV has a wide experimental host range, infecting about 100 species from 44 genera over nine families including Amaranthaceae, Chenopodiaceae, Fabaceae, Solanaceae, and Tetragoniaceae, although individual isolates may be much more restricted.

BCNMV naturally infects *P. vulgaris, P. lunatus, Centrosema pubescens, Crotalaria incana, Lablab purpureus, Senna bicapsularis, S. hirsuta, S. sophora*, and *V. vexillata*, and has been detected using a BCMNV-specific monoclonal antibody in naturally infected plants of *Albizia coriaria, Desmodium intortum, D. uncinatum, Rhynchosia resinosa, Tephrosia barbigera*, and *T. paniculata*. Experimental hosts include several other leguminous species, such as *Canavalia ensiformis, Crotalaria spinosa, Macroptilium lathyroides, Rhynchosia minima*, and *V. radiata*.

Both BCMV and BCMNV are transmitted by aphids in a nonpersistent manner typical of viruses in the genus *Potyvirus*; the most important vector species in many countries are *Acyrthosiphon pisum, Aphis fabae*, and *Myzus persicae*. *Aphis craccivora* may be the most important vector in India. Minimum acquisition access and inoculation access periods are less than 1 min; no latent period is required between acquisition and inoculation. Aphid transmission is the cause of secondary transmission within the crop, but the primary means of introduction to the crop is through infected seed. Aphid transmission and infection rates tend to be higher when beans are grown under irrigation in dry regions.

Both BCMV and BCMNV are also seed-transmitted. Seed transmission varies with isolates and host species, as well as the timing of infection, but rates of up to 93% of infected seed may result from diseased plants, with erratic distribution of infected seed within individual pods. Plants infected after flowering typically do not yield infected seed, probably because the virus is not found in the embryo and cotyledons, and the virus may not be able to spread into these locations if flowering occurs prior to infection of the mother plant. Infection of the seed, and perhaps of the mother plant, can occur through infected pollen. Infectivity of the virus is retained during prolonged storage of seed.

Dendrobium mosaic (a strain of BCMV) is transmitted by aphids and by vegetative propagation of *Dendrobium* orchids.

## Properties of the Virion and Genome

### Properties of Particles

BCMV and BCMNV are typical of viruses in the genus *Potyvirus* in that particle preparations contain a single sedimenting (154–158S) and buoyant density component ($1.31$–$1.32 \, \text{g cm}^{-3}$ in CsCl). Virions are nonenveloped flexuous filaments, 12–15 nm wide, and 847–886 nm (BCMV) or 810–818 nm (BCMNV) long. Virions are composed of one single-stranded RNA of *c.* 9600 nt for BCMNV and *c.* 10 000 nt for BCMV (accounting for 5% of the particle weight) encapsulated in about 1700–2000 monomer units of one CP polypeptide species with a molecular mass of *c.* 30 kDa for BCMNV, and *c.* 33 kDa for BCMV (comprising 95% of the particle weight). In gel electrophoresis the CP of BCMNV migrates with apparent $M_r$ 33 kDa, and that of BCMV of $M_r$ 34.5–35 kDa. Virus preparations that have undergone limited proteolysis also contain lower-molecular-weight peptides of apparent $M_r$ 29–34 kDa.

### Serological Relationships

BCMV and BCMNV particles are strongly immunogenic and can be easily distinguished using polyclonal antisera and monoclonal antibodies (McAbs). Polyclonal antisera show only a very distant serological relationship between strains of these two viruses. BCMV- and BCMNV-specific antisera are available commercially for use in standard serological tests such as enzyme-linked immunosorbent assay (ELISA). Early work with BCMV and homologous antisera showed that BCMV was closely related serologically to AzMV, BlCMV, PStV, and several soybean infecting-isolates from Taiwan. As noted above, these viruses are now recognized as strains of BCMV. BCMV is distantly related to BYMV, ClYVV, CABMV, PVY, PMV, PWV, SMV, TEV, and WVMV.

Several broad-spectrum McAbs have been identified that detect all of the strains of BCMV, including AzMV, BlCMV, DeMV, and PStV, as well as the BCMV subgroup member CABMV, but not with any strains of BCMNV. Other McAbs are reported which display various degrees of specificity for these viruses (e.g., can only detect CABMV or BCMV-PStV). While these McAbs can be used to discriminate among the BCMV isolates, biological diversity among these viruses may limit the use of many of these antibodies for diagnostic purposes. McAbs produced to BCMNV are specific only for strains of BCMNV (i.e., have no cross-reactivity with any strains of BCMV, nor other potyviruses).

Another series of McAbs were raised against three strains of BCMV (AzMV, BlCMV, and PStV). Results of ELISA, immunosorbent electron microscopy, and western-blot analyses indicated that these McAbs are specific for a series of epitopes located entirely on the virion surface. From western-blot analyses of untreated and trypsin- as well as endopeptidase Lys-C-treated CPs, these virus-specific epitopes appear to be located exclusively on the amino-terminus of the CP. These McAbs were found to discriminate between strains of BCMV. These and other N-terminal targeted antibodies often enable clear distinction of strains in mixed infections, but a BCMV-PStV-specific McAb that reacts well with most blotch isolates failed to detect a necrotic isolate from Taiwan and several blotch isolates from Georgia, while a BCMNV-specific McAb misdiagnosed a naturally occurring genomic recombinant between BCMV and BCMNV.

## Properties of the Genome and Replication

The single-stranded RNA genome, which has a VPg protein covalently linked to its 5′ end, is about 10 000 nt for BCMV and 9600 nt for BCMNV. The organization of the BCMV and BCMNV genomes is similar to other potyviruses, consisting of short untranslatable sequences at the 5′ and 3′ ends, and a single, long open reading frame (ORF). The ORF is translated into a single polyprotein: c. 3222 aa for BCMV and c. 3071 aa for BCMNV. The polyprotein undergoes co- and post-translational proteolytic processing by three viral-encoded proteinases to form ten individual gene products. Most of the polyprotein cleavage sites differ between BCMV and BCMNV, with the exception of the HC-Pro/P3 cleavage. The ten viral proteins (see **Figure 2**) include, in order from the 5′ end of the genome, P1 proteinase; helper

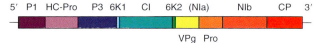

**Figure 2** Genome organization of BCMV and BCMNV. 5′, 5′-UTR; P1, P1 proteinase; HC-Pro, helper component-proteinase; P3, P3 protein; 6K1, a 6 kDa protein; CI, cylindrical inclusion; 6K2, a second 6 kDa protein; NIa, nuclear inclusion 'a', cleaved into VPg and Pro; VPg, genome-linked protein; Pro, proteinase; NIb, nuclear inclusion 'b'; CP, coat protein; 3′,3′-UTR.

component-proteinase (HC-Pro); P3; a 6 kDa protein; cylindrical inclusion (CI); a second 6 kDa protein; the nuclear inclusion 'a' (NIa)-VPg protein; NIa-proteinase; nuclear inclusion 'b' (NIb); and the CP. Genomic RNA replicates via production of a full-length negative-sense RNA.

## Pathogenicity, Pathology, and Resistance Genes

Isolates of BCMV and BCMNV can be differentiated into at least ten pathotypes (out of 16 theoretical pathotypes) based on their reactions on a series of differential bean cultivars. BCMV normally produces only mosaic symptoms in susceptible genotypes; BCMNV isolates induce lethal systemic necrosis and plant death in bean cultivars carrying the incompletely dominant 'I' gene. In most such cultivars the systemic necrosis develops at or above 20 °C; in hypersensitive cultivars the response is restricted to necrotic local lesions at normal temperatures, and systemic or 'black root' symptoms only at temperatures above 30 °C. Hypersensitivity results in field resistance, and is genetically dominant; no seed transmission of BCMV occurs in genotypes with the dominant 'I' gene. Other resistance genes ($bc$-$1$, $bc$-$1^2$, $bc$-$2$, $bc$-$2^2$, $bc$-$3$, and $bc$-$u$) condition resistance to particular pathotypes in various combinations; the combination of the dominant $I$ gene with the strain-specific recessive genes, $bc$-$1^2$, $bc$-$2^2$, or $bc$-$u$ protects against systemic infection and seed transmission of BCMNV. In the absence of the recessive strain-specific resistance genes effective against necrosis-inducing isolates of BCMNV, the $I$ gene alone is not sufficient to protect against systemic infection by BCMNV in many regions of Africa. The viruses carry various combinations of six pathogenicity determinants ($P0$, $P1$, $P1^2$, $P2$, $P2^2$, and $Px$), which to date have not been correlated with particular viral genes. Pathotype is not correlated with either CP or 3′-UTR phylogeny, nor is there evidence for determinants controlling systemic movement in the 3′-terminal region. Evidence from natural recombinants suggests that the P1 gene carries pathogenicity determinants.

The incompletely dominant $I$ gene was first identified in the cultivar 'Corbett Refugee', and the first recessive genes in 'Robust' and 'Great Northern 1' in the 1930s; strain-specificity of the recessive genes was identified in the 1950s, and the strain-unspecific gene bc-u shown in the 1970s to be necessary for expression of the strain-specific genes. No isolates have yet been identified that break resistance conferred by bc-3; however, the gene originated in a single European breeding line, and it took many years to introgress bc-3 into genotypes adapted to the tropics. Cultivars carrying the dominant I gene may have increased susceptibility to cowpea severe mosaic comovirus.

There is genetic linkage between certain seed colors and BCMV susceptibility, including red-mottled seed coat; the dominant I gene is closely linked to darker purple-mottled seed color. The lighter seed color associated with susceptibility is preferred in many cultures and commercial uses of the beans; this linkage can be partially avoided by selecting resistant types with commercially acceptable intermediate seed colors. Introgression of bc-3 can be problematic, as lines carrying the gene can be symptomless; serological or infectivity assays may be necessary to differentiate resistant and susceptible lines. Introgression of I and bc-3 has become simpler with development of molecular markers associated with each gene, allowing marker-assisted selection without virus challenge and phenotypic evaluation. The I gene is closely linked to a multigene family of TIR-NBS-LRR resistance genes conferring resistance to several related potyviruses and a comovirus.

In Latin America most black-seeded bean cultivars carry the I gene, and many additional bean lines bred at Centro Internacional de Agricultura Tropical and associated national programs also carry resistance. Many land races grown by traditional farmers, and particularly beans with light-colored seed coats, are susceptible; infection rates of up to 100%, and yield losses of 35–98% are reported. In North America many bean varieties have effective resistance to BCMV, although many pinto, navy, and red-seeded types have only strain-specific resistance. However, recent breeding has yielded pinto and navy bean cultivars carrying both I and bc-3. BCMNV was previously rare in the US, but combined resistance to BCMV and BCMNV is now available in some cultivars of most bean types.

Many of the beans grown by small farmers in eastern and southern Africa are susceptible to both BCMV and BCMNV. The presence of necrosis-inducing strains of BCMNV, also found in wild legumes and species grown for fodder, means that the I gene cannot be deployed independent of recessive resistance genes, as infection by necrosis-inducing isolates could result in death of a significant proportion of plants in a field. In India infections of up to 100% are reported in bean, urdbean, mungbean, and cowpea, resulting in mottle, mosaic, blistering, stunting, and poor pod set. Many Asian bean types also lack resistance, due to linkage between susceptibility and desirable seed colors.

Peanut stripe strains of BCMV causing symptoms described as blotch, blotch-CP-N, blotch stripe, chlorotic line pattern, chlorotic ring mottle, mild mottle, necrosis, and stripe have also been differentiated on the basis of disease reactions on specific host genotypes. Losses of 23–38% to peanut mild mottle have been documented in China. Peptide profiles of five symptom variants from Thailand that were not serologically distinct confirmed close relationships. No resistance was identified among over 8000 cultivated peanut genotypes in the germplasm collection at the International Crops Research Institute for the Semi-Arid Tropics. Some wild *Arachis* species have been identified as immune or highly resistant, including accession PI 475998 of *A. cardenasii*, which could not be infected.

Resistance genes have been identified and incorporated into resistant cultivars of *Pisum sativum, Lupinus angustifolius, Lablab purpureus,* and *Macroptilium lathyroides*.

Transgenic resistance has been demonstrated in peanut transformed with the CP gene of a peanut stripe isolate; a high level of resistance was observed in lines expressing either untranslatable or N-terminally truncated forms of the CP gene, suggesting that resistance resulted from RNA silencing. While beans have been transformed for resistance to bean golden mosaic geminivirus, and soybean for resistance to soybean mosaic potyvirus, there are no reports of transgenic BCMV- or BCMNV-resistant bean lines.

## Recombination and Variability

Although recombination between different isolates of BCMV had been reported for some years, no recombinants between BCMV and BCMNV were reported until recently. No recombinants were recovered from deliberate mixed infections of BCMV strain US-5 (pathogroup IV) and BCMNV strain NL-8 (pathogroup III) following many serial passages in either of two susceptible hosts. However, multiple recombinants were recovered within 28 days when the same strains were inoculated on opposite primary leaves of beans that were susceptible to one virus and resistant to the other; these recombinants fell into five different classes based on combinations of serotype (A, BCMNV; or B, BCMV) and pathogroup (IV, V, or VI). Subsequently, a naturally occurring strain of BCMNV (NL-3 K) with atypical responses on differential hosts was sequenced and shown to have a 5′-UTR and 5′ region of the P1 gene almost identical to BCMV strain RU1, while the rest of the genome was almost identical to BCMNV strain NL-3 D. The P1 gene of the recombinant was larger than that of BCMNV, and similar to the BCMV P1. These results suggest that the P1 gene may play a significant role in pathogenicity and virulence, as previously suggested for BYMV.

The observed phenotypes of recombinants indicate that serotype, pathogenicity, and symptom expression are independent traits with separate determinants in the genome. The occurrence of recombination in bean genotypes with only recessive, strain-specific resistance genes suggests that this may be the primary cause of emergence of new resistance-breaking strains.

Much of the variability between isolates of either BCMV or BCMNV appears to be within the 5′ portion of the genome, as has been shown both within and between other potyviruses. The P1 and P3 genes, and the N-terminal region of the CP gene are the most variable regions of the potyvirus genome, and may be involved in host–virus interactions. Differences in the 5′-UTR and N-terminus of P1 of both BCMV and BCMNV have been associated with symptom differences in common bean (*P. vulgaris*), and for BCMV in asparagus bean (*V. sesquipedalis*). Variability in the P1 gene is extensive, with as little as 61% identity at the amino acid level between BCMV strains, with most of the differences in the N-terminal domain. Differences in the CP gene result in clustering of BCMV isolates according to host species (bean, cowpea, peanut, asparagus bean, guar, *Dendrobium*, etc.) and to some extent by geographic origin within host clusters; peanut stripe isolates from China, Indonesia, and Thailand form separate clusters.

## Diagnosis

### Host Range and Symptomatology

Potyviruses can be detected and identified by a variety of techniques that are based on biological, cytological, biochemical, antigenic, and structural properties. With respect to BCMV and BCMNV, host range and symptomatology are not very useful because there are many viruses that infect bean, and significant differences in symptoms may be reproduced depending on the virus pathotype and host genotype. Morphology of cytoplasmic inclusions has been useful in the identification of many potyviruses since inclusions can be virus-specific. However, inclusion morphology is not reliable for detection and differentiation of BCMV and BCMNV. BCMNV induces pinwheel and scroll types of cytoplasmic inclusions (subdivision I) in infected tissue. Most strains of BCMV, including BCMV-AzMV and BCMV-BlCMV, also induce subdivision I type inclusions. However, PStV and Taiwan soybean isolates induce subdivision IV cylindrical inclusions.

### Serological Techniques

BCMV- and BCMNV-specific polyclonal antisera and McAbs are useful for the detection and differentiation of these seed-borne viruses, and several methods have been used for detection in bean seed. When indirect ELISA and a dot immunoassay (DIA) with both McAbs and polyclonal antisera were evaluated, ELISA with McAbs proved the most sensitive method. However, BCMV antigen was found to be erratically distributed within the seed, with more than 50% of infected samples having detectable levels of BCMV antigen only in the seed coat. Tests on individual seeds proved unreliable and it is recommended that flour from a bulked sample of eight seeds be used when testing seed lots. Other assays include immunosorbent electron microscopy to detect BCMV in bean seed; or double antibody sandwich (DAS)-ELISA and bioassay on *Chenopodium quinoa* to test hydrated mature cowpea seed for presence of BCMV-BlCMV. In these tests infectious virus was detected in cotyledons and sometimes in the embryo axis of infected seed. Viral antigen was found in or on the testae, but very little infectious virus was present. Others have used direct antigen-coated-ELISA to detect BCMV-BlCMV in cowpea seed.

### Reverse Transcription-Polymerase Chain Reaction

The sensitivity, capacity, and potential of the PCR technique has been applied to BCMV and BCMNV. Several sets of BCMV- and BCMNV-specific RT-PCR primers have been designed and utilized to sensitively and accurately detect, differentiate, and characterize the two viral species. In one study using virus-specific CP primers, the specific geographical distribution of these viruses in Mexico was determined. In addition, the alignment of nine nucleotide sequences from cloned amplicons from the RT-PCR reactions for each viral species confirmed the identities of the viruses and was helpful in assigning them tentatively to pathogroups. Other primer sets have been developed and utilized, for example, in the successful detection and differentiation of the necrotic and blotch isolates of BCMV-PStV, and in the differentiation of naturally occurring genomic BCMV-BCMNV recombinants.

Considering the seed-borne nature of these viruses, RT-PCR has potential in quarantine programs for screening imported seed material and germplasm.

## Prevention and Control

It is possible to eradicate BCMV and BCMNV from a particular production region, or at least to significantly reduce primary inoculum, by eliminating susceptible genotypes and planting only resistant varieties. There are few important noncrop reservoirs of inoculum, and infected seed are responsible for primary infections in the crop.

Rogueing of symptomatic plants is not recommended as a means of control, as it is highly likely that systemically infected plants without significant symptoms will remain to act as significant reservoirs for secondary infection by aphid transmission.

Exclusion of the virus is highly effective in areas where the virus is not present, but this may preclude introduction of new germplasm or cultivars without effective resistance. One or both viruses are almost universally present in seed lots of varieties that do not carry the *I* gene, and the majority of Latin American land race types do not possess any of the strain-specific recessive resistance genes. Most cultivars derived through extensive breeding programs carry the dominant *I* gene for resistance to BCMV and BCMNV. Increasingly *bc-3* is being combined with *I* to yield cultivars with effective resistance to both BCMV and BCMNV.

Certified seed can be an effective means of control. A certified seed program in California with a limit of <0.5% BCMV has aided in the control of the disease, as resistant varieties are not yet available for all bean classes currently grown in California. Monitoring of BCMV for certification also aided in identification and containment of an outbreak of BCMNV introduced in seed of a navy bean cultivar.

Vector exclusion may aid in control, but as aphids rarely colonize common beans, migratory aphids are responsible for most transmission in many areas. As BCMV and BCMNV can be acquired and transmitted in less than 1 min access to plants, insecticide treatment is unlikely to prevent, and may even encourage vector movement and transmission. Vector control through insecticide application may be effective in parts of Africa and Asia where aphids do colonize beans. Oil sprays are known to reduce transmission of stylet-borne viruses such as BCMV and BCMNV, but their lack of systemic action, and thus need for frequent reapplication to a growing crop, reduces the cost-effectiveness of their use.

*See also:* Potyviruses.

## Further Reading

Adams MJ, Antoniw JF, and Fauquet CM (2005) Molecular criteria for genus and species discrimination within the family *Potyviridae*. *Archives of Virology* 150: 459–479.

Berger PH, Wyatt SD, Shiel PJ, *et al.* (1997) Phylogenetic analysis of the *Potyviridae* with emphasis on legume-infecting potyviruses. *Archives of Virology* 142: 1979–1999.

Dijkstra J and Khan JA (1992) A proposal for a bean common mosaic subgroup of potyviruses. *Archives of Virology* (supplement 5): 389–395.

Drijfhout E (1978) Genetic interaction between *Phaseolus vulgaris* and bean common mosaic virus with implications for strain identification and breeding for resistance. *Agricultural Research Reports* No. 872. Wageningen: Center for Agricultural Publishing and Documentation.

Higgins CM, Hall RM, Mitter N, Cruickshank A, and Dietzgen RG (2004) Peanut stripe potyvirus resistance in peanut (*Arachis hypogea* L.) plants carrying viral coat protein gene sequences. *Transgenic Research* 13: 59–67.

Larsen RC, Miklas PN, Druffel KL, and Wyatt SD (2005) NL-3 K strain is a stable and naturally occurring interspecific recombinant derived from *Bean common mosaic necrosis virus* and *Bean common mosaic virus*. *Phytopathology* 95: 1037–1042.

McKern NM, Ward CW, and Shukla DD (1992) Strains of bean common mosaic virus consist of at least two distinct potyviruses. *Archives of Virology* (supplement 5): 407–414.

Milne RG (ed.) (1988) *The Plant Viruses, Vol. 4: The Filamentous Plant Viruses*. New York: Plenum.

Mink GI and Silbernagel MJ (1992) Serological and biological relationships among viruses in the bean common mosaic virus subgroup. *Archives of Virology* (supplement 5): 397–406.

Morales FJ (1998) Present status of controlling bean common mosaic virus. In: Hadidi A, Kheterpal RK, and Koganezawa H (eds.) *Plant Virus Disease Control*, pp. 524–533. St. Paul: APS Press.

Provvidenti R and Hampton RO (1992) Sources of resistance to viruses in the *Potyviridae*. *Archives of Virology* (supplement 5): 189–211.

Sengooba TN, Spence NJ, Walkey DGA, Allen DJ, and FemiLana A (1997) The occurrence of bean common mosaic necrosis virus in wild and forage legumes in Uganda. *Plant Pathology* 46: 95–103.

Shukla DD, Ward CW, and Brunt AA (1994) *The Potyviridae*. Wallingford: CAB International.

Vallejos CE, Astua-Monge G, Jones V, *et al.* (2006) Genetic and molecular characterization of the *I* locus of *Phaseolus vulgaris*. *Genetics* 172: 1229–1242.

Vetten HJ, Lesemann D-E, and Maiss E (1992) Serotype A and B strains of bean common mosaic virus are two distinct potyviruses. *Archives of Virology* (supplement 5): 415–431.

# Bean Golden Mosaic Virus

**F J Morales**, International Center for Tropical Agriculture, Cali, Colombia

© 2008 Elsevier Ltd. All rights reserved.

## Glossary

**3′ or 5′ end** Chemical convention of naming carbon atoms numerically in the nucleotide sugar ring.

**Bipartite** Viruses made up of two particles containing different nucleic acids.

**Complementary sense** Nucleic acid that has its sequence in the opposite direction in which nucleic acids usually replicate.

**DNA probes** DNA molecule converted into single strands and then labeled (radioactively with 32P or otherwise) that will hybridize with complementary nucleic acid strands, detecting their presence due to the label attached.

**ELISA** Enzyme-linked immunosorbent assay, a serological test that uses an enzyme to increase the sensitivity of the assay by inducing a colorimetric reaction.

**Gene silencing** Switching off a gene.
**GMO** Genetically modified organism that has received a foreign gene from other organism.
**Immunity** Lack of interaction between a pathogen and an organism.
**Infectious pseudo-recombinants** Mixture of A and B molecules from related strains of a begomovirus or begomovirus species, which can induce disease when inoculated in a susceptible plant.
**Inoculation** Bringing a pathogen into contact with a potentially susceptible organism.
**Isometric** Round particle with similar dimensions in all directions.
**MAS** Use of molecular markers that detect specific genes present in plants to select desirable plant genotypes under laboratory conditions.
**ORF** Any sequence of DNA or RNA that can be translated into a protein.
**PCR (polymerase chain reaction)** The reproduction of a nucleic acid segment in a DNA molecule, using short nucleic acid strands ('primers') that serve as the starting point for DNA replication.
**Polyadenylation** Addition of adenosine residues to the end of an mRNA.
**Polyclonal antiserum** An animal serum that has different antibodies produced against different antigenic determinants.
**Replication** A nucleic acid makes a copy of itself.
**Resistance** Relative ability of an organism to attenuate the effects of a pathogen.
**Sequence identity** Same nucleotide or amino acid sequence in a nucleic acid molecule.
**Sieve-tube cells** Cells that belong to phloem tissue; do not have a nucleus; have a few vacuoles and some organelles, such as ribosomes.
**Single-stranded** Viruses consisting of only one strand of viral RNA or DNA.
**Translational codons** Tri-nucleotide units coding for a single amino acid.
**Viroplasm** Amorphous aggregation of viral components.

## Introduction

Common bean (*Phaseolus vulgaris* L.) is one of the most important food legumes in the world. This species was domesticated in its center of origin, Latin America, where different races exist according to their origin (Mexico, Central America, and the Andean region of South America). Common bean has been one of the major staples of most Latin American countries since pre-Columbian times, and it has also become a vital food crop in East Africa. The largest producer of common beans in the world is Brazil, with total areas planted with this legume ranging from 4 to 5 million hectares. The second producer of common bean in the world is Mexico, with a total area of approximately 1 900 000 ha. Common bean is also a major staple in Central America (*c.* 700 000 ha) and the rest of South America (*c.* 600 000 ha), excluding Brazil.

Bean golden mosaic was first observed in 1961 by the pioneer Brazilian plant virologist Dr. Alvaro Santos Costa in the state of São Paulo, Brazil. Common bean (*P. vulgaris* L.) plants affected by this disease generally develop striking yellowing symptoms that affect the entire foliage in a truly systemic manner, unlike other partial yellowing diseases incited by different viruses in this legume species. The affected leaves of most common bean varieties also show downward curling and various degrees of malformation. Golden mosaic-affected common bean plants are usually stunted, and some highly susceptible common bean genotypes may not grow at all when infected at the early stages of plant development. Flower abortion has also been observed in golden mosaic-affected plants, particularly in hot environments. Pods formed by systemically infected plants are usually noticeably underdeveloped and malformed, and produce few small and badly damaged seeds.

Although Costa reported in 1965 that bean golden mosaic was 'not currently of sufficient economic importance', he noted its epidemiological potential. A decade later, Costa informed that bean golden mosaic was present in the two main common bean-producing states of Brazil, São Paulo and Parana. The role of *Bemisia tabaci* as a vector of the 'Abutilon mosaic virus' suspected to cause the 'infectious chlorosis of the Malvaceae' had already been demonstrated in Brazil in 1946, and Costa rapidly associated the occurrence of bean golden mosaic with the presence of *B. tabaci* in affected common bean plantings. He attributed the rapid dissemination of this disease to the noticeable increase in the population of the whitefly *B. tabaci* Gennadius (Homoptera: Aleyrodidae), as a result of the exponential growth in the area planted with soybean, a reproductive host of this whitefly species. Soybean plantings in Brazil had increased from 1.3 million hectares in the early 1970s, to almost 6 million hectares by 1975; and bean golden mosaic had disseminated to the remaining common bean-growing states: Minas Gerais, Goiás, and Bahia in the same period.

In Central America, a 'golden mosaic' of common bean had been observed to occur in the late 1960s from Guatemala to Panama. In 1966, a 'yellow mottle' severely affected common bean plantings in the Pacific coastal lowlands of Guatemala, El Salvador, and Nicaragua, mostly in regions where *B. tabaci* had previously been reported as a major pest of cotton in the 1950s. Based on the research conducted in Brazil on bean golden mosaic, and the similarities between the bright yellowing symptoms observed in

Central America and Brazil, it was assumed that the causal virus was the same. The golden mosaic disease of common bean became economically important in the mid-1970s, in southeastern Guatemala and the Pacific region of El Salvador. In Honduras, Nicaragua, and Costa Rica, the disease took longer to affect common bean production, and it was only in the 1980s that the disease became economically important in these countries.

In the Caribbean region, the name 'bean golden yellow mosaic' was used for the first time in 1973 to describe a disease of common bean in Puerto Rico. A 'golden mosaic' of common bean had also been reported in the Dominican Republic and Jamaica, in 1970 and 1975, respectively. Similar symptoms were observed in Cuba in the early 1970s, particularly in the province of Velasco, where farmers referred to this disease as 'amachamiento' (female sterility), because the pods of affected common bean plants did not produce seeds or the seeds were underdeveloped. The emergence of this disease was associated with an increase in the population of the whitefly *B. tabaci*. In Haiti, the first association between the presence of whiteflies and bean golden yellow mosaic in the lowlands was made in 1973. By 1978, the presence of the disease had been registered in all major bean-producing regions of the country. In Central America, other *Phaseolus* species, such as *P. acutifolius*, *P. coccineus*, and particularly *P. lunatus* (lima bean), are commonly found affected by this disease under field conditions.

In 1974, a 'yellow mosaic' was observed in common bean fields in the Valley of Culiacan, Sinaloa, north western Mexico. The disease was later observed in the Valley of Mochis, Sinaloa, and the Valley of Santo Domingo, Baja California Sur. The disease was referred to as 'bean golden mosaic', again, due to the similarities with the yellowing diseases of common bean in Brazil and Central America, and their association with the whitefly vector *B. tabaci*.

## Etiology

Until 1975, Costa in Brazil reported that "the causal agent of the golden mosaic disease of beans is assumed to be a virus and is considered to belong to the Abutilon mosaic virus complex. It probably represents an evolved variant of this complex that became adapted epidemiologically to certain species of the legume family." Attempts to visualize the causal agent in infected plant extracts or purified preparations by electron microscopy yielded negative results. However, ultrathin sections of vascular tissue obtained from golden mosaic-affected bean plants in Brazil revealed the presence of isometric particles (20–25 nm) in sieve tube elements. The causal agent of bean golden mosaic in Brazil could not be transmitted by mechanical means or sexual seed, and it did not infect species of *Sida* that are known hosts of the Abutilon mosaic virus.

The first successful isolation of a whitefly transmitted virus that induced golden mosaic-like symptoms in common bean was achieved in 1975 by Galvez and Castaño at CIAT, Colombia, using a bean golden yellow mosaic virus (BGYMV) isolate from El Salvador, Central America. The isolated virus had quasi-isometric twin particles $c.$ 32 nm × 19 nm (**Figure 1**). Using the same methodology, similar particles were isolated from common bean plants infected with a Puerto Rican isolate of the bean golden yellow mosaic pathogen. This isolate was used to characterize the viral nucleic acid in 1977, as a single-stranded DNA molecule. In 1978, the bean golden mosaic-like viruses and other similar single-stranded DNA viruses transmitted by *B. tabaci* or leafhoppers (Homoptera: Cicadellidae) were included in the newly created Geminivirus group. In 1981, the Puerto Rican bean golden yellow mosaic geminivirus isolate was shown to possess a divided ssDNA genome of approximate $M^r$ $(7-8 \times 10^5)$ ($\sim$2510 nt) consisting of two different molecules separately encapsidated in each paired particle. In 1989, 'bean golden mosaic virus' (BGMV) was considered a 'subgroup II geminivirus' (whitefly transmitted viruses with a bipartite ssDNA genome); and in 1995, a member of 'subgroup III of the Geminivirus group', to further differentiate the whitefly borne geminiviruses from similar viruses having different homopteran vectors.

In 1991, a Brazilian isolate of BGMV was cloned in order to prepare broad-spectrum DNA probes that detected all of the selected whitefly borne geminivirus isolates tested from Brazil, Colombia (Bean dwarf mosaic virus), Central America (Guatemala), and the Caribbean (Dominican Republic). Specific DNA probes were also prepared in this study, which detected only individual

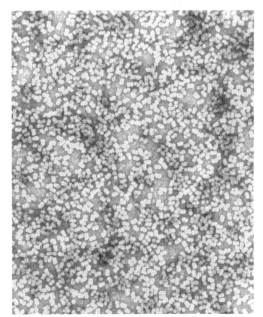

**Figure 1** Characteristic geminate particles of BGYMV.

viruses, with the exception of the Central American and Caribbean viruses that were detected by a single DNA probe. These assays demonstrated the genetic variability of the 'BGMV' isolates tested. In the same year, the polymerase chain reaction (PCR) method was used to produce nucleotide sequence of a hypervariable region (intergenic region between BC1 and the common region of the B component). Nucleotide sequence identities between different isolates from the Dominican Republic (95% sequence identities among them); 'BGMV' isolates from Guatemala (86%), Puerto Rico (75%); and BGMV from Brazil (46%), further suggested the existence of significant genetic variability, particularly between the Mesoamerican geminiviruses and the Brazilian BGMV isolate. In 1993, the complete nucleotide sequences of the Brazilian and Puerto Rican 'BGMV' isolates were available for comparison. The percent nucleotide identities between these two viruses ranged between 71% and 82%; the latter value corresponding to the coat protein gene (AV1), which is the most conserved gene among whitefly transmitted geminiviruses. These results clearly suggested that BGMV-Brazil (BR) was significantly different from BGMV-Puerto Rico (PR) and, consequently, BGMV-BR was considered as a type I BGMV, and BGMV-PR as a type II BGMV. In the same year, the Central American 'BGMV' isolates from Guatemala and Dominican Republic were sequenced and shown to belong to the BGMV type II cluster with the BGMV isolate from Puerto Rico.

In 2000, the family *Geminiviridae* was created to host four genera of single-stranded DNA plant viruses, including the genus *Begomovirus* (sigla from BGMV). However, the name of the type species from Puerto Rico was changed from BGMV to BGYMV, leaving BGMV to designate the Brazilian virus, thus recognizing these two viruses as distinct species of the genus *Begomovirus*.

The bean golden mosaic disease observed in north western Mexico was shown in 1992, to be caused by yet another distinct begomovirus originally isolated in 1981 from squash in southwestern United States: Squash leaf curl virus (SLCV). Some SLCV isolates sampled from diseased common beans in Mexico had already changed significantly and were thus considered a separate species named *Bean calico mosaic virus* (BcaMV). However, SLCV can still be found in northwestern Mexico causing bean calico mosaic, a severe disease that often progresses from the characteristic systemic yellowing of bean golden/yellow mosaic into severe foliar bleaching and plant death.

## Molecular Characterization of Bean Golden/Yellow Mosaic Viruses

The first nucleotide sequence was produced for the two genomic DNA molecules of the original BGYMV-PR isolate in 1985. DNA-A and DNA-B were 2646 and 2587 nt long, respectively. The only region that showed a high degree of homology in the two molecules, was the 'common region' (CR), approximately 205 nt long. The CR had direct and inverted repeats that could form a stem–loop structure. Other direct and inverted repeats were present in A+T-rich regions adjacent to G+C regions in DNA-A. The highly conserved CR contains critical sequences for the replication of begomoviruses. Six open reading frames (ORFs) were found in BGYMV-PR that could encode six proteins >10 kDa (15.6, 19.6, 27.7, 29.7, 33.1, and 40.2 kDa). By 1993, nine different begomoviruses had been fully sequenced, including a Brazilian isolate of BGMV. The DNA-A of BGMV-BR had 2617 base pairs (bp) and its DNA-B 2580 bp, with little sequence identity except for the CR of 181 nt. DNA-A had four ORFs encoding proteins >10 kDa: one in the viral sense (AV1) and three in the complementary sense (AC1, AC2, and AC3). The complementary sense ORFs were in different frames, and the 3′ end of AC1 overlapped the 5′ end of AC2, and AC2 overlapped all but the 3′ end of AC3. A shorter complementary sense ORF (AC4) that encodes a 9.5 kDa protein overlaps (nt 2227–2464) AC1. The four AC genes are involved in tasks associated with DNA replication (AC1), and possibly trans-activation and virus movement (AC2), virus replication enhancement (AC3), and gene silencing (AC4). AV1 and AC3 overlapped in opposite directions in an AT-rich region and shared four nucleotides at their C-termini, which contained their translational stop codons. AV1 encodes the coat protein, also necessary for whitefly transmission. On DNA B of BGMV-BR, two non-overlapping ORFs were detected: one in the viral sense (BV1), the nuclear shuttle protein (NSP), and one in the complementary sense (BC1), the cell-to-cell movement protein (MP). These genes code for proteins required for systemic movement of susceptible plants. All ORFs had one or more possible regulatory sequences (TATA boxes) within 100 nt of the putative start codon, except for the AC3 and BC1 ORFs. Putative polyadenylation signals (AATAAA motif) were located within the AT-rich regions at or near the 3′ ends of AV1 and AC3 ORFs and BV1 ORFs. BGMV ORFs AC1, AC2, and AC3 appeared to share a single polyadenylation signal located inside the 3′ end of AC3. Bipartite begomoviruses seem to utilize both rolling circle replication and recombination-dependent replication strategies. A specific intergenic region contains the origin of replication and the signature stem–loop structure that contains a nona-nucleotide motif involved in the rolling circle replication. The replication-associated protein (Rep) functions as a specific DNA binding protein that acts as both nuclease and ligase in the initiation and termination of the rolling circle replication process (**Figure 2**).

Sequence comparisons between BGYMV-PR and BGMV-BR showed nucleotide identities of 82%, 71%, and 72% for their respective AV1, AC1, and BV1 ORFs,

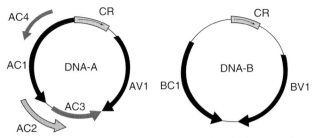

**Figure 2** Genomic structure of a typical bipartite begomovirus: CR, common region; BC1 and AC1–4, genes encoded in the complementary sense in the B and A components, respectively; AV1 and BV1, virion sense genes.

which indicates that these two viruses are genetically distinct. In 1993, the complete nucleotide sequences of BGMV isolates from Guatemala and Dominican Republic were obtained. Comparisons of the complete DNA-A sequence of the Guatemalan isolate with the Dominican Republic isolate, and with BGYMV-PR and BGMV-BR yielded identities of 97%, 96%, and 74%, respectively. For DNA-B, the nucleotide identities were 92%, 87%, and 65%, respectively. These results demonstrate that the Central American and the Caribbean viruses form one cluster (BGYMV), whereas the Brazilian BGMV belongs to a different clade. The genome organization of the BGMV and BGYMV isolates sequenced to date is quite similar in terms of number, location, and direction of their DNA-A and -B components, and presence of large intergenic regions between the CR and AV1 in DNA-A, and between the CR and BC1 and BV1 in DNA-B. Infectious pseudo-recombinants between the DNA-A and -B components were only produced between the BGYMV isolates, but not between BGYMV and BGMV. However, infectious pseudo-recombinants have been produced between distinct begomovirus species, and symptom expression depends on the components selected.

## Diagnosis

The fully systemic and intense yellowing induced by BGMV and BGYMV in most susceptible common bean genotypes is very characteristic of the diseases induced by these viruses. However, as mentioned above, other begomoviruses, such as SLCV and BCaMV, may also induce similar symptoms. Other viruses that may induce similar symptoms (intense yellowing) in common bean are alfalfa mosaic virus, bean yellow mosaic virus, and cowpea chlorotic mottle virus (which causes 'bean yellow stipple' in common bean). The most practical method available to detect the presence of BGMV or BGYMV is serology, using the enzyme-linked immunosorbent assay (ELISA) technique and suitable broad-spectrum polyclonal and/or monoclonal antibodies. Specific monoclonal antibodies have also been developed to differentiate between BGMV and BGYMV, but mutations in their coat protein genes have rendered the monoclonal antibodies useless. Nucleic acid hybridization tests have also been implemented to develop universal and specific probes for BGMV, BGYMV, and other bean begomoviruses. This technique, however, is more demanding in materials, time and facilities than serology. Finally, the PCR can be used to detect begomoviruses and produce nucleotide sequence for phylogenetic analyses and proper identification of begomovirus species and strains. A simple diagnostic characteristic of BGMV and BGYMV is that the former species is not manually transmissible, whereas BGYMV can be manually transmitted, although with some difficulty.

## Pathogenicity and Disease Resistance

Begomoviruses, including BGMV and BGYMV, induce major changes in the nuclear structure of infected cells. One of the earliest changes observed is the hypertrophy of the nucleolus, which eventually occupies most of the nuclear volume. The nucleolar components also segregate into granular and fibrillar regions composed of ribonucleoprotein. Characteristic, electron dense fibrillar rings appear next, which have been shown to contain DNA. Viroplasms and virus particle aggregates are only observed in the nucleus of infected phloem cells in the case of single infections of BGMV or BGYMV.

The first reports on BGMV in Brazil already suggested that BGMV was a highly pathogenic virus capable of systemically infecting a wide range of dry and snap common bean genotypes possessing different grain types and origins. BGMV was also observed to infect lima bean (*P. lunatus*). In the first evaluation of 28 different common bean (*P. vulgaris*) varieties conducted in Brazil in 1965, none of the genotypes inoculated by viruliferous *B. tabaci* individuals was immune to BGMV. The virus was not seed-borne in any of the common bean genotypes inoculated. The search for sources of resistance to BGMV and BGYMV continued in the 1970s, when over 4500 different common bean genotypes were evaluated and found to be susceptible to these viruses. In the late 1970s, Brazilian breeders made individual selections of symptomless common bean plants found under field conditions among BGMV-susceptible varieties, only to find out that their progenies were also susceptible. Some of these individual selections were further used to produce atomic mutants, with equally disappointing results. In 1975, a large common bean improvement project was launched in Central America to breed for resistance to BGYMV. Out of 7000 common bean accessions evaluated under natural virus pressure, only few black-seeded genotypes of Mesoamerican origin showed a moderate level of resistance to BGYMV, but none was immune to the virus. These materials were used as parents to increase the level of resistance of the main common bean cultivars grown in bean golden/yellow mosaic-affected regions of Latin America. The first BGYMV-resistant varieties obtained were derived from crosses made between the best parental materials selected (Porrillo Sintetico, ICA-Pijao, Turrialba 1). Of these resistant varieties, line DOR 41 (Porrillo Sintetico x ICA Pijao) became not only a cultivar in Central America, and in Argentina, South America, but one of the most frequently used sources of BGYMV and BGMV resistance in Latin America. In the 1980s, over 90% of all common bean genotypes possessing (partial) resistance to these viruses had DOR 41 or Porrillo Sintetico in its pedigree. The limitation with these sources of resistance was that the improved cultivars were all black-seeded, and many of the commercial common bean cultivars grown in bean golden/yellow mosaic virus-prone regions had other seed colors that were changed when crossed with the black-seeded parental sources of begomovirus resistance. Moreover, the level of virus resistance of the improved black-seeded varieties depended on the populations of viruliferous *B. tabaci*. Under high virus pressure, significant yield losses were to be expected. Consequently, a search for different sources of BGMV/BGYMV resistance was initiated in the mid-1980s.

The first breakthrough occurred when an advanced common bean line, A 429, developed for its superior growth habit (not for virus resistance), was observed to show low levels of BGYMV expression under field conditions in Guatemala, Central America. An evaluation of the parental materials involved in the development of this genotype revealed the presence of a 'pinto' (cream with brown stripes) genotype originally from Mexico (race 'Durango'). This genotype was fully susceptible and, in fact, severely affected by BGYMV in terms of arrested plant growth, plant malformation, and flower abortion, but it did not manifest the characteristic yellowing induced by BGYMV. The use of this genotype in crosses with virus susceptible common bean varieties led to the development of highly resistant genotypes possessing seed colors other than black. Eventually, the resistance was traced to a recessive gene known as *bgm-1*, for which a molecular marker is already available. The search for new sources of resistance eventually led to the identification of other useful mechanisms of resistance to BGMV and BGYMV. The most important mechanisms of resistance currently used to pyramid different bean golden/yellow mosaic resistance genes are: 'tolerance', defined as the ability of a genotype to yield acceptably despite its systemic infection and expression of characteristic disease symptoms. This mechanism is often found in genotypes belonging to race 'Nueva Granada' of Andean origin. Another recessive gene, *bgm-2*, was found to be associated with this trait. Some Andean genotypes also possessed the ability to grow normally despite their systemic infection by these viruses, that is, they showed resistance to the 'dwarfing' commonly induced by these viruses in susceptible common bean genotypes. The gene responsible for the dwarfing reaction in Andean genotypes was identified as *dwf*. Some varieties that possess this gene tend to escape infection under low-to-moderate disease pressure, but, in early infections, these genotypes do not grow at all. Nevertheless, some of these genotypes have been effectively used as source of resistance, as in the case of two Andean genotypes identified in the Dominican Republic as 'Pompadour J, and G', and line DOR 303. Highly susceptible cultivars possessing the *dwf* gene include the highly prized, large, white-seeded 'Alubia' grain type of Argentina. Another useful trait found in some Andean genotypes was the resistance to flower abortion and, more important, to pod and seed malformation, one of the most damaging effects of these begomoviruses in common bean. A dominant gene, *Bgp*, has been associated with this desirable trait. The resistance originally identified in Mesoamerican, black-seeded common bean genotypes is still useful and effective in combination with the genes mentioned above. The mechanism of resistance present in 'Porrillo Sintetico' is associated with its ability to 'escape infection' and, also, with a mechanism known as 'adult plant resistance'. That is, the longer plants grow in the field without being infected by these viruses, the greater the chance they have to escape disease. Most of the cultivated species of *Phaseolus* are also susceptible to BGMV and BGYMV, and those accessions that exhibit adequate levels of resistance have not been widely used in crossing programs due to the undesirable agronomic traits associated with the inter-specific hybridization of cultivated plants.

## Bean Golden/Yellow Mosaic Disease Management

Bean golden yellow mosaic is currently a severe disease of common bean in southern Mexico, Central America, the Caribbean (including Florida in the continental USA), and northern South America, mainly Colombia. Bean golden mosaic is restricted to Brazil, Argentina, and Bolivia in South America. Although these two viruses belong to different species, most of the sources of resistance found in *P. vulgaris* germplasm collections have proved to be equally effective as parental materials to increase the levels of resistance to either BGMV or BGYMV. Unfortunately, the recent emphasis on 'marker-assisted selection' (MAS) and 'genetically modified organisms' (GMOs) has greatly reduced the output of BGMV/BGYMV resistant cultivars possessing the necessary agronomic and commercial characteristics demanded by farmers and consumers alike. In the absence of a constant supply of new cultivars, the resistance present in the early common bean cultivars improved for their resistance to BGMV and

BGYMV starts to decline due to the gradual adaptation of begomoviruses to common bean cultivars exposed to these viruses for over a decade. The new generation of systemic insecticides currently used to control whitefly pests may prolong the life span of some of the golden/yellow mosaic resistant common bean varieties grown in Latin America. These systemic insecticides have some time to act in the plant against viruliferous whiteflies, because begomoviruses are transmitted in a semipersistent manner that require an average of 20 min for the viruses to be inoculated into the vascular (phloem) tissues of susceptible plants. However, in the presence of high populations of viruliferous *B. tabaci* individuals, the effectiveness of these new insecticides is significantly reduced. Hence, some cultural practices, such as avoidance of continuous planting of virus/vector susceptible crops, are also important components of integrated disease management programs designed to reduce the incidence of BGYMV and BGMV.

*See also:* Bean Common Mosaic Virus and Bean Common Mosaic Necrosis Virus.

## Further Reading

Bird J, Sanchez J, and Vakili NG (1973) Golden yellow mosaic of beans (*Phaseolus vulgaris*) in Puerto Rico. *Phytopathology* 63: 1435.
Brown JK, Ostrow KM, Idris AM, and Stenger DC (1999) Biotic, molecular, and phylogenetic characterization of Bean calico mosaic virus, a distinct begomovirus species with affiliation in the Squash leaf curl virus cluster. *Phytopathology* 89: 273–280.
Costa AS (1965) Three whitefly-transmitted diseases of beans in the State of São Paulo, Brazil. *FAO Plant Protection Buletin* 13: 121–130.
Costa AS (1975) Increase in the populational density of *Bemisia tabaci*, a threat of widespread virus infection of legume crops in Brazil. In: Bird J and Maramorosch K (eds.) *Tropical Diseases of Legumes*, pp. 27–49. New York: Academic Press.
Faria JC, Gilbertson RL, Hanson SF, *et al.* (1994) Bean golden mosaic geminivirus type II isolates from the Dominican Republic and Guatemala: Nucleotide sequences, infectious pseudorecombinants, and phylogenetic relationships. *Phytopathology* 84: 321–329.
Gamez R (1970) Los virus del frijol en centroamérica. I. Transmisión por moscas blancas (*Bemisia tabaci* Gen.) y plantas hospedantes del virus del mosaico dorado. *Turrialba* 21: 22–27.
Gilbertson RL, Faria JC, Ahlquist P, and Maxwell DP (1993) Genetic diversity in geminiviruses causing bean golden mosaic disease: the nucleotide sequence of the infectious cloned DNA components of a Brazilian isolate of Bean golden mosaic geminivirus. *Phytopathology* 83: 709–715.
Goodman RM and Bird J (1978) Bean golden mosaic virus. Commonwealth Mycological Institute/Association of Applied Biologists. *Descriptions of Plant Viruses* 192: 1–4.
Lazarowitz SG (1987) The molecular characterization of geminiviruses. *Plant Molecular Biology Reporter* 4: 177–192.
Morales FJ and Anderson PK (2001) The emergence and dissemination of whitefly-transmitted geminiviruses in Latin America. *Archives of Virology* 146: 415–441.
Morales FJ and Jones PG (2004) The ecology and epidemiology of whitefly-transmitted viruses in Latin America. *Virus Research* 100: 57–65.
Morales FJ and Niessen AI (1988) Comparative responses of selected *Phaseolus vulgaris* germplasm inoculated artificially and naturally with bean golden mosaic virus. *Plant Disease* 72: 1020–1023.
Morales FJ and Singh SP (1991) Genetics of resistance to bean golden mosaic virus in *Phaseolus vulgaris* L. *Euphytica* 52: 113–117.

# Beet Curly Top Virus

**J Stanley,** John Innes Centre, Colney, UK

© 2008 Elsevier Ltd. All rights reserved.

## Glossary

**Enation** Virus-induced swelling of the plant tissue.
**Endoreduplication** DNA replication in the absence of cell division.
**Etiology** The cause of a disease.
**Hemolymph** The insect circulatory system.
**Hyperplasia** Unregulated cell division.
**Hypertrophy** Cell enlargement.
**Nonpropagative transmission** The virus does not replicate in the insect.
**RFLP** Restriction fragment length polymorphism, an analytical tool to distinguish DNA viruses.
**Transovarial transmission** Transmission of virus through the insect egg.

## History

Curly top disease of sugar beet was first reported in the USA in 1888. With the establishment and growth of the sugar beet industry, it was soon realized that the disease was widespread throughout the western states where it also affected a variety of crops including tomato, bean, and potato. The disease spread eastward during the middle of the twentieth century and similar diseases were reported to occur in western Canada, northern Mexico, Brazil, Argentina, the Caribbean Basin, Turkey, and Iran, although in most cases the etiology still remains to be established. A relationship between the disease and the beet leafhopper *Circulifer tenellus* (Baker) (Homoptera: Cicadellidae) was established in 1909. Although the leafhopper transmission characteristics suggested viral etiology, the low level of accumulation of the pathogen in plants hampered its isolation, and it was not until 1974 that a virus was eventually isolated from tobacco and shown to adopt a twinned particle morphology characteristic of a geminivirus infection. The viral genomic component was eventually cloned from viral DNA isolated from symptomatic sugar beet and characterized by sequence analysis in 1986. Infectivity studies using the cloned genomic component and transmission of the clone progeny by *C. tenellus* confirmed that the disease was caused by the geminivirus beet curly top virus (BCTV).

## Taxonomic Classification and Phylogenetic Relationships

Members of the family *Geminiviridae* are divided into the genera *Begomovirus*, *Curtovirus*, *Topocuvirus*, and *Mastrevirus* on the basis of genome organization, host range, and insect vector characteristics. Originally collectively referred to as strains of *Beet curly top virus*, the genus *Curtovirus* currently includes BCTV as the type species and *Beet mild curly top virus*, *Beet severe curly top virus*, *Horseradish curly top virus*, and *Spinach curly top virus* (**Figure 1**). The names given to beet mild curly top virus (BMCTV) and beet severe curly top virus (BSCTV) reflect the symptom severity of these viruses in sugar beet although their phenotypes vary in other hosts. Species demarcation within the genus is based primarily on an 89% nucleotide sequence identity threshold for the entire genomic DNA while other biological factors such as host range, pathogenicity, and replication compatibility help to distinguish species and variants. The organization of BCTV complementary-sense genes resembles that of whitefly-transmitted begomoviruses while the virion-sense coat protein gene is more closely related to that of leafhopper-transmitted mastreviruses, suggesting that a curtovirus ancestor may have evolved by recombination between members of distinct genera. Phylogenetic analysis of horseradish curly top virus (HrCTV) suggests that a recombination event involving a region within the complementary-sense genes has also occurred between an ancestral curtovirus and begomovirus. Phylogenetic analysis of spinach curly top virus (SpCTV) indicates that its complementary-sense genes most closely resemble those of BMCTV and BSCTV while its virion-sense genes are more closely related to those of HrCTV, suggesting that it may have originated by recombination between members of the genus. However, because recombination frequently occurs between geminiviruses and undoubtedly plays a major role in their evolution, it is extremely difficult to establish an exact lineage for a particular virus and whether it represents the parent or progeny of a recombination event.

## Geographic and Seasonal Distribution

Curly top disease occurs widely throughout arid and semiarid regions of western USA that are favored by the leafhopper vector, and it is here that the epidemiology of the disease has been most closely monitored. Early investigations of host range, symptom induction, and virulence suggested that curly top disease was caused by a complex of viral pathogens, a notion borne out by the identification of distinct species and variants with overlapping geographic distributions. Adult leafhoppers overwinter on perennial weeds from which they acquire the virus. Viruliferous adults migrate to cultivated areas during spring where they undergo several generations while feeding on weeds and crops to which they transmit the disease. In autumn, the adult leafhoppers migrate back to their overwintering grounds. Using restriction fragment length polymorphism (RFLP) analysis, a field survey of sugar beet growing in Texas during 1994 showed BSCTV to be the predominant species. Both BMCTV and BSCTV were detected in a more comprehensive survey of sugar beet from California, Colorado, Idaho, New Mexico, Oregon, Washington, Wyoming, and Texas in 1995. Plants frequently contained genotypic variants of these species and occasionally maintained mixed infections of both

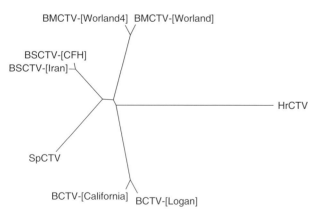

**Figure 1** Phylogenetic relationships between members of the genus *Curtovirus*. The unrooted tree is based on the alignment of the complete nucleotide sequences of isolates of beet curly top virus (BCTV), beet mild curly top virus (BMCTV), beet severe curly top virus (BSCTV), horseradish curly top virus (HrCTV), and spinach curly top virus (SpCTV).

species. Although BCTV was detected in samples maintained in laboratories and nurseries for analytical purposes, it was not recovered from field samples. This might reflect limitations of the sampling and screening procedures although a change in the population structure in which BCTV is no longer prevalent in the field seems likely. This is supported by a more recent survey of beet, pepper, and tomato as well as native weed species growing in California between 2002 and 2004, which again indicated the presence of BMCTV and BSCTV and the absence of BCTV. HrCTV was isolated in 1990 from horseradish originating from Illinois that exhibited brittle root disease, and SpCTV was isolated in 1996 from spinach growing in southwest Texas. The lack of RFLP patterns diagnostic of these two species in other surveys suggests they may have rather restricted geographic distributions.

A variant of BSCTV is also associated with curly top disease in the Mediterranean Basin, implying a common origin for the New and Old World viruses. Interbreeding experiments indicated that the leafhopper populations in these two parts of the world are closely related, although leafhopper genetic diversity in the Old World is much greater than in the USA. Coupled with the fact that sugar beet is native to Europe, it has been suggested that disease originated in the Old World and was introduced into the USA from the Mediterranean Basin as a result of movement of infected plants and accompanying viruliferous insects.

## Genome Organization and Gene Expression

In common with all geminiviruses, BCTV has a genome of circular single-stranded DNA (ssDNA) that is encapsidated in twinned quasi-isometric particles approximately $20 \times 30$ nm in size. The BCTV genome comprises a single component of $c$. 3000 nt that encodes seven genes distributed between the virion-sense (V) and complementary-sense (C) strands (also referred to as rightward (R) and leftward (L) strands in the literature), separated by an intergenic region (IR) that contains the origin of replication. A nonanucleotide motif (TAATATTAC) that is highly conserved in geminiviruses is located within the IR, between inverted repeat sequences with the potential to form a stem–loop structure. Short repeat sequences, termed iterons, occur upstream of the stem–loop (**Figure 2**).

The complementary-sense ORFs encode the replication-associated protein (Rep) required for the initiation of viral DNA replication (ORF C1), a replication-enhancer protein (REn) (ORF C3), and two proteins that contribute to viral pathogenicity (ORFs C2 and C4), all of which may be considered to be required during the early stages of infection. The virion-sense ORFs encode proteins that

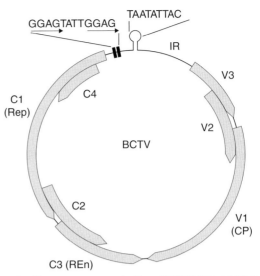

**Figure 2** The genome organization of BCTV. The position and orientation of virion-sense (V) and complementary-sense (C) ORFs are shown in relation to the intergenic region (IR) that contains the invariant nonanucleotide motif TAATATTAC and iterons (repeat sequences indicated by arrows) involved in the initiation of viral DNA replication. CP, coat protein; Rep, replication-associated protein; REn, replication-enhancer protein.

are required later in infection, namely the coat protein (ORF V1) and proteins involved in the regulation of the relative levels of viral ssDNA and double-stranded DNA (dsDNA) (ORF V2) and virus movement (ORF V3).

The identification of BCTV virion- and complementary-sense transcripts is consistent with a bidirectional transcription strategy. The two most abundant complementary-sense transcripts are similar in size to their begomovirus counterparts, the larger transcript mapping across all four ORFs and the smaller transcript across ORFs C2 and C3. However, the most abundant transcripts map across the virion-sense ORFs, downstream of two consensus eukaryotic promoter sequences, and are suitably positioned for the expression of ORFs V2 and V3, while a less abundant transcript maps across ORF V1. Precisely how the viral proteins are expressed from these transcripts is not yet understood although the overlapping nature of both the virion-sense and complementary-sense ORFs may provide a means for their temporal control during the infection cycle.

## Replication

Fractionation of viral DNA forms by a combination of chromatography and two-dimensional gel electrophoresis has identified BCTV intermediates consistent with both rolling-circle and recombination-dependent replication strategies. At the onset of infection, the viral ssDNA is uncoated and converted to a circular dsDNA intermediate by host factors. During rolling-circle replication, Rep interacts with iterons located in the IR and introduces a nick into the virion-sense strand of the dsDNA within the nonanucleotide motif (TAATATT↓AC). By analogy with the replication strategy of begomoviruses, Rep then forms a covalent bond with the 5′ terminus of the nicked virion-sense strand. However, as Rep does not have polymerase activity, the 3′ terminus must be extended by a host polymerase. The full-length nascent strand is nicked and religated by Rep to produce circular ssDNA that either reenters the replication cycle or is encapsidated. BCTV ORF C3 mutants produce severe symptoms and accumulate to wild-type levels in *Nicotiana benthamiana*, although they induce only a mild infection in sugar beet associated with significantly reduced levels of viral DNA accumulation, consistent with the proposed role for REn in viral DNA replication. Interestingly, HrCTV does not contain an intact ORF C3.

The interaction between Rep and the origin of replication, defined by the composition of the N-terminal region of Rep and the iteron sequence, is highly specific and determines whether distinct curtovirus species and variants are compatible for *trans*-replication. For example, BCTV Rep can only functionally interact with its own iteron (GGAGTATTGGAG; **Figure 2**) and not those of BMCTV and BSCTV (GGTGCTATGGGAG and GGTGCTTTGGGTG, respectively). Conversely, BMCTV and BSCTV Reps will not functionally interact with the BCTV iteron. However, the BMCTV and BSCTV iterons are sufficiently similar to allow mutual Rep recognition and, hence, *trans*-replication compatibility. Comparison of the Rep N-terminal sequences has identified amino acid residues conserved between BMCTV and BSCTV but differing in BCTV that may participate in iteron recognition, but this awaits experimental confirmation. While replication incompatibility between *cis*- and *trans*-acting elements serves to maintain the integrity of a particular species, this constraint may be overcome by recombination whereby the functional module comprising the 5′ terminus of the Rep ORF and the origin of replication is exchanged between incompatible viruses, as has been suggested to have occurred during the evolution of SpCTV.

Although replication intermediates consistent with a rolling-circle strategy for BCTV replication have been observed, the production of other viral DNA forms can be explained readily by a recombination-dependent replication mechanism. Recombination-dependent replication is generally considered to occur later in the infection cycle than rolling-circle replication whereby a ssDNA byproduct recombines with homologous sequences within a circular or multimeric dsDNA template to initiate replication, a process mediated by a recombination protein and often triggered by a double-stranded break. It remains to be established which, if any, viral proteins contribute to this process, although it is possible that Rep helicase activity could participate in both replication strategies. The nick site for rolling-circle replication has been shown to be a recombinational hot spot. However, recombination-dependent replication may explain the propensity of geminiviruses to undergo recombination throughout their genomes.

Subgenomic-sized viral DNAs of diverse size (ranging from 800 to 1800 nt) and complexity are rapidly produced from the cloned BCTV genomic component in sugar beet, tomato, and *N. benthamiana*. Deletions occur within all ORFs and only the IR is retained, consistent with its participation in viral DNA replication. Field-affected sugar beet frequently contains such defective DNAs, indicating that they are also produced under natural conditions. They occur as both single- and double-stranded DNA forms but it is not known if they are encapsidated or have the ability to move systemically in the plant. However, in view of their rapid appearance, it is likely that they are produced *de novo* in every infected cell. Short repeat sequences of 2–6 nt at the deletion boundaries may specify the deletion endpoints and suggest that the defective DNAs are produced by recombination or as a consequence of errors during replication. Many geminiviruses produce subgenomic-sized DNAs, several of which, including those associated with BCTV, have been shown to interfere with infection. Hence, they may have a biological role in modulating pathogenicity of the helper virus, thus conferring a selective advantage by slowing down infection of the plant to encourage leafhopper feeding and virus transmission.

## Virus Movement and Insect Transmission

The coat protein is a multifunctional protein that is targeted to the nucleus where virions accumulate, and is essential for systemic infection and leafhopper transmission. BMCTV coat protein mutants that retain the ability to form virions (mainly N-terminal mutations) are generally able to produce a systemic infection while those unable to form virions (mainly C-terminal mutations) cannot, despite being competent for replication. This serves to reinforce the proposal that long-distance movement of the virus throughout the plant occurs in the form of virions, consistent with the observation that high levels of virus-like particles accumulate in the phloem. It further suggests that C-terminal amino acids play an important role in determining the structure of the coat protein during virion assembly. ORF V2 mutants accumulate high levels of dsDNA and greatly reduced levels of ssDNA compared to the wild-type virus, implicating V2 protein in the regulation of the relative levels of these viral DNA forms. The function of V2 protein may be to ensure that ssDNA is available in sufficient amounts for encapsidation and systemic movement in the latter stages of the infection cycle. ORF V3 mutants remain competent for replication but produce only sporadic and asymptomatic infections in sugar beet and *N. benthamiana* associated with low levels of viral DNA accumulation, suggesting a role for V3 protein in virus movement.

All curtoviruses are transmitted in nature by the leafhopper *C. tenellus* (**Figure 3**) in a persistent circulative manner, and there is no evidence to suggest that there are any significant differences in transmission characteristics between curtovirus species. *Circulifer opacipennis* is reported to be an additional vector of the disease in the Mediterranean Basin, but this leafhopper does not occur in the USA. The leafhopper feeds primarily in the phloem from where it rapidly acquires the virus. Using a PCR-based approach, BMCTV has been detected in the digestive tract of the insect after an acquisition access period (AAP) of 1 h, in the hemolymph after 3 h, and the salivary glands after 4 h, from where it is reintroduced into plants through saliva during feeding. This is consistent with earlier reports estimating the minimum time between virus acquisition and the leafhopper becoming infective (the latent period) to be 4 h. The minimum feeding time necessary for transmission of the virus can be as short as 1 min. Depending on the length of the AAP, leafhoppers can remain infective for most of their lifetime, although the amount of virus they retain and their ability to transmit the disease decline with time when maintained on plants that are not hosts for the virus, implying that transmission is nonpropagative. There is no evidence for transovarial transmission of the virus.

Several lines of evidence demonstrate that coat protein composition defines leafhopper transmission specificity. First, BMCTV coat protein mutants that are unable to produce virions are not transmitted by *C. tenellus*. Second, recombinant virus in which the coat protein coding sequence of the whitefly-transmitted begomovirus African cassava mosaic virus has been replaced with that of BCTV produces virions that are transmissible by the leafhopper. Finally, the autonomously replicating nanovirus-like DNA-1 component, normally associated with the whitefly-transmitted begomovirus ageratum yellow vein virus, can be maintained in systemically infected sugar beet by BCTV and is encapsidated in BCTV coat protein, conferring on DNA-1 the ability to be transmitted by *C. tenellus*. This implies that surface features of the virion interact in a highly specific manner during circulative transmission, possibly by receptor-mediated endocytosis

**Figure 3** (a) Adult *Circulifer tenellus* (Baker) leafhopper. Symptoms of BCTV infection in (b) sugar beet and (c) *Nicotiana benthamiana*. Plants show upward leaf roll and vein-swelling symptoms typical of curly top disease.

during virus acquisition across the gut wall and movement into the salivary glands.

BCTV and other members of the genus are generally poorly transmitted by mechanical inoculation, reflecting their phloem limitation. Infection can be achieved using a fine-gauge needle (pin-pricking) to introduce virus or viral DNA inoculum directly into the phloem, although infection rates are dramatically improved using either biolistic delivery under pressure or *Agrobacterium tumefaciens*-mediated inoculation (agroinoculation) of cloned DNA. Agroinoculation exploits the ability of *A. tumefaciens* to introduce binary vector T-DNA, containing partially or tandemly repeated cloned genomic DNA, into the plant cell nucleus. To initiate an infection, circular viral DNA is resolved from the T-DNA, either by recombination between the repeat sequences or by a more efficient replicative mechanism if two copies of the origin of replication are present.

## Host Range and Pathogenesis

In contrast to most other geminiviruses which tend to have a limited host range, BCTV is reported to have an extremely wide host range confined to dicotyledonous plants, including over 300 species in 44 plant families, particularly members of the Chenopodiaceae, Compositae, Cruciferae, Leguminosae, and Solanaceae. Although the wide host range is not in doubt, these largely early observations were conducted using diseased plants of undefined etiology and, hence, the exact host range of BCTV remains to be confirmed. BMCTV, BSCTV, and SpCTV also have wide host ranges, although HrCTV has an atypically narrow host range. Curly top disease was originally described in association with agricultural crops showing severe symptoms, and virus tends to accumulate to higher levels in sugar beet and tomato compared to weeds. However, hosts include a wide variety of annual and perennial weed species which act as a reservoir for the disease and in which the virus frequently produces a mild or asymptomatic infection.

The products of BCTV complementary-sense ORFs C2 and C4 have been implicated in pathogenesis. ORF C2 mutants induce severe symptoms and accumulate to wild-type virus levels in sugar beet and *N. benthamiana*. However, plants infected with the mutants have a greater propensity for recovery from severe symptoms than those infected with wild-type virus, suggesting that C2 protein suppresses a plant stress or defense mechanism. Consistent with this, transgenic *N. benthamiana* and *N. tabacum* plants constitutively expressing C2 protein exhibit enhanced susceptibility to BCTV infection, manifested by a decrease in the inoculum concentration required to elicit an infection and a reduction in latent period before the onset of infection, although they do not develop more severe symptoms or accumulate higher levels of viral DNA than nontransformed plants. The transgenic plants are also more susceptible to the begomovirus tomato golden mosaic virus (TGMV) and the unrelated RNA virus tobacco mosaic virus, indicating that C2 protein impacts on a nonspecific host response. C2 protein has been shown to suppress post-transcriptional gene silencing (PTGS) in *N. benthamiana*, and in this respect resembles its positional counterpart (AC2) in the begomovirus TGMV. However, the two proteins share only limited homology and BCTV C2 protein does not appear to have transcriptional activator activity for virion-sense gene expression shown by its TGMV counterpart. Despite this, both BCTV C2 and TGMV AC2 proteins bind to, and inactivate, adenosine kinase (ADK), an enzyme required for 5'-adenosine monophosphate (AMP) synthesis. ADK activity is also reduced in BCTV-infected plants and transgenic plants expressing C2 protein. Regulation of adenosine levels by ADK plays a key role in the control of intermediates required for methylation, providing a possible link between C2 protein activity and suppression of transcriptional gene silencing (TGS) of the viral genome. Both BCTV C2 and TGMV AC2 proteins also bind to, and inhibit, SNF1-related nucleoside kinase. Reduction of SNF1 expression in *N. benthamiana* results in an enhanced susceptibility phenotype resembling that associated with the expression of C2 and AC2 proteins, while SNF1 overexpression produces plants exhibiting enhanced resistance. This activity is not directly linked to silencing suppression, suggesting that C2 protein inhibition of SNF-1 activity affects a distinct plant defense pathway.

BCTV typically induces upward curling of the leaves associated with vein clearing and the development of enations on the lower surface of veins (**Figure 3**). *In situ* localization studies have shown that BCTV is tightly phloem-limited in sugar beet and *N. benthamiana*. Although the virus cannot access apical meristematic tissues, it is possible that it exploits undifferentiated cambium cells in the vascular bundles. Infected tissues exhibit hyperplasia and hypertrophy within the phloem and adjacent parenchyma, and the dividing cells differentiate into sieve-like elements. In older infected tissues, the affected phloem cells eventually become necrotic and collapse. Pathogenic effects occur in developing tissues only after mature sieve elements have developed, suggesting that the virus moves from source to sink cells with the flow of metabolites. Infected sugar beet leaves accumulate enhanced levels of sucrose, attributed to impaired transport resulting from disruption of the phloem. This is associated with a reduction in chlorophyll content, reduced activity of key photosynthetic enzymes and concomitant reduction in the rate of photosynthesis, and altered turgor pressure that causes an increase in mesophyll cell size.

Geminiviruses control the plant cell cycle to produce an environment suitable for their replication. This is achieved by the interaction of Rep with the plant homolog of retinoblastoma-related tumor-suppressor protein (pRBR), which relieves the constraint imposed by pRBR binding to E2F transcription factor and allows the expression of E2F-responsive genes. In most cases, infected cells enter S phase and may undergo endoreduplication but do not divide. In contrast, BCTV infection induces cell division within vascular tissues, attributed to the action of C4 protein. Sugar beet infected with ORF C4 mutants remain asymptomatic and *N. benthamiana* shows an altered phenotype, and neither host develops hyperplasia of the phloem and enations. Furthermore, ectopic expression of C4 protein in transgenic *N. benthamiana* results in abnormal plant development and tumorigenic growths. Consistent with this phenotype, C4 protein interacts with BIN2, a negative regulator of transcription factors in the brassinosteroid signaling pathway that controls cell division and tissue development. The reason for the induction of cell division remains obscure, particularly as the hyperplastic tissues often do not contain detectable levels of virus.

The host range of BCTV includes *Arabidopsis thaliana* which represents an important resource for the study of virus–host interactions. Most susceptible ecotypes become stunted and develop enations on affected tissues although the Sei-O ecotype is hypersusceptible to BCTSV and develops callus-like structures containing high levels of the virus, suggesting virus-induced hormonal imbalance. Symptom development correlates with enhanced expression of the cell cycle gene *cdc2* and small auxin-upregulated RNA gene (*saur*). Hence, disruption of the phloem in infected tissues may affect auxin transport, causing localized increases that result in cell proliferation.

## Disease Control

As young plants are most susceptible to infection, a significant reduction in disease incidence may be achieved by protecting seedlings using either a physical barrier to prevent leafhopper access or by timing the emergence of the crop to avoid predicted leafhopper spring migration from the overwintering grounds. Breeding resistant varieties of sugar beet has been very successful although the basis of resistance is not understood. Plants remain susceptible to infection but are sufficiently productive, particularly when used in combination with an integrated pest-management scheme. Breeding program have provided better uniformity of resistance with improved yield and sugar content, which has reinvigorated the sugar beet industry in the western USA. Resistant bean varieties are also available, although breeding for resistance to curly top disease in tomato has not been so successful. Systemic insecticides, for example, soil treatment with imidacloprid and dimethoate foliar sprays, have been used to control the leafhopper vector on crops and on weeds located in the breeding areas, both with some success, particularly when applied to young susceptible plants, although this approach is both costly and damaging to the environment. Parasites and predators of the leafhopper are numerous, but it has proved difficult to assess their impact on leafhopper populations under natural conditions.

The subgenomic-sized DNAs produced during BCTV infection are known to adversely affect virus proliferation, for which reason they are termed defective interfering (DI) DNAs. Transgenic *N. benthamiana* plants containing an integrated tandem copy of the DI DNA are less susceptible to BCTV infection. The DI DNA is mobilized from the transgene and amplified to high levels during infection. Competition between the helper virus and DI DNA results in a significant reduction in both viral DNA accumulation and symptom severity. DI DNA amplification is dependent on *trans*-replication by the helper virus, governed by the specific interaction between Rep and the origin of replication. However, as BMCTV and BSCTV are the predominant species in the western USA and are mutually competent for *trans*-replication, genetic modification of sugar beet with the appropriate DI DNA transgene may provide a viable alternative for the production of resistant plants.

*See also:* Defective-Interfering Viruses; Plant Resistance to Viruses: Geminiviruses; Plant Virus Diseases: Fruit Trees and Grapevine; Plant Virus Diseases: Ornamental Plants; Replication of Viruses; Vector Transmission of Plant Viruses.

## Further Reading

Bennet CW (1971) *American Phytopathology Society Monograph No. 7: The Curly Top Disease of Sugar Beet and Other Plants*. St. Paul, MN: American Phytopathology Society.

Esau K and Hoefert LL (1978) Hyperplastic phloem in sugarbeet leaves infected with the beet curly top virus. *American Journal of Botany* 65: 772–783.

Hanley-Bowdoin L, Settlage SB, Orozco BM, Nagar S, and Robertson D (1999) Geminiviruses: Models for plant DNA replication, transcription, and cell cycle regulation. *Critical Reviews in Plant Sciences* 18: 71–106.

Stanley J, Bisaro DM, Briddon RW, et al. (2005) Geminiviridae. In: Fauquet CM, Mayo MA, Maniloff J, Desselberger U, and Ball LA (eds.) *Virus Taxonomy: Eighth Report of the International Committee on Taxonomy of Viruses*, pp. 301–326. San Diego, CA: Elsevier.

Stanley J, Markham PG, Callis RJ, and Pinner MS (1986) The nucleotide sequence of an infectious clone of the geminivirus beet curly top virus. *EMBO Journal* 5: 1761–1767.

# Benyvirus

**R Koenig,** Institut für Pflanzenvirologie, Mikrobiologie und biologische Sicherheit, Brunswick, Germany

© 2008 Elsevier Ltd. All rights reserved.

## Glossary

**Benyvirus** Siglum derived from *Beet necrotic yellow vein virus*.
**Rhizomania, root beardedness (root madness)** Extensive proliferation of often necrotizing secondary rootlets at the expense of the main tap root.

## History

*Beet necrotic yellow vein virus* is the type species of the genus *Benyvirus*. Originally it was classified as a possible member of the tobamovirus group, because its rod-shaped particles resemble those of tobamoviruses. In 1991, the fungus-transmitted rod-shaped viruses, all of which have multipartite genomes, were separated from the monopartite tobamoviruses to form a new group, named furovirus group. Soil-borne wheat mosaic virus (SBWMV) became the type member of this new group; beet necrotic yellow vein virus (BNYVV) and rice stripe necrosis virus (RSNV) were listed as possible members. Molecular studies performed in the following years revealed that the genome organization of many of these furoviruses greatly differed from that of SBWMV. Eventually, four new genera were created, that is, the genus *Furovirus* with *Soil-borne wheat mosaic virus* as the type species, the genus *Benyvirus* with *Beet necrotic yellow vein virus* as the type species, the genus *Pomovirus*, and the genus *Pecluvirus*. The genus *Benyvirus* presently comprises two definitive species, viz. *Beet necrotic yellow vein virus* and *Beet soil-borne mosaic virus*, and two tentative species, viz. RNSV and burdock mottle virus (BdMV).

## Host Ranges and Diseases

BNYVV is the causal agent of rhizomania, one of the most damaging diseases of sugar beet. Fodder beets, Swiss chard (*Beta vulgaris* var. *cicla*), red beets, and spinach may also become infected naturally. Infections of sugar beet are mainly confined to the root system. Susceptible sugar beet varieties show an extensive proliferation of nonfunctional, necrotizing secondary rootlets, a condition described by the names 'root beardedness', 'root madness', or rhizomania. The tap roots are stunted, their shape tends to be constricted, and their sugar content is low. A brownish discoloration of the vascular system is often seen. Leaves may become pale and have an upright position. Under dry conditions, wilting is often observed due to the disturbances in the root system. The upper parts of the plants are invaded only rarely by the virus. In infected leaves, the veins turn yellow, and occasionally become necrotic, a condition after which the virus has been named. In susceptible varieties, rhizomania may cause yield losses of 50% and more. Like other diseases caused by soil-borne organisms, rhizomania often occurs in patches, especially in recently infested fields. BNYVV can be mechanically transmitted to many species in the Chenopodiaceae and to some species in the Aizoacea (e.g., *Tetragonia expansa*), Amaranthaceae, and Caryophyllaceae. *Beta macrocarpa* becomes infected systemically. Recently obtained isolates also infect *Nicotiana benthamiana* systemically, causing slight mottling and growth reduction.

The symptoms caused by beet soil-borne mosaic virus (BSBMV) on its natural host *B. vulgaris* are more variable than those produced by BNYVV. Infected roots may remain symptomless or show rhizomania-like symptoms. Infections of the upper parts of the plants occur somewhat more frequently than with BNYVV. Leaves may develop vein banding or faint mottling. When plants are dually infected with BSBMV and BNYVV, foliar symptoms appear more frequently. In general, BSBMV causes much less damage to sugar beets than BNYVV. BSBMV can be transmitted mechanically to *Chenopodium quinoa*, *Chenopodium album*, and *Tetragonia tetragonioides*, all of which become infected locally, and to *Beta maritima*, which becomes infected systemically.

RSNV was first described in 1983 in West Africa (Ivory Coast) as the causal agent of a long-known disease named 'rice crinkling disease'. The percentage of infection was found to vary greatly according to the year and to the variety. In 1991, the virus was first noticed in Columbia where it causes a severe disease of rice named 'entorchamiento' that is characterized by seedling death, foliar striping, and severe plant malformation. The virus can be transmitted mechanically to *C. quinoa*, *C. album*, and *Chenopodium amaranticolor* where it produces local lesions, but not to rice. *Nicotiana benthamiana* does not become infected.

BdMV has been isolated from the leaves of naturally infected *Arctium lappa* L. Leaves as well as roots of this plant are common vegetables in Japan. The virus often occurs in mixed infections with other viruses and produces only mild symptoms which often become masked in older plants. BdMV can be transmitted mechanically to *C. quinoa*, *Chenopodium murale*, *Nicotiana clevelandii*, and *Nicotiana rustica*, which are infected systemically, and to

*B. vulgaris* var. *cicla* and var. *rapa*, *Spinacia oleracea*, *Cucumis sativus*, and *Tetragonia expansa*, which are infected only locally, some of them only with difficulty.

## Geographic Distribution and Epidemiology

BNYVV is now found worldwide in sugar beet-growing areas. Molecular analyses have revealed the existence of different genotypes (A type, B type, P type) that cannot be distinguished serologically, but their RNAs differ in *c.* 1–5% of their nucleotides. The A type is most common and occurs in Southern, Eastern, and parts of Northern Europe, in the USA, and – with a number of nucleotide exchanges – also in East Asia. The B type is prevalent in Germany, France, and other central European countries. B-type-like BNYVV is also found in East Asian countries. The P type, which usually has a fifth RNA species and causes especially severe symptoms, has only been found in limited areas of France (Pithiviers), of the UK, and in Kazakhstan. In the UK, a new RNA5-containing virus source (FF) has recently been described. Its RNA5 occupies an intermediate position between East Asian and European forms of RNA5. In the UK, where rhizomania was observed later than in most other European countries, a larger number of different BNYVV types have now been identified than in any other country in the world. Originally it has been assumed that rhizomania, which was first observed in Italy in 1952, has spread from that country to central and later to Northern Europe. It is unlikely, however, that all BNYVV types have originated in Italy where only the A type has so far been observed. It seems more likely that naturally infected local hosts in different geographic regions may harbor various types of BNYVV which may be transmitted to beets only with difficulty, perhaps due to vector/host compatibility problems or to their occurrence in different habitats. BNYVV is not commonly found in weeds in sugar beet fields. Once a transmission from a native host to sugar beet has been successful, further spread in beet-growing areas by machinery, irrigation, or infested soil may be very rapid.

BSBMV is widely distributed in the USA, but has so far not been found in other parts of the world. Single-strand conformation analyses of polymerase chain reaction (PCR) products indicated that this virus is genetically much more variable than BNYVV. This may explain the variability in the symptoms produced by this virus. RSNV occurs throughout West Africa and several countries in South America, that is, Colombia, Ecuador, Brazil, and Panama. It is assumed that it has been introduced to these countries from West Africa with seed contaminated at the surface with RSNV-harboring *Polymyxa graminis* resting spores. BdMV has been reported only from Japan.

## Transmission

In nature, the assigned species of the genus *Benyvirus*, viz. *Beet necrotic yellow vein virus* and *Beet soil-borne mosaic virus*, are transmitted by the zoospores of *Polymyxa betae*, a soil-borne ubiquitous plasmodiophorid protozoan. *P. betae* and the related *P. graminis* that has always been found in natural RSNV infections had formerly been considered to belong to the fungi. *Polymyxa*-transmitted viruses (or their RNAs?) are taken up by the plasmodia of the vector in infected root cells, but there is no evidence that they multiply in the vector. The multinucleate plasmodia which are separated from the host cytoplasm by distinct cell walls may either develop into zoosporangia from which secondary viruliferous zoospores are released within a few days. Alternatively, the plasmodia may form cystosori which act as resting spores and may survive in the soil for many years even under extreme conditions. They may be distributed on agricultural equipment, by irrigation, or even by blowing of wind, and upon germination they release primary zoospores transmitting the virus. Zoospores inject their contents into the cytoplasm of root cells where new plasmodia are formed. With the furovirus SBWMV, immunolabeling and *in situ* RNA hybridization tests have recently revealed the presence of viral movement protein and RNA but not of viral coat protein in the resting spores of *P. graminis*. This might suggest that the vector does not transmit virions but rather a ribonucleoprotein complex possibly formed by the viral movement protein and the viral RNAs. It remains to be shown whether benyviruses are transmitted in a similar manner. The RNA-binding ability of P42 movement protein and further properties of the other two benyviral movement proteins as well as the additional involvement in the transmission process of two transmembrane regions found in the coat protein readthrough proteins of benyviruses and other *Polymyxa*-transmitted viruses are described in the section 'Organization of the genome and properties of the encoded proteins'.

RSNV-harboring cystosori may be carried in soil adhering to the surface of rice seeds, but true seed transmission has not been observed. The natural mode of transmission of BdMV is unknown. *P. betae* and the aphid species *Myzus persicae* and *Macrosiphum gobonis* failed to transmit a BdMV isolate obtained from leaves of *Arctium lappa*.

Under experimental conditions, benyviruses are readily transmitted mechanically from test plants to test plants. However, attempts to transmit BNYVV mechanically from sugar beet rootlets to test plants may not always be successful.

## Control

*Polymyxa*-transmitted plant viruses may survive in soil in the long-living resting spores of the vector for many years, probably even decades. The diseases caused by these viruses are, therefore, more difficult to control than those caused by, for example, insect-transmitted viruses. Chemical control of these soil-borne diseases, for example, by soil treatment with methyl bromide, is neither efficient nor acceptable for economic and ecological reasons. Growing resistant or tolerant varieties currently represents the only practical and environmentally friendly means to lower the impact of theses diseases on yield. BNYVV-tolerant or partially resistant sugar beet varieties are now available which enable good yields also on infested fields. Their resistance is based mainly on the *Rz1* gene from a sugar beet line ('Holly' resistance) and the *Rz2* gene from the *Beta maritima* line WB42. In some locations, there are indications of a breakdown of the *Rz1*-mediated resistance. Genes which would confer immunity against BNYVV to sugar beet have not been found so far. Vector resistance has been detected in *Beta patellaris* and *Beta procumbens*, but attempts to develop agronomically acceptable sugar beet cultivars resistant to *Polymyxa* have so far failed. The degree of resistance or tolerance to RSNV differs in different rice cultivars. A high degree of resistance is found in *Oryza glaberrima*. The genes responsible for this resistance have been transferred to cultivated rice in breeding programs.

Genetically modified sugar beets expressing various portions of the BNYVV genome have been shown to be highly resistant to BNYVV, but have not yet been utilized commercially. Genome recombinations have not been observed when A-type BNYVV coat protein gene-expressing beets were grown in soil infested with B-type BNYVV-carrying *P. betae*.

## Particle Properties and Relations of Particles with Cells

Benyvirus virions are nonenveloped rods which have a helical symmetry (**Figure 1**). The diameter of the particles is *c.* 20 nm and they usually show several length maxima ranging from *c.* 80 to 400 nm depending on the RNA species encapsidated. Additional length maxima may be due to end-to-end aggregation or breakage of particles. The right-handed helix of BNYVV particles has a 2.6 nm pitch with an axial repeat of four turns involving 49 subunits of the *c.* 21 kDa major coat protein which consists of 188 amino acids. Each coat protein subunit occupies four nucleotides on the RNAs of BNYVV. The 75 kDa coat protein readthrough protein (**Figure 2**) has been detected by immunogold-labeling on one end of particles in freshly extracted plant sap. It is believed to act as a minor coat protein which initiates the encapsidation process.

Green fluorescent protein (GFP)-labeled particles of BNYVV were shown to localize to the cytoplasmic surface of mitochondria early during infection, but later they are relocated to semi-ordered clusters in the cytoplasm. In ultrathin sections, BNYVV particles are found scattered throughout the cytoplasm or occur in aggregates. More or less dense masses of particles arranged in parallel or angle-layer arrays may be formed. Membranous accumulations of endoplasmic reticulum may also be found.

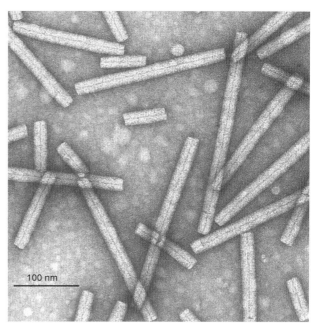

**Figure 1** Particles of *Beet necrotic yellow vein virus* in a purified preparation negatively stained with uranyl acetate. Courtesy D. E. Lesemann and J. Engelmann, BBA, Braunschweig, Germany.

## Antigenic Properties

Benyviruses are moderately to strongly immunogenic. Polyclonal antibodies have been obtained for BNYVV, BSBMV, and RSNV. BNYVV and BSBMV are only very distantly related serologically – some antisera may fail to detect this relationship. Particles of RSNV and BdMV do not react with antisera to BNYVV. Monoclonal antibodies to BNYVV have been used for diagnostic purposes and for determining the accessibility of stretches of the coat protein amino acid chain on the virus particles. The C-terminal amino acids 182–188 which are readily cleaved off by treatment with trypsin are exposed along the entire surface of the particles whereas amino acids 42–51 and 156–121 are accessible only on one end of the particles; amino acids 125–140 are located inside and are exposed only after disrupting the particles. Antibody single-chain fragments (scFv's) have been expressed in *Escherichia coli* and *N. benthamiana*.

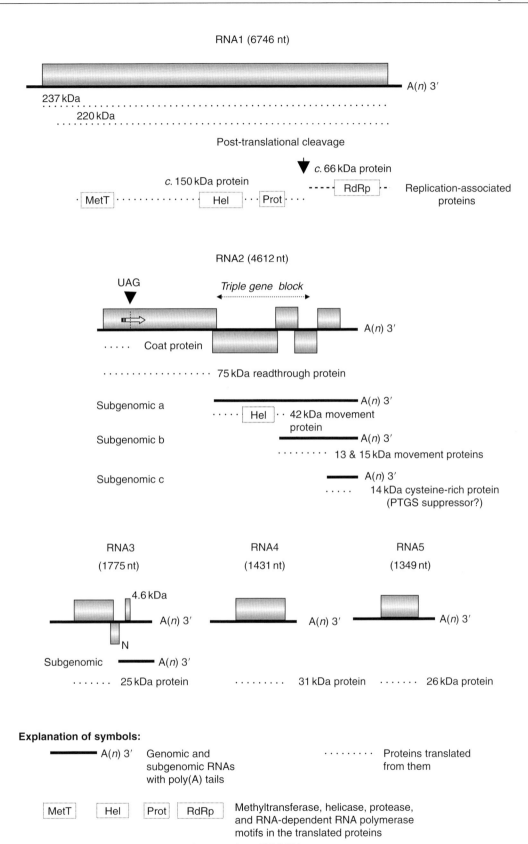

**Figure 2** Organization of the BNYVV genome and expression of BNYVV genes.

## Nucleic Acid Properties and Interrelationships between Benyviruses

Benyvirus genomes consist of two to five molecules of linear positive-sense single-stranded RNAs (ssRNAs) which terminate in a 3′ poly(A) tail. In naturally infected sugar beets, the BNYVV genome consists of four and in some isolates five RNAs of c. 6.7, 4.6, 1.8, 1.4, and 1.3 kb (**Figure 2**). Complete or partial nucleotide sequences have been determined for several isolates. After repeated mechanical transfers to local lesion hosts, BNYVV isolates often contain only partially deleted forms of their RNAs 3, 4, and/or 5, or they lose these small RNAs altogether. RNA1 and RNA2 contain all the genetic information necessary to enable multiplication, encapsidation, and cell-to-cell movement on local lesion hosts (see below); the additional presence of RNA3 and RNA4, however, is essential for vector-mediated transmission and disease development in sugar beet roots. RNA5 which is found only in limited areas increases the virulence of the virus.

Four genomic RNA species have also been identified for BSBMV and RSNV, but only two for BdMV. The BdMV isolate might possibly have lost additional small RNAs after prolonged cultivation on *C. quinoa* as do isolates of BNYVV. The complete sequence has been published for all four RNAs of an isolate of BSBMV. The complete sequences of RNA1 (c. 7.0 kb) and RNA2 (c. 4.3 kb) of BdMV and a partial sequence of RSNV RNA1 (2239 nt) have also been determined but not yet released.

Available data indicate that the genome organization is similar for all four viruses. Highest nucleotide sequence identities are found in the RNA1-encoded replication-associated proteins (c. 84% for the pair BNYVV/BSBMV) and the RNA2-encoded second triple gene block (TGB) proteins (c. 82% for the pair BNYVV/BSBMV; c. 50% for the pair BNYVV/BdMV), but the lowest ones in the RNA2-encoded cysteine-rich proteins, which are presumably RNA-silencing suppressors (c. 38% for the pair BNYVV/BSBMV; c. 20% for the pair BNYVV/BdMV).

## Organization of the Genome and Properties of the Encoded Proteins

The genome organization of BNYVV is outlined in **Figure 2**. BNYVV RNA1 contains one large open reading frame (ORF) for a replication-associated protein which is cleaved autocatalytically by a papain-like proteinase. In *in vitro* systems, its translation may start either at the first AUG at position 154 or at a downstream AUG at position 496. The resulting proteins of 237 and 220 kDa, respectively, contain in their N-terminal part methyltransferase motifs (MetT), in their central part helicase (Hel) and papain-like protease motifs (Prot), and in their C-terminal part RNA-dependent RNA polymerase (RdRp) motifs. BNYVV RNA2 contains six ORFs, viz. the coat protein gene which is terminated by a suppressible UAG stop codon, the coat protein readthrough protein gene, a TGB coding for proteins of 42, 13, and 15 kDa involved in viral movement, and a gene coding for a 14 kDa cysteine-rich protein, a putative RNA silencing suppressor. The N-terminal part of the 75 kDa coat protein readthrough protein is apparently necessary for initiating encapsidation (see section 'Particle properties and relations of particles with cells'), the C-terminal part for enabling transmission by *P. betae*. A KTER motif in positions 553–556 of the 75 kDa coat protein readthrough protein is essential for efficient transmission of the virus by *P. betae*. After prolonged cultivation on a local lesion host, the area containing this motif may be lost. Computer analyses have revealed the presence of two complementary transmembrane domains (TM1 and TM2) in the coat protein readthrough proteins not only of benyviruses, but also of the likewise *Polymyxa*-transmitted furoviruses and pomoviruses and in the P2 proteins of the *Polymyxa*-transmitted bymoviruses. The second domain is absent or disrupted in deletion mutants which are not vector transmitted. The TM helices are apparently tightly packaged with ridge/groove arrangements between the two helices and strong electrostatic associations. It has been suggested that they facilitate the movement of the virus across the membrane surrounding the plasmodia in plant cells. Benyvirus TM1 and TM2 helices are identical in length, each consisting of 23 amino acids. The E in the KTER motif is highly conserved in position −1 of TM 2.

Three subgenomic RNAs are derived from BNYVV RNA2. The first TGB protein (p42) is translated from subgenomic RNA2a, the second and third TGB proteins (P13 and P15) from the bicistronic subgenomic RNA2b, and the 14 kDa cysteine-rich protein from subgenomic RNA2c (**Figure 2**). The three TGB proteins can be functionally substituted in *trans* by the 30 kDa movement protein of tobacco mosaic virus or by the three TGB proteins of peanut clump virus when these are supplied together but not when they are substituted for their BNYVV counterparts one by one. This suggests that highly specific interactions among cognate TGB proteins are important for their function and/or stability *in planta*. The N-terminal part of the first (42 kDa) TGB protein has nucleic acid-binding activity; its C-terminal part contains helicase motifs. P42 labeled with GFP on its N-terminus is targeted by the second and third TGB proteins (P15 and P13) to punctuate bodies associated with plasmodesmata. It has been speculated that P15 and P13 provide a docking site for a P42–viral RNA complex at the plasmodesmata where P42 alters the plasmodesmatal size exclusion limit and potentiates transit of viral RNA.

The typical rhizomania symptoms in beet are produced only in the presence of BNYVV RNA3. The 25 kDa protein (P25) encoded by this RNA acts as a movement protein in beet roots and *Beta macrocarpa* and is also responsible for the production of bright yellow rather than pale green local lesions on *C. quinoa*. P25 has an amino acid tetrad in positions 67–70 which is highly variable in virus isolates originating from different geographical areas, especially in A-type BNYVV. A connection between the composition of this tetrad and the severity of symptoms in beets has been discussed, but not yet proven. P25 is found in the cytoplasm as well as in the nuclei of infected cells. A nuclear localization signal has been identified on its amino acids 57–62 and a nuclear export sequence within the region of amino acids 169–183. The four valine residues in positions 169, 172, 175, and 178 are all necessary for nucleocytoplasmic shuttling. Mutants that lack this shuttling ability fail to produce yellow lesions on *C. quinoa*. P25 has also a zinc-finger motif in positions 73–90. The small N gene (**Figure 2**) is not detectably expressed from full-length RNA3, but it is translationally activated by the deletion of upstream sequences which positions it closer to the 5' end of RNA3. Expression of the N gene induces the formation of necrotic lesions in BNYVV infections and also when it is expressed by another plant virus (i.e., cauliflower mosaic virus). The function of the RNA3-encoded 4.6 kbp ORF, which is presumably expressed from an abundantly present subgenomic RNA, is not known. RNA5 enhances symptom expression in sugar beet. Scab-like symptoms rather than root proliferation have been observed with a recombinant virus containing RNA5, but lacking RNA3. Addition of P-type RNA5 to RNA1 and -2 results in the formation of necrotic local lesions in *C. quinoa*. The RNA5-encoded P26 apparently occurs mainly in the nucleus as suggested by studies with fluorescent fusion proteins. BNYVV RNA4 greatly increases the transmission rate of the virus by *P. betae*. RNA3 and RNA4 or RNA5 and RNA4 may act in a synergistic way. Some authors have classified BNYVV RNA5 as a satellite RNA and RNAs 3 and 4 as satellite-like RNAs.

## Diagnosis

Symptoms caused by benyviruses in the field can easily be confused with those produced by other causes, for example, nematodes, soil-borne fungi, or nitrogen deficiency. Enzyme-linked immunosorbent assay, immunoelectron microscopy, and PCR techniques, such as immunocapture reverse transcription-PCR, are useful for diagnosing benyvirus infections in plant parts likely to be infected (e.g., root beards of sugar beets). Infestation of soil may be detected by means of bait plants which are tested for the presence of virus by the above techniques. PCR also allows the detection of genome deletions and specific primers can be designed to differentiate the various BNYVV types (A type, B type, etc.), which are undistinguishable serologically. Even more detailed information is obtained by sequence analyses of PCR products. The sensitivity of BNYVV detection has been increased by using nested primers and real-time PCR.

## Similarities and Dissimilarities with Other Taxa

The morphology of benyviruses resembles that of other rod-shaped viruses, that is, of furoviruses, pecluviruses, pomoviruses, hordeiviruses, tobraviruses, and tobamoviruses. The CPs of these viruses have a number of conserved residues, for example, RF and FE in their central and C-terminal parts, respectively, which are presumably involved in the formation of salt bridges. The benyvirus genomes – with the possible exception of the BdMV genome – consist of at least four RNA species, whereas the tobamoviruses have monopartite, the furoviruses, pecluviruses, and tobraviruses bipartite, and the pomoviruses and hordeiviruses tripartite genomes. The fact that the RNAs of benyviruses are polyadenylated differentiates them from the RNAs of the other viruses listed above. Benyviruses have a single large ORF on their RNA1: it codes for a polypeptide which is cleaved post-translationally to yield two replication-associated proteins (**Figure 2**). This also differentiates the benyviruses from the other viruses listed above which have their replication-associated proteins encoded by two ORFs located either on two different RNA species (hordeiviruses) or on the same RNA where an ORF-1 is terminated by a leaky stop codon and extends into an ORF-2 (tobamoviruses, tobraviruses, furoviruses, pecluviruses). Benyviruses like pomo-, peclu-, and hordeiviruses, but unlike furo-, tobamo-, and tobraviruses, have their movement function encoded on a TGB. Sequence identities in the first and second TGB-encoded proteins reveal affinities not only to pomo- and hordeiviruses, but also to potex- and carlaviruses. The methyltransferase, helicase, and RdRp motifs in the putative replication-associated proteins of benyviruses show a higher degree of similarity to those of hepatitis virus E (genus *Hepevirus*) and rubella virus (family *Togaviridae*) than to those of other rod-shaped plant viruses.

*See also:* Cereal Viruses: Rice; *Furovirus*; *Pomovirus*; Satellite Nucleic Acids and Viruses; Vector Transmission of Plant Viruses.

## Further Reading

Adams MJ, Antoniw JF, and Mullins JG (2001) Plant virus transmission by plasmodiophorid fungi is associated with distinctive transmembrane regions of virus-encoded proteins. *Archives of Virology* 146: 1139–1153.

Erhardt M, Vetter G, Gilmer D, *et al*. (2005) Subcellular localization of the triple gene block movement proteins of beet necrotic yellow vein virus by electron microscopy. *Virology* 340: 155–166.

Harju VA, Skelton A, Clover GR, et al. (2005) The use of real-time RT-PCR (TaqMan) and post-ELISA virus release for the detection of beet necrotic yellow vein virus types containing RNA 5 and its comparison with conventional RT-PCR. *Journal of Virological Methods* 123: 73–80.

Koenig R and Lennefors BL (2000) Molecular analyses of European A, B and P type sources of beet necrotic yellow vein virus and detection of the rare P type in Kazakhstan. *Archives of Virology* 145: 1561–1570.

Koenig R and Lesemann D-E (2005) Benyvirus. In: Fauquet CM, Mayo M, Maniloff J, Desselberger U, and Ball LA (eds.) *Virus Taxonomy: Eighth Report of the International Committee on Taxonomy of Viruses*, pp. 1043–1048. San Diego, CA: Elsevier Academic Press.

Lee L, Telford EB, Batten JS, Scholthof KB, and Rush CM (2001) Complete nucleotide sequence and genome organization of beet soil-borne mosaic virus, a proposed member of the genus Benyvirus. *Archives of Virology* 146: 2443–2453.

Link D, Schmidlin L, Schirmer A, et al. (2005) Functional characterization of the beet necrotic yellow vein virus RNA-5-encoded p26 protein: Evidence for structural pathogenicity determinants. *Journal of General Virology* 86: 2115–2125.

Meunier A, Schmit JF, Stas A, Kutluk N, and Bragard C (2003) Multiplex reverse transcription-PCR for simultaneous detection of beet necrotic yellow vein virus, beet soil-borne virus, and beet virus Q and their vector *Polymyxa betae* KESKIN on sugar beet. *Applied and Environmental Microbiology* 69: 2356–2360.

Ratti C, Clover GR, Autonell CR, Harju VA, and Henry CM (2005) A multiplex RT-PCR assay capable of distinguishing beet necrotic yellow vein virus types A and B. *Journal of Virological Methods* 124: 41–47.

Rush CM (2003) Ecology and epidemiology of benyviruses and plasmodiophorid vectors. *Annual Review of Phytopathology* 41: 567–592.

Rush CM, Liu HY, Lewellen RT, and Acosta-Leal R (2006) The continuing saga of rhizomania of sugar beets in the United States. *Plant Disease* 90: 4–15.

Tamada T (2002) Beet necrotic yellow vein virus. *AAB Descriptions of Plant Viruses* 391. http://www.dpvweb.net/dpv/showadpv.php?dpvno=391 (accessed July 2007).

Valentin C, Dunoyer P, Vetter G, et al. (2005) Molecular basis for mitochondrial localization of viral particles during beet necrotic yellow vein virus infection. *Journal of Virology* 79: 9991–10002.

Vetter G, Hily JM, Klein E, et al. (2004) Nucleo-cytoplasmic shuttling of the beet necrotic yellow vein virus RNA-3-encoded p25 protein. *Journal of General Virology* 85: 2459–2469.

Ward L, Koenig R, Budge G, et al. (2007) Occurrence of two different types of RNA 5-containing beet necrotic yellow vein virus in the UK. *Archives of Virology* 152: 59–73.

# Beta ssDNA Satellites

**R W Briddon and S Mansoor,** National Institute for Biotechnology and Genetic Engineering, Faisalabad, Pakistan

© 2008 Elsevier Ltd. All rights reserved.

## Glossary

**Koch's postulates** Criteria designed to establish a causal relationship between a causative microbe and a disease.

**Post-transcriptional gene silencing (PTGS)** An epigenetic process of gene regulation that results in the mRNA of a particular gene being destroyed. PTGS is believed to protect the organism's genome from, among other things, transposons and viruses.

**Rolling-circle replication** A process of DNA replication that synthesizes multiple copies of circular molecules of DNA. Rolling-circle replication is initiated by an initiator protein which nicks one strand of the double-stranded, circular DNA molecule. The initiator protein remains bound to the 5′-phosphate end of the nicked strand, and the free 3′-hydroxyl end is released to serve as a primer for DNA synthesis using the un-nicked strand as a template; replication proceeds around the circular DNA molecule, displacing the nicked strand as single-stranded DNA.

## Introduction

The DNA β components are a diverse group of symptom-modulating, single-stranded DNA (ssDNA) satellites associated with begomoviruses. These begomovirus–DNA β complexes cause some of the economically damaging diseases of crops across Asia including cotton leaf curl disease (CLCuD, an epidemic of which decimated cotton production in Pakistan and India during the 1990s, and is making a resurgence in recent times) and leaf curl diseases of tomato, tobacco, peppers, and papaya, as well as affecting numerous ornamental plants and weeds. Now, the DNA β components are classified as satellite nucleic acids in the satellite subviral agents.

## Satellites

Satellites are defined as viruses or nucleic acids that depend on a helper virus for their replication but do not have extensive nucleotide sequence similarity to their helper virus and are dispensable for its proliferation. Satellite viruses encode a structural protein that encapsidates its own nucleic acid while satellite nucleic acids rely on the helper virus structural protein for encapsidation and do not necessarily encode additional nonstructural proteins. A third type of agent, referred to as satellite-like nucleic acid, also depends on the helper virus for its replication but provides a function that is necessary for the biological success of the helper virus and is therefore considered as being part of the helper virus genome. The vast majority of these satellites consists of RNA and is associated with viruses which have genomes of RNA. Most satellites do not encode functional proteins but can nevertheless have a dramatic effect on the symptoms induced by their helper viruses, ranging from symptom amelioration to an increase in symptom severity.

## The DNA β Satellites of Begomoviruses

### Identification of the Begomovirus Satellites

The first begomovirus satellite discovered, referred to as tomato leaf curl virus-sat (ToLCV-sat), was identified in tomato plants infected with the monopartite begomovirus tomato leaf curl virus (ToLCV) originating from Australia. The component is a small (682 nt) circular ssDNA with no extensive open reading frames (ORFs) and sequence similarity to its helper virus limited to sequences within the apex of two stem–loop structures. The first contains the ubiquitous geminivirus nonanucleotide (TAATATTAC) motif and is thus similar, in both structure and sequence, to the origin of replication of geminiviruses. The second is unique to this satellite and contains a putative ToLCV replication-associated protein (Rep) binding motif. ToLCV-sat is not required for ToLCV infectivity and has no effect on the symptoms induced by the helper virus but is dependent on the helper begomovirus for its replication and encapsidation and hence has the hallmarks of a satellite DNA.

During the early 1990s, there was some astonishment within the geminivirus community with the revelation that some begomoviruses could apparently dispense with the second genomic component, DNA B. Two whitefly-transmitted viruses causing tomato leaf curl disease, a disease of significant economic importance and at the time confined to the Mediterranean Basin, were shown to have genomes consisting of only a single component. These viruses are now known as tomato yellow leaf curl virus (TYLCV) and tomato yellow leaf curl Sardinia virus (TYLCSV). Up to this point, all characterized begomoviruses – including African cassava mosaic virus, tomato golden mosaic virus, and bean golden mosaic virus (now known as bean golden yellow mosaic virus) – were shown to have genomes consisting of two components, both of which are essential for infection of hosts. There was some reluctance to accept this finding until Koch's postulates were conclusively satisfied, which ultimately they were.

Subsequent to the identification of TYLCV and TYLCSV, numerous apparently monopartite begomoviruses were identified. However, for a number of these viruses Koch's postulates could not be satisfied at the time. Although viruses such as ageratum yellow vein virus (AYVV) and cotton leaf curl Multan virus (CLCuMV) were infectious to experimental hosts, they were not, or only poorly, infectious to the hosts from which they were isolated (*Ageratum conyzoides* and *Gossypium hirsutum* (cotton), respectively) and did not induce the symptoms characteristic of the diseases with which they are associated. This inability to fulfill Koch's postulates was the first indication that, for some monopartite begomoviruses, a single component was not the entire infectious unit. All attempts to identify a DNA B component were unsuccessful and the hunt for additional components was initiated.

Plants infected with geminiviruses frequently contain less than unit length (less than 2800 nt) circular molecules derived from the virus component(s). These are deletion mutants which retain the origin of replication and are *trans*-replicated by the virus. Their presence in some cases leads to symptom attenuation, due to competition for cellular resources, and these components are for this reason known as defective interfering DNAs (diDNAs). diDNAs are typically either one-half or one-quarter the size of the virus' genomic component(s), due to stringent size selection for encapsidation or, more likely, movement within the host plant. Since they are virus derived and play no part in the etiology of the disease (other than to attenuate symptoms), less-than-unit-length molecules encountered in plants were frequently disregarded. However, in the late 1990s, a less-than-unit-length molecule was identified with some unusual characteristics. Associated with CLCuD, this molecule was half unit length, encoded a Rep (a rolling-circle replication initiator protein), and was capable of autonomous replication. Surprisingly, the molecule, named DNA 1, showed no significant sequence similarity to geminiviruses (other than the presence of a hairpin structure containing the nonanucleotide sequence) but was instead related to the Rep-encoding components of a second group (now family) of phytopathogenic DNA viruses, the nanoviruses. Nanoviruses are multicomponent, ssDNA viruses that are transmitted by aphids. Each virus is believed to have a genome consisting of six to eight components, the majority of which encode only a single product. All the *bona fide* virus components are *trans*-replicated by the Rep encoded on one component known as the master Rep (mRep) component. In addition to the mRep component, many nanovirus isolates are also associated with a varying number of other Rep-encoding components. These satellite-like Rep-encoding components (slReps) are not required by the virus, are sequence-distinct from the mRep component, and capable of autonomous replication. It is likely that DNA 1 evolved from a nanovirus-associated slRep by a process of component capture, following infection of a plant with both a geminivirus and a nanonovirus. The major differences between DNA 1 and the slRep are a slightly larger Rep-encoding gene and the presence of an adenine-rich (A-rich) region of sequence, which is likely required to raise the slRep from the size of a nanovirus component (~1000 nt) to half the size of a begomovirus component (~1400 nt). Subsequent to the identification of DNA 1 in CLCuD-affected cotton, this class of molecule was shown to be associated with many other monopartite begomoviruses, including AYVV. Unfortunately, DNA 1 proved to be an interesting 'red herring' in the search for the etiology of CLCuD and a number of other monopartite begomovirus-associated diseases. It was shown to play no part in symptom induction and to be dispensable to the helper begomovirus with which it is associated.

Encouraged by the identification of DNA 1, efforts were renewed to identify additional components among the less-than-unit-length DNAs associated with begomoviruses. The breakthrough came with the identification of a second ~1400-nt component associated with AYVV. This component was subsequently shown to be associated with numerous other monopartite begomoviruses that induce a wide range of symptoms across a diverse range of plant species (**Figure 1**). This group of molecules is now known as satellite DNA β.

## Structure of Satellite DNA β

DNA β molecules have a highly conserved structure (**Figure 2**). They are typically in the region of 1350 nt in length, roughly half the size of their helper begomoviruses. Their sequences encode multiple small ORFs in both the virion and complementary-sense orientations. Of these, only one is conserved in both position and

**Figure 1** Typical symptoms induced by begomovirus associated with DNA β satellites. These fall essentially into two types: the yellow vein type, represented here by (a) *Malvastrum coromandelianum;* (b) *Croton* sp.; (c) okra; and (d) *Ageratum conyzoides*, and the leaf curl type illustrated by (e) cotton (*G. hirsutum*); (f) tomato; (g) tobacco; and (h) papaya. Note the leaf-like enation on the cotton which is characteristic of CLCuD.

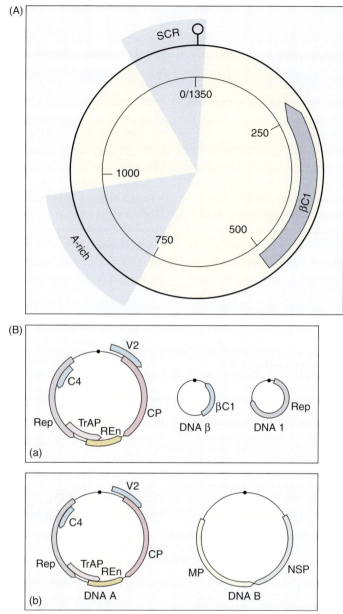

**Figure 2** Features of the DNA β satellites (A) and components of begomovirus DNA β complexes (B, a) and bipartite begomoviruses (B, b). Approximate nucleotide positions of the features are shown within the circle in (A). The position and orientation of genes are indicated with filled arrows. These are labeled as the DNA β C1 gene (βC1), replication-associated protein (Rep), transcriptional activator protein (TrAP), replication enhancer protein (REn), coat protein (CP), movement protein (MP), and nuclear shuttle protein (NSP). The functions of the V2 and C4 genes remain unclear. Also shown are the A-rich and satellite conserved region (SCR) of DNA β. The DNA-β-associated begomoviruses frequently are also associated with the satellite-like DNA 1 (B, a). Monopartite begomoviruses lack the DNA β component present in bipartite viruses (B, b). The black dot (B) and small black circle (A) indicate the position of the predicted stem–loop structure, containing the nonanucleotide motif (TAA/GTATTAC), that marks the origin of replication for the geminiviruses and likely also for DNA β and DNA 1 (although this is yet to be confirmed experimentally).

sequence. Transcript mapping of the AYVV and CLCuD-associated DNA β molecules has identified a single major transcript spanning the conserved ORF; this is now known as the βC1 gene.

A sequence of c. 80–90 nucleotides highly conserved between all identified DNA β components occurs adjacent to the putative stem loop. This is referred to as the satellite conserved region (SCR). A set of 'universal primers', designed to the sequence of the SCR, for polymerase chain reaction (PCR)-mediated amplification of the components, has provided a simple and reliable diagnostic test for the presence of DNA β. These primers have also, in the majority of cases, been shown to produce clones suitable for infectivity studies and are a major factor in the rapid identification of the large number of DNA β components. The SCR contains the stem–loop structure that, for the majority of the DNA β, contains the TAATATTAC motif. A small number have the nonanucleotide sequence TAGTATTAC, more typical of the nanoviruses. For the geminiviruses, the nonanucleotide marks the origin of virion-strand DNA replication that is nicked by virus-encoded Rep to initiate rolling-circle replication. For the DNA β components, this is yet to be confirmed.

DNA β components share, in addition to the nonanucleotide-containing hairpin structure and βC1, one further feature, a region of sequence rich in adenine. The function (if any) of this sequence to the satellite remains unclear. Deletion of the sequence does not unduly disable the component; it remains capable of trans-replication and maintenance by the helper begomovirus. The most plausible hypothesis for the presence of the A-rich sequence is that, like DNA 1, this may be required to increase the size of a preexisting component to half that of a begomovirus. This, in turn, has implications for the possible origins of DNA β, as discussed below.

## Functions Provided by Satellite DNA β to Its Helper Begomovirus

Early studies with the DNA β components associated with AYVV and CLCuMV showed that these satellites make an important contribution to the disease phenotype. Stable integration of a dimeric construct of the DNA β component associated with AYVV, and subsequently with that associated with CLCuMV, into *Nicotiana benthamiana* resulted in transgenic plants with severe developmental abnormalities, demonstrating that the components encode a pathogenicity determinant that is active in the absence of the helper virus. That the pathogenicity of the satellite is attributable to the βC1 ORF was initially provided by the analysis of naturally occurring mutants in which the entire βC1 coding sequence was deleted. The mutants were capable of trans-replication and systemic movement by the helper virus but did not induce the severe disease phenotype of the intact component. Transcript mapping of both the AYVV and CLCuMV DNA β satellites subsequently confirmed that only a single gene, on the complementary-sense strand, was present. More direct evidence for the activity of βC1 was provided by studies involving the disruption of the coding sequence and the production of transgenic *N. benthamiana* constitutively expressing the gene. For AYVV DNA β, replacement of the first in-frame methionine codon with a nonsense codon resulted in attenuated symptoms. It is likely that in this case initiation of translation occurs from a downstream methionine codon. Constitutive expression of βC1 resulted in transgenic plants with severe developmental abnormalities, resembling the plants transformed with dimeric DNA β constructs.

Significant advances have been made in determining the function of the only gene encoded by the DNA β components. A general feature of DNA β components is that they enhance the systemic accumulation of their helper begomoviruses. However, at least for AYVV DNA β, no significant increase in the accumulation of AYVV is seen in *N. benthamiana* leaf disk assays. This may suggest that βC1 is involved in virus movement, possibly in a host-specific manner, and is supported by the finding that CLCuMV DNA β can complement the DNA B functions of the bipartite tomato leaf curl New Delhi virus (ToLCNDV), as well as mutants of the V1 and C4 genes of ToLCV (the V1 and C4 genes have been implicated in virus movement). However, the nuclear localization of βC1, demonstrated by green fluorescent protein (GFP)-tagging the βC1 of the DNA β associated with tomato yellow leaf curl China virus (TYLCCNV), would seem to disagree with a role for βC1 in mobilizing viral DNA across the plasma membrane. βC1 does appear to have a role in overcoming host defenses. The βC1s encoded by the TYLCCN- and CLCuM-encoded DNA β components have been shown to suppress post-transcriptional gene silencing (PTGS). PTGS is a double-stranded RNA (dsRNA)-induced response that targets and degrades foreign nucleic acid sequences and is believed to be part of the plant's defense against viruses and other pathogens. The TYLCCNV DNA βC1 has additionally been shown to bind both ssDNA and double-stranded DNA (dsDNA) in a sequence-nonspecific manner.

## Diversity and Geographic Distribution of the DNA β Components

More than 200 full-length DNA β sequences have been deposited in the databases since they were first identified – an indication of the widespread nature, agricultural importance, and ease with which the components can be cloned and sequenced. So far, no DNA β components have been identified in association with viruses originating from the New World where monopartite begomoviruses are not known to occur. Similarly no DNA β components have been identified in Australia, although the identification of the DNA β remnant, ToLCV-sat, may indicate that at some time in the past they were present in this region. With the exception of these two areas, DNA β components have been identified in all tropical and subtropical areas, including Africa, the Middle East, Asia, and the Far East. Based on presently available numbers, the center of diversity for DNA β appears to lie either in South Asia (southern China and the Indian subcontinent) or in Southeast Asia, possibly pinpointing these regions as the center of origin.

The DNA β components fall into two major groups, those isolated from plant species of the family

Malvaceae and those associated with plants of other families. Although it is common to encounter malvaceous DNA βs in non-malvaceous hosts in the field, it is rare to identify non-malvaceous DNA β components in plants of the Malvaceae. The most prominent example of this is the DNA β associated with Sida yellow mosaic China virus in southern China. This is a typical non-malvaceous DNA β that occurs in the malvaceous species *Sida acuta* in the field (**Figure 3**). However, some doubt remains as to whether this begomovirus–DNA β pair can cause disease in *S. acuta*, since infectivity of the virus and DNA β clones to *S. acuta* was not shown experimentally.

The DNA β molecules identified from Africa differ significantly from their Asian counterparts. So far only satellites of the malvaceous type have been identified from this continent, possibly indicating that all DNA βs present in Africa originated from malvaceous plants. However, the relatively small number of DNA β sequences available from Africa may be misleading in this respect. Nevertheless, this finding does indicate that DNA β molecules have diversified due to geographical isolation.

It is likely that DNA β components also have a degree of host adaptation. For the DNA β components causing CLCuD on the Indian subcontinent, the satellite prevalent in *G. hirsutum* is distinct from that identified in *G. barbadense* (77% overall nucleotide sequence identity and 80.5% amino acid sequence similarity of the βC1 gene product). Neither of these cotton species is indigenous to the subcontinent, suggesting that the DNA β components and their helper begomoviruses evolved from other, possibly weed-infecting progenitors. The relatively close relationship between the DNA β components of *G. hirsutum* and *G. barbadense* to components isolated from *Malvastrum coromandelianum*, a pantropical weed, suggests that this species

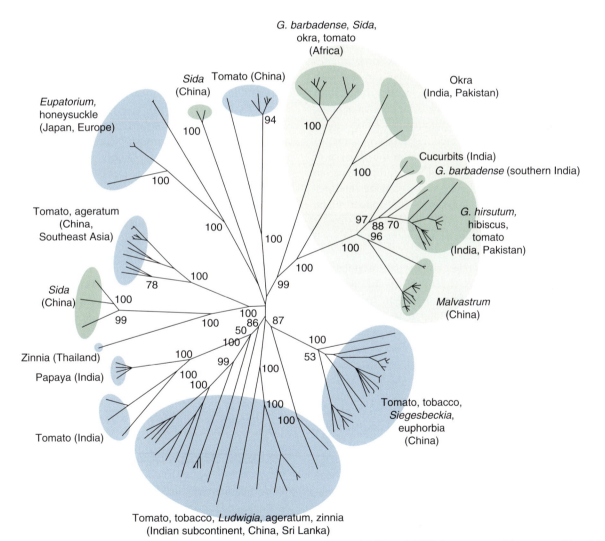

**Figure 3** Phylogenetic dendrogram based upon an alignment of 170 selected full-length DNA β sequences. The geographic origins and hosts from which the DNA β components were isolated are indicated. Clusters of DNA βs that include malvaceous hosts are highlighted in dark green, those that do not are highlighted in blue. The 'Malvaceae' group of DNA βs are shown in a light green oval. Numbers at major nodes indicate percentage bootstrap confidence scores (1000 replicates).

may have been the host for the satellite which adapted to infect cotton. It is also interesting to note that the species of cotton native to the subcontinent (*G. arboreum* and *G. herbaceum*) are immune to CLCuD, likely indicative of a long association with the disease.

## Complexity of DNA β Interaction with the Helper Begomoviruses

The first DNA β satellites to be isolated, those associated with AYVV and CLCuMV, have an obligate association with their helper begomoviruses. Although these viruses are capable of infecting plants in the absence of their satellite, in the field the virus is not found without the satellite. We can take this to indicate that these viruses are lacking in some way, and that the satellite provides an important, if not strictly essential, function to overcome this deficit in the competitive and demanding conditions of the 'wild'.

Recently, some begomoviruses have been identified which have a facultative association with their DNA β component. A recent study has shown that some isolates of tobacco curly shoot virus (TbCSV) are associated with a satellite whereas others are not. Cloned viruses from these isolates were shown to behave similarly, each being able to associate with DNA β component (yielding a more severe infection but not elevated viral DNA levels) as well as induce a symptomatic infection in the absence of the satellite. It is possible that TbCSV represents an evolutionary intermediate between the obligate satellite-associated begomoviruses and those that appear to be entirely solitary. However, it is unclear whether it is evolving to gain or lose the requirement for a DNA β component.

Even some strains of the archetypal monopartite begomovirus TYLCV have been shown to associate with DNA β in the field. The interaction in this case has not been analyzed in detail and it is as yet unclear whether this is a productive association – whether the DNA β concerned is contributing to the disease or is merely maintained as a nonfunctional satellite. Nevertheless, this is a disturbing finding. TYLCV is a global problem in tomatoes, having been spread by human hands from its probable place of origin (the Middle East/Mediterranean basin area) to the Far East, Australia, North Africa, Southern Europe, the Caribbean, and North and Central America. Few truly monopartite begomoviruses, for which no association with a DNA β has been shown, remain. These include TYLCSV and its close relatives in Southern Europe. It is likely that, in future, even these viruses will be shown, at some point during their evolution, to have had a relationship with this important group of satellites.

DNA β satellites, however, are not confined to monopartite begomoviruses. Isolates of the bipartite begomoviruses ToLCNDV and mungbean yellow mosaic India virus (MYMIV) have recently been identified in association with DNA β in the field. Both these viruses are prevalent in the Indian subcontinent, causing major losses to solanaceous (particularly tomato) and leguminous crops, respectively. Experimentally, the DNA A components have been shown to be complemented (for movement) by DNA β, and the presence of both the DNA B and DNA β results in a distinctive and more severe symptom phenotype.

A major question that remains to be answered is how DNA β satellites are *trans*-replicated by their helper begomoviruses. Current dogma maintains that Rep interaction with its cognate recognition sequences is highly specific. Thus the integrity of bipartite begomoviruses is maintained due to the presence, in both the DNA A and DNA B components, of a region of high sequence similarity (known as the 'common region' (CR)) which spans the origin of replication – encompassing both the nonanucleotide-containing hairpin structure and the Rep recognition sequences. However, DNA β components lack the Rep recognition sequences of their helper begomoviruses which raises the question as to how the recognition between helper virus-encoded Rep and DNA β genomic sequences functions. The presence of the SCR, which occurs in all DNA βs at a position equivalent to the CR of bipartite begomoviruses, may indicate that this has a function in Rep recognition. The specificity for *trans*-replication of DNA β components is also more relaxed than between cognate DNA A and DNA B components of a bipartite virus. The monopartite begomoviruses associated with CLCuD (eight identified to date), each having distinct predicted Rep-binding specificities, are all able to *trans*-replicate a single DNA β (CLCuD DNA β), and a single monopartite begomovirus, AYVV, has been shown to be able to *trans*-replicate the majority of DNA β in the model host *N. benthamiana*. Whether this is an adaptation by the DNA-β-associated viruses (by encoding a Rep with a more relaxed specificity) or an adaptation by DNA β (by encoding Rep-binding sequences recognized by a greater range of helper begomoviruses – a possible 'universal Rep recognition sequence') remains unclear. Whatever the underlying basis for the phenomenon, the ability to easily exchange components undoubtedly provides these pathogens with an evolutionary advantage.

## Potential Origins of Satellite DNA β

While begomovirus-associated DNA 1 components undoubtedly derive from nanoviruses, the evolutionary origin of DNA β remains unclear. Like DNA 1 components, DNA β components contain an A-rich region (**Figure 2**), suggesting that they may have originated as a component of another, as-yet-unidentified or extinct, pathogenic agent prior to being captured by a begomovirus. If the

A-rich sequence is a 'stuffer', required to increase the size of the component, this would suggest that the pathogenic agent from which DNA β was captured had circular ssDNA and a genome (or genomic components) <1400 nt in size. The putative pathogenic agent need not necessarily have originated in plants. For example, ToLCV has been shown to replicate in *Agrobacterium*, a soil-borne prokaryote that can transfer exogenous DNA into plants where it becomes integrated into the nuclear genome, an observation that prompted the suggestion that geminiviruses may have originated from prokaryotic episomal replicons that undergo rolling-circle replication. The genomic components of the bipartite begomovirus abutilon mosaic virus (AbMV) have been shown to be associated with plastids as well as the nucleus, an observation that may reflect a past functional relationship with these prokaryotic-like endosymbionts. Phylogenetic analysis provides compelling evidence to suggest that vertebrate circoviruses (a family of circular ssDNA viruses) may have originated from nanoviruses, possibly facilitated by arthropod vector intermediaries. Hence, it is not inconceivable that genetic material can also be transferred in the opposite direction, to plants from animals or arthropods that feed on plants.

Interestingly, further symptom-modulating satellite molecules have recently been identified in cassava mosaic disease-affected cassava in East Africa. These are apparently unrelated to DNA β but are yet to be fully characterized.

*See also:* African Cassava Mosaic Disease; Bean Golden Mosaic Virus; Cotton Leaf Curl Disease; Nanoviruses; Satellite Nucleic Acids and Viruses; Tomato Yellow Leaf Curl Virus; Tomato Leaf Curl Viruses from India.

## Further Reading

Briddon RW, Bull SE, Amin I, et al. (2003) Diversity of DNA β: A satellite molecule associated with some monopartite begomoviruses. *Virology* 312: 106–121.

Briddon RW and Stanley J (2006) Sub-viral agents associated with plant-infecting single-stranded DNA viruses. *Virology* 344: 198–210.

Dry IB, Krake LR, Rigden JE, and Rezaian MA (1997) A novel subviral agent associated with a geminivirus: The first report of a DNA satellite. *Proceedings of the National Academy of Sciences, USA* 94: 7088–7093.

Mansoor S, Zafar Y, and Briddon RW (2006) Geminivirus disease complexes: The threat is spreading. *Trends in Plant Science* 11: 209–212.

Tao X and Zhou X (2004) A modified satellite that suppresses gene expression in plants. *Plant Journal* 38: 850–860.

# Birnaviruses

**B Delmas**, INRA, Jouy-en-Josas, France

© 2008 Elsevier Ltd. All rights reserved.

## Glossary

**LexA repressor** Protease involved in the bacterial SOS response.
**Lon protease** ATP-dependent bacterial protease involved in cell homeostasis.
**Rotifera** Phylum of planktonic animals.

## Introduction

Birnaviruses form a distinct family of double-stranded (ds) RNA viruses infecting animals as different as vertebrates (birds, fish), mollusks, insects, and rotifers. No birnavirus has been reported to infect mammals, including humans. The name 'birna' highlights two important features of these viruses: the *bi*segmented nature of their genome and *rna* the ribonucleic acid they contain. Due to their economic impact, mainly two birnaviruses have been studied: the infectious bursal disease virus (IBDV) that is responsible for important losses in the poultry industry, and the infectious pancreatic necrosis virus (IPNV) that infects salmonids and represents a main concern for the aquaculture of North European and North American countries and Japan. The first characterizations of these two viruses were carried out in the early 1960s and the family was established as a distinct entity of classification in 1986. Several other viruses with bisegmented dsRNA genomes have been isolated from rats, rabbits, and humans in the 1980s, and due to their genomic organization and specific morphological features, they have recently (2006) been reclassified from a genus of the *Birnaviridae* into a family of their own, the *Picobirnaviridae*. Picobirnaviruses will be described at the end of this article.

## The *Birnaviridae*

### Taxonomy and Classification

The family *Birnaviridae* contains four genera (**Table 1**): *Aquabirnavirus* (type species: IPNV), *Avibirnavirus* (type species: IBDV), *Blosnavirus* (type species: blotched snakehead

**Table 1** Virus members in the family *Birnaviridae*

| Genus | Virus | Abbreviation | GenBank sequence accession number | PDB accession number |
|---|---|---|---|---|
| *Aquabirnavirus* | Infectious pancreatic necrosis virus | IPNV | U56907, M58757 | 2PNM |
| | Tellina virus 2 | TV-2 | AF342730 | |
| | Yellowtail ascites virus | YTAV | AB006783, AY129662 | |
| | Marine birnavirus[a] | MABV | AY123970 | |
| *Avibirnavirus* | Infectious bursal disease virus | IBDV | X92761, X92760 | 1WCE, 1WCD, 2DF7, 2GSY, 2PGG, 2IMU |
| *Blosnavirus* | Blotched snakehead virus | BSNV | AJ459383, AJ459382 | 2GEF |
| *Entomobirnavirus* | Drosophila X virus | DXV | U60650, AF196645 | |
| Unassigned viruses | Tellina virus 1 | TV1 | AJ920335, AJ920336 | |
| | Rotifer birnavirus | | | |

[a]Tentative member in the genus.

virus (BSNV) infecting blotched snakehead fish (*Channa lucius*)), and *Entomobirnavirus* (type species: drosophila X virus (DXV) of the fruit fly). While complete nucleotide sequences have been reported for a large number of IBDV and IPNV strains and isolates, only a single set of nucleotide sequences has been determined for the two genomic segments of each DXV and BSNV. Several representative GenBank and PDB accession numbers are listed in **Table 1**. Due to the lack of genome sequence information, several other tentative birnavirus species (tellina virus 1 (TV-1) and rotifer birnavirus (RBV)) have not yet been assigned to these genera. Cross-neutralization assays indicate that TV-1 and RBV are antigenically distant from other birnaviruses.

## Virion Structure and Composition

The birnaviruses have nonenveloped isometric particles, approximatively 60 nm in diameter, with icosahedral symmetry (**Figure 1**). The IBDV capsids are single shelled and consist of 260 VP2 trimers (441 amino acids × 3), arranged in $T=13$*laevo* lattices. The crystal structure of IBDV virions has been determined to a resolution of 7 Å. Purified particles also contain another abundant protein, VP3 (257 amino acids) and several peptides that are not visible in the icosahedral shell. Different ratios of VP3 over VP2 have been reported in virions from different birnavirus species. VP1, the viral RNA-dependent RNA polymerase (RdRp, 879 amino acids), is covalently linked to the two genomic segments. In addition, free VP1 molecules are present within viral particles. The buoyant density of particles in cesium chloride gradients is $1.33\,\mathrm{g\,ml^{-1}}$, a low value compared to 1.43 and $1.44\,\mathrm{g\,ml^{-1}}$ for the cores of other dsRNA viruses like reovirus and rotavirus, respectively, which enclose a similar volume for a genome three to four times bigger (see further ahead).

## Genome Organization and Gene Expression

The genome organization and expression strategy of the birnaviruses are similar; each virus genome contains two dsRNA segments (segments A and B) that range in size from 2.9 to 3.4 kbp (**Figure 2**). Segment B contains a single open reading frame that encodes the viral polymerase VP1. Segment A contains two open reading frames: a large one encoding a polyprotein precursor pVP2-VP4-VP3 (in N- to C-terminal order) and an overlapping (in IPNV and IBDV, designated VP5) or internal (in DXV or BSNV) small reading frame encoding a 17–27 kDa protein. For BSNV, an additional polypeptide, designated [X], is encoded in frame between pVP2 (the VP2 capsid protein precursor) and VP4. VP4 is a protease that cleaves its own N- and C-termini in the polyprotein, releasing pVP2 and VP3 within the infected cell. This primary cleavage occurs co-translationally. Subsequently, serial cleavages are catalyzed by VP4 at the C-terminus of pVP2, yielding the mature VP2 protein and several (three or four) peptides that remain associated with the virion. This cleavage can be incomplete; pVP2 may also be found in purified virus, although VP2 predominates.

## Virion Proteins and Peptides

### The RdRp VP1

All the RdRps and many of the DNA-dependent polymerases employ a fold whose organization has been compared to the shape of a cupped right hand with three domains, termed fingers, palm, and thumb. Only the palm domain, composed of a four-stranded antiparallel β-sheet with two α-helices packed beneath, is well conserved among all of these enzymes. This domain contains several ordered sequence motifs, with motifs A, B, and C being the most prominent. The C motif contains the Asp-Asp dipeptide signature often preceded by a Gly. The VP1 molecule of birnaviruses and of two positive-strand RNA ('tetra-like') insect viruses share a unique

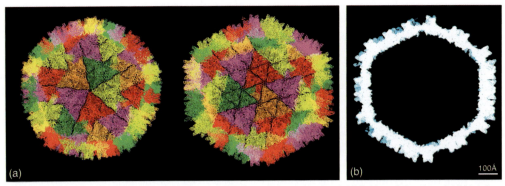

**Figure 1** The IBDV virion. (a) Half portion of an IBDV particle viewed down the fivefold (left) and the threefold (right) axes. Five different colors (yellow, orange, red, green, and magenta) were used for the 60 icosahedral asymmetric units, distributed such that immediate neighbors are colored differently. A flat face of the icosahedron is formed by three colors. For simplicity, the front half of the virion has been removed, showing the concave internal face of the particle. Threefold and fivefold axes are located where three and five colors meet, respectively. (b) Vertical 200 Å thick electron density slab through the center of the IBDV particle, viewed down the threefold axis. Reproduced from Coulibaly F, Chevalier C, Gutsche I, et al. (2005) The birnavirus crystal structure reveals structural relationships among icosahedral viruses. *Cell* 120: 761, with permission from Elsevier.

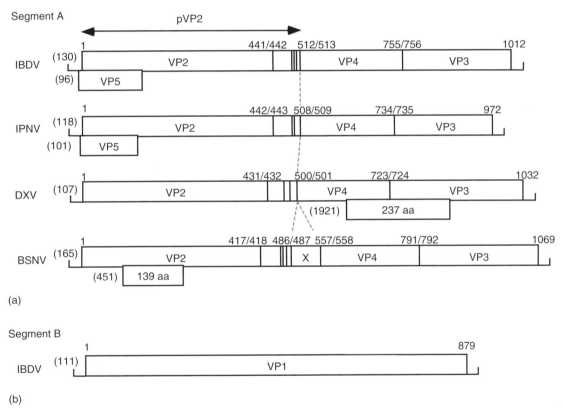

**Figure 2** Genome organization of birnaviruses. (a) Schematic representation of the gene arrangement of genomic segment A of infectious bursal disease virus (IBDV), infectious pancreatic necrosis virus (IPNV), drosophila X virus (DXV) and blotched snakehead virus (BSNV). (b) Schematic representation of the genomic segment B of IBDV. The B segments of others birnaviruses have the same organization.

characteristic polymerase motif rearrangement not found in other viral RdRps. Computer-assisted sequence analysis and resolution of the 2.5 Å structure of the IBDV VPI demonstrated that the birnavirus RdRp displays its catalytic motifs arranged in the permutated order C-B-A in the primary sequence of the palm domain. In addition, the conventional motif C (Gly-Asp-Asp) is missing and has been replaced by Ala-Asp-Asn in all birnavirus VP1s.

VP1 initiates RNA synthesis via a protein-priming strategy. Whilst the protein primer and the polymerase activity are carried by separate entities in many viruses (e.g., in picornaviruses and caliciviruses), the VP1 molecules of birnaviruses possess both RNA polymerase and protein priming functions. VP1 has a self-guanylyltransferase activity to produce VP1-pG and VP1-pGpG.

### The capsid protein VP2

The VP2 protein carries the immune determinants that control antibody-mediated neutralization and protection. While IBDV displays only two serotypes, nine different serotypes of IPNV have been described to reflect their antigenic diversity. The crystal structure of the IBDV VP2 trimer has been determined at 2.6 Å resolution. The VP2 subunit is folded into three distinct domains disposed radially in the virus particle. They are designated base (B), shell (S), and projection (P) (**Figure 3**). The B and P domains are both structured as barrels with the jelly roll topology, orientated in such a way that the strands, respectively, run tangentially and radially, to the spherical particle. There are ions bound at the trimer axis, a calcium ion at a distance of 11 Å at the inside of VP2, and a chloride ion at a distance of 17 Å from the calcium ion. The calcium ion is tightly bound by three pairs of symmetry-related conserved aspartic acid residues. Sequence alignment analysis of birnavirus VP2 molecules shows that domains S and B are relatively well conserved but that domain P is more variable. The amino acid differences between the two IBDV serotypes, and among the pathogenic serotype 1 isolates, map mainly in the P domain. Antibody neutralization–escape mutations are located at only a few residues, all in the outmost loops of the P jelly roll. Using the reverse-genetics approach, it was shown that IBDV adaptation to cell culture and virulence is also controlled by changes in these most exposed loops of domain P, suggesting that these residues may engage directly in contacts with a cell surface receptor which has not yet been identified.

Domains S and B together display a very high structural homology with the capsid protein of positive-strand RNA viruses of the families *Nodaviridae* and *Tetraviridae*. The coat proteins of these $T=3$ and $T=4$ icosahedral viruses exhibit the same topology features, including the α-helical N- and C-terminal extensions making up the domain B. In contrast, the domain P of birnavirus VP2 is most similar to the corresponding domain of the rotavirus VP6 protein, and its homolog forms the $T=13$ layer in members of the family *Reoviridae*, containing 10–12 segments of dsRNA as their genome.

### The peptides derived from pVP2

In the infected cells, the processing of pVP2 generates VP2 and three or four peptides that remain associated with the virus particles. In IBDV virions, the four peptides do not follow the icosahedral symmetry of the particles,

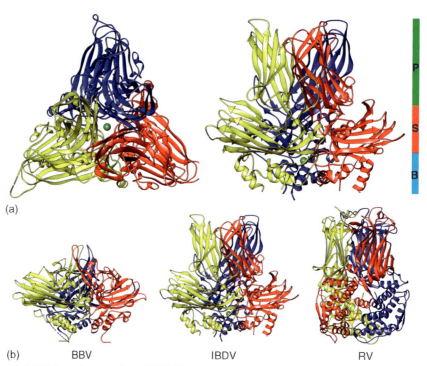

**Figure 3** The VP2 trimer. (a) Ribbon diagram of the IBDV VP2 trimer (left, top view; right, side view). Each subunit is colored differently. The sphere in green color indicates the calcium ion. The domains P, S, and B are indicated at the right. (b) The VP2 structure is a link between two categories of RNA viruses. Ribbon diagrams of the capsid proteins of black beetle virus (BBV, an alphanodavirus; protein β, PDB code 2bbv), IBDV (protein VP2), and rotavirus (RV, protein VP6, PDB code 1qhd). The DALI score between VP2 and BBV protein β, 15.3, and the rms is 3.3 Å for 211 α–carbons (out of 250 residues from domains B and S). The DALI score between VP2 and RV VP6 is 5.1, with 3.5 Å rms for 105 αcarbons (out of 140 residues from domain P).

and they are accessible to proteases or biochemical labeling. The largest peptide, pep46 (46 amino acids long), was shown to induce large structural rearrangements in liposomes and to destabilize biological membranes, suggesting its implication in cell entry.

### The internal viral protein VP3

There is only little information available on the three-dimensional structure of VP3. This abundant virion protein is able to form stable complexes with VP1 in the cytoplasm of IBDV-infected cells, suggesting that VP1 is incorporated in the virions through interaction with VP3. The VP3 domain interacting with VP1 was mapped to its 10 C-terminal amino acids. VP3 also interacts with viral dsRNA, suggesting a key role for VP3 in birnavirus assembly. In IPNV particles, VP3 and the genomic RNA have been shown to form thread-like ribonucleoprotein complexes.

## The Nonstructural Polypeptides

### The viral protease VP4

Using bioinformatics tools and mutagenesis analyses, it was demonstrated that the VP4s of IBDV and IPNV are proteases that utilize a serine/lysine dyad to process the birnavirus polyprotein. The catalytic residues of the VP4 of BSNV have also been experimentally ascertained. This catalytic dyad represents a so far unrecognized feature among viral proteases. On the other hand, this dyad has been described in several proteases of different origin, such as bacterial signal peptidase (SP), LexA repressor, and Lon protease. The VP4 thus belongs to the clan SJ and family S50 of proteases in the MEROPS databank. During polyprotein maturation, VP4 first autoprocesses the pVP2-VP4-VP3 polyprotein (*cis*-cleavage) and then cleaves pVP2 at its C-terminus (*trans*-cleavage) during virus assembly.

The crystal structure of the VP4 of BSNV has been solved at a resolution of 2.2 Å. These data confirmed the presence of a lysine in the catalytic site (**Figure 4**). The topology of the substrate's binding site is consistent with the substrate specificity and a nucleophilic attack from the *si*-face of the substrates scissile bond. Despite low scores of sequence identity, VP4 shows structural similarities in its active site with other serine/lysine proteases such as *E. coli* SP, LexA and Lon proteases (**Figure 4**). The binding pocket of VP4 is significantly larger and more extended than that of the *E. coli* Lon protease, the structure of which is most similar in the domain of the active site.

### The nonstructural protein encoded by the small segment A open reading frame

This protein has been designated VP5 for IBDV and IPNV. Reverse genetic studies on these two viruses have

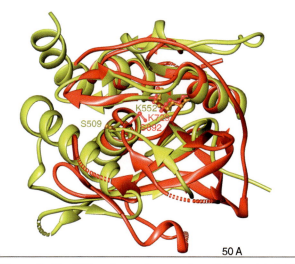

**Figure 4** The VP4 protease. A comparison of the protein folds of the BSNV VP4 protease (red) and the proteolytic domain of *Archaeoglobus fulgidus* Lon protease (yellow, PDB code 1zob). The catalytic Ser/Lys dyads are indicated.

shown that VP5 is dispensable for virus viability. VP5 plays a role in the modulation of apoptosis and pathogenesis. In contrast to its parental strain, an IBDV VP5 knockout mutant in which the initiation codon was mutated does not induce bursal lesions in susceptible infected chicken. Comparisons of the replication kinetics and markers for apoptosis indicate that the mutant induces increased DNA fragmentation, elevated caspase activity and NF-KB activation earlier in the replication cycle, allowing for the conclusion that VP5 has antiapoptotic functions at early steps of viral replication.

The VP5 of IBDV has been shown to be a class II membrane protein with a cytoplasmic N-terminal domain. VP5 is targeted to the plasma membrane and its expression without other viral components results in the alteration of the cell morphology and a reduction of the cell viability. VP5 has been proposed to play an important role in the release of the virions from the infected cell.

### Other proteins

The proteins encoded by the small segment A open reading frame of BSNV and DXV have not yet been characterized.

## Replication Cycle and Virus Assembly

There is still little information available about the early events of birnavirus replication. Investigations to identify their cellular receptors have so far not revealed the nature of the molecule(s) involved. Entry of IPNV does not seem to depend on endosomal acidification. Like all the other dsRNA viruses, birnaviruses have been shown to become transcriptionally active *in vitro* in the presence of nucleotides, extruding nonpolyadenylated mRNAs through channels in the capsid. It is presently not known if these mRNAs are capped. The single-layered capsid architecture raises questions concerning the particle translocation across the cell membrane for transcription. The two mRNAs can be detected by 3–4 h post infection and are synthesized at equal molar ratio throughout the replication cycle. Similarly, viral polypeptides are present in similar relative proportions until the end of the replication cycle.

Birnavirus morphogenesis is a complex process that is controlled by interactions of VP3 with both pVP2 and VP1 and by the regulated proteolytic activity of VP4. In IBDV, expression of the pVP2-VP4-VP3 polyprotein gene in the absence of VP1 results in the assembly of tubes displaying helical symmetry and composed of unprocessed pVP2. While expression of the gene segment corresponding to the mature VP2 (441 amino acids) in the absence of other viral components leads to the formation of $T=1$ subviral particles with diameter 25 nm, expression of the VP2 precursor, pVP2 (512 amino acids), results in the formation of irregular particles. Co-expression of the polyprotein and VP1 genes allows the sequential maturation of pVP2 to proceed and the concomitant pVP2/VP2/VP3/VP1 assembly into $T=13$ laevo virus-like particles (VLP) that are morphologically identical to the virions. Whereas the VLP shell is constituted by 260 VP2 trimers, the subviral particles contain only 20 trimers. VLP formation requires VP1 to screen for the negative charges at the C-terminus of VP3, as in the virus particle. Any exogenous peptide fused at the C-terminus of VP3 or simply a deletion of its last C-terminal residues similarly allows VLP assembly. Whereas VP3 plays a pivotal role in virus morphogenesis, it is not closely associated with VP2 in the virus particle. Moreover, the cleavage of pVP2 into VP2 and several peptides only occurs when the particle assembles. Two of the four structural peptides, pep11 and pep46 (11 and 46 amino acids long, respectively) also control virus assembly. Deletion of pep11 or substitution of most of its residues by other amino acids blocks capsid morphogenesis. Removal of pep46 also prevents capsid assembly but leads to the formation of $T=1$ particles which consist of an unprocessed VP2 species. Structural analysis of these particles showed that the presence of uncleaved pVP2 causes a steric hindrance at the vertices, blocking fivefold axis formation. Furthermore, addition of an exogenous sequence at the N-terminus of a VP2 form containing part of pep46 allows assembly of this species into $T=13$ and $T=7$ VLPs, in part mimicking the assembly controlled by VP3. Thus, the structural polymorphism observed with the various VP2 assemblies represents a paradigm in the analysis of nonequivalent interactions among identical subunits.

## Evolutionary Relationship between Birnaviruses and Other RNA Icosahedral Viruses

Due to the many similarities between birnaviruses and several positive-strand RNA viruses, it has been argued that birnaviruses represent an evolutionary link between dsRNA and positive-strand RNA viruses. The structural homologies among the coat proteins of nodaviruses, tetraviruses and VP2 (**Figure 3**) and the characteristic polymerase motif rearrangement not found in viruses other than birnaviruses and two positive-strand RNA viruses are the main features pointing to the existence of an evolutionary relationship among these viruses. Additional similarities among birna-, noda-, and tetraviruses are found in the organization of their bisegmented genomes, one RNA segment coding for the polymerase and the other for a capsid precursor, the latter undergoing proteolytic maturation at its C-terminus with a peptide of about 45 amino acids in length remaining associated with the virus particle. The mRNAs are not polyadenylated. The birnavirus VPg-linked genome replication strategy, which has been observed elsewhere only in a subset of positive-strand RNA viruses, including the picornaviruses, is a further common feature bringing positive-strand RNA viruses and birnaviruses together. Besides their dsRNA genomes, the only bridging element between birnaviruses with the other dsRNA viruses is the $T=13$ laevo protein layer and the structural homology of the VP2 P domain with the proteins forming the $T=13$ layer of the Reoviridae.

To better assess the evolutionary link between birnaviruses and positive-strand RNA viruses, it should be of interest to characterize additional dsRNA viruses having a bisegmented genome, such as picobirnaviruses and insect dsRNA and positive-strand RNA viruses.

## The Disease Induced by IBDV and IPNV

### IBDV

IBDV causes a highly contagious disease to young chickens, namely infectious bursal disease or 'Gumboro disease'. Due to its worldwide distribution, IBDV poses an important threat to the commercial poultry industry. Lymphoid cells in the bursa of Fabricius are the target cells of IBDV serotype-1 strains. Infection results in lymphoid depletion and the final destruction of the bursa. In the classical form of the outbreaks, the mortality rate ranges from 1% to 50%. In addition to mortality, IBDV is immunosuppressive and infected chicken become more susceptible to other pathogens and refractory to vaccines. In 1986, very virulent (vv) strains of IBDV, which can cause up to 70% mortality have emerged in Europe. These strains cause lesions typical of IBDV and are antigenically similar to 'classical' strains. Remarkably, vvIBDV can establish infection in the face of maternally derived antibodies that were previously protective against 'classical' strains. vvIBDV infections have been observed in Africa, Asia, and South America.

Using reverse-genetics systems, it was shown that IBDV virulence (and cell culture adaptation) is determined by changes in a few amino acid positions in the capsid protein VP2. These residues may engage directly in contacts with a receptor of the target cell. It is also believed that VP1, the RdRp, plays a role in virulence.

### IPNV

IPNV is the causative agent of infectious pancreatic necrosis in juvenile salmonid fish worldwide. Clinical signs observed in diseased fish are distended abdomen, aberrant swimming, darkened pigmentation and focal necrosis in the exocrine pancreatic tissue. Mortality observed in an outbreak varies considerably, thus probably revealing differences in genetic susceptibility of the host. Virulence determinism is also due to variations at discrete amino acid positions in the primary sequence of VP2.

## The *Picobirnaviridae*

Picobirnaviruses were first detected in fecal specimens from human and rats in 1988, and later also in other mammalian species and in chicken. They have been found in both healthy and diarrheic human patients. An association of their presence with diarrhea is particularly apparent in patients with immunodeficiencies. In a search for RNA viruses in feces of healthy humans, a metagenomic analysis has established that the most common animal viruses identified in RNA libraries generated from feces are picobirnaviruses. (This study also revealed that 97% of the virus-like sequences were homologous to plant virus sequences.) The study also revealed that the picobirnavirus population in the same individual can drastically change over time, highlighting the dynamic nature of picobirnavirus infections in humans. Picobirnaviruses have not been adapted to *in vitro* cell culture and are therefore mainly characterized by partial nucleotide sequence analyses of the RdRp gene.

Picobirnaviruses have recently been proposed to be reclassified from a genus of the *Birnaviridae*, based on the singularity of their genomic sequences and highly singular morphological features. Picobirnavirus particles are 35–41 nm in diameter with a triangulation number of 1, 3, or 4. They are nonenveloped and have a buoyant density of 1.38–1.40 g ml$^{-1}$. Their genome consists of two dsRNA segments of 2.3–2.6 and 1.5–1.9 kbp, respectively. While the smaller segment encodes the viral RdRp, the larger appears to code for the capsid protein. Only two sequences of the larger segment of a rabbit and a human picobirnavirus are available in the sequence database (**Table 2**). Two or three nonoverlapping open reading frames are present in the larger segment. Comparison of the amino acid sequences of the RdRp of human picobirnaviruses of different origins showed that they may differ by as much as 60%.

The RdRp of picobirnavirus is about 500 amino acids long, with the polymerase motifs of the palm domain arranged in the conventional A-B-C motifs order. Therefore, picobirnaviruses probably do not represent the missing link between +ssRNA viruses and birnaviruses.

**Table 2** Virus members in the family *Picobirnaviridae*

| Genus | Virus | GenBank sequence accession number |
|---|---|---|
| Picobirnaviruses | Human picobirnavirus | NC_007026, NC_007027 |
| | Rabbiit picobirnavirus | AJ244022 |
| Unassigned viruses | Rat picobirnavirus | |
| | Chicken picobirnavirus | |
| | Calf picobirnavirus | |

*See also:* Taxonomy, Classification and Nomenclature of Viruses; Reoviruses: General Features; Reoviruses: Molecular Biology; Nodaviruses; Viral Pathogenesis.

## Further Reading

Birghan C, Mundt E, and Gorbalenya AE (2000) A non-canonical Lon proteinase lacking the ATPase domain employs the ser–lys catalytic dyad to exercise broad control over the life cycle of a double-stranded RNA virus. *EMBO Journal* 19: 114.

Chevalier C, Galloux M, Pous J, *et al.* (2005) Structural peptides of a nonenveloped virus are involved in assembly and membrane translocation. *Journal of Virology* 79: 12253.

Coulibaly F, Chevalier C, Gutsche I, *et al.* (2005) The birnavirus crystal structure reveals structural relationships among icosahedral viruses. *Cell* 120: 761.

Delmas B, Kibenge FSB, Leong JC, *et al.* (2005) *Birnaviridae*. In: Fauquet C, Mayo M, Maniloff J, Desselberger U, and Ball LA (eds.) *Virus Taxonomy: Eighth Report of the International Committee on Taxonomy of Viruses*, pp. 561–569. San Diego, CA: Elsevier Academic Press.

Feldman AR, Lee J, Delmas B, *et al.* (2006) Crystal structure of a novel viral protease with a serine/lysine catalytic dyad mechanism. *Journal of Molecular Biology* 358: 1378.

Giordano MO, Martinez LC, Espul C, *et al.* (1999) Diarrhea and enteric emerging viruses in HIV-infected patients. *AIDS Research and Human Retroviruses* 15: 1427.

Gonzalez GG, Pujol FH, Liprandi F, *et al.* (1998) Prevalence of enteric viruses in human immunodeficiency virus seropositive patients in Venezuela. *Journal of Medical Virology* 55: 288.

Gorbalenya AE, Pringle FM, Zeddam J-L, *et al.* (2002) The palm subdomain-based active site is internally permuted in viral RNA-dependent RNA polymerases of an ancient lineage. *Journal of Molecular Biology* 324: 47.

Liu M and Vakharia VN (2006) Nonstructural protein of infectious bursal disease virus inhibits apoptosis at the early stage of virus infection. *Journal of Virology* 80: 3369.

Lombardo E, Maraver A, Espinosa I, et al. (2000) VP5, the nonstructural polypeptide of infectious bursal disease virus, accumulates within the host plasma membrane and induces cell lysis. *Virology* 277: 345.

Mundt E and Vakharia VN (1996) Synthetic transcripts of double-stranded birnavirus genome are infectious. *Proceedings of the National Academy of Sciences, USA* 93: 11131.

Pan J, Vakharia VN, and Tao YJ (2007) The structure of a birnavirus polymerase reveals a distinct active site topology. *Proceedings of the National Academy of Sciences, USA* 104: 7385.

Saugar I, Luque D, Ona A, et al. (2005) The structural polymorphism of the major capsid protein of a double-stranded RNA virus: An amphipathic helix as a molecular switch. *Structure* 13: 1007.

Tacken MG, Peeters BP, Thomas AA, et al. (2002) Infectious bursal disease virus capsid protein VP3 interacts both with VP1, the RNA-dependent RNA polymerase, and with viral double-stranded RNA. *Journal of Virology* 76: 11301.

Zhang T, Breitbart M, Lee WH, et al. (2006) RNA viral community in human feces: Prevalence of plant pathogenic viruses. *PLoS Biology* 4: 108.

## Relevant Website

http://merops.sanger.ac.uk – MEROPS, Peptidase Database.

# Bluetongue Viruses

**P Roy,** London School of Hygiene and Tropical Medicine, London, UK

© 2008 Elsevier Ltd. All rights reserved.

## Glossary

**Arthrogryposis** Persistent flexure or contracture of a joint.

**Campylognathia** A condition marked by misalignment of the jaws leading to a twisted appearance of the face.

**Hemoconcentration** An increase in the concentration of red blood cells in the circulating blood.

**Hemorrhagic** Related to bleeding.

**Hydranencephaly** Related to the abnormal buildup of cerebrospinal fluid in the ventricles of the brain.

**Hydropericardium** A noninflammatory accumulation of fluid within the linings surrounding the heart.

**Hydrothorax** An accumulation of serous fluid in the cavity of the chest.

**Hypotension** Abnormally low blood pressure.

**Mucosal edema** The presence of abnormally large amounts of fluid in the intercellular tissue spaces of the connective tissue lining the internal cavities of the body.

**Prognathia** A condition marked by abnormal protrusion of the lower jaw.

**Pulmonary edema** A state of increased interstitial fluid within the lung that leads to flooding of the lung alveoli with fluid.

**Serosal hemorrhages** Bleeding in the delicate membranes of connective tissue which line the internal cavities of the body.

**Thrombotic** Related to formation of thrombus; that is, an aggregation of blood factors, frequently causing vascular obstruction at the point of its formation.

**Triskelions** A shape consisting of three protruding branches radiating from a common center.

## History

Bluetongue disease (initially known as 'malarial catarrhal fever') was first observed in the late eighteenth century in sheep, goats, cattle, and other domestic animals, as well as in wild ruminants (e.g., white-tailed deer, elk, and pronghorn antelope) in Africa. A distinctive lesion in the mouths of the infected animals with severely affected, dark blue tongues was a characteristic symptom. That the disease was caused by a filterable agent was discovered in 1905. The first confirmed outbreak outside of Africa occurred in sheep in Cyprus in 1924 and this was followed by a major outbreak in 1943–44 with 70% mortality. The disease was recognized subsequently in the USA in 1948 and in Southern Europe in 1956 where approximately 75% of the affected animals died. The outbreaks of bluetongue disease in the Middle East, Asia, Southern Europe, and the USA in the early 1940s and 1950s led to its subsequent description as an 'emerging disease'. To date, based on serum-neutralization tests, 24 different serotypes have been isolated in tropical, semitropical, and temperate zones of the world including Africa, North and

South America, Australia, Southeast Asia, the Middle East, and, more recently, Southern and Central Europe. An important factor in the distribution of bluetongue virus (BTV) worldwide is the availability of suitable vectors, usually biting midges (gnats) of the genus *Culicoides*.

## Properties of the Virion

*Bluetongue virus* is the type species of the genus *Orbivirus* within the family *Reoviridae*. Details of the structure, genetics, and molecular properties of orbiviruses have been gleaned largely from studies of BTV. BTV and other orbiviruses are nonenveloped with two concentric protein shells and a genome consisting of ten double-stranded RNA (dsRNA) segments. BTV virions (550S) are architecturally complex icosahedral structures and are composed of seven discrete proteins that are numbered VP1 to VP7 in order of their decreasing size. When viewed by electron microscopy (EM), negative-stained, 550S BTV particles exhibit a poorly defined surface structure with a 'fuzzy' appearance. Complete virions are relatively fragile and the infectivity of BTV is lost easily in mildly acidic conditions. The outer capsid of BTV consists of two major proteins (VP2 and VP5) which constitute approximately 40% of the total protein content of the virus. Both proteins are removed shortly after infection to yield a transcriptionally active core (470S) particle (see schematic, **Figure 1**) which, in contrast to virions, is fairly robust. They are composed of two major proteins (VP7 and VP3), three minor proteins (VP1, VP4, and VP6), and the dsRNA genome. The cores may be further uncoated to form subcore particles (390S) that lack VP7. The 470S cores can be derived from virions *in vitro* by physical or proteolytic treatments that remove the outer capsid and cause activation of the BTV transcriptase. Considerable information on the three-dimensional (3-D) structures of BTV particles and proteins has been generated in recent years.

## Virion Structure and Outer Capsid Proteins

In contrast to negative staining, cryoelectron microscopy (cryo-EM) of intact BTV particles shows the icosahedral morphology of the mature particle, with a diameter of 86 nm. The outer layer consists of 60 sail-shaped, spike-like structures made up of VP2 (110 kDa) trimers and of 120 globular structures made up of VP5 (59 kDa) trimers (**Figure 2**). The most external part of the outer capsid is the propeller-shaped triskelion blade of VP2, the tip of which bends upward, perpendicularly to the plane of the virus. These bent tips give the entire virion a diameter of ~88 nm and extend from the main body of the particle by 3 nm. Interspersed between the triskelions and lying more internally are the globular VP5 trimers, each with a ~6 nm diameter. These are also entirely exposed in the virion, not covered by VP2, and both proteins make extensive contacts with the core VP7 layer underneath.

## Core Particle and Proteins

The core particles derived from purified virus by proteolytic treatment have been analyzed by cryo-EM and by X-ray crystallography. The icosahedral core has a diameter of 73 nm and a triangulation number of 13 ($T = 13$) with the surface layer made up of 260 VP7 (38 kDa) trimers. Trimers are arranged around 132 distinctive channels (three types: I, II, and III) as six-member rings, with five-member rings at the vertices of the icosahedrons (**Figure 3**). The five quasiequivalent trimers form a visible protomeric unit (P, Q, R, S, T). Each trimer consists of two distinct domains, 'upper' (an antiparallel β-sandwich) and 'lower' (mainly α-helical), which are twisted in such a

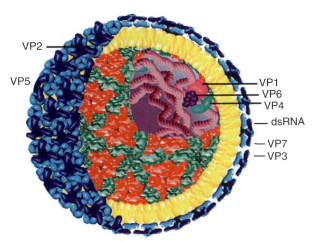

**Figure 1** A schematic diagram of bluetongue virus (BTV) showing the positions and structural organizations of BTV components. Reproduced courtesy of professor David Ian Stuart.

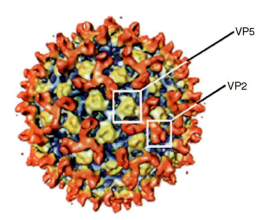

**Figure 2** Surface representations of the 3-D cryo-EM structures (22 Å resolution) of BTV. Whole particle showing sail-shaped triskelion propellers (VP2 trimers) in red and globular domains (VP5 trimers) in yellow.

way that the top domain of one monomer rests upon the lower domain of an adjacent monomer and the interaction between monomers is extensive. The lower domains are attached to an inner shell (59 nm in diameter) made up of 120 VP3 (103 kDa) molecules arranged as 60 dimers with $T=2$ symmetry (**Figure 4**). Each molecule consists of three distinct domains: a rigid 'carapace' domain, an 'apical' domain, and a 'dimerization' domain. Five of the VP3 dimers form a decamer with fivefold axes of symmetry, and 12 decamers, each a convex disk shape, are arranged together to form the complete VP3 shell.

Much of the genomic RNA can also be detected as an electron-dense region within the central core space. The dsRNA appears to be highly ordered, and approximately 80% of the entire genome can be modeled as four distinct concentric layers that have center-to-center spacing between RNA strands of 26–30 Å.

The arrangement of three internal minor proteins within the VP3 layer is not discernible in the core structure. However, cryo-EM of core-like particles (CLPs) consisting of VP3 and VP7 together with two minor proteins, VP1 (150 kDa) and VP4 (78 kDa), though lacking the genome, and VP6 has revealed a flower-shaped complex formed by VP1 and VP4, directly beneath the icosahedral fivefold axes. The exact position of the smallest minor protein VP6 (35 kDa) remains unclear although it forms a stable hexamer *in vitro* in the presence of BTV RNA, indicating that it is closely associated with the genomic RNA.

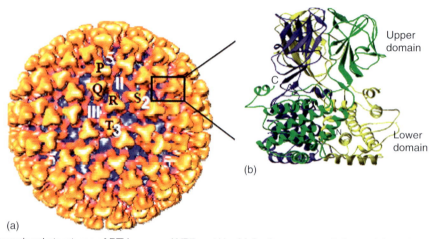

**Figure 3** Three-dimensional structures of BTV core and VP7 protein. (a) Surface representations of the 3-D cryo-EM structures of BTV core (700 Å in diameter) viewed along the icosahedral threefold axes showing the 260 trimers of VP7 (in blue). The five quasiequivalent trimers (P, Q, R, S, and T) and the locations of channels II and III are marked. (b) The trimer image of the VP7 atomic structure solved at 2.8 Å resolution. Two domains of the molecule are indicated. The view is shown from the side. Note the flat base of the trimer lies in a horizontal plane in this view.

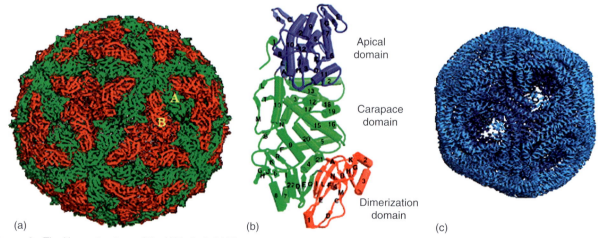

**Figure 4** The X-ray structure of the VP3 shell. (a) The structure of the VP3 layer showing the arrangement of 120 copies of two conformationally distinct types of VP3 molecules, 'A' shown in green and 'B' shown in red. (b) Structure of VP3 molecule shown as a thin triangular plate consisting of three domains as indicated. (c) The model diagram of genomic dsRNA that has been built into the four layers of electron density in the core. Reproduced courtesy of professor David Ian Stuart.

## The Virus Genome

The BTV genome comprises ten dsRNA segments in equimolar amounts which appear to be organized in an orderly fashion as four distinct layers within the core. The RNA segments, numbered 1–10 in order of migration by polyacrylamide gel electrophoresis (PAGE), were historically referred to as large, medium, and small segments (i.e., L1–L3, M4–M6, and S7–S10). Their relative order of mobility may vary according to the electrophoretic conditions used (agarose gel vs. PAGE) and the dsRNA profiles of members representing each orbivirus serogroup are generally distinctive and different from those of other members of the family *Reoviridae*.

The complete sequences of all ten dsRNA segments of a number of BTV serotypes (BTV-10 being completed as early as in 1989) are available. Each segment has six nucleotides at the 3' end and eight nucleotides at the 5' end conserved in each messenger sense RNA (mRNA) strand. For BTV-10, the genome is 19 218 bp in length with a molecular weight of $1.3 \times 10^7$ Da and the individual sizes of BTV-10 RNAs range from 3954 bp (segment 1, molecular weight $2.7 \times 10^6$ Da) to 822 bp (segment 10, molecular weight $5 \times 10^5$ Da). The 5' noncoding regions range from 8 nt (segment 4) to 34 nt in length (segment 6) while the 3' noncoding sequences are generally longer, that is, from 31 nt (segment 5) to 116 nt (segment 10) in length.

The 5' and 3' terminal sequences of the ten mRNA strands of BTV have inverted complementarity capable of forming intramolecular hydrogen bonds (i.e., end-to-end hydrogen bonding) and share some common features (e.g., a looped-out sequence proximal to the 3' termini). Apart from segment 2 and segment 5 (segment 6 in agarose gels) that code for the two outer capsid proteins (VP2 and VP5), all other eight RNA segments are highly conserved. Despite this sequence conservation, some genetic diversity exists for each RNA segment representing the various BTV serotypes as well as for various isolates within a single serotype. In addition, high-frequency segment reassortment occurs between different BTV serotypes in cell cultures, vertebrate hosts, or *Culicoides* vectors to generate new genotype combinations. Thus, both genetic drift and genetic shift contribute to BTV evolution.

Apart from segment 10, each of the AUG codons on the positive-sense mRNA strand of each of the segments initiates a single long open reading frame (see **Table 1**). There are two methionine codons in the same reading frame in the segment-10 RNA sequences, one at triplet 20–22 and the other at triplet 59–61, encoding two overlapping proteins. Thus, segment 10 codes for two nonstructural proteins, NS3 and NS3A. In addition to the seven structural proteins, two major nonstructural proteins (NS1, NS2) are also synthesized in infected cells. Table 1 summarizes the coding assignments of the ten genome segments of BTV-10.

## Viral Replication

The basic features of the BTV replication cycle, including the transcription process and the conservative mode of genome replication, are similar to those of reoviruses and rotaviruses. Unlike reoviruses or rotaviruses, however, BTV and other orbiviruses multiply in arthropods as well as in vertebrate hosts. Also, in view of the structural differences, it is likely that some stages of BTV replication and morphogenesis are unique.

## Attachment and Entry into Cells

BTV adsorbs rapidly to susceptible cells at both 4 and 37 °C. The cell receptor for BTV is not known although it binds to a sialoglycoprotein via the outer capsid protein VP2 in mammalian cells. VP2 is the serotype determinant and viral hemagglutinin protein, elicits serotype-specific virus-neutralizing antibody, and is the most variable protein among different serotypes. Although protease-treated particles with VP2 cleavage products attached are fully infectious, removal of VP2 results in loss of virus infectivity

**Table 1** Coding assignments and function of the BTV-10 viral RNA segments

| Genome segment[a] | Size (kbp) | Protein | Protein size (kDa) | Location | Function |
|---|---|---|---|---|---|
| Seg. 1 (L1) | 3.954 | VP1 | 149.5 | Core | Polymerase |
| L2 | 2.926 | VP2 | 111 | Outer shell | Attachment protein |
| L3 | 2.772 | VP3 | 103 | Core | Structural |
| M4 | 2.011 | VP4 | 764 | Core | Capping enzyme |
| M5 | 1.770 | NS1 | 644 | Tubules | Trafficking? |
| M6 | 1.639 | VP5 | 591 | Outer shell | Fusion protein |
| S7 | 1.156 | VP7 | 385 | Core | Insect cell attachment |
| S8 | 1.123 | NS2 | 409 | Nonstructural | RNA selection |
| S9 | 1.046 | VP6 | 357 | Core | Helicase |
| S10 | 0.822 | NS3 | 256 | Nonstructural | Egress |

[a]Genome segment nomenclature based on order of migration by electrophoresis. In some gel systems the migration of segments 5 and 6 is frequently, but not always, reversed.

in mammalian cells. Core particles, however, are infectious only for invertebrate cells.

The attachment of VP2 leads to receptor-mediated endocytosis by clathrin-coated vesicles which are subsequently lost when large endocytic vesicles form. Only the early endosomes are involved in BTV entry. BTV entry into the cytoplasm requires endosomal acidic pH which allows the second outer capsid protein VP5 to permeabilize the endosomal membrane. This occurs via the function of the N-terminal 40 hydrophobic residues which act as a 'pore-forming' peptide, analogous to the fusion peptides of envelope viruses. Thus, VP2 makes an initial contact with the host cell and VP5 mediates the penetration of the host cell membrane by destabilizing the endosomal membrane. Core particles lacking both VP2 and VP5 proteins then enter the cytoplasm.

## Transcription

In the cytoplasm, core particles do not disassemble further but initiate the transcription of the viral genome, and newly synthesized viral mRNAs, capped but not polyadenylated, are extruded into the cytoplasm. When intact cores, isolated from virus particles, are activated *in vitro* by the presence of magnesium ions and nucleoside triphosphate (NTP) substrates, distinct conformational changes can be seen around the fivefold axes of the core, forming pores in the VP3 and VP7 layers through which mRNAs are extruded. The role of each internal minor protein in genome replication activity has been established using individual purified proteins.

The smallest minor protein VP6, which is rich in basic amino acids (Arg, Lys, and His), binds both single-stranded RNAs (ssRNAs) and dsRNAs and, in isolation, has the ability to unwind dsRNA substrates *in vitro*. The protein exhibits physical properties characteristic of other helicases, including being hexameric, and has the ability to form ring-like structures in the presence of BTV RNAs. Mutation of amino acid residues in the active site destroys the catalytic activity of the protein.

The replication of viral dsRNA occurs in two distinct steps. First, plus-strand RNAs (mRNAs) are transcribed and extruded from the core particle. Second, the plus-strand RNAs serve as templates for the synthesis of new minus-strand RNAs. The largest core protein VP1 (150 kDa), in soluble form, has the ability to both initiate and elongate minus-strand synthesis *de novo*, but the catalytic activity is lost when a GDD motif (amino acids 287–289), the polymerase signature motif, is mutated.

The third minor protein VP4 is the mRNA-capping enzyme. The purified, soluble, recombinant VP4 alone has the ability to synthesize type 1-like 'cap' structures on uncapped BTV transcripts *in vitro*, and 'cap' structures are identical to those found on authentic BTV mRNAs. Thus, VP4 possesses methyltransferase, guanylyltransferase, and RNA triphosphatase activities. The recently resolved atomic structure has shown how distinct domains of this single protein perform each of these catalytic activities consecutively.

In summary, each of the three minor proteins of the BTV core has the ability to function on its own. Together they constitute a molecular motor that can unwind RNAs, synthesize ssRNAs of both polarities, and modify the 5' termini of the newly synthesized mRNA molecules. The transcripts are not produced in equimolar amounts from the ten segments of BTV; the smaller genome segments are generally the most frequently transcribed, although segment-6 RNA (encoding NS1) is synthesized more abundantly than segment-10 (the smallest) RNA (encoding NS3). The molar ratios of the ten different BTV mRNAs remain the same throughout the infection cycle. The ratios of mRNAs synthesized *in vivo* and *in vitro* are similar.

Much less is known about the *in vivo* RNA replication mechanisms of BTV. It is believed that, like reoviruses and rotaviruses, the packaged plus-strand RNA serves as a template for synthesis of a minus strand, and once the minus strand is synthesized, the dsRNA remains within the nascent progeny particle. As discussed, VP1 acts as the replicase enzyme but the roles of other proteins in minus-strand synthesis remain undefined.

## Protein Synthesis and Replication

In tissue culture, the first BTV-specific proteins are detectable 2 h post infection and the rate of protein synthesis increases rapidly until 12–13 h post infection, after which it slows down but continues until cell death. BTV infection of mammalian cells, in contrast to insect cells, leads to a rapid inhibition of cellular macromolecular synthesis and the induction of a robust apoptotic response triggered via multiple apoptotic pathways.

Two NS proteins, NS1 (64 kDa) and NS2 (41 kDa), are synthesized abundantly early in BTV infection and coincide with two virus-specific intracellular structures, tubules, and viral inclusion bodies (VIBs), respectively. By contrast, synthesis of NS3 and NS3A (26 and 25 kDa) varies from barely detectable to highly expressed, depending on the host cells, and may correlate with virus release.

Tubules are present in large numbers, mostly in perinuclear locations, and are made up of helically coiled NS1 dimers, on average 52.3 nm in diameter and ~1 μm long. The exact role of tubules or NS1 dimers in virus replication is not known. During BTV infection, VIBs are found in the infected cells, predominantly near the nucleus. VIBs act as the nucleation site for newly synthesized proteins that form the core structure and transcripts as well as the subsequent assembly of subviral particles. The major component of VIBs is the phosphoprotein NS2. Phosphorylation plays a key role in the formation of

VIBs and may be involved in stabilizing NS2 folding. The cellular protein kinase, casein kinase II (CKII), is responsible for NS2 phosphorylation via two serine residues.

Soluble NS2 and NS2 in VIBs have a strong affinity for ssRNA. NS2 preferentially binds BTV transcripts via specific hairpin structures, an indication of its role in the recruitment and selection of BTV mRNA during virus replication and RNA packaging. Since phosphorylation of NS2 is not necessary for recruitment of core components, but important for VIB formation, phosphorylation and dephosphorylation of NS2 is plausibly a dynamic process that controls the assembly and release of core particles.

## Capsid Assembly

The assembly of BTV capsids requires a complex and highly ordered series of protein–protein interactions. The use of CLPs and virus-like particles (VLPs), together with structure-based mutagenesis, has revealed the key principles that drive the assembly process. The VP3 shell appears to play the major role in the initiation of core assembly with formation of VP3 decamers and the complex formed with VP1 and VP4. These assembly intermediates subsequently recruit the viral RNA, and possibly VP6, prior to completion of the assembly of the VP3 subcore and the addition of VP7 trimers. Initially, multiple sheets of VP7 trimers form around different nucleation sites and thus it is likely that a number of strong VP7 trimer–VP3 contacts act as multiple equivalent initiation sites and that a second set of weaker interactions then 'fill the gaps' to complete the outer core layer. The VP7 layer gives increased rigidity and stability to the core particle. Assembly of the core takes place entirely within VIBs and assembled core is released from VIBs prior to the simultaneous addition of VP2 and VP5, which most likely occurs within the vimentin component of the cytoskeleton.

## Egress from Host Cells

The majority of mature virus particles remain cell-associated in mammalian cells, causing substantial cytopathic effect. However, some particles bud through the cell membrane or move in groups through a local disruption of the plasma membrane. NS3 (229 aa) and its shorter form NS3A (216 aa) are the only BTV glycoproteins and are associated with smooth-surfaced, intracellular vesicles. NS3/NS3A proteins have long N-terminal and shorter C-terminal cytoplasmic domains connected by two transmembrane domains and a short extracellular domain. A single glycosylation site is present in the extracellular domain of BTV NS3. The NS3/NS3A proteins are synthesized at very low levels in infected mammalian cells but at very high levels in invertebrate cells where there is nonlytic virus release. The NS3 protein is localized at the site of the membrane where viruses or VLPs are released. NS3 may cause local disruption of the plasma membrane, allowing virus particles to be extruded through a membrane pore without acquiring a lipid envelope. The N-terminal residues of NS3 interact with the calpactin light chain (p11) of the cellular annexin II complex, itself involved in membrane secretory pathways. The interaction of p11 with NS3 may direct NS3 to sites of active cellular exocytosis, or NS3 could become part of an active extrusion process. NS3 is also capable of interaction with Tsg101, a cellular protein implicated in the intracellular trafficking and release of a number of enveloped viruses. Interactions of both p11 and Tsg101 with NS3 appear to impact the nonlytic release of virions from infected cells. The significance of these interactions for BTV egress becomes more apparent in the light of the observation that the other cytoplasmic domain of the protein, situated at the C-terminal end, interacts specifically with the BTV outer capsid protein VP2. Current data suggest that NS3 makes use of host proteins and acts as an intermediate to facilitate the release of newly synthesized progeny virus across the cell membrane.

## Pathogenesis

BTV produces a spectrum of conditions from subclinical infection to severe and fatal disease, depending on virus strain and host species. Virus is transmitted to vertebrate hosts through blood-feeding insect vectors and infectious particles migrate to lymph nodes where they initially replicate and subsequently spread to spleen, thymus, and other lymph nodes. In the final phase of infection, the virus begins circulating in the bloodstream and can persist for several months. In certain vertebrate hosts (e.g., cattle), BTV can induce a prolonged viremia. BTV binds to glycophorins on the surface of bovine and ovine erythrocytes where it may remain in an infectious state as invaginations of the erythrocyte cell membrane for prolonged periods of time. This precludes contact with antibody and T cells and thus provides multiple opportunities for transmission by infection of blood-sucking midges.

The role of leukocytes in cell-associated viremia is less certain, although BTV has been recovered from bovine mononuclear cells during the early stage of infection. A characteristic feature of BTV pathogenesis is the ability of viruses to replicate in and to damage endothelial cells with tropisms for cells representing distinct anatomic sites. Infection of endothelial cells is followed by infection of vascular smooth muscle cells and pericytes. These then undergo lytic infections, resulting in virus-induced vascular injury and a cascade of pathophysiologic events characterized by capillary leakage, hemorrhage, and disseminated intravascular coagulation. Clinically, these events are manifested by mucosal edema, hemoconcentration, pulmonary edema, hydrothorax, hydropericardium, serosal hemorrhages, other hemorrhagic and thrombotic phenomena, hypotension, and shock. A curious feature of BTV infection in sheep is hemorrhage at the base of the pulmonary artery. Generally, sheep develop potentially fatal hemorrhagic disease and have extensive endothelial cell infections, whereas cattle typically develop subclinical infections and have only minimal endothelial infections. Thus, BTV and other related orbiviruses in their respective large animal hosts have pathophysiologic features resembling other viral hemorrhagic fevers.

Infection of pregnant cattle and sheep with BTV can result in maternal death, abortion and fetal death, or congenital anomalies including runting, blindness, deafness, hydranencephaly, arthrogryposis, campylognathia, and prognathia.

## Epidemiology

In many countries, the prevalence of BTV infection is high in ruminants although clinical disease is often not recorded. The epidemiology of bluetongue disease can be considered in the context of three major zones of infection: an endemic zone (generally subclinical infection), an epidemic zone (disease occurs at regular intervals), and an incursive zone (disease occurs usually only at extended intervals, but when it does occur, the epidemic is often extensive). Transmission of BTV is influenced by both the distribution and biology of insect vectors and the weather pattern, and different strains of the virus are perpetuated within distinct ecosystems and in separate geographic regions by different vector species. This has been emphasized by the application of molecular genotyping techniques which have given further insights into the epidemiology of BTV and the extent of BTV evolution. Indeed molecular epidemiology studies have demonstrated the basis of the recent European emergence of BTV and how changes in BTV epidemiology coincide with climate changes which appear to have increased both the distribution and size of vector insect populations in the region. It is also evident that bluetongue viruses have crossed from the original insect vector species (e.g., *Culicoides imicola*) and are now being transmitted by novel vectors (e.g., *C. obsoletus* and *C. pulicaris*) that are abundant across Central and Northern Europe. As a result, many of the serotypes associated with the initial outbreaks have persisted and spread across much of Southern and Central Europe.

## Immunity

Neutralizing antibodies, which play major role in protection against reinfection, appear within 8–14 days of virus infection. Although BTV infection results in the production of antibodies to all ten proteins, the outer capsid protein VP2 alone is sufficient to induce protective neutralizing antibodies. Cell-mediated immunity (CMI) also plays an important role in recovery from infection and protection against reinfection. Cytotoxic (CD8+) cells responsive *in vitro* to BTV antigens have been demonstrated in the blood of cattle and sheep during the first week of infection and observed to peak at 2 weeks post infection. Both cross-reactive and serotype-specific cytotoxic T-lymphocyte (CTL) responses occur in BTV-infected animals. The role of CMI in protection against BTV infection has been demonstrated in sheep in which possible transfer of specific CD8+ cells conferred protection on challenge with live virus. Immunity was major histocompatibility complex (MHC) restricted and thus was host specific.

## Prevention and Control

Despite the high morbidity and mortality associated with BTV infection, little attention has been given to the development of efficacious vaccines. Early vaccination attempts included the use of virus in the serum of sheep which had recovered from the disease. This mild strain of the virus, which had been serially propagated in sheep, was used as an attenuated vaccine for more than 50 years despite evidence that it was not entirely safe and the resultant immunity was not adequate.

The only vaccines currently in use are live-attenuated vaccines developed by serial passage of BTV in embryonated chicken eggs. These are administered as polyvalent vaccines consisting of a total of 15 serotypes. Sheep develop BTV antibodies by 10 days post-vaccination. The antibody response reaches a maximum at 4 weeks and may persist up to 1 year. There is a temporal relationship between the increase in neutralizing antibody titer and clearance of virus from the peripheral circulation. However, there are several problems associated with these vaccines including incomplete protection and reversion to virulence. Attenuated virus can also be transmitted by insects from vaccinated sheep to other animals. This poses risks since live virus vaccine is teratogenic in pregnant sheep and may cause fetal death and abnormalities.

To avoid the problems encountered by live virus vaccines, a number of inactivated vaccines have been developed in various countries and are used locally. Many experimental trials have also been undertaken using VP2 antigen and VLPs as candidate vaccines. VP2 administered at high dose (>100 μg per dose) alone or together with VP5 (~20 μg per dose) has been demonstrated to be protective against virulent virus challenge in sheep. VLPs (containing the four major proteins) appear to be highly protective at a lower dose (10 μg per dose) and have generated protective immunity for up to 15 months. However, despite their high protective efficacy and safety, no subunit vaccines are yet available commercially. One of the likely reasons for lack of interest in these vaccines is the lower demand of BTV vaccine in the Western countries, but this may change following the recent emergence of BTV in Europe.

## Future Perspectives

BTV has been a major focus for understanding of the molecular biology of the *Reoviridae* and has served as a model system for the other members of the genus *Orbivirus*. Significant recent advances have been made in understanding the structure–function relationships of BTV proteins and their interactions during virus assembly. By combining structural and molecular data, it has been possible to make progress on the fundamental mechanisms used by the virus to invade, replicate in, and escape from susceptible host cells. Evidence has been obtained for the role of cellular proteins in nonenveloped virus entry and egress.

Despite these advances, some critical questions remain unanswered. In particular, host–virus interactions during virus trafficking is one area needing intense attention in the future. Exactly how each genome segment is packaged into the progeny virus is another outstanding question. One of the major drawbacks of research with BTV and other members of *Reoviridae* has been the lack of availability of a suitable system for genetic manipulation of the virus and this has limited our understanding of the replication processes. However, in a recent major development in BTV research, live virus has been rescued by transfection of BTV transcripts. There is no doubt that this will soon be extended to establish *in vitro* manipulative genetic systems and will allow molecular and structural studies of individual BTV proteins to be placed in the context of the whole virus.

*See also:* Reoviruses: General Features; Reoviruses: Molecular Biology.

## Further Reading

Boyce M and Roy P (2007) Recovery of infectious bluetongue virus from RNA. *Journal of Virology* 81: 2179–2186.

Eaton BT, Hyatt AD, and Brookes SM (1990) The replication of bluetongue virus. In: Roy P and Gorman BM (eds.) *Current Topics in Microbiology, Vol. 162: Bluetongue Viruses*, pp. 89–118. Berlin: Springer.

Huismans H and Verwoerd DW (1973) Control of transcription during the expression of the bluetongue virus genome. *Virology* 52: 81–88.

MacLachlan NJ (1994) The pathogenesis and immunology of bluetongue virus infection of ruminants. *Comparative Immunology, Microbiology, and Infectious Diseases* 17: 197–206.

Mertens PPC, Maan S, Samuel A, and Attoui H (2005) *Reoviridae* – Orbivirus. In: Fauquet CM, Mayo MA, Maniloff J, Desselberger U, and Ball LA (eds.) *Virus Taxonomy: Eighth Report of the International Committee on Taxonomy of Viruses*, pp. 466–483. New York: Academic Press.

Noad R and Roy P (2006) Bluetongue virus assembly and morphogenesis. In: Roy P and Gorman BM (eds.) *Current Topics in Microbiology and Immunology, Vol. 309: Reoviruses: Entry, Assembly and Morphogenesis*, pp. 87–116. Berlin: Springer.

Parsonson IM (1990) Pathology and pathogenesis of bluetongue infections. In: Roy P and Gorman BM (eds.) *Current Topics in Microbiology and Immunology*, Vol. 162: *Bluetongue Viruses*, pp. 119–141. Berlin: Springer.

Roy P (2005) Bluetongue virus proteins and particles and their role in virus entry, assembly and release. In: Roy P (ed.) *Virus Structure and Assembly, Vol. 64*, pp. 69–114. New York: Academic Press.

Roy P (2006) Orbiviruses and their replication. In: Knipe DM and Howley PM (eds.) *Field's Virology*, 5th edn., pp 1975–1997. Philadelphia, PA: Lippincott–Raven Publishers.

Stott JL and Osburn BI (1990) Immune response to bluetongue virus infection. In: Roy P and Gorman BM (eds.) *Current Topics in Microbiology and Immunology, Vol. 162: Bluetongue Viruses*, pp. 163–178. Berlin: Springer.

Verwoerd DW, Huismans H, and Erasmus BJ (1979) Orbiviruses. In: Fraenkel-Contrat H and Wagener RR (eds.) *Comprehensive Virology*, vol. 14, pp. 163–178. New York: Plenum.

# Border Disease Virus

**P Nettleton and K Willoughby,** Moredun Research Institute, Edinburgh, UK

© 2008 Elsevier Ltd. All rights reserved.

## Glossary

**Cytopathic effect** A deleterious change in the microscopic appearance of cultured cells.

**Genotype** Individual or groups of viruses which form a significant branch in a phylogenetic tree generated using computer-assisted nucleotide sequence analysis.

**Persistent infection (PI)** A lifelong infection with noncytopathic BDV resulting from *in utero* infection during the first 60 days of gestation. Infected fetuses have an immune tolerance of the virus which lasts long after birth. PI sheep can live for at least 5 years and constantly excrete high levels of infectious virus.

## Introduction

Border disease virus (BDV) is one of the four species of pestivirus classified within the family *Flaviviridae*. The genus *Pestivirus* comprises BDV, classical swine fever virus (CSFV), and the two distinct species of bovine viral diarrhea virus (BVDV), BVDV 1 and BVDV 2. The viruses were named after the diseases they cause, so traditionally pestiviruses isolated from sheep or goats were termed BDV, those from swine CSFV, and those from cattle BVDV. It is now recognized, however, that domestic ruminants and pigs can all be infected by more than one pestivirus. Pestiviruses also have been recovered from wild animals of various species, and modern methods of antigenic and genetic analysis have identified BDV from sheep, goats, pigs, cattle, reindeer, European bison, and chamois.

BDV is principally recognized as a significant cause of congenital disease in sheep, resulting in barren ewes, abortion, stillbirths, and the birth of small weak lambs some of which show tremor and hairy fleece. Many such lambs are persistently infected (PI) with BDV all their lives, and because they constantly excrete virus they are very efficient spreaders of infection. The ability of BDV to produce PI animals following *in utero* infection is a fascinating biological phenomenon of virological and immunological interest. Control of the disease in a sheep flock relies on the identification of PI sheep and the prevention of infection of susceptible ewes during pregnancy.

## History

### First Reports of Border Disease

Although shepherds in the border region of England and Wales had been seeing weak, rough-coated, and trembling lambs for many years, it was not until 1959 that workers at the then newly opened Veterinary Laboratory at Worcester described a clinico-pathological condition they named border or 'B' disease. Recognition of the disease in New Zealand, Australia, and the United States soon followed and affected sheep were aptly named 'hairy-shaker' or 'fuzzy' lambs. In the early years, when the condition was only recognized in two closely related breeds of sheep, a genetic etiology was seriously considered. As more reports appeared in other sheep breeds, an infectious cause was thought more likely and the transmissibility of the condition was confirmed in 1967 by Dickinson and Barlow in Edinburgh and Shaw and colleagues at Worcester.

### Isolation of Virus

Evidence for a viral etiology came from workers at the Moredun Research Institute in Edinburgh who reported in 1972 that, antibiotic-treated, cell-free filtrates of tissue homogenates could cause infection. They showed the agent to be ether sensitive, to be destroyed by heating to 60 °C for 90 min, and to have an estimated size of about 27 nm. Similar experiments in Australia, Ireland, and the USA confirmed the transmissibility findings and added the valuable information that ewes receiving inocula derived from border disease (BD)-affected lambs developed antibodies against BVDV, so that by the mid-1970s the serological relationship between BDV, BVDV, and CSFV was established. Throughout the 1970s there were reports that BVDV or BDV from sheep could be isolated in cell culture. The first reports were from Australia, where a BVD virus caused abortion but no other signs of BD. The full clinical picture of BD was subsequently reproduced with noncytopathic cell culture isolates by Harkness at Weybridge, UK, Plant in Australia, and Terpstra in the Netherlands. Meanwhile, Vantsis at Moredun had isolated a virus in fetal lamb kidney cells, a virus that was cytopathic and reproduced BD at the 12th and 29th cell culture passages.

## Understanding the Pathogenesis of BD

In all the early transmission experiments clinical disease was rarely detected in recently infected animals. Slight fever and mild leukopenia was detected 1–2 weeks post infection, after which serum neutralizing antibodies were demonstrable. Inoculation of pregnant ewes, however, resulted in BD-affected offspring. In a series of experiments at Moredun, Barlow and Gardiner studied the effects on the ewe and the fetus of different inocula given at different stages of gestation. They showed that the main effect on the ewe was a necrotizing placentitis which varied from a mild focal lesion to a diffuse lesion, sometimes serious enough to contribute to fetal death and abortion. In affected fetuses and newborn lambs BDV infection interfered with pre- and postnatal development of the viscera, for example, thymus, spleen, skeleton, and muscles. The abnormal coarseness of the fleece was shown to be due to increased size and altered structure of the primary wool fibers (**Figure 1**), while the neurological signs predominantly resulted from a myelin deficiency throughout the central nervous system (CNS), with occasional brain malformations being seen. In all these outcomes the strain of virus and the timing of infection were critical.

### Persistent Viremia

Current understanding of the timing of infection hinges on the onset of immune competence by the fetal lamb. The ovine fetus can first respond to an antigenic stimulus between 60 and 85 days of its 150-day gestation period. In

**Figure 1** (a) A group of clinically affected BD lambs showing variable ability to stand. With careful nursing all these lambs survived and their neurological signs were no longer evident by 6 months old. (b) Hairy-shaker lamb in the foreground with breed matched normal lamb behind. The fleece of the hairy-shaker lamb is less curly with long fibers most prominent over the top of the head and neck. Reproduced by permission of Moredun Research Institute.

fetuses infected before the onset of immune competence, viral replication is uncontrolled and it is common to observe that half these fetuses die. Lambs that survive infection in early gestation are tolerant to the virus, which is widespread in all organs. Such PI lambs are born with a persistent viremia, so that a precolostral blood sample is virus positive and antibody negative. In such lambs there are no inflammatory changes. At all levels in the CNS there is a deficiency of myelin. This may be slight in lambs with only mild trembling of the head, ears, and tail, but can be pronounced in lambs unable to stand due to violent rhythmic contractions of their hind leg muscles. In lambs that suck colostrum containing anti-BDV antibodies, viremia may be masked for up to 2 months but after that lambs have high virus titers in their blood and continue to excrete virus, usually for the rest of their lives. With careful nursing a proportion of BD lambs can be reared although deaths may occur at any age. The neurological signs gradually become less noticeable and may have disappeared by 3–6 months. Weakness, swaying of the hind quarters, and fine trembling of the head may reappear at times of stress. Affected lambs can grow slowly and many die at or around weaning. However, a small proportion of viremic sheep appear normal and can survive to breeding age. Lambs born to viremic dams will always be PI. PI sheep are a continual source of infectious virus and their identification is a major factor in any control program. Sheep being traded for breeding should be screened for the absence of BDV viremia.

## Fetal Infection Later in Gestation

Fetal infections occurring just as the immune system is developing, around mid-gestation, can result in lambs with severe nervous signs, locomotor disturbances, and abnormal skeletons. Such lambs have destructive lesions of cerebellar hypoplasia and dysplasia, hydranencephaly, and porencephaly resulting from necrotizing inflammation. The severe lesions appear to be immune-mediated and such lambs usually have high levels of anti-BDV antibody at birth. Lambs infected in late gestation mount an immune response to the virus and are born apparently healthy, free of virus but with antibody to BDV. Such lambs, however, can be stillborn or weak and losses would be above average in early life.

## Late-Onset Disease in PI Sheep

In the early 1980s, experiments at Moredun investigating the nature of immune tolerance to BDV unexpectedly resulted in the deaths of PI sheep, which had been challenged 2–3 weeks earlier with live, cytopathic BDV. The sheep developed intractable scour, wasting, and excessive ocular and nasal discharges, sometimes with respiratory distress. At necropsy there was gross thickening of the distal ileum, cecum, and colon resulting from focal hyperplastic enteropathy. Cytopathic BDV was recovered from the guts of dying lambs and the syndrome had several similarities with bovine mucosal disease. Similar lesions also had been seen in PI sheep that had been housed away from other animals and from which noncytopathic BDV had been repeatedly recovered since birth. At postmortem of these spontaneous cases of late-onset disease, cytopathic virus was consistently recovered from gut tissues. With no obvious outside source of cytopathic virus, it was believed that such virus originated from the lamb's own virus pool.

Virtually, all pestivirus isolates from sheep and goats are noncytopathic in cell culture. The NS2–3 genes of two cytopathic sheep strains (Moredun and Cumnock) have been studied, and, in both these, it has been shown that they contain insertions of cellular sequences within the NS2–3 encoding region, which results in its cleavage to NS2 and NS3. This is analogous to the similar process in BVDV, which is associated with the development of mucosal disease in cattle. Such cattle are PI with noncytopathic (NCP) BVDV following *in utero* infection. Mutation of the persisting virus RNA in the region coding for NS2–3 can result in overproduction of NS3, which correlates with the development of mucosal disease. While it has proved possible to isolate and biologically clone pairs of noncytopathic and cytopathic viruses from individual cattle dying of mucosal disease, this has so far proved not to be the case with the analogous condition in sheep. Thus, no cytopathic isolate of BDV has ever been successfully biologically cloned and shown not to also contain noncytopathic virus.

## Physico-Chemical Properties of BDV

Pestiviruses are enveloped, spherical particles approximately 46 nm in diameter with a 20–25 nm core. They consist of an outer lipid envelope surrounding an inner capsid protein that contains the 12.3 kbp long, single-stranded, positive-oriented RNA viral genome, which comprises two terminal flanking untranslated regions and a single open reading frame (ORF), encoding approximately 4000 amino acids. The lipid envelope is pleiomorphic which makes it difficult to purify infectious particles by ultracentrifugation or to visualize virions by electron microscopy. The $M_r$ of a virion is estimated at $6 \times 10^7$ and the buoyant density in sucrose of BDV is $1.09–1.15 \text{ g ml}^{-1}$, with peak infectivity at $1.115 \text{ g ml}^{-1}$. Virions are stable within a pH range of 5.7–9.3, but are inactivated by lipid solvents (e.g., 5% chloroform for 10 min at room temperature), detergents, and treatment with 0.05% trypsin for 60 min at 37 °C. Viruses in cell culture fluids are inactivated by heating at 56 °C for 30 min, but this is not a reliable method for the inactivation of viral infectivity in fetal calf serum. At lower temperatures the virus is more stable; but even at 37 °C reductions in infectivity can be detected after 24 h. Inactivation of pestiviruses in fetal bovine serum is reliably achieved using gamma irradiation (20–30 kGy) or ethyleneimine (10 mM at 37 °C for 2 h).

## Antigenic and Genetic Relationships among Pestiviruses

The *Eighth Report of the International Committee on Taxonomy of Viruses* recognized four species of pestiviruses, namely BDV, CSFV, and BVDV-1 and -2 with a fifth virus originating from a giraffe as a tentative species. All pestiviruses are antigenically related but criteria for virus species demarcation include: nucleotide sequence relatedness, antigenic relatedness as measured in cross-neutralization or cross-protection assays, and host of origin. While nucleotide sequences can be generated quickly from new viruses and easily compared with sequences from known viruses, comprehensive antigenic relatedness studies are more demanding and often lacking. As pestiviruses are increasingly identified in wild animals, the natural host of origin becomes less certain. Among pestiviruses, rapid sequencing of nucleic acids coupled with computer-assisted phylogenetic analysis can be readily used to identify viruses of the four recognized species plus the giraffe pestivirus. Nearly all isolates can be fitted into these species, but as more isolates are examined, groups of viruses which form significant branches within each pestivirus species are becoming recognized. Thus, recent studies of ovine pestivirus isolates from Germany and Switzerland by Becher and colleagues at Giessen in Germany led to the proposal for three such genogroups among BDVs. BDV-1 contains many of the traditional well-characterized BDV isolates, BDV-2 contains recent German isolates plus a reindeer virus, while BDV-3 contains a single recent German isolate. Since that work a fourth genogroup BDV-4 has been proposed following analysis of Pyrenean chamois and Iberian sheep isolates. Furthermore, pestiviruses isolated from Tunisian sheep may represent an additional BDV genotype and an Italian goat isolate may represent a putative novel pestivirus close to BDV. Thus, it would appear that BDV may contain the widest spectrum of virus subtypes of all the pestiviruses.

## Molecular Biology of BDV

### Viral Genome

The complete genomic sequences of two sheep BDVs have been published. Strain BD31 (GenBank accession number U70263) is an American hairy-shaker lamb isolate which is available from the American Type Culture Collection. Strain X818 (GenBank accession number AF037405) is an Australian isolate also derived from a lamb clinically affected with BD. Both BD31 and X818 belong within genogroup BDV-1. The only other complete BDV genome sequence published is that of an isolate from a reindeer (GenBank accession number NC003678 or AF144618), a representative strain of the BDV-2 genotype. The structures of these three genomes are comparable and are similar to those of other pestiviruses. Thus, the 12.3 kbp single-stranded genome of BDV codes for one long ORF of 3895 codons bracketed by a 5′ untranslated region (UTR) of 356–372 bp and a 3′ UTR of 223–273 bp. When compared with other pestiviruses, BDVs unexpectedly have closer sequence homology to strains of CSFV than to some BVD viruses.

The 5' UTR is recognized as having the highest overall nucleic acid sequence identity among pestiviruses. Conservation of 5' UTR sequences suggests a strong functional pressure likely to be related to the formation of tertiary structures required for the internal ribosomal entry site (IRES) from which translation is initiated. Within the highly conserved 5' UTR, there are two short variable regions located between nucleotide positions 210–230 and 300–320. The 5' UTR region thus lends itself to the design of panpestivirus polymerase chain reaction (PCR) primers which allow the amplification of all pestiviruses, as well as primers which allow the speciation of pestiviruses.

The 3' UTR region of BDVs is of interest since it is longer than those found in other pestiviruses and contains a conserved sequence motif at more than one location, also found in CSFVs but absent from BVDV viruses.

## Viral Proteins

The identification of BDV-encoded proteins together with the predictions of conserved cleavage sites strongly suggests that the polyprotein of BDV is processed similarly as those of the other pestiviruses.

## Structural Proteins

The large ORF is translated as a single polyprotein with the order of individual proteins being $N^{pro}$, C, $E^{rns}$, E1, E2, p7, NS 2/3, NS4A, NS4B, NS5A, NS5B (**Figure 2**). The polyprotein is processed during and after translation by viral and host proteases. There are four structural proteins in the mature virion: C, $E^{rns}$, E1, and E2. C is the nucleocapsid protein and the other three are glycosylated envelope proteins. E2 is the outer immunodominant envelope glycoprotein against which the host produces a significant neutralizing antibody response. As E2 is the most variable protein among pestiviruses, monoclonal antibodies against E2 can be useful for discriminating pestivirus isolates. There is some evidence that $E^{rns}$ also stimulates protective host immune responses, but there is no evidence as yet that E1, the inner envelope protein, stimulates protective antibody. $E^{rns}$ is less conserved than is E1 and is a protein of fascination to virologists, since it has ribonuclease activity, has anti-interferon activity, and is present in the serum of PI animals at levels high enough for diagnostic detection.

## Nonstructural Proteins

The first viral protein of the pestivirus ORF is the nonstructural protein, $N^{pro}$. Like $E^{rns}$, this protein is unique to pestiviruses and is crucial in the establishment and maintenance of persistent infection. It has protease activity and cleaves itself from the polyprotein. The tiny nonstructural protein p7 follows E2 in the polyprotein and can remain associated with it, so that two forms of E2 with different C-termini (E2 and E2-p7) are found in infected cells. However, p7 alone or as E2-p7 is not found associated with infectious virions and its role is unknown. The next nonstructural protein, NS2–3, has attracted considerable attention because of its role in the cytopathogenicity of pestiviruses. The success of pestiviruses is due to the unique biological properties of the noncytopathic virus. Only the noncytopathic viruses are biologically able to ensure they are maintained. When cytopathic viruses evolve from noncytopathic viruses by nonhomologous RNA recombination, they signal the end of the life of the PI animal and the removal of that source of biologically competent noncytopathic virus. The NS2–3 in noncytopathic viruses contains a serine protease that is essential for the release of the NS proteins located downstream of NS3. In cells infected with cytopathic BDV, two antigenically related polypeptides of apparent $M_r$ 80 kDa and 130 kDa are detectable, whereas only a single polypeptide of 120 kDa is recognized in cells infected with noncytopathic BDV. The 120 kDa and the 130 kDa proteins represent NS2–3, with the polypeptide from the cells infected with cytopathic virus containing a cellular insertion. The 80 kDa polypeptide in the cells infected with cytopathic BDV is NS3. NS2 appears to be nonimmunogenic and little is known about its function. On the other hand, NS3 produced by cytopathic pestiviruses is of considerable interest. It is highly conserved across the pestiviruses and contains immunogenic epitopes. Purified NS3 possesses serine protease, RNA helicase, and RNA-stimulated NTPase activities. A catalytically active NS3 serine protease is essential for pestivirus replication.

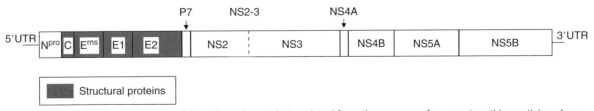

**Figure 2** Diagram of the proteins formed from the polyprotein translated from the genome of a noncytopathic pestivirus. In a cytopathic pestivirus a protease cleavage site results in NS2–3 being cleaved into NS2 and the conserved, immunodominant NS3. Reproduced by permission of Moredun Research Institute.

It appears to be produced in large amounts in cells infected with cytopathic viruses and its buildup coincides with cell death.

Less is known about the NS proteins downstream of NS3. The NS3 protease depends on a cofactor domain in the central region of NS4A and there is evidence that duplications of codons in NS4B and an insertion of viral sequences between NS4A and NS4B are well tolerated and do not interfere with efficient viral replication. This has led to the suggestion that the NS4A–NS4B region in pestiviruses and related viruses may be suited to the insertion of heterologous sequences.

The NS5A is the least conserved protein when BDV X818 is compared to other pestiviruses. No information is available on the role of this protein in BDVs. Work with BVDV, CSFV, and the related hepatitis C virus (HCV, genus *Hepacivirus*, family *Flaviviridae*) shows that pestivirus NS5A contains an essential zinc-binding site similar to that of HCV. Also structural conservations within NS5A point toward conserved roles of the terminal in-plane membrane anchor in the replication complex formation in these viruses.

The pestivirus NS5B protein possesses RNA-dependent RNA polymerase activity and is involved in the assembly of infectious viruses. It has attracted attention as a target for antiviral drug discovery.

## Viral Replication

BDV has rarely been used to study pestivirus replication. It is appropriate to assume that BDV replicates in the same way as do other pestiviruses. Studies on the replication of pestiviruses have received added impetus because their mechanisms of virion maturation and assembly seem to be equivalent to those of HCV. BVDV has consequently been used as a surrogate model of HCV for the evaluation of antiviral agents.

## Viral Attachment

The cell-surface receptors for pestiviruses require further clarification. There is good evidence that bovine CD46 serves as a cellular receptor for BVDV, and some evidence that heparin sulfate also does. E2 and $E^{rns}$ are the viral glycoproteins essential for attachment and they appear to use different cell-surface receptors. A chimeric BVDV with the E2 coding region of a BDV has been used to establish a role for E2 in determining different tropisms of a pestivirus for bovine or ovine cell cultures. It may be, however, that such results are virus strain and cell line dependent since another study showed that CSFV E2 and BVDV E2 share an identical receptor on porcine and bovine cells. In the same study several BVDV strains isolated from cattle were not inhibited by the presence of CSFV $E^{rns}$ suggesting that $E^{rns}$ also plays a role in host tropism.

## Virus Replication

After binding to a cellular receptor, BVDV is internalized via clathrin-dependent endocytosis. Virus-membrane fusion occurs in the endosomal compartment and release of viral RNA follows pH-dependent breakdown of the envelope glycoproteins. The viral RNA serves as mRNA and, using its own and cellular enzymes, undergoes translation into viral proteins. Viral RNA must also serve as the template for RNA replication through the synthesis of complementary negative strands. These, in turn, serve as templates for the synthesis of genome length positive strands for incorporation into new virions. Electron micrographs comparing BDV- and BVDV-infected cells showed mature virions of both viruses to be similar. Virus replication takes place totally within the cytoplasm in association with structures formed from modified endoplasmic reticulum (ER). Free virus-like particles were not seen in the cytoplasm, nor were any seen budding through internal host cell membranes or passing through the cytoplasmic membrane. Virus release is presumed to occur when remnants of ER are released or when cells disintegrate. Pestiviruses have a surprisingly large number of unique properties, for example, $N^{pro}$, $E^{rns}$, NS2–3 cleavage, and RNA recombination, all of which contribute to its replication and fascinating biology.

## Detection and Control

The presence of BDV in an infected flock is most easily determined by detection of the virus in dead or otherwise clinically affected lambs at birth, and by demonstrating the presence of antibody in ewes with BD-affected lambs. Virus detection can be by virus isolation, immunohistochemistry, viral antigen detection by the enzyme-linked immunosorbent assay (ELISA), or by nucleic acid detection by reverse transcriptase-polymerase chain reaction (RT-PCR). Antibody to BDV can be detected in sheep sera using virus neutralization or ELISA.

The control of BDV in a sheep flock has two essential requirements: the identification of PI sheep and the prevention of infection of susceptible pregnant ewes especially during the first half of gestation. One commercial vaccine has been produced, a killed adjuvanted vaccine containing representative strains of BDV and BVDV-1. It is administered to young animals before they reach breeding age in order to maximize their immunity during early pregnancy. An annual booster dose is required. Further vaccine development is required, with candidate vaccines being tested for efficacy in pregnant sheep.

*See also:* Bovine Viral Diarrhea Virus; Classical Swine Fever Virus.

## Further Reading

Barlow RM and Patterson DSP (1982) Border disease of sheep: A virus-induced teratogenic disorder. *Advances in Veterinary Medicine* (supplement to the *Journal of Veterinary Medicine*) No. 36: 1–90.

Nettleton PF, Gilray JA, Russo P, and Dlissi E (1998) Border disease of sheep and goats. *Veterinary Research* 29: 327–340.

Thiel H-J, Collett MS, Gould EA, et al. (2005) *Flaviviridae*. In: Fauquet CM, Mayo MA, Maniloff J, Desselberger U, and Ball LA (eds.) *Eighth Report of the International Committee on Taxonomy of Viruses*, pp. 981–998. San Diego, CA: Elsevier Academic Press.

Vilcek S and Nettleton PF (2006) Pestiviruses in wild animals. *Veterinary Microbiology* 116: 1–12.

# Bornaviruses

**L Stitz and O Planz,** Federal Research Institute for Animal Health, Tuebingen, Gemany
**W Ian Lipkin,** Columbia University, New York, NY, USA

© 2008 Elsevier Ltd. All rights reserved.

## History

Borna disease (BD) is one of the oldest known viral infections of animals. The name was adopted in 1895 after an epidemic among horses of a cavalry regiment in the town of Borna in Saxony near Leipzig. However, descriptions of natural infection in horses and sheep were reported in the veterinary literature as early as 1767. Synonyms such as disease of the head (Kopfkrankheit), brain fever, subacute meningitis, or hypersomnia of horses have been known for more than 200 years and reflect the restriction of the disease to the central nervous system (CNS). A viral etiology was proposed in 1925 by Zwick and Seyfried based on filtration experiments. The causative agent, Borna disease virus (BDV), was identified as a unique, nonsegmented negative-strand RNA virus in the late 1980s following the first successful application of subtractive cloning in microbiology.

## Properties of the Virion

Viral particles have not been visualized in infected tissues. Thus, morphological descriptions are based on observations in cultured cells. BDV particles are enveloped, spherical, 70–130 nm in diameter, and contain a 60 nm diameter helical ribonucleoprotein. BDV does not differ in its physical properties from other known enveloped viruses. It is sensitive to lipid solvents such as chloroform or acetone and to detergents. BDV is stable at low temperature and at pH 5–12; however, infectivity is destroyed by high temperatures and pH less than 5.

## Properties of the Genome

Data on the nature of the BDV genome have been obtained from complementary DNA clones using libraries prepared either from the brain of BDV-infected rats or from cell cultures. The genomic single-stranded RNA (ss RNA) of 8.9 kb with complementary 3′- and 5′-termini has been shown to be of negative polarity and represents the smallest genome of all *Mononegavirales* (**Figure 1**). Investigations of the organization of the genome have revealed six open reading frames (ORFs) on the antiviral genome (3′-N-X/P-M-G-L). Several unusual aspects for an ssRNA virus have been recognized in molecular biological studies, in particular an overlap of reading frames, a post-translational processing of subgenomic RNAs (RNA splicing), and nuclear phases indispensable for virus replication. Results of sequence analyses of viruses obtained from various animal species, man, or tissue culture, revealed a remarkable conservation of coding sequences.

## Properties of the Viral Proteins

Infected brains and cultured cells contain BDV proteins with an approximately relative molecular masses of 56, 38/40, 24, 14, and 10 kDa can be detected in cultured cells and brains after BDV infection. The most abundant viral proteins in infected cells as well as in viral particles have a molecular mass of 38/40 and 24 kDa. Together these comprise what is known as the soluble antigen (S-antigen). The BDV genome encodes six major BDV proteins. ORF I, which is the prominent 3′-ORF, encodes the nucleoprotein (NP) in its two isoforms (38 and 40 kDa proteins). ORF II codes a 24 kDa protein, a protein phosphorylated in its serine residues, representing the viral phosphoprotein (P). An additional ORF X has been identified that overlaps ORF II, encoding a nonglycosylated BDV protein designated X or p10. ORF III encodes the 14 kDa matrix protein (M). ORF IV encodes a 57 kDa precursor that is found in its N-glycosylated form at approximately 90 kDa (described as gp84 and gp94) and represents the glycoprotein of the virus (G); post-translational cleavage

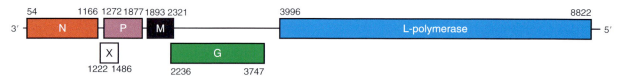

**Figure 1** Organization of the 8.9 kb BDV genome.

products of this G have been reported to result in molecular masses of approximately 40 kDa by the subtilisin-like cellular protease furin, corresponding to the N- and C-terminal regions of G. ORF V, which is localized at the 5′-end of the viral antigenome, has been identified as the 180 kDa RNA-dependent RNA polymerase (L) of BDV. It contains conserved regions characteristic for *Mononegavirales* and three RGD motifs which are generally important in protein–protein interactions.

Proteins X, P, and N are found mainly in the nucleus of infected cells. Colocalization of P and X on the one hand and N and X on the other hand might be explained by the fact that sequences within the X sequence have been identified which are typical for nuclear export trafficking. X itself has been demonstrated both in the nucleus and the cytoplasm. In addition, evidence has been presented that indicates the presence of a nuclear targeting sequence in the 40 kDa isoform of N which is not present in the 38 kDa isoform. This results in accumulation of the 40 kDa protein in the nucleus, whereas the 38 kDa isoform is found in both the nucleus and the cytoplasm. The heavily N-glycosylated gp84 has been seen both intracytoplasmically in the ER as well as on the cell surface, and its function is seen in context with receptor-mediated viral entry, assembly, and budding of BDV. The gp84 and gp94 forms of G as well as the M contain neutralization epitopes.

## Replication and Life Cycle

Both neural and non-neural cell lines support BDV replication. However, replication is more efficient in neural cell lines. Even in neural lines a single cell produces less than 10 infectious particles and titers rarely exceed $10^6$ TCID50/ml. This observation is unexplained as large amounts of viral antigen can be demonstrated immunohistologically.

After receptor-mediated endocytosis the BDV genome is replicated and transcribed in the nucleus. The cytoplasmic-nuclear shuttle is mediated by specific nuclear localization signals (NLSs) located in the N (p40), X, P, and L proteins. X contains a nuclear export signal (NES); although export activity has not formally been shown, X is proposed to participate in the subcellular localization of N and P. A variety of mechanisms are used to regulate gene expression. The use of cellular splicing machinery to process primary RNA transcripts is unique by BDV among the *Mononegavirales*. Other intriguing unusual aspects of BDV molecular biology include overlapping reading frames, overlapping transcription units, alternate RNA splicing, and leaking scanning of ribosomes during translation. The replication cycle seems to be completed within 24 h. Little, if any, infectious virus is released from infected cells. Additional information concerning the role of individual gene products in the virus life cycle will be forthcoming with the recent introduction of reverse genetic systems.

The distribution of BDV *in vivo* has been studied in rat models of experimental infection using immunohistochemical and molecular probes (**Figure 2**). Early in infection of adult immunocompetent animals, viral gene products are confined to the CNS and are found predominantly in neurons. At later phases, virus is found in astrocytes, oligodendroglia, ependymal cells, and Schwann cells. In animals that are immunocompromised or tolerant of the virus because of infection within 24 h of birth, BDV may also replicate in cells of peripheral organs. In recent years, it could be demonstrated that BDV interacts with intracellular signaling pathways and with the cell cycle of the host cell.

## Geographic Distribution

BD was originally described as a disease restricted to certain endemic areas in eastern and southern Germany and Switzerland. More recent seroepidemiological studies, however, have shown that natural infection may be more widely distributed than previously anticipated. Seropositive horses have been found throughout Germany, in other European countries, and in North America and Japan. The potential impact for agriculture of BDV infection has almost certainly led to under-reporting. One of the authors performed serological tests for an agricultural ministry that revealed evidence of infection in horses, sheep, and cattle but was not provided with the information needed to publish the results.

## Host Range and Virus Propagation

Natural infection with BDV was first described in horses, sheep, cattle, and rabbits. Later, cats, ostriches, and various

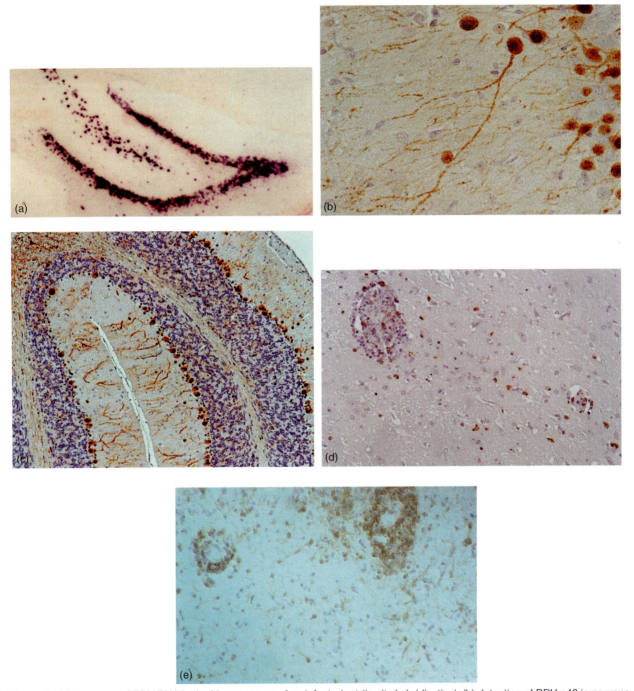

**Figure 2** (a) Presence of BDV RNA in the hippocampus of an infected rat (*in situ* hybridization); (b) detection of BDV p40 in neurons of rat brain cortex by specific monoclonal antibody (immunohistochemistry); (c) presence of BDV p40 in rat cerebellum (immunohistochemistry); (d) CD8+ T cells in the parenchyma of BDV-infected rat; and (e) presence of CD4+ T cells in perivascular cuffing of infected animal. Courtesy to Dr. H. J. Rziha, Tuebingen.

zoo ruminants were identified as natural hosts. Reports of infection in goats, deer, and donkeys have not been proved unequivocally, but are plausible in view of the broad experimental host range. The role of BDV in human disease remains controversial. Seropositive reactions have been reported in patients with psychiatric disorders from Germany, the US, and Japan. The presence of virus-specific nucleic acid has also been reported in postmortem brain samples of patients with schizophrenia, affective disorders, and hippocampal sclerosis. Virus has been recovered from the brain of one schizophrenic patient. Some investigators have also reported BDV nucleic acid in monocytes of patients with chronic fatigue syndrome, affective disorders, and schizophrenia. However, other investigators have been unable to independently replicate experiments showing human infection. The route for human infection (if true) is unknown as no convincing correlation has been established between the distribution of potential cases and infected individual animals or herds.

Experimental hosts comprise a wide variety of species ranging from birds to nonhuman primates. The most thoroughly investigated experimental model of BDV infection is the Lewis rat. This model has contributed considerably to the elucidation of BD pathogenesis, because the manifestations of infection are similar to those observed in naturally infected horses. Mice can be experimentally infected and have advantages with respect to genetic studies; however, the utility of murine models has been limited to immunopathology because behavioral consequences do not accurately represent what is observed in natural infection. Infection of adult mice can result in persistent CNS infection, but only rarely results in clinical disease. Infection of suckling mice can result in profound disease dependent on the mouse strain.

## Serological Relationship and Variability

Serologically, no distinction can be made between different virus isolates by using polyclonal antibodies. Variable detectability of viral antigens in brain specimens by monoclonal antibodies indicates, however, that some strain variation may exist. Using phylogenetic analyses virus isolates can be clustered corresponding to the geographical regions from which the viruses originated. In addition to the natural isolates and to the so-called laboratory strains, a BDV vaccine strain (Dessau) exists. Nevertheless, no conclusive data are available on the occurrence of biologically different virus strains in natural hosts. A recently described BDV strain (No/98) originated from an area in Austria where BD is nonendemic, is highly divergent to all field and laboratory strains, as well as to the vaccine strain.

It remains to be seen whether inapparent infections in horses and sheep might be due to variant strains. All virus strains used for infection of experimental animals originate from brain tissue of naturally infected diseased horses. These isolates mostly underwent serial passages in rabbits and were adapted thereafter to different animals. Different biological properties of virus isolated at various times after infection from rats could also point to the existence of variants. In this context it should be emphasized that virus isolates exist that cause extreme obesity syndrome or behavioral alterations in infected rats. Sequence analyses confirmed the general view of highly conserved regions in the BDV genome.

## Epidemiology

Originally, BD was considered an endemic disease occurring sporadically in Central Europe. Seroepidemiological investigations in horses have revealed virus-specific antibodies in 12% of randomly collected horse sera from Germany, Switzerland, and other European countries. Most of these animals were without clinical signs of BD, but some exhibited disease symptoms within an observation period of 1 year. Since the chance of contact between horses has increased over the past few decades through worldwide horse trading and international sport activities, it cannot be excluded that BD has been distributed worldwide. Investigations in endemic areas in Germany, Switzerland, Austria, and Liechtenstein have supported the notion of regional virus reservoirs. In this context, the finding that shrews harbored a virus with 99.9% sequence identity to a virus sequence derived from a diseased horse in the same area of Switzerland might be of great importance. However, it cannot be excluded that virus is shed during an inapparent infection of horses and sheep which also could lead to further propagation of the infection. Virus-specific RNA has been demonstrated in leukocytes, saliva, conjunctival and nasal fluids, and in urine of clinically healthy horses in stables with BD history. If one takes into account that persistent infections can be established in tolerant infected rats and that these animals continuously shed virus, this mode of infection – in addition to other reservoirs as mentioned – appears to also be a possibility for maintaining BDV in the horse population.

## Transmission and Tissue Tropism

Intranasal infection seems to be the most likely route of natural infection. Nerve endings in the nasal mucosa are readily accessible to the virus and pathological features observed in naturally infected horses such as edema of the bulbus olfactorius can be found in experimentally infected animals only after intranasal infection. An oral route via the trigeminal nerve is also plausible. In its natural hosts and in adult experimentally infected animals the virus shows a strict neurotropism. After inoculation in the periphery, BDV migrates centripetally to the brain where it replicates in neurons and astrocytes, especially in neurons of the hippocampus. Various routes of infection have been shown to result in clinical disease in susceptible animals. The most common route of experimental infection is intracerebral inoculation; however, intranasal, intraperitoneal, and intravenous routes are also effective. In rats infected as adults the virus shows the same strict neurotropism as in the natural hosts. Infectious virus is found in the brain as early as 7 days after infection and reaches titers of $10^6$ TCID50/ml. Infectious virus can also be detected in the spinal cord, in ganglia of the autonomic nervous system, in peripheral nerves, in the adrenal glands and neural cells of the retina although at considerably lower titers. No infectivity is found in the eyes late during the course of infection, possibly because degeneration of neurons in the retina eliminates cells in which the virus replicates most efficiently. In other nervous system tissues infectivity can be found throughout the life span. In sharp contrast to rats infected as adults, neonatally infected rats harbor virus not only in the CNS but also in peripheral organs, including salivary and lachrymal glands and the urinary bladder. Indeed, virus can be found in saliva, lachrymal fluid, and in urine. A similar pattern of distribution of BDV was found in long-term immunosuppressed adult rats.

## Pathogenicity

To date virus strains isolated from naturally infected animals have been equally pathogenic. However, because the isolates have been obtained from animals with active disease, it is conceivable that these strains do not reflect the full spectrum of pathogenicity. Support for the existence of less pathogenic virus variants can be found in experimentally infected rats, where some strains are associated with mild or inapparent disease. Some species appear to be infectible but resistant to disease. Although adult rabbits and rats develop severe meningoencephalitis following inoculation with virus isolated from diseased horses or sheep, adult mice, hamsters, and tree shrews can have inapparent infection or mild disease. The basis for differences in susceptibility to disease is still unknown. Some experimentally infected rats become obese. Although this has been linked to anatomical abnormalities in hypothalamus, the basis for this tropism is unknown.

## Immune Response

The immune response to BDV has been shown to represent the basis of the pathogenesis of the disease. Although experimentally infection of adult immunocompetent animals regularly results in disease, infection of newborn or athymic rats or animals rendered immunodeficient by drugs leads neither to encephalitis nor to overt disease despite the persistence of the virus. These findings in conjunction with the property of BDV being a noncytolytic virus that persists in the brain of immunocompromised hosts at titers comparable to those in immunocompetent animals provide all characteristics of an immunopathological disease. Rats infected as adults have both a T- and B-cell response to BDV (**Figure 2**). The role of the latter is unclear because BDV-specific antibodies adoptively transferred to immunoincompetent recipients do not induce pathological changes or disease. In contrast, adoptive transfer of T cells does induce BD in infected immunosuppressed recipients. The role of specific T-cell subsets has been investigated. Passive transfer of *in vitro* established homogeneous BDV-specific CD4+ T-cell lines can elicit BD in persistently infected asymptomatic rats; CD4+ T cells do not cause disease by themselves but rather induce activity of cytolytic CD8+ T cells. In the absence of CD8+ T cells, no disease or immunopathology is seen even if CD4+ T cells can be detected in the brain. One mechanism for the activity of CD8+ T cells may be their capacity to express perforin, the major molecule involved in cytolysis of target cells by CD8+ T cells. Other mechanisms are under investigation, such as pro-inflammatory cytokines or free radicals. Interestingly, virus-specific MHC class II-restricted T cells, when applied before infection, prevent the manifestation of clinical signs in rats. Antibodies to virus-specific antigens are synthesized early after infection in all BDV-infected animals; however, neutralizing antibodies do not appear until infection is well established in the CNS and never reach high titers.

## Clinical Features of Infection

Incubation time in natural infection ranges from weeks to months. If the infection becomes manifest, the disease is characterized during the early phase by hyperexcitability or somnolence, impairment of posture and balance, hyperesthesia, visual disturbance, anorexia, fever, and colic. Most naturally infected animals die 1–2 weeks after onset of the disease, but recoveries or recurrences of disease can also be observed. Manifestations of disease are similar in natural and experimental infection; however, the time course and features of experimental infection have been mapped in greater detail. In rats, disease starts with alertness and loss of fear. Later, rats show incoordination, occasionally with increasing hyperactivity and aggressiveness. Thereafter, most rats enter a stage of disease where passiveness and hypersomnia dominate. These symptoms are at least in part attributable to the onset of blindness. In the chronic stage of the disease, signs of dementia and chronic debility predominate.

## Pathology and Histopathology

Pathological changes in BDV are restricted to the brain, spinal cord, and the retina. In acute disease in naturally infected hosts, a massive, perivascular as well as parenchymal inflammation is found, composed of infiltrated macrophages, T and B cells (**Figure 2**). The sequence and development of pathological changes have been extensively studied in rats. BDV-infected rats develop a severe nonpurulent disseminated mononuclear meningoencephalitis with most intense inflammatory reaction restricted to the gray matter of cortex and diencephalon (**Figure 3**). Macrophages and T lymphocytes are the most numerous cell populations present in the inflammatory lesions during the earlier phases of the local reaction in both experimentally infected rats and in naturally infected horses and sheep; B lymphocytes predominate at later stages. After intracerebral infection in rats, the initial reaction is characterized by a focal accumulation of mononuclear cells in the leptomeninges at day 8–10. Thereafter, perivascular and parenchymal infiltrations are observed which intensify during the course of the disease and consist of massive accumulations of inflammatory cells in perivascular spaces and the neuropil between day 20 and 30 post infection. Beyond day 30 p.i. the intensity of the inflammatory reaction decreases and late after infection (>60 days) the number and intensity of the inflammatory infiltrates is significantly reduced although the level of infectious virus remains constant. Simultaneously with the development of the encephalitis, degenerative lesions can be found in hippocampal and cortex regions. One of the consequences of the necrotic process is a dilatation of the lateral ventricles resulting in a marked hydrocephalus *ex vaccuo*. Reactive astrogliosis is found throughout the brain. Infected neurons contain intranuclear acidophilic inclusion bodies (Joest–Degen inclusion bodies) that are regarded as pathognomonic for BD. In the brain the virus is predominantly found in neurons, but viral antigen is also detected in astrocytes, oligodendrocytes, and ependymal cells. Histopathological lesions in the eye of infected rats and rabbits are seen almost exclusively in the retina resulting in a degeneration of rods and cones and ultimately in a progressive disappearance of neurons from the inner and outer nuclear layers. The loss of photoreceptors results in blindness. Edema of the bulbus olfactorious is observed in naturally infected horses and in rats infected intranasally. These findings support the hypothesis that natural infection occurs via the nasal route.

## Prevention and Control

Although vaccines have shown efficacy in animal models, no vaccine is approved for use in veterinary medicine. Knowledge of the epidemiology of infection is critical to control viral diseases; nonetheless, studies in this vein have been initiated only recently. With the exception of

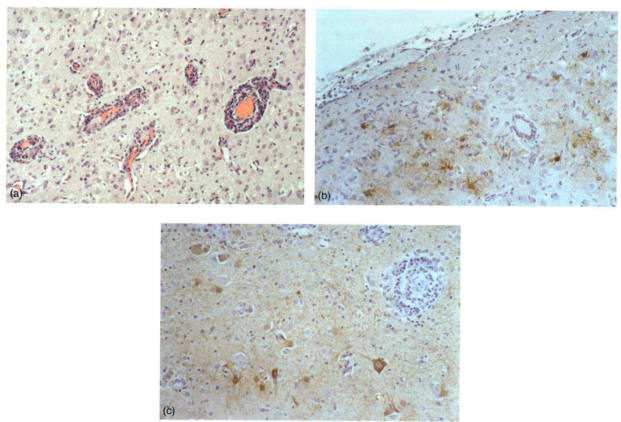

**Figure 3** Histology of natural cases of Borna disease. (a) Horse (severe inflammatory reaction; HE-staining); (b) sheep (inflammatory perivascular cuff and presence of viral p40 antigen; immunohistology counterstained); and (c) cattle (inflammatory reaction and presence of p40 antigen; immunohistology counterstained). Courtesy to Prof. T. Bilzer, Düsseldorf.

virus isolation or postmortem analysis of brain using immunohistochemical or *in situ* hybridization techniques, there are no standardized tests for BDV infection. Many investigators continue to use a labor-intensive and subjective immunofluorescence assay for serology in vitam diagnosis. However, more sensitive assays are gaining acceptance such as enzyme-linked immunosorbent assay (ELISA) in its various modifications (classical, capture, competitive, reverse-type sandwich) and Western blots. The advent of viral sequence information has allowed development of gene amplification technologies such as PCR and ligase chain reaction. While these methods are sensitive, results obtained have been complex to interpret. Analyses of some amplified sequences have revealed similarities to laboratory strains. Whether signal in these assays represents *bona fide* natural infection or contamination is unclear. Control of the disease will depend on identification of potential reservoirs of infection, including asymptomatic carriers.

## Future Perspectives

Of urgent importance is the refinement of reliable methods for routine diagnosis. These should include assays for the detection of BDV-specific antibodies as well as detection of the virus and viral gene products. Without such assays it is impossible to pursue rigorous epidemiology. Molecular biology characterization and pathogenesis also merit investment. This is a unique agent and investigating the mechanisms by which it maintains a broad host range, targets the nervous system, and establishes a persistent infection will almost certainly lead to seminal insights in microbiology, neuroscience, and medicine.

*See also:* Epidemiology of Human and Animal Viral Diseases.

## Further Reading

de la Torre JC (2002) Bornavirus and the brain. *Journal of Infectious Disease* 186(supplement 2): S241–S247.

Gonzalez-Dunia D, Volmer R, Mayer D, and Schwemmle M (2005) Borna disease virus interference with neuronal plasticity. *Virus Research* 111(2): 224–234.

Lipkin WI and Briese T (2006) Bornaviridae. In: Knipe DM and Howley PM (eds.) *Fields Virology*, 5th edn., pp. 1829–1851. Philadelphia: Lippincott Williams and Wilkins.

Lipkin WI and Hornig M (2004) Psychotropic viruses. *Current Opinion in Microbiology* 7: 420–425.

Planz O and Stitz L (2005) The role of Borna disease virus in neuropsychiatric disorders. In: Fatemi SH (ed.) *Infectious Etiologies of Neuropsychiatric Disorders*, pp. 123–134. Oxford: Taylor and Francis.

Staeheli P (2002) Bornaviruses. *Virus Research* 82(1–2): 55–59.

Stitz L, Bilzer T, and Planz O (2002) The immunopathogenesis of Borna disease virus infection. *Frontiers in Bioscience* 7: d541–d555.

# Bovine and Feline Immunodeficiency Viruses

**J K Yamamoto**, University of Florida, Gainesville, FL, USA

© 2008 Elsevier Ltd. All rights reserved.

## History

Since the isolation of human immunodeficiency virus (HIV) in 1983, the search for its counterpart in domestic and laboratory animals was initiated in a number of laboratories. Bovine immunodeficiency virus (BIV) was isolated in 1969 from a Louisiana dairy cow (called R29) during an intensive search for bovine leukemia virus (BLV). However, the viral sequencing of the frozen cell preparation from this cow was not performed until 1987, when this isolate ($BIV_{R29}$) was determined to be a lentivirus. Unlike the long history of BIV discovery, feline immunodeficiency virus (FIV) was isolated in 1986 from two laboratory cats inoculated with tissues derived from stray cats found in Petaluma, California. The stray cats were from a private cattery with high incidence of mortality and immunodeficiency-like syndrome. The isolated virus, called feline T-lymphotropic virus, resembled HIV in biochemical and morphological characteristics and was renamed FIV in 1988. Viral sequencing in 1989 confirmed this isolate (FIV-Petaluma, $FIV_{Pet}$) to be a lentivirus. BIV and FIV infections are prevalent worldwide and each has a distinctive impact on the survival of its host.

## Taxonomy and Classification

Based on nucleotide sequence analysis, both BIV and FIV belong to the genus *Lentiviruses* in the family *Retroviridae*. Phylogenetic tree analyses of Pol sequences indicate that BIV is related to caprine arthritis encephalitis virus (CAEV), whereas FIV is related to the equine infectious anemia virus (EIAV). Both BIV and FIV are distantly related to the primate lentiviruses, HIV and SIV. Serosurvey for BIV has identified a bovine lentivirus, called Jembrana disease virus (JDV), which causes an acute disease with 17% mortality in Balinese cattle. The full nucleotide sequence of JDV demonstrates that JDV is more closely related to BIV than to other lentiviruses.

FIV has been classified into five clades or subtypes (A–E) based on *env* and *gag* sequence analyses. Both serological and sequence analyses suggest that wild cats are infected with species-specific FIV. Eighteen of 37 nondomestic feline species, including lions, pumas, ocelots, leopards, cheetah, and Pallas cats (**Figure 1(b)**), are infected with species-specific FIV. Lion FIV-Ple and puma FIV-Pco are currently classified into three subtypes (A–C) and two subtypes (A, B), respectively. Based on *pol-RT* sequence, domestic FIV (FIV-Fca) is more closely related to puma FIV-Pco and ocelot FIV-Lpa than to other nondomestic FIV. Recent serological and *pol-RT* sequence analyses demonstrated species-specific FIV infection in Hyaenidae species (spotted hyena FIV-Ccr), the closest relative to the feline family.

## Geographic Distribution

The global distribution of BIV infection was determined by serological survey. BIV infection was found in Europe; North, Central, and South America; Asia; Middle East; Australia; New Zealand; and Africa (**Figure 1(a)**). In the USA, BIV infection was more prevalent in cattle herds from Southern states than those from Northern states. Standard serological survey has difficulty distinguishing BIV infection from JDV infection due to antibody cross-reactivity in the capsid protein (CA). JDV infection was reported on the island of Bali and neighboring districts in Indonesia.

A recent survey concluded that cattle in Australia have antibodies to a lentivirus antigenically more closely related to JDV than BIV.

The worldwide prevalence of FIV infection in domestic cats was determined by serosurvey and confirmed by sequence analysis (**Figure 1(b)**). In general, FIV subtypes are distributed according to the geographic location. Subtype B FIV was the predominant global subtype with higher prevalence in USA, Canada, eastern Japan, Argentina, Italy, Germany, Austria, and Portugal. Subtype A was the next predominant subtype with prevalence in western USA, western Canada, Argentina, Nicaragua, northern Japan, Australia, Germany, Italy, Netherlands, France, Switzerland, United Kingdom, and South Africa. Subtype C was detected predominantly in Vietnam,

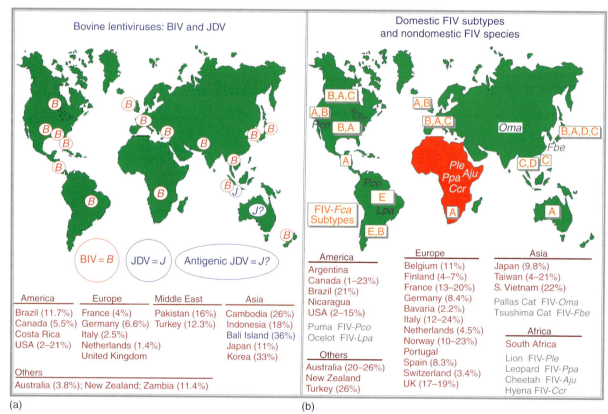

**Figure 1** Global distribution of (a) bovine lentiviruses and (b) domestic FIV subtypes and nondomestic FIV species. Seroprevalence rates of BIV, JDV, and domestic FIV-Fca (prevalence in high risk group) are described in the parentheses next to the corresponding country. Nomenclatures of the nondomestic FIV in wild cats and hyena are shown in the prevalent country or continent.

northern Taiwan, Japan, southern Germany, and British Columbia; subtype D in southern Japan and Vietnam; and subtype E in Brazil and Argentina.

## Virion and Genome Structure

BIV, JDV, and FIV are RNA viruses with two identical single-stranded RNA molecules of approximately 8.5, 7.7, and 9.5 kb per genome, respectively. The viral genomes, viral enzymes, and nucleoproteins (NC) involved in viral assembly are all packaged within the nucleocapsid core (**Figure 2(a)**). The mature core is shaped like a cone, typical of primate lentiviruses. Viral enzymes are protease (PR), reverse transcriptase (RT), and integrase (IN) encoded by the *pol* gene. FIV also produces a deoxyuridine triphosphatase (dUTPase or DU), which is unique to certain ungulate lentiviruses and important for replication in nondividing macrophages or resting T lymphocytes. The virus core, composed of CA, is surrounded by the matrix protein (MA), which coats the inner surface of viral envelope membrane. FIV has a myristoylated MA, similar to that of HIV and SIV. In contrast, BIV has a nonmyristoylated MA, typical of ungulate lentiviruses (EIAV, maedi-visna virus (MVV)). Myristoylation is important in targeting Gag to the inner surface of plasma membrane. The viral envelope (Env) consists of surface (SU) and transmembrane (TM) glycoproteins, projecting from the lipid bilayer derived from host plasma membrane.

BIV, JDV, and FIV have a genomic organization of $5'$-LTR-gag-pol-env-LTR-$3'$, like all retroviruses. The viral RNA is reverse-transcribed into double-stranded DNA in the cell cytoplasm, which are transported into the cell nucleus and then integrated into the host genome as viral provirus using viral IN. A genome-length mRNA, transcribed from the provirus, is translated into the Gag and Gag-Pol precursor polyproteins. These polyproteins are cleaved by the viral protease into MA, CA, and NC from Gag and PR, RT, DU (only FIV), and IN from Pol (**Figure 2(b)**). Similarly, the spliced *env* mRNA is translated into the Env precursor polyprotein, which is subsequently cleaved by a cellular protease into SU and TM. Like other lentiviruses, the BIV, JDV, and FIV genomes are more complex in organization than other retroviruses by containing regulatory genes that are found in the open reading frames (*orfs*) overlapping or flanking the $5'$ and $3'$ ends of the *env* gene. These viruses do not possess a *nef* gene or produce Nef protein. BIV *orfs* contain *tat, rev, vpy, vpw, vif,* and *tmx* genes but JDV *orfs* have no *vpy* and *vpw*. Tat is a spliced gene product that binds to viral transcripts to enhance transcribing polymerase activity. Rev is a spliced gene product that is important in production and transport of viral RNA transcripts. Vpy and Vpw are speculated to have similar function as HIV-1 Vpu and Vpr. Vpu is involved in increasing virion release, while

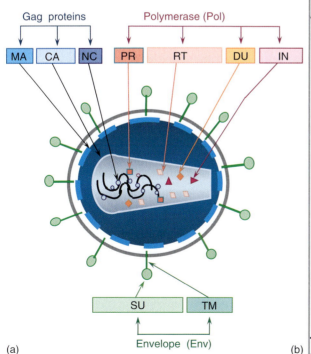

| Viral components | FIV | BIV |
|---|---|---|
| *Gag proteins* | | |
| Capsid (CA) | p24 | p26 |
| Matrix (MA) | p15 | p16 |
| Nucleoprotein (NC) | p7 | p13 |
| *Pol proteins* | | |
| Protease (PR) | p13* | p11* |
| Reverse Transcriptase (RT) | p61* | p64 |
| Integrase (IN) | p31 | p32* |
| *Env glycoproteins* | | |
| Surface Env (SU) | gp95–gp100 | gp100 |
| Transmembrane Env (TM) | gp39 | gp45 |
| *Regulatory proteins* | | |
| Regulator of expression of viral protein (Rev) | p23 | p23 |
| Virus infectivity factor (Vif) | p29 | Vif** |
| Transactivator (Tat) | ?*** | p14 |
| Virus protein y (Vpy) | None | p10* |
| Virus protein w (Vpw) | None | p7* |
| Transmembrane x (Tmx) | None | p19 |

\* Predicted value; ** Predicted; *** 79 aa Tat-like protein

**Figure 2** General structure of (a) BIV and FIV with (b) the size designations of structural proteins, enzymes, and regulatory proteins.

Vpr promotes the transport of the DNA pre-integration complex into the cell nuclei. Tmx has been detected in infected cells but its function remains unknown. FIV has four main *orfs* (*orf 1–4*). FIV *orf 1* appears to be analogous to the HIV-1 *vif* gene. Vif acts late in the viral life cycle and is needed for optimal production of virions. FIV *orf 2* (also called *orf A*) was speculated to be the first exon of the *tat* gene (encoding a 79 aa Tat-like protein) but recently was reported to be related to the *vpr* gene found in primate lentiviruses. FIV *orf 3* (3′ end of *orf A* or *orf 2*) and *orf 4* (*orf H*) are analogous to the exons of the *rev* gene found in HIV and SIV.

## Host Range and Virus Propagation

BIV has been isolated from dairy and beef cattle, and buffaloes. In addition, rabbits, goats, and sheep can be experimentally infected with BIV. BIV infects primary cells from bovine embryonic tissues (lung, brain, thymus, and spleen) and established cell lines of epithelial bovine trachea, canine fetal thymus, and embryonic rabbit epidermis, but does not infect human cell lines. Monocytes and macrophages are the major cell types infected by BIV. Recent studies show low-level infection in T cells and B cells from $BIV_{R29}$-infected cattle.

Domestic FIV has limited host range with productive infection found only in felid species. In contrast, nondomestic FIV sequence was recently identified in Hyaenidae species. Puma FIV-pco and lion FIV-Ple cause persistent infection in laboratory cats without clinical disease. Domestic FIV has been found in the wild cat population, Tsushima cat (*Felis bengalensis euptilura*), and may present problems for the conservation of this endangered species. FIV of domestic cats infects macrophages, endothelial cells, glial cells, B cells, $CD8^+$ T cells, and $CD4^+$ T cells. *In vitro* infection has been demonstrated in feline astrocytes, fibroblastic Crandell feline kidney cells, feline T cell lines, macaque peripheral blood mononuclear cell (PBMC), and low-level-to-defective infection in human cell lines. Experimental inoculation with domestic FIV grown in macaque PBMC resulted in FIV infection of macaques. However, there has been no evidence of zoonotic transmission and productive infection of humans with FIV.

## Genetics and Evolution

Due to the difficulty of isolating BIV from infected cattle, only a few full sequences, all from the USA (Louisiana, Florida, and Oklahoma), are available. Genetic variability of different USA strains shows as much as 10% variation in the *pol* and considerably more at the *env*. Genetic variability was also assessed by analyzing polymerase chain reaction (PCR)-amplified short sequences of *env* and *pol*. Phylogenetic analysis of the *SU* gene showed clustering of Asian isolates distinct from USA isolates. Moreover, inter- and intra-animal *env* variations were observed in $BIV_{R29}$-infected rabbits, suggesting the potential development of quasispecies. Only one Bali JDV isolate has been fully sequenced and therefore, little is known about the genetics of JDV.

Due to the wide genetic variability between worldwide isolates, FIV is classified into five subtypes (A–E). Studies on mutation rates in FIV *env* and *gag* show positive selective pressure for *env* mutations, consistent with reports that HIV *env* has a high mutation rate compared to *gag* and *pol*. PCR analyses of FIV variants isolated within individual cats over a 3 year period indicate that sequence variation in *env* increases over time, with later isolates showing divergence of 0.5–1.5%. Hence, the divergence of the variants within individual cats appears to be twofold less than those variants from an HIV-1-positive individual. Superinfections and recombinations have been reported for FIV much like those reported for HIV-1. Phylogenetic analyses suggest domestic FIV originated from a lineage associated with the wild cat lentivirus (see the section titled 'Taxonomy and classification').

## Serologic Relationships and Variability

Sera from BIV-infected cattle reacted to CA from HIV-1, SIV, EIAV, and JDV, and vice versa. Strongest serum cross-reactivity to CA was observed between BIV- and JDV-infected cattle. Sera from $BIV_{R29}$-infected cattle contained virus-neutralizing antibodies (VNAs) to $BIV_{R29}$ throughout 5 years post infection (pi), but these VNAs did not cross-neutralize Florida $BIV_{FL112}$. The *env* sequences between BIV and JDV and among BIV, HIV-1, SIV, and EIAV are less conserved. Consequently, cross-neutralization of HIV-1, SIV, and EIAV with sera from BIV-infected cattle is not expected.

Sera from cats infected with different FIV strains reacted to prototype $FIV_{Pet}$. The cross-reactivity of the infected sera has been the basis for the commercial FIV diagnostic kit. Sera from wild cats infected with nondomestic lentivirus cross-reacted predominantly to CA p24 and lesser degree to MA p15 of domestic FIV, and vice versa. Sera from HIV-1-positive subjects cross-reacted to domestic FIV p24 and vice versa. Rabbit polyclonal antibodies to ungulate lentiviruses, MVV and CAEV, also reacted to domestic FIV at p24 and p15, demonstrating the evolutionarily conserved epitopes on p24 and p15. In general, FIV-specific VNAs can neutralize closely related strains but not divergent strains. Preliminary studies suggest that a loose correlation exists between the genotype based on *env* sequence and the VNAs elicited by infected cats.

## Epidemiology

Global distribution of BIV infection in dairy and beef cattle has a seroprevalence of 1.4–33% (**Figure 1(a)**). Clinical signs were more frequently observed in BIV-infected dairy cattle than in BIV-infected beef cattle. This may be due to fewer stress factors in beef cattle than in dairy cattle. In some countries, BIV infection was higher in the dairy cattle than the beef cattle, because of the management practices (hand-feeding pooled colostrums/milk to dairy calves) and the longer lifespan (more risk of exposure) of the dairy cattle than the beef cattle. The global prevalence of FIV infection is 1–26% in the high-risk populations (symptomatic cats) (**Figure 1(b)**) and 0.7–16% in the healthy (minimal to no-risk) populations. FIV infection is found more frequently in cats >5 years of age than in younger cats and rarely in cats <1 year of age. Furthermore, free-roaming cats have the highest incidence of infection as compared to indoor cats. Male cats are 2–3 times more likely to be infected than female cats. Since male cats are territorially aggressive and have higher incidence of wounds and bite abscesses, the high prevalence of FIV infection in male free-roaming cats is consistent, with the major mode of FIV transmission being via biting.

## Transmission and Tissue Tropism

BIV has been detected in the spleen, liver, brain, lymph nodes, and PBMC of infected cattle. Based on PCR analysis of purified cell populations from $BIV_{R29}$-infected cattle, BIV is pantropic and infects $\alpha\beta$ T cells, $\gamma\delta$ T cells, B cells, and monocytes/macrophages. The cellular receptor for BIV is still unknown. Transmission of BIV is by exposure to contaminated blood, contaminated milk/colostrum, and possibly by sexual contact. *In utero* vertical transmission was reported. BIV was detected by PCR in embryo and occasionally in semen from infected cattle. Controversy exists about the BIV PCR detection in the semen and on the viability of BIV from the PCR-positive frozen semen and embryos, which are the sources for artificial insemination. Transmission by arthropod vector has been suggested but the evidence for this mode is lacking. Transmission by ingesting contaminated colostrum/milk was also suspected since BIV was detected in milk. Moreover, iatrogenic transmission (reused hypodermic syringe and equipment for pregnancy examination, de-horning, artificial insemination, and castration) of contaminated blood is suspected to be the main source of passing the infection to large numbers of uninfected cattle.

Although significant $CD4^+$ T-cell loss occurs during FIV infection, the feline CD4 molecule is not the receptor for FIV. The primary cellular receptor for domestic FIV was recently identified to be feline CD134, also called OX40. CD134 is a member of tumor necrosis factor receptor family that is transiently expressed on activated T cells. FIV cannot use human CD134 as receptor to infect human cells, but can use both feline and human CXCR4, a chemokine receptor, as co-receptor. It has been speculated that CXCR4 can also serve as primary cellular receptor for FIV. Based on experimental transmission studies, the major route of FIV transmission is through bites from infected cats. This transmission route is consistent with the epidemiological studies and with the fact that cats shed significant amounts of virus in saliva. FIV transmission by ingestion of virus via grooming and licking bleeding wounds of an infected cat cannot be excluded, since oral administration of infected blood can cause infection. FIV infection was demonstrated by experimental vaginal and rectal inoculation, and FIV was isolated from vaginal swab and semen of infected cats. Another route of transmission is by ingesting contaminated colostrum/milk of infected queens. Experimental studies suggest the possibility of transplacental infection and transmission during birth through a contaminated birth canal. In conflict with these experimental observations is the low incidences of FIV infection in cats <1 years of age and in kittens born to chronically infected queens, suggesting that vertical transmission of FIV may be rare in nature.

## Clinical Features and Infection

Experimental $BIV_{R29}$ infection of dairy cattle caused lymphocytosis, hypertrophic regional lymph nodes, and hypertrophic hemal lymph nodes, which was identical to the lymphoid changes observed in cow R29. Seroconversion and virus recovery also demonstrated BIV infection. Unlike cow R29, clinical disease was not observed in the experimentally infected cattle. Naturally infected dairy cows had lymphoid changes similar to the experimentally infected cattle, and a portion of the population displayed postparturition problems. These included foot problems, mastitis, diarrhea, pneumonia, neuropathy, and decreased milk production. Clinical signs in beef cattle were less frequent when compared to dairy cattle. Infected beef cattle grew normally and gained weight but displayed dullness, lumbering gait, and enlarged subcutaneous hemal lymph nodes. The low pathogenicity of BIV has been described in a number of studies. In contrast, JDV infection of Bali (*Bos javanicus*) cattle had pronounced clinical disease with mortality of 17% and was more pathogenic than BIV or JDV infection of Ongole (*Bos taurus*) and Friesian (*Bos indicus*) cattle.

The immunological hallmark of FIV infection is depletion of peripheral $CD4^+$ T cells and reduced CD4:CD8 ratios, leading to B- and T-cell dysfunctions and hypergammaglobulinemia. The clinical stages of FIV infection are similar to human AIDS in several ways. The acute stage of experimental FIV infection was characterized by immunological abnormalities followed by depression, fever, diarrhea, neutropenia, and persistent generalized lymphadenopathy. FIV was primarily detected in lymphoid tissues followed by dissemination of the virus into nonlymphoid organs. Both antibodies against FIV and virus recovery from PBMC persisted throughout infection. The FIV load in the blood was lower and $CD4^+$ T-cell decline was slower at the asymptomatic stage than acute stage. By late symptomatic stage, the animals were severely immunosuppressed and displaying wasting syndrome, neurological disorders, and persistent secondary opportunistic infections. The virus load was extremely high at this stage, and FIV was readily isolated from nonlymphoid tissues and organs, such as kidney, saliva, and central nervous system (CNS). Lymphomas with and without FIV proviral integration were reported in naturally and experimentally infected cats, including B-cell lymphoma of more unusual extranodal forms (predominantly in the neck and head). The major clinical manifestations observed in naturally infected cats were a progressively degenerative immune disorder, neurological disorders, wasting syndrome, and persistent secondary opportunistic infections. The specific signs were chronic oral diseases, chronic upper respiratory tract disease, chronic enteritis, chronic conjunctivitis, anorexia, fever of unspecified origin, and recurrent cystitis. Abnormal behavioral problems, lymphosarcoma, and myeloproliferative disease were observed in a small proportion of affected cats.

## Pathology and Histopathology

Experimental $BIV_{R29}$ infection of dairy cattle caused B-cell lymphocytosis, hypertrophic regional lymph nodes with lymphoid hyperplasia, and primary hypertrophic hemal lymph nodes with follicular lymphoid hyperplasia. The lymph nodes and hemal lymph nodes during natural infection showed hyperplasia followed by follicular exhaustion and dysfunction, and follicular hypoplasia with atrophy and central follicular depletion especially of T-dependent zones in severe cases. Infected cattle developed early atypical proliferation of lymphocytes in lymphoid tissue followed by recurrent opportunistic infections, loss of circulating monocytes, poor body condition, and weight loss. Encephalitis was diagnosed with lesions of lymphocytic infiltration into the meninges and perivascular spaces, and foci of microglia and astrocytosis. Overall, the primary lesions were in the brain, lymphoid tissue, and feet (lymphocyte infiltration of hoof tissues). Secondary diseases included nerve paralysis, persistent mastitis, septicemia, laminitis, secondary pododermatitis (plasmacytic inflammation), gastrointestinal diseases, bronchopneumonia, and subcutaneous and intramuscular abscesses.

Major histopathological changes during acute stage of experimental FIV infection were observed in the lymphoid tissues. In the first three weeks pi, lymphoid hyperplasia was observed in the lymph nodes, tonsils, spleens, and gut-associated lymphoid tissues. A majority of the FIV-infected cells were found in the germinal centers of lymphoid follicles of these tissues. Some infected cells were observed in the paracortex and medullary cords of lymph nodes, and in periarterial lymphoid sheaths and red pulp of the spleen. Shortly after, cats developed myeloid hyperplasia in the bone marrow and cortical involution, thymitis, and follicular hyperplasia of the medulla in the thymus. Both T cells and monocytes/macrophages were infected with FIV at an early phase of the acute stage, followed by infection of B cells. By early symptomatic stage, lymph nodes displayed follicular hyperplasia, involution, and lymphoid depletion, and by late phase, destruction of nodal architecture with involution and depletion of lymphoid follicles.

The CNS disease of FIV resembles those induced by HIV, and includes perivascular mononuclear cell infiltrates, glial nodules and diffuse gliosis of gray and white matter, and neuronal loss. Similar to HIV, neurotropic FIV strains infect microglia, astrocytes, and brain microvascular endothelial cells, but do not infect neuronal cells. Both anti-FIV antibodies and FIV virions have been isolated from cerebral spinal fluid (CSF) of infected cats, in addition to the elevated IgG index detected in the CSF. Like HIV, the level of FIV infection in the CNS cannot account for the cognitive/motor function abnormalities observed in these cats, suggesting that cytokines induced during CNS infection may play a key role in FIV neuropathogenesis. Neurological abnormalities included limb paresis, delayed righting and pupillary reflexes, behavioral changes, delayed visual and auditory evoked potentials, decreased spinal and peripheral nerve conduction velocities, and sleep abnormalities (e.g., increased awake time with decreased rapid eye movement) similar to sleep disturbances described in AIDS patients.

## Immune Response

Antibodies to p26 followed by antibodies to Env developed during early $BIV_{R29}$ infection, while a transient CD4/CD8 ratio decrease developed 2–7 weeks pi. This decrease was attributed to the greater $CD8^+$ cell increase than the slight $CD4^+$ cell increase. No significant $CD4^+$ cell or CD4/CD8 ratio decreases were observed throughout 5 years pi, even though persistent infection was demonstrated by the continued presence of TM-specific antibodies and VNAs. High VNA titers resulted in faster detection of virus, indicating high viral load. The role that VNAs play during natural pathogenesis of BIV is still unclear. It has been speculated that cellular immunity such as BIV-specific cytotoxic T lymphocytes (CTLs) may control the infection from progressing into clinical disease. However, decreased proliferation responses to T-cell mitogen and depressed T-dependent antibody responses to recall antigens at 4 and 5 years pi suggest loss in functional T-cell responses. In another study, B-cell lymphocytosis but no changes in $CD4^+$ cells and $CD8^+$ cells were observed in cattle infected with slightly pathogenic $BIV_{FL112}$. Although BIV infection does not cause $CD4^+$ cell loss, signs of T-cell dysfunction appear to develop upon prolonged infection.

Both the exposure dose and the FIV strain infecting the cats determine the nature of humoral and cellular immunity generated against the virus. Some strains do not elicit high or even significant VNAs even though high antibody titers to SU and TM are produced. Upon experimental infection with a moderate dose of FIV, anti-FIV antibodies were detected as early as 3 weeks pi, followed by VNAs starting 6–9 weeks pi. A decrease in primary proliferative response to only foreign antigen (none to T-cell mitogens and recall antigen) was observed at 5 weeks pi, the earliest time point tested. Hence, selective defects in primary antigen-specific response of naive $CD4^+$ T-helper cells are the early signs of T-cell dysfunction. Meanwhile, PBMC developed CTL responses to FIV Gag and Env at 7–9 and 16 weeks pi, respectively. The time of CD4/CD8 inversion depended greatly on the FIV strain used, with a moderate dose of pathogenic strains causing CD4/CD8 inversion as early as 4–6 weeks pi. In general, the $CD4^+$ T-cell loss accounted for the CD4/CD8 ratio inversion with most strains, but $CD8^+$ T-cell increases also contributed to this inversion in a number of strains. A defect in proliferative response of memory T cells to recall antigen developed at about 19 weeks pi. Decreased T-cell mitogen responses developed upon prolonged infection when considerable CD4/CD8 ratio inversion prevailed. B-cell dysfunctions were milder and were decreased primary antibody responses to T-dependent antigens and increased serum IgG levels, indicative of virus-specific B-cell hyperactivity. In addition, functional abnormalities in macrophages, neutrophils, and natural killer cells were observed in the FIV-infected cats.

## Prevention and Control

The control of BIV infection in animal food relies on the practice of testing and preventing the spread of BIV infection by decreasing iatrogenic transmission through improved management practices. Public health concern appears to be minimal, since there is no evidence of BIV zoonosis and the virus in milk is readily inactivated by pasteurization. The importance of BIV infection to the cattle industry depends on the severity of the economic losses resulting from lower milk production and poor beef quality. Similarly, the need for developing a BIV vaccine will hinge on the perception of the cattle industry rather than the significance this virus has as an animal model for AIDS.

Experimental FIV vaccine trials, ranging from inactivated single-subtype virus vaccine to proviral DNA vaccine, were performed with minimal-to-no success against homologous (identical to vaccine strains) and heterologous strains. A commercial dual-subtype FIV vaccine, consisting of inactivated subtype A and D viruses, was released in USA in 2002. This vaccine was effective against homologous strains, heterologous subtype A strains, and subtype B strains. This vaccine induced VNAs to closely related FIV strains from subtypes A and D. The vaccine protection against subtype B viruses was achieved in the absence of VNAs to challenge viruses, suggesting the importance of vaccine-induced cellular immunity. Both the prototype and commercial dual-subtype vaccines induced strong FIV-specific $CD4^+$ T-helper and $CD8^+$ CTL responses. Since there is no evidence of FIV zoonosis, the efforts to develop FIV-specific antiretroviral drugs have been limited. Only few antiretroviral drugs for HIV-1 therapy have been tested in infected cats. These include azidothymidine (AZT), 9-(2-phosphonylmethoxyethyl) adenine, dideoxycytidine 5′-triphosphate, dideoxycytidine, lamivudine (3TC), cyclosporine A, interferon-α, and commercial HIV-1 protease inhibitors. Prophylactic 2 week therapy with a high-dose AZT/3TC combination, started either 3 days before or on the day of FIV inoculation, resulted in 100% and 67% protection of cats from infection, respectively. Although the prophylaxis with nucleoside analogs was remarkable, the therapeutic use of the aforementioned drugs, including nucleoside analogs, was somewhat disappointing. Like HIV-1 drug therapy, FIV therapy will require multiple drug combination with each drug inhibiting different stages of FIV replication cycle.

## Future

The importance of BIV infection to the cattle industry is still unclear. The current strategy is to contain BIV infection by methodical testing and imposing management practices that prevent the spread of BIV. More information about the genetic and pathogenic evolution of BIV is needed to assess whether fatal pathogenic strains can evolve from current low-pathogenic strains. Overall, the policies set by cattle industry and government agencies will influence the extinction or survival of BIV infection in cattle.

FIV infection causes an important disease in domestic pet cats. The commercial vaccine was effective against strains from global subtypes A and B, and should be able to contain the global spread of FIV. However, the inability of current FIV diagnostics (enzyme-linked immunosorbent assay (ELISA) and immunoblot assay) to distinguish vaccinated cats from FIV-infected cats has caused a dilemma in the use of this vaccine. This problem can be resolved by developing sensitive molecular diagnostics or a vaccine that does not conflict with current FIV diagnostics. Identifying the protective vaccine epitopes should assist in designing a vaccine that is devoid of diagnostic epitopes. Moreover, FIV research on vaccines will provide new insights to HIV vaccine development for humans. The recent discovery of fatal pathogenic FIV strains (10% acute mortality) demonstrates the pathogenic evolution of FIV similar to HIV-1 immunopathogenesis. Hence, FIV infection is not only important for feline medicine but serves as an important small animal model for testing novel antiretroviral drugs, immunotherapy, and vaccine approaches for HIV/AIDS.

*See also:* Equine Infectious Anemia Virus; Human Immunodeficiency Viruses: Antiretroviral agents; Human Immunodeficiency Viruses: Molecular Biology; Human Immunodeficiency Viruses: Origin; Human Immunodeficiency Viruses: Pathogenesis; Simian Immunodeficiency Virus: Animal Models of Disease; Simian Immunodeficiency Virus: General Features; Simian Immunodeficiency Virus: Natural Infection; Visna-Maedi Viruses.

## Further Reading

Bachmann MH, Mathiason-Dubard C, Learn GH, *et al.* (1997) Genetic diversity of feline immunodeficiency virus: Dual infection, recombination, and distinct evolutionary rates among envelope sequence clades. *Journal of Virology* 71: 4241–4253.

Burkhard MJ and Dean GA (2003) Transmission and immunopathogenesis of FIV in cats as a model for HIV. *Current HIV Research* 1: 15–29.

Evermann JE, Howard TH, Dubovi EJ, *et al.* (2000) Controversies and clarifications regarding bovine lentivirus infections. *Journal of the American Veterinary Medical Association* 217: 1318–1324.

Podell M, March PA, Buck WR, and Mathes LE (2000) The feline model of neuroAIDS: Understanding the progression towards AIDS dementia. *Journal of Psychopharmacology* 14: 205–213.

Snider TG, Hoyt PG, Jenny BF, *et al.* (1997) Natural and experimental bovine immunodeficiency virus infection in cattle. *The Veterinary Clinics of North America. Food Animal Practice* 13: 151–176.

St-Louis M-C, Cojocariu M, and Archambault D (2004) The molecular biology of bovine immunodeficiency virus: A comparison with other lentiviruses. *Animal Health Research Reviews* 5: 125–143.

Troyer JL, Pecon-Slattery J, Roelke ME, *et al.* (2005) Seroprevalence and genomic divergence of circulating strains of feline immunodeficiency virus among Felidae and Hyaenidae species. *Journal of Virology* 79: 8282–8294.

Uhl EW, Heaton-Jones TG, Pu R, and Yamamoto JK (2002) FIV vaccine development and its importance to veterinary and human medicine: A review. *Veterinary. Immunology and Immunopathology* 90: 113–132.

Wilcox GE, Chadwick BJ, and Kertayadnya G (1995) Recent advances in the understanding of Jembrana disease. *Veterinary Microbiology* 46: 249–255.

# Bovine Ephemeral Fever Virus

**P J Walker,** CSIRO Livestock Industries, Geelong, VIC, Australia

© 2008 Elsevier Ltd. All rights reserved.

## Glossary

**Epicardium** The outer layer of heart tissue.
**Hypocalcemia** A low level of calcium in the circulating blood.
**Leucopenia** A decreased total number of white blood cells in the circulating blood.
**Pericardial fluid** Fluid within a double-walled sac that contains the heart and the roots of the great blood vessels.
**Sternal recumbency** Reclined in a position of comfort on the chest bone.
**Synovial membranes** Connective tissue membranes lining the cavities of the freely movable joints.
**Thoracic fluid** Fluid in the chest cavity.

## Introduction

Bovine ephemeral fever virus (BEFV) is an arthropod-borne rhabdovirus that causes a disabling and sometimes lethal disease of cattle (*Bos taurus, Bos indicus,* and *Bos javanicus*) and water buffaloes (*Bubalus bubalis*). Unapparent infections can also occur in cape buffalo, hartebeest, waterbuck, wildebeest, deer, and possibly goats. Bovine ephemeral fever (BEF) was first recorded in East Africa and Egypt during the late nineteenth century. However, BEF (which is also variously called three-day sickness, bovine enzootic fever, bovine influenza, and stiffsiekte) is thought to have been endemic since antiquity in much of tropical and subtropical Africa, and Asia. As the name suggests, BEF is often characterized by the rapid onset of and recovery from clinical signs that can include a bi- or multiphasic fever, ocular and nasal discharge, muscle stiffness, anorexia, rumenal stasis, lameness, and sternal recumbency. Although mortality rates rarely exceed 1–2%, a particularly severe outbreak in Taiwan in 1996 has been reported to have resulted in the death or culling of 11.3% of a population of 110 247 dairy cattle on 516 farms. Severe infections commonly occur in larger, more valuable animals. Morbidity rates may approach 100% with significant economic impacts including loss of milk production, temporary infertility in bulls, abortion, loss of condition in beef herds, and disablement of draft animals at the time of harvest. The economic consequences of BEF can be significant. An outbreak of BEF in Israel in 1999 has been estimated to have cost, on average $280 per lactating cow and $112 per nonlactating cow. In Australia, sweeping epidemics in the 1970s have been estimated to have caused industry-wide losses exceeding $200 million in today's values. Due to limitations on the importation of livestock and semen from animals with evidence of BEFV infection, the disease can also have significant impact on international trade.

## Taxonomic Classification

BEFV is classified in the order *Mononegavirales*, family *Rhabdoviridae* as the type species of the genus *Ephemerovirus*. Ephemeroviruses have morphological and genetic characteristics common to all rhabdoviruses, including an enveloped, bullet-shaped virion containing a nonsegmented, negative-sense, single-stranded (ss) RNA genome. However, unlike viruses classified in other established rhabdovirus taxa, ephemeroviruses share the unusual characteristic of two type 1 transmembrane glycoproteins (G and $G_{NS}$) that are related in amino acid sequence and appear to have arisen by gene duplication. Other recognized species in the genus include *Berrimah virus* (BRMV) and *Adelaide River virus* (ARV). These viruses have each been isolated from cattle in northern Australia and also appear to be transmitted by biting insects. Tentative species in the genus include Kimberley virus (KIMV), Malakal virus (MALV), and Puchong virus (PUCV), each of which is closely related serologically to BEFV. KIMV has been isolated from *Culex annulirostris* mosquitoes and *Culicoides brevitarsis* midges in Australia, and from healthy cattle. MALV was isolated in Sudan and PUCV in Malaysia, each from *Mansonia uniformis* mosquitoes. Potential vertebrate hosts for these viruses are yet to be identified. Although BEFV is the only ephemerovirus known to be associated with disease, Kotonkan virus (KOTV), which was originally isolated from *Culicoides* spp. in Nigeria, does cause an ephemeral fever-like illness in cattle. Recent phylogenetic studies using the N gene and segments of the L gene suggest that KOTV, as well as Obodhiang virus (OBOV) which was isolated from *Mansonia uniformis* mosquitoes in Sudan, should also be classified as ephemeroviruses. It has also been suggested that all arthropod-borne rhabdoviruses, including the ephemeroviruses, vesiculoviruses, and a large number of other unclassified rhabdoviruses share a common ancestor and should form a larger taxonomic group for which the name 'dimarhabdovirus' (dipteran-mammalian rhabdovirus) has been proposed (**Figure 1**).

## Geographic Distribution

BEFV is known to occur in most tropical and subtropical regions of Africa, Asia, and Australia. The disease occurs throughout much of Africa and in Asian countries south of a line that includes Israel, Iraq, Iran, Syria, India, Pakistan, Bangladesh, southern and central China, and east to Taiwan, and southern Japan. It occurs throughout Southeast Asia, parts of New Guinea, and in most of northern and eastern Australia. It does not occur in New Zealand or the islands of the Pacific, or in the Americas, or Europe, where it is considered an important exotic pathogen. There is reported serological evidence of infection in southern Russia but the disease has not been described. The distribution of BEFV is determined by the geographic range of available insect vectors and is limited by international trade restrictions on live animals and semen showing evidence of infection.

## Epizootiology

Bovine ephemeral fever is a seasonal disease. It occurs principally in the summer and early autumn, and with the onset of the monsoon season in Asia. Outbreaks are usually associated with periods of high rainfall which precipitate the emergence of insect vectors in large numbers. BEFV can also spread in epizootics that follow the pattern of prevailing winds with a general southward movement in the Southern Hemisphere and northward movement above the Equator. Wind-borne movement of insect vectors is the likely mechanism of spread. Vectors include biting midges (*Culicoides* spp.) and mosquitoes, in which the virus replicates. BEFV has been isolated from *Culicoides brevitarsis*, *Culicoides coarctatus*, *Anopheles bancroftii*, a mixed pool of mosquito species in Australia, and from a mixed pool of biting midges in Africa. The virus has also been recovered from several other species of biting midge and mosquito following experimental infection. The abundance and distribution of insects from which BEFV has been isolated suggests that several major vectors may be involved in transmission. There is no evidence of direct transmission of BEFV between cattle, even when encouraged by smearing nasal or ocular discharges on mucosal surfaces.

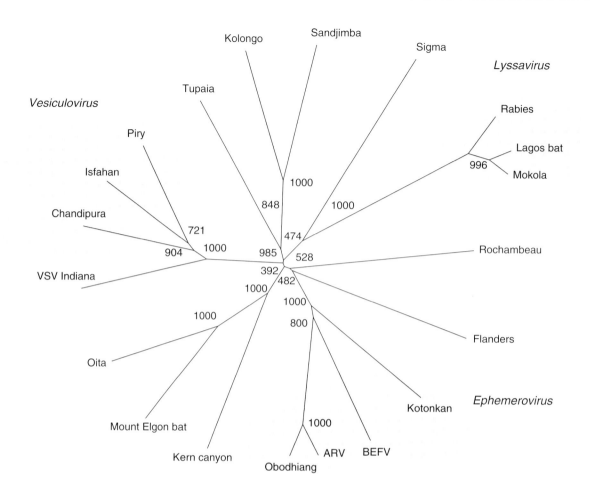

**Figure 1** Phylogenetic tree of partial N gene sequences of 20 rhabdoviruses infecting mammals and/or insects. The tree was generated from a Clustal X alignment of the sequences by the neighbor-joining method and presented graphically using Treeview software. The viruses include representatives of three rhabdovirus genera: *Ephemerovirus* (BEFV, ARV, Obodhiang virus, and Kotonkau virus); *Vesiculovirus* (VSV Indiana, Chandipura, Isfahan, and Piry viruses); and *Lyssavirus* (rabies, Mokola, and Lagos bat viruses). Unclassified rhabdoviruses include Sandjimba, Rochambeau, Flanders, Kern Canyon, Mount Elgon bat, Oita, Tupaia and Kolongo, and sigma viruses. Confidence in branch nodes was determined by bootstrap analysis on 1000 replicates (indicated as numbers).

## Pathology and Pathogenesis

Bovine ephemeral fever is principally an inflammatory disease. The incubation period is normally 2–4 days. Viremia usually persists for 1–3 days and peaks approximately 24 h before the onset of fever. The initial sites of infection are not known but the virus has been isolated from neutrophils and reticuloendothelial cells of the lungs, spleen, and lymph nodes. There is also evidence of infection in synovial membranes, epicardium and aorta, and in cells derived from synovial, pericardial, thoracic, and abdominal fluids. There is not widespread tissue damage. The primary lesion is a vasculitis affecting the endothelium of small vessels of synovial membranes, tendon sheaths, muscles, facia, and skin. The onset of fever and other clinical signs is accompanied by marked leucopenia, relative neutrophilia, elevated plasma fibrinogen, and elevated levels of cytokines including interferon α, interleukin 1, and tissue necrosis factor. There is also a significant hypocalcemia that is thought to be responsible for sternal recumbency. The major clinical signs can be treated very effectively with anti-inflammatory drugs.

## Virion Structure and Morphogenesis

BEFV virions are enveloped, bullet-shaped particles (approximately $70 \times 180$ nm) containing a precisely coiled, helical nucleocapsid with 35 cross-striations at an interval of 4.8 nm. Virions have a prominent axial channel intruding from the base and typically are more cone-shaped than commonly observed for viruses in other genera of animal rhabdovirus (e.g., lyssaviruses and vesiculoviruses). The envelope contains a single 81 kDa class

1 transmembrane glycoprotein (G) that forms visible projections on the virion surface. The G protein mediates cell attachment and entry, is the target for virus-neutralizing antibodies, and induces protective antibodies in cattle. Nucleocapsids contain a negative-sense ssRNA genome, a 52 kDa nucleoprotein (N) which is tightly bound to the genome, a large 250 kDa replicase protein (L), and a highly charged 43 kDa replicase cofactor (P). Virions also contain a 29 kDa matrix protein (M) which is a major structural component and appears to lie between nucleocapsids and the inner surface of the lipid envelope (**Figure 2**).

Viral replication is cytoplasmic and morphogenesis occurs primarily at the plasma membrane in association with accumulations of a filamentous, granular, intracytoplasmic matrix. However, late in infection, there is a proliferation of plasma membrane, cells become highly vacuolated, and virions are observed both at the plasma membrane and within cytoplasmic vacuoles. Following budding as cone-shaped extrusions, virions accumulate in extracellular spaces. The general characteristics of viral morphogenesis appear to be similar in infected mammalian cell cultures and mouse neurons.

## BEFV Genome Organization and the Encoded Proteins

The 14 900 nt BEFV genome is the largest known for any rhabdovirus and one of the largest and most complex genomes for any nonsegmented, negative-sense RNA virus. The genome comprises 12 genes, flanked by terminal, partially complementary leader ($l$) and trailer ($t$) sequences, and arranged in the order $3'$-$l$-N-P/C-M-G-$G_{NS}$-$\alpha 1/\alpha 2/\alpha 3$-$\beta$-$\gamma$-L-$t$-$5'$. By analogy with other rhabdoviruses, the $3'$ leader (50 nt) and $5'$ trailer (70 nt) sequences are likely to have important roles in the initiation and control of replication and transcription, and in nucleoprotein assembly and packaging. The structural protein genes (N, P, M, G, and L) are arranged in the same order as for all other known rhabdoviruses and encode proteins with similar functional characteristics (**Figure 3**).

### The N Protein

The BEFV N gene encodes the 431 amino acid nucleoprotein (N). The N protein is a highly hydrophilic, RNA-binding protein containing 14.4% basic residues

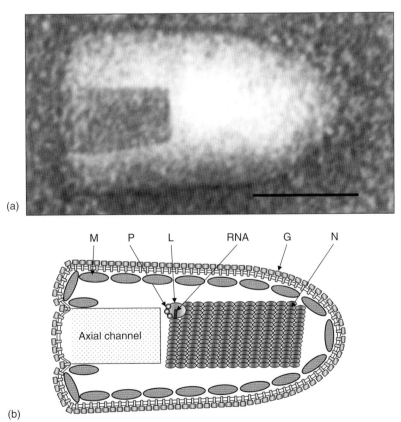

**Figure 2** (a) Negative-contrast electron micrograph of a purified BEFV virion. (b) Schematic representation of the BEFV virion. Structural proteins N, P, M, G, and L, and the negative-sense ssRNA genome are indicated. The axial channel is also depicted. The size and relative quantities of the proteins do not accurately reflect the content in virions. Scale = 50 nm (a).

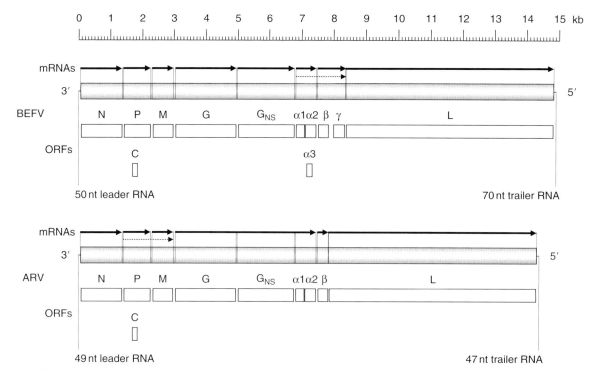

**Figure 3** Illustration of the genome organization and transcription strategy of BEFV and ARV. Solid arrows indicate the major transcriptional products. Minor transcripts are indicated by dotted arrows. The length of the ARV L protein is estimated from the alignment of available sequences with those of BEFV.

(lysine, arginine, and histidine) distributed relatively evenly throughout the molecule. Seven of these residues are highly conserved in rhabdoviruses and are involved in RNA-binding and stabilizing the interaction. The N protein also contains 14.6% acidic residues (glutamate and aspartate) that are less evenly distributed with significant clustering in a short domain near the C terminus. A similar acidic domain in the rabies N protein is a phosphorylation site involved in binding P protein to nucleocapsids. The BEFV N protein is also phosphorylated when packaged in virions and contains a similar phosphorylation site in this acidic domain.

### The P Protein

The P gene encodes the 278 amino acid, highly hydrophilic P protein. It corresponds to the polymerase-associated phosphoproteins of rabies virus (RV) and vesicular stomatitis virus (VSV) which are components of nucleocapsids and act as essential cofactors to the L protein during transcription and replication. The BEFV P protein has not been observed to be phosphorylated when extracted from virions but is phosphorylated when expressed from a recombinant baculovirus in insect cells. The BEFV P gene also contains an alternative open reading frame (ORF) encoding a 48 amino acid, highly basic 5.8 kDa polypeptide. This protein has not been detected in BEFV-infected cells and it is not known if it is expressed. However, alternative ORFs in the P gene occur in vesiculoviruses and are a common feature of many viruses in the *Mononegavirales*. As in VSV, the BEFV C protein has two potential initiation codons, suggesting it could be expressed in two different forms.

### The M Protein

The M gene encodes the 691 amino acid, basic, hydrophilic protein that corresponds to the matrix protein (M) of rabies and VSV. The M protein is a major component of rhabdovirus virions and has been shown to have important functions in regulation of viral replication and transcription, inhibition of host cell protein synthesis and induction of apoptosis, and in budding of nucleocapsids at cytoplasmic membranes. The BEFV M protein has been shown to be phosphorylated when extracted from virions but not when expressed from a recombinant baculovirus in insect cells. This is consistent with observations that phosphorylation of the VSV M protein occurs at a late stage in viral assembly. The BEFV M protein also contains a 'late domain' sequence motif (PPSY) which, in VSV and several other RNA viruses, is essential for efficient budding from infected cells.

### The G Protein

The BEFV G gene encodes the 623-amino-acid virion transmembrane glycoprotein (G). The G protein is a class 1 membrane protein. It shares significant amino acid sequence identity with other animal rhabdovirus G proteins and contains a core of conserved cysteine residues suggesting preservation of fundamentally similar secondary structure. The G protein contains five potential N glycosylation sites, three of which appear to align with similar sites in VSV and/or rabies virus. The G protein is responsible for cell docking and entry, and is the target of virus-neutralizing antibodies for which the major binding sites have been defined (see below).

## The L Protein

The L gene encodes the 2144-amino-acid RNA-dependent RNA polymerase (L protein). The L protein is a structural component of nucleocapsids that, in cooperation with the N and P proteins, forms the ribonucleoprotein (RNP) complex that is responsible for replication and transcription of the viral genome. The BEFV L protein shares a high level of sequence similarity with other rhabdovirus L proteins and contains all of the conserved sequence motifs associated with the major functional domains; these include the polymerase catalytic site and other regions involved in replication, transcription, initiation, elongation, and termination, 3′ polyadenylation, 5′ capping, and cap methylation.

## The $G_{NS}$ Protein

The BEFV nonstructural protein genes are located in a 3442 nt region of the genome between the G gene and L gene. Immediately downstream of the G gene is a second gene encoding a class 1 transmembrane glycoprotein ($G_{NS}$). The 90 kDa $G_{NS}$ protein is abundant in infected cells but has not been detected in virions. It is related in structure and sequence to the BEFV virion G protein and other rhabdovirus G proteins and the evidence suggests that it has arisen by gene duplication. The $G_{NS}$ protein contains eight potential N-glycosylation sites and, as the size by sodium dodecyl sulfate-polyacrylamide gel electrophoresis (SDS-PAGE) is approximately 21 kDa in excess of the calculated molecular weight of the unmodified polypeptide, it appears to be highly glyscosylated. Sequence alignments indicate that it shares 10 of 12 cysteine residues that appear to form the core of secondary structure common to all animal rhabdoviruses. However, the $G_{NS}$ protein does not share antigenic sites with the BEFV G protein and antibody to $G_{NS}$ does not neutralize the infectivity of BEFV produced in either mammalian or insect cells. It has been shown to accumulate at the cell surface in association with amorphous structures but not with budding or mature virions. The function of the $G_{NS}$ is currently unknown.

## The Small Nonstructural Proteins

Downstream of the $G_{NS}$ gene is a complex region of the genome-encoding proteins that appear to be unique to BEFV and other ephemeroviruses. None of these proteins has yet been detected in virions or BEFV-infected cells. The α-gene coding region contains three long ORFs (α1, α2, and α3). The α1 ORF encodes an 88 amino acid, 10.6 kDa protein. It features a central transmembrane domain comprising 16 hydrophobic amino acids bounded by arginine residues, and a highly basic C-terminal domain in which 12 of 18 amino acids are lysine or arginine residues. This structure suggests α1 may function as a viroporin, a class of proteins that causes cytopathic effects by increasing membrane permeability. The BEFV α1 protein has been shown to be cytotoxic when expressed in insect cells from a recombinant baculovirus vector. The α2 ORF encodes a 116 amino acid, 13.7 kDa polypeptide. It overlaps the α3 ORF which encodes a 51 amino acid, 5.7 kDa polypeptide that contains an unusual triple repeat of isomers of the sequence KLMEE at intervals of four residues. The β-gene encodes a 107 amino acid, 12.3 kDa polypeptide. The γ-gene encodes a 114 amino acid, 13.5 kDa polypeptide. The α2, α3, β, and γ products share no sequence homology with known viral proteins (other than ARV, see below) and their functions are yet to be determined.

## BEFV Transcription

The RNP complex, comprising the RNA genome and the N, L, and P proteins, is the active replication and transcription unit of the virus. As for other rhabdoviruses, BEFV transcription from the (−) RNA genome generates 5′ methylated, capped, and polyadenylated mRNAs by a progressive mechanism that initiates and terminates at short, conserved sequences flanking each gene. For each of the structural protein genes (N, P, N, G, and L), and the nonstructural glycoprotein gene ($G_{NS}$), transcription initiates at the sequence UUGUCC and terminates at the polyadenylation signal GUAC $[U]_7$. Transcription of the α-, β-, and γ-coding regions is more complex. The α-coding region is translated as an α1-α2-α3 polycistronic mRNA that initiates at UUGUCC but terminates at the variant polyadenylation signal GUUC $[U]_7$. This variant signal appears to cause incomplete termination and partial read-through of a longer tri-cistronic α-β-γ mRNA. The β gene is also immediately followed by a variant polyadenylation signal (GUAC $[U]_6$). However, the truncated ($U_6$) palindrome does not allow transcription termination, and reinitiation does not occur at a UUGUCC sequence located immediately in advance of the γ gene. As a result, the β- and γ-coding regions are transcribed as a bi-cistronic mRNA that initiates at UUGUCC upstream of the β ORF and terminates at the functional GUAC $[U]_7$

polyadenylation signal downstream of the γ ORF. This polyadenylation signal overlaps the L gene initiation sequence by 21 nt, requiring an upstream repositioning of the polymerase to commence L gene transcription. A similar arrangement for L gene transcription has been observed for several other nonsegmented (−) RNA viruses.

## ARV Genome Organization and Transcription

The genome organization and transcription strategy of ARV are similar to those of BEFV. However, there are subtle differences that reveal aspects of genome evolution and the control of gene expression in ephemeroviruses. The ARV genome includes the five structural protein genes that are common to all rhabdoviruses (N, P, M, G, and L). Each of the proteins encoded in these genes is similar in size and shares a high level of sequence identity with the corresponding BEFV proteins. Like BEFV, the ARV P gene also contains an alternative ORF that encodes a basic protein of similar size (7.4 kDa) to the BEFV C protein. The ARV 3′ leader sequence (49 nt) is similar in size to the BEFV leader RNA and shares a high level of sequence identity (21/22 nt) in the U-rich terminal domain. The ARV 5′ trailer sequence (47 nt) is shorter than the BEFV trailer RNA (70 nt), primarily due to the absence of a 26 nt direct repeat of the BEFV leader sequence that occurs in the BEFV trailer RNA. The function (if any) of this direct repeat is not known. The A-rich 5′ terminal region of the ARV trailer RNA shares only moderate sequence identity (15/21 nt) with the BEFV trailer, and is partially complementary (18/21 nt) to the U-rich ARV 3′ leader. This complementarity reflects the specificity of interaction of the polymerase with both the (−) RNA genome and (+) RNA antigenome during replication.

Like BEFV, the ARV genome encodes a second, class 1 transmembrane glycoprotein ($G_{NS}$) immediately downstream of the G gene. ARV $G_{NS}$ is also nonstructural and shares significant amino acid sequence identity with the BEFV $G_{NS}$ protein, as well as the G proteins of BEFV and other animal rhabdoviruses. ARV $G_{NS}$ has eight potential N-glycosylation sites, four of which appear to align with sites in BEFV $G_{NS}$. There is a high level of preservation of cysteine and proline residues with BEFV $G_{NS}$, suggesting a similar folded secondary structure. The ARV and BEFV $G_{NS}$ glycoproteins also appear to have preserved a core of cysteine residues conserved in rhabdovirus G proteins and is crucial for maintaining a fundamentally similar secondary structure. A proline-rich motif that, in VSV, forms a crucial 'P' helix in the membrane fusion domain, is also present in the BEFV and ARV G proteins. The 'P' helix motif is also present in BEFV $G_{NS}$ but it is absent from the ARV $G_{NS}$ protein. However, unlike the BEFV G protein, BEFV $G_{NS}$ does not induce cell fusion when expressed in insect cells from a recombinant baculovirus vector. The biological significance of these observations will not be clear until further studies are conducted to better define the functions of ephemerovirus $G_{NS}$ proteins.

Between the $G_{NS}$ and L genes, ARV also contains a complex region encoding several proteins of uncertain function. The genes in this region are arranged in the order −$G_{NS}$-α1/α2-β-L-t-5′. The α1 ORF encodes a membrane-spanning, nonstructural protein with a highly basic C-terminal domain which is similar to the BEFV α1 protein and may well also function as a viroporin. The ARV α2 and β ORFs encode proteins similar in size to the corresponding BEFV proteins but, although the overall sequence similarity is relatively high, there is no significant sequence identity. A 17 kDa protein reported in purified ARV virions is similar in size to that predicted for the β protein. The α3 and γ ORFs are not present in the ARV genome.

The ARV transcription strategy is somewhat different from that of BEFV. Only the N-gene, L-gene, and β-gene are transcribed solely as monocistronic mRNAs. For the N-gene and L-gene, transcription initiates and terminates at standard UUGUCC and GUAC $[U]_7$ signals flanking each gene. Transcription of the β-gene initiates at the variant UUGUCU sequence and terminates at GUAC $[U]_7$. The P-gene and the M-gene are transcribed both as high abundance monocistronic mRNAs and a lower abundance (approximately 10%) bicistronic P/M mRNA. Each initiates at UUGUCC and the M-gene terminates at GUAC $[U]_7$. However, the P-gene terminates at the leaky variant signal GCAC $[U]_7$. The G-, $G_{NS}$-, and α-genes are transcribed primarily as a long polycistronic mRNA that initiates at UUGUCG upstream of the G-gene and terminates at GUAC $[U]_7$ following the α-gene. Corrupted termination/polyadenylation signals following the G-gene (GUAC $[U]_4$C $[U]_2$) and the $G_{NS}$-gene (GUGC $[U]_2$C $[U]_4$) appear to allow a very low level of termination and transcription initiation at UUGUCC signals immediately preceding both the $G_{NS}$- and α-genes. As for the BEFV γL junction, there is an overlap (22 nt) of the β-L gene junction in ARV, highlighting the importance of polymerase repositioning in the control of L gene expression. There is also a high level of nucleotide sequence identity between the α-β and β-γ gene junctions in BEFV, and between the ARV β-L gene junction and the BEFV γL gene junction. This suggests that the BEFV γ-gene may have evolved as a consequence of β-gene duplication. It appears, therefore, that gene duplication may have an important role in ephemerovirus evolution.

## Antigenic Variation

As defined either by cross-protection experiments in cattle, or by cross-neutralization tests in mice, or in cell cultures, BEFV exists as a single serotype globally. The relative antigenic stability of BEFV is most likely due to the occurrence of viremia and vector-borne transmission several days prior to the appearance of significant levels of virus-neutralizing antibodies. The BEFV virion G protein is the target of neutralizing antibody and four major neutralization sites have been identified and mapped to the amino acid sequence. Antigenic site G1 is a linear site that maps as two minimal B-cell epitopes at each end of the sequence spanning amino acids $Y^{487}$ to $K^{503}$ in the 'stem' domain of the G protein. This domain appears to be a unique feature of ephemeroviruses. Antigenic site G2 is conformational. It is located adjacent to two cysteine residues ($C^{172}$ and $C^{182}$) that appear to form a disulfide bridge linking a tight glycosylated loop structure in the folded G protein. Site G3 is the major conformational site comprising two partially overlapping elements (G3a and G3b). The site encompasses three different domains of the cysteine-rich 'head' structure of the folded G protein spanning $Q^{49}$ to $D^{57}$, $K^{215}$ to $E^{229}$, and $Q^{265}$. Similarly complex antigenic sites map to corresponding regions of other animal rhabdoviruses, again supporting the view that essential elements of G protein secondary structure are preserved. Site G4 is a linear site. It has not yet been mapped to the G-protein sequence but it is known to be conserved in BRMV and KIMV which are also neutralized by site G4 monoclonal antibodies. (Amino acid residues are numbered here to include the N-terminal signal peptide that is cleaved during maturation of the G protein; **Figure 4**).

Limited natural antigenic variation has been reported between BEFV isolates. Variations in sites G3a and G3b have been identified among 70 Australian BEFV isolates collected from diverse locations between 1956 and 1992. There appears to be a temporal basis for the shift in site G3a which is present in most strains isolated after 1973. Comparison of prototype Australian and Chinese BEFV isolates has also indicated variations in site G3a. In Taiwan, variations have been reported in sites G1 and G3. The pattern of amino acid substitutions indicates that the isolates cluster into those which included the 1984 Taiwanese vaccine strain (Tn73) and those which were isolated after 1986. It is possible that incomplete protection provided by available BEF vaccines is contributing to antigenic instability in the G protein.

## Immune Response and Vaccination

Natural BEFV infection induces a strong neutralizing antibody response and apparently durable immunity. Following experimental infection, neutralizing IgG antibody appears 4–5 days after the onset of clinical signs and peaks within 1–4 weeks. Although there are some reports that cattle with high levels of neutralizing antibody can be susceptible to experimental challenge, other evidence

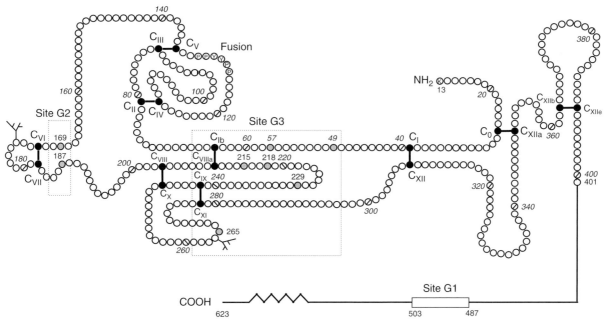

**Figure 4** Schematic illustration of the structure of the BEFV G protein showing the locations of the major neutralization sites (G1, G2, and G3) and the predicted fusion domain including the highly conserved poly-proline helix (PPYYPP). Amino acids known to be located in the major antigenic sites are indicated as shaded circles. Disulfide bridges are assigned according to previous predictions from sequence alignments with other rhabdoviruses and from the known crystallographic structure of the low-pH form of the VSV G protein. Amino acids are numbered from the first residue of the translated protein, prior to removal of the N-terminal signal peptide.

suggests a good correlation between protection and neutralizing antibody. Colostral antibody has also been shown to protect calves against experimental challenge. High levels of cytokines circulate during the acute phase of infection but little is known of the role of innate or adaptive cell-mediated immunity in recovery from infection or protection against natural or experimental challenge.

Several forms of live-attenuated, inactivated, subunit, and recombinant BEFV vaccines have been reported and vaccines of varying format are produced for commercial use. Live-attenuated vaccines have been produced in mice and in cell cultures. In general, live vaccines are relatively effective in inducing protection but require at least two doses in adjuvant to generate durable immunity. Inactivated vaccines have been produced by treatment of BEFV with formalin or β-propiolactone, but have generally poor efficacy. Consecutive vaccinations with live-attenuated and killed preparations have also been used with some success. A purified G-protein subunit vaccine delivered in Quil A adjuvant has been shown to provide reliable protection following a two-dose treatment at an interval of 21 days. Recombinant BEFV vaccines employing the BEFV G protein delivered in vaccinia and capripox viral vectors have also been trialed.

*See also:* Animal Rhabdoviruses; Chandipura Virus; Fish Rhabdoviruses; Rabies Virus; Vesicular Stomatitis Virus.

## Further Reading

Inaba Y, Kurogi H, Takahashi A, et al. (1974) Vaccination of cattle against bovine ephemeral fever with live attenuated virus followed by killed virus. *Archiv fur die Gesamte Virusforschung* 44: 121–132.

Kirkland PD (2002) Akabane and bovine ephemeral fever virus infections. *Veterinary Clinics of North America: Food Animal Practice* 18: 501–514.

Kongsuwan K, Cybinski DH, Cooper J, and Walker PJ (1998) Location of neutralizing epitopes on the G protein of bovine ephemeral fever rhabdovirus. *Journal of General Virology* 79: 2573–2578.

Kuzmin IV, Hughes GJ, and Rupprecht CE (2006) Phylogenetic relationships of seven previously unclassified viruses within the family Rhabdoviridae using partial nucleoprotein gene sequences. *Journal of General Virology* 87: 2323–2331.

McWilliam SM, Kongsuwan K, Cowley KA, Byrne KA, and Walker PJ (1997) Genome organization and transcription strategy in the complex $G_{NS}$-L intergenic region of bovine ephemeral fever rhabdovirus. *Journal of General Virology* 78: 1309–1317.

Nandi S and Negi BS (1999) Bovine ephemeral fever: A review. *Comparative Immunology, Microbiology, and Infectious Diseases* 22: 81–91.

St. George TD (1990) Bovine ephemeral fever virus. In: Dinter Z and Morein B (eds.) *Virus Infections of Vertebrates, Vol. 3: Virus Infections of Ruminants*, pp. 405–415. Amsterdam: Elsevier.

Theodoridis A, Giesecke WH, and Du Toit IJ (1973) Effects of ephemeral fever on milk production and reproduction of dairy cattle. *The Onderstepoort Journal of Veterinary Research* 40: 83–92.

Tomori O, Fagbami A, and Kemp G (1974) Kotonkan virus: Experimental infection of Fulani calves. *Bulletin of Epizootic Diseases of Africa* 22: 195–200.

Tordo N, Benmansour A, Calisher C, et al. (2005) Rhabdoviridae. In: Fauquet CM, Mayo MA, Maniloff J, Desselberger U, and Ball LA (eds.) *Virus Taxonomy: Eighth Report of the International Committee on Taxonomy of Viruses*, pp. 623–644. San Diego, CA: Elsevier Academic Press.

Uren MF, St. George TD, and Zakrzewski H (1989) The effect of anti-inflammatory agents on the clinical expression of bovine ephemeral fever. *Veterinary Microbiology* 19: 99–111.

Venter GJ, Hamblin C, and Paweska JT (2003) Determination of the oral susceptibility of South African livestock-associated biting midges, Culicoides species, to bovine ephemeral fever virus. *Medical and Veterinary Entomology* 17: 133–137.

Walker PJ (2005) Bovine ephemeral fever in Australia and the World. In: Fu Z and Kaprowski H (eds.) *The World of Rhabdoviruses. Current Topics in Microbiology and Immunology*, vol. 292, pp. 57–80. Berlin: Springer.

Walker PJ, Byrne KA, Riding GA, et al. (1992) The genome of bovine ephemeral fever rhabdovirus contains two related glycoprotein genes. *Virology* 191: 49–61.

Wang YH, McWilliam SM, Cowley JA, and Walker PJ (1994) Complex genome organization in the $G_{NS}$-L intergenic region of Adelaide River rhabdovirus. *Virology* 203: 63–72.

# Bovine Herpesviruses

**M J Studdert,** The University of Melbourne, Parkville, VIC, Australia

© 2008 Elsevier Ltd. All rights reserved.

## History

Although some of the clinical diseases caused by herpesviruses in members of the family Bovidae have been recognized for centuries, it was not until the first and probably most important alphaherpesvirus now called bovine herpesvirus 1 (BHV1) was isolated in the late 1950s from the genital disease coital exanthema (also called infectious pustular vulvovaginitis (IPV) in the female) and from the respiratory disease infectious bovine rhinotracheitis (IBR) that any of these diseases was confirmed to be caused by a herpesvirus. Historically, IPV and its male counterpart infectious pustular balanoposthitis (collectively the male and female diseases are termed coital exanthema or *blaschenausschlag*) were commonly described diseases in central Europe throughout the nineteenth century. It was common for a single bull in a village to serve all the female cattle in that village and, where distances were small, also in nearby villages, and *blaschenausschlag* was a frequently observed sequel to mating. The isolation of IBR virus in 1957 and IPV virus in 1958, and subsequent work that established that the two viruses were essentially identical, led to the designation of BHV1.

Around 1970, a distinctly different, but BHV1-related, alphaherpesvirus was recognized as a cause of encephalitis. This virus is designated BHV5 pending identification of its definitive host, which may not be European cattle.

The alphaherpesvirus BHV2 was recognized as a cause of pseudolumpy skin disease and, as an independent syndrome, mammilitis, in about 1960, although both diseases were known clinically well before this time.

An alphaherpesvirus with a natural history similar to that of BHV1 was isolated from goats in the mid-1960s. An increasing number of alphaherpesviruses, again assumed to have a natural history similar to BHV1, have been isolated from various deer and other wild ruminant species. Curiously, no alphaherpesvirus has been isolated from sheep.

A slowly growing, highly cell-associated gammaherpesvirus, uncertainly associated with a number of disease syndromes in cattle, is designated BHV4.

The disease bovine malignant catarrhal fever (MCF) has been described for at least a century and the causative gammaherpesvirus of the African form of the disease, formerly designated BHV3 (sometimes BHV4), was first isolated in 1968. However, the natural host for the best-characterized MCF gammaherpesvirus, acquired by European cattle, as originally reported in southern Africa, is the wildebeest (*Connochaetes gnu*) and this virus is now termed alcelaphine herpesvirus 1 (AlHV1). A poorly characterized gammaherpesvirus (OvHV1) associated with ovine pulmonary adenomatosis has been described and a second ovine gammaherpesvirus (OvHV2) is the cause of sheep-associated bovine MCF.

An increasing number of gammaherpesviruses have been identified in normally free-ranging ruminant species when they are farmed or held in zoological collections. Several of these viruses have caused MCF-like syndromes when transmitted to other in-contact ruminant species.

## Classification

Members of the family *Herpesviridae* (in the proposed order *Herpesvirales*) that infect certain members of the family *Bovidae* are classified in the subfamilies *Alphaherpesvirinae* or *Gammaherpesvirinae*. Some of the viruses are listed in **Table 1**, which also lists their respective genera, together with some other salient properties, where known, including nucleotide composition and genome size. All of the ruminant alphaherpesviruses are classified in the genus *Varicellovirus* with the notable exception of BHV2, which is classified in the genus *Simplexvirus* because it is most closely related to human herpes simplex virus. All of the ruminant gammaherpesviruses have been proposed as members of the new genus *Macavirus* (*maca* is short for MCF) with the exception of BHV4, which remains in the genus *Rhadinovirus*.

## Structure

Each of the viruses listed in **Table 1** has a typical herpesvirus morphology. Virions are enveloped and about 150 nm in diameter. The double-stranded DNA genome is spooled within the capsid. There is an icosahedral nucleocapsid 100 nm in diameter composed of 162 hollow capsomers (150 hexamers and 12 pentamers). The nucleocapsid is surrounded by a layer of globular material called the tegument that is enclosed by a typical bilayer lipoprotein envelope in which are embedded glycoproteins, which generally appear as projecting spikes in negatively stained electron micrographs. There are about 12 distinct glycoproteins associated with the envelope spikes. Though the size of the DNA genome varies (**Table 1**), there is evidence that there are up to 76 open reading frames (genes) minimally coding for a corresponding number of individual proteins. About 40 of these proteins are structural (i.e., associated with the virion), while the remainder are nonstructural, being found only in infected cells. Repeat DNA sequences are found in the genomes of all bovine herpesviruses. For the alphaherpesviruses a set of two inverted repeats bracket the so-called short region of the genome, and in the case of BHV2 only a second set of inverted repeats bracket the so-called long region. The gammaherpesviruses have a set of terminal repeat structures, within each of which a variable number of tandemly repeated sequences is found.

## Replication

Virus replication occurs in the nucleus of cells, and in the case of alphaherpesviruses typically results in the production of a rapid cytopathic effect, with characteristic large intranuclear, eosinophilic inclusion bodies present in appropriately stained preparations. Some of the gammaherpesviruses can be cultivated in monolayer cell cultures, where they produce a cytopathic effect, but others have not been isolated in cell cultures. The replication cycle involves at least three classes of genes termed $\alpha$, $\beta$, and $\gamma$ or immediate early, early, and late, the synthesis of which is coordinately regulated in a cascade manner during the replication cycle.

Herpesviruses transcribe sets of micro (mi)RNAs that add complexity to understanding the replication cycle and the host–virus relationship and change views of the antiviral roles of RNA interference (RNAi), also known as gene silencing. Rather than being inhibited, many herpesviruses appear to be able to usurp or divert

**Table 1** Herpesviruses of the family Bovidae and some other ruminant species

| Virus | Abbreviation | Disease names/synonyms | Subfamily | Genus[a] | Genome Size (kbp) | G+C (mol.%) |
|---|---|---|---|---|---|---|
| Bovine herpesvirus 1 | BHV1 | Infectious bovine rhinotracheitis; infectious pustular vulvovaginitis; coital exanthema | Alphaherpesvirinae | Varicellovirus | 135 | 72 |
| Bovine herpesvirus 2 | BHV2 | Bovine mammillitis; Allerton virus; pseudolumpy skin disease | Alphaherpesvirinae | Simplexvirus | 133 | 64 |
| Bovine herpesvirus 4 | BHV4 | Movar virus | Gammaherpesvirinae | Rhadinovirus | 160[b] | 41 |
| Bovine herpesvirus 5 | BHV5 | Bovine encephalitis | Alphaherpesvirinae | Varicellovirus | 135 | 74 |
| Bovine herpesvirus 6 | BHV6 | Bovine lymphotropic HV | Gammaherpesvirinae | Macavirus | | |
| Bubaline herpesvirus 1 | BuHV1 | Buffalo HV | Alphaherpesvirinae | Varicellovirus | | |
| Elk herpesvirus 1 | ElHV1 | Elk HV related to BHV1 | | | | |
| Caprine herpesvirus 1 | CapHV1 | Goat HV1 | Alphaherpesvirinae | Varicellovirus | | |
| Caprine herpesvirus 2 | CapHV2 | Goat HV2 | Gammaherpesvirinae | Macavirus | | |
| Ovine herpesvirus 1 | OvHV1 | Sheep pulmonary adenomatosis-associated HV | | | | |
| Ovine herpesvirus 2 | OvHV2 | Cause of sheep-associated bovine malignant catarrhal fever | Gammaherpesvirinae | Macavirus | 160[b] | 52[b] |
| Cervid herpesvirus 1 | CerHV1 | Red deer HV | Alphaherpesvirinae | Varicellovirus | | |
| Cervid herpesvirus 2 | CerHV2 | Reindeer HV | Alphaherpesvirinae | Varicellovirus | | |
| Alcelaphine herpesvirus 1 | AlHV1 | Wildebeest HV, cause of malignant catarrhal fever of European cattle | Gammaherpesvirinae | Macavirus | 155–160[b] | 46[b] |
| Alcelaphine herpesvirus 2 | AlHV2 | Barbary red deer HV, cause of malignant catarrhal fever of Jackson's hartebeest | Gammaherpesvirinae | Macavirus | | |

[a]Viruses listed in proposed genus *Macavirus* are currently in genus *Rhadinovirus* or are unclassified.
[b]Sizes shown include the terminal repeats, which vary in total size around 25–30 kbp. G + C contents exclude the terminal repeats.

the host RNA silencing machinery to their advantage. Herpesvirus-encoded miRNAs can act in *cis* to ensure accurate expression of viral genomes or in *trans* to modify the expression of host RNA transcripts.

## Geographic Distribution

In general, each of the bovine alphaherpesviruses occurs worldwide, paralleling the distribution of the host species. The bovine alphaherpesviruses, with minor exceptions, have a restricted host range. None is known to infect nonbovine species; most are restricted to the primary host species.

Increasing numbers of gammaherpesviruses have been identified in normally free-ranging (exotic) ruminant species that have been farmed or held in zoological collections, and several of these viruses have caused MCF-like syndromes when transmitted to other in-contact ruminant species. At least six members of the MCF virus group of ruminant gammaherpesviruses have been identified thus far. Four of these viruses are clearly associated with clinical disease: alcelaphine herpesvirus 1 (AlHV-1) carried by wildebeest (*Connochaetes* spp.); ovine herpesvirus 2 (OvHV-2), ubiquitous in domestic sheep; caprine herpesvirus 2 (CapHV-2), endemic in domestic goats; and the virus of unknown origin that caused classic MCF in white-tailed deer (*Odocoileus virginianus*, MCFV-WTD). Gammaherpesviruses in the MCF virus group have been found in musk ox (*Ovibos moschatus*), Nubian ibex (*Capra nubiana*), and gemsbok (South African oryx, *Oryx gazella*). Gammaherpesviruses have also been found in bighorn sheep, bison, black-tailed deer, mule deer, fallow deer, elk, and addax.

## Antigenic Relationships

Bovine herpesviruses are genetically stable, with only a single antigenic type described for each species and no major changes in antigenicity over time recognized. Some intratypic (within species) differences are detectable using restriction endonuclease DNA fingerprinting, but these differences have not been correlated with major antigenic differences in the proteins coded for by the regions where variable sequences have been identified.

## Epidemiology

In general, transmission of the alphaherpesviruses requires close contact, particularly the kinds of physical contact that bring moist epithelial surfaces into apposition (e.g., coitus, or licking and nuzzling as between mother and offspring). In large, closely confined populations such as cattle feedlots or zoo collections, short-distance aerosol is an important mode of transmission.

Most of the bovine gammaherpesviruses are recognized when they are transmitted to heterologous hosts. For example, MCF in European cattle is acquired from wildebeest, sheep, or other nondomestic (exotic) ruminant species. Where exotic species are the source of infection, more often than not the transmissions occur outside of the natural geographic habitat of the transmitting host.

The gammaherpesviruses are probably transmitted primarily via nasal secretions in both the natural host and to the heterologous hosts that develop MCF. The latter transmission cycle is probably much less efficient, at least for some of the viruses, since many cases of MCF are sporadic. It is only recently that cell-free OvHV2 virions have been demonstrated in nasal secretions of sheep, which is a reflection of the highly cell-associated nature of gammaherpesviruses.

## Pathogenesis

Alphaherpesviruses typically cause localized lesions, particularly of mucosal surfaces of the respiratory and genital tracts or, less commonly, the skin. Progression is characterized by the sequential production of vesicles, pustules, and shallow ulcers that become covered by a pseudomembrane and heal after 10–14 days, usually without scar formation (**Figure 1**).

Generalized alphaherpesvirus infections may occur in very young calves or in a fetus prior to abortion. Encephalitis produced by bovine encephalitis herpesvirus (BHV5) occurs as a consequence of spread of virus from the nasal cavity to the brain, via trigeminal nerve branches.

The gammaherpesvirus BHV4 is associated with low-grade clinical infection. MCF is a uniformly fatal disease associated with mucosal erosions, ophthalmia, and encephalitis that appear to be immune mediated. Lesions are characterized by infiltration and proliferation of lymphocytes. It is still not clear which lymphocyte population is the site of latency; both B and T lymphocytes have been implicated for different gammaherpesviruses. Immune complexes of viral antigen are probably also produced and contribute to the pathology.

Latency is a hallmark of bovine herpesviruses. The genome, probably as a circularized episome, persists in ganglion cells, typically the trigeminal and sciatic in the case of alphaherpesviruses, and in white blood cells in the case of gammaherpesviruses. From these sites of latency, virus is periodically shed to give rise to recurrent disease, shedding, and transmission to in-contact animals.

BHV1 establishes latency in sensory neurons of trigeminal ganglia, and in germinal centers of pharyngeal tonsil and similar sites related to the genital tract. BHV1 reactivates

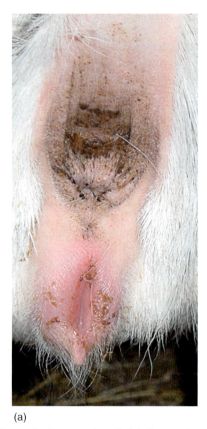 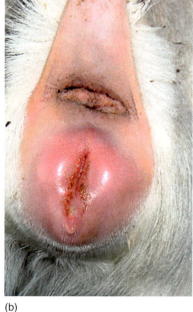 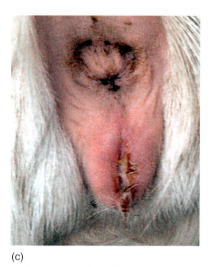

(a) (b) (c)

**Figure 1** As examples of alphaherpesvirus genital disease, three naturally occurring cases (a (early case), b, c) of acute, primary, infectious pustular vulvovaginitis in 8-month-old does caused by caprine herpesvirus 1 are shown. The extent of the individual vesicular/pustular lesions on the vaginal mucosa is not well shown because of the swelling, and the pain associated with closer examination. Reproduced from Piper KL, Fitzgerald CJ, Ficorilli N, and Studdert MJ (2008) Isolation of caprine herpesvirus 1 from a major outbreak of infectious pustular vulvovaginitis in goats. *Australian Veterinary Journal* 86: 136–138, with permission.

periodically from latency, virus is shed, and consequently virus transmission occurs. Two RNA transcripts, the latency-related RNA and ORF-E RNA, are abundantly expressed in trigeminal ganglia of latently infected cattle, and these transcripts probably regulate the BHV1 latency–reactivation cycle.

## Clinical Diseases

The alphaherpesvirus BHV1 causes coital exanthema and IBR. Both diseases are a consequence of lesions of the mucous membrane at the two sites. The extent and severity of the lesions may vary from mild, subclinical disease to acute, complicated and severe clinical disease that is more likely to occur in the case of IBR in feedlot cattle, where the disease is complicated by secondary bacterial infections. Complicated cases of IBR extending to pneumonitis in feedlot cattle may be fatal.

BHV1 may occasionally cause enteritis in calves. Encephalitis caused by BHV1 has not been confirmed; all cases of alphaherpesvirus encephalitis in cattle have been caused by the distinctly different virus BHV5. In groups of young calves the mortality caused by BHV5 encephalitis may approach 100%.

Mammilitis caused by BHV2 may be acute, leading to loss of skin from the teats, udder, and perineal regions following vesicle and pustule formation. Pseudolumpy skin disease caused by BHV2 appears to be a consequence of viremic spread, possibly as a cell-associated viremia with localization of the virus in the skin resulting in large golf-ball-sized subcutaneous swellings. These eventually resolve after a course of 3–4 weeks.

BHV6 was identified in bovine B-lymphoma cells and peripheral blood mononuclear cells.

The natural history of caprine herpesvirus 1 is similar to that of BHV1, with disease characterized by a variety of clinical signs including conjunctivitis and lesions of the genital tracts and sometimes the respiratory and gastrointestinal tracts. Abortion may occur.

The red deer and reindeer alphaherpesviruses probably cause clinical disease and have a similar natural history to BHV1.

Bovine MCF in European cattle is caused by alcelaphine herpesvirus 1 or 2 or ovine herpesvirus 2. The disease follows an incubation period of 3 weeks and is characterized by fever, depression, leucopenia, profuse nasal and ocular discharge, generalized lymphadenopathy, extensive mucosal erosions, central nervous system signs, and bilateral ophthalmia that begins as a keratoconjunctivitis and extends to a panophthalmitis. Death, which is invariable, occurs about 1 week after the onset of clinical signs.

Caprine herpesvirus 2 causes clinical MCF when transmitted to farmed white-tailed deer. The transmission pattern of caprine herpesvirus 2 in goats is similar to that of ovine herpesvirus 2 in sheep, with nasal secretions believed to be the major mode of transmission.

Ovine herpesvirus 2 is a noncultivable, lymphotropic gammaherpesvirus that asymptomatically infects most sheep, but causes MCF in cattle, bison, and, somewhat surprisingly, pigs and deer. A uniformly fatal enteric form of MCF caused by ovine herpesvirus 2 was identified in American bison (*Bison bison*) at a large feedlot in the American Midwest in 1998. An estimated 150 bison died. Clinical onset was acute, and most affected bison died within 1–3 days following the onset of clinical disease.

## Immune Response

Both antibody and cell-mediated immune responses are generated during herpesvirus infections. Neutralizing antibody primarily directed against envelope glycoproteins is probably important in long-term immunity. Viral antigens, some of which may be nonstructural immediate early and early proteins, are incorporated into the cell membrane and serve as targets for cytotoxic T lymphocytes. The immune response associated with infection does not prevent the establishment of latency and its role in regulating reactivation of latent virus and recurrent disease and shedding is debated. A central contradiction of herpesvirus immunity is that following natural infection immune animals are also animals that are infected for life.

## Prevention and Control

BHV1 genital disease can be controlled by eliminating carrier cattle identified either serologically or by reactivation and isolation of virus following the administration of corticosteroids such as dexamethasone. Alternatively, where it is important to do so, such as for bulls in artificial breeding centers, a two-herd system may be established. IBR is often associated with stress of transport, intercurrent disease, overcrowding and the mixing together of cattle from different sources, all of which are typically associated with feedlot operations. Awareness and minimization of these predisposing factors can reduce the severity of clinical disease. In an increasing number of countries, test and slaughter programs have achieved total eradication of BHV1 from national herds.

In many parts of the world, including North America, control of BHV1 is achieved by vaccination with conventional live attenuated or inactivated vaccines. With parts of Europe being BHV1 free, the ability to differentiate infected from vaccinated animals has become critical for trade. Live and killed glycoprotein E-deleted marker vaccines are now widely used in Europe, in combination with glycoprotein-based diagnostic tests to monitor cattle. There is debate about the cost and sustainability of eradication programs other than in limited settings such as artificial insemination centers or small countries. Conventional inactivated and attenuated vaccines are less efficacious in neonates because of interference by virus-specific, passively derived maternal antibody. Alternative vaccine types, such as those incorporating CpG oligonucleotides as adjuvant for recombinant protein vaccines or DNA vaccines, are being explored.

Vaccines are not generally available for the control of other bovine or other ruminant herpesvirus diseases.

The epidemiology of BHV2 is not well understood and a possible approach to prevention and control would be to consider removal of known infected cattle.

Since MCF is acquired from a heterologous host (wildebeest, sheep, goats, or other ruminant species), it is clearly preventable by avoiding such contacts. The often-sporadic nature of the disease and the lack of detailed knowledge of the putative sheep-associated virus make avoidance difficult. In zoological collections, bovid species known to harbor alcelaphine herpesviruses 1 and 2 or any of the other gammaherpesviruses should not be cohabited with those species known to be susceptible to MCF.

## Future

Continued progress in understanding the molecular biology of the bovine herpesviruses, including full genome sequencing such as already reported for some viruses including BHV1, BHV4, BHV5, and AlHV1, will occur. Characterization of the transcripts and proteins of the viruses will be taken forward. Progress in developing better vaccines and diagnostic reagents for BHV1 including those based on recombinant DNA technologies, including DNA vaccines, will continue. How effective these new vaccines and diagnostic tests will be in national and international eradication programs is a question of will and financial commitment. The unusual epidemiologies of BHV2 and BHV5 are matters for future inquiry. Further progress in the unusual pathogenesis of MCF and the characterization of the sheep-associated virus responsible for many cases of MCF in the Western world are a part of ongoing work. It may be anticipated that the number of new heterologous host transmission cycles of gammaherpesviruses leading to highly fatal MCF outbreaks will continue to increase, and definition of these should lead to better control measures where ruminant species cohabit. There are many members of the family *Bovidae* for which neither alphaherpesviruses nor gammaherpesviruses have been identified, and over time it may be expected that more of these viruses will be isolated and characterized.

*See also:* Herpes Simplex Viruses: General Features; Herpes Simplex Viruses: Molecular Biology; Herpesviruses: Discovery; Herpesviruses: Latency; Pseudorabies Virus; Taxonomy, Classification and Nomenclature of Viruses.

## Further Reading

Jones C, Geiser V, Henderson G, *et al.* (2006) Functional analysis of bovine herpesvirus 1 (BHV-1) genes expressed during latency. *Veterinary Microbiology* 113: 199–210.

Li H, Gailbreath K, Flach EJ, *et al.* (2005) A novel subgroup of rhadinoviruses in ruminants. *Journal of General Virology* 86: 3021–3026.

Murphy FA, Gibbs EPJ, Horzinek MC, and Studdert MJ (eds.) (1999). Herpesviridae. In: *Veterinary Virology*, 3rd edn., ch.18, pp. 301–325. New York: Academic Press.

O'Toole D, Li H, Sourk C, Montgomery DL, and Crawford TB (2002) Malignant catarrhal fever in a bison (*Bison bison*) feedlot, 1993–2000. *Journal of Veterinary Diagnostic Investigation* 14: 183–193.

Pfeffer S, Sewer A, Lagos-Quintana M, *et al.* (2005) Identification of microRNAs of the herpesvirus family. *Nature Methods* 2: 269–276.

Piper KL, Fitzgerald CJ, Ficorilli N, and Studdert MJ (2008) Isolation of caprine herpesvirus 1 from a major outbreak of infectious pustular vulvovaginitis in goats. *Australian Veterinary Journal* 86: 136–138.

Rovnak J, Quackenbush SL, Reyes RA, Baines JD, Parrish CR, and Casey JW (1998) Detection of a novel bovine lymphotropic herpesvirus. *Journal of Virology* 72: 4237–4242.

Taus NS, Oaks JL, Gailbreath K, Traul DL, O'Toole D, and Li H (2006) Experimental aerosol infection of cattle (*Bos taurus*) with ovine herpesvirus 2 using nasal secretions from infected sheep. *Veterinary Microbiology* 116: 29–36.

Thiry J, Keuser V, Muylkens B, *et al.* (2006) Ruminant alphaherpesviruses related to bovine herpesvirus 1. *Veterinary Research* 37: 169–190.

Thonur L, Russell GC, Stewart JP, and Haig DM (2006) Differential transcription of ovine herpesvirus 2 genes in lymphocytes from reservoir and susceptible species. *Virus Genes* 32: 27–35.

van Drunen Littel-van den Hurk S (2006) Rationale and perspectives on the success of vaccination against bovine herpesvirus-1. *Veterinary Microbiology* 113: 275–282.

# Bovine Spongiform Encephalopathy

**R G Will,** Western General Hospital, Edinburgh, UK

© 2008 Elsevier Ltd. All rights reserved.

## Introduction

Bovine spongiform encephalopathy (BSE) is a prion disease of cattle, which was first identified in 1986, and has subsequently become a source of widespread concern for policymakers and public health. To date, more than 184 000 cases have been identified in the UK and more than 5000 cases in other countries, primarily, but not exclusively, in Europe. The origin of BSE is unknown, but may have been related to scrapie contamination of cattle feed, with amplification of the epidemic through within-species recycling of infection. Legislative measures, including restrictions on the feeding of ruminant protein to ruminants, has led to a decline in annual numbers of identified cases in most countries, and there is a possibility that the introduction and implementation of appropriate control measures will lead to the eradication of BSE.

In 1996, a new form of human prion disease, variant Creutzfeldt–Jakob disease (vCJD), was identified in the UK, and epidemiological and laboratory data indicate that this disease is a zoonosis caused by infection with BSE, probably through past dietary exposure to infection. The human population in the UK and many European countries were exposed to significant titers of BSE infectivity over a period of years from about 1980, but the possibility of an extensive epidemic of vCJD has not, as yet, materialized. There has been a relatively limited and declining annual mortality rate in the UK and, with the exception of France, only isolated cases in other, mainly European, countries. However, future outbreaks of vCJD, perhaps related to polymorphisms in the human prion protein gene (*PRNP*), cannot be excluded and concern for public health has increased with the demonstration of transmission of vCJD through blood transfusion. Accurate predictions of future numbers of cases are hampered by many uncertainties including the mean incubation period of human BSE infection and the prevalence of sub- or preclinical infection in exposed populations.

## BSE

### Clinical and Subclinical Infection

All prion diseases are degenerative conditions of the central nervous system and present with progressive and fatal neurological disorders. The clinical features of BSE include weight loss, reduced milk yield, ataxia and hyperesthesia, progressing to recumbency and death. Although there is a wide age range of affected cattle, the majority of cases are aged 4–6 years. Identification of clinically affected animals is critical to analysis of the epidemiology of BSE and for protection of public health, but depends on recognition of the clinical phenotype or active testing in abattoirs. As in other prion diseases, BSE has a protracted incubation period, in BSE a mean of about 5 years prior to the onset of clinical signs, and infectivity may be present in some tissues, particularly in the pre-terminal stages. However, the tissue distribution of infectivity in BSE is relatively restricted in comparison to other prion diseases such as sheep scrapie. Effective protection of public health depends on accurate case identification in the field or testing for the presence of disease in the abattoir to prevent clinically unrecognized cases entering the human food chain.

Passive surveillance for BSE has proved to be a relatively inefficient strategy for case identification and varies by country according to available skills and resources and the size of the cattle population. Active testing for BSE in abattoirs by examination of the obex region of the brain stem for disease-associated prion protein ($PrP^{Sc}$) has proved to be a reliable method of identifying infected animals. This has allowed more precise information on the course of the BSE epidemic, although the costs of systematic testing of cattle populations are significant.

Infectivity in prion diseases is not restricted to the central nervous system and may involve peripheral tissues, particularly in the lymphoreticular system, in which the agent replicates during the incubation period prior to neuroinvasion and clinical disease. In order to minimize human exposure to infection from preclinical or unrecognized clinical cases, many countries have introduced a ban on certain bovine tissues from entering the human food chain from apparently healthy cattle. A 'specified bovine offal' ban was introduced in the UK in 1989 and an extended list of tissues, the 'specified risk materials' in other European countries in 2000. Implementation and enforcement of these measures is essential to protect public health in countries with a significant risk of human exposure to BSE.

The original list of proscribed tissues was based on information from previous studies in sheep scrapie, but experimental pathogenesis studies have subsequently provided information on the tissue distribution of infectivity in BSE during the incubation period and in the clinical phase. Infectivity in BSE can be identified in tonsil at 10 months after challenge, in terminal ileum after 6–18 months, in the dorsal root ganglia at 32 months, with clinical onset and involvement of brain at 35 months. Many tested tissues have been negative by bioassay in mice and a restricted range of tissue by similar studies in cattle, indicating that the anatomical distribution of BSE is relatively restricted in comparison to other prion diseases such as sheep scrapie. A more extensive tissue involvement in BSE, including involvement of sciatic nerve, has been suggested by the development of more sensitive techniques for the identification of either $PrP^{Sc}$ or infectivity.

## The Origin of BSE

BSE was first identified in the UK in 1986 and epidemiological investigation indicated that the disease was a common source epidemic caused by infection in cattle feed in the form of meat and bone meal. This hypothesis has been strongly supported by the decline in the BSE epidemic in the UK about 5 years after a ban on feeding ruminant protein to cattle and the effectiveness of similar measures in other countries.

The original hypothesis was that the initial source of infection was sheep scrapie, which had been inadvertently included in cattle feed, and that sufficient levels of infection had been present to cross the 'species barrier' between sheep and cattle. Circumstantial evidence of a change in the production methods for meat and bone meal in the 1970s provided an explanation for the timing of the initial cases and the subsequent extensive epidemic was attributed to subsequent recycling of infected cattle tissues to cattle. An alternative hypothesis for the origin of BSE is that this was due to the development of spontaneous disease in a single animal, which was used in the production of meat and bone meal and recycling of infection within the cattle population resulted in an epidemic.

The true origin of BSE is unknown and will probably never be established with certainty. However, any hypothesis must be consistent with the origin of BSE in the UK rather than any other country. The UK had a large sheep population, a high incidence of scrapie, and a practice of feeding meat and bone meal to calves. If spontaneous BSE actually occurs, by analogy with sporadic CJD, the probability of a spontaneous case will be proportionate to the size of the cattle population and that in the UK was smaller than some other countries such as the USA, Australia, and New Zealand, and the latter two countries are believed to be free of scrapie and BSE.

## Epidemiology

The first cases of BSE in the UK probably occurred in the early 1980s and the subsequent epidemic peaked in 1992 and then declined as a result of the ban on feeding ruminant protein to ruminants, introduced in the UK in 1988 (**Figure 1**). At its peak, more than 30 000 clinical cases were identified annually by passive surveillance, although it is likely that there was under-ascertainment of cases, particularly in the early years of the epidemic and before the introduction of an active abattoir testing program. In addition, mathematical models suggest that at least 1 000 000 preclinically infected cattle may have entered the human food chain in the

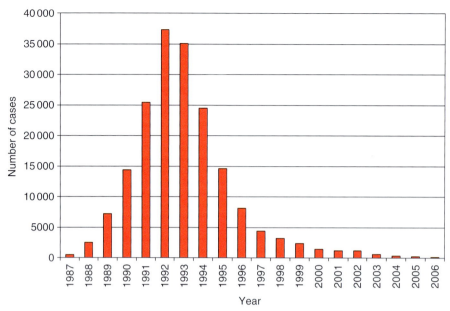

**Figure 1** Annual number of cases of BSE identified in the UK. Data from OIE Website.

1980s prior to the introduction of measures to minimize human exposure to BSE.

To date, more than 40 000 affected cattle born after the introduction of the ruminant feed ban have been identified, indicating that this measure was not fully enforced. There is no good evidence of alternative vertical or lateral routes of transmission in BSE. A likely explanation for these cases is that meat and bone meal was still being fed to other species such as pigs and poultry and that cross contamination in feed mills resulted in continuing cattle exposures. A reinforced feed ban, prohibiting the feeding of meat and bone meal to any farmed species, was introduced in 1996 and only 140 cases of BSE born after this date have been identified, possibly linked to importation of animal feed. An experimental challenge study in BSE has shown that as little as 1 mg of infected brain is sufficient to cause infection by the oral route. It is of note, however, that the BSE epidemic continues to decline and in 2006 there were only 114 cases in the UK.

Live cattle and bovine products, including cattle feed, were exported from the UK in the 1980s and early 1990s and from other European countries in later years. It is likely that the risk of cattle exposure to BSE has a widespread geographical distribution. BSE was identified in Ireland in 1989, in Portugal and Switzerland in 1990, in France in 1991, and has subsequently been identified in all original member states of the European Union (EU). A ban on feeding ruminant protein to ruminants was introduced in the EU in 1994 and an SRM ban in 2000, although some countries introduced these measures earlier. In contrast to the UK, the passive surveillance system appears to have been relatively inefficient in identifying cases of BSE in most countries (with the exception of Switzerland), and it was with the introduction of a mandatory abattoir testing program in 2000–01 that some countries first identified indigenous BSE (e.g., Denmark, Germany, Italy, and Spain), with resulting extensive public concern. It is likely that cases of BSE may not have been identified in preceding years and the true size of BSE outbreaks in some counties, although limited in relation to the size of the UK epidemic, is unknown. In recent years, there has been a decline in the number of cases in almost all European countries (**Figure 2**), underlining the importance of introducing and enforcing measures to prevent the recycling of infection within cattle populations.

Cases of BSE have been found in small numbers in non-European countries, including Canada (9), Israel (1), Japan (31), and the USA (2, including 1 that originated in Canada). Despite the limited numbers of these cases, their identification has had important implications for trade. The possibility that risk of BSE may have been widely disseminated has resulted in a recommendation that all countries carry out a risk assessment, taking into account the possibility of importation of relevant risk materials and the possibility of recycling infection within cattle populations. Active abattoir testing for BSE can be an efficient means of identifying cases of BSE, but the precise populations to be tested, that is, normal slaughter, fallen stock, casualty animals, etc., and the numbers of required tests in specific populations are controversial. Over 10 000 000 cattle are currently tested per annum in the EU at a cost of 45 euros per test, with 561 positives in 2005.

New and atypical forms of BSE have been identified through the active testing programs. Cases of a novel form of BSE defined by a differential neuropathology and biochemical prion protein characteristics, bovine

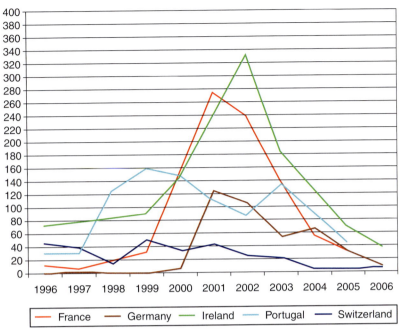

**Figure 2** Annual number of cases of BSE identified in some European countries. Data from OIE Website.

amyloidotic spongiform encephalopathy, were first recognized in Italy, and subsequently a small number of cattle with brain prion protein characteristics different from both the Italian cases and BSE itself have been found, mainly in Europe. The total number of atypical BSE cases identified is currently 18 and the majority are in the older age groups (>10 years) and most had no clinical signs. The origin of these cases is unknown (curiously no cases have yet been identified in the UK) and the implications for human health, if any, are uncertain.

## Variant CJD

### Clinical and Subclinical Infection

vCJD presents clinically with psychiatric symptoms, including depression and withdrawal, followed after a mean of 6 months by progressive ataxia and cognitive impairment, associated with involuntary limb movements. The mean survival is 14 months. vCJD affects younger age groups than in sporadic Creutzfeldt–Jakob disease (sCJD) with a mean age at death of 29 years (range 16–74 years), and it has been proposed that this may be related to either age-related susceptibility or variation in dietary exposure by age. All tested clinical cases have been methionine homozygotes at codon 129 of the human prion protein gene (*PRNP*).

The pathogenesis of vCJD is distinct from other human prion diseases as there is evidence of significant involvement of peripheral tissues and in particular the lymphoreticular system, including lymph nodes, spleen, appendix, and tonsil, in addition to the central nervous system. Infectivity is also present in peripheral nerves and large intestine and $PrP^{Sc}$ in enteric plexus, adrenal, ileum, and skeletal muscle. Infectivity may be present in some peripheral tissues during the incubation period and act as a source of potential secondary iatrogenic infection, for example, through blood transfusion.

The prevalence of sub- or preclinical infection has not been established with certainty in any population, but anonymized screening of appendectomy and tonsillectomy specimens in the UK has led to estimates that there may be a minimum prevalence of infection of 237 per million, translating to about 4000 individuals in the age group 10–30 years who are currently infected, taking account of the age distribution of those from whom specimens were sourced. Two out of the three positive appendix specimens were analyzed for codon 129 genotype and both were valine homozygotes, suggesting that individuals with this genetic background may be susceptible to infection with BSE.

### The Origin of vCJD

The hypothesis of a causal link between BSE and vCJD is supported by a range of evidence. The clinical and pathological phenotypes are remarkably consistent and distinct from previous experience. The characteristic neuropathological findings, including widespread deposition of florid plaques of $PrP^{Sc}$, have not been recognized previously in human prion disease, and review of archive tissues in a number of countries has failed to identify any case with the typical pathological phenotype prior to the identification of vCJD in the UK. Retrospective review of deaths certified under a range of rubrics and review of neuropathology in a limited number of these cases have failed to identify past cases of unrecognized vCJD. This evidence strongly suggests that vCJD is a new disease.

Laboratory studies have demonstrated that the infectious agent in vCJD is almost identical to the BSE agent in terms of incubation period and brain lesion distribution in experiments carried out on wild-type and transgenic mice. The biochemical characteristics of the $PrP^{Sc}$ deposited in the brain in vCJD are similar to BSE and distinct from other human prion diseases. Macaque monkeys inoculated experimentally with BSE develop florid plaques similar to those in vCJD. These studies indicate that the BSE agent is the cause of vCJD.

It has been proposed that BSE-infected humans through past dietary exposure, probably to high-titer bovine tissues and, in particular, spinal cord, dorsal root ganglia, and products containing mechanically recovered meat. Direct evidence of this hypothesis is lacking, not least because of the difficulties in investigating exposures that may have taken place years or even decades in the past. Furthermore, details of dietary history are necessarily obtained from surrogate witnesses because of the cognitive impairment that develops in vCJD. A case-control study comparing dietary exposures in cases of vCJD and age-matched population controls is consistent with increased risk through past oral intake of food products likely to have contained high levels of BSE infectivity, but the potential biases in this study compromise any firm conclusions. It is however of note that the mortality rate of vCJD is approximately double in the north when compared to the south of the UK, and this may reflect regional differences in past dietary exposures. Neither descriptive analyses nor case-control studies have provided evidence of any plausible alternative route of BSE exposure in vCJD cases, including past occupation or previous surgery.

The occurrence of a novel form of human prion disease, vCJD, in a country with a potentially new risk factor, BSE, first raised the possibility that these conditions were linked. Importantly, data from a harmonized system for surveillance of CJD in Europe indicated, in 1996, at the time vCJD was first found in the UK, that similar cases had not been identified in other countries. Improved efficiency of surveillance in the UK was therefore unlikely to explain the identification of this new disease. Subsequently, cases of vCJD have been found in other countries, but the fact that some of the cases occurred in countries with a very limited risk of exposure to indigenous BSE and had a history of residence in the UK during the time of maximal human exposure to BSE supports the concept that BSE is indeed the cause of vCJD.

## Epidemiology

Up to February 2007, 165 cases of vCJD have been identified in the UK, all but three of which are presumed to be related to past dietary exposure to BSE. The annual number of deaths from vCJD in the UK peaked in 2000 with 28 cases and has subsequently declined to 5 deaths in both 2005 and 2006 (**Figure 3**). Fears of a large epidemic have receded, but there remains the possibility of further outbreaks of cases related to BSE infection in individuals with a heterozygous or valine homozygous genotype at codon 129 of *PRNP* and it is possible that such cases may occur with an extended incubation period and perhaps with a different clinical and pathological phenotype. It is also likely, by analogy with other human prion diseases such as kuru, that there will be an extended tail to the epidemic with a low annual number of deaths for years or even decades.

Cases of vCJD have been found in a number of other countries, mainly, but not exclusively, in Europe (**Table 1**). To date, 21 cases have been identified in France with a peak in annual deaths some 5 years later than in the UK, consistent with a mathematical model which attributes the French cases to exposure to exports of BSE-infected materials from the UK rather than to indigenous BSE.

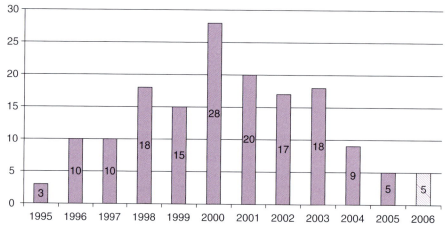

**Figure 3** Number of vCJD deaths per annum (UK).

**Table 1** Number of vCJD cases per country (Feb. 2007)

| Country | Total number of primary cases (number alive) | Total number of secondary cases: blood transfusion (number alive) | Residence in UK >6 months during period 1980–96 |
|---|---|---|---|
| UK | 162 (6) | 3 (1) | 165 |
| France | 21 (1) | | 1 |
| Rep. of Ireland | 4 (1) | | 2 |
| Italy | 1 (0) | | 0 |
| USA | 3[a] (0) | | 2 |
| Canada | 1 (0) | | 1 |
| Saudi Arabia | 1 (1) | | 0 |
| Japan | 1[b] (0) | | 0 |
| Netherlands | 2 (0) | | 0 |
| Portugal | 1 (1) | | 0 |
| Spain | 1 (0) | | 0 |

[a]The third US patient with vCJD was born and raised in Saudi Arabia and has lived permanently in the United States since late 2005. According to the US case report, the patient was most likely infected as a child when living in Saudi Arabia.
[b]The case from Japan had resided in the UK for 24 days in the period 1980–96.

Cases of vCJD are attributed by country according to the country of normal residence at the time of the onset of clinical symptoms. This does not necessarily correlate with the country in which exposure to BSE took place and it is of note that 2/3 US cases, 2/4 Irish cases, and the single Canadian case all had a history of extended residence in the UK during 1900–96 and were probably exposed to BSE in the UK rather than the country of attribution. The third US case had lived for most of his life in the country of origin, Saudi Arabia, and a further case has been identified from this country, which is not known to have BSE. One possibility is that human BSE exposure was related to exports from the UK, a matter of concern as exports from the UK and other European countries were distributed worldwide. The single case of vCJD in Japan could be related to this factor, although the individual had also spent a short period of time in the UK. It is of note that, although Italy did import bovine material from the UK and has indigenous BSE, only a single Italian case of vCJD has been identified and that was in 2001, with no new cases for 6 years. Although measures to minimize human exposure to high-titer bovine tissues were introduced in continental Europe more than 10 years after the UK, it is likely that the total number of vCJD cases in other countries will be significantly less than in the UK.

Concerns about the public health implications of BSE and vCJD have increased with the identification of transfusion transmission of vCJD. Four recipients of non-leucodepleted red cells, donated by individuals who later developed vCJD, have either developed vCJD ($n = 3$) or have become sub- or preclinically infected ($n = 1$). In the latter case, an individual who died of an intercurrent illness was found to have immunostaining for $PrP^{Sc}$ in spleen and one lymph node. The three clinical cases were methionine homozygotes and the preclinical case a heterozygote at codon 129 of *PRNP*, thus indicating that methionione homozygotes are not the only ones susceptible to secondary infection. In all four instances of transmission of infection by blood transfusion, the donation had been given months to years prior to clinical onset in the donor, indicating that infection is present in blood during the incubation period. The three clinical cases developed symptoms between 6 and ~9 years after transfusion, and it is of note that the four infections developed out of a total cohort of 26 individuals who survived at least 5 years after transfusion, indicating that this route is an efficient mechanism of transmitting vCJD infection from person to person. Although transfusion transmission of vCJD has only been identified in the UK, individuals with vCJD who had previously donated blood have been found in France, Spain, Ireland, and Saudi Arabia.

There is no evidence, to date, of secondary transmission of vCJD through plasma-derived products, contaminated surgical instruments, or vertically from mother to child and, although risk assessments suggest that the risks by some of these routes are limited, the period of observation is currently too short to exclude the possibility of alternative routes of transmission in the future, taking account of the potentially extended incubation periods in these diseases. A range of measures have been introduced in many countries to limit the risks of secondary transmission of vCJD, including, for example, deferral of blood donors with a history of extended residence in the UK.

*See also:* Viral Pathogenesis.

## Further Reading

Bradley R (1998) An overview of the BSE epidemic in the UK. *Developments in Biological Standard* 93: 65–72.

Brown P, McShane LM, Zanusso G, and Detwiler L (2007) On the question of sporadic or atypical bovine spongiform encephalopathy and Creutzfeldt–Jakob disease. *Emerging Infectious Diseases* 12(12): 1816–1821.

Collee JG and Bradley R (1997) BSE: A decade on – Part 1. *Lancet* 349: 636–641.
Collee JG and Bradley R (1997) BSE: A decade on – Part 2. *Lancet* 349: 715–721.
Cousens S, Everington D, Ward HJT, Huillard J, Will RG, and Smith PG (2003) The geographical distribution of variant Creutzfeldt–Jakob disease in the UK: What can we learn from it? *Statistical Methods in Medical Research* 12: 235–246.
Hewitt PE, Llewelyn CA, Mackenzie J, and Will RG (2006) Creutzfeldt–Jakob disease and blood transfusion: Results of the UK Transfusion Medicine Epidemiology Review study. *Vox Sanguins* 91: 221–230.
Hilton DA, Ghani AC, Conyers L, et al. (2004) Prevalence of lymphoreticular prion protein accumulation in UK tissue samples. *Journal of Pathology* 203: 733–739.
Kimberlin RH (1996) Speculations on the origin of BSE and the epidemiology of CJD. In: Gibbs CJ, Jr. (ed.) *Bovine Spongiform Encephalopathy: The BSE Dilemma*, pp. 155–175. New York: Springer.
Valleron A-J, Boelle P-Y, Will R, and Cesbron J-Y (2001) Estimation of epidemic size and incubation time based on age characteristics of vCJD in the United Kingdom. *Science* 294: 1726–1728.
Ward HJT, Everington D, Cousens SN, et al. (2006) Risk factors for variant Creutzfeldt–Jakob disease: A case-control study. *Annals of Neurology* 59: 111–120.
Wells GAH, Scott AC, Johnson CT, et al. (1987) A novel progressive spongiform encephalopathy in cattle. *Veterinary Record* 121: 419–420.
Wilesmith JW, Ryan JBN, and Atkinson MJ (1991) Bovine spongiform encephalopathy: Epidemiological studies of the origin. *Veterinary Record* 128: 199–203.
Wilesmith JW, Wells GAH, Cranwell MP, and Ryan JB (1988) Bovine spongiform encephalopathy: Epidemiological studies. *Veterinary Record* 123: 638–644.
Will RG, Ironside JW, Zeidler M, et al. (1996) A new variant of Creutzfeldt–Jakob disease in the UK. *Lancet* 347: 921–925.
World Health Organisation, Food and Agricultural Organisation, Office International des Epizooties (2002) Technical Consultation on BSE: Public health, animal health, and trade.

## Relevant Website

http://www.oie.int – OIE website.

# Bovine Viral Diarrhea Virus

**J F Ridpath**, USDA, Ames, IA, USA

Published by Elsevier Ltd.

## Glossary

**Biotype** Designation based on expression of cytopathic effect in cultured epithelial cells.
**Cytopathic effect** Alteration in the microscopic appearance of cultured cells following virus infection.
**Genotype** Group designation determined by genetic sequence comparison.
**Hemorrhagic syndrome** Form of severe acute bovine viral diarrhea (BVD) characterized by high morbidity and mortality as high as 50%. Clinical signs include hemorrhages throughout digestive system, high fevers, and bloody diarrhea. While similar in clinical presentation to mucosal disease it differs in that it is the result of an uncomplicated acute infection, only one biotype is present (noncytopathic) and it is not 100% fatal.
**Mucosal disease (MD)** Form of BVD characterized by low morbidity and 100% mortality. Clinical signs include severe and bloody diarrhea, sores in the mouth, and rapid wasting. MD occurs in animals born persistently infected with a noncytopathic BVDV that are subsequently superinfected with a cytopathic bovine viral diarrhea virus (BVDV). While similar in clinical presentation to hemorrhagic syndrome it differs in that it only occurs in persistently infected animals, two biotypes of virus are present, and it is 100% fatal.
**Persistent infection** Results from infection with noncytopathic virus during the first 125 days of gestation. The infected fetus develops an immune tolerance to the virus and sheds virus throughout its subsequent lifetime. Cytopathic BVDV are not able to establish persistent infections.
**Species** A fundamental category of taxonomic classification, ranking below a genus or subgenus and consisting of related organisms grouped by virtue of their common attributes and assigned a common name.

## Introduction

Bovine viral diarrhea viruses (BVDVs) present the researcher and veterinary clinician with arguably the most complicated combination of clinical presentation, pathogenesis, and basic biology of all bovine viral pathogens (**Figure 1**). There are two distinct species of BVDV (BVDV1 and BVDV2), two distinct biotypes (cytopathic and noncytopathic), two states of infection (acute and persistent), five recognized forms of clinical acute presentation (acute BVDV infection, severe acute BVDV infection, hemorrhagic BVDV infection, acute BVDV infection/bovine respiratory disease (BRD), and acute BVDV infection/immunosuppression), and one clinical

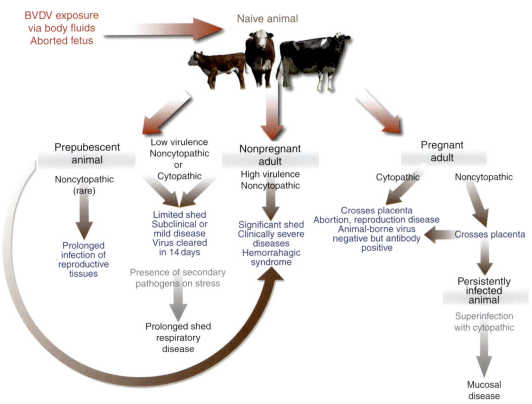

**Figure 1** BVDV infections. Clinical presentation following BVDV infection is dependent on viral strain, viral biotype, viral virulence, age of host, reproductive status of host, and presence of other pathogens. Acute uncomplicated BVDV infections are always accompanied by a loss of immune tissue and circulating lymphocytes.

presentation that is the sequeale of a persistent infection with one biotype followed by an acute infection with the other biotype. While BVDV infections are most commonly associated with cattle, they also occur in a wide variety of domesticated and wild ruminants including white tail and mule deer, bison, elk, and sheep.

In the last 10 years a host of advances in BVDV research and diagnostics have led to the development of improved diagnostics and burgeoning eradication/control programs. While the economic impact of BVDV is largely due to affects of acute infections, persistently infected animals are the most frequent vector. Thus, identification and removal of persistently infected animals is key to effective control strategies.

## History

### First Reports of BVD and MD

Bovine viral diarrhea (BVD) was first reported as a 'new' disease of cattle observed in New York dairies in 1946. The first report, by Dr. Francis Fox of Cornell University, described a 'rinderpest like' disease characterized by leukopenia, high fever, depression, diarrhea and dehydration, anorexia, salivation, nasal discharge, gastrointestinal erosions, and hemorrhages in various tissues. In the five initial herds in which it was observed, morbidity rates ranged from 33% to 88% and mortality rates ranged from 4% to 8%. In addition, fetal abortions were observed 10 days to 3 months following infection. It was shown that this disease could be transmitted experimentally. Rinderpest virus was ruled out as a causative agent because sera from convalescent animals did not neutralize rinderpest virus and cattle that had recovered from BVD were not resistant to rinderpest virus infection.

In 1953 another disease was reported in the US. Given the name mucosal disease (MD), it was characterized by severe diarrhea, fever, anorexia, depression, profuse salivation, nasal discharge and gastrointestinal hemorrhages, erosions, and ulcers. The gut-associated lesions were similar to, but more severe, than those reported for BVD. However, unlike BVD, MD could not be transmitted experimentally. In addition, while BVD outbreaks were marked by high morbidity but low mortality, MD usually only infected a small number of animals in the herd but once contracted was invariably fatal. Based on differences in lesions, transmissibility, and morbidity/mortality rates MD and BVD were initially thought to have different causative agents.

## Isolation of Virus

In 1957 Dr. James Gillespie of Cornell University isolated a noncytopathic virus as the causative agent of BVD. Three years later, in 1960, he isolated a cytopathic virus from a MD case. Cross-neutralization studies demonstrated that the viral agents associated with BVD and MD were the same and led to the realization that BVD and MD were different disease manifestations of infection with the same agent. The agent was termed bovine viral diarrhea virus (BVDV). While BVDV was identified as the causative agent, the etiology of MD, now referred in the literature as BVD-MD, remained a puzzle. In the late 1960s several research groups reported that animals succumbing to MD had persistent BVDV infections. It was also noted that persistently infected animals did not mount a serological immune response to the virus that they carried. The observations that fetal bovine serum was frequently contaminated with BVDV and that BVDV infections could be detected in newborn and one-day-old calves suggested that persistent infection might arise from *in utero* exposure.

## Release of First MLV Followed by Reports of pvMD

Because cytopathic strains could be more easily detected, quantitated, and studied in tissue culture than noncytopathic strains, the discovery of cytopathic BVDV was a boon to the study of BVDV. In 1964, the first cytopathic BVDV strain discovered (Oregon C24 V) was incorporated into a modified live multivalent vaccine. Soon it was reported that subsequent to use of this vaccine a minority of animals became sick with MD-like symptoms and died. These cases were referred to as postvaccinal MD (pvMD).

Further investigation revealed that the vaccinated animals that succumbed to pvMD responded with serum antibodies to the other components of the vaccine but did not respond to the BVDV component. This suggested that the susceptibility to pvMD might be correlated with failure of the immune system to recognize BVDV. Questions raised by pvMD lead to the elucidation of the etiology of MD.

## Unraveling the Etiology of MD

By the early 1970s the prevailing wisdom was that calves with persistent BVDV infections were uniformly unthrifty and usually died within the first few months of life. In a series of papers published in the late 1970s and early 1980s, Arlen McClurkin of the USDA's National Animal Disease Center (then known as the National Animal Disease Laboratory or NADL) reported persistent infection and immune tolerance in apparently healthy adult animals. Calves born to persistently infected (PI) cows were persistently infected at birth indicating maternal transmission. Further, McClurkin was able to generate persistently infected calves by exposing seronegative cows to noncytopathic BVDV between 42 and 125 days gestation. He followed the fate of the PI calves he generated and observed the following:

1. while many PI animals appeared weak and had congenital malformations, some appear apparently normal;
2. while that majority of PI animals died soon after birth, some lived to breeding age;
3. PI lines of cattle could be generated by breeding PI animals; and
4. PI animals spontaneously developed MD.

Using the McClurkin studies as a springboard, Joe Brownlie (Institute for Animal Health, Compton Laboratory, UK) and Steve Bolin (NADC/ARS/USDA) in separate but nearly concurrent studies experimentally reproduced MD in PI cattle. In the Brownlie study a cytopathic virus isolated from an animal that died from MD was inoculated into healthy PI herd mates. Both animals succumbed to MD. In the Bolin study, PI cattle were infected with noncytopathic and cytopathic strains of BVDV. Only those animals receiving cytopathic BVDV developed MD. Both studies concluded that cattle born persistently infected with a noncytopathic BVDV succumb to MD later in life when they are superinfected with a cytopathic BVDV. Follow-up studies revealed that the noncytopathic and cytopathic viruses isolated from individual MD cases were antigenically similar. At this point the origin of the cytopathic virus remained a mystery.

## Discovery of the Molecular Basis for Biotype

The first two BVDV strains to be sequenced were the cytopathic strains Osloss (European origin, sequence published in 1987 by L. De Moerlooze) and NADL (North American origin, sequence published in 1988 by M. Collett). Studies of proteins associated with BVDV replication in cultured cells, done during this same time period, demonstrated that cytopathic BVDV could be distinguished from noncytopathic BVDV by the production of an extra nonstructural protein (now known as NS3). This protein is a smaller version of a nonstructural now known as NS2-3 (**Figure 2**). Comparison of the sequences of the cytopathic viruses Osloss and NADL to noncytopathic viruses from the genus *Pestivirus* revealed insertions in the region of the genome coding for the NS2-3. Subsequent studies, in which the NS2-3 coding region of cytopathic and noncytopathic viral pairs isolated from MD cases were sequenced, revealed that nearly all

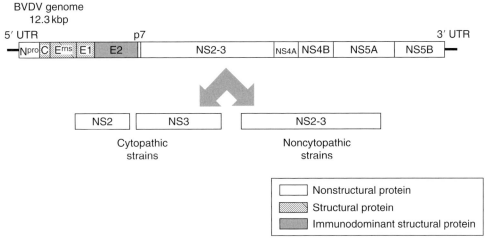

**Figure 2** BVDV genome. The organization of the single-stranded RNA genome of BVDV is shown. The cleavage of the NS2-3 protein to NS2 and NS3 is observed with cytopathic, but not noncytopathic, isolates.

cytopathic viruses had host-cell genetic sequences or duplicated BVDV genetic sequences inserted into the NS2-3 region when compared to their noncytopathic counterparts. These studies suggested that cytopathic viruses arise from noncytopathic viruses by a recombinational event.

## Segregation of BVDV into Two Genotypes

Studies done in the late 1980s and early 1990s, comparing vaccination cross-protection and monoclonal antibody binding, revealed antigenic variability among BVDV strains. While these studies indicated that there was considerable variation among BVDV strains, no standard means of grouping viruses based on these variations was generated. Meanwhile, based on hybridization analysis and sequence comparison, several groups evaluated the 5′ untranslated region (UTR) as a target for polymerase chain reaction (PCR) tests designed to detect the wide range of BVDV strains or to differentiate BVDV strains from other pestiviruses.

Concurrent to this research a highly virulent form of BVD, termed hemorrhagic syndrome, was reported in Canada and the US. While the first report of BVDV in 1946 described a severe acute disease, the most commonly reported form of acute uncomplicated BVD between the 1950s and late 1980s was a mild or subclinical infection. Acute BVDV infections came to be regarded as clinically unimportant and textbooks of the day stated that the transmission of the virus between healthy immunocompetent cattle was probably insignificant. The potential for infection with BVDV to result in clinical disease was downplayed and the research focused on the consequences of transplacental infections and the pathogenesis of BVD-MD. However, the recognition of hemorrhagic syndrome in the late 1980s and early 1990s brought the concept of severe acute BVD once more to the forefront. Case records of cattle admitted to the Cornell College of Veterinary Medicine for the years 1977–87 revealed that 10% of clinically acute BVD infections in adult cattle were associated with thrombocytopenia. During this period an outbreak in the state of New York resulted in 50 of 100 animals in a milking herd becoming ill and 20 of them subsequently dying. Clinical signs included high temperatures, bloody diarrhea, hemorrhages, and prolonged bleeding from venipuncture sites. This disease came to be considered as a distinct form of severe acute BVD termed hemorrhagic syndrome. Severe acute BVD cases were reported with increasing frequency in North America in the early 1990s. These outbreaks were particularly devastating in the Canadian provinces of Quebec and Ontario. The disease in some ways resembled MD, but differed in that it was transmissible to normal non-PI cattle and that only one virus, from the noncytopathic biotype, was present. In 1994, in separate studies, Charles Pellerin (Institut Armand-Frappier, University of Quebec, Canada) and Julia Ridpath (NADC/ARS/USDA) performed phylogenetic analysis of BVDV strains, isolated from animals suffering from hemorrhagic syndrome and arrived at the same conclusion. The BVDV strains associated with the outbreaks of hemorrhagic syndrome belong to a genetic group (genotype) clearly distinct from the BVDV strains commonly used, at that time, in vaccine production, diagnostic tests, and research. The newly recognized group of BVDV was designated BVDV genotype II, while the group containing the strains used in vaccines, detection, and research was termed BVDV genotype I. The names of these two genotypes were later modified to BVDV1 and BVDV2 in keeping with taxonomic conventions in use with other viruses. It was further noted that viruses from the BVDV2 genotype were also isolated from PI calves born to dams that had been vaccinated with vaccines based on BVDV1 isolates. In 2005 the Eighth Report of the International Committee on Taxonomy of Viruses officially classified BVDV1 and BVDV2 as separate and distinct species within the genus *Pestivirus*.

In the late 1990s there was some speculation that BVDV2 strains represented newly emerging viruses that originated in the US as a result of use of vaccines by US producers and were then transferred to Europe. While BVDV2 strains were first recognized in 1994, retrospective characterization of strains collected from BVDV outbreaks in Ontario that occurred between 1981 and 1994 demonstrated that BVDV2 were present in North America at least since the early 1980s. However, the first isolate, retrospectively identified as BVDV2, described in the literature was isolated in Europe prior to 1979. Interestingly, this strain was isolated from a pig and was referred to as an atypical classical swine fever virus (CSFV).

## Emergence of BVDV Reduction/Eradication Programs

In the decades after the first description of BVD and the discovery of its causative agent, systematic reduction/eradication programs were not considered. This was partially due to an underestimation of the economic impact of BVDV infections and the lack of suitable diagnostics. In the mid-1990s studies began to appear showing the sizable economic impact BVDV infections have on beef and dairy industries worldwide. Initially, veterinarians and producers assumed that vaccination alone could substantially reduce the incidence of BVDV infections. However, by the late 1990s it became apparent that four decades of vaccination had not reduced the incidence of BVDV. At this time, better diagnostics, particularly for the detection of PI animals, began to be developed. Concurrently, in the Scandinavian countries, programs designed around a strict testing and removal program for PI accompanied by movement restrictions for infected herds resulted in near eradication of BVDV in those countries by 2005. The success of these efforts encouraged other European countries and US producers to implement reduction/eradication programs. The Scandinavian programs, which were conducted in countries with a relatively low incidence of both BVD and cattle densities, eschewed the use of vaccines and were based solely on the test and removal of PI cattle. In areas of the world in which cattle densities are high and BVDV is endemic, current control programs are focused on a combination of test and removal of PI's, systematic vaccination, and consistent biosecurity.

## Characteristics of BVDV

### Characteristics Common to Both BVDV1 and BVDV2

All BVDV belong to the genus *Pestivirus* within the family *Flavivirus*. The BVDV virion is an enveloped, spherical particle 40–50 nm in diameter that consists of an outer lipid envelope surrounding an inner protein shell or capsid that contains the viral genome. The capsid appears as an electron-dense inner core with a diameter of approximately 30 nm. The lipid envelope that surrounds the virion is pleomorphic which impedes purification of infectious particles by banding in sucrose gradients and identification by electron microscopy. The $M_r$ of the virion is estimated as $6.0 \times 10^7$ and the buoyant density in sucrose is $1.10–1.15 \text{ gm cm}^{-3}$.

The viral genome consists of a single strand of positive-sense RNA, that in the absence of insertions is about 12.3 kbp long and codes for a single open reading frame (ORF). The ORF is preceded and followed by relatively long UTRs on the $5'$ and $3'$ ends of the genome (**Figure 2**). Similar to other members of the genus *Pestivirus*, BVDV1 and BVDV2 viruses encode two unique proteins, $N^{pro}$ and $E^{rns}$. The nonstructural protein $N^{pro}$ is encoded at the very beginning of the ORF and is a proteinase, whose only known function is to cleave itself from the viral polypeptide. The $E^{rns}$ is an envelope glycoprotein that possesses an intrinsic RNase activity.

All *Pestivirus* species, including BVDV1 and BVDV2, are antigenically related. However, neutralizing antibody titers found in convalescent sera are typically several-fold higher against viruses from the same species as compared to viruses from other *Pestivirus* species. Both BVDV1 and BVDV2 viruses may exist as one of two biotypes, cytopathic and noncytopathic. The noncytopathic biotype is the predominant biotype in both BVDV species. Noncytopathic viruses from both the BVDV1 and the BVDV2 species can cross the placenta and establish persistent infections. All BVDV strains, regardless of species or biotype, are lymphotrophic and acute infection always results in destruction of immune tissues. The extent of the loss of immune tissue and the accompanying immunosuppression is dependent on viral strain.

Virions are stable within a pH range of 5.7–9.3. Infectivity is not affected by freezing but decreases at temperatures above 40 °C. Like other enveloped viruses, BVDV are inactivated by organic solvents and detergents. Other methods of inactivation include trypsin treatment (0.5 mg ml$^{-1}$, 37 °C, 60 min), ethylenimine (reduction of 5 log10 units using 10 mM at 37 °C for 2 h), electron beam irradiation (4.9 and 2.5 kGy needed to reduce virus infectivity 1 log10 unit for frozen and liquid samples, respectively), and gamma irradiation (20–30 kGy).

## Differences between BVDV1 and BVDV2

While severe disease and death loss have been reported in association with both BVDV1 and BVDV2 strains, severe acute BVD and hemorrhagic syndrome have only been reproduced experimentally with BVDV2 strains. Aside from severe acute BVDV, which is only caused by a small proportion of BVDV2 strains, it is difficult to distinguish BVDV1 infections from BVDV2 infections based on clinical signs. Further, while virulence is strain dependent, clinical presentation may also be affected by immune status, reproductive status, stress, and the presence of secondary pathogens.

BVDV1 and BVDV2 strains are antigenically distinct as demonstrated by serum neutralization using polyclonal sera and monoclonal antibody binding. The practical significance of antigenic differences is indicated by the birth of BVDV2 PI animals to dams that had been vaccinated against BVDV1 strains. While modified live BVDV1 vaccines may induce antibodies against BVDV2 strains, the titers average one log less than titers against heterologous BVDV1 strains. These observations have lead to the inclusion of both BVDV species in BVDV vaccines.

While the first studies segregating BVDV strains into two different genotypes were based on comparison of the 5′ UTR, differences between BVDV1 and BVDV2 strains are consistently found throughout the genome. Based on complete genomic sequence comparisons, the genetic sequences of BVDV1 and BVDV2 differ from each other as much as other member species of the genus *Pestivirus*, such as CSFV and border disease virus, differ from each other. The level of the genetic difference was the basis for declaring BVDV1 and BVDV2 to be two separate and distinct species.

## Molecular Biology of BVDV

### Viral Genome

The single-stranded RNA genome of BVDV codes for one long ORF (approximately 4000 codons in the absence of insertions). The ORF is bracketed by relatively large 5′ (360–390 bp) and 3′ (200–240 bp) UTRs. The 5′ terminus does not contain a cap structure and there is no poly(A) tract present at the 3′ end. Similar to the genomes of other pestiviruses, both BVDV1 and BVDV2 genomes terminate at the 3′ end with a short poly(C) tract. Sequence identity is highest between BVDV1 and BVDV2 strains in the 5′ UTR region. It is thought that conservation of 5′ UTR sequences is related to formation of tertiary structures required for internal ribosomal entry-mediated initiation of translation. While sequence conservation between BVDV1 and BVDV2 is high in the 5′ UTR, there are two short regions that are notable for their variability. These are located between nucleotides 208–223 and nucleotides 294–323 (nucleotide position numbers based on the sequence of BVDV1-SD-1). (Although BVDV1a-NADL and BVDV2-890 are the type virses for genotypes BVDV1 and BVDV2, respectively, both genomes have inserted sequences. Insertions can cause confusion when indicating genomic location based on nucleotide number. For this reason BVDV1a-SD-1 is used as the reference for nucleotide position. It was the first noncytopathic BVDV1 sequenced and does not have an insertion. The accession number for BVDV1a-SD-1 is M96751.) Sequence variations in these regions have been exploited in PCR-based tests designed to differentiate BVDV1 strains from BVDV2 strains.

### Viral Proteins

#### Structural proteins

The large ORF is translated as a polyprotein. The order of the individual viral proteins within the polyprotein is as follows: $N^{pro}$-C-$E^{rns}$-E1-E2-p7-NS2/3-NS4A-NS4B-NS5A-NS5B (**Figure 2**). The polyprotein is processed co- and post-translationally by host and viral proteases. The proteins associated with the mature virion (structural proteins) are C, $E^{rns}$, E1, and E2. C is the virion nucleocapsid protein. $E^{rns}$, E1, and E2 are associated with the outer envelope of the BVDV virion. These three proteins are highly glycosylated and possess the antigenic determinants of the virus. It is not known whether the $E^{rns}$ and E1 possesses neutralizing epitopes that are important in disease control. The E2 protein is the immunodominant structural protein and possesses neutralizing epitopes that function in disease control. Protective antibodies induced by killed vaccines are predominantly against the E2. Monoclonal antibodies (Mab's) produced against the E2 have been used to differentiate between BVDV1 and BVDV2 strains.

#### Nonstructural proteins

The first viral protein encoded by the BVDV ORF is the nonstructural protein, $N^{pro}$. This protein, as discussed above, is unique to the genus *Pestivirus*. Its only known function is to cleave itself from the polyprotein. The next nonstructural protein, p7, follows the structural protein E2 in the polyprotein. While the role of this cell-associated protein is unknown, it is hypothesized that it is required for production of infectious virus but not for RNA replication. The p7 protein is inefficiently cleaved from the E2 during processing of the polyprotein. This leads to two intracellular forms of E2 with different C termini (E2 and E2-p7). However, neither p7 or E2-p7 are found associated with infectious virus.

Following p7 is the serine protease, NS2-3. As discussed above, in BVDV strains from the cytopathic biotype the NS2-3 is cleaved to NS2 and NS3 (**Figure 2**). Both the uncleaved NS2-3 and the cleaved NS3 act as serine proteases that cleave the remaining nonstructural proteins from the polyprotein. The function of the NS2 is unknown. It is not required for RNA replication and its cleavage from the NS2-3 does not affect serine protease activity. Purified BVDV NS3 also possesses RNA helicase and RNA-stimulated NTPase activities and all three activities (serine protease, RNA helicase, and RNA-stimulated NTPase) are essential to virus viability. While antibodies to the NS2-3 and NS3 do not neutralize infectivity, these proteins possess immunodominant epitopes. The NS2-3 and NS3 (but not the NS2), are strongly recognized by polyclonal convalescent sera and animals vaccinated with modified live vaccines have as nearly a strong antibody response to the NS2-3 and/or NS3 protein as to the E2 structural protein. In contrast, animals vaccinated with inactivated (killed) vaccines primarily react with structural proteins and not the NS2-3 or NS3. The difference in recognition of NS2-3 or NS3 may be useful in differentiating between immune responses to inactivated vaccines and immune responses to natural infection.

The NS4A and NS4B proteins are similar in size, composition, and hydrophobicity to the NS4A and NS4B proteins of other flaviviruses. NS4A acts as a cofactor for the NS2-3 and NS3 serine protease activity. NS4B and NS5A probably are replicase complex components. RNA polymerase activity has been demonstrated for the NS5B protein.

## Viral Replication

### Viral uptake

Uptake of virus appears to be a multistep process that occurs by endocytosis. In the initial step, the virus attaches to the cell surface through interaction of $E^{rns}$ envelope protein and a docking glycosaminoglycan receptor molecule. The next step is mediated by attachment of the E2 envelope protein to the low-density lipoprotein receptor (LDLR) followed by internalization via endocytosis.

### Release of genomic RNA, translation, and replication

The mechanism of release of genomic RNA into the cell cytoplasm is unknown but probably involves acidification of endocytic vesicles. Following release, the genomic RNA must act as mRNA, directing the translation of viral proteins. The translated viral proteins provide functions necessary for RNA replication, protein processing (protease cleavages), and protein trafficking, but are insufficient to perform all or perhaps even most of the functions required. Thus, the virus relies on host cell machinery to provide many functions required for virus replication. The most important of these host-provided functions is protein synthesis. After translation to produce viral proteins, RNA replication begins with the synthesis of complementary negative strands. It has been proposed that a secondary structure motif in the 5' UTR enables the switch of viral RNA from a template for translation to a template for replication. Using these negative strands as templates, genome-length positive strands are synthesized by a semiconservative mechanism involving replicative intermediates and replicative forms. Because viral proteins are not detected on the surface of infected cells, it is thought that virions mature in intracellular vesicles and are released by exocytosis. A substantial fraction of the infectious virus remains cell associated.

## Detection and Control

BVDV diagnostics have focused on the detection of PI animals. Virus isolation on cultured bovine cells remains the gold standard. However, due to ease and lower expense, antigen detection by either immunohistochemistry or antigen capture ELISA or nucleic acid detection by RT-PCR are gaining favor. Both killed and modified live vaccines are available for the prevention of BVD. Control by vaccination alone is compromised by the heterogeneity observed among BVDV strains, lack of complete fetal protection afforded by vaccination, and the failure to remove PI animals from cattle populations.

*See also:* Border Disease Virus; Classical Swine Fever Virus; Flaviviruses: General Features.

## Further Reading

Dubovi ED, Brownlie J, Donis R, et al. (eds.) (1996) *International Symposium on Bovine Viral Diarrhea Virus: A 50 Year Review*. Ithaca, NY: Cornell University.

Goyal SM and Ridpath JF (eds.) (2005) *Bovine Viral Diarrhea Virus: Diagnosis, Management and Control*. Ames, IA: Blackwell Publishing.

Houe H, Brownlie J, and Steinar Valle P (eds.) (2005) *Bovine Virus Diarrhea Virus (BVDV) Control, Vol. 72: Preventive Veterinary Medicine*. Special Issue. Amsterdam: Elsevier Academic Press.

Smith RA (consulting ed.) and Brock KV (guest ed.) (2004) *Veterinary Clinics of North America Food animal practice. Vol. 20: Bovine Viral Diarrhea Virus: Persistence is the Key*. Philadelphia, PA: Saunders.

Thiel H-J, Collett MS, Gould EA, et al. (2005) Family *Flaviviridae*. In: Fauquet CM, Mayo MA, Maniloff J, Desselberget U, and Ball LA (eds.) *Virus Taxonomy Classification and Nomenclature of Viruses*, 8th edn., pp. 981–998. Amsterdam: Elsevier Academic Press.

# Brome Mosaic Virus

**X Wang and P Ahlquist,** University of Wisconsin – Madison, Madison, WI, USA

© 2008 Elsevier Ltd. All rights reserved.

## Introduction

Brome mosaic virus (BMV) is a positive-strand RNA virus that infects cereal plants, causing mosaic symptoms and stunting. BMV is the type member of the genus *Bromovirus* in the family *Bromoviridae*, which belong to the alphavirus-like superfamily of human, animal, and plant viruses. BMV has been used as a model for studying gene expression, RNA replication, host–virus interactions, recombination, and encapsidation by positive-strand RNA viruses. Results produced in these areas by many researchers have revealed insights and principles that extend beyond BMV to many other viruses and to general cellular biology. This article reviews selected aspects of this work.

## Genome Structure, Expression, and Sequence

In 1971, Lane and Kaesberg used buoyant density gradients to separate BMV virions into three classes having identical capsids but different RNAs (**Figure 1**). Heavy virions contain a single copy of RNA1 (3.2 kbp), medium-density virions contain one copy each of RNA3 (2.1 kbp) and RNA4 (0.9 kbp), and light virions contain a single copy of RNA2 (2.9 kbp). Productive infections require all three virions, but not all four RNAs: infectivity is abolished by omitting RNA1, RNA2, or RNA3, but not RNA4.

Interest in characterizing the protein(s) encoded by each of the four BMV virion RNAs motivated early *in vitro* translation studies with purified BMV virion RNAs. Shih and Kaesberg found that RNAs 1, 2, and 4 are monocistronic while RNA3 is dicistronic. Although early infectivity studies showed that RNA3 encodes coat protein (CP), trypsin degradation analysis showed that the principal translation product of RNA3 was unrelated to CP, while RNA4 served as an excellent template for CP. Moreover, when mixtures of all BMV RNAs were added to wheat germ extracts in increasing amounts, RNA4 inhibited translation of genomic RNAs 1–3. These results showed that RNA4 is a subgenomic CP mRNA derived from RNA3, and implied an elegant system of gene regulation by translational competition: early in infection when viral RNA concentrations are low, all viral proteins including nonstructural RNA replication factors are translated, while after the virion RNAs are sufficiently amplified, CP is preferentially translated to encapsidate these RNAs.

By the early 1970s, RNA bacteriophage studies had provided valuable information on prokaryotic translation initiation sites, including the finding that the first AUG initiation codon usually was 100 or more nucleotides from the RNA 5′-end. In 1975, Dasgupta and Kaesberg characterized the first eukaryotic translation initiation site by isolating two fragments of BMV RNA4 that were efficiently bound by wheat germ ribosomes. RNA sequencing (a challenging, chromatography-based process at the time) revealed that the two fragments were 5′-terminal, overlapping, and encoded the first 4 and 14 amino acids of CP, respectively. The most distinguishing feature was that the initiating AUG codon for CP began only 10 nucleotides (nt) from the $m^7G^{5'}ppp^{5'}Gp$-capped RNA4 5′-end, presaging the now well-known mechanistic linkage between most eukaryotic translation initiation and 5′-mRNA caps.

The ~8.2 kbp BMV genome sequence was completed in 1984. In good agreement with *in vitro* translation results, RNA1 and RNA2 encode single proteins 1a (109 kDa) and 2a (94 kDa), respectively, the 5′-half of RNA3 encodes protein 3a (35 kDa), and the 3′-half of RNA3 encodes RNA4, the subgenomic mRNA for CP (20 kDa) (**Figure 1**). Later work discussed below showed that 3a is required for infection movement in plants. Comparisons with other emerging viral RNA and protein sequences quickly revealed that the BMV 1a and 2a proteins, already implicated in RNA replication by protoplast experiments, shared extensive amino acid sequence similarities with proteins encoded by an outwardly diverse set of plant and animal positive-strand RNA viruses. These similarities initially were recognized between BMV, alfalfa mosaic virus, tobacco mosaic virus, and Sindbis virus, and subsequently were found to extend to many other viruses now grouped together as the alphavirus-like superfamily. Similarities shared by these viruses include a polymerase domain in BMV 2a (hereafter $2a^{pol}$) and RNA capping and RNA helicase-like domains in 1a (**Figure 1**).

## Infectious *In Vitro* Transcripts and Foreign Gene Expression

Also in 1984, BMV was used to produce the first infectious transcripts from cloned RNA virus cDNA. Specially designed BMV cDNA clones were transcribed to produce capped *in vitro* transcripts of genomic RNAs 1–3, each with the natural viral RNA 5′-end and only a few extra nucleotides at the 3′-end. Mixtures of all three BMV

RNA transcripts, but not their parent cDNA clones, were infectious to barley plants, a natural BMV host.

This ability to engineer the expression of infectious transcripts provided a means to manipulate the viral RNA genome at the cDNA level using recombinant DNA technology, which has subsequently proved applicable to many other RNA viruses. In one of the first applications of these new reverse genetics approaches, French and co-workers demonstrated that foreign genes could be inserted into the viral genome while retaining the ability to replicate and express genes. Using a transcribable BMV RNA3 cDNA clone, the CP open reading frame (ORF) was replaced with the ORF of the bacterial reporter gene chloramphenicol acetyltransferase (CAT). When *in vitro* transcribed and inoculated onto barley protoplasts with RNA1 and RNA2 transcripts, this RNA3 derivative was replicated and produced CAT activity at higher levels than previously achieved by DNA-based transformation. This first demonstration that RNA viruses can be engineered at the cDNA level showed that the viral RNA genome functions in a sufficiently flexible and modular fashion to tolerate even large changes such as whole gene replacements without substantial optimization, which has significant implications for virus evolution, basic research and biotechnology applications such as development of additional gene expression vectors.

transcripts from cloned viral cDNAs. For BMV, barley protoplast systems developed and refined by the groups of Okuno and Furusawa, Hall and others in the late 1970s have allowed studies of all aspects of BMV RNA replication, subgenomic RNA synthesis, progeny RNA encapsidation, and the like. The highly synchronized infections obtained also allowed detailed kinetic studies.

In the early 1990s, it was demonstrated that BMV also would direct RNA replication, subgenomic RNA synthesis, selective viral RNA encapsidation, RNA recombination and the like in the well-studied yeast, *Saccharomyces cerevisiae*. This host provides some of the same advantages as plant protoplasts together with rapid growth, a particularly small genome, well-characterized cell biology, and powerful classical and molecular genetic tools including genome-wide arrays of isogenic yeast strains with each gene systematically modified by deletion, GFP-tagging, etc. Yeast expressing BMV RNA replication factors 1a and $2a^{pol}$ support the replication of BMV genomic RNAs introduced by transfection or DNA-dependent transcription from plasmids or chromosomally integrated expression cassettes, duplicating nearly all features of replication in plant cells. To facilitate yeast genetic studies, BMV RNA replicons can express selectable or screenable markers in yeast and are transmitted to yeast daughter cells during cell division with 85–90% efficiency, rivaling the transmission of yeast DNA plasmids.

## *In Vitro* and *In Vivo* Replication Studies

Positive-strand RNA viruses like BMV replicate their genomes in a completely RNA-dependent manner, producing a negative-strand RNA replication intermediate for each genomic RNA. Studies on BMV RNA replication were greatly advanced by the development and subsequent use of such tools as *in vitro* RNA-dependent RNA polymerase (RdRp) extracts and cultured plant protoplasts. In 1979, Hall and colleagues developed a virus-specific *in vitro* RdRp extract from BMV-infected barley leaves that synthesized full-length negative-strand RNAs using added BMV virion RNAs as templates. This was the first eukaryotic *in vitro* RdRp preparation exhibiting a high level of template specificity, with other viral RNAs having less than 20% of the template activity of BMV RNAs. As noted in part below, this and similar BMV *in vitro* systems have been utilized by the groups of Hall, Quadt and Jaspars, Kao, and others to make many advances regarding promoters for positive- and negative-strand RNA synthesis, initiation mechanisms, and other issues.

In parallel to *in vitro* systems, cultured plant protoplasts provided a valuable substrate for *in vivo* replication studies due to their ability to be infected or transfected with nearly 100% efficiency with virions, virion RNAs, or *in vitro*

## BMV Proteins in RNA Replication

Protoplast studies showed that BMV RNA replication and subgenomic RNA4 synthesis require the viral 1a and $2a^{pol}$ proteins but not 3a or CP. As mentioned above, 1a and $2a^{pol}$ share sequence similarity with proteins encoded by other viruses in and beyond the alphavirus-like superfamily. The conserved central domain of $2a^{pol}$ shows similarity to RdRps encoded by picornaviruses and some other RNA viruses. The N-terminal 1a protein domain is related to alphavirus protein nsp1 and contains $m^7G$ methyltransferase and $m^7GMP$ covalent binding activities required for capping viral RNA *in vivo*. The C-terminal 1a domain has sequence similarity to superfamily I NTPase/helicases and NTPase activity that is required for RNA template recruitment and RNA synthesis.

In plant and yeast cells, BMV RNA replication occurs on endoplasmic reticulum (ER) membranes, predominantly in the perinuclear region. In both cell types, 1a localizes to ER membranes in the absence of other viral factors. This localization depends on the N-terminal 1a domain and brings 1a to the cytoplasmic face of ER membranes as a peripheral, not transmembrane, protein. The C-terminal 1a NTPase domain recruits $2a^{pol}$ to ER membranes by interacting with $2a^{pol}$ N-proximal sequences.

In yeast cells replicating BMV RNAs, the ER membrane is modified by numerous 50–60 nm spherular invaginations into the ER lumen. Similar membrane invaginations, designated spherules, are seen in plant cells infected by any of several bromoviruses and in animal cells infected by viruses in and beyond the alphavirus-like superfamily. Immunogold labeling and electron microscopy localized 1a, 2a$^{pol}$, and nascent BMV RNAs to the spherule interiors, which remain connected to the cytoplasm by a narrow neck. Protein 1a is sufficient to induce these spherules, and immunogold and biochemical studies show that each spherule contains from one to a few hundred 1a proteins and ~25-fold fewer 2a$^{pol}$ proteins. These and other features imply that the structure, assembly, and operation of these spherule replication complexes have functional and perhaps evolutionary links to the replicative cores of retrovirus and dsRNA virus virions, which sequester viral RNA replication templates and their polymerases in a protein shell.

## Noncoding Regions and cis-Signals

The first region of BMV RNAs to attract attention as a potential regulatory element was the 3'-end. Synergistic work by multiple groups showed that the 3'-noncoding regions of BMV RNAs are highly conserved, multifunctional domains that direct negative-strand RNA synthesis, contribute to RNA encapsidation, translation, and stability, and possess multiple tRNA-like features and functions.

Limited early sequence data showed that BMV RNAs 1–4 share a tRNA-like CCA$_{OH}$ 3'-end. In 1972, Hall and colleagues showed that all four BMV RNAs were selectively aminoacylated *in vitro* with tyrosine. Their further studies showed that tyrosylated BMV RNAs interacted with translation elongation factor 1a and that BMV RNAs were tyrosylated *in vivo* during infection of barley protoplasts. BMV RNA 3'-ends were also found to interact with (ATP, CTP):tRNA nucleotidyl transferase, which can add 3'-CCA$_{OH}$ ends to mature or maintain incomplete BMV RNAs, thus acting as a primitive telomerase. Beginning in the 1970s with work by Dasgupta and Kaesberg, further sequencing, enzymatic structure probing, and three-dimensional computer modeling showed that the 3' ~200 nt of all BMV RNAs were strongly conserved and folded into an extended, tRNA-like structure with at least two alternate forms that differed in pairing nt near the 3' with local or distal partners. Similarly conserved, highly structured 3'-regions with related alternate forms were also found in other members of the family *Bromoviridae*.

*In vitro* and *in vivo* studies using RNA fragments and mutations showed that sequences within the 3'-terminal tRNA-like structure direct negative-strand RNA initiation for RNA replication. Recent results have also implicated the tRNA-like sequence in translation and encapsidation (see below). Involvement in all of these functions led to early and continuing suggestions that the 3'-region mediates co-regulation of the varied uses of BMV positive-strand RNAs to minimize conflicts between multiple essential processes.

A second class of elements combining tRNA-like features, replication signals, and possible interaction with translation are the BMV template recruitment elements, first recognized in RNA3. Deletion analysis revealed that, in addition to 3'- and 5'-sequences required for negative- and positive-strand RNA initiation, BMV RNA3 replication *in vivo* requires a segment of ~150 nt in the 5'-half of the intergenic noncoding region between the 3a and CP ORFs (**Figure 1**). Subsequent studies

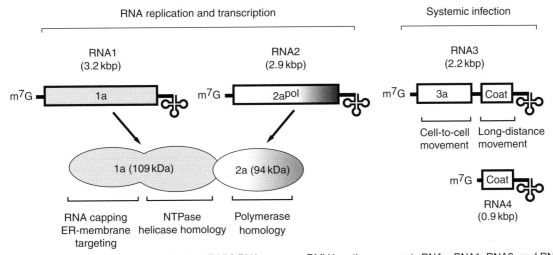

**Figure 1** Organization of the brome mosaic virus (BMV) RNA genome. BMV has three genomic RNAs, RNA1, RNA2, and RNA3, and a subgenomic mRNA, RNA4, encoded by the 3'-portion of RNA3 as shown. Each of these four BMV positive-strand RNAs bears a 5' m$^7$G cap and 3' tRNA-like structure (cloverleaf). Open boxes represent the ORFs for the viral proteins 1a and 2a$^{pol}$, which are required for RNA replication, and 3a and CP, which are required for systemic movement in host plants. Some specific functions and features of each viral protein are listed.

showed that this region is required for a step prior to negative-strand RNA synthesis, and is necessary and sufficient for 1a to recognize and recruit an RNA to a membrane-associated, translationally inaccessible, nuclease-resistant state that appears to be the spherule interior. Structure probing studies show that this intergenic RNA3 sequence folds into a long, bulged stem loop, which presents at its apex the invariant sequence and structure of tRNA TΨC stem loops. In plant and yeast cells, the appropriate BMV residues in this consensus are modified to T and Ψ, showing that, like the 3′-end, this sequence interacts *in vivo* with tRNA-specific enzymes. Moreover, any mutations to this TΨC stemloop mimicry abolish 1a-mediated template recruitment. Similar stem loops with apical TΨC stem loop regions are found at the extreme 5′-ends of BMV RNA1 and RNA2, where they similarly direct 1a-mediated template recruitment.

As one important aspect of *cis*-signals in BMV replication, the Kao group in particular has defined and dissected minimal core promoters for negative-strand, positive-strand, and subgenomic RNA synthesis, using a variety of approaches. Among other results, their mutational studies imply that the BMV RdRp–promoter interaction has the characteristics of an induced fit, wherein the RdRp has some tolerance to adjust its binding to a range of promoter variants as long as some key sequence features remain. This model potentially reconciles the specificity of BMV RNA synthesis with the ability of the RdRp to synthesize different forms of viral RNA from separate promoters with distinct primary sequences and secondary structures. Kao and associates further found that DNA or DNA/RNA hybrid templates containing the key BMV promoter sequences can be recognized *in vitro* by BMV RdRp extracts and copied into RNA. Although the efficiency of copying DNA templates is ~15-fold lower than for BMV RNA templates, these results have significant potential implications for virus evolution.

## Subgenomic mRNA Synthesis

Early observations that the nature of CP in BMV infections is dictated by RNA3 rather than RNA4, and that RNA4 was regenerated when omitted from BMV inoculum, were partially explained when sequencing revealed that RNA4 was encoded by the 3′-portion of RNA3. Nevertheless, whether RNA4 was produced from RNA3 by cleavage or any of several possible RNA synthesis pathways remained uncertain. In 1984, Miller, Hall, and colleagues showed that a BMV RdRp extract produced subgenomic RNA4 *in vitro* when supplied with negative-strand RNA3 templates, and that the product RNA4 could be labeled by $\gamma$-$^{32}$P-GTP incorporation, demonstrating *de novo* initiation. This first elucidation of a pathway for subgenomic mRNA synthesis appears to provide a meaningful precedent for similar subgenomic mRNA synthesis by many positive-strand RNA viruses, and an important foundation for understanding the diversity of alternate mechanisms that has begun to emerge with the demonstration of distinctly different subgenomic mRNA synthesis pathways used by coronaviruses, nodaviruses, and some other positive-strand RNA viruses.

*In vivo* and *in vitro* analyses of the BMV subgenomic mRNA promoter have complemented well to reveal a core promoter within the 20 nt immediately upstream of the RNA4 start site, which directs low level but accurate initiation of subgenomic mRNA. *In vivo*, the activity of this core promoter is greatly enhanced by upstream sequences that include an oligo(A) tract of variable, ~16–22 nt length in the viral population as well as upstream, partially conserved repeats of core promoter sequences. The important role of the oligo(A) tract suggests that while negative-strand ssRNA can serve as a subgenomic mRNA template *in vitro*, the natural *in vivo* template might be a dsRNA within which the oligo(A) provides a melting site to facilitate internal initiation.

## Host Factors in RNA Replication

As for many other viruses, the small size of the 8.2 kbp BMV genome relative to the complexity of BMV replication suggests that many, if not most, steps in BMV RNA replication involve contributions from host factors. Since most of the viral genome is devoted to RNA replication functions, host factors appear particularly likely to be involved in that process as well. Biochemical and genetic results support these ideas and have begun to elucidate some of the relevant host factors and contributions. In 1990, for example, Quadt and Jaspars used anti-1a antiserum to precipitate a complex of BMV 1a, 2a$^{pol}$, and approximately five host proteins from an active BMV RdRp extract from barley cells. Subsequently, the 41 kDa subunit of cellular translation initiation factor-3 (eIF-3) was found tightly associated with and capable of stimulating BMV RdRp activity.

The ability of BMV to direct RNA replication in yeast further facilitated analysis of host contributions to these processes. Classical yeast genetic approaches identified a number of host genes important to BMV RNA replication, such as *DED1*, a translation factor helicase involved in selective regulation of 2a$^{pol}$ translation; multiple components of the *LSM1–7/PAT1/DHH1* complex, which facilitates recruitment of BMV genomic RNA templates into translation and from translation into RNA replication; *YDJ1*, encoding a chaperone required to activate the RNA replication complex, likely through action on 2a$^{pol}$; and *OLE1*, encoding a fatty acid desaturase required to produce a membrane lipid profile compatible with membrane-associated viral RNA replication. More recently, systematic screening of a genome-wide array of yeast knockout strains identified nearly 100 host genes that, when deleted, inhibited or enhanced BMV RNA replication by 3- to >25-fold. These include host genes in RNA, protein, or membrane modification pathways and genes of unknown function, which are shedding further light on BMV replication, virus–host interactions, and cell processes.

## RNA Recombination

In 1986, Bujarski and Kaesberg used BMV to provide an early demonstration of RNA recombination in a plant virus. Subsequent work by the Bujarski group and others demonstrated many forms and features of inter- and intramolecular, homologous, and nonhomologous RNA recombination in BMV. Among other findings, Bujarski and collaborators showed that mutations in BMV RNA replication factors 1a and 2a$^{pol}$ could alter the frequencies and distributions of crossover sites, implying that at least a significant portion of such recombination was a byproduct of RNA replication, as by template switching. These and other results show that RNA recombination is a major force for repairing BMV genomes damaged by the high mutation rates of viral RNA replication and other events, thereby contributing to BMV survival and adaptability.

## Virion Structure and RNA Encapsidation

BMV forms nonenveloped virions ~28 nm in diameter. The outer capsid is composed of 180 copies of CP arranged with $T=3$ quasi-icosahedral symmetry. Cryo-electron microscope reconstructions by Baker and collaborators and subsequently X-ray crystallography by McPherson and co-workers showed that the BMV capsid structure is extremely similar to that of the related bromovirus, cowpea chlorotic mottle virus (CCMV), whose capsid structure was solved by Johnson and colleagues in 1995. Intriguingly, some features of these capsids are dissimilar from other known isometric RNA virus capsids, but similar to capsids of the DNA-containing papovaviruses. These features include orientation of the core β-barrel nearly perpendicular to the capsid surface to form distinctively prominent hexameric and pentameric capsomeres, and linking of adjacent capsomere clusters by exchange of invading C-terminal arms. These shared features suggest that BMV CP and the CPs of polyoma- and papillomaviruses likely share a common ancestor. BMV virion RNA is arranged as an interior subshell inside the capsid, leaving a hollow virion center. The N-terminal 26 amino acids of the CP, which are highly basic and required for RNA packaging, interact with the RNA to neutralize its charge.

In addition to the predominant 180-subunit capsid, BMV CP can also assemble *in vivo* into a 120-subunit capsid, composed of 60 CP dimers, first discovered in yeast and subsequently confirmed in infected plants. This assembly polymorphism is controlled *in vivo* by the RNA packaged, with BMV RNA2 packaged in 180-subunit capsids, while a small chimeric mRNA containing the CP ORF as its only BMV sequence is packaged in 120-subunit capsids. Structural features shared by 120- and 180-subunit capsids imply that a common pentamer of CP dimers is an important intermediate in BMV virion assembly.

*In vitro* and *in vivo* encapsidation studies by the Rao group and by Mise, Okuno, Furusawa, and colleagues identified portions of the BMV 3a coding region whose deletion blocked RNA3 encapsidation and interfered with normal co-encapsidation of RNA3 and RNA4. Studies by the Rao group also implicated the 3′ tRNA-like structure of BMV RNAs as a facilitator of encapsidation in *cis* or *trans*, possibly by nucleating CP interactions into productive assemblies such as pentamers of dimers, and showed that BMV RNA replication promotes selective encapsidation of viral RNAs, possibly by inducing coupled synthesis of viral RNA and CP in close proximity.

## Infection Spread and Host Range

BMV replicates and encapsidates its RNAs in directly inoculated cells from a wide variety of plants, but has a fairly restricted host range for systemic infection of whole plants. The effective host range for BMV infection thus appears to be determined at the level of initiating or sustaining infection spread from the sites of primary infection. Work by De Jong, Mise, Okuno, Furusawa, Rao, their colleagues, and others have elucidated many features of such spread and its viral determinants.

BMV 3a and CP are dispensable for RNA replication but required for systemic spread. Disruption of the 3a gene blocks cell-to-cell movement, limiting infection to individual, directly inoculated cells. The 3a protein shares multiple properties with cell-to-cell movement proteins of other viruses, including cooperative binding to single-stranded RNA, localization to the plasmodesmatal connections, and induction of virion-containing tubules from the surface of BMV-infected protoplasts. Disruption of the CP gene stops virus spread to noninoculated leaves. Whether local cell-to-cell spread occurs in the absence of CP depends on several factors including the 3a allele and the host plant.

Exchanging genomic RNAs, individual genes, and gene segments among BMV strains and between BMV and other viruses shows that adaptation for infection spread in particular host plants depends not only on 3a and CP but also on features of RNA1 and RNA2. Host adaptation of 3a generally exerts the predominant effects on infection spread, and only a few amino acid changes in 3a are required to extend BMV host range from monocotyledonous to dicotyledonous plants. However, changes modulating the efficiency of systemic spread also map to 1a and $2a^{pol}$. Such changes may alter systemic spread through host-specific effects on RNA replication, as by influencing the ability of the virus to replicate and spread faster than host defense responses. Alternatively, 1a and $2a^{pol}$ may possess additional functions, as for some C-terminal $2a^{pol}$ sequences that are dispensable for RNA replication but required for efficient systemic infection.

## Future Perspectives

Through a variety of intrinsic features and the work of many investigators, research on BMV not only has advanced understanding of bromoviruses, but also has contributed significantly to general virology and molecular biology. Some of the challenges for the future include improved definition and analysis of distinct substeps in viral RNA synthesis including initiation, elongation, termination, and capping; better characterization at molecular, cellular, and tissue levels of the pathways and mechanisms involved in infection spread and the interplay of virus-directed processes and host defenses; and improved understanding of the linkages between different infection processes, including regulated gene expression, RNA replication, encapsidation, and spread.

*See also:* Alfalfa Mosaic Virus; Bromoviruses; Cucumber Mosaic Virus; *Ilarvirus*; Recombination.

## Further Reading

Bujarski JJ and Kaesberg P (1986) Genetic recombination between RNA components of a multipartite plant virus. *Nature* 321: 528–531.

Dreher TW, Bujarski JJ, and Hall TC (1984) Mutant viral RNAs synthesized *in vitro* show altered aminoacylation and replicase template activities. *Nature* 311: 171–175.

French R, Janda M, and Ahlquist P (1986) Bacterial gene inserted in an engineered RNA virus: Efficient expression in monocotyledonous plant cells. *Science* 231: 1294–1297.

Hardy SF, German TL, Loesch-Fries LS, and Hall TC (1979) Highly active template-specific RNA-dependent RNA polymerase from barley leaves infected with brome mosaic virus. *Proceedings of the National Academy of Sciences, USA* 76: 4956–4960.

Johnson JE and Speir JA (1997) Quasi-equivalent viruses: A paradigm for protein assemblies. *Journal of Molecular Biology* 269: 665–675.

Lane LC and Kaesberg P (1971) Multiple genetic components in bromegrass mosaic virus. *Nature New Biology* 232: 40–43.

Lucas RW, Larson SB, and McPherson A (2002) The crystallographic structure of brome mosaic virus. *Journal of Molecular Biology* 317: 95–108.

Miller WA, Dreher TW, and Hall TC (1985) Synthesis of brome mosaic virus subgenomic RNA *in vitro* by internal initiation on (−)-sense genomic RNA. *Nature* 313: 68–70.

Nagano H, Okuno T, Mise K, and Furusawa I (1997) Deletion of the C-terminal 33 amino acids of cucumber mosaic virus movement protein enables a chimeric brome mosaic virus to move from cell to cell. *Journal of Virology* 71: 2270–2276.

O'Reilly EK and Kao CC (1998) Analysis of RNA-dependent RNA polymerase structure and function as guided by known polymerase structures and computer predictions of secondary structure. *Virology* 252: 287–303.

Rao ALN (2006) Genome packaging by spherical plant RNA viruses. *Annual Review of Phytopathology* 44: 61–87.

Schwartz M, Chen J, Janda M, Sullivan M, den Boon J, and Ahlquist P (2002) A positive-strand RNA virus replication complex parallels form and function of retrovirus capsids. *Molecular Cell* 9: 505–514.

Shih DS and Kaesberg P (1976) Translation of the RNAs of brome mosaic virus: The monocistronic nature of RNA1 and RNA2. *Journal of Molecular Biology* 103: 77–88.

Siegel R, Bellon L, Beigelman L, and Kao CC (1999) Use of DNA, RNA and chimeric templates by a viral RNA-dependent RNA polymerase: Evolutionary implications for the transition from the RNA to the DNA world. *Journal of Virology* 73: 6424–6429.

# Bromoviruses

**J J Bujarski,** Northern Illinois University, DeKalb, IL, USA and Polish Academy of Sciences, Poznan, Poland

© 2008 Elsevier Ltd. All rights reserved.

## Introduction

The family *Bromoviridae* represents one of the most important families of plant viruses, infecting a wide range of herbaceous plants, shrubs, and trees. Several of them are responsible for major epidemics in crop plants. The *Bromoviridae* include the spherical icosahedral viruses with a tripartite positive-sense RNA genome. Since these viruses usually accumulate to a high level in the infected tissue, they have been a convenient subject of molecular studies. The type members of different genera, such as cucumber mosaic virus (CMV), brome mosaic virus (BMV), and alfalfa mosaic virus (AMV), constitute excellent molecular models for basic research on viral gene expression, RNA replication, virion assembly, and the role of cellular genes in basic virology.

The genus *Bromovirus* (**Table 1**) comprises not only the best-characterized RNA viruses of the family such as

**Table 1** Main characteristics of the RNA genome in members of the genus *Bromovirus*

| Species | Particle size (nm) | RNA1[a] | RNA2 | RNA3 | 3'end[b] | sgRNAs[c]/DI RNAs[d] |
|---|---|---|---|---|---|---|
| BMV | 26 | 3234 | 2865 | 2117 | tRNA | 2/yes |
| BBMV | 26 | 3158 | 2799 | 2293 | tRNA | 1/yes |
| CCMV | 26 | 3171 | 2774 | 2173 | tRNA | 1/n.a. |
| CYBV | 26 | 3178 | 2720 | 2091 | n.a. | 1/n.a. |
| MYFV | 25 | n.a. | n.a. | n.a. | n.a. | 1/n.a. |
| SBLV | 28 | 3252 | 2898 | 2213 | n.a. | 1/n.a. |

[a]Sequences of RNAs 1 through 3 are available in GenBank.
[b]For these bromoviruses tRNA-like structures were analyzed.
[c]Shows the number of identified sgRNAs.
[d]n.a. – not analyzed.

brome mosaic virus (BMV), broad bean mottle virus (BBMV), or cowpea chlorotic mottle virus (CCMV), but also melandrium yellow fleck virus (MYFV), spring beauty latent virus (SBLV), and cassia yellow blotch virus (CYBV). BMV infects cereal grains and numerous members of the Graminae, and is commonly distributed throughout the world. BBMV and CCMV infect legume species, while other bromoviruses infect selected members of several plant families. Interestingly, BMV, MYFV, and SBLV were found to infect *Arabidopsis thaliana*, a model plant host.

Bromoviruses are mostly related to cucumoviruses, the agriculturally important genus of *Bromoviridae*. Both bromo- and cucumoviruses share such properties like the molecular and genetic features of their tripartite RNA genome, the number of encoded proteins, and similar virion structure. The computer-assisted comparisons of amino acid sequences revealed similarity among their RNA replication proteins. More broadly, the replication factors share amino acid sequence similarity within the alphavirus-like replicative superfamily of positive-strand RNA viruses, which includes numerous plant and important animal/human viruses.

## Phylogeny and Taxonomy of *Bromoviridae*

The current taxonomy divides the family *Bromoviridae* into five genera, named after their best-known members: *Alfamovirus* (one member, type virus: *Alfalfa mosaic virus*, AMV), *Bromovirus* (five members, type virus: *Brome mosaic virus*), *Cucumovirus* (three members, type member: *Cucumber mosaic virus*), *Ilarvirus* (16 viruses, type virus: *Tobacco streak virus*), and *Oleavirus* (one member: *Olive latent virus 2*, OLV-2). However, the evaluation of the phylogenetic relationships based on integrating information on the entire proteome of viruses of the *Bromoviridae* revealed that (1) AMV should be considered a true ilarvirus instead of forming a distinct genus *Alfamovirus*; (2) one virus, pelargonium zonate spot virus (PZSV), should probably constitute a new genus; (3) the genus *Ilarvirus* should be divided into three subgroups; and (4) the exact location of OLV-2 within the *Bromoviridae* remains unresolved.

## Virion Properties and Structure

The bromovirus virions are nonenveloped, possess a $T = 3$ icosahedral symmetry, measure about 28 nm in diameter, and are composed of 180 molecules of a 20 kDa coat protein (CP) clustered into 12 pentamers and 20 hexamers. Three types of icosahedral particles of identical diameters (28 nm) and similar sedimentation coefficients (*c.* 85 S) encapsidate different RNA components: RNA1 (mol. wt. *c.* $1.1 \times 10^6$), RNA2 (mol. wt. *c.* $1.0 \times 10^6$), and RNA3 plus sgRNA4 (mol. wt. *c.* $0.75 \times 10^6$ and $0.3 \times 10^6$). The crystal structures of both BMV and CCMV have been resolved showing very similar organization (**Figure 1**). The structure of CCMV serves as a model bromovirus structure. Similar to other icosahedral viruses, its CP subunits, folded into a β-barrel core, are oriented vertically to the capsid surface and organized within both the protruding pentameric and hexameric capsomers. The interactions among hydrophobic amino acid residues stabilize the capsomers, and the hexameric subunits are further stabilized via interactions between N-terminal portions, where six short β-strands form a hexameric tubule called β-hexamer. Mutational analysis demonstrated that this structure was not required for virion formation but modulated virus spread *in planta*. In addition, the capsomers are held together by interactions through C-terminal portions that extend radially from the capsid. The C-termini are anchored between the β-barrel core and the N-proximal loop, and this interaction might be responsible for initiation of assembly of CCMV capsids.

The molecular replacement using the CCMV structure as a model revealed that distinct portions of the BMV capsids can also form the pentameric and the hexameric capsomers. The CP has the canonical β-barrel topology with extended N-terminal polypeptides where a significant fraction of the N-terminal peptides is cleaved. Overall, the virion appears to assemble loosely among the hexameric capsomers. This is likely responsible for virion swelling at neutral pH. The structure also coordinates metal ($Mg^{2+}$) ions. Interestingly, $T = 1$ imperfect icosahedral particles of BMV virions can be created by treatment of $T=3$ virus, with particles composed of a loose arrangement of reoriented pentameric capsomers.

The single-stranded viral RNA is located inside the capsid, as a separate torus-shaped subshell, where the basic N-terminal amino acids of the CP interact with the RNA and neutralize the phosphate groups. Other sites of RNA interaction localize to the internally proximal basic amino acids of the CP subunits. CCMV capsid undergoes well-studied reversible structural transitions where shifting pH from 5.0 to 7.0 causes capsid expansion. The swollen forms can be completely disassembled at 1 M salt concentration. Some CP mutations can further stabilize the capsids by a new series of bonds that are resistant to high ionic strength. Other mutations in the CP subunits show that capsid geometry is flexible and may adapt to new requirements as the virus evolves. Capsids of bromoviruses are also stabilized by metals. In CCMV, there are 180 unique metal-binding sites that coordinate five amino acids from two adjacent CP subunits. Some metals have less affinity for the binding sites than the others. The biological significance of capsid swelling is not completely understood.

The detailed knowledge about the structure of bromoviral virions has found applications in nanotechnology. The pH-dependent structural transitions of CCMV capsids provide opportunities for the reversible pH-dependent gating, useful during the regulation of size-constrained biomimetic mineralization. The interior surface of CCMV capsids can be genetically engineered to act as a ferritin surrogate that spatially constrains the formation of iron oxide nanoparticles, whereas the exterior surfaces have been functionalized for ligand presentation. The latter can be tuned by asymmetric reassembly of differentially functionalized CCMV CP subunits. The electrostatically driven adsorption behavior of CCMV on Si and amine-functionalized Si as well as the fabrication of multilayer CCMV films have been reported.

## Genome Organization and Expression

High-resolution density gradients as well as gel electrophoresis under denaturing conditions reveal three classes of BMV virion particles, each encapsidating separate plus-sense components of the viral genome. One class encapsidates one copy of RNA1, another class one copy of RNA2, and the third class one copy of RNA3 and one copy of subgenomic RNA4 (**Figure 2**). All three virion

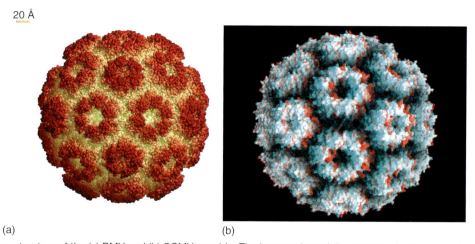

**Figure 1** Surface structure of the (a) BMV and (b) CCMV capsids. The hexameric and the pentameric structural elements are visible. See also text for further description (a, b). From the virion picture collection at the web-site of the Institute for Molecular Virology at the University of Wisconsin–Madison). Reproduced by permision of Institute for Molecular Virology.

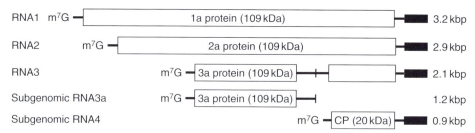

**Figure 2** Molecular organization of a typical bromovirus RNA genome. The RNA components 1 and 2 encode two replicase polypeptides (1a and 2a) while RNA3 encodes the movement protein (3a) and the coat protein (CP). The open reading frames are boxed and labeled. The 3′ terminal sequences, which are common among all four RNA components, are marked as black solid boxes on the right. The oligo(A) tract is shown as a small vertical rectangle within the intercistronic region of RNA3 or at the 3′ end of sgRNA3a. The existence of sgRNA3a has been observed, so far, only for BMV.

classes must infect a cell to initiate viral infection. All BMV RNAs are 5′-capped with RNA1 and RNA2 coding for viral replicase proteins 1a and 2a, respectively, while RNA3 encodes for two proteins, the 5′ movement protein 3a and the 3′ CP. The CP is translated from the 3′ subgenomic RNA4. Recent studies reveal that 3a may be translated from another 5′ subgenomic RNA3a (sgRNA3a). *In vitro* data demonstrate that sgRNA3a is a more efficient RNA template for translation of 3a than full-length RNA3, suggesting a separation of translation (sgRNA3a and RNA4) from replication (whole RNA3) functions.

The purified bromovirus genomic RNAs are directly infectious. By using recombinant DNA technology, the *in vitro* transcripts from complete cDNA clones of RNAs 1, 2, and 3 can be synthesized, and such combined RNAs are also infectious to the bromovirus hosts. In addition, transient expression of DNA constructs carrying bromovirus cDNAs adjacent to the transcription promoter can directly induce bromovirus infection.

## RNA Replication

Replication of bromovirus RNAs completely relies on RNA templates without known DNA intermediates. Both positive and negative RNA components accumulate in the infected tissue but the plus strands reach a 100-fold excess over the minus strands. In addition to whole plants, bromovirus RNAs can replicate in single cells (protoplasts) isolated from various plants (including nonhosts) or even in yeast cells. The complete BMV RNA replication cycle occurs in yeast, demonstrating the presence of compatible host factors.

Only viral proteins 1a and 2a, but not proteins 3a or CP, are required from bromovirus RNA replication cycle, as demonstrated for BMV. These proteins form the active RNA replicase complex that localize to the endoplasmic reticulum membranes called spherules. Computer-assisted comparisons reveal two large domains in protein 1a and a central domain in protein 2a that are conserved with viral proteins of other members of the alphavirus-like superfamily (**Figure 2**). Protein 2a is actual RdRp enzyme whereas 1a has both helicase and methyltransferase activities, and both function cooperatively inside the spherular structures.

Bromoviral RdRp enzyme preparations can be extracted from virus-infected plants or from yeast cells. These activities are more or less virus RNA-specific and in case of BMV, the enzyme can copy full-length bromoviral RNAs and can generate *in vitro* both sgRNAs (sgRNA4 and sgRNA3a) on (−)-RNA3 templates. The well-studied promoter of (−)-strand synthesis comprises the last 200 nt of the 3′ noncoding region in all genomic BMV RNAs. In addition, it participates in several tRNA-specific activities, including adenylation or aminoacylation of the 3′ CAA terminus. The promoters of (+)-strand synthesis have also been mapped to the 5′ proximal noncoding region in BMV RNAs.

The sgp promoter in (−)-strand of BMV RNA3 of CCMV RNA3 includes a core region that binds the RdRp complex, an upstream polyU tract, and an A/U-rich enhancer, both increasing the level of RNA4 synthesis (transcription). Apparently the sgp is responsible not only for the initiation of sgRNA4 but also for premature termination of sgRNA3a (as demonstrated recently for BMV).

The RNA encapsidation signals have been mapped for BMV RNAs. The 3′ tRNA-like structure has been shown to function as a nucleating element of CP subunits. In addition, a *cis*-acting, position-dependent 187 nt region present in the 3a open reading frame (ORF) is essential for efficient packaging of RNA3. The co-packaging of sgRNA4 is contingent upon both RNA replication and translation of CP.

## Homologous and Nonhomologous Recombination of Bromoviral RNAs

Genetic RNA recombination in plant RNA viruses has been first demonstrated by noting the crossovers between BMV RNAs during infection. Both homologous and nonhomologous crossovers have been reported to occur during BMV replication cycle and the restoration of CCMV infectious virus from genomic RNA3 fragments has been reported.

Most importantly, homologous crossing-over has been demonstrated between the same BMV RNA components, and a high-frequency hot spot was observed within the sgp region in BMV RNA3. This type of crossing-over is analogous to the well-studied meiotic DNA crossing-over in DNA-based organisms.

Bromoviruses are capable of generating the defective-interfering (DI) RNAs. In particular, strains of BBMV accumulate RNA2-derived deletion variants that tend to exacerbate the severity of symptoms. The *de novo* generation of DI-like RNAs was demonstrated during serial passages of BBMV in broad bean, and the importance of sequence features for BBMV DI RNA accumulation has been demonstrated. In BMV, both replicating and nonreplicating truncated RNA2-derived artificial DI RNAs have been shown to interfere with accumulation of BMV RNAs.

## Host Factors Involved in Replication of Bromoviral RNAs

Numerous results, including the role of tRNA-specific enzymes, the composition of BMV RdRp preparations, or the dependence of bromovirus infectivity on the host plant, all suggest possible functions of cellular host factors during replication of bromovirus RNAs. More recently, the use of powerful yeast genetics has demonstrated a direct involvement of several host genes in BMV replication, including those related to RNA, protein, or membrane modification pathways. Also, some of the bromoviruses, including BMV or SBLV, have been shown to infect *Arabidopsis thaliana*, and preliminary data reveal the role of host genes in BMV replication in *Arabidopsis*.

## Transmission and Virus–Host Relationship

Mechanical inoculation has been used to efficiently transmit bromoviruses from plant to plant. Also in the field, mechanical transmission by human activity can spread bromoviruses among plant hosts. In general, seed transmission has not been reported in case of bromoviruses, although there is one report of seed transmission of BBMV. Beetles can reproducibly but inefficiently transmit bromoviruses. Aphids are generally negative in transmission whereas inefficient transmission of BMV by nematodes has been reported.

Bromoviruses have restricted host range for systemic spread, with BMV generally infecting monocotyledonous plants whereas BBMV and CCMV infect dicotyledonous plants. BMV requires the CP for both the cell-to-cell and long-distance movement in barley. The interactions between BMV CP and certain host genes (e.g., an oxidoreductase) have been reported during viral infection in barley. The use of interstrain pseudorecombinants has demonstrated the involvement of RNA3 genetic information in virus spread and in symptom formation for BMV and CCMV, and during the breakage of CCMV resistance in soybean. Certain changes in RNAs 1 and 2 also confirmed their roles as determinants of systemic spread and symptom formation in BMV and CCMV. Similarly, the genome exchange experiments suggest that the adaptations of bromoviruses to their hosts rely on movement protein and on both the replicase proteins. Also, various strains of BBMV cause different symptom intensity in their host plants. However, there is no direct correlation between bromovirus yield and symptom severity, but rather the symptoms seem to be associated with changes in the CP gene or in the subgenomic promoter.

*See also:* Alfalfa Mosaic Virus; Brome Mosaic Virus; Cucumber Mosaic Virus; *Ilarvirus*; Recombination.

## Further Reading

Fauquet CM, Mayo MA, Maniloff J, Desselberger U, and Ball LA (eds.) (2005) *Virus Taxonomy: Eighth Report of the International Committee on Taxonomy of Viruses,* 2nd edn., 1162pp. San Diego, CA: Elsevier: Academic Press.

Figlerowicz M and Bujarski JJ (1998) RNA recombination in brome mosaic virus, a model plus strand RNA virus. *Acta Biochimica Polonica* 45(4): 847–868.

Johnson JE and Speir JA (1997) Quasi-equivalent viruses: A paradigm for protein assemblies. *Journal of Molecular Biology* 269: 665–675.

Noueiry AO and Ahlquist P (2003) Brome mosaic virus RNA replication: Revealing the role of the host in RNA virus replication. *Annual Review of Phytopathology* 41: 77–98.

Wooley RS and Kao CC (2004) Brome mosaic virus. Descriptions of Plant Viruses, No. 405. http://www.dpvweb.net/dpv/showdpv.php?dpvno=405 (accessed Jul. 2007).

# Bunyaviruses: General Features

**R M Elliott,** University of St. Andrews, St. Andrews, UK

© 2008 Elsevier Ltd. All rights reserved.

## Glossary

**Ambisense** Coding strategy used by some bunyaviruses in which a genome segment encodes proteins in both positive and negative sense orientations. This strategy also is employed by arenaviruses.

**Bunyavirus** Any member of the family *Bunyaviridae*.

**Cap snatching** Cleavage of the capped 5′-end of a cellular mRNA to produce short oligonucleotides that are used to prime viral mRNA transcription. The cellular sequences are incorporated into the viral mRNAs.

**Reassortment** Exchange of genome segments during the course of a mixed infection; generates new viruses.

**Reverse genetics** Rescue of infectious virus from cloned cDNA copies of the viral genome.

**Vector competence** Ability of an arthropod to transmit a pathogen.

## Introduction

Currently more than 300 mainly arthropod-transmitted viruses are classified within the family *Bunyaviridae*. The majority of these viruses are classified into one of five genera: *Orthobunyavirus, Hantavirus, Nairovirus, Phlebovirus,* and *Tospovirus*, though more than 40 remain unassigned. The basic features that unite these viruses are similar virion morphology, negative-sense tripartite single-stranded RNA genome, and cytoplasmic site of viral replication with intracellular maturation in the Golgi. However, considering the large number of viruses within this one taxonomic grouping, it is not surprising that considerable diversity occurs in terms of biological behavior, and in genome coding and replication strategies. In this article, genus-specific traits are indicated with the terms orthobunyavirus, hantavirus, etc., while the term bunyavirus is reserved for familial traits. Bunyaviruses in all five genera impinge on human health and well-being (**Table 1**), either directly in causing diseases ranging from encephalitis (e.g., La Crosse virus) to hemorrhagic fever (e.g., Crimean-Congo hemorrhagic fever virus), or indirectly by causing disease in animals (e.g., Cache Valley and Akabane viruses) or crop plants (e.g., tomato spotted wilt virus). Furthermore, this family contains prime examples of emerging viruses such as Sin Nombre and related hantaviruses that have been identified in the Americas since 1993.

## Bunyavirus Biology

Viruses in each genus are associated with a principal arthropod vector, with the exception of hantaviruses, which have no arthropod involvement in their life cycle. Gross generalizations can be made that orthobunyaviruses are mainly transmitted by mosquitoes, nairoviruses primarily by ticks, and phleboviruses by phlebotomine flies, ticks, and, notably for Rift Valley fever virus, mosquitoes as well. Tospoviruses are plant-infecting bunyaviruses that are transmitted by thrips. In all cases, the virus can replicate efficiently in the vector without doing overt damage to it, though behavioral changes to mosquito feeding patterns have been noted with some of them. Hantaviruses cause persistent infections in rodents and usually one particular hantavirus is associated with rodents of one particular species. Humans are infected by inhaling aerosolized rodent excretions. Human-to-human transmission is rare but has been reported for Andes hantavirus.

**Table 1** Selected important pathogens in the family *Bunyaviridae*

| Genus/ virus | Disease | Vector | Distribution |
| --- | --- | --- | --- |
| *Orthobunyavirus* | | | |
| Akabane | Cattle: abortion and congenital defects | Midge | Africa, Asia, Australia |
| Cache Valley | Sheep, cattle: congenital defects | Mosquito | N. America |
| La Crosse | Human: encephalitis | Mosquito | N. America |
| Oropouche | Human: fever | Midge | S. America |
| Tahyna | Human: fever | Mosquito | Europe |
| *Hantavirus* | | | |
| Hantaan | Human: severe hemorrhagic fever with renal syndrome (HFRS), fatality 5–15% | Field mouse | Eastern Europe, Asia |
| Seoul | Human: moderate HFRS, fatality 1% | Rat | Worldwide |
| Puumala | Human: mild HFRS, fatality 0.1% | Bank vole | Western Europe |
| Sin Nombre | Human: hantavirus cardiopulmonary syndrome, fatality 50% | Deer mouse | N. America |
| *Nairovirus* | | | |
| Crimean–Congo hemorrhagic fever | Human: hemorrhagic fever, fatality 20–80% | Tick, culicoid fly | Eastern Europe, Africa, Asia |
| Nairobi sheep disease | Sheep, goat: fever, hemorrhagic gastroenteritis, abortion | Tick, culicoid fly, mosquito | Africa, Asia |
| *Phlebovirus* | | | |
| Rift Valley fever | Human: encephalitis, hemorrhagic fever, retinitis, fatality 1–10% | Mosquito | Africa |
| | Domestic ruminants: necrotic hepatitis, hemorrhage, abortion | | |
| Sandfly fever | Human: fever | Sandfly | Europe, Africa |
| *Tospovirus* | | | |
| Tomato-spotted wilt | Plants: over 650 species, various symptoms | Thrips | Worldwide |

The maintenance and transmission cycles of orthobunyaviruses have been studied in more detail than for other bunyaviruses, and involve both horizontal and vertical transmission between a restricted number of mosquito and vertebrate hosts (**Figure 1**). The integrity of the cycles is maintained even when the geographic distributions of the different hosts overlap in nature. Horizontal transmission usually refers to transmission to susceptible vertebrate hosts via the biting arthropod, with the resulting vertebrate viremia a source for infection for another feeding arthropod. Humans are often dead end hosts in that they rarely function as a source for further vector infection. Horizontal transmission can also refer to venereal transmission between mosquitoes, and has been reported for orthobunyaviruses. Vertical transmission is a major maintenance mechanism, particularly in periods of unfavorable climatic conditions. Virus can be maintained in mosquito eggs over winter or during periods of drought for months if not years, and when conditions allow, mosquito larvae emerge infected with the virus.

Vector competence is the term used to describe the ability of a particular arthropod species to transmit a virus, and is governed by both host and viral factors. After ingestion of virus in a blood meal, the virus infects the midgut cells, and then escapes into the hemocoel to allow disseminated infection of all organs. High level replication in the salivary glands provides the viral inoculum for transmission at the next feeding occasion. For orthobunyaviruses in the California serogroup, experiments with reassortant viruses (see later) determined that the viral factors associated with efficient infection of midgut cells and subsequent transmission mapped to the medium (M) RNA genome segment.

## Virion Proteins and RNA Segments

The spherical bunyavirus particle is about 100 nm in diameter and electron microscopy reveals a fringe of spikes on the surface (**Figure 2**). The virion consists of four structural proteins: two internal proteins (N and L) and two external glycoproteins, termed Gn and Gc, which are inserted in the Golgi-derived viral membrane (**Figure 3**). The N protein (2100 copies per particle) encapsidates the RNA genome segments to form ribonucleoprotein complexes termed RNPs or nucleocapsids, to which the L protein (25 copies per particle) associates. There is no equivalent of a matrix protein to stabilize the virion structure, and it is presumed that interaction between the cytoplasmic tail(s) of either or both of the glycoproteins and the N protein in the RNP is important for structural integrity of the virion.

The three genomic RNA segments, which are designated L (large), M (medium), and S (small), characteristically have complementary terminal sequences that are similar for the three segments of viruses within a genus, but differ between genera; the genus-specific consensus sequences are shown in **Table 2**. A consequence of the terminal complementarity is that the ends of the RNAs may base-pair, and circular or panhandle bunyavirus RNAs have been seen in the electron microscope. Bunyavirus nucleocapsids are also circular, and the ends of the RNA segments are base-paired within the RNP. It is probable

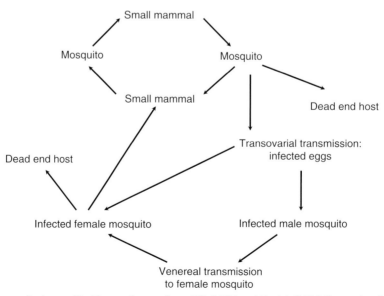

**Figure 1** Life cycle of mosquito-transmitted bunyaviruses. From Elliott RM and Koul A (2004) *Bunyavirus/Mosquito Interactions*, symposium 63, pp. 91–102. Society for General Microbiology.

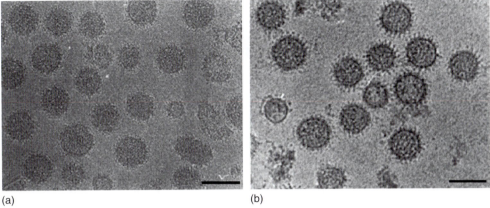

**Figure 2** Electron micrographs of vitrified-hydrated La Crosse orthobunyavirus virions. (a) Small defocus value which demonstrates the membrane bilayer; (b) large defocus value which demonstrates the glycoprotein spikes (see Elliott (1990) for further details). Scale = 100 nm. The photographs were generously provided by Dr. B. V. V. Prasad. From Elliott RM (1990) Molecular biology of Bunyaviridae. Journal of General Virology 71(Pt. 3): 501–522.

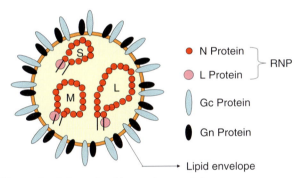

**Figure 3** Schematic of bunyavirus particle.

**Table 2** Consensus 3' and 5' terminal nucleotide sequences of bunyaviral genome RNAs

| | |
|---|---|
| Orthobunyavirus | 3' UCAUCACAUGA...UCGUGUGAUGA 5' |
| Hantavirus | 3' AUCAUCAUCUG..........AUGAUGAU 5' |
| Nairovirus | 3' AGAGUUUCU............AGAAACUCU 5' |
| Phlebovirus | 3' UGUGUUUC................GAAACACA 5' |
| Tospovirus | 3' UCUCGUUAG............CUAACGAGA 5' |

that the encapsidation signal for the N protein is at the 5'-end of the RNA.

## Coding Strategies of the Viral Genomes

Complete nucleotide sequences have been determined for at least one representative of each genus in the *Bunyaviridae* which has allowed the coding strategies of the individual genome segments to be elucidated. These are shown schematically in **Figure 4**.

The L RNA encodes the L protein using a conventional negative-strand strategy, that is, in a complementary positive-sense mRNA. The L protein contains motifs found in all RNA polymerases, and expression of recombinant L proteins, via a variety of systems, demonstrated the L had RNA synthesis activity. The L protein is therefore at least a component of the virion-associated transcriptase.

The M segment encodes in the complementary sense mRNA the virion glycoproteins in the form of a precursor polyprotein. The M segment gene products have been implicated in many biological attributes of the virus including hemagglutination, virulence, tissue tropism, neutralization, and cell fusion. By convention, the glycoprotein of greater molecular weight (or slower electrophoretic mobility in sodium dodecyl sulfate–polyacrylamide gels) was termed G1, and the smaller glycoprotein G2. However, functional analyses of glycoproteins of viruses in different genera have revealed some commonality between the glycoproteins according to whether they are encoded at the N- or C-terminal region on the polyprotein. Therefore, the glycoproteins are now referred to as Gn or Gc. In the case of orthobunyaviruses, hantaviruses, and phleboviruses, the precursor is co-translationally cleaved to yield the mature proteins, which are characteristically rich in cysteine residues.

Orthobunyaviruses encode an additional, nonstructural protein called NSm between Gn and Gc, while some phleboviruses encode an NSm protein upstream of the glycoproteins, also as part of the precursor. Only the N-terminal domain of NSm is required for orthobunyaviruses to replicate in cell cultures, while the entire NSm region of Rift Valley fever phlebovirus can be deleted without compromising replication in cultured cells. However, the function of these NSm proteins remains unknown.

Tospoviruses encode an NSm protein, using an ambisense strategy. The NSm coding region is contained in the 5' terminal part of the viral genomic RNA, but is translated from a subgenomic mRNA. This mRNA is transcribed from the full-length complement of the

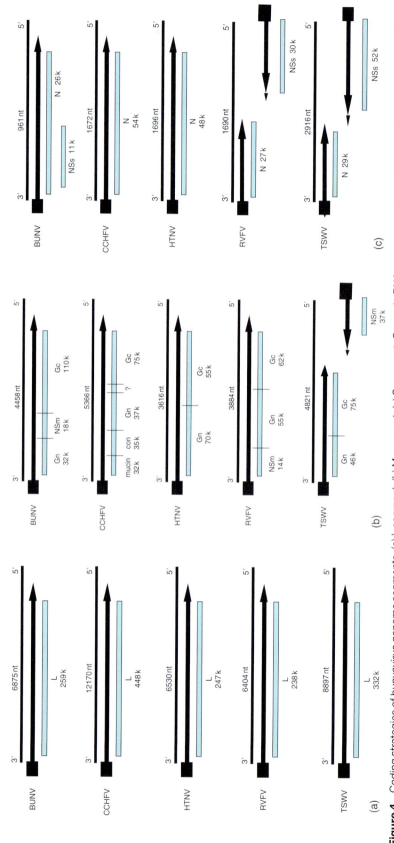

**Figure 4** Coding strategies of bunyavirus genome segments. (a) L segment; (b) M segment; (c) S segment. Genomic RNAs are represented by thin lines (the length in nucleotides is given above each segment) and the mRNAs are shown as arrows (filled square indicates host-derived sequences at 5′-end). Gene products, with their apparent $M_r$, are represented by light colored boxes. Abbreviations: BUNV, Bunyamwera orthobunyavirus; CCHFV, Crimean-Congo hemorrhagic fever nairovirus; HTNV, Hantaan hantavirus; k, mole weight of protein 2; RVFV, Rift Valley fever phlebovirus; TSWV, tomato spotted wilt tospovirus. In (b), CCHFV M segment, mucin represents mucin-like region; con the connector region; and ? unidentified protein product.

genomic RNA. The tospovirus NSm functions as a movement protein by forming tubular structures that penetrate through plasmadesmata and allow cell-to-cell transport of viral nucleocapsids.

Processing of the nairovirus M segment-encoded polyprotein is more complex, occurs over a period of several hours and involves the production of a number of intermediate proteins. Upstream of Gn are two domains, a highly variable region at the N-terminus (resembling the mucin-like domain observed in other viral glycoproteins) followed by a second domain (connector domain), both of which are cleaved in separate events from Gn, while a further domain is cleaved from between mature Gn and Gc (**Figure 3**). The function(s) of the released domains awaits further investigation.

The NSs protein of orthobunyaviruses and phleboviruses is dispensable for replication and is regarded as an accessory virulence factor. Although these NSs proteins are of different sizes, show no obvious sequence similarity, and are encoded by different mechanisms, they appear to have a similar function in antagonizing the host interferon (IFN) response. Studies on Bunyamwera and La Crosse orthobunyaviruses, engineered by reverse genetics (see below) to ablate the NSs open reading frame, and on an attenuated variant of Rift Valley fever phlebovirus that expresses a truncated NSs protein, showed that these mutant viruses induce IFN, whereas wild-type viruses do not. Further work revealed that the NSs proteins inhibited RNA polymerase II-mediated host transcription, thus preventing synthesis of IFN mRNA.

The S RNA segments also show great diversity in their coding strategies. The bunyavirus S segment encodes two proteins, N, and a nonstructural protein, NSs, in different overlapping reading frames in the complementary-sense RNA. The two proteins are translated from the same mRNA species, the result of alternative initiation at different AUG codons. For hantaviruses and nairoviruses, a single open reading frame, encoding the N protein, is found in the S segment complementary-sense RNA. The S segments of phleboviruses and tospoviruses employ an ambisense coding strategy; the N protein is encoded in the complementary-sense RNA corresponding to the 3′ half of the genomic S segment, whereas the coding sequence for the NSs protein is contained in the 5′ half of the genomic RNA. The proteins are translated from separate subgenomic mRNAs.

## Genome Segment Reassortment

In common with other viruses with segmented genomes, bunyaviruses can reassort their genome segments during co-infection (**Figure 5**), and this phenomenon has been extensively studied for orthobunyaviruses in both cell cultures and in mosquitoes. There are barriers to the extent to which reassortment occurs in that it is restricted to antigenically closely related viruses within a genus, and certain segment combinations seem favored over others. In mosquitoes, reassortment of orthobunyavirus RNA segments occurs following ingestion of two viruses in a single blood meal or in separate feedings (mimicking interrupted feeding due to the host's physical reaction to the mosquito) provided the two events are temporally close. If the time between the infections is greater than 2 or 3 days, there is resistance to infection with the second virus. Reassortant viruses were used to map viral proteins and biological functions to these proteins with individual genome segments before the advent of nucleotide sequencing. Reassortment has been detected in nature, for instance, among various orthobunyaviruses, and of significance is the discovery that Ngari (originally called Garrissa) virus, an orthobunyavirus associated with human hemorrhagic fever, is a reassortant between the relatively innocuous viruses Bunyamwera and Batai.

## Viral Replication Cycle

### Attachment, Entry, and Uncoating

Infection of the target cell is mediated by one or both of the virion glycoproteins interacting with the cellular receptor. Receptors have not been formally identified for any bunyavirus, though cellular entry of hantaviruses involves β1-integrins (apathogenic hantaviruses) or β3-integrins (pathogenic hantaviruses). The relative importance of either of the glycoproteins in attachment has not been fully elucidated and may differ between the genera. For orthobunyaviruses, it has been suggested that Gc is the major attachment protein for vertebrate cells, whereas Gn may contain the major determinants for attachment to mosquito cells. Neutralization and hemagglutination-inhibition sites have been mapped to both glycoproteins encoded by hantaviruses and phleboviruses, suggesting that for these viruses both Gn and Gc may be involved in attachment. In common with many other enveloped viruses bunyaviruses can fuse cells at acidic pH; at least for orthobunyaviruses, this is accompanied by a conformational change in Gc. Based on electron microscopic studies of phleboviruses, entry into cells is by endocytosis. It is probable that uncoating occurs when endosomes become acidified, thus initiating fusion of the viral membrane and endosomal membrane, followed by release of the nucleocapsids into the cytoplasm.

### Transcription

The classical scheme for replication of a negative-strand RNA virus is that the infecting genome is first transcribed into mRNAs by the virion-associated RNA polymerase or transcriptase (**Figure 6**). This process, termed primary

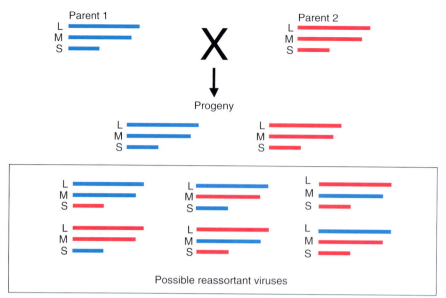

**Figure 5** Generation of reassortant bunyaviruses following coinfection of the same cell. From Elliott RM and Koul A (2004) *Bunyavirus/Mosquito Interactions*, symposium 63, pp. 91–102. Society for General Microbiology.

transcription, is independent of ongoing protein synthesis. Following translation of the primary transcripts into viral proteins, the genome is replicated via a complementary full-length positive-strand RNA, the antigenome, and then further mRNA synthesis (secondary transcription) ensues. Transcriptase activity has been detected in detergent-disrupted virion preparations of representatives of most *Bunyaviridae* genera. The enzymatic activity was weak compared to, for example, the transcriptase of vesicular stomatitis virus, which has hampered extensive biochemical characterization of the enzyme. However, the bunyavirus transcriptase was shown to be stimulated by oligonucleotides of the (A)nG series, cap analogs (e.g., mGpppAm) and natural mRNAs, such as alfalfa mosaic virus RNA 4. These appeared to act as primers for transcription. Further support for this notion was provided by sequencing studies of the 5′-ends of both *in vivo* and *in vitro* synthesized mRNAs, which showed they contained an additional 10–18 nontemplated nucleotides; a cap structure was present at the 5′ terminus. *In vitro*, an endonuclease activity which specifically cleaved methylated capped mRNAs was detected. Taken together, these data indicate that bunyavirus transcription is markedly similar to that of influenza viruses in using a cap-snatch mechanism to prime transcription. In contrast to influenza viruses, bunyavirus transcription is not sensitive to actinomycin D or α-amanitin, and occurs in the cytoplasm of infected cells. The apparent reiteration of viral terminal sequences at the junction between the primer and viral sequence itself suggests that the polymerase may slip during transcription; further analysis of hantavirus RNAs suggests that this may occur during replication as well, and has been dubbed prime-and-realign.

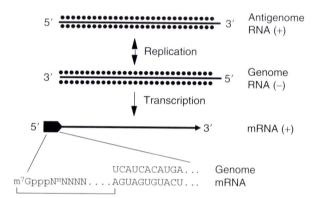

**Figure 6** Transcription and replication scheme of negative-sense bunyavirus genome segments. The genome RNA and the positive-sense complementary RNA known as the antigenome RNA are only found as ribonucleoprotein complexes and are encapsidated by N protein (filled square). The mRNA species contain host-derived primer sequences at their 5′-ends (filled square) and are truncated at the 3′-end relative to the vRNA template; the mRNAs are neither polyadenylated nor encapsidated by N protein. The sequence at the 5′-end of an orthobunyavirus mRNA is shown. From Elliott RM (2005) Negative strand RNA virus replication. In: Mahy B and ter Meulen V (eds.) *Topley and Wilson's Microbiology and Microbial Infections*, 10th edn., vol. 1, pp. 127–134. London: Hodder Arnold.

Analysis of bunyavirus primary transcription *in vivo*, however, appeared initially to produce results incompatible with the presence of a virion transcriptase, in that no mRNA synthesis could be detected in the presence of protein synthesis inhibitors in certain virus-cell systems. Further work showed that only short transcripts were produced in the absence of protein synthesis *in vivo*; subsequent gel electrophoresis analyses of the *in vitro* transcriptase products showed that these too were short transcripts. If the *in vitro* reaction was supplemented with rabbit reticulocyte lysate, however, full-length RNAs were synthesized. The translational requirement was not at the level of mRNA initiation, but rather during elongation or, more precisely, to prevent the transcriptase from terminating prematurely. A model to account for these observations proposes that in the absence of ribosome binding and protein translation the nascent mRNA chain and its template can base-pair, thereby preventing progression of the transcriptase enzyme. This translational requirement is not ubiquitous, however, since concurrent translation is not needed for efficient readthrough of premature termination sites in some strains of BHK cells or in C6/36 mosquito cells. Reconstitution and mixing experiments suggest that the translational requirement is mediated by a host cell factor, present in some BHK cells, which may promote interaction between the nascent mRNA and its template.

The 3′-ends of the bunyavirus mRNAs are not co-terminal with their genome templates. For the nonambisense segments, mRNAs terminate 50–100 nucleotides before the end of the template, though there does not appear to be a universal termination signal in the bunyaviruses. The pentanucleotide sequence UGUCG has been mapped as the transcription-termination signal in the Bunyamwera orthobunyavirus S segment, and the same or highly related motifs have been identified in the Bunyamwera virus L segment and in the S segments of other orthobunyaviruses. However, the motif is not present in orthobunyavirus M segments nor in the genomes of viruses in other genera. The subgenomic mRNAs transcribed from ambisense S segments terminate within the noncoding intergenic sequences in the RNA (**Figure 7**); for some but not all viruses, the intergenic region has the potential to form a stable hairpin structure, though the role of secondary structure in transcription termination is unclear. Bunyavirus mRNAs are not demonstrably 3′ polyadenylated, but many have the potential to form stem–loop structures which may confer stability.

## Genome Replication

In order to replicate the negative-sense genome RNA, a full-length complementary, positive-sense RNA, the antigenome, must be synthesized (**Figures 6** and **7**). This molecule differs from the positive-sense mRNA in that it does not have the 5′ primer sequences and the 3′-end extends to the 5′ terminus of the genomic RNA template. Experiments with minigenome systems show that only the viral L and N proteins are required for replication, but it is not known what controls the switch from transcriptive to replicative mode of the polymerase. Two events differ between transcription and replication: initiation, which does not require a primer, and readthrough of the mRNA termination signal. The difference in initiation may be because the RNA polymerase is modified by a cellular protein. In the infected cell antigenomes are only found assembled into nucleocapsids; therefore, encapsidation of the nascent antigenome RNA may prevent its interaction with the template, thereby overcoming the mRNA termination signal.

## Assembly and Release

Maturation of bunyaviruses characteristically occurs at the smooth membranes in the Golgi apparatus, and hence is inhibited by monensin, a monovalent ionophore. The viral glycoproteins accumulate in the Golgi complex and cause a progressive vacuolization. However, the morphologically altered Golgi complex remains functionally active in its ability to glycosylate and transport glycoproteins destined for the plasma membrane. Using recombinant expression systems, it has been shown that the targeting of the bunyavirus glycoproteins to the Golgi is a property of the glycoproteins alone, and does not require other viral proteins or virus assembly. Electron microscopic studies revealed that viral nucleocapsids condense on the cytoplasmic side of areas of the Golgi vesicles, whereas viral glycoproteins are present on the luminal side. The absence of a matrix-like protein in bunyaviruses, which for other viruses may function as a bridge between the nucleocapsid and the glycoproteins, suggests that direct transmembrane interactions between the bunyavirus nucleocapsid and the glycoproteins may be a prerequisite for budding. After budding into the Golgi cisternae, vesicles containing viral particles are transported to the cell surface via the exocytic pathway, eventually releasing their contents to the exterior.

There are important exceptions, however, to the above maturation scheme; Rift Valley fever phlebovirus has been observed to bud at the surface of infected rat hepatocytes, and it appears that a characteristic of the newly described American hantaviruses which cause hantavirus pulmonary syndrome is assembly and maturation occurring at the plasma membrane.

## Persistent Infections

Arboviruses share a common biological property in their capacity to replicate in both vertebrate and invertebrate cells. The outcomes of these infections can be markedly different; whereas infection of vertebrate cells is often lytic, leading to cell death, infection of invertebrate cells is often asymptomatic, self-limiting, and leads to a persistent infection. For bunyaviruses this has been demonstrated both at the organismic level and in cultured cells. Studies of persistent infections of mosquito cells with orthobunyaviruses showed no inhibition of host cell

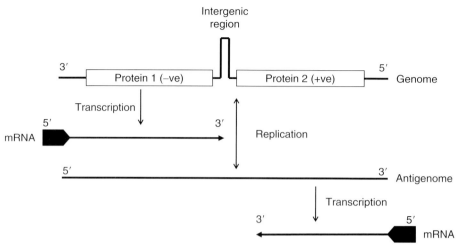

**Figure 7** Transcription and replication scheme of ambisense-sense bunyavirus genome segments. The genome RNA encodes proteins in both negative- and positive-sense orientations, separated by an intergenic region that can form a hairpin structure. The proteins are translated from subgenomic mRNAs, with the mRNA encoding protein 2 transcribed from the antigenome RNA after the onset of genome replication. From Elliott RM (2005) Negative strand RNA virus replication. In: Mahy B and ter Meulen V (eds.) *Topley and Wilson's Microbiology and Microbial Infections*, 10th edn., vol. 1, pp. 127–134. London: Hodder Arnold.

protein synthesis, in sharp contrast to replication in mammalian cells. A feature of the persistently infected cells was the excess amount of S segment RNA they contained, but although the level of S mRNA remained high the amount of N protein translated declined. Blockage of N protein synthesis was shown to be because N was able to encapsidate its own mRNA, thereby preventing its translation. Defective L segment RNAs were also found in the persistently infected cells, but these were not packaged into virions. In contrast, a novel type of defective interfering particle was produced which contained only S segment RNA. It is highly probable that host cell factors contribute to these events, but the identity of these factors awaits further investigation.

## Reverse Genetics

Reverse genetic approaches to the study of bunyavirus RNA synthesis have been described, in which transiently expressed recombinant N and L proteins transcribe and replicate synthetic RNA transcripts containing a reporter gene. A system to recover infectious Bunyamwera orthobunyavirus entirely from cloned cDNA copies of the three genome segments was developed in 1996, and improvements to this system paved the way for the subsequent recovery of La Crosse virus and Rift Valley fever viruses. These represent significant accomplishments in bunyavirology and will allow future in-depth studies of the functions of all the viral proteins as well as detailed investigation of biological properties such as virulence, tissue tropism, and vector competence. It is expected that similar systems will be developed for viruses in the other *Bunyaviridae* genera. In the longer term, the design and recovery of specifically modified viruses having potential as vaccines may be feasible. For instance, a recombinant Bunyamwera virus containing a modified L RNA segment (L coding region flanked by M segment noncoding sequences) was found to be attenuated for replication in cell cultures and in mice.

*See also:* Akabane Virus; Crimean-Congo Hemorrhagic Fever Virus and Other Nairoviruses; Hantaviruses.

## Further Reading

Billecocq A, Spiegel M, Vialat P, et al. (2004) NSs protein of Rift Valley fever virus blocks interferon production by inhibiting host gene transcription. *Journal of Virology* 78: 9798–9806.

Bridgen A and Elliott RM (1996) Rescue of a segmented negative-strand RNA virus entirely from cloned complementary DNAs. *Proceedings of the National Academy of Sciences USA* 93: 15400–15404.

Bridgen A, Weber F, Fazakerley JK, and Elliott RM (2001) Bunyamwera bunyavirus nonstructural protein NSs is a non-essential gene product that contributes to viral pathogenesis. *Proceedings of the National Academy of Sciences USA* 98: 664–669.

Elliott RM (ed.) (1996) *The Bunyaviridae*, 337pp. New York: Plenum Press.

Elliott RM (1990) Molecular biology of *Bunyaviridae*. *Journal of General Virology* 71: 501–522.

Elliott RM (2005) Negative strand RNA virus replication. In: Mahy BWJ and ter Meulen V (eds.) *Topley and Wilson's Microbiology and Microbial Infections*, 10th edn., vol. 1, pp. 127–134. London: Hodder Arnold.

Elliott RM and Koul A (2004) *Bunyavirus/Mosquito Interactions*, symposium 63, pp. 91–102. Society for General Microbiology.

Lowen AC, Boyd A, Fazakerley JK, and Elliott RM (2005) Attenuation of bunyavirus replication by rearrangement of viral coding and non-coding sequences. *Journal of Virology* 79: 6940–6946.

Lowen AC, Noonan C, McLees A, and Elliott RM (2004) Efficient bunyavirus rescue from cloned cDNA. *Virology* 330: 493–500.

Nichol ST, Beaty BJ, Elliott RM, et al. (2005) Bunyaviridae. In: Fauquet CM, Mayo MA, Maniloff J, Desselberger U, and Ball LA (eds.) Virus Taxonomy: Eighth Report of the International Committee on Taxonomy of Viruses, pp. 695–716. San Diego, CA: Elsevier Academic Press.

Nichol ST, Spiropoulou CF, Morzunov S, et al. (1993) Genetic identification of a hantavirus associated with an outbreak of acute respiratory illness. Science 262: 914–917.

Sanchez AJ, Vincent MJ, Erickson BR, and Nichol ST (2006) Crimean-Congo hemorrhagic fever virus glycoprotein precursor is cleaved by Furin-like and SKI-1 proteases to generate a novel 38-kilodalton glycoprotein. Journal of Virology 80: 514–525.

Schmaljohn C and Hooper JW (2001) Bunyaviridae: The viruses and their replication. In: Knipe DM and Howley PM (eds.) Fields Virology 4th edn., pp. 1581–1602. Philadephia: Lippincott Williams and Wilkins.

Weber F, Bridgen A, Fazakerley JK, et al. (2002) Bunyamwera bunyavirus non-structural protein NSs counteracts the induction of type I interferon. Journal of Virology 76: 7949–7955.

Won S, Ikegami T, Peters CJ, and Makino S (2006) NSm and 78-kilodalton proteins of Rift Valley fever virus are nonessential for viral replication in cell culture. Journal of Virology 80: 8274–8278.

# Bunyaviruses: Unassigned

**C H Calisher,** Colorado State University, Fort Collins, CO, USA

© 2008 Elsevier Ltd. All rights reserved.

## Glossary

**Arbovirus** A virus transmitted to vertebrates by hematophagous (blood-feeding) insects.
**Arthralgia** Joint pain.
**Myalgia** Muscle pain.
**Orthobunyavirus, hantavirus, phlebovirus, nairovirus, tospovirus** Any virus in the genus Orthobunyavirus, Hantavirus, Phlebovirus, Nairovirus, or Tospovirus.
**Polyarthritis** Simultaneous inflammation of several joints.
**Taxon** A taxonomic category or group, such as a phylum, order, family, genus, or species.
**Unassigned bunyavirus** A virus placed in a family Bunyaviridae but not in a genus within that family.
**Unclassified** A virus not yet placed in a taxon.
**Ungrouped** A virus not yet shown to be related to any other virus.
**Viremia** Virus in the blood.

## Introduction

Although most viruses in the family Bunyaviridae have been placed in one of the five established genera (Orthobunyavirus, Phlebovirus, Nairovirus, Hantavirus, or Tospovirus), some have not. The original studies of all these viruses were done by serologic or other methods and were meant only to determine whether a virus was a newly recognized one or simply an additional isolate of a known virus. When two or more viruses were shown to be related antigenically, they were considered members of a 'group', nothing more. However, when results of further molecular studies provided definitive genetic information, it became possible to place a virus, or a group of viruses, in a particular taxon. Still other viruses, for which we lack data that would allow placement in a genus, were shown by electron microscopy (morphology of the virus and morphogenesis), or by other physical or biochemical characteristics to be analogous to recognized members of the family Bunyaviridae and so were placed in the family as 'unassigned' members. Many of these viruses are known to be related to each other and so are assigned to groups of unassigned virus and the rest are considered as ungrouped and unassigned. The unassigned viruses of the family Bunyaviridae (i.e., 'bunyaviruses') are listed in **Table 1**.

Certain of these viruses cause human illness but have not been adequately characterized nor, with few exceptions, have their epidemiologies been studied methodically and we do not have information regarding their prevalences.

## Bhanja Virus

This virus, first isolated from Haemaphysalis intermedia, was collected from a paralyzed goat in India. It was subsequently isolated from ticks of other species collected in Senegal, Central African Republic, Nigeria, Cameroon, Somalia, Armenia, Bulgaria, Croatia (former Yugoslavia), and Italy. Infection of a laboratory worker provided the first evidence that this virus is pathogenic for humans. The patient had a mild illness characterized by myalgias and arthralgias, moderate frontal headache, slight photophobia, slight elevation of temperature, all lasting less than 2 days. Subsequently, a few additional human infections, including one fatal case, were recorded. Clearly, this virus is widespread geographically and its significance underreported.

**Table 1** Unassigned viruses in the family *Bunyaviridae*

| Group | Viruses |
|---|---|
| Bhanja | Bhanja, Forecariah, Kismayo |
| Kaisodi | Kaisodi, Lanjan, Silverwater |
| Mapputta | Gan Gan, Mapputta, Maprik, Trubanaman |
| Tanga | Okola, Tanga |
| Resistencia | Antequera, Barranqueras, Resistencia |
| Upolu | Aransas Bay, Upolu |
| Yogue | Kasokero, Yogue |

Ungrouped: Bangui, Belem, Belmont, Bobaya, Caddo Canyon, Chim, Enseada, Keterah (= Issyk-Kul), Kowanyama, Lone Star, Pacora, Para, Razdan, Salanga, Santarem, Sunday Canyon, Tai, Tamdy, Tataguine, Wanowrie, Witwatersrand, Yacaaba.

## Kasokero Virus

Kasokero virus was first isolated from the bloods of Egyptian rousette bats (*Rousettus aegyptiacus*) in Uganda. It subsequently was isolated from bloods of three people working in the laboratory at the time the bats and the virus were being manipulated and from the blood of a driver who only occasionally entered the laboratory, suggesting the possibility of aerosol transmission of the virus. Patients suffered from fever, headache, abdominal pain, diarrhea, and from some or all of nausea, abdominal pain, chest pain, hyperactive reflexes, coughing, and severe myalgias and arthralgias, lasting 5–14 days, but followed by complete recovery.

## Tataguine Virus

Tataguine virus was first isolated from mosquitoes aspirated from dwelling places near Tataguine, Senegal, and subsequently from mosquitoes collected in Nigeria, Central African Republic, Burkina Faso, Cameroon, and Senegal. Clinical manifestations of Tataguine virus infection in humans are mild, characterized by fever and rash, and can be confused with malaria and fevers of unknown origin in West Africa. When active surveillance for this virus was done in Nigeria and Senegal, it was found to be one of the most prevalent viruses in blood samples taken from febrile humans.

## Wanowrie Virus

Wanowrie virus, first isolated from *Hyalmomma marginatum* ticks in India, has been isolated from other ticks in Egypt and Iran and from mosquitoes in India. It has also been isolated from the brain of a child in Sri Lanka, suggesting that further studies of this ungrouped pathogen are warranted.

## Keterah Virus (Also Known As Issyk-Kul Virus)

This virus was first isolated from *Argas pusillus* ticks collected from bats in Malaysia and from the bloods of those bats (*Scotophilus temmencki*), and has been isolated from ticks and mosquitoes that had fed on infected bats. It was subsequently isolated from pooled brain, liver, spleen, and kidney tissues of other bats in Kyrgyzstan and Tadzhikistan. The virus also was isolated from a staff member of a virology institute in Tadzhikistan who had contracted the infection during field work with bats. Sporadic human cases of this disease have been recognized for more than 20 years in central Asia, particularly in Tadzhikistan, and in Malaysia, and the virus may occur in parts of Iran, Afghanistan, India, and Pakistan. The nonfatal illness caused by this virus is characterized by fever, headache, and myalgias, which are sufficiently nonspecific to be generally undiagnosed or mistaken for diseases caused by other pathogens.

Experimental infections of African green monkeys (*Cercopithecus aethiops*), golden hamsters (*Mesocrictus auratus*), and laboratory mice with the Issyk-Kul strain of this virus demonstrated virus in blood and organs of all animals. Histological studies revealed inflammatory and dystrophic changes in the central nervous system, lungs, liver, and kidneys and pronounced morphological changes in the spleen. The virus is pantropic, causing generalized infection in all animals, irrespective of the route of infection. In monkeys, asymptomatic infection was accompanied by marked organ damages and viremia.

## Other Unassigned Orthobunyaviruses Causing Disease in Humans

Gan Gan virus, isolated thus far only from Australian mosquitoes, has been associated with a few cases of acute epidemic polyarthritis-like illness, and may be confused with other viruses causing this syndrome in Australia, such as Ross River virus and Barmah Forest viruses (*Togaviridae, Alphavirus*). Similarly, Trubanaman virus, also isolated only from Australian mosquitoes, is suspected of being pathogenic. These viruses may employ marsupials as principal vertebrate hosts. Finally, Tamdy virus, isolated from *Hyalomma* spp. ticks in Uzbekistan and Turkmenistan, also has been shown to be a human pathogen. Clearly, many more studies are needed, if we are to understand the natural cycles of these viruses, and their prevalences in humans and in wild animals.

*See also:* Baculoviruses: Molecular Biology of Granuloviruses; Hepatitis B Virus: Molecular Biology.

## Further Reading

Calisher CH and Goodpasture HC (1975) Human infection with Bhanja virus. *American Journal of Tropical Medicine and Hygiene* 24: 1040–1042.

Dandawate CN, Shah KV, and D'Lima LV (1970) Wanowrie virus: A new arbovirus isolated from *Hyalomma marginatum isaaci*. *Indian Journal of Medical Research* 58: 985–989.

Fagbami AH, Monath TP, Tomori O, Lee VH, and Fabiyi A (1972) Studies on Tataguine infection in Nigeria. *Tropical and Geographic Medicine* 24: 298–302.

Fauquet CM, Mayo MA, Maniloff J, Desselberger U, and Ball LA (eds.) (2005) *Virus Taxonomy: Eighth Report of the International Committee on Taxonomy of Viruses*, pp. 713–714. San Diego, CA: Elsevier Academic Press.

Kalunda M, Mukwaya LG, Mukuye A, et al. (1986) Kasokero virus: A new human pathogen from bats (*Rousettus aegyptiacus*) in Uganda. *American Journal of Tropical Medicine and Hygiene* 35: 387–392.

L'vov DK, Sidorova GA, Gromashevskii VL, Skvortsova TM, and Aristova VA (1984) Isolation of Tamdy virus (*Bunyaviridae*) pathogenic for man from natural sources in Central Asia, Kazakhstan and Transcaucasia (in Russian). *Voprosy Virusologii* 29: 487–490.

L'vov DK, Terskikh II, Abramova LN, Savosina NS, Skvortsova TM, and Gromashevskii VL (1991) An experimental infection caused by the Issyk-Kul arbovirus (in Russian). *Meditsinskaia Parazitolologiia I Parazitarnye Bolezni* 4: 15–16.

Mackenzie JS, Lindsay MD, Coelen RJ, Broom AK, Hall RA, and Smith DW (1994) Arboviruses causing human disease in the Australasian zoogeographic region. *Archives of Virology* 136: 447–467.

Moore DL, Causey OR, Carey DE, et al. (1975) Arthropod-borne viral infections of man in Nigeria, 1964–1970. *Annals of Tropical Medicine and Parasitology* 69: 49–64.

Pavri KM, Anandarajah M, Hermon YE, Nayar M, Wikramsinghe MR, and Dandawate CN (1976) Isolation of Wanowrie virus from brain of a fatal human case from Sri Lanka. *Indian Journal of Medical Research* 64: 557–561.

# Cacao Swollen Shoot Virus

**E Muller,** CIRAD/UMR BGPI, Montpellier, France

© 2008 Elsevier Ltd. All rights reserved.

## Introduction

Cacao swollen shoot virus (CSSV) is a member of the genus *Badnavirus*, family *Caulimoviridae*. The importance and diversity of badnaviruses has only been recognized relatively recently and particularly on tropical plants, due to progress made in molecular diagnostic techniques.

In addition to CSS3V, other badnaviruses commonly reported are banana streak virus (BSV) on banana, dioscorea alata bacilliform virus (DaBV) on yam, taro bacilliform virus (TaBV) on taro, sugarcane bacilliform virus (SCBV) on sugarcane, citrus mosaic bacilliform virus (CMBV) on citrus species, and pineapple bacilliform virus (PBV) on pineapple, but these viruses also infect various ornamental plants such as Aucuba, Commelina, Kalanchoe, Yucca, Mimosa, and Schefflera. Some of the above are only tentative species of the genus *Badnavirus* due to insufficient molecular data. There are regularly new reports of badnaviruses on other plants.

CSSV is naturally transmitted to cacao (*Theobroma cacao*) in a semipersistent manner by several mealybug species, the vector of most badnaviruses. Cacao swollen shoot disease occurs in all the main cacao-growing areas of West Africa, where it has caused enormous damage. It was highlighted for the first time in Ghana in 1922 but was described and named in 1936, then in Nigeria in 1944, in Ivory Coast in 1946, in Togo in 1949, and in Sierra Leone in 1963. The disease was also described in Trinidad and Tobago but currently seems to have disappeared. Cacao swollen shoot disease is present in Sri Lanka and in Indonesia (Java and North Sumatra).

## Host Range and Symptomatology

Experimental host range is limited to species of the families of Sterculiaceae, Malvaceae, Tiliaceae, and Bombaceae but the principal host of the virus is *Theobroma cacao*. Symptoms are mostly seen in leaves, but stem and root swellings, as well as pod deformation also occur. In some varieties of cocoa, particularly Amelonado cocoa, reddening of primary veins and veinlets in flush leaves is characteristic (**Figure 1(a)**). This red vein banding later disappears. There can be various symptoms on mature leaves, depending on the cocoa variety and virus strain. These symptoms can include: yellow clearing along main veins; tiny pin-point flecks to larger spots; diffused flecking; blotches or streaks. Chlorotic vein flecking or banding is common and may extend along larger veins to give angular flecks.

Stem swellings may develop at the nodes, internodes, or shoot tips. These may be on the chupons, fans, or branches (**Figure 1(a)**). Many strains of CSSV also induce root swellings. Stem swellings result from the abnormal proliferation of xylem, phloem, and cortical cells.

Infected trees may suffer from partial defoliation initially due to the incompletely systemic nature of the infection. Ultimately, in highly susceptible varieties, severe defoliation and dieback occurs.

Smaller, rounded to almost spherical pods may be found on trees infected with severe strains. Occasional green mottling of these pods is seen and their surface may be smoother than the surface of healthy pods. Various isolates were described in Ghana as in Togo with a variability and a gradation in the type of symptoms observed.

A few avirulent strains occur in limited, widely scattered outbreaks, usually inducing stem swellings only, and having little effect, if any, on growth or yield. Moreover, there are periods of remission during which symptoms are not visible.

The natural host range of the virus includes *Ceiba pentandra*, *Cola chlamydantha*, *Cola gigantea* var. *glabrescens*, *Sterculia tragacantha*, and *Adansonia digitata* with associated symptoms of transient leaf chlorosis or conspicuous leaf chlorosis.

**Figure 1** (a) Symptoms of red-vein banding observed on young flush leaves of cocao tree in Togo, Kloto area. (b) Symptoms of swellings on chupons of cocao tree, in Togo, Litimé area. (c) Symptoms of swollen shoot observed on *Theobroma cacao* plantlets four months after agroinoculation with the Togolese Agou 1 isolate. From left to right: A plantlet inoculated with the wild strain *Agrobacterium tumefaciens* LBA4404 and two plantlets inoculated with the recombinant *A. tumefaciens* bacteria LBA4404 (pAL4404, pBCPX2) containing CSSV insert.

## Transmission

Fourteen species of mealybugs (*Pseudococcidae* spp.), including *Planococcoides njalensis, Planococcus citri, Planococcus kenyae, Phenacoccus hargreavesi, Planococcus* sp. *Celtis, Pseudococcus concavocerrari, Ferrisia virgata, Pseudococcus longispinus, Delococcus tafoensis,* and *Paraputo anomalus,* have been reported to transmit CSSV. Only the nymphs of the first, second, and third larval stages and the adult females are able to transmit the virus. The virus does not multiply in the vector and is not transmitted to its progeny. CSSV is transmitted neither by seed nor by pollen. CSSV can infect cocoa at any stage of plant growth. The virus is transmitted experimentally to susceptible species by grafting, particle bombardment, by agro-infection using transformed *Agrobacterium tumefaciens* and with difficulty by mechanical inoculation. Seedlings usually produce acute red vein banding within 20–30 days and, 8–16 weeks later, swellings on shoots and tap roots (**Figure 1(c)**).

As this disease appeared soon after the introduction of cacao to West Africa, it is likely that CSSV came from indigenous hosts. The species *Adansonia digitata, Ceiba pentandra, Cola chlamydantha,* and *Cola gigantea* were identified as reservoir hosts in Ghana.

## Economic Importance

CSSV is a serious constraint to cocoa production in West Africa, particularly in Ghana. Severe strains of this virus can kill susceptible cocoa trees within 2–3 years. They affect Amelonado cocoa, widely considered to give the best-quality cocoa, more seriously than Upper Amazon cocoa and hybrids which have been selected for resistance to the virus.

The disease was first recognized in 1936 but almost certainly occurred in West Africa in the 1920s. Estimates of annual yield losses due to this virus vary from about 20 000 tonnes to approximately 120 000 tonnes of cocoa from the Eastern Region of Ghana alone. The average annual loss between 1946 and 1974 in Ghana was estimated to be worth over £3 650 000.

Attempts at CSSV control in Ghana have required substantial financial and manpower inputs. The 'cutting-out' policies in place in Ghana since the early 1940s which attempted to control cocoa swollen shoot disease resulted in the removal of over 190 million trees up to 1988. Over ten million infected trees which still required 'cutting out' were identified in the field by 1990.

CSSV is currently confined to West Africa, Sri Lanka, and Sumatra. The disease does not pose a real economic problem for the culture of cocoa in Sri Lanka and Indonesia. However, the devastation which has occurred in West Africa has serious implications for germplasm movement. International attempts at crop improvement are hampered by the need to index cocoa germplasm for this virus, particularly if the germplasm is to be moved to where highly susceptible varieties are grown. Although seed transmission of this virus is not known to occur, there is often the need to move germplasm as stem cuttings, which must then pass through intermediate quarantine for indexing.

## Molecular Characterization

### Characteristics of the Virions

The viral origin was shown in 1939. CSSV possesses small nonenveloped bacilliform particles and a double-stranded DNA genome of 7–7.3 kbp. The bacilliform particles measure 130 nm × 28 nm in size and have been shown by dot–blot hybridization to occur in the cytoplasm of phloem companion and xylem parenchyma cells.

## Description of Full-Length Sequence

Molecular characterization of CSSV had a boost in 1990, thanks to the improvement of the techniques of purification.

The first complete sequence of a CSSV isolate (Agou1 from Togo) was determined in 1993. Five putative open reading frames (ORFs) are located on the plus strand of the 7.16 kbp CSSV genome. ORF1 encodes a 16.7 kDa protein whose function is not yet determined. The ORF2 product is a 14.4 kDa nucleic acid-binding protein. ORF3 codes for a polyprotein of 211 kDa which contains, from its amino- to carboxyl-terminus, consensus sequences for a cell-to-cell movement protein, an RNA binding domain of the coat protein, an aspartyl proteinase, a reverse transcriptase (RTase), and an RNase H. The last two ORFs X (13 kDa) and Y (14 kDa) overlap ORF3 and encode proteins of unknown functions. CSSV, CMBV, and TaBV are the only badnaviruses which, to date, are known to code for more than three ORFs, CMBV, and TaBV encode six and four ORFs respectively.

## Variability of the CSSV

A more extensive study of the molecular aspects of CSSV variability is relevant for three reasons. First, the knowledge of molecular variability will allow the improvement and the validation of a PCR diagnostic test for better virus-indexing procedures. Second, the variability of the virus must be taken into account for resistance screening of new cocoa varieties to CSSV. Finally, a better understanding of the genetic diversity of CSSV in West Africa and elsewhere will in turn help to provide a better understanding of the development of the epidemics and their eradication, and of the evolution of viral populations.

Badnaviruses are highly variable at both the genomic and serological level, a feature which complicates the development of both molecular and antibody-based diagnostic tests. Moreover, CSSV isolates were for a long time classified according to the variability of the symptoms expressed on *T. cacao*, but it is not known if there is a correspondence between this variability of symptoms and the intrinsic molecular variability of the virus.

A first study analyzed the molecular variability of the area of the ORF3 coding for N-terminal coat protein for 1A-like isolates from Ghana and recently the analysis of new whole sequences made it possible to have a better idea of the variability of the CSSV.

**Figure 2** presents the circular genome maps of the six isolates of CSSV sequenced to date, and these isolates come from Togo (Agou1, Nyongbo2 from Kloto area, and Wobe12 from Litimé area) and Ghana. The size of the total genome varies from 7024 bp for isolate N1A to 7297 bp for the isolate Wobe12. The overall organization of ORFs is the same for all the isolates but differs by the partial disappearance of the ORFX for the three Ghanaian isolates, and by the existence of an additional ORF for the Peki isolate, which codes for a putative protein of 11 kDa. **Table 1** shows the nucleotide and amino acid sequence percentage identity and similarity between the different ORFs of isolates. ORF1 is the most conserved ORF coding region (81 synonymous mutations versus 101 mutations among the six isolates) and the amino acid dissimilarity of pairwise combinations of CSSV isolates ranges from 1.4% to 8.4%. The maximum amino acid dissimilarity between isolates is 25.5% for ORF2, 23.8% for ORF3, and slightly higher for ORFY (33.6%). The value of overall variability calculated for ORF3 corresponds to a succession of more or less variable regions that can be identified on alignment. In agreement with the division of the ORF3 polyprotein in three regions, we observed that the first region (amino acids 1–350), which corresponds to the movement protein region, is highly conserved as already observed for other pararetroviruses. Region 2 is far less conserved (particularly amino acids 370–500 and 1010–1070). Region 3 has an intermediary level of variability. ORFX is the least conserved ORF.

Maximum nucleotide sequence variability between pairwise combinations of complete genomic sequences of CSSV isolates was 29.4% (between Wobe12-CSSV and Peki-CSSV).

The alignments of the nucleotide and protein sequences of the ORFs (**Table 1**) and the phylogenetic trees built from these sequences make it possible to separate the isolates into three groups according to their geographical origin, rather than their aggressiveness (CSSV-N1A are CSSV-Peki are considered as mild isolates).

A study of sequence variability was made on the level of the first region of the ORF3 on isolates taken in two areas different from Togo having distinct epidemiologic histories. The swollen shot disease was observed for the first time in 1949 in Togo in the area of Kloto, and then spread only in this area. It is only toward the end of the 1990s that the disease started to be observed in Litimé. The phylogenetic tree built from the sequences of isolates of these two areas distinguishes three groups (A, B, and C), as does the analysis based on the whole sequences (**Figure 3**). One of these groups, A, containing Wobe12 is more distantly related to the others and contains only isolates coming from the area of Litimé, Togo. Moreover, the amplified sequence of isolates from group A (724 bp instead of the 721 bp amplified for the other isolates) code for an additional amino acid. Group A of isolates cannot therefore originate from the area of Kloto, infected before and probably originate from Ghana. In Kloto, only one group of isolates is present (C), whereas in the area of Litimé the three groups of diversity coexist.

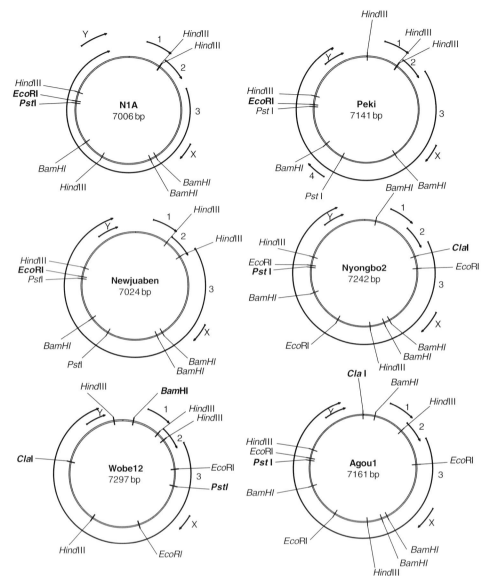

**Figure 2** Organization of the circular genomes of new CSSV isolates compared to Agou 1. Arrows indicate the deduced ORFs 1–4, X and Y capable of encoding proteins larger than 9 kDa. *Bam*HI, *Cla*I, *Eco*RI, *Hind*III, and *Pst*I restriction sites are shown. Unique restriction sites are in bold. Reprinted from Muller E and Sackey S (2005) Molecular variability analysis of five new complete cacao swollen shoot virus genomic sequences. *Archives of Virology* 150: 53–66, with permission from Springer Wien.

It is probable that the group B of isolates of Kloto, that are present also in Litimé, come from a contamination of Litimé by isolates of Kloto.

## Homology with Other Viruses

A phylogenetic tree was built from all the available full-length sequences of badnaviruses (**Figure 3**). CSSV isolates are closer to each other than to badnaviruses and to members of the genus *Tungrovirus*. In addition, among the badnaviruses sequenced to date, CSSV has the closest relationship with CMBV whose host is also a tree crop. The majority of the hosts of the badnavirus are indeed plants with vegetative multiplication.

However, one group of BSV-like sequences recently detected in Uganda are closer to CSSV sequences than to other BSV sequences and one sequence found integrated in *Musa acuminata* germplasm has good homology with CSSV.

The maximum variability level observed between the six CSSV isolates is slightly higher than the one observed between the three rice tungro bacilliform virus (RTBV) isolates but much lower than between the various isolates of BSV sequenced until now.

**Table 1** Nucleotide sequence identity and amino acid sequence similarity of pairwise combinations of CSSV isolates for the ORFs 1, 2, 3, X, and Y

percentage of nucleotide sequence identity
percentage of amino acid sequence similarity

|  | ORF1 | | | | | | ORF2 | | | | | | ORF3 | | | | | | ORFX | | | | | | ORFY | | | | | |
| --- | --- | --- | --- | --- | --- | --- | --- | --- | --- | --- | --- | --- | --- | --- | --- | --- | --- | --- | --- | --- | --- | --- | --- | --- | --- | --- | --- | --- | --- | --- |
|  | Ag1 | Nb2 | NJ | Peki | N1A | w12 | Ag1 | Nb2 | NJ | Peki | N1A | w12 | Ag1 | Nb2 | NJ | Peki | N1A | w12 | Ag1 | Nb2 | NJ | Peki | N1A | w12 | Ag1 | Nb2 | NJ | Peki | N1A | w12 |
| Agou1 | 100 | 98.8 | 80 | 80.9 | 80.7 | 95.6 | 100 | 88.3 | 73.3 | 73.6 | 73.6 | 70.7 | 100 | 98.2 | 85.5 | 84.7 | 81.8 | 74.4 | 100 | 97.1 | 58.7 | 58.4 | 59.5 | 40.1 | 100 | 98.5 | 89.3 | 87.3 | 89.1 | 71 |
|  | 100 | 98.6 | 93.7 | 94.4 | 93 | 97.2 | 100 | 89 | 82.8 | 82.1 | 82.1 | 74.5 | 100 | 98.2 | 90.6 | 90.2 | 86.5 | 79.8 | 100 | 92.9 | 41.9 | 45.3 | 42.7 | 22.4 | 100 | 97.7 | 92.4 | 87.8 | 89.3 | 69.5 |
| Nyongbo2 |  | 100 | 80.2 | 81.1 | 80.9 | 94.9 |  | 100 | 81.8 | 81.8 | 81.8 | 77.6 |  | 100 | 85.1 | 84.4 | 81.5 | 74.2 |  | 100 | 58.7 | 58.1 | 60.1 | 41.3 |  | 100 | 88.3 | 86.3 | 88 | 71 |
|  |  | 100 | 93.7 | 94.4 | 93 | 97.2 |  | 100 | 89.7 | 88.3 | 88.3 | 76.5 |  | 100 | 90.3 | 89.8 | 86.3 | 79.7 |  | 100 | 41.9 | 43.6 | 44.4 | 22.4 |  | 100 | 90.1 | 85.5 | 87 | 71 |
| New Juaben |  |  | 100 | 97.2 | 96.7 | 80.9 |  |  | 100 | 96.6 | 96.3 | 76.7 |  |  | 100 | 96.4 | 93.3 | 73.3 |  |  | 100 | 96 | 96.3 | 47.3 |  |  | 100 | 92.9 | 94.9 | 71.8 |
|  |  |  | 100 | 97.9 | 97.9 | 92.3 |  |  | 100 | 96.6 | 96.6 | 77.9 |  |  | 100 | 96.2 | 93.6 | 79.5 |  |  | 100 | 89 | 91.2 | 14.3 |  |  | 100 | 90.1 | 92.4 | 69.5 |
| Peki |  |  |  | 100 | 97.7 | 81.4 |  |  |  | 100 | 96.6 | 76.5 |  |  |  | 100 | 91.8 | 73.5 |  |  |  | 100 | 94.1 | 46.9 |  |  |  | 100 | 93.4 | 69.7 |
|  |  |  |  | 100 | 97.2 | 93.7 |  |  |  | 100 | 95.9 | 77.2 |  |  |  | 100 | 92 | 80 |  |  |  | 100 | 89 | 20.5 |  |  |  | 100 | 88.5 | 65.6 |
| N1A |  |  |  |  | 100 | 80.7 |  |  |  |  | 100 | 77 |  |  |  |  | 100 | 70.4 |  |  |  |  | 100 | 51.3 |  |  |  |  | 100 | 70.7 |
|  |  |  |  |  | 100 | 91.6 |  |  |  |  | 100 | 78.5 |  |  |  |  | 100 | 76.2 |  |  |  |  | 100 | 20.5 |  |  |  |  | 100 | 66.4 |
| Wobe12 |  |  |  |  |  | 100 |  |  |  |  |  | 100 |  |  |  |  |  | 100 |  |  |  |  |  | 100 |  |  |  |  |  | 100 |
|  |  |  |  |  |  | 100 |  |  |  |  |  | 100 |  |  |  |  |  | 100 |  |  |  |  |  | 100 |  |  |  |  |  | 100 |

Ag1, Agou1; Nb2, Nyongbo2; NJ, New Juaben; w12, Wobe12.
Reproduced from Muller and a Sockey (2005) Molecular variability analysis of five new complete cacao swollen shoot virus genomic sequences. *Archives of Virology* 150: 53–66, with permission from Springer Wien.

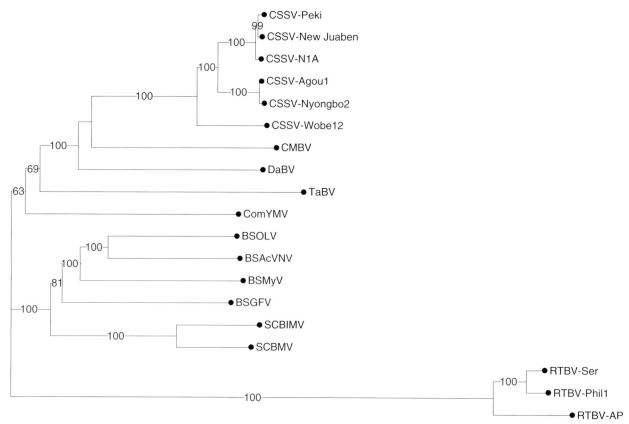

**Figure 3** Neighbor-joining tree generated by the Darwin 4 program based on complete nucleotide sequences of badnaviruses and rice tungro bacilliform virus (RTBV). Numbers at the nodes of the branches represent percentage bootstrap values (1000 replicates) when superior to 60. The GenBank accession numbers of sequences are L14546 (CSSV-Agou1), AJ534983 (CSSV-Nyongbo2), AJ608931 (CSSV-New Juaben), AJ609019 (CSSV-Peki), AJ609020 (CSSV-NIA), AJ781003 (CSSV-Wobe12), AJ002234 (banana streak OL virus – BSOLV), AY805074 (banana streak mysore virus – BSMyV) AY750155 (banana streak acuminate vietnam virus – BSAcVNV), AY493509 (banana streak GF virus – BSGFV), X52938 (commelina yellow mottle virus – comYMV), AF347695 (citrus mosaic bacilliform virus – CMBV), X94576 (dioscorea alata bacilliform virus – DaBV), M89923 (sugarcane bacilliform Mor virus – SCBMV), AJ277091 (sugarcane bacilliform IM virus – SCBMIV), AF357836 (taro bacilliform virus – TaBV), AJ292232 (RTBV-AP), O76470 (RTBV-Ser), X57924 (RTBV-Phill).

## Molecular Epidemiology

A systematic molecular characterization of viral isolates should make it possible to study the molecular epidemiology of CSSV. Moreover, studies on CSSV in adventitious plants or insect vectors would make it possible to better understand the epidemiologic factors that lead to the rapid expansion of the disease in some plots.

## Diagnosis

### Serological Diagnostics

The virus is not strongly immunogenic; however, several antisera have been raised and shown to react with CSSV. Enzyme-linked immunosorbent assay (ELISA) has been used to detect CSSV but high background values were obtained and difficulties were found in detecting CSSV in plants suspected of having only a low virus titer. Immunosorbent electron microscopy has been used for detection and comparison of some isolates of CSSV in Ghana. Both of these techniques cannot detect latent infection.

The virobacterial agglutination (VBA) test has been found to be a useful test for detecting CSSV in leaf tissue from infected trees. Using this assay, CSSV can be detected in trees showing symptoms as well as in infected, but symptomless, trees. The immunocapture polymerase chain reaction (IC-PCR) technique was adapted to the only detection of CSSV-1A isolates from Ghana.

There is considerable strain variation among the many recognized CSSV isolates, some of which react only weakly with certain antisera. Using monoclonal antibodies, four serotypes of CSSV have been distinguished by ELISA analysis of 31 samples of the virus from different geographical locations in Ghana. The efficiency of serological diagnosis depends on the use of polyvalent antiserum able to detect all serotypes.

### PCR Diagnosis

For a reliable PCR diagnosis, it is necessary to design primers from conserved regions of the genome. Until recently, only few badnaviruses were sequenced in full, and only the end of the ORF3 which contains the conserved motifs coding for RTase and RNaseH made it possible to obtain primers useful for diagnosis. PCR-based diagnosis is able to detect CSSV not only in symptomless leaves of symptomatic plants, but also in symptomless plants as early as one week post-inoculation.

The alignment of the six full-length sequences of CSSV allowed conserved regions to be identified and polyvalent primers to be designed for the diagnosis of CSSV. These primers can detect all the isolates tested to date from Togo and Ghana. These primers are located in the first part of the ORF3 and amplify a fragment from 721 to 724 bp, including from isolates having only 70% nucleotide identity (CSSV-Wobe12 and CSSV-Peki). The extraction method is based on a buffer containing MATAB (mixed alkyltrimethylammonium bromide), which reduces the co-extraction of PCR-inhibitory substances, as well as being less costly and generates a higher DNA yield.

These primers for CSSV diagnosis may require modifications as more sequence data from other CSSV isolates are generated. As the sensitivity of diagnosis is better when young symptomatic host plant leaves are tested, it is recommended that validation of the diagnostic test is done using this type of material in the first instance, especially when testing isolates from different areas.

### Control of the Disease

Control of the swollen shoot disease based only on 'cutting-out' campaigns has not been successful due to several factors including political and socioeconomic problems. The better strategy for dealing with the disease is to develop a combination of control measures in an integrated approach. Moreover, this approach should implicate cocoa farmers as much as possible.

Intermediate quarantine facilities are at present hampered by the lack of suitable indexing methods for the virus. The early PCR diagnosis could be easily tested in intermediate quarantine facilities and compared with the grafting procedure used for indexation on Amelonado cocoa seedlings.

Mild strain protection is a possibility which is being investigated. Mild strains which appear to confer some protection against the severe strains are available and are being tested in the field in Ghana. However, the degree of protection afforded so far is not sufficient and further research is necessary.

The possibility of isolating new cocoa plantings from infected cocoa by using barriers of CSSV-immune crops could be considered. These crops would form a barrier to the movement of vectors. Examples of possible barrier crops include oil palm, coffee, cola, and citrus. The 'cutting out' of the adventitious plants of the type Commelina and Taro could be a possible strategy if it is confirmed that these plants harbor CSSV.

The use of resistant cocoa is advocated as many of the new hybrids available in West Africa do have some resistance to CSSV and because it seems to be the most sustainable method. Replanting with resistant cacao trees, however, requires the installation of a protocol of effective screening for resistance. Severe isolates representative of the different molecular groups should be used as well as a suitable screening method. However, a standardized inoculation method is not yet available because particle bombardment is difficult to develop on a large scale and agroinoculation needs biosafety confinement. The screening for CSSV resistance for two types of severe isolates has been initiated in 2003 in Togo.

*See also:* Caulimoviruses: General Features; Caulimoviruses: Molecular Biology; Virus Classification by Pairwise Sequence Comparison (PASC).

### Further Reading

Brunt AA (1970) *CMI/AAB Descriptions of Plant Viruses No 10. Cacao Swollen Shoot Virus.* Wellesbourne: Association of Applied Biologists.

CABI, Crop Protection Compendium(2002) *Cacao Swollen Shoot Virus.* Wallingford: CABI Publishing.

Castel C, Amefia YK, Djiekpor EK, Partiot M, and Segbor A (1980) Le swollen shoot du cacaoyer au Togo. Les différentes formes de viroses et leurs conséquences économiques. *Café, Cacao, Thé* 24(2): 131–146.

Fauquet C, Mayo M, Maniloff J, Desselberger U, and Ball L (eds.) (2005) *Virus Taxonomy: Eighth Report of the International Committee on Taxonomy of Viruses.* San Diego, CA: Elsevier Academic Press.

Hagen LS, Jacquemond M, Lepingle A, Lot H, and Tepfer M (1993) Nucleotide sequence and genomic organization of cacao swollen shoot virus. *Virology* 196(2): 619–628.

Hagen LS, Lot H, Godon C, Tepfer M, and Jacquemond M (1994) Infection of *Theobroma cacao* using cloned DNA of cacao swollen shoot virus and particle bombardment. *Molecular Plant Pathology* 84: 1239–1243.

Lot H, Djiekpor E, and Jacquemond M (1991) Characterization of the genome of cacao swollen shoot virus. *Journal of General Virology* 72: 1735–1739.

Muller E and Sackey S (2005) Molecular variability analysis of few new complete cacao swollen shoot virus genomic sequence. *Archives of Virology* 150: 53–66.

# Caliciviruses

**M J Studdert and S J Symes,** The University of Melbourne, Parkville, VIC, Australia

© 2008 Elsevier Ltd. All rights reserved.

## Glossary

**Chimeric virus** A virus containing genetic material from at least two genetically different parents.

**Genetic recombination** The molecular process of forming a chimeric virus and in the caliciviruses is likely to involve a copy-choice mechanism, whereby the viral RNA polymerase switches template during replication.

**Histo-blood group antigens** Molecules with carbohydrate epitopes, first described on the surface of red blood cells but subsequently shown to be expressed on the surface of many cells and some of which are receptors for some caliciviruses.

**Pinniped** Marine mammals including sea lions, walruses, fur seals, and true seals.

**Splenomegaly** Enlargement of the spleen.

**Vesicular disease** A disease characterized by a fluid-filled sac arising within an epithelial surface (blister).

## Introduction

The members of the family *Caliciviridae* are classified in four genera: *Vesivirus, Lagovirus, Norovirus*, and *Sapovirus* (**Table 1**). Caliciviruses produce diseases that are important in both human and veterinary medicine.

## History

### Vesiviruses

In 1932 there occurred in southern California, in pigs fed uncooked garbage, a vesicular disease that was provisionally diagnosed as foot and mouth disease. The disease was eradicated by slaughter and quarantine. However, in 1933 the disease recurred and at that time it was realized that it was not foot and mouth disease but a new disease that was named vesicular exanthema of swine. Between 1932 and 1951, the disease continued to occur in southern California and more than 2.5 million pigs representing 21% of the total pig population in southern California were involved in the outbreaks. There was a clear linkage between the disease and the feeding of raw, that is, uncooked garbage containing, it was assumed, pork scraps. Outbreaks of the disease continued to occur despite the introduction of laws requiring that all garbage fed to pigs be cooked. In late 1951 a train left San Francisco and off-loaded in Wyoming, to a pig rancher, garbage containing pork scraps that were fed to his pigs. When a few of the pigs in the herd developed vesicular lesions, the owner shipped the pigs for sale. Following this, between 1952 and 1956, outbreaks of vesicular exanthema occurred in 40 of the 48 states. The last outbreak occurred in New Jersey in 1951. By 1956 rigid enforcement of laws that required all garbage fed to pigs be cooked, or a total ban on feeding garbage, resulted in the disappearance of the disease from the national swineherd. In 1959 vesicular exanthema of swine virus was declared eradicated from the United States, the only country in which it has occurred.

In 1972 a calicivirus was isolated from sea lions during an investigation of an outbreak of abortion on San Miguel Island off the southern Californian coast. This virus, called San Miguel sea lion virus, closely resembled vesicular exanthema virus and was transmissible to pigs. It is believed that the origin of vesicular exanthema virus of swine was a consequence of feeding uncooked sea lion carcasses that were washed up on the beaches of southern California, to pigs. Subsequently, caliciviruses have been isolated from other pinniped species including Northern fur seals inhabiting the Pribilof Islands, Alaska.

Acute upper respiratory disease of cats is common and in 1957 the first feline calicivirus, originally identified as a picornavirus, was isolated. It was shown to be one of the two major viral causes of respiratory disease in cats, the other virus being feline herpesvirus 1.

### Lagoviruses

In 1984, a new, highly infectious disease of the European rabbit, *Oryctolagus cuniculus*, was identified in China. It was characterized by hemorrhagic lesions, particularly affecting the lungs and liver. The virus was eventually called rabbit hemorrhagic disease virus. It killed some 470 000 rabbits in the first 6 months and by 1985 had spread throughout China. By 1988, it had spread throughout eastern and western Europe and had reached North Africa. In December 1988, cases occurred in Mexico City. Both wild and domestic *O. cuniculus* were affected, but all other species of mammals, except the European hare, appear to be resistant to infection. The disease was unknown in Europe before 1984; however, a very similar disease called European brown hare syndrome had been recognized in the early 1980s affecting *Lepus europaeus* and subsequently some other *Lepus* species.

**Table 1** Overview of the family *Caliciviridae*

| Genus | Virus | Acronym | Primary host | Cultivable | Disease |
|---|---|---|---|---|---|
| *Vesivirus* | Vesicular exanthema of swine virus | VESV | Pig | Yes | Vesicular lesions of mouth, snout, hooves |
| | San Miguel sea lion virus | SMSV | San Miguel sea lion | Yes | Vesicular lesions of flippers, abortions |
| | Feline calicivirus | FCV | Cat | Yes | Upper respiratory tract and oral cavity disease |
| *Lagovirus* | Rabbit hemorrhagic disease virus | RHDV | Rabbit | No | Generalized hemorrhagic disease, liver necrosis, disseminated intravascular coagulation |
| | European brown hare syndrome virus | EBHSV | European brown hare | No | Hemorrhagic disease, liver necrosis |
| *Norovirus* | Norwalk virus | NV | Human | No | Gastroenteritis |
| *Sapovirus* | Sapporo virus | SV | Human | No | Gastroenteritis |

Rabbit hemorrhagic disease virus was imported into Australia in 1991 to a high security laboratory. Australian native animal species susceptibility studies were conducted prior to determining whether the virus would be a safe and effective biocontrol agent. In Australia, rabbits are in plague numbers, perhaps as many as 100 million. Rabbits are estimated to cause $600 million in annual losses; some of these losses, including native species and habitat destruction, may be permanent. The virus was transferred from the high security laboratory to Wardang Island for further pen trials. During these trials it escaped to the mainland, possibly by insect vector transmission (mosquitoes, bush flies) or carrion-eating birds (crows, hawks, eagles), and subsequently spread to many areas prior to 'official' release as a biocontrol agent. In New Zealand in mid-1997, the virus was illegally introduced and spread, probably, by farmers, irate, following a decision of the government not to allow legal importation of the virus until more was known about its host range, including native animal species susceptibility, particularly of the kiwi.

### Noroviruses and Sapovirues

Noroviruses are a major cause of epidemic nonbacterial gastroenteritis outbreaks in humans. Norwalk virus, the prototype strain of the genus *Norovirus*, was identified in 1972 by negative stain electron microscopy of fecal specimens obtained from human volunteers. These volunteers were exposed to filtrates of feces collected during an outbreak of gastroenteritis at a school in Norwalk, Ohio in 1968. Noroviruses lack the typical calicivirus morphology and were initially and awkwardly referred to as 'small round-structured viruses'.

Noroviruses have not been cultivated in conventional monolayer cell cultures. Noroviruses have been cultured in a physiologically relevant three-dimensional (3D) organoid model of human small intestinal epithelium. Determination of the genomic sequences of Norwalk virus and of Southampton virus (a norovirus detected in the United Kingdom) in the early 1990s allowed the development of diagnostic tests, including reverse transcription–polymerase chain reaction for virus detection and enzyme-linked immunosorbant assays using recombinant proteins for antibody detection. The detection and sequencing of a large number of norovirus genomes revealed significant genetic variability.

Sapporo virus, the prototype of the genus *Sapovirus*, was identified as a cause of gastroenteritis in human infants following an outbreak in a children's home in Japan in 1982. Unlike the noroviruses, sapoviruses possess the classic calicivirus morphology when viewed by electron microscopy. Historically, the noroviruses and sapoviruses have taken their name from the location of the outbreak.

Some caliciviruses of veterinary importance are still to be officially classified. Bovine enteric caliciviruses including Newbury agent 2 and Jena viruses, and a murine virus are proposed members of the genus *Norovirus*, while a porcine enteric calicivirus is a proposed member of the genus *Sapovirus*. The recent genomic characterization of another bovine enteric calicivirus, Newbury agent 1, shows that phylogenetically this virus does not fall within any of the four defined calicivirus genera and may represent a fifth genus, within the family *Caliciviridae*.

### Classification and Properties

Caliciviruses were initially classified as picornaviruses with which they share a number of properties. However, the distinctive properties of caliciviruses led to the creation of a new family *Caliciviridae*. Human hepatitis E virus was originally considered to be a calicivirus because of apparent similarity in morphology but is now classified in the genus *Hepevirus* in an undefined family.

The family name *Caliciviridae* derives from the cup-shaped (*calix* = cup) surface depressions that give the virion its unique appearance. By cryoelectron microscopy, virions are 40.5 nm with 32 cup-shaped surface structures comprising 90 arch-like capsomers arranged in $T=3$ icosahedral symmetry (**Figure 1**). The individual capsomers are dimers of the $\sim 60$ kDa capsid protein and have two main structural domains: the protruding (P) domain, which consists of the P1 and P2 subdomains, which are joined to the shell (S) domain by a flexible binge (**Figure 1**). The capsid protein has six regions (A–F) that probably form hypervariable, loop-like structures on the surface of the folded protein and are sites of antigenic variation. Noroviruses are exceptional in that they lack the characteristic cup-shaped depressions and have a fuzzy edge.

Virion buoyant density is 1.33–1.41 g$^{-1}$ ml in CsCl. Physicochemical properties have not been fully established for all members of the family and may not be common to all. Rabbit hemorrhagic disease virus is reported to be stable over a wide range of pH values (4.5–10.5). Noroviruses are acid, ether, and relatively heat stable. Vesiviruses are inactivated at pH 3–5, thermal inactivation is accelerated in high concentrations of Mg$^{2+}$ ions, and virions are insensitive to treatment with ether, chloroform, or mild detergents.

Complete genomic sequences have been determined for a least one member of each genus. The phylogenetic

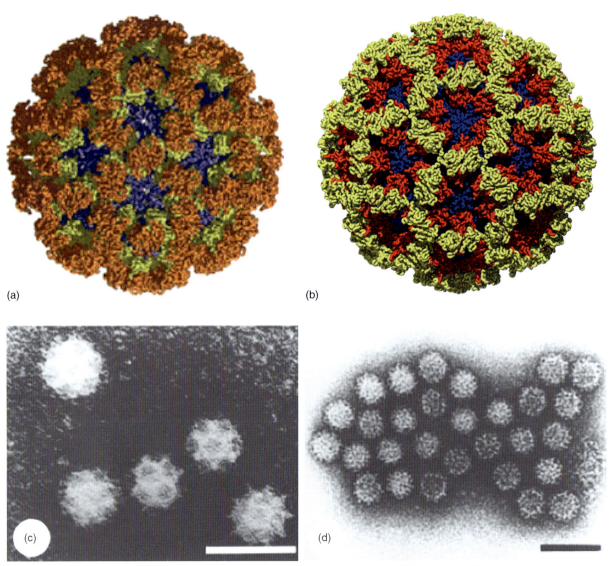

**Figure 1** (a) X-ray structure of the capsid of San Miguel Sea Lion virus. The S, P1, and P2 domains of the capsid protein are colored in blue, yellow, and darkish orange respectively. (b) X-ray structure of the capsid of Norwalk virus. The S, P1, and P2 domains of the capsid proteins are colored in blue, red, and yellow respectively. (c) Negative contrast electron micrograph of Norwalk virus in a human fecal sample. (d) Negative contrast electron micrograph of rabbit hemorrhagic disease virus in a rabbit liver homogenate. Scale = 100 nm (c,d). Images (a, b) provided by Dr. B. V. Venkataram Prasad, Department of Biochemistry and Molecular Biology, Baylor College of Medicine. Images (c, d) from Dr. A. Z. Kapikian (Norwalk) and M. König and H.-J. Thiel (RHDV), respectively and are adapted from the *seventh Report, International Committee on the Taxonomy of Viruses*, Academic Press, 2000.

relationships among the genera and viruses are shown in (**Figure 2**). The RNA genomes (**Figure 3**) consist of a linear, positive-sense, ssRNA molecule of 7.3–8.6 kb. A 10–15 kDa protein, designated VPg, is covalently attached to the 5′-end of the genomic RNA and the 3′-end is polyadenylated. Viruses of each genus have a characteristic open reading frame (ORF) arrangement but differences are found between the genera. In general, the nonstructural proteins are encoded at the 5′-end of the genome and the structural proteins at the 3′. For lagoviruses and sapoviruses the nonstructural proteins and the major structural protein are expressed as a single polyprotein encoded by ORF1, while for the vesiviruses and noroviruses the nonstructural polyproteins and the structural capsid protein are separately encoded by ORF1 and ORF2, respectively.

The nonstructural proteins 2C NTPase, 3C trypsin-like serine protease, and 3D RNA-dependent RNA polymerase are somewhat similar in both sequence identity and relative position, to similar proteins found in picornaviruses. There are helicase motifs at the 5′-end of the genome but no helicase activity has been detected. The individual nonstructural proteins are released from the polyprotein by specific viral protease cleavage in a cascade similar to that of picornaviruses. Viruses from all four genera also possess an ORF at the 3′-end of the genome, which is ORF3 for vesiviruses and noroviruses and ORF2 for lagoviruses and sapoviruses. The 3′ encoded protein is a small basic protein believed to be a minor structural protein. Some sapoviruses also possess an additional ORF (designated ORF3) contained completely within the capsid-encoding gene. However, the biological significance of this ORF is unclear.

Vesiviruses have two polyadenylated RNA species, one corresponding to the genome of 7.6 kb and the second, a subgenomic species of 2.4 kb, that is bicistronic encoding ORF2 and ORF3. The subgenomic RNA, like the full length genomic RNA, is VPg-linked. Feline calicivirus subgenomic RNA can be packaged into virions. Similar polyadenylated RNA species of 7.5 and 2.2 kb are recognized for rabbit hemorrhagic disease virus and both species are packaged into virions, either the same or separate virions. The 5′-ends of the genomic and subgenomic RNAs show high levels of sequence conservation. The 5′ untranslated region (UTR) of the caliciviruses is relatively short comprising 4 to 21 nucleotides.

## Geographic Distribution

As noted, vesicular exanthema was eradicated in 1956 from the United States. Other marine vesiviruses, including San Miguel sea lion virus, are endemic in pinniped species and have been particularly described as occurring

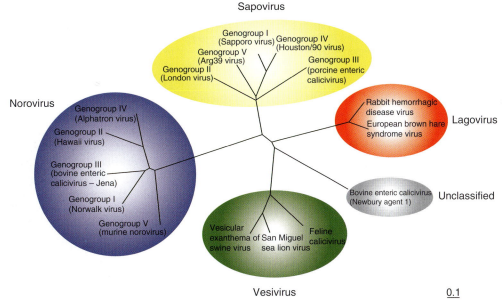

**Figure 2** Phylogenetic relationships among members of the family *Caliciviridae*. Phylogenetic analysis was performed using nucleotide sequences of the full capsid encoding region of representative caliciviruses. This unrooted maximum-likelihood tree is drawn with branch lengths to scale, and the scale bar indicating 0.1 substitutions per site. Shading indicates genera. Assignment into genera is supported by high bootstrap values (not shown). The following viruses were used in the analysis: vesicular exanthema of swine virus (A48, GenBank accession U76874), San Miguel sea lion virus (serotype 1, M87481), feline calicivirus (F9, Z11536), rabbit hemorrhagic disease virus (FRG, M67473), European brown hare syndrome virus (Z69620), Norwalk virus (M87661), Hawaii virus (U07611), bovine enteric calicivirus (Jena, AJ011099), Alphatron virus (AF195847), murine norovirus (AY228235), Sapporo virus (U65427), London virus (U95645), porcine enteric calicivirus (AF182760), Houston/90 virus (U95644), Arg/39 virus (AY289803), bovine enteric calicivirus (Newbury agent 1, NC_007916).

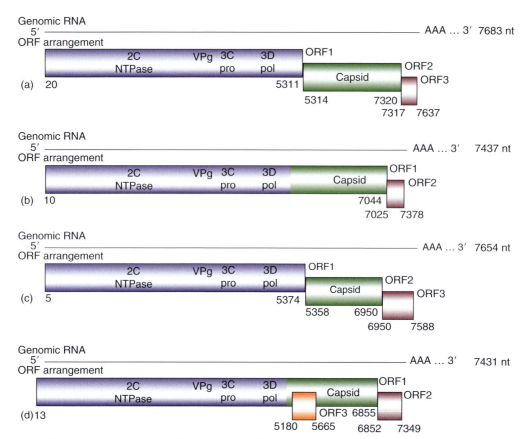

**Figure 3** Comparative genome organization of the caliciviruses. The four calicivirus genera illustrated are: (a) *Vesivirus* represented by feline calicivirus (Urbana, GenBank accession L40021); (b) *Lagovirus* represented by rabbit hemorrhagic disease virus (FRG, M67473); (c) *Norovirus* represented by Norwalk virus (M87661); and (d) *Sapovirus* represented by Manchester virus (X86560). Numbers below the ORFs correspond to the first and last nucleotides. Amino acid motifs in the nonstructural polyprotein correspond to the 2C NTPase, VPg, 3C protease (pro), and 3D RNA-dependent RNA polymerase (pol). The regions of the genomes are shown as follows: blue region – nonstructural polyprotein; green region – capsid protein; purple region – 3′ terminal ORF; orange region – sapovirus ORF3.

along the western seaboard of North America. Feline calicivirus is a very common cause of respiratory disease of cats throughout the world. Rabbit hemorrhagic disease has occurred in China, Eastern and Western Europe, North Africa, Korea, Mexico, Cuba and is now endemic in Australia and New Zealand.

Human noroviruses have a worldwide distribution and have been detected in association with outbreaks and sporadic cases of gastroenteritis on all permanently inhabited continents. Sapoviruses also have a global distribution but are less frequently detected than the noroviruses and are often linked to gastroenteritis in pediatric patients.

## Host Range and Virus Propagation

Caliciviruses infect a broad range of animals that include pigs, rabbits, hares, cats, pinnipeds, cetaceans, mice, cattle, reptiles, skunks, chimpanzees, and humans. Although individual calicivirus species generally have a natural host restriction, vesicular exanthema of swine virus appears an exception and has been isolated from several marine animal species (including fish), birds, reptiles, and land mammals. The geographic distribution of each calicivirus species usually reflects the host distribution. A probable calicivirus has been isolated from dogs with a vesicular genital disease.

Pigs are a significant alternative host for at least some of the pinniped caliciviruses. The extent to which pinniped and perhaps other marine species share caliciviruses as hosts has not been fully defined. Feline calicivirus is restricted to members of the family *Felidae* although most commonly, infections are recognized in domestic cats. Rabbit hemorrhagic disease virus, in addition to infecting rabbits, *O. cuniculus* and other *Oryctolagus* species, is known to infect and cause disease in the European brown hare, *Lepus europaeus* and some other *Lepus* species. Noroviruses and Sapoviruses are primarily associated with human gastroenteritis, but infections of other animal species including pigs and cattle occur. It is unlikely that filter-feeding bivalve marine species (mollusks) such as oysters and clams, which are common sources of human infection, usually as a consequence of contamination with human feces, support virus replication.

The pig, pinniped, and feline viruses grow readily and rapidly in monolayer cell cultures derived from pig and feline tissues respectively; most strains of vesicular exanthema virus grow in Vero cells. Rabbit hemorrhagic disease virus, human noroviruses, and human sapoviruses have not been cultivated in conventional cell cultures. However, the porcine enteric sapovirus has been cultivated in cell cultures when gnotobiotic pig intestinal contents are added to the growth medium and a 3D model of human intestinal epithelium has been shown to support norovirus replication.

The construction of an infectious cDNA clone of Norwalk virus has been described. When Norwalk virus cDNA and a replication-deficient vaccinia virus were co-transfected into mammalian cells, Norwalk viral RNA was replicated and packaged into virus particles. Norwalk virus subgenomic RNA was transcribed from genomic RNA using nonstructural proteins expressed from genomic RNA and was translated into the major capsid protein VP1. An infectious clone for rabbit hemorrhagic disease virus has also been described.

## Serologic Relationships and Variability

Based on serum neutralization assays and cross-protection studies in pigs, vesicular exanthema virus has a very high number of antigenic types. It was not uncommon for more than one type to be isolated during a single outbreak of disease or indeed for more than one virus type to be isolated from a single pig. The exact number of types cannot be recorded with certainty since some of the early collections of viruses were lost, although at least 13 distinct antigenic types exist in one collection of viruses.

A similar pattern of antigenic variation (types) was initially recognized among feline caliciviruses. When rabbit antiserums were used in serum neutralization assays, a large number of different antigenic types were identified in collections examined by the few individual laboratories that attempted these studies; no single set of typing criteria has been used to examine all feline caliciviruses. Curiously, however, when antisera raised in specific pathogen-free cats were used in virus neutralization assays to examine reasonably large collections of feline caliciviruses, while considerable antigenic variation was recognized, extensive cross-reactions were identified essentially between all viruses. It was concluded that feline caliciviruses were related as a single antigenic type and these findings paved the way for the development of monotypic, later bivalent, vaccines.

Antigenic variation among rabbit hemorrhagic disease virus isolates has not been much studied primarily because of the lack of a cell culture system; however, monotypic vaccines appear to be fully protective. Rabbit hemorrhagic disease and European brown hare syndrome viruses are antigenically distinct.

Several attempts have been made to define antigenic types for noroviruses and sapoviruses using methods including cross-challenge studies in human volunteers, immune electron microscopy, and ELISA using antisera made to virus-like particles (VLPs). Originally, four antigenic types of human noroviruses were proposed: Norwalk, Hawaii, Snow Mountain, and Taunton viruses; other schemes have also been proposed, but a unified system for serotyping based on neutralization is not available.

The most comprehensive information about variability of noroviruses and sapoviruses comes from the sequencing of various regions of genomic RNA from a large number of strains. As might be expected for a single-stranded RNA virus, much sequence variation is observed. Available data suggest the division of both genera into several distinct 'genogroups'. Noroviruses have been divided into at least five genogroups, with the human viruses in genogroups I, II, and IV, the bovine noroviruses in genogroup III, and murine norovirus in genogroup V. Sapoviruses fall into at least five genogroups. Within the genogroups, viruses have been further divided into 'clusters'. Because of the lack of appropriate cell culture systems or standardized serological reagents, the correlation between the genogroup and cluster designation of a virus and its antigenic type has not been completely established.

Genetic recombination resulting in chimeric viruses has been observed for noroviruses, sapoviruses, and feline calicivirus. For noroviruses the most commonly identified recombination site is the ORF1–ORF2 junction. There are two features of this site that predispose it to recombination. First, the site is highly conserved among different viruses, increasing the likelihood of homologous recombination between separate viruses. Second, it is the putative beginning of subgenomic RNA, which may itself play a role in recombination. Recombination is believed to occur by a copy-choice mechanism. A similar form of recombination was originally described for picornaviruses, specifically poliovirus. Copy choice occurs during viral replication when the viral RNA-dependent RNA polymerase switches templates from the RNA of one parent virus to the RNA of a second parent.

## Epidemiology

Vesicular exanthema was initially and on an ongoing basis transmitted to pigs by the feeding of uncooked pinniped carcass meat and additional transmission occurred by feeding uncooked garbage containing pork scraps. Within a herd, pig-to-pig transmission occurred since ruptured vesicles shed large quantities of virus into the environment such that transmission would readily occur by contact or via fomites.

Feline caliciviruses are transmitted by contact and particularly by sneezing when cats are closely confined as in multiple cat households, breeding and boarding establishments, cat shelters, and veterinary hospitals. Recovered cats remain persistently infected with virus for many months or years presumably as a consequence of low-grade infection in pharyngeal tonsillar tissues. True long-term persistent infection in individual cats is relatively rare. Progressive genetic variation following primary infection and perhaps even more importantly, in persistently infected cats leads to a gradual increase in the diversity of a given strain of virus, both in the individual and in the population, until eventually new strains emerge.

Rabbit hemorrhagic disease virus is readily transmitted between rabbits via a fecal–oral route and over longer distances apparently by fomites including contaminated vehicles. Insects were believed to have played a role as mechanical vectors in the escape and subsequent dissemination of rabbit hemorrhagic disease virus from an offshore island location to the mainland of Australia. Infected rabbit carcass meat could be carried over considerable distances by predatory and carrion-eating mammals and birds.

Noroviruses and sapoviruses are transmitted via a fecal–oral route. Transmission often results from consumption of fecally contaminated water or food; bivalve filter-feeding mollusks (oysters, clams) are commonly implicated. However, transmission from person to person in semiclosed communities such as nursing homes, hospitals, and cruise ships is also common. Short-distance aerosol transmission following vomiting has been suggested. Common source outbreaks associated with food and water contamination have involved up to 2000 individuals. In more recent years, cruise ships have emerged as sites for many outbreaks, causing major problems for the cruise ship industry and ruining holidays for passengers. Hospital ward closures following norovirus outbreaks cause major disruptions.

Currently, noroviruses belonging to genogroup II, especially cluster 4, are the predominant type circulating worldwide. However, other types continue to circulate concurrently. Several variants of the genogroup II cluster 4 viruses, correlated with an increase in the detection of norovirus-associated gastroenteritis in various regions around the world, have been described.

## Pathogenesis

Vesicular exanthema virus gains entry via abrasions usually around the snout and mouth or on the feet. Secondary vesicles may occur as a result of direct local spread or following viremia. Abortion may occur and death of baby pigs from agalactia in their dams also occurs.

Feline calicivirus produces vesicular lesions on the muzzle, and within the oral cavity and the respiratory tract. These tend to rupture quickly. During the past decade a more severe disease termed virulent-systemic (VS) feline calicivirus disease has been recognized. No genetic or *in vitro* diagnostic methods to distinguish viruses isolated from cases of VS disease from other isolates have been described. Phylogenetic analyses and alignments of the capsid and protease–polymerase regions of the genome did not reveal any conserved changes that correlated with virulence and the VS isolates did not segregate into a distinct cluster. These results suggest that VS isolates arise independently and at multiple sites. VS isolates spread more efficiently in cell culture than other isolates when cultures are inoculated at low multiplicity, although the overall growth kinetics of both standard strains and VS isolates are very similar.

Rabbit hemorrhagic disease is a generalized infection in which viremic spread results in lesions in a wide range of tissues. At postmortem lesions are particularly evident as hemorrhagic necrosis of the liver and lung. The key to the pathogenesis is massive liver necrosis leading to disseminated intravascular coagulation.

For noroviruses the incubation period is 24–48 h (range 10–77 h) and illness usually lasts 24–48 h. Virus shedding as detected by immune electron microscopy coincides with the onset of symptoms and stops at about 72 h. By reverse transcription polymerase chain reaction (RT-PCR) virus shedding has been detected for up to 7 days, although, in one case, an otherwise healthy child excreted norovirus for 60 days. Excretion times are even longer in immunocompromized individuals, 156 days in one instance. Asymptomatic virus excretion occurs in some individuals.

Virions are acid stable and hence safely traverse the stomach to reach the site of primary replication, which is assumed to be the upper intestinal tract. Based on human volunteer studies biopsy samples from the jejunum show that there is a broadening and blunting of the intestinal villi although curiously the epithelial cells remain intact; virus has not been detected in these cells by thin section electron microscopy. There is a shortening of microvilli, infiltration of mononuclear cells, and cytoplasmic vacuolation.

There is evidence to indicate that carbohydrate histo-blood group antigens are receptors for noroviruses. The histo-blood group antigens, first described on the surface of red blood cells, are also expressed on the surface of epithelial cells. First, a correlation has been observed between the ABO histo-blood group and susceptibility. Second, norovirus particles, either native virions or VLPs, directly bind to the blood group antigens *in vitro*. Different norovirus strains may utilize histo-blood group antigens encoded for by different alleles as receptors. Therefore, susceptibility alleles for infection with one strain are likely to be different to susceptibility alleles for infection with another strain, unless the strains are closely related.

There is evidence that rabbit hemorrhagic disease virus may also use histo-blood group antigens as receptors. Hemagglutination of human erythrocytes and binding to synthetic blood group oligosaccharides was demonstrated for rabbit hemorrhagic disease virus using both native virions and recombinant VLPs. Rabbits less than 2 months of age are reported to express negligible amounts of histo-blood group antigens on the surface of epithelial cells. This observation may explain the resistance of young rabbits to fatal infection. Feline junctional adhesion molecule 1 (JAM-1), a glycoprotein belonging to the immunoglobulin superfamily, was shown to act as a receptor for feline calicivirus.

## Clinical Signs

Vesicular exanthema of pigs is clinically indistinguishable from the other three, so-called vesicular diseases that affect pigs, that is, foot and mouth disease, vesicular stomatitis, and swine vesicular disease. Following an incubation period of 12–48 h there is a marked febrile response, anorexia, and listlessness. Primary vesicles are blanched, raised areas of epithelium up to 3 cm in diameter and up to 1 cm high filled with a serous virus-rich fluid. They easily rupture leaving raw bleeding ulcers that subsequently become covered with a fibrinous pseudomembrane. Secondary vesicles appear 48–72 h post infection. Notably these appear on the soles of the feet and in the interdigital space and at the coronary band. There is severe four-footed lameness. Secondary bacterial infection of lesions, particularly of the feet, occurs and prolongs recovery where slaughter and eradication are not pursued.

Feline calicivirus infection may produce subclinical, acute, or subacute disease. Symptoms include conjunctivitis, rhinitis, tracheitis, pneumonia (usually in young kittens) and vesiculation/ulceration of the epithelium of the oral cavity and muzzle. There is fever, anorexia, lethargy, stiff gait, and usually a profuse ocular and nasal discharge. Morbidity is high and mortality in untreated cases may reach 30%. The virulent systemic disease is characterized by high, persistent fever, anorexia, depression, facial and limb edema, sores or alopecia on the face, pinnae and paws, pulmonary edema, coagulation abnormalities, pancreatitis, and hepatic necrosis.

Intriguingly, rabbit hemorrhagic disease does not generally affect rabbits less than 2 months of age even where there is no maternal antibody protection. As indicated earlier, there is evidence that rabbits less than 2 months of age do not express histo-blood group antigens that are believed to be the receptors for the virus. The disease is often peracute, characterized by sudden death following a 6–24 h period of depression and fever. Infection is via a fecal–oral route. Morbidity rates of 100% and mortality rates of 90% are observed in rabbits older than 2 months. At postmortem there is congestion and hemorrhage in the lungs, with accentuated lobular markings, necrosis of the liver, and splenomegaly. Massive blood clots are present throughout the vasculature.

Norovirus and sapovirus gastroenteritis usually lasts 24–48 h although more severe episodes requiring medical intervention occur. Clinical signs may include nausea, vomiting, diarrhea, abdominal cramps, headache, fever, chills, and myalgia.

## Immunity

Recovered pigs are immune to the particular antigenic types of vesicular exanthema virus with which they were infected but not to other types. Since slaughter and eradication policies are pursued, the questions of long-term immunity and vaccine development are not at issue.

Cats recovered from feline calicivirus infection or immunized with feline calicivirus vaccine appear to remain relatively free of disease following subsequent challenges. This appears to be the case despite the considerable degree of antigenic variability recognized among feline caliciviruses. It is practice to recommend annual vaccination.

Vaccines have been developed for rabbit hemorrhagic disease and are widely used particularly in those many countries such as China and Italy where rabbit farming for meat and pelt production are major industries.

The basic features of calicivirus immunity have been best studied for feline calicivirus. Antibody and cell-mediated immune responses, including the generation of cytotoxic T-lymphocyte responses, have been described. Chromium release assays for cytotoxic T lymphocytes in which autologous feline fetal kidney monolayer cell cultures were used as target cells have been reported. Using these assays the nature and kinetics of the cytotoxic T-cell responses following vaccination and challenge in kittens were described.

Immunity to norovirus and sapovirus is not well understood since there is no cell culture system for assaying neutralizing antibody. Much of the early understanding of immunity to these viruses comes from volunteer studies. There is evidence that immunity may not be durable and that the presence of serum antibody does not invariably correlate with protective immunity. As the putative histo-blood group antigen receptors are variable within the human population and different virus strains bind to different allelic forms of these molecules, innate resistance to infection with certain strains may occur. It has been suggested that the distribution of susceptibility alleles among human populations may go some way to explain the recent predominance of the genogroup II, cluster 4 noroviruses globally. Due to their preference for carbohydrate attachment molecules, it appears that these viruses have a wider host range among humans and may therefore spread more efficiently.

The recently identified murine norovirus has been used as a model for studying norovirus immunity. These studies point to a critical role of the innate immune response, an observation that may help explain the relatively short clinical course of a norovirus disease. However, clearance of the virus requires an adaptive immune response. Shedding norovirus may occur for some time after clinical symptoms have resolved.

## Prevention and Control

Vesicular exanthema is effectively controlled by slaughter and is now considered an extinct virus disease. A real time PCR that detects a wide range of available mammalian and pinniped vesiviruses has been described.

Feline calicivirus infection is most difficult to control in large open cat populations. Vaccination is an important means of control. Clinically ill cats should be isolated and incoming cats of uncertain status should be held in isolation for at least a week before being introduced into the general colony. Although recovered cats remain persistently infected, the amount of virus spread is not usually large, so they do not pose the same threat to in-contact cats as cats with obvious clinical disease.

Preventing entry of rabbit hemorrhagic disease virus into commercial rabbitries, either via fomites or via infected wild rabbits, creates a major challenge in control. Where feed and other supplies are delivered to multiple farms in a common delivery run, special care is required. Feeding pellets that have been sterilized should minimize transmission. Vaccines for the control of the disease are prepared as inactivated, adjuvanted homogenates of infected rabbit tissues. VLPs produced by recombinant DNA technology in baculovirus, yeast, or plant expression systems are effective as vaccines following parenteral or oral administration but are not available commercially.

As noted, norovirus and sapovirus produce mild self-limiting gastroenteritis that normally resolves without complications or the need for treatment. Oral rehydration and electrolyte replacement delivered orally may be required. Complications, sometimes with fatal outcomes, are seen in elderly or immunocompromised patients. A key to prevention is meticulous attention to hygiene, which should include effective hand washing and disposal or disinfection of contaminated materials. Hygiene in food preparation is important. Depuration of oysters is not adequate to clear them of norovirus. Therefore, prevention of contamination of oyster beds with human feces or vomitus is of critical importance. More simply, oysters and other shellfish should be cooked before human consumption and detailed protocols for such procedures have been developed by the US Food and Drug Administration. Vaccines based on VLPs are in various stages of development but none is at present licensed for human use. Such vaccines would have important applications in settings such as the cruise ship industry, hospitals, nursing homes, military and various other institutions as well as in reducing the incidence of diarrhea in young children in both the developed and developing world.

## Future

Vigilance is required to avoid the reemergence of vesicular exanthema of swine either from marine sources or from laboratory escape sources. The host range of the virus strains currently known requires constant monitoring, as does the emergence of variant strains. A clearer understanding of the molecular basis for virulence remains a major challenge. The molecular basis for antigenic variation and variation in virulence among feline caliciviruses remain important objectives. There is a sense that currently used vaccines appear effective but require ongoing validation. The nature of the carrier state and its distinction from reinfection requires further understanding.

Further details of the replication cycle of caliciviruses in general should continue to emerge in the coming few years. Further comparative sequence analysis among caliciviruses will be as interesting and fascinating as it has been thus far. The taxonomic status of the unclassified bovine Newbury agent 1, which may require the establishment of a fifth calicivirus genus, should be resolved. In addition, a universally accepted classification scheme for the noroviruses and sapoviruses needs to be established. Such a scheme would need to reconcile the phylogenetic ambiguities resulting from recombinant viruses.

The need to cultivate rabbit hemorrhagic disease virus is a high priority although this has been somewhat abrogated by the availability of an infectious cDNA clone.

For those countries such as Australia and New Zealand where wild rabbits are a plague upon the nations, reducing profits from farming and degrading the land, continuing assessment and enhancement of the effectiveness of rabbit hemorrhagic disease virus to control rabbits should be an ongoing priority. The emergence of a 'smooth' virion, possibly nonvirulent, form of rabbit hemorrhagic disease virus in Europe, presumably as mutation rather than a phenotypic change due to enzymatic digestion of outer peptide residues of the virion surface, may be one of many factors that in the long term diminish the effectiveness of rabbit hemorrhagic disease virus as a biocontrol agent. In Australia in many areas rabbit numbers have declined precipitously (>60%) and there is evidence for the restoration of original habitats and species in unfarmed areas. It is too early to assess the long-term effectiveness and benefits following the introduction of the virus into Australia and New Zealand but it is likely that it, together with other established methods of control including myxomatosis (in Australia), will bring long-term benefit. The use of rabbit hemorrhagic disease virus as a biocontrol agent has caused conflict among farming groups and animal welfare groups on the one hand and those other greens among us concerned with the preservation and return of the environment to its more natural state as was before the arrival of the rabbit in Australia and New Zealand. The paradox created by the use of the virus as a lethal biocontrol agent and the recovery of some environments to near pristine state with the reappearance of native plant and animal species that were either much depleted or considered extinct will continue to be discussed. As in much of the rest of the world it is now essential in Australia and New Zealand to have effective vaccination programs in place to protect farmed, pet, and laboratory rabbits. In contrast to the situation in the Antipodes, the preservation of wild rabbit populations for hunting purposes throughout all of Europe has been a contentious political issue and the mass immunization of these wild populations will be considered.

The availability of infectious cDNA clones to study the replication of the noncultivable norovirus, sapovirus, and rabbit hemorrhagic disease virus will bring a better understanding of the biology of these viruses.

## Further Reading

Asanaka M, Atmar RL, Ruvolo V, Crawford SE, Neill FH, and Estes MK (2005) Replication and packaging of Norwalk virus RNA in cultured mammalian cells. *Proceedings of the National Academy of Sciences, USA* 102: 10327–10332.

Chen R, Neill JD, Estes MK, and Prasad BV (2006) X-ray structure of a native calicivirus: Structural insights into antigenic diversity and host specificity. *Proceedings of the National Academy of Sciences, USA* 103: 4048–4053.

Coyne KP, Gaskell RM, Dawson S, Porter CJ, and Radford AD (2006) Evolutionary mechanisms of persistence and diversification of a calicivirus within endemically infected natural host populations. *Journal of Virology* 81: 1961–1971.

Dingle KE, Lambden PR, Caul EO, and Clarke IN (1995) Human enteric *Caliciviridae:* The complete genome sequence and expression of virus-like particles from a genetic group II small round structured virus. *Journal of General Virology* 76: 2349–2355.

Green KY, Chanock RM, and Kapikian AZ (2001) Human caliciviruses. In: Knipe DM and Howley PM (eds.) *Fields Virology,* 4th edn., pp. 841–874. New York: Lippincott Williams and Wilkins.

Guo M, Chang KO, Hardy ME, Zhang Q, Parwani AV, and Saif LJ (1999) Molecular characterization of a porcine enteric calicivirus genetically related to Sapporo-like human caliciviruses. *Journal of Virology* 73: 9625–9631.

Jiang X, Wang M, Wang K, and Estes MK (1993) Sequence and genomic organization of Norwalk virus. *Virology* 195: 51–61.

Karst SM, Wobus CE, Lay M, Davidson J, and Virgin HW (2003) STAT1-dependent innate immunity to a Norwalk-like virus. *Science* 299: 1575–1578.

Koopmans MK, Green KY, Ando T, *et al.* (2005) Family *Caliciviridae.* In: Fauquet CM, Maniloff J, D'esselberger U, and Ball DA (eds.) *Virus Taxonomy: Eigth Report of the International Committee on Taxonomy of Viruses*, pp. 843–851. San Diego, CA: Elsevier Academic Press.

Oliver SL, Asobayire E, Dastjerdi AM, and Bridger JC (2006) Genomic characterization of the unclassified bovine enteric virus Newbury agent-1 (Newbury1) endorses a new genus in the family *Caliciviridae*. *Virology* 350: 240–250.

Ossiboff RJ, Sheh A, Shotton J, Pesavento PA, and Parker JS (2007) Feline caliciviruses (FCVs) isolated from cats with virulent systemic disease possess in vitro phenotypes distinct from those of other FCV isolates. *Journal of General Virology* 88: 506–527.

Prasad BV, Hardy ME, Dokland T, Bella J, Rossmann MG, and Estes MK (1999) X-ray crystallographic structure of the Norwalk virus capsid. *Science* 286: 287–290.

Studdert MJ (1978) Caliciviruses. Brief review. *Archives of Virology* 58: 157–191.

Tham KM and Studdert MJ (1987) Antibody and cell-mediated immune responses to feline calicivirus following inactivated vaccine and challenge. *Zentralbl Veterinarmed B* 34: 640–654.

Wirblich C, Thiel HJ, and Meyers G (1996) Genetic map of the calicivirus rabbit hemorrhagic disease virus as deduced from *in vitro* translation studies. *Journal of Virology* 70: 7974–7983.

# *Capillovirus, Foveavirus, Trichovirus, Vitivirus*

**N Yoshikawa**, Iwate University, Ueda, Japan

© 2008 Elsevier Ltd. All rights reserved.

## Glossary

***Alfavirus*** A genus in the family *Togaviridae*. The type species is *Sindbis virus*.

**Helical symmetry** A form of capsid structure found in many RNA viruses in which the protein subunits which interact with the nucleic acid form a helix.

**Methyltransferase** Enzyme activity involved in capping of viral mRNAs.

**Movement protein** A virus-encoded protein which is essential for the cell-to-cell movement of the virus in plant tissues.

**Semipersistent manner** The relationship between a plant virus and its arthropod vector which is intermediate between nonpersistent manner and persistent manner. It has the features of short acquisition feed and no latent period found in nonpersistent manner, but the vector remains able to transmit the virus for periods of hours to days which is longer than the nonpersistent manner.

**Stem grooving** Deformation of the normally smooth surface of a trunk cased by its furrowing.

**Stem pitting** A plant disease characterized by the formation of larger or smaller depression in the old wood, between the phloem and the xylem of the tree trunk.

**Subgenomic RNA** A species of RNA less than genomic length found in infected cells. Viral genomic RNA codes for several proteins but all the 3' open reading frames will effectively be closed for translation. The formation of subgenomic RNAs overcomes this problem as each species has a different cistron at its 5' end, thus opening it for translation.

## Genus *Capillovirus*

The genus *Capillovirus* contains three species – *Apple stem grooving virus* (ASGV, the type species), *Cherry virus A* (CVA), and *Lilac chlorotic leafspot virus* (LiCLV) – and a tentative species – Nandina stem pitting virus (NSPV) (**Table 1**). Citrus tatter leaf virus (CTLV) from citrus and lily is indistinguishable from ASGV from Rosaceae fruit trees biologically, serologically, in genome organization, and in nucleotide sequence, and these days CTLV is regarded as an isolate of ASGV.

## Biological Properties

ASGV occurs worldwide in Rosaceae fruit trees, including apple, European pear, Japanese pear, Japanese apricot and cherry, and it is usually symptomless. However, the virus causes stem grooving, brown line, and graft union abnormalities in Virginia Crab, and it causes topworking disease of apple trees grown on Mitsuba kaido (*Malus sieboldii*) in Japan. ASGV is also widespread in citrus, and it induces bud union abnormalities of citrus trees on trifoliate orange. It also infects lily. CVA occurs in cherry trees in Germany and Japan, but it probably is not associated with any disease. LiClV and NSPV occur in England and the United states, respectively.

No vectors of any of these viruses have been reported. ASGV has been known to be transmitted through seeds to progeny seedlings of lily (1.8%) and *Chenopodium quinoa* (2.5–60%).

## Particle Structure

Virions are flexuous filamentous particles 670–700 nm long and 12 nm in diameter, with obvious cross-banding and helical symmetry and a pitch of *c.* 3.8 nm (**Figure 1**). Virus particles are composed of a linear positive-sense ssRNA, 6.5–7.4 kbp, and a single polypeptide species of $M_r$ 24–27 kDa. The 3' terminus of the RNA has poly A-tail and the 5' terminus probably has a cap structure.

## Genome Organization and Replication

The complete nucleotide sequences (6496 bases) of the single RNA genome of three ASGV isolates have been determined: isolate P-209 from apple, and isolates L and

**Table 1** Virus species in the genus *Capillovirus, Foveavirus, Trichovirus, Vitivirus* in the family *Flexiviridae*

| Genus | Species | Sequence accession numbers |
| --- | --- | --- |
| Capillovirus | Apple stem grooving virus (ASGV) | D14995, D16681, D16368, D14455, AB004063 |
| | Cherry virus A (CVA) | X82547 |
| | Lilac chlorotic leafspot virus (LiCLV) | |
| | Nandina stem pitting virus (NSPV)[a] | |
| Foveavirus | Apple stem pitting virus (ASGV) | D21829, AB045731, D21828 |
| | Apricot latent virus (ApLV) | AF057035 |
| | Rupestris stem pitting-associated virus (RSPaV) | AF026278, AF057136 |
| Trichovirus | Apple chlorotic leaf spot virus (ACLSV) | M58152, D14996, X99752, AJ243438 |
| | Cherry mottle leaf virus (CMLV) | AF170028 |
| | Grapevine berry inner necrosis virus (GINV) | D88448 |
| | Peach mosaic virus (PMV) | |
| Vitivirus | Grapevine virus A (GVA) | X75433, AF007415 |
| | Grapevine virus B (GVB) | X75448 |
| | Grapevine virus D (GVD) | Y07764 |
| | Heracleum latent virus (HLV) | X79270 |
| | Grapevine virus C (GVC)[a] | |

[a]Tentative species.

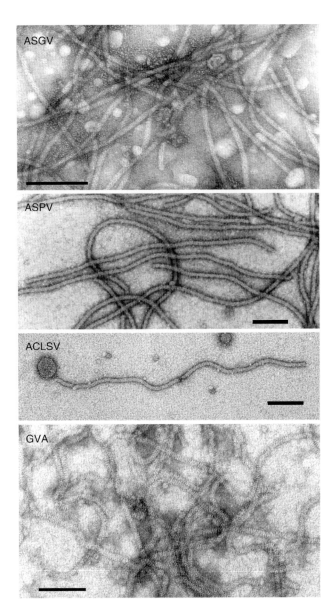

**Figure 1** Electron micrographs of particles of *Apple stem grooving virus* (ASGV), the type species of the genus *Capillovirus*; *Apple stem pitting virus* (ASPV), the type species of the genus *Foveavirus* (courtesy of H. Koganezawa); *Apple chlorotic leaf spot virus* (ACLSV), the type species of the genus *Trichovirus*; and *Grapevine virus A* (GVA), the type species of the genus *Vitivirus* (courtesy of J. Imada). Scale = 100 nm.

Li-23 from lilies. Identities of the nucleotide sequences are 82.9% (P-209/L), 83.0% (P-209/Li-23), and 98.4% (L/Li-23). The genomic RNA has the same structural organization and two overlapping open reading frames (ORFs) in the positive strand (**Figure 2**). ORF1 (bases 37–6341) encodes a 241–242 kDa polyprotein (2105 aa) containing the consensus motifs of methyltransferase (Met), papain-like protease (P-pro), nucleotide triphosphate-binding helicase (Hel), RNA polymerase (Pol), and coat protein (CP) in the C-terminal region. The protein has homologies with putative polymerase of the 'alphavirus-like' supergroup of RNA viruses. ORF2 (bases 4788–5747) encodes a 36 kDa putative movement protein (320 aa). A region (amino acid (aa) position 1585–1868) of the ORF1-encoded protein between the polymerase and the CP, that encodes ORF2 in another frame has none of the other functional motifs found in other known plant virus genomes. This region (designated the V-region) shows high variability among isolates and sequence variants. The genome organization of CVA is composed of a 266 kDa polyprotein (ORF1) and a 52 kDa protein (ORF2) located within ORF1. ORF1 encodes the CP (24 kDa) in the C-terminal region. The overall nucleotide sequence identity between CVA and ASGV is 57.6%.

Although the ASGV-CP is located in the C-terminal region of the ORF1-encoded polyprotein and genomic RNA directs the synthesis of a polypeptide of c. 200 kDa as a major product in *in vitro* translation, and is immunoprecipitated by antiserum to virus particle preparations, the following evidence suggests that the CP is expressed from a subgenomic RNA. Analysis of double-stranded (ds) RNA from infected *C. quinoa* tissues indicates that all ASGV isolates tested contain five virus-specific dsRNAs (6.5, 5.5, 4.5, 2.0, and 1.0 kbp). The 6.5 kbp species represents the double-stranded form of the full-length genome, whereas the 2.0 and the 1.0 kbp species may be the double-stranded forms of subgenomic RNAs coding for the putative movement protein and the CP, respectively. The 5.5 and 4.5 kbp species are thought to be 5′ co-terminal with the genome. The size of the *Escherichia coli*-expressed protein corresponding to the C-terminal region of the ORF1-encoded protein, which starts with the methionine at aa position 1869, agrees with that of the CP. The single-stranded subgenomic RNAs for movement protein (MP) and CP have also been reported in infected tissues.

In infected *C. quinoa* leaves, the particles occur singly or as aggregates in the cytoplasm of mesophyll and phloem parenchyma cells, suggesting that the replication of the genome and the assembly of the particles may occur in the cytoplasm, although no virus-specific inclusion bodies, such as pinwheels, viroplasms, or vesicles, have been observed.

## Serology

Polyclonal antisera were prepared in rabbits against purified virus or CP expressed in *E. coli*. Enzyme-linked immunosorbent assay was used to detect the virus in fruit trees. Monoclonal antibodies were produced in mice against an isolate from citrus and three selected monoclonal antibodies reacted with all isolates tested, including nine isolates from citrus trees in Japan, four isolates from citrus in the USA, six isolates from Chinese

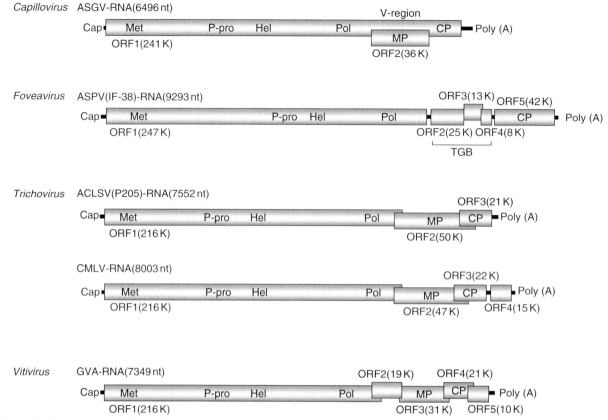

**Figure 2** Gemone organization of virus species in the genera *Capillovirus*, *Foveavirus*, *Trichovirus*, and *Vitivirus*. ASGV, apple stem grooving virus; ASPV, apple stem pitting virus; ACLSV, apple chlorotic leaf spot virus; CMLV, cherry mottle leaf virus; GVA, grapevine virus A; Met, methyltransferase; P-pro, papain-like protease; Hel, nucleotide triphosphate-binding helicase; Pol, RNA-dependent RNA polymerase; CP, coat protein; MP, movement protein; and V-region, variable region.

citrus, and isolates from lily, apple, and Japanese apricot. ASGV is serologically unrelated to all known virus species in the genus *Capillovirus*.

## Strains and Genome Heterogeneity

Many isolates have been reported from apple, Japanese pear, European pear, Japanese apricot, lily, and citrus plants, but most have not been characterized. Some isolates have been differentiated only on symptomatology. Virus isolates from apple, Japanese pear, and European pear trees comprise at least two to four variants that differ considerably from each other in nucleotide sequence. The composition of sequence variants within a tree differs among leaves from different branches, showing that each sequence variant is distributed unevenly within an individual tree.

## Genus *Foveavirus*

The genus *Foveavirus* consists of three species: *Apple stem pitting virus* (ASPV, the type species), *Apricot latent virus* (ApLV), and *Rupestris stem pitting-associated virus* (RSPaV) (**Table 1**).

## Biological Properties

ASPV is one of the causative agents of apple topworking disease in Japan and induces lethal decline in apple trees grown on *Malus sieboldii* (Regel) Rehd. rootstock. The virus is usually latent in apple cultivars and is distributed widely in many apple trees. Pear vein yellows is also thought to be caused by this virus. ApLV infects apricot, peach, and sweet cherry, and may be the causal agents of the peach asteroid spot disease and of the peach sooty ringspot diseases. RSPaV is thought to be an agent of rupestris stem pitting, that is probably the most common component of the Rugose wood complex on grapevines. No vectors have been reported for viruses in the genus *Foveavirus*, and the viruses are probably spread in nature by graft-transmission.

## Particle Structure

ASPV (isolate B-39) has flexuous filamentous particles, approximately 800 nm in length and 12–15 nm in width (**Figure 1**). Virus particles readily form end-to-end aggregates with four prominent peaks appearing at 800, 1600, 2400, and 3200 nm in length. ASPV is comprised of a single species of RNA of $M_r\ 3.1 \times 10^6$ and a major CP of $M_r$ 48 kDa.

## Genome Organization and Replication

The genomes of two ASPV isolates (PA66 and IF 38) have been completely determined and found to consist of 9306 and 9237 nucleotide (nt), respectively, excluding the 3′ poly A-tail. The base composition of ASPV genome is 27.6% A, 20.0% C, 23.4–23.8% G, and 28.6–29.0% T. Analysis of the putative ORFs of the nucleotide sequence in both positive and negative strands showed that ASPV genome contains five ORFs in the positive-strand, encoding proteins with $M_r$'s of 247K (ORF1), 25K (ORF2), 13K (ORF3), 7–8K (ORF4), and 42–44K (ORF5) (**Figure 2**). The 5′-noncoding regions of the genome have been reported to be 33 nt (PA66) and 60 nt (IF38). ORF1 encodes a protein (247K) with motifs associated with Met, P-pro, Hel, and Pol (**Figure 2**). ORF2 encodes a protein (25K) containing a helicase motif (GSGKT, aa positions 31–35). ORF3 and ORF4 encode proteins with 13K and 7–8K, respectively. The ORFs 2, 3, and 4 constitute the triple gene block (TGB) found in allexi-, carla-, potex-, and mandariviruses. ORF5 encodes a protein with an $M_r$ of 42–44K, and the sequence of this protein contains the conserved amino acids (R and D) potentially involved in a salt bridge formation that is typical of the CPs of rod-shaped and filamentous plant viruses. The 3′-noncoding region of both genomes of PA66 and IF38 isolates consists of 132 nucleotides, excluding the poly A-tail. The comparison of the nucleotide sequence of the PA66 genome with that of the IF38 genome showed a high level of divergence (76% identity) between the two isolates. Comparisons of aa sequences of five proteins between PA66 and IF38 show the identities of 87% (247K), 94% (25K), 87% (13K), 77% (7–8K), and 81% (42–44K). A hypervariable region was found between the MET and P-Pro domains in the 247K protein. This hypervariable region between MET and P-Pro domains of the 247K protein was also found in the apple chlorotic leaf spot virus (ACLSV) 216K protein, indicating that this region has not undergone severe evolutionary constraint. The 25K protein was mostly conserved between the two isolates. Another striking variability between PA66 and IF38 was found in the N-terminal region of ASPV-CP, in which there were many deletions compared with PA66. This resulted in an aa number of IF38-CP which is 18 aa fewer than that of PA66.

The genome of RSPaV consists of 8726 nt excluding the 3′ poly A-tail and has five potential ORFs on its positive strand which have the capacity to code for the replicase (ORF1; 244 K), the TGB (ORF2–4; 24K, 13K, and 8K), and the CP (ORF5; 28K). The identities of aa sequences of five proteins between RSPaV and ASPV (PA66) are 39.6% (ORF1), 38.0% (ORF2), 39.3% (ORF3), 27.1% (ORF4), and 31.3% (ORF5). Partial sequences of the 3′-terminal regions of ApLV have been reported.

In Northern blot hybridization analysis of virus-specific dsRNAs and ssRNAs in ASPV (IF38)-infected tissues by use of three negative-sense RNA probes complementary to nt positions 6207–6683 (ORF1 region), nt positions 6678–7447 (ORF2 region), nt positions 8717–9293 (CP and the 3′-noncoding regions) of the IF38 genome showed the presence of five dsRNAs (9, 7.5, 6.5, 2.6, and 1.6 kbp) and three ssRNAs (9, 2.6, and 1.6 kbp) in infected tissues (**Figure 3**). The slowest migrating RNA (c. 9 kbp) was equivalent to that of the ASPV genome, and other two RNAs (2.6 and 1.6 kbp) are thought to be subgenomic RNAs of ORF2–ORF4 proteins and CP (ORF5), respectively.

In electron microscopy of infected leaves, virus particles were found in mesophyll, epidermal, and vascular

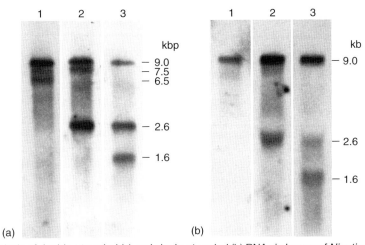

**Figure 3** Northern blot analysis of double-stranded (a) and single-stranded (b) RNAs in leaves of *Nicotiana occidentalis* infected with apple stem pitting virus (IF-38) using three different RNA probes specific to different sequences within the genomic RNA. Lane 1, a probe complementary to nt positions 6207–6683 (ORF1 region); lane 2, nt positions 6678–7447 (ORF2 region); and lane 3, nt positions 8717–9293 (CP and the 3′-noncoding regions).

parenchyma cells. The particles were observed often as large aggregates in the cytoplasm, but not in the vacuole, the nucleus, or in other cellular organelles.

### Strains and Genome Heterogeneity

Restriction fragment length polymorphism (RFLP) analysis of the cDNA clones from four ASPV isolates from apple (B39, B12, IG39, and IF38) showed that three different patterns were found in cDNA clones from isolate B39, two patterns from isolates B12 and IG39, and one pattern from isolate IF38. Sequence analysis of the 3′-terminal 600 nt of the genomes indicated that sequence identities among cDNA clones showing the same RFLP pattern were more than 99%, and in contrast, considerable sequence variations were found among the cDNA clones showing a different RFLP pattern. These results indicated that isolate B39 is composed of at least three sequence variants (SVs), and isolates B12 and IG39 at least two SVs. The nucleotide sequence identity among nine isolates and SV including PA66 was 77.9–95.7% in the 3′-terminal 600 nt regions. The aa number of CP varied from 396 to 415 among nine isolates and SV, and their identities ranged from 69.4% to 92.7%.

## Genus *Trichovirus*

The genus *Trichovirus* consists of four viral species: *Apple chlorotic leaf spot virus* (ACLSV, the type species), *Cherry mottle leaf virus* (CMLV), *Grapevine berry inner necrosis virus* (GINV), and *Peach mosaic virus* (PcMV) (**Table 1**).

### Biological Properties

ACLSV occurs in woody plants of the family Rosaceae including apple, pear, peach, plum, cherry, apricot, and prune. Though ACLSV infection is symptomless in most commercial apple varieties, the virus causes topworking disease of apple grown on Maruba kaido (*Malus prunifolia* var. *ringo*) rootstocks in Japan. ACLSV may cause plum bark split, plum pseudopox, pear ring pattern mosaic, and apricot pseudopox disease. CMLV occurs naturally in sweet cherry, peach, and apricot, and causes mottle leaf disease of cherry in some regions of North America. GINV is the causal virus of grapevine berry inner necrosis disease, one of the most important virus diseases of grapevine in Japan.

The viruses can be transmitted by mechanical inoculation, grafting, and through propagating materials. No vectors have been reported for ACLSV. On the other hand, CMLV and GINV are transmitted by the peach bud mite *Eriophyes insidiosus* and the grape erineum mite *Colomerus vitis*, respectively.

### Particle Structure

Virus particles are very flexuous filaments 680–780 nm long and 9.25–12 nm in width (**Figure 1**), with obvious cross-banding and helical symmetry; the pitch of the helix is about 3.47–3.8 nm. ACLSV has a buoyant density of 1.27 g ml$^{-1}$ in CsSO$_4$ gradients, and the particles are degraded in CsCl gradients. The $A_{260}/A_{280}$ of purified ACLSV preparation is 1.85–1.89. The particles require the presence of divalent cations to maintain the integrity of the quaternary structure.

Virus particles are composed of a linear positive-sense ssRNA, 7.2–8.0 kbp, and a single polypeptide species of $M_r$ 21–22 kDa. The genomic RNA is about 5% of particle weight. The 3′ terminus of the RNA has a poly A-tail, and the 5′ terminus probably has a cap structure. Nucleotide base ratios are 32% A, 18% C, 23% G, and 27% U for ACLSV-RNA.

### Genome Organization and Replication

The complete nucleotide sequences of the genomes of ACLSV (four isolates from plum, apple, and cherry), CMLV, and GINV have been determined. The genomes of ACLSV consist of 7549–7555 nt in length excluding the 3′ poly A-tail. The complete genomes of CMLV and GINV are 8003 nt and 7243 nt long excluding the poly A tail, respectively.

The genomic RNAs of ACLSV and GINV have three slightly overlapping ORFs in the positive strand (**Figure 2**). ORF1 encodes a replication-associated protein of $M_r$ 214–216 kDa containing the consensus motifs of Met, P-pro, Hel, and Pol of the alphavirus-like supergroups of ssRNA. ORF2 encodes a putative MP of $M_r$ 39–50 kDa. The 50 kDa protein encoded by ORF2 of ACLSV genome is an MP, which has the following characteristics: (1) it is localized to plasmodesmata of infected and transgenic plant cells; (2) it can spread from the cells that initially produce it into neighboring cells; (3) it enables cell-to-cell trafficking of green fluorescent protein (GFP) when the 50 kDa protein and GFP are co-expressed in leaf epidermis; (4) it induces the production of tubular structures protruding from the surface of protoplast; and (5) it binds to single-stranded nucleic acids. Additionally, transgenic plants expressing the 50 kDa protein can complement the systemic spread of mutants of an infectious cDNA clone that are defective in ORF2. ORF3 encodes a 21–22 kDa CP.

The genomic RNA of CMLV has four putative ORFs encoding 216 kDa (ORF1), 47 kDa (ORF2), 22 kDa (ORF3), and 15 kDa (ORF4) proteins (**Figure 2**). The ORF4 protein that is not present in the genomes of ACLSV and GINV may be a nucleic acid-binding protein because similarities have been found between ORF4 of CLMV and ORF5 of grapevine virus B.

In analysis of the dsRNAs of ACLSV-infected plants, six dsRNA species of c. 7.5, 6.4, 5.4, 2.2, 1.1, and 1.0 kbp were found in infected tissues. The 7.5 kbp species corresponding to the full-length genome may be a replicative form of the ACLSV genome, and the 2.2 and 1.1 kbp species are thought to be the double stranded forms of subgenomic RNAs coding for the MP and the CP, respectively. The 6.4 and 5.4 kbp species were found to be 5′ co-terminal with the genomic RNA. In the model for the expression of the genome of ACLSV, only the 5′-proximal ORF (216 kDa) coding for the putative viral replicase is translated directly, and the two other ORFs encoding MP and CP are expressed through two subgenomic RNAs.

In infected plant leaves, ACLSV and GINV particles occur as aggregates in the cytoplasm. ACLSV was also observed in nucleoplasm of a mesophyll cell in the infected *C. quinoa*. Replication is presumed to occur in the cytoplasm although no virus-specific inclusions, such as vesicles and viroplasm, were observed.

## Serology

No serological relationships were found among ACLSV, CMLV, and GINV. Polyclonal and monoclonal antibodies (MAbs) of CMLV cross-reacted with all isolates of PcMV tested, indicating that CMLV and PcMV are closely related viruses. MAbs against ACLSV particles were produced and used to investigate the antigenic structure of the virus. Epitope studies using these MAbs identified seven independent antigenic domains in ACLSV particles.

## Strains

Many isolates have been reported from apple, cherry, peach, plum, and prune. Several strains could be differentiated serologically as well as by symptoms in indicator plants. Sequence comparisons among isolates indicate large molecular variability, that is, sequence conservation rates vary between 77.4% and 99.4%, with most of the isolates differing by 10–20% from any other given isolates.

## Genus *Vitivirus*

The genus *Vitivirus* consists of four viral species – *Grapevine virus A* (GVA, the type species), *Grapevine virus B* (GVB), *Grapevine virus D* (GVD), and *Heracleum latent virus* (HLV) – and a tentative species – Grapevine virus C (**Table 1**).

## Biological Properties

GVA, GVB, and GVD naturally infect only grapevines and induce severe diseases, the rugose wood complex of grapevine, characterized by pitting and grooving of the wood. HLV occurs in hogweed (*Heracleum* sp. family Apiaceae) without causing obvious symptoms. Herbaceous plant species, *Nicotiana benthamiana* and *N. occidentalis*, are used for propagation hosts of GVA and GVB, respectively, and *Chenopodium* species are used for diagnostic and propagation species of HLV.

All virus species are transmitted by mechanical inoculation. Transmission by grafting and dispersal through propagating materials is common with infected grapevines. GVA and GVB are transmitted in nature by several species of the pseudococcid mealybug genera *Planococcus* and *Pseudococcus* in a semipersistent manner, whereas HLV is transmitted from naturally infected hogweed plants by the aphids in a semipersistent manner, which depends on a helper virus present in naturally infected plants. No seed transmission of HLV is found in *C. quinoa*, chervil, coriander, or hogweed.

## Particle Structure

Virus particles are flexuous filaments 725–825 nm long and 12 nm in width, showing obvious cross-banding and helical symmetry with a pitch of 3.5 nm. Particles sediment as a single peak in sucrose or $Cs_2SO_4$ density gradients with a sedimentation coefficient ($S_{20w}$) of about 96S and a buoyant density in $Cs_2SO_4$ of 1.24 g cm$^{-3}$ for HLV. The $A_{260}/A_{280}$ of purified virus preparation is 1.5–1.52 for GVB and HLV.

Virus particles are composed of a linear positive-sense ssRNA, 7.3–7.6 kbp, and a single polypeptide species of $M_r$ 18–21.5 kDa. The genomic RNA is about 5% of the particle weight. The 3′ terminus of the RNA has a poly A tail, and the 5′ terminus probably has a cap structure. The overall A+U and G+C content of GVB-RNA is 53.5% and 46.5%, respectively.

## Genome Organization and Replication

The complete nucleotide sequences of the genomes of GVA and GVB have been determined. The genomes of GVA and GVB consist of 7349 and 7598 nt in length excluding the 3′-poly A tail, respectively.

The genomic RNAs of GVA and GVB have five slightly overlapping ORFs in the positive strand (**Figure 2**). ORF1 of both viruses encodes a replication-associated protein of $M_r$ 194–195 kDa containing the consensus motifs of methyltransferase, papain-like protease, nucleotide triphosphate-binding helicase, and RNA polymerase of the alphavirus-like supergroups of ssRNA. ORF2 encodes a protein of $M_r$ 19 kDa (GVA) or 20 kDa (GVB) that does not show any significant sequence homology with protein sequences from the databases. The biological function of these proteins has not been determined yet. ORF3 encodes an MP with $M_r$ 31 kDa (GVA) and 36.5 kDa (GVB), possessing the G/D motif of the '-30K' superfamily movement protein. ORF4 encodes

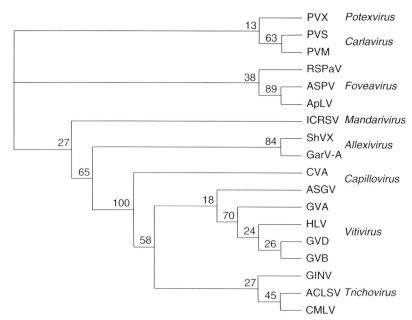

**Figure 4** Dendrogram showing the relationships among virus species in the family *Flexiviridae* using the aa sequences of the coat protein. The tree was produced and bootstrapped using CLUSTAL W. PVX, potato virus X; PVS, potato virus S; PVM, potato virus M; RSPaV, ruspestris stem pitting-associated virus; ASPV, apple stem pitting virus; ApLV, apricot latent virus; ICRSV, Indian citrus ringspot virus; ShVX, shallot virus X; GarV-A, garlic virus A; CVA, cherry virus A; ASGV, apple stem grooving virus; GVA, grapevine virus A; HLV, heracleum latent virus; GVD, grapevine virus D; GVB, grapevine virus B; GINV, grapevine berry inner necrosis virus; ACLSV, apple chlorotic leaf spot virus; CMLV, cherry mottle leaf virus.

a CP with $M_r$ 21.5 kDa (GVA) and 21.6 kDa (GVB). ORF5 encodes a protein with $M_r$ 10 kDa (GVA) and 14 kDa (GVB). A 14 kDa protein encoded by GVB-ORF5 shares weak homologies to proteins with nucleic acid-binding properties.

The gene expression strategy may be based on proteolytic processing of ORF1 protein and subgenomic RNA production for translation of ORF2 to ORF5 proteins. In analysis of the dsRNAs of infected plants, four major dsRNA bands with a size of 7.6, 6.48, 5.68, and 5.1 kbp for GVA and GVD, and 7.6, 6.25, 5.03, and 1.97 kbp for GVB were found in infected tissues. In *N. benthamina* plants infected with GVA, three 3′-terminal subgenomic RNAs of 2.2, 1.8, and 1.0 kbp were also detected and thought to serve for the expression of ORF2, ORF3, and ORF4, respectively.

In electron microscopy of leaves infected with GVA and GVB, tonoplast-associated membranous vesicles containing finely fibrillar materials were found in phloem parenchyma cells, and virus particles occurred only in phloem tissues. Thus, replication may occur in the cytoplasm, possibly in association with vesicles protruding from the tonoplast of phloem cells.

## Serology

In immunoelectron microscopy, the antiserum against GVB-NY clearly decorated homologous (GVB-NY) and heterologous (GVB-CAN) virus particles, but did not decorate GVA particles, indicating no serological relationships between GVA and GVB.

## Phylogenetic Relationships

Sequences from viruses within the genus *Foveavirus* clustered into a branch different from other TGB-containing viruses (allexi, carla, potex, and mandariviruses) (**Figure 4**). Sequences for viruses within the genera *Capillovirus*, *Trichovirus*, and *Vitivirus* also clustered into branches different from each other (**Figure 4**).

*See also:* Allexivirus; Carlavirus; Flexiviruses; Plant Virus Diseases: Fruit Trees and Grapevine; Potexvirus.

## Further Reading

Adams MJ, Antoniw JF, Bar-Joseph, *et al.* (2004) The new plant virus family *Flexiviridae* and assessment of molecular criteria for species demarcation. *Archives of Virology* 149: 1045–1066.

Adams MJ, Accotto GP, Agranovsky AA, *et al.* (2005) Family *Flexiviridae*. In: Fauquet CM, Mayo MA, Maniloff J, Desselberger U, and Ball LA (eds.) *Virus Taxonomy Eighth Report of the International Committee on Taxonomy of Viruses*, pp. 1089–1124. San Diego, CA: Elsevier Academic Press.

Boscia D, Savino V, Minafra A, *et al.* (1993) Properties of a filamentous virus isolated from grapevines affected by corky bark. *Archives of Virology* 130: 109–120.

German S, Candresse T, Lanneau M, et al. (1990) Nucleotide sequence and genome organization of apple chlorotic leaf spot closterovirus. *Virology* 179: 1104–1112.

Jelkmann W (1994) Nucleotide sequence of apple stem pitting virus and of the coat protein of a similar virus from pear associated with vein yellows disease and their relationship with potex- and carlaviruses. *Journal of General Virology* 75: 1535–1542.

Koganezawa H and Yanase H (1990) A new type of elongated virus isolated from apple trees containing the stem pitting agent. *Plant Disease* 74: 610–614.

Magome H, Yoshikawa N, Takahashi T, Ito T, and Miyakawa T (1997) Molecular variability of the genomes of capilloviruses from apple, Japanese pear, European pear, and citrus trees. *Phytopathology* 87: 389–396.

Martelli GP and Jelkmann W (1998) *Foveavirus*, a new plant virus genus. *Archives of Virology* 143: 1245–1249.

Martelli GP, Minafra A, and Saldarelli P (1997) *Vitivirus*, a new genus of plant viruses. *Archives of Virology* 142: 1929–1932.

Meng B, Pang S-Z, Forsline PL, McFerson JR, and Gonsalves D (1998) Nucleotide sequence and genome structure of grapevine rupestris stem pitting associated virus-1 reveal similarities to apple stem pitting virus. *Journal of General Virology* 79: 2059–2069.

Minafra A, Saldarelli P, and Martelli GP (1997) Grapevine virus A: Nucleotide sequence, genome organization, and relationship in the *Trichovirus* genus. *Archives of Virology* 142: 417–423.

Yanase H (1974) Studies on apple latent viruses in Japan. *Bulletin of the Fruit Tree Research Station, Japan, Series* C1: 47–109.

Yoshikawa N (2000) Apple stem grooving virus. AAB Descriptions of Plant viruses 376. http://www.dpvweb.net/dpv/showdpv.php?dpvno=376 (accessed January 2008).

Yoshikawa N (2001) Apple chlorotic leaf spot virus. AAB Descriptions of Plant viruses 386. http://www.dpvweb.net/showdpv.php?dpvno=386 (accessed January 2008).

Yoshikawa N, Sasaki E, Kato M, and Takahashi T (1992) The nucleotide sequence of apple stem grooving capillovirus geneme. *Virology* 191: 98–105.

# Capripoxviruses

**R P Kitching**, Canadian Food Inspection Agency, Winnipeg, MB, Canada

© 2008 Elsevier Ltd. All rights reserved.

## Glossary

**Abomasal** Pertaining to the fourth stomach of ruminants.
**Agalactia** Shortage of milk supply.
**Myiasis** Infestation with maggots.
**Hydropic** Accumulating water.

## History

Sheeppox and goatpox are malignant pox diseases of sheep and goats easily recognizable by their characteristic clinical signs, and described in the earliest texts on animal diseases. Lumpy skin disease (Neethling) of cattle (LSD), however, was first described in 1929 in Northern Rhodesia (Zambia), having apparently been absent from domestic cattle until that time. From Zambia, LSD spread south to Botswana and Zimbabwe, and by 1944 the disease appeared in South Africa, where it caused a major epizootic, affecting over 6 million cattle. In 1957, LSD was first diagnosed in Kenya, and was thought at the time to be associated with the introduction of a flock of sheep affected with sheeppox on the farm. Since then LSD has been present in most of the countries of sub-Saharan Africa, often associated with large epizootics followed by periods in which the disease is only rarely reported. In 1988, LSD caused a major outbreak in Egypt, and in 1989 it spread from Egypt to a village in Israel. This was the first time that a diagnosis of LSD outside of Africa had been supported by laboratory confirmation.

## Taxonomy and Classification

The viruses that cause sheeppox, goatpox, and LSD are all members of the genus *Capripoxvirus*, in the subfamily *Chordopoxvirinae* of the family *Poxviridae*, and have morphological, physical, and chemical properties similar to vaccinia virus. Originally, the viruses were classified according to the species from which they were isolated, but comparisons of their genomes indicate that the distinction between them is not so clear, and that recombination events occur naturally between isolates from different species. This is reflected in the ability of some strains to cause disease in both sheep and goats and in experimental results which show that all the sheep isolates examined could infect goats, and that goat isolates could infect sheep.

The epidemiological relationship between sheeppox and goatpox isolates and cattle isolates is less clear, apparent in differences in the geographical distribution of sheeppox and goatpox and LSD (see below). However, some isolates recovered from sheep and goats in Kenya have genome characteristics very similar to cattle isolates. It has been proposed that confusion can be reduced by referring to the malignant pox diseases of sheep, goats, and cattle, including Indian goat dermatitis and Kenyan sheep and goatpox, as capripox. It is envisaged that when sufficient isolates have been examined biochemically, no clear distinction will be possible between sheep, goat, and cattle isolates, but a spectrum will emerge in which some strains have clear host preferences while others will be less defined and will naturally infect the host with which they come into contact.

## Geographical and Seasonal Distribution

Capripox of sheep and goats is enzootic in Africa north of the equator, the Middle East and Turkey, Iran, Afghanistan, Pakistan, India, Nepal, and parts of the People's Republic of China, and, in 1986, Bangladesh. Sheeppox was eradicated from Britain in 1866, and from France, Spain, and Portugal in 1967, 1968, and 1969, respectively. Sporadic outbreaks still occur in Europe, for instance in Italy in 1983 and Greece and Bulgaria both in 1995 and 1996, and Greece in 1997 and 2000. In 2005, goatpox was first reported in Vietnam, following its introduction from China.

LSD is enzootic in the sub-Saharan countries of Africa and is still present in Egypt. The single outbreak in Israel was eradicated by slaughter of affected and in-contact cattle.

There is no clear seasonality to outbreaks of capripox in sheep and goats. In enzootic areas, lambs and kids are protected against infection with capripoxvirus for a variable time dependent on the immunity of the mother. However, the spread of LSD is related to the density of biting flies and consequently major enzootics have been associated with humid weather when fly activity is greatest.

## Host Range and Virus Propagation

Among domestic species, capripoxvirus is restricted to cattle, sheep, and goats. Experimentally, it is possible to infect cattle, sheep, or goats with isolates derived from any of these three species, although clinically the reaction following inoculation may be indiscernible. Viral genome analysis using restriction endonucleases has identified fragment size characteristics by which it is possible to classify strains into cattle, sheep, or goat isolates. However, the identification of strains that have intermediate characteristics between typical sheep and goat isolates does suggest the movement of strains between these species. Analysis of some Kenyan isolates derived from sheep and goats shows very close homology with cattle LSD isolates.

The involvement of the African buffalo (*Syncerus caffer*) in the maintenance of LSD has not been clearly established. Some surveys have shown the presence of capripoxvirus antibody in buffalo, while others have failed to show its presence. Buffaloes clinically affected with LSD have not been described. Experimental infection of giraffe (*Giraffe camelopardalis*), impala (*Aepyceros melampus*), and gazelle (*Gazella thomsonii*) has resulted in the development of clinical disease.

*Bos indicus* cattle are generally less susceptible to LSD and develop milder clinical disease than *Bos taurus*, of which the fine-skin Channel Island breeds are particularly susceptible. Similarly, breeds of sheep and goats indigenous to capripoxvirus enzootic areas appear less susceptible to severe clinical capripox than do imported European or Australian breeds.

Capripoxvirus will grow on the majority of primary and secondary cells and cell lines of ruminant origin. Primary lamb testes cells are considered the most sensitive system for isolation and growth of capripoxvirus. The virus produces a characteristic cytopathic effect (cpe) on these cells which can take up to 14 days for field isolates, but can be as short as 3 days for well-adapted strains.

Isolates of capripoxvirus derived from cattle have been adapted to grow on the chorioallantoic membrane of embryonated hens' eggs, although attempts to grow isolates from sheep and goats in eggs have been unsuccessful. Vaccine strains of capripoxvirus have been adapted to grow on Vero cells. Capripoxvirus will not grow in any laboratory animals.

## Genetics

Less is known concerning the specific genetics of capripoxvirus than is known about the orthopoxvirus genome. Studies on field isolates taken from cattle suggest that the virus is very stable, as *Hin*dIII restriction endonuclease digest patterns of isolates from the 1959 Kenya outbreak of LSD are identical to those obtained from 1986 LSD isolates. However, recombination has been shown to occur between cattle and goat isolates and this could be the natural method by which the virus evolves. By analogy with the orthopoxviruses, it is also likely that sequences are deleted or repeated within the genome in the normal replicative cycle.

The genomes of those capripoxvirus isolates that have been sequenced, representing isolates from cattle, sheep, and goats, have a 96% nucleotide homology along their entire length of approximately 150 kbp.

Sheep and goat isolates have 147 putative genes and LSD isolates an additional nine. However, the sheep and goat isolates have these nine in a disrupted form, suggesting that the LSD virus is the more ancient progenitor. This is clearly not consistent with the apparent first appearance of LSD in 1929. Nevertheless, the published sequences have shown a range of genes coding for host cell protein homologs, in common with many identified in the orthopoxviruses. In fact, in the central region of the genome, there is a high degree of similarity (*c.* 65%) with amino acid sequences found in other poxvirus species, in particular suipoxvirus, yatapoxvirus, and leporipoxvirus. These sequence studies and those reported from India suggest that distinction can be made between sheep and goat isolates, but limited numbers were studied; however, some isolates examined from the Middle East show less evidence of host-species-specific sequences.

## Evolution

The capripoxviruses have evolved into specific cattle, sheep, and goat lines, but, as has been described above, intermediate strains exist, particularly those with cattle and goat genome characteristics. In Kenya, there is evidence of movement of strains between all three species, but the absence of sheeppox or goatpox in LSD enzootic areas in southern Africa, and the absence of LSD outside of Africa, would suggest that host-specific strains are being maintained and presumably are continuing to evolve.

## Serologic Relationships and Variability

Polyclonal sera fail to distinguish in the virus neutralization test between any of the isolates of capripoxvirus so far examined. Sheep, goats, or cattle that have been infected with any of the isolates are totally resistant to challenge with any of the other isolates. On this basis, it has been possible to use the same vaccine strain to protect all three species. No monoclonal antibodies are as yet available against capripoxvirus, but it can be expected that differences will emerge between strains using these reagents.

Capripoxviruses share a precipitating antigen with parapoxviruses, but no cross-immunity has been shown between these two genera, or between capripoxvirus and any other poxvirus genera.

## Epizootiology

In sheeppox and goatpox enzootic areas the distribution of disease is frequently a reflection of the traditional form of husbandry. For instance, in the Yemen Arab Republic, the sheep and goat flocks kept on the grassland of the central plateau and better irrigated regions of the coastal plain move about in search of food, frequently mixing with flocks from neighboring villages at water holes, and in this situation disease is restricted to the young stock. Animals over 1 year of age have a solid immunity. The animals belonging to villages in the more mountainous regions and the arid areas of the coastal plain are isolated by terrain or semidesert from mixing with animals from other villages. It is not known what is the critical number of animals required to maintain capripoxvirus within a single population but it is over a thousand adult animals, which is the approximate village sheep and goat populations. In these villages, disease is usually only seen following the introduction of new animals, typically from market, and generally affects animals of all age groups. The disease spreads through the village, usually within 3–6 months, and then disappears in the absence of more susceptible animals. Occasionally, even within areas of high sheep and goat density, it is possible to encounter animals that have been kept totally isolated in the confines of a domestic residence, and these may remain susceptible to infection until adult.

In Sudan, large numbers of sheep and goats are trekked from the west to the large collecting yards and markets of Omdurman, outside Khartoum. Here also, it is possible to see capripox infection in adult animals. Many of the flocks originate in villages which, like in the Yemen, are isolated from their neighbors. Capripoxvirus does not persist in these villages, and as a result the animals acquire no resistance, and are fully susceptible when they first encounter disease on the long journey across Sudan. Animals being exported from countries that are free of capripoxvirus may suffer a similar fate when they arrive in a capripoxvirus enzootic area, as often seen in Australian or New Zealand sheep imported into the Middle East.

In a study of 49 outbreaks of capripox in the Yemen, only 8 were reported to affect both sheep and goats, the remaining 41 causing clinical disease in either sheep or goats. It is possible that both sheep and goats could have been involved in more than the eight outbreaks, but that the disease was inapparent in one species; whether, therefore, the species in which the disease was inapparent could transmit virus and become a vector for disease has not been determined. In Kenya, capripox is frequently encountered in both sheep and goats within the same flock, and there is the possibility that the same strain of capripoxvirus could also cause LSD in cattle.

The epidemiology of sheeppox, goatpox, and LSD is similar; the severity of outbreaks depends on the size of the susceptible population, the virulence of the strain of capripoxvirus, the breed affected (indigenous animals tending to be less susceptible to clinical disease than imported), and, with LSD, the presence of suitable insect vectors. Morbidity rates vary from 2% to 80%, and mortality rates may exceed 90%, particularly if the infection is in association with other disease or bad management.

## Transmission and Tissue Tropism

Under natural conditions, capripoxvirus is not transmitted very readily between animals, although there are circumstances when transmission appears very rapid; for example, in association with factors that damage the mucosae, such as peste des petits ruminants infection or feeding on abrasive forage. Animals are most infectious soon after the appearance of papules and during the 10-day period before the development of significant levels of protective antibody. High titers of virus are present in the papules, and those papules on the mucous membranes quickly ulcerate and release virus in nasal, oral, and lachrymal secretions, and into milk, urine, and semen. Viremia may last up to 10 days, or in fatal cases until death. Those animals that die of acute infection before the development of clinical signs and those that develop only very mild signs or single lesions rarely transmit infection, while those that develop generalized lesions produce considerable virus and are highly infectious. Aerosol infection over a few meters only, as with other poxvirus infections, is probably the usual form of transmission. Contact transmission of LSD virus under experimental conditions in the absence of insect vectors has only rarely been reported. Biting flies are significant in the mechanical transmission of LSD, and *Stomoxys calcitrans* and *Biomyia fasciata* have been implicated. There are probably a number of insects capable of mechanically transmitting LSD virus, but insects such as mosquitoes, which preferentially feed on hyperemic sites such as papules and if interrupted inoculate a new host intravenously, are considered the most likely to be involved in outbreaks characterized by large numbers of affected animals with generalized infections. Experimentally, *S. calcitrans* has also been shown to be capable of transmitting sheeppox and goatpox, and mosquitoes have transmitted LSD virus under experimental conditions.

During the recovery phase following infection, the papules on the skin become scabs. It is relatively easy to demonstrate virions in the scab, but difficult to isolate virus on tissue culture, probably because of the complexing of antibody and virus within the scab. Capripoxvirus is reported to remain viable in wool for 2 months and in contaminated premises for 6 months, and is reported to remain infectious in skin lesions of cattle for 4 months. The true epidemiological significance of the virus within the scab, and ultimately the environment, is not clear. It has been suggested that the protein material that envelops the virus within the type A intracytoplasmic inclusion bodies of infected cells protects the virus in the environment.

There is no evidence for the existence of animals persistently infected with capripoxvirus. Transplacental transmission of capripoxvirus may be possible in association with simultaneous pestivirus infection, as may occur with pestivirus-contaminated capripox vaccine.

Capripoxvirus can be isolated from the leukocytes during viremia, and has been isolated from lesions in the liver, urinary tract, testes, digestive tract, and lungs; however, the cells of the skin and skin glands and the internal and external mucous membranes appear to be the major sites of virus replication.

## Pathogenicity

There is considerable variation in the pathogenicity of strains of capripoxvirus. Little is known concerning the genes responsible in the capripoxvirus genome for virulence or host restriction; some preliminary results have been published.

## Clinical Features of Infection

The incubation period of capripox infection, from contact with virus to the onset of pyrexia, is approximately 12 days, although it frequently appears longer as transmission is often not immediate between infected and susceptible animals. Following experimental inoculation of virus, the incubation period is approximately 7 days, and this is similar to that shown experimentally using biting flies to transmit virus.

The clinical signs of malignant disease are similar in sheep, goats, and cattle. Twenty-four hours after the development of pyrexia of between 40 and 41 °C, macules (2–3 cm diameter areas of congested skin) can be seen on the white skin of sheep and goats, particularly under the tail. Macules are not seen on the thicker skin of cattle, and are frequently missed on skin of pigmented sheep and goats. After a further 24 h, the macules swell to become hard papules of between 0.5 and 2 cm diameter, although they may be larger in cattle. In the generalized form of capripox, papules cover the body, being concentrated particularly on the head and neck, axilla, groin, and perineum, and external mucous membranes of the eyes, prepuce, vulva, anus, and nose. In cattle, these papules may exude serum, and there may be considerable edema of the brisket, ventral abdomen, and limbs. The papules on the mucous membranes quickly ulcerate, and the secretions of rhinitis and conjunctivitis become mucopurulent (**Figure 1**). Keratitis may be associated with the conjunctivitis.

All the superficial lymph nodes, particularly the prescapular, are enlarged. Breathing may become labored as the enlarged retropharyngeal lymph nodes put pressure on the trachea. Mastitis may result from secondary infection of the lesions on the udder.

The papules do not become vesicles and then pustules, typical of orthopoxvirus infections. Instead, they become necrotic, and if the animal survives the acute stage of the disease, change to scabs over a 5–10 day period from the first appearance of papules. The scabs can persist for up to

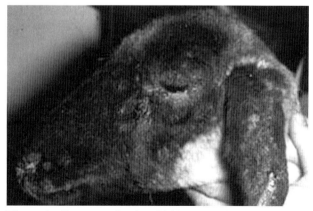

**Figure 1** Sheeppox showing rhinitis and conjunctivitis.

a month in sheep and goats, whereas in cattle the necrotic papules that penetrate the thickness of the skin may remain as 'sitfasts' for up to a year.

Severe disease is accompanied by significant loss of condition, agalactia, possibly secondary abortion, and pneumonia. Eating, drinking, and walking may become painful, and death from dehydration is not uncommon. Secondary myiasis is also a major problem in tropical areas.

## Pathology and Histopathology

The lesions of capripox are not only restricted to the skin, but also may affect any of the internal organs, in particular the gastrointestinal tract from the mouth and tongue to the anus, and the respiratory tract. In generalized infections, papules are prominent in the abomasal mucosa, trachea, and lungs. Those in the lungs are approximately 2 cm in diameter, and papules may coalesce to form areas of gray consolidation (**Figure 2**).

In affected skin, there is an initial epithelial hyperplasia followed by coagulation necrosis as thrombi develop in the blood vessels supplying the papules. Histiocytes accumulate in the areas of the papules, and the chromatin of the nuclei of infected cells marginates. The cells appear stellate as their boundaries become poorly defined, and many undergo hydropic degeneration with the formation of microvesicles. Intracytoplasmic inclusion bodies are present in infected cells of the dermis and also in the columnar epithelial cells of the trachea where frequently gross lesions may not be apparent. These are initially type B inclusions at the sites of virus replication (**Figure 3**), but later in infection they are replaced by type A inclusions (see above). The maximum titer of virus is obtained from papules approximately 6 days after their first appearance.

## Immune Response

Capripoxvirus, like orthopoxvirus, is released from an infected cell within an envelope derived from modified cellular membrane. The enveloped form of the virus is more infectious than the nonenveloped form, which can be obtained experimentally by freeze–thawing infected tissue culture. By analogy with orthopoxvirus, antigens on the envelope and on the tubular elements of the virion surface may stimulate protective antibodies. Animals immune to nonenveloped virus are still fully susceptible to the enveloped form. Passively transferred antibody, either colostral or experimentally inoculated, will protect susceptible animals against generalized infection; however, in the vaccinated or recovered animal, there is no direct correlation between serum levels of neutralizing antibody and immunity to clinical disease. Antibody may limit the

**Figure 2** Sheeppox showing severe lung lesions.

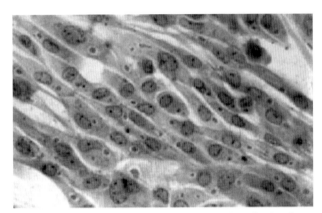

**Figure 3** Capripoxvirus growing in lamb testis cells showing many intracytoplasmic inclusion bodies. Magnification ×400.

spread of capripoxvirus within the body, but it is the cell-mediated immune response that eliminates infection. In sheep, major histocompatibility complex-restricted cytotoxic T-lymphocytes are required in the protective immune response to orthopoxvirus infection, and therefore probably also capripoxvirus infection.

Immune animals challenged with capripoxvirus by intradermal inoculation develop a delayed-type hypersensitivity reaction at the challenge site. This may not be apparent in animals with high levels of circulating antibody. It has been suggested that the very severe local response shown by some cattle at the site of vaccination against LSD may be a hypersensitivity reaction due to previous contact with the antigens of parapoxvirus.

There is total cross-immunity between all strains of capripoxvirus, whether derived from cattle, sheep, or goats.

## Prevention and Control

In temperate climates, capripox can be effectively controlled by slaughter of affected animals, and movement control of all susceptible animals within a 10 km radius for

6 months. In tropical climates, particularly in humid conditions when insect activity is high, movement restrictions are not sufficient and vaccination of all susceptible animals should be considered. In outbreaks of LSD, it is not considered necessary to vaccinate sheep and goats, although theoretically cattle strains of virus could infect them. Similarly, in outbreaks of capripox in sheep and goats, cattle are not normally vaccinated.

Countries in which capripoxvirus is absent can maintain freedom by preventing the importation of animals from infected areas. There is always a possibility that skins from infected animals could introduce infection into a new area, although there have been no proven examples of this. The insect transmission of capripoxvirus into Israel from Egypt over a distance of between 70 and 300 km would indicate that it is impossible for countries neighboring enzootic areas to totally secure their borders.

In enzootic areas, annual vaccination of susceptible animals with a live vaccine will control the disease. Calves, kids, and lambs up to 6 months of age may be protected by maternal antibody, but this would only occur if the mother had recently been severely affected with capripox. Although maternal antibody will inactivate the vaccine, it is advisable to vaccinate all stock over 10 days of age. No successful killed vaccines have been developed for immunization against capripoxvirus infection, other than those that give only very short-term immunity.

## Future Perspectives

Capripox of sheep and goats is present in most of Africa and Asia, whereas LSD is restricted to Africa. There is no good explanation as to why LSD has not spread into the Middle East and India, carried by the considerable trade in live cattle. Unless there is a reservoir host in Africa which is required for the maintenance of the cattle-adapted capripoxvirus, it can be anticipated that LSD will spread out of Africa, with major economic consequences.

While considerable attention has been given to vaccinia virus as a vector of other viral genes for development as a recombinant vaccine, little attention has been given to capripoxvirus as a potential vector vaccine. Although its use would be restricted to the not inconsiderable capripoxvirus enzootic area, it would have the advantage of not being infectious to humans, and be a useful vaccine in its own right.

The high degree of homology between the genomes of capripoxvirus isolates from different species and the apparent differences between their virulence and host preference would make them good candidates for studying the genetic basis of virulence and host specificity.

*See also:* Leporipoviruses and Suipoxviruses; Poxviruses.

## Further Reading

Black DN, Hammond JM, and Kitching RP (1986) Genomic relationship between capripoxvirus. *Virus Research* 5: 277–292.

Carn VM and Kitching RP (1995) An investigation of possible routes of transmission of lumpy skin disease (Neethling). *Epidemiology and Infection* 114: 219–226.

Gershon PD, Ansell DM, and Black DN (1989) A comparison of the genome organization of capripoxvirus with that of the orthopoxviruses. *Journal of Virology* 63: 4703–4708.

Gershon PD, Kitching RP, Hammond JM, and Black DN (1989) Poxvirus genetic recombination during natural virus transmission. *Journal of General Virology* 70: 485–489.

Kitching RP, Bhat PP, and Black DN (1989) The characterization of African strains of capripoxvirus. *Epidemiology and Infection* 102: 335–343.

Kitching RP, Hammond JM, and Taylor WP (1987) A single vaccine for the control of capripox infection in sheep and goats. *Research in Veterinary Science* 42: 53–60.

Tulman ER, Afonso CL, Lu Z, Zsak L, Kutish GF, and Rock DL (2001) Genome of lumpy skin disease virus. *Journal of Virology* 75: 7122–7130.

Tulman ER, Afonso CL, Lu Z, et al. (2002) The genomes of sheeppox and goatpox viruses. *Journal of Virology* 76: 6054–6061.

# Capsid Assembly: Bacterial Virus Structure and Assembly

**S Casjens**, University of Utah School of Medicine, Salt Lake City, UT, USA

© 2008 Elsevier Ltd. All rights reserved.

## Glossary

**Coat protein** The protein which is the structural building block for the nucleic acid container part of the virion.

**DNA encapsidation or DNA packaging** The process of enclosing a nucleic acid inside the protein coat of a virion.

**Head** The icosahedrally symmetric portion of *Caudovirales* virions that contains the nucleic acid.

**Portal protein** The *Caudovirales* protein through which DNA enters and exits the coat protein shell; it is present at only one of the 12 otherwise identical icosahedral vertices.

**Procapsid or prohead** The virion precursor particle that contains no nucleic acid; it is the preformed container into which the nucleic acid is packaged.

**Scaffolding protein** A protein that directs the proper assembly of the coat protein, but is not present in the completed virion.

**Tail** The portion of *Caudovirales* virions that binds to cells bearing the correct surface receptor and through which DNA traverses during delivery from the virion into the cell. The tail is attached to the portal protein of the head in the complete virion.

**Triangulation (*T*) number** Formally, *T* is the number of smaller equilateral triangles (facets) into which each of the 20 icosahedral faces is subdivided in an icosadeltahedron. In viruses, this is directly related to the number of coat protein subunits ($S$) per virion ($S = 60T$), since there are typically three subunits per facet and 20 faces in icosahedral viruses.

**Virion** The stable virus particle that is released from infected cells and is capable of binding to and infecting other sensitive cells.

## History and Overview

The structure of bacterial viruses (bacteriophages or phages) remained a mystery for over a quarter of a century after their original discovery in about 1910, and even then early electron micrographs of phage T4 showed only a vague outline of its complex features. The invention in the late 1950s of negative staining with heavy metal salts and high-resolution metal evaporation techniques allowed electron microscopy to be used to visualize virions with a resolution of a few nanometer – a big step forward, but most individual protein components could still not be resolved. Meanwhile, viruses were shown to be made up primarily of protein on the outside and nucleic acid inside. It is worth noting that phage T2 was a major participant in the proof that DNA is in fact the genetic material, when Hershey and Chase used it to show that during an infection the virion's DNA entered the cell and the bulk of the protein remained outside. After their discovery that DNA is the genetic material, in 1953, Crick and Watson also recognized that virus chromosomes could not encode a single protein molecule large enough to enclose their nucleic acid molecules, and they correctly postulated that virions had to be constructed with repeating arrays of protein molecules. They also noted that there are only two ways of building virions with symmetric arrays of coat proteins that enclose a space suitable for a virion's nucleic acid, and those are arrays built with cubic or helical symmetry (the former includes icosahedral, octahedral, and tetrahedral arrays which would be built from 60, 24, and 12 identical protein subunits, respectively). In 1962, Caspar and Klug deduced that many viruses must be built from icosahedral shells that contain more than 60 subunits, and developed a geometric theory for such arrangements if the subunits were given some flexibility in bonding to their neighbors (referred to as 'quasi-equivalence', since the flexibility gives rise to subunits that are not quite identical in their conformation and/or bonding properties). Some viruses, especially the tailed phages (*Caudovirales*), which are in large part the subject of this article, were found not to be this 'simple' and their repeating arrays of proteins are embellished with parts that are not icosahedrally symmetric. These asymmetric parts can be very complex; for example, phage T4 virions contain over 40 different protein species, of which a substantial majority are not icosahedrally arranged like the coat protein. Recent determination of the complete atomic structure of virions, procapids, or parts of them by X-ray diffraction (usually 2–4 Å resolution) and three-dimensional (3-D) reconstruction by superposition of electron micrographs of many particles and averaging of the resulting structure about the icosahedral rotational symmetry axes (current best is 6–8 Å resolution) has given rise to much detailed information about the structure of protein arrays in viruses in general and phages in particular, and very recent advances have allowed structural determination of virions' asymmetric parts to below 20 Å in advantageous cases (**Figure 1**).

As structural information about virus particles and other macromolecular assemblies accumulated, scientists naturally wondered how such structures were built. Animal and plant viruses were difficult to study inside infected cells, and so the assembly of the bacterial viruses in the easier-to-study bacterial cells became an important model for understanding the mechanisms utilized in the assembly of many macromolecular structures in addition to the viruses themselves. Critical in these studies was the determination of 'parts lists' for the bacteriophages under study, and these were determined both biochemically by delineation of the different protein components and genetically by identification of the genes that encode the protein components; both approaches were crucial in understanding the true nature of phage virions and how they assemble. Although some eukaryotic virions do include host encoded proteins, this has not been found to be the case for phages; all of their protein components are encoded by their own genomes. Although small organic molecules (e.g., putrescine) have been found in phage virions, none have been shown to be essential structural components.

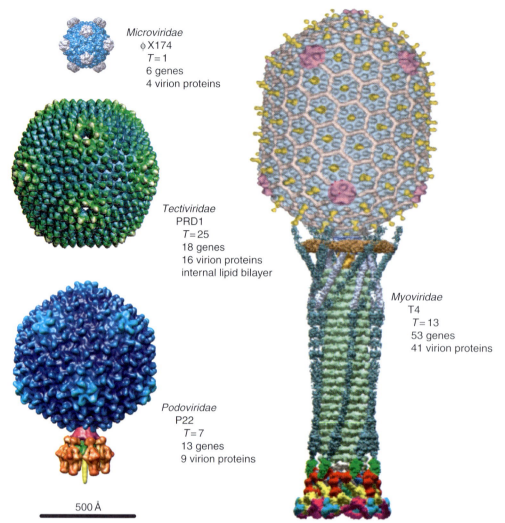

**Figure 1** Bacterial virus structures. The structures of representatives from four bacteriophage families are shown at the same scale. The images are shown at $20 \pm 5$ Å resolution. The 'number of genes' indicated in the figure is the number of genes currently known to be involved in virion assembly and the 'virion proteins' is the number of these present in the completed virion. In φX174, DNA packaging into the virion is dependent upon DNA replication but the genes that encode the replication machinery are not included. In each case, different structural proteins visible from the outside are shown as different colors. The PRD1 structure shown is not the actual virion but only the arrangement of coat proteins. The T4 virion is shown in its 'tail fiber up' conformation, in which the tail fibers (dark green) are folded up and the whiskers are folded down, and both are lying along the body of the tail. The T4 image was provided by V. Kostyuchenko, A. Fokine, and M. Rossmann, the P22 image was provided by G. Lander and J. Johnson, and the other two images are from the Virus Particle Explorer website (http://viperdb.scripps.edu/index.php).

Phages are known that contain any one of the four types of nucleic acid, single-stranded DNA (ssDNA), double-stranded DNA (dsDNA), single-stranded RNA (ssRNA), and double-stranded RNA (dsRNA), and they are built with icosahedral symmetry, helical symmetry, and a combination of the two. **Table 1** lists the eight formally named phage orders which appear to be largely unrelated to one another, and one of these, the *Caudovirales*, has been divided into three families with different tail types. Note that there are poorly studied phages that infect the bacteria class Mollicutes that have not yet been formally classified, and viruses that infect the Archeas are not discussed here. Given this great diversity of phage structures and assembly mechanisms, it is not possible to describe them all in detail here, and the reader is urged to consult other articles in this volume for more specific information on the various bacteriophage systems. In this article, we explore phage virion structure and the strategies and mechanisms by which virions assemble. Focus is largely upon the *Caudovirales*, since their assembly is the best understood, but other phage phyla will be mentioned as they are relevant.

## Assembly Pathways

A major theme that emanates from the early study of phage assembly is the idea of an 'assembly pathway'.

**Table 1** Types of bacteriophages

| Phage type | Best-studied phages | Major virion symmetry | Virion nucleic acid | Comments |
|---|---|---|---|---|
| Microviridae | ϕX174 | Icosahedral; protein fold A[a] | Circular ssDNA; 4–6 kbp | $T=1$ |
| Inoviridae | f1, M13 | Helical | Circular ssDNA; 5–9 kbp | Special proteins at both ends |
| Caudovirales Myoviridae | T4, P2 | Icosahedral head; protein fold B[a]; long helical contractile tail | Linear dsDNA; 30–500 kbp[b] | Caudovirales heads range from $T=3$ to $T=52$ and can be isometric or elongated |
| Caudovirales Siphoviridae | λ, HK97, SPP1 | Icosahedral head; protein fold B[a]; long noncontractile tail | Linear dsDNA; 22–140 kbp[b] | Heads can be isometric or elongated |
| Caudovirales Podoviridae | P22, T7, ϕ29 | Icosahedral head; protein fold B[a]; short tail | Linear dsDNA; 18–75 kbp[b] | Heads can be isometric or elongated |
| Tectiviridae | PRD1 | Icosahedral; protein fold A[a] | Linear dsDNA; 13–15 kbp | $T=25$; contains internal lipid bilayer |
| Corticoviridae | PM2 | Icosahedral | Circular dsDNA; 10 kbp | $T=21$; contains internal lipid bilayer |
| Plasmaviridae | L2 | Pleomorphic | Circular dsDNA; 2 kbp | Contains lipid bilayer |
| Leviviridae | MS2, R17 | Icosahedral; protein fold C[a] | Linear ssRNA; 3.5–4.3 kbp | $T=3$ |
| Cystoviridae | ϕ6 | Icosahedral; protein fold D[a] | Linear dsRNA; 13–14 kbp | $T=13$; three different dsRNA molecules in virion; contains external lipid bilayer |

[a]The protein folds (named arbitrarily in the table) of typical Microviridae, Caudovirales (Siphoviridae), Leviviridae, and Tectiviridae coat proteins have been determined by X-ray diffraction; that of Cystoviridae is suspected from its arrangement in virions which is similar to other dsRNA viruses.
[b]Approximate size ranges – Caudovirales phages are extremely diverse and the limits of genome sizes are not fully known.

The complex phages (Caudovirales), which can have as many as 50 different proteins in their virions, typically do not synthesize these different proteins in the temporal order of their assembly. Instead, the different proteins are all synthesized simultaneously, and the inherent properties of the proteins dictate their order of assembly. Thus, virion assembly proceeds via a specific 'pathway' in which the components assemble in an orderly manner. How is such a program of assembly achieved? The generally accepted mechanism for this type of control, which is now known to occur in the construction of many macromolecular structures, not just phages, is that the component proteins change conformation upon binding to the growing structure. Thus, once an initiator complex (of say protein A) is generated, protein B binds to it and changes its conformation so that protein C can bind, protein C in turn changes its conformation so protein D can bind, etc. In some cases, the precise juxtaposition of proteins B and C on the growing structure may also contribute to the creation of a binding site for D. In such a scheme, the unassembled proteins B, C, and D have no affinity for each other and so do not assemble in alternate ways. Thus, an ordered 'assembly pathway' of protein A first creating an initiation complex, then B, then C and then D binding in this defined order is determined solely by the properties of A, B, C, and D proteins. Because of these properties, when one component is missing, in most cases the assembly intermediate to which the missing component would have bound accumulates, and downstream components remain unassembled. This has allowed the examination of many such intermediates in the characterization of phage assembly pathways. The isolation in relevant genes of conditionally lethal nonsense mutations, which are particularly tractable in phage systems, made this approach extremely productive. Several issues regarding the assembly pathway strategy remain poorly understood, and these include such questions as: What controls initiation of the assembly process? Control could simply be the synthesis and proper folding of a limiting amount of protein A in the example above, but in some situations such as procapsid assembly (below) it appears to be more complex. And what is the exact nature of the conformational changes that occur upon binding? These remain fertile ground for research.

## Assembly Strategies and Mechanisms

### Branched Pathways, Subassemblies, and Quality Control

Phages have apparently evolved ways of maintaining quality-control systems in virion assembly. One such system is assembly though 'subassemblies' or discrete assembly intermediates that are unable to join with other subassemblies unless they are properly assembled. Thus, if an incorrectly built subassembly (say missing one of 10 protein molecules) is defective and unusable, it is much less costly than if a complete virion (containing hundreds to thousands of protein molecules) were rendered defective due to one missing or defective protein molecule. Thus in the *Caudovirales*, for example, head precursors called procapsids are built first and most likely only package DNA if they are correctly assembled, tails are built independently (though a series of smaller subassemblies), and DNA-filled heads and tails only join when both are successfully completed. A second quality-control system may be that phage proteins have evolved to minimize assembly problems by ensuring that the participating proteins do not assemble unless their C-terminus is intact. Such a strategy ensures that prematurely terminated polypeptides (arguably the most probable translation error) are unable to participate, since they are missing their assembly site. Although this is an evolutionary argument, the fact that among the many phage assembly genes and proteins that have been studied, very few of the N-terminal fragments have 'dominant negative' genetic characteristics or are incorporated into the assembling virion.

## Assembly of Symmetrical Protein Arrays

### Assembly and maturation of icosahedral arrays

The use of an icosahedral protein shell to enclose phages' virion chromosomes appears to have evolved independently at least four times. Phage coat proteins are sufficiently ancient that even those with common ancestors are often not recognizably similar in amino acid sequence. We are thus left with comparison of their folded (secondary and tertiary) structures to deduce common or separate ancestries, and such structural information is not known for most phages. Nonetheless, high-resolution atomic structures are known by X-ray diffraction for a few members of several groups and those from the *Caudovirales*, *Tectiviridae*, *Microviridae*, and *Leviviridae* form three very different coat protein folding groups that are certainly not related. But do coat proteins from all members within each of these groups have a common ancestor? Our current limited knowledge suggests that this may well be true. High-resolution 3-D reconstructions of electron micrographs of several members of the *Podoviridae* and *Myoviridae* strongly suggest that all *Caudovirales* have the same unusual protein fold that has been determined for phage HK97. Finally, the *Cystoviridae* are thought to have a fourth type of subunit that is related to dsRNA viruses that infect eukaryotes.

Although much is understood about the actual structures of the phage (and other virus) icosahedral arrays that encase the nucleic acid in so many virions, in different viruses the size and shape of these arrays can vary greatly and control of these variations is poorly understood. No octahedral or tetrahedral viruses have been discovered. The geometry of icosahedral arrays is well understood, and if strict conformational and bonding identity is maintained among the participating chemically identical coat proteins, only exactly 60 subunits can be accommodated in the shell. However, if some flexibility (quasi-equivalence, above) is allowed, then the number of subunits ($S$) allowed in an isometric icosahedral shell is equal to $60T$ where $T = h^2 \times hk \times k^2$, where $h$ and $k$ are any integer including zero. This gives rise to the infinite series $T = 1, 3, 4, 7, 9, 12, \ldots$ and $S = 60, 180, 240, 420, 540, 720, \ldots$, respectively. Such shells can be thought of as having pentamers of the coat protein at each of the 12 icosahedral vertices and different numbers of hexamers making up the 20 triangular faces in different-sized shells. Phage heads are known with $T$ numbers ranging from 1 to 52. Little is understood about exactly how the assembly of such shells is initiated or how their size is controlled. In many cases, the coat protein by itself will assemble into several sizes of shell as well as long tubes and irregular structures. In some cases (notably phages T4 and $\phi 29$), the icosahedral shell is elongated along the tail axis, and evidence from the systems under study suggests that both portal protein and scaffolding protein (see below) can be involved in this control, so factors beyond the innate coat protein properties clearly participate in the control of the spatial aspect of shell assembly.

Coat protein shells of the dsDNA phages assemble into a metastable 'immature' procapsid shell which undergoes a maturation that consists of a conformational transition to its final more stable state at about the time of DNA packaging. This maturation transition is typically accompanied by a significant expansion of the shell, the release of the scaffolding protein, and any proteolytic cleavage of the head proteins that might occur. This transition is surprisingly complex, and in the case of phage HK97, where it has been studied at the highest resolution, it proceeds through at least five separable coat protein conformational/organizational stages, in which there is little or no refolding of the main structural domain of coat protein, but there are significant changes in the orientations of the coat protein and the conformations of surface parts of the protein that are not within the central domain (**Figure 2**). The trigger(s) for this transition is not fully understood, but it clearly requires the successful completion of procapsid assembly. DNA entry and/or head protein proteolysis may be the natural signal for the transition to occur. This transition is universal in the assembly of tailed phages, and it has been suggested that topological difficulties in the construction of a closed shell with building units that have uneven surfaces might demand building a loosely bonded shell first, which subsequently rearranges into a more tightly bonded structure. The transition could also be part of a quality-control mechanism to ensure the structure is correctly built, and the concomitant expansion increases the available internal space for occupation by the DNA.

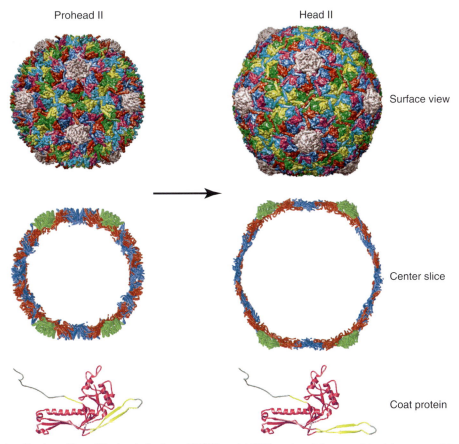

**Figure 2** The maturation transition of the bacteriophage HK97 head. HK97 coat protein assembles into procapsids, its N-terminal 102 residues are removed by proteolytic cleavage, and the resulting structure is called prohead II. This coat protein shell undergoes a several-stage transition to become head II, which has expanded in diameter and has a thinner shell. In the surface view, the seven different 'quasi-equivalent' coat subunit locations are given different colors. In the slice view, the penton subunits are green and hexon subunits red or blue. In addition to subunit reorientation during the transition, the loop (called the E loop) shown at the lower right of the coat protein ribbon diagrams (the bottom tier) undergoes a major change in conformation during the transition. The images were provided by Lu Gan, Jack Johnson, and Roger Hendrix.

### Symmetry disruptions and symmetry mismatches

The tails of the tailed phages (*Caudovirales*) and the portal of the *Tectiviridae* disrupt the icosahedral symmetry of coat protein shells of these virions, by being present at only one of the 12 otherwise identical vertices. How does this occur? Where it is known (phages P22, T4, T7, ɸ29), this 'disruption' replaces exactly one of the coat pentamers and it is present in the procapsid before the tail joins to the head. In procapsids, a dodecameric (12-mer) ring of portal protein occupies the position of the coat protein pentamer and tails will eventually bind to this portal protein. A single unique vertex is most simply explained if the portal protein forms the site that initiates coat shell assembly, and, in agreement with this idea, it has been shown that the absence of portal protein affects the spatial aspects of coat protein assembly in several phages (e.g., T4, λ, ɸ29). However, in some other phages, such as P22, coat assembles quite normally in the absence of portal protein. In all these systems, other factors, especially scaffolding protein (below), clearly play an as yet imprecisely defined role in initiation of coat protein assembly and in creating the unique portal vertex. The 12-fold symmetry of the portal ring and the fivefold symmetry of the capsid at the portal vertex means that there is a 'symmetry clash' between these two parts of the virion, and thus there cannot be identical interactions between the individual coat and portal subunits; it has been suggested that this could allow rotation of the portal with respect to the coat shell, but this has not yet been shown experimentally to be the case.

### Helical arrays

Cylindrically shaped helical and stacked-ring protein arrays are present in the long tails of *Myoviridae* and *Siphoviridae*. In both types of viruses, these arrays assemble (in the cases studied) from the tail tip up to the head proximal end to give rise to the tail subassembly. Both types of tails have a complex baseplate (or tail tip) structure at their head-distal ends. Baseplates assemble first, and when they are complete, they form the site at which the helical tail shaft begins to assemble. But helical arrays can be indefinitely long; so how are the discrete tail lengths programmed? Phage tails use a protein template (discussed in more detail below) to determine tail length, and when this length is attained, other proteins bind to the growing tip to block further growth. There are three types of helix (or stacked ring) building proteins, the *Myoviridae* tail tube (inner) and sheath (outer) proteins and the *Siphoviridae* tail shaft protein. Of these three, the sheath proteins are most highly conserved, and in many (but not all) tailed phages they are recognizably related in amino acid sequence. The other two are extremely variable, and with no structural information it is impossible to know whether or not they form discrete groups of distantly related proteins. Thus, there is no convincing evidence that the *Myoviridae* tail core and sheath and the *Siphovirdiae* tail shaft subunits have a common ancestor, and their different functionalities and roles suggests that they may have independent origins.

The *Inoviridae* or filamentous phages, unlike most other bacterial viruses, do not escape the host cell by lysing it. Their helical coat is assembled from subunits embedded in the cell membrane as the DNA is extruded through a complex assembly machine built into the membranes of the infected cell. Since they assemble in a such a different manner from the other phages under discussion here, and the detailed mechanism of their assembly is not understood in detail, they will not be considered in depth here.

## Catalysis and Assembly

### Scaffolds, templates, and jigs

Scaffolding proteins are proteins that assemble in fairly large numbers into procapsids but are not present in the completed virion. They act transiently to help in the proper assembly of coat protein shells and in the tailed-phage cases in the determination of the one unique vertex (above). The precise mechanisms by which scaffolding proteins perform these functions is unknown, but typically the interior space of *Caudovirales* procapsids contains several hundred molecules of scaffolding protein that are essential for proper procapsid assembly. They are then either proteolytically destroyed or released intact before DNA is packaged. Among the latter type, the scaffold of phage P22 has been shown to participate in an average of five rounds of procapsid assembly and so clearly acts catalytically in the assembly process. Stable oligomeric complexes between coat and scaffold molecules have not been found, yet scaffold 'nucleates' coat assembly (in the absence of all other proteins) at concentrations at which coat fails to assemble by itself. Scaffold remains as small oligomers in the absence of coat protein. It is not clear whether scaffold is forming a template whose surface guides coat assembly or is transiently binding and modifying the coat conformation so that it assembles correctly. *Microviridae*, in spite of their 'simple' $T = 1$ coat protein shell, have both an internal and external scaffolding protein. In addition to roles in both temporal and spatial control of coat protein assembly, scaffolds may also function to fill the interior and thus exclude cellular macromolecules from the procapsid interior.

Perhaps the best example of an assembly template is the assembly of the *Caudovirales* tail shafts. *Siphoviridae* tail shaft subunits require two things to assemble, an initiation site on the completed baseplate and a template along which to assemble. This template has been best studied in phage λ, but even there tail shaft assembly is complex and not yet fully understood. It has been shown that the phage λ gene *H* (or 'tape measure') protein serves as a template for the assembly of the shaft subunit outward from the distal tip structure, and it is the length of the extended tape measure protein that determines the length of the tail shaft; shortening the tape measure protein causes the formation of a commensurately shorter tail. Interestingly, in these virions, the tape measure is thought to occupy the lumen of the tail, through which DNA must pass during injection, and the tape measure protein is ejected from the virion along with the DNA during the infection process. A second type of tape measure has been suggested for assembly of the phage PRD1 $T = 25$ coat protein shell. The protein product of PRD1 gene *30* is stretched along the face edges of the icosahedral structure and may determine the length of the face and thus the size of the icosahedron.

A third type of spatial control of assembly is the 'jig', of which the phage T4 whisker is the prime example. This fiber protrudes from the head–tail junction, and its distal end binds to the elbow of the bent tail fiber, holding it in the correct position for joining of the tail fiber to the baseplate. The use of such a jig here may reflect some inherent difficulty in successfully attaching the tail fibers (see below).

### Assembly enzymes and chaperonins

Aside from the scaffolding proteins, there are other well-studied examples of proteins that act catalytically during phage virion assembly. Perhaps the most curious of these is the gene *63* product of phage T4. This protein is required for the addition of tail fibers to the otherwise completed virion, but it is not present in virions. Its exact mechanism of action is not known, but it has been suggested that it and the jig described above are required to allow the creation of a flexible ball-and-socket joint between the fiber and the rest of the virion, which in turn might allow the fiber to 'search' more efficiently for its binding site on the surface of bacterial cells. In addition, T4 gene *38* protein appears to bind to and catalyze efficient trimerization of the subunit that forms the outer half of the bent tail fiber, but then it is released and not found in virions. Chaperonins are required in the folding of many proteins and so can be critical to phage assembly in that the proteins must be properly folded to participate in the assembly process. It is interesting to note that the chaperonin protein-folding function was discovered during the genetic analysis of the host's role in phage λ head assembly, and that some phages (e.g., T4) encode replacement subunits for the GroE chaperonin to ensure that their proteins are folded correctly.

Other catalytic actions during phage assembly are the ATP cleavage-dependent action of DNA translocases that move the DNA into the procapsid through the portal ring and the cleavage of overlength replicated DNA to virion length by the phage-encoded enzyme called 'terminase' in the *Caudovirales* (the nucleic acid packaging process is less well understood in other phages). Both of these enzymatic functions are essential for tailed-phage assembly, and the reader is referred to the article in this volume on nucleic acid encapsidation for a more detailed discussion. In addition, enzymatic proteolytic modification of virion proteins is often essential during assembly (below).

## Nucleic Acid Encapsidation

The packaging of nucleic acid within virions is covered in more depth in other articles of this volume, but it is noted here that different phages utilize very different strategies to build a virion with nucleic acid on the inside and protein (and lipid in some cases) surrounding that nucleic acid. The packaging of an ssRNA within the virion molecule by members of the *Leviviridae* family occurs by concomitant assembly of coat protein molecules into a $T = 3$ icosahedral structure and condensation of the RNA (which is predicted to be rather compact due to extensive secondary structure) by a process that is not understood in detail. Here, as apparently in all phages, recognition of a particular sequence in the viral nucleic acid is required to initiate packaging, thereby ensuring that only virus-specific nucleic acid is encapsidated. The other phage types either assemble a procapsid into which the nucleic acid is inserted by an energy-consuming nucleic acid translocase (*Caudovirales*, *Tectiviridae*, *Cystoviridae*) or assemble the virion around the ssDNA as it is extruded from the cell (*Inoviridae*).

## Covalent Protein Modifications during Assembly

### Protein cleavage

A common, but not universal, feature of phage assembly is the controlled proteolytic cleavage of some protein participants. This is best studied in the *Caudovirales*, and so they will be discussed here. It was discovered early that an N-terminal portion of all of phage T4's coat protein molecules are removed by proteolytic cleavage during assembly, and that this cleavage is dependent upon proper assembly of the coat protein shell. It has been suggested that the role of such coat protein trimming, which is common, is to make assembly irreversible, allow coat protein to find a more stable folded state, to alter coat protein function between virion assembly and DNA delivery, to help create room inside for the DNA, to be a kind of quality control, and/or to simply to remove part of the virion that is no longer needed after successful assembly. These of course are not mutually exclusive roles. Assembly-dependent cleavage of other virion assembly proteins such as portal proteins, scaffolding proteins and tape measure proteins, as well as the phage-encoded protease itself, have also been observed. Where it has been studied, a phage-encoded protease co-assembles into the procapsid and subsequently proteolyzes the head proteins, including itself. Tail protein cleavage is less well understood, and the protease that trims, for example, the C-terminus of phage λ tape measure protein has not been identified, but a host protease is responsible for the essential but non-assembly-dependent removal of the N-terminal region of a phage P22 virion protein that is injected with the DNA.

### Protein cross-linking

*Escherichia coli* phage HK97 typifies a subset of *Siphoviridae* that are known to covalently cross-link all of their coat proteins as the final step in head shell maturation. Other covalent cross-links have been described between a few of the coat protein molecules of phage λ and a fragment of its putative head protease, but their role in assembly is not known. In HK97, each coat subunit is joined to its neighbors through lysine–asparagine side-chain isopeptide bonds. The topology of these cross-links is such that they form covalent rings of five and six subunits that are interlocked, thus making a sort of molecular 'chain mail'. This cross-linking is catalyzed by the coat protein itself after the shell has expanded, and the cross-links contribute to the stability of the head.

## Lipid Membrane Acquisition

Several bacterial virus families have lipid bilayers in their virions (**Table 1**). These have been studied most in the *Tectiviridae* phage PRD1, which has a lipid bilayer between the icosahedral capsid shell and the internal dsDNA. This layer may aid in the protection of the intravirion nucleic acid, but also participates in DNA delivery into sensitive cells. Members of the *Tectiviridae*, *Cystoviridae*, and *Corticoviridae* all appear to acquire their lipid membrane in the cytoplasm from the host's membrane by as yet poorly understood mechanisms, after virus-encoded proteins have been inserted into the host membrane. Information is available on some of the virion proteins that interact with the lipid bilayers, but little is known about the detailed mechanism of the assembly of the membrane into these phages' virions. The *Plasmaviridae* phage L2, which infects a member of the bacterial *Mycoplasma* family, appears to bud from the host cell membrane. This kind of release may only be possible in the mycoplasmas, since they are the only bacteria that have no cell wall, but the morphogenesis of L2 has not been studied in detail.

## Summary

Clearly, there is no single 'mechanism of bacteriophage capsid assembly'. Bacterial viruses are varied and complex, and they utilize many different mechanisms on the road to assembling completed virions. Two major take home lessons regarding phage assembly are (1) the ubiquitous use of obligate pathways in assembly processes and (2) the use of proteins that are essential for assembly but which are not present in the completed virion. The latter means that assembly cannot be understood by simply examining the properties of the components of the completed virion. In addition, the protein and lipid parts of virions are not simple containers designed solely for the protection of the nucleic acid inside. They are also sophisticated molecular devices that are designed to deliver their nucleic acid payloads into sensitive cells by mechanisms that are as varied as the ways in which they are assembled. No doubt, this requirement that virions be metastable 'spring-loaded' structures, that can spontaneously release their nucleic acid when the right external signal is received, is largely responsible for the complexity of phage virion structure and assembly. Because of this diversity and complexity, the study of phage assembly has shed considerable light on the mechanisms of many other macromolecular assembly processes.

## Acknowledgments

The author thanks Lu Gan, Roger Hendrix, Jack Johnson, Victor Kostyuchenko, and Michael Rossmann for material for the figures.

*See also:* Assembly of Viruses: Nonenveloped Particles; Filamentous ssDNA Bacterial Viruses; Genome Packaging in Bacterial Viruses; Icosahedral dsDNA Bacterial Viruses with an Internal Membrane; Icosahedral Enveloped dsRNA Bacterial Viruses; Icosahedral ssDNA Bacterial Viruses; Icosahedral ssRNA Bacterial Viruses; Inoviruses; Virus Particle Structure: Nonenveloped Viruses; Virus Particle Structure: Principles.

## Further Reading

Abrescia N, Cockburn J, Grimes J, et al. (2004) Insights into assembly from structural analysis of bacteriophage PRD1. *Nature* 432: 68–74.

Calendar R (ed.) (005) *The Bacteriophages*, 2. New York: Oxford University Press.

Casjens S (1985) An introduction to virus structure and assembly. In: Casjens S (ed.) *Virus Structure and Assembly*, pp. 1–28. Boston: Jones & Bartlett.

Casjens S and Hendrix R (1988) Control mechanisms in dsDNA bacteriophage assembly. In: Calendar R (ed.) *The Bacteriophages*, vol. 1, pp. 15–91. New York: Plenum.

Caspar D and Klug A (1962) Physical principles in the construction of regular viruses. *Cold Spring Harbor Symposium on Quantitative Biology* 27: 1–24.

Chiu W, Burnett R, and Garcea R (eds.) (1997) *Strucutral Biology of Viruses*. New York: Oxford University Press.

Fane B and Prevelige P , Jr. (2003) Mechanism of scaffolding-assisted viral assembly. *Advances in Protein Chemistry* 64: 259–299.

Fokine A, Chipman P, Leiman P, Mesyanzhinov V, Rao V, and Rossmann M (2004) Molecular architecture of the prolate head of bacteriophage T4. *Proceedings of the National Academy of Sciences, USA* 101: 6003–6008.

Hendrix R (2005) Bacteriophage HK97: Assembly of the capsid and evolutionary connections. *Advances in Virus Research* 64: 1–14.

Horn W, Tars K, Grahn E, et al. (2006) Structural basis of RNA binding discrimination between bacteriophages Qβ and MS2. *Structure* 14: 487–495.

Kostyuchenko V, Chipman P, Leiman P, Arisaka F, Mesyanzhinov V, and Rossmann M (2005) The tail structure of bacteriophage T4 and its mechanism of contraction. *Nature Structural and Molecular Biology* 12: 810–813.

Lander G, Tang L, Casjens S, et al. (2006) A protein sensor for headful viral chromosome packaging is activated by spooled dsDNA. *Science* 312: 1791–1795.

Poranen M, Tuma R, and Bamford D (2005) Assembly of double-stranded RNA bacteriophages. *Advances in Virus Research* 64: 15–43.

Steven A, Heymann J, Cheng N, Trus B, and Conway J (2005) Virus maturation: Dynamics and mechanism of a stabilizing structural transition that leads to infectivity. *Current Opinion in Structural Biology* 15: 227–236.

Wikoff W, Conway J, Tang J, et al. (2006) Time-resolved molecular dynamics of bacteriophage HK97 capsid maturation interpreted by electron cryo-microscopy and X-ray crystallography. *Journal of Structural Biology* 153: 300–306.

# Cardioviruses

**C Billinis,** University of Thessaly, Karditsa, Greece
**O Papadopoulos,** Aristotle University, Thessaloniki, Greece

© 2008 Elsevier Ltd. All rights reserved.

## Glossary

**Artiodactyls** Even-toed hoofed plant-eating mammals, including pigs, deer, antelope, cows, camel, giraffe, and hippopotamus.

**Autogenous vaccine** A vaccine manufactured from virus isolated from the same animal or herd of animals.

**Myocardiotropic** Virus tropism to the myocardium.

**Myocarditis** Inflammation of the myocardium.

**Neurotropic** Virus tropism to the nervous system.
**Phylogeny** The evolutionary relationships between groups of organisms.

## History

The genus *Cardiovirus* contains two species, *Encephalomyocarditis virus* and *Theilovirus* (**Table 1**). The first report of a cardiovirus was by Theiler in 1934. He reported that a neurological disease of mouse could be transmitted by intracerebral or intranasal inoculation, and was caused by a virus, which persisted in the central nervous system (CNS). This virus is known as Theiler's murine encephalomyelitis virus (TMEV). The disease was described as a polio-like infection with destruction of neurons of the anterior horns of spinal cord. Later, similar agents were isolated from the intestinal contents of normal mice and other rodents. Initially, some of these viruses were confused with poliomyelitis virus. In the early 1950s, Daniels *et al.* reported that a recently isolated strain of Theiler's virus, the DA strain, was responsible for a chronic white matter disease characterized by focal demyelination. A comprehensive picture was finally given in the mid-1970s by Lipton who showed that the DA strain, as well as the strains originally isolated by Theiler, cause a biphasic disease characterized by an initial polio encephalomyelitis with destruction of neurons followed by a chronic inflammatory and demyelinating disease of white matter resembling multiple sclerosis.

The first strain of encephalomyocarditis virus (EMCV), Columbia SK, was isolated from cotton rats (*Sigmodon hispidus*) by Jungeblut and Sanders in 1940. Immunological studies showed that this agent, which caused a flaccid paralysis of the hind limbs of mice followed by death, was a new and different neurotropic virus. This first description was followed by the isolation of MM virus in 1943, of EMCV in 1945, and Mengo virus in 1946. A last member of this serological group was isolated in 1949 by Gronnert, from the intestinal tissues of a sick mouse; this was called mouse-encephalomyelitis or maus-Elberfeld (ME) virus. A close relationship between all these viruses was soon recognized. Warren and Smadel in 1948 reported that sera from humans or animals containing neutralizing antibodies against EMCV would also neutralize Columbia SK and MM viruses. In 1949, Dick demonstrated cross-neutralization between Mengo and EMC viruses. Since the late 1960s, EMCV infection in laboratory animals has been used extensively as a research tool and model for human diseases. Extensive work has

**Table 1** Cardiovirus strains

| Serotype | Isolate | Origin | Species | Year of isolation | Pathogenicity | Accession no. for P1 |
|---|---|---|---|---|---|---|
| *Encephalomyocarditis virus* | | | | | | |
| EMCV | VR-129B | USA | Chimpanzee | 1945 | Myocardial | AJ617356 |
| EMCV | Rueckert | USA | Chimpanzee | 1945 | Myocardial | M81861 |
| EMCV | Mengo | Uganda | Rhesus monkey | 1946 | Paralysis | L22089 |
| EMCV | B | Panama | Pig | 1958 | Myocardial | M22457 |
| EMCV | MN-30 | USA | Pig | 1987 | Reproductive | AY296731 |
| EMCV | GRE-424 | Greece | Pig | 1990 | Myocardial | AJ617362 |
| EMCV | BEL-2887A | Belgium | Pig | 1991 | Reproductive | AF356822 |
| EMCV | CYP-108 | Cyprus | Pig | 1995 | Myocardial | AJ617359 |
| EMCV | BEL-279 | Belgium | Pig | 1995 | Myocardial | AJ617361 |
| EMCV | BEL-440 | Belgium | Pig | 1995 | Reproductive | AJ617360 |
| EMCV | ITL-136 | Italy | Pig | 1986 | Reproductive | AJ617358 |
| EMCV | ITL-001 | Italy | Pig | 1996 | Reproductive | AJ617357 |
| *Theilovirus* | | | | | | |
| TMEV | GDVII | USA | Mouse | 1937 | Neurovirulent | M20562 |
| TMEV | GDVII | USA | Mouse | 1937 | Neurovirulent | X56019 |
| TMEV | FA | USA | Mouse | 1937 | Neurovirulent | U32924 |
| TMEV | TO4 | USA | Mouse | 1937 | Biphasic | U33045 |
| TMEV | TO Yale | USA | Mouse | 1943 | Biphasic | U33047 |
| TMEV | WW | USA | Mouse | 1977 | Biphasic | U33046 |
| TMEV | DA | USA | Mouse | 1952 | Biphasic | M20301 |
| TMEV | BeAn8386 | Brazil | Mouse | 1957 | Biphasic | M16020 |
| TMEV | M2 | Brazil | BeAn variant | 1957 | Biphasic | AF030574 |
| TLV | NGS910 | | Rat | 1991 | | AB090161 |
| VHEV | Siberia-55 | Siberia | Human | 1963 | Encephalitis | M94868 |

EMCV, encephalomyocarditis; TMEV, Theiler's murine encephalomyelitis virus; TLV, Theiler's-like virus of rats; VHEV, Vilyuisk human encephalomyelitis virus.

been carried out using animal models for diabetes mellitus, cardiovascular diseases, polymyositis, and diseases of the CNS. More recently, EMCV has been recognized as a viral pathogen of domestic animals, especially swine. Clinical outbreaks of EMCV infection in pig herds were first encountered in Panama in 1958 and in Florida in 1960. Subsequently, clinical outbreaks in pigs have been reported in North America, New Zealand, Australia, Brazil, Cuba, Greece, Italy, Belgium, and Cyprus. In addition, EMC virus or antibodies have been detected in several species, including rodents, domestic and wild mammals, birds, and humans, often without evidence of clinical disease.

## Properties of the Virion and Genome

The virion of all cardioviruses contains single-stranded positive-sense RNA of $2.6 \times 10^6$ Da enclosed in an icosaedral capsid. The genome varies in length from 7800 to 8000 nt and is composed of the leader protein followed by the four structural proteins (VP1–4) and seven nonstructural proteins (2A, 2B, 2C, 3A, 3B, 3C, 3D). The coding region is surrounded by two untranslated regions (UTRs) at the 5′ and 3′ ends. The 5′ UTR comprises an internal ribosome entry site (IRES) from which viral protein translation is initiated. EMC viruses, but not theiloviruses, contain an unusual polypyrimidine (poly C) tract within the distal region of their 5′ UTRs. This tract could be involved in viral virulence, although its exact role remains unclear, and can vary in length from 60 to 420 residues, depending upon the strain of virus. In all cardioviruses, the 3′ UTRs terminate with a heterogeneous poly (A) tail, being involved in the binding process of the viral RNA-dependent RNA polymerase, whose encoding gene is located upstream to the 3′ UTR.

Cardioviral proteins and their precursors take their names (L, P1, P2, P3) from their sequential locations within the polyprotein (**Figure 1**). The leader or L proteins are present only in the EMC viruses and the persistent strains of TMEV. The EMC L proteins play an undefined role in host or tissue tropism and may also be involved in translational regulation of the IRES. The four P1 peptides are the capsid proteins 1A, 1B, 1C, and 1D. The middle portion of the polyprotein (P2) contains peptides 2A, 2B, and 2C. Protein 2C is an ATPase, not a polymerase, and its contribution to the replication cycle remains unclear. The P3 peptides, 3A, 3B$^{VPg}$, 3C$^{pro}$, and 3D$^{pol}$ are more closely associated with genome replication. Protein 3B is VPg, the peptide covalently linked to the 5′ end of the genome. VPg sequences are rich in basic, hydrophilic amino acids and have only one tyrosine residue (the attachment site) at position 3 from the amino end of the peptide. Protease 3C$^{pro}$ is the central enzyme in the viral cleavage cascade.

## Taxonomy and Classification

The cardioviruses constitute a separate genus, *Cardiovirus*, within the family *Picornaviridae*. They are distinguished from other picornaviruses by their pathological properties, genome organization, and the dissociability of their virions at pHs between 5 and 7 (in 0.1 M NaCl). The genus *Cardiovirus* consists of two species, *Encephalomyocarditis virus* and *Theilovirus*. Comparative analysis of RNA sequencing of different isolates has shown a clear difference between the two species (**Table 1**; **Figures 2** and **3**).

*Encephalomyocarditis virus* is represented by a single serotype of the same name. Mengovirus, Columbia SK virus, and Maus Elberfeld virus are strains of EMCV, based on serological cross-reaction and sequence identity. The species *Theilovirus* consists of three viruses (which are probably distinct serotypes): TMEV, Vilyuisk human encephalomyelitis virus (VHEV), and rat Theiler's-like virus (TLV). A fourth virus, rat encephalomyelitis virus (REV), is probably a strain of TMEV, although conclusive sequence data are awaited. Theiloviruses are also divided into two subgroups, based on neurovirulence (see the section titled 'Pathogenicity').

## Virus Propagation

Cardioviruses replicate in primary or continuous cell lines originating from a variety of species, including murine, bovine, porcine, human, primate, guinea pig, and hamster. Baby hamster kidney (BHK-21) and Vero cells are most commonly used. The virus also replicates in baby mice and chicken embryos and is pathogenic to many laboratory animals. EMC virus hemagglutinates guinea pig, rat, horse, and sheep erythrocytes. Serial passages of EMC viruses in cell culture can alter *in vitro* growth characteristics, reduce virulence, and affect hemagglutinating activity.

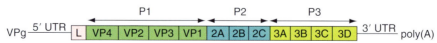

**Figure 1**  Diagram of the cardiovirus RNA genome, showing the genome-linked VPg at the 5′ end, the 5′ untranslated region (5′ UTR), the leader protein (L), the three viral protein coding regions P1 (VP4, VP3, VP2, VP1), P2 (2A, 2B, 2C), and P3 (3A, 3B, 3C$^{pro}$, 3D$^{pro}$), the 3′ untranslated region (3′ UTR), and the poly (A) tail.

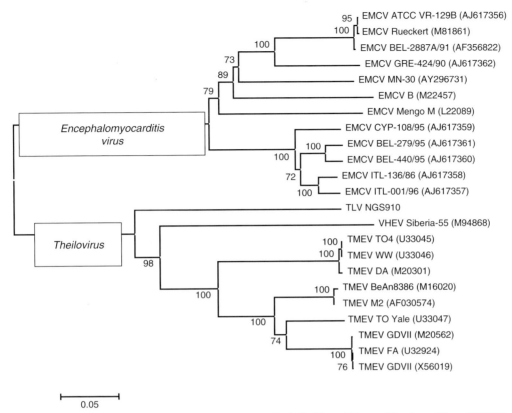

**Figure 2** Neighbor-joining tree based on a comparison of nucleotide similarities of P1 capsid polypeptides of EMCVs and theiloviruses. The tree was generated with the MEGA 3.1 program. Neighbor-joining trees were constructed from a different matrix employing the Kimura two-parameter correction. One thousand bootstrap pseudoreplicates were used to test the branching (shown as percentages). For abbreviations, see **Table 1**. Data provided by Dr. N. J. Knowles, Institute for Animal Health, Pirbright Laboratory, UK.

## Replication

The virus particle attaches to a cell receptor and is taken into the cell. The RNA genome is then uncoated and translated into a single polyprotein which is cleaved in a cascade of events catalyzed by the viral protease 3C$^{pro}$, and possibly by protein 2A in the case of the 2A/2B cleavage. The viral RNA is replicated by the RNA polymerase 3D$^{pol}$ through a complementary negative strand. Notably, proteins L and 2A are not required for viral replication in BHK-21 cells, although they may play important roles *in vivo*. Provirions are assembled by packaging RNA within capsid pentamers to form infectious particles. The replication takes place in the cell cytoplasm (**Figure 7**). However, in the case of EMCV, dense and empty viral particles have also been found in the nucleus of cardiac muscle cells (**Figure 8**).

## Evolution

The complete genome sequences of P1 capsid-coding region (CCR) of 12 EMC viruses, nine TMEVs, a Theiler's-like virus of rats, and a Vilyuisk human encephalomyelitis virus have been deposited in GenBank databases. Phylogenetic and molecular evolutionary analyses using MEGA 3.1 confirmed that two distinct species exist: EMCVs and theiloviruses. The 12 EMCVs formed one cluster, while the second cluster comprised all theiloviruses. These clusters differed by an average of approximately 42%. The maximum nucleotide variation among EMCVs was 21.70%, whereas within Theilovirus group it was 29.70%. The tree constructed from amino acids shows how viruses within a serotype are more closely related than they appear on the tree constructed from nucleotides, while members of different serotypes are still more distantly related (**Figures 2** and **3**).

Genetic analysis of European pig EMC viruses, in comparison with Mengo virus, have been performed with sequences of the 3D$^{pol}$ gene and the VP3/VP1 junction. These studies showed clustering of all the European isolates in two lineages, A and B, with a similar tree topology for both genome regions. A correlation between the phylogeny and the geographic origin was demonstrated, but no link was established with the clinical picture induced in pigs. Phylogenetic analyses of nucleotide sequences of the coding regions (the Leader gene, the CCR, and the end of 3D$^{pol}$ gene) also confirmed the A and B subgroups of

**Figure 3** Neighbor-joining tree based on a comparison of amino acid similarities of P1 capsid polypeptides of EMCVs and theiloviruses. The tree was generated with the MEGA 3.1 program. Neighbor-joining trees were constructed from a different matrix employing the Kimura two-parameter correction. One thousand bootstrap pseudoreplicates were used to test the branching (shown as percentages). For abbreviations, see **Table 1**. Data provided by Dr. N. J. Knowles, Institute for Animal Health, Pirbright Laboratory, UK.

EMCV. In addition, the topology of the tree constructed on the 5′ UTR showed that a Greek isolate, with complex clinical picture, was different, and most closely related to Mengo, in comparison to the others. The divergence of this isolate was also observed for the sequence of the 3′ UTR reduced by the 14 nt upstream to the poly (A) tail. This short adenosine-rich region involved in RNA polymerase binding influenced considerably the topology of the tree. It also appeared that the length of the poly (C) tract was not related to the difference in virulence observed between various EMCV isolates as it has been described for other EMCV strains in the past.

Serial passages *in vivo* of an EMCV pig isolate had a limited impact on the nt and aa sequence of the regions targeted, with an exception of the 2A gene and VP1, but it had a critical effect on the virulence of the virus by inducing its attenuation. A large deletion of 115 aa was found within the 2A gene which encodes a protein able to affect a co-translational break in the polyprotein chain at its own C-terminus releasing P1–2A. The N-terminal cleavage of 2A from P1 CCR is achieved by the virus-encoded proteinase 3C$^{pro}$. The EMCV 2A protein also contains a nuclear localization signal, which allows it to migrate to nuclei and nucleoli where it is involved in inhibition of cellular mRNA transcription and inhibition of cap-dependent mRNA translation, respectively. The deletion, which had been reported in the *in vitro*-passaged EMCV, removed all of 2A except the 18 aa necessary for the N- and C-terminal protein processing; thus, this deletion could, at least in part, be responsible for the attenuation observed experimentally. In addition, the mutation located at position 62 of VP1, which is adjacent to the aa responsible for the ability of certain EMCV strains to induce diabetes in mice, could also be involved in attenuation. Moreover, the low variability that has been detected after *in vivo* passages revealed that EMCV is stable in pigs.

## Epidemiology (Geographic Distribution, Host Range, and Transmission)

Cardioviruses have a worldwide distribution and are associated with a wide variety of species. Rodents, especially rats, are considered to be the natural host and reservoir of EMCV. In wild rodents, the virus usually persists without causing disease, whereas infection in laboratory rodents, especially baby mice, commonly produces fatal encephalitis and/or myocarditis. In pigs, natural infections with EMCV have different clinical characteristics, probably depending on different EMCV strains, which act with different pathogenicity. The infection has been recognized either as a cause of mortality in young pigs, due to acute myocarditis, or reproductive failure in sows. Clinical outbreaks of EMCV infection in pig herds were reported in Europe, Australia, New Zealand, and America. In Europe, the myocardial form has been reported in Greece, Italy, Cyprus, and Belgium. The reproductive form has only been reported in Belgium. In addition, apparently subclinical infections were reported in British and Austrian pigs, with a serological prevalence of 28% and 5%, respectively.

Two major routes of infection are suggested for the introduction and/or subsequent spread of the virus within a pig farm. The first infection route is ingestion of feces or carcasses of infected rodents, as many outbreaks have been linked with rodent plagues. The second route is horizontal pig-to-pig transmission through virus excretion by acutely infected pigs for 1–4 days. The virus can also be transiently excreted from persistently infected pigs, after reactivation of the infection. Transplacental infection has also been described as a potential route of virus spread. Other factors, such as infectious dose, route of infection, and age of pigs, have been found to be important in the spread of the virus under experimental conditions.

Clinical outbreaks in zoological parks and research institutions, involving several animal species, including elephants, artiodactyls, marsupials, mongooses, porcupines, and primates, have also been reported. In addition, EMCV virus or antibodies have been detected in domestic and wild mammals, birds, and humans, often without evidence of clinical disease.

There have been reports that humans are susceptible to EMCV. However, most of the evidence has been indirect, based on the presence of antibodies. In the 1950s, only a few documented cases of EMCV infection in humans were associated with fever, neck stiffness, lethargy, delirium, and headache. Human cases have been reported in Australia, in an area with a high incidence of EMCV in pigs. An EMCV outbreak in a zoo in the US, involving multiple animal species, did not result in illness in humans, although a zoo attendant who cared for EMCV-infected primates demonstrated a high antibody titer. Recent serological surveys indicate that human EMCV infections are relatively common in certain areas of the world (i.e., Austria, Greece), but generally are asymptomatic or unrecognized. Recent advances toward using pig tissues as a means of overcoming the acute shortage of transplantation tissues and organs for humans have made it necessary to determine the risk of transmitting this zoonotic virus from pigs to humans. The experimental infection of mice by transplanting pig organs infected with EMCV demonstrates that the risk is real.

TMEV is a natural enteric pathogen of mice that causes asymptomatic infection, but on rare occasions the virus will invade the CNS. TMEV shares many biological characteristics in common with poliovirus, another member of the picornavirus family, that may include spread from host to host via the fecal–oral route or the oropharyngeal–oral route.

## Pathogenicity, Clinical Features, and Pathology

### Encephalomyocarditis Virus

The expression of clinical disease in animals depends upon both viral and host factors. In susceptible species, strains of EMCV vary in virulence and/or tissue tropism. In the natural hosts, wild rodents, the virus usually persists without causing disease. Among laboratory animals, clinical manifestations and pathogenesis of EMCV are variable. Certain virus strains cause predominantly fatal encephalitis, widespread myocardial damage, or specific destruction of pancreatic β-cells. The course of infection varies in rats, mice, guinea pigs, hamsters, gerbils, rabbits, and monkeys, depending on the age of the animals and the virus strain used.

On the basis of organ tropism in mice, Graighead in 1966 classified EMCV into two variants, E (neurotropic) and M (myocardiotropic). EMCV-M also shows tropism for the β-cells of the pancreas, producing a disease syndrome similar to insulin-dependent diabetes mellitus. In 1980, Yoon *et al.* established a highly diabetogenic D variant and a nondiabetogenic B variant by repeated purification of EMCV-M. In addition, EMCV strains vary in pathogenicity and tissue tropism. In adult rats and mice experimentally infected with two myocardial EMCV strains of pig origin, no clinical signs or gross lesions were observed. Regardless, virus was isolated from several tissues from the third day post infection until the end of the observation, at 62 days post infection. EMCV was most frequently isolated from Peyer's patches and thymus. The results suggested that these tissues represent a site of persistence. Furthermore, EMCV antigen was detected in various organs but consistently in the heart and thymus.

In zoo animals, the disease is characterized by sudden death due to acute myocarditis. With the exception of pigs, domestic animals are generally resistant to overt clinical EMCV infection. In pigs, disease due to EMCV may take one of two main forms, depending on the tropism of the viral strain: an acute myocarditis (usually in piglets) or reproductive failure in sows. Each form of the disease in pigs (myocardial or reproductive) seems to be restricted to certain geographical areas, probably reflecting the character of viral strains originating from local rodent populations. Pig age at the time of infection is an important determinant of clinical severity. Extremely high mortality, approaching 100%, is limited to pigs of pre-weaning age. Infections in pigs from post-weaning age to adulthood are usually subclinical, although some mortality may be observed even in adult pigs. The myocardial form is most commonly characterized by acute disease with sudden death due to myocardial failure. Other clinical signs may be observed, including lethargy, inappetence, trembling, staggering, paralysis, vomiting, and dyspnea. Persistent infection has been demonstrated, the virus persisting in lymphoid tissues and transiently excreted in the feces following reactivation. Macroscopical lesions vary in severity and consist of multiple white-gray linear or circular areas, which are visible in the wall of the ventricles, especially those of the right ventricle (**Figure 4**). Histopathological lesions are characterized mainly by interstitial mononuclear cell infiltration, degeneration, and necrosis of cardiac muscle cells and occasionally calcium deposits (**Figures 5** and **6**). Electron microscope examination reveals presence of the virus intracytoplasmically in cardiac muscle cells, Purkinje fibers, and endothelial cells of the capillaries and intranuclearly in cardiac muscle cells (**Figures 7** and **8**). In breeding females, clinical signs may vary from no obvious illness to various forms of reproductive failure, including poor conception rates, embryo resorption, mummification, stillbirths, abortions, and neonatal death.

## Theiler's Virus

Theiloviruses are divided into two subgroups, based on neurovirulence following intracerebral inoculation of mice: GDVII and Theiler's Original (TO). The GDVII group includes GDVII and FA strains; it is extremely neurovirulent and induces acute fatal polioencephalomyelitis without demyelination. The TO group is less

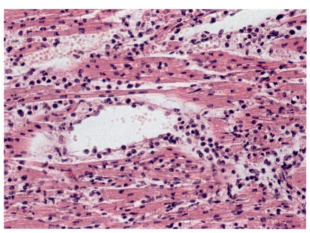

**Figure 5** Pig myocardium. Interstitial mononuclear cell infiltration, consisting of lymphocytes and macrophages. Courtesy of Dr. V. Psychas, Faculty of Veterinary Medicine, Aristotle University, Thessaloniki, Greece.

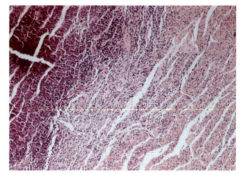

**Figure 6** Pig myocardium. Diffuse interstitial mononuclear cell infiltration accompanied by calcium deposits (toward left). Courtesy of Dr. V. Psychas, Faculty of Veterinary Medicine, Aristotle University, Thessaloniki, Greece.

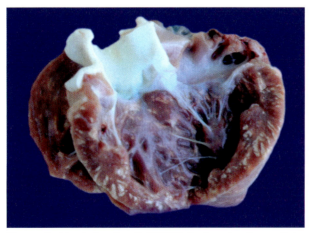

**Figure 4** Heart of a pig with EMCV infection showing multiple white-gray linear and circular calcified necrotic areas in the myocardium also visible under the endocardium. Courtesy of Dr. V. Psychas, Faculty of Veterinary Medicine, Aristotle University, Thessaloniki, Greece.

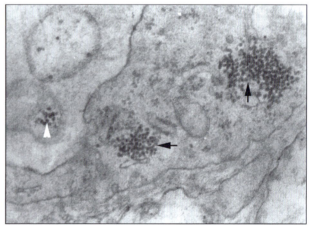

**Figure 7** Electron photomicrograph from myocardium of a pig infected with EMCV. Notice the numerous virus particles (arrows) in the cytoplasm of an endothelial cell and the membranous vesicles (arrowhead) containing virus particles free in the lumen of the capillary. Courtesy of Dr. V. Psychas, Faculty of Veterinary Medicine, Aristotle University, Thessaloniki, Greece.

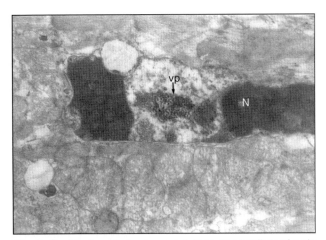

**Figure 8** Electron photomicrograph from myocardium of a pig infected with EMCV. Notice a pyknotic nucleus (N) with aggregations of dense virus particles (vp). Courtesy of Dr. V. Psychas, Faculty of Veterinary Medicine, Aristotle University, Thessaloniki, Greece.

neurovirulent and includes the TO4, TO-Yale, BeAn, WW, and DA strains. These strains cause a biphasic disease in which the early phase is characterized by an acute encephalomyelitis. If the virus persists, the animals will develop a chronic demyelinating disease in the CNS.

A large number of detailed pathogenesis studies in mice have been reported and reviewed. Generally, no clinical signs are observed, but on rare occasions the virus may invade the CNS and cause neurological disease. After ingestion, the virus replicates in the pharynx or the lower gastrointestinal tract, spreads to local and regional lymphatic tissue, and then invades the blood stream before penetrating and replicating in the nervous system, resulting in unilateral or bilateral flaccid paralysis of the hind limbs and rarely other neurological signs. It is postulated that invasion of the CNS is taking place via the circulation and crossing the blood–brain barrier (BBB) or through lymphatic channels. The BBB is composed of specialized cerebrovascular endothelial cells, which restrict vesicular transport and protect the CNS from harmful substances circulating in the blood. Hematogenous virus entry into the CNS may involve endothelial cells and/or macrophages. TMEV viral RNA has been detected in vascular endothelial cells, neurons, and glial cells in the brain and spinal cord. Moreover, mononuclear cells isolated directly from the CNS inflammatory infiltrates of TMEV-infected mice were found to contain infectious viral antigens, and TMEV was found to replicate in brain macrophages. Several studies also suggest that the main viral dissemination route for Theiler's virus may occur by infection of peripheral organs, such as muscle, followed by viral entry from these secondary sites via peripheral neural routes.

## Diagnosis

### Encephalomyocarditis Virus

In piglets, the disease is often characterized by sudden death without any prior clinical signs. The most striking lesions are white-gray linear or circular areas under the endocardium and epicardium. All age categories may be affected, but mostly pigs 20–40 days are involved. In sows, a clinical history of reproductive failure and pre-weaning piglet mortality is suggestive of EMCV infection. EMCV-induced reproductive problems should be differentiated from other pathogens, such as porcine reproductive and respiratory syndrome, Aujeszky's disease, and porcine parvovirus.

Histopathological examination may contribute to diagnosis. A variable degree of nonsuppurative interstitial myocarditis is characteristic of EMCV infection. The presence of virus antigen both in the cytoplasm and in the nucleus of cardiac muscle cells by immunohistochemical examination with the use of monoclonal antibodies further strengthens the diagnosis. A conclusive diagnosis of EMCV should be demonstrated by virus isolation in sensitive cell culture (e.g., BHK-21) and identification by cross-serum neutralization with a reference antiserum. Virus isolation should be attempted mainly from heart and spleen tissues of acutely infected piglets. Experimental infection in piglets demonstrated that the virus could be easily isolated during the period of viremia, that is, from 1 to 5 days post infection. Nucleic acid probes and reverse transcription-polymerase chain reaction (RT-PCR) may provide sensitive and specific methods of diagnosis.

The commonly used serological test is virus-neutralization test. Enzyme-linked immunosorbent assay (ELISA) may also be used, if available. However, serology should not be used alone, as EMCV antibodies have been detected in areas with no clinical disease, presumably due to nonpathogenic EMCV strains.

### Theiler's Virus

Molecular detection by PCR is highly sensitive and specific for TMEV. It is used for monitoring laboratory mouse colonies. Serological detection of the virus is inadequate, especially in nude or immunocompromised mice.

## Prevention and Control

There is no specific treatment for EMC. In pigs, avoiding stress or excitement in affected animals may minimize mortality. EMCVs appear to persist in rodents and are most likely to affect pigs and zoo animals when rodent populations are high. Elimination or reduction of feral rodent populations is recommended in the prevention and control of EMCV in order to minimize environmental contamination with virus. Prompt and proper disposal of animals that have died of the disease is also recommended. EMCV is inactivated by the use of disinfectants labeled for use against nonenveloped viruses. Autogenous and commercial killed vaccines or a genetically engineered vaccine have been used for the prevention of the myocardial form in pigs or zoo animals.

Recent studies demonstrated that the vaccination of mice with cDNA encoding the capsid proteins of TMEV could alter the course of TMEV-induced demyelinating disease. A single vaccination with cDNA encoding VP1 led to increased pathology in the CNS and increased clinical expression of disease. In another study, vaccinating mice three times with cDNA encoding the capsid proteins VP2 and VP3 reduced TMEV-induced disease, as demonstrated by a decrease in clinical expression of disease and reduce CNS pathology.

*See also:* Phylogeny of Viruses; Picornaviruses: Molecular Biology; Theiler's Virus.

## Further Reading

Brahic M and Bureau JF (1998) Genetics of susceptibility to Theiler's virus infection. *BioEssays* 20: 627–633.

Denis P, Liebig HD, Nowotny N, *et al.* (2006) Genetic variability of encephalomyocarditis virus (EMCV) isolates. *Veterinary Microbiology* 113: 1–12.

Koenen F (2006) Encephalomyocarditis virus. In: Straw BE, Zimmerman JJ, D'Allaire S, and Taylor DJ (eds.) *Diseases of Swine*, p. 331. Ames, IA: Blackwell.

Lipton HL and Jelachich ML (1997) Molecular pathogenesis of Theiler's murine encephalomyelitis virus-induced demyelinating disease in mice. *Intervirology* 40: 143–152.

Lipton HL, Kumar MAS, and Trottier M (2005) Theiler's virus persistence in the central nervous system of mice is associated with continuous viral replication and a difference in outcome of infection of infiltrating macrophages versus oligodendrocytes. *Virus Research* 111: 214–223.

Racaniello V (2001) *Picornaviridae*: The viruses and their replication. In: Knipe DM and Howley PM (eds.) *Fields Virology,* 4th edn., pp. 685–722. Philadelphia: Williams and Wilkins.

Scraba DG and Palmenberg A (1999) Cardioviruses (*Picornaviridae*). In: Webseter RG and Granoff A (eds.) *Encyclopedia of Virology*, pp. 229–238. London: Academic Press.

Zimmerman JJ (1994) Encephalomyocarditis. In: Beran GW and Steele JH (eds.) *Handbook of Zoonoses*, pp. 423–436. Boca Raton, FL: CRC Press.

# *Carlavirus*

**K H Ryu and B Y Lee,** Seoul Women's University, Seoul, South Korea

© 2008 Elsevier Ltd. All rights reserved.

## Glossary

**Symptomless** Having no apparent symptoms of disease.

**Triple gene block (TGB)** A specialized evolutionarily conserved gene module of three partially overlapping ORFs involved in the cell-to-cell and long-distance movement of some plant viruses.

## Introduction

The genus *Carlavirus* derived from the name of the type species *Carnation latent virus* is one of nine genera in the family *Flexiviridae* and contains a large number of members and tentative members. The carlavirus-infected natural host plants usually have very mild symptoms or remain symptomless, and thus the term 'latent' appears in the names of many species of the genus *Carlavirus*. The natural host range of individual species is restricted to a few plant species. This tendency toward mild or latent (symptomless) infection, a characteristic feature of carlaviruses, has led to carlaviruses and carlavirus-associated diseases failing to attract any special attention from phytopathologists and phytovirologists. However, many members of the genus have been identified in the last few decades and most of them have been associated with more serious diseases when plants are co-infected with other viruses. Most species are transmitted by aphids in a nonpersistent manner but two species (*Coupea mild mottle virus* and *Melon yellowing-associated virus*, CPMMV and MYaV) are transmitted by whiteflies (*Bemisia tabaci*).

## Taxonomy and Classification

The genus *Carlavirus* belongs to the family *Flexiviridae*. The type species of the genus is *Carnation latent virus*. The name carlavirus was derived from the type species. The genus contains a large number of members. In total, 68 members of the genus *Carlavirus* (39 definitive species and 29 tentative species) are listed in **Table 1**. Some members listed as species and all members listed

**Table 1** Virus species in the genus *Carlavirus*

| Mode of transmission | Virus species name | Abbreviation | Accession number |
|---|---|---|---|
| *Species* | | | |
| Aphid-transmitted | *American hop latent virus* | AHLV | |
| | *Blueberry scorch virus* | BlScV | NC_003499 |
| | *Cactus virus 2* | CV-2 | |
| | *Caper latent virus* | CapLV | |
| | *Carnation latent virus* | CLV | X52627, AJ010697 |
| | *Chrysanthemum virus B* | CVB | NC_009087 |
| | *Cole latent virus* | CoLV | AY340584 |
| | *Dandelion latent virus* | DaLV | |
| | *Daphne virus S* | DVS | AJ620300 |
| | *Elderberry symptomless virus* | ESLV | |
| | *Garlic common latent virus* | GarCLV | AB004805 |
| | *Helenium virus S* | HVS | D10454 |
| | *Honeysuckle latent virus* | HnLV | |
| | *Hop latent virus* | HpLV | NC_002552 |
| | *Hop mosaic virus* | HpMV | AB051109 |
| | *Hydrangea latent virus* | HdLV | |
| | *Kalanchoe latent virus* | KLV | AJ293570-1 |
| | *Lilac mottle virus* | LiMoV | |
| | *Lily symptomless virus* | LSV | NC_005138 |
| | *Mulberry latent virus* | MLV | |
| | *Muskmelon vein necrosis virus* | MuVNV | |
| | *Nerine latent virus* | NeLV | DQ098905 |
| | *Passiflora latent virus* | PLV | NC_008292 |
| | *Pea streak virus* | PeSV | AF354652, AY037925 |
| | *Potato latent virus* | PotLV | AY007728 |
| | *Potato virus M* | PVM | NC_001361 |
| | *Potato virus S* | PVS | NC_007289 |
| | *Red clover vein mosaic virus* | RCVMV | |
| | *Shallot latent virus* | SLV | NC_003557 |
| | *Sint-Jan's onion latent virus* | SJOLV | |
| | *Strawberry pseudo mild yellow edge virus* | SPMYEV | |
| Whitefly-transmitted | *Cowpea mild mottle virus* | CPMMV | DQ444266 |
| | *Melon yellowing-associated virus* | MYaV | AY373028 |
| Unknown vector | *Aconitum latent virus* | AcLV | NC_002795 |
| | *Narcissus common latent virus* | NCLV | NC_008266 |
| | *Poplar mosaic virus* | PopMV | NC_005343 |
| | *Potato rough dwarf virus* | PRDV | AJ250314 |
| | *Sweetpotato chlorotic fleck virus* | SPCFV | NC_006550 |
| | *Verbena latent virus* | VeLV | AF271218 |
| *Tentative species* | | | |
| Aphid-transmitted | Anthriscus latent virus | AntLV | |
| | Arracacha latent virus | ALV | |
| | Artichoke latent virus M | ArLVM | |
| | Artichoke latent virus S | ArLVS | |
| | Butterbur mosaic virus | ButMV | |
| | Caraway latent virus | CawLV | |
| | Cardamine latent virus | CaLV | |
| | Cassia mild mosaic virus | CasMMV | |
| | Chicory yellow blotch virus | ChYBV | |
| | Coleus vein necrosis virus | CVNV | DQ915963 |
| | Cynodon mosaic virus | CynMV | |
| | Dulcamara virus A | DuVA | |
| | Dulcamara virus B | DuVB | |
| | Eggplant mild mottle virus | EMMV | |
| | Euonymus mosaic virus | EuoMV | |
| | Fig virus S | FVS | |
| | Fuchsia latent virus | FLV | |
| | Gentiana latent virus | GenLV | |
| | Gynura latent virus | GyLV | |
| | Helleborus mosaic virus | HeMV | |

Continued

**Table 1** Continued

| Mode of transmission | Virus species name | Abbreviation | Accession number |
|---|---|---|---|
| | Hydrangea chlorotic mottle virus | HCMV | DQ412999 |
| | Impatiens latent virus | ILV | |
| | Lilac ringspot virus | LiRSV | |
| | Narcissus symptomless virus | NSV | NC_008552 |
| | Plantain virus 8 | PIV-8 | |
| | Potato virus P | PVP | DQ516055 |
| | Prunus virus S | PruVS | |
| | Southern potato latent virus | SoPLV | |
| | White bryony mosaic virus | WBMV | |

as tentative species have no sequence data, and therefore the taxonomic status of these members is yet to be verified.

Carlaviruses are either closely or distantly related to each other and these relationships have been confirmed by phylogenetic analysis of some viral proteins. The phylogenetic tree analysis of sequenced carlaviruses is presented in **Figure 1**.

Species demarcation criteria in this genus include the following:

1. distinct species have less than about 72% nucleotides or 80% amino acids identical between their coat protein (CP) or replicase genes;
2. distinct species are readily differentiated by serology, and strains of individual species are often distinguishable by serology;
3. distinct species do not cross-protect in infected common host plant species, and each distinct species usually has a specific natural host range and distinguishable experimental host ranges.

## Virus Structure and Composition

Virions are slightly flexuous filaments that are not enveloped, and measure 610–700 nm in length and 12–15 nm in diameter. Nucleocapsids appear longitudinally striated. Virions have helical symmetry with a pitch of 3.3–3.4 nm. Virion $M_r$ is about $60 \times 10^6$ with a nucleic acid content of 5–7%. Virions contain a monopartite, linear, single-stranded, positive-sense RNA that has a size range of 7.4–9.1 kbp in length. The 3′ terminus has a poly(A) tract and the 5′ terminus occasionally has a methylated nucleotide cap, or a monophosphate group. CP subunits are of one type, with size 31–36 kDa. The genome is encapsidated in 1600–2000 copies of CP subunit. Purified preparations of carlaviruses such as potato virus S and helinium virus S (PVS, HVS) contain small amounts of encapsidated subgenomic RNAs.

## Physicochemical Properties

Virus particles in purified preparations have a sedimentation coefficient of 147–176 S. Isoelectric point of virions is about pH 4.5. The UV absorbance spectra of carlaviruses have maxima at 258–260 nm and minima at 243–248 nm, with $A_{max}/A_{min}$ ratios of 1.1–1.3, and the $A_{260}/A_{280}$ ratio of purified preparation is 1.08–1.40. Physical properties of viruses in this genus are: thermal inactivation point (TIP) 50–85 °C, longevity *in vitro* (LIV) 1–21 days, and dilution endpoint (DEP) $10^{-2}$–$10^{-7}$. Infectivity of sap does not change on treatment with diethyl ether. Infectivity is retained by the virions despite being deproteinized by proteases and by phenols or detergents.

The CPs of the carlaviruses are partly degraded during purification and storage of virus preparations. The carlavirus virions are susceptible to and can be broken down by denaturing agents such as sodium dodecyl sulfate (SDS), urea, guanidine hydrochloride, acetic acid, and alkali.

## Genome Structure and Gene Expression

The genome is a single-stranded RNA (ssRNA), 7.4–9.1 kbp in size, which contains six open reading frames (ORFs), encoding the replicase, three putative protein components of movement proteins called triple gene block (TGB), the CP, and a putative nucleic acid-binding regulatory protein, from the 5′–3′-end in that order (**Figure 2**). In genome organization, the genus *Carlavirus* particularly resembles the genera *Allexivirus, Foveavirus, Mandarivirus,* and *Potexvirus* but is distinguished from them by its six ORFs and a large replication protein. The genome RNA is capped at the 5′ end and polyadenylated at the 3′ end. The 223 kDa viral replicase ORF is translated from the full-length genomic RNA. For expression of its 3′-proximal viral genes, the virus utilizes at least two subgenomic RNAs. The genome structure of potato virus M (PVM) (8553 nt) is represented in **Figure 2**. PVM RNA has 75 nt 5′ untranslated region (UTR) at the 5′ terminus and 70 nt 3′ UTR followed by

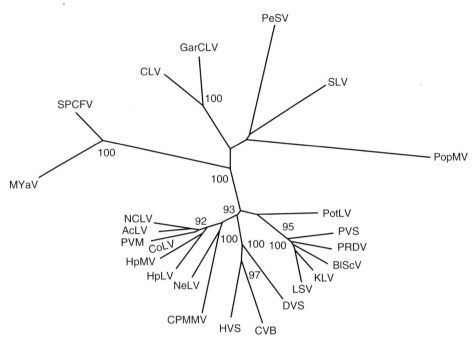

**Figure 1** Phylogenetic tree of carlaviruses derived from multiple-aligned deduced amino acid sequences of CP. See **Table 1** for abbreviations of carlavirus names.

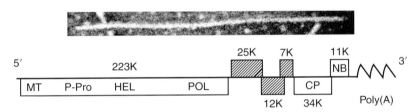

**Figure 2** Particle morphology and genome organization of potato virus M (PVM), showing a typical genome structure for the genus Carlavirus in the family Flexiviridae. The 5′-proximal one large ORF encodes an RNA-dependent RNA polymerase (viral replicase), three overlapping ORFs encode the putative MPs (TGBs), and two ORFs encode CP and nucleic acid binding protein (NB). Motifs in the replicase are methyltransferase (MT), papain-like protease (P-Pro), helicase (HEL), and RNA-dependent RNA polymerase (POL).

a poly(A) tail at the 3′ terminus and intergenic UTRs of 38 and 21 nt between three large blocks of coding sequences. Carlavirus-infected plants contain double-stranded RNA (dsRNA) whose molecular mass corresponds to that of genomic RNA.

ORF1 encodes a 223 kDa polypeptide that is presumed to be the viral replicase. It contains motifs of methyltransferase, a papain-like protease, helicase, and RNA-dependent RNA polymerase. With blueberry scorch virus (BLScV) and probably other carlaviruses, the 223 kDa viral replicase is proteolytically processed by papain-like protease activity, which results in about 30 kDa being removed. ORFs 2–4 form the TGB encoding polypeptides of 25, 12, and 7 kDa, respectively, which facilitate virus cell-to-cell movement. ORF5 encodes the 34 kDa CP, and ORF6 encodes a cysteine-rich 11 kDa protein. The function of the 3′-proximal 11 kDa polypeptide has yet to be elucidated, but its ability to bind nucleic acid and agroin-filtration-mediated transient expression studies suggest that it may facilitate vector transmission, may be involved in viral RNA transcription, or may be a viral pathogenicity determinant for plant defense system.

Putative promoters for the subgenomic RNA synthesis have been identified for several carlaviruses, C/UUUAGGU, 19–43 residues upstream from both putative subgenomic RNA initiation sites.

Complete genomic sequences have recently been obtained for some carlaviruses: BLScV (8512 nt), chrysanthemum virus B (CVB) (8870 nt), daphne virus S (DVS) (8739 nt), hop latent virus (HpLV) (8612 nt), lily symptomless virus (LSV) (8394 nt), PVM (8553 nt), PVS (8478 nt), shallot latent virus (SLV) (8363 nt), aconitum latent virus (AcLV) (8657 nt), poplar mosaic virus (PopMV) (8741 nt), passiflora latent virus (PLV) (8386 nt), narcissus symptomless virus (NSV) (8281 nt), sweetpotato chlorotic fleck virus (SPCFV) (9104 nt), and narcissus common latent virus (NCLV) (8539 nt).

Infectious cDNA clones have been reported for two carlaviruses, BlScV and PopMV.

## Viral Transmission

Carlaviruses are spread by mechanical contact, by horticultural and agricultural equipment, and transmitted in a nonpersistent or semipersistent manner by aphids with varying efficiencies, or by whiteflies. Some members are not transmitted by mechanical inoculation. Seed transmission may occur with some legume-infecting carlaviruses, but is not common. Transmission by root grafts may occur in the case of PopMV. Those viruses that infect vegetatively propagated hosts persist in the propagated parts.

## Cytopathology

The distribution of carlaviruses in the infected plant is not tissue specific. Virions are usually found in the cytoplasm, or sometimes in chloroplasts or in mitochondria of infected tissues, or may occur in membrane-associated bundle-like or plate-like aggregates. Inclusions are present in infected cells of some members. They occur as crystals in the cytoplasm, as amorphous X-bodies, membranous bodies, and viroplasms and they sometimes contain virions.

## Host Range

Infection by most members of this genus is symptomless in the natural host. The individual natural host ranges are, usually narrow, although a few species can infect a wide range of experimental hosts.

## Symptomatology

Carlavirus infection often results in no apparent symptoms in natural hosts. Symptoms vary cyclically or seasonally or may disappear soon after infection. If symptoms are evident, more severe symptoms such as mosaics appear in early stages of infection. Some carlaviruses such as PVM, PVS, and BlScV cause diseases that are of economic importance on their own; however, most of them are associated with more serious diseases when the plants are co-infected with other viruses.

## Serology

Carlavirus virions are good immunogens. Serological relationships among carlaviruses may be close or distant or sometimes undetectable: some species are serologically interrelated but others apparently distinct. A number of carlaviruses are more or less closely interrelated serologically, with serological differentiation indices (SDIs) ranging from about 3.5 to 6.5.

## Geographical Distribution

Some carlaviruses are found wherever their natural hosts are grown, but the geographical distribution of many species is restricted to only certain parts of the world. Those infecting vegetatively propagated crops are usually widely distributed. Most species commonly occur in temperate regions, but whitefly transmitted carlaviruses are restricted to tropical and subtropical regions.

## Viral Epidemiology and Control

Most carlavirus-associated diseases are usually very mild or symptomless and few efforts are made to control them. However, crops such as potatoes, certain legumes, and blueberries that may contain more damaging carlaviruses require suitable control measures. Seed potatoes and vegetatively propagated materials must be screened continuously to certify a virus-free status. Rapid removal of infected plants is particularly important for plants associated with aphid- or whitefly-vectored carlaviruses. PVS is the only carlavirus for which transgenic plants (potato and *Nicotiana debney*), which are resistant to virus infection, have been reported so far.

See also: Allexivirus; *Capillovirus, Foveavirus, Trichovirus, Vitivirus*; Flexiviruses; Plant Virus Vectors (Gene Expression Systems); *Potexvirus*; Vector Transmission of Plant Viruses.

## Further Reading

Adams MJ, Antoniw JF, Bar-Joseph M, et al. (2004) The new plant virus family *Flexiviridae* and assessment of molecular criteria for species demarcation. *Archives of Virology* 149: 1045–1060.

Fauquet CM, Mayo MA, Maniloff J, Desselberger U, and Ball LA (eds.) (2005) *Virus Taxonomy, Classification and Nomenclature of Viruses, Eighth Report of the International Committee on the Taxonomy of Viruses*, p. 1101. San Diego, CA: Elsevier Academic Press.

Hillman BI and Lawrence DM (1994) Carlaviruses. In: Singh RP, Singh US, and Kohmoto K (eds.) *Pathogenesis and Host Specificity in Plant Diseases*, vol. 3, 35pp. New York: Plenum.

Lee BY, Min BE, Ha JH, Lee MY, Paek KH, and Ryu KH (2006) Genome structure and complete sequence of genomic RNA of Daphne virus S. *Archives of Virology* 151: 193–200.

Martelli GP, Adams MJ, Kreuze JF, and Dolja VV (2007) Family Flexiviridae: A case study in virion and genome plasticity. *Annual Review of Phytopathology* 45. 73–100.

# Carmovirus

**F Qu and T J Morris,** University of Nebraska, Lincoln, NE, USA

© 2008 Elsevier Ltd. All rights reserved.

## Taxonomy, Classification and Evolutionary Relationships

*Carmovirus* is one of eight genera in the family *Tombusviridae*. Members of this family all have icosahedral virions of about 30 nm in diameter with $T = 3$ symmetry that consists of 180 coat protein (CP) subunits of about 38–43 kDa and a single-stranded (ss) RNA genome ranging in size from 4.0 to 4.8 kbp. Carmoviruses share recognizable yet varied sequence similarity with members of other genera of *Tombusviridae*.

Carmoviruses contain a single-component positive-sense genome of about 4.0 kbp. The genome, as exemplified by turnip crinkle virus (TCV) in **Figure 1**, consists of five definitive open reading frames (ORFs) which encode proteins of about 28, 88, 8, 9, and 38 kDa from the 5′ to the 3′ end, respectively. Some carmoviruses such as hibiscus chlorotic ringspot virus (HCRSV) have additional ORFs of unknown functions. The virions are icosahedral in symmetry and consist of 180 CP subunits of approximately 38 kDa. The genus name is derived from the first member of the genus to be sequenced, carnation mottle virus (CarMV). Much more detailed knowledge about virus structure and genome function is, however, known for TCV because its crystal structure has been determined and it was the first carmovirus for which infectious transcripts were produced from a cDNA clone of the genome.

To date, the nucleotide sequences of 12 definitive carmoviruses have been determined (see **Table 1**). These sequenced members share similar morphological and physicochemical properties with about a dozen other viruses listed in **Table 1** that are recognized as species or tentative species depending on the level of detail of the molecular characterization of the viral genomes. Various carmoviruses are sufficiently distant from each other that they do not cross-react in standard RNA hybridization or serological tests.

Carmoviruses share properties with viruses belonging to other genera of the family *Tombusviridae*. Their particle structure and CP sequences are closely related to tombus-, aureus-, diantho-, and avenaviruses. Their RNA-dependent RNA polymerase (RdRp) genes share significant homology with viruses of the following genera: *Machlomovirus*, *Panicovirus*, *Necrovirus*, *Aureusvirus*, and *Tombusvirus*. Carmovirus RdRp genes also share similarity with more distantly related viruses outside of *Tombusviridae*, such as umbraviruses and luteoviruses. In a broader context, phylogenetic comparisons of viral RNA polymerase genes have identified the *Tombusviridae* as a representative plant virus cluster for one of three RNA virus supergroups with relatedness to animal viruses of *Flaviviridae* and small RNA phage (*Leviviridae*).

## Distribution, Host Range, Transmission, and Economic Significance

Carmoviruses occur worldwide and are generally reported to cause mild or asymptomatic infections on relatively restricted natural host ranges. Most accumulate to high concentrations in infected tissues and are mechanically transmitted. Beetle transmission has been reported for some carmoviruses as has transmission in association with soil and/or irrigation water, and in some cases in association with fungal zoospores.

A number of carmoviruses have been identified in association with ornamental hosts and have been widely distributed in such hosts by vegetative propagation. CarMV is the most noteworthy, being widespread in cultivated carnations, and recognized as one of the more important components of viral disease complexes in this crop worldwide. It accumulates to high concentrations without producing severe symptoms and spreads primarily by contact transmission and vegetative propagation. It has a broad experimental host range that includes over 30 species in 15 plant families. Pelargonium flower break virus (PFBV) is widespread in vegetatively propagated *Pelargonium* species causing disease in association with other viruses. The incidence of narcissus tip necrosis virus (NTNV) in narcissus cultivars and HCRSV in hibiscus primarily reflects distribution of infected nursery stock.

Numerous small RNA viruses have been reported to naturally infect cucurbits causing significant disease problems. Several of these viruses are recognized tombusviruses while others such as melon necrotic spot virus (MNSV) and cucumber soil-borne virus (CSBV) have been identified as carmoviruses based on sequence and genome organization properties. MNSV occurs worldwide in greenhouse cucurbits and is both soil and seed transmitted, while CSBV has been primarily restricted to infrequent outbreaks around the Mediterranean. Both have been reported to be transmitted in association with the fungus *Olpidium bornovanus*.

Several carmoviruses have been discovered in natural leguminous hosts, with glycine mottle virus (GMoV) being potentially the most important, causing serious disease losses in legumes in Africa. Bean mild mosaic virus (BMMV) has been reported to be a latent virus

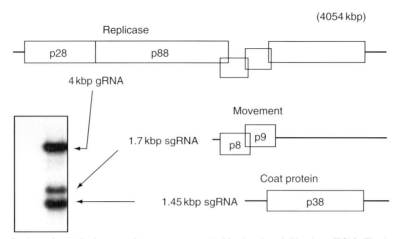

**Figure 1** Genome organization of a typical carmovirus as represented by turnip crinkle virus (TCV). The boxes represent open reading frames with the sizes of the encoded proteins indicated within the boxes, in kilodaltons. The two proteins required for replication (p28 and p88) are translated from the 4 kbp genomic RNA with the p88 protein translated by readthrough of an amber codon at the end of the p28 gene. Two small proteins involved in cell-to-cell movement (p8 and p9) are translated from a 1.7 kbp subgenomic RNA. The viral coat protein (p38) is translated from a 1.45 kbp subgenomic RNA. The panel on the left is a Northern blot showing the typical pattern of accumulation of genomic and subgenomic RNAs in carmovirus-infected cells.

**Table 1** Virus members in the genus *Carmovirus*

| | |
|---|---|
| *Sequenced viruses* | |
| Angelonia flower break virus | (AnFBV) |
| Cardamine chlorotic fleck virus | (CCFV) |
| Carnation mottle virus | (CarMV) |
| Cowpea mottle virus | (CPMoV) |
| Galinsoga mosaic virus | (GaMV) |
| Hibicus chlorotic ringspot virus | (HCRSV) |
| Japanese iris necrotic ring virus | (JINRV) |
| Melon necrotic spot virus | (MNSV) |
| Pea stem necrosis virus | (PSNV) |
| Pelargonium flower break virus | (PFBV) |
| Saguaro cactus virus | (SCV) |
| Turnip crinkle virus | (TCV) |
| *Unsequenced viruses* | |
| Ahlum water-borne virus | (AWBV) |
| Bean mild mosaic virus | (BMMV) |
| Cucumber soil-borne virus | (CSBV) |
| Weddel water-borne virus | (WWBV) |
| *Viruses assigned to tentative species* | |
| Blackgram mottle virus | (BMoV) |
| Calibrachoa mottle virus | (CbMV) |
| Elderberry latent virus | (ELV) |
| Glycine mottle virus | (GMoV) |
| Narcissus tip necrosis virus | (NTNV) |
| Plaintain virus 6 | (PlV-6) |
| Squash necrosis virus | (SqNV) |
| Tephrosia symptomless virus | (TeSV) |

It has a relatively wide experimental host range in some 20 plant families including experimentally useful species such as *Arabidopsis* and *Brassica* in which it accumulates to extremely high concentrations, often approaching a level equivalent to 0.5% of the fresh weight of the plant tissue. Cardamine chlorotic fleck virus (CCFV) was first discovered in the Mount Kosiusko alpine region of Australia in *Cardamine lilacina*, a wild perennial of the family Brassicaceae. It has also been shown to infect *Arabidopsis* and other Brassicaceae species.

Other carmoviruses have been isolated worldwide from natural hosts with little apparent disease and are presumably of little agricultural concern. These include tephrosia symptomless virus (TeSV) from legumes in Kenya, GMoV from glycine in Australia, saguaro cactus virus (SCV) from saguaro cactus in Arizona galinsoga mosaic virus (GaMV) from potato weed in Australia, and plaintain virus 6 (PlV-6) from plantain weed in England. The infrequent isolation of these genetically similar viruses in remote locations around the world has prompted the speculation that ancestor carmoviruses may have been introduced into their natural hosts well before the last Ice Age and have since co-evolved in isolation in their diverse host plants.

widely distributed in bean cultivars in El Salvador, and backgram mottle virus (BMoV) has been found in *Vigna* species in Asia. Beetle vectors have been identified for these viruses, but seed transmission may also be an important factor in their distribution.

TCV is neither common nor widespread in nature in spite of the fact that it is reportedly beetle transmitted.

## Virion Structure and Assembly

Four carmoviruses (TCV, CarMV, cowpea mottle virus (CPMoV), and HCRSV) have now been analyzed in structural detail, with TCV being the first and most thoroughly studied by high-resolution X-ray crystallography. The detailed information about CP structure and intersubunit interactions established that TCV and tomato bushy stunt tombusvirus (TBSV) show marked structural conservation. In this regard, the common structural features shared by other members of the family *Tombusviridae* have been primarily deduced from alignment of the amino acid sequence of the coat proteins of TBSV and TCV. TCV consists of a $T=3$ icosahedral capsid of 180 subunits of the 38 kDa CP. The individual CP subunit folds into three distinct domains typical of CP subunit of tombusviruses. The relatively basic N-terminal R domain extends into the interior of the virus particle and presumably interacts with viral RNA. The R domain is connected by an arm to the S domain which constitutes the virion shell. The S domain is attached through a hinge to the P domain which projects outward from the virion surface. The protein subunits are believed to form dimers in solution and during assembly.

TCV is also the only carmovirus on which detailed *in vitro* assembly studies have been performed. The virion has been shown to dissociate at elevated pH and ionic strength to produce a stable RNA–CP complex (rp-complex) and free CP subunits. Reassembly under physiological conditions in solution could be demonstrated using the isolated rp-complex and the soluble CP subunits. This rp-complex, consisting of six CP subunits tightly attached to viral RNA, could be generated *in vitro* and was shown to be important in selective assembly of TCV RNA. A model for assembly was proposed in which three sets of dimeric CP interact with a unique site on the viral RNA to form an initiation complex to which additional subunit dimers could rapidly bind. Preliminary characterization of the 'origin of assembly' for this virus identified two possible sites based on the identification of RNA fragments in the rp-complex protected from RNase digestion by CP. Further *in vivo* studies narrowed the assembly origin site to a bulged hairpin-loop of 28 nt within a 180 nt region at the 3' end of the CP gene.

## Genome Structure

Complete nucleotide sequences have been determined for 12 carmoviruses as listed in **Table 1**. Comparative studies of the deduced ORFs revealed that all of these viruses encode a similar set of genes that are closely related and in the same gene order as illustrated for TCV in **Figure 1**. The genome organization of the carmoviruses is quite compact with most of the identified ORFs overlapping each other. Both the product of the most 5' proximal ORF (26–28 kDa) and its readthrough product (86–89 kDa) are essential for replication of the TCV genome. The 3' proximal gene encodes the viral CP which varies from 37 to 42 kDa for the different viruses. All of the sequenced carmoviruses characteristically encode two small ORFs in the middle of the genome that both have been shown in TCV to be indispensable for cell-to-cell movement (movement proteins or MP). Although the genome organizations of all sequenced carmoviruses are quite similar, there are some unique features evident in the individual carmoviruses. For example, the two small central ORFs in MNSV (p7a and p7b) are connected by an in-frame amber codon that could result in the production of a 14 kDa fusion protein of the two ORFs by a readthrough mechanism. The CPMoV as well as HCRSV are predicted to encode a sixth ORF nested within the 3' proximal CP gene. HCRSV has also been reported to encode another novel protein that is nested within the RdRp gene.

The 5' end of the genome is not capped. The 5' noncoding region varies from 34 nt in CPMoV to 88 nt in MNSV. No extensive sequence homology was observed within this region. The 3' noncoding region of carmoviruses varies from about 200 to 300 nt in length and possesses neither a poly-A tail nor a tRNA-like structure.

## Replication and Gene Expression

Carmoviruses replicate to very high concentrations in protoplasts, with the genomic RNA accumulating to levels approaching that of the ribosomal RNAs. Upon infection of susceptible plants, carmoviruses transcribe two 3' coterminal subgenomic RNAs (sgRNAs) for expression of the MP and CP genes. The smaller sgRNA (*c.* 1.5 kbp) is the mRNA for CP. The larger sgRNA (*c.* 1.7 kbp) presumably functions as the mRNA for the two MP genes utilizing a leaky scanning mechanism. Results involving transgenic expression of the p8 and p9 gene products of TCV in *Arabidopsis* plants have demonstrated that both of them are essential for viral cell-to-cell movement and that they function by in *trans*-complementation in the same cell.

Carmoviruses are thought to replicate through a (−) strand intermediate because virus-specific double-stranded RNAs (dsRNAs) corresponding in size to the genomic RNA and sgRNAs characteristically accumulate in infected plant tissue. The product of the 5' proximal ORF (p28 in TCV) and its readthrough product (p88 in TCV) are the only virus-encoded components of the polymerase complex. When expressed from two separate mRNAs, p28 and p88 complemented *in trans* to enable the genome replication. The only host factor found to augment TCV replication so far is the eukaryotic translation initiation factor 4G (eIF4G) of *Arabidopsis*, presumably through more efficient translation of viral genes. Recently, a membrane-containing extract prepared from evacuolated protoplasts of uninfected *Arabidopsis* plants has been shown to faithfully produce both genomic and subgenomic RNAs from a full-length TCV RNA template. Such an extract should be useful in characterizing viral as well as host elements required for virus replication.

Recent studies have demonstrated that, besides the virus-encoded proteins, a plethora of structural elements in the viral RNA play critical roles in the genome replication of carmoviruses. These elements are located throughout the entire viral genome and occur in both the (+) and (−) strands. Their roles range from promoters, enhancers, or repressors for RNA replication and transcription, enhancers for translation, and specificity determinants for virus assembly. The current knowledge of these RNA structures suggests that different secondary structural motifs, some of them mutually exclusive, are formed at the different stages of virus multiplication, and the highly coordinated nature of their formation ensures optimal utilization of the compact viral genome.

## Satellites, Defective-Interfering RNAs

TCV is the only carmovirus in which replication of associated small subviral RNAs in infected plants has been characterized, and the situation for this virus is curiously complex. TCV infections are associated with defective interfering (DI) RNAs derived totally from the parent genome (e.g., RNA G of 342–346 nt), satellite RNAs of nonviral origin (e.g., RNAs D, 194 nt and F, 230 nt), and chimeric RNAs (e.g., RNA C of 356 nt) with a 5′ region derived from sat RNA D and a 3′ region derived from the 3′ end of the TCV genome. All three types of small RNAs depend on the helper virus for their replication and encapsidation within the infected plant. The different satellite and DI RNAs have been shown to affect viral infections in different ways. Both RNA C and G intensify viral symptoms while interfering with the replication of the helper virus, while RNAs D and F seem to produce no detectable effects on either expression of symptoms or helper virus replication.

## Virus–Host Interaction

Considerable progress has been made in the last decade in our understanding of the molecular mechanisms of virus–plant interactions with significant contributions coming from studies utilizing TCV and its host plant *Arabidopsis*. Most notably, one ecotype of *Arabidopsis thaliana* has been determined to be resistant to TCV infection and the resistance gene (*HRT*) and its encoded protein have been characterized. Additional host factors that contribute to the resistance response have also been elucidated including the identification of a novel transcription factor (TIP) whose interaction with TCV CP is critical for the initiation of the resistance response. The determination that TCV CP is targeted by HRT resistance protein suggests that TCV CP plays a key role in combating antiviral defense mechanisms of the plant host, in addition to being required for virus assembly. Interestingly, recent studies have also established that TCV CP is a strong suppressor of RNA silencing, a host defense mechanism that targets invading RNA. RNA silencing is a potent defense mechanism that is conserved in nearly all eukaryotic organisms. Accordingly, TCV CP is an effective silencing suppressor in both plant and animal cells. Silencing suppressor activity has also been associated with several other carmovirus CPs, establishing this activity as a conserved feature of the structural subunit.

TCV CP is also important in satellite RNA C interactions in the host plant. Normally, the presence of RNA C results in symptom intensification in TCV infections. However, when the TCV CP ORF is either deleted or replaced by the CCFV CP ORF, RNA C attenuates symptoms caused by the helper virus suggesting that CP either downregulates the replication of RNA C or enhances its own competitiveness.

Finally, the replicase gene has also been implicated in the symptom modification by satellite RNA C by two independent groups. The 3′ end of the TCV genome, a sequence common in TCV RNA, RNA C, and DI RNA G, was also suggested to be a symptom determinant. Environmental conditions also affect the extent of resistance of *Arabidopsis* plants to TCV. In conclusion, it is clear that recent intensive studies of TCV have established this small RNA virus as an ideal model for unraveling the complicated processes involved in viral pathogenesis.

*See also:* Legume Viruses; Luteoviruses; Machlomovirus; *Necrovirus*; Plant Virus Diseases: Ornamental Plants; Tombusviruses.

## Further Reading

Hacker DL, Petty ITD, Wei N, and Morris TJ (1992) Turnip crinkle virus genes required for RNA replication and virus movement. *Virology* 186: 1–8.

Kachroo P, Yoshioka K, Shah J, Dooner HK, and Klessig DF (2000) Resistance to turnip crinkle virus in *Arabidopsis* is regulated by two host genes, is salicylic acid dependent but NPR1, ethylene and jasmonate independent. *Plant Cell* 12: 677–690.

Komoda K, Naito S, and Ishikawa M (2004) Replication of plant RNA virus genomes in a cell-free extract of evacuolated plant protoplasts. *Proceedings of the National Academy of Sciences, USA* 101: 1863–1867.

Lommel SA, Martelli GP, Rubino L, and Russo M (2005) *Tombusviridae*. In: Fauquet CM, Mayo MA, Maniloff J, Desselberger U and Ball LA (eds.) *Virus Taxonomy: Eighth Report of the International Committee on Taxonomy of Viruses*, pp. 907–936. San Diego, CA: Elsevier Academic Press.

Morris TJ and Carrington JC (1988) Carnation mottle virus and viruses with similar properties. In: Koenig R (ed.) *The Plant Viruses: Polyhedral Virions with Monopartite RNA Genomes*, vol. 3, pp. 73–112. New York: Plenum.

Nagy PD, Pogany J, and Simon AE (1999) RNA elements required for RNA recombination function as replication enhancers *in vitro* and *in vivo* in a plus-strand RNA virus. *EMBO Journal* 18: 5653–5665.

Qu F and Morris TJ (2005) Suppressors of RNA silencing encoded by plant viruses and their role in viral infections. *FEBS Letters* 579: 5958–5964.

Ren T, Qu F, and Morris TJ (2000) HRT gene function requires interaction between a NAC protein and viral capsid protein to confer resistance to turnip crinkle virus. *Plant Cell* 12: 1917–1925.

Russo M, Burgyan J, and Martelli GP (1994) Molecular biology of *Tombusviridae*. *Advances in Virus Research* 44: 381–428.

Yoshii M, Nishikiori M, Tomita K, *et al.* (2004) The *Arabidopsis cucumovirus multiplication 1* and *2* loci encode translation initiation factors 4E and 4G. *Journal of Virology* 78: 6102–6111.

# Caulimoviruses: General Features

**J E Schoelz,** University of Missouri, Columbia, MO, USA

© 2008 Elsevier Ltd. All rights reserved.

## Glossary

**Agroinoculation** An inoculation technique in which *Agrobacterium tumefaciens* is used to deliver a full-length infectious clone of a virus into a plant cell.

**Plasmodesmata** A narrow channel of cytoplasm that functions as a bridge between two plant cells to facilitate movement of macromolecules.

**Reverse transcriptase** An enzyme that utilizes an RNA template for synthesis of DNA.

**Ribosome shunt** A translational mechanism in which ribosomes enter the RNA at the 5′ end and scan for a short distance before being translocated to a downstream point.

**Semipersistent transmission** Vector acquires the virus in minutes to hours, and can transmit to other plants for hours after the initial feeding.

**Transgene** A gene introduced into an organism through any one of a number of genetic engineering techniques.

## Introduction

The members of the family *Caulimoviridae* are plant viruses that replicate by reverse transcription of an RNA intermediate and whose virions contain circular, double-stranded DNA (dsDNA). They replicate by reverse transcription, but unlike the true retroviruses, integration into the host chromosomes is not required for completion of their replication cycle. The circular DNA encapsidated in virions is not covalently closed, as it contains at least one discontinuity in each DNA strand, and these discontinuities occur as a consequence of the replication strategy of the virus. There are six genera in this family and they can be divided into two groups based on virion morphology; the members of the genera *Caulimovirus*, *Petuvirus*, *Cavemovirus*, and *Soymovirus* contain viruses that form icosahedral particles that are largely found within amorphous inclusion bodies in the cell (**Figures 1(a)** and **1(b)**). In contrast, the members of the genera *Badnavirus* and *Tungrovirus* form bacilliform particles that are not associated with inclusion bodies (**Figure 1(c)**). Their virions are found in the cytoplasm either individually or clustered in palisade-like arrays.

Cauliflower mosaic virus (CaMV), a member of the type species *Cauliflower mosaic virus* of the genus *Caulimovirus*, was the first of the plant viruses to be shown to contain dsDNA in its icosahedral virion by Shepherd in 1968. This discovery led to an extended investigation into its replication strategy throughout the 1970s and early 1980s, culminating in Pfeiffer and Hohn's study in 1983 that showed that CaMV replicates through reverse transcription of an RNA intermediate. Perhaps because the genome of CaMV is composed of dsDNA, it was also the first of the plant viruses to be completely sequenced and cloned into bacterial plasmids in an infectious form.

In the early 1980s, CaMV was thought to have some promise as a vector for foreign genes in plants. However, the effort to convert CaMV into a vector was scaled back when it was shown that the virus genome could tolerate only small insertions of up to a few hundred basepairs of DNA. A few small genes, such as dihydrofolate reductase and interferon, were eventually expressed in plants via a CaMV vector, but other viruses have been shown to be much more versatile as vectors for foreign genes. Although the caulimoviruses have had only limited utility as plant virus vectors, they continue to have a great impact on plant biotechnology. The 35S promoter of CaMV is capable of directing a high level of transcription in most types of plant tissues. This promoter was used to drive expression of one of the first transgenes introduced into transgenic plants, and it is still widely used for expression of transgenes for both research and commercial applications. The promoter regions of several other

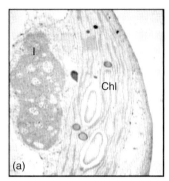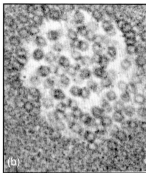

**Figure 1** Inclusion bodies and virions of the family *Caulimoviridae*. (a) Amorphous inclusion body of CaMV (I) adjacent to a chloroplast (Chl). Individual virions can be seen in the vacuolated regions of the inclusion body. (b) Icosahedral virions of CaMV visualized within an inclusion body. (c) Purified bacilliform virions of a badnavirus. (a) Reproduced with permission from the *Encyclopedia of Virology*, 2nd edn. (c) Photo courtesy of Ben Lockhart (University of Minnesota).

caulimoviruses have also been evaluated for expression of transgenes in both monocots and dicots, and they can be good alternatives to the CaMV 35S promoter.

The feature that was used initially to characterize the caulimoviruses was the presence of circular dsDNA encapsidated into icosahedral virions of approximately 50 nm diameter. However, as icosahedral viruses continued to be isolated from a variety of hosts and their genomes sequenced, it was discovered that their genome structure diverged from that of CaMV. These differences were significant enough for the International Committee on Taxonomy of Viruses (ICTV) to create three new genera in its VIII ICTV Report. Consequently, *Soybean chlorotic mottle virus*, *Cassava vein mosaic virus*, and *Petunia vein clearing virus* became the type species of the genera *Soymovirus*, *Cavemovirus*, and *Petuvirus*, respectively.

For many years, only plant viruses that had icosahedral virions of 50 nm in diameter were thought to have genomes composed of dsDNA. However, in 1990, Lockhart showed that the bacilliform virions of commelina yellow mottle virus (CoYMV) contained circular, dsDNA, and *Commelina yellow mottle virus* became the type species for the genus *Badnavirus*. Lockhart showed that nucleic acid isolated from the bacilliform virions was resistant to RNase and degraded by DNase. Furthermore, DNA treated with S1 nuclease revealed that the CoYMV genome contained at least two single-stranded discontinuities, a hallmark of the plant viruses that replicate by reverse transcription. In addition to CoYMV, Lockhart also purified DNA from banana streak virus, kalanchoe top-spotting virus (KTSV), and canna yellow mottle virus, providing evidence that they should also be placed in the new genus *Badnavirus*. There are now 18 species in the genus *Badnavirus* (**Table 1**), many of which cause economically important diseases in the tropics. However, new badnaviruses have been characterized since the publication of the VIII Report of the ICTV, so this number will almost certainly be revised upward in the near future.

Soon after the discovery that CoYMV was a DNA virus, the bacilliform component of rice tungro disease was also shown to contain circular, dsDNA. Rice tungro is the most important virus disease of rice in South and Southeast Asia, with annual losses approaching $1.5 billion dollars. The disease is caused by a complex of an RNA virus, called rice tungro spherical virus (RTSV) coupled with the bacilliform dsDNA virus called rice tungro bacilliform virus (RTBV). The genus *Tungrovirus* is distinguished from the genus *Badnavirus* because RTBV has one more open reading frame (ORF) than CoYMV and overall, the RTBV genome has only 20–25% nucleotide sequence identity with members of the genus *Badnavirus*. There is only one species in the genus *Tungrovirus*, although several isolates of the species *Rice tungro bacilliform virus* have been collected and sequenced.

## Taxonomy and Classification

The family *Caulimoviridae* consists of six genera (**Table 1**), and they can be conveniently divided into two groups based on virion morphology. Viruses with icosahedral virions (**Figure 1(b)**) include the genera *Caulimovirus*, *Petuvirus*, *Soymovirus*, and *Cavemovirus*. Viruses with bacilliform virions (**Figure 1(c)**) include the genera *Badnavirus* and *Tungrovirus*. Genera can be further distinguished because of differences in genome organization and nucleotide differences in common genes such as the reverse transcriptase. All members of the family *Caulimoviridae* infect only plants. There are no animal or insect viruses in this family.

## Virion Structure and Composition

The viruses in the genera *Caulimovirus*, *Cavemovirus*, *Soymovirus*, and *Petuvirus* form nonenveloped, isometric particles that vary in size from 43 to 50 nm. The virion is

**Table 1** Virus members in the family *Caulimoviridae*

| Genus | Virus | Abbreviation | Host (family) | Geographic distribution | Representative accession number |
|---|---|---|---|---|---|
| *Caulimovirus* | Carnation etched ring virus | CERV | Carnation (Caryophyllaceae) | Worldwide | X04658 |
| | **Cauliflower mosaic virus**[a] | **CaMV** | **Brassica sp. (Crucifereae)** | **Worldwide** | **V00140** |
| | Dahlia mosaic virus | DMV | Dahlia (Compositae) | Worldwide | |
| | Figwort mosaic virus | FMV | Figwort (Scrophulariaceae) | USA | X06166 |
| | Horseradish latent virus | HRLV | Horseradish (Crucifereae) | Denmark | |
| | Mirabilis mosaic virus | MiMV | Mirabilis (Nyctaginaceae) | USA | AF454635 |
| | Strawberry vein banding virus | SVBV | Strawberry (Rosaceae) | Worldwide | X97304 |
| | Thistle mottle virus | ThMoV | Thistle (Compositae) | Europe | |
| | Aquilegia necrotic mosaic virus[b] | ANMV | Columbine (Ranunculaceae) | East Asia, Japan | |
| | Plantago virus 4[b] | PlV-4 | Plantain (Plantaginaceae) | United Kingdom | |
| | Sonchus mottle virus[b] | SMoV | *Sonchus* sp. (Asteraceae) | | |
| *Petuvirus* | **Petunia vein clearing virus** | **PVCV** | **Petunia (Solanaceae)** | **Worldwide** | **U95208** |
| *Soymovirus* | Blueberry red ringspot virus | BRRSV | Blueberry (Ericaceae) | USA | AF404509 |
| | Peanut chlorotic streak virus | PCSV | Peanut (Leguminosae) | India | U13988 |
| | **Soybean chlorotic mottle virus** | **SbCMV** | **Soybean (Leguminosae)** | **Japan** | **X15828** |
| | Cestrum yellow leaf curling virus[b] | CmYLCV | *Cestrum* sp. (Solanaceae) | Italy | AF364175 |
| *Cavemovirus* | **Cassava vein mosaic virus** | **CsVMV** | **Cassava (Euphorbiaceae)** | **Brazil** | **U59751** |
| | Tobacco vein clearing virus | TVCV | *Nicotiana* sp. (Solanaceae) | Worldwide | AF190123 |
| *Badnavirus* | Aglaonema bacilliform virus | ABV | *Aglaonema* sp. (Araceae) | Southeast Asia | |
| | Banana streak GF virus | BSGFV | Banana (Musaceae) | Worldwide | AY493509 |
| | Banana streak MYsore virus | BSMyV | Banana (Musaceae) | Worldwide | AY805074 |
| | Banana streak OL virus | BSOLV | Banana (Musaceae) | Worldwide | AJ002234 |
| | Cacao swollen shoot virus | CSSV | *Theobroma* sp. (Sterculiaceae) | Africa | NC001374 |
| | Canna yellow mottle virus | CaYMV | Canna (Cannaceae) | Japan, USA | |
| | Citrus mosaic virus | CMBV | *Citrus* sp. (Rutaceae) | India | AF347695 |
| | **Commelina yellow mottle virus** | **ComYMV** | **Commelina (Commelinaceae)** | **Caribbean** | **X52938** |
| | Dioscorea bacilliform virus | DBV | Yam (Dioscoreaceae) | Africa | X94576 |
| | Gooseberry vein banding-associated virus | GVBAV | *Ribes* sp. (Grossulariaceae) | Worldwide | AF298883 |
| | Kalanchoe top-spotting virus | KTSV | Kalanchoe (Crassulaceae) | UK, USA | AY180137 |
| | Piper yellow mottle virus | PYMoV | Pepper (Piperaceae) | Brazil, India, Asia | |
| | Rubus yellow net virus | RYNV | Raspberry (Rosaceae) | Eurasia, USA | AF468454 |
| | Schefflera ringspot virus | SRV | *Schefflera* (Araliaceae) | Worldwide | |
| | Spiraea yellow leaf spot virus | SYLSV | Spiraea (Rosaceae) | | AF299074 |
| | Sugarcane bacilliform IM virus | SCBIMV | Sugarcane (Poaceae) | USA, Cuba, Morocco | AJ277091 |
| | Sugarcane bacilliform Mor virus | SCBMV | Sugarcane (Poaceae) | USA, Cuba, Morocco | M89923 |
| | Taro bacilliform virus | TaBV | Taro (Araceae) | South Pacific | AF357836 |

*Continued*

**Table 1** Continued

| Genus | Virus | Abbreviation | Host (family) | Geographic distribution | Representative accession number |
|---|---|---|---|---|---|
| | Aucuba bacilliform virus[b] | AuBV | *Aucuba* sp. (Garryaceae) | Japan, UK | |
| | Mimosa bacilliform virus[b] | MBV | *Mimosa* sp. (Fabaceae) | USA | |
| | Pineapple bacilliform virus[b] | PBV | Pineapple (Bromeliaceae) | Australia | Y12433 |
| | Stilbocarpa mosaic bacilliform virus[b] | SMBV | Stilbocarpa (Araliaceae) | Subantarctic islands | AF478691 |
| | Yucca bacilliform virus[b] | YBV | Yucca (Agavaceae) | South America, Italy | AF468688 |
| *Tungrovirus* | **Rice tungro bacilliform virus** | **RTBV** | **Rice (Poaceae)** | **East Asia, China** | **X57924** |

[a]The entry for type member of each genus is highlighted in bold font.
[b]Tentative member.

composed of 420 subunits with a $T = 7$ structure. The viruses in the genera *Badnavirus* and *Tungrovirus* form nonenveloped, bacilliform particles that are 30 nm in width, but can vary in length from 60 to 900 nm. The length most commonly observed is 130 nm. Their structure is based on an icosahedron, in which the ends are formed from pentamers and the tubular section is made up of hexamers.

The virions of the family *Caulimoviridae* contain a single, circular dsDNA that is 7.2–8.3 kbp in length. Complete and partial nucleotide sequences of members of the family *Caulimoviridae* are listed in **Table 1**. Only a single accession number is given for CaMV and RTBV, although multiple strains of each have been sequenced.

## Genome Organization and Expression

Several genome features are common to all members of the family *Caulimoviridae*, in addition to their circular dsDNA genomes. For example, the dsDNA is not covalently closed; it has at least two discontinuities, but may have up to four. Furthermore, all of the ORFs are found on only one of the DNA strands and all of the viruses have at least one strong promoter that drives the expression of a terminally redundant mRNA. In addition, all of the viruses have a reverse transcriptase, and in most cases it is located downstream from the coat protein. Several active sites can be identified within the reverse transcriptase protein, including a protease, the core reverse transcriptase, and RNaseH activity. The reverse transcriptase of PVCV is distinguished from all other members of the family *Caulimoviridae* because it has the core features of an integrase function, in addition to other functions. The most significant difference among the caulimoviruses concerns the arrangement and number of ORFs, and this is illustrated in an examination of the genome organization and expression strategies of CaMV and CoYMV (**Figure 2**).

The genome of CaMV is approximately 8000 bp in size and it contains three single-stranded discontinuities (**Figure 2(a)**). One discontinuity occurs in the negative-sense DNA strand and, by convention, this is the origin of the DNA sequence. Two other discontinuities occur in the positive-sense strand, one at nucleotide position 1600 and a second at approximately nucleotide position 4000. The virus genome consists of seven ORFs. ORF1 encodes a protein necessary for cell-to-cell movement (P1). ORF2 and -3 (proteins P2 and P3) are both required for aphid transmission. P2 is responsible for binding to the aphid stylet, whereas P3 is a virion-associated protein. To complete the bridge for aphid transmission of virions, the C-terminus of P2 physically interacts with the N-terminus of P3. Recent evidence also indicates that P3 may have an additional role in cell-to-cell movement. ORF4 encodes the coat protein, and ORF5 encodes a reverse transcriptase that also has protease and RNaseH domains. With the exception of PVCV, none of the reverse transcriptases of the *Caulimoviridae* have evidence for an integrase function. The ORF6 product (P6) was originally described as the major inclusion body protein, but it also has a function as a translational transactivator (TAV). It physically interacts with host ribosomes to reprogram them for reinitiation of translation of the polycistronic 35S RNA. Furthermore, P6 is an important symptom and host range determinant. No protein product has been found for ORF7. Its nucleotide sequence appears to play a regulatory role in aligning ribosomes for translation of other CaMV gene products.

Two transcripts are produced from the CaMV genomic DNA. The 19S RNA serves as the mRNA for P6, whereas the terminally redundant 35S RNA serves as a template for reverse transcription and as a polycistronic mRNA for ORFs 1–5. One feature common to many of the caulimoviruses is the ribosomal shunt mechanism of translation. The ribosomal shunt has undoubtedly evolved over time to compensate for the complexity and length of the leader sequence of the genomic RNA. In the case of CaMV, this

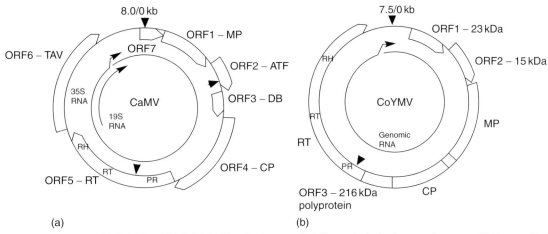

**Figure 2** Genomic maps of (a) CaMV and (b) CoYMV. The single-stranded discontinuity in the negative-sense DNA strand is indicated by the triangle outside of the circle, whereas single-stranded discontinuities in the positive-sense DNA strand are indicated by triangles inside the circle. The mRNAs for each virus are represented by the inner circles and the 3′ end of the RNA is indicated by the arrowhead. The functions for each of the ORFs are MP, cell-to-cell movement protein; ATF, aphid transmission factor; DB, DNA-binding protein, which also has role in aphid transmission; CP, coat protein; RT, reverse transcriptase; TAV, translational transactivator and major inclusion body protein. The domains within the reverse transcriptase are: PR, protease; RT, reverse transcriptase domain; and RH, ribonucleaseH activity. ORF3 of CoYMV encodes a 216 kDa polyprotein that is cleaved to produce the MP, CP, and RT proteins.

leader sequence is approximately 600 bp in length and it contains up to nine short ORFs that vary in size from 9 to 102 nt. The complexity of the CaMV 35S RNA leader sequence is bypassed through the formation of a large stem–loop structure, which allows ribosomes to bypass most of the leader and to initiate translation at ORF7. CaMV utilizes several other strategies for expression of the 35S RNA, including splicing and reprogramming of host ribosomes by the TAV for reinitiation of translation.

The genome of CoYMV is approximately 7500 bp in size, and the circular DNA contains two single-stranded discontinuities, one in each strand (**Figure 2(b)**). As with CaMV, the discontinuity in the negative-sense DNA strand serves as the origin of the DNA sequence, whereas the discontinuity in the positive-sense strand is found at approximately nucleotide position 4800. CoYMV encodes a single transcript that is terminally redundant, with 120 nucleotides reiterated on the 5′ and 3′ ends. The virus genome consists of three ORFs, encoding proteins of 23, 15, and 216 kDa, respectively. Both the 23 and 15 kDa proteins are associated with the virions. The 23 kDa protein contains a coiled-coil motif and by analogy with the P3 protein of CaMV, may be necessary for cell-to-cell movement. The 216 kDa P3 protein is a polyprotein that contains motifs for a movement protein, coat protein, aspartate protease, reverse transcriptase, and RNase H.

## Replication

Caulimoviruses replicate by reverse transcription of an RNA intermediate, but integration of the viral DNA into host chromosomes is not required to complete the replication process. **Figure 3** illustrates the replication of an icosahedral caulimovirus, but the same steps are applicable for the bacilliform viruses. After virions enter a plant cell, the viral DNA becomes unencapsidated and is transported into the nucleus. The viral DNA contains two to four single-stranded discontinuities, which form as a consequence of the reverse transcription process. Once in the nucleus, the single-stranded discontinuities are covalently closed and the DNA is associated with histones to form a minichromosome. The host RNA polymerase II is responsible for synthesizing an RNA that is terminally redundant; the sequence on the 5′ end is reiterated on the 3′ end. In the case of CaMV, the terminal redundancy is 180 nt in size, whereas the terminal redundancy in the CoYMV RNA is 120 nt. The terminally redundant RNA is transported out of the nucleus into the cytoplasm where it can either serve as a template for translation of viral proteins or as a template for reverse transcription. Reverse transcription is thought to occur in nucleocapsid-like particles.

First strand DNA synthesis is primed by a methionine (Met) tRNA that binds to a complementary sequence near the 5′ end of the terminally redundant RNA. In the case of CaMV, the Met tRNA binds to a sequence approximately 600 nt from the 5′ end of its 35S RNA. The virally encoded reverse transcriptase synthesizes DNA up to the 5′ end of the terminally redundant RNA and the RNase H activity of the reverse transcriptase degrades the 5′ end of the RNA. The terminal redundancies present in the genomic RNA provide a mechanism for a template switch, as the reverse transcriptase is able to switch from the 5′ end of the genomic RNA to the same sequences on the 3′ end of the genomic RNA and continue to synthesize the first strand of DNA.

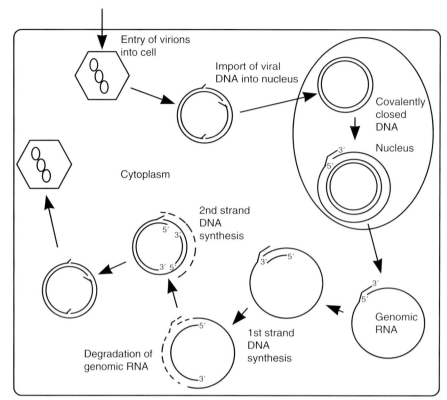

**Figure 3** Replication strategy of the family *Caulimoviridae*.

The RNase H activity of the reverse transcriptase degrades the viral RNA template and small RNA fragments serve as the primers for second strand DNA synthesis. The RNA fragments bind to guanosine-rich tracts present in the first DNA strand, and these priming sites determine the positions of the discontinuities in the second DNA strand. A second template switch is required to bridge the gap from the 5′ to the 3′ end of the first DNA strand. The completion of the second DNA strand results in the formation of a dsDNA molecule that contains the characteristic single-stranded discontinuities.

## Integration of Some Petuviruses, Cavemoviruses, and Badnaviruses into Host Chromosomes

Although caulimoviruses do not integrate into the host as part of their replication strategy, the sequences of several viruses have been detected in the genomes of their hosts. Furthermore, the integrated copies of several of these viruses, the banana streak viruses, TVCV and PVCV, can be activated to yield episomal infections. All three viruses can form virions once episomal infections are initiated, but they differ in their capacity to be transmitted to other plants. Banana streak virus can be transmitted by mealybugs, whereas PVCV is transmitted only by grafting. TVCV and PVCV are transmitted vertically, through seed, but this probably involves only the integrated copies of the virus rather than the episomal forms.

The sites of integration are complex, as the viral sequences have undergone rearrangements, and multiple copies are arranged in tandem. Banana streak virus is integrated at two loci in *Musa* chromosomes, TVCV is integrated at multiple loci in *Nicotiana edwardsonii* chromosomes, and PVCV is integrated into four loci in *Petunia hybrida* chromosomes. The mechanism that results in episomal infections remains to be elucidated, but there are some features common to all three viruses. First, the infections arise in interspecific hybrids. Consequently, each parent plant must contribute some factor to activate the integrated virus. Second, the infections arise after the hybrid has been exposed to some sort of stress. PVCV is activated when plants are exposed to water or nutrient stress, or the plants are wounded. TVCV infections arise in *N. edwardsonii* in the winter months in the greenhouse and are thought to be related to changes in light quality or duration. Integrated banana streak virus sequences are activated in otherwise healthy *Musa* species when they are subjected to tissue culture. In each case, the virus is likely released through a series of recombination events or through reverse transcription of an RNA template.

Portions of other DNA viruses have been detected in plant chromosomes, but they have not been associated with episomal infections. This has led to speculation that plant DNA viruses might be capable of recombination in every infected plant, likely through a mechanism involving nonhomologous recombination. However, since episomal forms of the caulimoviruses and badnaviruses have not been found in tissues fated for seed formation, the integrated forms would not appear in the next generation of plants. Consequently, the banana streak viruses, TVCV, and PVCV must have gained access to germline cells at some point such that their integrants would be passed through seed. This is likely to have been a relatively rare event.

## Transmission and Host Range

Most members of the genus *Caulimovirus* are transmitted by as many as 27 species of aphids in a semipersistent manner. The virions are acquired rapidly by the aphid and can be transmitted immediately upon acquisition. Virions can be maintained by the aphid for as little as 5 h up to 3 days, but are not retained after the aphid molts and are not passed on to aphid progeny. The protein products encoded by ORF2 (P2) and ORF3 (P3) are both required for aphid transmission. P3 forms a tetramer that binds to both the virion and to the C-terminus of P2. P2 is responsible for the binding of this complex to the aphid, as the N-terminus of P2 binds to a site in the aphid foregut. No insect vectors have been identified for species in the genera *Petuvirus*, *Soymovirus*, or *Cavemovirus*.

None of the virions of the icosahedral viruses in the genera *Caulimovirus*, *Petuvirus*, *Soymovirus*, or *Cavemovirus* are transmitted through seed or pollen. Most are transmitted after mechanical inoculation, but there are a few exceptions, as PVCV, TVCV, blueberry red ringspot virus, and strawberry vein banding virus cannot be mechanically inoculated. In addition, most of these viruses are transmitted by grafting. In particular, dahlia mosaic virus and blueberry red ringspot virus infections in the field may be initiated through vegetative propagation or through grafting of infected plant material.

Badnaviruses are transmitted primarily by mealybugs, but a few are transmitted by aphids. Badnaviruses transmitted by mealybugs include the banana streak virus species, the sugarcane bacilliform viruses, cacao swollen shoot virus, dioscorea bacilliform virus, kalanchoe top spotting virus (KTSV), piper yellow mottle virus, taro bacilliform virus, and schefflera ringspot virus. These viruses are transmitted in a semipersistent manner and can be retained after molts, but do not multiply in the mealybug and are not transmitted to progeny. The badnaviruses shown to be transmitted by aphids include gooseberry vein banding-associated virus, rubus yellow net virus, and spiraea yellow leafspot virus.

Other modes of transmission of the badnaviruses vary with the species. For example, CoYMV and KTSV can be mechanically inoculated, cacao swollen shoot virus and piper yellow mottle virus are mechanically inoculated with some difficulty, and rubus yellow net virus and taro bacilliform virus have not been shown to be mechanically inoculated. The preferred method for inoculation of infectious badnavirus clones is agroinoculation. Some of the badnaviruses are transmitted through seed, including the banana streak virus, KTSV, and mimosa bacilliform virus. KTSV is very efficiently transmitted by seed, with transmission rates from 60% to 90%, and is also transmitted in pollen.

RTBV is dependent on the RNA virus, RTSV, for its transmission. RTSV is vectored by a leafhopper, but causes only very mild symptoms. RTBV is only transmitted by the leafhopper in the presence of RTSV, but is responsible for the severe symptoms associated with the rice tungro disease. RTBV is not mechanically transmitted or carried in seed or pollen.

The host range of most caulimoviruses is fairly narrow, as in nature they generally infect plants within a single family (**Table 1**). Their experimental host range may extend to members of one or two other families, but in many instances, they may only infect a single genus of plants. There are a few exceptions. For example, the sugarcane bacilliform viruses have a broader host range than most badnaviruses, as they can infect *Sorghum*, *Rottboellia*, *Panicum*, rice, and banana. A limited host range can also be associated with similar limitations in the geographic distribution of the virus. For example, soybean chlorotic mottle virus has only been recovered from a few samples in Japan. Perhaps the virus with the smallest geographic distribution is stilbocarpa mosaic bacilliform virus, which has only been found on a single, small island in the Subantarctic, midway between Tasmania and Antarctica. Other viruses, such as CaMV, carnation etched ring virus, and dahlia mosaic virus are found worldwide, wherever their hosts are grown. Furthermore, the distribution of viruses that originate from integrated copies in their host's genomes, PVCV, TVCV, and the banana streak virus species, are also closely aligned with the locations of their hosts. Interestingly, the icosahedral viruses of the caulimoviruses tend to infect hosts in temperate climates, whereas the bacilliform viruses of the badnavirus group are more likely to infect hosts in tropical or subtropical climates.

## Virus–Host Relationships

Caulimoviruses induce a variety of systemic symptoms in their hosts, from chlorosis, streaking, and mosaics, to necrosis. The best-characterized pathogenicity determinant is the P6 protein of CaMV, as it has been shown to play a key role in the formation of chlorotic symptoms in turnips. This virulence function was first associated with P6 through gene-swapping experiments between CaMV isolates. It was confirmed when P6 was transformed into several species of plants, and in most cases, they exhibited virus-like symptoms. The P6 protein is also responsible for triggering systemic cell death in *Nicotiana clevelandii*, as well as a non-necrotic resistance response in *N. glutinosa*, and a hypersensitive resistance response in *N. edwardsonii*.

One feature that distinguishes the icosahedral viruses from the bacilliform viruses is that the former have the capacity to aggregate into amorphous inclusion bodies (**Figure 1(a)**), whereas the latter do not form inclusions. The inclusion bodies formed by the icosahedral viruses are not bound by a membrane, can range in size from 5 to 20 µm, and occur in virtually all types of plant cells. The inclusions can be visualized with a light microscope in strips of epidermal tissue that has been stained with phloxine B. Close examination of CaMV inclusion bodies by electron microscopy reveals that there are actually two types. One type contains many vacuoles and consists of an electron-dense, granular matrix that is composed primarily of the P6 protein (**Figure 1(a)**). A second, electron translucent type is made up of the P2 protein, a protein required for aphid transmission. Both types of inclusions are thought to have a role in the biology of the virus. The vacuolated inclusion bodies may be considered pathogen organelles, as they are thought to serve as the sites for replication of the viral nucleic acid, as well as translation of the 35S RNA and assembly of the virions. The electron translucent inclusions are considered to have a role in aphid transmission.

A second feature characteristic of the caulimoviruses is that the plasmodesmata of infected cells are enlarged enough to accommodate the 50 nm virions, as electron micrographs have revealed the presence of CaMV and CoYMV virions in the enlarged plasmodesmata. For both CaMV and CoYMV, the alteration in size is mediated by their proteins required for cell-to-cell movement. In infected protoplasts, the CaMV P1 protein has been shown to induce the formation of tubular structures that extend away from the protoplast surface. It is hypothesized that virions are assembled in the cell and then are escorted to the enlarged plasmodesmata by the cell-to-cell movement proteins.

*See also:* Caulimoviruses: Molecular Biology.

## Further Reading

Calvert LA, Ospina MD, and Shepherd RJ (1995) Characterization of cassava vein mosaic virus: A distinct plant pararetrovirus. *Journal of General Virology* 76: 1271–1278.

Geering ADW, Olszewski NE, Harper G, et al. (2005) Banana contains a diverse array of endogenous badnaviruses. *Journal of General Virology* 86: 511–520.

Harper G, Hull R, Lockhart B, and Olszewski N (2002) Viral sequences integrated into plant genomes. *Annual Review of Phytopathology* 40: 119–136.

Hasagewa A, Verver J, Shimada A, et al. (1989) The complete sequence of soybean chlorotic mottle virus DNA and the identification of a novel promoter. *Nucleic Acids Research* 17: 9993–10013.

Hay JM, Jones MC, Blakebrough ML, et al. (1991) An analysis of the sequence of an infectious clone of rice tungro bacilliform virus, a plant pararetrovirus. *Nucleic Acids Research* 19: 2615–2621.

Hohn T and Fütterer J (1997) The proteins and functions of plant pararetroviruses: Knowns and unknowns. *Critical Reviews in Plant Sciences* 16: 133–161.

Hull R (1996) Molecular biology of rice tungro viruses. *Annual Review of Phytopathology* 34: 275–297.

Hull R, Geering A, Harper G, Lockhart BE, and Schoelz JE (2005) Caulimoviridae. In: Fauquet CM, Mayo MA, Maniloff J, Desselberger U, and Ball LA (eds.) *Virus Taxonomy: Eighth Report of the International Committee on Taxonomy of Viruses*, pp. 385–396. San Diego, CA: Elsevier Academic Press.

Lockhart BEL (1990) Evidence for a double-stranded circular DNA genome in a second group of plant viruses. *Phytopathology* 80: 127–131.

Pfeiffer P and Hohn T (1983) Involvement of reverse transcription in the replication of cauliflower mosaic virus: A detailed model and test of some aspects. *Cell* 33: 781–789.

Qu R, Bhattacharya M, Laco GS, et al. (1991) Characterization of the genome of rice tungro bacilliform virus: Comparison with Commelina yellow mottle virus and caulimoviruses. *Virology* 185: 354–364.

Richert-Pöggeler KR and Shepherd RJ (1997) Petunia vein-clearing virus: A plant pararetrovirus with the core sequences for an integrase function. *Virology* 236: 137–146.

Schoelz JE, Palanichelvam K, Cole AB, Király L, and Cawly J (2003) Dissecting the avirulence and resistance components that comprise the hypersensitive response to cauliflower mosaic virus in *Nicotiana*. In: Stacey G and Keen N (eds.) *Plant–Microbe Interactions,* vol. 6, pp. 259–284. St. Paul, MN: The American Phytopathological Society.

Shepherd RJ, Wakeman RJ, and Romanko RR (1968) DNA in cauliflower mosaic virus. *Virology* 36: 150–152.

Skotnicki ML, Selkirk PM, Kitajima, et al. (2003) The first subantarctic plant virus report: Stilbocarpa mosaic bacilliform badnavirus (SMBV) from Macquarie Island. *Polar Biology* 26: 1–7.

# Caulimoviruses: Molecular Biology

**T Hohn,** Institute of Botany, Basel university, Basel, Switzerland

© 2008 Elsevier Ltd. All rights reserved.

## Viruses of the Genus *Caulimovirus*

Several viruses have been assigned to the genus *Caulimovirus*, characterized by their content of open circular DNA, their icosahedral capsid, and their arrangement of seven open reading frames (ORFs) (**Table 1**). They have a narrow host range, usually restricted to one of the plant families. The genus *Caulimovirus* is one of six genera of the family *Caulimoviridae*, or plant pararetroviruses, which include three more genera of icosahedral viruses,

**Table 1** Caulimoviruses species

| Species | Abbr. | Host range | Sequence |
| --- | --- | --- | --- |
| Cauliflower mosaic virus | CaMV | Cruciferae | V00141; X02606; J02046 |
| Carnation etched ring virus | CERV | Caryophyllaceae | X04658 |
| Figwort mosaic virus | FMV | Scrophulariaceae | X06166 |
| Strawberry vein banding virus | SVBV | Rosaceae | X97304 |
| Horseradish latent virus | HRLV | Cruciferae | AY534728 - 33 |
| Dahlia mosaic virus | DMV | Compositae | |
| Mirabilis mosaic virus | MiMV | Nyctaginacea | NC_004036 |
| Thistle mottle virus | ThMoV | Cirsium arvense | |

Note that *Blueberry ringspot virus* and *Cestrum yellow leaf curl virus*, originally classified as caulimoviruses, are not included because they belong to the genus *Soymovirus*.

*Soymovirus*, *Cavemovirus*, and *Petuvirus*, and two genera of bacilliform viruses, *Badnavirus* and *Tungrovirus*. Besides the differences in capsid structure of the two categories, the genera are distinct in the details of their genome arrangement and expression strategies (**Figure 1**).

Although members of the family *Caulimoviridae* (plant pararetroviruses), unlike retroviruses, do not integrate obligatorily into the host genome, 'illegitimately' integrated sequences have been found for several genera to date, that is, caulimo, petu-, cavemo-, badna-, and tungroviruses. These have been named commonly 'endogenous plant pararetroviruses' (EPRVs). EPRVs can be found in the pericentromeric region of chromosomes, are passively replicated together with the host DNA, and are inherited from generation to generation.

## Properties of the Virion and Inclusion Bodies

Members of the genus *Caulimovirus* are icosahedral ($T = 7$), 45–50 nm in diameter and sediment between 200 and 250S. The main capsid components of Cauliflower mosaic virus (CaMV) are proteins with mobilities of 37 and 44 kDa on sodium dodecyl sulfate polyacrylamid electrophoresis. They are derived from ORF IV. The capsid is further decorated with 420 subunits of loosely bound 15 kDa virus-associated protein (VAP) derived from ORF III. In addition, CaMV particles contain minor amounts of other polypeptides: protease and reverse transcriptase derived from ORF V, and host casein kinase II. In cells infected with CaMV or a number of other caulimoviruses large numbers of virus particles accumulate in typical stable inclusion bodies, the matrix of which is the virus-encoded transactivator/viroplasmin (TAV) protein of 62 kDa, derived from ORF VI. Inclusion bodies are visible in the light microscope and can be stained with phloxin.

## Properties of the Genome

The genome of the members of the family *Caulimoviridae* is built from 7200 to 8200 bp. It exists (1) in infected plant nuclei as supercoiled DNA bound to nucleosomes (minichromosome), (2) in infected cytoplasm as RNA with a ~180 nt redundancy, and (3) in virus particles as open circular dsDNA. The open form is due to nicks at specific sites in both strands with short 5' overlaps (**Figure 2**), remnants of the reverse transcription process, which are apparently removed in the infected nucleus. The sequences of several of the caulimoviruses are known (**Table 1**). They contain seven ORFs in the order VII (dispensable unknown function), I (cell-to-cell movement), II (aphid transmission factor), III (VAP), IV (capsid protein precursor), V (protease, reverse transcriptase, RNase H), and VI (multifunctional protein, transactivator of translation/viroplasmin).

## Properties of the Proteins

### Movement Protein

CaMV and probably all the other members of the family *Caulimoviridae* with icosahedral capsid move from cell to cell as particles through tubular structures spanning the cell walls between adjacent cells. In CaMV the 37 kDa movement protein (MOV), coded for by ORF I, is responsible for the formation of these tubules and also provides their main component.

### Aphid Transmission Factor

In general, aphids transmit viruses of the genus *Caulimovirus* and probably other genera of icosahedral (but not bacilliform) Caulimoviruses. In the case of CaMV, ORF II codes for the 18 kDa aphid transmission factor (ATF). Some of the viral inclusion bodies in CaMV-infected cells consist mainly of ATF. ATFs interact with the virus particles via their VAP and with tubulin and also with the cuticulum lining the tip the aphid's stylet. ATF mutants can still be transmitted by mechanical inoculation and by aphids that had previously be in contact with wild-type ATF either by feeding on plants infected with a related virus or on nutrient solutions supplied with artificially produced ATF.

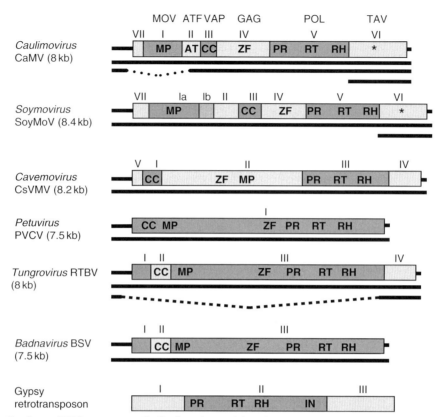

**Figure 1** Open reading frame (ORF) arrangement in *Caulimovirus* and the other genera of the family *Caulimoviridae*. Examples of each of the six genera are shown, as well as the retrotransposon 'Gypsy' for comparison. For CaMV, the ORFs code for dispensable unknown function (VII), movement protein (I, MOV), aphid transmission factor (II, ATF), virion-associated protein (III, VAP), capsid protein precursor (IV, GAG), Pol polyprotein (V, POL), and transactivator/viroplasmin (VI, TAV). Although the number of final proteins might be similar for the other genera, the number of original ORFs differs, with soymoviruses having the most ORFs and petuviruses the least, namely only one original ORF, meaning more work for the protease to produce individual proteins. Key motifs are conserved in the family *Caulimoviridae*, movement protein motif (MP), coiled-coil domain (CC), zinc-finger (ZF), protease domain (PR), reverse transcriptase domain (RT), RNase H domain (RH), transactivator domain (*). In contrast to retroviruses and retrotransposons, an integrase domain (IN) is missing in members of the family *Caulimoviridae*.

## Virion-Associated Protein

The VAP provides for coupling between virions and the movement and insect transmission factors and is therefore required for both insect transmission and cell-to-cell movement.

MOV, ATF, and VAP have no counterpart in retroviruses, reflecting the differences in cell-to-cell movement and infection routes used by animal and plant viruses. However, the movement proteins of plant pararetroviruses are related to a corresponding class of genes in most other plant viruses. In animals, viruses usually move from cell to cell by endocytosis and budding, while in plants, passage through modified plasmodesmata is used.

## Capsid Proteins and Their Precursor

The product of CaMV ORF IV has a molecular weight of 56 kDa and an electrophoretic mobility of 80 kDa. It is flanked by very acidic regions, which are removed from the mature capsid proteins, which have a mobility corresponding to proteins of 44 and 37 kDa, although their molecular weight is probably smaller. The sequence of p37 is included within p44 and thus antibodies against p37 react also with p44. p44, but not p37, is phosphorylated. It is not known whether both proteins are required for full infection, or whether only one of them is the true functional capsid protein. It might be that p44 is the mature form and p37 a degradation product. Alternatively, p37 might be the mature form and p44 still a precursor. Both these proteins include a Zn-finger motif, which is implicated in RNA binding. Furthermore, large stretches of basic amino acids at the C-termini of both proteins constitute DNA- and RNA-binding motifs.

The precapsid protein has a nuclear localization signal which is masked by the N-terminal acidic domain. Upon virus assembly and removal of the acidic regions, the NLS becomes exposed and virus particles are transported to the nuclear pore where they release the DNA into the nucleus.

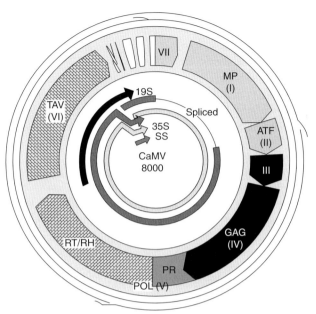

**Figure 2** Map of cauliflower mosaic virus. On the outside the open circular double-stranded viral DNA with its overhangs is symbolized. The main circle shows the arrangement of the open reading frames (ORFs). ORF I encodes movement protein (MP or MOV), ORF II aphid transmission factor (ATF), ORF III the VAP, ORF IV the capsid protein precursor (GAG, in analogy to the corresponding protein in retroviruses), ORF V the POL polyprotein consisting of protease (PR), reverse transcriptase (RT), and RNase H (RH), and ORF VI the transactivator/viroplasmin (TAV). On the inside, the primary transcripts and the essential spliced RNAs are shown.

## POL Polyprotein, Protease, Reverse Transcriptase/RNase H

CaMV ORF V corresponds to the POL ORF of retroviruses. It produces a polyprotein which is cleaved during virus production into a 15 kDa aspartic proteinase and a 60 kDa reverse transcriptase/RNase H. It is not known whether the RNase H functions as part of the 60 kDa protein or whether it is released by another cleavage reaction. One has to assume that the proteinase also functions as part of the polyprotein, that is, by releasing itself. Members of the family *Caulimoviridae* do not have an integrase and consequently their DNA is not or only accidentally integrated into the host chromatin.

## Transactivator of Translation/Viroplasmin

The 62 kDa transactivator of TAV, encoded by ORF VI, has no significant homologies to any other known viral or host genes. It has been implicated in virus assembly, reverse transcription, host range determination, symptom severity, control of polycistronic translation, and seclusion of virus functions and silencing suppression. Perhaps as a consequence of the role of the ORF VI protein in translation, all CaMV proteins including foreign proteins carried by a CaMV vector can be found within the inclusion bodies. A subdomain in its center has been assigned to the translational transactivation and other domains have been found to unspecifically bind single- and double-stranded RNA, respectively. Furthermore, the protein interacts with capsid protein. TAV also has properties of a nuclear shuttle protein and might be involved in viral RNA transport.

## Virus Stability

Both, the CaMV inclusion bodies and the virus particles, are very stable. Dissolution of the inclusion bodies to obtain virus particles requires treatment with 1 M urea over a long period. Virus particles aggregate at acidic pH and disintegrate at very high pH (0.1 M NaOH). Virus particles can also be disintegrated by boiling in sodium dodecyl sulfate/dithiothreitol or by treatment with proteinase K and phenol extraction. During these treatments, VAP is released and therefore usually escapes detection as virus component.

## Genome Replication

As in retroviruses, replication occurs by production of a terminally redundant RNA by transcription in the nucleus and its reverse transcription in the cytoplasm. Also as in retroviruses, and unlike in hepadnaviruses, a tRNA primer (met-initRNA) is used as the primer for (−)-strand DNA synthesis from the RNA template. The RNA template is digested by RNase H following the reverse transcriptase. However, oligo(G) stretches are resistant to this digestion and remain bound to the (−)-strand DNA where they act as primers for (+)-strand DNA synthesis. The number of (+)-strand synthesis initiation events varies between one and three major events in different members of the family *Caulimoviridae* and even within strains from a particular virus. Minor initiation events also occur. The synthesis of both the strands overshoots, creating short overhangs that can only be repaired by repair exonuclease and ligase after transport into the nucleus. Accordingly, the packaged viral DNA is open circular, while the nuclear one is closed and supercoiled. The latter interacts with histones and forms minichromosomes. Caulimovirus DNA does not integrate into the host chromatin obligatorily. The template for transcription is the supercoiled circular DNA.

## Transcription

Transcription in caulimoviruses is unidirectional. All caulimoviruses produce a transcript covering the total genome plus about 180 nt, such that the RNA is terminally redundant. This RNA is called 35S RNA due to its sedimentation behavior. The terminal redundancy is caused by the polyadenylation signal located on the circular DNA 180 bp downstream of the transcription start site being ignored by the polymerase at its first passage (see below). CaMV-like caulimoviruses produce a second transcript, the 19S RNA, covering ORF VI and encoding the inclusion body protein/translation transactivator.

The 35S promoter of CaMV, FMV, and probably other caulimoviruses is very strong and quasi-constitutive, that is, it is expressed in nearly all types of cells and at all developmental stages. This constitutivity is caused by a number of different enhancer elements, each with some specificity for certain cell types. Some of the corresponding transcription factors have been identified, for example, ASF-1, ASF-2, and CAF recognizing a TGACG motif, a GATA motif, and CA-rich region, respectively.

## RNA Processing

Full-length RNA made from the circular DNA template includes a polyadenylation signal. If this is used on the first encounter, a short-stop RNA of 180 nt is formed; if used on the second encounter full-length terminally redundant RNA is formed. The role of the short-stop RNA, if any, is not known.

The CaMV polyadenylation signal consists of an AAUAAA sequence, which determines the cleavage site 13 nt downstream of it. Occasionally cryptic signals are used in addition. In contrast to the animal system, single point mutations in the AAUAAA signal are partially tolerated by the plant system. Polyadenylation enhancers are located upstream of the AAUAAA signal and not downstream as in most animal cases. In CaMV and FMV, these elements are tandem repeats of UAUUUGUA.

In addition to the primary CaMV 35S transcript, alternatively spliced versions of it have been detected. By removing an intron extending from close to the end of the leader to a position within ORF II, an mRNA is created in which ORF III is the first ORF and ORF IV the second. Additional splicing events were found using the same splice acceptor but using donors within ORF I. These led to ORF I–ORF II in-frame fusions, the function of which is not known. Whether the ratio of spliced to unspliced RNA is controlled, for example by nuclear export of unspliced RNA as in human immunodeficiency virus, is not known.

## Translation Mechanism

The 35S RNA of caulimoviruses and also its spliced derivatives serve as polycistronic mRNAs for the viral proteins. The ORFs of these viruses closely follow each other, are opened by efficiently recognized start codons, and usually also contain internal start codons. In plant protoplasts and in transgenic plants most of the downstream ORFs are poorly expressed in the absence of TAV interestingly the only ORF that is translated from a monocistronic mRNA. Transactivation activity has been demonstrated for CaMV and FMV, and probably resides also in the corresponding ORFs of the other CaMV-like and PCSV-like caulimoviruses. Transactivation is thought to be based on a reinitiation mechanism.

Another special feature of the caulimovirus (and also the badnavirus) 35S RNAs is their 600–700-nt-long leader, which is rather large for eukaryotic RNAs. These leaders contain several small ORFs and include a large hairpin structure. These are features that usually make an RNA a poor messenger, since initiation factors and/or 40S ribosomes initiate scanning at the cap of an RNA moving in the 3′ direction until an AUG codon is encountered and translation proper begins, whereas translation from AUGs further downstream is precluded. To overcome this problem caulimoviruses employ a shunting mechanism for translation initiation, whereby scanning is initiated at the cap as usual, but the scanning complex bypasses (shunts) most of the leader to reach the first longer ORF. This mechanism is unlike the internal initiation mechanism used by enteroviruses.

## Protein Processing

Many of the caulimovirus proteins are processed, that is, in CaMV-like (and by analogy probably the other) caulimoviruses the precapsid protein is cleaved at both ends removing its very acidic termini. This process might be coupled to virus assembly and maturation. The capsid protein is both methylated and glycosylated. The Pol protein is cleaved to yield the aspartic proteinase and reverse transcriptase/RNase H. At least some of these cleavages occur by the action of the viral proteinase. In contrast, the ORF III product is cleaved by a host cysteine proteinase.

## Silencing and Silencing Suppression

Another level of expressional control is provided by transcriptional and post-transcriptional silencing, which lead to strong inhibition of transcription and destabilization of the transcript, respectively. Silencing is viewed as a plant defense mechanism directed against viruses and transposons. On the other hand, viruses have developed silencing suppressors to counteract this plant defence strategy. The response of *Brassica napus* to systemic infection with cauliflower mosaic virus that results first in enhancement followed by subsequent suppression of viral gene expression in parallel with changes in symptom formation can be explained by the battle being waged between silencing suppression and silencing.

Silencing is autocatalytic and systemic, and hence the silencing of CaMV can also lead to the silencing in *trans* of transgenes driven by the CaMV 35S promoter. In a special case, herbicide resistance in oilseed rape conferred by expression of a 35S promoter-driven bialaphos tolerance transgene can be silenced due to the host response to CaMV infection.

*See also:* Caulimoviruses: General Features; Endogenous Retroviruses; Legume Viruses; Plant Virus Vectors (Gene Expression Systems); Rice Tungro Disease; Viral Suppressors of Gene Silencing; Virus Induced Gene Silencing (VIGS).

## Further Reading

Blanc S (2002) Caulimoviruses. *Virus–Insect–Plant Interactions.* New York: Academic Press.

Covey SN and Al Kaff NS (2000) Plant DNA viruses and gene silencing. *Plant Molecular Biology* 43(special issue): 307–322.

Goldbach R and Hohn T (1996) Plant viruses as gene vectors. In: Bryant JA (ed.) *Methods in Plant Biochemistry 10b: Molecular Biology,* pp. 103–120. San Diego: Academic Press.

Haas M, Geldreich A, Bureau M, *et al.* (2005) The open reading frame VI product of cauliflower mosaic virus is a nucleocytoplasmic protein: Its N terminus mediates its nuclear export and formation of electron-dense viroplasms. *Plant Cell* 17: 927–943.

Hohn T (2007) Plant virus transmission from the insect point of view. *Proceedings of the National Academy of Sciences, USA* 104: 17905–17906.

Hohn T, Park HS, Guerra-Peraza O, *et al.* (2001) Shunting and controlled reinitiation: The encounter of cauliflower mosaic virus with the translational machinery. *Cold Spring Harbor Symposia on Quantitative Biology* 66: 269–276.

Hohn T and Richert-Poeggeler K (2006) Caulimoviruses. In: Heffron KL (ed.) *Recent Advances in DNA Virus Replication,* pp. 289–319. Kerala: Research Signpost Transworld.

Kiss-László Z and Hohn T (1996) Pararetro- and retrovirus RNA: Splicing and the control of nuclear export. *Trends in Microbiology* 4: 480–485.

Lam E (1994) Analysis of tissue specific elements in the CaMV 35S promoter. In: Nover L (ed.) *Results and Problems in Cell Differentiation, #20,* pp. 181–196. Berlin: Springer.

Plisson C, Uzest M, Drucker M, *et al.* (2005) Structure of the mature P3-virus particle complex of cauliflower mosaic virus revealed by cryo-electron microscopy. *Journal of Molecular Biology* 346: 267–277.

Rothnie HM, Chapdelaine Y, and Hohn T (1994) Pararetroviruses and retroviruses: A comparative review of viral structure and gene expression strategies. *Advances in Virus Research* 14: 1–67.

Ryabova L, Pooggin MM, and Hohn T (2005) *Translation Reinitiation and Leaky Scanning in Plant Viruses. Virus Research (Translational Control during Virus Infection).* Oxford: Elsevier.

Staginnus C and Richert-Poggeler KR (2006) Endogenous pararetroviruses: Two-faced travelers in the plant genome. *Trends in Plant Science* 11: 485–491.

# Central Nervous System Viral Diseases

**R T Johnson and B M Greenberg,** Johns Hopkins School of Medicine, Baltimore, MD, USA

© 2008 Elsevier Ltd. All rights reserved.

## Introduction

Most viral infections of the nervous system represent serious and potentially life-threatening complications of systemic viral infections. With the possible exception of rabies, viruses are not neurotropic in the literal sense of having a specific affinity for the nervous system. Some viruses frequently invade the nervous system yet seldom cause serious disease; for example, mumps virus may cause meningitis but, even during uncomplicated mumps parotitis, cerebrospinal fluid (CSF) changes in over 50% of patients indicate probable nervous system infection. Other viruses, such as herpes simplex virus, rarely infect the central nervous system (CNS), but when they do, they often cause fatal disease. Thus, both mumps and herpes simplex viruses are regarded as neurotropic; mumps is highly neuroinvasive but has limited neurovirulence; herpes simplex has low neuroinvasiveness but is highly neurovirulent. Such variations are dependent on the structural and functional determinants of nervous system invasion, on the particular neural cells that specific viruses infect, on the effect of this infection on the host cells, and on the immune response or immunopathologic responses to the infection. Several terms are defined as follows:

- Neurotropic: able to infect neural cells.
- Neuronotropic: able to infect neurons in contrast to other nervous system cells.
- Neuroinvasive: able to enter the nervous system.
- Neurovirulent: able to cause neurologic disease.

## Anatomic and Physiologic Considerations

Both structural and functional features of the CNS present a unique milieu for viral replication. The blood–brain barrier and the compact structure of the brain and cord pose formidable barriers to the entry or dissemination of viruses within the nervous system. Yet, the multitude of dense dendritic connections among neurons provide a unique environment for cell-to-cell spread of pathogens. Furthermore, neurons are unique cells with high metabolic rates, intense membrane specialization, and no regenerative capacity. The same barriers that exclude viruses also limit access of immunocompetent cells and antibodies, and the nervous system lacks an intrinsic lymphatic system and has a paucity of phagocytic cells. Thus, the barriers that inhibit virus invasion also deter viral clearance. Therefore, many persistent infections involve the CNS.

The blood–brain barrier was originally conceptualized from the observation that dyes, such as Trypan Blue, stain all tissues except the brain and spinal cord after injection into the systemic circulation. The barrier at the cerebral capillary level consists of tight junctions between the capillary endothelial cells (beyond which most dyes do not pass), a dense basement membrane around the cells and tightly opposed astrocytic footplates. In the choroid plexus, the blood–CSF barrier is structurally different. The capillaries of the plexus are fenestrated, lack a basement membrane, and are surrounded by a loose stroma. Dyes and particles readily pass into the choroid plexus but are prevented from entering the CSF by tight junctions located at the apices of the secretory epithelial cells of the choroid plexus. Tight junctions between the arachnoid cell over the surface of the brain complete the barrier.

There is no comparable barrier between the brain and the spinal fluid. The ependymal cells are not joined by tight junctions and, therefore, there is a free exchange between the extracellular space of the brain and the CSF. However, the intracellular gap between neural cells measures only 10–15 nm, less than the diameter of even the smallest virus, so that free movement of virus particles or inflammatory cells within the extracellular space is relatively restricted.

Neurons have specialized membranes for the transmission and receipt of specific messages; they also have axonal extensions to carry signals to and from distant neuronal populations, motor endplates, and sensory endings. In humans these cytoplasmic extensions may exceed a meter in length. These features are important in viral infections, since different subpopulations of neurons have different receptors usurped by viruses to permit entry into cells. Furthermore, viruses in some cases can be carried by axoplasmic transport systems either into the nervous system or within the nervous system where axonal processes link functionally related neurons.

Antibodies found in the normal CNS are derived entirely from the serum. Antibody levels of immunoglobulin G (IgG) are approximately 0.4% of the serum levels. Since diffusion of macromolecules across the barrier is largely size-dependent, immunoglobulin M (IgM) is present in even lower levels. Complement is largely excluded. There is also no lymphatic system in the usual sense and few phagocytic cells. When inflammation disrupts the blood–brain barrier, antibody molecules leak into the nervous system along with other serum proteins. When a mononuclear inflammatory response is mounted against infection, T lymphocytes usually enter the nervous system first followed by macrophages and B lymphocytes. These B cells from the peripheral circulation move into the perivascular space and can generate immunoglobulins intrathecally.

The postmitotic nature of neurons is one the most important features when considering viral infections of the nervous system. Unlike most organ systems, the fundamental component of the nervous system lacks the ability to regenerate. Thus, by definition, infections that cause cell death (directly or indirectly via the elicited immune response) create irreversible damage. In this setting, neurons have developed various strategies for suppressing viral replication, clearing virus infections, and for creating environments suitable for latent infections.

## Pathways of the CNS Invasion

Viruses have been shown to enter the nervous system both along nerves and from the blood. The first experimental studies of viral invasion employed rabies, herpes simplex, or polioviruses, all of which, under experimental circumstances, can penetrate the nervous system along peripheral nerves. The precise mechanisms of neural spread remained a mystery for many years, since it was thought that the axoplasm slowly oozed in an anterograde direction. It was proposed that virus might move in perivascular lymphatics, by ascending infection of the supportive cells within the peripheral nerve, or even by replication in axons, a speculation that is now untenable because of the observed lack of ribosomes or protein synthesis within axons. In the 1960s active anterograde and retrograde axon transport systems were found. Viruses or other particles can be taken up in vesicles at the nerve terminals and transported to the cell body of the sensory or motor neuron (**Figure 1**). This neural route of entry is important in primary viral infections such as rabies and possibly poliomyelitis. Retrograde transport also moves herpes simplex and varicella-zoster viruses from mucous membranes or skin into sensory ganglia at the time of primary infection. Subsequently, anterograde transport carries the reactivated virus from the ganglia to the periphery during exacerbations. Anterograde transport of herpes simplex virus by nerves innervating the dura from the trigeminal ganglia may explain the unique temporal lobe localization of herpes simplex virus encephalitis.

The olfactory spread of virus is a variation of neural spread. In the olfactory mucosa, neural fibers provide a

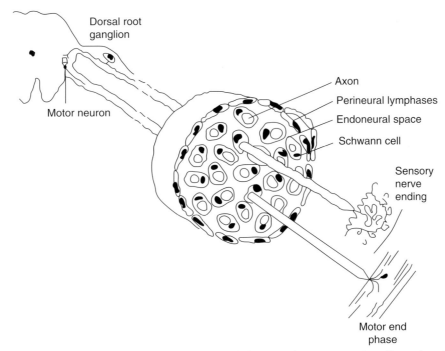

**Figure 1** Schematic diagram of possible routes of neural spread of viruses to the nervous system. Reproduced with permission from Johnson RT (1982) *Viral Infections of the Nervous System*. New York: Raven Press.

unique pathway; the apical processes of receptor cells extend beyond the free surface of the epithelium as olfactory rods and the central processes synapse in the olfactory bulb. These are the only nerve cells with processes that link the CNS and ambient environment. Indeed, some colloidal particles placed on the olfactory mucosa can be found in the olfactory bulbs within 1 h. Experimental studies show that viruses can enter through this route, and this may occur in some aerosol infections in humans such as laboratory accidents or rabies virus infections in bat-infested caves. Also, the olfactory pathway has been postulated as a possible route of herpes simplex virus entry into the nervous system as an alternative explanation for the orbital–frontal and medial–temporal lobe localization of herpetic encephalitis. Nevertheless, despite the apparent ease of spread along this route, it appears to be a rare route of natural infection.

In most experimental and natural infections, viruses invade the brain from the blood. Historically, the blood–brain barrier was believed to be impervious to viruses. This belief was based in part on the fact that viruses experimentally inoculated directly into brain cause disease after a brief inoculation period, whereas the incubation period after intravenous inoculation is longer and comparable to that following cutaneous or peritoneal inoculation. The reason for this delay in infection is that virus in the blood is rapidly removed by the reticuloendothelial system; therefore, intravenous inoculation is, in fact, an inoculation primarily of the Kupffer cells of the liver and other reticuloendothelial cells. Therefore, virus must establish a nidus of peripheral replication that can effectively seed a viremia of sufficient magnitude and duration to allow invasion across the blood–brain barrier (**Figure 2**). Thus, some viruses grow in lymphatics and seed into the blood directly via the thoracic duct, others grow in vascular epithelial cells, and others replicate in highly vascular tissue such as muscle.

A persistent viremia can be achieved by several mechanisms. Rate of clearance is dependent upon particle size; small particles such as togaviruses and flaviviruses can maintain high-titer plasma viremias with sufficient rapid replication in peripheral tissue. Other viruses adsorb to red blood cells and thus evade clearance. Many large viruses such as measles and herpes viruses infect white blood cells thus evading clearance and replicating at the same time.

Some viruses enter the nervous system either across the capillary endothelium and others across the choroid plexus. Some viruses infect the capillary endothelial cells and simply grow into the brain while others are able to transit across endothelial cells despite their paucity of endopinocytotic vesicles. Entry in infected leukocytes is a theoretical possibility but leukocyte traffic into the nervous system is limited, although trauma or inflammation due to other causes may predispose the nervous system to infection with viruses that infect white blood cells. Although there are areas of increased blood–brain barrier permeability, no viral infection has been shown to infect these areas selectively. Other viruses, such as mumps

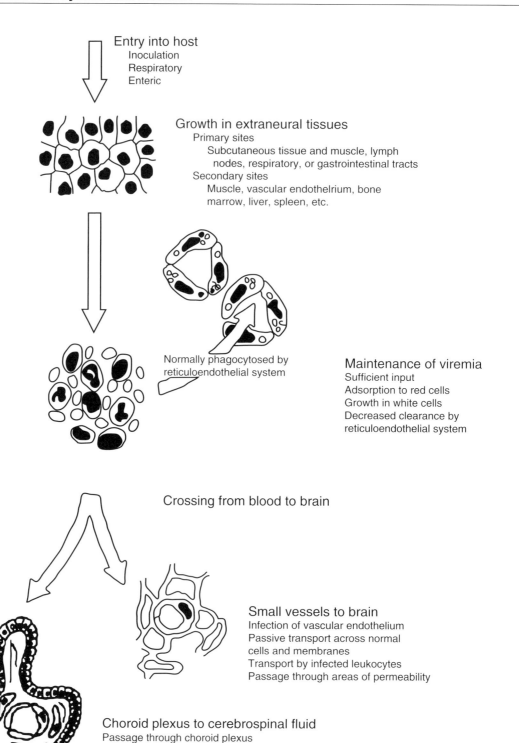

**Figure 2** Schematic diagram of steps in hematogenous spread of viruses to the nervous system. Reproduced with permission from Johnson RT (1982) *Viral Infections of the Nervous System*. New York: Raven Press.

virus, grow in choroid plexus epithelium and seed into CSF. Thus, the clearance of particles by the reticuloendothelial system, barriers of nonsusceptible extraneural cells, production of interferon and other nonspecific inhibitors, and the physical barriers of the nervous system itself probably explain why viral infections of the brain are rare, even though systemic infections with the potential pathogens are very common.

## Infections of Neural Cells

Once a virus has penetrated into the nervous system it must contact a susceptible cell and spread through the compact neuropil which is a theoretical problem. The fact that some viruses can be neutralized by extracellular antibody even after CNS invasion shows that viruses such as togaviruses and flaviviruses do spread in extracellular space, but this is not true of larger viruses. Conversely, the compact neuropil may facilitate the contiguous cell-to-cell spread of viruses. For example, in subacute sclerosing panencephalitis, a chronic brain infection of humans with measles, extracellular enveloped virus is never seen, and there are enormous titers of extracellular antibody. Apparently, the fusion protein allows measles virus to move from cell to cell through the brain.

Cell-to-cell spread may also involve axoplasmic flow causing infection of functionally linked cells; for example, in poliovirus infections the virus is rapidly spread through the motor system. Some viruses infect only neuronal populations such as rabies, polioviruses, and the arthropod-borne viruses (arboviruses), and some infect selective neuron populations. Other viruses such as herpes simplex virus appear to infect neurons and glial populations with little selectivity.

Infection limited to vascular endothelial cells is found with rickettsial infections but is not recognized in any viral infection of the nervous system. Infection limited to choroid plexus and meningeal cells appears to occur with those viruses that cause benign meningitis. In experimental studies with a number of viruses, widespread lytic infection of ependymal cells can lead to closure or stenosis of the aqueduct of Sylvius and resultant hydrocephalus. Similar aqueductal stenosis and hydrocephalus have been described in children after mumps virus meningitis.

The selective infection of oligodendrocytes has been recognized in nature in the disease progressive multifocal leukoencephalopathy caused in humans by the JC virus and in monkeys by SV-40 virus. In the course of immunosuppression, now seen most frequently with acquired immune deficiency syndrome (AIDS), a selective lytic infection of oligodendrocytes causes multifocal areas of demyelination in the brain. This usually fatal condition has also been seen in patients treated with immunomodulating medical regimens that included natalizumab (a monoclonal antibody against alpha 4 integrin). This drug causes a release of lymphocytes from lymph nodes and prevents the trafficking of lymphocytes across the blood–brain barrier. Thus, this medication may have interfered with the normal immune surveillance of the CNS, leading to the unabated emergence of JC virus within the brain.

With changes in age, the specificity of infection and vulnerability of neural cells may change. For example, bluetongue virus infection of fetal sheep destroys the precursors of neurons and glia of the subependymal plate which leads to hydranencephaly or porencephaly dependent on the age of fetal development, whereas the virus fails to infect the mature postmigratory cells in the late gestational or postnatal animal. Similarly, the external granular cells of the cerebellum in fetal or newborn animals are selectively infected by parvoviruses, and destruction of these mitotic cells leads to the granuloprival cerebellar degeneration seen in both natural and experimental animal infections. Alphaviruses which cause encephalomyelitis have more profound effects in young hosts including mice and humans.

## Mechanisms of Cell Damage

Lytic infections of neural cells cause regional destruction of brain or lysis of specific cell populations. Noncytopathic infections of neural cells also occur and lead to persistent infection with no disease or disorders without obvious histological changes. For example, neuroblastoma cells in culture infected with noncytopathic viruses such as rabies can show normal morphology growth rates and protein synthesis, but reduced synthesis of specific neurotransmitters or receptors. These have been termed 'luxury functions', although *in vivo* the ability of neurons to synthesize transmitters or receptors would hardly be considered a luxury. Analogous noncytopathic infection has been demonstrated in mice congenitally infected with lymphocytic choriomeningitis virus. Congenitally infected mice are usually 'runts', but recent studies have shown selective infection of cells of the anterior pituitary which normally generates growth hormone. The animals are actually pituitary dwarfs responsive to growth hormone therapy.

Alternatively, the infected cell may not be damaged by virus replication but destroyed by the immune responses, as seen in adult mice infected with lymphocytic choriomeningitis virus. Indirect cell damage can occur in viral infections that leads to sensitization of the host to neural antigens. This has recently been demonstrated in rats infected with coronaviruses, where the infection of neural cells leads to a cell-mediated autoimmune response to myelin proteins and to subsequent demyelination. In postmeasles encephalomyelitis of humans, autoimmune demyelination appears to occur in the absence of infection of neural cells. Infection of lymphoid tissue leads to disruption of normal immune regulation, and about 1 per 1000 persons develop a symptomatic autoimmune reaction to myelin basic protein.

Other indirect mechanisms of neural cell damage have been postulated to explain the diverse neurological diseases seen in the course of human immunodeficiency virus (HIV) infection. The virus does not appear to cause significant or readily documented infections of neuronal or glial cells. The primary cells that are infected are macrophages and the microglia of the brain derived from macrophage populations. Possibly viral proteins produced by these cells or cytokines released by these cells interfere with neuronal function or are toxic to neurons or glial cells. For example, tumor necrosis factor, a lymphokine often increased in brains of neurologically affected AIDS patients has been shown *in vitro* to induce demyelination.

## Clinical Features

### Acute Infections

The varied clinical features of viral infections of the nervous system can be explained in large part by the factors discussed above. Thus, a virus may infect only meningeal or ependymal cells, and cause a clinical syndrome known as viral meningitis or acute aseptic meningitis. This clinical syndrome is characterized by fever, headache, and nuchal rigidity secondary to meningeal irritation but without clinical signs suggesting parenchymal disease. The commonest causes of acutely viral meningitis are enteroviruses and mumps virus.

Encephalitis is a clinical syndrome in which in addition to fever, headache, and stiff neck, there is paralysis, altered mental status, seizures, or other evidence of parenchymal disease of the brain. The commonest causes of severe encephalitis in man are herpes simplex virus and the arboviruses. The former infects neurons and glia and causes diffuse necrotizing encephalitis in the newborn but focal encephalitis in the immune adult presumably because diffuse virus spread is contained by immune responses. Focal signs, hallucinations, behavioral abnormalities, and aphasia are more common with herpes encephalitis because of the localization to temporal lobes. Arboviruses have a propensity to infect neurons, and some flaviviruses tend to infect basal ganglia and brainstem neurons causing movement disorders and sudden respiratory failure. If signs of spinal cord involvement accompanies encephalitis, the term encephalomyelitis is often used. However, the term encephalomyelitis is also used to distinguish an acute postinfectious demyelinating disease of assumed autoimmune origin from acute viral encephalitis. Postinfectious encephalomyelitis (or acute disseminated encephalomyelitis) usually occurs 3–14 days after exanthems (measles, varicella, or rubella) or respiratory infections (mumps, influenza, and others) and clinically is characterized by the abrupt onset of fever, obtundation, seizures, and multifocal neurological signs.

The clinical syndromes of rabies and poliomyelitis are the most distinctive of viral infections because of the selective infection of specific populations of neurons. Polioviruses selectively infect motor neurons which lead to flaccid paralysis. Rabies causes early infection of the limbic system neurons with a relative sparing of cortical neurons which leads to behavioral abnormalities. Rabies infections represent a diabolical adaptation of virus to animal host, causing the animal to remain alert but to lose timidity and develop aggressive behavior to transmit the virus. The advantage of this selectivity is evident considering that strains of rabies that cause the so-called 'dumb' or passive rabies are seldom transmitted in nature.

### Slow Infections

Slow infections are characterized by long incubation periods of months to years followed by an afebrile progressive disease. The term was originally coined in veterinary medicine to describe several transmissible diseases in sheep. The prototype slow infections are scrapie, a chronic noninflammatory spongiform encephalopathy due to a transmissible agent in which no nucleic acid has been identified (a prion), and visna, a chronic inflammatory encephalomyelitis caused by a lentivirus. The first slow infection identified in man was 'kuru', a progressive ataxia of a tribal group in New Guinea where the agent, resembling the agent of scrapie, was apparently transmitted by ritual cannibalism. Creutzfeldt–Jakob disease, a subacute dementia with myoclonus, is a worldwide disease due to prion agents. In some cases, transmission is genetic, in others, transmission is by transplants or administration of contaminated growth hormone, but in the vast majority the means of transmission is unknown. In both of these human spongiform encephalopathies, the clinical disease progresses irrevocably to death in about 6 months, but without fever or other clinical or histological findings to suggest infection.

Dementia, a chronic deterioration of cognitive function, can also be caused by several conventional viruses. Subacute sclerosing panencephalitis is a chronic dementing illness caused by measles virus. One per million otherwise healthy children develop this chronic illness at an average of 7 or 8 years after uncomplicated measles. Dementia evolves slowly, associated with myoclonic movements and massive levels of measles antibody in serum and CSF. Children usually die a year or less after infection, but survival may vary from 6 weeks to 6 years. The disease is due to a subacute, slowly progressive diffuse infection of neurons and glial cells. Another viral cause of dementia, HIV, has become a common problem as patients are surviving longer while on highly active antiretroviral therapy.

Clinical symptoms associated with HIV are very diverse (**Figure 2**). This virus now represents the commonest viral infection of the nervous system. From prospective studies of CSF changes, it appears that the majority of people infected with the virus have early invasion of the nervous system, that is, the virus is highly neuroinvasive. However, early during this infection neurological disease is rare. A presumed autoimmune disorder occasionally causes a demyelinating peripheral neuropathy (Guillian–Barre syndrome); acute meningitis is occasionally seen at the time of seroconversion or during early asymptomatic infection. An asymptomatic pleocytosis is often found. Therefore, the virus is, at this stage, not highly neurovirulent. However, after the onset of immunodeficiency the infection is neurovirulent; at least 50% of the patients develop progressive dementia with cerebral involvement, myelopathies, or painful sensory neuropathies. The pathogenesis of these complications is unknown, but they are thought to be due to some viral protein or lymphokine incited by the virus because of the relative paucity of the virus in the nervous system and its localization to the microglial and macrophage populations.

Tropical spastic paraparesis complicating human T-cell leukemia virus type 1 infections is a recently recognized slow nervous system infection. Less than 1% of the patients infected with this virus develop either acute T-cell leukemia or neurological disease. Since many of those infected are infected by breast milk and the onset of tropical spastic paraparesis is usually in the fourth or fifth decade of life, the incubation period is extraordinarily long. A subacute disease develops with progressive paralysis of the lower extremities associated with impotence, incontinence, and sensory symptoms, but usually minimal sensory findings. Disease progresses slowly until the patients are wheelchair bound, but the disorder remains primarily at the level of the thoracic spinal cord. The involvement of the upper extremities is minimal with hyperreflexia but usually good function, and there is usually little, if any, indication of cerebral involvement. Early pathology found in a biopsy of the spinal cord in a single case showed vasculitis. In late cases, hyalinization of vessels with necrosis and demyelination of spinal cord is found, and findings are most prominent in the thoracic cord. Whether virus replicates in any cells other than T lymphocytes is unknown. These observations lead to the questions of why less than 1:100 who are infected develop disease; why the incubation period is as long as 40 years; why the disease localizes to the thoracic spinal cord; and why over years the disease becomes relatively quiescent despite the fact that there is ongoing high levels of intrathecal antibody synthesis, suggesting that there is still antigenic stimulation by virus within the nervous system.

Identification of viruses or virus-like agents (prions) in a variety of chronic neurological diseases has led to speculation of a viral etiology for multiple sclerosis, Parkinson's disease, amyotrophic lateral sclerosis, Alzheimer's disease, schizophrenia, and other illnesses. Experimental evidence for viruses in these chronic diseases is still tenuous.

*See also:* Herpesviruses: Latency; Measles Virus; Mumps Virus; Persistent and Latent Viral Infection; Rabies Virus; Viral Receptors; Visna-Maedi Viruses.

## Further Reading

Johnson RT (1982) *Viral Infections of the Nervous System.* New York: Raven Press.

# Cereal Viruses: Maize/Corn

**P A Signoret,** Montpellier SupAgro, Montpellier, France

© 2008 Elsevier Ltd. All rights reserved.

## Introduction

Maize is the main cereal crop in the world regarding total yield; it is grown on about 120 Mha. Maize has been grown for millennia in Central America. From a plant mainly used as human food, maize is now a main component for feeding animals but is still a major staple food crop in sub-Saharan Africa and America.

Viruses can cause important diseases of maize worldwide. While some viruses are widespread, others are localized. Some maize viruses can be sporadic and devastating causing severe yield losses, others may occur each year but losses are relatively minor. The average yield of maize in several African countries is only about a third of the world's average, and some viral diseases of maize are one of the major factors responsible for this low productivity. When available, resistant or tolerant hybrids provide the most effective means to control maize viruses. A pathogen-derived resistance strategy was developed for some viruses such as maize dwarf mosaic virus (MDMV) and maize streak virus (MSV). The major viruses affecting maize are transmitted by leaf- or planthoppers (16) but four viruses have aphids, two have mites, and one has beetles as vector. Populations of leafhopper and planthopper vectors can be controlled by systemic insecticides applied as seed dressing or at sowing time. Herbicide eradication of Johnson grass, an overwintering host of some viruses, had a significant impact on disease control. In some cases, crop rotation and selection of planting date result in significant decreases in virus infection.

## Maize Virus Diseases

### Barley Yellow Dwarf (MAV-PAS-PAV), Cereal Yellow Dwarf (GPV-RPS-RPV), RMV-SGV

All the virus species of the family *Luteoviridae* in this subsection cause similar symptoms (see **Table 1** for a list of relevant virus names). Usually infected cultivars express discoloration of the leaves, reddening and yellowing, but some present no symptoms. More or less red or yellow bands are observed mainly on the borders of the older leaves. Compared to the healthy plants, possible grain yields are lower by 15–20%. Seeds coated with a systemic insecticide have to be used in areas where the disease is frequently encountered. A few maize inbred lines tolerant to barley yellow dwarf virus are available. Maize hybrids are in general tolerant.

### Chloris Striate Mosaic Virus

This virus found in 1963 in Australia causes striate mosaic in *Chloris gayana* Kunth and other grass species and in cereals including barley, maize, and oats. It is transmitted by the leafhopper *Nesoclutha pallida* (Evans). Natural hosts are

**Table 1**  Taxonomic position of maize viruses

| Virus names | Genus | Family |
| --- | --- | --- |
| Barley yellow dwarf-MAV virus | *Luteovirus* | *Luteoviridae* |
| Barley yellow dwarf-PAS virus | *Luteovirus* | *Luteoviridae* |
| Barley yellow dwarf-PAV virus | *Luteovirus* | *Luteoviridae* |
| Barley yellow striate mosaic virus | *Cytorhabdovirus* | *Rhabdoviridae* |
| Brome mosaic virus[b] | *Bromovirus* | *Bromoviridae* |
| Cereal yellow dwarf-GPV virus | *Polerovirus* | *Luteoviridae* |
| Cereal yellow dwarf-RPS virus | *Polerovirus* | *Luteoviridae* |
| Cereal yellow dwarf-RPV virus | *Polerovirus* | *Luteoviridae* |
| Chloris striate mosaic virus | *Mastrevirus* | *Geminiviridae* |
| Corn lethal necrosis complex | | |
| Corn stunt complex | | |
| Cynodon chlorotic streak[a] | | |
| High Plains virus[a] | | |
| Indian peanut clump virus | *Pecluvirus* | |
| Johnson grass mosaic virus | *Potyvirus* | *Potyviridae* |
| Maize chlorotic dwarf virus | *Waikavirus* | *Sequiviridae* |
| Maize chlorotic mottle virus | *Machlomovirus* | *Tombusviridae* |
| Maize dwarf mosaic virus | *Potyvirus* | *Potyviridae* |
| Maize fine streak virus | *Nucleorhabdovirus* | *Rhabdoviridae* |
| Maize Indian Fiji-like virus[a] | | |
| Maize Iranian mosaic virus[a] | | |
| Maize mosaic virus | *Nucleorhabdovirus* | *Rhabdoviridae* |
| Maize mottle/chlorotic stunt virus[a] | | |
| Maize necrotic streak virus[a] | | |
| Maize rayado fino virus | *Marafivirus* | *Tymoviridae* |
| Maize rough dwarf virus | *Fijivirus* | *Reoviridae* |
| Maize sterile stunt virus[a] | | |
| Maize streak virus | *Mastrevirus* | *Geminiviridae* |
| Maize stripe virus | *Tenuivirus* | *Tenuiviridae* |
| Mal de Rio Cuarto virus | *Fijivirus* | *Reoviridae* |
| Pennisetum mosaic virus[a] | | |
| Rice black-streaked dwarf virus | *Fijivirus* | *Reoviridae* |
| RMV | | *Luteoviridae* |
| SGV | | *Luteoviridae* |
| Sugarcane mosaic virus | *Potyvirus* | *Potyviridae* |
| Wheat American striate mosaic virus | *Cytorhabdovirus* | *Rhabdoviridae* |
| Wheat streak mosaic virus | *Tritimovirus* | *Potyviridae* |

[a]Unassigned virus.
[b]Not described, virus with low or no agronomical importance.

several grasses, and barley, oats, maize, and wheat. Infected plants develop chlorotic grayish white striation on leaves; notching and curling of leaves may also occur. Maize is occasionally infected but most hybrids are resistant.

## Cynodon Chlorotic Streak

Cynodon chlorotic streak virus (CCSV) can cause a serious disease in the Mediterranean region. Symptoms on maize are characterized by the development on the youngest leaves of yellow streaks which become larger and can coalesce to form yellow stripes (**Figure 1(e)**). Finally the whole leaf turns yellow with some parts becoming reddish. If contamination occurs at a young stage, the plants are dwarfed and die early. The virus is transmitted by planthoppers and occurs naturally in Bermuda grass (*Cynodon dactylon* (L.) Pers.) and maize. The level of CCSV infection in maize is directly related to the level of CCSV infection and planthopper vector populations in Bermuda grass. As many as 70–84% virus-infected plants have been observed for some years in southern France. Seed dressing with imidachloprid allowed a good control of the vector population.

## High Plains Virus

High Plains virus (HPV), a previously unknown pathogen, was found in 1993 to infect maize in US High Plains states. It has also been found in other countries. The virions are flexible, thread-like particles. Analysis of the nucleic acid shows four bands of double-stranded RNA (dsRNA) after electrophoresis. A great variety of symptoms may be expressed in maize, including spots or flecks along vascular bundles, purpling or reddening of the leaf margins or sectors of the leaf, and stunting. The virus is transmitted obligately by the wheat curl mite *Aceria tosichella* (Keifer) that also transmits wheat streak mosaic virus (WSMV). HPV can be seed transmitted. The host range includes barley, maize, and wheat in addition to several weed species. Three dominant genes for resistance have been identified and mapped in maize, and they provide a high degree of resistance.

## Johnson Grass Mosaic Virus

Johnson grass mosaic virus (JGMV) was first reported in Australia in maize and sorghum and later in the US and South America. Johnson grass (*Sorghum halepense* (L.) Pers.), maize, and some other plants of the Poaceae are natural hosts. The virus causes mosaic, mottling, and ringspot symptoms on maize and grass hosts, which is particularly evident on young leaves. Since the symptoms of JGMV and MDMV are very similar, serology or polymerase chain reaction (PCR) may be necessary to distinguish between the two viruses. The type strain from Australia can infect oats, which is not susceptible to the other graminaceous potyviruses. The perennial host Johnson grass allows virus survival between seasons in rhizomes. Resistant or tolerant hybrids provide the best control option.

## Maize Chlorotic Dwarf Virus

Confirmed reports of maize chlorotic dwarf virus (MCDV) infection are limited to the US, where the virus was discovered in 1969. The extent of distribution of MCDV in the US is second to that of MDMV. The virus is associated with the occurrence of indigenous Johnson grass (*S. halepense* (L.) Pers.), its overwintering host, and the black-faced leafhopper, *Graminella nigrifrons* (DeLong and Mohr), its principal vector. Symptoms on susceptible maize are shortening of the upper internodes or proportionate stunting of the internodes, chlorotic clearing or banding of the tertiary leaf veins, and red or yellow discoloration of upper leaves. Infected leaves are also more difficult to tear off a plant. On resistant maize, MCDV induces mainly chlorotic clearing or banding of tertiary veins without marked stunting or discoloration. The diagnostic symptom of MCDV infection is vein banding. MCDV's host range is limited to species of the Poaceae. MCDV may complex with MDMV in infections of maize and Johnson grass in the same field or plant and with sugarcane mosaic virus strain MDB in maize. Biological tests for MCDV include transmission by the leafhopper in a semipersistent manner and vascular-puncture inoculation (VPI) to susceptible plants with expression of the diagnostic symptoms of vein clearing or banding of the tertiary leaf veins. Confirmation by serological methods is often necessary. No molecular tests for MCDV have been developed.

Agronomical control of MCDV is possible by early planting of maize to escape peak vector populations and by herbicide eradication of overwintering Johnson grass. However, the latter treatment may result in increased MCDV incidence in susceptible hybrids due to mass migration of viruliferous leafhoppers from dying Johnson grass. Maize inbreds with high levels of tolerance to MCDV have been developed. Virus resistance appears inherited mostly as a dominant to partially dominant trait. Resistant Caribbean germplasm has been identified and the inbred line Oh1VI highly resistant to MCDV has been developed. Commercial hybrids with various degrees of resistance or tolerance and with agronomically satisfactory characteristics are available. The systemic insecticide carbofuran which remains toxic to the leafhopper for up to 55 days after application at planting time is very effective and can control the disease. As a result the economic impact of the virus has decreased and it no longer poses a significant threat to US maize production.

**Figure 1** (a) Long continuous stripes along the veins of sweet corn caused by maize mosaic virus. (b) Severe dwarfing, stripping of upper leaves and 'leek aspect' of maize rough dwarf virus-infected maize plant. (c) Symptoms of maize streak virus on maize. (d) Maize plants infected with Mal de Rio Cuarto virus showing severe dwarfing and leaves distorted and ragged. (e) Yellow stripe and dwarfing on maize infected by cynodon chlorotic streak virus.

## Maize Chlorotic Mottle Virus

Maize chlorotic mottle virus (MCMV) was first reported from Peru in 1973, and was later found in central USA, Hawaii, Mexico, and Argentina. Symptoms vary from a mild chlorotic mottling to severe yellowing, necrosis, and plant death depending on the maize variety and plant age when infected. Plants may be stunted with short, malformed, partially filled ears, and male inflorescences may be shortened with sparse spikes. Under natural infection, 10–15% yield reductions can be observed. Maize is the only natural host of MCMV, and the experimental host range is limited to Poaceae. Laboratory transmission of MCMV can be obtained with six species of chrysomelid beetles, but only *Diabrotica* spp., which have a New World distribution and are mostly tropical, are natural vectors. In areas of South America where maize is grown continuously, MCMV spreads from older plants to younger plants via adult and larval beetles. Control would require a halt in continuous cultivation or management of the beetles. Most maize cultivars are susceptible to MCMV but a new variety (N211) is resistant. Transgenic maize shows milder symptoms from MCMV infection. MCMV is sometimes found as part of the corn lethal necrosis complex.

## Maize Dwarf Mosaic Virus

MDMV was first reported on corn in 1965 from Ohio, USA. It is considered to be distributed worldwide. Symptoms of MDMV in corn vary with the stage of host development at time of infection, genotype, and virus strain. Initial leaf symptoms are chlorotic spots and short streaks followed by typical mosaic or mottle. When the plants develop, flecks, streaks, and rings appear on new leaves. Ear formation and development are arrested, resulting in losses in grain yield. MDMV is widely adapted to species of the Poaceae infecting over 66% of 293 species tested. Maize and sorghum are important agricultural hosts. More than 15 species of aphids transmit this virus in a nonpersistent manner. MDMV can be mechanically transmitted; it survives by overwintering in the rhizomes of infected Johnson grass. Crop loss due to MDMV ranges from 9% to 72% for susceptible dent corn hybrids. Losses are generally important in seed production or sweet corn hybrid fields.

Many commercial corn MDMV-resistant hybrids are available. A major gene for resistance to MDMV is a dominant gene (*mdm1*) located on chromosome 6. The control of Johnson grass helps to control disease.

## Maize Fine Streak Virus

Maize fine streak virus (MFSV) was isolated in 1999, from sweet corn grown in Georgia, USA. Symptoms in the field include fine chlorotic streaks along the major leaf veins, similar to symptoms associated with maize rayado virus (**Figure 1(c)**). Maize is so far the only identified natural host of MFSV. The virus is transmitted by the leafhopper *Graminella nigrifons* (DeLong & Mohr) in a persistent manner. MFSV is readily transmitted mechanically by kernel vascular-puncture inoculation. Maize lines resistant to maize mosaic virus (MMV: Hi31 and Hi34) are also resistant to MFSV.

## Maize Indian Fiji-Like Virus

Symptoms include severe plant dwarfing, dark green leaves, enations on the lower leaf surface, and small malformed ears. All of these symptoms are similar to those produced by Mal de Rio Cuarto virus (MRCV) and maize rough dwarf virus (MRDV) on maize. This unreported reovirus was found in southern India, where its incidence ranged from 4% to 61%.

## Maize Iranian Mosaic Virus

First observed in the Fars Province of Iran, it is now distributed in other parts of the country. Maize Iranian mosaic virus (MIMV) symptoms consist of long chlorotic lines and stripes along the veins and sheaths, stunting of the plants, and abortion of ears when infected early. MIMV is serologically distinct from others rhabdoviruses infecting Poaceae plants. In polyacrylamide gel electrophoresis, six proteins corresponding to proteins L, G, N, NS, M, and M2 were detected. MIMV is a new virus in the genus *Nucleorhabdovirus* that may be distantly related to MMV. Delphacids are vectors of MIMV. The natural host range includes maize, oat, rice, and wheat. The disease incidence in maize fields can be as high as 80% depending on sowing date and variety. Late planting effectively controls the virus.

## Maize Mosaic Virus

MMV causes a serious disease of maize in tropical and subtropical areas of Africa, the Americas, and Australia. It was first reported in 1914 in Hawaii. MMV can occur in mixed infections with maize stripe virus (MSpV). Initial leaf symptoms on sweet corn inoculated with MMV are long stripes, colored light green to yellow, along the midrib (**Figure 1(a)**). These stripes elongate to form distinct, chlorotic stripes between and along the veins extending on the whole leaf. Stripes can appear on sheath and husk. Stunting of plants is common. Mosaic patterns do not occur despite the disease name. Only species of the Poaceae are known to be hosts of MMV. It occurs naturally in itchgrass, maize, and *Setaria vulpiseta* (Lam.) Roem. & Schult. MMV is transmitted in a persistent-propagative manner by the corn planthopper *Peregrinus maidis* (Ashmead), the sole vector. Transmission is by both nymphs and adult males and females. MMV has been shown to be inoculated by vascular puncture of maize kernels (VPI). Several contact and systemic insecticides can control *P. maidis*. Host-plant resistance is likely to be the most effective means of control. Tropically adapted parental inbred lines of maize resistant to MMV have been developed and sweet corn cultivars have been bred and released, both by the University of Hawaii. From Caribbean and Mascarene germ plasm, high resistance was found in line Hi40 and resistance to some levels was found in lines 37–2, A211, and Mp705.

## Maize Mottle Chlorotic Stunt Virus

This virus was first reported, in 1937, in East Africa and called maize mottle virus (MMotV). Maize mottle chlorotic stunt virus (MMCSV) is restricted to tropical African countries and the adjacent islands. The virions are spherical, *c.* 40 nm in diameter. MMCSV can be found in mixed infections with maize mosaic and maize streak viruses. Symptoms on maize appear as chlorotic mottling with mild to severe chlorosis of tertiary veins and stunting. Young mottled leaves fail to support themselves in a normal upright position. Tassel abortion can occur in severe cases. Maize is the only known natural host of the virus which is transmitted by several *Cicadulina* spp. in a persistent manner. Disease control can be achieved by vector control or by growing resistant cultivars, which are now available. Most maize streak virus-resistant cultivars have also moderate to high levels of resistance to MMCSV.

## Maize Necrotic Streak Virus

Maize necrotic streak virus (MNeSV) was discovered in 2000 in maize samples from the US state of Arizona. The virus is of little agronomic importance. Virions are isometric, *c.* 32 nm in diameter. MNeSV appears to be a member of the family *Tombusviridae*. Distinctive symptoms of MNeSV infection are initially chlorotic spots and streaks that become spindle shaped. They later coalesce into long chlorotic bands that become translucent and necrotic around the edges. Plant stalks show a chlorosis that becomes necrotic, a symptom especially distinctive for MNeSV. There is no known aerial vector – the virus is transmitted through the soil. Maize lines Oh 1VI OSU23i and Mo17 are highly resistant to MNeSV.

## Maize Rayado Fino Virus

Maize rayado fino virus (MRFV) was first described in 1969 in Costa Rica and El Salvador. MRFV is widespread and becoming increasingly important in tropical areas of the Americas. It is the type member of the genus *Marafivirus*, and is the only known indigenous virus of maize in Mesoamerica. Initially, infection causes fine chlorotic dots or stipples, which coalesce into chlorotic stripes at the base and along the veins of young leaves. As the plant grows older, the symptoms become less conspicuous and may disappear when the plant reaches maturity. In locally adapted Central American varieties, the virus incidence varies from 0% to 20%, but may reach 100% in more susceptible foreign or newly developed cultivars. Maize, teosinte, and the perennial *Zea* spp. are the only known natural hosts. In nature, MRFV is transmitted exclusively by the corn leafhopper vector *Dalbulus maidis* (DeLong and Wolcott) in a persistent manner. The natural host range of MRFV may be limited by the ability of the vector to feed on different species. It has been proposed that MRFV, its insect vector, and the maize host co-evolved in a triad in which the parasitic members (insect and virus) displayed highly specialized interactions. It has been suggested that MRFV originated in Mexico or Guatemala, from where it spread in the region.

MRFV is one of the three major components of the corn stunt disease complex. MRFV is endemic in many areas of Central and South America, where the virus and *D. maidis* reach epidemic levels. The use of systemic insecticides does not reduce virus incidence. Crop rotation, mixed plantings, and selection of planting dates result in significant decreases in virus infection. The Guatemalteco variety introduced in Ecuador in 1962 is used as a source of virus tolerance. Production of transgenic plants resistant to MRFV and their evaluation for their susceptibility are in progress in Costa Rica.

## Maize Rough Dwarf Virus

This virus was discovered in Europe when corn hybrids were introduced from the US after World War II. In 1949, in Italy, a severe outbreak threatened the maize cultivation, lowering the yield by ~40%. Later, MRDV was reported in several European countries, where it had the potential to be economically damaging. In China, a similar disease was found to be caused by rice black-streaked dwarf virus (RBSDV), not by MRDV. In young field-grown corn, symptoms caused by MRDV are dark green color of leaves, stunting and irregular swellings of veins (enations) along the lower surfaces of leaves and sometimes also of leaf sheaths, ligules, and husks. The enations are rough to the touch, hence the disease name. Plants are stunted with increased girth giving the plant a 'leek' aspect (**Figure 1(b)**). Short chlorotic streaks develop on mature leaves and coalesce into yellowish green stripes parallel to the veins. Later the leaves can turn reddish. Tassels are sterile. The root system is reduced, roots develop swellings, resulting in their frequent splitting. MRDV has a fairly wide experimental host range including only a few species of Poaceae, among which maize is the only one of economic importance. Oats, wheat, and several grasses can be infected naturally. The planthopper *Laodelphax striatellus* (Fallén) is the natural vector and the only known winter host of MRDV. However, in northern Italy, oats were reported as the first overwintering plant. Planthoppers can be controlled either by spraying grasses around the fields with an insecticide in the spring, by seed treatment with imidachlopride, or by row application of an insecticide at sowing time (e.g., carbofuran). Some maize hybrids are less susceptible to MRDV, but the resistance can be overcome if the vector population is very high.

## Maize Sterile Stunt Virus

Maize sterile stunt virus (MSSV) was first reported from Australia in 1977. It is a strain of barley yellow striate mosaic virus (BYSMV). The virus causes severe stunting, sterility, purple coloration, and top necrosis in a small number of maize genotypes. MSSV is transmitted in a persistent manner by the delphacid planthopper *Sogatella longifurcifera* Esaki and Ishihara and inefficiently by *Sogatella kolophon* (Kirk) and *P. maidis* (Ashm.). Natural hosts of MSSV are barley, maize, triticale, and wheat.

## Maize Streak Virus

Maize streak was first recorded in 1901 in South Africa and is the most widespread and important disease of maize in sub-Saharan Africa and adjacent islands. Maize streak symptoms are characterized by the development of chlorotic spots and streaks in longitudinal lines on leaves. There is a progressive increase in the number and length of streaks that occur on new leaf tissue. The streaks often fuse laterally, resulting in narrow broken chlorotic stripes, which can in some cases cover the entire leaf. In very susceptible maize cultivars, severe chlorosis occurs leading to stunted plant and premature death. MSV is transmitted by several species of leafhoppers in the genus *Cicadulina*, with *C. mbila* (Naudé) being the most common vector. It transmits MSV in a circulative nonpropagative manner. Streak symptoms have been described in numerous species of wild grasses. Late plantings are generally more severely infected than early ones. The major source of MSV infection on maize is a previously infected maize field. Therefore, planting of maize close to a previous maize field must be avoided. Carbamate insecticides (such as carbofuran) can effectively prevent MSV transmission but prices of chemicals and equipment for spraying are often prohibitive for smallholder farmers. The development of insect resistant plants is under study.

## Maize Stripe Virus

MSpV has often been confused with MMV because of their similarity in vector specificity, mode of transmission, symptomatology, and host range. It often occurs in mixed infection with MMV. The virus is commonly found in most parts of the subtropics and the tropics where the vector occurs. MSpV is found to infect maize, sorghum, and itchgrass (*Rottboellia exaltata* L.). It can be transmitted experimentally to barley, oats, rye, and triticale. Initial symptoms on sweet corn are fine chlorotic stipplings and narrow stripes similar to that of MMV. Later, continuous chlorotic stripes develop with varying width and intensity which show a 'brushed out' appearance, and apical bending occurs. Infected plants are often stunted and die prematurely. The virus is transmitted by *P. maidis* (Ashm.), also a vector of MMV, in a persistent-propagative manner. Both nymphs and adults transmit the virus. Control of the maize stripe disease is largely dependent on chemical control of *P. maidis*.

## Mal de Rio Cuarto Virus

The Mal de Rio Cuarto was first reported in maize fields of the Rio Cuarto County in Argentina, at the end of the 1960s. MRCV is a major constraint to maize production in Argentina. It spread in the endemic area of the province of Cordoba, but also appeared in the central region. MRCV initially reported as a strain of MRDV but is now considered to be a separate species within the genus *Fijivirus*. In field-infected maize, the symptoms show up within 4–5 weeks. Plants are dwarfed, showing fine chlorotic flecks on secondary and tertiary veins. Stems are usually flattened with shortened internodes and leaves from the upper third may be stiff, distorted, and ragged (**Figure 1(d)**). The root system is fragile, and greatly reduced. The ears are reduced and malformed. Natural dispersion of the virus occurs by the planthopper *Delphacodes kuscheli* Fennah in a propagative manner. Apart from maize, MRCV can infect several monocots, winter small grains (barley, oats, rye, and wheat), spring–summer grains (millet and sorghum), and several spring–summer annual and perennial weeds. The initial inoculum source is its vector that feeds on winter cereals like oat and wheat, where it acquires the virus and migrates to maize when these cereals become senescent. Early sowing of maize helps to prevent high virus infection. Systemic insecticides (carbofuran, imidachlopride) applied as seed coating protect maize seedlings from the vector. Some commercial hybrids with high tolerance levels are now available.

## Pennisetum Mosaic Virus

A potyvirus was isolated from whitegrass (*Pennisetum flaccidum* Griseb) in North China, in the early 1980s and was later named pennisetum mosaic virus (PenMV). This virus infects maize and sorghum naturally. Symptoms on maize are not specific when compared to other members of the sugarcane mosaic virus (SCMV) subgroup in the genus *Potyvirus*. The virus is less important, in terms of yield loss, than the other prevalent viruses of the SCMV subgroup.

## Rice Black-Streaked Dwarf Virus

The causal agent of characteristic symptoms on maize (stunting and rough white line veins of leaves) was initially identified as MRDV in China in the late 1970s. Subsequently, molecular studies demonstrated that rice black-streaked virus was the cause of the disease. These two viruses are quite similar in host range, serology, morphology, and the genome sequence. Reverse transcription-polymerase chain reaction can discriminate between RBSDV and MRDV. Host range is limited to species in the Poaceae: oats, hordeum, oryza, triticum, and zea. This virus is transmitted by the leafhopper *L. striatellus* in a persistent manner. RBSDV is managed mainly by control of leafhoppers.

## Sugarcane Mosaic Virus

SCMV was first detected in sugarcane in 1919 and in maize in 1963, both in the US. SCMV occurs throughout the world. Infected maize plants develop distinct mosaic symptoms which are especially clear on the lower part of the younger leaves. Very susceptible cultivars react with a strong chlorosis, sometimes together with a red striped pattern. It is often impossible to distinguish SCMV and MDMV infections on the basis of symptoms. In Southern Europe, MDMV is the most prevalent virus. Maize inbred lines resistant to the virus have been described and transgenic maize expressing the coat protein gene of SCMV is also available.

## Wheat Streak Mosaic Virus

The symptoms consist of linear arrays of small, yellow-ringed eyespots. On some germplasm, the spots and streaks later form mosaics, or mottle patterns. WSMV is mite transmitted. Three independent genes controlling resistance to WSMV have been identified in maize. Resistance in certain maize lines is not effective against all strains of WSMV. Late planting of corn adjacent to maturing wheat should be avoided.

## Maize Diseases Caused by Complex of Pathogens

### Corn Lethal Necrosis Complex

This disease is caused by a synergistic interaction between MCMV and certain members of the family *Potyviridae* (MDMV-A, SCMV-MDB, JGMV, and WSMV). Maize plants infected at early stages develop leaf chlorosis followed by necrosis. Plants are stunted, produce small deformed ears, and die prematurely. Only terminal leaves show symptoms with a late infection. Crop losses of up to 90% have been reported.

### Corn Stunt Complex

Corn stunt disease complex, also known by the names achaparramiento, maize stunt, and red stunt, is caused by two or more of the following agents: MRFV, corn stunt spiroplasma (CSS), and maize bushy stunt phytoplasma (MBSP). The corn stunt complex pathogens appear to be restricted to the Americas. Early symptoms of CSS consist of yellowish streaks in the youngest leaves. Later, much of the leaf turns purple. Infected plants are stunted and will often have more ears, but the ears are smaller. Maize plants infected with MBSP show stunted growth and reduced grain production, ear proliferation, and greater tillering. The leafhopper vector *D. maidis* (DeLong & Wolcott) can simultaneously transmit CSS, MBS, and MRFV. Although *D. maidis* is the primary vector, other leafhoppers transmit one or more of the corn stunt complex pathogens under experimental conditions.

*See also:* Cereal Viruses: Rice; Cereal Viruses: Wheat and Barley; Maize Streak Virus.

## Further Reading

Lapierre H and Signoret PA (2004) *Viruses and Virus Diseases of Poaceae Gramineae,* 857pp. Versailles, France: INRA.

Lübberstedt T, Ingvardsen C, Melchinger AE, *et al.* (2006) Two chromosome segments confer multiple potyvirus resistance in maize. *Plant Breeding* 125: 352–356.

Redinbaugh MG, Jones MW, and Gingery RE (2004) The genetics of virus resistance in maize (*Zea mays* L.). *Maydica* 49: 183–190.

# Cereal Viruses: Rice

**F J Morales,** International Center for Tropical Agriculture, Cali, Colombia

© 2008 Elsevier Ltd. All rights reserved.

## Glossary

**Ambisense** A viral nucleic acid that can be 'read' in both directions to be translated into viral encoded proteins.

**Capsid** Protein envelope of the viral genome (nucleic acid).

**Circulative** A virus that passes through the intestine of an insect vector into the hemolymph, and eventually reaches the salivary glands of the insect to be injected back into a plant.

**Etiology** Studies the cause of disease.

**Inclusion** Virus particles or viral components found in infected plant cells.

**Incubation** Period of time that a virus needs inside its vector before it can be transmitted to a susceptible host.

**ORF** Any sequence of DNA or RNA that can be translated into a protein.

**Persistent** A virus that is not lost by a vector after feeding in a series of healthy plants, without having access to a virus-infected plant (virus source).

**Pinwheel** A cytoplasmic protein inclusion induced by potyviruses.
**Propagative** A virus that multiplies in its vector.
**Semi-persistent** Virus is lost after a vector feeds on a series of virus-free plants.
**Single-stranded** Viruses consisting of only one strand of viral RNA or DNA.
**Transovarial transmission** Of a virus from a female insect vector to her progeny.
**Transtadial** A virus that is retained by an insect vector after each developmental stage until it reaches adulthood.
**Virion** A mature virus particle (capsid protein, nucleic acid, and any other constituents).
**Viroplasm** Amorphous aggregation of viral components.

## Introduction

Rice (*Oryza sativa* L.) is the second most extensive food crop grown in the world (c. 154 million ha) after wheat, and is the main food crop grown in the tropics. Half of the world's population (over 3 billion people) depends on rice, particularly in Asia, where annual per capita rice consumption may be as high as 170 kg in countries such as Vietnam, as compared to 9 kg in the United States. Asia produces almost 90% of all the rice cultivated in the world (136 million ha), mainly China and India, where rice cultivation and domestication seemed to have started over 5000 years ago. Africa cultivates approximately 9 million ha, and Latin America 6.8 million ha of rice. Nigeria and Brazil are the main producers of rice in Africa and Latin America, respectively, with average cultivated areas of 3.8 million ha, each.

Given the history and extensive cultivation of rice around the world, including Europe (565 000 ha) and the United States (1 355 000 ha), it is not surprising to note that this crop has several pest and disease problems of significant socioeconomic importance. Among the various biotic constraints of rice production, viruses constitute an important group of plant pathogens, particularly in Asia. Viral diseases are also considered major constraints to rice production in Africa and Latin America, even though only two viruses are considered pathogens of economic importance outside Asia.

Rice is also affected in the tropics by numerous insect pests, particularly by several species of leafhoppers and planthoppers (Homoptera:Auchenorrhyncha) that cause direct feeding damage and, more important, transmit economically important plant viruses. This article describes the main viral diseases of rice in the world; their causal viruses and vectors; and the different disease management strategies implemented to date.

## Rice Dwarf

Rice dwarf was first described in Japan, in 1883. A Japanese rice grower was the first person to demonstrate in 1894, the relation between this disease and the presence of leafhoppers in affected rice fields. Rice dwarf is also known to occur in Korea, China, Nepal, and the Philippines. In 1885, K. Takata reported that the leafhopper *Recilia dorsalis* was 'responsible' for the disease. In 1900, the leafhopper *Nephotettix cincticeps* was thought to be the cause of rice dwarf. In 1899, *N. cincticeps* individuals captured in a disease-free area could not induce disease in healthy rice plants, unless these individuals had been previously allowed to feed on rice dwarf-affected rice plants. Despite this evidence suggesting the existence of an unknown causal agent, it was not until 1960 that the first electron micrographs of an isometric virus c. 70 nm in diameter were obtained. The suspected virus was shown in 1966–70, to contain double-stranded (ds) RNA, divided in 12 different segments (4.354–0.828 kb) with a total genome size of 25.13 kb. The causal virus was classified in 1972 as *Rice dwarf virus*, a species of the genus *Phytoreovirus*, in the family *Reoviridae*. The virus replicates in the cytoplasm of infected cells, and the dsRNA genome is encapsidated in super-coiled form.

Rice dwarf virus (RDV) is transmitted by the leafhoppers *N. cincticeps* (main vector in Japan, Korea and China), *Nephotettix nigropictus* (formerly *N. apicalis*; the main vector in Nepal and the Philippines), *Nephotettix virescens*, and *R. dorsalis* (formerly *Inazuma dorsalis*). Few individuals of these species are active or potential vectors, and they acquire the virus by feeding for up to a day on infected plants. The incubation period of the virus in a potential vector ranges from 12 to 25 days ('propagative' virus), and the leafhopper vector remains viruliferous for life ('persistent' mode of transmission). Viruliferous individuals require from a few minutes to half an hour to transmit the virus to a healthy plant, and there is transovarial transmission of RDV to the progeny of viruliferous female vectors. Some nymphs infected through the egg die prematurely. The average lifespan of viruliferous *N. cincticeps* females is 12.1 days, and of virus-free females is 16.6 days. RDV is not mechanically transmissible and it is not transmitted by sexual seed or pollen.

RDV induces stunting and chlorotic specks in infected rice plants. RDV's host range is restricted to the Gramineae, including the genera *Alopecurus*, *Avena*, *Echinochloa*, *Oryza*, *Panicum*, *Paspalum*, *Poa*, and some cultivated species, namely, barley, millet, oats, rye, and wheat. The disease usually appears in the field after transplanting, suggesting an early infection in the seedbed stage. Early infection generally results in severe stunting, shortening of internodes, and small, rosette-like tillers. The root system of diseased plants does not develop well either, and infected plants usually do not produce panicles or seed. Diseased plants remain green in contrast to virus-free plants, which naturally turn yellow when mature. Virions are found in phloem, cytoplasm, and in cell vacuoles. Viroplasms can be observed in infected cells. Genetic resistance to RDV has been observed in some rice varieties, but current efforts are directed toward the development of genetically modified (transgenic) rice plants possessing resistance to the virus. Complementary cultural practices and other integrated disease management strategies are also recommended.

## Rice Black Streaked Dwarf

Rice black streaked dwarf was first reported to affect rice in Japan in the early 1950s, and it also spreads in China and Korea. This disease was shown to be different from rice dwarf, and to be associated with the planthopper *Laodelphax striatellus*. The causal agent was isolated in 1969 and further characterized in 1974 as a spherical virion c. 60 nm in diameter, belonging to the reovirus group. Rice black streaked dwarf virus (RBSDV) is currently classified as a member of a species of the genus *Fijivirus*, family *Reoviridae*. It has 10 dsRNA segments coding for the RNA polymerase, core protein, nonstructural components, spikes, major outer shell, and other proteins of unknown function. Three major proteins (130, 120, and 56 kDa) and three minor proteins (148, 65, and 51 kDa) have been detected in RBSDV virions. The 56 kDa protein is the main component of the outer capsid shell. Most of the genome segments are monocistronic, and replication occurs in the cytoplasm of phloem cells. Inclusion bodies of unknown nature are observed in malformed tissue. Some degree of genomic variability among RBSDV isolates has been observed.

RBSDV is transmitted by the delphacid vector *L. striatellus*, and occassionally by *Unkanodes sapporona* and *Unkanodes albifascia*, in a persistent, circulative, and propagative manner. The proportion of active transmitters of *L. striatellus* is about 30%. Shortest acquisition feeding period was 30 min; and incubation periods ranging from 4 to 35 days in the vector *L. striatellus* have been reported. The virus is not transmitted congenitally. *Laodelphax striatellus* breeds on rice, wheat, and barley, but not on maize; whereas *U. sapporona* favors maize, wheat, and barley, but does not breed on rice. Besides rice, RBSDV infects oats, barley, wheat, maize, rye, and species of *Agrostis, Alopecurus, Digitaria, Echinochloa, Eragrostis, Glyceria, Lolium, Panicum,* and *Poa* in the Gramineae. Perpetuation of the virus between seasons occurs through overwintering planthoppers. Main symptoms associated with this disease are: stunting, darkening of leaves, leaf malformation, and waxy swellings along veins of the abaxial leaf surface, which later form dark tumors. Infected plants usually produce excessive tillers.

Resistance to RBSDV has been observed under field conditions. There is at least one dominant gene involved in the resistance to RBSDV, but some modifying genes may also be involved in the outcome of the interaction between the plant genotype and the amount of virus inoculated. Some wild rice species have exhibited moderate resistance to the virus. Cultural practices aimed at reducing vector populations, such as sowing dates that do not coincide with peak populations of the vector, are recommended.

## Rice Gall Dwarf

Rice gall dwarf was first observed in Thailand, in the early 1980s. This disease is also present in China, Malaysia, and Thailand. The causal virus belongs to a species in the genus *Phytoreovirus*, family *Reoviridae*. Virions are icosahedral (60–70 nm) and possess 12 dsRNA genome segments and up to seven structural proteins (45–160 kDa). Rice gall dwarf virus (RGDV) is confined to phloem tissues of the plant host. Largest genome segment has 4.4 kb, and the smallest, 0.6 kb. Virions are found in the phloem, cytoplasm, and cell vacuoles in low concentrations, together with viroplasms containing immature virions.

RGDV is transmitted by the leafhoppers *N. nigropictus, R. dorsalis, N. cincticeps, N. malayanus,* and *N. virescens,* in a persistent, circulative, and propagative manner. RGDV is transmitted transovarially, is acquired rapidly from infected rice plants, and has an incubation period of 10–20 days in the insect vector. Main symptoms in rice are stunting, translucent gall formation on the underside of the leaves and on leaf sheaths (enations), and permanent dark green leaves. Infected plants produce few tillers, and form poorly developed panicles. Other grasses susceptible to this virus include oats, barley, rye, wheat, Italian grass (*Lolium* sp.), and *Oryza rufipogon*. Japanese grass (*Alopecurus aequalis*) is an important reservoir of RGDV; and maize also seems to be a host of RGDV in China.

Cultural practices aimed at breaking the cycle of the leafhopper vectors that move from spring to autumn plantings; and the elimination of ratoon plantings from over-wintered rice are recommended.

## Rice Ragged Stunt

Rice ragged stunt was first reported from the Philippines and Indonesia in 1976. It is currently distributed in Bangladesh, China, India, Japan, Malaysia, Sri Lanka, Taiwan, and Thailand. The causal virus was isolated in the early 1980s, and shown to belong to a species of the genus *Oryzavirus*, family *Reoviridae*. Rice ragged stunt virus (RRSV) has icosahedral, double-shelled particles (75–80 nm) with conspicuous spikes. Oryzaviruses have a genome of 10 linear dsRNA segments with a total size of

26.6 kb. RRSV particles are composed of five major structural proteins (33–120 kDa) and three nonstructural proteins (31, 63, and 88 kDa). The dsRNA segments contain a single ORF each, except S4, which contains two ORFs. RRSV is serologically related to Echinochloa ragged stunt virus.

RRSV is transmitted by the delphacid planthoppers *Nilaparvata lugens* (brown planthopper) and *N. bakeri*, in a persistent, circulative, and propagative manner. RRSV is not transovarially transmitted. RRSV infects rice and other *Oryza* species, causing stunting, enations, ragged (serrated) leaves, and suppression of reproductive structures. Infected rice plants remain green when healthy plants have senesced. RRSV also infects maize and wheat. Virions are found in phloem enations; and infected cells show the presence of viroplasms and cell wall proliferation.

RRSV persists in rice and in the brown planthopper, and, consequently, continuous rice planting perpetuates the virus and its vectors. Planthoppers migrate from the tropics to temperate areas in the summer time. Resistance to RRSV has been identified in some breeding lines; but molecular breeding techniques have been preferred to genetically modify rice genotypes for resistance to the virus.

## Rice Tungro

*Tungro* means 'degenerate growth' in Ilocano, a Philippine language. Tungro-like symptoms were first observed in the Philippines, Indonesia, Malaysia, and Thailand in the early 1960s, under different names, such as 'mentek', 'penyakit habang', 'penyakit mera', and 'yellow orange leaf'. Severe outbreaks of tungro disease occurred in the late 1960s, following the introduction of early maturing, fertilizer-responsive, and high-yielding rice varieties. The disease has been reported from other Asian countries, namely, India, Bangladesh, and China. Tungro was first associated with the presence of leafhoppers in affected rice fields, but one of the causal agents could only be visualized in 1967, as isometric particles, 30–35 nm in diameter. Between 1967 and 1970, at least four different strains of the tungro 'virus' were identified, and this number increased to eight strains by 1976. In 1978, Indonesian scientists reported the observation of both isometric (30 nm) and bacilliform (35 × 350 nm) particles in rice plants affected by the 'penyakit habang' disease (tungro) in Indonesia. Severe symptoms were associated with both particles, whereas rice plants infected only by the bacilliform virus showed moderate symptoms. Plants infected only by the isometric virus remained symptomless. The isometric particles were transmitted by the leafhopper *N. virescens*, but the bacilliform virus could not be transmitted in the absence of the isometric virus. This constituted the first demonstration that tungro is a synergistic disease, and that the isometric virus acts as an 'assistor' or 'helper' virus for the bacilliform virus that causes main symptoms associated with tungro disease. In 1979, *N. nigropictus* was also shown to be an occasional vector of the isometric virus, but not of the bacilliform virus. The shortest acquisition feeding for *Nephotettix impicticeps* is 5–30 min, and there is no incubation period. Therefore, the virus does not persist in the vector; and there is no trans-stadial or transovarial passage of the virus in the insect vector.

Rice tungro disease is thus caused by two different viruses: rice tungro spherical virus (RTSV), a positive-sense RNA virus, and rice tungro bacilliform virus (RTBV), a circular dsDNA virus. The nucleotide sequence of rice tungro baciliform badnavirus (RTBV) DNA was determined in 1991. The circular genome has 8.3 kb and one strand contains four open reading frames (ORFs). One ORF encodes a protein of 24 kDa (P24). The other three ORFs potentially encode proteins of 12, 194, and 46 kDa. Comparative analyses with retroviruses suggested that the 194 kDa polyprotein is proteolytically cleaved to yield the virion coat protein, a protease, and replicase (reverse transcriptase and RNase H), characteristic of retro-elements. The DNA sequence of RTBV suggests that RTBV is a pararetrovirus. RTBV is currently classified as a member of a species of the genus *Badnavirus*, family *Caulimoviridae*.

The 12 433 nucleotide (nt) sequence of RTSV contains a large ORF, capable of encoding a viral polyprotein of 390.3 kDa. Two viral subgenomic RNAs of *c.* 1.2 and 1.4 kb, respectively, were detected in RTSV-infected leaf tissues. There are at least three capsid protein subunit cistrons near the N-terminus of the large ORF. The C-terminal half of the large ORF revealed conserved protein sequence motifs for a viral RNA polymerase, proteinase, and a putative nucleoside triphosphate (NTP)-binding protein. The sequence motifs are arranged in a manner that resembles those of picorna-like viruses. RTSV is the type species of the genus *Waikavirus*, family *Sequiviridae*.

Rice tungro bacilliform virus isolates from Bangladesh, India, Indonesia, Malaysia, and Thailand were compared with the type isolate from the Philippines. Restriction endonuclease maps revealed differences between the isolates, and cross-hybridization showed that they formed two groups, one from the Indian subcontinent and a second one from Southeast Asia.

Diseased leaves become yellow or orange-yellow, and malformed, starting with the tip of the lower leaves. Young infected leaves may show mottling and chlorotic stripes parallel to the veins. Root development is poor and the reproductive stage may be delayed. Affected panicles are usually small and sterile. Symptom expression varies considerably depending on the variety affected and environmental conditions. Yield losses depend upon plant age at the time of infection, and the plant genotype. *Eleusine indica*, *Echinochloa colonum*, *E. crusgalli*, *Oryza* spp., *Paspalum* sp.,

Setaria sp., sorghum, and wheat have also shown to be susceptible to tungro disease.

Several sources of resistance to RTSV have been identified at the International Rice Research Institute (IRRI). Promising lines have been identified and some of the superior lines resistant to tungro have been released as varieties. IR20, IR26, and IR30 were the first tungro-resistant cultivars released by IRRI followed by IR28, IR29, IR34, IR36, IR38, and IR40 during the 1970s. Some cultivars resistant to *N. virescens*, such as IR54, IR56, IR60, and IR62, were released in the 1980s, followed by PSBRc4, PSBRc10, PSBRc18, and PSBRc28 during the 1990s. Recently, five tungro-resistant lines showing resistance to RTSV were released as cultivars in Indonesia and the Philippines. Utri Merah, Balimau Putih, Habiganj DW8, and *O. rufipogon* served as donors of resistance genes.

## Rice Stripe

Rice stripe is the second oldest viral disease of rice, having been first observed in the 1890s, in Japan. In 1931, the disease was associated with the presence of the leafhopper *L. striatellus*, but its causal agent could not be isolated. The disease continued to spread in Japan in the early 1950s, due to the introduction of new cultural practices (seeding of rice in wheat fields and early plantings) that expanded the temporal availability of young rice plants to the leafhopper vector. Rice stripe eventually emerged in Korea, Taiwan, China, and southwestern regions (Vladivostok) of the former USSR. The causal agent eluded detection until 1975, when Japanese scientists isolated thin (3–8 nm), coiled, filamentous particles of different lengths (2110–510 nm) from rice stripe-affected plants. In 1989, rice stunt virus (RSV) was shown to consist of four ssRNA components, similar to the genome of maize stripe virus (MSpV) isolated in 1981 in the United States. These viruses, together with rice hoja blanca virus (RHBV), isolated in Colombia in 1982, formed the 'rice stripe virus group' in 1988. Further molecular work led to the creation of the genus *Tenuivirus* in 1995. Tenuiviruses consist of up to six segments of linear, negative-sense, and ambisense ssRNA. RSV is believed to have four to five segments, depending on the isolate. Sizes are *c.* 9 kb for RNA 1 (negative polarity); 3.3–3.6 kb for RNA 2 (ambisense), 2.2–2.5 kb for RNA 3 (ambisense); 1.9–2.2 kb for RNA 4 (ambisense); and 1.2 kb for RNA 5 (negative polarity). The viral genome encodes structural (32–35 kDa) and nonstructural proteins (230 kDa), both of which are useful for diagnostic purposes. The nucleocapsid protein is encoded by the 5′ proximal region of the virion-complementary sense strand of RNA 3. Some proteins are translated from subgenomic RNAs.

RSV is transmitted by *L. striatellus*, the 'smaller brown planthopper', *Terthron albovittatum*, *Unkanodes sapporonus*, and *U. albifascia* (formerly *Ribautodelphax albifascia*). The virus multiplies in the insect vector (propagative) and there is transovarial passage of the virus to the progeny. The proportion of active vectors of *L. striatellus* can be as high as 54%. The shortest virus acquisition feeding period may be less than 30 min, and the incubation period in the vector is 5–10 days, although longer incubation periods have been reported for *L. striatellus* (21 days) and *U. albifascia* (26 days). The virus persists in its planthopper vectors for life ('persistent' transmission). RSV is not transmitted manually or through seed. The main hosts of *L. striatellus* are wheat, barley, and Italian rye in spring, and upland rice, *Echinochloa crusgalli*, and *Digitaria* spp. in summer.

Initial symptoms induced by RSV are failure of emerging leaves to unfold normally. Affected leaves may show chlorosis, and chlorotic stripes running parallel to the leaf veins. Affected leaves may eventually develop necrosis and die. In infected, older leaves, a chlorotic mottle may appear. The tillering capacity is often reduced, and panicles may fail to produce seed. RSV also infects at least 35 other species of plants in the Gramineae, including oats, barley, rye, sorghum, wheat, and maize.

All the Japanese paddy rice cultivars tested as seedlings proved susceptible, whereas most Japanese upland rice cultivars are highly resistant or immune. Outside Japan, most *indica* rice cultivars were resistant, but *japonica* rice cultivars were susceptible. Resistance in Japanese upland varieties was found to be controlled by two complementary dominant genes ($St_1$ and $St_2$). Multiple alleles of the latter gene may be responsible for the resistance associated with *indica* cultivars. Partial dominant resistance has been identified, linked to another dominant gene ($St_3$). Resistance to the virus is not necessarily linked to resistance to the vector. Transgenic rice cultivars (coat protein mediated resistance) possessing resistance to RSV have already been developed.

## Rice Grassy Stunt

Rice grassy stunt was first observed in 1963 in the Philippines, where the characteristic stunting was initially referred to as 'yellow dwarf' or 'rice rosette'. This disease is now present in China, India, Indonesia, Japan, Malaysia, Sri Lanka, Taiwan, and Thailand. The association of this disease with the brown planthopper, *N. lugens*, was made in 1964. The causal agent of rice grassy stunt was isolated in the early 1980s, as a member of a distinct species of the genus *Tenuivirus*. Hence, rice grassy stunt virus (RGSV) shares most of its physicochemical characteristics with RSV. However, RGSV has an additional (sixth) ssRNA segment (ambisense), and its RNA 3 (3.1 kb) and RNA 4 (2.9 kb) are different from the corresponding segments of other tenuiviruses. RGSV

RNA 1 (9.8 kb), 2 (4.1 kb), 5 (2.7 kb), and 6 (2.6 kb) are homologous to RNAs 1, 2, 3, and 4, respectively, of other tenuiviruses. Both the capsid and non-structural proteins of RSV and RGSV are serologically related. A tentative tenuivirus, rice wilted stunt virus, originally reported in 1981 from Taiwan, is often considered a synonym of RGSV.

RGSV is transmitted by the planthoppers *Nilaparvata bakeri*, *N. lugens*, and *N. muiri*, in a persistent manner. RGSV multiplies in the vector, but it is not transmitted congenitally. Main symptoms consist of severe stunting, excessive tillering, erect growth, and short, narrow chlorotic leaves. Young leaves may show mottling and/or stripes. Inclusions are present in infected cells. Other hosts of RGSV include: *E. colonum*, *Cynodon dactylon*, *Cyperus rotundus*, *Leersia hexandra*, and *Monochoria* sp.

Genetic resistance to RGSV has been identified in some rice varieties. A single dominant gene conferring resistance to RGSV was identified in *Oryza nivara*. Some sources of RGSV resistance are susceptible to the planthopper vectors.

## Rice Hoja Blanca

Rice hoja blanca ('white leaf') was first observed in the Cauca Valley of Colombia, South America, in 1935. By 1940, the disease was present in most rice-producing regions of Colombia. The disease was detected in Panama in 1952, in Cuba in 1954, in Venezuela in 1956, in Costa Rica in 1958, in El Salvador in 1959, in Guatemala in 1960, and in Nicaragua, Honduras, and the Dominican Republic in 1966. In South America, rice hoja blanca also occurs in Ecuador and Peru. The virus was detected in southern United States in the late 1950s, but it is not a problem there. Until the end of the twentieth century, hoja blanca was the only known viral disease of rice in the Americas.

The etiology of rice hoja blanca remained elusive until 1983, when virus-like particles isolated from infected rice plants in Colombia, were shown to be similar to those described in Japan for RSV. The causal virus was finally characterized as a member of the rice stripe virus group created in 1988, which also include MSpV. In 1995, these viruses became species of the genus *Tenuivirus*. The main physicochemical characteristics of these viruses have been described above, but RHBV virions consist of four species of ssRNA, with a total genome of 17.6 kb. RNA 1 is 9.8 kb and encodes the viral polymerase. RNA 2 (3.5 kb) has the capacity to code for two proteins in an ambisense manner, and RNA 3 (2.3 kb) encodes two proteins in an ambisense manner, including the capsid protein (35 kDa). RNA 4 (1.9 kb) has the same basic arrangement as RNAs 2 and 3, and codes for a nonstructural protein (23 kDa). Echinochloa hoja blanca virus (EHBV) was initially considered a strain of *Rice hoja blanca virus*, because it coexists with this virus in most affected rice fields, but it is now considered a distinct, serologically related tenuivirus.

RHBV is transmitted in a persistent, circulative, and propagative manner by the planthopper *Tagosodes orizicolus* (previously *Sogatodes oryzicola*). The virus also infects the insect vector and passes transovarially to the progeny of viruliferous females (80–95% efficiency). However, only 5–15% of the wild *T. orizicolus* population has the ability to transmit the virus in nature. Incubation periods of 6–36 days have been reported for RHBV in its planthopper vector. *Tagosodes cubanus*, the planthopper vector of EHBV, has been reported to transmit RHBV from rice to *E. colonum* but not from rice to rice, or from *Echinochloa* to rice, unless the viruliferous insects are forced to feed on the test plants. Colonies containing over 90% active vectors can be developed by crossing known male and female vectors. If left unattended, these colonies revert to their natural 5–15% of active vectors, suggesting that viruliferous individuals are at a biological disadvantage with respect to nontransmitters. The acquisition feeding period observed for at least 50% of the potential vectors to transmit RHBV is 15 min, with an optimum of 1 h. The incubation period is 6–9 days for individuals who acquire the virus transovarially, but may be up to 36 days for individuals acquiring the virus as nymphs or adults (males usually have a shorter life span and must acquire the virus transovarially). Nonviruliferous *T. orizicolus* females lay more eggs than viruliferous females, which suggested a deleterious effect of RHBV in its insect vector.

RHBV induces small chlorotic spots at the base of infected leaves, which then show longitudinal chlorotic stripes. The following leaves may be completely chlorotic (hence the name 'white leaf'). Plants affected at an early stage are usually stunted, and may eventually die. Malformation of grains or sterility of the panicle can be observed in systemically infected rice plants that reach the reproductive stage. RHBV has been shown to infect *Digitaria* spp., *Leptochloa* spp., oats, barley, rye, and wheat. Massive screenings of rice cultivars took place in Cuba, Venezuela, and Colombia in the late 1950s and early 1960s. Most commercial long-grain rice varieties are susceptible. Resistance to RHBV was identified in short-grain *japonica* varieties and hybrids with *indica* types. These hybrids have been used to develop resistant rice cultivars, but their level of resistance ultimately depends on virus pressure. Some varieties possessing high levels of RHBV resistance have been developed in recent years in Colombia, which seem to be the result of combining different mechanisms of resistance, not necessarily against the virus. Transgenic plants that show moderate levels of resistance to the causal virus have been developed. RHBV epidemics are cyclic, often separated by several years of low disease incidence. The deleterious effect of the virus in viruliferous females may account for this cyclical phenomenon.

## Rice Yellow Stunt

Rice yellow stunt (previously known as 'rice transitory yellowing') was first reported from southern Taiwan in 1960. The disease was initially believed to be caused by the lack of aeration in paddy fields. The characteristic foliar yellowing, stunting, and root rot were once blamed for the destruction of over 2000 ha of rice in the 1960s. In 1965, a virus was first suspected as the causal agent, and the leafhopper *N. nigropictus* (previously *N. apicalis*) was implicated as the vector. The name 'transitory yellowing' was given to this disease because affected rice plants often recovered from the yellowing symptoms at later stages of growth. The causal virus was isolated in 1986 and shown to be a bullet-shaped rhabdovirus approximately $190 \times 95$ nm. *Rice yellow stunt virus* is currently characterized as a species of the genus *Nucleorhabdovirus*, family *Rhabdoviridae*. Nucleorhabdoviruses multiply in the nucleus of infected cells, forming large granular inclusions. Virus maturation occurs at the inner nuclear envelope, where rice yellow stunt viruses (RYSVs) acquire their lipid envelope. The virus contains a single molecule of negative-sense ssRNA about 13.5 kb in size. A modular organization of the genome shows three highly conserved blocks, encoding structural and nonstructural proteins, membrane proteins, movement protein, and the polymerase.

RYSV is transmitted by the leafhoppers *N. nigropictus*, *N. cincticeps*, and *N. virescens* in a persistent and propagative manner. The virus does not pass congenitally to the progeny. Warm winters and continuous rice plantings increase virus incidence. *Japonica* varieties are more susceptible to RYSV than *indica* types. Highly resistant rice varieties have been selected since the early 1960s to manage this disease.

## Rice Yellow Mottle

Rice yellow mottle was first observed in 1966 in Kenya and is currently the most important viral disease of rice in sub-Saharan Africa. The disease also affects rice in Madagascar. Yield losses ranging from 25% to 100% are not uncommon in West Africa. The causal agent was described in 1974 as an isometric virus about 30 nm in diameter, possessing positive-sense ssRNA ($1.4 \times 10^6$ M). The RNA has 4450 nt organized into four ORFs, three of which overlap (except ORF 1). ORF4 codes for the 26 kDa capsid protein. This protein is required for cell-to-cell and long-distance movement. Rice yellow mottle virus (RYMV) is currently classified as a member of the species *Rice yellow mottle virus* in the unassigned genus *Sobemovirus*.

RYMV is transmitted by chrysomelid beetles (Coleoptera): *Sesselia pusilla, Chaetocnema pulla, C. dicladispa, Trichispa sericea, Dicladispa (Hispa) viridicyanea*, and *D. gestroi*, in a semi-persistent manner. The virus is lost when beetles molt; it does not multiply in the vector and is not congenitally transmitted. RYMV is transmitted by mechanical means but is not seed-borne. Rice yellow mottle disease is characterized by stunting, crinkling, mottling, yellow streaks, malformation of panicles, reduced tilllering, and sterility. RYMV also infects other *Oryza* species, and the grasses *Acroceras zizanioides, Dinebra retroflexa, Eragrostis aethipica, E. ciliaris, E. namaquensis, E. tenella, Echinochloa colona, Ischaemum rugosum, Panicum repens, P. subalbidum, Phleum arenarium, Sacciolepsis africana*, and *Setaria longiseta*.

Most African rice cultivars are highly susceptible to RYMV. Two types of resistance have been identified in *Oryza*: polygenic resistance in *O. sativa*, *japonica*, which slows down virus dissemination; and monogenic resistance in *Oryza glaberrima*, which prevents infection. Transgenic rice plants showing 'enhanced' resistance to RYMV have also been developed. The epidemiology of RYMV is poorly understood, specially regarding the identification of primary inoculum sources, and propagation of the virus in the field. Resistance-breaking isolates of RYMV have been recently detected in West and Central Africa.

## Rice Stripe Necrosis

Rice stripe necrosis first emerged in Ivory Coast, West Africa, in 1977. The disease induces chlorotic striping, malformation, and necrosis of leaves, besides stunting and reduced tillering. The causal virus was shown to consist of rod-shaped virions of two predominant lengths (110–160 and 270–380 nm), and 20 nm wide. The virus can be mechanically transmitted to *Chenopodium amaranticolor* (local lesions), but is not seed-borne. The virus was transmitted through contaminated soil, which led to the identification of the fungus *Polymyxa graminis* as the vector. The causal virus, rice stripe necrosis virus (RSNV), was tentatively classified as a member of the genus *Furovirus*. However, its relatively low incidence and economic importance, when compared to rice yellow mottle virus, relegated RSNV to the condition of a minor pathogen. In the meantime, RSNV slowly disseminated into the main rice-growing areas of West Africa, where farmers and scientists alike mistook the disease for a soil or physiological problem. It was not until 1991, when a new disease of rice emerged in the Eastern Plains of Colombia, South America, that research conducted at CIAT, Palmira, Colombia, demonstrated that the new disease was in fact 'rice stripe necrosis', probably introduced from West Africa in contaminated rice seed. The Colombian isolate of RSNV had rod-shaped particles with a bi-modal length of 260 and 360 nm, and a particle width of 20 nm. These particles were observed in the cytoplasm of infected rice cells, affecting mainly mitochondria and the endoplasmic reticulum.

The virus was isolated as four components of ssRNA (6.3, 4.6, 2.7, and 1.8 kb), and a capsid protein subunit of 22.5 kDa. At least two distinct replication-associated domains of RSNV RNA1 show sequence identities of >78% with the corresponding domains of Beet necrotic yellow vein virus (BNYVV-RNA 1). As in the case of BNYVV, the RNAs of RSNV terminate in a 3′-poly(A) tail. RSNV's particle length, number of RNA segments, polyadenylation of the RNAs 3′-end, and partial homology to BNYVV suggest that RSNV is a species of the genus *Benyvirus*. Partial sequence obtained from the Colombian isolate's RNA 1 segment also showed a 51% similarity to the corresponding region of BNYVV, member of the type species of the genus *Benyvirus*. The vector of the Colombian RSNV isolate was molecularly characterized as a type II isolate of the fungus *P. graminis*.

Since the 1970s, when RSNV was first reported from West Africa, it was evident that rice varieties reacted differentially to RSNV. Some IRAT lines (8, 9, and 13) showed low virus incidences or escaped disease. Screening of the main rice varieties grown in Colombia showed that most of these cultivars were susceptible. Some breeding lines and some Asian varieties exhibited moderate resistance levels. *Oryza glaberrima*, a related species planted for human consumption in West Africa, was observed to possess a high level of resistance to RSNV in that region and in Colombia. The disease seems to spread more rapidly in mechanized and irrigated rice plantings, but the main virus-vector dissemination mechanism is undoubtedly the use of seed produced in affected fields, where the fungus and the virus persist for many years. Rice stripe necrosis is currently found throughout West Africa, and in four neighboring countries of Latin America: Colombia, Ecuador, Panama, and Costa Rica.

## Rice Necrosis Mosaic

Rice necrosis mosaic was first observed in Japan in 1959. The causal virus was observed in 1968 to have filamentous particles of two different lengths (275 and 550 nm) and 13 nm in width. Rice necrosis mosaic virus (RNMV) is currently classified as a member of the species *Rice necrosis mosaic virus* of the genus *Bymovirus*, family *Potyviridae*. Virions contain two molecules of positive-sense ssRNA. RNA 1 is 7.5–8.0 kb, and RNA 2 is 3.5–4.0 kb, and both RNAs are polyadenylated at their 3′-ends. The capsid protein is located in the 3′-proximal region of RNA 1. Virions have a single capsid protein subunit of *c.* 33 kDa. As other species in the family *Potyviridae*, RNMV induces cylindrical (pinwheels) inclusions in the cytoplasm of infected cells.

RNMV is transmitted by mechanical means and through soil by the vector fungus *P. graminis*. The main symptoms are initial mottling of the lower leaves, which gives rise to streaks and irregular foliar patches. Affected leaves turn yellow. Although susceptible rice plants are not markedly stunted, the number of tillers is usually reduced. Necrotic lesions can appear on leaf sheaths and the base of culms. Grain production is significantly reduced in diseased plants. The virus seems to be limited to *O. sativa*.

Viruses transmitted by *P. graminis* are generally disseminated through contamination of agricultural tools and machinery; irrigation water; and seed contaminated with soil particles containing viruliferous fungal propagules (cystosori). These propagules can remain viable in contaminated soils for over 10 years. Soil used in seedbeds should be kept free of the vector. Resistance to this disease has been identified in some rice varieties.

*See also:* Cereal Viruses: Maize/Corn; Cereal Viruses: Wheat and Barley; Plant Reoviruses; Plant Rhabdoviruses; Rice Tungro Disease; Rice Yellow Mottle Virus; *Tenuivirus*; Vector Transmission of Plant Viruses.

## Further Reading

APS (1992) *Compendium of Rice Diseases*, 62pp. St. Paul, MN: APS Press.

Brunt AA, Crabtree K, Dallwitz MJ, et al. (1996) *Viruses of Plants*, 1484pp. Wallingford: CAB International.

Fauquet CM, Mayo MA, Maniloff J, Desselberger U, and Ball LA (eds.) (2005). *Virus Taxonomy: Eighth Report of the International Committee on Taxonomy of Viruses*, 1259pp. San Diego, CA: Elsevier Academic Press.

IRRI (1969) *The Virus Diseases of the Rice Plant*, 354pp. Los Baños, The Philippines: The International Rice Research Institute.

Ling KC (1972) *Rice Virus Diseases*, 142pp. Los Baños, The Philippines: The International Rice Research Institute.

Morales FJ and Niessen AI (1983) Association of spiral filamentous virus-like particles with rice hoja blanca. *Phytopathology* 73: 971–974.

Morales FJ, Ward E, Castaño M, Arroyave J, Lozano I, and Adams M (1999) Emergence and partial characterization of rice stripe necrosis virus and its fungus vector in South America. *European Journal of Plant Pathology* 105: 643–650.

Wilson MR and Claridge MF (1991) *Handbook for the Identification of Leafhoppers and Planthoppers of Rice*, 142pp. Bristol, UK: Natural Resources Institute and C.A.B. International.

# Cereal Viruses: Wheat and Barley

**H D Lapierre and D Hariri,** INRA – Département Santé des Plantes et Environnement, Versailles, France

© 2008 Elsevier Ltd. All rights reserved.

## Glossary

**Guttation** Phenomenon linked to the natural movement of the fluid from the apoplastic areas to the leaf surface through stomata under conditions that allow gutate to form.

**Pathotype** A pathotype is a virus strains that exhibits a differential interaction with the host, usually increased virulence.

**Pyramiding genes** A combination of different resistance genes introduced in one line or cultivar.

**Strain** A virus isolate called strain exhibits specific interactions in front of several plant species. A strain may also be charaterized by particular properties of the viral particles.

## Introduction

Most of the viruses that infect barley also infect wheat. Of the 57 viruses infecting wheat and or barley (**Table 1**), ten have been discovered less than 30 years ago and new strains or pathotypes are regularly reported. A few apparently nonpathogenic viruses belonging to the genera *Metavirus*, *Pseudovirus*, and *Endornavirus* have been characterized but are not described here. Only the most damaging viruses and some emerging viruses are described.

## Agropyron Mosaic Virus

This virus of low agronomical importance on wheat is believed to be transmitted by the cereal rust mite *Abacarus hystrix* (Nalepa). In USA, agropyron mosaic virus (AgMV) may occur in wheat fields infected with wheat streak mosaic virus (WSMV). In the presence of the two viruses, a yield loss of 85% has been estimated.

## Arabis Mosaic Virus

In Switzerland, arabis mosaic virus (ArMV) has been detected in fields of winter barley. Partial stunting, without yellowing, can occasionally be observed. Barley infection is achieved by the nematode *Xiphinema diversicaudatum* (Micoletzky).

## Aubian Wheat Mosaic Virus

This emergent virus was observed in some winter wheat fields from northern and southern France. A similar virus is present in the UK. The rod-shaped particles of this virus are detected by enzyme-linked immunosorbent assay (ELISA). *Polymyxa graminis* cystosori are detected in the roots of infected plants but aubian wheat mosaic virus (AWMV) is not serologically related to the *Polymyxa*-transmitted viruses already known. The only known natural hosts of this virus belong to the genus *Triticum*. AWMV is mechanically transmitted to wheat and to some eudicots but mosaic symptoms were observed only on wheat. All the cultivars of bread and durum wheat inoculated were infected. In wheat fields, yellow patches of diseased plants are observed from the beginning of the year till heading stage. Most of the cultivars are slightly stunted showing variable mosaic symptoms. Nucleic acid sequences of the genome and natural transmission of the virus are under investigation.

## Barley Stripe Mosaic Virus

This virus is no longer a global agronomical problem to barley. In the past, up to 60% yield loss has been reported. Pollen transmission of barley stripe mosaic virus (BSMV) is the main way of contamination. Immunological control of the seeds allows growing virus-free cultivars. BSMV is less frequently found on wheat. Its recent detection in Turkey means this virus should be monitored closely. With its tripartite genome and the high seed transmission, BSMV remains a model for studies of virus/plant interactions.

## Barley Yellow Streak Mosaic Virus

This virus has been detected in several US states and Canada. The agronomical importance of barley yellow streak mosaic virus (BaYSMV) has not been evaluated. Nearly all plants in a field may be infected resulting in yield loss up to 100%. Infected plants are stunted and show chlorotic streaks and stripes parallel to the leaf veins. Virus particles are filamentous $c.$ 64 nm in diameter and of varying lengths between 127 and 4000 nm. Particles have an outer lipid-like envelope. From purified particles, a dominant protein of $\sim$32 Da and several single-stranded RNA (ssRNA) species of $\sim$13–11 kbp have been isolated. This unassigned virus which resembles some insect and animal viruses probably belongs to a new genus.

**Table 1** Viruses infecting naturally wheat and barley in different parts of the world

| Viruses | Genera | Families |
|---|---|---|
| Agropyron mosaic virus [a] | Rymovirus | Potyviridae |
| Arabis mosaic virus [b] | Nepovirus | Comoviridae |
| Aubian wheat mosaic virus [a, l] | | |
| Barley mild mosaic virus [b, c] | Bymovirus | Potyviridae |
| Barley stripe mosaic virus [d] | Hordeivirus | |
| Barley yellow dwarf virus-MAV [d, e] | Luteovirus | Luteoviridae |
| Barley yellow dwarf virus-PAS [d, e] | Luteovirus | Luteoviridae |
| Barley yellow dwarf virus-PAV [d, e] | Luteovirus | Luteoviridae |
| Barley yellow mosaic virus [b, c] | Bymovirus | Potyviridae |
| Barley yellow streak mosaic virus [d, l] | | |
| Barley yellow striate mosaic virus [d] | Cytorhabdovirus | Rhabdoviridae |
| Brazilian wheat spike virus [a, l] | | |
| Brome mosaic virus [d] | Bromovirus | Bromoviridae |
| Brome streak mosaic virus [f] | Ttritimovirus | Potyviridae |
| Cereal mosaic virus [g, l] | | |
| Cereal pseudorosette virus [g, l] | | |
| Cereal yellow dwarf virus-GPV [d, e] | Polerovirus | Luteoviridae |
| Cereal yellow dwarf virus-RPS [d, e] | Polerovirus | Luteoviridae |
| Cereal yellow dwarf virus-RPV [d, e] | Polerovirus | Luteoviridae |
| Chinese wheat mosaic virus [a, h] | Furovirus | |
| Chloris striate mosaic virus [d] | Mastrevirus | Geminiviridae |
| High Plains virus [d, l] | | |
| Indian peanut clump virus [d] | Pecluvirus | |
| Iranian wheat stripe virus [a, l] | | |
| Maize Iranian mosaic virus [a, l] | | |
| Maize rough dwarf virus [a, l] | Fijivirus | Reoviridae |
| Maize sterile stunt virus [d, l] | | |
| Maize streak virus [d] | Mastrevirus | Geminiviridae |
| Maize white line mosaic virus [a, l] | | |
| Mal de Rio Cuarto virus [d] | Fijivirus | Reoviridae |
| Northern cereal mosaic virus [d] | Cytorhabdovirus | Rhabdoviridae |
| Peanut clump virus [i] | Pecluvirus | |
| Rice black streaked dwarf virus [d] | Fijivirus | Reoviridae |
| RMV [d, e] | | Luteoviridae |
| SGV [d, e] | | Luteoviridae |
| Soil-borne cereal mosaic virus [a, h] | Furovirus | |
| Soil-borne wheat mosaic virus [a, h] | Furovirus | |
| Soil-borne wheat mosaic virus Marne strain [b, h] | Furovirus | |
| Triticum mosaic virus [a, l] | | |
| Wheat American striate mosaic virus [a] | Cytorhabdovirus | Rhabdoviridae |
| Wheat dwarf virus B and W strains [a, b, l] | Mastrevirus | Geminiviridae |
| Wheat Eqlid mosaic virus [a, l] | | |
| Wheat rosette stunt virus [g, l] | | |
| Wheat spindle streak mosaic virus [a, j] | Bymovirus | Potyviridae |
| Wheat spot mosaic [d, k, l] | | |
| Wheat streak mosaic virus [a] | Tritimovirus | Potyviridae |
| Wheat yellow head virus [a, l] | | |
| Wheat yellow leaf virus [d] | Closterovirus | Closteroviridae |
| Wheat yellow mosaic virus [a, j] | Bymovirus | Potyviridae |
| Winter wheat Russian mosaic virus [g, l] | | |

[a] Virus infecting wheat.
[b] Virus infecting barley.
[c] See section on bymoviruses of barley.
[d] Virus infecting wheat and barley.
[e] See section on yellow dwarf viruses.
[f] Virus detected in barley volunteers.
[g] See section on Northern cereal mosaic virus.
[h] See section on furoviruses.
[i] See section on Indian peanut clump virus.
[j] See section on bymoviruses of wheat.
[k] See section on High Plains virus.
[l] Unassigned virus.

BaYSMV is transmitted by the brown wheat mite *Petrobia latens* and is retained in eggs of mites which are the overwintering reservoir. No resistant barley cultivars are known.

## Barley Yellow Striate Mosaic Virus

The distribution of this virus was first described in the Mediterranean areas. Recently, barley yellow striate mosaic virus (BYSMV) has been detected in East European countries. More than 10% yield loss in susceptible cultivars of durum wheat is suspected. Symptoms consist of leaf chlorotic striations, dwarfing, and excess tillering, and heads may be confined in the boot. Awns are short or broken and show mosaic. Wheat and barley plants die in spring in the case of an early infection in the autumn. BYSMV is transmitted in a persistent manner mainly by the planthopper *Laodelphax striatellus* (Fallén). The virus can be transmitted transovarially. Limited molecular tools have been used to compare BYSMV with other serologically related viruses (digitaria striate virus, maize sterile stunt virus, northern cereal mosaic virus (NCMV), wheat rosette stunt virus (WRSV)). Small grain species and several grasses are hosts of BYSMV. Durum wheat is generally more susceptible to BYSMV than bread wheat. Genetic basis of susceptibility/tolerance in wheat is under study.

## Brazilian Wheat Spike Virus

This virus found in Brazil seems to be of low incidence on wheat crops. Young leaves of infected plants are entirely chlorotic, older leaves show bright stripes. Spikes of affected plants are pale yellow to whitish, containing empty grains. Particle morphology and cytology indicate that Brazilian wheat spike virus (BWSpV) could be a tenuivirus.

## Brome Mosaic Virus

Brome mosaic virus (BMV) disease has been reported on wheat and triticale. Infected plants show yellow streak and stunting. The virus is present in the guttation fluid of the leaves. Beetles are considered to be the main vectors. Mechanical inoculation by animal and machinery traffic probably also occur.

## Bymoviruses of Barley

### Barley Yellow Mosaic Virus, Barley Mild Mosaic Virus

Bymoviruses have bipartite ssRNA genomes and each segment (RNA1 and RNA2) carries a single open reading frame (ORF) which encodes a polyprotein. The coat protein (CP) gene is located at the C-terminus of the RNA1.

Barley yellow mosaic virus (BaYMV) and barley mild mosaic virus (BaMMV) are major pathogens of winter barley in Europe and East Asia. Yellow discolored patches of different sizes are often the first sign of field infection in early spring. First symptoms are chlorotic or pale green spots and streaks along the leaf veins and can be observed in December or after snow break in February or March. Generally, the third leaves show as rolled young leaves. The bymoviruses of barley significantly reduce height and number of fertile tillers and losses caused by these viruses vary from 10% to 90%. When temperature rises, the new leaves remain green but show distinct pale green streaks. BaYMV and BaMMV can be found, either separately or together. In general, high yield reductions can be anticipated when the winter is very severe (**Figures 1** and **2**).

Different biological or/and serological variants of BaYMV, BaMMV, and wheat spindle streak mosaic virus (WSSMV) have been reported in Europe. In Europe, the resistant genes *rym4* and *rym5* have been overcome by

**Figure 1** Symptoms of barley yellow mosaic virus pathotype 2 (BaYMV 2) in winter barley.

**Figure 2** Symptoms of barley mild mosaic virus Sillery pathotype (BaMMV-Sil) in barley.

**Figure 3** Symptoms of wheat spindle streak mosaic virus in winter bread wheat.

BaYMV 2 and BaMMV Sillery strain (BaMMV-Sil). In France, the presence of several pathotypes of BaYMV and BaMMV able to overcome at least seven of 15 known genes has been reported. These pathotypes, in slow progression, do not pose an agronomic problem for the moment. In Japan, seven strains of BaYMV and two strains of BaMMV have been described on the basis of pathogenicity toward barley cultivars. A Korean strain of BaMMV (BaMMV-Kor) differing biologically and serologically from the Japanese and European isolates and several biological variants of BaYMV in China have also been recognized.

The vector of BaYMV/BaMMV is *P. graminis* Ledingham, a soil organism belonging to the family Plasmodiophoraceae.

The only effective means of controlling these viruses is through the use of resistant cultivars. Resistant genes have been described and localized in the barley genome. Some of them confer complete immunity against BaYMV and/or BaMMV while others only delay the appearance of symptoms. Pyramiding genes have been introduced in barley cultivars and other genes are studied in various *Hordeum* species.

## Bymoviruses of Wheat

### Wheat Spindle Streak Mosaic Virus, Wheat Yellow Mosaic Virus

WSSMV and wheat yellow mosaic virus (WYMV) were reported in America/Europe and Asia, respectively. These two viruses are major pathogens of bread and durum wheat, rye, and triticale (**Figure 3**). In the case of WSSMV, it was demonstrated that its genome associated to viral CP is internalized by *P. graminis*. In France, a new serotype (sII) of the WSSMV has been reported. Depending on the wheat cultivars, WSSMV resistance is controlled by a single dominant gene or by two pairs of alleles, which show complementary effects.

## Furoviruses

### Chinese Wheat Mosaic Virus, Soil-Borne Cereal Mosaic Virus, Soil-Borne Wheat Mosaic Virus

Several wheat furoviruses infecting cereal species have been described in the worldwide: soil-borne wheat mosaic virus (SBWMV) in wheat, rye, and barley; soil-borne cereal mosaic virus (SBCMV) in wheat and rye (**Figure 4**); and Chinese wheat mosaic virus (CWMV) in wheat. These viruses are transmitted by *P. graminis*. SBWMV, the type member of the genus *Furovirus*, occurs mostly on winter wheat in America, Europe, Asia, and Africa. Based on genomic organization and biological characteristics, it was proposed that the American SBWMV, CWMV, European SBCMV, and Japanese SBWMV are four strains of the same virus species. SBWMV is internalized by *P. graminis* in the form of viral RNAs associated to movement proteins. Genomes of these viruses are divided into two ssRNA species that are individually encapsidated. The furoviruses of wheat often causes a 'rosetting' disease, meaning that the plants are severely stunted. The number of tillers and kernel weight are also reduced. The disease usually results in a 10–30% yield loss, but may cause up to 80% yield loss in seriously infested fields. Resistance to SBCMV is controlled by a single locus *Sbm* on 5DL chromosome. Recently a bulk segregant analysis demonstrated that the SBWMV resistance gene was also on chromosome 5DL. The relationship between the two loci is being investigated. This resistance limits the movement of the viruses from the roots to leaves.

Soil-borne wheat mosaic virus Marne strain (SBWMV-Mar) was isolated from barley cv. Esterel in France (Marne department). This furovirus, in contrast to common French isolates of SBCMV, is mechanically transmitted to barley and to oats. Nucleotide and amino acid sequence analyses revealed that the French virus infecting barley is closely related to a Japanese isolate of SBWMV (SBWMV-JT), which was originally isolated from barley.

**Figure 4** Symptoms of soil-borne cereal mosaic virus in winter wheat.

## High Plains Virus

This virus, discovered in USA in 1993, has been identified in two other continents. High Plains virus (HPV) is similar or identical to wheat spot mosaic virus described since 1952. Severe symptoms are shown by wheat plants when infected by HPV. The host range of HPV is large including barley, maize, and many grasses. The virus is transmitted obligatorily by the mite *Aceria tritici* (Shevtchenko) that also transmits WSMV. Two sources of resistance to WSMV *RonL* and *Wsm1* are not effective on HPV.

## Indian Peanut Clump Virus

Found in five states of India and in Pakistan, Indian peanut clump virus (IPCV) is one of the few viruses infecting both Poaceae and Eudicot species. Several Poaceae species are important reservoirs of this virus which induces a severe disease in peanut. Like barley and maize, wheat is severely affected by IPCV. Early infected plants are dark green, stunted, and generally die. Spikes of late infected plants are malformed and produce shriveled seeds. IPCV is transmitted by the soil-borne vector *P. graminis* Ledingham and at a low level by wheat seeds. Peanut clump virus from West Africa is closely related to IPCV. Its agronomical importance on wheat and barley is unknown.

## Iranian Wheat Stripe Virus

The agronomical importance of this virus was not estimated but is probably not high in the Fars Province of Iran where it was discovered. Chlorosis, whitening, and striping are observed in infected plants which are dwarfed. Iranian wheat stripe virus (IWSV) is transmitted in a persistent-propagative manner by the planthopper *Unkanodes tanasijevici* (Dlabola). From the comparison of genomic nucleic acid sequences, it has been suggested that a common ancestor between IWSV and RHBV/EHBV tenuiviruses may exist in the New World.

## Maize Streak Virus – B Strain

Three strains of MSV have been distinguished on the basis of nucleic acid sequences. Strain A is associated to maize. Strain B (MSV-B) has been characterized in wheat, barley, and grasses in South Africa and in Tasmania. MSV-B is more aggressive on wheat and barley than on maize. Recombinants between strain A (from maize) and strain B have been characterized. Agronomical importance and epidemiology of strain B have to be investigated.

## Northern Cereal Mosaic Virus

Several viruses such as WRSV from China, cereal mosaic virus, and winter wheat Russian mosaic virus (WWRMV) from east Russia are probably isolates or strains of the NCMV described in Japan. On winter wheat associated with conditions of high viral inocula in the autumn, yield losses may reach 80%. The infected wheat plants show stunting and prolific tillers. The leaves develop longitudinal yellow and green stripes. Plants infected at the seedling stage die during winter. When infection occurs later, the plant develops thickened stems and foliage. Grains are frequently shriveled. The virus is transmitted in a persistent-propagative manner mainly by *L. striatellus*. Small grains and several grasses are natural hosts of NCMV. Maize has been reported to be resistant to NCMV. Late sowing in autumn and eradication of volunteer plants are recommended.

Particle size, antigenic properties, and vector specificity of these cytorhabdoviruses are very similar or identical. Cereal pseudorosette virus (CPV) is transmitted also by *L. striatellus*. Since its particles are shorter than those of NCMV, CPV is considered a distinct species.

## Triticum Mosaic Virus

This virus, discovered in 2006 in Kansas, has been found in areas where WSMV and HPV are present. The gene RonL giving resistance to WSMV is not efficient on triticum mosaic virus (TriMV). The Wsm1 source of resistance is active on TriMV at 19 °C but not at 25 °C. Symptoms induced by TriMV are similar to those of WSMV. In the future, serological diagnosis may become useful to distinguish TriMV from other viruses commoly found in wheat.

## Wheat American Striate Mosaic Virus

Reported from Canada and the north-central regions of US, this virus has a limited importance because it is usually present only on the borders of wheat fields. Wheat American striate mosaic virus (WASMV) has a relatively narrow natural host range including barley and maize. Fine chlorotic parallel streaks, brown necrotic streaking of culms and glumes are relatively good indicators of WASMV infection. Susceptible cultivars are stunted and most ears are sterile. *Endria inimica* (Say), the main leafhopper vector, transmits WASMV in a persistent-propagative manner. Early planting favors autumn infection. Several cultivars such as TAM107 are tolerant.

## Wheat Dwarf Virus

In continental areas of Europe and its borders, this virus may be very severe on winter wheat, durum wheat, triticale, rye, and barley. Two strains of wheat dwarf virus (WDV) have been characterized. Overall nucleic acid sequences of the wheat-adapted strain (WDV-W) which does not infect barley showed 16% divergence with the barley-adapted strain (WDV-B) which does not infect wheat. The monoclonal antibody MAB 3C10 developed in Germany detects only WDV-B. The two strains have several common hosts. In winter wheat, the leaves show large yellowing bands, the plants are stunted, and an excess of tillering is frequent in contrast to infection by yellow dwarf viruses (YDVs). In winter, barley plants are dwarfed, leaves are shortened, and start yellowing. As in wheat, early infected plants die. *Psammotettix* is the only known leafhopper vector of this virus. Both nymphs and adults of *Psammotettix alienus* (Dahlbom) transmit WDV in a persistent-circulative manner. Only local populations of *P. alienus* carry the viral inoculum to new crops. Depending on the climatic conditions, one, two, or three vector generations may be observed during the warm active periods. In autumn, flying insects are not seen when the maximum temperature is below 10 °C. *Psammotettix alienus* survives the winter in the egg stage on several Poaceae species. Fallows, regrowths, and volunteers play a major role in the constitution of the autumn viral inoculum. Herbicide sprays and tillage of fallows and of preceding susceptible crops reduce the size of *P. alienus* populations. Sowing as late as possible is also recommended. Insecticide-dressed seed offers good protection. Depending on the evolution of *P. alienus* populations, protection of crops can be achieved using one or two pyrethroid sprays.

## Wheat Eqlid Mosaic Virus

This emergent virus found in the Fars Province of Iran is transmitted by the root aphid *Forda marginata* (Koch). It has been suggested that wheat Eqlid mosaic virus (WEqMV) should be included in a new genus of the family *Potyviridae*.

## Wheat Streak Mosaic Virus

Agronomical importance of this virus is high mainly in European continental areas where analysis of the genome sequence suggested that this virus probably originated in the Fertile Crescent. WSMV-infected plants show mosaic symptoms and develop a rosette appearance. Symptoms on winter wheat usually appear only in spring at temperatures above 10 °C. Infected plants are stunted with sterile heads. Yield losses can be as high as 100%. WSMV is transmitted by the wheat curl mite (WCM) *Aceria tosichella* (Keifer). The virus is acquired during nymphal stages and is retained up to several weeks even following molting. Prevailing winds play an important role in long-distance dispersion of mites. Seed transmission of WSMV in wheat was estimated from 0.5% to 1.5%. The host range of WSMV is very wide including maize. Prevention of the infection of WSMV in winter wheat in the autumn is achieved by elimination of volunteers 2 weeks before planting.

*Wsm1* is a gene of *Thinopyrum intermedium* translocated in wheat. Other partial amphiploids (Zhong1, Zhong 2) are resistant to WSMV and WCM. Recently, two wheat origins from Iran (Adl Cross and 4004) have been reported to carry two distinct genes providing complete resistance to WSMV symptoms. Five genes of resistance to WCM detected in *Aegilops* have been transferred to wheat. Introduction of WSMV replicase (NIa) or CP gene into transgenic wheat confers virus resistance (**Figure 5**).

## Wheat Yellow Head Virus

This virus detected in USA is of limited agronomical importance. The infected plants show yellow heads and mosaic symptoms of flag leaves. Its vector is not known but wheat yellow head virus (WYHV) can be transmitted by vascular puncture inoculation to maize and to *Nicotiana clevelandii* plants. In infected plants, a major protein of 32–34 kDa is detected. Analysis of the amino acid sequences of this protein demonstrated that WYHV is closely related to rice hoja blanca virus, a tenuivirus.

## Wheat Yellow Leaf Virus

Agronomical importance of this virus, detected in Japan, China, and Italy, is limited. The virus causes diffuse chlorotic flecks or interveinal chlorosis and severe yellowing. All small grain crops are susceptible to wheat yellow leaf virus (WYLV). In Japan, the weed *Agropyron*

**Figure 5** Symptoms of wheat streak mosaic virus in winter wheat.

*tsukushiense* (Ohwi) is probably an important reservoir of the virus. *Rhopalosiphum maidis* (Fitch) and *Rhopalosiphum padi* (L.) transmit WYLV in a semi-persistent manner. Resistant Chinese cultivars have been reported.

## Yellow Dwarf Viruses

### Barley Yellow Dwarf-MAV, -PAS, -PAV; Cereal Yellow Dwarf-GPV, -RPS, -RPV; RMV; SGV

Different viruses of the family *Luteoviridae* infecting Poaceae (barley yellow dwarf-MAV, -PAS, -PAV; cereal yellow dwarf-GPV, -RPS, -RPV) and two other viruses not assigned to a genus are collectively called YDs. The distribution of these viruses is dependent on environmental conditions. BYDV-PAV is probably the most widespread. In fields mainly infected by BYDV-PAV, yield reduction from 5% to 20% and nearly 40% have been reported in susceptible wheat and barley cultivars, respectively. Symptoms in infected fields occur in patches. Yellowing (mostly in barley) or reddening (mostly in wheat) are characteristic of YDV infection. These symptoms appear at the two- to three-leaf stage in barley and often after heading in wheat. YDVs are transmitted in a circulative-persistent manner by one or several aphid species. The capsid of YDVs includes a CP and a read-through protein, both needed for aphid transmission.

In temperate areas, YDV vectors have two main flight periods. At the beginning of autumn, winged aphids leave summer hosts (ripening maize, perennial grasses, regrowths, and volunteers). These aphids are attracted to low plant density areas in young winter cereal fields. Depending on the temperature, one to several cycles of virus infection occur before aphids are killed by low temperatures. In oceanic and in other warm areas, low concentration of apterous aphids is maintained during winter. In spring, aphids leave their hosts and infect spring crops and then maize crops. Perennial grasses maintain a low but permanent reservoir of YDVs and aphids.

Risk assessment systems have been developed. In Australia, forecasts of vector incidence are based on the temperature and rainfall in late summer/early autumn. In France, a decision-support system based on temperature-driven simulations of aphid populations was proposed. In UK, a computer-based decision-support system associating the number of aphids found in suction traps and the numbers of foci of YDV infection per unit area of crop was developed. Using these systems, a risk index allows rationalization of foliar pesticide usage (usually synthetic pyrethroids). Treating seeds with an aphicide is necessary in some areas when early sowing of susceptible cultivars is decided or when the pressure of YDV inocula is known to be regularly high. Imidacloprid (nicotinic agonist) or fipronil (chloride channel agonist) gives a good protection during several weeks.

Only minor sources of resistance are known in wheat. Several sources of tolerance have been detected in other Triticinae genera. Several lines derived from crosses between wheat and *Thinopyrum* sp. are very tolerant to YDVs. The basis for *Thinopyrum*-derived resistance to CYDV-RPV is associated to resistance via inhibition of viral systemic infection.

In barley, a major gene of resistance mapped to chromosome 3H, Ryd2, has been introduced into present-day cultivars. The level of protection conferred by this gene varies according to strains of BYDV-PAV and has a low efficiency for CYDV-RPV. Recently, a new major gene Ryd3 mapped to chromosome 6 has been characterized. In resistant plants, these genes tend to increase seed yield compared to healthy controls.

Recent data on the mechanism and genetics of transmission by aphids and evidence of the glycosylation of virus CP open new avenues to control these viruses in the future.

*See also:* Barley Yellow Dwarf Viruses; Brome Mosaic Virus; Cereal Viruses: Maize/Corn; Cereal Viruses: Rice; *Furovirus*; *Hordeivirus*; Maize Streak Virus; *Nepovirus*; *Pecluvirus*; Plant Reoviruses; Plant Resistance to Viruses: Engineered Resistance; Plant Resistance to Viruses: Natural Resistance Associated with Dominant Genes; Plant Resistance to Viruses: Natural Resistance Associated with Recessive Genes; Plant Rhabdoviruses; *Tenuivirus*.

## Further Reading

Chain F, Riault G, Trottet M, and Jacquot E (2007) Evaluation of the durability of the barley yellow dwarf virus-resistant Zhong ZH and TC14 wheat lines. *European Journal of Plant Pathology* 117: 35–43.

Hariri D, Meyer M, and Prud'homme H (2003) Characterization of a new barley mild mosaic virus pathotype in France. *European Journal of Plant Pathology* 109: 921–928.
Lapierre H and Signoret P (eds.) (2004) *Viruses and Virus Diseases of Poaceae* (*Gramineae*), 857pp. Versailles, France: INRA.
Li H, Conner R, Chen Q, et al. (2005) Promising genetic resources for resistance to wheat streak mosaic virus and the wheat curl mite in wheat – *Thinopyrum* partial amphiploids and their derivatives. *Genetic Resources and Crop Evolution* 51: 827–835.
Malik R, Smith CM, Brown-Guedira GL, Harvey TL, and Gill BS (2003) Assessment of *Aegilops tauchii* for resistance to biotypes of wheat curl mite (Acari: Eriophyidae). *Journal of Economic Entomology* 96: 1329–1333.
Niks RE, Habeku ß A, Bekele B, and Ordon F (2004) A novel major gene on chromosome 6H for resistance of barley against barley yellow dwarf virus. *Theoretical and Applied Genetics* 109: 1536–1543.
Seddas P and Boissinot S (2006) Glycosylation of beet western yellows proteins is implicated in the aphid transmission of the virus. *Archives of Virology* 151: 967–984.
Stein N, Perovic D, Kumlehn J, et al. (2005) The eukaryotic translation initiation factor 4E confers multiallelic recessive bymovirus resistance in *Hordeum vulgare* (L.). *Plant Journal* 42: 912–922.
Wiangjun H and Anderson JM (2004) The basis for *Thinopyrum*-derivative resistance to cereal yellow dwarf virus. *Phytopathology* 94: 1102–1106.

# Chandipura Virus

**S Basak**, University of California, San Diego, CA, USA
**D Chattopadhyay**, University of Calcutta, Kolkata, India

© 2008 Elsevier Ltd. All rights reserved.

## Introduction

Vesicular stomatitis has been known for more than a century as a disease that causes severe mouth lesions in cattle, with ruptured vesicles in the gum or tongue and reddish ulcerations. The first report of vesicular stomatitis described the prevalence of the disease in South Africa as early as in 1897, whereas reports of infected horses with symptoms of vesicular stomatitis in the New World date back to an outbreak in the United States in 1916. In spite of multiple outbreaks of the disease in domestic animals that resulted in huge economic loss in different parts of the world, it was only in 1942 that the causative agent was identified as a virus and named vesicular stomatitis virus (Indiana serotype). In the subsequent year, the New Jersey serotype of vesicular stomatitis virus was isolated.

Chandipura virus (CPHV) closely resembles vesicular stomatitis virus but it was readily distinguished by its ability to infect humans. CHPV was first isolated in 1965 in Nagpur, India, from serum samples collected from two febrile patients. This virus isolate was shown to produce cytopathic effects when inoculated into cells in culture. Subsequently, CHPV was isolated from the serum of an encephalopathy patient in 1980 to further confirm its existence and its ability to propagate in the human population. Subsequent serological studies established the existence of this virus in many parts of India in both humans and in a variety of animals.

## Taxonomy and Classification

Vesiculoviruses are assigned to the genus *Vesiculovirus* in the family *Rhabdoviridae*, order *Mononegavirales*. Vesiculoviruses are enveloped and contain a nonsegmented, single (mono) strand of genomic RNA of negative sense. Virions display bullet-shaped morphology, which is typical of viruses included in the family *Rhabdoviridae* (rhabdo in Greek means rod-shaped) (**Figure 1(a)**). Other genera of the family *Rhabdoviridae* include *Lyssavirus*, *Ephemerovirus*, *Novirhabdovirus*, *Cytorhabdovirus,* and *Nucleorhabdovirus*, and these include viruses that are important pathogens of ruminants, fish and plants, and human pathogens such as rabies virus. *Vesicular stomatitis Indiana virus* (VSIV), *Vesicular stomatitis New Jersey virus* (VSNJV), and *Chandipura virus* (CPHV) have been formally recognized as distinct species within the genus *Vesiculovirus*. Virus species that have been classified in the genus *Vesiculovirus* are presented in **Table 1**. Other viruses that have been placed tentatively in the genus *Vesiculovirus* include important viruses such as spring viremia of carp virus, Tupaia virus, and Calchaqui virus.

## Viral Epidemiology

Although CHPV was first isolated in 1965, it was only in 2003 that the first evidence for its association in human epidemic was obtained during acute encephalitis outbreaks. One outbreak took place in the state of Andhra Pradesh and affected 329 children, resulting in 183 deaths. Simultaneously, another encephalitis outbreak was reported in the state of Maharashtra, resulting in about 400 encephalitis cases with 115 deaths. Subsequently, another outbreak of encephalitis associated with CHPV infection was reported in 2004 in the eastern state of Gujarat with a fatality rate of more than 75%. These recent outbreaks indicate the emergence of CHPV as a deadly human pathogen in the Indian subcontinent. Phylogenetic analysis has indicated

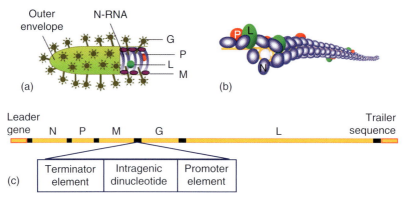

**Figure 1** Vesiculoviruses: a general overview. A general overview of the vesiculoviruses with particular reference to Chandipura virus (a) Schematic presentation of bullet-shaped CHPV. Outer membrane envelope and relative organization of N-RNA nucleocapsid, glycoprotein G, phosphoprotein P, large protein L, and matrix protein M within virion structure have been indicated; (b) Nucleocapsid protein enwrapped genome RNA along with associated phosphoprotein P and large protein L forms a helical structure; (c) CHPV genome RNA encoding leader gene, five protein-coding genes, and gene junctions. Relative arrangement of the *cis*-acting elements within the gene junction has been indicated.

**Table 1** Species within the genus *Vesiculovirus*

| Species name | Abbreviation | Original isolation | Natural host(s) |
| --- | --- | --- | --- |
| Vesicular stomatitis Indiana virus | VSIV | USA, 1942; *North Am. Veterinarian* 26: 726–730 | Horses, cattle, swine, sandflies, |
| Vesicular stomatitis New Jersey virus | VSNJV | USA, 1943; *North Am. Veterinarian* 26: 726–730 | Horses, cattle, swine, sandflies, mosquitoes |
| Piry virus | PIRYV | Brazil, 1960; USPHS Publication No. 1760 Government Printing Office 1967 | Human, cattle, mosquitoes |
| Cocal virus | COCV | Trinidad, 1964; *Am. J. Vet. Res.* 1964 Jan; 25: 236–42 | Horses, cattle, insects |
| Chandipura virus | CHPV | India, 1965; *Indian J. Med. Res.* 1967 Dec; 55(12): 1295–305 | Human, cattle, sandflies, |
| Vesicular stomatitis Alagoas virus | VSAV | Brazil, 1967; *Res. Vet. Sci.* 1967, 8; 103–117 | Human, sandflies, mosquitoes |
| Isfahan virus | ISFV | Iran, 1975; *Am. J. Trop. Med. Hyg.* 1977 Mar; 26(2): 299–306 | Human, gerbils, sandflies |
| Maraba virus | MARAV | Brazil, 1984; *Am. J. Trop. Hyg.* 1984 Sep; 33(5): 999–1006 | Sandflies |
| Carajas virus | CJSV | Brazil, 1984; *Am. J. Trop. Hyg.* 1984 Sep; 33(5): 999–1006 | Sandflies |

circulation in the population during these recent epidemics of multiple strains of CHPV which are subtly different from the original Nagpur strain.

The symptoms of CHPV encephalitis include high fever, generalized convulsions, vomiting, diarrhea, decerebrate posture leading to coma, neurological deficit, and acute encephalitis. During the outbreaks Chandipura encephalitis affected children below 15 years of age and resulted in a high mortality rate. Electron microscopy, complementation fixation and neutralization tests, as well as molecular biological techniques, were performed to examine the existence of Chandipura virus in the sera obtained from patients. The presence of anti-CHPV IgM antibodies and viral RNA was confirmed in the serum to implicate CHPV as an important etiological agent in the outbreak. Subjects infected with CHPV were shown to have an elevated level of cytokines in the blood that is thought to result in the damage to the central nervous system, and CHPV antigen and RNA was found in the brain tissue of the infected patients.

The prevalence of anti-CHPV antibodies in animal sera indicates that domestic animals might serve as a natural reservoir for this pathogen. Indeed, the presence of neutralizing antibodies to Chandipura virus was confirmed in 30.6% of pig, 14.3% of cattle, 17.9% of buffalo, 9.3% of goat, and 7.7% of sheep serum samples collected from the epidemic affected areas. The sandfly has been long recognized as a potential vector for the transmission of CHPV. The virus was isolated from a pool of 253 sandflies of the genus *Phlebotomus* in 1969. Transovarial transmission has also been established in *Phlebotomus papatasi*. However, investigations carried out in the outbreak-hit areas suggest the involvement of sandflies of the genus *Sergentomyia* in the transmission of the pathogen.

## Chandipura Virus Structure and Composition

At the molecular level, VSIV is the most extensively studied member of the family *Rhabdoviridae*. However, research on VSNJV and CHPV has also allowed us to gain further molecular insights into vesiculovirus' replication. The recent association with epidemics have emphasized the necessity to develop better understanding of the mechanism of Chandipura virus pathogenesis. Below, we summarize our current knowledge on the structure and replication of CHPV at the molecular level.

The CHPV virion is a bullet-shaped particle that is ~150 nm long and ~60 nm wide with a ~10 nm distinct surface projection. It has two compartments, the outer envelope and the inner nucleocapsid structure. The helical nucleocapsid is composed of a 11 119-nucleotide (nt) single-stranded genome RNA that is tightly wrapped with ~1200 molecules of nucleocapsid protein N (mol. wt. 50 kDa). In addition, ~50 molecules of large protein L (mol. wt. 241 kDa) and ~500 molecules of phosphoprotein P (mol. wt. 32.5 kDa) remain associated with the N-RNA complex. The envelope is a membranous structure that surrounds the nucleocapsid and contains an embedded transmembrane glycoprotein G that protrudes externally from the membrane surface. The matrix protein M lies at the inner side of the membrane and tethers the nucleocapsid to the envelope (**Figures 1(a)** and **1(b)**).

## Genetic Makeup and Gene Structure

Vesiculoviruses exemplify one of the simplest genome architectures among the *Mononegavirales*. The entire CHPV genome has been sequenced and comparison with other vesiculoviruses has confirmed similar genetic makeup among vesiculoviruses. The Chandipura virus RNA genome comprises a 49-nt long leader sequence (l), followed by five transcriptional units coding for viral polypeptides separated by intergenic, nontranscribed spacer regions, and a short nontranscribed trailer sequence (t) arranged in the order 3′ l-N-P-M-G-L-t 5′ (**Figure 1(c)**). The CHPV gene junctions encompass a 23-nt long sequence that is highly conserved among different vesiculoviruses; this conserved sequence is comprised of a 11-nt gene end signal that regulates transcription termination and polyadenylation of viral mRNA, a 2 nt nontranscribed spacer region, and a 10 nt conserved promoter element that has been implicated in the transcription initiation. Comparative sequence analysis has indicated CHPV to be evolutionarily equidistant from the New World vesiculoviruses VSIV and VSNJV and closely related to the Asian vesiculovirus Isfahan.

## Replication Cycle

The vesiculovirus replication cycle can be divided into nine discrete steps, namely: (1) adsorption of the virus particle; (2) viral entry into the host cell via receptor-mediated endocytosis; (3) membrane fusion and release of the core RNP into the cytosol from endosomal vesicles; (4) transcription of the genome by viral polymerase; (5) translation of viral mRNA; (6) post-translational modifications of viral proteins; (7) replication of viral genome; (8) assembly of progeny particles; and finally, (9) budding of the mature virion (**Figure 2**). The entire vesiculovirus replication cycle is cytosolic. Genomic RNA enwrapped with nuleocapsid protein N acts as a template during transcription to produce short leader RNA (l) and five monocistronic capped and polyadenylated viral mRNAs. Viral RNA-dependent RNA polymerase (RdRp) is composed of large protein L, which is the catalytic subunit and phosphoprotein P that acts as a regulatory subunit. Translation of viral mRNA results in the accumulation of viral polypeptides within the infected cell to prepare for the onset of genome replication. During replication, the same polymerase copies the entire genomic template into an exact polycistronic complement that acts as replication intermediate to produce multiple copies of negative-sense genome upon further cycles of replication. Subsequently, progeny of negative-sense genomic material is either packaged within the mature virion or subjected to further transcription, often referred to as secondary transcription. Notably, virus-specific genomic analogs always remain encapsidated by N protein; it is believed that the N-mediated enwrapment of the genomic RNA protects RNA from cellular RNases (**Figure 2**).

## Viral Invasion

The receptor utilized by Chandipura virus to invade into the host cell has not yet been characterized. However, studies on VSIV have provided valuable information on the entry mechanism used by vesiculoviruses, in general. Consistent with its broad host range, the receptor utilized by VSIV is thought to be phosphatidyl serine, a ubiquitously present membrane lipid. Subsequent to the attachment, clathrin-mediated endocytosis triggers entry of the virus particle into the host cell through the endosomal pathway, which is then followed by G protein-mediated fusion of the virus particle with the endosomal membrane and release of the core nucleocapsid into the host cytoplasm.

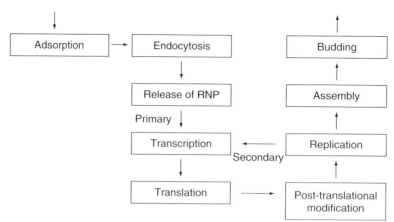

**Figure 2** Vesiculovirus replication cycle. Graphical presentation of nine steps of the vesiculovirus replication cycle with particular reference to Chandipura virus. Transcription from the nucleocapsid particle released from the infecting virion, also referred to as primary transcription, as well as transcription from nucleocapsid particles synthesized *de novo* within infected cells, also referred to as secondary transcription, have been indicated in the figure.

## Role of Glycoprotein G

Glycoprotein G is the sole spike protein that has been implicated in adsorption as well as budding of Chandipura virus. Studies on CHPV G protein expressed from a cloned DNA have revealed the presence of an N-terminal cleavable signal peptide, an N-terminal ectodomain with two N-linked glycosylation sites, a membrane anchor domain, and a cytosolic domain at the C-terminus. It has been proposed that a low pH-induced conformational change in G protein within the endosomal environment subsequent to viral entry results in membrane fusion to release core nucleocapsid particle into the host cytoplasm. The ectodomain of glycoprotein G mediates critical steps in membrane fusion. It is interesting to note that almost all the mutations in the G protein sequences identified in epidemic-associated virus strains were within this ectodomain.

## Role of Matrix Protein M

Matrix protein M glues the core nucleocapsid particle to the outer membrane within the mature virion. The highly basic N-terminal domain of the M protein is primarily responsible for membrane association, although additional domains have been proposed to contribute in membrane binding.

Chandipura virus makes use of the fact that host mRNA is synthesized in the nucleus, whereas viral mRNA synthesis occurs in the cytoplasm to promote preferential translation of viral mRNAs. To achieve this, CHPV utilizes M protein to inhibit nuclear export of host mRNAs and thus channel host machinery towards translating viral mRNAs. Furthermore, M proteins of both VSIV and CHPV have been shown to shut off RNA polymerase I- and II-mediated transcription within infected cells. Not surprisingly, M protein alone was shown to account for cytotoxicity in virus-infected cells.

## Gene Expression

### Sequential and Polar Transcription

Vesiculovirus gene expression illustrates two distinct well-conserved characteristics; namely sequential and polar transcription. Vesiculovirus mRNAs are synthesized in an obligatory sequential manner after polymerase entry at a single site at the 3' end of the genome, that is, at the beginning of leader RNA. Also, measurement of molar ratios of different VSIV mRNAs indicated that their abundance decreases with increasing distance from the polymerase entry site in the order $N > P > M > G > L$. To explain this polar and sequential mRNA synthesis, a stop–start model has been postulated for vesiculovirus transcription (**Figure 3**). In this model, subsequent to the entry into the N-RNA template, viral RdRp sequentially transcribes the genome with progressive attenuation at each gene boundary, which results in decreasing amounts of transcripts for genes that are distant from the entry site. Each termination event may cause the polymerase to fall off the template or may allow for re-initiation at the downstream promoter. Therefore, transcription of the downstream genes depends on the termination of the upstream gene transcription and re-initiation processes. Genomic RNA enwrapped with N protein into a nucleocapsid structure and not the naked viral RNA is recognized by the polymerase as a template for RNA synthesis. *In vitro* transcription reconstitution as well as *in vivo* studies have suggested that the proposed mechanism for vesiculovirus transcription is mostly conserved in CHPV system (**Figure 3(a)**).

### Viral RdRp

CHPV RdRp is composed of the L and P proteins. Catalytic activities for RNA polymerization, capping, and polyadenylation reside within the L protein. The entire ORF

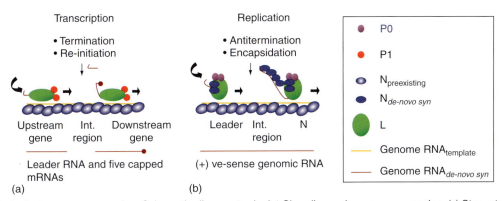

**Figure 3** Vesiculovirus gene expression. Schematic diagram to depict Chandipura virus gene expression. (a) Stop–start model to describe vesiculovirus transcription, where, polymerase terminates and re-initiates at each gene junction to sequentially synthesize leader RNA and five mRNAs. (b) Viral replication that is characterized by the antitermination at each gene junction with concomitant encapsidation of the nascent replication product by nucleocapsid protein N. A role for unphosphorylated P protein (P0) has been postulated during replication.

encoding the CHPV L gene has been sequenced. The L protein exhibits a distinct structural module similar to that of a right hand, with palm, thumb, and finger domains. The palm domain structure is particularly conserved among almost all known RdRps. However, L protein only in association with the phosphorylated form of the P protein, and not L alone, acts as viral transcriptase.

## Phosphoprotein P in Transcription

The P protein is packaged within the mature virion particle and remains associated with the viral nucleocapsid. Subsequent to the release of core nucleocapsid into host cytoplasm, the P protein plays an essential role in viral transcription. The P protein through its C-terminal domain interacts with both N-RNA template and the L protein, thus assembling transcriptase complex on the genomic template. The P protein utilizes its N-terminal acidic domain to interact with and activate polymerase function during transcription, similar to the activation mediated by classical eukaryotic transcription activators. The CHPV P protein exists in the virus-infected cells in two distinct states; the unphosphorylated form has been designated as P0, while its phosphorylated counterpart is referred to as P1. Phosphorylation of P at a distinct site is required for P to act as a transcriptional activator. Further investigations identified casein kinase II (CKII) as the host kinase responsible for phosphorylating P at specific residues. Biochemical and mutational analysis identified ser62 in the CHPV P protein as a single phosphorylation site both *in vitro* and *in vivo*.

How does phosphorylation regulate P protein function as a transcriptional activator? Studies on CHPV have revealed that the phosphorylated P protein forms a dimer. The N-terminal 46 amino acids were implicated in the dimerization of phosphorylated protein. It has been proposed that the unphosphorylated P protein exists as a monomer, whereas, dimerization upon phosphorylation enables P to act as a transcriptional activator. However, the unphosphorylated form of CHPV P protein has also been shown to form a dimer in a concentration-dependent manner. Recent structural analysis also indicated that a central domain that is conserved among different vesiculoviruses mediates dimer assembly of the VSIV-P protein in its unphosphorylated form. Therefore, dimerization alone is inadequate to explain transcriptional activation by phosphorylated P. Interestingly, phosphorylation at ser62 residue has been shown to promote a major conformational change in the P protein that exposes the N-terminal coiled-coil motif on the surface of protein tertiary structure. In light of the available information, it seems possible that the central domain assembles P into a dimer, whereas, additional interactions mediated through the N-terminal domain may stabilize such a dimer assembly. Conformational changes within the N-terminal domain of P upon phosphorylation and net negative charge attributed by the incorporated phosphate have been proposed to facilitate precise polymerase (L) contact by P1, resulting in optimal transcription.

## mRNA Modification by Viral Transcriptase

One unique feature of vesiculovirus gene expression is the utilization of a single protein (L) for mRNA synthesis, 5′ end modification, and 3′ end polyadenylation of the mRNA. With the exception of leader RNA, all other viral transcripts possess a 5′ end cap structure, similar to that observed in cellular mRNAs. Although certain host factors have been postulated to participate in the capping reaction, primary enzymatic activity that mediates viral mRNA capping resides within the L protein. Viral transcriptase is also involved in the polyadenylation of viral mRNA. Polymerase slippage during transcription termination at the U7 tract present in the gene boundaries results in addition of a poly(A) tail at the 3′ end of viral mRNAs. However, our current knowledge of mRNA modifications mostly relies upon contemporary studies that have been conducted in the VSIV system, due to the lack of a recombinant CHPV L protein expression system.

## Replication of the Genome RNA

### Transcription–Replication Switch

During transcription, the viral polymerase obeys the termination signals present at the gene boundaries to produce discrete viral mRNAs, whereas the same polymerase antiterminates at the gene boundaries during the replication phase to produce a positive-sense polycistronic complement of the entire genome. Subsequent rounds of replication that utilize the positive-sense replication intermediate as template produce many more copies of the progeny, negative-sense genomic RNA, which, after packaging buds out as a mature virion (**Figure 2**). The switch in polymerase function that allows it to ignore termination signals during replication is known as the transcription–replication switch. Although the same L protein is involved in RNA polymerization during both transcription and replication the phosphorylation state of the regulatory subunit P is proposed to be different in these two processes.

### Role of Unphosphorylated P Protein

Recent studies in both CHPV and VSIV systems have indicated a role for unphosphorylated P protein in transcription–replication switch. A tripartite complex composed of a phosphorylation-defective mutant of P, N, and L was shown to efficiently synthesize VSIV replication product *in vitro*. It was postulated that the formation of L–N–P0 complex enables L protein to recognize promoter elements for replication and suppress the transcription termination signal present at the gene boundaries during VSIV replication. Similarly, *in vivo* analysis revealed that the N-terminal phosphorylation of P protein, although necessary for transcription, was dispensable for CHPV or VSIV replication. Indeed, a preexisting pool of phosphorylation-defective mutant of CHPV P protein (Ser62Ala) was shown to boost viral replication with a two-log increase in the virus yield upon subsequent infection with the wild-type virus. Therefore, the form of P protein (P1) engaged in transcription appears to be distinct from the form (P0) that has been implicated in replication.

However, another intriguing observation connected the proposed role of P0 in viral replication with the events on the nascent leader RNA chain. In an *in vitro* RNA-binding assay, it has been shown that P protein in its unphosphorylated form binds to leader RNA, whereas, CKII-mediated phosphorylation abrogated its RNA-binding ability. Leader RNA–P0 interaction was observed to be distinct from the N-mediated encapsidation. Similar phosphoregulation of RNA-binding ability of P protein has also been reported for rinderpest virus, a paramyxovirus. Accordingly, a refined model has been proposed to explain Chandipura virus genome replication. In this proposal, recruitment on the nascent leader chain brings P0 to the vicinity of the transcribing polymerase to promote P0 association with the polymerase. Such P0–L interaction alters the polymerase conformation in a manner that enables read-through at the gene boundaries. Nonetheless, progressive encapsidation of nascent viral RNA by CHPV N has been proposed to maintain processivity of such a replicase complex (L–P0) to allow for the synthesis of ∼11.1 kb positive-sense genome (**Figure 3(b)**). Curiously, recent analysis identified a conserved Glu64Asp mutation in the CKII consensus site within the N-terminal region of P protein in all epidemic isolates. In the light of the proposed function of P0 in viral replication, functional consequences of such mutations need to be further explored.

## Genome Encapsidation and Assembly

### Genome Encapsidation

Viral replication intermediates or replication products always remain encapsidated by N protein within infected cells and enjoy protection from cellular RNases. Consistently, vesiculovirus replication critically relies upon *de novo* synthesis of N protein. However, N protein is prone to aggregation and this aggregated form is unable to appropriately encapsidate viral RNA. Phosphoprotein P, by virtue of its N-specific chaperone-like activity, prevents self-aggregation of N protein and maintains it in an encapsidation-competent state. *Cis*-acting elements present within the first 20 nucleotides from the 5′ end of the positive-sense anti-genome and also at the 5′ half of the trailer region of the negative-sense genome, have been implicated in the initiation of encapsidation. CHPV N protein, in its monomeric form, possesses intrinsic RNA-binding specificity and is able to recognize *cis*-elements present in the genome RNA. However, this RNA-binding specificity has been shown to be compromised upon homooligomerization of N protein. Accordingly, a model has been proposed to provide mechanistic insight into the encapsidation process. In this model, monomer N initiates encapsidation by recognizing specific sequence element present in the nascent viral RNA. Subsequently, N–N interactions allow oligomer N to continue the elongation phase of the encapsidation process to enwrap diverse RNA sequences into the nucleocapsid structure. It is believed that the ensuing encapsidation of nascent RNA by N protein may also facilitate antitermination during replication by opposing polymerase pausing at the gene boundaries. Nevertheless, the major role of

N protein in viral replication cycle remains to protect viral RNA from cellular RNases and present genomic RNA in a specialized helical form that is recognized by viral polymerase.

## Assembly and Budding

Progeny genomic RNA molecules enwrapped in the nucleocapsid structures associate with matrix protein M and are subjected to further M-mediated condensation. Subsequently, M protein participates in recruiting mature nucleocapsid particles to the membrane during assembly and budding in a manner that is dependent on the cellular ubiquitin–proteasome system, at least in the VSIV system. *De novo*-synthesized G protein is organized in the plasma membrane of virus-infected cells into microdomains that serve as assembly sites for the viral components.

## Concluding Remarks

Our current understanding of the discrete steps of the CHPV replication cycle relies on the studies being conducted on the CHPV as well as on other vesiculoviruses. Remarkable similarities between CHPV and VSIV in genetic makeup, gene structure, polypeptide composition as well as function allow the postulation of a generalized model to describe vesiculovirus transcription at the molecular detail. While the proposed mechanism for CHPV replication that involves RNA binding by the phosphoprotein needs to be tested in the VSIV system, existence of distinct mechanisms in different serotypes of vesiculoviruses is possible. Nevertheless, information obtained from the molecular analysis of the epidemic-associated strains in conjunction with the large body of information that is available on N–P, N–leader RNA, or P0–leader RNA interactions should be utilized to further understand CHPV pathogenesis. While retaining general academic interest to understand RNA synthesis in rhabdoviruses, future CHPV research should also focus on development of the potential antiviral therapeutic interventions that will allow for specific inhibition of virus multiplication.

*See also:* Rabies Virus.

## Further Reading

Banerjee AK (1987) Transcription and replication of rhabdoviruses. *Microbiological Reviews* 51(1): 66–87.

Barr JN, Whelan SP, and Wertz GW (2002) Transcriptional control of the RNA-dependent RNA polymerase of vesicular stomatitis virus. *Acta Biochimica et Biophysica Hungarica* 1577(2): 337–353.

Basak S, Mondal M, Polley S, Mukhopadhyay S, and Chattopadhyay DJ (2007) Reviewing Chandipura: A vesiculovirus in human epidemics. *Bioscience Reports* 27(4–5): 275–298.

Mishra AC (2007) Chandipura encephalitis: A newly recognized disease of public health importance in India. In: Scheld WH, Hooper DC, and Hughes JM (eds.) *Emerging Infections*, pp. 121–137. Washington, DC: ASM Press.

# Chrysoviruses

**S A Ghabrial,** University of Kentucky, Lexington, KY, USA

© 2008 Elsevier Ltd. All rights reserved.

## Glossary

**Hyphal anastomosis** The union of a hypha with another resulting in cytoplasmic exchange.

**Mycoviruses** Viruses that infect and multiply in fungi.

**Viruses with multipartite genomes** The essential genome is divided among several genomic segments (segmented genome) that are either separately encapsidated in identical capsids (multicomponent viruses, i.e., chrysoviruses) or jointly enclosed in a single particle (e.g., reoviruses).

## Introduction

The discovery in the late 1960s and early 1970s of polyhedral virus particles in many of the industrial strains of *Penicillium chrysogenum* used for penicillin production generated considerable interest in the study of *Penicillium* viruses. It was surmised then that virus infection might be responsible for the instability of some of these strains. *Penicillium chrysogenum virus* (PcV), the type species of the genus *Chrysovirus*, was one of the first mycoviruses to be extensively studied at the biochemical, biophysical, and ultrastructural levels. Although the results of these earlier studies on PcV properties were mostly in agreement, they differed in their explanation of the nature of genome complexity. Although PcV and the related *Penicillium* viruses penicillium brevicompactum virus (PbV) and penicillium cyaneo-fulvum virus (Pc-fV) have similar isometric particles, 35–40 nm in diameter, and are serologically related, there was confusion as to whether they contain three or four double-stranded RNA (dsRNA) segments. Because none of these viruses was characterized at the molecular level until recently, these three *Penicillium* viruses were originally grouped under the genus *Chrysovirus* and provisionally placed in the family *Partitiviridae* with the assumption that their genomes are bipartite, with dsRNA1 encoding the RNA-dependent RNA polymerase (RdRp) and dsRNA2 encoding the major capsid protein (CP). The additional dsRNAs (dsRNAs 3 and/or 4), like those of some partitiviruses, were presumed to be defective or satellite dsRNAs. The classification of the genus *Chrysovirus* was recently reconsidered because the complete nucleotide sequence and genome organization of each of the four monocistronic dsRNA segments associated with PcV virions and with the chrysovirus Helminthosporium victoriae 145S virus were recently reported. Based on the consistent simultaneous presence of their four dsRNA segments, the existence of extended regions of highly conserved terminal sequences at both ends of all four segments, sequence comparisons and phylogenetic analysis, it became clear that PcV and related viruses should not be classified with the family *Partitiviridae*. This led to the creation of the new family *Chrysoviridae* to accommodate the isometric dsRNA mycoviruses with multipartite genomes. The name *Chryso* is derived from the specific epithet of *Penicillium chrysogenum*, the fungal host of the type species.

## Virion Properties

The buoyant densities of virions of members in the family *Chrysoviridae* are in the range of 1.34–1.39 g cm$^{-3}$ and their sedimentation coefficients $S_{20,w}$ (in Svedberg units) are in the range of 145S to 150S. Generally, each virion contains only one of the four genomic dsRNA segments. However, purified preparations of PcV and Pc-fV can contain minor distinctly sedimenting components that include empty particles and replication intermediates.

Chrysovirus particles possess virion-associated RNA-dependent RNA polymerase (RdRp) activity, which catalyzes the synthesis of single-stranded RNA (ssRNA) copies of the (+) strand of each of the genomic dsRNA molecules. The *in vitro* transcription reaction occurs by a conservative mechanism, whereby the released ssRNA represents the newly synthesized plus strand.

## Genome Organization

The virions of members of the family *Chrysoviridae* contain four unrelated linear, separately encapsidated, monocistronic dsRNA segments (2.4–3.6 kbp in size; **Table 1**). the largest segment, dsRNA1, codes for the RdRp and

**Table 1**  List of members and tentative members in the family *Chrysoviridae*

| Virus species | Abbreviation | DsRNA segment no. (length in bp, encoded protein; size in kDa) | GenBank accession no. |
|---|---|---|---|
| Helminthosporium victoriae 145S virus | Hv145SV | 1 (3612; RdRp,125) | NC_005978 |
| | | 2 (3134; CP, 100) | NC_005979 |
| | | 3 (2972; chryso-P4, 93) | NC_005980 |
| | | 4 (2763; chryso-P3, 81) | NC_005981 |
| Penicillium brevicompactum virus | PbV | Four dsRNA segments; no molecular data | |
| Penicillium chrysogenum virus | PcV | 1 (3562; RdRp,129) | NC_007539 |
| | | 2 (3200; CP, 109) | NC_007540 |
| | | 3 (2976; chryso-P3 101) | NC_007541 |
| | | 4 (2902; chryso-P4, 95) | NC_007542 |
| Penicillium cyaneo-fulvum virus | Pc-fV | Four dsRNA segments; no molecular data | |
| Tentative members | | | |
| Agaricus bisporus virus 1 or LaFrance isometric virus | AbV-L1 | L1 (3396; RdRp, 122) | X94361 |
| | LFIV | L5 (2455; unknown, 82) | X94362 |
| Amasya cherry disease associated chrysovirus | ACDACV | 1 (3399; RdRp, 124) | AJ781166 |
| | | 2 (3128; CP, 112) | AJ781165 |
| | | 3 (2833; chryso-P4, 98) | AJ781164 |
| | | 4 (2498; chryso-P3, 77) | AJ781163 |
| Cherry chlorotic rusty spot associated chrysovirus | CCRSACV | 1 (3399; RdRp, 124) | AJ781397 |
| | | 2 (3125; CP, 112) | AJ781398 |
| | | 3 (2833; chryso-P4, 98) | AJ781399 |
| | | 4 (2499; chryso-P3, 77) | AJ781400 |

dsRNA2 codes for the major CP. The dsRNA segments 3 and 4 code for proteins of unknown function. The genomic structure of PcV, the type species of the genus *Chrysovirus*, comprising four dsRNA segments, is schematically represented in **Figure 1**. The earlier conflicting reports on whether PcV contains three or four segments were recently explained when studies on cDNA cloning and sequencing of the viral dsRNAs were completed. Although the dsRNA extracted from purified virions is resolved into three bands by agarose gel electrophoresis (**Figure 2**, lane EB), northern hybridization analysis using cloned cDNA probes representing the four dsRNA segments shows clearly that each of the four segment has unique sequences (**Figure 2**). Because dsRNAs 3 and 4 differ in size by only 74 bp (**Table 1**), they co-migrate when separated by agarose gel electrophoresis. Previous studies on sequencing analysis and *in vitro* coupled transcription–translation assays showed that each of the four dsRNAs is monocistronic, as each dsRNA contains a single major open reading frame (ORF) and each is translated into a single major product of the size predicted from its deduced amino acid sequence. Thus, the fact that PcV virions contain four distinct dsRNA segments has clearly been established.

Unlike PcV, dsRNAs 3 and 4 from other chrysoviruses, helminthosporium victoriae 145SV (Hv145SV), amasya cherry disease associated chrysovirus (ACDACV) and cherry chlorotic rusty spot associated chrysovirus (CCRSACV) are clearly resolved from each other when purified dsRNA preparations are subjected to agarose gel electrophoresis. As shown in **Table 1**, dsRNAs 3 and 4 from these viruses are significantly different in size. Assignment of numbers 1–4 to PcV dsRNAs was made according to their decreasing size. Following the same criterion used for PcV, the dsRNAs associated with Hv145SV, CCSRACV, and ACDACV were accordingly assigned the numbers 1–4. Sequence comparisons, however, indicated that dsRNAs 3 of Hv145SV, CCSRACV, and ACDACV are in fact the counterparts of PcV dsRNA4 rather than dsRNA3. Likewise, dsRNA4 of these three chrysoviruses are the counterparts of PcV dsRNA3. Since PcV was the first chrysovirus to be characterized at the molecular level and to avoid confusion, the protein designations P3 and P4 as used for PcV will be adopted and referred to as

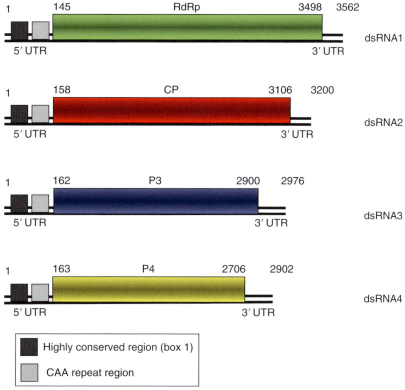

**Figure 1** Genome organization of penicillium chrysogenum virus (PcV), the type species of the genus *Chrysovirus*. The genome consists of four dsRNA segments; each is monocistronic. The RdRp ORF (nt positions 145–3498 on dsRNA1), the CP ORF (nt positions 158–3106 on dsRNA2), the p3 ORF (nt positions 162–2900 on dsRNA3), and the p4 ORF (nt positions 163–2706 on dsRNA4) are represented by rectangular boxes. Adapted from Ghabrial SA, Jiang D, and Castón RJ (2005) *Chrysoviridae*. In: Fauquet CM, Mayo MA, Maniloff J, Desselberger U, and Ball LA (eds.) *Virus Taxonomy: Eighth Report of the International Committee on Taxonomy of Viruses*, pp. 591–595. London: Academic Press, with permission from Elsevier.

chryso-P3 and chryso-P4. Thus, whereas the chryso-P3 protein represents the gene product of PcV dsRNA3, it comprises the corresponding gene product of Hv145SV dsRNA4, and so on.

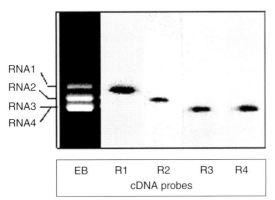

**Figure 2** Northern hybridization analysis of PcV dsRNA segments. Virion dsRNAs were separated on a 1.5% agarose gel, transferred onto Hybond-N$^+$ membrane and hybridized with $^{32}$P-labeled probes prepared by random-primer labeling of cloned cDNA to PcV dsRNA1 (R1), dsRNA2 (R2), dsRNA3 (R3), and dsRNA4 (R4). EB, ethidium bromide-stained virion dsRNAs separated on a 1.5% agarose gel. Reproduced from Jiang D and Ghabrial SA (2004) Molecular characterization of penicillium chrysogenum virus: Reconsideration of the taxonomy of the genus chrysovirus. *Journal of General Virology* 85: 2111–2121, with permission from Society for General Microbiology.

Except for CCRSACV and ACDACV dsRNAs, the 5' UTRs of chrysovirus dsRNAs are relatively long, between 140 and 400 nucleotides in length. In addition to the strictly conserved 5'- and 3'-termini, a 40–75 nt region with high sequence identity is present in the 5' UTR of all four dsRNAs (box 1; **Figure 3**). A second region of strong sequence similarity is present immediately downstream from 'box 1' (**Figure 3**). This consists of a stretch of 30–50 nt containing a reiteration of the sequence 'CAA'. The (CAA)$_n$ repeats are similar to the enhancer elements present at the 5' UTRs of tobamoviruses. Although the 5' UTR of CCRSACV and ACDACV contain the 'CAA' repeat region upstream of the translation initiation codon, the 'box 1' region is significantly shortened. Furthermore, these viruses do not share with PcV and Hv145SV the strictly conserved terminal 8 and 7 nucleotides, respectively at the 5' and 3' ends (**Table 2**). The discrepancies in the length and features of the 5' and 3' UTRs of CCRSACV and ACDACV dsRNAs compared to those of PcV and Hv145SV could be due to the cloning procedure used, which may not have allowed for the exact termini to be cloned. This, however, seems unlikely because of the strong similarities of the termini and internal sequences of the 5' and 3' UTRs of all four CCRSACV and ACDACV dsRNA segments. Alternatively, the divergence in the features of the 5' and 3' UTRs of ACDACV and CCRSACV may be because

**Figure 3** Comparison of the 5' and 3' UTRs of the four dsRNA segments of PcV. Multiple sequence alignments were obtained using CLUSTAL X (and some manual adjustments) with the nucleotide sequences of the 5' UTR (a) and the 3' UTR (b). The (CAA) repeats are underlined in (a). Asterisks signify identical bases at the indicated position (shaded) and colons specify that three out of four bases are identical at the indicated positions. Reproduced from Jiang D and Ghabrial SA (2004) Molecular characterization of penicillium chrysogenum virus: Reconsideration of the taxonomy of the genus chrysovirus. *Journal of General Virology* 85: 2111–2121, with permission from Society for General Microbiology.

**Table 2** Comparison of the nucleotide sequences at the 5' and 3' termini of chrysovirus dsRNAs[a]

|  | 5'-terminus | 3'-terminus |
|---|---|---|
| Hv145SV | | |
| dsRNA1 | 5'-GAUAAAAAGAAAAA-U.. | ..UUAGGACUUUAAGUGU-3' |
| dsRNA2 | 5'-GAUAAAAACAAAAAU.. | ..UUCGGACUUUAAGUGU-3' |
| dsRNA3 | 5'-GAUAAAAACAGAAAU.. | ..UUCGGACUUUAAGUGU-3' |
| dsRNA4 | 5'-GAUAAAAACAGAAAAU.. | ..UGCGGACUUUAAGUGU-3' |
| PcV | | |
| dsRNA1 | 5'-GAUAAAAAAAGAAUAA.. | ..GCUUUAAAAUAAGUGU-3' |
| dsRNA2 | 5'-GAUAAAAAACAAUAA.. | ..GCUUUAAAAUAAGUGU-3' |
| dsRNA3 | 5'-GAUAAAAAACGAUAA.. | ..GCUUUAAAAUAAGUGU-3' |
| dsRNA4 | 5'-GAUAAAAAACGAUAA.. | ..GUUUUAAAAUAAGUGU-3' |
| CCRSACV | | |
| dsRNA1 | 5'-GAAAUUAUGGUUUUUG.. | ..AUUGUCAAUAAUAUGC-3' |
| dsRNA2 | 5'-GAAAUUAUGGUUUUUG.. | ..GUGUUGAUAUAUAUGC-3' |
| dsRNA3 | 5'-GAAAUUAUGGUUUUUG.. | ..GGUUAUAACUAUAUGC-3' |
| dsRNA4 | 5'-GAAAUUAUGGAUUUUG.. | ..AUGUGUAACUAUAUGC-3' |

[a]Identical nucleotides in the same position are shaded.

they represent plant rather than fungal chrysoviruses (see section on 'Biological properties').

## Genome Expression and Replication

### Chrysovirus dsRNA1s Code for RdRp

The largest dsRNA segment (dsRNA1) of the chrysoviruses so far sequenced contains a single large ORF coding for RdRp. The calculated molecular mass of chrysovirus RdRps ranges from 122 to 129 kDa (**Table 1**). These values are consistent with those estimated by SDS-PAGE of the *in vitro* translation products of full-length transcripts derived from cDNAs to dsRNA1s of PcV and Hv145SV. Examination of the deduced amino acid sequence of the RdRp ORF reveals the presence of the eight conserved motifs characteristic of RdRps of dsRNA viruses of simple eukaryotes (**Figure 4**). A comparison of the conserved motifs of chrysovirus RdRps with those of totiviruses and partitiviruses reveals that the RdRps of chrysoviruses are more closely related to those of totiviruses than to those of partitiviruses (**Figure 4**). This conclusion is also supported by published results of phylogenetic analysis of RdRp conserved motifs and flanking sequences of chrysoviruses and viruses in the families *Totiviridae* and *Partitiviridae* (see section on 'Evolutionary relationships').

### Chrysovirus dsRNA2s Code for CP

The second largest dsRNA segment (dsRNA2) of chrysoviruses so far sequenced contains a single large ORF coding for CP. The calculated molecular mass of chrysovirus CPs ranges from 100 to 112 kDa (**Table 1**). The predicted size of PcV CP (109 kDa) is similar to that estimated by SDS-PAGE of purified PcV virions as well as that determined for the *in vitro* translation product of full-length transcript of dsRNA2 cDNA. Direct evidence that PcV dsRNA2 encodes CP was provided by amino acid sequencing of a tryptic peptide derived from a gradient-purified PcV capsid.

### Chryso-P3 Shares a 'Phytoreo S7 Domain' with Core Proteins of Phytoreoviruses

PcV dsRNA3 codes for its chryso-P3 protein, whereas Hv145SV, ACDACV, and CCRSACV dsRNA4s encode the corresponding chryso-P3s. Although the function of chryso-P3 is not known, sequence analysis and database searches offer some clues. ProDom database searches reveal that chryso-P3 sequences share a 'phytoreo S7 domain' with a family consisting of several phytoreovirus P7 proteins known to be viral core proteins with nucleic acid binding activities. The consensus for the three chrysoviruses is [X(V/I)V(M/L)P(A/M)G(C/H)GK(T/S)T-(L/I)]. Phytoreovirus P7 proteins bind to their corresponding P1 (transcriptase/replicase) proteins, which bind to the genomic dsRNAs. It is of interest, in this regard, that the N-terminal regions of all chryso-P3s (encompassing the amino acids within positions 1–500) share significant sequence similarity with comparable N-terminal regions of the putative RdRps encoded by chrysovirus dsRNA1s. A multiple alignment of a portion of the N-terminal region sequences of chrysovirus P3s and RdRps is shown in **Table 3** to demonstrate the level of similarity among the N-terminal sequences of these proteins. The regions in the dsRNA1-encoded proteins with high similarity to chryso-P3 occur upstream of the eight highly conserved motifs characteristic of RdRps of dsRNA viruses of simple eukaryotes. The significance of this sequence similarity to the function of chryso-P3 is not known for certain, but one may speculate that the N-terminal region of these proteins may play a role in viral RNA binding and packaging.

|  | 1 | 2 | 3 | 4 | 5 |
|---|---|---|---|---|---|
| SsRV1 | LLGRA(61) | WCVNGSQND(47) | KL-EHG-KTRAIFACDTRSY (47) | LDFDDFNSHHS(45) | TLPSGHRGTTIVNSVLNAAYI(14) |
| Hv190SV | LQGRY(61) | WCVNGSQNA(42) | KL-ENG-KDRAIFACDTRSY (47) | LDYDNFNSQHS(45) | TLMSGHRATTFTNSVLNAAYI 14) |
| SsRV2 | LQGRA(64) | WAVNGSQSG(46) | KL-EHG-KTRAIFACDTLNY (47) | LDYDDFNSHHS(46) | TLMSGRRGTTYISSVLNEVYL(14) |
| UmVH1 | LYGRG(66) | WLVSGSSAG(55) | KLNETGGKARAIYGVTLWHY (47) | YDYPDFNSMHT(64) | GLYSGDRDTTLINTLLNIAYA(20) |
|  | * ** | * *:** | ** * * *** :::: * | :*:::*** *: | :* ** * ** :::** :*: |
| Hv145SV | LLGRR(73) | WMTKGSLVS(56) | KLNENGHKDRVLLPGGLLHY (44) | YDWANFNVQHS (49) | GLYSGWRGTTWDNTVLNGCYM(20) |
| PCV | LVGRG(74) | WLTKGSLVY(60) | K-YEVGKKDRTLLPGTLVHF (44) | YDWADFNEQHS (49) | GLYSGWRGTTWINTVLNFCYV(19) |
| CCRSACV | LVGRR(73) | WLTKGSTVY(65) | KLNECGYKDRTLLPGSLFHY (44) | FDWANFNAFHS (49) | GLYSGWRGTSFLNSVLNSCYT(19) |
|  | *:**: | *:****:*: | *:* * ***:**** * *: | :***:** :** | ********** :: *:*** ** |
| FpV1 |  |  | SDRDGILKQRPVYAVDDLFL (47) | IDWSGFDQRLP(72) | GVPSGMLNTQFLDSFGNLFLL(19) |
| RhsV-717 |  |  | SKRDGTLKVRPVYAVDELFL (47) | IDWSGYDQRLP(71) | GVPSGMLLTQFLDSFGNLYLI(19) |
| AhV1 |  |  | SKRD-NLKVRPVYNAPMIYI (47) | IDWSRFDHLAP(92) | GVPSGILMTQFIDSFVNLTIL(19) |
| HaV |  |  | SKIT-KLKVRPVYNAPMLFL (47) | FDYSRFDQLAP(101) | GVPSGIFMTQILDSFVNLFIF(19) |
| BCV3 |  |  | ADLREKTKVRGVWGRAFHYI (48) | LDWSSFDSSVT(50) | GIPSGSYYTSIVGSVVNRLRI(15) |
| FsV1 |  |  | SPRD-DPKTRLAWIYPSEML (47) | LDFSSFDTKVP(61) | GVPSGSWWTQLVDSVVNWILV(14) |
|  |  |  | : :: :*:*::: : | :*:* :* : | *:*** : *: :*::*: |

|  | 6 | 7 | 8 |
|---|---|---|---|
| SsRV1 | LHTGDDVYIRA (18) | RINPAKQSVGFGTGEFLRM (8) | GYLARSVASFVSGNW |
| Hv190SV | LHAGDDVYLRL (18) | RMNPTKQSIGYTGAEFLRL (8) | GYLCRAIASLVSGSW |
| SsRV2 | IHVGDDVYLGV (18) | RMNPMKQSVGHTSTEFLRL (8) | GYLARAVASTISGNW |
| UmV-H1 | LCHGDDIITVH (18) | KGQESKLMIDHKHHEYLRI (9) | GCLARCVATYVNGNW |
|  | :: ***::: | : * *:::: *:**: | *:*:* :*: ::*:* |
| Hv145SV | DQGGDDVDQEF (18) | EATKSKQMIG-RNSEFFRV (8) | ASPVRGLATFVAGNW |
| PCV | DHGGDDIDLGL (18) | KANKWKQMFGTR-SEFFRN (8) | ASPTRALASFVAGDW |
| CCRSACV | DHGGDDIDGGI (18) | EAQKIKQMIGID-SEFFRI (8) | GSATRALARFVSGNW |
|  | *:*****:* : | :* * ***:* : *****: | :*:::*:** **:*:* |
| FpV1 | FIMGDDNSAFT (26) | SKTKSIITTLRHKIETLSY (8) | RPIGKLVAQLCFPER |
| RhsV-717 | FIMGDDNSIFT (26) | SKTKSVITTLRSKIETLSY (8) | RDVEKLIAQLVYPEH |
| AhV1 | FIMGDDNVIFT (26) | NISKSAVTSIRRKIEVLGY (8) | RSISKLVGQLAYPER |
| HaV | FIQGDDNLVFY (26) | SPDKSWITRLRTKIEVLGY (8) | RDVSKLIATLAYPER |
| BCV3 | YTQGDDSLIGE (20) | NPDKTEYSTDPGYVTFLGR (8) | RSLDKCLRLLMFPEY |
| FsV1 | RVLGDDS-AFM (21) | SDEKSISVEDATELKLLGV (8) | RETEEWFKLALYPEG |
|  | :: ***: : | : *: : : ::: *:: | * ::: : :** |

**Figure 4** Comparison of the conserved motifs of RdRps of selected isometric dsRNA viruses. Numbers 1–8 refer to the eight conserved motifs characteristic of RdRps of dsRNA mycoviruses. The amino acid positions corresponding to conserved motifs 1 and 2 for the RdRps of viruses in the family *Partitiviridae* are not well defined and, therefore, they are not presented. Multiple sequence alignments were obtained using the CLUSTAL X program with RdRp amino acid sequences of the following viruses. Upper set: viruses in the family *Totiviridae*: sphaeropsis sapinea RNA virus 1 (SsRV-1), SsRV-2, helminthosporium victoriae 190S virus (Hv190SV), and ustilago maydis virus H1 (UmV-H1). Middle set: viruses in the family *Chrysoviridae*: Hv145SV PcV and CCRSACV. Lower set: viruses in the family *Partitiviridae*: fusarium poae virus 1 (FpV1), rhizoctonia solani virus-717 (RhsV-717), atkinsonella hypoxylon virus 1 (AhV1), heterobasidion annosum virus (HaV), beet cryptic virus 3 (BCV3), and fusarium solani virus 1 (FsV1). Asterisks signify identical residues (shaded) at the indicated positions; colons signify highly conserved amino acid residue within a column; numbers in parentheses correspond to the number of amino acid residues separating the motifs.

## Chryso-P4 is Virion Associated as a Minor Protein

Present evidence, based on amino acid sequencing of a tryptic peptide derived from gradient-purified PcV virions, strongly supports the conclusion that PcV chryso-P4 is virion-associated as a minor protein. The chryso-P4 encoded by chrysoviruses contains the motif PGDGXCXXHX. This motif (I), along with motifs II (with a conserved K), III, and IV (with a conserved H), form the conserved core of the ovarian tumor gene-like superfamily of predicted cysteine proteases. Multiple alignments showed that motifs I–IV are also present in other viruses including AbV-1, a tentative member of the family *Chrysoviridae*. Whether the RNAs of these viruses indeed code for the predicted proteases remains to be investigated.

## Replication of Chrysoviruses

There is very limited information on how chrysoviruses replicate their dsRNAs. The virion-associated RdRp catalyzes *in vitro* end-to-end transcription of each dsRNA to produce mRNA by a conservative mechanism. Purified virions containing both ssRNA and dsRNA have been isolated from *Penicillium* spp. infected with PcV or Pc-fV, which may represent replication intermediates.

## Virion Structure

Virions are isometric, nonenveloped, 35–40 nm in diameter (**Figure 5**). The capsid structure of PcV was recently determined at relatively moderate resolutions ($\sim$2.5 nm) using cryotransmission electron microscopy combined with three-dimensional image reconstruction. The outer surfaces of full particles of PcV, viewed along a five-, three-, and twofold axis of symmetry are shown in **Figure 6**. The capsid comprises 60 protein subunit monomers arranged on a $T=1$ icosahedral lattice. The outer diameter of the capsid is 406 Å and the average thickness of the capsid shell is 44 Å. At this low to moderate resolution, some features are distinguished on the relatively smooth

**Table 3** Comparison of the amino acid sequences of the N-terminal regions of chryso-P3 and corresponding regions of RdRp proteins

| Viral protein | Amino acid sequence[a] |
|---|---|
| PcV-chryso-P3 | (102) LYGVVMPMGHGKTTLAQEEGWIDCDSLI (129) |
| Hv145S-chryso-P3 | (82) LYTVVMPAGCGKTTIANEFNCIDVDDLA (109) |
| ACDACV-chryso-P3 | (63) LFAIVLPAGCGKSTLCRKYGYLDIDECA (90) |
| CCRSACV-chryso-P3 | (63) LFAIVLPAGCGKSTLCRKYGYLDIDECA (90) |
| PcV-RdRp | (84) LFAVIMPSGCGKTTLARTYGMVDVDELV (111) |
| Hv145S-RdRp | (60) LFAIILPAGTGKTYLAKKYGFIDVDKCV (87) |
| ACDACV-RdRp | (62) LFAIVMPGGTGKTRWAREYGLVDVDELV (89) |
| CCRSACV-RdRp | (62) LFAIVMPGGTGKTRWAREYGLVDVDELV (89) |
| | **:**:*.*.***.:**.**.****... |

[a]Asterisks signify identical or similar residues (shaded) at the indicated position; colons signify at least six identical residues within a column; single dots signify 50% identical residues at the indicated position.

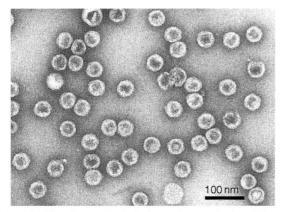

**Figure 5** Negative contrast electron micrograph of particles of penicillium chrysogenum virus, the type species of the genus *Chrysovirus* in the family *Chrysoviridae*.

The PcV capsid with its genuine $T=1$ lattice is the exception among dsRNA viruses whose capsids have '$T=2$' layers, which is the typical architecture for dsRNA viruses. PcV has the largest coat protein making a $T=1$ shell. Examination of the PcV coat protein sequence reveals a possible relationship with the canonical $T=2$ layers. Sequence analysis indeed suggests that the PcV capsid subunit falls into two similar domains that are likely to build a similar fold. Even though they are covalently linked, this unusual building unit could resemble the regular 120-subunit capsid. In this situation, PcV capsid might be considered a pseudo-($T=2$) structure. The coat protein of the chrysoviruses Hv145SV and CCRSACV share some of the repeated segments only in their amino-terminal half. The structures of the Hv145SV and CCRSACV capsids have yet to be elucidated.

## Biological Properties

There are no known natural vectors for the recognized chrysoviruses PcV, PbV, Pc-fV, and Hv145SV. They are transmitted intracellularly during cell division and sporogenesis (vertical transmission), and following hyphal anastomosis (cell fusion) between compatible fungal strains (horizontal transmission).

Unlike the *Penicillium* chrysoviruses, which are associated with latent infections of their hosts, all other known chrysoviruses occur in mixed infections with other mycoviruses (or possibly plant viruses) and are associated with disease phenotypes of their hosts. The chrysovirus Hv145SV, which together with the totivirus Hv190SV co-infect the plant pathogenic fungus *Helminthosporium* (*Cochliobolus*) *victoriae*, is associated with a debilitating disease of the fungal host. The role of Hv145SV in disease development, however, is not yet clear.

ACDACV and CCRSACV are associated with two diseases of cherry, the Amasya cherry disease (ACD) and

topography of full and empty particles. The $T=1$ full capsid is formed by 12 slightly outward protruding pentamers making an underlying cavity. The pentamers are rather complex; on the outer surface, they are formed by five connected, elongated, ellipsoid-like structures surrounded by another five smaller similar structures that are neither connected to each other, nor intercalated with inner ellipsoid-like structures. No holes are evident in the shell. The outer surface of the empty capsid is essentially identical to that of the full capsid except for the presence of five pores around the inner, elongated, ellipsoid-like structures at the fivefold axis, and three small pores around the threefold axis (not shown). There are significant differences between empty and full capsids on the inner surface around the five- and threefold positions, suggesting conformational changes. Although conformational changes have been characterized in related cores, the structural changes that are observed between empty and full PcV particles are considered unique.

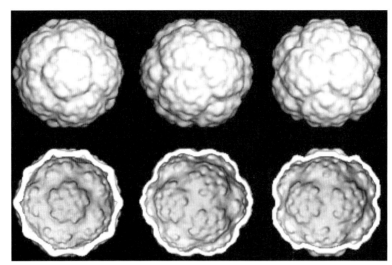

**Figure 6** Three-dimensional structures of full PcV capsids. Surface-shaded representations of the outer surfaces of full PcV capsids viewed along a five-, three-, and twofold (left to right, upper row) axis of icosahedral symmetry. Models with the front half of the protein shell removed viewed along the five-, three-, and twofold axis are shown (left to right, lower row). Courtesy of R. J. Castón.

cherry chlorotic rusty spot (CCRS) disease. Both ACD and CCRS diseases are associated with a complex pattern of virus-like dsRNAs. Symptomatologically, both diseases are indistinguishable, a conclusion that was further supported by similar PAGE profiles for their associated dsRNAs. Furthermore, the sequences of at least six of these dsRNAs are essentially identical. Four of the ACD- and CCRS-associated dsRNAs comprise the genomic dsRNAs of the chrysoviruses ACDACV and CCRSACV, respectively. The etiology of these two cherry diseases is unknown and it has yet to be determined whether the dsRNAs associated with the diseases represent the genomes of plant or fungal viruses. In addition to the chrysovirus dsRNAs, the mixture of dsRNAs associated with the disease also contain a partitivirus and and larger dsRNAs with similar sizes to totiviruses. Whether the chrysoviruses alone or in combination with other dsRNA viruses play a role in disease development has yet to be elucidated.

The tentative chrysovirus AbV-1, also designated La France isometric virus (LFIV) induces a serious disease of cultivated mushroom (named La France disease). The AbV1 virions, isolated from diseased fruit bodies and mycelia, are isometric 34–36 nm in diameter and co-purify with nine dsRNA segments (referred to as disease-associated dsRNAs). The sizes of the dsRNA segments vary from 3.6 to 0.78 kbp, at least three of which are believed to be satellites. AbV1 represents a multiparticle system in which the various particle classes appear to have similar densities. Interestingly, phylogenetic analysis of RdRp conserved motifs of AbV1, encoded by dsRNA segment L1, and other dsRNA mycoviruses showed that AbV1 is closely related to the multipartite chrysoviruses. Although there is convincing evidence that infection with AbV-1 alone is essential and sufficient for induction of the La France disease, another mushroom virus, the mushroom bacilliform virus (MBV), the type species of the family *Barnaviridae*, is commonly found in mixed infections with AbV1 in diseased mushrooms. MBV is apparently not essential for disease development since no obvious phenotypic changes were observed in an *Agaricus* culture singly infected by MBV. However, synergistic interactions between AbV1 and MBV in doubly infected mushrooms cannot be ruled out.

## Evolutionary Relationships among Chrysoviruses

The *Penicillium* chrysoviruses PcV, Pc-fV, and PbV are serologically related and have similar biochemical and biophysical properties. Although molecular data is only available for PcV, the three viruses could be considered as strains of the same virus for all practical purposes. The fact that these closely related viruses occur in different fungal species suggested that transmission by means other than hyphal anastomosis may occur in nature, since hyphal fusion between different fungal species is doubtful. Horizontal transmission of fungal viruses in nature, however, has yet to be demonstrated, and in the case of viruses of *Penicillium* species may not need to occur since the viruses replicate in parallel with their hosts and are carried intracellularly during the vegetative growth of the host (vertical transmission). Furthermore, the viruses are efficiently disseminated via the asexual spores (conidia) of *Penicillium* species. It seems feasible, however, that virus infection arose early in the phylogeny of *P. brevi-compactum*, *P. chrysogenum*, and *P. cyaneo-fulvum* before they diverged and that the resident virus remained associated with them during their subsequent evolution.

BLAST searches of PcV RdRp amino acid sequence showed that it has significantly high sequence similarity (40% identity and 57% aa sequence similarity) to the RdRps encoded by the chrysoviruses Hv145SV, ACDACV, and CCRSACV. High similarities (BLAST hits of $e^{-16}$ or lower) were also found to the RdRps of the tentative chrysovirus agaricus bisporus virus 1 (AbV-1) as well as to the totivirus ustilago maydis virus H1 (UmV-H1) and the giardiavirus trichomonas vaginalis virus (TVV). Still high similarity hits can be obtained with the RdRps of several members of the family *Totiviridae*. Interestingly, no significant hits were evident with any of the viruses in the family *Partitiviridae*, another validation for the removal of chrysoviruses from the family *Partitiviridae* and their placement in the newly created family *Chrysoviridae*. The conclusion that chrysovirus RdRps are more closely related to those of totiviruses than to those of partitiviruses is consistent with the results shown in **Figure 4**, where the RdRp conserved motifs of members of the three families are compared. Phylogenetic analysis (**Figure 7**), based on the complete nucleotide sequences of the RdRps of members of the three families, provides further confirmation to this conclusion.

BLAST searches of the deduced amino acid sequence of dsRNA2 ORF showed significant high similarity hits ($3e^{-62}$) to Hv145SV CP (29% identity and 50% amino acid sequence similarity) and ACDACV and CCRSACV CPs (24% identity and 46% similarity). It is interesting that the region of high sequence similarity between the CPs of these chrysoviruses is limited to the N-terminal half of the proteins (aa 19–560 of PcV CP; data not shown). This finding may have implications when considering the structural organization of chrysovirus capsids.

## Future Perspectives

### Functions of Chryso-P3 and Chryso-P4

Present evidence suggests that both chryso-P3 and chryso-P4 are virion associated and that they may play a role in RNA transcription, RNA binding, and packaging. Because infectivity assays are not amenable for chrysoviruses, no direct evidence is currently available in support of the conclusion that chrysovirus segments 3 and 4 and their gene products are essential for virus infection and replication. However, based on the consistent co-presence of the four dsRNA segments, the existence of extended regions of highly conserved terminal sequences at both ends of all four segments, sequence comparisons and phylogenetic analysis, it is abundantly clear that the structural features of chrysoviruses are typical of RNA viruses with multipartite and multicomponent genomes. The dsRNA pattern of PcV virions isolated from different strains of *Penicillium chrysogenum* has remained unchanged throughout the years since PcV was first isolated. This is also true for other chrysoviruses isolated from different *Penicillium* species and from various strains of *Helminthosporium victoriae*. The co-presence of all four segments in different fungal species and strains harboring chrysoviruses and the stability of the dsRNA patterns support the contention that all four segments are essential for infection and that none is defective or satellite in nature.

An alternative approach to infectivity assays that should be applicable to chrysoviruses is to transform virus-free fungal isolates with full-length cDNA clones of viral dsRNAs. In a recent preliminary study involving transformation of virus-free *H. victoriae* protoplasts with individual as well as different combinations of full-length cDNAs of the four dsRNAs of Hv145SV, productive yield of ssRNA transcripts corresponding to all four dsRNA segments was only generated in transformants containing DNA copies of all four dsRNAs. Although dsRNA synthesis was not launched in these transformants, the results suggest that dsRNA segments 3 and 4 and their gene products are necessary for accumulation and stability of the viral transcripts.

### Etiology of the CCRS Disease and Potential Occurrence of Plant Chrysoviruses

The etiology of the CCRS disease remains a mystery. Although there is circumstantial evidence for a fungal etiology for the disease, no fungal pathogen was ever isolated from diseased trees. Although it is true that no plant chrysoviruses have been identified to date, there is no reason to exclude this possibility. As a matter of fact, it is not yet determined whether the partitivirus cherry chlorotic rusty spot associated partitivirus (CCRSAPV), which is co-isolated along with CCRSACV from diseased cherry trees, is a plant or a fungal partitivirus. Phylogenetic analysis based on RdRp conserved motifs placed CCRSAPV in a cluster with a mixture of plant and fungal partitiviruses. This finding raised the interesting possibility of horizontal transfer of members of the family *Partitiviridae* between fungi and plants. This possibility is reasonable because some of the viruses in this phylogenetic cluster have fungal hosts that are pathogenic to plants. As more chrysoviruses from a wider range of fungal (and possibly plant) hosts are isolated and characterized, the reality of plant chrysoviruses may become apparent as well as the need to reconsider the taxonomy of the family *Chrysoviridae*.

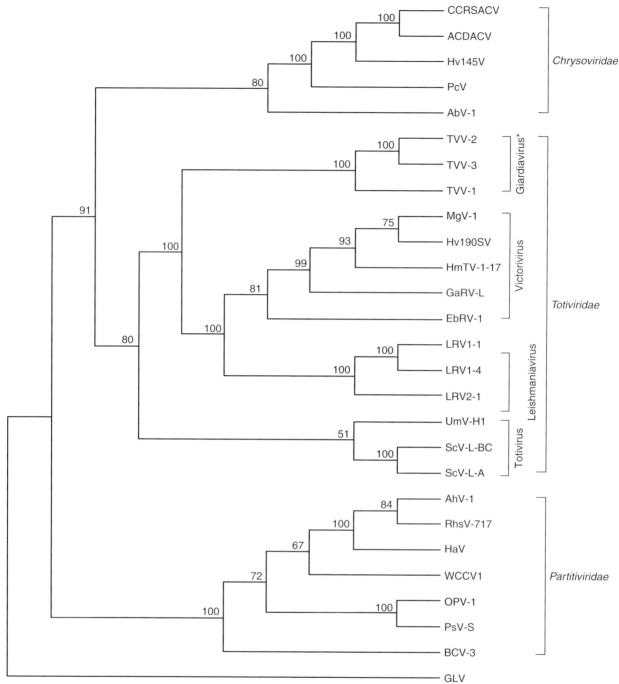

**Figure 7** Neighbor-joining phylogenetic tree constructed based on the complete amino acid sequences of RdRps of selected isometric dsRNA viruses. The RdRp sequences were derived from aligned deduced amino acid sequences of members of the families *Chrysoviridae*, *Partitiviridae*, and *Totiviridae* using the program CLUSTAL X. See **Table 1** for names and abbreviations of chrysoviruses. The following viruses in the family *Totiviridae* were included in the phylogenetic analysis (abbreviations in parenthesis): trichomonas vaginalis virus-1 (TVV-1), TVV-2, and TVV-3, magnaporthe grisea virus 1 (MgV-1), helminthosporium victoriae 190S virus (Hv190SV), Helicobasidium mompa *totivirus 1-17* (HmTV-1-17) gremmeniella abietina RNA virus L (GaRV-L), eimeria brunetti RNA virus 1 (EbRV-1), leishmania RNA virus 1-1 (LRV1-1), LRV1-4, LRV2-1, ustilago maydis virus H1 (UmV-H1), saccharomyces cerevisiae virus L-A (ScV-L-A), and ScV-L-BC. The following viruses in the family *Partitiviridae* were included: atkinsonella hypoxylon virus 1 (AhV-1), rhizoctonia solani virus-717 (RhsV-717), heterobasidion annosum virus (HaV), white clover cryptic virus 1 (WCCV-1), ophiostoma partitivirus 1 (OPV-1), penicillium stoloniferum virus S (PsV-S), and beet cryptic virus 3 (BCV-3). The phylogenetic tree was generated using the program PAUP*. Bootstrap numbers out of 1000 replicates are indicated at the nodes. The tree was rooted with the RdRp of Giardia lamblia virus, the type species of the genus *Giardiavirus* in the family *Totiviridae*, which was included as an outgroup. * Note that TVV-1, TVV-2, and TVV-3 are tentative members of the genus *Giardiavirus*.

*See also:* Fungal Viruses; Partitiviruses: General Features.

## Further Reading

Castón JR, Ghabrial SA, Jiang D, *et al.* (2003) Three-dimensional structure of Penicillium chrysogenum virus: A double-stranded RNA virus with a genuine *T* = 1 capsid. *Journal of Molecular Biology* 331: 417–431.

Covelli L, Coutts RHA, Di Serio F, *et al.* (2004) Cherry chlorotic rusty spot and Amasya cherry diseases are associated with a complex pattern of mycoviral-like double-stranded RNAs. Part I: Characterization of a new species in the genus *Chrysovirus*. *Journal of General Virology* 85: 3389–3397.

Ghabrial SA, Jiang D, and Castón RJ (2005) *Chrysoviridae*. In: Fauquet CM, Mayo MA, Maniloff J, Desselberger U, and Ball LA (eds.) *Virus Taxonomy: Eighth Report of the International Committee on Taxonomy of Viruses*, pp. 581–590. London: Academic Press.

Ghabrial SA, Soldevila AI, and Havens WM (2002) Molecular genetics of the viruses infecting the plant pathogenic fungus *Helminthosporium victoriae*. In: Tavantzis S (ed.) *Molecular Biology of Double-Stranded RNA: Concepts and Applications in Agriculture, Forestry and Medicine*, pp. 213–236. Boca Raton, FL: CRC Press.

Jiang D and Ghabrial SA (2004) Molecular characterization of Penicillium chrysogenum virus: Reconsideration of the taxonomy of the genus *Chrysovirus*. *Journal of General Virology* 85: 2111–2121.

# Circoviruses

**A Mankertz,** Robert Koch-Institut, Berlin, Germany

© 2008 Elsevier Ltd. All rights reserved.

## Glossary

**Apoptosis** Programmed self-induced cell death.
**Botryoid** Grape-like appearance.
**Bursa of Fabricius** Specialized organ in birds, that is necessary for B-cell development.
**Chimera** Virus created from two or more different genetic sources.
**Koch's postulates** Four criteria that must be fulfilled in order to establish a causal relationship between an agent and a disease.
**Mono/polycistronic mRNA** mRNA is termed polycistronic when it contains the genetic information to translate more than one protein, monocistronic if only one protein is encoded.
**Phylogenetic tree** Depicts the evolutionary interrelationships among species that are believed to have a common ancestor.
**Rolling-circle replication (RCR)** Mechanism of replication which takes its name from the characteristic appearance of the replicating DNA molecules, a special feature of RCR is the uncoupling of the synthesis of the two DNA strands.

## Circovirus Taxonomy

Circoviruses contain a covalently closed circular single-stranded DNA (ssDNA) genome with sizes between 1759 and 2319 nt. The circular nature of their genomes, which are the smallest possessed by animal viruses, has led to the family being termed *Circoviridae*. Differences in organization of the viral genomes and capsid morphology led to their classification into two different genera.

*Chicken anemia virus* is the only species member of the genus *Gyrovirus*, while the genus *Circovirus* currently comprises *Porcine circovirus type 1* and *Porcine circovirus type 2*, *Psittacine beak and feather disease virus*, *Pigeon circovirus*, *Canary circovirus*, and *Goose circovirus*. Duck circovirus (DuCV), finch circovirus (FiCV), and gull circovirus (GuCV) are members of tentative species in the genus, while the circoviruses of raven (RaCV) and starling (StCV) have been discovered only recently and are therefore not yet included in the taxonomic classification.

The reported size range for the chicken anemia virus (CAV) virion is 19.1–26.5 nm, the genome is of (−) polarity, and the open reading frames (ORFs) are overlapping. Only one mRNA molecule is produced from a promoter/enhancer region; it encodes three partially overlapping ORFs. CAV shows homology to the newly identified ssDNA viruses in humans, torque teno virus (TTV) and the related torque teno mini virus (TTMV), which are members of the unassigned genus *Anellovirus*. The noncoding regions of CAV and TTV are G/C-rich and show a low level of nucleotide homology. In addition, CAV and TTV specify structural proteins that contain two amino acid motifs with putative roles in rolling-circle replication (RCR) and nonstructural proteins that exhibit protein phosphatase activity. CAV and TTV are separately classified, since their sequence homology is limited and only one mRNA is produced from CAV, while splicing has been detected in TTV and TTMV. Phylogenetic investigation of the family *Circoviridae* revealed that CAV has no close relationship to the genus *Circovirus* (**Figure 1**).

Members of the genus *Circovirus* differ in several aspects from CAV, because they display a smaller particle size (12–20.7 nm, **Figure 2**), their ambisense genomes (**Figure 3**) are divergently transcribed, and splicing has been reported. The viruses of the genus *Circovirus* show homology in genome organization and protein sequences and function to the plant-infecting viruses of the family *Nanoviridae* and *Geminiviridae*.

## Virion Structure

The size for the CAV virion has been reported as 19.1–26.5 nm and 12–20.7 nm for circoviruses (**Figure 2**).

The virions of CAV, porcine circovirus type 1 (PCV1) and porcine circovirus type 2 (PCV2), are each comprised of one structural protein, for which sizes of 50 (CAV), 30 (PCV1), and 30 kDa (PCV2) have been estimated, respectively. Psittacine beak and feather disease virus (PBFDV) is reported to contain three proteins (26, 24, and 16 kDa). The protein composition of the other avian circoviruses is not known, but putative structural proteins have been identified by homology searches. All capsid proteins have a basic N-terminal region containing several arginine residues, which is expected to interact with the packaged DNA. Virions do not possess an envelope. Investigation of the structures of CAV, PCV1, and PCV2 revealed that all have an icosahedral $T = 1$ structure containing 60 capsid protein molecules arranged in 12 pentamer clustered units. PCV2 and BFDV show similar capsid structures with flat pentameric morphological units, whereas chicken anemia virus displayed protruding pentagonal trumpet-shaped units.

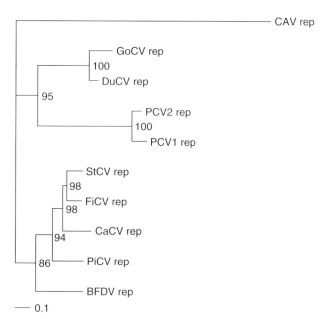

**Figure 1** Phylogenetic tree of members of the family *Circoviridae*. The amino acid sequences of the Rep proteins of the members of the genus *Circovirus* (GoCV, DuCV, BFDV, PiCV, StCV, FiCV, CaCV, PCV1, PCV2) and of the only member of the genus *Gyrovirus* (CAV) were compared. Analysis was performed using the MacVector program package by analyzing 100 data sets.

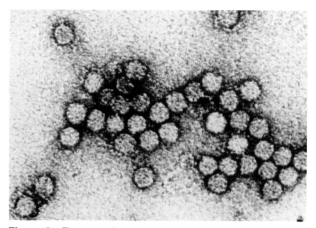

**Figure 2** Electron microscope picture of particles of PCV1. PCV1 particles in an immunoaggregation with a PCV1 hyperimmuneserum (180 000-fold magnification, negative contrasting with 1% UAc).

## Pathogenesis of Circoviruses

### Circoviruses and Gyroviruses Induce Diseases

Circoviruses are supposed to be host specific or to have narrow host ranges. The fecal–oral route of transmission is likely, but vertical transmission has been reported in some cases. With the exception of PCV1, all known circoviruses are pathogens, which cause immune suppression and damage in the lymphoreticular tissues.

CAV infections occur mainly in young chicken. The main targets of CAV replication are cells in the bone marrow (hemocytoblasts) and precursor lymphocytes in the thymus. Characteristic symptoms are aplastic anemia and hemorrhagic lesions, watery blood, pale bone marrow, lymphoid depletion, atrophy of thymus and bursa, and swollen and discolored liver. Since macrophages recovered from infected birds produce less interleukin 1 (IL-1) and the pathogenicity of co-infecting viruses such as Marek's disease virus, infectious bursal disease virus, and Newcastle disease virus are enhanced, immune suppression is thought to play a role in CAV-induced pathogenesis.

Another intensively studied circoviral disease is psittacine beak and feather disease (PBFD). PBFD is the most common disease in cockatoos and parrots and is typically detected in young birds. Deformation of beak, claws, and

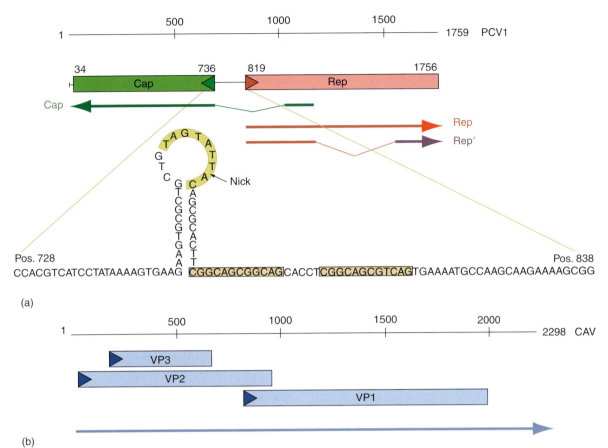

**Figure 3** Genomic organization of PCV1 member of the type species of the genus *Circovirus* and CAV member of the type species of the genus *Gyrovirus*. ORFs are shown in open boxes, the direction of transcription is indicated by triangles. Transcripts are indicated by arrows, splice processes by dotted lines. (a) PCV1, the species type of the genus *Circovirus*, is shown. The ambisense genomic organization is outlined, that is, both strands of the replicative form are coding for proteins. Therefore, the two major genes Rep and Cap are divergently transcribed. Between the start points of Rep and Cap, the origin of viral replication is located. This element is drawn to a larger scale, displaying its characteristic features, a putative hairpin, and adjacently located repeats. An arrow indicates the position where Rep and Rep' restrict the replicative intermediate to initiate the replication. (b) CAV is a member of the type species of the genus *Gyrovirus*. Its genome is of the 'negative-sense' type, that is, the viral strand does not encode genetic information, it has to be converted into a dsDNA version, from which one polycistronic mRNA is produced. Splice processes have not been observed.

feathers, lethargy, depression, weight loss, and severe anemia are the most prominent symptoms.

Young pigeon disease syndrome (YPDS) is a multifactorial disease in which PiCV is assumed to induce immunosuppression in young birds, which suffer from ill-thrift, lethargy, anorexia, and poor race performance. Depletion of splenic and bursal lymphocytes was seen and bacterial agents as *Escherichia coli* and *Klebsiella pneumoniae* were isolated more frequently from PiCV-infected birds. Inclusion bodies were present in various organs, especially the bursa of Fabricius.

In CaCV-infected neonatal canary birds, a condition known as 'black spot' has been reported. It is associated with abdominal enlargement, gall bladder congestion, failure to thrive, dullness, anorexia, lethargy, and feather disorder. Histological changes as lymphofollicular hyperplasia, lymphoid necrosis, cellular depletion, and cystic atrophy are observed in the thymus and the bursa of Fabricius. A general feature of circovirus infection is the formation of globular or botryoid, basophilic inclusion bodies in the cytoplasm, in which the virus may form paracrystalline arrays.

PCV1 and PCV2 seem to be restricted to pigs. PCV2 is the etiological agent of a new disease in swine, the so-called post-weaning multisystemic wasting syndrome (PMWS), and may be involved in several other porcine circoviral diseases (PCVDs) like porcine dermatitis and nephropathy syndrome (PDNS) or porcine respiratory disease complex. PMWS was first recognized in Canada in 1991. Since then it has been described as a major economic concern in virtually all pig-producing areas of the world. PMWS primarily occurs in pigs between 60 and

80 days old. Maternal antibodies confer titer-dependent protection against PCV2 infection – higher titers are generally protective, but low titers are not. PMWS is characterized by wasting, respiratory signs, enlargement of superficial inguinal lymph nodes, diarrhea, paleness of the skin or icterus, but the clinical signs are often variable. The most consistent feature of PMWS is a generalized depletion of lymphocytes. Secondary infections with opportunistic organisms are common. This indicates that the immune system is involved in the pathogenesis of PMWS. On affected farms, mortality may reach up to 40%, but it can be reduced, if special management plans are implemented. In the first attempts, experimental reproduction of the disease according to Koch's postulates has led to an amazing variety of results, since no symptoms were seen as well as histopathological lesions or a full-blown PMWS. The symptoms of the disease were aggravated when piglets were infected in which the immune system had been stimulated either by a prior vaccination or by a co-infection with porcine parvovirus (PPV) or porcine reproductive and respiratory syndrome virus (PRRSV). These and other findings indicate that PMWS is a multifactorial disease, in which not only factors such as the status of the immune system and genetic predisposition but also practical aspects such as nutrition and vaccination policy may influence the onset of the disease and the severity of the symptoms.

## Diagnosis of Circoviral Diseases

In general, diagnosis of circoviral diseases is based on detection of the virus by culture, polymerase chain reaction (PCR), immunohistochemistry or *in situ* hybridization, or detection of antibodies against the circovirus by serology. PBFD can also be diagnosed on the basis of feathering abnormalities. PCV2 is a ubiquitous virus and also prevalent in healthy pigs; therefore, diagnosis of PMWS must concurrently meet three criteria: (1) the presence of compatible clinical signs, (2) the presence of moderate to severe characteristic microscopic lymphoid lesions, and (3) the presence of moderate to high amount of PCV2 within these lesions.

## PCV1 and PCV2

A striking difference is seen in the pathogenicity of PCV1 and PCV2. No disease is attributed to PCV1, while PCV2 is the etiological agent of PMWS, a new emerging disease of swine. What may be the molecular basis for this distinct feature? The genomes of the two strains are highly conserved, especially the origin of replication (80% sequence homology) and the Rep gene (82%). Exchange of replication factors between PCV1 and PCV2 did not reveal differences, since the Rep protein of PCV1 (Rep/PCV1) replicated its cognate origin as well as the heterotype origin of PCV2 (and vice versa), suggesting that pathogenesis may not be linked to the replication factors. A higher degree of sequence deviation is found in the Cap genes with less than 62% homology between PCV1 and PCV2. Chimeras of PCV1 and PCV2 have been produced and tested for their potential to induce PMWS and to stimulate the immune answer. The chimera PCV2/1, containing the PCV1 capsid gene cloned into the backbone of the pathogenic PCV2 genome, was compared to a chimera PCV1/2, containing the PCV2 capsid gene in the nonpathogenic PCV1 genome. Both variants displayed similar growth characteristics *in vitro*. Gross lesions significant for PMWS were not observed, but PCV1/2 induces protective immunity to wild-type PCV2 challenge in pigs, indicating that it may be an effective vaccine candidate. ORF3 is comprised by the rep gene and may also contribute to the pathogenesis of PCV2, because its sequence differs significantly in PCV1 and PCV2 and induction of apoptosis has been reported.

## Interaction with the Immune System

PCV2 provides a valuable model for gaining insight into how ssDNA viruses interact with the host immune system and for understanding their pathogenesis. PCV2 is intriguing in its ability to persist in macrophages and dendritic cells without replication although its infectivity is retained. When natural interferon (IFN)-producing cells responded to an inducer of cytokine synthesis, their co-stimulatory function, which induces myeloid dendritic cell maturation, was clearly impaired in case of a concurrent PCV2 infection. Stimulation of the porcine immune system with IFN-$\alpha$ and IFN-$\gamma$ causes increased replication of PCV2 *in vivo*, while no changes were observed in IL-1-, IL-6-, tumor necrosis factor alpha (TNF-$\alpha$)-, or IL-10-treated cells. With the circumstantial evidence compiled over the last years, one may assume that PMWS can be considered as an acquired immunodeficiency syndrome of pigs although direct evidence for this hypothesis is still missing.

## Molecular Biology of Circoviruses

### Genome Organization of Circoviruses

The genomes of all circoviruses are composed from a circular ssDNA molecule with a size between 1759 and 2319 nt and therefore display the smallest genomes possessed by mammalian viruses. Nevertheless, members of the genera *Circovirus* and *Gyrovirus* show remarkable differences in their genome organization.

The genomes of CAV isolates are either 2298 or 2319 nt in size. Part of the noncoding region of the genome is

G–C-rich and able to form putative hairpin structures. The three genes are encoded by the viral (−)strand, therefore CAV has a negative-sense genome organization. One major polycistronic mRNA (2.0 kbp) is transcribed from the circular double-stranded (ds) replication form (RF), which is produced after infection. The nontranscribed region of the genome contains transcription initiation and termination signals and a tandemly arranged array of four or five 19 nt repeats with which promoter–enhancer activity is associated. Within this sequence, estrogen response element consensus half-sites were found, resembling the arrangement that can be recognized by the nuclear receptor superfamily. Since expression from the CAV promoter was significantly increased with estrogen treatment, members of the nuclear receptor superfamily may provide a mechanism to regulate CAV activity.

The hypothesis that the circular ssDNA genome replicates using the RCR mechanism is supported by the presence of the conserved nonanucleotide motif within the CAV genome, at which RCR is initiated in other circular ssDNA replicons, but it is not located at the apex of a putative hairpin. The presence of two amino acid motifs typical for enzymes involved in RCR within VP1 also suggests that this structural protein possesses DNA replication function, while its basic N-terminus implies that this protein is involved in capsid formation, too. This would be highly unlike the genus *Circovirus*, where two distinctly encoded proteins perform the two most elementary functions of replication and packaging of the genome. Coding regions of the avian and porcine circoviruses are arranged divergently resulting in an ambisense genome organization and creating two intergenic regions, a larger one between the 5′ ends of the two major ORFs rep and cap and a shorter one between their 3′ ends. In case of PCV1 and PCV2, the non-coding regions between the ATGs of the rep and cap gene comprise the origin of viral genome replication. Similar genomic structures are found in members of the families *Geminiviridae* and *Nanoviridae*.

## Viral ORFs and Proteins

Synthesis of three virus proteins is directed from the CAV genome. VP1 (52 kDa) is encoded by ORF1 and may combine the function of a structural protein as well as the initiator of replication. VP2 (26 kDa) is a protein phosphatase encoded by ORF2. VP2 protein phosphatase activity is required for efficient replication. It may also have a role as a scaffolding protein during virion assembly. Co-expression of both VP1 and VP2 is necessary for the induction of neutralizing antibodies. VP3 is a 14 kDa virulence factor known to induce apoptosis in transformed cell lines and has been called 'apoptin'.

Two major ORFs are encoded by the genomes of PCV1 and PCV2, encoding the viral functions for replication (Rep and Rep′) and a structural protein (Cap). A similar genomic structure is seen in the avian circoviruses. Several smaller ORFs have been found by computer analyses, but with the exception of ORF3 of PCV, which seems to be involved in pathogenesis and apoptosis, their expression has not been studied yet. The largest ORF of the ambisense organized circoviruses is located on the viral plus-strand (V1). It encodes the Rep protein (312 or 314 aa). Three motifs conserved in enzymes mediating replication in the RCR mode and a dNTP-binding domain have been identified. Both Rep proteins reside in the nucleus of infected cells. Phylogenetic analyses suggest that circovirus Rep proteins may have evolved by a recombination event between the Rep protein of nanoviruses and an RNA-binding protein encoded by picorna-like viruses or a helicase of prokaryotic origin. The second largest ORF of all circoviruses is located on the complementary strand (C1) and encodes the major structural protein Cap protein (234 aa), which displays a basic N-terminus rich in arginine residues. This suggests that this region is involved in binding to viral DNA. Some avian circoviruses use alternative start codons for Cap translation. After expression in bacteria and insect cells, Cap of PCV assembled into virus-like particles when viewed by electron microscopy. Cap has been shown to reside mostly in the nucleoli, but shuttling to the cytoplasm occurs during the infectious cycle. ORF3 is encoded counterclockwisely by ORF1. It encodes a protein that is not essential for PCV replication but has been reported to induce apoptosis.

## Transcription of PCV

The promoter of the cap gene of PCV has been mapped to a fragment at 1168–1428, that is, $P_{cap}$ is located within the rep gene. $P_{cap}$ is not regulated by virus-encoded proteins. The cap transcript starts at nucleotide 1238 with an untranslated leader sequence of 119 nt (1238 to 1120) joined to exon 2 of the ORF1 transcript at nucleotide 737, immediately adjacent to the start point of translation. Processing of this RNA has presumably evolved to avoid synthesis of another protein initiated at an internal start codon in the intron. The start of the rep transcript of PCV1 has been mapped to nucleotide $767 \pm 10$. The promoter of the rep gene, $P_{rep}$, is comprised within a fragment, nucleotides 640–796. $P_{rep}$ overlaps the intergenic region and the origin of replication. $P_{rep}$ is repressed by the Rep protein by binding to hexamers H1 and H2; these elements are involved in initation of replication, too. Mapping the rep mRNA revealed synthesis of several transcripts in PCV1 and PCV2. A full-length transcript directs synthesis of the Rep protein (312 aa, 35.6 kDa). In a spliced transcript, removal of an intron (nucleotides 1176 to 1558) results in synthesis of a truncated protein, which has been termed Rep'. Rep' is truncated to 168 aa (19.2 kDa) and, due to a frameshift, the last 49 aa are expressed in a different reading frame. Comparison of the ratio of Rep and Rep' transcript with a real-time PCR discriminating between the two transcripts indicated a variation of the ratio of the two transcripts in correlation to time. Replication of PCV is dependent on expression of both proteins. Splicing of Rep proteins is a well-known feature in other small ssDNA viruses (e.g., *Mastrevirus*), but, in contrast to PCV, one Rep protein is sufficient for replication of these viruses.

## Replication of Circoviruses

Although the main target for viral replication still remains unknown, PCV2 was seen *in vivo* in a variety of cell types including hepatocytes, enterocytes, epithelial and endothelial cells, lymphocytes, smooth muscle cells, and fibroblasts. This broad range of cells that support PCV2 infection indicates that PCV2 does not enter the cell via a rarely expressed receptor. When binding of PCV2 to monocytic cells was investigated, it became evident that surface proteins and glycosaminoglycans heparan sulfate and chondroitin sulfate B are attachment receptors for PCV2. This result is supported by the finding that the heparan sulfate binding motif (XBBXBX; B = basic amino acid, X = neutral/hydrophobic amino acid) is present on the PCV2 capsid protein. PCV2 enters the cells predominantly via clathrin-mediated endocytosis and requires an acidic environment for infection.

Replication of PCV has been studied in detail. PCVs are supposed to replicate their genomes using a circular, ds RF intermediate, which is produced by host cell DNA polymerases during the S phase of cell division. The origin of replication of PCV has been mapped to a fragment comprising the intergenic region between the start points of the two major ORFs (**Figure 3**).

Replication of these fragments cloned into a vector was observed after co-transfection of porcine kidney cells with plasmids expressing the rep gene. By sequence alignment, analogous elements can be identified for all other circoviruses with the exception of CAV. The origin of replication is characterized by a potential stem–loop structure with a nonamer (5'-TAGTATTAC; **Figure 3**) in its apex. Mutagenesis of the nonamer resulted in inactivation of PCV replication. Adjacent to the nonamer, short repeats are located, which serve as the binding site for the rep gene products. Nonamer, stem–loop, and adjacent short repeats are conserved in all other circoviruses, in the families *Nanoviridae* and *Geminiviridae*, and, although to a lesser extent, in many replicons replicating via RCR. The conserved elements in the *cis*-acting origin of replication as well as RCR signatures in the *trans*-acting Rep protein amino acid (aa) sequence (see below) indicate an RCR-like replication mechanism for the circoviruses.

## Functional Analyses of PCV-Encoded Proteins

Truncation of the rep gene as well as site-directed mutagenesis of the four conserved motifs abrogated replication of PCV, indicating that the rep gene products are indispensable. The roles of Rep and Rep' of PCV have been analyzed in detail.

*Binding to DNA.* Rep and Rep' bind *in vitro* to fragments of the origin of replication containing the stem–loop structure plus the conserved nonamer and the four hexamer (H) repeats (5'-CGGCAG; H1 to H4). Proteins bind either the two inner (H1/H2) or the two outer (H3/H4) hexamers. A minimal binding site (MBS) has been identified for Rep and Rep' protein using truncated substrates: the Rep MBS was mapped to the right leg of the stem–loop plus the two inner hexamer repeats H1/H2, while the MBS of Rep' was composed of only the two hexamer repeats H1/H2. Gel shift assays also revealed presence of several complexes, indicating that variable amount of proteins may be bound.

*Replication.* The covalently closed, ssDNA genome of PCV replicates via a dsDNA replicative intermediate. The replication occurs by RCR whereby a single-stranded break is introduced by Rep or Rep' leading to a free 3'-hydroxyl group serving as a primer for subsequent DNA synthesis. Replication does occur when Rep plus Rep' protein are expressed in the cells, indicating that both proteins are essential for replication from the PCV origin. This is in contrast to other ssDNA viruses, for example, AAV2 or members of the genus *Mastrevirus*, in which spliced Rep proteins are produced, but are not essential for viral replication. Rep and Rep' cleave the viral strand between nucleotides 7 and 8 *in vitro* within the conserved nonanucleotide located at the apex of a putative stem–loop structure. In addition, Rep and Rep' join viral ssDNA fragments, implying that these proteins also play a role in the termination of virus DNA replication. This joining activity is strictly dependent on preceding substrate cleavage and the close proximity of origin fragments accomplished by base pairing of the stem–loop structure. This dual 'nicking/joining' activity associated with Rep and Rep' are pivotal events underlying the RCR-based replication of porcine circoviruses in mammalian cells. Although presence of the palindrome plus a single H sequence is sufficient for PCV replication, a tandem repeat arrangement is more stable. Within the H sequence, selected nucleotides at specific positions are critical for Rep-associated protein recognition and for viral DNA replication.

*Repression of the Rep gene promoter $P_{rep}$.* When the influence of virus-encoded proteins upon $P_{rep}$ was investigated, it became apparent that $P_{rep}$ is repressed by its own gene product Rep but not by Rep′ or Cap. This finding illustrates that Rep protein initiates replication and controls its own transcription by binding to hexamers H1/H2. Interestingly, Rep′ also binds to H1/H2, but this does not result in repression of $P_{rep}$ activity. Since mutagenesis of H1/H2 decreases but does not inactivate $P_{rep}$ transcription, features other than binding of Rep may be necessary for repression of $P_{rep}$, for example, interaction of Rep protein with transcription factors. $P_{cap}$ is not regulated by Cap, Rep, and Rep′.

*Interaction.* Studies investigating the interaction of PCV-encoded proteins are bemusing, because the results were depending on the system used for expression. While two hybrid analysis in bacteria revealed interaction of Rep and Cap, this was not observed in yeast cells, suggesting that post-translational modifications of Rep and Cap may significantly modulate their function. Rep and Rep′ have been observed to interact in yeast cells and this observation was reproduced by immunoprecipitation in mammalian cells, the natural target of PCV.

*Localization.* Rep and Rep′ protein co-localize in the nucleoplasm of infected cells, but no signal was seen in the nucleoli. The localization did not change during the infection cycle. Rep and Rep′ carry three potential nuclear localization signals (NLSs) in their identical N-termini. Proteins mutated in their NLSs demonstrated that NLS1 and NLS2 mediate the nuclear import, whereas NLS3 enhances the nuclear accumulation of the replication proteins. In contrast to Rep and Rep′, the localization of the Cap protein was restricted to the nucleoli in plasmid-transfected cells. In PCV-infected cells, Cap was localized in the nucleoli in an early stage, while it was seen later on in the nucleoplasm and the cytoplasm. This signifies that Cap is shuttling between distinct cellular compartments during the infection cycle. Since Rep, Rep′, and Cap are all located in the nucleus, this points out that DNA replication and encapsidation of the circular closed ssDNA probably occur in the nucleus and not in cytoplasmic compartments. The biological function of the early localization of Cap to the nucleoli remains unclear. Nucleolar localization has been described for proteins of many DNA and RNA viruses and it has been proposed that virus proteins enter the nucleoli to support viral transcription or influence the cell cycle.

## Conclusion

Although the family *Circoviridae* comprises only a relatively small number of viruses, the increasing number of publications demonstrates rising interest. This may not only be related to the fact that circoviruses induce severe multifactorial diseases, which compromise and unbalance the immune system, but also to the fact that the apparent simplicity of the circovirus genome contrasts highly with the complex and poorly understood pathogenesis. Hopefully, this will induce many question-solving studies, enabling us to improve our understanding of these intriguing viruses in the future.

*See also:* Anellovirus; Nanoviruses; Plant Resistance to Viruses: Geminiviruses.

## Further Reading

Cheung AK (2004) Palindrome regeneration by template strand-switching mechanism at the origin of DNA replication of porcine circovirus via the rolling-circle melting-pot replication model. *Journal of Virology* 78: 9016–9029.

Clark EG (1997) Post-weaning wasting syndrome. *Proceedings of the American Association of Swine Practitioners* 28: 499–501.

Crowther RA, Berriman JA, Curran WL, Allan GM, and Todd D (2003) Comparison of the structures of three circoviruses: Chicken anemia virus, porcine circovirus type 2, and beak and feather disease virus. *Journal of Virology* 77: 13036–13041.

Darwich L, Segales J, and Mateu E (2004) Pathogenesis of postweaning multisystemic wasting syndrome caused by porcine circovirus 2: An immune riddle. *Archives of Virology* 149: 857–874.

Fenaux M, Opriessnig T, Halbur PG, Elvinger F, and Meng XJ (2004) A chimeric porcine circovirus (PCV) with the immunogenic capsid gene of the pathogenic PCV type 2 (PCV2) cloned into the genomic backbone of the nonpathogenic PCV1 induces protective immunity against PCV2 infection in pigs. *Journal of Virology* 78: 6297–6303.

Finsterbusch T, Steinfeldt T, Caliskan R, and Mankertz A (2005) Analysis of the subcellular localization of the proteins Rep, Rep′ and Cap of porcine circovirus type 1. *Virology* 343: 36–46.

Gibbs MJ and Weiller GF (1999) Evidence that a plant virus switched hosts to infect a vertebrate and then recombined with a vertebrate-infecting virus. *Proceedings of the National Academy of Sciences, USA* 96: 8022–8027.

Krakowka S, Ellis JA, McNeilly F, Ringler S, Rings DM, and Allan G (2001) Activation of the immune system is the pivotal event in the production of wasting disease in pigs infected with porcine circovirus-2 (PCV-2). *Veterinary Pathology* 38: 31–42.

Mankertz A, Mueller B, Steinfeldt T, Schmitt C, and Finsterbusch T (2003) New reporter gene-based replication assay reveals exchangeability of replication factors of porcine circovirus types 1 and 2. *Journal of Virology* 77: 9885–9893.

Miller MM, Jarosinski KW, and Schat KA (2005) Positive and negative regulation of chicken anemia virus transcription. *Journal of Virology* 79: 2859–2868.

Misinzo G, Delputte PL, Meerts P, Lefebvre DJ, and Nauwynck HJ (2006) Porcine circovirus 2 uses heparan sulfate and chondroitin sulfate B glycosaminoglycans as receptors for its attachment to host cells. *Journal of Virology* 80: 3487–3494.

Segales J, Allan GM, and Domingo M (2005) Porcine circovirus diseases. *Animal Health Research Reviews* 6: 119–142.

Steinfeldt T, Finsterbusch T, and Mankertz A (2006) Demonstration of nicking/joining activity at the origin of DNA replication associated with the Rep and Rep′ proteins of porcine circovirus type 1. *Journal of Virology* 80: 6225–6234.

Tischer I, Gelderblom H, Vettermann W, and Koch MA (1982) A very small porcine virus with circular single-stranded DNA. *Nature* 295: 64–66.

Todd D, Bendinelli M, Biagini P, et al. (2005) Circoviridae. In: Fauquet C, Mayo MA, Maniloff J, Desselberger U, and Ball LA (eds.) *Virus Taxonomy: Eighth Report of the International Committee on Taxonomy of Viruses*, pp. 327–334. San Diego, CA: Elsevier Academic Press.

# Citrus Tristeza Virus

**M Bar-Joseph,** The Volcani Center, Bet Dagan, Israel
**W O Dawson,** University of Florida, Lake Alfred, FL, USA

© 2008 Elsevier Ltd. All rights reserved.

## Glossary

**Cross-protection** Prevention of the symptomatic phase of a disease by prior inoculation with a mild or nonsymptomatic isolate of the same virus.

**Defective RNA** Subviral RNA molecules lacking parts of the genome while maintaining the signals enabling their synthesis by the viral replication system.

**Genome** The complete genetic information encoded in the RNA of the virus, including both translated and nontranslated sequences.

**Subgenomic RNA** Shorter than full-length genomic RNA molecules produced during the replication process sometimes to allow translation of open reading frames (ORFs).

**Transgenic pathogen-derived resistance** Plants genetically transformed to harbor viral or other pathogen-derived sequences expected to confer resistance against related virus isolates.

## History

Over the last 70 years citrus tristeza virus (CTV) has killed, or rendered unproductive, millions of trees throughout most of the world's citrus-growing areas and hence it is rightfully considered as the most important virus of citrus, the world's largest fruit crop, hence the name 'tristeza' which means 'sadness' in Spanish and Portuguese. However, as with many other disease agents, the actual damages of CTV infections and their timings varied considerably at different periods and geographical regions. The origins of CTV infections remain unknown; the virus however existed for centuries in Asia as an unidentified disease agent, but growers in these areas adapted citrus varieties and rootstock combinations with resistance or tolerance to CTV infections. Out of Asia, citrus production moved to the Mediterranean region, primarily through the introduction of fruit and propagation of seed which does not transmit CTV to the resulting plantings; hence, the new cultivation areas starting from seed sources remained free of the virus for centuries. However, the improvement in maritime transportation, allowing long-distance transport of rooted citrus plants, led to the outbreak of the deadly citrus root rot disease caused by *Phytophthora* sp., which was managed by adapting the more tolerant sour orange rootstock. The considerable horticultural advantages offered by this rootstock coincided with the large-scale expansion of plantings throughout the Mediterranean and American countries, and as a result citrus production in many areas was almost entirely based on a single rootstock. Consequently, this decision had grave effects when CTV pandemics swept throughout the world, causing 'quick decline' (death) of trees on this rootstock. Although the CTV problems were first noticed in South Africa and Australia where by the end of the nineteenth century the growers had found that sour orange was an unsuitable rootstock, it was only in the 1930s that the extent of this deadly disease problem manifested itself first in Argentina and shortly later in Brazil and California, with the death of millions of trees.

The virus-like nature of the tristeza-related diseases was demonstrated experimentally by graft and aphid transmission of the disease agent in 1946; however, description of CTV properties began with the seminal findings of Kitajima *et al.* of thread-like particles (TLPs) associated with tristeza-infected trees. These unusually long and thin particles presented a challenging problem for isolation (purification), which was an essential step toward virus characterization. Development of effective purification methods and biophysical characterization of TLP and similar viruses led to their assignment to the closteroviruses, a group of elongated viruses. The association of infectivity with the TLP-enriched preparations was demonstrated first by Garnsey and co-workers; however, the unequivocal completion of Koch's postulates for TLPs was completed only in 2001 with the mechanical infection of citrus plants with TLPs obtained from RNA transcripts of an infectious CTV cDNA clone amplified through serial passaging in *Nicotiana benthamiana* protoplasts.

Developments of TLP purification methods paved the way to antibody preparations and to improved CTV diagnosis by enzyme-linked immunosorbent assay (ELISA), which revealed considerable serological identity among CTV isolates. Later, RNA extracts from CTV-enriched particle preparations were used for molecular cloning of DNA molecules, which when used as probes demonstrated considerable genomic variation among CTV isolates. Nucleic acid probes also demonstrated that plants infected with CTV contained many defective RNAs (dRNAs) in addition to the normal genomic and subgenomic RNAs. Infected plants also contain unusually large amounts of dsRNAs corresponding to the genomic and subgenomic RNAs and the dRNAs. Because of the ease of purifying these abundant dsRNAs, they often have been the template of choice for producing cDNAs. The advent of cDNA cloning of CTV led to the sequencing of its genome and to description of dRNAs.

## Taxonomy and Classification

CTV belongs to the genus *Closterovirus*, family *Closteroviridae*. The family *Closteroviridae* contains more than 30 plant viruses with flexuous, filamentous virions, with either mono- or bipartite (one tripartite) single-stranded positive-sense RNA genomes. A recent revision of the taxonomy of the *Closteroviridae* based on vector transmission and phylogenetic relationships using three proteins highly conserved among members of this family (a helicase, an RNA-dependent RNA polymerase, and a homolog of the HSP70 proteins) led to the demarcation of three genera: the genus *Closterovirus* including CTV and other aphid-borne viruses with monopartite genomes; the genus *Ampelovirus* comprising viruses with monopartite genomes transmitted by mealybugs; and the genus *Crinivirus* that includes whitefly-borne viruses with bipartite or tripartite genomes. Viruses of this family are all phloem-limited viruses. Among the hallmarks of the group is the presence of a conserved five-gene module that includes the four proteins involved in assembly of virions, the major (CP) and minor (CPm) coat proteins, p61, and p64, a homolog of the HSP70 chaperon. In addition, infected phloem-associated cells have clusters of vesicles considered specific to the cytopathology of closteroviruses. Although the evolution of the three genera of the family *Closteroviridae* is unknown, its is interesting to note that dRNAs of CTV analogous to each of the two RNA segments of the crinivirus genome were found in infected plants raising the possibility that criniviruses arose from a closterovirus ancestor.

## Geographic Distribution

Citrus is a common fruit crop in areas with sufficient rainfall or irrigation from the equator to about 41° of latitude north and south. CTV is now endemic in most of the citrus-growing areas, with only a few places in the Mediterranean basin and Western USA still remaining free of CTV infections. The spread during the last decade of the brown citrus aphid (*Toxoptera citricida*) to Florida and its recent spread to parts of Portugal and Spain are now threatening most of the remaining CTV-free areas of North America and in the Mediterranean basin. One important aspect of CTV's geographic distribution is that some areas that have endemic isolates of CTV still do not have the most damaging stem-pitting isolates. With the spread of the vector comes the threat of extremely severe CTV isolates which so far have not spread to North America and Mediterranean countries, where only the milder CTV isolates were established in the past.

## Host Range and Cytopathology

CTV infects all species, cultivars, and hybrids of *Citrus* sp., some citrus relatives such as *Aeglopsis chevalieri*, *Afraegle paniculata*, *Fortunella* sp., *Pamburus missionis*, and some intergeneric hybrids. Species of *Passiflora* are the only nonrutaceous hosts, infected both naturally and experimentally. The CTV decline strains are associated with the death of the phloem near the bud union, resulting in a girdling effect that may cause the overgrowth of the scion at the bud union, loss of feeder roots, and thus drought sensitivity, stunting, yellowing of leaves, reduced fruit size, poor growth, dieback, wilting, and death. However, other virulent and damaging CTV strains cause stem-pitting (SP), which results in deep pits in the wood under depressed areas of bark and are often associated with severe stunting and considerably reduced fruit production. The seedling-yellow reaction (SY) which includes severe stunting and yellowing on seedlings of sour orange, lemon, and grapefruit is primarily a disease of experimentally inoculated plants but might also be encountered in the field in top-grafted plants.

CTV infection is closely restricted to the phloem tissues which often show strongly stained cells, termed as chromatic. Electron microscopy of infected cells shows that they are mostly filled with fibrous inclusions, consisting of aggregates of virus particles and of membranous vesicles differing in tonicity and containing a fine network of fibrils.

## The CTV Virions

The CTV virions are long flexuous particles, 2000 nm long and 10–12 nm in width. The virions have a helical symmetry with a pitch of 3–4 nm, about 8–9 capsids per helix turn, and a central hole of 3–4 nm. Unlike the virions of other elongated plant viruses that possess cylindrical nucleocapsids made of a single coat protein (CP), the CTV virions consist of bipolar helices with a long body and a short tail. Immunoelectron microscopy showed antibodies to CPm attached to only one end of the virions, with the major part (>97%) of the virion encapsidated by CP. Interestingly, the 'tail' corresponds to 5′ end region of the viral genome and the particle tails of other closteroviruses have been associated with small amounts of p61 and p65.

## The CTV Genome

Figure 1 shows a schematic presentation of the 19.3 kbp RNA of CTV. Generally, the CTV genome is divided almost equally into two parts, the 5′ part consisting of ORF1a and -1b harboring the viral replication machinery and the 3′ half harboring ten ORFs encoding a range of structural proteins and other gene products involved in virion assembly and host and vector interactions. CTV replicons containing only ORF1a and -1b plus the 5′ and 3′ nontranslated ends (Figure 1) fulfill all the requirements for efficient replication in protoplasts.

Interestingly, while the sequences of the 3′ half of all CTV isolates that have been sequenced are 97% and 89% identical when comparing the 3′ nontranslated regions (NTRs) and the rest of the 3′ halves, respectively, the 5′ half sequences often differ considerably. For instance, the isolates T36 and VT show only 60% and 60–70% identities for their 5′ NTR and the remaining 5′ halves, respectively. The considerable deviation of strain T36 from that of the VT group suggests that T36 might have resulted from a recent recombination event involving the 3′ part of a VT-like isolate and the 5′ half derived from a different closterovirus. The common finding of numerous recombination events both within and between CTV isolates and multiple dRNAs supports such a possibility.

## The Nontranslated Regions

The remarkable feature of CTV isolates are the close identities (97%) of their 3′ NTR primary sequences and the considerable divergence of their 5′ NTR sequences. Computer-assisted calculations, however, suggest that not only the 3′ NTRs of different isolates fold into similar predicted structures, but surprisingly the dissimilar sequences of the 5′ NTRs also are predicted to fold into similar secondary structures. The 3′ replication signal of several CTV isolates was mapped to 230 nt within the NTR and predicted to fold into a secondary structure composed of ten stem-and-loop (SL) structures. Three of these SL regions and a terminal 3′ triplet, CCA, were essential for replication. Replacement of T36 3′ NTR with 3′ NTRs from other strains allowed replication, albeit with slightly less efficiency, confirming the significance of the primary sequence of this part. The plus-strand sequence of the 5′ NTRs from different strains are predicted to form similar secondary structures consisting of two stem loops (SL1 and SL2) separated by a short spacer region, that were essential for replication. These structures were shown to be *cis*-acting elements involved in both replication and initiation of assembly by CPm.

## Genome Organization and Functions

In infected cells, the 12 ORFs of CTV (Figure 1) are expressed through a variety of mechanisms, including proteolytic processing of the polyprotein, translational frameshifting, and production of ten 3′-coterminal subgenomic RNAs. The first two mechanisms are used to express proteins encoded by the 5′ half of the genome while the third mechanism is used to express ORFs 2–11. The ORF1a encodes a 349 kDa polyprotein containing

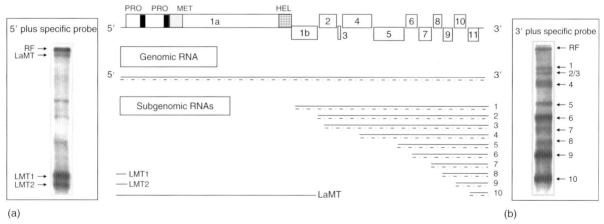

**Figure 1** A schematic presentation of the citrus tristeza virus (CTV) genomic (g) and subgenomic (sg) RNAs. The putative domains of papain-like proteases (PRO), methyltransferase (MT), helicase (HEL), and the various open reading frames (ORFs) with their respective numbers are indicated. Lines shown on the left and right side below the genomic map indicate 5′ large molecular single stranded (ss) subgenomic (sg) RNA (LaMT) and two low molecular weight ssRNAs (LMT1 and LMT2) and a nested set of 3′ co-terminal sgRNAs, respectively. Inset (a), shows Northern blot hybridization of dsRNA-enriched extracts from a citrus plant infected with the VT strain of CTV using riboprobes specific to the 5′ end of the viral RNA. Note the presence of the large replicative form (RF) molecules (upper band), LaMT, and two abundant LMT molecules. Inset (b), Hybridization of dsRNA with the 3′ probe. Note the hybridization bands of the different 3′ sgRNAs indicated by arrows.

two papain-like protease domains plus methyltransferase-like and helicase-like domains. Translation of the polyprotein could also continue through the polymerase-like domain (ORF1b) by a +1 frameshift. These proteins along with the signals at the 5′ and 3′ ends of the genome are the minimal requirements for replication of the RNA. The function of p33 (ORF2) is unknown, but is required for infection of a subset of the viral host range. The next five gene products from the 3′ genes include the unique signature block characteristic of closteroviruses, which consist of the small, 6 kDa hydrophobic protein (ORF3), 65 kDa cellular heat-shock protein homolog (HSP70h, ORF4), 61 kDa protein (ORF5), and the tandem pair of p27 (CPm, ORF6) followed by p25 (CP, ORF7). The four latter proteins are required for efficient virion assembly. The small hydrophobic p6 is a single-span transmembrane protein not required for virus replication or assembly but for systemic invasion of host plants. CP is also a suppressor of RNA silencing. The function of p18 (ORF8) and p13 (ORF9) is unknown. Protein p20 (ORF10) accumulates in amorphous inclusion bodies of CTV-infected cells and has been shown to be a suppressor of RNA silencing. p23 (ORF11) has no homolog in other closteroviruses but is a multifunctional protein that (1) binds cooperatively both single-stranded and dsRNA molecules in a non-sequence-specific manner; (2) contains a zinc-finger domain that regulates the synthesis of the plus- and minus-strand molecules and controls the accumulation of plus-strand RNA during replication; (3) is an inducer of CTV-like symptoms in transgenic *C. aurantifolia* plants; (4) is a potent suppressor of intracellular RNA silencing in *Nicotiana tabacum* and *N. benthamiana*; and (5) controls the level of genomic and subgenomic negative-stranded RNAs.

## The CTV Subgenomic (sg) RNAs

The replication of CTV involves the production of a large number of less than full-length RNAs. These include ten 3′ coterminal sg mRNAs, and ten negative-stranded sgRNAs corresponding to the ten 3′ sg mRNAs, plus ten 5′-coterminal sgRNAs that apparently are produced by termination just 5′ of each of the ten ORFs (**Figure 1**). The amounts of different sg mRNAs vary with the highest levels for sg mRNAs p23 and p20, located at the distal 3′ end. Infected cells also contain abundant amounts of two other small 5′ coterminal positive-stranded sgRNAs of ∼600 and 800 nt designated as low molecular weight tristeza (LMT).

## Defective RNAs

Most CTV contain dRNAs, which consist of the two genomic termini, with extensive internal deletions. CTV dRNAs accumulate abundantly even when their genomes contain less than 10% of the viral genome (**Figure 2**).

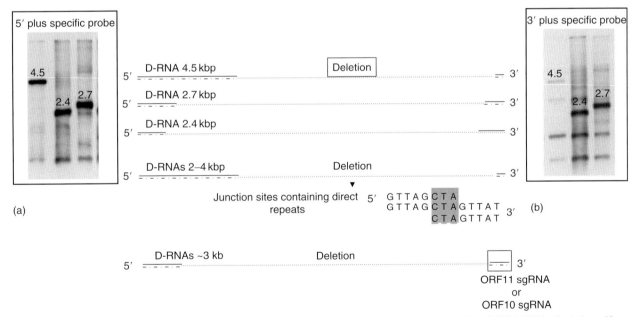

**Figure 2** A diagram of class I CTV defective (d) RNAs (=D-RNA) with three different sizes and class II CTV-dRNAs that show 3′ moieties of the size and structure of the full-length sgRNA of ORF11. Insets (a) and (b) show Northern hybridization of class I dsRNAs from three different CTV-VT subisolates, hybridized with riboprobes specific to the 5′ and 3′ ends, respectively, of the genomic RNA of CTV. Note the intense bands resulting from the hybridization of the two probes with the abundantly present dRNAs. Lower part is a diagrammatic presentation of direct repeats at the junction sites of some class I dRNAs. The bottom part shows a schematic presentation of a class II dRNA, with a 3′ terminus corresponding to the ORF11 sgRNA of CTV-VT.

For convenience, the dRNAs were divided into six groups, class I molecules contained different sizes of 5′ and 3′ sequences. Some dRNAs of this class contained direct repeats of 4–5 nt flanking or near their junction sites, supporting the possibility that they were generated through a replicase-driven template-switching mechanism. Class II molecules showed 3′ moieties of the size and structure of the full-length sgRNA of ORF11 (**Figure 2**, lower part). An extra C at the junction sites of several dRNAs of this class corresponded for the extra G reported for 3′ of minus strands from sgRNAs and RF molecules. Class III molecules are large (c. 12 kbp) encapsidated dRNAs which are infectious when used to inoculate *N. benthamiana* protoplasts. Class IV consists of dRNAs that retained all or most of the ten 3′ ORFs, analogous to crinivirus RNA 2. Class V includes double recombinants with identical 5′ regions of 948 nt followed with an internal ORF2 sequence and 3′ parts of different sizes, and class VI molecules are closely similar to class I with variable regions from the 5′ and 3′ with junction sites comprised of non-CTV inserts of 14–17 nt.

The biological roles of CTV dRNAs remain obscure as no specific associations have been established for any of the dRNA. Examination of the dRNAs from Alemow plants infected with SY and non-SY inducing isolates revealed mostly a major single dRNA of 4.5 or 5.1 kbp in non-SY plants and two different dRNAs of 2.4 or 2.7 kbp in SY plants. These results suggested the possibility that large dRNAs might play a role in suppression of SY symptoms.

## Transmission

Although the virus is phloem limited, mechanical inoculation can be done experimentally at relatively low efficiency by slashing the trunks of small citrus trees with blades containing sap extracts. In commercial groves, the virus is spread naturally by aphid vectors and by vegetative propagation in infected budwood. Long-distance spread of the virus, particularly from country to country, had been through the long-distance transfer of plant propagation material.

Several aphid species including *Aphis gossypii*, *A. spiraecola*, *A. craccivora*, and *T. citricida* transmit CTV semipersistently. The rate of transmissibility varies considerably between different virus isolates and different aphid species. The brown citrus aphid (*T. citricida*) is the most efficient vector, followed by *A. gossypii*. Despite the less efficient transmission by *A. gossypii* and the absence of brown citrus aphid in the Mediterranean region and California, mild strains are spread in these areas. Recently, the brown citrus aphid invaded Central America and Florida, and was also reported to be in the island of Madeira and Portugal.

## CTV Control Measures

Strategies to control CTV have varied at different locations and periods and included (1) quarantine and budwood certification programs to prevent the introduction of CTV; (2) costly and ambitious eradication programs to contain situations of virus spread; (3) the use of CTV-tolerant rootstocks and mild (or protective) strain cross-protection, often also named preimmunization; (4) breeding for resistance; and (5) attempts to obtain resistance by genetic engineering. Mild-strain cross-protection has been widely applied for millions of citrus trees in Australia, Brazil, and South Africa to protect against stem pitting of sweet orange and grapefruit trees. However, mild-strain cross-protection has not provided field protection against CTV isolates causing quick decline of trees on the sensitive sour orange rootstock.

Useful resistance to CTV has been found in a citrus relative, *Poncirus trifoliate*. This resistance has been mapped and shown to be controlled primarily by a single dominant gene, *Ctv*. Although the region containing this gene has been identified and sequenced, a specific gene has not been identified by transformation into susceptible citrus with the resistance phenotype.

## CTV Diagnosis

The outcome of CTV infections varies considerably depending on the virulence of the prevailing virus isolates and the sensitivity of infected citrus varieties and root stock combinations. Hence, the need of effective means of CTV diagnosis are of utmost importance. Biological indexing of the disease by grafting sensitive citrus indicators has been the definitive assay, although it is costly and time consuming. It has largely been replaced with more rapid immunoassays. ELISA has been widely practiced for almost 30 years for a variety of CTV sanitation programs. Development of recombinant antigens considerably advanced diagnostic possibilities and production of monoclonal antibodies has allowed for more precise differentiation of isolates. PCR and combinations of immunocapture PCR allow more sensitive CTV diagnosis. However, despite the progress in development of better diagnostic tools, none is effective for predicting the biological properties of new CTV isolates.

## Economic Costs of CTV

Control strategies varied at different geographic regions and periods, depending on the extent to which CTV was spreading within the newly infested areas, the sensitivity of the specific local varieties, and the economical availability of alternative crops to replace diseased groves. Most citrus-growing countries that managed to remain free of CTV did so by enforcing sanitation practices to prevent virus spread. These included certification schemes aimed to ensure that citrus nurseries propagate and distribute to growers only CTV-free planting material. Other areas had far more costly and ambitious programs of eradication. These efforts were mostly only temporarily successful, mainly because of lack of long-term grower and governmental commitment to such costly operation. In areas where CTV was endemic, alternative strategies to enable continued commercial citrus production in the presence of CTV were developed. Indeed, long before the viral nature of the tristeza disease was realized, Japanese citrus growers were grafting CTV-tolerant Satsuma mandarins on the cold-tolerant and CTV-resistant trifoliate orange *(P. trifoliate)* rootstocks, thus enabling them to produce quality fruits despite endemic CTV infections. Similarly, the change of rootstocks from the 'quick declining' sour orange rootstocks to rough lemon rootstocks, and to a less extent to mandarin rootstocks, allowed the South African growers to produce good crops of oranges despite the presence of the most efficient aphid vector and infections with severe CTV strains. Similarly, half a century later the Brazilian citrus industry that was completely decimated during 1940s, when all trees were grafted on sour orange rootstocks was saved by adopting CTV-tolerant rootstocks and mild-strain cross-protection.

Historically, the costs of CTV epidemics were estimated to be of the order of tens and even hundreds of millions of dollars. These estimates, however, varied depending on actual market value of the lost production capacity, alternative uses of the land, and the time needed for the CTV-tolerant replants to enter production. There were also indirect costs resulting in poorer performance and/or sensitivity of some CTV-tolerant rootstocks used to replace the widely adapted sour orange rootstocks to other citrus diseases such as citrus blight and citrus sudden death.

## Attempts to Apply Transgenic Resistance to CTV

Conventional breeding for CTV resistance is a long, difficult, and inconvenient process mainly because most citrus varieties are hybrids of unknown parentage. Hence, the considerable interest in applying pathogen-derived resistance (PDR) to render both citrus rootstocks and varieties tolerant to stem-pitting isolates and decline causing CTV isolates. Yet, literally hundreds of independent transformations with a range of different types of configuration of CTV sequences have resulted in failure to obtain durable protection against CTV. These failures are especially frustrating since the control of severe stem- pitting isolates is needed in several major citrus production areas. The lack of RNA silencing against CTV in citrus might be due to more effective suppression by the combination of the three suppressors of this virus.

*See also:* Bromoviruses; Plant Virus Diseases: Fruit Trees and Grapevine.

## Further Reading

Bar-Joseph M, Garnsey SM, and Gonsalves D (1979) The *Closteroviruses*: A distinct group of elongated plant viruses. *Advances in Virus Research* 25: 93–168.

Bar-Joseph M, Marcus R, and Lee RF (1989) The continuous challenge of citrus tristeza virus control. *Annual Review of Phytopathology* 27: 291–316.

Dawson WO (in press) Molecular genetics of *Citrus tristeza virus*. In: Karasev AV and Hilf ME (eds.) *Citrus Tristeza Virus and Tristeza Diseases*. St. Paul, MN: APS Press.

Karasev AV (2000) Genetic diversity and evolution of closteroviruses. *Annual Review of Phytopathology* 38: 293–324.

Martelli GP, Agranovsky AA, Bar-Joseph M, *et al.* (2002) The family *Closteroviridae* revised. *Archives of Virology* 147: 2039–2044.

# Classical Swine Fever Virus

**V Moennig and I Greiser-Wilke,** School of Veterinary Medicine, Hanover, Germany

© 2008 Elsevier Ltd. All rights reserved.

## Glossary

**CPE** Viruses can have a cytopathic effect on infected cells, that is, they either kill cells or change their properties.

**DI** Defective interfering particles lack part(s) of their genome and they depend on complete helper virus for replication; they interfere with the replication of the helper virus by competing for essential enzymes.

**ELISA** Enzyme-linked immunosorbent assays are widely used for virological and serological diagnosis. Samples can be processed automatically and results are read using optical methods.

**IFT** Immunofluorescence tests utilize specific antibodies conjugated with fluorescent dye. Viral antigens can be visualized after binding of conjugated antibodies.

**MLV** Modified live vaccines consist of attenuated viruses that have lost their pathogenic properties and cause an infection without significant clinical signs in the vaccine.

**PLA** Enzymes like peroxidase can be conjugated to virus-specific antibodies. Binding of these antibodies to viral antigens located in infected cells can be visualized by adding substrate to the antigen–antibody–enzyme complex.

**qPCR** Real-time PCR allows monitoring of the polymerase chain reaction in real time on a computer screen. In contrast to gel-based PCR systems qPCR yields semi-quantitative results.

**VNT** Virus neutralization tests are tissue-culture-based assays for the detection and quantification of virus neutralizing antibodies. Of all antibody detection tests they are generally considered the 'gold standard'.

## Introduction

Classical swine fever (CSF), formerly known as hog cholera, is a highly contagious viral disease of swine with high morbidity and mortality especially in young animals. Due to its economic impact it is notifiable to the Office International des Epizooties (OIE (World Organisation for Animal Health)). First outbreaks of the disease were observed in 1833 in Ohio, USA. The aetiological agent was long assumed to be a bacterium (hog cholera bacillus), until de Schweinitz and Dorset demonstrated in 1903 that the agent was filterable. CSF has been eradicated in Australia, Canada, the US, and almost all member states of the European Union (EU). Vaccination is banned in these countries. However, outbreaks of CSF still occur intermittently in several European countries in either wild boar or domestic pigs. The latter can cause large economic losses. During the last 16 years in the EU, close to 20 million pigs were euthanized and destroyed because of control measures imposed to combat CSF epizootics, causing total costs of about 5 billion euros. In many countries worldwide, CSF is still a major problem. In 2004, a total of 38 member states of the OIE reported CSF outbreaks.

## Virus Properties

### Genome Organization and Protein Expression

CSF virus (CSFV) is an enveloped virus with a diameter of c. 40–60 nm. The single-stranded RNA genome has a size of 12.3 kb. It has positive polarity with one open reading frame (ORF) flanked by two nontranslated regions (NTRs). The 5′ NTR functions as an internal ribosomal entry site (IRES) for cap-independent translation initiation of the large ORF that codes for a polyprotein of about 3900 amino acids. The polyprotein is co- and post-translationally processed by viral and cellular proteases. From 11 viral proteins, four constitute the structure of the particle, namely three envelope glycoproteins ($E^{rns}$, E1, and E2) and one core (C) protein. The remaining seven non-structural proteins are $N^{pro}$, p7, NS2-3, NS4A, NS4B, NS5A, and NS5B. The main target for neutralizing antibodies is the viral envelope glycoprotein E2. To a lesser extent, the host's immune system produces neutralizing antibodies to $E^{rns}$. This viral envelope glycoprotein occurs as a disulfide-bonded homodimer in the virus particle, and is also secreted by CSFV-infected cells. $E^{rns}$ was shown to be a potent ribonuclease specific for uridine. $N^{pro}$ has an autoproteolytic activity that achieves cleavage from the downstream nucleocapsid protein C. *In vitro*, it had been shown that $N^{pro}$ interferes with the induction of interferon $\alpha/\beta$, and *in vivo* CSFV with a deletion in the $N^{pro}$ gene was able to infect pigs but had lost its pathogenicity.

### Replication

CSFV replicates in cell lines from different species, for example, pigs, cattle, sheep, and rabbit. For most purposes, the viruses are cultivated in porcine cells. Virus replication is restricted to the cytoplasm of the cell and normally does not result in a cytopathic effect (CPE). The first progeny virus is released from the cells at 5–6 h post infection. Virion assembly occurs on membranes of the endoplasmic reticulum, and full virions appear within cisternae. They are released via exocytosis or cell lysis.

Cytopathic isolates of CSFV occur only sporadically, and most of them are defective interfering (DI) particles. Their genomes have large internal deletions, consisting mainly of the genomic regions coding for the structural proteins. The DI particles are strictly dependent on a complementing helper virus for replication.

### Taxonomy and Classification

CSFV belongs together with viruses in the species *Bovine viral diarrhea virus 1* (BVDV-1), BVDV-2, and viruses in the species *Border disease virus* (BDV) to the genus *Pestivirus*, in the family *Flaviviridae*. Classification of CSFV strains and isolates is currently performed by genetic typing. For this purpose, three regions of the CSFV genome are used, mainly 190 nt of the E2 envelope glycoprotein gene, but also 150 nt of the 5′ NTR, and 409 nt of the NS5B polymerase gene. The phylogenetic tree shows that the CSFV isolates can be divided into three groups with three or four subgroups: 1.1, 1.2, 1.3; 2.1, 2.2, 2.3; 3.1, 3.2, 3.3, 3.4. CSFV groups and subgroups show distinct geographical distribution patterns, which is important for epidemiologists. Whereas isolates belonging to group 3 seem to occur solely in Asia, all CSF virus isolates of the last 15 years isolated in the EU belonged to one of the subgroups within group 2 (2.1, 2.2, or 2.3) and were clearly distinct from historic CSFV of group 1, that are still being used as CSFV laboratory reference viruses. In contrast, current representatives of group 1 viruses continue to cause outbreaks in South America and Central America, and in the Caribbean. Interestingly, CSFVs of all groups have been found in various locations in Asia (**Figure 1**).

## Antigenic Relationships and Variability

All pestiviruses were originally classified according to their host origin and the disease they cause. Later, it was found that they are not restricted to their original hosts and that they are serologically related, for example, BVDV and BDV have the capacity to naturally infect many other ruminant species and pigs. The same applies for CSFV, although its extended host range has only been determined experimentally and does not seem to reflect natural conditions. Detailed studies on the antigenic relationship between CSFV and ruminant pestiviruses have been performed using monoclonal antibodies and neutralization studies.

In recent years, progress in molecular biology has allowed a more precise assessment of their relationship. Several genomic regions have been used to study genetic diversity, for example, the 5′ NTR and genes coding for proteins ($N^{pro}$, C, and E2). Phylogenetic analysis confirmed the classification of the genus *Pestivirus*, comprising CSFV, BVDV-1, BVDV-2, and BDV, and has

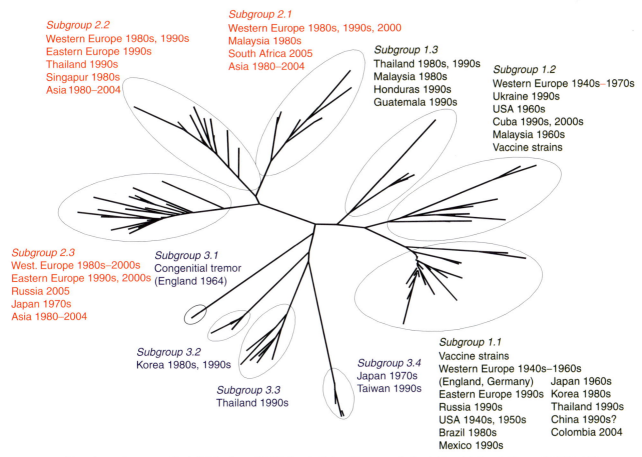

**Figure 1** Diversity and geographical distribution of CSF virus isolates. The unrooted neighbor-joining phylogenetic tree was calculated using 190 nt of E2 sequences from 108 CSF viruses. The nomenclature of virus subgroups is as suggested by Paton DJ *et al*. Adapted from Paton DJ, McGoldrick A, Greiser-Wilke I, *et al.* (2000) Genetic typing of classical swine fever virus. *Veterinary Microbiology* 73: 137–157.

revealed at least one additional species, a pestivirus isolated from giraffe.

There is as yet little knowledge concerning the evolution of CSFV, although some analyses have been performed, albeit using short regions of the genome only. These results are difficult to interpret as, in contrast to other RNA viruses, CSFV has little tendency to accumulate mutations. Extensive genomic analyses of isolates obtained between February 1997 and March 1998 in the Netherlands emphasized the genetic stability of CSFV even in the highly variable antigenic region of the E2 gene during a major epidemic that lasted longer than 1 year.

## Transmission and Host Range

Members of the family Suidae, in particular domestic pigs (*Sus scrofa domesticus*) and European wild boar (*Sus scrofa scrofa*), are the natural hosts of CSFV. Blood, tissues, secretions, and excretions from an infected animal contain the virus. Transmission occurs mainly by the oral–nasal route, though infection through the conjunctiva, mucous membrane, by skin abrasion, insemination, and percutaneous blood transfer (e.g., common needle, contaminated instruments) may occur. Airborne transmission is not thought to be important in the epizootiology of CSFV, but such transmission could occur between mechanically ventilated units within close proximity to each other.

Experimental transmission of CSFV to goats, sheep, cattle, peccaries (*Tayassu tajacu*), and rabbits was successful, while other vertebrates, for example, racoons, mice, and pigeons did not support the propagation of the virus. Of all species, the rabbit has been of major importance because it was, and in some countries still is, used for the attenuation of CSFV and for large-scale production of live vaccine virus.

## Clinical Features

The incubation period in individual animals is 3–10 days. Since transmission from animal to animal may be slow, in large holdings symptoms of CSF may only become evident several weeks after virus introduction ('herd incubation'). The severity of clinical signs mainly depends on the age of the animal and the virulence of the virus. In older breeding pigs, the course of infection is often mild or even subclinical. The virulence of a CSFV isolate is difficult to determine on a rational basis, as the same CSFV isolate can cause different forms of disease depending on age, breed, and immune status of the host animal. Basically acute, chronic, and prenatal forms of CSF can be distinguished, and there is no classical pattern of symptoms that is invariably associated with the disease.

### Acute CSF

Most piglets up to 12 weeks of age develop a severe form of acute CSF with high fatalities, whereas older breeding animals may only show mild symptoms. A constant finding in young animals is pyrexia higher than 40 °C. Initial signs are anorexia, lethargy, huddling together, conjunctivitis, respiratory symptoms, and constipation followed by diarrhea. Neurological signs are frequently seen, such as a staggering gait with weakness of hind legs, incoordination of movement, and convulsions. Hemorrhages of the skin are frequently observed on the ear, tail, abdomen, and the lower part of the limbs during the second and third week after infection until death. Virus is shed from the infected animal by all secretions and excretions. CSFV causes severe leukopenia and immunosuppression, which often leads to secondary enteric or respiratory infections. The signs of these secondary infections can mask or overlap the most typical signs of CSF and may mislead the veterinarian. With increasing age of the infected pigs (fattening and breeding animals), clinical signs are less specific and recovery is frequent. The infection is terminated by a strong, predominantly humoral immune response. First neutralizing antibodies against CSFV become detectable 2–3 weeks post infection.

### Chronic CSF

The chronic form of CSF is always fatal. It develops in a low percentage of infected animals, when pigs are not able to mount an effective immune response to overcome the virus. Initial signs are similar to those characteristic for the acute infection. Later, predominantly nonspecific signs are observed, for example, intermittent fever, chronic enteritis, and wasting. Animals usually survive for 2–4 months before they die. Until death, CSFV is constantly shed from the onset of clinical signs. Antibodies may be temporarily detected in serum samples, as the immune system begins to produce antibodies, although they are not able to eliminate the virus. Consequently, the antibodies are complexed by circulating virus and cease to be detectable. Since clinical signs of chronic CSF are rather nonspecific, a broad range of other diseases must be considered for differential diagnosis.

### Prenatal and Late Onset CSF

Although the course of infection in sows is often subclinical, CSFVs, as do other pestiviruses, cross the placenta of pregnant animals, thereby infecting fetuses during all stages of pregnancy. The outcome of transplacental infection depends primarily on the time of gestation and on viral virulence. Abortions and stillbirths, mummification, and malformations are observed after infection during early pregnancy. In breeding herds, this leads to a reduction in the fertility index of the affected pig herd. Infection of sows from about 50 to 70 days of pregnancy may lead to the birth of persistently viremic piglets, which may be clinically normal at birth and survive for several months. After birth, they usually show poor growth ('runt'), wasting, or occasionally congenital tremor. This course of infection is referred to as 'late onset CSF' and the outcome is fatal. During their lifetime, these animals constantly shed large amounts of virus and are dangerous virus reservoirs, spreading and maintaining the infection within the pig population. This feature of CSFV infection is comparable to cattle or sheep persistently infected with BVDV and BDV, respectively.

## Pathology and Histopathology

After infection, the tonsils are the location of primary virus replication. Thereafter, the agent progresses to neighboring lymphoreticular tissues. Through lymph channels the virus reaches regional lymph nodes, from where it spreads to the blood vascular system. Massive secondary virus replication takes place in spleen, bone marrow, and visceral lymph nodes. Major targets for the virus are cells of the immune system. In the peripheral blood, main target cells for the virus appear to be monocytes. In later stages of the disease, infected lymphocytes as well as granulocytes are found. Early events that are not completely understood play a significant role in the pathogenesis and manifestation of CSF. As early as 24 h post infection with virulent virus, that is, when animals are still asymptomatic and 2–4 days before virus can be detected in the peripheral blood, a progressive lymphopenia is observed resulting in severe immunosuppression. The reason for the massive cell death is not clear; however, a direct interaction with the virus can be ruled out. Viral E$^{rns}$ has been shown to induce apoptosis in lymphocytes of different species. Since this protein is secreted in large quantities from infected cells, it is conceivable that it causes massive destruction of lymphocyte populations. An additional or an alternative mechanism of cell death, respectively, might be triggered by cytokines that are activated shortly after infection, for example, tumor necrosis factor-$\alpha$ (TNF-$\alpha$), interleukin-1$\beta$ (IL-1$\beta$) and IL-1$\alpha$, as well as IL-6.

A severe thrombocytopenia develops once infected animals develop fever and virus is detectable in peripheral blood. Currently, there are two explanations for thrombocyte depletion: (1) abnormal peripheral consumption of thrombocytes may be responsible for thrombocytopenia; and (2) progressive degeneration of megakaryocytes, which is observed to begin at day 1 p.i., can result in cell death and shortage of thrombocyte production. It is not clear whether direct or indirect effects are responsible for the latter phenomenon. In analogy to the clinical picture, the severity of pathological lesions depends on time of infection, age of the animal, and virulence of the strain.

In acute cases, pathological changes visible on postmortem examination are observed most often in lymph nodes, spleen, and kidneys. Kidney parenchyma may display a yellowish brown colour. Infarctions of the spleen are considered pathognomonic; however, they are rarely observed. Lymph nodes are swollen, edematous, and hemorrhagic. Hemorrhages of the kidney may vary in size from petechial to ecchymotic. Petechiae can also be observed in the mucous membranes of other organs, for example, urinary bladder, larynx, epiglottis, and heart, and may be widespread over the serosae of the abdomen and chest. Inflammation in the respiratory, gastrointestinal, and urinary tract often are sequelae of secondary infections. Severe pneumonia sometimes is complicated by interstitial edema. Tonsils may display necrotic foci caused by micro infarcts followed by secondary infection. A nonpurulent encephalitis is often present.

In chronically infected animals, pathological changes are less pronounced, especially an absence of hemorrhages on organs and serosae. Instead, thymus atrophy, lymphocyte depletion in peripheral lymphatic organs, and hyperplasia of the renal cortex are often observed. In animals displaying chronic diarrhea, necrotic and ulcerative lesions on the ileum, the ileocaecal valve, and the rectum are found. 'Button' ulcers in the large intestine are pathognomonic, though rare. In cases of congenital infection, a proportion of piglets may show incomplete development of the cerebellum or other developmental abnormalities, such as atrophy of the thymus.

## Epizootiology

Primary outbreaks in CSF-free regions usually occur as a consequence of the feeding of swill containing infected pork. Although this practice is officially banned in almost all CSF-free and many other countries, it is still the major risk factor for the importation of the infection into CSF-free populations of domestic pigs or wild boar. The most common route for the spread of the infection among pigs is virus excretion by infected pigs via saliva, feces and urine, and oronasal virus uptake by uninfected animals. Further sources of infection are natural breeding and artificial insemination, since the virus is also excreted in sperm. Trade of live pigs, including at auction sales, has been shown to be the most frequent cause of virus spread from herd to herd. Other farming activities, for example, livestock shows, visits by feed dealers, and rendering trucks are also high-risk factors. Infected pregnant sows ('carrier sows') may give birth to persistently infected piglets that play an important role in the spread of the infection. CSF in small holdings, commonly called 'backyard holdings', is particularly difficult to control. Swill feeding, lack of animal registration and movement control, poor hygiene and lack of education of pig owners, as well as lack of awareness of private and government veterinarians facilitate the further spread and persistence of the infection in backyard pig populations.

CSF outbreaks in wild boar or feral pigs can be caused by contaminated garbage or 'spillover' from CSFV-infected domestic pigs. The outcome of these outbreaks mainly depends on the size and density of the wild boar populations affected. Outbreaks in small populations that live within natural confines, such as valleys, tend to be self-limiting. In contrast, infections leading to outbreaks in large areas and dense populations often become endemic. Most of the older animals survive the infection and become immune. Piglets become susceptible with waning maternal immunity and then can serve as reservoirs for the perpetuation of the infection. Most fatalities are registered in the young age class. CSFV in wild boar or feral pigs is a threat to any local domestic pig holding and strict measures have to be taken to avoid the spread to domestic pigs.

## Diagnosis

The majority of outbreaks of CSF are diagnosed tentatively on clinical grounds, especially in the severe acute form of disease. However, a number of diseases must be considered for differential diagnosis: acute African swine fever leads to a very similar clinical and pathological picture. Erysipelas, porcine reproductive and respiratory syndrome, cumarin poisoning, purpura hemorragica, postweaning multisystemic wasting syndrome, porcine dermatitis and nephropathy syndrome, *Salmonella* or *Pasteurella* infections, or any enteric or respiratory syndrome with fever not responding to antibiotic treatment may display features resembling CSF. Infections of pigs with related ruminant pestiviruses occasionally cause similar symptoms and cross-reactions in diagnostic laboratory tests. In conditions of reduced fertility, CSF, next to parvovirus infections, porcine reproductive and respiratory syndrome, leptospirosis, and Aujeszky's disease (pseudorabies), should be considered in the differential diagnosis. In any case, suspected CSF outbreaks need to be verified using laboratory diagnostic methods. Techniques for the detection of CSFV and virus-specific antibodies are well established and they are described in detail in the Manual of Standards of the OIE as well as in the Diagnostic Manual attached to Decision 2002/106/EC.

### Virus Isolation in Cell Culture

Virus isolation using susceptible porcine cell cultures is still considered the standard method for the direct diagnosis of CSFV infection. Suitable samples are whole blood, buffy coat, plasma, serum or clarified organ suspensions from tonsils, spleen, kidney, or gut lymph nodes. Since CSFV does not cause a CPE, the infection must be visualized by fixing and staining the cells. Antigen can then be detected either by direct or indirect immunofluorescence tests (IFTs), or by immunoperoxidase assays (PLAs), using conjugated virus-specific polyclonal or monoclonal antibodies. Virus isolation is time consuming, but the virus isolates can be stored and used for further analyses, for example, genotyping, and it allows the establishment of strain collections.

### Direct Antigen Detection

Antigen detection may be carried out on fixed cryosections of organs, using IFT or PLA with polyclonal or monoclonal antibodies. The tests yield results quickly and they are often used for a first laboratory investigation in a suspected case. However, due to limited sensitivity, a CSF suspicion cannot be ruled out in case of a negative result. The correct interpretation of results requires well-trained and experienced personnel.

Commercially available antigen capture enzyme-linked immunosorbent assays (ELISAs) are used for analyzing blood, buffy coat, organ suspension, plasma, or serum samples. Although the method yields quick results (4 h), its usefulness is limited due to its low sensitivity. It is only suitable for herd diagnosis and not for individual animals. In the near future, the ELISA will most probably be replaced by polymerase chain reaction after reverse transcription of the genome (reverse transcription-polymerase chain reaction, RT-PCR) using pools of samples.

### Detection of Viral Nucleic Acid by PCR

RT-PCR is becoming an increasingly important tool for the diagnosis of CSFV. Evaluation of RT-PCR can be performed either by agarose gel electrophoresis, or by real-time techniques (RT-qPCR). As laboratory equipment is becoming reliable and also more affordable, coupling of liquid handling robotics for nucleic acid isolation and RT-qPCR is becoming practicable. Another advantage of both the standard gel-based RT-PCR and the RT-qPCR is that, due to their high sensitivities, pooled samples can be tested. In particular, the use of RT-qPCR allows rapid and reliable testing of herds at the perimeter of an outbreak in order to avoid preemptive slaughter. Despite the advantages that RT-PCR methodology may have over conventional diagnostic tests, it is extremely vulnerable to false negative or false positive results. False negative results can arise when the nucleic acid is degraded, or when the reaction mixture contains inhibitors. Due to its high sensitivity, false positive results may arise from contaminations, either from sample to sample or from other sources. This implies that before diagnostic laboratories can replace any test, their

RT-PCR protocols have to be validated, and regular participation in proficiency testing must prove that performance of the methods used is accurate. In addition, specification concerning sensitivity of the detection must be defined. This is important when samples are pooled. In summary, analytical performance must be equal to or better than that of the standard method, that is, isolation of CSFV in permissive cells.

### Antibody Detection

CSFV-specific antibodies in pig populations are sensitive indicators for the presence of the infection. Hence, serological tests are valuable tools for diagnosis and surveillance. CSFV infection mainly induces antibodies against viral proteins E2, $E^{rns}$, and NS3. Detectable levels of antibodies appear 2–3 weeks post infection and persist lifelong in recovered animals. The virus neutralization test (VNT) is the most sensitive and versatile assay. It is very useful for quantifying neutralizing antibodies as well as for discriminating between infections with CSFV or ruminant pestiviruses. This often becomes necessary since, irrespective of its clinical status, a pig herd or single animal seropositive for CSFV is considered as CSFV infected, unless an involvement of CSFV is ruled out. In contrast, in the case of BDV or BVDV infections in pigs, no disease control measures are taken. The choice of test viruses should take into account the local epidemiological situation. While the above-mentioned differentiation is often possible, the VNT is not able to discriminate between antibody titers due to CSFV field infection and antibody titers resulting from immunization with modified live CSFV vaccines. As it relies on cell culture technology, the test is labor intensive and time consuming, and it is not suitable for mass screening of samples. Therefore, it is mainly used for cases where an accurate quantitative and discriminatory assessment of antibody levels is required.

For routine serological investigation, ELISAs are suitable. These tests are either designed as blocking or as indirect ELISAs. They are widely used for screening of antibodies during and after outbreaks, for monitoring of CSFV infections in wild boar, and to test coverage of immunization of wild boars after vaccination. In general, ELISAs are less sensitive than VNTs. However, although ELISAs have some limitations considering specificity and/or sensitivity, they yield results quickly and they are well suited for mass screening of animals, and on a herd basis they are suitable to detect field virus infections.

### Immune Response

Live CSFV infection induces a B-cell as well as a T-cell response, while inoculation of inactivated virus elicits only a B-cell response. The cellular immune response has not been thoroughly investigated. Antibodies are produced against viral proteins NS3, E2, and $E^{rns}$. Antibodies against E2 and to a lesser extent $E^{rns}$ are protective. Neutralizing antibodies against CSFV are regarded as the most important specific defence against infection and disease. This is in accord with the immune responses to other pestivirus infections. There is some antigenic variation among CSFV isolates, but not to the extent observed with ruminant pestiviruses. Therefore, convalescent animals have a stable and long-lasting, if not lifelong, immunity against all variants of CSFV.

### Vaccination

First attempts to vaccinate against swine fever date back to the beginning of the last century, when pigs were infected with live virus and simultaneously treated with serum from immune pigs. This high-risk practice was replaced in the 1940s by the use of inactivated and modified live vaccines (MLVs). While the inactivated vaccines proved rather inefficient, MLV turned out to be highly efficacious and safe in pigs of any age, for example, the $GPE^-$ and the lapinized Chinese strain (C-strain) of CSFV. The latter is probably the most popular MLV against CSF. When properly used, MLVs are powerful tools for prophylactic protection of domestic pigs against CSF. In countries still struggling with endemic CSF, vaccines are being used in order to limit economical damage. The systematic use of these vaccines often was and is the first step in the eradication of CSF. In countries that have eradicated the infection, prophylactic vaccination is usually banned, mainly because animals vaccinated with conventional MLV (e.g., the C-strain) cannot be distinguished serologically from animals that have recovered from field infection and thus the CSF-free status of the country would not be regained. However, in emergency situations, vaccination may be used, followed by international trade restrictions for vaccinated animals and their products for at least 6 months. For EU member states, provisions have been made in directive 2001/89/EC for limited vaccination of domestic pigs in cases of severe outbreak emergencies.

In order to overcome the severe restrictions after vaccination, novel marker vaccines have been developed. They are based on the concept that the pattern of antibodies against CSFV from a vaccinated animal can be distinguished from that of an animal which has recovered from a field virus infection. The recently developed E2 subunit CSFV marker vaccines induce neutralizing antibodies against the E2 glycoprotein only. Consequently, CSFV antibodies which are not directed against the E2 glycoprotein are indicative of an infection with wild-type CSFV. With the availability of E2 glycoprotein-based subunit marker vaccines against CSFV, two discriminatory

ELISAs were developed as companion tests that detect antibodies directed against the E$^{rns}$ glycoprotein. Positive results are indicative of the exposure to wild-type CSFV. The development of subunit marker vaccines was a major step forward. Compared to MLV, a few shortcomings have to be taken into account when using subunit marker vaccines. In comparison with conventional MLV, full immunity after subunit vaccination is slow (c. 21 days). Sometimes, a booster vaccination is needed, and it does not induce sterile immunity, for example, transplacental transmission of CSFV cannot be completely prevented. However, it was shown that the E2 subunit vaccine was able to stop virus spread among pigs that were vaccinated 10 days earlier. The discriminatory tests are suitable for herd diagnosis but they must not be used to assess the serological status of individual animals. Most likely the limitations of the first generation of marker vaccines may be overcome by a second generation of live marker vaccines. Candidates are being developed. Viral vectors carrying the E2 gene of CSFV or chimeras using the genomes of the CSFV C-strain and BVDV are promising candidates.

## Prevention and Control

There is no specific treatment for CSF. In countries where CSF occurs endemically, infected animals are killed and destroyed and vaccination is used to prevent further spread of the virus. Countries free of CSF usually implement measures to avoid outbreaks of the disease. The most effective sanction to prevent introduction of CSFV into a free pig population is the ban of swill feeding and the control of trade. Attempts must be made to prevent the illegal importation of meat and meat products. Professional farms must comply with standard biosecurity rules. In order to eradicate the disease, a system for the registration and identification of all holdings and pigs should be in place. This greatly facilitates the traceability of animal movements. In case CSFV has been introduced, eradication programs are based principally on the destruction of infected and serologically positive animals. In order to avoid trade restrictions, vaccination is usually prohibited. Despite continued efforts to control CSF, outbreaks have occurred intermittently in several European countries. In areas with high pig densities very high numbers of pigs had to be culled in order to stop virus spread, and in these cases direct and indirect economical damage was very high. For example, in the course of the 1997 CSF epidemic in the Netherlands, approximately 12 million pigs had to be destroyed and total economic losses amounted to more than 2 billion euro. In severe outbreak situations, an emergency vaccination option is available. Thus far, emergency vaccination has never been used in Western Europe.

Epidemics of CSF in wild boar populations may be long-lasting and difficult to control. Oral vaccination campaigns using MLV sometimes accompanied by specific hunting measures have been shown to be suitable tools to shorten the duration of these epidemics.

## Perspectives

The fascinating pathogenesis of CSFV displays similarities with other hemorrhagic diseases. Further elucidation of determinants of viral virulence and virus interaction with the host animal will contribute to the understanding of hemorrhagic diseases.

Control and eradication of CSFV will remain a challenge for many years to come. In CSF-free countries with a highly developed pig industry and densely populated livestock areas, the control of CSF outbreaks will change from the presently practiced excessive culling of pigs to a more sophisticated disease control strategy using sensitive and specific diagnostic tools (e.g., RT-PCR), for the tracing of virus at the perimeters of outbreaks, combined with emergency marker vaccination, preferably with a live marker vaccine, which is yet to be developed.

In countries where CSF is endemic and where a sizable proportion of pigs are held in backyards, extensive vaccination with MLV, in combination with movement controls and epidemiological surveillance, might help to control this pestilence. Of prime importance is thorough education in order to increase knowledge and awareness of all parties involved.

*See also:* Vaccine Safety; Vaccine Strategies.

## Further Reading

Anonymous (2001) Council Directive 2001/89/EC on community measures for the control of classical swine fever. *Official Journal of the European Communities* L 316/5.

Anonymous (2002) Commission Decision 2002/106/EC approving a Diagnostic Manual establishing diagnostic procedures, sampling methods and criteria for evaluation of the laboratory tests for the confirmation of classical swine fever. *Official Journal of the European Communities* L 39/71.

Armengol E, Wiesmüller K-H, Wienhold D, et al. (2002) Identification of T-cell epitopes in the structural and nonstructural proteins of classical swine fever virus. *Journal of General Virology* 83: 551–560.

Bouma A, de Smit AJ, de Kluijver EP, Terpstra C, and Moormann RJ (1999) Efficacy and stability of a subunit vaccine based on glycoprotein E2 of classical swine fever virus. *Veterinary Microbiology* 66: 101–114.

Meyers G and Thiel HJ (1996) Molecular characterization of pestiviruses. *Advances in Virus Research* 47: 53–118.

Moennig V, Floegel-Niesmann G, and Greiser-Wilke I (2003) Clinical signs and epidemiology of classical swine fever: A review of new knowledge. *Veterinary Journal* 165: 11–20.

Paton D and Greiser-Wilke I (2003) Classical swine fever – An update. *Research Veterinary Science* 75: 169–178.

Paton DJ, McGoldrick A, Greiser-Wilke I, et al. (2000) Genetic typing of classical swine fever virus. *Veterinary Microbiology* 73: 137–157.
Reimann I, Depner K, Trapp S, and Beer M (2004) An avirulent chimeric pestivirus with altered cell tropism protects pigs against lethal infection with classical swine fever virus. *Virology* 322: 143–157.
Ruggli N, Bird BH, Liu L, Bauhofer O, Tratschin D-J, and Hofmann MA (2005) N$^{pro}$ of classical swine fever virus is an antagonist of double-stranded RNA-mediated apoptosis and IFN-$\alpha/\beta$ induction. *Virology* 340: 265–276.
Stegeman A, Elbers A, de Smit H, Moser H, Smak J, and Pluimers F (2000) The 1997/1998 epidemic of classical swine fever in the Netherlands. *Veterinary Microbiology* 73: 183–196.
Thiel H-J, Plagemann PGW, and Moennig V (1996) Pestivirus. In: Fields B (ed.) *Virology,* 3rd edn., pp. 1059–1073. Philadelphia: Lippincott-Raven Publishers.

# Coltiviruses

**H Attoui,** Faculté de Médecine de Marseilles, Etablissement Français Du Sang, Marseilles, France
**X de Lamballerie,** Faculté de Médecine de Marseille, Marseilles, France

© 2008 Elsevier Ltd. All rights reserved.

## Glossary

**Lagomorph** Gnawing herbivorous mammals including rabbits and hares.
**Myocarditis** Inflammation of the heart muscle.
**Nosological** From nosology, which is a branch of medical science which deals with disease classification.
**Nymph** Immature insect or tick.
**Orchitis** Inflammation of the testicles.
**Pericarditis** Inflammation of pericardium.
**Sarcolemma** Thin transparent homogenous sheath enclosing a striated muscle fiber.
**Trans-ovarial** From parental ticks to progeny ticks through the eggs.
**Trans-stadial** In-between developmental stage forms of the ticks.

## Introduction

The family *Reoviridae* includes 12 recognized and 3 proposed genera. The genera *Coltivirus, Seadornavirus,* and *Orbivirus* are those which include causative agents of animal arboviral diseases (including human diseases). In 1991, the International Committee on Taxonomy of Viruses (ICTV) officially created the genus *Coltivirus* (sigla from *Colorado tick fever virus*) for the classification of 12-segmented dsRNA animal viruses. Until recently, this genus contained both tick-borne and mosquito-borne viruses. The tick-borne viruses included two species which are *Colorado tick fever virus* (CTFV) and *Eyach virus* (EYAV). The former species encompasses three serotypes that are the Florio isolate (CTFV-Fl, isolated from humans and ticks), California hare coltivirus (CTFV-Ca, isolated from a hare in northern California), and Salmon River virus (isolated from humans in Idaho). Many mosquito-borne viruses were also considered as tentative species including *Banna virus* (BAV, isolated from humans) and several isolates from mosquitoes. In the Seventh Report of the ICTV, the genus *Coltivirus* was subdivided into coltivirus group A which contained the various strains of CTFV and EYAV (tick-borne isolates), and coltivirus group B which contained the Southeast Asian mosquito-borne isolates. The tick-borne and mosquito-borne viruses of this group have many distinctive features and sequence data led to a re-evaluation of their taxonomic status. In the Eighth Report of the ICTV (2005), the genus *Coltivirus* encompasses only the tick-borne viruses, while the mosquito-borne viruses were assigned to a new genus designated *Seadornavirus.*

Coltiviruses have been implicated in a variety of pathological manifestations including flu-like illness, neurological disorders, and some severe complications as further described in the text below.

## Historical Overview of Coltiviruses

Coltiviruses have been isolated from rodents, humans, and various ticks in the family *Ixodidae.* Colorado tick fever virus is endemic in the northwestern part of the American continent. CTFV is responsible for Colorado tick fever (CTF), a human disease that was initially confused with a mild form of Rocky Mountain spotted fever (RMSF), caused by a *Rickettsia* species. The description of fever forms with CTF manifestations was made as early as 1850 and clinical studies of these fevers were initiated in 1880s. In 1930, CTF was recognized as a distinct nosological entity. The causative agent was isolated from human serum in 1944 by Florio using inoculation in adult hamsters, and the disease was reproduced in human volunteers by inoculation with serum from infected patients or egg-adapted CTFV. Between 1946 and 1950, the virus was adapted to mice, and suckling mice became the routine isolation system for CTFV. This considerably progressed the elucidation of the clinical features, epidemiology, and ecology of CTF. Up to 200 cases of CTF have been reported annually in Colorado State in the USA alone, in the past 10 years. Three isolates of Eyach virus were obtained from ticks in Europe. In 1976, isolate EYAV-Gr was obtained from *Ixodes ricinus* in Eyach village near Stugart, Germany. In 1981, two isolates of the virus designated EYAV-Fr577 and EYAV-Fr578 were obtained from *Ixodes ricinus* and *Ixodes ventalloi* in Mayenne, France. The antigenic relationship of EYAV to CTFV was established by a complement fixation assay.

The recent analysis of the complete genome sequences of both CTFV and EYAV has shown that these two viruses are closely related and confirmed the antigenic observations. The relatedness of EYAV to a well-known human pathogen and the identification of anti-EYAV antibodies in patients' sera indirectly have incriminated EYAV in human neurological diseases. Salmon River virus was isolated from a patient with a moderately severe CTF-like illness in Idaho.

## Distribution and Epidemiology

CTF is considered as the one of the most important arboviral diseases of humans in the USA. CTF occurs in mountain forest habitats at an altitude of 1200–3000 m in the Rocky Mountain region of the USA and in Canada (**Figure 1**). The virus distribution closely matches that of its vector, *Dermacentor andersoni*. This region covers British Columbia, Alberta, and Saskatchewan in Canada, and Washington, Montana, Oregon, Idaho, Wyoming, Nevada, Utah, and Colorado in the USA. However, CTFV was also isolated from the dog tick *Dermacentor variabilis* in Long Island, New York, in the late 1940s. California hare coltivirus (CTFV-Ca), identified previously as strain S6-14-03, was isolated from the blood of the white hare *Lepus californicus* in California (outside the range of *Dermacentor andersoni*). Antibodies have been found in hares in Ontario but without any virus isolation or occurrence of human cases. CTFV infection has been contracted from infected ticks between March and September. Interestingly, the presence of antibodies reacting with CTFV antigen have also been detected in sera of humans in South Korea.

Between 1980 and 1988, over 1400 cases were reported in the USA, mainly in Colorado State. The disease is under-reported since it has always been confused with other infections. More than 70% of CTF cases occur in adults and a few fatal cases have been reported, mainly in children.

The EYAV isolates EYAV-Gr, EYAV-Fr577, and EYAV-Fr578 were obtained from ticks in Europe. EYAV-Fr577 and EYAV-Fr578 were isolated in France from two ticks, *Ixodes ventalloi* and *Ixodes ricinus*, respectively, feeding on a rabbit (**Figure 1**). However, attempts to isolate these viruses from the rabbit failed. Serological surveys conducted in France identified antibodies to EYAV in rodents. It was found that 0.92% of the tested European rabbit (*Oryctolagus cunniculus*) population in Camargue and Vaucluse had antibodies to EYAV. Similar to CTFV whose principal hosts are rodents, European rabbits could represent the major reservoir for the EYAV. Other studies detected the presence of anti-EYAV antibodies in wild mice in the Sarre region and

**Figure 1** The known world distribution of coltiviruses (shaded areas): CTFV, CTFV-Ca, and Salmon River virus in North America and EYAV in Europe (France, Germany, Czech Republic and Slovakia). The geographical localization of the various coltiviruses is within the same range of latitude. The question mark indicates South Korea where antibodies to CTFV have been detected but the virus has not been formally identified by isolation.

in deer, mountain goats, and sheep in the Pyrénées and southern Alps regions of France. Serological surveys in higher mammals, including ovines, deer, and caprines, in the southern half of France, identified anti-EYAV antibodies in 1.35% of tested animals. Following the original isolation of EYAV in France, the serum from a French farmer (0.2% of the tested population) was found to be positive. The presence of the virus is also suspected in Eastern Europe as anti-EYAV antibodies have been identified in patients with neurological disorders. However, the natural cycle of the virus is still unclear and it is not known whether it circulates continuously in Europe. In 2003, the virus was re-isolated from ticks in Baden Wuerttemberg, Germany.

## Vectors, Host Range, and Transmission

The vectors of coltiviruses are ticks. Colorado tick fever virus is transmitted by the adult wood tick *Dermacentor andersoni*, but other ticks such as *Dermacentor occidentals*, *Dermacentor albopictus*, *Dermacentor parumapertus*, *Haemaphysalis leporispalustris*, *Otobius lagophilus*, *Ixodes sculptus*, and *Ixodes spinipalpis* have been found to transmit the virus. Eyach virus was isolated from the ticks *Ixodes ricinus* and *Ixodes ventalloi*. *Dermacentor* ticks are three-host ticks, requiring different hosts at the larval, nymph, and adult stages. Ticks become infected with CTFV on ingestion of a blood meal from an infected vertebrate host. Infected nymphs and adults transmit the virus to the host via saliva during feeding.

CTFV is transmitted *trans*-stadially but there is no evidence for *trans*-ovarial transmission. However, it is noteworthy that in other tick-borne viruses in the family *Reoviridae*, such as the orbivirus St. Croix River virus, there is strong suspicion that *trans*-ovarial transmission occurs. CTFV-infected larvae and nymphs can hibernate and both adult and nymphal ticks become persistently infected, providing an overwintering mechanism for the virus. A threshold of $10^2$ pfu/ml of blood is required for a tick to become infected. Some rodent species become viremic (as high as $10^{6.5}$ pfu/ml) and this could last more than 5 months, which may also facilitate overwintering and virus persistence in the field. The prevalence of viremic rodents in an endemic area ranges from 3.5% to 25%, producing a tick infection frequency ranging from 10% to 25%. Humans represent accidental hosts and usually become infected with CTFV when bitten by the adult wood tick *Dermacentor andersoni* but most probably do not represent a source of reinfection for other ticks. CTFV has a wide host range including ground squirrels, chipmunks, wild mice, wood rats, wild rabbits and hares, porcupines, marmots, deer, elk, sheep, and coyotes.

Transmission from person to person has been recorded as a result of blood transfusion. In the USA, the virus figures on the list of agents to be screened before bone marrow transplantation. The prolonged viremia observed in humans and rodents is due to the intra-erythrocytic location of virions, protecting them from immune clearance. The maintenance cycle of EYAV in the nature is not known, though the European rabbit *Oryctolagus cunniculus* is suspected to be the primary host.

## Virion Properties, Genome and Replication

Coltivirus particles have a total diameter of approximately 80 nm and a core that is about 50 nm in diameter. Electron microscopic studies, have revealed particles with a relatively smooth surface capsomeric structure and icosahedral symmetry (**Figure 2**). The majority of the

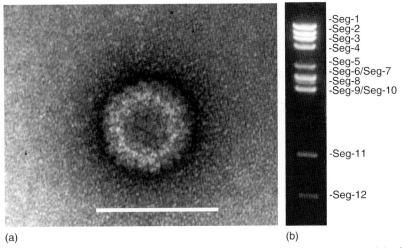

**Figure 2** Electron microscopy and the genome of CTFV. (a) Negative contrast electron micrograph of the CTFV Florio strain. The bar represents 100 nm. (b) The agarose gel (1.2%) electopherotype of CTFV genome.

viral particles are non-enveloped, but a few acquire an envelope structure during passage through the endoplasmic reticulum. The buoyant density of CTFV in CsCl is 1.36–1.38 g cm$^{-3}$. The virus is stable between pH 7.0 and 8.0 and there is total loss of infectivity at pH 3.0. CTFV can be stored at 4 °C for 2–3 months in presence of 50% fetal calf serum in 0.2 M Tris– HCl pH 7.8, or for years at –80 °C. Upon heating to 55 °C, CTFV rapidly loses infectivity. The virus is stable when treated with nonionic detergents (such as Tween 20) or with organic solvents such as Freon 113 or its ozone-friendly substitute Vertrel XF, but viral infectivity is abolished by treatment with sodium deoxycholate or sodium dodecyl sulfate.

The genome of coltivirus consists of 12 dsRNA segments that are named Seg-1 to Seg-12 in order of reducing molecular weight as observed by agarose and polyacrylamide gel electrophoresis. The genome comprises approximately 29 000 bp and the segment length ranges between 4350 bp and 675 bp. The genomic RNA of CTFV migrates as ten distinct bands during 1% agarose gel electrophoresis (AGE). Two of these bands contain co-migrating segments (**Figure 2**). The genome of the North American California hare coltivirus (CTFV-Ca) has shown a similar electropherotype by AGE. The dsRNA genome of EYAV has not been analyzed by AGE, since this virus is not cultivable. Sequence analysis has shown that the G+C content for coltiviruses ranges between 48% and 51%.

CTFV produces cytophatic effect (CPE) and plaque formation in many mammalian cells including KB (human carcinoma cells), BGM (monkey kidney cells), Vero (monkey kidney cells), BHK-21 (hamster kidney cells), and L-929 (mouse fibroblast). The virus has been found to replicate in a *Dermacentor variabilis* cell line with yields of $10^{6.5}$ pfu ml$^{-1}$. In cells infected by CTFV, granular matrices, which contain virus-like particles, are produced in the cytoplasm. These structures appear similar to viral inclusion bodies (VIBs) produced during orbivirus infections. In addition, bundles of filaments (tubules) characterized by cross-striations, and kinky threads, are found in the cytoplasm and, in some cases, in the nucleus of infected cells. These may also be comparable to the 'tubules' found in orbivirus infected cells. There is no evidence for virus release prior to cell death and disruption, after which more than 90% of virus particles remain cell-associated. Immunofluorescence has shown that viral proteins accumulate in the cytoplasm and can be detected from 12 h post infection.

The full-length genome sequence of Colorado tick fever virus has been characterized and found to be 29 210 nt long; that of Eyach virus was found to be 29 210 nt in length. In comparison to viruses classified in to other genera of the *Reoviridae*, the EYAV genome is the longest of all members sequenced (genomes range between 18 500 and 28 500 bp). All 12 segments of CTFV and EYAV have conserved sequences which are located at the termini. The conserved motifs 5′-$^G/_C$ACAUUUG-3′ and 5′-UGCAGU$^G/_C$-3′ have been found in the 5′ NCR and the 3′ NCR of CTFV, respectively, while in EYAV, these motifs are 5′-GACA$^A/_U$UU-3′ and 5′-UG$^C/_U$AGUC-3′. For both viruses, the 5′ and 3′ terminal tri-nucleotides of all segments are inverted complements.

The organization of the genome is well documented. Each genome segment contains a single ORF spanning almost the whole length of the segment (**Figure 3**), except genome segment 9 which has been found to contain a stop codon at position 1054 that is in-frame to a terminal stop codon at position 1846. The in-frame stop codon has been shown to belong to the category of so-called 'leaky stop codons' which allow a partial readthrough, generating two proteins. The first is the short form of the protein designated VP9 (ORF 41–1054) which is structural, and the second, which is the readthrough product designated VP9′ (ORF 41–1846), is nonstructural. The readthrough phenomenon (reading through a stop codon) is well characterized in some plant viruses, retroviruses, and aplhaviruses. None of the members of family *Reoviridae* had been known to use this translational regulatory feature until the identification of readthrough in genome segment 9 of coltiviruses (both CTFV and EYAV). VP9 is supposed to interfere with the action of eukaryotic peptide chain release factors (ERF1 and ERF3), thus facilitating readthrough of the stop codon and synthesis of VP9′.

Segment 6 of CTFV (encoding VP6) is homologous to segment 7 of EYAV (encoding VP7). The amino acid sequence (residues 370–490) of EYAV VP7 shows 50% similarity to the sarcolemmal-associated protein of the European rabbit *Oryctolagus cunniculus*, which is suspected to be its primary host. By comparison, the cognate protein VP6 of CTFV showed no match with this rabbit protein. This is probably due to the insertion of a sequence encoding a lagomorph protein into segment 7 of EYAV.

## Antigenic and Genetic Relationship among Coltiviruses

Antigenic variation between CTFV strains has been investigated using immune sera, and antigenic heterogeneity has been observed, especially between strains recovered from humans. However, the variation is not enormous and durable immunity to reinfection appears to be the rule. CTFV from North America and EYAV from Europe show little cross-reaction in neutralization assays. CTFV-Ca cross-reacts with CTFV-Fl but not with EYAV and is considered to be a serotype of CTFV-Fl. Three antigenic variants of EYAV are recognized: EYAV-Gr, EYAV-Fr577, and EYAV-Fr578. RNA cross-hybridization analysis

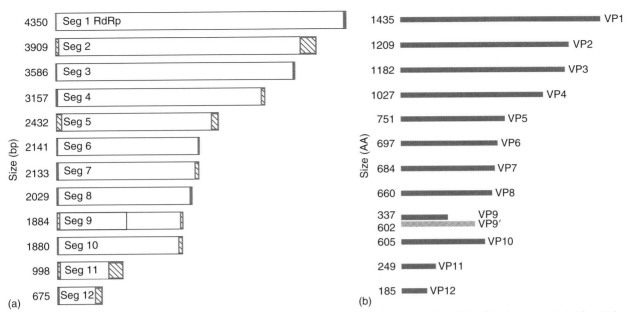

**Figure 3** (a) Organization of the genome segments of CTFV and (b) their putative encoded proteins. Shaded areas in A: 5' and 3' NCRs. Segment 9 is interrupted by an in-frame stop codon (position 1054) which has been shown to be leaky allowing readthrough, and therefore generating two proteins: a truncated short form designated VP9 (a structural component) and a long readthrough form designated VP9' (a nonstructural component).

shows that CTFV isolates have remained relatively homogenous (some variation has been reported based on cross-hybridization in genome segments 4 and 6, compared to the prototype strain CTFV-F1). This level of genetic divergence in CTFV is low compared to that found, for example, within the various species in the genus *Orbivirus*, suggesting that the CTFV is sheltered from immunological pressures causing genetic drift. The overall identities between nucleotide sequences of segments 9, 10, 11, and 12 of CTFV strains range between 90% and 100% : 97–100% (segment 9), 96–99% (segment 10), 90–94% (segment 11), and 94–96% (segment 12). The degree of identity between nucleotide sequences of segment 1–12 of CTFV and EYAV isolates ranges from 55% to 88%.

Within family *Reoviridae*, the most conserved gene between the various genera is that encoding the RNA-dependent RNA polymerase. Values lower than 30% amino acid sequence identity distinguish members of distinct genera. A tree constructed from an alignment of polymerase sequences from the various representative members of the *Reoviridae* (**Table 1**) is shown in **Figure 4**. Calculations of the amino acid sequence identity based on this alignment indicate that coltiviruses show only 15% identity with seadornaviruses, confirming the status of coltiviruses and seadornaviruses as members of distinct genera. Amino acid identities as high as 28% occur between coltiviruses and mycoreoviruses (reoviruses of fungi containing 11 or 12 genome segments).

Genome characterization has helped to shed more light on the origin of EYAV. Analysis of the most variable genes (segments 6, 7, and 12) indicates as high as 45% amino acid sequence divergence (37% nucleic acid divergence) as shown in **Figure 5**. Trees displayed in **Figure 5** show that it is possible to distinguish coltivirus species by sequence analysis of their variable genes (segment 6, 7, or 12). It has been proposed that EYAV was derived from CTFV following the introduction into Europe, in 1953 and 1972, of the North American cottontail rabbit *Sylvilagus floridanus* (following the myxoma disease that ravaged the European rabbit *Oryctolagus cuniculus*). EYAV may have then evolved from CTFV under the selective pressure of adaptation to the European rabbit, *Oryctolagus cuniculus*, and European ticks. It follows that the observed genetic divergence of 37% (in segments 6, 7, and 12) occurred over a very short period of time (less than 50 years). This implies a molecular evolutionary rate in the order of $0.5 \times 10^{-2}$ mutations/nt/year.

Alternatively, it has been also proposed that EYAV might have been derived from an ancestral virus introduced into Europe with the migration of ancestral lagomorphs (hares, rabbits) from North America through Asia. Lagomorph ancestors first appeared during the Eocene epoch (57.8–36.6 MYA) in what was then North America. They are thought to have first migrated into Asia during the Oligocene epoch (34–23 MYA) and by the high Miocene epoch (23–5 MYA) they were common in Europe. This hypothesis implies a molecular evolutionary rate in the order of $10^{-8}$ to $10^{-9}$ mutations/nt/year, a rate similar to that of dsDNA genomes. Genetic findings support the second hypothesis.

**Table 1** Sequences of RdRp used in phylogenetic analysis of various members of family Reoviridae

| Species | Isolate | Abbreviation | Accession number |
| --- | --- | --- | --- |
| Genus *Seadornavirus* (12 segments) | | | |
|   Banna virus | Ch | BAV-Ch | AF168005 |
|   Kadipiro virus | Java-7075 | KDV-Ja7075 | AF133429 |
| Genus *Coltivirus* (12 segments) | | | |
|   Colorado tick fever virus | Florio | CTFV-Fl | AF134529 |
|   Eyach virus | Fr578 | EYAV-Fr578 | AF282467 |
| Genus *Orthoreovirus* (10 segments) | | | |
|   Mammalian orthoreovirus | Lang strain | MRV-1 | M24734 |
| | Jones strain | MRV-2 | M31057 |
| | Dearing strain | MRV-3 | M31058 |
| Genus *Orbivirus* (10 segments) | | | |
|   African horse sickness virus | Serotype 9 | AHSV-9 | U94887 |
|   Bluetongue virus | Serotype 2 | BTV-2 | L20508 |
| | Serotype 10 | BTV-10 | X12819 |
| | Serotype 11 | BTV-11 | L20445 |
| | Serotype 13 | BTV-13 | L20446 |
| | Serotype 17 | BTV-17 | L20447 |
|   Palyam virus | Chuzan | CHUV | Baa76549 |
|   St. Croix River virus | SCRV | SCRV | AF133431 |
| Genus *Rotavirus* (11 segments) | | | |
|   Rotavirus A | Bovine strain UK | BoRV-A/UK | X55444 |
| | Simian strain SA11 | SiRV-A/SA11 | AF015955 |
|   Rotavirus B | Human/murine strain IDIR | Hu/MuRV-B/IDIR | M97203 |
|   Rotavirus C | Porcine Cowden strain | PoRV-C/Co | M74216 |
| Genus *Aquareovirus* (11 segments) | | | |
|   Golden shiner reovirus | GSRV | GSRV | AF403399 |
|   Grass carp reovirus | GCRV-873 | GCRV | AF260511 |
|   Chum salmon reovirus | CSRV | CSRV | AF418295 |
|   Striped bass reovirus | SBRV | SBRV | AF450318 |
| Genus *Fijivirus* (10 segments) | | | |
|   Nilaparvata lugens reovirus | Izumo strain | NLRV-Iz | D49693 |
| Genus *Phytoreovirus* (10 segments) | | | |
|   Rice dwarf virus | Isolate China | RDV-Ch | U73201 |
| | Isolate H | RDV-H | D10222 |
| | Isolate A | RDV-A | D90198 |
| Genus *Oryzavirus* (10 segments) | | | |
|   Rice ragged stunt virus | Thai strain | RRSV-Th | U66714 |
| Genus *Cypovirus* (10 segments) | | | |
|   Bombyx mori cytoplasmic polyhedrosis virus 1 | Strain I | BmCPV-1 | AF323782 |
|   Dendrlymus punctatus cytoplasmic polyhedrosis 1 | DsCPV-1 | DsCPV-1 | AAN46860 |
|   Lymantria dispar cytoplasmic polyhedrosis 14 | LdCPV-14 | LdCPV-114 | AAK73087 |
| Genus *Mycoreovirus* (11 or 12 segments) | | | |
|   Rosellinia anti-rot virus | W370 | RaRV | AB102674 |
|   Cryphonectria parasitica reovirus | 9B21 | CPRV | AY277888 |
| Genus *Mimoreovirus* (11 segments) | | | |
|   Micromonas pusilla reovirus | MPRV | MPRV | DQ126102 |

The antigenic and genetic relation between CTFV and EYAV is further corroborated by morphological characteristics. Electron microscopic analysis has shown that EYAV and CTFV are morphologically identical.

## Clinical Features and Diagnostic Assays

CTF is characterized in humans by the abrupt onset of fever, chills, headache, retro-orbital pains, photophobia, myalgia, and generalized malaise. Abdominal pain occurs in about 20% of patients. Rashes appear in less than 10% of patients. A diphasic, or even triphasic, febrile pattern has been observed, usually lasting for 5–10 days. Severe forms of the disease, involving infection of the central nervous system, or hemorrhagic fever, pericarditis and orchitis have been infrequently observed (mainly in children under 12 years of age). A small number of such cases are fatal. Severity is sufficient to result in hospitalization in about 20% of patients. Congenital infection with CTFV may occur, although the risk of abortion and congenital defects remains uncertain. It has been shown that there is

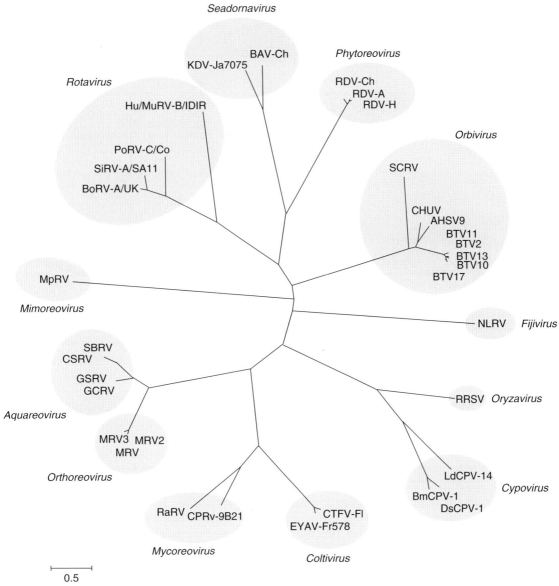

**Figure 4** Phylogenetic relations among members of the *Reoviridae* based on the sequences of their putative RdRps. Neighbor-joining phylogenetic tree built with available polymerase sequences (using the Poisson correction or gamma distribution algorithms) for representative members of 12 genera of family *Reoviridae*. The abbreviations and accession numbers are those provided in **Table 1**.

an incidence of teratogenic effects in the offspring of mice experimentally infected with CTFV. Congenital mother to infant transmission has been reported in pregnant women. According to different reports, the incidence of complications has reached up to 7%. CTFV causes leucopoenia (65% of infected humans) with the mean leukocyte count reaching 900 to 3900 mm$^{-3}$ and thrombocytopenia with platelet counts between 20 000 and 95 000. Patients with neurological disorders (meningitis, meningo-encephalitis, and encephalitis) have lymphocyte infiltration of cerebrospinal fluid (CSF) from which the virus has been isolated. Other complications such as pericarditis, myocarditis, and orchitis have been reported. The presence of CNS signs can confuse CTF with other causes of viral meningitis and encephalitis, including enteroviral infections, St. Louis encephalitis and Western equine encephalitis. CTFV infections have also been confused with other tick-borne infections such as Rocky Mountain spotted fever (RMSF: a rickettsial disease), tularemia, relapsing fever, and Lyme disease. Nevertheless, the character of the rash and the presence of leukocytosis distinguish RMSF from CTF.

The virus can be isolated from blood, since it is present in circulating erythrocytes for as long as 4 months after onset. This is because the virus infects hematopoetic progenitor cells and remains sheltered in erythrocytes after maturation. The presence of virus within erythrocytes can be demonstrated by immunofluorescence. Intracerebral

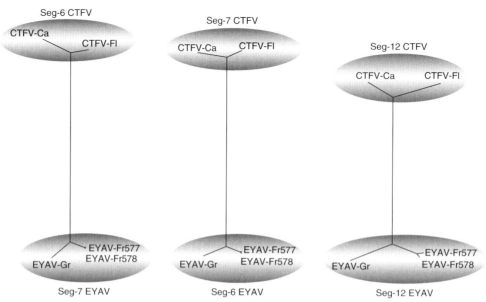

**Figure 5** Phylogenetic trees for the most variable genome segments (segments 6, 7, and 12) of different coltiviruses. Unrooted neighbor-joining tree created using MEGA program from Clustal W alignments. The groupings for the deduced amino acid sequence of homologous genes of different coltiviruses. The two species CTFV and EYAV that are currently recognized are labeled. The tree branching is almost identical using the three segments 6, 7, and 12. The genetic distance between species is expressed in percent. These have been calculated for the amino acid sequence of segment 12 and are between species CTFV/EYAV (40%) among isolates of EYAV (<8%); among isolates of CTFV (< 8%).

inoculation of blood in suckling mice has been considered for years to be the most sensitive isolation system. Reverse transcriptase polymerase chain reaction (RT-PCR) has now been developed for the diagnosis of CTFV infection, which permits the detection of human and tick virus isolates as early as 3 days post infection in experimentally infected mice. As little as one genome copy can be detected using the PCR assay.

Serological methods, which utilize infected cell cultures as antigens, have also been developed. These include a complement fixation test, which is relatively insensitive since 25% of patients do not develop CF antibodies, and in 75% of patients CF antibodies appear late. Another test is the virus neutralization test. Neutralizing antibodies appear 14–21 days after onset. An immunofluorescence assay using CTFV-infected BHK-21 or Vero cells offers an easy and rapid means to detect anti-CTFV antibodies. IgM and IgG antibodies can also be detected by enzyme-linked immunosorbent assay (ELISA) and appear concurrently or a few days after neutralizing antibodies (peak at 30–40 days). IgM antibodies decline sharply after day 45. A Western blot assay, which detects antibodies in the acute phase of infection, has also been developed. An ELISA based on the recombinant VP7 of CTFV has been developed and proved efficient and sensitive in detecting various antibodies to CTFV.

A complement fixation assay has been developed for EYAV. In the former Czechoslovakia, a population of 158 patients suffering from encephalitis diagnosed as tick-borne encephalitis (TBE) was tested for TBEV (genus *Flavivirus*), Kemerovo, Lipovnik, or Tribec viruses (all three are tick borne orbiviruses) and EYAV antibodies. Seventeen sera (11%) were found to contain solely anti-EYAV antibodies. Forty-seven patients with polyradiculoneuritis and meningopolyneuritis were tested for the same viruses and eight (17%) were found to have solely anti-EYAV antibodies. Recently, an ELISA based on the recombinant VP6 of EYAV was developed for serological diagnosis of the infection. Antibodies from mice immunized with EYAV and CTFV were used as controls for the assay of the specificity and sensitivity. This test selectively identifies anti-EYAV antibodies. An RT-PCR assay has also been developed for the specific detection of EYAV genome, based on the sequence of genome segment 12.

## Treatment, Immunity, and Prevention

There is no specific treatment for CTFV infection. Symptomatic treatment includes relief of fever and pain with paracetamol. Salicylates should be avoided because of thrombocytopenia and occurrence of bleeding disorders. An experimental vaccine (formalin-inactivated, purified CTFV) developed in the 1960s (stopped in 1970s) has been used for people at high risk of exposure to infection, and produced long-lasting immunity. After three doses, neutralizing antibodies persist for 5 years. Acaricides or repellents such as permethrin, DEET (*N,N*-diethyl-meta-toluamide), and picaridin may be used to control ticks in rodent niches and as repelling agents to ticks.

See also: Reoviruses: General Features; Reoviruses: Molecular Biology.

## Further Reading

Attoui H, Charrel R, Billoir F, et al. (1998) Comparative sequence analysis of American, European and Asian isolates of viruses in the genus Coltivirus. *Journal of General Virology* 79: 2481–2489.

Attoui H, de Lamballerie X, and Mertens PPC (2005) Coltivirus, Reoviridae. In: Fauquet CM, Mayo MA, Maniloff J, Desselberger U, and Ball LA (eds.) *Virus Taxonomy: Eighth Report of the International Committee on Taxonomy of Viruses*, pp. 497–503. San Diego, CA: Elsevier Academic Press.

Attoui H, Mohd Jaafar F, Biagini P, et al. (2002) Genus Coltivirus (family Reoviridae): Genomic and morphologic characterization of Old World and New World viruses. *Archives of Virology* 147: 533–561.

Chastel C, Main AJ, Couatarmanac'h A, et al. (1984) Isolation of Eyach virus (Reoviridae, Colorado tick fever group) from Ixodes ricinus and I. ventalloi ticks in France. *Archives of Virology* 82: 161–171.

Emmons RW (1988) Ecology of Colorado tick fever. *Annual Review of Microbiology* 42: 49–64.

Florio L, Miller MS, and Mugrage ER (1950) Colorado tick fever. Isolation of the virus from *Dermacentor andersoni* in nature and laboratory study of the transmission of the virus in the tick. *Journal of Immunology* 64: 257–263.

Karabatsos N, Poland JD, Emmons RW, et al. (1987) Antigenic variants of Colorado tick fever virus. *Journal of General Virology* 68: 1463–1469.

Lane RS, Emmons RW, Devlin V, Dondero DV, and Nelson BC (1982) Survey for evidence of Colorado tick fever virus outside of the known endemic area in California. *American Journal for Tropical Medicine and Hygiene* 31: 837–843.

Le Lay-Rogues G, Arthur CP, Vanderwalle P, Hardy E, and Chastel C (1990) Lapin de Garenne, *Oryctolagus cuniculus* L. et arbovirus dans le sud-est de la France résultats de deux enquêtes sérologiques. *Bulletins de la Société de Pathologies Exotiques* 83: 446–457.

Malkova D, Holubova J, Kolman JM, et al. (1980) Antibodies against some arboviruses in persons with various neuropathies. *Acta Virologica* 24: 298.

Monath TP and Guirakhoo F (1996) Orbiviruses and coltiviruses. In: Fields BN, Knipe DM, Howley PM, et al. (eds.) *Fields Virology*, 3rd edn., pp. 1735–1766. Philadelphia, PA: Lippincott–Raven.

Murphy FA, Coleman PH, Harrison AK, and Gary GW (1968) Colorado tick fever virus: An electron microscopic study. *Virology* 35: 28–40.

Rehse-Küpper B, Casals J, Rehse E, and Ackermann R (1976) Eyach, an arthropod-borne virus related to Colorado tick fever virus in the Federal Republic of Germany. *Acta Virologica* 20: 339–342.

# Common Cold Viruses

**S Dreschers and C Adams**, University of Duisburg–Essen, Essen, Germany

© 2008 Elsevier Ltd. All rights reserved.

## Glossary

**ASM** Acid sphingomyelinase (EC 3.1.4.12), enzyme cleaving sphingomyelin to ceramide and phosphatidylcholine under low pH conditions.

**ICAM-1** Intercellular adhesion molecule-1, belongs to the immunoglobulin superfamily.

**IRES** Internal ribosomal entry site; the sequence enables ribosomes to start translation within a mRNA.

**ITAM** Immunoreceptor-tyrosine based activation motif; enables proteins to bind to non-receptor tyrosine kinases.

**NPDA** Niemann–Pick disease type A; patients lack a functional ASM gene and suffer from a malfunctioning lipid storage due to the accumulation of sphingomyelin.

**MBCD** Methyl-β-cyclodextrin; drug depleting cholesterol out of membranes.

## Introduction

Everyone knows the first signs of a common cold. Symptoms usually begin 2–3 days after infection and often include obstruction of nasal breathing, swelling of the sinus membranes, sneezing, cough, and headache. Although the symptoms are usually mild, this (probably the most common illness known to man) exhibits an enormous economic impact due to required medical attention or restriction of activity. According to estimations of the National Center for Health Statistics (NCHS), the common cold caused 157 million days of restricted activity and 15 million days lost from work in 1992. The progress of molecular biology in the twentieth century allowed the isolation and *in vitro* culture of rhinoviruses as well as their ultrastructural and molecular analysis. From the beginning of recorded history to our time people have been interested in how the common cold is spread. Famous ancient investigators, among them Hippocrates and Benjamin Franklin to name two, were the first to suggest inhalation of contaminated air as the origin of infection. This observation still holds its validity, but touching infectious respiratory secretions on skin and on environmental surfaces (desktops, doorknob, etc.) and subsequently touching the eyes or nose is now believed the most common mode of transmission. This article aims to summarize the knowledge about rhinoviral infections and discusses mechanisms of internalization, activation of host-cell receptor molecules, and details of possible therapeutic strategies.

## Taxonomic Identification, Genome Organization, and Replication Cycle of Rhinoviruses

The establishment of cell and tissue culture techniques as well as nucleotide sequencing methods allowed for taxonomic classification of rhinoviruses. An overall RNA sequence compilation revealed a close genetic similarity to polio- and coxsackiviruses; hence, rhinoviruses were identified as members of the family *Picornaviridae*.

### Genome Organization

The genomic organization is similar in all genera of *Picornaviridae* (**Figure 1**). The 5′ end of the RNA is joined to a small viral protein called VPg, which is a prerequisite for the synthesis of the negative RNA strand as a template for replication. The genomes of *aphtho-, erbo-, kobu-, tescho-, cardioviruses* (e.g., the foot and mouth disease virus) contain an additional gene (leader) at the 5′ end of the RNA. The 5′ end also contains the internal ribosomal entry site (IRES) sequence and allows binding to the ribosomal subunits of the host translational machinery. Following the IRES sequence is a single long open reading frame (ORF) encoding the structural (or capsid forming) proteins designated as the P1 precursor and nonstructural proteins (proteases and RNA-dependent RNA polymerase) designated as the P2/P3 precursors(s) (see **Figure 1**). The RNA genome is completed by the 3′ nontranslated region and a poly-A tail.

## Translation of the Polyprotein, Processing, and Capsid Maturation

The protease 2A$^{pro}$ coded by the P2 precursor was found to have multiple functions. It is required for the inital proteolytic separation of the P1 and P2/P3 precursors in the nascent rhinoviral polyprotein, but also inactivates the cellular *eLF-4F* complex, thereby inhibiting the translation of the cellular mRNA and initiating the virus host shutoff (VHS – the translational machinery of the host cell is producing viral proteins only). Further steps of the post-translational maturation are catalyzed by the protease 3C$^{pro}$/3CD$^{pro}$ (see below). Products of the latter proteolytic activity are the VPg protein and the RNA polymerase. The capsid is composed of four proteins (viral proteins, VP1–VP4, see **Figure 1**), which arise from a single polyprotein generated from the viral RNA genome functioning as an mRNA. The very first cleavage occurs while translation is still in process. This first proteolytic maturation *in cis* is carried out by the protease 2A$^{pro}$, which separates structural proteins from those necessary for the replication (designated as P1 and P2/P3, see **Figure 1**). A series of further cleavages form VP1, VP3, and VP0; the latter is finally processed to result in VP2 and VP4. Many picornaviruses undergo VP0 cleavage during virion maturation. Apparently, intact VP0 is necessary for correct assembly of the protomers, whereas processing of VP0 is necessary for the final maturation of the virion. Both viral proteases are targets for an inhibitor screen (see below).

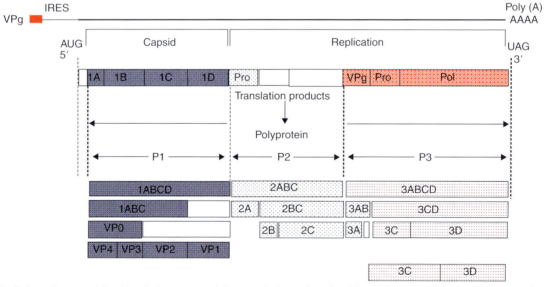

**Figure 1** Schematic map of the rhinoviral genome and the encoded proteins. The rhinoviral genome consists of the 5′ IRES sequence (thin black line), followed by one open reading frame (bold black line) and the polyadenylation site. At the 5′ end the VPg protein is linked covalently to the RNA. Below, the regions representing the rhinoviral polyprotein are given. Compounds forming the capsid (structural proteins P1, blue boxes) and nonstructural proteins (P2–P3) such as proteases and the polymerase (light-blue dotted boxes and red dotted boxes, respectively) are translated into one polyprotein. The polyprotein becomes (auto-) proteolytically cleaved until the mature proteins are ready to form new viral particles.

## Genome Replication

The following steps of rhinoviral reproduction probably do not differ from other members of the *Picornaviridae*. The RNA genome can be instantly translated; however, for replication of the genome a complementary (−) strand must be synthesized first. Essential for this step is a priming reaction, which is started by the VPg precursor 3AB. The hydrophobic portion of this protein is anchored in endomembranes and is linked to a poly-U tail. After formation of an RNA:RNA duplex at the 3′ end of the (+) strand, the polymerase synthesizes new (−) RNA strands. Only 5–10% of all rhinoviral RNAs are (−) strands. The synthesis of a complete genome takes 45 s to a minute, underlining that the reproduction cyle of picornaviruses in general and of rhinoviruses in particular is a fast process.

Synthesis of rhinoviral RNA in nasal washings could be detected 12 h post infection by quantitative reverse transcription-polymerase chain reaction (RT-PCR). The production of viral RNA and viral particles peaks within 48–72 h and declines thereafter.

## Attachment and Entry

Studying any viral reproduction prompts the question of how viruses attach and invade cells. Therefore, the identification of cellular receptors binding to rhinoviruses was a landmark in examining how the infection cycle works. Intercellular adhesion molecule-1 (ICAM-1; CD54) was first detected as the receptor for major group rhinoviruses and later very low density lipoprotein (V)LDL as the receptor for minor group rhinoviruses (**Figure 2**). The receptors were isolated by using monoclonal antibodies directed against cellular surface proteins (see below). The receptors for the minor and major group rhinoviruses are structurally nonrelated, and the molecular reason why rhinoviruses as well as other members of the *Picornaviridae* (e.g., coxsackieviruses) use different receptors is not well understood. One strain (CAV-A21) of the rhinovirus-related family of coxsackieviruses needs to engage both ICAM-1 and decay-accelerating factor (DAF; CD55), whereas other coxsackieviruses require DAF or integrins for binding. HRV binding to ICAM-1 allows a conformational change of the rhinoviral capsid and is a prerequisite for the uncoating of viral RNA. Although binding to (V)LDL receptors is also important for the internalization, the acidification of the endosome seems to be indispensable for proper entry, in particular for minor group rhinoviruses. (V)LDL receptors are well-described molecules, and the uptake of their natural ligands via clathrin-coated vesicles suggested that minor group rhinoviruses enter the host cell via the same pathway. Microscopical studies have detailed the uptake of the minor group rhinovirus HRV2. This viral serotype was shown

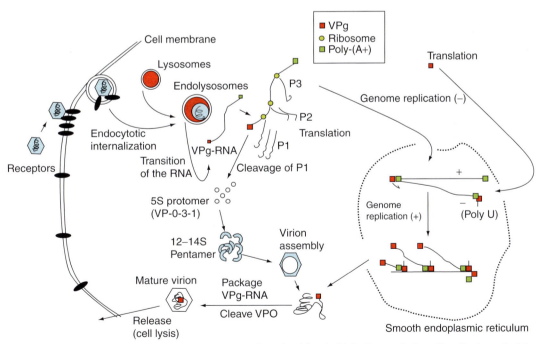

**Figure 2** Infection cycle of rhinoviruses. The scheme summarizes the rhinoviral infection cycle from the attachment of the capsid to the release of newly synthesized capsids. Shown are the internalization of the capsids via endocytosis, the transport to the endolysosmal compartment, transition of the RNA to the cytoplasm, synthesis of viral RNA within the smooth endoplasmic reticulum, maturation of the capsid proteins, and self-assembly.

to cluster to membrane domains positive for clathrin, supporting the notion of an uptake mechanism depending on clathrin-coated vesicles. In line with this finding is the blockade of HRV2 uptake by overexpression of dynamin-K44, a transdominant negative isoform of dynamin, which is required for pinching off clathrin-coated vesicles from the plasma membrane. Uptake of rhinoviruses via clathrin-coated vesicles is very fast and does not take longer than 30 min. Moreover, preincubation with methyl-β-cyclodextrin (MBCD) interrupts the uptake of HRV2 at an early stage. Rhinoviral capsids pile up close to the plasma membrane and are not transported to the endolysosomal compartment in cells treated with MBCD. MBCD depletes cholesterol from cellular membranes and leads to destabilization of membrane domains which are thought to play a central role in signal transduction and endocytosis. Recent work indicates that another subgroup of membrane domains termed ceramide-enriched membrane platforms plays a crucial role in the infection cycle of rhinoviruses. These studies demonstrate that pharmacological or genetic inhibition of ceramide-enriched platforms reduces rhinoviral titers in epithelial cells. Membrane domains may not only be involved in the first steps of internalization, but also play a role in transport of vesicles and fusion of transport vesicles, endosomes, and endolysosomal vesicles, finally triggering acidification of the vesicle which is required for uncoating of rhinoviruses and transition of the RNA to the cytoplasm (see below).

## Virion/Particle Structure and Phylogenetic Considerations

Human rhinoviruses (HRVs) represent the serologically most diverse group of picornaviruses (over 100 serotypes have already been identified). The reason for this diversity is not known.

### The Canyon Hypothesis

Ultrastructural analysis of rhinoviruses by X-ray crystallography resulted in a structural model of rhinoviruses (**Figure 3**). Along the fivefold icosahedral axis of the virion a 2.5 nm depression, the so-called 'canyon', was detected. The genomic region encoding the peptide residues that form this structure is more conserved than regions encoding any other structure on the virion surface. Studies using mutated viral strains showed that point mutations in the genomic region covering the capsid protein VP1 changed the affinity of radiolabeled mutant viruses to purified host cell membranes (**Figure 4** and **Table 1**). Alterations of amino acids at position 103 (Lys),

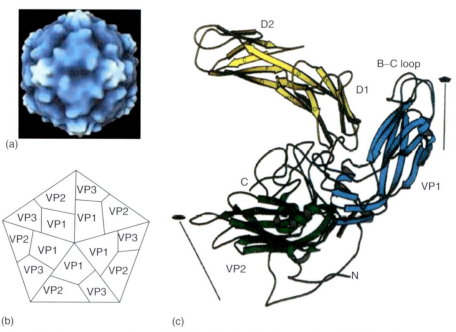

**Figure 3** Structural model of rhinoviruses and receptor interaction. (a) The structural model of rhinoviruses was evaluated from cryo-electron microscopic data, the schematic drawing (b) represents the position of the capsid proteins VP1–VP3 in the rhinoviral capsid. (c) Schematic drawing of the viral canyon structure (represented by the capsid compounds VP1 and VP2 given in blue and green, respectively) and the D1 and D2 domains of the receptor ICAM-1. The antiparallel β-strand structures of VP1, VP2, and ICAM-1 are listed in alphabetical order.

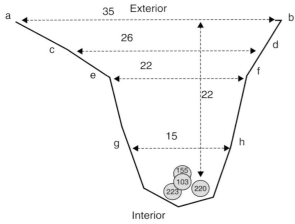

**Figure 4** Mutations in the rhinoviral protein VP1 interfere with attachment and growth. The figure displays a scheme of the HRV-14 canyon. Distances are shown in angstrom and circled numbers represent the four amino acids targeted in the studies. Coordinate points used in determining distances are as follows: a, $O^{\delta 1}$ of Asp-91; b, $O^{\epsilon 2}$ of Glu-210; c, $O^{\delta 1}$ Asn–92; d, O Glu–210; e, $O^{\epsilon 2}$ Glu–95; f, $N^{\epsilon 2}$ Gln–212; g, $O^{\delta 1}$ Asp–101; h, $C^{\delta 1}$ Ile–215. Modified from Colonno RJ, Condra JH, Mizutani S, Callahan PL, Davies ME, and Murcko MA (1988) Evidence for the direct involvement of the rhinovirus canyon in receptor binding. *Proceedings of the National Academy of Sciences, USA* 85(15): 5449–5453.

**Table 1** Mutations in the rhinoviral protein VP1 interfere with attachment and growth

| Position of wt amino acids | Substitutions | Virus yield (pfu/cell) |
|---|---|---|
| 103 Lys | — | 160 |
|  | Ile | 84 |
|  | Arg | 54 |
|  | Asn | 140 |
| 155 Pro | — | 160 |
|  | Gly | 1 |
| 220 His | — | 160 |
|  | Ile | 10 |
|  | Trp | 14 |
| 223 Ser | — | 160 |
|  | Ala | 8 |
|  | Thr | 70 |
|  | Asn | 88 |

155 (Pro), 220 (His), or 223 (Ser), had the strongest effect regarding the binding affinity of rhinoviral capsids. The structural analysis revealed a juxtaposition of these amino acids at the 'bottom of the canyon'. It is noteworthy that most of the substitutions reduced the binding of capsids to host cell membranes; however, the biochemical property of one amino acid (Pro 155) is of special importance because a change to glycine led to a higher binding affinity and an enhanced yield of mutated virus. Most publications concerning rhinoviral structure described members of the major group rhinoviruses, while only few publications dealt with the structure of minor group rhinoviruses. It was possible to define structural differences between HRV2 (representing minor group rhinoviruses) and HRV14 (representing major group viruses) by analyzing a special subportion of the 'canyon', the 'pocket' region within the VP1 protein. This region binds in particular to hydrophobic compounds – for example, fatty acids supplemented by membranes of the host cell. The interaction of the virus with the fatty acid seems to stabilize the virus during its spread from cell to cell, but it must be removed prior to the uncoating process. For major group viruses the fatty acid is displaced by ICAM-1. The minor group viruses exhibit a higher affinity for the fatty acids and binding of LDL-R is insufficient to expel the fatty acid in the 'pocket' of VP1. The conformational change of the viral capsid is obligatory for the transition of the viral RNA into the cytoplasm. Acidification of the endolysosomal compartment transfers the native virus to a so-called C-antigenic subviral particle in which VP1 forms a channel with a diameter around 10 A – large enough for the RNA to escape. Concomitantly, the N-terminus of VP1 and the entire VP4 are extruded and interact with the endolysosomal membrane. Studies with rhinovirus-related poliovirus identified a 3 kDa peptide from the N-terminus of VP1 capable of integrating into artificial liposomes.

## Rhinoviral Antigenicity

Antibodies that neutralize rhinoviruses *in vitro* and *in vivo* permitted the identification of the epitopes in the viral capsid critically important for binding to the receptors and/or release of the RNA into the cytoplasm. Antigenicity in HRVs has been extensively studied in serotypes HRV2 and HRV14. In HRV14 four antigenic sites were identified, designated NIm-IA, NIm-IB, NIm-II, and NIm-III. The sites could be accurately located at the external surface of the virus. In HRV2 three antigenic sites, called A, B, and C could also be mapped. Antigenic site A is located within the B–C loop of VP1 that flanks the rim of the canyon structure, thus, corresponding to the NIm-IA, NIm-IB site of HRV14. Site B encompasses residues from VP1, VP2, and VP3. VP2 residues of site B comprise a continuous epitope that is equivalent to the NIm-II site of HRV14 and mainly defined by the VP2 loop between the β-barrel strands E and F. Antigenic site C includes two amino acids of VP2 which do not correspond to any of the immunogenic sites of major group rhinoviruses.

## Phylogenetic Considerations

In addition to the classification based on serotyping, rhinoviruses can be grouped in different ways, on the basis of shared structural and biological properties (e.g., receptor specificity), sensitivity to antiviral agents or at least genetic similarity. The phylogenetic analysis of the genomic VP4/VP2 interval of 97 different rhinoviral serotypes resulted in two phylogenetic clades, A and B. Surprisingly, some of the 76 serotypes of class A were identified as minor group rhinoviruses such as HRV 1A, 2, etc., whereas class B consists of 25 serotypes, among them major group HRV 14. According to the phylogenetic analysis, two rhinoviral serotypes (HRV 23 and HRV 25) which bind to ICAM-1 (major group) constitute an 'intermediate' class between A and B, respectively. Further studies showed that soluble ICAM-1 could not compete with viral binding to cellular ICAM-1. Furthermore, the phylogenetic analysis identified one serotype, namely HRV 87, as closely related to another *Picornaviridae* genus the enteroviruses, namely enterovirus 70. HRV 87 does not bind to either ICAM-1 or (V)LDL receptors, but to sialoproteins, which also bind enterovirus 70. Although HRV 87 is acid-labile like all other rhinovirus serotypes, it appears to belong to the enterovirus species D. In general, the analysis shows a greater phylogenetic heterogeneity for minor group rhinoviruses than for major group rhinoviruses, which may indicate that minor group serotypes evolved more recently.

## Rhinovirus Infection Models

Rhinoviruses show an almost strict specificity for human cells. Most studies, therefore, have to be carried out in human volunteers or, alternatively, in primates, which necessitates special requirements regarding biosafety of the rhinoviral stock preparations, and raises ethical questions. Some publications reported that host restriction could be overcome by using a method to select rhinovirus strains which are able to infect murine cells. Another possible technique to overcome the host range problem is by expression of human/murine chimeric rhinoviral receptors. Engineered ICAM-1 molecules, stably expressed in murine cell lines, bind rhinoviruses and permit the uptake in murine respiratory cells and L-cells. The latter study underlined that mutations in noncapsid proteins (genomic region P2) are necessary to adapt viral proteins to the new host cell and finally allow an appropriate replication in murine cells. Other researchers developed human ICAM-1 transgenic mice that were recently used in infection studies with coxsackieviruses, and might also be utilized to study rhinoviral infections. Interestingly, a common finding of the field is noteworthy: receptors such as ICAM-1, VCAM-1, and also the (V)LDL play a predominant role in rhinovirus binding, uptake, and associated signal transduction processes. However, these data were determined from infection experiments employing poorly differentiated epithelial cells in culture. In contrast, these molecules are only infrequently expressed *in vivo* on the apical surface, leading to the question of how viruses access these receptors *in vivo*. In addition, the issue whether epithelial cell layers are the primary targets of rhinoviruses remains to be determined, since infections do not result in cytopathic effects in the ciliated epithelial cells as shown for many other respiratory viruses. Many studies focused primarily on the detection of rhinovirus RNA in rhinopharyngeal tissues by means of RT-PCR or *in situ* hybridization showed that cells and tissues (e.g., the germinal layer in the adenoid tissue below the epithelial layer) are positive for rhinoviral RNA. Rhinoviruses can be cultured in WI-38 fibroblasts suggesting that fibroblasts in the rhinopharyngeal tissue can be targets for rhinoviral infection. It is also possible that rhinoviruses attach to and invade the epithelial layer by a mechanism different from that observed in cultured epithelial cells, that is, binding to ICAM-1 and (V)LDL receptors. Further studies are required to address these questions.

## Host Cell Reactions upon Rhinoviral Infections – The Molecular Origin of Symptoms

Infection of the upper respiratory tract by HRV16 has been demonstrated to occur at localized portions of the epithelia and does not cause widespread lysis of the infected epithelia. This observation has suggested that the pathology induced by rhinovirus may be due in part to cytokine dysregulation rather than extensive epithelial necrosis observed in other viral infections.

HRV, as well as other respiratory pathogens, have been shown to induce production of many cytokines and chemokines, among them IL-4, IL-6, IL-8, and granulocyte macrophage colony stimulating factor (GMCSF), to name the most frequently published. The induction of chemokines triggers the inflammatory response, which is considered to enhance exacerbations like asthma and chronic obstructive pulmonary disease (COPD). A common feature of asthma and COPD exacerbations is a high level of IL-8 and an increased number of neutrophils in the sputum and nasal secretions of patients suffering an HRV infection.

Little is known, however, about the primary signaling pathways initiated by the rhinoviral infection cycle. Because signal transduction pathways of the LDL receptor family are difficult to understand, the experimental design concentrated on ICAM-1, the receptor for major group rhinoviruses. ICAM-1-inhibiting antibodies abrogated IL-8 production, but infection with UV-inactivated rhinovirus failed to do so, indicating that binding and entry events induced the chemokine response rather than causing effects resulting from viral replication.

HRV ligation to ICAM-1 provokes recruitment of the tyrosine kinase syk to membrane domains in human airway epithelial cells. The tyrosine kinase syk belongs to the family of src receptor tyrosine kinases. The immunoreceptor tyrosine-based activation motif (ITAM) of syk recruits another protein, ezrin, and the syk–ezrin complex finally binds to ICAM-1. Autophosphorylation of syk activates the MAP-kinase p38-K in a biphasic manner, that is, after 1 h and 8–12 h after infection with major group rhinovirus HRV14. This finding is especially interesting because the induction of chemokines is temporally correlating with the second p38-K activation peak. The PI-3 kinase is related to the family of MAP kinases. Again, subsequent studies concentrated on very early phases (5–60 min post infection) of rhinoviral infection. Active PI-3 kinase could be precipitated from lysates of infected cells and isolated plasma membranes showed the presence of metabolic products (i.e., PI $(3,4,5)P_3$) of the PI-3 kinase. The same studies identified Akt and nuclear factor kappa B (NF-κB) as downstream targets of PI-3 kinase and thus completing this signal transduction pathway showed the binding of NF-κB to the IL-8 promoters.

Interestingly, the activation of PI-3 kinase is functionally connected to the HRV entry. Inhibition of PI-3 kinase by application of the drug Wortmannin reduced the uptake of radiolabeled HRV, due to the retention of HRV2 in endosomal compartments, which results in a reduced transition of the RNA and, therefore, a reduced viral titer.

Inflammatory reactions require effector cells such as macrophages and monocytes. A recent publication addressed the question if the production of chemokines is restricted to the nasal epithelial tissues or if chemokines are synthesized and secreted by lymphocytes. Therefore, alveolar macrophages and monocytes were infected with rhinovirus. After infection with HRV16 the production of the macrophage attractive protein-1 (MCP-1) was shown to be increased in monocytes and macrophages. Surprisingly, HRV16 induces the same signal transduction pathways in lymphocytes as in epithelial cells. Again, the p38-K is activated and engages the nuclear factor ATF-2 involving NF-κB. The signal transduction could be attenuated by preincubation of the lymphocytes with ICAM-1, suggesting that HRV can bind to and enter lymphocytes. However, propagation of rhinoviruses in lymphocytes was not an aim of this study. Finally, these results present the novel idea that cell types other than epithelial cells contribute to viral exacerbations of asthma as well as causing the symptoms of the common cold.

## Antirhinoviral Strategies

Although the common cold is usually mild, with symptoms lasting a week or less, it is a leading cause of physician visits and of school/job absence in addition to several more severe exacerbations like asthma (see above). Initial studies that employed monoclonal antibodies to define candidates of viral receptors demonstrated a surprising 'immunological resistance' of rhinoviruses. On the one hand this is due to the fact that over 100 different rhinoviral serotypes have been isolated and antibodies raised against one serotype after immunization are not necessarily protective against a different serotype. However, the identification of the so-called 'canyon' allowed an additional explanation: if this 'canyon' represents the predominant site of receptor interactions, it is difficult to block with antibodies since the canyon is too narrow to enable interactions with the antigen. Both findings seem to rule out vaccination as a standard antirhinoviral treatment. Hence, the canyon structure dominated the development of therapeutic strategies (**Table 2**). Researchers were able to

Table 2  Substances interfering with the rhinoviral propagation

| Tested compound | Functions in or interferes with |
| --- | --- |
| WIN factors | Interaction with the canyon structure, viral adhesion is suppressed |
| AG 7088 | Blocks protease 3 C irreversibly and stops capsid maturation |
| Soluble ICAM-1 | Competition with ICAM-1 |
| Cytochalasin B | Interference with microfilaments and inhibition of vesicle maturation |
| Bafilomycin | Blocking endosomal acidification |
| Erythromycin | Reduction of ICAM-1 receptor, blocking endosomal acidification |
| Impramine | Blocking ASM and inhibiting ceramide-enriched platforms |
| Desipramine | Blocking ASM and inhibiting ceramide-enriched platforms |
| Nocodazole | Interference with macrofilaments and inhibiting vesicle maturation |
| zVAD | Antiapoptotic agent blocking caspases and release of new viruses |
| Wortmannin | PI-3-kinase inhibitor, blocks vesicle maturation |
| MBCD | Depletes cholesterol, destroys functional membrane domains, transport of virus is blocked |

design chemical structures designated as 'pocket factors', interacting with amino acid residues in the canyon. Because of their chemical similarity to the primarily described drug WIN51711, an isoxazole derivative, such chemicals are generally termed WIN compounds. WIN compounds seem to function both as competitors for binding to the receptor and in a steric blockage interfering with the conformational change of the capsid necessary for the release of the genomic RNA. While WIN compounds have a significant effect in reducing the viral titer *in vitro*, the *in vivo* application was difficult. Structurally related substances like flavinoids, for example 4–6 Dichlorflavane, exhibited a better feasibility regarding application, but caused only little reduction of rhinovirus titers. The chemokines interferon-$\alpha/\beta$ are produced after viral infection. Host cells detect the presence of double-stranded RNA via activation of the toll-like receptor 3 (TLR-3), and the main function of interferons is to induce IFN-stimulated genes, which attenuate viral replication. Some groups tested the application of interferons and interferon-inducers, respectively, and their interference with rhinoviral replication and production. Although high doses of interferons applied intranasally exhibited a prophylactic effect, these high doses of interferons required for the protective effect caused local erosions and nasal stuffiness.

Therefore, the interest shifted to alternative methods to attenuate or even to block rhinoviral infections and their symptoms. Even though vaccination seemed to be impossible (as pointed out before), some studies dealt with the question of how the human immune system responds to a rhinoviral infection. Rhinovirus inoculation induced both IgG and IgA presence in the serum and the airway, respectively, which neutralized the identical serotype upon reinfection. In studies using several different rhinovirus serotypes the generation of cross-reacting antibodies could be observed. Individuals with preexisting cross-reacting antibodies or cross-reacting antibodies derived from rabbits exhibited a partial immunity when challenged with any new rhinovirus serotype. However, a complete immunity is barely to be expected, because many studies showed that antirhinovirus-protecting IgA titers decreased already 2 months after infection. Nevertheless, a vaccination could attenuate the exacerbations of the infection, that is, reduce asthma and COPD.

Recent work indicates that lipid domains in the plasma membrane of infected cells are functional in signal transduction events caused by rhinoviral infections. Moreover, the lipid composition could influence the interaction of viral proteins with vesicular membranes during the transition of the RNA and replication. For example, rhinovirus uptake is sensitive to treatment with MBCD and blocks p38-K signaling, thus interrupting the release of chemokines and attenuate symptoms. The acid sphingomyelinase (ASM), which was shown to be crucial for generation of ceramide-enriched platforms, a subentity of lipid domains, might be an additional target to prevent HRV infections. Fibroblasts isolated from NPDA patients (Niemann-Pick disease type 1) lacking functional ASM were shown to be resistant to HRV infections. The ASM can be inactivated pharmaceutically by tricyclic antidepressants such as amitryptiline and imipramine. Studies in which the ASM was inhibited by these substances showed a dramatic decrease of viral reproduction, suggesting that the ASM is critically involved in the propagation of rhinoviruses in human cells. Drugs derived from the structure of tricyclic antidepressants might therefore be effective against rhinovirus infections.

Aside from antiviral drugs targeting the attachment and entry processes, antirhinovirus therapy could benefit from the identification of chemicals inhibiting viral maturation. As pointed out earlier, activation of protease 3C is a crucial step in the assembly of the capsid. Tripeptidyl alpha-ketoamides were identified as human rhinovirus protease 3C inhibitors. The protease 3C-inhibiting drug AG7088 (*ruprintivir*) showed a 100-fold reduction of rhinoviral titers and abrogated inflammatory responses as well. Clinical trials indicated that AG7088 could be administered to volunteers without adverse reactions. Subsequent studies are in progress, aiming at another rhinoviral protease (i.e., 2A) as a target for antirhinoviral therapy.

*See also:* Enteroviruses: Human Enteroviruses Numbered 68 and Beyond; Enteroviruses of Animals; Picornaviruses: Molecular Biology.

## Further Reading

Colonno RJ, Condra JH, Mizutani S, Callahan PL, Davies ME, and Murcko MA (1988) Evidence for the direct involvement of the rhinovirus canyon in receptor binding. *Proceedings of the National Academy of Sciences, USA* 85(15): 5449–5453.

Olson NH, Kolatkar PR, Oliveira MA, et al. (1993) Structure of human rhinovirus complexed with its receptor molecule. *Proceedings of the National Academy of Sciences USA* 90: 86–93.

Rossman MG (1989) The canyon hypothesis. Hiding the cell receptor attachment site on a viral surface from immune surveillance. *Journal of Biological Chemistry* 264: 14587–14590.

Rossman MG, Arnold E, Erickson JW, et al. (1985) Structure of human common cold virus and functional relationship to other picornaviruses. *Nature* 317: 145–153.

Rossman MG, Bella J, Kolatkar PR, et al. (2000) Cell recognition and entry by rhino- and enteroviruses. *Virology* 269: 239–247.

Verdaguer N, Blaas D, and Fita I (2000) Structure of human rhinovirus serotype 2 (HRV2). *Journal of Molecular Biology* 300: 1179–1194.

Wang X, Lau C, Wiehler S, et al. (2006) Syk is downstream of intercellular adhesion molecule-1 and mediates human rhinovirus activationof p38 MAPK in airway epithelial cells. *Journal of Immunology* 177: 6859–6870.

# Coronaviruses: General Features

**D Cavanagh and P Britton,** Institute for Animal Health, Compton, UK

© 2008 Elsevier Ltd. All rights reserved.

## Glossary

**Infectious clone** A full-length DNA copy of an RNA virus genome from which full-length viral RNA can be generated, leading to production of infectious virus.

**Nidovirales (nidoviruses)** An order comprising positive-sense RNA coronaviruses, toroviruses, arteriviruses, and roniviruses that have a common genome organization and expression, similar replication/transcription strategies, and form a nested set of 3′ co-terminal subgenomic mRNAs (*nidus*, Latin for nest).

**Ribosomal frameshifting** Movement (shift) backward by one nucleotide of a ribosome that is on an RNA, caused by particular RNA structures and sequences. Subsequent continuation of the progress of the ribosome is in a different open reading frame.

## Introduction

Coronaviruses are known to cause disease in humans, other mammals, and birds. They cause major economic loss, sometimes associated with high mortality, in neonates of some domestic species (e.g., chickens, pigs). In humans, they are responsible for respiratory and enteric diseases. Coronaviruses do not necessarily observe species barriers, as illustrated most graphically by the spread of severe acute respiratory syndrome (SARS) coronavirus among wild animals and to man, with lethal consequences. As a group, coronaviruses are not limited to particular organs; target tissues include the nervous system, immune system, kidney, and reproductive tract in addition to many parts of the respiratory and enteric systems. A great advance in recent years has been the development of systems ('infectious clones') for modifying the genomes of coronaviruses to study all aspects of coronavirus replication, and for the development of new vaccines.

## Taxonomy and Classification

The genus *Coronavirus* together with the genus *Torovirus* form the family *Coronaviridae*; members of these two genera are similar morphologically. The *Coronaviridae*, *Arteriviridae*, and *Roniviridae* are within the order *Nidovirales*. Members of this order have a similar genome organization and produce a nested set of subgenomic mRNAs (*nidus*, Latin for nest). To date, coronaviruses have been placed into one of three groups (**Table 1**). Initially, this was on the basis of serological relationships which subsequently have been supported by gene sequencing.

## Virion Properties

Virions have a buoyant density of approximately $1.18 \text{ g ml}^{-1}$ in sucrose. Being enveloped viruses (**Figure 1(a)**), they are destroyed by organic solvents such as ether and chloroform.

## Virion Structure and Composition

All coronaviruses have four structural proteins in common (**Figure 1(b)**): a large surface glycoprotein (S; *c.* 1150–1450 amino acids); a small envelope protein (E; *c.* 100 amino acids, present in very small amounts in virions); integral membrane glycoprotein (M; *c.* 250 amino acids); and a phosphorylated nucleocapsid protein (N; *c.* 500 amino acids). Group 2a viruses have an additional structural glycoprotein, the hemagglutinin-esterase protein (HE; *c.* 425 amino acids). This is not essential for replication *in vitro* and may affect tropism *in vivo*.

Virions are *c.* 120 nm in diameter, although they can be up to twice that size, and the ring of S protein spikes is approximately 20 nm deep. When present, the HE protein forms a layer 5–10 nm deep. In some species, the S protein is cleaved into two subunits, the N-terminal S1 fragment being slightly smaller than the C-terminal S2 sequence. The S protein is anchored in the envelope by a transmembrane region near the C-terminus of S2. The functional S protein is highly glycosylated and exists as a trimer. The bulbous outer part of the mature S protein is formed largely by S1 while the stalk is formed largely by S2, having a coiled-coil structure. S1 is the most variable part of the S protein; some serotypes of IBV differ from one another by 40% of S1 amino acids. S1 is the major inducer of protective immune responses. Variation in the S1 protein enables one strain of virus to avoid immunity induced by another strain of the same species.

The M glycoprotein is the most abundant protein in virions. In most cases, only a small part (~20 amino acids) at the N-terminus protrudes at the surface of the virus. There are three membrane-spanning segments and the C-terminal half of the M protein is within the lumen of

**Table 1** Species of coronavirus[a]

| Group 1 | Group 2 | Group 3 |
|---|---|---|
| *Subgroup 1a* | *Subgroup 2a* | |
| *Transmissible gastroenteritis virus* (TGEV) | *Murine hepatitis virus* (MHV) | *Infectious bronchitis virus* (IBV) |
| *Feline coronavirus* (FCoV) | *Bovine coronavirus* (BCoV) | Turkey coronavirus (TCoV) |
| *Canine coronavirus* (CCoV) | *Porcine haemagglutinating encephalomyelitis virus* (HEV) | Pheasant coronavirus (PhCoV) |
| | Equine coronavirus (EqCoV) | |
| Ferret coronavirus (FeCoV) | Canine respiratory coronavirus (CRCoV) | |
| | Human coronavirus HKU1 | |
| *Subgroup 1b* | | Duck coronavirus[b] (DCoV) |
| *Human coronavirus* (HCoV) *229E* | Human coronavirus (HCoV) OC43 | Goose coronavirus (GCoV) |
| *Porcine epidemic diarrhoea virus* (PEDV) | Human enteric coronavirus (HECoV) | Pigeon coronavirus (PiCoV) |
| | Rat coronavirus (RCoV) | |
| | Puffinosis coronavirus | |
| Bat coronavirus-61 | | |
| Bat coronavirus-HKU2 | | |
| Human coronavirus NL63 | *Subgroup 2b* | |
| | *Severe acute respiratory syndrome coronavirus* (SARS-CoV) | |
| | Bat-CoV-HKU3-1 | |

[a]Recognized virus species.
[b]The viruses duck coronavirus, goose coronavirus, and pigeon coronavirus have been recommended by the Coronavirus Study Group for recognition as species by the ICTV.
Official virus species names are in italics. Coronaviruses that have not yet been recognized as distinct species have names that are not italicized. Rabbit coronavirus is considered as a tentative member of the genus. A coronavirus has been isolated from a parrot. On the basis of very limited sequence data, it is not clear in which, if any, of the three groups that this virus would be placed.

the virus. In transmissible gastroenteritis virus (TGEV), a proportion of M molecules have four membrane-spanning segments, resulting in the C-terminus also being exposed on the outer surface of the virus (M′ in **Figure 1(b)**). The E protein is anchored in the membrane by a sequence near its N-terminus.

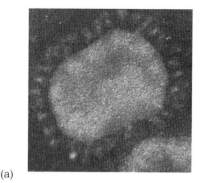

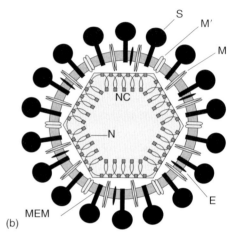

**Figure 1** (a) Electron micrograph of an IBV virion, showing the bulbous S protein. (b) Diagrammatic representation of the composition and structure of a coronavirus virion: S, spike glycoprotein; M, M′, integral membrane glycoprotein; E, small envelope protein; N, nucleocapsid protein; NC, nucleocapsid (nucleoprotein) comprising the RNA genome and N protein. Cryoelectron microscopy of TGEV has indicated a core structure comprising the NC and the M protein. Two forms of M protein (M, M′) have been observed for TGEV (see main text). The coronavirus membrane proteins, S, E, M, and M′, are inserted into a lipid bilayer (MEM) derived from internal cell membranes. (b) Reproduced from González JM, Gomez-Puertas P, Cavanagh D, Gorbalenya AE, and Enjuanes L (2003) A comparative sequence analysis to revise the current taxonomy of the family *Coronaviridae*. *Archives of Virology* 148: 2207–2235, with permission from Springer-Verlag.

## Genome Organization and Expression

Coronaviruses have the largest known RNA genomes, which comprise 28–32 kb of positive sense, single-stranded RNA. The overall genome organization is being 5′ UTR–polymerase gene–structural protein genes–3′ UTR, where the UTRs are untranslated regions (**Figure 2**). The first 60–90 nucleotides at the 5′ end form a leader sequence. The structural protein genes are in the same order in all coronaviruses: (HE)–S–E–M–N. Interspersed among these genes are one or more gene (depending on the species; SARS-CoV has four) that encode small proteins of unknown function. Some of these genes encode two or three proteins. In some cases (e.g., gene 3 of IBV and gene

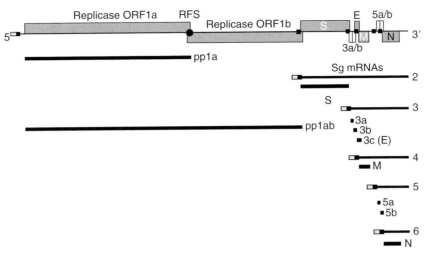

**Figure 2** Schematic diagram representing the genomic expression of the avian coronavirus IBV. The upper part of the diagram shows the IBV genomic RNA, with the various genes highlighted as boxed regions. The black boxes represent the transcription regulatory sequences (TRSs) found upstream of each gene and direct the synthesis, via negative-sense counterparts, of the sg mRNAs (2–6 for IBV). The leader sequence, represented by a gray box, is at the 5′ end of the genomic RNA and at the 5′ ends of the sg mRNAs. The genomic RNA is translated to produce two polyproteins, pp1a and pp1ab, that are cleaved by virus-encoded proteases to produce the replicase proteins. The structural proteins, S, E, M, and E, and the accessory proteins, 3a, 3b, 5a, and 5b, produced from IBV genes 3 and 5, respectively, are translated from the sg mRNAs. The proteins produced by the sg mRNAs are represented by lines below the corresponding sg mRNA. All of the sg mRNAs, except the smallest species, are polycistronic but only produce a protein from the 5′-most gene. The ribosome frameshift (RFS) region, denoted as a black circle on the genomic RNA, directs the −1 frameshift event for the synthesis of pp1ab. Translation of the genomic RNA results in the production of pp1a. However, the translating ribosomes undergo the −1 frameshift about 30% of the time resulting in pp1ab. The 5′ and 3′ UTR sequences are represented as single lines downstream of the leader and N gene sequences, respectively.

5 of murine hepatitis virus (MHV)), translation of the third and second open reading frame (ORF), respectively, is effected by the preceding ORFs acting as internal ribosome entry sites. The proteins encoded by these small ORFs are mostly not required for replication *in vitro*; some of them might function as antagonists of innate immune responses, though this has not yet been demonstrated.

Following entry into a cell and the release of the virus ribonucleoprotein (genome surrounded by the N protein) into the cytoplasm, ribosomes translate gene 1, which is approximately 20 kb, into two polyproteins (pp1a and pp1ab). These are cleaved by gene 1-encoded proteases, to generate 15 or 16 proteins (**Figure 3**). Translation of ORF 1b involves ribosomal frameshifting, which has two elements, a slippery site followed by an RNA pseudoknot. At the slippery site (UUUAAAC in IBV), the ribosome slips one nucleotide backward and then moves forward, this time in a −1 frame compared with translation ORF 1a, resulting in the synthesis polyprotein 1ab.

Proteins, including the RNA-dependent RNA polymerase, from gene 1 associate to form the replicase complex, which is membrane associated. Coronavirus subgenomic mRNAs are generated by a discontinuous process. At the beginning of each gene is a common sequence (CUUAACAA in the case of IBV) called a transcription regulatory sequence (TRS). It is believed that when the polymerase producing the nascent negative sense RNA, reaches a TRS, RNA synthesis is attenuated, followed by continuation at the 5′ end of genomic RNA. This results in the addition of a negative copy of the leader sequence to the negative-sense RNA, resulting in a negative-sense copy of an sg mRNA. Of course, progress of the polymerase is not always halted at a TRS. Rather, it sometimes continues, producing a nested set of negative-sense sg mRNAs. These are the templates for the generation of the positive-sense sg mRNAs (**Figure 2**). The amount of each sg mRNA does not necessarily decrease in a linear fashion; the efficiency of termination by a TRS is dependent on adjacent sequences, which are different for each gene. The leader sequence is found at the very 5′ end of the genomic RNA and at the 5′ ends of each sg mRNA.

## Replication Cycle

The N-terminal (S1) part of the S protein mediates that mediates attachment to cells. It is a determinant of host species specificity and, in some cases, pathogenicity, by determining susceptible cell range (tissue tropism) within a host. The C-terminal S2 part triggers fusion of the virus envelope with cell membranes (plasma membrane or endosomal membranes), which can occur at neutral or slightly acidic pH, depending on species or even strain. The virus glycoproteins (S, M, and HE, when present) are synthesized at the endoplasmic reticulum. Both subunits

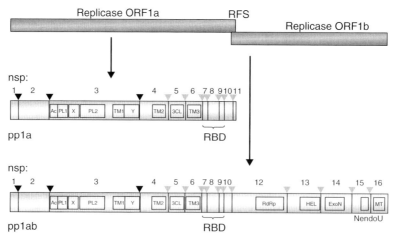

**Figure 3** Organization of the coronavirus replicase gene products. Translation of the coronavirus replicase ORF 1a and ORF 1b sequences results in pp1a and pp1ab; the latter is a C-terminal extension of pp1a, following a programmed −1 frameshift event (see legend to **Figure 2**). The two polyproteins are proteolytically cleaved into 10 (pp1a; nsp1–11) and 16 (pp1ab; nsp1–16) products by the papain-like proteinases (PL1$^{pro}$ and PL2$^{pro}$) and the 3C-like (3CL$^{pro}$) proteinase. The PL$^{pro}$ proteinases cleave at the sites indicated with a black triangle and the 3CL$^{pro}$ proteinase cleaves at the sites indicated with a gray triangle. The nsp11 product of pp1a is produced as a result of the ribosomes terminating at the ORF 1a translational termination codon, a −1 frameshift results in the generation of nsp12, part of the pp1ab replicase gene product. Various domains have been identified within some of the replicase products: Ac is a conserved acidic domain; X = ADP-ribose 1′-phosphatase (ADRP) domain; PL1 and PL2 the two papain-like proteinases; Y is a conserved domain; TM1, TM2, and TM3 are conserved putative transmembrane domains; 3CL = 3CL$^{pro}$ domain; RdRp, RNA-dependent RNA polymerase domain; HEL, helicase domain; ExoN, exonuclease domain; NendoU, uridylate-specific endoribonuclease domain; MT, 2′-O-ribose methyltransferase domain. nsp's 7–9 contain RNA-binding domains (RBDs).

of the S protein are multiply glycosylated, while the M protein has one or two glycans close to its N-terminus. Interestingly, glycosylation of the M protein can be either N- or O-linked, depending on the type of coronavirus, although experiments using reverse genetics showed that conversion of an O-linked glycosylated M protein to an N-linked version had no effect on virus growth.

Early and late in infection, formation of virus particles can occur in the endoplasmic reticulum–Golgi intermediate compartment (ERGIC) and endoplasmic reticulum, but most assembly occurs in the Golgi membranes. The M protein is not transported to the plasma membrane; its location at internal membranes determines the sites of virus particle formation. It interacts with the N protein (as part of the RNP) and C-terminal part of the S protein, retaining some, though not all, of the S protein at internal membranes. The E protein is essential for virus particle formation, though it is not known how it functions. It has a sequence that determines its accumulation at internal membranes, and its interaction with the M protein. The latter interacts with the C-terminus of the S protein, retaining some of it at internal membranes, and with the N protein (itself part of the ribonucleoprotein structure), enabling the formation of virus particles with spikes.

## Genome Replication and Recombination

Following infection of a susceptible cell, the coronavirus genomic RNA is released from the virion into the cytoplasm and immediately recognized as an mRNA for the translation of the replicase pp1a and pp1ab proteins. These proteins are cleaved by ORF1a-encoded proteases, after which they become part of replicase complexes for the synthesis of either complete negative-sense copies of the genomic RNA or negative-sense copies of the sg mRNAs. The negative-sense RNAs are used as templates for the synthesis of genomic RNA and sg mRNAs (**Figure 2**). Following synthesis of the sg mRNAs, the structural proteins are produced for the assembly and encapsidation of the *de novo*-synthesized genomic RNA, resulting in the release of new infectious coronavirus virions. The release of new virions starts 3–4 h after the initial infection. As indicated above, the synthesis of the sg mRNAs is the result of a discontinuous process in which the synthesis of a negative-sense copy of an sg mRNA is completed by the addition of the negative-sense leader sequence by a recombination mechanism. If a cell is infected with two related coronaviruses, the polymerase may swap between two RNA templates, in a similar way to addition of the leader sequence. This 'copy-choice' mechanism of genetic recombination results in a chimeric RNA. Such RNAs may give rise to new viruses with modified genomes with a capacity to infect a different cell and, in some cases, new host species.

## Evolutionary Relationships among Coronaviruses

Phylogenetic analyses of the structural proteins have resulted in the grouping of coronavirus species in accordance with earlier antigenic groups (**Table 1** and **Figure 4**).

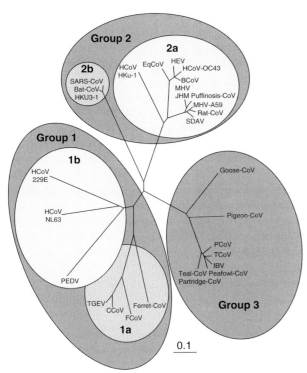

**Figure 4** Phylogenetic relationship of aligned coronavirus-derived nucleoprotein amino acid sequences. The complete N protein sequences represent coronaviruses from each of the three groups (**Table 1**). The tree is unrooted and the three main coronavirus groups, 1–3, are highlighted as dark gray ellipsoids. Groups 1 and 2 are divided into two subgroups, a and b, representing some divergence of the sequences within their corresponding groups. Similar relationships are observed when comparing other structural proteins and replicase-derived proteins.

Members of subgroups have higher amino acid sequence identities to each other ($\geq 60\%$) than to members of another group in the same group (with which they share $\leq 40\%$ identity). Comparing one group with another, protein sequence identities are generally in the range 25–35%. Unlike other members of group 2, SARS-CoV does not have an HE glycoprotein. Phylogenetic analysis using all the encoded proteins indicates that recombination has been a feature of coronavirus evolution. For example, some group 1a viruses are clearly recombinants between a feline and canine group 1 coronavirus.

## Diseases and Host Range

Probably all coronaviruses replicate in epithelial cells of the respiratory and/or enteric tracts, though not necessarily producing clinical damage at those sites. Avian IBV not only causes respiratory disease but can also damage gonads in both females and males, and causes serious kidney disease (dependent on the strain of virus, and to some extent on the breed of chicken). IBV is able to replicate at virtually every epithelial surface in the host. Some coronaviruses have their most profound effect in the alimentary tract (e.g., porcine TGEV causes $\geq 90\%$ mortality in neonatal pigs). Human coronaviruses are known to be associated with enteric and respiratory diseases (e.g., diarrhea), in addition to respiratory disease. SARS-CoV was also associated with diarrhea in humans, in addition to serious lung disease. Other coronaviruses, for example, MHV and porcine HEV, spread to cells of the central nervous system, producing disease, for example, acute or chronic demyelination in the case of MHV.

Coronavirus replication and disease are not necessarily restricted to a single host species. Canine enteric CoV and feline CoV can replicate and cause disease in pigs; these two viruses have proteins with very high amino acid identity to those of porcine TGEV. Canine respiratory CoV has proteins, including the S protein (which is the attachment protein and a determinant of host range), with very high amino acid identity ($\geq 95\%$) to other group 2 viruses Hu CoV-OC43 and BCoV. This raises the possibility of co-infection in these hosts. Bovine CoV causes enteritis in turkeys following experimental oral infection. There is evidence that pheasant CoV can infect chickens, and IBV infect teal (a duck), though without causing disease. The most dramatic demonstration that coronaviruses can have a wide host range was provided by SARS-CoV. This may have had its origin in bats, was transferred to various other species (e.g., civet cat) that were captured for trade, and then caused lethal disease in humans.

Persistent infections *in vivo* are well known for MHV, and less well known for other coronaviruses (e.g., IBV). Following infection of very young chickens, IBV is re-excreted when hens start to lay eggs. The trigger for release is probably the stress of coming into lay.

The S protein is a determinant of both tissue tropism within a host and host range. This has been elegantly demonstrated by genetic manipulation of the genome of MHV, which is unable to attach to feline cells. Replacement of the MHV S protein gene with that of CoV from feline coronavirus resulted in a recombinant virus that was able to attach, and subsequently replicate in, feline cells. However, other proteins can also affect pathogenicity. Research with genetically modified coronaviruses, using targeted recombination or 'infectious clones', has shown that modifications to proteins encoded in ORF1 and the small genes interspersed among the structural protein genes, result in attenuation of pathogenicity. Although the roles of these 'accessory proteins' are not known, this may offer a route to the development of a new generation of live vaccines. Currently, the most widely used prophylactics for control of IBV in chickens include killed vaccines and live vaccines attenuated by passage in embryonated eggs. However, disease control is complicated by extensive variation in the S1 protein which is the inducer of protective immunity.

*See also:* Nidovirales.

## Further Reading

Britton P and Cavanagh D (2007) Avian coronavirus diseases and infectious bronchitis vaccine development. In: Thiel V (ed.) *Coronaviruses: Molecular and Cellular Biology* pp. 161–181. Norfolk, UK: Caister Academic Press.

Britton P and Cavanagh D (2007) Nidovirus genome organization and expression mechanisms. In: Perlman S Gallagher T, and Snijder EJ (eds.) *Nidoviruses*, pp. 29–46. Washington, DC: ASM Press.

Cavanagh D (2003) SARS vaccine development: Experiences of vaccination against avian infectious bronchitis coronavirus. *Avian Pathology* 32: 567–582.

Cavanagh D (2005) *Coronaviridae*: A review of coronaviruses and toroviruses. In: Schmidt A and Wolff MH (eds.) *Coronaviruses with Special Emphasis on First Insights Concerning SARS*, pp. 1–54. Basel: Birkhäuser.

Cavanagh D (2005) Coronaviruses in poultry and other birds. *Avian Pathology* 34: 439–448.

Enjuanes L, Almazán F, Sola I, and Zuñiga S (2006) Biochemical aspects of coronavirus replication: A virus–host interaction. *Annual Reviews in Microbiology* 60: 211–230.

González JM, Gomez-Puertas P, Cavanagh D, Gorbalenya AE, and Enjuanes L (2003) A comparative sequence analysis to revise the current taxonomy of the family *Coronaviridae*. *Archives of Virology* 148: 2207–2235.

Masters PS (2006) The molecular biology of coronaviruses. *Advances in Virus Research* 66: 193–292.

Siddell S, Ziebuhr J, and Snijder E (2005) Coronaviruses, toroviruses and ateriviruses. In: Mahy BWJ and ter Meulen V (eds.) *Topley and Wilson's Microbiology and Microbial Infections, Virology*, pp. 823–856. London: Hodder Arnold.

# Coronaviruses: Molecular Biology

**S C Baker**, Loyola University of Chicago, Maywood, IL, USA

© 2008 Elsevier Ltd. All rights reserved.

## Glossary

**Cell tropism** Process that determines which cells can be infected by a virus. Factors such as receptor express can influence the cell type that can be infected.

**Discontinuous transcription** Process by which the coronavirus leader sequence and body sequence are joined to generate subgenomic RNAs.

**Double membrane vesicles (DMVs)** Vesicles that are generated during coronavirus replication when viral replicase proteins sequester host cell membranes. These vesicles are the site of coronavirus RNA synthesis.

**Transcriptional regulatory sequences (TRSs)** Sequences that are recognized by the coronavirus transcription complex to generate leader-containing subgenomic RNAs.

## Introduction

Coronaviruses (CoVs) were first identified during the 1960s by using electron microscopy to visualize the distinctive spike glycoprotein projections on the surface of enveloped virus particles. It was quickly recognized that CoV infections are quite common, and that they are responsible for seasonal or local epidemics of respiratory and gastrointestinal disease in a variety of animals. CoVs have been named according to the species from which they were isolated and the disease associated with the viral infection. Avian infectious bronchitis virus (IBV) infects chickens, causing respiratory infection, decreased egg production, and mortality in young birds. Bovine coronavirus (BCoV) causes respiratory and gastrointestinal disease in cattle. Porcine transmissible gastroenteritis virus (TGEV) and porcine epidemic diarrhea virus (PEDV) cause gastroenteritis in pigs. These CoV infections can be fatal in young animals. Feline infectious peritonitis virus (FIPV) and canine coronavirus (CCoV) can cause severe disease in cats and dogs. Depending on the strain of the virus and the site of infection, the murine CoV mouse hepatitis virus (MHV) can cause hepatitis or a demyelinating disease similar to multiple sclerosis. CoVs also infect humans. Human coronaviruses (HCoVs) 229e and OC43 are detected worldwide and are estimated to be responsible for 5–30% of common colds and mild gastroenteritis. Interestingly, HCoV-OC43 and BCoV share considerable sequence similarity, indicating a likely transmission across species (either from cows to humans or vice versa) and then adaptation of the virus to its host. In contrast to the relatively mild infections caused by HCoV-229e and HCoV-OC43, the CoV responsible for severe acute respiratory syndrome (SARS-CoV) causes atypical pneumonia with a 10% mortality rate. Two additional HCoVs, HCoV-NL63 and HCoV-HKU1, have been recently identified using molecular methods and are associated with upper and lower respiratory tract infections in children, and elderly and immunosuppressed patients. CoVs are grouped according to sequence similarity. CoVs that infect mammals are assigned to group 1 and group 2, whereas CoVs that infect birds are in group 3.

To date, the most infamous example of zoonotic transmission of a CoV is the outbreak of SARS in 2002–03. We now know that the outbreak started with cases of atypical pneumonia in the Guangdong Province in southern China in the fall of 2002. The infection was spread to tourists visiting Hong Kong in February, 2003, resulting in the dissemination of the outbreak to Hong Kong, Vietnam, Singapore, and Toronto, Canada. After attempting to treat cases of atypical pneumonia in Vietnam and acquiring the infection himself, Dr. Carlo Urbani alerted the World Health Organization (WHO) that this disease of unknown origin may be a threat to public health. The WHO rapidly organized an international effort to identify the cause of the outbreak, and within months a novel CoV was isolated from SARS patients and identified as the causative agent. Sequence analysis revealed that the virus was related to, but distinct from, all known CoVs. This led to an intensive search for an animal reservoir for this novel CoV. Initially, the masked palm civet and raccoon dog were implicated in the chain of transmission, since a SARS-CoV-like virus could be isolated from some animals found in wild animal markets in China. However, SARS-CoV-like viruses were never detected in animals captured from the wild, indicating that the civets may have only served as an intermediate host in the chain of transmission. Further investigation revealed that the likely reservoir for SARS-CoV is the Chinese horseshoe bat (*Rhinolophus* spp.), which is endemically infected with a virus, named bat-SARS-CoV, that is closely related to SARS-CoV. The existence of an animal reservoir presents the possibility of re-emergence of this significant human pathogen. By improving our understanding of the molecular aspects of CoV replication and pathogenesis, we may facilitate development of appropriate antiviral agents and vaccines to control and prevent diseases caused by known and potentially emerging CoV infections.

## Molecular Features of CoVs

CoV virions (**Figure 1(a)**) are composed of a large RNA genome, which combines with the viral nucleocapsid protein (N) to form a helical nucleocapsid, and a host cell-derived lipid envelope which is studded with virus-specific proteins including the membrane (M) glycoprotein, the envelope (E) protein, and the spike (S) glycoprotein. CoV particles vary somewhat in size, but average about 100 nm in diameter. The genomic RNA (gRNA) inside the virion, which ranges in size from 27 to 32 kb for different CoVs, is the largest viral RNA identified to date. CoV gRNAs have a broadly conserved structure which is illustrated by the SARS-CoV genome shown in **Figure 1(b)**. The gRNA is

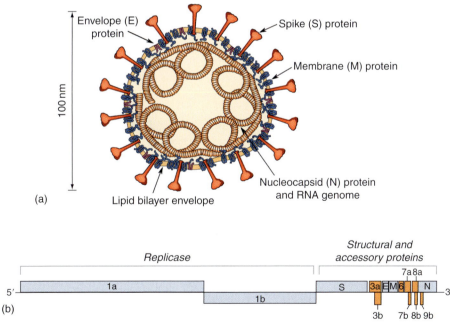

**Figure 1** CoV virion and the genome of SARS-CoV. (a) Schematic diagram of a CoV virion with the minimal set of four structural proteins required for efficient assembly of the infectious virus particles: S, spike glycoprotein; M, membrane glycoprotein; E, envelope protein; and N, nucleocapsid phosphoprotein which encapsidates the positive-strand RNA genome. (b) Schematic diagram of the gRNA of SARS-CoV. Translation of the first two open reading frames (ORF1a and ORF1b) generates the replicase polyprotein. ORFs encoding viral structural and accessory (orange) ORFs are indicated at the 3′ end of the genome. (a) Reprinted from Masters PS (2006) The molecular biology of coronaviruses. *Advances in Virus Research* 66: 193–292, with permission from Elsevier.

capped at the 5′ end, with a short leader sequence followed by two long open reading frames (ORFs) encoding the replicase polyprotein. The remaining part of the genome encodes the viral structural and so-called accessory proteins. The structural protein genes are always found in the order S–E–M–N, but accessory protein genes may be interspersed at various sites between the structural genes. SARS-CoV has the most complex genome yet identified, with eight ORFs encoding accessory proteins. The expression of these ORFs is not required for viral replication, but they may play a role in the pathogenesis of SARS. In addition, the products of accessory genes may be incorporated into the virus particle, potentially altering the tropism or enhancing infectivity. For SARS-CoV, the proteins encoded in ORFs 3a, 6, 7a, and 7b have been shown to be incorporated in virus particles, but the exact role of these proteins in enhancing virulence is not yet clear.

The features of CoV structural proteins are shown in **Figure 2**. For each structural protein, a schematic diagram of the predicted structure of the protein is shown on the left and a linear display of the features is shown on the right. The CoV spike glycoprotein is essential for attachment of the virus to the host cell receptor and fusion of the virus envelope with the host cell membrane. CoV spike glycoproteins assemble as trimers with a short cytoplasmic tail and hydrophobic transmembrane domain anchoring the protein into the membrane. The spike glycoprotein is divided into the S1 and S2 regions, which are sometimes cleaved into separate proteins by cellular proteases during the maturation and assembly of virus particles. S1 contains the receptor-binding domain (RBD) and has been shown to provide the specificity of attachment for CoV particles. The cellular receptors and corresponding RBDs in S1 have been identified for several CoVs. MHV binds to murine carcinoembryonic

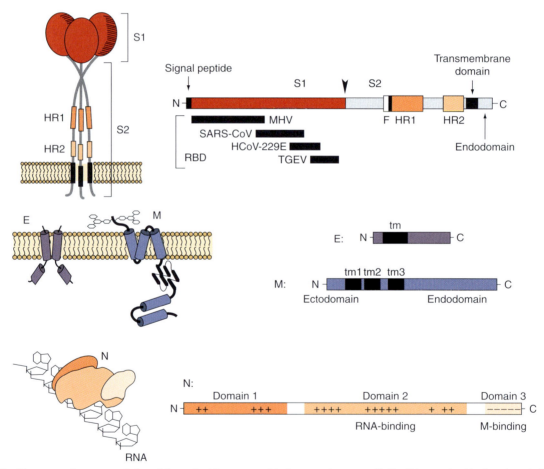

**Figure 2** Diagrammatic representation of the spike trimer assembled on membranes, with the S1 receptor binding domain (RBD) and S2 fusion domain indicated. The linear map of spike indicates the location of the RBDs for three CoVs, and the relative location of the heptad repeat domains 1 and 2 (HR1 and HR2) which mediate the conformational changes required to present the fusion peptide (F) to cellular membranes. The membrane (M), envelope (E), and nucleocapsid (N) proteins represented in association with membranes or viral RNA. The linear map of each protein highlights the transmembrane domains of M and E and the RNA-binding and M protein-binding domains of N. Domains 1 and 2 of N are rich in arginine and lysine (indicated by +). Reprinted from Masters PS (2006) The molecular biology of coronaviruses. *Advances in Virus Research* 66: 193–292, with permission from Elsevier.

antigen-related cell adhesion molecules (mCEACAM1 and MCEACAM2); TGEV, FIPV, and HCoV-229e bind to species-specific versions of aminopeptidase N. Interestingly, both HCoV-NL63 and SARS-CoV have been shown to bind to human angiotensin-converting enzyme 2 (ACE2). ACE2 is expressed in both the respiratory and gastrointestinal tracts, consistent with virus replication at both these sites.

Once the S1 portion of the spike has engaged the host cell receptor, the protein undergoes a dramatic conformational change to promote fusion with the host cell membrane. Depending on the virus strain, this can occur at the plasma membrane on the surface of the cell, or in acidified endosomes after receptor-mediated endocytosis. The critical elements in the conformational change are the heptad repeats, HR1 and HR2, and the fusion peptide, F. After engaging the receptor, there is a dissociation of S1 which likely triggers the rearrangement of S2 so that HR1 and HR2 are brought together to form an antiparallel, six-helix bundle. This new conformation brings together the viral and host cell membranes and promotes the fusion of the lipid bilayers and introduction of the nucleocapsid into the cytoplasm. During infection, the spike glycoprotein is also present on the surface of the infected cell where it may (depending on the virus strain) promote fusion with neighboring cells and syncytia formation. The spike glycoprotein is also the major antigen to which neutralizing antibodies develop. The spike protein is a target for development of therapeutics for treatment of CoV infections. Monoclonal antibodies directed against the spike neutralize the virus by blocking binding to the receptor; synthetic peptides that block HR1-HR2 bundle formation have also been shown to block CoV infection.

The membrane (M) and envelope (E) proteins are essential for the efficient assembly of CoV particles. M is a triple-membrane-spanning protein that is the most abundant viral structural protein in the CoV virion. The ectodomain of M is generally glycosylated, and is followed by three transmembrane domains and an endodomain which is important for interaction with the nucleocapsid protein and packaging of the viral genome. The E protein is present in low copy numbers in the virion, but is important for efficient assembly. In the absence of E protein, few or no infectious virus particles are produced. The exact role of the E protein in the assembly of virus particles is still unknown, but recent studies suggest that E may act as an ion channel. The nucleocapsid protein (N) is an RNA-binding protein and associates with the CoV gRNA to assemble ribonucleoprotein complexes. The N protein is phosphorylated, predominantly at serine residues, but the role of phosphorylation is currently unknown. The N protein has three conserved domains, each separated by highly variable spacer elements. Domains 1 and 2 are rich in arginine and lysine residues, which is typical of many RNA-binding proteins. Domain 3 is essential for interaction with the M protein and assembly of infectious virus particles. The N protein has been shown to be an important cofactor in CoV RNA synthesis and is proposed to act as an RNA chaperone to promote template switching, as described below.

## Replication and Transcription of CoV RNA

The replication and transcription of CoV RNA takes place in the cytoplasm of infected cells (**Figure 3**). The CoV virion attaches to the host cell receptor via the spike glycoprotein and, depending on the virus strain, the spike mediates fusion directly with the plasma membrane or the virus undergoes receptor-mediated endocytosis and spike-mediated fusion with endosomal membranes to release the viral gRNA into the cytoplasm. Once the positive-strand RNA genome is released, it acts as a messenger RNA (mRNA) and the 5' end (ORF1a and ORF1b) is translated by ribosomes to generate the viral RNA-dependent RNA polymerase polyprotein, termed the viral replicase. Translation of ORF1b is dependent on ribosomal frameshifting, which is facilitated by a slippery sequence and RNA pseudoknot structure present in all CoV gRNAs. The replicase polyprotein is processed by replicase-encoded proteases (papain-like proteases and a poliovirus 3C-like protease) to generate 16 mature replicase products. These viral replicase proteins sequester host cell membranes to generate distinctive double-membrane vesicles (DMVs) that have been shown to be the site of CoV RNA synthesis. The replicase complex on the DMVs then mediates the replication of the positive-strand RNA genome to generate full-length and subgenomic negative-strand RNAs, and the subsequent production of positive-strand gRNAs and sgmRNAs. The sgmRNAs are translated to generate viral structural and accessory proteins, and virus particles assemble with positive-strand gRNA in the endoplasmic reticulum-Golgi intermediate compartment (ERGIC) and bud into vesicles, with subsequent release from the cell. Depending on the virus strain, this replication can be robust and result in destruction of the host cell or a low-level, persistent infection that can be maintained in cultured cells or infected animals.

A hallmark of CoV transcription is the generation of a nested set of mRNAs, with each mRNA having the identical 'leader' sequence of approximately 65–90 nt at the 5' end (**Figure 4(a)**). The leader sequence is encoded only once at the 5' end of the gRNA. Each subgenomic mRNA (sgmRNA) has the identical leader sequence fused to the 5' end of the body sequence. How are the leader-containing mRNAs generated during CoV transcription? Current evidence supports a model of discontinuous transcription, whereby the replicase complex switches templates during the synthesis of negative-strand RNA

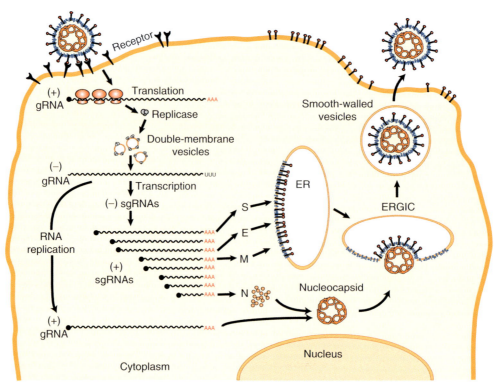

**Figure 3** Replication cycle of CoVs. The spike glycoprotein on the virus particle interacts with host cell receptors to mediate fusion of the virus and host cell membranes and release of the positive-strand RNA genome into the cytoplasm. The 5′-proximal open reading frames (ORF1a and ORF1b) are translated from the gRNA to generate the replicase polyprotein. The replicase polyprotein is processed by viral proteases into 16 nonstructural proteins which assemble with membranes to generate double-membrane vesicles (DMVs) where RNA synthesis takes place. A nested set of 3′ co-terminal subgenomic (sg) RNAs is generated by a discontinuous transcription process. The sgRNAs are translated to generate the viral structural and accessory proteins. Viral gRNA is replicated and associates with nucleocapsid protein and viral structural proteins in the endoplasmic reticulum-Golgi intermediate compartment (ERGIC), where virus particles bud into vesicles before transport and release from the cell. Reprinted from Masters PS (2006) The molecular biology of coronaviruses. *Advances in Virus Research* 66: 193–292, with permission from Elsevier.

(**Figure 4(b)**). The key sequence element in this process is the transcriptional regulatory sequence (TRS). The TRS is a sequence of approximately 6–9 nt (5′-ACGAAC-3′ for SARS-CoV) which is found at the end of the leader sequence and at each intergenic region (the sites between the open reading frames encoding the viral structural and accessory proteins). Site-directed mutagenesis and deletion analysis has revealed the critical role of the TRS in mediating transcription of sgmRNAs. Deletion of any intergenic TRS results in loss of production of the corresponding sgmRNA. In addition, the CoV leader TRS and the intergenic TRS sequences must be identical for optimal production of the sgmRNAs. A three-step working model for template switching during negative-strand RNA synthesis has been proposed to describe the process for the generation of CoV leader-containing sgmRNAs (**Figure 4(b)**). In this process, the 5′ end and 3′ end of the gRNA form a complex with host cell factors and the viral replication complex. The 3′ end of the positive strand is used as the template for the initiation of transcription of negative-strand RNA. Negative-strand RNA synthesis continues up to the point of the TRS. At each TRS, the viral replicase may either read through the sequence to generate a longer template, or switch templates to copy the leader sequence. The template switch allows the generation of a leader-containing sgmRNA. In this model, alignment of the leader TRS, the newly synthesized negative-strand RNA, and the genomic TRS is critical for the template switching to occur. Disruption of the complex, or loss of base-pairing within the complex, will result in the loss of production of that sgmRNA. Further studies of the CoV replication complex may yield new insights into the role of the viral helicase and endoribonuclease in the generation of the leader-containing CoV RNAs.

Another hallmark of CoV replication is high-frequency RNA recombination. RNA recombination occurs when a partially synthesized viral RNA dissociates from one template and hybridizes to similar sequences present in a second template. Viral RNA synthesis continues and generates a progeny virus with sequences from two different parental genomes. This RNA recombination event is termed copy-choice recombination. Copy-choice RNA recombination can be demonstrated experimentally when two closely related CoV strains (such as MHV-JHM and

# Coronaviruses: Molecular Biology

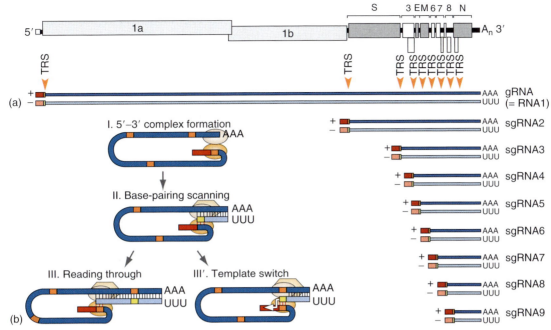

**Figure 4** Model of SARS-CoV gRNA and sgRNAs, and a working model of discontinuous transcription. (a) Diagram of gRNA and the nested set of sgRNAs of SARS-CoV. The 5′ leader sequence, the transcriptional regulatory sequences (TRSs), and the positive- and negative-sense sgRNAs are indicated. (b) A working model of CoV discontinuous transcription. I. 5′–3′ complex formation. Binding of viral and cellular proteins to the 5′ and 3′ ends of the CoV gRNA is represented by ellipsoids. The leader sequence is indicated in red, the TRS sites are in orange. II. Base-pair scanning step. Minus-strand RNA (light blue) is synthesized from the positive-strand template by the viral transcription complex (hexagon). At the TRS site, base-pairing may occur between the template, the nascent negative-strand RNA, and the leader TRS sequence (dotted lines). III. The synthesis of negative-strand RNA can continue to make a longer sgRNA III, or a template switch can take place III′ to generate a leader-containing subgenomic negative-strand RNA, which could then serve as the template for leader-containing positive-strand sgRNAs. Modified from Enjuanes L, Almazán F, Sola I, and Zunia S (2006) Biochemical aspects of coronavirus replication: A virus–host interaction. *Annual Reviews in Microbiology* 60: 211–230.

MHV-A59) are used to coinfect cells. Recombinant viruses with cross-over sites throughout the genome can be isolated, although sequences within the spike glycoprotein may be a 'hot spot' for recombination due to the presence of RNA secondary structures that may promote dissociation and reassociation of RNA. It has been proposed that copy-choice recombination is also the mechanism by which many CoVs have acquired accessory genes, and it has been exploited experimentally for the deletion or insertion of specific sequences in CoV genomes to assess their role in virus replication and pathogenesis.

## CoV Accessory Proteins

Sequence analysis of CoVs isolated from species ranging from birds to humans has revealed that all CoVs encode a core canonical set of genes, replicase (rep), spike (S), envelope (E), membrane (M), and nucleocapsid (N), and additional, so-called accessory genes (**Table 1**). The canonical genes are always found in the same order in the genome: rep-S-E-M-N. Reverse genetic studies (see below) have shown that this is the minimal set of genes required for efficient replication and assembly of infectious CoV particles. However, the genomes of all CoVs sequenced to date encode from one to eight additional ORFs, which code for accessory proteins. As the

**Table 1** Coronavirus **canonical** and *accessory* proteins

| Virus | Proteins: **canonical (rep-S-E-M-N)** and *accessory* |
|---|---|
| *Group 1* | |
| TGEV | **rep-S**-*3a,3b*-**E-M-N**-*7* |
| FIPV | **rep-S**-*3a,3b,3c*-**E-M-N**-*7a,7b* |
| HCoV-229E | **rep-S**-*4a,4b*-**E-M-N** |
| PEDV | **rep-S**-*3*-**E-M-N** |
| HCoV-NL63 | **rep-S**-*3*-**E-M-N** |
| *Group 2* | |
| MHV | **rep**-*2a*, *HE*-**S**-*4-5a*,**E-M-N**,*7b* |
| BCoV | **rep**-*2a*, *HE*-**S**-*4a,4b-5*,**E-M-N**,*7b* |
| HCoV-OC43 | **rep**-*2a*, *HE*-**S**-*5*,**E-M-N**,*7b* |
| HCoV-HKU1 | **rep**-*HE*-**S**-*4*-**E-M-N**,*7b* |
| SARS-CoV | **rep-S**-*3a,3b*-**E-M**-*6-7a,7b-8a,8b*-**N**,*9b* |
| Bat-SARS-CoV | **rep-S**-*3*-**E-M**-*6-7a,7b-8*-**N**,*9b* |
| *Group 3* | |
| Avian IBV | **rep-S**-*3a,3b,3c*-**E-M**-*5a,5b*-**N** |

name implies, these accessory proteins are not required for CoV replication in tissue culture cell lines, but they may play important roles in tropism and pathogenesis *in vivo*. How were these additional genes acquired? Current evidence indicates that these additional sequences may have been acquired by RNA recombination events between co-infecting viruses. For example, the hemagglutinin-esterase (HE) glycoprotein present in four different CoVs (MHV, BCoV, HCoV-OC43, and HCoV-HKU-1) was likely acquired by recombination of an ancestral CoV with the HE glycoprotein gene of influenza C. Interestingly, the expression of the HE gene has no effect on replication of the virus in cultured cell lines, but has been shown to enhance virulence in infected animals. Other CoV accessory genes may have been acquired through recombination with host cell mRNA or other viral mRNAs. The specific role of the accessory proteins in CoV replication and pathogenesis is under investigation. For SARS-CoV, accessory protein 6 has been implicated as an important factor in viral pathogenesis. Researchers have shown that mice infected with murine CoV expressing SARS-CoV protein 6 rapidly succumb to the infection, indicating that the protein 6 enhances virulence. In addition, recent studies suggest that SARS-CoV accessory proteins may play a role in blocking host cell innate immune responses, which may enhance viral replication and virulence. Other accessory proteins, such as SARS-CoV 3a and 7a, have been shown to be packaged into virus particles, where they may enhance infectivity or alter cell tropism. Future studies will be aimed at elucidating how CoV accessory proteins may modulate the virulence of CoV infection.

## Manipulating CoV Genomes Using RNA Recombination and Reverse Genetics

Genetic manipulation of CoV sequences is challenging because of the large size (27–32 kbp) of the RNA genomes. However, two approaches have been developed to allow researchers to introduce mutations, deletions, and reporter genes into CoV genomes. These approaches are (1) targeted RNA recombination and (2) reverse genetics using infectious cDNA constructs of CoV. The first approach exploits high-frequency copy-choice recombination to introduce mutations of interest into the 3′ end of the CoV gRNA. In the first step of targeted RNA recombination, a cDNA clone encoding the region from the spike glycoprotein to the 3′ end of the RNA is generated. These sequences can be easily manipulated in the laboratory to introduce mutations or deletions, or for the insertion of reporter or accessory genes, into the plasmid DNA. Next, RNA is transcribed from the plasmid DNA and the RNA is transfected into cells coinfected with the CoV of interest. RNA recombination occurs between the replicating CoV and the transfected substrate RNA, and viruses with the 3′ end sequences derived from the transfected substrate RNA will be generated. The recombinant viruses are generated by high-frequency copy-choice recombination, but the challenge is to sort or select for the recombinant virus of interest from the background of wild-type virus. To facilitate selection of recombinant viruses, Masters and Rottier introduced the idea of host range-based selection. They devised a clever plan to use a mouse hepatitis virus (MHV) that encodes the spike glycoprotein from a feline CoV as the target for their recombination experiments. This feline-MHV, termed fMHV, will infect only feline cell lines. Substrate RNAs that encode the MHV spike and mutations of interest in the 3′ region of the genome can be transfected into feline cells infected with fMHV, and progeny virus can be collected from the supernatant and subsequently selected for the ability to infect murine cell lines. Recombinant CoVs that have incorporated the MHV spike gene sequence (and the downstream substrate RNA with mutations of interest) can be selected for growth on murine cells, thus allowing for the rapid isolation of the recombinant virus of interest. This host range-based selection step is now widely used by virologists to generate recombinant viruses with specific alterations in the 3′ end of the CoV genome.

The second approach for manipulating CoV sequences, generating infectious cDNA constructs of CoV, has been developed in several laboratories. Full-length CoV sequences have been cloned and expressed using bacterial artificial chromosomes (BACs), vaccinia virus vectors, and from an assembled set of cDNA clones representing the entire CoV genome. The generation of a full-length cDNA and subsequent generation of a full-length CoV gRNA allows for reverse genetic analysis of CoV sequences. Successful reverse genetics systems are now in place to study the replication and pathogenesis of SARS-CoV, MHV, HCoV-229e, and IBV. These reverse genetics systems have allowed researchers to introduce mutations into the replicase gene and identify sites that are critical for enzymatic activities of many replicase products such as the helicase, endoribonuclease, and the papain-like proteases. Reverse genetic approaches are also being used to investigate the role of the TRSs in controlling the synthesis of CoV mRNAs. Interestingly, the SARS-CoV genome can be 're-wired' using a novel, nonanonical TRS sequence, which must be present at both the ends of the leader sequence and at each intergenic junction. This 're-wired' SARS-CoV may be useful for generating a live-attenuated SARS-CoV vaccine. An important feature of this 're-wired' virus is that it would be nonviable if it recombined with wild-type virus, since the leader TRS and downstream TRS would no longer match in a recombinant virus. The development of reverse genetics systems for CoVs has opened the door to investigate how replicase gene products function in the complex mechanism of CoV discontinuous transcription, and provides new opportunities to generate novel CoVs as potential live-attenuated or killed virus vaccines to reduce or prevent CoV infections in humans and animals.

## Vaccines and Antiviral Drug Development

Because of the economic importance of CoV infection to livestock and domestic animals, a variety of live-attenuated and killed CoV vaccines have been tested in animals. Vaccines have been developed against IBV, TGEV, CCoV, and FIPV. However, these vaccines do not seem to provide complete protection from wild-type virus infection. In some cases, the wild-type CoV rapidly evolves to escape neutralization by vaccine-induced antibodies. In studies of vaccinated chickens, a live-attenuated IBV vaccine has been shown to undergo RNA recombination with wild-type virus to generate vaccine escape mutants. Killed virus vaccines may also be problematic for some CoV infections. Vaccination of cats with a killed FIPV vaccine has been shown to exacerbate disease when cats are challenged with wild-type virus. Therefore, extensive studies will be required to carefully evaluate candidate vaccines for SARS-CoV. A variety of approaches are currently under investigation for developing a SARS-CoV vaccine, including analysis of killed virus vaccines, live-attenuated virus vaccines, DNA immunization, and viral vector vaccines (such as modified vaccine virus Ankara, canarypox, alphavirus, and adenovirus vectors). The development of improved animal models for SARS will be essential for evaluating SARS-CoV candidate vaccines. Transgenic mice expressing human ACE-2 may be an appropriate small animal model. Initial studies suggest that Syrian hamsters and ferrets develop pneumonia and lung pathology similar to that seen in humans after infection with SARS-CoV, and therefore may be appropriate animal models for viral pathogenesis. CoV vaccine studies will benefit from an improved understanding of conserved viral epitopes that can be targeted for vaccine development.

The use of neutralizing monoclonal antibodies directed against the SARS-CoV spike glycoprotein is another approach that may provide protection from severe disease. The success in the development and use of humanized monoclonal antibodies against respiratory syncytial virus (family *Paramyxoviridae*) to protect infants from severe disease indicates that this approach is certainly worth investigating. Preliminary studies have indicated that patient convalescent serum and monoclonal antibodies directed against the SARS-CoV spike glycoprotein efficiently neutralize infectious virus. Further studies are essential to evaluate any concerns about potential antibody-mediated enhancement of disease and to determine if neutralization escape mutants arise rapidly after challenge with infectious virus. Studies evaluating monoclonal antibodies directed against a variety of structural proteins, and monoclonal antibodies directed against conserved sites in the spike glycoprotein will provide important information on the efficacy of passive immunity to protect against SARS.

Currently, there are no antiviral drugs approved for use against any human CoV infection. With the potential for the emergence or re-emergence of pathogenic CoV from animal reservoirs, there is considerable interest in identifying potential therapeutic targets and developing antiviral drugs that will block viral replication and reduce the severity of CoV infections in humans. Two promising targets for antiviral drug development are the SARS-CoV protease domains, the papain-like protease (PL$^{pro}$) and the 3C-like protease (3CL$^{pro}$, also termed the main protease, M$^{pro}$) (**Figure 5**). These two protease domains

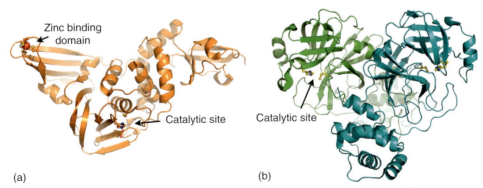

**Figure 5** CoV proteases are targets for antiviral drug development; X-ray structures of the two SARS-CoV protease domains encoded in the replicase polyprotein. (a) The SARS-CoV papain-like protease (PL$^{pro}$) with catalytic triad cysteine, histidine, and aspartic acid residues, and zinc-binding domain indicated. (b) The 3C-like protease (3CL$^{pro}$, also termed main protease, M$^{pro}$) dimer with catalytic cysteine and histidine residues indicated.

are encoded within the replicase polyprotein gene and protease activity is required to generate the 16 replicase nonstructural proteins (nsp1–nsp16) that assemble to generate the viral replication complex. The crystal structure of the 3CL$^{pro}$ was determined first from TGEV and then from SARS-CoV. Rational drug design, much of which was based on our knowledge of inhibitors directed against the rhinovirus 3C protease, has provided promising lead compounds for 3CL$^{pro}$ antiviral drug development. Interestingly, these candidate antivirals have been shown to inhibit the replication of SARS-CoV and other group 2 CoVs such as MHV, and the less related group 1 CoV, HCoV-229e. This indicates that the active site of 3CL$^{pro}$ is highly conserved among CoVs and that antiviral drugs developed against SARS-CoV 3CL$^{pro}$ may also be useful for inhibiting the replication of more common human CoVs such as HCoV-229e, HCoV-OC43, HCoV-NL63, and HCoV-HKU1. Further studies are needed to determine if these inhibitors can be developed into clinically useful antiviral agents.

Analysis of SARS-CoV papain-like protease led to the surprising discovery that this protease is also a viral deubiquitinating (DUB) enzyme. The SARS-CoV PL$^{pro}$ was shown to be required for processing the amino-terminal end of the replicase polyprotein and to recognize conserved cleavage site (-LXGG). The LXGG cleavage site is also the site recognized by cellular DUBs to remove polyubiquitin chains from proteins targeted for degradation by proteasomes. Analysis of the X-ray structure of the SARS-CoV PL$^{pro}$ has revealed that it has structural similarity to known cellular DUBs. These studies suggest that CoV papain-like proteases have evolved to have both proteolytic processing and DUB activity. The DUB activity may be important in preventing ubiquitin-mediated degradation of viral proteins, or the DUB activity may be important in subverting host cell pathways to enhance viral replication. PL$^{pro}$ inhibitors are now being developed using structural information and by performing high-throughput screening of small molecule libraries to identify lead compounds. Additional CoV replicase proteins, particularly the RNA-dependent RNA polymerase, helicase, and endoribonuclease, are also being targeted for antiviral drug development.

## Future Perspectives

The development of targeted RNA recombination and reverse genetics systems for CoVs has provided new opportunities to address important questions concerning the mechanisms of CoV replication and virulence, and to design novel CoV vaccines. In the future, improved small animal models for testing vaccines and antivirals, and the availability of additional X-ray crystallographic structure information for rational drug design will be critical for further progress toward development of effective vaccines and antiviral drugs that can prevent or reduce diseases caused by CoVs.

*See also:* Coronaviruses: General Features; Nidovirales; Severe Acute Respiratory Syndrome (SARS); Torovirus.

## Further Reading

Baker SC and Denison M (2007) Cell biology of nidovirus replication complexes. In: Perlman S, Gallagher T, and Snijder E (eds.) *The Nidoviruses*. Washington, DC: ASM Press.

Baric RS and Sims AC (2005) Development of mouse hepatitis virus and SARS-CoV infectious cDNA constructs. *Current Topics in Microbiology and Immunology* 287: 229–252.

Enjuanes L, Almazán F, Sola I, and Zuñiga S (2006) Biochemical aspects of coronavirus replication: A virus–host interaction. *Annual Reviews in Microbiology* 60: 211–230.

Lau YL and Peiris JS (2005) Pathogenesis of severe acute respiratory syndrome. *Current Opinion in Immunology* 17: 404–410.

Li W, Wong SK, Li F, et al. (2006) Animal origins of the severe acute respiratory syndrome coronavirus: Insights from ACE2-S-protein interactions. *Journal of Virology* 80: 4211–4219.

Masters PS (2006) The molecular biology of coronaviruses. *Advances in Virus Research* 66: 193–292.

Masters PS and Rottier PJM (2005) Coronavirus reverse genetics by targeted RNA recombination. *Current Topics in Microbiology and Immunology* 287: 133–160.

Perlman S and Dandekar AA (2005) Immunopathogenesis of coronavirus infections: Implications for SARS. *Nature Reviews Immunology* 5: 917–927.

Ratia K, Saikatendu K, Santarsiero BD, et al. (2006) Severe acute respiratory syndrome coronavirus papain-like protease: Structure of a viral deubiquitinating enzyme. *Proceedings of the National Academy of Sciences, USA* 103: 5717–5722.

Shi ST and Lai MMC (2005) Viral and cellular proteins involved in coronavirus replication. *Current Topics in Microbiology and Immunology* 287: 95–132.

Stadler K, Masignani V, Eickmann M, et al. (2003) SARS – Beginning to understand a new virus. *Nature Reviews Microbiology* 1: 209–218.

Thiel V and Siddell S (2005) Reverse genetics of coronaviruses using vaccinia virus vectors. *Current Topics in Microbiology and Immunology* 287: 199–228.

Wang L, Shi Z, Zhang S, Field H, Daszak P, and Eaton BT (2006) Review of bats and SARS. *Emerging Infectious Diseases* 12: 1834–1840.

Yang H, Xie W, Xue X, et al. (2005) Design of wide-spectrum inhibitors targeting coronavirus main proteases. *PLoS Biology* 3: 1742–1751.

Yount B, Roberts RS, Lindesmith L, and Baric RS (2006) Rewiring the severe acute respiratory syndrome coronavirus (SARS-CoV) transcription circuit: Engineering a recombinantion-Resistant genome. *Proceedings of the National Academy of Sciences, USA* 103: 12546–12551.

## Relevant Websites

http://patric.vbi.vt.edu – Coronavirus bioinformatics resource, PATRIC (PathoSystems Resource Integration Center).

http://www.cdc.gov – Severe acute respiratory syndrome (SARS) resource, Centers for Disease Control and Prevention.

# Cotton Leaf Curl Disease

**S Mansoor, I Amin, and R W Briddon,** National Institute for Biotechnology and Genetic Engineering, Faisalabad, Pakistan

© 2008 Elsevier Ltd. All rights reserved.

## Glossary

**Host plant resistance** Refers to the ability of a natural host species to control population of the parasite. This is the most desired way to limit damage caused by the causative agent.

**Koch's postulates** The criteria conceived by Robert Koch in 1890 to establish a causal relationship between a microbe and a disease. These conditions are that the pathogen must be found in all symptomatic hosts, be grown in pure culture, and should be able to cause the disease when reintroduced into a healthy host. The pathogen must then be reisolated from the experimentally infected host.

**Pathogen-derived resistance** An engineered type of resistance to pathogens created by expression of pathogen-derived genes, or sequences, that interfere with infection by the homologous and closely related pathogens.

## Introduction

Cotton leaf curl disease (CLCuD) is a serious disorder of cotton and several other malvaceous hosts that is transmitted by the whitefly *Bemisia tabaci*. CLCuD is endemic across most of Pakistan and northwestern India. The disease has also been reported from Egypt, Sudan, Nigeria, Malawi, and South Africa. Affected plants exhibit very distinctive symptoms; they consist of vein swelling, upward or downward curling of the leaves, and the formation of enations on the main veins on the undersides of leaves. The enations frequently develop into cup-shaped, leaf-like structures (**Figure 1**). Conspicuously CLCuD-infected cotton plants are greener than noninfected plants due to the proliferation of chloroplast-containing tissues. Symptoms are highly variable between cotton varieties, and they also depend on the age of the plant at infection. Plants infected late in the season show only mild symptoms and do not suffer a significant yield reduction. Plants infected early are severely stunted, with tightly rolled leaves, and they usually yield no harvestable lint.

Cotton has been grown in tropical and subtropical regions of the world since prehistoric times and is increasingly being adapted to more temperate climates as well.

The genus *Gossypium* encompasses *c*. 50 species, of which five are tetraploid and the remainder are diploid. Of these, four produce spinable fiber and are cultivated as cotton. These include two diploid species, *G. arboreum* and *G. herbaceum*, and two tetraploid species, *G. hirsutum* and *G. barbadense*. More than 90% of world cotton production is obtained from *G. hirsutum*. Diploid cotton species were grown throughout Asia and Africa before the introduction of tetraploid cotton from the New World. The diploid cotton species grown in the Old World are completely immune to CLCuD.

## History of CLCuD

### CLCuD in Africa

The first report of CLCuD was from Nigeria in 1912, where the disease was a sporadic but minor problem, with a second outbreak in this country occurring in 1924. In North Africa (mainly Egypt and Sudan), *G. barbadense* is the main cotton species cultivated. In these areas, CLCuD is endemic, although it is only sporadically a problem. This was not the case during the early part of the twentieth century when the disease caused major crop losses, particularly in Sudan. An epidemic in Sudan in 1927–28 stimulated interest, and it was during this time that the disease was shown to be transmitted by the whitefly, *B. tabaci*. Extensive work to understand the disease showed that it was also graft transmissible. Although the causative agent was not identified, a virus-like agent was suspected. The disease in this case was brought under control by imposing a cotton-free cultivation period and the introduction of virus-tolerant *G. barbadense* varieties. The disease continues to occur sporadically throughout the region but does not cause major losses.

### CLCuD in Southern Asia

CLCuD was noted infrequently across the Indian subcontinent prior to the 1980s. In Pakistan, cotton production, the main foreign exchange earner for the country, suffered heavily from an epidemic of CLCuD which initiated in the vicinity of the city of Multan in the mid-1980s and spread to virtually all cotton-growing areas, as well as into western India. The epidemic has been attributed to the introduction of high-yielding, but also highly susceptible, cotton varieties such as S-12 and CIM70. The cotton

**Figure 1** Typical symptoms of CLCuD-affected cotton (*G. hirsutum*) plants. Symptoms include upward curling of the leaves, vein swelling, vein darkening, and enations on the main veins on the undersides of leaves which frequently develop into cup-shaped, leaf-like structures (left). A severely infected plant showing a reduction in leaf size, downward leaf curling, leaf crumpling, and severe stunting (right).

species native to this region, *G. arboreum*, is immune to CLCuD, but unfortunately this does not produce the high-grade cotton lint desired by the processing industry. Nevertheless, *G. arboreum* continues to be grown on a small scale. Although CLCuD remained endemic, losses due to the disease were reduced in both India and Pakistan during the 1990s by the gradual replacement of susceptible varieties with locally developed, tolerant, and resistant cotton varieties. By the late 1990s, cotton production was again at record levels, exceeding the output achieved prior to the epidemic. Then, in 2001, there was a change in the prevalent virus population affecting cotton in Pakistan, as will be described later.

## Etiology of CLCuD

The etiology of CLCuD from two regions (North Africa (Egypt and Sudan) and the Indian subcontinent) has been determined. In both regions, the disease is caused by begomovirus complexes consisting of monopartite begomoviruses (genus *Begomovirus*, family *Geminiviridae*) and a recently identified single-stranded DNA satellite termed DNA β. Invariably, begomovirus-DNA β infections of cotton are also associated with a third component, known as DNA-1. This is a satellite-like molecule that plays no essential part in the etiology of CLCuD.

## Elucidation of the Etiology of CLCuD

During the early 1990s, the losses to the Pakistan and Indian cotton industries due to the newly emergent CLCuD initiated an urgent investigation into the etiology of the disease. A number of distinct begomoviruses were initially cloned and fully sequenced but infectivity of these was not shown to cotton; thus, Koch's postulates were not fulfilled. Later efforts to introduce one of these viruses, now known as cotton leaf curl Multan virus (CLCuMV), into cotton resulted in mild, atypical symptoms. This indicated that either CLCuMV was not involved in causing the disease or that some other essential component was required to do so.

The genomes of many begomoviruses consist of two components. Component DNA-A encodes all viral functions required for replication, control of gene expression, and encapsidation. The second, DNA-B, encodes two products involved in movement of the virus between and within plant cells. A small number of begomoviruses, mostly isolated from tomato, have monopartite genomes, appearing to be able to dispense with the functions encoded by DNA-B, their genomes consisting of only a homolog of the DNA-A component of bipartite viruses. For these viruses, the products of other genes are able to take over the functions of the DNA-B products in movement. One possible effect of the absence of DNA-B is that monopartite begomoviruses are limited to phloem-associated cells, whereas the majority of bipartite begomoviruses can spread to tissues outside the phloem. The monopartite begomoviruses include tomato yellow leaf curl virus, a major cause of losses to tomato crops globally.

The absence of a detectable DNA-B component associated with CLCuMV suggested that this was a monopartite virus. However, the inability to infect cotton and induce typical disease symptoms was troubling and mirrored the situation with several other begomoviruses at the time. Further analysis of plants identified an additional, novel, subviral component in CLCuD-affected plants. Similar components have been shown to be associated with a number of other monopartite begomoviruses and this group of molecules has been named DNA-1. However, DNA-1 was found not to play a part in the etiology of CLCuD. Nevertheless, this finding caused a renewed analysis of subviral DNA molecules associated with the monopartite begomoviruses. The breakthrough finally came with the identification of a second subgenomic DNA associated with the monopartite begomovirus, ageratum yellow vein virus, and soon thereafter with CLCuMV, as well as numerous other viruses. This group of molecules has become known as DNA β. Inoculation of CLCuMV with CLCuD DNA β induced full CLCuD symptoms in cotton and established the etiology of the disease as well as fulfilled Koch's postulates.

## Components of the CLCuD Complex

### CLCuD-associated DNA-1 component

DNA-1 components are satellite-like molecules that are associated with most (but not all) begomovirus-DNA β complexes (**Figure 2**). Their function remains unclear, since they are not required for infectivity or symptom induction in host plants. They are small (~1380 nt), circular, single-stranded DNA molecules that contain a single gene which encodes a rolling-circle replication initiator protein (known for plant-infecting, single-stranded DNA viruses as the replication-associated protein (Rep)). These molecules are capable of autonomous replication in the cells of host plants; hence they are described as satellite-like since, by definition, satellites depend upon their helper virus for replication. However, these components require the helper begomovirus for movement in plants and for transmission between plants, presumably by *trans*-encapsidation in the helper virus' coat protein.

Surprisingly, the DNA-1 components are closely related to the Rep-encoding, satellite-like components of nanoviruses. Nanoviruses are a second family of plant-infecting, single-stranded DNA viruses which are transmitted by aphids. As well as the so-called master Rep (mRep) component, the component which encodes the Rep responsible for replicating all the *bona fide* virus components, some nanovirus infections are also associated with additional Rep-encoding components. These satellite-like molecules are distinct from the mRep in both sequence and in the organization of the genome (the positioning of the promoter being different from that of the mRep). The DNA-1 components associated with begomovirus complexes are believed to have evolved from the nanovirus satellite-like components; most probably by component exchange during co-infection.

The major difference between DNA-1 molecules and nanovirus satellite-like components is size. The components of nanoviruses are typically 1000–1100 nt in length, whereas DNA-1 components are on average 1380 nt – almost exactly half the size of the genomes of their helper begomoviruses. The majority of this size increase is due to the presence, in DNA-1, of ~200 nt region of sequence which is rich in adenine. The function of this A-rich sequence is unclear; it is not essential for the molecule, since DNA-1 components with the A-rich sequence deleted remain capable of replication and movement in plants in the presence of a begomovirus. It is likely that there is a very subtle selection mechanism for maintenance of these additional sequences, most likely for efficient encapsidation by the helper virus' coat protein.

The presence of a DNA-1 component with the CLCuD complex in North Africa has not been investigated.

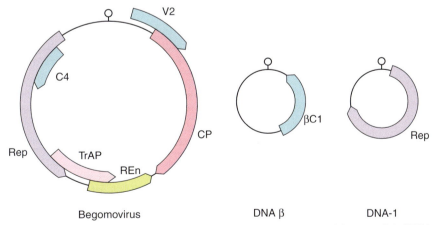

**Figure 2** Components associated with CLCuD. Shown are the begomovirus genome and the associated DNA β and DNA-1 components. Only the begomovirus and DNA β component are required to cause the disease. The position and orientation of genes encoded on these are indicated as arrows and labeled as coat protein (CP), replication-associated protein (Rep), transcriptional-activator protein (TrAP), replication enhancer protein (REn), and DNA βC1 protein (βC1). The function of the begomovirus-encoded V2 and C4 genes remain unclear. The ubiquitous stem–loop structure containing the nonanucleotide loop sequence TAATATTAC (which for geminiviruses is the origin of virion-strand, rolling-circle DNA replication) is indicated by the small circle for each component.

### CLCuD-associated DNA β components

DNA β components are a recently identified class of single-stranded DNA satellites that are associated with monopartite begomoviruses (family *Geminiviridae*). DNA β depends on the begomovirus for its proliferation and is responsible for the accumulation of the helper virus to levels normally found in their original hosts in the field, suggesting its involvement in either replication, systemic movement, or the suppression of a host defense mechanism. They are typically half the length of their helper begomoviruses (~1350 nt) and share no sequence homology with their helper virus other than the presence of a potential stem–loop structure containing the ubiquitous nonanucleotide sequence TAATATTAC. DNA β components encode a single gene, the product of which (known as βC1) is a pathogenicity determinant (determines the symptoms of the infection), a suppressor of post-transcriptional gene silencing (overcoming host plant defenses), upregulate viral DNA levels *in planta*, and bind DNA. DNA β satellites depend upon their helper viruses for replication, movement in plants, and transmission between plants, presumably by *trans*-encapsidation in the helper virus' coat protein. DNA β components have been cloned from a diverse range of hosts and DNA sequence comparison shows them to be highly diverse. Despite their highly diverse nature, only a single type of DNA β is associated with CLCuD and is replicated by several distinct begomoviruses, both experimentally and in the field. Expression of βC1 of CLCuD in tobacco, either stably transformed or from a virus vector, induces virus-like symptoms, demonstrating that DNA β is the most important pathogenicity determinant of the disease. However, the DNA β associated with CLCuD from Sudan is distinct from the one associated with the disease on the Indian subcontinent.

### Monopartite begomoviruses associated with CLCuD

On the Indian subcontinent, seven distinct begomovirus species (CLCuMV, cotton leaf curl Alabad virus (CLCuAV), cotton leaf curl Kokhran virus (CLCuKV), cotton leaf curl Rajasthan virus (CLCuRV), papaya leaf curl virus (PaLCuV), tomato leaf curl Bangalore virus (ToLCBV), and cotton leaf curl Burewala virus (CLCuBV)), either singly or as dual/multiple infections, have been shown to be involved in causing CLCuD (**Figure 3**). Some of these viruses have been shown experimentally to induce CLCuD in cotton when co-inoculated with CLCuD DNA β. Among these begomoviruses, only ToLCBV is infectious to tomato, in the absence of a DNA β, causing tomato leaf curl disease. An eighth virus, cotton leaf curl Bangalore virus (**Figure 3**), is associated with CLCuD in southern India (isolated from *G. barbadense*) and is thus unlikely to be part of the epidemic of CLCuD affecting the north of the country; its infectivity to cotton has not been shown experimentally. The viruses shown to be associated with CLCuD are not monophyletic (**Figure 3**), and thus are unlikely to have a common recent ancestor. This suggests that the viruses have been drawn into the disease by gaining an interaction with CLCuD DNA β, likely from other malvaceous hosts since a host range including cotton is a prerequisite for causing CLCuD.

The majority of begomoviruses associated with CLCuD show features both of monopartite as well as bipartite begomoviruses. These viruses replicate to high level in experimental hosts such as *Nicotiana benthamiana*, a situation that differs from bipartite begomoviruses which accumulate to only very low levels, or not at all, in the absence of DNA-β. Experimentally CLCuD DNA β can be replicated and maintained by other begomoviruses that are not found in cotton, including the bipartite begomovirus tomato leaf curl New Delhi virus (ToLCNDV; a virus that occurs widely in solanaceous crops across the Indian subcontinent). However, ToLCNDV has not been reported to infect cotton, suggesting that begomoviruses mobilized into cotton are also limited by their host range. The presence of so many viruses causing a single disease in cotton suggests that they were mobilized into this host after gaining an interaction with CLCuD DNA β, most probably following co-infection of a common host such as tomato. The presence of multiple begomoviruses may help the disease complex to adapt to other hosts or help in overcoming resistance. Nevertheless, there are other monopartite begomoviruses found on the Indian subcontinent that are not found in cotton, suggesting that some begomoviruses are better adapted to cotton. Experimentally some begomoviruses that are not found in the region (such as ageratum yellow vein virus originating from Southeast Asia and tomato leaf curl virus originating from Australia) are capable of supporting replication of CLCuD DNA β. The ability of CLCuD DNA β to interact with begomoviruses from other parts of the world indicates that CLCuD is a global threat to cotton cultivation. No study has been carried out to assess the ability of bipartite begomoviruses from the New World to support replication of CLCuD DNA β. No native monopartite begomoviruses occur in the New World, although TYLCV has recently been introduced and now affects crops in the Caribbean, southern North America, and Central America.

In North Africa, a single begomovirus, cotton leaf curl Gezira virus, has been shown to be associated with CLCuD. This virus is only distantly related to the Asian CLCuD-associated begomoviruses, being instead more closely related to other begomoviruses infecting plants of the Malvaceae originating from Africa and the Mediterranean (**Figure 3**). This is typical of the relationship of begomoviruses within an area (viruses being more related geographically than by the host from which they are

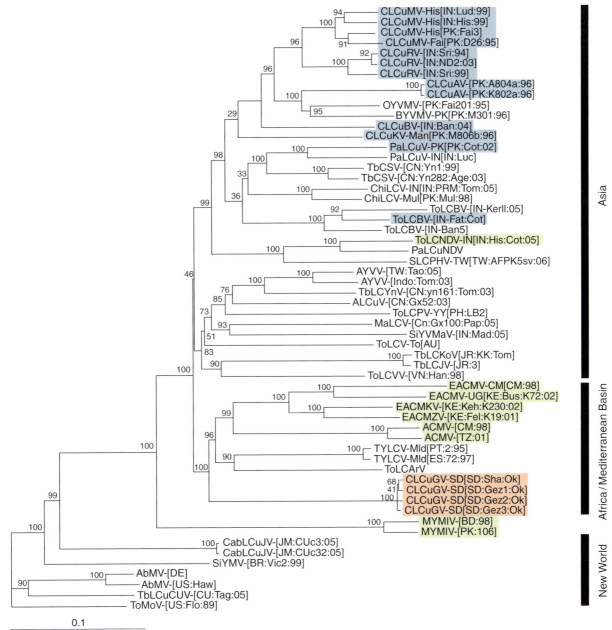

**Figure 3** Phylogenetic dendrogram derived from an alignment of 58 selected full-length sequences of the genomes (or DNA A genomic components) of begomoviruses. Indicated on the right are the viruses originating from the New World, Africa/Mediterranean Basin, and Asia. Viruses associated with CLCuD on the Indian subcontinent are highlighted in blue whereas those associated with CLCuD in North Africa are highlighted in brown. Viruses known to be bipartite are highlighted in yellow. The viruses are labeled as abutilon mosaic virus (AbMV), African cassava mosaic virus (ACMV), ageratum leaf curl virus (ALCuV), ageratum yellow vein virus (AYVV), Bhendi yellow vein mosaic virus (BYVMV), cabbage leaf curl Jamaica virus (CabLCuJV), chilli leaf curl virus (ChiLCV), cotton leaf curl Bangalore virus (CLCuBV), cotton leaf curl Gezira virus (CLCuGV), cotton leaf curl Kokhran virus (CLCuKV), cotton leaf curl Multan virus (CLCuMV), cotton leaf curl Rajasthan virus (CLCuRV), cotton leaf curl Alabad virus (CLCuAV), East African cassava mosaic Kenya virus (EACMKV), East African cassava mosaic virus (EACMV), East African cassava mosaic Zanzibar virus (EACMZV), malvastrum leaf curl virus (MaLCV), okra yellow vein mosaic virus (OYVMV), mungbean yellow mosaic India virus (MYMIV), papaya leaf curl New Delhi virus (PaLCuNDV), papaya leaf curl virus (PaLCuV), sida yellow mosaic virus (SiYMV), sida yellow vein Madurai virus (SiYVMaV), squash leaf curl Philippines virus (SLCPHV), tobacco curly shoot virus (TbCSV), tobacco leaf curl Cuba virus (TbLCuCUV), tobacco leaf curl Japan virus (TbLCJV), tobacco leaf curl Kochi virus (TbLCKoV), tobacco leaf curl Yunnan virus (TbLCYnV), tomato leaf curl Bangalore virus (ToLCBV), tomato leaf curl New Delhi virus (ToLCNDV), tomato leaf curl Pune virus (ToLCPV), tomato leaf curl virus (ToLCV), tomato leaf curl Vietnam virus (ToLCVV), tomato mottle virus (ToMoV), tomato yellow leaf curl virus (TYLCV), and tomato leaf curl Arusha virus (ToLCArV). Strain and isolate designation follow established convention. Vertical lines are arbitrary; horizontal lines are proportional to calculated mutation distances. The numbers at nodes indicate percentage bootstrap values (1000 replicates). The tree was rooted on the sequence of ToMoV originating from North America.

isolated or the disease they cause) and indicates that the CLCuDs occurring in Asia and Africa likely have distinct evolutionary origins and are not, as some have claimed, the result of accidental introduction of the virus to Pakistan/India from North Africa.

## Control of CLCuD

### Natural Resistance to the CLCuD Complex

Resistance to the Asian CLCuD complex is available in the form of G. arboreum. This cotton species is immune to CLCuD (in fact, no geminiviruses are reported that infect it) and is grown commercially on a small scale. However, the cotton fiber it produces is not of a sufficient quality or quantity to make it a viable alternative to G. hirsutum. Introgression of resistance from G. arboreum into G. hirsutum is also technically not a simple task (G. arboreum being diploid whereas G. hirsutum is tetraploid), although some efforts in that direction are in progress.

Limited resistance/tolerance is available in G. hirsutum germplasm. The cultivars LRA 511 and CP15/2 were the basis of much of the conventional resistance used to control CLCuD in Pakistan/India during the late 1990s. Cultivation of these cultivars was successful in alleviating losses to CLCuD in endemic areas and during this period cotton production in Pakistan showed an upward trend. However, recently, leaf curl symptoms were noted on previously resistant cotton varieties, indicating a change in the prevalent CLCuD complex.

### Breakdown of resistance to CLCuD

During 2001, in the vicinity of the town of Burewala (Punjab, Pakistan), previously resistant cotton varieties began to exhibit the symptoms typical of CLCuD infection. Since then there has been a resurgence of the disease with all previously resistant varieties and virtually all cotton-growing areas of Pakistan and neighboring India affected. Initial analysis of the resistance-breaking strain of CLCuD shows it to be a typical begomovirus-DNA β complex (although the presence of a DNA-1 component has not been confirmed). Only a single begomovirus, which has since been named cotton leaf curl Burewala virus (CLCuBV), a recombinant virus consisting of sequences derived from the previously occurring CLCuMV and CLCuKV, has been identified in all resistant varieties infected. Similarly, the DNA β associated with resistance breaking is recombinant. In this case, the molecules consist for the most part of sequences derived from the original CLCuD DNA β, but with a small fragment of the satellite conserved region (SCR), ~80 nt, derived from a DNA β first identified from tomato exhibiting tomato leaf curl disease symptoms. Although these components are infectious to experimental hosts, it has thus far not been possible to infect resistant cotton varieties with clones of CLCuBV and the recombinant DNA β; the precise molecular basis for resistance breaking thus remains unclear.

### Engineered Resistance to the CLCuD Complex

With the absence of any diversity in the natural resistance to CLCuD in the G. hirsutum germplasm, efforts have been made to obtain genetically engineered resistance to the disease complex prevalent on the Indian subcontinent. The major challenge to obtaining transgenic CLCuD resistance is the diversity of viruses which cause the disease. It is essential to introduce a broad-spectrum resistance, which is effective against all the viruses present in the field, if the approach is to stand any chance of durability. The strategies under investigation all rely on post-transcriptional gene silencing (PTGS) or transcriptional gene silencing (TGS) – homology-dependent, RNA-mediated phenomena which stimulate the plant's own defenses to target the invading virus. Since the one 'target' present in all CLCuD-affected plants is CLCuD DNA β, initial studies attempted to induce PTGS/TGS against this molecule, with little success. More fruitful have been the studies that have targeted the Rep and AV2 genes, by anti-sense expression as either full-length (AV2) or truncated (Rep) coding sequences. Both these strategies are presently being assessed in cotton under field conditions. However, it remains to be seen whether the sequences being used provide a sufficiently broad-spectrum resistance to all the CLCuD-associated begomoviruses to be effective and durable under field conditions.

*See also:* Banana Bunchy Top Virus; Beta ssDNA Satellites; Nanoviruses; Plant Resistance to Viruses: Geminiviruses; Plant Resistance to Viruses: Natural Resistance Associated with Recessive Genes; Satellite Nucleic Acids and Viruses; Tomato Leaf Curl Viruses from India; Tomato Yellow Leaf Curl Virus; Virus Induced Gene Silencing (VIGS).

## Further Reading

Asad S, Haris WAA, Bashir A, et al. (2003) Transgenic tobacco expressing geminiviral RNAs are resistant to the serious viral pathogen causing cotton leaf curl disease. *Archives of Virology* 148: 2341–2352.

Briddon RW (2003) Cotton leaf curl disease, a multicomponent begomovirus complex. *Molecular Plant Pathology* 4: 427–434.

Briddon RW, Bull SE, Amin I, et al. (2004) Diversity of DNA-1; A satellite-like molecule associated with monopartite begomovirus-DNA β complexes. *Virology* 324: 462–474.

Briddon RW and Markham PG (2000) Cotton leaf curl virus disease. *Virus Research* 71: 151–159.

Fauquet CM and Stanley J (2005) Revising the way we conceive and name viruses below the species level: A review of geminivirus taxonomy calls for new standardized isolate descriptors. *Archives of Virology* 150: 2151–2179.

Idris AM, Briddon RW, Bull SE, and Brown JK (2005) Cotton leaf curl Gezira virus-satellite DNAs represent a divergent, geographically isolated Nile Basin lineage: Predictive identification of a satDNA REP-binding motif. *Virus Research* 109: 19–32.

Mansoor S, Briddon RW, Bull SE, *et al.* (2003) Cotton leaf curl disease is associated with multiple monopartite begomoviruses supported by single DNA β. *Archives of Virology* 148: 1969–1986.

Mansoor S, Khan SH, Bashir A, *et al.* (1999) Identification of a novel circular single-stranded DNA associated with cotton leaf curl disease in Pakistan. *Virology* 259: 190–199.

Sanjaya VVS, Prasad V, Kirthi N, Maiya SP, Savithri HS, and Sita GL (2005) Development of cotton transgenics with antisense AV2 gene for resistance against cotton leaf curl virus (CLCuD) via *Agrobacterium tumefaciens*. *Plant Cell, Tissue and Organ Culture* 81: 55–63.

# Cowpea Mosaic Virus

**G P Lomonossoff,** John Innes Centre, Norwich, UK

© 2008 Elsevier Ltd. All rights reserved.

## Introduction

Cowpea mosaic virus (CPMV) is the type member of the genus *Comovirus* which includes 13 additional members in the family *Comoviridae*. CPMV was first isolated from an infected cowpea (*Vigna unguiculata*) plant in Nigeria in 1959. Subsequently, it has been found to occur in Nigeria, Kenya, Tanzania, Japan, Surinam, and Cuba. While its natural host is cowpea, it can infect other legumes, and *Nicotiana benthamiana* has proven to be an extremely valuable experimental host. In nature, CPMV is usually transmitted by leaf-feeding beetles, especially by members of the Chrysomelidae. CPMV has also been reported to be transmitted by thrips and grasshoppers. The beetle vectors can acquire the virus by feeding for as little as one minute and can retain and transmit the virus for a period of days or weeks. The virus does not, however, multiply in the insect vector. Experimentally, CPMV is mechanically transmissible.

In Nigeria, infection of cowpeas with CPMV causes a considerable reduction in leaf area, flower production, and yield. Infected plant cells show a number of characteristic cytological changes. These include the appearance of viral particles, a proliferation of cell membranes and vesicles in the cytoplasm, and a variety of modifications to plasmodesmata.

## Physical Properties of Viral Particles

Viral particles can reach a yield of up to $2 \text{ g kg}^{-1}$ of fresh cowpea tissue and can be readily purified by polyethylene glycol precipitation and differential centrifugation. The particles are very stable with a thermal inactivation point in plant sap of 65–75 °C and a longevity in sap of 3–5 days at room temperature. Once purified, the particles can be stored for prolonged periods at 4 °C. The ease with which virus particles can be propagated, purified, and stored has undoubtedly contributed to the early popularity of CPMV as an object of study.

CPMV preparations consist of nonenveloped isometric particles, 28 nm in diameter, which can be separated on sucrose density gradients into three components, designated top (T), middle (M), and bottom (B), with sedimentation coefficients of 58S, 95S, and 115S, respectively. The three components have identical protein compositions, containing 60 copies each of a large (L) and small (S) coat protein, with sizes 42 and 24 kDa, respectively, as calculated from the nucleotide sequence. The discovery, in 1971, that CPMV particles contained equimolar amounts of two different polypeptides suggested that the capsids had an architecture more similar to the animal picornaviruses than to other plant viruses of known structure. This provided an early clue as to the common origins of plant and animal viruses.

The difference in the sedimentation behavior of the three centrifugal components of CPMV lies in their RNA contents. Top components are devoid of RNA, while middle and bottom components each contain single molecules of positive-strand RNA of 3.5 and 6.0 kbp, respectively. The two RNA molecules were originally termed middle (M) and bottom (B) component RNA after the component from which they were isolated. However, more recently they have been referred to as RNA-2 and RNA-1, respectively. The three-component nature of CPMV preparations is summarized in **Figure 1**. The determination of the component structure of the virus, and particularly the relationship between this and infectivity, was important in establishing the principle that plant viruses frequently have divided genomes, the individual components of which are separately encapsidated.

Because of their differing RNA contents, the three components of CPMV also differ in density and can hence be separated by isopycnic centrifugation on cesium chloride gradients. While T and M components give single bands of densities of 1.30 and $1.41 \text{ g ml}^{-1}$, respectively,

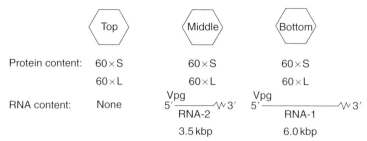

**Figure 1** The three-component nature of CPMV indicating the protein and RNA content of each component.

B component gives two bands of 1.43 and 1.47 g ml$^{-1}$, the proportion of the denser band increasing under alkaline conditions. This increase in density results from an increase in capsid permeability, which allows the exchange of the polyamines present in B components (where they serve to neutralize the excess negative charges from RNA-1) for cesium ions.

CPMV preparations are not only centrifugally heterogeneous but can also be separated in two forms, fast and slow, electrophoretically. Both electrophoretic forms contain all three centrifugal components. The proportion of the two electrophoretic forms in a given virus preparation varies both with the time after infection at which the virus was isolated and the age of the preparation itself. Conversion of one form to the other is caused by loss of 24 amino acids from the C-terminus of the S protein.

## Viral Structure

X-ray crystallographic studies on CPMV, as well as the related comoviruses bean pod mottle virus (BPMV) and red clover mottle virus (RCMV), have provided a detailed picture of the arrangement of the two viral coat proteins in the three-dimensional structure of the particle. Overall, the virions are icosahedral, with 12 axes of fivefold and 20 axes of threefold symmetry, and resemble a classic $T=3$ particle. The two coat proteins taken together consist of three distinct β-barrel domains, two being derived from the L and one from the S protein. Thus, in common with the $T=3$ viruses, each CPMV particle is made up of 180 β-barrel structures. The S protein, with its single domain, is found at the fivefold symmetry axes and therefore occupies a position analogous to that of the A-type subunits in $T=3$ particles (**Figure 2**). The N- and C-termini domains of the L protein occur at the threefold axes and occupy the positions equivalent to those of the C- and B-type subunits of a $T=3$ particle, respectively (**Figure 2**). This detailed analysis confirmed the earlier suggestion that CPMV particles are structurally homologous to those of picornaviruses, with the N- and C-terminal domains of the L protein being equivalent to viral protein VP2 and viral protein VP3, respectively, and the S protein being equivalent to viral protein VP1 (**Figure 2**). However, CPMV particles are structurally less complex than those of picornaviruses. The L and S subunits lack the extended N- and C-termini found in VP2, VP3, and VP1 of picornaviruses and there is no equivalent of VP4. Moreover, CPMV subunits lack the relatively large insertions between the strands of β-sheet, sequences that form the major antigenic determinants of picornaviruses. No RNA is visible in either the M or B components of CPMV, in contrast to the situation found with BPMV where segments of ordered RNA could be detected in middle components.

## Genome Structure

Both M and B (but not T) components of a virus preparation are essential for infection of whole plants. As CPMV is a positive-strand RNA virus, a mixture of the genomic RNAs within the particles can also be used to initiate an infection. However, RNA-1 is capable of independent replication in individual plant cells but this leads to the establishment of gene silencing, rather than a productive infection, in the absence of RNA-2. Both genomic RNAs have a small basic protein (VPg) covalently linked to their 5′ termini and both are polyadenylated at their 3′ ends. The elucidation of the overall structure of the RNA segments once more underscored the similarity between CPMV and picornaviruses. However, unlike picornaviruses, the VPg is linked to the viral RNA via the β-hydroxyl group of its N-terminal serine residue rather than via a tyrosine. The VPg is not required for the viral RNAs to be infectious.

The complete nucleotide sequences of both genomic RNAs were reported in 1983, making CPMV one of the first RNA plant viruses to be completely sequenced. The length of the RNAs are 5889 and 3481 nucleotides, for RNA-1 and RNA-2, respectively, excluding the poly(A) tails and the full sequences appear in GenBank under accession numbers NC_003549 and NC_003550. The two genomic RNAs have no sequence homology apart from that at the 5′ and 3′ termini. Full-length infectious cDNA clones of both RNAs of CPMV have been

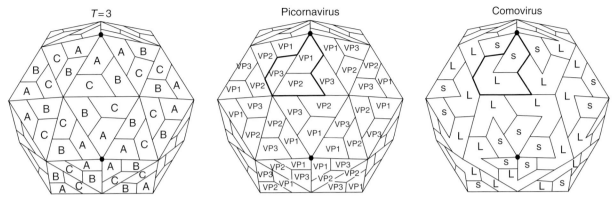

**Figure 2** Arrangement of the coat protein subunits of comoviruses (including CPMV) compared with those of simple $T=3$ viruses and picornaviruses. The asymmetric unit of $T=3$ viruses contains three β-barrels contributed by three coat protein subunits with identical amino acid sequences (labeled A, B, and C). The asymmetric unit of a picornavirus also contains three β-barrels, but in this case each is contributed by a different coat protein (labeled viral protein VP1, viral protein VP2, and viral protein VP3). The comovirus capsid is similar to that of a picornavirus except that two of the β-barrels (corresponding to viral protein VP2 and viral protein VP3) are fused to give the L protein. Reproduced from Lomonossoff GP, Shanks M, et al. (1991) Comovirus capsid proteins: Synthesis, structure and evolutionary implications. *Proceedings of the Phytochemical Society of Europe* 32: 76.

constructed, allowing the genome to be manipulated. This was an important development since it allowed both reverse genetic experiments to be undertaken and for the virus to be used for biotechnological applications.

## Expression of the Viral Genome

Both genomic RNAs contain a single long open reading frame, which occupies over 80% of the length of the RNA. A combination of *in vitro* translation and protoplast studies has unraveled the basic mechanism of gene expression of the virus. Both RNAs of CPMV are expressed through the synthesis and subsequent cleavage of large precursor polyproteins. This was the first example of a plant virus using this strategy for the expression of its genome, a strategy that was subsequently shown to be used by a number of other plant viruses. On RNA-1, initiation of translation occurs at the first AUG encountered on the sequence (at position 207) and results in the synthesis of a protein of approximately 200 kDa (the 200K protein). This initial product undergoes rapid cotranslational autoproteolysis to give proteins with apparent sizes of 32 and 170 kDa (the 32K and 170K proteins). The 170K protein undergoes further cleavages to give the range of virus-specific proteins shown in **Figure 3**. *In vitro* translation studies using mutant RNA-1 molecules have shown that all the cleavages occur most efficiently in *cis*. The 170K product can initially be cleaved at three different sites to give three different combinations of secondary cleavage products, 58K + 112K, 60K + 110K, and 84K + 87K. *In vitro*, and probably also *in vivo*, the 60K and 110K are stable and do not undergo further cleavage reactions. This is particularly curious in the case of the 110K protein as it contains both the 24K proteinase domain and a cleavage site. By contrast, the 112K and 84K proteins do undergo further cleavages. The end products of the cleavage pathway of the 170K protein are, from N- to C-terminus, the 58K protein, the VPg, the 24K proteinase, and the 87K protein.

Initiation of translation of RNA-2 occurs at two different positions on the RNA and results in the synthesis of two carboxy coterminal proteins, the 105K and 95K proteins (**Figure 3**). This double initiation phenomenon, which occurs as a result of 'leaky scanning', is found with the RNA-2 molecules of all comoviruses. In the case of CPMV, synthesis of the 105K protein is initiated from an AUG at position 161 while initiation from an AUG at position 512 directs the synthesis of the 95K protein. CPMV RNA-2 has an additional AUG (position 115) upstream of both these initiation sites but this feature is not conserved in the RNA-2 molecules of other comoviruses. Both RNA-2-encoded primary translation products are cleaved by the RNA-1-encoded proteolytic activity to give either the 58K or the 48K protein (depending on whether it is the 105K or 95K protein that is being processed) and the two viral coat proteins. Processing of the RNA-2-encoded polyproteins, at least at the site between the 48K and L coat protein, has been shown to require the presence of the 32K protein as well as the 24K proteinase.

## Functions of the Viral Proteins

Functions have been ascribed to most of the regions of the polyproteins encoded by both RNA-1 and RNA-2 of CPMV. In most cases, however, it is not certain at what stage(s) in the cleavage pathway they manifest their activity.

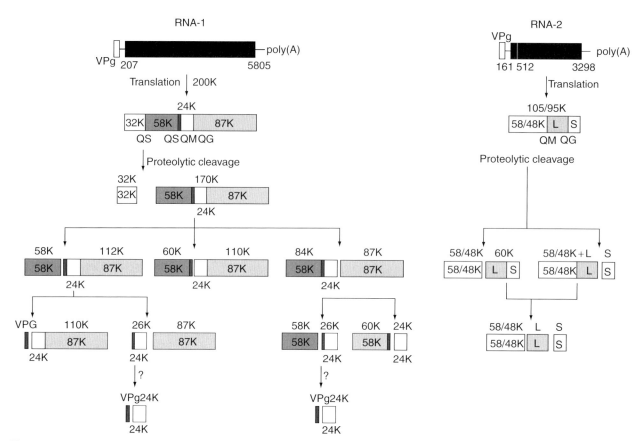

**Figure 3** Expression of CPMV RNA-1 and RNA-2. Both RNAs contain a single long open reading frame, which is processed to yield a series of proteins. The positions of the initiation and termination codons and the dipeptides at the proteolytic processing sites are shown.

In the case of RNA-1, the 32K protein, which is rapidly cleaved from the N-terminus of the 200K primary translation product, is a cofactor which modulates the activity of the virus-encoded protease. As described earlier, the presence of the 32K protein is required for the cleavage of the RNA-2-encoded 105K and 95K proteins but is not essential for the cleavage of the RNA-1-encoded 170K protein. It does, however, seem to play a role in determining the rate at which cleavage of the 170K protein occurs. When mutant RNA-1 molecules carrying deletions in the region encoding the 32K protein are translated *in vitro*, the rate of processing of the 170K protein is greatly increased, indicating that the 32K protein acts as an inhibitor of processing. This inhibition may be achieved through the interaction of the 32K with the 58K domain of the 170K protein. The mechanism by which the 32K protein enables the 24K proteinase to cleave in *trans* is unclear.

The RNA-1-encoded 58K protein is associated with cell membranes and contains a nucleotide-binding motif. The 60K protein (**Figure 3**), containing the amino acid sequence of the 58K protein linked to VPg, is involved in rearrangements in the endoplasmic reticulum of CPMV-infected cells and acts in concert with the 32K protein. The 24K protein is the virus-encoded proteinase that carries out all the cleavages on both the RNA-1- and RNA-2-encoded polyproteins. Its proteolytic activity has been shown to be expressed in a number of the processing intermediates that contain its sequence. Indeed, it is not known whether the free form of the protein has any biological significance. Although the proteinase contains a cysteine at its active site, it is structurally related to serine proteases, such as trypsin, rather than cellular thiol proteases, such as papain. In this regard, it is similar to the 3C proteinases of picornaviruses. All comoviral cleavage sites identified so far have glutamine (Q) residue at the −1 position. The enzyme encoded by a given comovirus is specific for the polyproteins encoded by that virus and is unable to cleave the polyproteins from other comoviruses either in *cis* or in *trans*.

The 87K protein is believed to contain the virus-encoded RNA-dependent RNA polymerase (RdRp) activity since it contains the G-D-D sequence motif found in all such enzymes. It also has amino acid sequence homology to the 3D$^{Pol}$ polymerases encoded by picornaviruses. However, when replication complexes capable of elongating nascent RNA chains were isolated from CPMV-infected cowpea plants, they were found to contain the 110K protein (**Figure 3**), consisting of the sequence of 87K protein linked to the 24K proteinase, rather than the free 87K protein.

In the case of RNA-2, the 48K protein, derived from processing of the 95K protein, is involved in potentiating the spread of the virus from cell to cell. This protein is found in tubular structures that are formed in the plasmodesmata of infected cells. Tubules extending into the culture medium can also be seen in protoplasts either infected with CPMV or transiently expressing the 48K protein. Virus particles can be seen within these tubules when protoplasts are infected with CPMV but not when only 48K protein is expressed. At present, no definite role has been assigned to the 58K protein, which is produced by processing of the 105K protein. Mutants in which translation of the 105K protein is disrupted replicate poorly, if at all. In light of these observations, it has been suggested that the 105K protein may play a role in the replication of RNA-2. Apart from containing many hydrophobic and aromatic amino acids, the approximately 10 kDa of protein present in the 58K but absent from the 48K protein is not conserved between comoviruses. The viral coat proteins are required to enable capsids to be formed. As well as protecting the genomic RNAs, capsid formation is essential for the virus to be able to spread from cell to cell through modified plasmodesmata and long-distance movement also requires capsid formation. An additional function in suppressing gene silencing is also provided by the C-terminal region of the S protein.

## Replication

CPMV replicates to high level in infected cells. Replication is believed to involve the initial transcription of the incoming positive-sense RNA into minus-strands followed by initiation and synthesis of new plus-strands from the recently formed minus-strands. It has been shown that the 5′ ends of both the plus- and minus-strands are covalently linked to the VPg, suggesting that this protein has an essential role in the initiation of RNA synthesis. There also appears to be a tight linkage between the translation of the viral RNAs and their replication.

Replication of the viral RNAs has been shown to occur in the membraneous cytopathological structures, which are formed in the cytoplasm of cells during infection through the action of the RNA-1-encoded 32 and 60K proteins. Both CPMV-specific double-stranded replicative form (RF) RNA and an enzyme activity capable of completing nascent RNA strands can be isolated from such structures. Purified preparations of the enzyme activity contain the RNA-1-encoded 110K protein and two host-encoded proteins of 68 and 57 kDa. However, at present, no enzymatic activity capable of initiating RNA synthesis *in vitro* has been described.

## Relationships with Other Viruses

Together with the genera *Nepovirus* and *Fabavirus*, the genus *Comovirus* belongs to the family *Comoviridae*. Within the family, the greatest affinity is between the genera *Comovirus* and *Fabavirus*. On a wider scale, consideration of genome structure and organization, translational strategy, and amino acid homologies between the virus-encoded proteins has led to grouping the family *Comoviridae* with the families *Potyviridae* and *Picornaviridae* as members of picorna-like superfamily of viruses. Members of this superfamily are all nonenveloped positive-strand RNA viruses with 3′ polyadenylated genomic RNAs, which have a protein (VPg) covalently linked to their 5′ ends. All members of the supergroup have a similar mode of gene expression, which involves the synthesis of large precursor polyproteins and their subsequent cleavage by a virus-encoded proteinase. The members of the superfamily all contain similar gene order, membrane-bound protein-VPg-proteinase-polymerase (see **Figure 3**) and share significant amino acid sequence homology in the membrane-bound proteins, the proteinases, and polymerase coding regions. Comovirus capsids are also clearly structurally related to those of picornaviruses (**Figure 2**).

## Use in Biotechnology

CPMV has been extensively used as a vector for the expression of foreign peptides and proteins in plants. To date, all vectors have involved modifications to RNA-2 (**Figure 4**). In the first instance, antigenic peptides (epitopes) were genetically fused to exposed loops on the surface of the viral capsids. The resulting chimeric virus

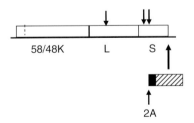

**Figure 4** Structure of CPMV RNA-2-based vectors used to express heterologous peptides and proteins in plants. Positions where epitopes have been inserted into the sequence of the coat proteins are shown by black arrows. The position where foreign proteins (shown hatched) have been inserted into RNA-2 is indicated. The FMDV 2A sequence is shown as a black box at the N-terminus of the foreign protein.

particles (CVPs) could be propagated in plants and the modified virions purified. When injected into experimental animals, CVPs can elicit the production of antibodies against the inserted epitope and in a number of instances can confer protective immunity against the pathogen from which the epitope was derived. This was a significant breakthrough as it represented the first instance where protection against an animal pathogen was conferred by material produced from a plant virus-based vector.

In an alternative approach, the sequence encoding an entire heterologous polypeptide has been fused to the C-terminus of the RNA-2-encoded polyprotein via a 2A catalytic peptide derived from foot-and-mouth disease virus (FMDV). The inclusion of the 2A sequence promotes efficient release of the foreign polypeptide from the polyprotein (**Figure 4**). This system has been used to express antibody derivatives in cowpea plants and crude plant extracts containing the antibodies have been shown to be capable of passively immunizing newborn pigs against challenge with the porcine coronavirus, transmissible gastroenteritis virus.

Recent developments in the use of comoviruses in biotechnology include the creation of combined transgene/viral vector systems based on CPMV, and the use of CPMV particles in bionanotechnology.

*See also:* *Nepovirus*; Plant Virus Diseases: Economic Aspects; Poliomyelitis; Virus Particle Structure: Nonenveloped Viruses.

## Further Reading

Cañizares MC, Lomonossoff GP, and Nicholson L (2005) Development of cowpea mosaic virus-based vectors for the production of vaccines in plants. *Expert Review of Vaccines* 4: 687–697.

Gergerich RC and Scott HA (1996) Comoviruses: Transmission, epidemiology, and control. In: Harrison BD and Murant AF (eds.) *The Plant Viruses 5: Polyhedral Virions and Bipartite RNA Genomes*, 77p. New York: Plenum.

Goldbach RW and Wellink J (1996) Comoviruses: Molecular biology and replication. In: Harrison BD and Murant AF (eds.) *The Plant Viruses 5: Polyhedral Virions and Bipartite RNA Genomes*, 35p. New York: Plenum.

Lin T and Johnson JE (2003) Structure of picorna-like plant viruses: Implications and applications. *Advances in Virus Research* 62: 167–239.

Lomonossoff GP, Shanks M, et al. (1991) Comovirus capsid proteins: Synthesis, structure and evolutionary implications. *Proceedings of the European Phytochemical Society* 32: 76.

# Cowpox Virus

**M Bennett,** University of Liverpool, Liverpool, UK
**G L Smith,** Imperial College London, London, UK
**D Baxby,** University of Liverpool, Liverpool, UK

© 2008 Elsevier Ltd. All rights reserved.

## History

The first published account of cowpox in man and cattle is probably that of Edward Jenner in his *Inquiry* published in 1798, although others, such as Benjamin Jesty, performed immunization of humans with cowpox material earlier. Jenner described the clinical signs of cowpox in both hosts, and how infection in man with *Variolae vaccinae* ('known by the name of the cowpox') provided protection against smallpox. At that time, smallpox was responsible for between 200 000 and 600 000 deaths each year in Europe and about 10% of all deaths in children. Jenner's discovery, despite the concern of some over the consequences of inoculating bovine material into man, soon led to the establishment of smallpox vaccination schemes around the world. However, not until Pasteur's work *c.* 100 years later was the principle of immunization used again. In fact, it was Pasteur who suggested that all such immunizations be called vaccines in honor of Jenner's work.

Although Jenner's first vaccines probably came indirectly from cattle, later vaccine material was often derived from horses, and the origin(s) of modern vaccinia virus (VACV) the smallpox vaccine, remain unknown. That cowpox virus (CPXV) and VACV are different was first published in 1939, since when further biological and genetic studies have confirmed that VACV represents a species in its own right, and is not simply a mutant of CPXV or a recombinant of variola virus (VARV) and CPVX.

Even Jenner seems to have had difficulty finding cowpox cases, and CPXV is not endemic in cattle. Rather, it is endemic in rodents, and cattle and man are merely accidental hosts. The domestic cat is the animal diagnosed most frequently with clinical cowpox in Europe.

## Taxonomy and Classification

CPXV represents the species *Cowpox virus*, a member of the genus *Orthopoxvirus* in the family *Poxviridae*, and the international reference strain, Brighton Red, was isolated from farm workers in contact with infected cattle in 1937. CPXV can be differentiated from other orthopoxviruses

(OPVs) by a combination of biological tests, including the ability to produce hemorrhagic pocks on chorioallantoic membranes, the production of A-type inclusions (ATIs) in infected cells, and its ceiling temperature for growth (40 °C, the highest temperature at which the virus will replicate), by minor antigenic differences, by restriction enzyme digestion of the entire genome, particularly with *Hin*dIII, by sequencing of certain genes, and by polymerase chain reaction (PCR). There is considerably more variation between some CPXV isolates than between strains of other OPV species, such that some CPXV strains might be reclassified as separate species rather than strains of the same species. This variation is seen in biological properties (e.g., ceiling temperature of growth, and possibly virulence in different hosts) and genome (restriction enzyme fragment polymorphism, gene content, and nucleotide sequence). It is not yet known whether these differences reflect variation in geographic range or reservoir host.

## Properties of the Virion

CPXV has a typical OPV morphology, and is indistinguishable from VACV by electron microscopy. Virions are brick shaped, approximately 300 nm × 200 nm × 200 nm in size, and are enveloped. There are two types of virion. The simpler form, termed intracellular mature virus (IMV), consists of a biconcave core, and within each concavity lies a lateral body. The whole is surrounded by a lipid membrane, and an outer layer of protein. The core contains DNA and proteins, many of which are virus-encoded enzymes. The second form is called extracellular enveloped virus (EEV), and these virions are surrounded by an additional lipid envelope that is fragile and is derived from either the *trans*-Golgi network or endosomes.

The life history of CPXV within the cell is very similar to that of VACV, except that intracellular nonenveloped, yet infectious, virions can become incorporated into large intracytoplasmic ATIs (**Figure 1**). It is thought that these inclusions help protect the virus after cell lysis, and so are important in survival in the environment and spread from animal to animal. In contrast, EEV is more important for spread within individual hosts.

## Properties of the Genome

The genome consists of linear, double-stranded DNA with covalently linked inverted repeats at the termini. The CPXV genome is the largest of all the OPV genomes and for strain Brighton Red is 224 501 bp. Restriction endonuclease mapping and nucleotide sequencing demonstrated that the middle portion of the CPXV genome (approximately 100 kbp) is highly conserved between OPVs, but more variation occurs toward either

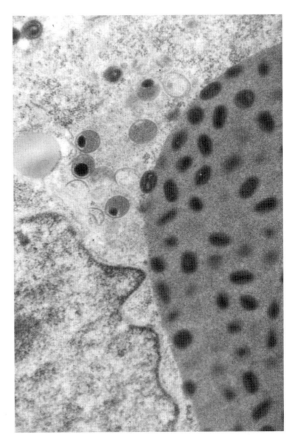

**Figure 1** A CPXV-infected cell, showing intracytoplasmic virion synthesis and an ATI containing intracellular mature virions.

end. Digestion with *Hin*dIII usually differentiates CPXVs from other known OPVs, but isolates, particularly geographically distinct strains, do vary in profile, and digestion of genomic DNA with other enzymes often reveals much greater differences between strains. However, overall the genome is fundamentally very stable. The main exception to this is the deletion of 32–39 kbp of DNA from the right end of the genome and its replacement by DNA from the left end: these terminal transpositions are the cause of the 1% white pocks observed on infected chorioallantoic membranes. The size of the transposed fragment varies (5–50 kbp) but the net effect is a change in the length of the inverted terminal repeat (ITR), the duplication of some genes, and the loss of others.

The nucleotide sequences of two CPXV genomes, and of particular genes for a much greater number of isolates, have been determined. CPXV not only has the largest OPV genome, but it also encodes the greatest number of complete protein-coding open reading frames (ORFs) and has relatively few genes that are broken by mutation into fragments. Such broken ORFs are more common in other OPVs such as VARV, camelpox virus (CMLV), and taterapox virus (TAPV). These observations suggest that CPXV might be closest in its genetic complement to the ancestral poxvirus from which OPVs have evolved. PCR-based diagnostic methods are now used to distinguish CPXV from other OPVs.

## Properties of Proteins

The genome of CPXV Brighton Red is predicted to encode more than 200 proteins. Like other OPVs, the proteins encoded in the central part of the genome are mostly essential for virus replication, and in this region 89 genes have counterparts in all sequenced chordopoxviruses. Generally, these CPXV proteins are very closely related to their counterparts in VACV, VARV, and other OPVs. This is certainly true for the proteins that are present in the infectious virion, and the finding that the proteins on the surface of IMV and EEV are very closely related provides a plausible explanation why VACV and CPXV were effective vaccines against smallpox. In contrast, genes located toward the genome termini are nonessential for virus replication in cell culture and these include approximately half of all CPXV genes. Notably, CPXV encodes many genes that are not found in VACV strains.

Interest has focused on proteins concerned with immune modulation, and these are encoded mostly near the left and right ends of the genome. Experimentation has demonstrated that several of these genes (or their orthologs in VACV) are associated with virulence in accidental hosts, but their functions in natural rodent hosts may be more subtle. Possibly their role in natural infection, where clinical disease is not seen and presumably not relevant to transmission, is to moderate the host's immune response, ensuring efficient contact between individuals and therefore transmission.

Other proteins expressed by CPXV but not VACV have been studied, such as those affecting the ability to grow on Chinese hamster ovary (CHO) cells and produce ATIs. Growth of CPXV in CHO cells is dependent on production of a 77 kDa protein and the insertion of the gene encoding this protein into VACV enables that virus to replicate in CHO cells. Formation of ATIs involves the production of two late proteins. One, of 160 kDa, is the major component of the inclusion itself. The other is required for occlusion of virions into the ATI.

A red pock character is associated with the *crmA* gene, which encodes a 38 kDa protein that is one of the most abundant early gene products and a serine protease inhibitor that inhibits apoptosis and the cleavage of pro-interleukin (IL)-1β to IL-1β. This gene is lost in the frequent deletion mutants referred to above. The essential difference between the wild-type and mutant white pock is the greater hemorrhage in the former and the massive leukocyte infiltration in the latter. Although a protein with 93% amino acid sequence identity is encoded by VACV strain Western Reserve, pocks formed by this virus are white irrespective of whether or not the protein (B13) is expressed.

## Physical Properties

Little work has been done directly on the physical properties of CPXV, which are usually assumed to be similar to those of VACV and ectromelia virus (ECTV). The outer lipid envelope of extracellular forms of CPXV is labile and readily lost by mechanical stress, but is not essential for virus infectivity because its removal releases an infectious IMV particle. The virus is generally very hardy, and can survive for months in dry scab material at room temperature, and indefinitely at −70 °C. It is inactivated by moist heat at 60 °C for 10 min, and by hypochlorites, phenolics, and detergents, but less effectively by alcohols. The significance of survival in the environment to the epidemiology of cowpox is unknown. The transmission dynamics of natural infections suggest that most transmission is direct. However, in populations of rodents on small islands, infection occurs as small epidemics followed by apparent extinction: whether these epidemics result from immigration of infected individuals, or survival of the virus in the environment with rare reintroduction into naive populations, has not been resolved.

## Replication

The mechanisms of DNA replication and temporally regulated transcription and translation of CPXV mRNA are assumed to be very similar to those in VACV. Virus replication takes place in the cytoplasm, and virus assembly occurs in areas known as B-type inclusions. Like VACV, CPXV virions can leave the cell as IMV by lysis of the cell membrane, or as EEV by exocytosis following fusion between the outer membrane of intracellular enveloped virus (IEV) and the plasma membrane. CPXV differs from VACV in that its genome also encodes for large proteinaceous inclusions known as ATIs, into which virions are incorporated (**Figure 1**). These inclusions are released by cell lysis, and are thought to act as protective packets that aid survival of the virus outside of the animal and therefore increase the chance of spread to another host on fomites.

## Geographic and Seasonal Distribution

CPXV has been isolated, or detected by PCR and sequencing, throughout much of Northern Europe and as far east as Kazakhstan. Although cases of human and feline cowpox may be seen at any time of year, infection is most common in the late summer and autumn, which probably reflects increased infection in the reservoir hosts, which in turn reflects increased numbers of hosts at that time of year.

## Host Range and Propagation

CPXV has a wide host range both *in vivo* and *in vitro*. It has been isolated from cattle, humans, domestic cats (perhaps the most common source of human infection), dogs, a horse, and a variety of zoo animals including cheetahs, ocelots, panthers, lynx, lions, pumas, jaguars, anteaters, elephants, rhinoceroses, and okapis. All of these are, however, accidental hosts, and the virus circulates mainly in wild rodents. The main reservoir hosts appear to be voles (*Clethrionomys* spp. and *Microtus* spp.) and wood mice (*Apodemus* spp.) throughout the virus' range. House mice (*Mus musculus*) and rats (*Rattus norvegicus*) are infected rarely and are probably accidental hosts (explaining the limited geographic range of the virus), although they may, like cats, act as liaison hosts and transmit the infection onward to man. Other rodents may also act as reservoirs toward the eastern range of the virus. CPXV has been isolated from wild susliks and gerbils (*Rhombomys* spp., *Citellus* spp., and *Meriones* spp.) in Turkmenia. Guinea pigs and rabbits have been infected experimentally with CPXV.

CPXV can be isolated and propagated on the chorioallantoic membrane of hens' eggs, but, unlike VACV, it does not grow in feather follicles of adult chickens. It can be propagated in a variety of cell cultures derived from human, simian, bovine, feline, murine, and rabbit tissues.

## Genetics

CPXV produces red hemorrhagic pocks on the chorioallantoic membrane with about 1% white pock mutants. The virus that produces the white pocks breeds true and the change in pock character reflects deletion and transposition events near the termini of the genome. White pock mutants are better able to grow on arginine-deprived cells than is the parent virus. Other properties that are inherited independently, and vary between individual strains, include production of the hemagglutinin, incorporation of virions into ATIs, resistance to heat inactivation, ceiling temperature for growth, and virulence for newborn mice and chick embryos. OPVs undergo genetic recombination in dual-infected cells, for example producing hybrids of CPXV and VARV.

## Evolution

The genetic relationships of CPXV and other OPVs have been studied by bioinformatics. Nucleotide sequencing demonstrated that the OPVs are very closely related as a whole (although the North American OPVs appear more distantly related), but each species can be readily differentiated by biological properties and genome structure. Bioinformatic analyses demonstrated that the two CPXV genomes sequenced are quite divergent and it was proposed they should be reclassified as separate species. As noted above, the presence of the greatest number of intact genes and fewest broken genes suggests that CPXV is closest to the ancestral OPV. With the exception of terminal transposition events (see above), CPXV isolates are genetically stable *in vitro* and *in vivo*. Modern isolates often have near-identical restriction enzyme profiles to isolates made many years ago, and the Brighton Red strain behaves the same now as when first isolated over 60 years ago. However, isolates of CPXV do differ in restriction enzyme pattern and minor biological characteristics, and generally greater differences occur between geographically distinct isolates. Furthermore, the greatest variation appears to be seen among isolates in central Europe, and this may reflect a central European origin for CPXV and divergence as individual strains spread out into different host reservoirs and different geographical areas.

CPXV isolates are all clearly different from other OPVs, such as VACV and VARV, but the genome sequence of horsepox virus (from Central Asia) shows that it is intermediate between VACV and CPXV and shares some properties of each virus.

## Serologic Relationships and Variability

Within the genus *Orthopoxvirus*, there is extensive antigenic cross-reactivity in all serologic tests, although minor differences can allow differentiation of species using monoclonal antibodies. No significant serologic differences have been reported among CPXV strains.

## Epidemiology

CPXV is rarely isolated from cattle, and serologic surveys show that cattle are not the reservoir host. Most human infection cannot be traced to contact with infected cattle, but about half of the recent human cases in Britain were traced to contact with an infected cat. The domestic cat, although the species in which clinical cowpox is most frequently diagnosed in Europe, is not the reservoir host of CPXV either. Although cat-to-cat transmission can occur, antibody to CPXV is uncommon in surveys of healthy cats. There is no evidence that CPXV can become endemic in any of the zoo animals that have been infected, but with increasing reliance on zoo populations for the survival of many species, these outbreaks may cause conservation problems.

Rather, CPXV is endemic in rodent populations. Transmission has been studied in most detail in several populations of voles (*Clethrionomys glareolus* and *Microtus agrestis*) and wood mice (*Apodemus sylvaticus*) in northern England, where it has been used as a model system for investigating the ecology of endemic infections in wildlife populations. Transmission dynamics appear linked to host population dynamics, with the highest incidence of infections occurring in the autumn when population sizes are at their greatest. This is reflected in the highest incidence of infection in cats and human beings also at this time of year. Although cross-species transmission may occur, empirical evidence combined with mathematical models suggests that most transmission occurs among members of the same rodent species, even where two rodent host species share the same habitat.

## Transmission and Tissue Tropism

Both natural rodent hosts and laboratory rodents can be infected with CPXV by the oral and respiratory routes as well as skin inoculation: natural hosts can often be infected with less virus than is required to infect cell culture, as is the case with ECTV where the lethal dose for some strains of mice may be less than one plaque-forming unit. However, it is not known how the virus is transmitted in the wild, and transmission has not occurred among laboratory-housed rodent hosts. This may be because a particular behavior, not elicited in the laboratory, is involved in transmission. Transmission rates in the wild can be very high, and longitudinal studies of naturally infected populations suggest that much transmission, although not all, is frequency, rather than density, dependent, which might itself suggest an important role for particular behaviors in transmission.

Among the occasional hosts of CPXV, the most frequent route of infection appears to be through the skin, probably through a cut or abrasion. Domestic cats, however, can be infected experimentally by oronasal inoculation, and limited respiratory spread sometimes occurs in domestic or zoo cat colonies.

Virus replication in cattle and man is mainly limited to the epidermis at the site of entry, and possibly also to draining lymph nodes. In cats, virus can be isolated not only from skin lesions but also from lymphoid, lung, and turbinate tissue. Skin inoculation of cats is followed by virus replication both at the site of entry and in draining lymph nodes, which leads to the development of a viremia, and virus can be isolated from the white cell fraction of blood, from the spleen, and other lymphoid organs. After about 7–10 days, virus can be detected in the epidermis, leading to the development of secondary skin lesions. The viremia in cats appears to last 1–8 days, and no virus has been isolated from cats after the skin lesions have healed, which may take 5–6 weeks.

Oral, nasal, and skin inoculation of various rodents with CPXV also causes systemic infection. Virus has been isolated from lung, kidney, liver, and lymph nodes of susliks and gerbils, and detected by PCR in a similar range of tissues, and the cellular fraction of blood from naturally and experimentally infected British wild rodents. Laboratory infection and studies of naturally infected wild voles and wood mice suggest that they remain infected for around 4 weeks before clearing the infection.

## Virulence

It is not known whether different strains of CPXV vary in virulence for most accidental hosts, including man. In a small-scale experiment, no differences in the ability to cause infection or a primary lesion in domestic cats could be detected between the Brighton Red strain and isolates from a cheetah and a domestic cat. Differences in virulence of different strains do exist for newborn laboratory mice and chick embryos, but these differences are not associated with the ability to infect various accidental hosts and it is not known whether they have any significance in the maintenance of different strains of CPXV in different reservoir tests.

Wild-type CPXV outgrows the white pock mutants and is more virulent for laboratory animals.

## Clinical Features of Infection

In cattle, CPXV causes teat lesions, but little apparent systemic disease. Human infection is characterized usually by a single skin lesion, often on a hand or the face. Spread of skin lesions in man is usually the result of direct transmission, for example, from hand to face, but multiple lesions may also occur if there is a preexisting skin condition. Cowpox in man is often accompanied by systemic signs such as nausea, fever, and lymphadenopathy, and children are often hospitalized. Death is rare, and usually the result of an underlying condition, such as immunodeficiency, which increases the severity of disease.

CPXV infection in domestic cats is usually a more severe disease than in cattle or man. There is often a history of a single primary lesion, especially on the head or a forelimb, but by the time the cat is presented for veterinary attention widespread skin lesions have usually developed. The primary lesions vary enormously in character, and secondary bacterial infection is common. The widespread secondary lesions first appear as small erythematous macules, which develop into papules and ulcers over several days. These scab over, and the cat usually recovers within 6–8 weeks. Cats may be slightly pyrexic in the early stages of the disease and some show signs of mild upper respiratory disease. More severe illness such as large nonhealing lesions or pneumonia usually, but not always, results from secondary bacterial infection or immunosuppression. In some zoo-kept cats, such as cheetahs, pneumonia is more common and is associated with a high mortality rate.

No obvious clinical signs are observed in either naturally or experimentally infected rodents (unless the experimental dose used is high). However, longitudinal studies of naturally infected wild voles and wood mice, at both the individual and population levels, as well as experimental infections, have shown an effect on rodent fecundity. Infected voles and mice delay reproduction compared to uninfected animals in the wild by perhaps a whole season.

## Pathology and Histopathology

Skin lesions associated with cowpox in most accidental hosts are typical of those expected of an OPV infection, developing through papule, vesicle, pustule, ulcer, and healing stages, although macroscopic vesicles often ulcerate quickly because of abrasion or, in the case of domestic cats, because the epidermis is too thin in most areas to support a vesicle. Microscopic examination reveals hypertrophy and hyperplasia of infected cells, multilocular vesicle formation, large, intracytoplasmic eosinophilic inclusion bodies (ATIs) (**Figure 1**) in epithelial cells, and a vigorous polymorph infiltration of the dermis. Immunostaining demonstrates virus antigen in epithelial cells of the skin, hair follicles, and sebaceous glands, and in dermal macrophages.

Even in lungs and turbinates of cats from which large amounts of virus can be isolated, there are often no gross lesions, and microscopic lesions may be difficult to find. In cats showing clinical signs of pneumonia, there is often an interstitial pneumonia, and, again, eosinophilic inclusions can be seen in infected cells. The tonsils and lymph nodes of infected cats contain many large reactive follicles, and immunostaining may demonstrate antigen in macrophage-like cells. Some follicles have large necrotic centers, suggesting that virus replication is occurring here.

After 3 days of growth on the chorioallantoic membrane of 12 day chick embryos, CPXV causes hemorrhagic pocks approximately 2 mm in diameter, with a few (usually 1%) white pocks. Microscopically, the red pocks consist of ectodermal proliferation and hypertrophy with many cells containing ATIs, and extensive edema and hemorrhage into the mesoderm. The histopathology of white pocks is similar but consists of more inflammatory infiltration and no hemorrhage.

## Immune Response

Relatively little is known about the immune response to CPXV in naturally infected hosts. Antibody can be detected by enzyme-linked immunosorbent assay (ELISA), immunofluorescence, virus neutralization (VN) (usually done with the IMV form of virus) with or without complement, complement-fixation, and hemagglutination-inhibition (HAI) tests. HAI antibody can be detected before VN antibody, and in cattle and cats begins to decline after about 6 months. HAI antibody is therefore more useful for the diagnosis of acute infections than VN antibody, and in epidemiological studies indicates more recent infection than VN antibody alone.

As in VACV and ECTV infections, cell-mediated immunity plays an important role in protection against CPXV disease, but less work has been published specifically on the cell-mediated immunity to CPXV. A possible delayed-type hypersensitivity response has been reported in cats.

## Prevention and Control

Infection in man and domestic animals is relatively uncommon and so measures to prevent infection are generally not warranted. Vaccination against smallpox is no longer routine, and might not protect against the development of a skin lesion, but would reduce the severity of any systemic illness. VACV infection of cattle and man often causes lesions and disease similar to cowpox. VACV does not grow well in cats, and its efficacy as a vaccine (although not necessarily as a vaccine vector) in felids is uncertain, although it has been recommended for some zoo animals.

Although CPXV can cause quite severe disease in man, it does not appear to be very infectious. Human-to-human spread of cowpox has not been reported (in contrast to VACV). Many human cases of cowpox have been traced to contact with infected cats, but we know of no cases of cat-to-human transmission after cowpox was diagnosed in the cat. Simple hygiene – washing hands after handling the cat, keeping the cat or scab material away from cuts and the eyes – seems adequate to prevent transmission to man, although special measures might be taken for the very young, elderly, or immunosuppressed. Similarly, if an outbreak occurs in cattle, the main route of spread among the cows is through milking equipment, and simple hygiene should suffice to control spread.

## Future Perspectives

CPXV is one of several OPVs that have wildlife reservoirs; others include monkeypox virus, raccoonpox virus, and Californian volepox virus. Buffalopox virus is now regarded as a variant of VACV and may also have a wild animal reservoir in India. Occasionally, these and other OPVs may infect other, accidental, hosts, such as domestic animals: there are reports of uncharacterized OPVs being isolated from horses, for example, in Africa, North America, and Australia, and a recent report of raccoonpox virus infection in a cat in Canada. Study of the ecology of CPXV is therefore useful as a model for OPV maintenance in a wildlife reservoir host and the mechanisms of transmission from animals to man. It is also being studied in wild rodents as an ecological model of transmission of endemic infections, and the interactions between host and parasite dynamics.

*See also:* Adenoviruses: Pathogenesis; Mousepox and Rabbitpox Viruses; Smallpox and Monkeypox Viruses; Vaccinia Virus.

## Further Reading

Baxby D, Bennett M, and Getty B (1994) Human cowpox; A review based on 54 cases, 1969–93. *British Journal of Dermatology* 131: 598–607.
Bennett M, Gaskell RM, and Baxby D (2005) Feline poxvirus infection. In: Greene C (ed.) *Infectious Diseases of the Dog and Cat,* 3rd edn., pp. 158–160. Philadelphia: WB Saunders.
Carslake D, Bennett M, Hazel S, Telfer S, and Begon M (2006) Inference of cowpox virus transmission rates between wild rodent host classes using space–time interaction. *Proceedings of the Royal Society Series B* 272: 775–782.
Essbauer S and Meyer H (2007) Genus *Orthopoxvirus*: Cowpox virus. In: Mercer AA, Schmidt A and Weber O (eds.) *Poxviruses*, pp. 75–88. Basel: Birkhäuser Verlag.
Fenner F, Wittek R, and Dumbell KR (1989) *The Orthopoxviruses.* London: Academic Press.
Gubser C, Hué S, Kellam P, and Smith GL (2004) Poxvirus genomes: A phylogenetic analysis. *Journal of General Virology* 85: 105–117.
Moss B (2001) *Poxviridae*: The viruses and their replication. In: Fields BN, Knipe DM, Howley PM, *et al.* (eds.) *Virology,* 4th edn., pp. 2849–2883. Philadelphia: Lippincott-Raven Publishers.
Pickup DJ, Ink BS, Hu W, Ray CA, and Joklik WK (1986) Hemorrhage in lesions caused by cowpox virus is induced by a viral protein that is related to plasma protein inhibitors of serine proteases. *Proceedings of the National Academy of Sciences, USA* 83: 7698–7702.
Shchelkunov SN, Safronov PF, Totmenin AV, *et al.* (1998) The genomic sequence analysis of the left and right species-specific terminal region of a cowpox virus strain reveals unique sequences and a cluster of intact ORFs for immunomodulatory and host range proteins. *Virology* 243: 432–460.
Telfer S, Bennett M, Bown K, *et al.* (2002) The effects of cowpox virus on survival in natural rodent populations: Increases and decreases. *Journal of Animal Ecology* 71: 558–568.
Telfer S, Bennett M, Bown K, *et al.* (2005) Infection with cowpox virus decreases female maturation rates in wild populations of woodland rodents. *Oikos* 109: 317–322.
Tulman ER, Delhon G, Afonso CL, *et al.* (2006) Genome of horsepox virus. *Journal of Virology* 80: 9244–9258.

## Relevant Website

http://www.poxvirus.org – Poxvirus Bioinformatics Research Center.

# Coxsackieviruses

**M S Oberste and M A Pallansch,** Centers for Disease Control and Prevention, Atlanta, GA, USA

Published by Elsevier Ltd.

## Glossary

**Chemosis** Edema of the bulbar conjunctiva.
**Conjunctival hyperemia** Increased blood flow to the conjunctiva.
**Cytopathic effect** Degeneration of cultured cells due to virus infection and replication, often characteristic of a given virus.
**Exanthem** Illness characterized by skin eruption (rash).
**Myalgia** Nonspecific muscle pain or tenderness.
**Nuclear pyknosis** Condensation and reduction in size of the nucleus, often the result of a pathogenic process.
**Pleurodynia** Illness characterized by chest pain due to inflammation of the pleura, the membrane surrounding the lungs and chest cavity.
**Protomer** Basic unit of a virus capsid, containing one or more viral proteins.

## History

The coxsackieviruses (genus *Enterovirus*, family *Picornaviridae*) were discovered in the late 1940s, as a result of intense efforts to develop better systems to propagate and study poliovirus. At the time, nonhuman primates were the model system for poliovirus and poliomyelitis, as cell culture and inoculation of suckling mice were just being introduced. Investigators at the New York State Department of Health inoculated suckling mice with fecal suspensions from two suspected poliomyelitis cases. The mice became paralyzed, but the viruses were not polioviruses – the virus was later named 'coxsackievirus' (CV) for the patient's home town, Coxsackie, New York. Subsequent studies identified related viruses, some causing spastic paralysis rather than the flaccid paralysis observed with the initial isolates. Based on these differences in pathogenicity, the viruses were classified as group A coxsackieviruses (flaccid paralysis) or group B coxsackieviruses (spastic paralysis), with the individual virus types being numbered sequentially within each group (e.g., coxsackievirus A2 or coxsackievirus B5). Once cell culture became established as a standard laboratory technique, additional viruses were discovered that were able to replicate in culture but caused no disease in mice. Since many of these were derived from stools of healthy individuals and they were not yet known to cause disease in humans, they were termed enteric cytopathic human orphan viruses or 'echoviruses' – these were also named using consecutive numbers (e.g., echovirus 9). The early coxsackieviruses had been isolated from cases of 'nonparalytic poliomyelitis' (aseptic meningitis), showing that this disease was not necessarily caused by polioviruses. Other illnesses, including herpangina, rash, pleurodynia, and myocarditis, were also shown to be associated with coxsackievirus infection; later, the echoviruses were later shown to be associated with many of these same illnesses.

## Taxonomy and Classification

As mentioned, the term 'coxsackievirus' was applied to enteroviruses that caused paralysis upon intracerebral inoculation in suckling mice, with the type of paralysis – flaccid versus spastic – being used to place strains in group A or group B. When it was discovered that antigenically related viruses could have varying degrees of pathogenicity in mice, the original naming convention was abandoned and since 1967 all new enteroviruses (whether coxsackievirus-like or echovirus-like) have been named 'enterovirus' followed by a sequential number, starting with enterovirus 68. Enteroviruses have traditionally been classified antigenically, using well-characterized, standardized antisera in a neutralization assay, either in cell culture or in suckling mice (for viruses that replicate poorly in culture). Using this approach, 30 coxsackievirus serotypes were defined, with 24 in group A and six in group B. Coxsackievirus A23 was later discovered to be antigenically identical to echovirus 9, so CVA23 was dropped as a distinct serotype.

It was recognized early on that there were many limitations to the antigenic typing approach. In addition to the labor-intensive nature of the neutralization assay and the requirement for extensive characterization of new antisera, different serotypes may cross-react in complex combinations and individual strains of a given type may react differently with different preparations of homotypic antisera. Once nucleotide sequences became available for a large number of serotypes, it became clear that the sequences of certain parts of the capsid-coding region could serve as a surrogate for antigenic type. In the late 1990s and early 2000s, several investigators developed typing systems based on polymerase chain reaction (PCR) amplification of a portion of the capsid region, followed by sequencing and analysis of the sequence to determine type. Eventually, VP1 was settled upon as the most reliable region for molecular virus typing. Sequence comparisons have shown that CVA15 is a strain of CVA11 and CVA18 is a strain of CVA13.

Sequence relationships are now recognized as a primary discriminating characteristic in enterovirus taxonomy. There are currently eight species in the genus *Enterovirus*: *human enterovirus* (HEV) A–D, *Poliovirus*, *Bovine enterovirus*, *Porcine enterovirus*, and *Simian enterovirus A*. There are current proposals to merge *Poliovirus* with HEV-C and to move the two human rhinovirus species from their own genus into *Enterovirus*. The coxsackieviruses are distributed among three species: HEV-A (CVA2–8, CVA10, CVA12, CVA14, and CVA16), HEV-B (CVA9, CVB1–6), and HEV-C (CVA1, CVA11, CVA13, CVA17, CVA19–22, and CVA24) (**Table 1**).

## Host Range and Virus Propagation

Coxsackieviruses are primarily human pathogens but many can also infect certain nonhuman primate species and all serotypes are able to infect and cause disease in mice (by definition), though there is some variability among strains of any given type. Infection of nonhuman primates often fails to induce clinical disease, but CVA7

**Table 1** Coxsackievirus taxonomy

| Species | Serotype |
| --- | --- |
| Human enterovirus A[a] | CVA2–8, CVA10, CVA12, CVA14, CVA16 |
| Human enterovirus B[b] | CVA9, CVB1–6 |
| Human enterovirus C[c] | CVA1, CVA11, CVA13, CVA17, CVA19–22, CVA24 |

[a]Also includes EV71, EV76, EV89–92.
[b]Also includes all echoviruses, EV69, EV73–75, EV77–88, EV93, EV97–98, EV100–101.
[c]Also includes polioviruses, EV95–96, EV99, and EV102.

can induce polio-like paralysis in monkeys. Swine vesicular disease virus, an important livestock pathogen because symptoms of illness are similar to those of foot and mouth disease, is closely related genetically and antigenically to CVB5. Genetic and epidemiologic studies strongly suggest that human CVB5 was introduced into swine decades ago, with subsequent adaptation and diversification in the new host species.

The best specimens for virus isolation are stool specimens or rectal swabs, throat swabs or washings, and cerebrospinal fluid, in that order. Isolation from stool is most sensitive because virus is usually present at higher titer and it is present in stool longer than in any other specimen. Nonfecal specimens are most likely to yield virus isolates if they are obtained early in the acute phase of illness. For cases of acute hemorrhagic conjunctivitis, the best specimens are conjunctival swabs and tears. A number of different primary cell cultures and continuous cell lines, both generally of human or monkey origin, have been used for isolation and propagation of coxsackieviruses. These cell systems include primary monkey kidney cells; monkey kidney cell lines such as BGM, Vero, and LLC-MK2; and human cell lines such as HeLa, Hep2, KB, and RD. Virus infection and replication results in a characteristic cytopathic effect which is observed microscopically 1–7 days after inoculation. The cells become rounded and refractile, with nuclear pyknosis and cell degeneration. Ultimately, the cells are lysed and become detached from the surface of the culture vessel.

Attempts at virus isolation in cell culture may sometimes be unsuccessful, necessitating the inoculation of suckling mice. For example, CVA1, CVA19, and CVA22 are rarely, if ever, isolated in culture, but grow readily in suckling mice. If the virus titer is extremely low, blind passage in mice may be necessary; blind passage may also be needed to allow the virus to adapt to growth in mice. The two groups of coxsackieviruses can be distinguished by the distinct pathology that they cause in mice. With CVA infection, newborn mice develop flaccid paralysis and severe, extensive degeneration of skeletal muscle (sparing the tongue, heart, and CNS), and they may have renal lesions. Death usually occurs within a week. CVB infection proceeds more slowly and is characterized by spastic paralysis and tremors associated with encephalomyelitis, focal myositis, brown fat necrosis, myocarditis, hepatitis, and acinar cell pancreatitis.

## Epidemiology, Geographic Distribution, and Seasonality

The coxsackieviruses are distributed worldwide, but there can be significant geographic and temporal variation in prevalence of individual serotypes. These differences are attributed to differences in climate and public hygiene, overall population immunity (e.g., an increase in the number of susceptible individuals in the years following an outbreak), and other factors. In the United States, for example, CVA9 is among the most commonly reported enteroviruses, but circulation tends to peak every 3–5 years. CVB2, CVB4, and CVB5 are also common, with cyclic peaks of activity, and CVB5 has been associated with large outbreaks, while CVB1 and CVB3 circulate at low, relatively constant levels, and CVB6 is rarely reported. Like the other enteroviruses, most coxsackievirus infections tend to occur during the warmer months in areas with a temperate climate, peaking in late summer or fall. In the tropics, peak activity generally correlates with the rainy season.

Coxsackieviruses are isolated in the highest titer and for the longest time in stool specimens but can also be isolated from respiratory secretions. Fecal-oral transmission and spread by contact with respiratory secretions (person-to-person, fomites, and possibly large particle aerosol) are the most important modes of transmission. The relative importance of the different mode probably varies with the virus and the environmental setting. Viruses that cause a vesicular exanthem are also presumably spread by direct or indirect contact with vesicular fluid, which contains infectious virus. Exceptions to the usual modes of enterovirus transmission are the agents of acute hemorrhagic conjunctivitis, coxsackievirus A24 variant (CA24v), and enterovirus 70 (a member of HEV-D). These two viruses are seldom isolated from respiratory or fecal specimens and are probably transmitted primarily by direct or indirect contact with eye secretions.

## Virion Structure and Host Cell Receptors

Coxsackievirus virions, like those of other picornaviruses, are approximately 30 nm in diameter, with little discernable fine-scale structure (**Figure 1**). Sixty copies of each of the mature virion proteins, VP1–VP4, form an icosahedral virion particle with pseudo $T=3$ symmetry. The proteins VP1–VP3 combine to form the virion protomer, with VP4 internal to the particle.

Virus infection is dependent on the presence of specific receptors. At least four distinct receptors are used by coxsackieviruses for entry into human cells. CVA9 uses integrin $\alpha_v\beta_3$, CVB1–6 use the 'coxsackievirus–adenovirus receptor' (CAR) and some may also use decay-accelerating factor (DAF), and some of the HEV-C coxsackieviruses use intracellular adhesion molecule 1 (ICAM-1). The receptor(s) used by most viruses in HEV-A, and by many of the viruses in HEV-C remain unknown.

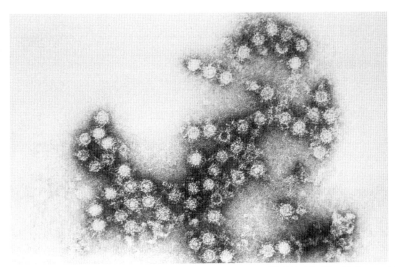

**Figure 1** Electron micrograph of coxsackievirus B4 virions. Image courtesy of CDC Public Health Image Library.

## Genetics, Genetic Diversity, and Evolution

Like the other enteroviruses, the coxsackievirus genome is a single-stranded, polyadenylated, positive-sense RNA of approximately 7.4 kbp, with a 22-amino-acid virus-encoded protein ($3B^{VPg}$) covalently linked to the 5' end. Flanked by 5'- and 3'-nontranslated regions (NTRs), the single long open reading frame encodes a polyprotein of approximately 2200 amino acids that is processed during and following translation by viral proteases to yield the mature viral polypeptides (**Figure 2**). The P1 region encodes the capsid proteins 1A-1D (VP4, VP2, VP3, and VP1, respectively). The P2 and P3 regions encode proteins involved in polyprotein processing, RNA replication, and shutdown of host cell protein synthesis.

Coxsackieviruses, like other picornaviruses, evolve extremely rapidly, because the viral RNA-dependent RNA polymerase is error-prone and lacks a proofreading function (like other RNA virus replicases). Most of the nucleotide substitutions are translationally silent, because most substitutions that become fixed in a virus population are in the third ('wobble') position of codons. This phenomenon is likely due to intense selection against amino acid substitutions, especially in the proteins that are required for replication, protein processing, and shutdown of host cell synthesis. Most viable amino acid substitutions occur in surface loops of the external capsid proteins, VP1, VP2, and VP3, while residues that contribute to the proteins' beta-barrel structure tend to be highly conserved within a serotype and, to a large extent, within a species. Within a serotype, VP1 nucleotide sequence can vary by up to 25%, representing near-saturation of synonymous sites, as well as multiple nonsynonymous substitutions – the amino acid sequence can vary by as much as 15% with a type.

The different enterovirus species each form a distinct phylogenetic group throughout the coding region, as well as in the 3'-NTR. In the 5'-NTR, the human enteroviruses form only two clusters: cluster I contains HEV-C and HEV-D, while cluster II contains HEV-A and HEV-B. The nonhuman species each form distinct 5'-NTR clusters. The coxsackie B virus capsid sequences cluster together as a group, probably reflecting their shared, but unique, use of CAR as a receptor. In other parts of the coding region, the CVB are not phylogenetically coherent; rather, they interdigitate with other viruses in HEV-B, due to the high frequency of RNA recombination within any given enterovirus species.

## Pathogenesis and Immunity

Much of the disease associated with coxsackievirus infection is thought to be a direct result of tissue-specific cell destruction, analogous to the cytopathic effect in cultured cells (**Figure 3**). For the most part, the detailed mechanisms of virus-induced disease have not been well-characterized. The primary site of infection is typically the respiratory or gastrointestinal epithelium, leading to viremia that may result in a secondary site of tissue infection. Such secondary spread of virus to the CNS can result in aseptic meningitis or, rarely, encephalitis or paralysis. Other tissue-specific infection can result in pleurodynia or myocarditis. Disseminated infection can lead to exanthems, nonspecific myalgias, or severe multiple-organ disease in neonates. Some disease manifestations, enteroviral exanthems and myocarditis, for example, are thought to result from the host immune response to the infection.

Enterovirus infection elicits a strong humoral immune response. Often the response is heterotypic; that is, infection with one serotype induces a broad immune response

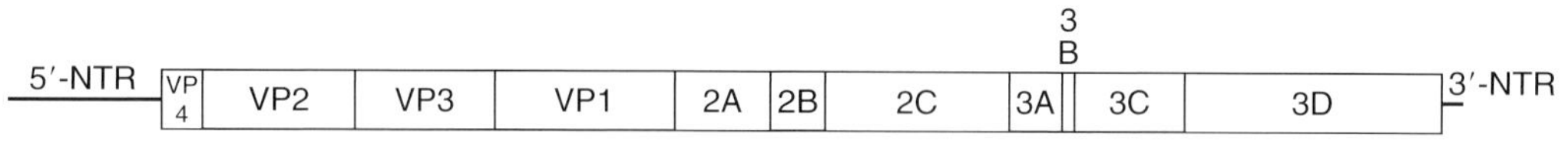

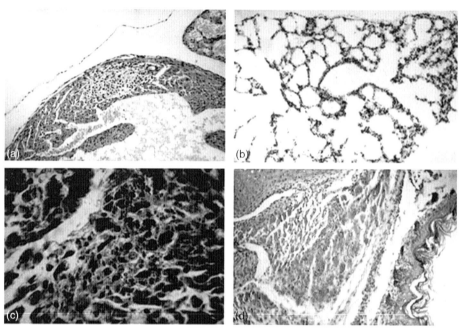

**Figure 2** Coxsackievirus genome, indicating the locations of the mature viral proteins, 1A–1D (VP4, VP2, VP3, and VP1, respectively), 2A–2C, and 3A–3D, flanked by the 5'- and 3'-nontranslated regions (NTRs).

**Figure 3** Tissue damage caused in baby mice by a freshly isolated strain of coxsackievirus B (a–c) or coxsackievirus A (d). (a) A large necrotic focus in the atrial myocardium ($\times 100$); (b) steatitis in the interscapular brown fat pad ($\times 250$); (c) a ventricular focus of necrotic cardiomyocytes ($\times 450$); (d) intercostal muscles with diffuse hyalinization of myocytes and mild inflammatory infiltrate ($\times 250$). Reproduced from Pozzetto B and Gaudin OG (1999) Coxsackieviruses (*Picornaviridae*). In: Granoff A and Webster R (eds.) *Encyclopedia of Virology*, 2nd edn. San Diego: Academic Press, with permission from Elsevier.

that cross-reacts with several other serotypes. Young children develop a more homotypic response, whereas older children and adults tend to develop a more heterotypic response. This age difference in the specificity of the antibody response probably reflects exposure to a greater number of serotypes with increasing age. The heterotypic response may reflect the presence of epitopes that are shared among multiple serotypes, but the actual mechanism is unknown. The presence of antibody does not prevent infection or primary virus replication, but it is sufficient to protect from disease, probably by limiting spread of the virus to secondary sites such as the CNS. Infection also elicits a T-cell response which helps clear the virus, but the cell-mediated immune response is not required for protection from disease.

## Clinical Manifestations

Coxsackievirus infections can result in a wide variety of disease syndromes (**Table 2**). Most infections are asymptomatic or result in only mild upper respiratory tract symptoms (common cold). Other mild enteroviral illness, such as fever, headache, malaise, and mild gastrointestinal symptoms, may also occur. Serious illness that brings the patient to the attention of a physician is much less frequent. Inapparent infections and prolonged excretion of virus, especially in stools, are common. These properties of enterovirus infection, and the fact that enterovirus infection is extremely common, make it difficult to establish a definitive link between infection and specific disease unless the virus can be isolated from a nonsterile site that is linked to the observed pathology (e.g., isolation from cerebrospinal fluid in the case of aseptic meningitis). Often, the association between infection and disease is based on studies of outbreaks in which a large number of persons with the same clinical signs and symptoms have evidence of infection with the same serotype. Such studies have clearly demonstrated that enterovirus infection can cause aseptic meningitis, pericarditis, pleurodynia, myocarditis, and encephalitis.

**Table 2** Diseases commonly associated with coxsackievirus infection

| Illness or syndrome | Group A | Group B |
|---|---|---|
| Acute hemorrhagic conjunctivitis | X[a] | |
| Aseptic meningitis | X | X |
| Encephalitis[b] | X | X |
| Exanthema | X | X |
| Hand, foot, and mouth disease | X[c] | |
| Hepatitis | X | X |
| Herpangina | X | |
| Infantile diarrhea | X | |
| Myocarditis/pericarditis | | X |
| Paralysis[b] | X | X |
| Pleurodynia | | X |
| Respiratory illness | X | X |
| Severe systemic infection in infants | | X |
| Undifferentiated febrile illness | X | X |

[a]CVA24 variant.
[b]Rare.
[c]Primarily CVA16 and EV71; also CVA6, CVA10.

## Central Nervous System Disease

The most commonly recognized serious manifestation of enterovirus infection is CNS disease, usually aseptic meningitis, but sometimes, encephalitis or paralysis. Aseptic meningitis is the most common CNS infection in the United States, and enteroviruses are the main recognized cause of aseptic meningitis in both children and adults in developed countries. Other than poliovirus, echoviruses and group B coxsackieviruses are most commonly isolated from cases of enterovirus-associated CNS disease, but this may be largely an artifact of low-efficiency isolation of group A coxsackieviruses in culture.

## Febrile Rash Illnesses

Herpangina is a common illness in school-age children, characterized by vesicular inflammation of the oral mucosa, including throat, tonsils, soft palate, and tongue. Herpangina has commonly been associated with CVA2–6, CVA8, and CVA10, as well as with some of the echoviruses. Hand, foot, and mouth disease (HFMD) is characterized by vesicular rash on the palms, soles, and oral mucosa, with frequent involvement of the limbs and trunk. CVA 16 and EV71 are the most common viruses isolated from HFMD cases, but CVA4, CVA6, and CVA10 are also frequent causes. Many of the coxsackieviruses (both groups A and B) may also be associated with undifferentiated rash, often in connection with other symptoms such as meningitis.

## Prenatal, Perinatal, and Neonatal Infection

A number of enterovirus serotypes have been associated with severe illness in neonates, including generalized disseminated infection that can mimic some aspects of bacterial sepsis. CVB4 infection is associated with a higher risk for severe disease in infants less than 1 month old, but other CVB serotypes and other enteroviruses can also cause similarly severe disease in this age group. Neonatal disease can also include other typical enterovirus syndromes, such as meningitis, encephalitis, pneumonia, hepatitis, myocarditis, and pancreatitis. One systematic study in Nassau County, New York, estimated that one of every 2000 infants in that area was hospitalized during the first 3 months of life for CVB sepsis-like disease, with significant mortality. This probably underestimates the true rate, since virus isolation studies may be insensitive and other enteroviruses were not included in the estimate.

Although there are few studies that examine the relationship between enterovirus infection and adverse effects on the fetus, one study found serologic evidence of CNS infection with CVB in ventricular fluid from 4 of 28 newborns with congenital neural tube defects. The infants had neutralizing antibody to only one CVB serotype in the ventricular fluid, but to several in serum. The unique distribution of antibodies in the ventricular fluid compared to that in serum supports the purported association. The mothers had antibodies to the same serotype as well as some other CVB serotypes. No virus was isolated from infants or mothers. Two other studies have documented an association between enterovirus infection and miscarriages and stillbirths. Further studies are needed to assess the possibility of enterovirus infection of the fetus.

## Acute Hemorrhagic Conjunctivitis

Coxsackievirus A24 variant and enterovirus 70 are associated with acute hemorrhagic conjunctivitis. This disease is different from other enteroviral illnesses, having occurred in global pandemics since its introduction around 1969. The incubation period for these agents is shorter than for other enteroviruses (24–72 h); systemic illness is much less common and conjunctival replication the rule. The disease is characterized by acute onset of lacrimation, severe pain, chemosis and periorbital edema, photophobia, conjunctival hyperemia, and mild-to-severe subconjunctival hemorrhages. The disease is usually bilateral. It is generally self-limiting, but may lead to secondary bacterial infection.

## Other Acute Illnesses

Group B viruses have been associated with pleurodynia, also known as 'Bornholm disease' or 'epidemic myalgia'. Pleurodynia is characterized by sudden onset of chest pain, due to inflammation of the diaphragm, accompanied by general malaise, headache, fever, and sore throat. Enteroviruses, primarily the coxsackieviruses, are also a common cause of respiratory illness. Most of these infections are relatively mild – even subclinical – and restricted to the upper respiratory tract (e.g., common cold, croup, and epiglottitis). Occasionally, the lower respiratory tract becomes infected, resulting in more serious illness, such as bronchiolitis or pneumonia, especially in very young children.

### Cardiac Disease

Although the association between myocarditis and pericarditis and enterovirus infection is clearly established, it is not yet clear how often enterovirus infections are responsible for the disease syndromes. One study has shown that CVB IgM in a group of patients with acute myocarditis is significantly higher than in controls. Enterovirus RNA has also been detected in myocardial biopsy specimens from patients with myocarditis. These and other studies suggest, but do not clearly show, that CVB infection may be associated with a large fraction of cases of acute myocarditis. By contrast, different studies have failed to show conclusive evidence for the involvement of enterovirus infection in idiopathic dilated cardiomyopathies.

### Diabetes

As with myocarditis, numerous studies have suggested a link between enterovirus infection and the onset of type I diabetes mellitus. In some cases, enteroviruses, such as CVB4, have been isolated from pancreas of fulminant diabetes cases, but most of the evidence is indirect, through serologic studies. Several mechanisms have been hypothesized to explain the possible link between infection and diabetes onset, including direct, lytic destruction of insulin-producing beta cells in the pancreas; molecular mimicry of self-antigens, resulting in induction of prediabetic autoimmunity that eventually leads to beta-cell destruction; and indirect damage mediated through activation of a nonspecific inflammatory response in the pancreas ('bystander effect'). A number of studies are attempting to address various steps in virus-induced diabetes pathogenesis, in human patients and in animal model systems.

### Treatment and Prevention

Antiviral therapy is not presently available for enterovirus infections, so treatment is directed toward ameliorating symptoms. Drugs have been identified that exhibit antiviral activity against several enteroviruses in tissue culture and experimental animals, and some have been tested in human clinical trials; however, no enterovirus drugs are currently licensed in the United States or elsewhere. Interferon has been proposed for treatment of acute hemorrhagic conjunctivitis, but further studies are needed to evaluate its effectiveness. Chronic enterovirus infections in patients with agammaglobulinemia have been treated with $\gamma$-globulin, and in some cases this has controlled the infection.

No vaccines are available to prevent coxsackievirus infection or disease. General preventive measures include enteric precautions and good personal hygiene. Nosocomial infections are most serious in newborns and persons with compromised immune systems, but others may also be affected. Hospital staff can inadvertently carry the virus between patients or become infected themselves and spread the virus. The main strategy to prevent nosocomial infections is to manage patients with suspected enterovirus infection using enteric precautions. During outbreaks, patients and staff can be cohorted. In some cases, neonatal nurseries were closed to new admissions during newborn outbreaks to prevent further spread of the virus.

*See also:* Echoviruses; Enteroviruses: Human Enteroviruses Numbered 68 and Beyond; Enteroviruses of Animals; Evolution of Viruses; Human Eye Infections; Picornaviruses: Molecular Biology; Poliomyelitis; Quasispecies; Recombination; Rhinoviruses; Viral Receptors.

### Further Reading

Bergelson JM, Cunningham JA, Droguett G, et al. (1997) Isolation of a common receptor for coxsackie B viruses and adenoviruses 2 and 5. *Science* 275(5304): 1320–1323.

He Y, Bowman VD, Mueller S, et al. (2000) Interaction of the poliovirus receptor with poliovirus. *Proceedings of the National Academy of Sciences, USA* 97(1): 79–84.

Huber S and Ramsingh AI (2004) Coxsackievirus-induced pancreatitis. *Viral Immunology* 17: 358–369.

Hyöty H and Taylor KW (2002) The role of viruses in human diabetes. *Diabetologia* 45: 1353–1361.

Hyypiä T and Stanway G (1993) Biology of coxsackie A viruses. *Advances in Virus Research* 42: 343–373.

Khetsuriani N, LaMonte A, Oberste MS, and Pallansch MA (2006) Enterovirus Surveillance – United States, 1970–2005. *Morbidity and Mortality Weekly Report Surveillance Summaries* 55(SS-8): 1–20.

Kim K-S, Hufnagel G, Chapman N, and Tracy S (2001) The group B coxsackieviruses and myocarditis. *Reviews in Medical Virology* 11: 355–368.

Muckelbauer JK, Kremer M, Minor I, et al. (1995) The structure of coxsackievirus B3 at 3.5 Å resolution. *Structure* 3: 653–667.

Pallansch MA and Oberste MS (2004) Coxsackievirus, echovirus, and other enteroviruses. In: Gorbach SL, Bartlett JG, and Blacklow NR (eds.) *Infectious Diseases*, 3rd edn, pp. 2047–2051. Philadelphia, PA: Lippincott Williams and Wilkins.

Pallansch MA and Roos R (2006) Enteroviruses: Polioviruses, coxsackieviruses, echoviruses, and newer enteroviruses. In: Knipe DM, Howley PM, Griffin DE, et al. (eds.) *Fields Virology*, 5th edn Philadelphia, PA: Lippincott Williams and Wilkins.

Pozzetto B and Gaudin OG (1999) Coxsackieviruses (*Picornaviridae*). In: Granoff A and Webster R (eds.) *Encyclopedia of Virology*, , 857pp San Diego: Academic Press.

Rossmann MG, He Y, and Kuhn RJ (2002) Picornavirus–receptor interactions. *Trends in Microbiology* 10(7): 324–331.

Semler BL and Wimmer E (eds.) (2002) *Molecular Biology of Picornaviruses*. Washington, DC: ASM Press.

Stanway G, Brown F, Christian P, et al. (2005) Picornaviridae. In: Fauquet CM, Mayo MA, Maniloff J, Desselberger U, and Ball LA (eds.) *Virus Taxonomy: Eighth Report of the International Committee on Taxonomy of Viruses*, pp. 757–778. San Diego, CA: Elsevier Academic Press.

Tracy S, Oberste MS, and Drescher K (eds.) (2008) *Coxsackie B viruses. Current Topics in Microbiology and Immunology*. Berlin: Springer.

# Crenarchaeal Viruses: Morphotypes and Genomes

**D Prangishvili and T Basta,** Institut Pasteur, Paris, France
**R A Garrett,** Copenhagen University, Copenhagen, Denmark

© 2008 Elsevier Ltd. All rights reserved.

## Glossary

**Hyperthermophile** An organism that grows optimally at temperature of 80 °C or higher.

## Introduction

The susceptibility to viral infection is as common for the domain Archaea as it is for the other two domains of life, the Bacteria and the Eukarya. Initially, the knowledge about archaeal viruses was restricted to head–tail viruses infecting members of the kingdom Euryarchaeota. However, the isolation of viruses from the other archaeal kingdom, the Crenarchaeota, has demonstrated that head–tail phages constitute only a part of viral diversity associated with this domain.

The double-stranded (ds) DNA viruses of Crenarchaeota isolated by Wolfram Zillig and colleagues were so unusual in their morphotypes that for their classification four novel viral families had to be introduced: *Fuselloviridaae, Lipothrixviridae, Rudiviridae,* and *Guttaviridae*.

Further studies on viral diversity in geothermally heated hot habitats, where Crenarchaeota predominate demonstrated that unusual morphotypes are a general feature of crenarcheal viruses and resulted in isolation of more viruses with unique morphologies. They have been classified into three novel viral families the approval of which is pending at the ICTV: Globuloviridae, Ampullaviridae, and Bicaudaviridae.

In this article we summarize our present knowledge on viruses infecting hyperthermophilic crenarchaea (**Table 1**). Virion morphotypes and constituents are described and data regarding different stages of replication cycle presented. Sequences of dsDNA genomes of all viruses are discussed jointly, in accord with the idea that crenarchaeal viruses form a distinctive group, unrelated to any other viruses, with a small pool of shared genes and a unique origin, or more likely, multiple origins.

## Morphotypes

The vast majority of the viruses replicated by Crenarchaeota have morphotypes that have not been observed among dsDNA viruses infecting Bacteria and Eukarya. In this section we describe different morphotypes of crenarchaeal viruses grouped according to their gross shape features.

### Spindle-Shaped Viruses

Spindle-shaped virions, single or two-tailed, constitute a large fraction of the known archaeal viruses and are abundant in hot acidic springs where crenarchaea predominate.

Virions of the family *Fuselloviridae*, represented by *sulfolobus* spindle-shaped viruses (SSVs), are enveloped with a short tail attached to one of the two pointed ends and measure about $(55–60) \times (80–100)$ nm (**Figure 1(a)**). The tail carries fibers which serve for adsorption to the host membrane. The structure of the inner core of the enveloped virions which generates the unusual shape is unknown. The mature virions were observed intracellulary (**Figure 1(a)**) indicating that the budding process is not required for the coating by a lipid-containing membrane.

The virions of *Acidianus* two-tailed virus (ATV) are most unusual because they undergo extensive morphological changes outside the host cell (**Figures 1(c)–1(g)**). When released from the host cell, the particles are spindle-shaped with an overall length of 250–300 nm and a diameter of 110–120 nm (**Figure 1(c)**). Then, specifically at temperatures above 75 °C, a tail-like appendage protrudes from each of the pointed ends (**Figures 1(d)–1(g)**). The particles with fully developed tails are about 1 μm in length. This process is linked with a reduction of the volume of the virion by a factor of 2, concurrent with a slight expansion of its surface area.

The tails are tube-like and terminate with an anchor-like structure formed by two furled filaments (**Figure 1(g)**). This structure is employed for adsorption on the host cell surface. The tube contains a filament of an unknown nature,

**Table 1** Crenarchaeal viruses

| Morphotype | Species | Virus family | Origin | Host | Lipids | Genome form and size (bp) | G+C content (%) | Genome integration | Accession no. |
|---|---|---|---|---|---|---|---|---|---|
| Linear | AFV1 | Lipotrixviridae | Y | Acidianus | + | linear, 20 869 | 37 | − | AJ567472 |
|  | AFV2 | Lipotrixviridae | P | Acidianus | − | linear, 31 787 | 36 | − | AJ854042 |
|  | TTV1 | Lipotrixviridae | I | Thermoproteus tenax | + | linear, 15 900 | 37 | ND | X14855 (85%)[a] |
|  | TTV2 | Lipotrixviridae | I | T. tenax | ND | ND | ND | ND | ND |
|  | TTV3 | Lipotrixviridae | I | T. tenax | ND | ND | ND | ND | ND |
|  | SIFV | Lipotrixviridae | I | Sulfolobus | + | linear, 40 852 | 33 | − | AF440571 (97%)[a] |
|  | SIRV1 | Rudiviridae | I | Sulfolobus | − | linear, 32 308 | 25 | − | AJ414696 |
|  | SIRV2 | Rudiviridae | I | Sulfolobus | − | linear, 35 450 | 25 | − | AJ344259 |
|  | ARV1 | Rudiviridae | Y | Acidianus | − | linear, 24 655 | 39 | − | AJ875026 |
| Spindle-shaped | SSV1 | Fuselloviridae | J | Sulfolobus | ND | ccc, 15 465 | 40 | + | XO7234 |
|  | SSV2 | Fuselloviridae | I | Sulfolobus | ND | ccc, 14 796 | 39 | + | AY370762 |
|  | SS-K1 | Fuselloviridae | R | Sulfolobus | ND | ccc, 17 385 | 39 | + | AY423772 |
|  | SSVRH | Fuselloviridae | Y | Sulfolobus | ND | ccc, 16 473 | 39 | + | AY388628 |
|  | ATV | "Bicaudaviridae"[b] | P | Acidianus | ND | ccc, 62 730 | 41 | + | AJ888457 |
|  | STSV1 | unclassified | C | Sulfolobus | + | ccc, 75 294 | 35 | − | AJ783769 |
| Spherical | PSV | "Globuloviridae"[b] | Y | Pyrobaculum | + | linear, 28 337 | 48 | − | AJ635162 |
|  | TTSV1 | "Globuloviridae"[b] | K | Thermoproteus | + | linear, 20 933 | 50 | ND | AY722806 |
|  | STIV | unclassified | Y | Sulfolobus | ND | ccc, 17 663 | 36 | ND | AY569307 |
| Bottle-shaped | ABV | "Ampullaviridae"[b] | P | Acidianus | + | linear, 23 900[c] | 48 | ND | ND |
| Droplet-shaped | SDNV[d] | "Guttaviridae" | I | Sulfolobus | ND | circular, 20 000[c] | 41 | ND | ND |

[a] Percentage of the determined sequence.
[b] Proposed families.
[c] Approximate value.
[d] The virus do not exist any more in laboratory collections.
J, Japan; I, Iceland; R, Kamchatka, Russia; Y, Yellowstone National Park, USA; P, Pozzuoli, Italy; K, Korea; C, China; ND, not determined.

**Figure 1** Electron micrographs of spindle-shaped viruses. (a) SSV1 and its extrusion from the host cell; (b) STSV1; (c–g) ATV at different stages of extracellular morphogenesis (in inserts, a horizontal slice through the 3D reconstruction of terminal structure by electron tomography); negative stain with 3% uranyl acetate. Scale: 100 nm (a); 200 nm (b, e–g); 500 nm (c).

which exhibits a structural periodicity. Protrusion of tails does not require the presence of host cells, an exogenous energy source or any cofactors. The only requirement is temperature in the range of host growth, above 75 °C.

The virus *Sulfolobus tengchongensis* spindle-shaped virus 1 (STSV1) is the largest of the known spindle-shaped viruses (107 × 230 nm) and has a single short tail (**Figure 1(b)**). The tail length varies in the range from 0 to 133 nm suggesting that the virus may also undergo an extracellular morphogenesis similar to ATV.

## Rod-Shaped and Filamentous Viruses

The combination of rod or filamentous shape and dsDNA genome occurring among members of *Rudiviridae* and the *Lipothrixviridae* is unique in the viral world.

The three characterized rudiviruses *Sulfolobus* rod-shaped virus 1 (SIRV1), *Sulfolobus* rod-shaped virus 2 (SIRV2), and *Acidianus* rod-shaped virus 1 (ARV1) are nonenveloped stiff rods about 23 nm wide and 610–900 nm long (**Figure 2(a)**). The particle length is proportional to the size of the dsDNA genome (**Table 1**). The rods are formed by a tube-like superhelical nucleoprotein complex that consists of linear dsDNA and copies of a single DNA-binding protein. The tubes are closed at each end by plug-like structures about 50 nm in length. Each plug carries three tail fibers (**Figure 2(a)**, inset) that are probably involved in adsorption to host cell membranes.

The lipothrixviruses include the highest number of known species among the crenarchaeal viral families. Virions are filamentous and mostly flexible, except for *Thermoproteus tenax* virus 1 (TTV1), and enveloped (**Figure 2(b)**).

The structures of the virion termini, as well as of the inner core are diverse, and owing to this diversity, as

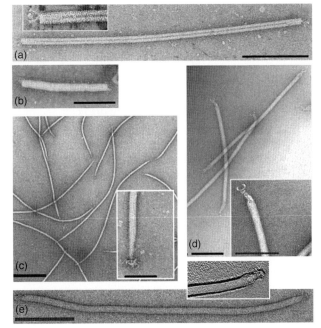

**Figure 2** Electron micrographs of linear viruses. (a) SIRV1; (b) TTV1; (c) SIFV; (d) AFV1; (e) AFV2; terminal structures are shown in inserts. Negative stain with 3% uranyl acetate. Scale: 200 nm (a–e) (in insets, 100 nm).

well as to presumed different replication strategies, four genera have been established in the family.

The body of *Sulfolobus islandicus* filamentous virus (SIFV) particles tapers and ends in identical mop-like structures (**Figure 2(c)**), the *Acidianus* filamentous virus 1 (AFV1) has claw-like terminal structures (**Figure 2(d)**), and *Acidianus* filamentous virus 2 (AFV2) carries a complex collar with two sets of filaments, resembling a bottle brush with a solid round cap at each end (**Figure 2(e)**, inset). The virion termini of TTV1 have not been analyzed in detail.

The terminal structures of lipothrixviruses serve for the adsorption onto the host cell. Both termini of a given virus may be involved in this process. The mop-like terminal structures of SIFV have been shown to unfold like spiders legs prior to attachment to cell membranes. Most lipothrixviruses adsorb to host cell membrane with exception of AFV1 which uses the claw-like ends for attachment to host pili.

The architecture of virion cores differs between the lipothrixviruses although they all carry linear dsDNA genomes. The cores of AFV1 and TTV1 are helical, and the latter was shown to contain equimolar amounts of two DNA-binding proteins. The DNA of SIFV is wound around a zipper-like array of subunits of a single protein. Such a core arrangement is reminiscent of nucleosomes. In contrast, no regular structure was detected in the core of AFV2.

### Spherical Viruses

The spherical viruses infecting hyperthermophilic Crenarchaeota encompass *Pyrobaculum* spherical virus, PSV, *Thermoproteus tenax* spherical virus 1, TTSV1, and *Sulfolobus* turreted icosahedral virus, STIV (**Figure 3**). This is the only morphotype of crenarchaeal viruses that is also observed in the domains Bacteria and Eukarya.

Virions of PSV are around 100 nm in diameter and they carry an envelope which encases tightly packed nucleoprotein in a superhelical conformation (**Figures 3(a)** and **3(b)**). Such core structure is unique for a DNA virus but resembles the core structure of single-stranded RNA viruses of the family *Paramyxoviridae*. The virion surface is covered by a variable number of spherical protrusions, about 15 nm in diameter, most likely involved in adsorption to the host cell.

The properties of the virion structure and genome of TTSV1 are very similar to those of PSV suggesting that these viruses belong to the same family.

The virions of the STIV are nonenveloped icosahedra with a diameter of about 74 nm (**Figure 3(c)**). The capsid most likely contains an internal lipid envelope. Single-particle reconstruction revealed a unique virus architecture including complex, turret-like appendages at the virion surface. The crystal structure of the major capsid protein of STIV is very similar to those of the bacterial tectivirus PRD1 and the eukaryal phycondnavirus PBCV-1, suggesting a common ancestry, despite the fact that the corresponding protein sequences show no significant similarity.

### Bottle- and Droplet-Shaped Viruses

The extraordinary shapes of crenarchaeal viruses are exemplified by two viruses with most unusual virion structures, *Acidianus* bottle-shaped virus, ABV, and *Sulfolobus neozealandicus* droplet-shaped virus, SNDV (**Figure 4**).

The virions of ABV resemble in their shape a bottle with an overall length of 230 nm and a width varying from 75, at the broad end, to 4 nm, at the pointed end (**Figure 4(a)**). The broad end of the virion exhibits 20 ($\pm 2$) thin rigid filaments, which appear to be inserted into a disk, or ring, and are interconnected at their bases. The 9-nm-thick envelope encases cone-shaped core that is formed by a toroidally supercoiled nucleoprotein filament (**Figures 4(b)** and **4(c)**). The narrow end of virions is likely involved in cellular adsorption and in channeling of viral DNA into a host cell. The function of filaments at the opposite end remains unclear.

The SNDV virions are droplet-shaped, 110–185 nm long and 70–95 nm wide. The pointed end is densely covered by thin fibers (**Figure 4(d)**). The core is protected by a beehive-like structure, the surface of which appears to be built up of components stacked in a helical manner.

### Virion Constituents

The protein content of particles of crenarchaeal viruses is often complex. The number of the major protein components present in the virions varies from two proteins for rudiviruses, fuselloviruses, and lipothrixviruses up to 11 proteins for the ATV. Moreover, in the majority

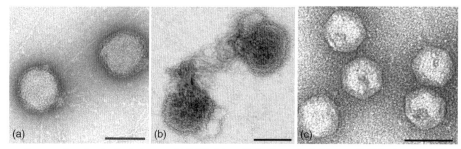

**Figure 3** Electron micrographs of spherical viruses. (a) PSV; (b) partially disrupted virions of PSV releasing helical nucleoprotein filament; (c) STIV. Negative stain with 3% uranyl acetate. Scale = 100 nm.

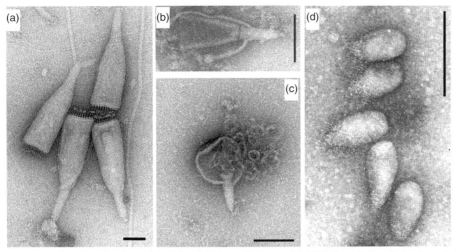

**Figure 4** Electron micrographs of bottle-shaped and droplet-shaped viruses. (a) ABV; (b and c) partially disrupted ABV virions, revealing an inner cone-shaped core and its constituent nucleoprotein filament; (d) SNDV. Negative stain with 3% uranyl acetate. Scale: 50 nm (a); 100 nm (b, c); 200 nm (d).

of viruses a few minor protein components are also present. Several structural proteins have been identified (**Figure 5**).

Virion proteins have an important role in protecting the viral DNA from the harsh conditions of the host's habitat. For example, the single DNA-binding protein of nonenveloped rudiviruses can partially protect the virion structure during autoclaving at 120 °C for 10 min. The particles are completely disrupted only after a further 50 min.

Some of the structural proteins can form structures of higher order. This is exemplified by the 90 kDa protein from ATV that is rich in coiled-coil motifs and can generate long filaments. They may contribute to the protrusion of tail-like appendages during the extracellular development of ATV. The virions of PSV exhibit three major protein components the largest of which can build dimers and higher aggregates. This protein also contains two-thirds of hydrophobic amino acid residues, about 15% of which are aromatic, and no cysteine residues, properties which are likely to contribute to high thermostability.

Almost all crenarchaeal viruses are enveloped. The only exception are the members of the *Rudiviridae*. The envelopes of the PSV, and of all lipothrixviruses, except AFV2, have been shown to contain host lipids that have been modified probably by virus-encoded enzymes. The STIV particles have an internal host lipids-containing envelope. The presence of lipids in virions of other crenarchaeal viruses has not yet been reported.

## Virus–Host Interactions

Little is known about the modes of cellular entry and the assembly mechanisms of the crenarchaeal viruses. Presumably, after attachment to the host cell surface either the whole virion is taken up or the viral DNA is injected through the host membrane. The latter mechanism could be employed by the virions of ABV that were shown to release a nucleoprotein complex through the pointed end (**Figure 4**(c)). Another mechanism is possible for the AFV1 which attaches with its claw-like terminal structures onto the host pili. The virions could be pulled into the cell by retraction of the pili.

Hosts of cultured crenarchaeal viruses are extreme acidophiles (*Sulfolobus, Acidianus*), or neutrophiles and obligate anaerobes (*Pyrobaculum, Thermoproteus*). Infectivity of the latter organisms with viruses is unaffected by exposure to oxygen. For all of the hosts which grow optimally at 80 °C or above, viral infection occurs most effectively at the optimal growth temperature of the host such that viruses can also be considered as hyperthermophiles.

Viruses of the Crenarchaeota are rather exceptional with respect to virus–host interactions because they generally do not kill their host during release of progeny virions. Instead, the viruses persist in host cells and are not segregated even after prolonged growth and several dilutions of infected cell cultures. This is the result of an equilibrium that is established between the viral genome replication and cell division. It is usually argued that such inoffensiveness of crenarchaeal viruses is beneficial, helping them to avoid direct, and possibly prolonged, exposure to the harsh environmental conditions of the host habitat.

The only two examples of crenarchaeal viruses that cause lysis of their hosts are TTV1 and ATV.

The ATV exhibits the characteristics of a true temperate virus. At the optimal temperature of host growth, 85 °C, infection with ATV results in integration of the viral genome into the host chromosome and lysogenization of the host cell. The lysogeny can be interrupted by

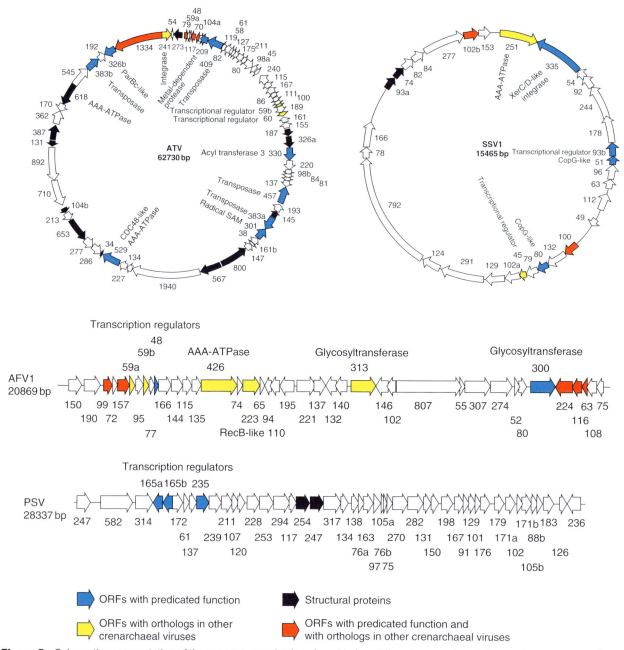

**Figure 5** Schematic representation of the genome organizations found in four different crenarchaeal viruses. Arrows represent predicted viral ORFs and their size and orientation. Numbers assigned to the ORFs correspond to number of amino acids in the encoded proteins. The genomes are not in scale to each other. ORFs have been assigned functions according to sequence similarities with genes in public databases.

subjecting the host cell to stress conditions, for example, UV-irradiation, treatment with mytomycin C, freezing–thawing cycles, or host growth at suboptimal temperatures. Induction of virus replication eventually leads to cell lysis.

The integration of the ATV genome into the host chromosome is facilitated by a virus-encoded integrase. The genes encoding these enzymes have been detected also on circular genomes of the fuselloviruses and STSV1. The former insert generally within specific host tRNA genes and the integrated form of the latter has not been detected. The fuselloviruses are present in the infected host cell cultures in episomal and integrated form but it is not known whether the two forms exist in the same cell. The replication of the SSV1 can be induced by ultraviolet irradiation leading to temporary growth inhibition, but not lysis, of host cells. Subsequently, virus proliferation decreases and the host cells assume their earlier growth rate.

The linear genomes of crenarchaeal viruses do not encode integrases and they have not been detected in chromosomes of their hosts. Moreover, their replication is not affected by UV-irradiation or other stress factors.

## Genome Structure

The crenarchaeal viruses investigated so far all carry dsDNA genomes. The genomes of the fuselloviruses ATV and STSV1 are circular, 15–75 kbp in size (**Table 1**). Known rudiviruses, lipothrixviruses, and PSV contain linear genomes in the size range 25–45 kbp.

For the fuselloviruses and some lipothrixviruses, about half of each genome, including genes for viral structural proteins, is always highly conserved in gene content and order with the remainder of the genome being more variable.

Modifications have been detected in two of the viral genomes. The SNDV carries a *dam*-like N(6) methylation of adenine residues in GATC sequences while the DNA of the STSV1, which encodes three putative modifying enzymes, undergoes methylation of specific cytosine residues within the sequence CCGG. The latter modifications are probably absent from the host chromosomal DNA.

Most of the linear genomes carry inverted terminal repeats (ITRs) that can vary significantly in their size between viruses. The rudiviral genomes exhibit large ITRs (600–2000 bp), the PSV carries a 190 bp ITR, and the lipothrixvirus AFV1 has a very short 11 bp ITR.

Genomic termini of some crenarchaeal viruses are modified. In the rudiviruses SIRV1 and SIRV2 the two DNA strands are covalently linked at the ends, generating a 4 bp hairpin loop which can be degraded by *Bal*31 endonuclease. A similar structure may occur at the termini of the PSV genome.

Restriction fragments containing the terminal sequences of the TTV1 and AFV1 produce low molar yields on phenol extraction suggesting the presence of tightly bound proteins at the ends of genomic DNA.

## Replication of Viral Genomes

The genome replication of crenarchaeal viruses has not been studied experimentally. Consequently, the knowledge about the replication mechanisms employed and participating enzymes is very scarce. Nevertheless, some clues about possible replication strategies of crenarchaeal viruses could be obtained from the sequences of the genomes and the genome arrangements.

Linear genomes of some lipothrixviruses exhibit an internal, repeat-rich, non-protein-coding region. The AFV2 carries a 1008 bp element bordered by an ITR that is rich in repeat sequences. Such regions could constitute internal origins of replication.

Even more remarkable are the genomic features of the STSV1. Its circular genome is highly asymmetric and divides into equal halves with respect to gene orientation. Between the two halves of the genome was located the 1.4 kbp long intergenic region with an unusually high AT content, including two sets of tandem repeats and two sets of inverted repeats. This region constitutes the putative origin of replication. Most likely, genome replication proceeds bidirectionally in the theta-mode from the proposed origin.

The genomic termini of lipotrixvirus AFV1 do not resemble any other virus. The two very short (11 bp) ITRs are preceded by about a 300 bp region consisting of many direct repeats of the pentanucleotide TTGTT, or close variants thereof. Such structures resemble telomeric ends of eukaryotic chromosomes and raise the possibility of a primitive telomeric mechanism operating in AFV1 genome replication.

Rudivirus SIRV1 replicates its genome via head-to-head and tail-to-tail linked intermediates. This suggested a self-priming mechanism of replication, similar to that proposed for large eukaryotic dsDNA viruses including poxviruses. Consistent with this proposal are the similarities in the structures of linear genomes of these archaeal and eukaryal viruses, including long ITRs and covalently closed termini. Moreover, each of these viruses encodes a Holliday junction resolvase which is likely to resolve the replicative intermediates into single-progeny genomes. The large ITRs of the rudiviruses carry highly conserved sequence located 100–150 bp from the genomic termini that could serve as signal for the initiation of replication.

## Transcription of Viral Genes

The studies on transcription of the fusellovirus SSV1 resulted in identification of archaeal promoter sequences and were crucial for our seminal understanding of mechanisms of transcriptional regulation in archaea. SSV1 transcripts were mapped after UV induction of virus production in the lysogenes. Eight constitutive and one UV-inducible transcript were identified.

The first detailed analysis of transcription over the complete replication cycle was performed on rudiviruses SIRV1 and SIRV2. Already 30 min post infection all the ORFs, with one exception, are transcribed. Nevertheless, some ORFs have different transcription patterns during the replication cycle. One example is the ORF coding for coat protein that is initially transcribed as polycistronic mRNA but its monocistronic transcript was detected in the later stage of virus infection. The rather simple and ordered transcriptional pattern of the rudiviruses fits with fairly unsophisticated virus–host relationship (stable carrier state) which probably does not require much transcriptional control.

SIRV promoters, like those of their hosts, carry a TATA-like box and adjacent to it a purine-rich region, corresponding to the transcription factor B binding site. In addition, virus-specific consensus element, the trinucleotide GTC is located immediately downstream from the TATA-like box. The GTC motif is also present in one-third of the promoter regions of another rudivirus, the ARV1.

Many viral transcripts for single genes and the first genes of SIRV operons are leaderless, and lack the consensus Shine–Dalgarno motif GGTG.

## Genomics of Crenarchaeal Viruses

The most prominent feature of genomes of known crenarchaeal viruses is the extremely low number of gene products with homologs in the public sequence databases (**Figure 5**).

### Genes with Confirmed Functions

Putative functions of only few genes (with homologs in public databases) have been confirmed experimentally. They encode the dUTPase and the Holliday junction resolvase of the rudiviruses SIRV1 and SIRV2 and the integrase/recombinase of the fusellovirus SSV1. In addition genes for structural proteins were identified.

### Genes with Putative Functions Unique for Individual Viruses

A few genes of crenarchaeal viruses have homologs in archaea and/or bacteria. These include DNA-modification methylases of the STSV1, superfamily 2 helicases of the lipothrixvirus SIFV, ABC-class ATPase of the rudivirus ARV1, and thioldisulfide isomerase of the globulovirus TTSV1.

### Genes with Putative Functions Shared by Different Viruses

Many crenarchaeal viruses share a repertoire of genes with putative functions including transcription regulators, ATPases, enzymes of DNA precursor metabolism, and RNA modification, and glycosylases.

The most common gene products in crenarchaeal viruses are small proteins containing ribbon–helix–helix (RHH) domain. These proteins are transcription regulators, common in archaea, and are nearly as abundant as classical helix–turn–helix (HTH) domains. The RHH proteins of crenarchaeal viruses are rather heterogeneous, and apparently have a complex history, probably involving multiple independent acquisitions as well as horizontal gene transfers.

Other putative transcription regulators encompass HTH-domains, looped-hinge helix domains, and Zn-fingers. Remarkable is the presence of C2H2 Zn-finger proteins with moderate similarity to a variety of eukaryotic Zn-fingers but no obvious homologs in prokaryotes.

Each of the crenarchaeal viruses encodes at least one of the P-loop ATPases, which are known to be the most abundant protein domain in prokaryotes and the great majority of viruses, and typically are involved in viral replication, transcription, or packaging. Other predicted enzymes with probable functions in DNA replication are the RecB-family nucleases in the *Lipothrixviridae*, *Rudiviridae*, and *Fuselloviridae*, the XerC/D-like integrase/recombinase in the Bicaudaviridae, and the STSV1. As for the Holliday junction resolvases of the *Rudiviridae*, these enzymes may be involved in intermediate resolution during viral genome replication.

The *Rudiviridae* and STSV1 encode a dUTPase and/or the flavin-dependent thymidylate synthase (ThyX) which apparently are involved in DNA precursor metabolism, a function widely represented in DNA viruses of Bacteria and Eukarya. Another function common for bacterial viruses, RNA modification, is also represented in the members of *Rudiviridae*: they encode predicted tRNA-ribosyltransferase and S-adenosylmethionine-dependent methyltransferase. The former enzyme is encoded also by STSV1 and the latter enzyme by the lipothrixvirus SIFV.

All known members of the *Rudiviridae* and *Lipothrixviridae* encode glycosyltransferases that may be involved in modification of virions proteins and/or the host cell wall during viral entry and/or release. In members of other viral families in this process maybe implicated such enzymes as a membrane-associated acyltransferase (in ATV) and nucleoside-diphosphate-sugar epimerase (in STSV1).

The gene pool shared by crenarchaeal viruses includes also two protein families with unknown functions. One of these, exemplified by the AFV1 protein 03 of 99 amino acids, is found exclusively in the *Rudiviridae* and *Lipothrixviridae*. Another family of small proteins has a single bacterial representative, the uncharacterized protein YddF of *Bacillus subtilis*, with all other members found in crenarchaeal viruses: in the *Rudiviridae*, *Lipothrixviridae*, *Fuselloviridae*, the ATV, and STIV.

### Orthologous Genes

Results of identification of orthologous genes among the genes shared by crenarchaeal viruses are in astounding accord with evolutionary relationships postulated on the basis of viral morphotypes. (Because of the uncertainty regarding the very existence of a unique common ancestor virus, orthologs were defined only conditionally, as genes that are more closely related to each other in a given set of viruses than they are to any homologs that may exist outside that set of genomes.)

A significant number of orthologs (18–44) are observed among viral species assigned to the same genus, which probably derived from relatively recent common ancestors. These are four members of the *Fuselloviridae*, three members of the *Rudiviridae*, and two members of the Globuloviridae.

A smaller number of orthologous genes (5–9), representing moderately related groups which probably have evolved from a single ancestral virus in a more distant past, were observed for members of different genera in the family *Lipothrixviridae*: the betalipothrixvirus SIFV, gammalipothrixvirus AFV1, and deltalipothrixvirus AFV2. A comparable number of orthologous genes were observed also between some members of different families, the *Rudiviridae* and *Lipothrixviridae*, suggesting a common ancestry for these two families.

In those cases when the viruses share only 1–4 orthologous genes, it remains unclear, whether a common ancestral virus ever existed. The existence of a common viral ancestor is even less likely, when no orthologs are detected among viruses, as it occurs for the two members of the Globuloviridae and the rest of the crenarchaeal viruses.

## Conclusions and Future Prospects

Known dsDNA viruses infecting hyperthermophilic crenarchaea represent a remarkable collection of unique morphotypes, not encountered among dsDNA viruses of the Bacteria and Eukarya. It is now clear that crenarcheal viruses are not just an oddity but represent a general picture of viral diversity in hot environments. Therefore, the observed variety of viral morphotypes in hot environments is probably only a tip of an iceberg and the isolation of new viruses should remain an important part of future studies.

Analysis of the sequences of crenarchaeal viral genomes demonstrated that they form a distinct group, evolutionary unrelated to dsDNA viruses of the other two domains of life. Moreover, the majority of predicted genes of crenarchaeal viruses have no homologs in public sequence databases and only a few of the viral genes could be assigned functions. Consequently, this hinders our understanding of the fundamental processes that take place during the replication cycle of the crenarchaeal viruses. It is now time to initiate studies of the mechanisms of these processes, notably attachment, virion assembly and release, transcription, and genome replication. The instrumental discoveries made during studies of bacteriophages in the past argue that the studies of crenarchaeal viruses have potential to provide the archaeal research community with robust genetic tools.

The discovery of the unique group of crenarchaeal viruses provides a new perspective on events related to the origin of diverse groups of DNA viruses and their evolution. It is a challenge for future studies to reveal evolutionary aspects of involvement of sets of nonorthologous genes in establishment of similar lifestyles in the three unrelated groups of dsDNA viruses.

*See also:* Viruses Infecting Euryarchaea.

## Further Reading

Häring M, Rachel R, Peng X, Garrett RA, and Prangishvili D (2005) Viral diversity in hot springs of Pozzuoli, Italy, and characterization of a unique archaeal virus, *Acidianus* bottle-shaped virus, from a new family, the *Ampullaviridae*. *Journal of Virology* 79(15): 9904–9911.

Häring M, Vesterdaard G, Rachel R, Chen L, Garrett R, and Prangishvili D (2005) Virology: Independent virus development outside a host. *Nature* 436(7054): 1101–1102.

Kessler A, Brinkman AB, van der Oost J, and Prangishvili D (2004) Transcription of the rod-shaped viruses SIRV1 and SIRV2 of the hyperthermophilic archaeon *Sulfolobus*. *Journal of Bacteriology* 186(22): 7745–7753.

Nadal M, Mirambeau G, Forterre P, Reiter WD, and Duget M (1986) Positively supercoiled DNA in a virus-like particle of an archaebacterium. *Nature* 321: 256–258.

Prangishvili D (2003) Evolutionary insights from studies on viruses of hyperthermophilic archaea. *Research in Microbiology* 154(4): 289–294.

Prangishvili D and Garrett RA (2005) Viruses of hyperthermophilic crenarchaea. *Trends in Microbiology* 13: 535–542.

Prangishvili D, Garrett RA, and Koonin EV (2006) Evolutionary genomics of archaeal viruses: Unique viral genomes in the third domain of life. *Virus Research* 117: 52–67.

Prangishvili D, Stedman K, and Zillig W (2001) Viruses of the extremely thermophilic archaeon *Sulfolobus*. *Trends in Microbiology* 9(1): 39–43.

Rachel R, Bettstetter M, Hedlund BP, et al. (2002) Remarkable morphological diversity of viruses and virus-like particles in hot terrestrial environments. *Archives of Virology* 147(12): 2419–2429.

Rice G, Stedman K, Snyder J, et al. (2001) Viruses from extreme thermal environments. *Proceedings of the National Academy of Sciences, USA* 98: 13341–13345.

Rice G, Tang L, Stedman K, et al. (2004) The structure of a thermophilic archaeal virus shows a double-stranded DNA viral capsid type that spans all domains of life. *Proceedings of the National Academy of Sciences, USA* 101: 7716–7720.

Wiedenheft B, Stedman K, Roberto F, et al. (2004) Comparative genomic analysis of hyperthermophilic archaeal *Fuselloviridae* viruses. *Journal of Virology* 78: 1954–1961.

Xiang X, Chen L, Huang X, Luo Y, She Q, and Huang L (2005) *Sulfolobus tengchongensis* spindle-shaped virus STSV1: Virus–host interactions and genomic features. *Journal of Virology* 79: 8677–8686.

Zillig W, Prangishvili D, Schleper C, et al. (1996) Viruses, plasmids and other generic elements of thermophilic and hyperthermophilic Archaea. *FEMS Microbiology Reviews* 18: 225–236.

# Crimean–Congo Hemorrhagic Fever Virus and Other Nairoviruses

**C A Whitehouse,** United States Army Medical Research Institute of Infectious Diseases, Frederick, MD, USA

Published by Elsevier Ltd.

## Glossary

**Argasid ticks** Any member of the family *Argasidae*, which are soft-shelled; commonly called soft ticks.
**Ecchymosis** The escape of blood into the tissues from ruptured blood vessels.
**Ixodid ticks** Any member of the family *Ixodidae*, which have a hard outer covering called scutum; commonly referred to as hard ticks.
**Petechiae** Minute reddish or purplish spots containing blood that appear in the skin or mucous membrane, especially in some infectious diseases.

## Introduction

The genus *Nairovirus* within the family *Bunyaviridae* includes 34 predominantly tick-borne viruses. These viruses are enveloped, with a tripartite negative-sense, single-stranded RNA genome. Among the members of this genus, the most important pathogens are the Crimean–Congo hemorrhagic fever virus (CCHFV), which causes severe and often fatal hemorrhagic fever in humans, Crimean–Congo hemorrhagic fever (CCHF), and Nairobi sheep disease virus (NSDV), which circulates in Africa and Asia and causes acute hemorrhagic gastroenteritis in sheep and goats.

CCHF was first recognized during a large outbreak among soldiers and agricultural workers in the mid-1940s in the Crimean peninsula but the etiologic agent was not isolated at that time. In 1956 a virus named Congo virus was isolated from a febrile patient in the Belgian Congo (now the Democratic Republic of the Congo) and it was later recognized that Congo virus was the same as Crimean hemorrhagic fever virus, isolated in 1967 from a patient with this disease in (now) Uzbekistan. The disease now occurs sporadically throughout much of Africa, Asia, and Europe and results in an approximately 30% case–fatality rate. CCHF is characterized by a sudden onset of high fever, chills, severe headache, dizziness, and back and abdominal pains. Additional symptoms can include nausea, vomiting, diarrhea, neuropsychiatric disorders, and cardiovascular changes. In severe cases, hemorrhagic manifestations, ranging from petechiae to large areas of ecchymosis, develop. Ixodid ticks of numerous genera serve both as vector and reservoir for CCHFV. Ticks in the genus *Hyalomma* are particularly important in the ecology of this virus and exposure to these ticks represents a major risk factor for contracting disease. Other important risk factors include direct contact with blood and/or body fluids from infected patients or animals. The highly pathogenic nature of CCHFV has restricted research on the virus to biosafety level 4 (BSL-4) laboratories and has led to the fear that it might be used as a bioweapon.

NSDV, which is an important cause of veterinary disease, was originally isolated from sheep from Nairobi, Kenya in 1910. Dugbe virus was originally isolated from adult male *Amblyomma variegatum* ticks collected from cattle in Ibadan, Nigeria in 1964. Both Dugbe and Ganjam viruses (a variant of NSDV) have been isolated repeatedly from ticks removed from domestic animals and have both caused febrile illnesses in humans. Most other nairoviruses have been isolated from ixodid ticks or from argasid ticks, which are ectoparasites of seabirds and other birds, and their medical or veterinary significance is not known.

## History

### Crimean–Congo Hemorrhagic Fever Virus

A disease now considered to be CCHF was described by a physician in the twelfth century from the region that is presently Tadzhikistan. The description was of a hemorrhagic disease with the presence of blood in the urine, rectum, gums, vomitus, sputum, and abdominal cavity. The disease was said to be caused by a louse or tick, which normally parasitizes a blackbird. The arthropod described may well have been larvae of a species of *Hyalomma* ticks, which are frequently found on blackbirds. CCHF has also been recognized for centuries under at least three names by the indigenous people of southern Uzbekistan: *khungribta* (blood taking), *khunymuny* (nose bleeding), or *karakhalak* (black death). Various other names, including acute infectious capillarotoxicosis and Uzbekistan hemorrhagic fever, have been used for centuries in Central Asia to refer to CCHF.

In modern times, CCHF (then known as Crimean hemorrhagic fever (CHF)) was first described as a clinical entity in 1944–45 when about 200 Soviet military personnel and civilian farmers were infected during an epidemic in war-torn Crimea. Shortly thereafter, a viral etiology was suggested by reproducing a febrile syndrome in psychiatric patients who were inoculated with a filterable agent from the blood of CHF patients. Further evidence of a viral etiology and of a suspected tick-borne route of infection was demonstrated by inducing a mild form of disease in healthy human volunteers after their inoculation with filtered suspensions of nymphal *H. marginatum* ticks. In 1967, the Russian virologist M. P. Chumakov and his colleagues at the Institute of Poliomyelitis and Viral Encephalitides in Moscow were the first to use newborn white mice for CHF virus isolation.

In 1956, Dr. Ghislaine Courtois at the Provincial Medical Laboratory in what was then Stanleyville, Belgian Congo, isolated a virus, referred to as Congo virus, from the blood of a 13-year-old boy who had fever, headache, nausea, vomiting, and generalized joint pains. Interestingly, a second isolation of the virus was made from the blood of Dr. Courtois, who subsequently became ill with symptoms similar to those of the boy. Subsequent work showed that Crimean hemorrhagic fever virus was antigenically indistinguishable from Congo virus and were in fact, the same virus, leading to the new name, Crimean–Congo hemorrhagic fever virus.

### Other Important Members of the Genus *Nairovirus*

NSDV, an important cause of livestock disease, was originally isolated from sheep in Nairobi, Kenya in 1910. Dugbe virus was originally isolated from *A. variegatum* ticks from Ibadan, Nigeria in 1964. Ganjam virus (**Figure 1**), which is now considered to be a strain of NSDV, was originally isolated from *Haemaphysalis intermedia* ticks collected from goats in the Ganjam district of India in 1954. Most other nairoviruses were first isolated in the 1960s or 1970s from ticks parasitizing seabirds and other birds (**Table 1**).

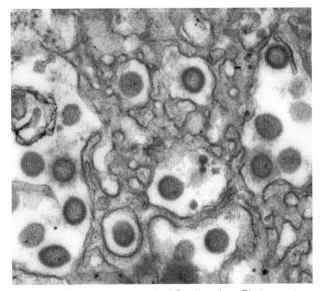

**Figure 1** Electron micrograph of Ganjam virus. Photo courtesy of Frederick A. Murphy, Centers for Disease Control and Prevention, Atlanta, Georgia.

## Taxonomy and Classification

According to the Eighth Report of the International Committee on the Taxonomy of Viruses, the species *Crimean–Congo hemorrhagic fever virus* is classified in the genus *Nairovirus*. There are seven recognized species in the genus, containing a total of 34 viruses, most of which are transmitted by either ixodid or argasid ticks (i.e., hard or soft ticks, respectively) (**Table 1**).

## Geographical Distribution

The known distribution of CCHFV covers the greatest geographic range of any tick-borne virus and there are reports of viral isolation and/or disease from more than 30 countries in Africa, Asia, southeast Europe, and the Middle East. Evidence for its presence in France, Portugal, Egypt, and India is based only on limited serologic observations. Interestingly, after several decades of only serologic evidence of the existence of CCHFV in Turkey, a large outbreak of disease in that country began in 2002, is ongoing, and has resulted in more than 900 cases. This highlights the ability of CCHF to emerge in geographic regions that have previously been devoid of the disease.

## Ecology and Epidemiology

As do other tick-borne zoonotic agents, CCHFV generally circulates in nature unnoticed in an enzootic tick-vertebrate-tick cycle. CCHFV has been isolated from numerous domestic and wild vertebrates, including cattle and goats, sheep, hares, hedgehogs, and even domestic dogs. Sera from wild mammals of several species have been shown to have antibodies to CCHFV and seroepidemiological studies have also detected antibodies to CCHFV in domestic cattle, horses, donkeys, sheep, goats, and pigs from various parts of Europe, Asia, and Africa. Interestingly, there has been only one report of antibody to CCHFV detected from a reptile, a tortoise from Tadzhikistan.

Although many domestic and wild vertebrates are infected with CCHFV, as evidenced by development of viremia and/or antibody response, birds, in general, appear to be refractory to infection with CCHFV. One interesting exception is ostriches, which become infected with CCHFV and have been the source of several cases of CCHF associated with slaughtering ostriches in South Africa.

The natural cycle of CCHFV includes transovarial (i.e., passed through the eggs) and trans-stadial (i.e., passed directly from immature ticks to subsequent life stages) transmission among ticks in a tick–vertebrate–tick cycle. CCHFV has been isolated from ticks in at least 31 species. Viral isolations have been made from ticks of two species

**Table 1** List of currently accepted species and other viruses within the genus Nairovirus[a]

| Virus | Original source | Year of isolation | Abbreviation |
|---|---|---|---|
| *Crimean–Congo hemorrhagic fever virus* | | | |
|    Crimean–Congo hemorrhagic fever virus | Human | 1967 | CCHFV |
|    Hazara virus | *Ixodes* sp. | 1964 | HAZV |
|    Khasan virus | *Haemaphysalis* sp. | 1971 | KHAV |
| *Dera Ghazi Khan virus* | | | |
|    Dera Ghazi Khan | *Hyalomma* sp. | 1966 | DGKV |
|    Abu Hammad virus | *Argas hermanni* | 1971 | AHV |
|    Abu Mina virus | *Argas streptopelia* | 1963 | AMV |
|    Kao Shuan virus | *Argas* sp. | 1970 | KSV |
|    Pathum Thani virus | *Argas* sp. | 1970 | PTHV |
|    Pretoria virus | *Argas* sp. | 1970 | PREV |
| *Dugbe virus* | | | |
|    Dugbe virus | *Hyalomma* sp. | 1966 | DUGV |
|    Nairobi sheep disease virus | Sheep | 1910 | NSDV |
|    Ganjam virus[b] | *Haemaphysalis* sp. | 1954 | GANV |
| *Hughes virus* | | | |
|    Hughes virus | *Ornithodoros* sp. | 1962 | HUGV |
|    Farallon virus | *Carios capensis* | 1964 | FARV |
|    Fraser Point virus | ND[c] | ND | FPV |
|    Great Saltee virus | *Ornithodoros* sp. | 1976 | GRSV |
|    Puffin Island virus | *Ornithodoros maritimus* | 1979 | PIV |
|    Punta Salinas virus | *Ornithodoros* sp. | 1967 | PSV |
|    Raza virus | *Carios denmarki* | 1962 | RAZAV |
|    Sapphire II virus | *Argas cooley* | ND | SAPV |
|    Soldado virus | *Ornithodoros* sp. | 1963 | SOLV |
|    Zirqa virus | *Ornithodoros* sp. | 1969 | ZIRV |
| *Qalyub virus* | | | |
|    Qalyub virus | *Ornithodoros erraticus* | 1952 | QYBV |
|    Bakel virus | ND | ND | BAKV |
|    Bandia virus | *Mastomys* sp. | 1965 | BDAV |
|    Omo virus | *Mastomys* sp. | 1971 | OMOV |
| *Sakhalin virus* | | | |
|    Sakhalin virus | *Ixodes* sp. | 1969 | SAKV |
|    Avalon virus | *Ixodes* sp. | 1972 | AVAV |
|    Clo Mor virus | *Ixodes* sp. | 1973 | CMV |
|    Kachemak Bay virus | *Ixodes signatus* | 1974 | KBV |
|    Taggert virus | *Ixodes* sp. | 1972 | TAGV |
|    Tillamook virus | *Ixodes uriae* | 1970 | TILLV |
| *Thiafora virus* | | | |
|    Thiafora virus | *Crocidura* sp. (shew) | 1971 | TFAV |
|    Erve virus | *Crocidura russula* (shew) | 1982 | ERVEV |

[a]Species names are in italics. Under each virus species name is listed other viruses of undetermined taxonomic status.
[b]Considered to be an Asian variant of NSDV.
[c]ND, not determined.

in the family *Argasidae* and from ticks of seven genera of the family *Ixodidae*. However, the virus appears to be most efficiently transmitted by ticks of the genus *Hyalomma* (**Figure 2**) and, in general, the known occurrence of CCHFV in Europe, Asia, and Africa coincides with the geographic distribution of these ticks. Although *Hyalomma* spp. ticks are considered the most important in the epidemiology of CCHF, the virus has been isolated from ticks in other genera (i.e., *Rhipicephalus*, *Boophilus*, *Dermacentor*, and *Ixodes* spp.) as well, which may contribute to its wide geographical distribution.

The principal vector of NSDV is the ixodid tick *Rhipicephalus appendiculatus*, which occurs throughout East and Central Africa. Most other nairoviruses have been isolated from ticks parasitizing birds and their ecology has not been well studied.

## Properties of the Virion and Genome

The morphology and structure of the CCHFV virion was first described in the early 1970s from the brains of infected newborn mice. It is now known that CCHFV and nairoviruses in general, are typical of other members of the family *Bunyaviridae* in terms of their basic structure, morphogenesis, replication cycle, and physicochemical

**Figure 2** Dorsal view of a *Hyalomma marginatum marginatum* female tick. Photo courtesy of Dr. Zati Vatansever, Ankara University Faculty of Veterinary Medicine, Ankara, Turkey.

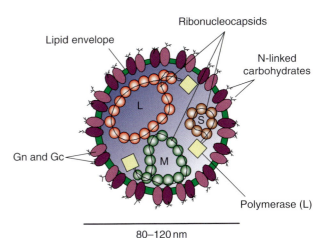

**Figure 3** Schematic cross-section of a *Bunyaviridae* virion. The three RNA genome segments (S, M, and L) are complexed with nucleocapsid protein to form the ribonucleocapsids. The nucleocapsids and RNA-dependent RNA polymerase are packaged within a lipid envelope that contains the viral glycoproteins, Gn and Gc. Adapted from Schmaljohn CS and Hooper JW (2001) Bunyaviridae: The viruses and their replication. In: Knipe DM and Howley PM (eds.) *Fields Virology*, 4th edn., pp. 1447–1472. Philadelphia: Lippincott Williams and Wilkins, with permission from Lippincott Williams and Wilkins.

properties. Virions are spherical, approximately 100 nm in diameter, and have a host cell-derived lipid bilayered envelope approximately 5–7 nm thick, through which protrude glycoprotein spikes 8–10 nm in length (**Figure 3**). Virions of members of the family *Bunyaviridae* contain three structural proteins: two envelope glycoproteins (Gn and Gc (previously termed G2 and G1)), named in accordance with their relative proximity to the amino or carboxy terminus of the M segment-encoded polyprotein, respectively, and a nucleocapsid protein (N), plus a large polypeptide (L), which is the viron-associated RNA-dependent RNA polymerase (**Figure 3**).

The genome is typical of those of other members of the family and is composed of three negative-strand RNA segments, S, M, and L, encoding the N nucleocapsid, Gn and Gc glycoproteins, and the L polymerase, respectively. The RNA segments are complexed with N to form individual S, M, and L nucleocapsids, which appear to be circular or loosely helical. The M segment of nairoviruses is 30–50% larger than the M segments of members of other genera in the *Bunyaviridae* family and has a potential coding capacity of up to 240 kDa of protein. At least one of each of the S, M, and L ribonucleocapsids must be contained in a virion for infectivity; however, equal numbers of nucleocapsids may not always be packaged in mature virions. Recent data show that the N protein is targeted to the perinuclear region of infected cells in the absence of native RNA segments and that this targeting is actin filament dependent. The first 8–13 nucleotide bases at the 3′-termini of all three RNA segments have a sequence (3′-AGAGUUUCU-) that is conserved within viruses of the genus, with a complementary consensus sequence at the 5′-termini. Base-pairing of the terminal nucleotides is predicted to form stable panhandle structures and noncovalently closed circular RNAs, which have been directly observed by electron microscopy of RNA extracted from the bunyavirus *Uukuniemi Virus*.

The principal stages of the replication process for viruses in the *Bunyaviridae* are similar to those of many other enveloped viruses. The viral glycoproteins are believed to be responsible for recognition of receptor sites on susceptible cells. Viruses that attach to receptors on susceptible cells are internalized by endocytosis, and replication occurs in the cytoplasm. Virions mature by budding through endoplasmic reticulum into cytoplasmic vesicles in the Golgi region, which are presumed to fuse with the plasma membrane to release virus.

Recently, a reverse genetics system was developed for CCHFV. The development of such a system was a major step forward in efforts to understand the biology of the virus. Developing an infectious clone for CCHFV will allow for more extensive studies of its biology and pathogenesis, and may ultimately lead to better therapeutic and prophylactic measures against CCHFV infections.

## Phylogenetic Relationships

Many early studies, based on serological testing, suggested that there are very few significant differences among strains of CCHFV. Recent data based on nucleic acid sequence analyses have revealed extensive genetic diversity. For example, analysis of the S RNA segment reveals the existence of seven distinct virus groups (i.e., genotypes) (**Figure 4**). These genotypes show a strong correlation with the geographical area of viral isolation. Furthermore, studies have discovered similar genotypes in distant geographical locations. Movement of CCHFV-infected livestock or uninfected livestock carrying infected ticks may explain some of the movement of viral genetic lineages within diverse regions. Other explanations include the movement of virus via migratory animals or birds that are either infected or are carrying virus-infected ticks. It is interesting that the Greek strain AP92 differed greatly from other European strains, and therefore is in a group by itself (**Figure 4**). AP92 strain was originally isolated in Greece from a *Rhipicephalus bursa* tick and has not yet been associated with disease in humans.

Sequence information on the L RNA segment has lagged behind those of the S and M RNA segments; nevertheless, there is evidence that the S and L RNA segments have the same evolutionary history. Thus, essentially identical S and L tree topologies are seen when analyzing all the available segment sequences. This, however, is not the case for the M RNA segment. This is generally taken for evidence of RNA segment reassortment. RNA viruses with segmented genomes have the capacity to reassort their genomic segments into new, genetically distinct viruses if the two or more viruses co-infect the same target cell.

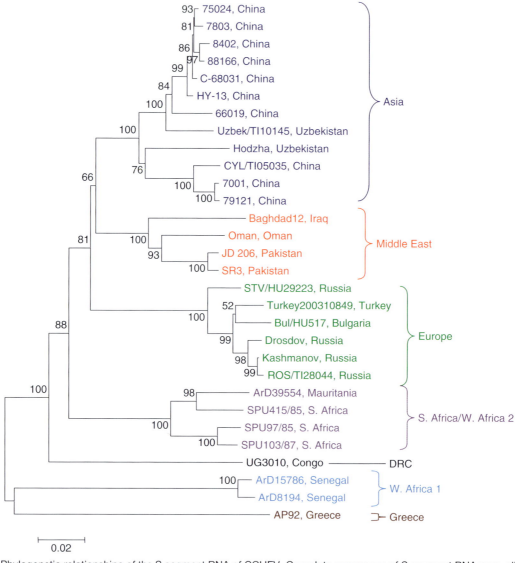

**Figure 4** Phylogenetic relationships of the S segment RNA of CCHFV. Complete sequences of S segment RNA were aligned and analyzed by a neighbor-joining method with Kimura two-parameter distances by using MEGA software (version 3.1). The lengths of the horizontal branches are proportional to the number of nucleotide differences between taxa. Bootstrap values above 50%, obtained from 500 replicates of the analysis, are shown at the appropriate branch points. CCHFV strains are described as strain designation, country of origin. Scale = 2% divergence.

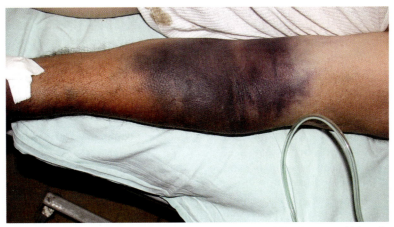

**Figure 5** Ecchymosis seen on the arm of a CCHF patient. Photo courtesy of Dr. Miro Petrovec, University of Ljubljana, Slovenia.

Indeed, several examples of RNA segment reassortment have been documented among viruses of the family *Bunyavirdae*, including CCHFV. Furthermore, reassortment appears to be much more frequently observed among CCHFV M RNA segments than among S and L RNA segments.

## Clinical Features

Except for newborn mice, humans appear to be the only host of CCHFV in which disease is manifested. In contrast to the inapparent infection in most other vertebrate hosts, human infection with CCHFV often results in severe hemorrhagic disease. The typical course of CCHF is noted by some authors as progressing through four distinct phases, that is, 'incubation', 'prehemorrhagic', 'hemorrhagic', and 'convalescence'; it is noteworthy that the duration and associated symptoms of these phases can vary greatly. In general, the incubation period after a tick bite can be as brief as 1–3 days, but can be much longer depending on several factors including route of exposure. After the incubation period, the prehemorrhagic period is characterized by sudden onset of fever, chills, severe headache, dizziness, photophobia, and back and abdominal pains. Additional symptoms such as nausea, vomiting, diarrhea, and an accompanying loss of appetite are common. Fever is often very high (39–41 °C) and can remain continually elevated for 5–12 days or may be biphasic. It is interesting that neuropsychiatric changes have been reported in some CCHF patients. These have included sharp changes in mood, with feelings of confusion and aggression and even some bouts of violent behavior. Cardiovascular changes can also be seen and include bradycardia and low blood pressure. In severe cases, 3–6 days after the onset of disease, hemorrhagic manifestations develop. These can range from petechiae to large areas of ecchymosis and often appear on the mucous membranes and skin, especially on the upper body and/or extremities (**Figure 5**). Bleeding in the form of melena, hematemesis, and epistaxis is also commonly seen by day 4 or 5 and can often be characterized by dark 'coffee grounds' vomitus and tarlike stools resulting from intestinal hemorrhages. Bleeding from other sites including the vagina, gingival bleeding, and, in the most severe cases, cerebral hemorrhage have been reported. About 50% of the patients develop hepatomegaly. Not surprisingly, poor prognosis is associated with cerebral hemorrhage and massive liver necrosis. Mortality rates for the various CCHF epidemics and outbreaks have varied greatly. The average mortality rate is often cited at 30–50%; however, rates as high as 72.7% and 80% have been reported from the United Arab Emirates and China, respectively. During the recent outbreak in Turkey, of the 500 cases reported to the Turkish Ministry of Health, 26 (5.2%) were fatal. Mortality rates of nosocomial infections are often much higher than those acquired naturally through tick bite. The exact reasons for this phenomenon are not known, but may simply relate to viral dose.

## Pathogenesis

Because of the difficulties of working with CCHFV (e.g., the need for specialized containment laboratories) and the lack of an animal model of disease, the pathogenesis of CCHF is poorly understood. Capillary fragility is a common feature of CCHF, suggesting infection of the endothelium. This is surely where the alternative term 'capillary toxicosis', given to CCHF by the early Soviet workers, was derived. Localization of CCHFV in tissues by immunohistochemistry has shown that mononuclear phagocytes and endothelial cells are major sites of viral infection. Endothelial damage would account for the characteristic rash and contribute to hemostatic failure by stimulating platelet aggregation and degranulation, with consequent activation of the intrinsic coagulation cascade. Thrombocytopenia appears to be a consistent feature of CCHF infection and platelet counts can often be extremely low beginning at the early stage of illness in fatal cases.

Some have argued that the characteristic endothelial damage seen in CCHF is not necessarily the result of direct infection of the endothelial cells by CCHFV. Evidence is mounting that for viral hemorrhagic fever, caused by the Ebola virus, much of the cellular damage and resulting coagulopathy results from multiple host-induced mechanisms. These include massive apoptosis of lymphocytes both intravascularly and in lymphoid organs; induction of pro-inflammatory cytokines, including tumor necrosis factor (TNF)-α; and the dysregulation of the coagulation cascade leading to disseminated intravascular coagulation (DIC). Indeed, many of these same features are seen in CCHF, including DIC, vascular dysfunction, and shock. Clearly, more work needs to be done in this area to completely understand the pathogenesis of CCHF.

## Diagnosis

A diagnosis of CCHF should be considered when severe flu-like symptoms with a sudden onset are seen in patients with a history of tick bite, travel to an endemic area, and/or exposure to blood or other tissues of livestock or human patients. Early diagnosis is essential, both for the outcome of the patient and because of the potential for nosocomial infections. The differential diagnosis should include rickettsiosis (tick-borne typhus or African tick bite fever caused by *Rickettsia conorii* and *R. africae*, respectively), leptospirosis, and borreliosis (relapsing fever). Additionally, other infections, which present as hemorrhagic disease such as meningococcal infections, hantaviral hemorrhagic fever, malaria, yellow fever, dengue, Omsk hemorrhagic fever, and Kyasanur Forest disease should be considered. In Africa, Lassa fever and infection with the filoviruses, Ebola and Marburg, must also be included in the differential diagnosis.

The diagnosis of CCHF is confirmed by detecting viral nucleic acid by reverse transcription-polymerase chain reaction (RT-PCR), demonstration of viral antigen by enzyme-linked immunosorbent assay (ELISA), or isolation of the virus. Any attempts at isolating and culturing the virus should be performed only under BSL-4 containment. The traditional method and 'gold standard' for CCHFV isolation has been by intracranial or intraperitoneal inoculation of a sample into newborn mice. However, isolation in cell culture is far simpler and safer and provides a more rapid result. Virus can be isolated from blood and organ suspensions in a wide variety of susceptible cell lines including LLC-MK2, Vero, BHK-21, and SW-13 cells with maximal viral yields after 4–7 days of incubation. In some cases however, depending on the cell line and strain, the virus may induce little or no cytopathic effect (CPE). In these cases, the presence of virus can be identified by performing an immunofluorescence assay (IFA) with specific antibodies to CCHFV or by RT-PCR.

The neutralizing antibody response to CCHFV, as well as to other nairoviruses, is weak and difficult to demonstrate. Both IgG and IgM antibodies are detectable by IFA by about day 5 of illness and are present in the sera of survivors by day 9. The IgM antibody declines to undetectable levels by the fourth month after infection, and IgG titers may also begin to decline gradually at this time, but remain demonstrable for at least 5 years. An antibody response is rarely detectable in fatal cases and diagnosis is usually confirmed by isolating the virus from the serum or from liver biopsy specimens or by demonstrating the presence of CCHFV antigen by immunohistochemical techniques of paraffin-embedded liver sections or by IFA of liver impression smears.

Molecular-based diagnostic assays, such as RT-PCR, provide a useful complement to serodiagnosis and now often serve as the front-line tool both in the diagnosis of CCHF and in epidemiological studies of the disease.

## Prevention and Control

Several groups of individuals are considered to be at risk of contracting CCHF – specifically, people from endemic areas who are liable to be fed upon by ticks, particularly *Hyalomma* spp. ticks. These include individuals who work outdoors, particularly those who work with large domestic animals. Exposures such as crushing infected ticks and butchering infected animals have also been a frequent source of CCHFV infection. Other groups who are at risk include those caring for CCHF patients. In fact, the risk of nosocomial infection in healthcare workers is well documented and can be extremely high, especially during the hemorrhagic period of disease.

Avoiding or minimizing exposure to the virus is the best means of preventing CCHF. Persons in high-risk occupations (i.e., slaughterhouse workers, veterinarians, sheep herders, etc.) should take every precaution to avoid exposure to virus-infected ticks or virus-contaminated animal blood or other tissues. For example, wearing gloves and limiting exposure of naked skin to fresh blood and other tissues of animals are effective practical control measures. Likewise, medical personnel who care for suspected CCHF patients should practice standard barrier-nursing techniques. Tick control may not be practical in many regions of the world where *Hyalomma* ticks are most prevalent. However, acaricide treatment of livestock in CCHFV-endemic areas is effective in reducing the population of infected ticks. Applying commercially available insect repellents (i.e., meta-$N$,$N$-diethyl toluamide (DEET)) to exposed skin and using clothes impregnated with permethrin can provide some protection against tick bites. As for other tick-borne viruses, inspecting one's body and clothes for ticks, and their prompt removal can minimize the risk of infection. An inactivated vaccine, prepared from the brains of infected suckling mice, was used in Eastern Europe and the former Soviet Union in the past, but no vaccines are currently available.

## Treatment

Treatment options for CCHF are limited and primarily consist of supportive and replacement therapies. Standard treatment consists of monitoring patients with replacement of red blood cells, platelets, and other coagulation factors. Immunotherapy has been attempted by passive transfer of CCHF survivor convalescent plasma, but the efficacy of this treatment is not clear. There is currently no specific antiviral therapy for CCHF approved for use in humans by the US Food and Drug Administration. An antiviral drug, ribavirin, is effective against CCHFV in culture and has shown the most promise in treating CCHF patients over the years. However, for the best patient outcome, treatment should be started early, ideally, before day 5 of illness.

## Disclaimer

The views, opinions, and findings contained herein are those of the author and should not be construed as an official Department of the Army position, policy, or decision unless so designated by other documentation.

*See also:* Orthobunyaviruses; Rift Valley Fever and Other Phleboviruses.

## Further Reading

Flick R and Whitehouse CA (2005) Crimean–Congo hemorrhagic fever virus. *Current Molecular Medicine* 5: 753–760.

Hoogstraal H (1979) The epidemiology of tick-borne Crimean–Congo hemorrhagic fever in Asia, Europe, and Africa. *Journal of Medical Entomology* 15: 307–417.

Linthicum KJ and Bailey CL (1994) Ecology of Crimean–Congo hemorrhagic fever. In: Sonenshine DE and Mather TN (eds.) *Ecological Dynamics of Tick-Borne Zoonoses*, pp. 392–437. New York: Oxford University Press.

Montgomery E (1917) On a tick-borne gastro-enteritis of sheep and goats occurring in British East Africa. *Journal of Comparative Pathology and Therapy* 30: 28–57.

Schmaljohn CS and Hooper JW (2001) *Bunyaviridae:* The viruses and their replication. In: Knipe DM and Howley PM (eds.) *Fields Virology*, 4th edn., pp. 1447–1472. Philadelphia: Lippincott Williams and Wilkins.

Swanepoel R (1995) Nairovirus infections. In: Porterfield JS (ed.) *Exotic Viral Infections*, pp. 285–293. London: Chapman and Hall.

Watts DM, Ksiazek TG, Linthicum KJ, and Hoogstraal H (1989) Crimean–Congo hemorrhagic fever. In: Monath TP (ed.) *The Arboviruses: Epidemiology and Ecology*, vol. II, pp. 177–222. Boca Raton: CRC Press.

Whitehouse CA (2004) Crimean–Congo hemorrhagic fever. *Antiviral Research* 64: 145–160.

# Cryo-Electron Microscopy

**W Chiu and J T Chang,** Baylor College of Medicine, Houston, TX, USA
**F J Rixon,** MRC Virology Unit, Glasgow, UK

© 2008 Elsevier Ltd. All rights reserved.

## Introduction

X-ray crystallography has yielded atomic models of icosahedral viruses of up to 700 Å diameter (e.g., bluetongue virus, reovirus, and bacteriophage PRD1). However, crystallography has not been successful with larger viruses, is not yet able to yield high-resolution details of nonicosahedral components in intact virus particles, and is of limited use in examining partially purified or low-concentration specimens or samples with mixtures of different conformational states. Many of these limitations can be overcome by using cryoelectron microscopy (cryoEM) and image reconstruction approaches. Furthermore, cryoEM has been useful not only to virologists for understanding virus assembly and virus–antibody and virus–receptor interactions, but also to electron microscopists for driving developments in electron imaging technology.

In the early 1970s, DeRosier, Crowther, and Klug introduced the method for icosahedral particle reconstruction, which marked the beginning of three-dimensional (3-D) electron microscopy for spherical viruses. Initially, these studies used stained and dried specimens. However, in the early 1980s, Dubochet and co-workers developed vitrification methods that use rapid freezing of solutions to preserve the intact structures of biological macromolecules for electron imaging. They demonstrated the feasibility of reconstructing the 3-D structure of a virus particle in its native conformation without using stain or fixative.

Both the quality of the electron images and the resolving power of the single-particle reconstruction algorithms have been continuously improved in the last 20 years to the extent that it is now possible to routinely obtain virus structures at subnanometer resolutions (6–10 Å). Although not yet equal to X-ray crystallographic resolution, this does allow identification of secondary structure elements such as long α-helices and large β-sheets in protein subunits of virus particles.

All the subnanometer resolution cryoEM structures of viruses solved so far, have made use of the icosahedral properties of the particle. However, the assembly and infection processes in many viruses depend on structures which break the icosahedral symmetry of the particle. Details of the organization of these nonicosahedral components are absent from almost all crystal structures of virions due to the icosahedral averaging used in the data processing. Similarly, cryoEM reconstructions from single-particle images that are dependent on icosahedral averaging do not show the nonicosahedral components.

Recently, image reconstruction techniques have been introduced to compute a density map without imposing icosahedral symmetry. This approach has been successfully used to visualize all the structural components of bacteriophages T7, Epsilon15, and P22 at moderate resolution (17–20 Å).

Another imaging method called electron cryotomography (cryoET) has been used to reconstruct large enveloped viruses such as human immunodeficiency virus (HIV), vaccinia virus, and herpes simplex virus (HSV) at 100–60 Å resolution. Subsequent single-particle averaging of certain computationally extracted components has yielded their 3-D structures to 50–30 Å resolution. This approach is ideal for obtaining low-resolution information from virus particles containing both conformationally constant and variable components.

## CryoEM of Single Virus Particles

CryoEM of single particles consists of four basic steps: cryo-specimen preparation, imaging in an electron cryomicrosope, image reconstruction, and structural interpretation. Because there are excellent reviews that describe the detailed procedures involved, in this article the focus is on the fundamental concepts of this methodology, including some practical considerations where appropriate.

## Cryo-Specimen Preparation

Specimens in the electron microscope have to be examined in a high vacuum ($10^{-6}$ torr) environment, which would dehydrate and thus destroy the native structure of any normal biological specimen. Fortunately, the rapid freezing methods developed by Dubochet and co-workers can preserve specimens embedded in a thin layer of vitreous ice. Using this procedure, the virus particles are maintained in their original solution state. Vitrification is the process by which water solidifies without undergoing crystallization and it requires extremely rapid cooling. This can be achieved by applying the sample to a microscope grid that has been overlaid with a thin film of holey carbon, blotting away excess sample, and plunging the very thin aqueous film into liquid ethane that has been cooled to liquid nitrogen temperature. An ideal situation is for the ice to be slightly thicker than the virus particles. In practice, there is always variation in ice thickness even across a single microscope grid. For any new specimen, a process of systematic trial and error will eventually identify freezing conditions that yield optimal ice thickness in at least some parts of a grid.

An important practical consideration in cryo-specimen preparation is the concentration of the sample, which ideally should be in the order of $10^{10}$ particles ml$^{-1}$. A subnanometer-resolution icosahedral reconstruction generally needs images of several thousands of particles in different orientations. Therefore, a high particle concentration will reduce the number of micrographs that have to be collected, which will, in turn, reduce the number of data sets to be processed. An added benefit is that having a large number of particles per micrograph facilitates the determination of the defocus value of each micrograph, which is necessary for subsequent data processing. However, such concentrations can be difficult to obtain with some virus samples, in which case the necessary data collection and processing can become very time consuming.

## Low-Dose Single-Particle Imaging

Following sample preparation, the grids containing the ice-embedded virus particles can be stored in liquid nitrogen until placed into the electron microscope for imaging. Throughout all stages of the process, the frozen-hydrated specimens must be handled extremely carefully to avoid any ice contamination from moisture in the atmosphere. The specimen is kept at or below liquid nitrogen temperature during the electron microscopy session. Frozen-hydrated biological specimens are extremely sensitive to radiation and must be scrutinized using a very low electron dose. They are examined initially at low magnification (100–1000×) to find a suitable area on the grid and

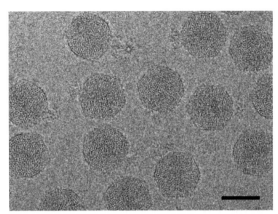

**Figure 1** CryoEM imaging. Electron image of frozen-hydrated Epsilon15 phage taken at 300 kV displays the capsids as dark circles. The viral DNA inside is observed as a fingerprint pattern. Scale = 500 Å. Courtesy of Joanita Jakana, Baylor College of Medicine.

then images are recorded at a higher magnification (40 000–80 000×) from the chosen area (**Figure 1**).

The total electron dose that can be tolerated by a specimen depends only slightly on the specimen temperature and the accelerating electron voltage. For most practical purposes, a typical dose is about 15–25 electrons Å$^{-2}$ for single-particle imaging aimed at a resolution of better than 20 Å. A major variable in recording any electron micrograph is the defocus setting for the objective lens, which will vary according to the targeted resolution of the experiment and the type of microscope used. Higher defocus increases image contrast while smaller defocus yields higher-resolution information. Typically, the data for reconstruction are collected with defocus values between 1 and 3 μm in order to produce visible image contrast in the raw micrograph while preserving the higher-resolution features retrievable through image processing.

There are two ways to record an electron micrograph: on photographic film or with a charge-coupled device (CCD) camera. The advances in CCD camera technology now enable most medium-resolution data (down to 6–8 Å) to be recorded directly in digital format ready for subsequent steps of image processing. Given the rapid growth in CCD technology, it is likely that photographic film will eventually be replaced completely.

## Icosahedral Reconstruction

Many viruses are composed of spherical protein shells arranged icosahedrally. **Figure 1** shows a representative cryoEM image of this type of particle. It is important that the particles are oriented randomly in the matrix of embedding ice because the task of image reconstruction involves combining particles from different

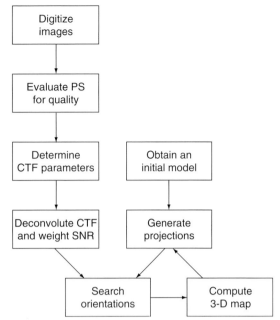

**Figure 2** A 3-D image reconstruction scheme for icosahedral particles. The quality of digitized images is evaluated by analyzing their power spectrum (PS). Next, the contrast transfer function (CTF) parameters are determined for each image and used to deconvolute and weigh them according to their signal-to-noise ratio (SNR). The orientations of the deconvoluted images are determined by comparing them against projections from a 3-D map generated from a starting model. The oriented images are used to reconstruct a better 3-D map, which is subsequently used to generate better projections. The new projections are used to improve the accuracy of the search for the image orientations. The 3-D map improves after each iteration until it converges and the processing can stop.

orientations into a 3-D density map. The separate steps in image reconstruction are outlined in **Figure 2** and described briefly below. To simplify the technical aspects of the procedure, numerous image-processing software packages that can be used for icosahedral reconstruction have been developed.

Initially, each micrograph captured from a CCD camera or digitized from photographic film is evaluated for its image quality based on the power spectrum derived from the average of multiple Fourier transforms of computationally isolated particle images. This power spectrum contains information about the structure of the particles, the imaging conditions used, and the image quality. The ring pattern in the power spectrum (**Figure 3(a)**) reflects the imaging condition known as the contrast transfer function (CTF). It can vary in appearance according to factors such as defocus, astigmatism, specimen drift, and signal decay as a function of resolution, all of which are attributable to various instrumental factors. The signal-to-noise ratio (SNR) in each micrograph can be computed from the circularly averaged power spectrum. The SNR plot (**Figure 3(b)**) indicates the potential signal strength

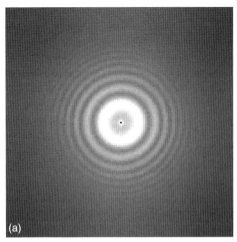
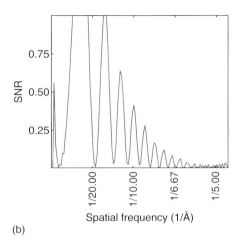

**Figure 3** Evaluating image quality. (a) The power spectrum derived from averaging the Fourier transforms of particles boxed out from electron images such as those in **Figure 1**. (b) SNR plot of the boxed particle images as a function of spatial frequency (resolution) computed from the circularly averaged power spectrum. For a conservative measure of 10% SNR, the usable data extends beyond 8 Å resolution. Courtesy of Donghua Chen, Baylor College of Medicine.

at different resolutions and gives an indication of the ultimately achievable resolution in the reconstruction.

In order to obtain a correct structure, each raw particle image has to be deconvoluted by its CTF and be compensated for the signal decay at high resolution (**Figure 2**). The image deconvolution and signal-to-noise weighing are usually carried out for each raw particle image before the orientation of the particles is determined. Following orientation determination, the corrected particle images are merged to synthesize the 3-D map. Alternatively, the deconvolution can be carried out on the preliminary 3-D maps computed from each micrograph, with the deconvoluted maps being merged subsequently to produce the final map.

The most computationally demanding aspect of image processing is the estimation and refinement of the five spatial parameters (three for rotations around $x$-, $y$-, and $z$-axes, and two for center in $x$- and $y$-directions) for each particle image (**Figure 2**). There are a number of different algorithms and software packages that can be used to perform this critical task. Most of these programs require an initial model of the particle, from which they generate multiple projection images showing the particle model in different orientations. These projection images are then compared with each of the raw particle images. The projection image that gives the closest match is assumed to have the same orientation as the virus particle that produced the raw image. Each algorithm uses different criteria for comparing the reference and raw particle images to derive the spatial parameters. Throughout a series of iterative refinement steps, the model is continuously updated and its resolution is gradually increased (**Figure 2**).

During the steps of orientation and center determination, and 3-D reconstruction, the algorithms assume that the particle has icosahedral symmetry. This means that the particle is made up of 60 repeated (asymmetric) units, which are related to each other by five-, three-, and twofold symmetry operations. If the particle does not have the expected symmetry, the algorithm will not yield a consistent map and will not be able to derive its structure. On the other hand, it is possible that in some particles the icosahedral symmetry is preserved only up to a certain resolution, which would limit the ultimate resolution of the reconstruction. This is analogous to a protein crystal only diffracting up to some resolution limit beyond which the crystal structure cannot be resolved.

## Interpreting CryoEM Single-Particle Density Maps

The challenge in interpreting a cryoEM density map varies according to its resolution. Maps of different resolutions reveal different structural features of the icosahedral shell proteins (**Figure 4**). A low-resolution map (down to 30 Å) can yield the size, shape, and organization (triangulation number) of the icosahedral particle. In fact, cryoEM is probably the most reliable technique for providing such information. In a moderate resolution map (20–10 Å), the subunit boundaries may begin to be resolved. The segmentation of the map into subunits can be done qualitatively by visual inspection of the structure in front of a graphics terminal or quantitatively by a software program. The accuracy of the segmentation is ultimately dictated by the resolution of the map.

When a cryoEM density map reaches subnanometer resolution, long α-helices and large β-sheets become visually recognizable and can be assessed quantitatively (**Figure 5**). In intermediate resolution (7–9 Å) density maps, it has been shown that it is possible to correctly

(a) (b) (c)

**Figure 4** CryoEM maps of Epsilon15 phage at different resolutions. The surface-rendered images of the whole capsid (top row) and a close-up of the asymmetric unit (bottom) show that (a) at 40 Å resolution, the overall morphology and triangulation number can be distinguished; (b) at 20 Å resolution, it is possible to visualize the subunit separation in the hexons and pentons; and (c) at 9.5 Å resolution, α-helices are discernible as rod-like densities (EMD-1176). The display was colored radially.

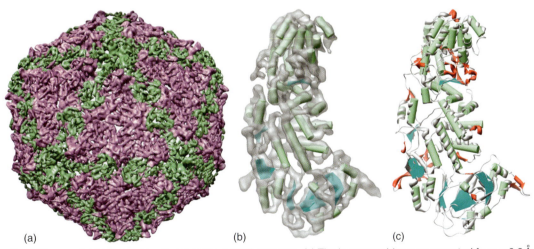

(a) (b) (c)

**Figure 5** CryoEM map of rice dwarf virus matches the crystal structure. (a) The inner capsid map segmented from a 6.8 Å resolution cryoEM reconstruction of the entire virus, with the two different conformations of the capsid protein P3A and P3B colored green and violet, respectively (EMD-1060). (b) Segmented inner capsid shell protein P3A with predicted secondary structure elements annotated as cylinders (α-helices) and surfaces (β-sheets). (c) Overlaying the crystal structure (ribbons, 1UF2) onto the secondary structure prediction shows a good match. (a) Reproduced from Zhou ZH, Baker ML, Jiang W, et al. (2001) Electron cryomicroscopy and bioinformatics suggest protein fold models for rice dwarf virus. *Nature Structural Biology* 8: 868–873. (c) Reproduced from Baker ML, Ju T, and Chiu W (2007) Identification of secondary structure elements in intermediate-resolution density maps. *Structure* 15: 7–19, with permission from Elsevier.

identify and annotate nearly all of the helices with $\geq 2.5$ turns and sheets with $\geq 2$ strands. For example, the match (**Figure 5(c)**) between the secondary structural elements determined from the 6.8 Å cryoEM map of the rice dwarf virus and the crystal structure that was obtained later demonstrated the reliability of the cryoEM reconstruction and of the structural feature identification.

## Asymmetric Reconstruction of an Icosahedral Particle with a Unique Vertex

Though icosahedral reconstruction methods have been very successful in providing structural information on capsid shell proteins, they reveal nothing about any non-icosahedral components. Recently, however, several image-processing procedures have emerged for reconstructing spherical virus particles with tail structures without imposing symmetry. Roughly speaking, the reconstruction procedure uses an initial icosahedral reconstruction of the particle as a starting model for the asymmetric reconstruction. Because only one vertex of the virus particle has the additional mass of the tail, a crude model can be generated by computationally adding a cylinder of density at one of the fivefold vertices of the icosahedral reconstruction. This tailed model is projected in each of the 60 icosahedrally equivalent views and compared with the raw particle image to identify the view that produces the best match. Once the orientations of all the images have been determined, they are used to reconstruct the map without imposing any symmetry. This process of projection matching is iterated using a constantly updated reconstruction, until the cylindrical symmetry of the tail is broken, revealing its true structure. **Figure 6** shows the asymmetric reconstruction of Epsilon15 phage where, in addition to the capsid shell, multiple nonicosahedrally ordered molecular components and the internal DNA can be visualized.

## Modeling Virus Particles with CryoEM Density Map Constraints

It is not uncommon for components of a virus particle to be crystallized and structurally determined while only a low- or medium-resolution cryoEM map of the entire virus particle is available. In these cases, a pseudo-atomic model of the virus particle can often be built by docking the crystal structures of the solved components into the cryoEM map. The docking can be done visually or quantitatively using a fitting program. Examples of this are provided by adenovirus and herpesvirus. In neither case has the intact capsid been crystallized, but inserting the crystal structures of the adenovirus hexon and penton base proteins, or a domain of HSV-1 major capsid protein

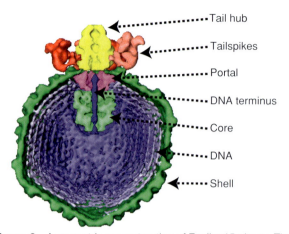

**Figure 6** Asymmetric reconstruction of Epsilon15 phage. The map reveals all the structural components of the capsid shell, DNA, core, portal, tailspikes, and tail hub (EMD-1175). Reproduced from Jiang W, Chang J, Jakana J, Weigele P, King J, and Chiu W (2006) Structure of Epsilon15 phage reveals organization of genome and DNA packaging/injection apparatus. *Nature* 439(7076): 612–616.

into the respective cryoEM structures, has provided us with pseudo-atomic models that reveal important features of the capsids. An outstanding example of how cryoEM and crystallographic data can be merged to establish a pseudo-atomic model is provided by the T4 bacteriophage baseplate (**Figure 7**). Here, the crystal structures of six separate molecular components have been fitted into the reconstruction of the baseplate, leading to a proposal for the mechanism of DNA injection into the host cell.

Where there is no crystal structure available, it may still be possible to construct a set of structural models using either comparative modeling or *ab initio* modeling based on the amino acid sequence alone. In these cases, a cryoEM density map can be used as a template to select the most native-like model. A domain of the VP26 capsid protein of HSV, which has a molecular mass of $\sim 10$ kDa, has been modeled *ab initio* using the cryoEM structure of the HSV capsid as a selection constraint. Although the validity of this approach has not yet been confirmed by crystal data, it is encouraging that the model fitting the cryoEM data was consistent with published genetic and biochemical data for VP26.

## Low-Dose Cryoelectron Tomographic Imaging of Virus Particles

Particles of some viruses have components that assume different conformations in different particles, making it impossible to generate a 3-D density map by averaging these particles. For this type of particle, the only option available is tomography. While the reconstruction techniques described above work by combining views of multiple particles in different orientations, tomographic reconstruction combines multiple views of the same particle as its orientation is changed by rotating it between consecutive exposures using a tilting microscope holder. Therefore, this technique allows a low-resolution map to be generated by collecting a tilt series from any cryoEM specimen, and it has been used successfully with a range of viruses including HSV virions and capsids, vaccinia virus, and HIV. Because of the constraints on the number of images that can be acquired due to the effects of radiation damage, cryotomography yields only low-resolution data. Therefore, the strategy is to record multiple images of the specimens at low magnification and under low-dose condition. Typically, a magnification of 10 000–25 000× and a dose of 1–2 electrons $\text{Å}^{-2}$ per micrograph are used (**Figure 8**).

Once captured, the tilt series images are aligned and merged to generate a 3-D tomogram. Because the grid supporting the specimen obscures the electron beam, the range of tilts that can be used is limited to about ±70°. As a result, there is a missing wedge of data, which means that each tomogram is subject to distortion and must be interpreted cautiously.

Although tomographic reconstructions are generally of low resolution (100–60 Å), post-tomographic averaging of components with constant conformations can yield higher resolutions. For example, post-tomographic data sorting, mutual alignment, and averaging have been used to generate a model of the herpesvirus capsid with an averaged view of the single portal vertex (**Figure 9(a)**). Similarly, the spike of the simian immunodeficiency virus (SIV) particle has been aligned and averaged to produce a map that permitted docking of the crystal structures of the spike proteins (**Figure 9(b)**). In this type of approach, the final resolution is limited by a number of factors, including the accuracy of the alignment, the number of independent views of the components included in the

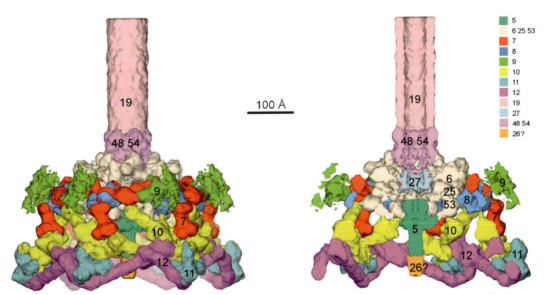

**Figure 7** Pseudo-atomic models. Side and cut-away view of the T4 phage tube–baseplate complex. Six crystallized components of the T4 baseplate were docked into the cryoEM map, which permitted identification of the baseplate components (color-coded gene products with legend on right) (EMD-1086). Courtesy of M. G. Rossmann at Purdue University and reproduced with permission from Rossmann MG, Mesyanzhinov VV, Arisaka F, and Leiman PG (2004) The bacteriophage T4 DNA injection machine. *Current Opinion in Structural Biology* 14(2): 171–180.

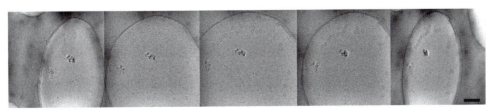

**Figure 8** Tomography images. Selected CCD frames recorded at −60°, −30°, 0°, +30°, and +60° tilts (left to right) show herpesvirus capsids forming an arc in the top half of each image. Scale = 5000 Å.

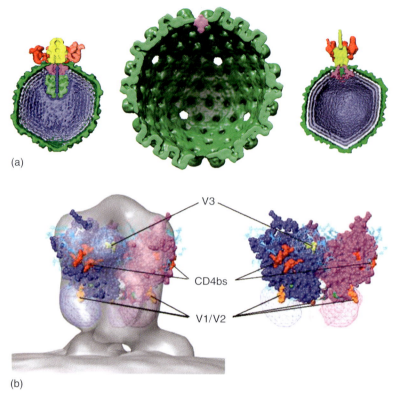

**Figure 9** Increasing resolution by averaging subvolumes from tomographic reconstructions. (a) Structure of the herpes simplex virus-1 capsid (center) shown after removal of the pentons to reveal the unique portal vertex (purple). This structure was derived by averaging 13 subvolumes in different tomographic reconstructions (EMD-1308). The arrangement of the portal in the capsid floor is similar to that seen in the tailed double-stranded DNA (dsDNA) phages Epsilon15 (left, EMD-1175) and P22 (right, EMD-1222). (b) Averaging membrane spikes from tomographic reconstructions of SIV particles and then imposing threefold symmetry improves the resulting map, thereby allowing the crystal structure of the spike protein to be fitted. The V1/V2 (orange) and V3 (yellow) loops and CD4 binding sites (red) are labeled in this model. (a) Reproduced from Chang JT, Schmid MF, Rixon FJ, and Chiu W (2007) Electron cryotomography reveals the portal in the herpesvirus capsid. *Journal of Virology* 81(4): 2065–2068, with permission from American Society for Microbiology. (b) Courtesy of K. Taylor and K. H. Roux at Florida State University and reproduced with permission from Zhu P, Liu J, Bess J, *et al.* (2006) Distribution and three-dimensional structure of AIDS virus envelope spikes. *Nature* 441(7095): 847–852.

averaging, the symmetry of the components, the proper image deconvolution, the correct compensation for the missing wedge data, and the conformational uniformity of the components.

## Biological Insights from CryoEM Maps

The extent of biological information that can be obtained from cryoEM reconstruction varies according to the resolution of the map as well as the design of the experiment. As shown in **Figures 4** and **5**, different structural features can be derived from the cryoEM maps at varying resolutions, ranging from shape, size, triangulation number, capsomere morphology and number, quaternary, tertiary, and secondary structural features of individual protein subunits. Here, a small and not exhaustive set of examples is provided to demonstrate the usefulness of cryoEM maps in the context of virology.

## Conservation of Structural Motifs among Viruses

Questions regarding the origin of viruses and their distant evolutionary relationships are handicapped by the low sequence conservation generally found in these small and rapidly evolving entities. Therefore, structural comparison has been used as an alternative indicator of their relatedness. The best-known example of structural analysis revealing a link between supposedly unrelated viruses is that of bacteriophage PRD1 and human adenovirus. Although early cryoEM analysis had shown that the PRD1 capsid structure was very similar to that of adenovirus, it was not until the crystal structures of their respective major capsid proteins were compared that the evolutionary relationship between these viruses was firmly established.

However, in cases where crystal structures are not available, cryoEM maps determined at subnanometer resolution can provide sufficient information on the pattern of secondary structure elements and thus of the protein folds. **Figure 10** shows a structural comparison of the capsid proteins of the eukaryote-infecting herpesvirus (*Herpesviridae*) and the prokaryote-infecting tailed DNA bacteriophages (*Caudovirales*). Although these proteins have no detectable sequence homology, it is clear from the similar spatial distributions of secondary structural elements that they share a characteristic fold. CryoEM-derived structural data have also revealed conserved features in the unique portal vertices of the two virus types, including a characteristic subunit arrangement and a similar location within the capsid shell (**Figure 9(a)**). These observations have led to the conclusion that these very distinct, extant viruses must have arisen from the same primordial progenitors.

## Structural Polymorphism during Virus Maturation

Many virus particles undergo structural changes during their morphogenesis. For example, tailed bacteriophages typically form their capsid shell around an internal scaffold of proteins. This initial assembly product (the procapsid) then undergoes a maturation process during which the DNA is packaged, the internal scaffold is lost, and the capsid shell undergoes extensive structural reconfiguration. **Figure 11** demonstrates the structural changes in P22 bacteriophage as revealed from reconstructions of the procapsid and of the mature virion. Not only can changes in the overall size and shape of the capsid be observed but also correlative changes in the quaternary and tertiary structures of individual subunits can be

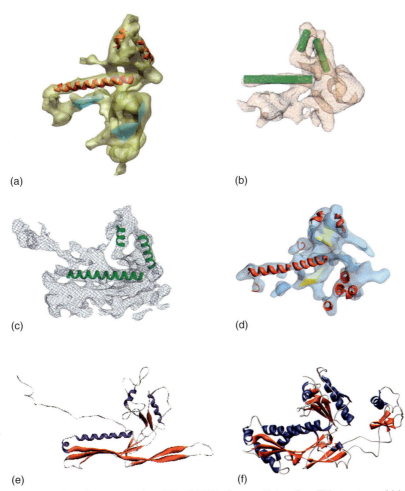

**Figure 10** Evolutionary conservation of a structural motif in dsDNA virus particles. CryoEM structure of (a) Epsilon15 phage shell protein, with annotated helices and sheets (EMD-1176), (b) P22 phage coat protein gp5 (EMD-1101), (c) Phi29 phage head protein gp8 (EMD-1120), (d) HSV-1 major capsid protein VP5 floor domain, and crystal structures of (e) HK97 phage head protein gp5 (1OGH) and (f) T4 phage head protein gp24 (1YUE) all have similar structural signatures, which suggest a common ancestry even though they have little sequence similarity. Reproduced from Jiang W, Chang J, Jakana J, Weigele P, King J, and Chiu W (2006) Structure of Epsilon15 phage reveals organization of genome and DNA packaging/injection apparatus. *Nature* 439(7076): 612–616.

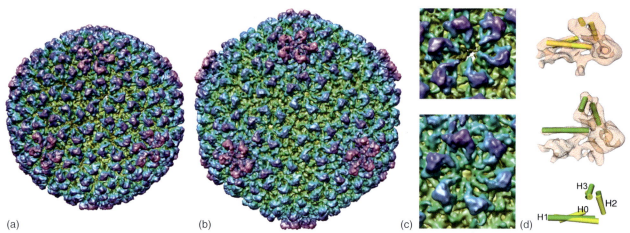

**Figure 11** Structural changes during virus maturation. (a) The spherical procapsid of P22 is ~600 Å in diameter, and its hexons are elongated, with a hole in the center. (b) The angular mature capsid is ~700 Å in diameter, and its hexon subunits are arranged hexagonally with more lateral interactions, resulting in a smaller hole (EMD-1101). (c) A close-up view of the procapsid (top) and mature (bottom) hexon. (d) The P22 capsid protein gp5 rearranges when maturing from the procapsid (top, annotated secondary structures in yellow) to the capsid (middle, annotated secondary structures in green), resulting in changes to secondary, tertiary, and quaternary structure, as shown by overlapping the annotated secondary structures (bottom). (d) Reproduced from Jiang W, Li Z, Zhang Z, Baker ML, Prevelige PE, and Chiu W (2003) Coat protein fold and maturation transition of bacteriophage p22 seen at subnanometer resolutions. *Nature Structural Biology* 10:131–135.

noted. For instance, the organization of subunits in the hexon capsomere changes from a skewed shape to a six-fold symmetrical arrangement. In addition to the overall domain movements in the hexon subunit that bring about this change, the cryoEM structures also reveal more detailed changes such as the refolding of an N-terminal α-helix. Large-scale structural changes have also been visualized during maturation of the HSV capsid. In addition to comparing relatively stable forms of a particle, cryoEM can be used to follow rapid structural changes (e.g., by changing the pH of the buffer immediately before freezing) as observed in Semliki forest virus, Sindbis virus, and La Crosse virus.

## Viral Genome Organization

To date, most information on how viral nucleic acids are arranged within particles has come from cryoEM reconstructions. Low-resolution cryoEM reconstruction has revealed that small sections of the RNA genomes in some RNA viruses are bound icosahedrally to the capsid shell proteins, although the bulk of the genome was not resolved even in these cases. However, in general, viral genomes are not packaged icosahedrally within particles and so cannot be visualized by icosahedral reconstruction techniques.

Those double-stranded DNA viruses that package their genomes into preformed capsids pose a particularly interesting dilemma because this must be accomplished without organizing proteins (such as histones) to counter the electrostatic charge of the DNA and the organization must allow for efficient entry and ejection without forming knots or tangles. The arrangement of the nucleic acids in such capsids has been extensively studied and debated. Proposed models include the co-axial spool, spiral fold, liquid crystal, and folded toroid. CryoEM has led the way in shedding light on this question. Analysis of a bacteriophage T7 mutant that was found to orient preferentially in vitreous ice showed the capsids having either concentric rings (axial view) of DNA spaced about 25 Å apart or punctate arrays (side view) suggesting that the DNA was organized in concentric layers. Subsequently, it became possible to examine the DNA in more detail using asymmetric reconstruction algorithms. In Epsilon15 phage, for example, DNA strands are packed as concentric layers in the capsid cavity suggesting a co-axial spool arrangement (**Figure 12(a)**), and a slice normal to the DNA strands and parallel to the tail axis shows hexagonal features (**Figure 12(b)**). A DNA strand connecting each layer has not yet been observed, but this could be the result of errors in the alignment of the particle images during reconstruction, or due to heterogeneity in where the DNA winding commences. The goal of such studies is to trace the path of the entire nucleic acid molecule within the virion. Asymmetric reconstruction methods at higher resolution are well poised for these determinations.

## Virus–Antibody and Cell Receptor Complexes

There is increasing interest in using cryoEM to analyze aspects of virus biology involving interactions with other molecules such as receptors or antibodies. A good

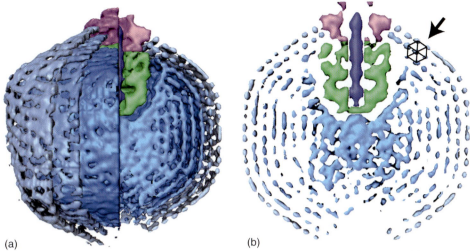

**Figure 12** Viral genome organization in tailed dsDNA phages. (a) In Epsilon15 phage the dsDNA (blue) is arranged in concentric layers surrounding a protein core (green) and portal (purple) (EMD-1175). (b) A slice parallel to the tail axis and normal to the dsDNA strands shows hexagonal features (arrow over polygon).

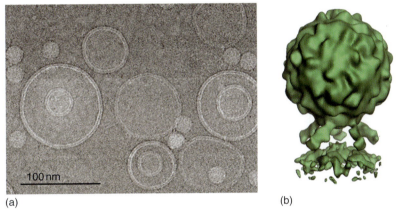

**Figure 13** Virus–receptor interactions. (a) CryoEM image of poliovirus attached to liposomes with embedded receptors. (b) The reconstruction shows the icosahedral capsid binding to five receptors on the membrane, which orients the capsid so that a fivefold vertex is directed toward the membrane. Courtesy of J. M. Hogle at Harvard Medical School and reproduced with permission from Bubeck D, Filman DJ, and Hogle JM (2005) Cryo-electron microscopy reconstruction of a poliovirus–receptor–membrane complex. *Nature Structural and Molecular Biology* 12: 615–618.

example of the value of this approach is provided by work on virus–receptor interactions in the small icosahedral picornaviruses, notably poliovirus and the common cold virus (rhinovirus). In order to recognize and enter their target cells, viruses must usually bind to specific receptor molecules on the cell surface. When the crystal structure of the rhinovirus particle was first determined, it revealed the existence of a depression or 'canyon' around each fivefold axis of the particle, which was proposed as the binding site for the receptor. However, direct confirmation of this proposal was not obtained until icosahedral reconstruction was carried out on cryoEM images of virus particles bound to their receptors. Although virus receptors are components of the cell membrane, these initial binding studies were carried out using soluble fragments of receptor molecules.

Typical eukaryote cells are too large to freeze rapidly by normal methods making it difficult to study virus binding *in situ*. However, by embedding receptor molecules in artificial liposomes that were small enough to vitrify (**Figure 13**(a)), the interaction between the poliovirus particle and its receptor could be visualized in the context of a membrane. Comparison of the micrographs with projection images of an existing icosahedral reconstruction allowed the orientation of the virus particles bound to the liposome to be determined. Then, in a process roughly analogous to that used in the asymmetric reconstruction of tailed bacteriophage particles (described earlier), an artificial density was introduced at the point on the membrane closest to the virus particle and the reconstruction was repeated without imposing icosahedral symmetry. The structure that resulted from this procedure showed the virus particle, receptor molecules, and adjacent parts of the liposome membrane (**Figure 13(b)**), thereby demonstrating that the particle adopts a consistent alignment with respect to the lipid membrane. This not only provided information on how the receptor molecules interact with the virus particle in the context of a membrane, but also revealed the presence of a perturbation to the membrane induced by virus binding. This perturbation might represent an early stage in the formation of the pore, which is necessary for release of the viral genome into the cell.

## Concluding Remarks

CryoEM is an important technique that has been used to determine the structures of many viruses. In many cases, achieving subnanometer resolutions has revealed new insights into the structural organization and biological properties of viruses. CryoEM results have been complemented by docking crystal structures or by molecular modeling to provide additional information. Furthermore, conformational changes, structural transformations, and virus–receptor interactions have been observed. Even in cases where the sample is pleomorphic, electron cryotomography has revealed important information. Overall, the study of viruses has been enormously enriched by the use of cryoEM.

## Acknowledgment

Research support has been provided by National Institutes of Health, National Science Foundation, and the Robert Welch Foundation.

*See also:* Electron Microscopy of Viruses; Virus Particle Structure: Nonenveloped Viruses; Virus Particle Structure: Principles.

## Further Reading

Adrian M, Dubochet J, Lepault J, and McDowall AW (1984) Cryo-electron microscopy of viruses. *Nature* 308(5954): 32–36.

Baker ML, Ju T, and Chiu W (2007) Identification of secondary structure elements in intermediate-resolution density maps. *Structure* 15(1): 7–19.

Baker TS, Olson NH, and Fuller SD (1999) Adding the third dimension to virus life cycles: Three-dimensional reconstruction of icosahedral viruses from cryo-electron micrographs. *Microbiology and Molecular Biology Reviews* 63(4): 862–922.

Chang JT, Schmid MF, Rixon FJ, and Chiu W (2007) Electron cryotomography reveals the portal in the herpesvirus capsid. *Journal of Virology* 81(4): 2065–2068.

Crowther RA, Amos LA, Finch JT, DeRosier DJ, and Klug A (1970) Three dimensional reconstructions of spherical viruses by Fourier synthesis from electron micrographs. *Nature* 226(5244): 421–425.

Dubochet J, Adrian M, Chang JJ, et al. (1988) Cryo-electron microscopy of vitrified specimens. *Quarterly Reviews of Biophysics* 21(2): 129–228.

Jiang W, Chang J, Jakana J, Weigele P, King J, and Chiu W (2006) Structure of Epsilon15 phage reveals organization of genome and DNA packaging/injection apparatus. *Nature* 439(7076): 612–616.

Jiang W and Chiu W (2006) Electron cryomicroscopy of icosahedral virus particle. In: Kuo J (ed.) *Methods in Molecular Biology*, pp. 345–363. Totowa, NJ: The Humana Press.

Johnson JE and Rueckert R (1997) Packaging and release of the viral genome. In: Chiu W, Burnett RM, and Garcea RL (eds.) *Structural Biology of Viruses*, pp. 269–287. New York: Oxford University Press.

Lucic V, Forster F, and Baumeister W (2005) Structural studies by electron tomography: From cells to molecules. *Annual Review of Biochemistry* 74: 833–865.

Rossmann MG, He Y, and Kuhn RJ (2002) Picornavirus–receptor interactions. *Trends in Microbiology* 10(7): 324–331.

Rossmann MG, Mesyanzhinov VV, Arisaka F, and Leiman PG (2004) The bacteriophage T4 DNA injection machine. *Current Opinion in Structural Biology* 14(2): 171–180.

Smith TJ (2003) Structural studies on antibody–virus complexes. *Advances in Protein Chemistry* 64: 409–454.

Zhou ZH, Baker ML, Jiang W, et al. (2001) Electron cryomicroscopy and bioinformatics suggest protein fold models for rice dwarf virus. *Nature Structural Biology* 8: 868–873.

Zhou ZH, Dougherty M, Jakana J, He J, Rixon FJ, and Chiu W (2000) Seeing the herpesvirus capsid at 8.5 Å. *Science* 288(5467): 877–880.

# Cucumber Mosaic Virus

**F García-Arenal,** Universidad Politécnica de Madrid, Madrid, Spain
**P Palukaitis,** Scottish Crop Research Institute, Dundee, UK

© 2008 Elsevier Ltd. All rights reserved.

## Glossary

**Chlorotic** Yellowing or light green symptoms induced by virus infection affecting chlorophyll accumulation.

**Cross-protection** Inhibition of systemic virus accumulation and disease by prior inoculation of plants with a mild or symptomless strain of the same virus.

**Filiformism** Narrowing of the leaf blade, often leading to symptoms referred to as shoestring.
**Pseudorecombination** Reassortment of the genomic RNAs of two or more strains of a multipartite RNA virus to generate novel combinations of the full complement of genomic RNAs.
**Satellite RNA** A subviral RNA genome dependent on a helper virus for both replication and encapsidation.
**Tonoplast** The membrane surrounding the central vacuole of plant cells.

## History

Cucumber mosaic virus (CMV) was first described as a disease of cucurbits in 1916 by Doolittle in Michigan and Jagger in New York. The virus can infect a large number of indicator plant species and has been isolated from over 500 naturally infected species. Cross-protection was used in the 1930s to discriminate isolates of CMV with differences in phenotypes or host range (strains). CMV was not purified reliably until the middle 1960s. Later serology and hybridization technology were used to detect and differentiate two major subgroups of CMV. The nucleotide sequence and the genome organization of one strain of each CMV subgroup were determined between 1984 and 1990, while biologically active cDNA clones of several CMV strains were developed in the early 1990s. The major functions of each of the five encoded proteins have been assigned, although each protein is also involved in other host–virus relationships.

## Taxonomy and Classification

Isolates of CMV are heterogeneous in symptoms, host range, transmission, serology, physicochemical properties, and nucleotide sequence of the genomic RNAs. On the basis of different criteria (e.g., serological typing, peptide mapping of the coat protein, sequence similarity of their genomic RNA) CMV isolates can be classified into two major subgroups, now named subgroup I and subgroup II. The percentage identity in the nucleotide sequence between pairs of isolates belonging to each of these subgroups ranges from 69% to 77%, depending on the pair of isolates and the RNA segment compared, dissimilarity being highest for RNA2. Nucleotide sequence identity among isolates within a subgroup is above 88% for subgroup I and above 96% for subgroup II, indicating a higher heterogeneity of subgroup I. Analysis of the open reading frames (ORFs) and 5′ noncoding regions of RNA3 of subgroup I isolates shows a group of closely related isolates forming a monophyletic cluster, named subgroup IA; the rest of subgroup I isolates are included in the nonmonophyletic group IB. Analyses of RNA2 show that subgroups IA and IB constitute monophyletic groups, while analyses of RNA1 show no clear division into groups IA and IB. Hence, the different genomic segments have followed different evolutionary histories. Cross-protection occurs between strains from all subgroups. Isolates of subgroup I and II can be distinguished using monoclonal antibodies, and isolates from subgroup IA, IB, and II can be distinguished by reverse transcriptase polymerase chain reaction (RT-PCR). CMV isolates differ from isolates of the other two cucumovirus species, *Tomato aspermy virus* (TAV) and *Peanut stunt virus*, having only 50–67% nucleotide sequence identity, depending on the RNA and isolates being compared.

## Geographic Distribution

CMV isolates have a worldwide distribution, having been reported from both temperate and tropical regions. Most reported isolates belong to subgroup I. Subgroup II isolates are found more frequently in cooler areas or seasons of temperate regions. This has been associated with lower temperature optima for *in planta* virus accumulation shown for the few isolates characterized for this property. Most isolates in subgroup IB have been reported from East Asia, which is presumed to be the origin of this subgroup. Subgroup IB isolates also have been reported from other areas, for example, the Mediterranean region, California, Brazil, and Australia. Those in the Mediterranean could have been introduced recently from East Asia.

## Host Range and Propagation

The host range of the collective isolates of CMV is over 1300 species in more than 500 genera of over 100 families, with new hosts reported each year. Some recently described strains from new hosts have lost the ability to infect many of the typical hosts of CMV. This may be a general feature for adaptation to unusual hosts. CMV infects most of the major horticultural crops as well as many weed species; the latter act as reservoirs for the virus. Infection of various indicator plants was used to differentiate CMV from other viruses, since unlike most other viruses of cucurbitaceous or solanaceous hosts, CMV could infect representative species of both families. These include cucumber (*Cucumis sativus*), tomato (*Solanum lycopersicum*), and tobacco (*Nicotiana tabacum*), all systemic hosts of CMV, as well as cowpea (*Vigna unguiculata*) and *Chenopodium quinoa*, which limit CMV infection to the inoculated leaves, although there are legume strains that will infect cowpea systemically. Most isolates of CMV are best propagated in squash (marrow) (*Cucurbita pepo*), tobacco, *N. clevelandii*, or *N. glutinosa*.

## Virus Structure and Properties

CMV has icosahedral particles 29 nm in diameter, which sediment as a single component with an $S_{20,w}$ of 98.6–104$c$ ($c$ being virus concentration in mg ml$^{-1}$). Particles are built of 180 capsid protein subunits arranged with $T = 3$ quasisymmetry, contain about 18% RNA, and have an extinction coefficient at 260 nm (1 mg ml$^{-1}$, 1 cm light path) of 5.0. RNA1 and RNA2 are encapsidated in different particles, whereas RNA3 and RNA4 are probably packaged together in the same particle, some particles may contain three molecules of RNA4. Thus, virus preparations contain at least three different types of particles, but with similar morphology and sedimentation properties. Virus particles also contain low levels of the RNA species designated RNA4A, RNA5, and RNA6. There is a limit to the size of encapsidated RNAs, those larger than RNA1 are not encapsidated *in vivo*. CMV particles are stabilized by RNA–protein interactions, and no empty particles are formed. Particles disrupt at high neutral chloride salt concentrations or at low sodium dodecyl sulfate concentrations; biologically active particles can be reassembled by lowering the salt concentration or removing the sodium dodecyl sulfate. Particles are stable at pH 9.0 and do not swell at pH 7.0, an important difference with bromoviral particles.

The structure of CMV particles has been resolved at 3.2 Å by X-ray crystallography. The $T = 3$ lattice of capsids is composed of 60 copies of three conformationally distinct subunits designated A, B, and C that form trimers with quasi-threefold symmetry. There are 20 hexameric capsomers of B and C subunits with quasi-sixfold symmetry and 12 pentameric capsomers formed by A subunits with fivefold symmetry. The exterior radius along the quasi-sixfold axes is 144 Å, the RNA is tightly packaged against the protein shell and leaves a hollow core of about 110 Å along the threefold axes. The protein subunit has a β-barrel structure, with the long axis of the β-barrel domain oriented roughly in a radial direction. The N-terminal 22 amino acids of the capsid protein are needed for particle assembly. This region is positively charged and, in the B and C subunits, forms amphipatic helices that run parallel to the quasi-sixfold axes. There is an external region of negative electrostatic potential that surrounds the fivefold and quasi-sixfold axes and locates above regions of positive potential which extend to cover nearly homogeneously the inner surface of capsids, where interaction with encapsidated RNA occurs. Electrostatic distributions in CMV particles explain the physicochemical conditions required for particle stability.

## Genome Organization

The genome of CMV consists of five genes distributed over three, single-stranded, positive-sense, capped, genomic RNAs: RNA1 (3.3–3.4 kb) encodes the *c*. 111 kDa 1a protein. RNA2 (3.0 kb) encodes the 98 kDa 2a protein, as well as the 13–15 kDa 2b protein, which is translated from a 630–702 nt subgenomic RNA designated RNA 4A that is co-terminal with the 3′ end of RNA2. The ORF expressing the 2b protein overlaps with the ORF encoding the 2a protein, but in a +1 reading frame. RNA3 encodes the 30 kDa 3a protein, as well as the 25 kDa 3b protein, which is expressed from a 1010–1250 nt subgenomic RNA designated RNA4 that is co-terminal with the 3′ end of RNA3. The 224–338 nt 3′ nontranslated regions of all three genomic and both subgenomic RNAs are highly conserved, forming a tRNA-like structure as well as several pseudoknots. The 5′ nontranslated regions of RNA1 (95–98 nt) and RNA2 (78–97 nt) are more conserved in sequence with each other than with those of RNA3 (96–97 nt or 120–123 nt). CMV also produces an RNA5 of unknown function, which is co-terminal with the 3′ nontranslated regions of RNA1 and RNA2. RNA4A and RNA5 are only encapsidated by subgroup II strains. CMV particles also encapsidated a low level of tRNAs, which have been reported in the literature as CMV RNA6. Although rarely reported, some strains of CMV can also encapsidate defective RNAs derived from CMV RNA3.

## Satellite RNAs

CMV can also support satellite RNAs varying in size from 333 to 405 nt. These satellite RNAs are dependent upon CMV as the helper virus for both their replication and encapsidation, but have sequence similarity to the CMV RNAs limited to no more than 6–8 contiguous nt. More than 100 satellite variants have been found associated with over 65 isolates of CMV from both of the CMV subgroups. These satellite RNAs usually reduce the accumulation of the helper viruses and on most hosts also reduce the virulence of CMV. However, this attenuation of disease is not due to competition between the helper virus and the satellite RNA for a limited amount of replicase or capsid protein. Some CMV satellite RNAs can also be replicated and packaged by strains of the cucumovirus TAV, although these satellite RNAs do not attenuate the symptoms induced by TAV. Certain satellite RNAs in some selected hosts can enhance the disease induced by CMV. In the case of tomato plants infected by CMV and certain satellites, this has led to systemic necrosis observed in the fields of several Mediterranean countries. This necrosis is actually caused by sequences of the complementary-sense satellite RNA produced in large quantities during satellite RNA replication.

## Genetics

The CMV 1a and 2a proteins encode proteins that replicate the virus, but which also function in promoting virus movement in several host species. The 2b protein is an RNA-silencing suppressor protein that antagonizes the salicylic acid defense pathway and also influences virus movement in some hosts. The 3a protein is the major movement protein of the virus and is essential for both cell-to-cell as well as long-distance (systemic) movement. The 3b protein is the sole viral capsid protein and is also required for cell-to-cell and long-distance movement, although the ability to form virions is not a requirement for movement. All of the CMV-encoded proteins are RNA-binding proteins. Viral RNAs from different strains and subgroups can be exchanged to form novel viruses, allowing mapping of some phenotypes to specific RNAs. Higher-resolution mapping requires the use of biologically active cDNA clones for generating chimeras and site-directed mutants. By these approaches, the following functions have been delimited to specific RNAs, with some mapped to specific nucleotide changes: hypersensitive response in tobacco, rapid local and systemic movement in squash, seed transmission in legumes, temperature-sensitive replication in melon; and replication of satellite (all in RNA1); hypersensitive response versus systemic infection in cowpea, suppression of gene silencing, and host range and pathology responses (all in RNA2); limited movement between epidermal cells, systemic movement in cucurbits, virion assembly, hypersensitive response on *Nicotiana* sp., local and systemic infection in cucurbits or in maize, symptom responses, and aphid transmission (all in RNA3).

## Replication

CMV replication takes place on the vacuole membrane (the tonoplast). Replication involves the 1a and 2a proteins of the virus and presumably several host proteins. The purified CMV replicase contains the 1a and 2a proteins, as well as a host protein of *c.* 50 kDa of unknown function. The 1a protein of CMV contains a putative N-terminal proximal methyltransferase domain believed to be involved in capping of the RNAs, as well as a putative C-terminal proximal helicase domain, presumed to be required for the unwinding of the viral RNAs during replication. The 1a protein has been found to be able to bind to several tonoplast intrinsic proteins, although what roles these have in virus replication has not been established. The N-terminal region of the 2a protein interacts with the C-terminal region of the 1a protein *in vivo* and *in vitro*. Phosphorylation of the 2a protein prevents interaction with the 1a protein. The C-terminal half of the 2a protein contains conserved domains found in RNA polymerases and therefore together with the 1a protein forms the core of the CMV replicase.

CMV replication is initiated by the binding of the tonoplast-associated replicase to the tRNA-like structure and various pseudoknots present in the 3′ nontranslated region of the positive-sense CMV RNAs. Minus-sense RNA is then synthesized from each of the genomic RNAs and the synthesized minus-sense RNA acts as a template for synthesis of new plus-stranded genomic RNAs. The minus-sense RNA2 and RNA3 also serve as the templates for the synthesis of the two plus-sense subgenomic RNAs (4 and 4A), through recognition of the subgenomic promoters present on the minus-sense RNAs. The subgenomic RNAs are not themselves replicated, but defective RNAs and satellite RNAs are replicated by the CMV replicase. Differences in the relative levels of accumulation of the various CMV genomic RNAs and satellite RNAs have been observed in different host species, which probably is due to host-specific differences in template copying.

## Movement

CMV moves cell to cell via plasmodesmata between cells until it reaches the vasculature, when the virus moves systemically via the phloem. The viral RNAs move as a nucleoprotein complex between cells involving the 3a movement protein and some involvement by the capsid protein. This movement does not involve interactions with microtubules. No specific plant proteins have yet been identified as being involved in cell-to-cell movement. CMV appears to move between epidermal cells as well as from epidermal cells down to mesophyll cells toward vascular cells. The 2b protein also influences the path of virus movement, since without the 2b protein, the virus moves preferentially to and between mesophyll cells. The virus replicates in all of these cell types, but not in the sieve elements of the vasculature. Virion assembly may take place inside sieve elements from RNAs and capsid protein moving from neighboring vascular cells. Virion assembly is necessary in some, but not all, species for systemic infection. A 48 kDa phloem protein (PP1) from cucumber phloem exudates interacts with CMV particles *in vitro* and increases virus particle stability. How the virus moves from the vasculature back to mesophyll and epidermal cells is unknown. The virus moves from plant to plant either by transmission by the aphid vectors of CMV, or in some cases via seed transmission, at a low but variable frequency.

## Pathology

The symptoms induced by CMV are not generally specific to CMV, but rather reflect sets of host responses to viral pathogens. Therefore, symptoms such as light green–dark green mosaics, generalized chlorosis, stunting, leaf filiformism, and local chlorotic or necrotic lesions associated with various strains of CMV are not specific to CMV, but can also be elicited by other viruses in the same plant species. Some strains of CMV can induce a bright yellow chlorosis in some *Nicotiana* species. This can be due to either specific amino acid changes in the viral capsid protein, or the presence of a chlorosis-inducing satellite RNA. A white-leaf disease of tomato also was due to the effects of a particular satellite RNA, as well as systemic necrosis and lethal necrosis disorders seen in the field of several Mediterranean countries. Some pathogenic responses, such as local lesions versus systemic infection in cowpea have been mapped to two amino acid changes in RNA 2, while other pathogenic responses have been mapped to sequences present in more than one viral RNA molecule, and to different viral sequences for different strains, indicating that several interactions are involved in the elicitation of some symptom responses.

Cytopathology associated with infection by CMV includes viral inclusions within the cytoplasm and vacuoles, and as membrane-bound clusters in sieve elements. In some cases, angular inclusions corresponding to virus crystals can be seen in vacuoles by staining and light microscopy. CMV infection usually also leads to proliferation of cytoplasmic membranes, which originate from the plasma membrane, the endoplasmic reticulum, or the tonoplast. Effects on the nucleus or nucleolus (e.g., vacuolation), usually due to virion accumulation, have been observed with some strains of CMV. Similarly, some strains have caused effects on mitochondria or chloroplasts. Yellowing strains, in particular, show effects on chloroplasts development leading to smaller and rounded chloroplasts that have fewer grana and starch granules. These various effects appear to be host specific. How virus–plant interactions lead to cytopathic effects remains unknown; however, as the virus expands from the initial site of infection, rings or zones of responses occur, in which the expression patterns of numerous plant genes are altered in a spatial- and temporal-specific manner.

## Transmission

Seed transmission of CMV has been reported in many plant species, with efficiencies varying from less than 1% up to 50%. Virus may be present in the embryo, endosperm, and seminal integuments, as well as in pollen. RNA1, and possibly protein 1a, affects the efficiency of seed transmission.

Horizontal transmission of CMV is vectored by aphids in a nonpersistent manner. Over 80 species of aphids have been reported to transmit CMV, *Aphis gossypii* and *Myzus persicae* being two efficient and most studied vectors. Transmission efficiency depends on several factors, particularly the specific combination of virus isolate and aphid species, and the accumulation of particles in the source leaf. Differences in transmissibility of various isolates by different aphid species are determined solely by the virus coat protein and amino acid determinants for transmission have been mapped. The amino acid positions that determine transmission are either exposed on the outer surface of capsids (e.g., amino acid position 129, on the β strand H-I loop) or lay in the inner surface (e.g., position 162). Hence, the effect could be through direct interaction with the aphid mouth parts or by affecting particle stability. Loss of transmissibility upon repeated mechanical passage seems to be rare, perhaps because of tradeoffs with particle stability, but particle stability is not always correlated with transmissibility. Transmission efficiency also depends on the host plant, as shown by the resistance to transmission on melon genotypes containing the *Vat* gene, which does not confer resistance to the aphid or the virus, but impairs aphid transmission.

## Ecology and Epidemiology

CMV infects a wide range of food crops, ornamentals, and wild plant species. Economic losses in crops are highest in field-grown vegetables and ornamentals, pasture legumes, and banana. In recent times, CMV has caused severe epidemics in many crops, including necrosis of tomato in Italy, Spain, and Japan; mosaic and heart rot of banana worldwide; mosaic of melons in California and Spain; mosaic of pepper in Australia and California; and mosaic of lupins and other legumes in Australia and the USA. In crops in which seed transmission is effective (e.g., pasture or fodder legumes), the primary inoculum for epidemics may be the seedlings from infected seeds. In most vegetable and ornamental crops, seed transmission does not occur or is negligible, and the primary inoculum must come from outer sources as other crops or weeds, which should be near the crop as aphid transmission is nonpersistent. In the absence of crops during unfavorable seasons, infected perennial weeds or crops, and infected seeds from weeds act as reservoir inoculum. Seed transmission has been shown to be important in several weed species from different regions. In spite of its general broad host range, there is evidence of host adaptation or preference for some CMV strains, which might have important consequences for inoculum flows among host species. Also, the dynamics of virus infection may differ largely in weeds and crops within a region, indicating that the relevant reservoirs and inoculum sources for crops need not be the most frequently infected weeds. In banana, secondary spread of infection within the crop is ineffective for most strains, and alternative hosts are both primary and secondary inoculum sources for epidemics.

## Variation and Evolution

In agreement with the high mutation rates of RNA genomes, populations of CMV derived from biologically active cDNA clones were found to be genetically diverse. When the cDNA-derived population was passaged in different hosts, the amount of genetic diversity depended on the host species. Genetic diversity has been shown to be countered by genetic drift associated to population bottlenecks during systemic colonization of the host and, probably also, during host-to-host transmission. Sequence analyses have shown different evolutionary constraints for the different viral proteins, which show different evolutionary dynamics. A second source of genetic variation is the exchange of RNA sequences by recombination or by reassortment of genomic segments. Experimentally, recombination has been shown to occur between the 3′ nontranslated region of the genomic RNAs, with recombinants being up to 11% of the population. Recombinants in the 3′ nontranslated region may have an increased fitness in some hosts, as shown for isolates infecting alstroemeria. Recombination in the RNA3 also was frequent in mixed infection between CMV and TAV. Recombination between CMV strains or CMV and TAV strains seems to be facilitated by stem–loop structures in the RNA. In spite of abundant evidence for frequent recombination, analyses of the genetic structure of field populations of CMV show that recombinant RNAs are not frequent, and that selection operates against most recombinants. A second mechanism of genetic exchange is reassortment of genomic segments also called pseudorecombination. Reassortants exchanging any genomic segment have been obtained between different CMV strains, which multiply efficiently under experimental conditions. Natural reassortants also have been described, and reassortment may have played an important role in the evolution of CMV, as suggested by the different phylogenies obtained for each genomic RNA. In field population, reassortant isolates are rare and seem to be selected against, as is the case for recombinants. Evidence for selection against genotypes originating by genetic exchange suggests co-adaptation of the different viral genes that, when disrupted, results in a decreased fitness. Analyses of the population structure of CMV in Spain and California indicate a metapopulation structure, with local extinctions and recolonizations, which suggests that population bottlenecks occur, probably associated with unfavorable seasons for the host plants and/or the aphid vectors. Interestingly, this is not the case for the population structure of the satellite RNA. Analyses in Italy and Spain during epidemics of CMV plus satellite RNAs have shown that the satellite RNAs have an undifferentiated population. The different population structure for CMV and its satellite RNAs indicates that the satellite RNAs expanded as a molecular parasite on the CMV population, rather than satellite expansion was linked to a particular CMV isolate.

## Control

Control of CMV can be achieved by planting of resistant crops, but resistance in many crop species is often not available to a broad range of CMV strains. The use of insecticides to control the aphid vectors of the virus has met with only limited success, since the virus is transmitted in a nonpersistant manner and thus before the aphids would have been killed by the insecticide, it would have transmitted the virus. Rather, insecticides are used to reduce aphid numbers and thus reduce the progressive spread of infections. Since many species of weeds act as reservoirs for the virus and many of these are asymptomatic hosts, it is important to remove these from the borders of fields to eliminate the source of infectious material that could then be spread by aphids. This also applies to removal of infected crop plants during the growing season. Others sources of resistance include the use of transgenic plants expressing either protein-mediated or RNA silencing-mediated resistance. Although most of these approaches have led to resistance to only members of one of the two major subgroups, the use of pyramiding of viral segments from different subgroups offers the promise of obtaining a broad spectrum resistance to CMV together with other viruses infecting the same crop species. Transgenic expression of satellite RNAs has also been used to confer resistance to CMV. This has met with success, but has raised concerns about using a virulent pathogen, as did the use of mild strains of CMV for cross-protection against severe strains.

*See also:* Brome Mosaic Virus; *Ilarvirus*.

## Further Reading

Edwardson JR and Christie RG (eds.) (1991) Cucumoviruses. *CRC Handbook of Viruses Infecting Legumes*, pp. 293–319. Boca Raton, FL: CRC Press.

Kaper JM and Waterworth HE (1981) Cucumoviruses. In: Kurstak E (ed.) *Handbook of Plant Virus Infections and Comparative Diagnosis*, pp. 257–332. New York: Elsevier/North-Holland.

Palukaitis P and Garcia-Arenal F (2003) Cucumoviruses. *Advances in Virus Research* 62: 241–323.

Palukaitis P and García-Arenal F (2003) Cucumber mosaic virus. In: Antoniw J and Adams M (eds.) *Description of Plant Viruses. DPV400*, Rothamstead Research, UK: Association of Applied Biologists. http://www.dpvweb.net/dpv/showdpv.php?dpvno=400 (accessed September 2007).

Palukaitis P, Roossinck MJ, Dietzgen RG, and Francki RIB (1992) Cucumber mosaic virus. *Advances in Virus Research* 41: 281–348.

# Cytokines and Chemokines

**D E Griffin,** Johns Hopkins Bloomberg School of Public Health, Baltimore, MD, USA

© 2008 Elsevier Ltd. All rights reserved.

## Glossary

**Chemokine** Chemotactic cytokine that regulates trafficking of cells during immune responses.
**Cytokine** Soluble protein produced by a variety of cells that mediates cell-to-cell communication important for innate and adaptive immune responses.
**Inflammasome** A multiprotein complex of more than 700 kDa that is responsible for the activation of caspases 1 and 5, leading to the processing and secretion of the pro-inflammatory cytokines IL-1β and IL-18.
**Lymphotoxin** A cytokin secreted by activated $T_H1$ cells, fibroblasts, endothelial and epithelial cells.

## Introduction

Cytokines are small soluble proteins produced by a wide variety of cells and are important mediators of cell-to-cell communication. Production of these biologically potent proteins is particularly associated with generation of the innate and acquired immune responses. Cytokines regulate both the initiation and maintenance of immune responses and some also induce production of cellular antiviral molecules. Chemokines (chemotactic cytokines) regulate trafficking of leukocytes during immune responses and infiltration of leukocytes into infected tissues. Both cytokines and chemokines share the properties of pleiotropy and redundancy. Each cytokine and chemokine has more than one function and most functions can be performed by more than one of these soluble factors.

Cytokines can act at a distance, but their effects are usually most potent in the areas in which they are produced. They mediate their effects by binding to small numbers of high-affinity receptors on responding cells. Cytokine receptors have a modular design consisting of two or more chains, each with a single transmembrane-spanning domain. Individual chains may participate in formation of the receptor for more than one cytokine. Chemokine receptors are single G-protein-coupled molecules with seven-transmembrane domains. The biologic function of many cytokines and chemokines is regulated by a short half-life for both the protein and the mRNA and by the production of circulating cytokine inhibitors such as soluble forms of the cytokine receptor or biologically inactive forms of the cytokine itself.

A number of viruses, particularly in the herpesvirus and poxvirus families, encode inhibitory homologs of cytokines, chemokines, or their receptors, attesting to their importance for antiviral immunity.

## Overview of Cytokine Networks

Cytokines associated with the innate immune response are those induced early after infection by virus replication *per se*. One of the most important of these early immunologically nonspecific responses are type 1 ($\alpha$ and $\beta$) interferons (IFNs) that are generally synthesized and released by infected cells. Some cells, for example, plasmacytoid dendritic cells, produce IFN rapidly after infection. IFN-$\alpha/\beta$ can induce an antiviral state in surrounding cells and can also influence the characteristics of the subsequent adaptive immune response. In addition, phagocytic cells, responsible for processing and presentation of viral antigens to initiate the antigen-specific immune response, produce a variety of cytokines, including IFN-$\alpha/\beta$. These antigen-presenting cells, generally dendritic cells and monocyte/macrophages, may or may not be infected or produce infectious virus. Important cytokines and chemokines produced by antigen-presenting cells in the initial phases of the response to many viral infections are tumor necrosis factor (TNF), monocyte chemotactic proteins (MCP), and interleukin (IL)-1, IL-6, IL-12, and IL-18. These soluble factors promote the activation of T cells and recruitment of inflammatory cells into areas of virus replication. IL-12 stimulates natural killer (NK) cells to produce IFN-$\gamma$, promotes differentiation of $CD8^+$ cytotoxic T lymphocytes, and influences the early development and differentiation of CD4 T cells toward a delayed type hypersensitivity cellular immune response (type 1).

$CD4^+$ and $CD8^+$ T lymphocytes are the effectors of the specific cellular immune response and function primarily through synthesis of cytokines. T cells do not produce cytokines constitutively, but upregulate expression in response to stimulation through the T-cell receptor. $CD8^+$ T cells are stimulated by viral antigens presented in association with class I major histocompatibility (MHC) antigens. This type of stimulation is most likely to occur if the virus replicates in the antigen-presenting cell because antigens presented in association with MHC class I must be processed by proteasomes in the cytosol. $CD8^+$ T cells often have cytotoxic activity through production of granzymes and perforin or Fas ligand, but also produce substantial amounts of IFN-$\gamma$ and lymphotoxin (LT). Viral antigens are presented to $CD4^+$ T cells as peptides in association with class II MHC antigens. Processing of antigens for presentation in association with MHC class II occurs through an endosomal pathway; therefore, virus replication is not required. Activated CD4$^+$ T cells produce IL-2, a T-cell growth factor that supports proliferation of both CD4$^+$ and CD8$^+$ T cells.

With further development of the virus-specific immune response, CD4$^+$ T cells tend to differentiate into different types of T-helper (Th) cells. Type 1 CD4$^+$ T (Th1) cells develop under the influence of IFN-γ and produce primarily IFN-γ and TNF-β (LT-α) and are associated with classic delayed type hypersensitivity or cellular immune responses. Type 2 CD4$^+$ T cells develop under the influence of IL-4 and produce primarily IL-4, IL-5, and IL-13 and are associated with strong antibody responses to viral antigens. A third type of CD4$^+$ effector cells (Th17) develop under the influence of transforming growth factor (TGF)-β and produce IL-17 and IL-6 and are associated with inflammation and autoimmune disease. Cytokines produced by CD4$^+$ T cells are necessary for B-cell development and for the switch from B cells producing IgM to B cells producing more mature forms of immunoglobulin such as IgG, IgA, and IgE and for B cell maturation into long-lived antibody-secreting cells. Both CD4$^+$ and CD8$^+$ T-cell responses are associated with mononuclear infiltrates into areas of virus replication. Regulatory T cells control and dampen ongoing immune responses, in part, through production of IL-10 or TGF-β.

## Brief Description of Cytokine Families

### TNF

The TNF family has three members: TNF-α, TNF-β/LT-α, and LT-β. These cytokines have approximately 30% homology and interact with cellular receptors in the TNF receptor (TNFR) superfamily. Activated macrophages are the major source of TNF-α and T cells are the major source of LT. TNF-α is synthesized as a proprotein (26 kDa) which is membrane bound and cleaved by a specific multidomain cell-surface metalloprotease to yield a 17 kDa protein. The active form of the soluble protein is a cone-shaped trimer that can bind to either of two TNFRs (TNFR1, p55 or TNFR2, p75). Binding of TNF to TNFR1 induces nuclear factor kappa B (NF-κB) and can trigger apoptosis. Homotrimers of LT-α also bind to TNFR1 and TNFR2, but LT-β reacts with a separate member of the TNFR family, the LT-β receptor. TNF interacts with a wide variety of cells systemically and locally and induces pro-inflammatory cellular responses such as expression of adhesion molecules, release of cytokines, chemokines and procoagulatory substances, synthesis of acute phase proteins, and production of fever. TNF also has some antiviral activity and is an important early participant in the immune response to infection. In bacterial infections, TNF is one of the mediators of endotoxin-induced (septic) shock. TNF and LT also play an important role in development of lymphoid tissues because lymphoid architecture in mice with deletions of these genes is abnormal.

### IFN-γ

IFN-γ is a glycosylated protein of 25 kDa that is produced by NK cells and by type 1 CD4 and CD8 T cells (immune IFN). Although IFN-γ induces antiviral activity, it is structurally unrelated to IFN-α/β and uses a distinct receptor. However, IFN-γ shares some components of the intracellular signaling pathways for type 1 IFN. The biologically active form of IFN-γ is a dimer. A primary role for IFN-γ is the activation of macrophages to increase phagocytosis, tumoricidal properties, and intracellular killing of pathogens, particularly bacteria and fungi. IFN-γ induces macrophage production of a variety of inflammatory mediators and reactive oxygen and nitrogen intermediates. IFN-γ increases expression of high-affinity immunoglobulin F$_c$ receptors on phagocytes, which increases recognition of opsonized microorganisms by these cells. IFN-γ also increases expression of MHC antigens by macrophages and this facilitates antigen presentation to T cells. The IFN-γ cell-surface receptor is composed of two chains, IFN-γR1 (α) and IFN-γR2 (β). The α (R1) chain is sufficient for binding, but the β (R2) chain is required for signaling and receptor complex formation. T-cell responsiveness is regulated by receptor expression. Both Th1 and Th2 cells express IFN-γR1, but only Th2 cells express IFN-γR2. Therefore, Th1 cells produce, but do not respond to, IFN-γ while Th2 cells respond to, but not produce this cytokine. Macrophages express both receptor chains and are a primary target of IFN-γ activity.

### IL-1

The IL-1 family has three members: IL-1α, IL-1β, and IL-1 receptor antagonist (IL-1RA). IL-1α and IL-1β have only distantly related amino acid sequences, but similar structure, and recognize the same receptors. IL-1RA binds the IL-1 receptor without transducing a signal and blocks the activities of IL-1α and IL-1β. Activated monocytes and macrophages are the major source of IL-1, although many other cells can produce this cytokine. IL-1β is synthesized as a precursor protein of 31 kDa that is processed by the caspase interleukin-1 β-converting enzyme (caspase-1) within the 'inflammasome' to the active secreted 17 kDa form. IL-1 acts systemically, as well as locally, and can produce fever, sleep, and anorexia, frequent symptoms of viral infection. Hepatocytes are among the cells that produce IL-1RA as a part of the acute-phase response to inflammation and infection, presumably to control the effects of IL-1. There are two IL-1 receptors but only IL-1R1 transduces an activation signal to responding cells. IL-1R2 appears to function as a decoy receptor that regulates IL-1 activity by binding the cytokine without transducing a signal.

IL-18, also known as IFN-γ-inducing factor, is an 18 kDa cytokine produced by macrophages that shares structural features with the IL-1 family of proteins. It is synthesized as a proprotein that is cleaved by caspase 1 to its biologically active form. IL-18 stimulates IFN-γ production by T cells, NK cytotoxicity, and T-cell proliferation.

## IL-2

IL-2 is a unique 15 kDa cytokine produced primarily by activated type 1 $CD4^+$ T cells. IL-2 plays an important role in supporting proliferation of activated helper T cells, cytotoxic T cells, B cells, macrophages, and NK cells. For stimulated $CD4^+$ T cells, IL-2 acts in an autocrine fashion because Th1 cells can produce IL-2 and upregulate high-affinity IL-2 receptors as well. The high-affinity IL-2R consists of three chains – IL-2Rα, β, and γ – which are expressed on the surface of activated T cells. IL-2Rα is a unique protein, whereas IL-2Rβ and IL-2Rγ are members of the hematopoietic growth factor receptor superfamily. Members of this superfamily share a 210-amino-acid domain in the extracellular region of the molecule that contains a distinct cysteine and tryptophan motif at the N-terminus and a Trp-Ser-X-Trp-Ser motif at the C-terminus. The whole domain is folded into two barrel-like structures with the ligand-binding region between. In addition to the IL-2 receptor, this receptor superfamily includes components of receptors for cytokines IL-3, IL-4, IL-5, IL-6, IL-7, IL-9, granulocyte macrophage-colony stimulating factor (GM-CSF), and granulocyte-colony stimulating factor (G-CSF). Signal transduction is through the IL-2Rβ chain. Intermediate-affinity IL-2R βγ receptors are expressed on monocytes and NK cells. IL-2Rγ (γ constant) also participates in the formation of the multichain IL-4, IL-7, IL-9, and IL-15 receptors. Defects in expression of this molecule lead to severe combined immunodeficiency in humans. Binding of IL-2 to the high-affinity receptor leads to T-cell proliferation.

## IL-4

There are five members of the IL-4 family of cytokines: IL-3, IL-4, IL-5, IL-13, and GM-CSF. These cytokines have approximately 30% protein sequence homology, have similar α-helical structures, are linked in the same chromosomal region, and are produced by type 2 $CD4^+$ T cells. IL-3 and GM-CSF are two of the four major human myeloid growth factors that also include G-CSF and M-CSF. IL-3 is produced by activated T cells and binds to a heterodimeric receptor. The β subunit of the IL-3R is shared with the receptors for IL-5 and GM-CSF.

IL-4 and IL-13 have similar effects on B cells and monocytes. Both can serve as a costimulating factor for B-cell proliferation, drive activated B cells into immunoglobulin secretion, and, in human B cells, induce class switching to IgG4 and IgE. IL-4 is regarded as an anti-inflammatory cytokine because it antagonizes the effects of IFN-γ on macrophages and downregulates type 1 $CD4^+$ T-cell responses. IL-13 induces monocyte proliferation and monocytosis in vivo. IL-4 is important for promoting the differentiation of $CD4^+$ T cells to type 2 cytokine production. IL-4 and IL-13 are produced by type 2 $CD4^+$ T cells. IL-4 is also produced by a subset of $CD4^+$ T cells that are $NK1.1^+$ (NK T cells), by type 2 $CD8^+$ T cells and by mast cells, basophils, and eosinophils.

## IL-5

IL-5, along with IL-3 and GM-CSF, acts as an eosinophil stimulating factor and its receptor shares a common β chain with these cytokines. The high-affinity IL-5R has a unique α chain. IL-5 induces proliferation and differentiation of eosinophil progenitors, whereas IL-3 and GM-CSF probably act at earlier stages of development. IL-5 is produced by type 2 $CD4^+$ and $CD8^+$ T cells and by NK cells. IL-5 also induces antigen-stimulated B cells to differentiate into immunoglobulin-secreting plasma cells and enhances secretion of IgA.

## IL-6

IL-6 is a member of a family of cytokines and neuronal differentiation factors that includes leukemia inhibitory factor, oncostatin M, IL-11, and ciliary neurotrophic factor. The receptors for these cytokines share the transmembrane protein gp130 that transduces the cytokine signal. IL-6 is a functionally diverse cytokine produced by macrophages and monocytes early in the response to infection. It induces hepatic synthesis of acute phase proteins, supports proliferation of B cells, and has an important role in production of IgA.

## IL-7

IL-7 is a 15 kDa protein that is produced primarily by epithelial cells and supports the growth and development of immature B and T cells. IL-7 is produced by bone marrow stromal cells, thymic epithelial cells, keratinocytes, and intestinal epithelial cells and supports lymphocyte development in these locations. In the thymus, IL-7 induces rearrangement of germline T-cell receptor genes by upregulating expression of the RAG-1 and RAG-2 recombinases. The α chain of the IL-7 receptor is a member of the hematopoietic growth factor family and associates with the IL-2Rγ chain to form a functional high-affinity IL-7R.

## IL-10

IL-10 is an 18 kDa polypeptide that forms homodimers. It is predicted to be a member of the 4 α–helix bundle family of cytokines. It is produced by a wide variety of cells, including type 2 $CD4^+$ and $CD8^+$ cells, B cells, mast cells, macrophages, and keratinocytes. Epstein–Barr virus encodes a viral homolog of IL-10 that is functionally active. The activities of IL-10 are diverse, but its principal function appears to be limitation of inflammatory responses. IL-10 inhibits cytokine production by Th1 cells, NK cells, and macrophages and has been associated with viral persistence. It suppresses the synthesis of TNF-α, IL-1α, IL-6, IL-8, GM-CSF, and G-CSF by macrophages while increasing the production of IL-1RA. IL-10 stimulates proliferation and differentiation of activated B cells and T cells and of mast cells and is important for differentiation of regulatory T cells. The functional IL-10R consists of two subunits, IL-10R1 and IL-10R2, both of which belong to the class II cytokine receptor family. Expression of both subunits is required for signal transduction.

## IL-12

IL-12 is a structurally unique heterodimeric cytokine composed of two disulfide-linked 35 and 40 kDa polypeptides. The 40 kDa protein conveys receptor-binding activity, is shared with IL-23 and is constitutively produced by many cells. Only the heterodimer is biologically active, so regulation of synthesis of the 35 kDa chain controls IL-12 activity, while synthesis of the IL-23 19 kDa chain controls IL-23 activity. IL-12 is produced by macrophages and dendritic cells and stimulates the production of IFN-γ by NK and T cells. IL-12 is important for differentiation of CD4$^+$ T cells into Th1 cells and IL-23 promotes differentiation into Th17 cells. At least two classes of the IL-12 receptor exist. The IL-12R β1 chain is expressed in both Th1 and Th2 cells while the β2 chain is expressed only in Th1 cells. The IL-23 receptor is composed of IL-12β1 and a unique IL-23R subunit. For IL-12 responsiveness and for T-cell differentiation, both β1 and β2 subunits are necessary so only potential Th1 cells respond to this cytokine. Both IL-12R β1 and β2 subunits are related to the gp130 group of cytokine receptors. Homodimers of p40 bind with high affinity to the IL-12R, transduce no signal and block the effect of the heterodimer.

## IL-15

IL-15 is a novel cytokine functionally related to IL-2, but without significant IL-2 sequence homology. It binds to a receptor with the β and γ chains of the IL-2R, but has its own unique α chain. IL-15 induces T-cell proliferation, enhances NK cell function, is a potent T-lymphocyte chemoattractant, and stimulates production of IL-5. Unlike IL-2, it is produced primarily by monocyte/macrophages and endothelial cells.

## IL-17

IL-17 (IL-17A) is the founding member of a family of six cytokines (IL-17A-F). IL-17 is a 155-amino acid glycoprotein that functions as a disulfide-linked 35 kDa homodimer. IL-17 is produced by a separate lineage of activated CD4$^+$ (Th17) and CD8$^+$ T cells that have been implicated in autoimmune disease.

## Chemokines and Chemokine Families

The chemokine superfamily includes a large number of structurally related, basic, heparin-binding, small molecular weight (8–10 kDa) proteins that have potent chemotactic and immunoregulatory functions. Chemokines attract specific, but overlapping, populations of inflammatory cells, induce integrin activation, and can be divided into four families with different biologic activities based on the spacing of the first two of four conserved disulfide-forming cysteine residues in the amino-terminal part of the molecule. The largest groups are the CXC or α-chemokines and the CC or β-chemokines. In general, CXC chemokines are the primary attractants for neutrophils, whereas CC chemokines modulate responses of monocytes, macrophages, lymphocytes, basophils, and eosinophils. C chemokines have only one set of cysteines and are chemotactic for lymphocytes. The CX3C chemokine fractalkine is unusual in that it is a type 1 transmembrane protein with the chemokine domain on top of an extended mucin-like stalk. Herpesviruses and poxviruses encode viral chemokine binding proteins (vCKBPs) that are important regulators of the host immune response.

### α-Chemokines

The first CXC chemokine to be identified was platelet factor 4 (PF4, CXCL4) that is stored in the granules of platelets that also contain neutrophil-activating protein (NAP-2, CXCL7) ready to be released at sites of tissue injury. The best-characterized α-chemokine is IL-8 (CXCL8) that can be rapidly induced in most cells and is also chemotactic for neutrophils. Other α-chemokines with chemotactic activity for neutrophils include GRO-α (CXCL1), GRO-β (CXCL2), GRO-γ (CXCL3)γ, epithelial cell-derived neutrophil activating protein (CXCL5), and granulocyte chemotactic protein-2 (CXCL6). Sequence identities between these various family members range from 24% to 46%. IP-10 (CXCL10) is a structurally similar, but functionally distinct, CXC chemokine that is an attractant for monocytes and T cells rather than granulocytes.

### β-Chemokines

The CC chemokines have two large subgroups: the MCPs and the macrophage inflammatory proteins (MIPs). MCP-1 (CCL2), produced by endothelial cells in response to IL-1β and TNF-α, functions as a dimer, attracts and activates monocytes, NK cells, CD4$^+$, and CD8$^+$ T cells, promotes

Th2 responses and is a potent histamine-releasing factor for basophils. MCP-2 (CCL8) and MCP-3 (CCL7) are structurally and functionally related chemokines important in Th2 responses, whereas MIP1α (CCL3) and MIP1β (CCL4) are important in Th1 responses. Other CC chemokines commonly increased during viral infection are RANTES (CCL5) and I-309 (CCL1). Sequence identities between family members are 29–71%. These chemokines activate integrins, thereby inducing or increasing adhesion to endothelial cells or extracellular matrix proteins to promote development of an inflammatory response. Two novel, related β chemokines, MCP-4 (CCL13) and eotaxin (CCL11), are preferential chemoattractants for eosinophils. CCL11 is produced by epithelial and endothelial cells, often in concert with IL-5.

## Chemokine Receptors

Chemokine receptors are seven-transmembrane, G-protein-coupled molecules that signal through the pertussis toxin-sensitive Gαi subfamily of hetereotrimeric G proteins. In general, the α chemokine receptors (CXCRs) are distinct from the β receptors (CCRs). Each receptor can bind multiple chemokines within its class and each chemokine can bind multiple receptors. Therefore, the specificity of cellular attraction to a site and subsequent activation is probably achieved by selective mixtures of chemokines and cellular receptors. These molecules serve as secondary receptors for entry of human immunodeficiency virus into T cells and monocytes.

CXCR5 and CCR7 and their ligands (CXCL13, CCL19, and CCL21) are major regulators of dendritic cell and lymphocyte trafficking to secondary lymphoid tissue. Th1 cells more frequently express CXCR3, CXCR6, CCR2, CCR5, and CX3CR1 than Th2 cells which more frequently express CCR3, CCR4, and CCR8. Tissue-specific homing is chemokine dependent. For instance, T cells that home to skin preferentially express CCR4 and CCR10, whereas their ligands CCL2 and CCL27 are produced by dermal macrophages, dendritic cells, and keratinocytes. B cells producing IgG upregulate CXCR4 which promotes homing.

*See also:* Entomopoxviruses; Epidemiology of Human and Animal Viral Diseases; Epstein–Barr Virus: General Features; Epstein–Barr Virus: Molecular Biology; Immune Response to viruses: Antibody-Mediated Immunity; Immune Response to viruses: Cell-Mediated Immunity; Immunopathology.

## Further Reading

Boehm U, Klamp T, Groot M, and Howard JC (1997) Cellular responses to interferon-gamma. *Annual Review of Immunology* 15: 747–795.

Brown MA and Hural J (1997) Functions of IL-4 and control of its expression. *Critical Reviews in Immunology* 17: 1–32.

Del Prete G (1998) The concept of type-1 and type-2 helper T cells and their cytokines in humans. *International Reviews of Immunology* 16: 427–455.

Dinarello CA (1996) Biologic basis for interleukin-1 in disease. *Blood* 87: 2095–2147.

Kolls JK and Linden A (2004) Interleukin-17 family members and inflammation. *Immunity* 21: 467–476.

Moore KW, de Waal Malefyt R, Coffman RL, and O'Garra A (2001) Interleukin-10 and the interleukin 10 receptor. *Annual Review of Immunology* 19: 683–765.

Murphy PM (2003) Chemokines. In: Paul WM (ed.) *Fundamental Immunology,* 5th edn., pp. 801–840. Philadelphia: Lippincott Williams and Wilkins.

Sugamura K, Asao H, Kondo M, et al. (1996) The interleukin-2 receptor gamma chain: Its role in the multiple cytokine receptor complexes and T cell development in XSCID. *Annual Review of Immunology* 14: 179–205.

Taga T and Kishimoto T (1997) Gp130 and the interleukin-6 family of cytokines. *Annual Review of Immunology* 15: 797–819.

Trinchieri G (1998) Immunobiology of interleukin-12. *Immunologic Research* 17: 269–278.

Weaver CT, Harrington LE, Mangan PR, Gavriell M, and Murphy KM (2006) Th17: An effector CD4 T cell lineage with regulatory T cell ties. *Immunity* 24: 677–688.

# Cytomegaloviruses: Murine and Other Nonprimate Cytomegaloviruses

**A J Redwood, L M Smith, and G R Shellam,** The University of Western Australia, Crawley, WA, Australia

© 2008 Elsevier Ltd. All rights reserved.

## Glossary

**Cytomegalia** From Greek *kytos*, cell, and *megas*, large. Refers to the cellular enlargement or swelling (cytomegalia) seen in cytomegalovirus-infected cells. This swelling is typically accompanied by the presence of intranuclear inclusion bodies, which are sometimes called 'owl eye' inclusion bodies because the dense staining inclusion body is surrounded by a cleared halo.

## Classification of Cytomegaloviruses

Cytomegaloviruses (CMVs) are large, enveloped, double-stranded DNA viruses with an icosahedral capsid that belong to subfamily *Betaherpesvirinae* of the family *Herpesviridae*. There are three genera. Genus *Cytomegalovirus* contains human cytomegalovirus (HCMV; species *Human herpesvirus 5*) and a number of other primate CMVs. Genus *Muromegalovirus* contains murine cytomegalovirus (MCMV; *Murid herpesvirus 1*) and rat cytomegalovirus (RCMV; *Murid herpesvirus 2*). Genus *Roseolovirus* contains human herpesviruses 6 and 7 (HHV-6 and HHV-7; *Human herpesvirus 6* and *Human herpesvirus 7*). Other CMVs that have not yet been fully classified within the *Betaherpesvirinae* are guinea pig cytomegalovirus (GPCMV; *Caviid herpesvirus 2*), tree shrew herpesvirus (THV; *Tupaiid herpesvirus 1*), swine cytomegalovirus (SuHV-2; suid herpesvirus 2), European ground squirrel cytomegalovirus (ScHV-1; sciurid herpesvirus 1), and American ground squirrel cytomegalovirus (ScHV-2; sciurid herpesvirus 2).

The recognized nonprimate CMVs are described in **Table 1**. With the exception of MCMV, RCMV, and GPCMV, little is known about the life cycle and pathogenesis of nonprimate CMVs. Nevertheless, general characteristics include strict species specificity, an ability to induce cytomegalia in infected cells, cell-associated replication in cell culture, a slow replication cycle and the establishment of persistent and latent, lifelong infection in the natural host. Tropism for secretory glands, particularly the salivary gland, is also a common feature. Infection is generally asymptomatic unless the host is immunosuppressed or has an immature immune system. In such hosts, infection may result in morbidity or even mortality.

## Murine Cytomegalovirus

*Murid herpesvirus 1* is the type species of the genus *Muromegalovirus*. The term murine cytomegalovirus (MCMV) is more commonly used for the virus than murid herpesvirus 1. The natural host for MCMV is the house mouse, *Mus musculus domesticus*. Because of the strict species specificity of CMVs, MCMV is widely used as an animal model of HCMV infection. Consequently, more is known about MCMV than any other nonprimate CMV.

### Virion Structure and Morphology

CMV virions are spherical, *c.* 230 nm in diameter, and comprise four morphologically distinct elements: the core, capsid, tegument, and envelope. The core encompasses the double-stranded DNA viral genome, which is packaged as a single linear molecule into the protein capsid.

The viral capsid is composed of 162 capsomers comprising hexons and pentons in a $T = 16$ icosahedral lattice structure. The capsids of CMVs are larger and incorporate a larger genome than other herpesviruses. Unlike other CMVs, MCMV preparations may contain a high proportion of multicapsid virions, which contain a number of capsids enclosed within a common membrane. The capsid of MCMV is composed of five proteins: the major capsid protein (MCP), the minor capsid protein, the minor capsid binding protein, the smallest capsid protein, and an assembly/protease. These are encoded by genes *M86*, *M85*, *M46*, *M48.2*, and *M80*, respectively.

The tegument of MCMV is a proteinaceous layer of material between the capsid and envelope, and resembles the matrix of other viruses. By electron microscopy the tegument is seen to have an ordered structure, particularly proximal to the capsid. The tegument proteins of MCMV have been defined by their homology to known HCMV tegument proteins. In HCMV there are at least 25 proteins associated with the tegument. Typically, tegument proteins are phosphorylated (and have the prefix pp). At least nine MCMV tegument proteins with homologs in HCMV have been detected in MCMV virions. These are the upper and lower matrix phosphoproteins (encoded by *M82* and *M83*), large tegument protein (*M48*), pp150 (*M32*), and the gene products from *M25*, *M47*, *M51*, *M94*, and *M99*. The role of most tegument proteins is unknown, but the function of some can be inferred from the function of HCMV homologs. For example, the upper matrix protein of HCMV, also known as pp71, is a transcriptional transactivator that regulates immediate early gene expression, possibly by

**Table 1** Classification of known and putative nonprimate CMVs by the International Committee on Taxonomy of Viruses

| | | |
|---|---|---|
| Family | 00.031 | Herpesviridae |
| Subfamily | 00.031.2 | Betaherpesvirinae |
| Genus | 00.031.2.02 | Muromegalovirus |
| Examples | 00.031.2.02.001 | Murid herpesvirus 1 (murine cytomegalovirus) |
| | 00.031.2.02.002 | Murid herpesvirus 2 (rat cytomegalovirus) |
| Unassigned in the subfamily | 00.031.2.00.004 | Caviid herpesvirus 2 (guinea pig cytomegalovirus) |
| | 00.031.2.00.049 | Tupaiid herpesvirus 1 (tree shrew herpesvirus) |
| Unassigned in the family | 00.031.0.00.062 | Suid herpesvirus 2 (swine cytomegalovirus) |
| | 00.031.0.00.063 | Sciurid herpesvirus 1 (European ground squirrel cytomegalovirus) |
| | 00.031.0.00.046 | Sciurid herpesvirus 2 (American ground squirrel cytomegalovirus) |

inhibiting the effects of the host cell transcriptional inhibitor hDaxx. Other tegument phosphoproteins are likely to play similar roles in the transcriptional regulation of viral genes.

The envelope of MCMV is composed predominantly of lipids obtained from the intracellular membranes of the host cell and contains a considerable number of virus-encoded glycoproteins. The envelope of CMVs is more pleiomorphic, and contains more glycoproteins, than envelopes of other herpesviruses. The major envelope glycoprotein, a product of the *M55* gene, is the highly conserved glycoprotein B (gB) and is a dominant B-cell antigen in CMV-infected animals and humans. It has been found in every mammalian herpesvirus and is one of the most highly conserved herpesvirus proteins. The glycoproteins of CMVs form three distinct complexes that mediate viral attachment and entry into host cells (**Table 2**).

## Genome Architecture and Coding Potential

The MCMV genome comprises a single unique sequence with short terminal direct repeats and several short internal repeats. Unlike that of HCMV, the linear genome of MCMV does not have an isomeric structure because it lacks internal repeat sequences related to the terminal repeats. The nomenclature devised for MCMV genes numbers them left to right along the genome. MCMV genes with homologs in HCMV are assigned the uppercase prefix '*M*', while genes with no sequence identity with HCMV genes are identified by the lowercase prefix '*m*'.

The genome of the Smith strain of MCMV is 230 278 bp in size, and has a coding potential estimated to be between 170 and 204 open reading frames (ORFs), including newly defined splice variants of previously described ORFs. MCMV shares with other members of the herpesvirus family a number of evolutionarily conserved proteins, which are involved in processes such as DNA replication and virion maturation and structure, and are located within the central region of the genome. The two terminal regions of the genome contain genes that are unique to MCMV and include the *m02* and *m145* gene families, respectively (**Figure 1**). Many of the genes in these unique regions encode proteins that are involved in modulation of the host's immune response.

## Replication

Viral replication is initiated when infectious virions enter a susceptible host cell (**Figure 2**). A series of receptors have been implicated in the binding, fusion, and entry of CMVs. Most information has come from studies of HCMV, in which infection is initiated by loose tethering of the virus via viral gCI and gCII complexes and heparan sulfate proteoglycans (HSPGs) on the cell surface. Tethering promotes a stronger binding reaction between the viral gCI complex and host cellular receptors. These receptors have not been fully defined, although for HCMV the epidermal growth factor receptor (EGFR) is thought to be one of the host receptors important in viral binding. This receptor binding promotes receptor colocalization with other host proteins, in particular integrins, into lipid rafts. The β1 and β3 chains of integrins have been implicated in the entry of MCMV into host cells. Integrin-binding domains have been localized to the N-terminus of gB in all betaherpesviruses and most gammaherpesviruses. Finally, receptor binding and/or colocalization triggers intracellular signaling and fusion between the host cell and viral membranes. Cellular signaling appears important for translocation of the capsids to the nucleus.

**Table 2**  Structural glycoproteins of MCMV

| Glycoprotein (gene) | Complex | Essential | Function |
|---|---|---|---|
| gB (*M55*) | gCI | Yes | Binds to cellular receptors (e.g., EGFR) and triggers intracellular signaling. *May activate host TLR2. Major B-cell epitope* |
| gN (*M73*) | gCII | Yes | With gM binds HSPG |
| gM (*M100*) | gCII | Yes | With gN binds HSPG |
| gH (*M75*) | gCIII | Yes | As a component of the gCIII complex is involved in binding and membrane fusion. *May activate host TLR2* |
| gL (*M115*) | gCIII | Yes | Required for transport of gH to the gCIII complex. As a component of the gCIII complex is involved in binding and membrane fusion |
| gO (*m74*) | gCIII | No; deletion virus has small plaque size | Enhances cell-to-cell spread of MCMV. As a component of the gCIII complex is involved in binding and membrane fusion |
| gp24 (*m73.5*) | | Unknown | Unknown |

The data are a compilation of information from HCMV and MCMV. Text in italics denotes the host response to glycoproteins. In HCMV, gB forms the disulfide-linked homodimeric gCI complex. The gCII complex is a disulfide-linked heterodimer between gM (*UL100*) and gN (*UL73*). The HCMV glycoproteins gL, gH, and gO (encoded by the genes *UL115, UL75,* and *UL74,* respectively) form the noncovalently associated heterotrimeric complex gCIII. All three complexes are essential for viral replication. MCMV homologs are believed to serve a similar function, although gN has only been demonstrated in the virions of HCMV. Note that gp24 has only been demonstrated in MCMV. In HCMV, gB and gH appear to bind TLR2, but this has not been demonstrated for MCMV.

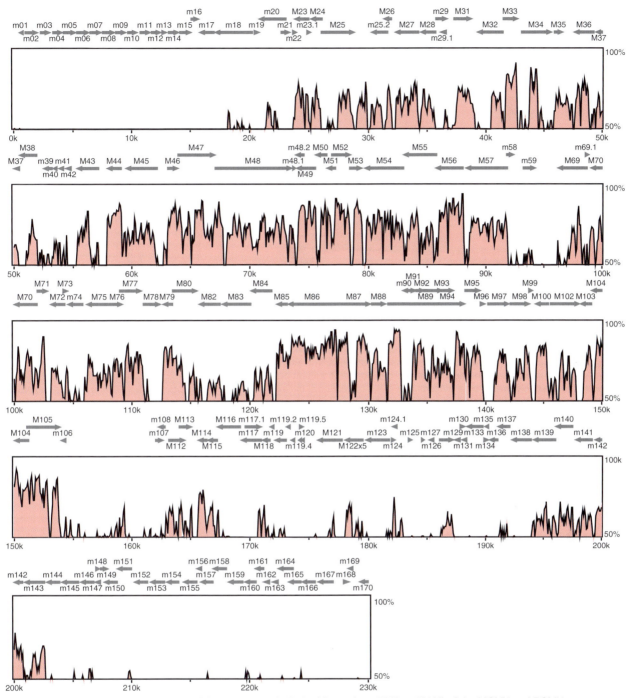

**Figure 1** Moving window comparison of the sequence similarity (shown from 50% to 100%) of the MCMV and RCMV genomes. Selected MCMV genes are marked with an arrow above the plot. The central regions of the genome are highly conserved and represent the herpesvirus-conserved genes. The left and right terminal sequences contain species-specific genes and therefore show little or no sequence similarity. Image created using LAGAN and VISTA (http://genome.lbl.gov/vista/index.shtml).

The exact mechanism by which viral DNA enters the nucleus is unknown. However, it is believed that viral DNA migrates to nuclear domain 10 (ND10) complexes within the nucleus where viral gene transcription occurs. All betaherpesviruses have three gene sets, α, β, and γ, that are temporally regulated. These genes are expressed in the immediate early (IE), early (E), and late (L) phases of viral replication, respectively. The IE phase occurs immediately after viral DNA enters the nucleus and is controlled by the major IE promoter (MIEP) in MCMV. The MIEP controls expression of the transcriptional activator genes *ie1* (*m123*) and *ie3* (*M122*). While both the *ie1* and *ie3* genes are essential for replication, the former is required specifically at low multiplicities of

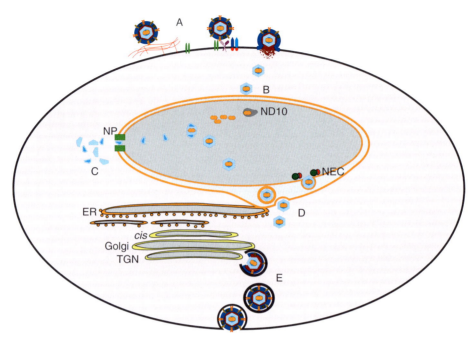

**Figure 2** CMV life cycle. (A) Virus is loosely tethered to the host cell by binding to HSPG via either gB (gCI complex) or gM/gN (gCII complex). Virus is bound more tightly by association with cell receptors such as EGFR (HCMV). Binding causes receptor clustering and the interaction of other receptors such as host cell integrins. The interaction between host receptors and viral glycoproteins induces membrane fusion and the delivery of viral capsid and tegument proteins into the host cell cytoplasm. (B) Viral capsid is transported to the nucleus and, by a process that is not understood, the viral DNA enters the nucleus. Viral DNA is localized at ND10 complexes where viral gene expression is initiated. (C) Viral proteins, including capsid proteins, are produced in the cytoplasm. DNA encapsidation and capsid assembly occur in the nucleus. Capsid proteins either diffuse into the nucleus via nuclear pores or are transported due to the presence of nuclear localization motifs. Large capsid proteins, such as MCP, which lack these motifs, are transported to the nucleus in association with capsid proteins that possess them. (D) Viral capsids acquire a primary envelope as they migrate through the inner nuclear membrane (INM). Viral proteins p35 and p38 (*M50* and *M53*) form the nuclear egress complex (NEC) and serve to disassociate the INM. Transport of the capsid through the outer nuclear membrane results in de-envelopment. (E) The final viral envelope is derived from cytoplasmic organelles, possibly in the *trans*-Golgi network (TGN). Fully formed virions enter the secretory pathway and exit the cell by exocytosis.

infection. A third IE gene, *ie2*, not present in HCMV, is transcribed from a different promoter and in the opposite direction. The *ie2* gene is not essential for MCMV replication either *in vitro* or *in vivo*.

There is an absolute requirement for IE gene expression prior to E gene expression. Genes transcribed during the E phase of viral replication include those required for entry into the L phase of viral replication, and other genes, such as the immune evasion genes. L phase genes mostly encode structural proteins and are expressed after the start of viral DNA replication, which occurs in the nucleus approximately 16 h after infection. During DNA replication, the sequences at the genomic termini fuse via a 3′ nucleotide extension to form the intermediates for MCMV replication, which occurs by a rolling-circle mechanism. Maturation of the MCMV genome involves the processing and cleavage of newly synthesized concatemeric viral DNA into genome-length monomers prior to packaging into preformed nucleocapsids in the cell nucleus. Herpesvirus-conserved *pac1* and *pac2* DNA sequence motifs are required for cleavage and packaging of the MCMV genome.

The origin of DNA replication of MCMV is the *ori* Lyt region between *M57* and *M69* (**Figure 1**), which extends over 1.7 kbp and is extremely rich in repeat sequences that act as binding elements for various transcription factors. MCMV-encoded proteins required for origin-dependent replication include the DNA polymerase (*M54*), a polymerase accessory protein (*M44*), the single-stranded DNA-binding protein (*M57*), and a helicase-primase complex encoded by *M70*, *M102*, and *M105*. All of these proteins have been detected within purified MCMV virions.

The formation of capsids and the packaging of viral DNA occur in the nucleus of infected cells. Capsid proteins are produced in the cytoplasm and are transported back to the nucleus across the nuclear membrane. The transport of capsid proteins is a result of either their small size, which allows diffusion across the nuclear pore complex, or the presence of nuclear localization signals. Large capsid proteins such as the MCP, which do not contain nuclear localization signals, are transported in association with those that do. Viral DNA is packaged into complete capsids and transported to the cytoplasm via the nuclear membrane where the capsids acquire their primary envelope as they bud through the inner nuclear membrane (INM) as shown in **Figure 2**. In MCMV infection, the virus penetrates the INM with the aid of *M50*/p35 and a partner gene, *M53*/p38. The proteins p35 and p38 form the nuclear egress complex (NEC) and recruit cellular kinases to the INM, specifically to the nuclear lamina. Recruitment of cellular kinases results in phosphorylation and degradation of the lamina, and facilitates egress of the virus from the nucleus.

The mechanism of egress of CMVs from infected cells is incompletely understood. Studies from other herpesviruses suggest that there are two phases of envelopment. A primary envelopment involving the host's INM precedes a final secondary envelopment from another host compartment, possibly the *trans*-Golgi network (TGN), where the virions also gain their tegument. The mature enveloped virions, now present in secretory vesicles, are transported to the plasma membrane, where they are released into the extracellular space by exocytosis (**Figure 2**).

## Pathogenesis

During acute experimental intraperitoneal infection, MCMV replicates predominantly in the spleen and liver and to a lesser extent in the lungs. However, the virus persists in the lungs for longer than in the spleen or liver. Other organs infected during the acute phase include the adrenal glands, kidneys, heart, and ovaries. During the chronic or persistent phase, the virus replicates predominantly in the salivary gland. Intranasal inoculation, which may mimic natural infection, results in viral replication predominantly in the lungs and salivary gland. Histologically, MCMV-infected cells exhibit typical swelling or cytomegalia (from which the virus derives its name) with intranuclear inclusion bodies. In MCMV-infected mice, mononuclear cell infiltration may be observed in inflammatory responses in the heart, lung, adrenals, and other organs.

The seroprevalence of MCMV in free-living mice can be up to 100%, particularly when mouse population densities are high. Transmission of the virus is presumed to be via saliva, possibly as a result of biting and grooming. However, MCMV is found in urine and breast milk, as well as the reproductive tract of male and female mice. This suggests additional modes of transmission, such as from mother to pup during feeding and by sexual activity. Interestingly, infection with multiple strains of MCMV is common, suggesting that mice are either infected simultaneously with multiple MCMV strains or that the lack of sterilizing immunity allows sequential reinfection of mice.

A number of diseases induced by HCMV in humans have been modeled with MCMV in mice. These include myocarditis, hepatitis, adrenalitis, and interstitial pneumonitis. However, natural congenital infection has not been modeled in mice, since MCMV, unlike HCMV, does not cross the placenta in immunocompetent hosts. Mice have also been used to investigate host resistance to CMV. A number of factors affect the capacity of MCMV to cause disease in mice. These include the dose and route of inoculation of the virus as well as the age and genetic constitution of the host. Several genetically linked innate resistance mechanisms have been identified in mice, including *Cmv1*, *Cmv3*, and *Cmv4*. These loci control resistance to MCMV via innate immune responses. Other genetic resistance mechanisms have been demonstrated in New Zealand white mice (*Cmv2*) and in a mouse strain carrying an *N*-ethyl-*N*-nitrosourea-induced mutation of the protein Unc93b1 in which signaling via Toll-like-receptors (TLRs) 3, 7, and 9 is deficient. TLRs 3, 7, and 9 recognize single-stranded RNA, double-stranded RNA (dsRNA), and unmethylated DNA, respectively. Previous studies have highlighted the importance of TLRs 2, 3, and 9 in resistance to MCMV.

## Latency

Persistent infection is a common feature of the betaherpesviruses. The infection may be either chronic, in which infectious virus is produced at very low levels for long periods in particular organs or tissues such as the salivary gland, or latent, in which infectious virus is no longer detected, although the viral genome is present in certain cells in the body. Reactivation from latency usually occurs during immunosuppression to yield infectious virus, which may induce disease and be transmitted to susceptible hosts.

Regulation of transcription is tightly controlled during latency. In MCMV, regulation of IE gene expression is an early checkpoint on the way from latency to recurrence. The IE1/3 transcriptional unit gives rise to IE1 and IE3 mRNAs by differential splicing, which is driven by the $P^{1/3}$ promoter with a strong upstream enhancer that serves as a molecular switch, connecting IE1/3 transcription to the cellular environment. External stimuli, such as the proinflammatory cytokine tumor necrosis factor alpha (TNF-$\alpha$), act as the first signal in the reactivation pathway, inducing transcription factors which activate the enhancer by binding to defined sequence motifs. IE3 is believed to be the major transactivator of E gene expression. MCMV latency is controlled after the initiation of IE1/3 transcription and this is the second checkpoint in the pathway of molecular reactivation. The full mechanism of reactivation has not been elucidated, but models postulate a multistep system of MCMV reactivation involving many checkpoints before the production of infectious virus.

## Host Immune Responses and Viral Evasion Strategies

The host immune response to MCMV and the many countermeasures employed by MCMV reflect a dynamic host and pathogen interaction and co-evolution. A range of host immune responses involving antibody, CD4+ T cells, CD8+ T cells, and natural killer (NK) cells help control MCMV infection, and are summarized in **Table 3**.

Innate intracellular defense mechanisms, such as interferon production (see below), apoptosis, and the dsRNA-dependent protein kinase R (PKR)-mediated shutdown of protein synthesis, are all targets of MCMV genes. There are three genes encoded by MCMV whose products inhibit apoptosis and facilitate tropism in specific cell types: apoptosis is inhibited in macrophages by the product of *M36*, in endothelial cells and fibroblasts by the product of *M45*, and in endothelial cells by the product of *m41*. An additional gene, *m38.5*, appears to be a homolog of the HCMV viral mitochondria-localized inhibitor of apoptosis (vMIA) encoded by *UL37*. Finally, the products of MCMV genes *m142* and *m143* form a dsRNA-binding complex that inhibits host PKR-mediated shutdown of protein synthesis.

NK cells are the primary host cell involved in the early innate response to MCMV. NK cells limit the severity, extent, and duration of acute infection and also affect the subsequent acquired immune response. MCMV infection can be lethal in their absence. NK cell control of MCMV infection is mediated by direct lysis of infected cells and by the production of cytokines. Direct NK cell killing is mediated predominantly by perforin, particularly in the spleen. Cytokine control of infection is most evident in the liver and is predominantly mediated by interferon gamma (IFN-γ). MCMV encodes at least six genes that

**Table 3** Immune evasion strategies of MCMV

| Host control | Effect | MCMV immune subversion | | |
| --- | --- | --- | --- | --- |
| | | Gene | Function | Mechanism |
| *Innate immunity* | | | | |
| NK cells | Control of MCMV replication in acute infection | m145 | Inhibits NK cell killing | Downregulates MULT-1, a ligand for the NK cell activating receptor, NKG2D, on infected cells |
| | Cytokine production enhances later immune responses such as T-cell activation and may affect DC/NK cell cross talk | m152 | Inhibits NK cell killing | Downregulates RAE-1, a ligand for the NK cell activating receptor, NKG2D, on infected cells |
| | | m155 | Inhibits NK cell killing | Downregulates H60, a ligand for the NK cell activating receptor, NKG2D, on infected cells |
| | | m144 | Inhibits NK cell killing | MHC homolog, ligand unknown |
| | | m138 | Inhibits NK cell killing | Downregulates MULT-1 and H60 ligands for the NK cell activating receptor, NKG2D, on infected cells |
| | | m157 | Activates NK cells | Binds the NK cell activating receptor Ly49H (Cmv1) |
| Monocyte/ macrophages | Phagocytosis of infected cells and cytokine production | Nil | MCMV infects monocytes and macrophages, increasing IL-10 expression and reducing MHC class II expression | Infected monocytes disseminate virus to other organs such as the salivary gland. Differentiation in the tissues to macrophages allows for productive infection |
| Cytokines | Various effects on both innate and acquired immune responses | M27 | Inhibits innate intracellular resistance to MCMV | Downregulates STAT-2, induces type I and type II IFN resistance |
| | | M33 | Migration of cells, including smooth muscle cells | Agonist independent GPCR, functional homolog of HCMV US28 |
| | | m129/131 | Promotes inflammation | Chemokine homolog, macrophage chemoattractant. May aid dissemination of MCMV to the salivary gland within macrophages |

**Table 3** Continued

| | | MCMV immune subversion | | |
|---|---|---|---|---|
| Host control | Effect | Gene | Function | Mechanism |
| *Acquired immunity* | | | | |
| DCs | Priming of T cells<br>Cytokine production<br>Activation of NK cells | *m147.5* and other unidentified genes | Inhibition of T-cell priming<br>Inhibits NK cell function<br>Reduction in IL-2 and IL-12 production | Failure of DC maturation, possibly due to multiple MCMV genes affecting MHC expression, co-stimulatory molecule expression, and cytokine production. *m147.5* specifically downregulates CD86 |
| CD8 T cells | Direct killing of infected cells | *m04* | Blocks CTL function | Binds MHC class I and remains associated on cell surface |
| | Control of reactivation from latency | *m06* | Blocks CTL function | Targets MHC class I to lysosome for degradation |
| | | *m152* | Blocks CTL function | Retains MHC class I in ERGIC |
| CD4 T cells | Control of viral replication in the salivary gland | Nil | | |
| Antibody | Reduces viral dissemination<br>Reduces MCMV titer after reactivation<br>Passive transfer protects mice from MCMV | *m138* | Fc receptor homolog (fcr-1). No known role in evasion of antibody responses | |

MCMV has multiple mechanisms for evading host control. NK cells and CD8[+] CTLs are the primary effector cells in host control and this is reflected in the number of immune evasion strategies MCMV employs to evade these mechanisms. Dendritic cells (DCs) also play a major role in the acquired immune systems control of MCMV infection by priming MHC class I and II restricted responses and, in the innate response, by cytokine release and by activation of NK cells. MCMV causes functional paralysis of DC and inhibits the expression of MHC and co-stimulatory molecules on the surface of DCs. These effects presumably serve to reduce T-cell priming as well as the activation of NK cells and may alter the type of immune response generated by perturbing cytokine responses. Macrophages are a major host control mechanism, but MCMV subverts this role by infecting monocytes and using the cells as a means of dissemination to other organs such as the salivary gland. Once in the tissues, monocytes differentiate into macrophages and become permissive for MCMV production. MCMV-encoded chemokine (MCK-2) and chemokine receptor (*M33*) may serve to recruit inflammatory cells to the site of infection to aid viral dissemination.

affect NK cell responses (**Table 3**). The complexity of this interaction is demonstrated in **Figure 3**.

CD8[+] cytotoxic T lymphocytes (CTLs) are crucial for the resolution of acute MCMV infection in BALB/c mice. CTLs lyse MCMV-infected fibroblasts *in vitro* and are found in the spleen of infected mice within 3 days. Sensitized CTLs recognize both structural and nonstructural viral antigens. This has been best studied in BALB/c mice where almost 50% of CTLs are directed against the nonstructural pp89 (*IE1*) protein of the Smith strain of MCMV. Epitope mapping has identified the immunodominant H-2L[d]-restricted pp89 peptide as the nonomer YPHFMPTNL. Other peptides recognized in BALB/c mice are encoded by genes *m04*, *M45*, *m164*, *M83*, and *M84*. Recent studies in C57BL/6 mice have also identified a broad range of CD8 T-cell epitopes.

MCMV encodes a number of genes that are believed to affect CTL-mediated control of infection. Some, such as *m04*/gp34, *m06*/gp48, and *m152*/gp40, affect major histocompatibility complex (MHC) class I molecule expression on the surface of infected cells and/or inhibit CTL recognition and killing (**Table 3** and **Figure 3**). Their role *in vivo* is yet to be fully explained, but MCMV mutants in which all three genes are deleted exhibit reduced growth in the salivary glands of mice. MCMV also inhibits T-cell priming by causing functional paralysis of dendritic cells (DCs). The MCMV genes responsible for this effect are largely unknown, although *m147.5* is known to downregulate the co-stimulatory molecule B7.2 on the surface of infected DCs.

Antibody and CD4[+] T cells play lesser roles in the resistance of mice to MCMV infection. Antibody reduces dissemination of the virus within the host and reduces viral loads after reactivation from latency. As yet, there are no known MCMV genes that target antibody function directly. The MCMV-encoded (*m138*) Fc receptor inhibits NK cell function but does not appear to play a role in inhibiting host-antibody-mediated protection. Similarly, no MCMV-encoded gene has been identified that directly targets host CD4[+] T-cell responses. This perhaps reflects the moderate role of host CD4[+] T cells in resistance to MCMV, although CD4[+] T cells are important in the control of MCMV replication in the salivary gland.

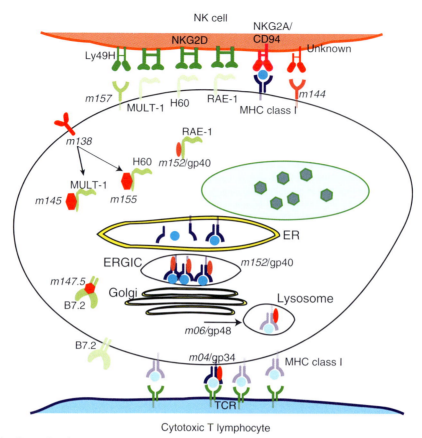

**Figure 3** MCMV NK cell and CD8+ CTL immune evasion strategies. MCMV has multiple methods of moderating host immune responses. Receptors or genes that inhibit immune responses are shown in red; those that activate immune responses are shown in green. The MHC class I molecules activate immune responses when ligated by host TCR (T-cell receptor) but inhibit responses when ligated by host NK cell NKG2A/CD94 heterodimer and are therefore shown in blue. When downregulated by an MCMV gene, host proteins are shown on the surface of the cell with reduced shading. MCMV targets NK cells by multiple mechanisms. MCMV downregulates host stress ligands that activate NK cells via the NK cell activating receptor NKG2D. MCMV genes *m138*, *m145*, *m152*, and *m155* affect the host proteins MULT-I and H60, MULT-1, H60, and RAE-1, respectively. These interactions have been depicted to occur in the cytoplasm because their actual locations are unknown. MCMV *m144* is an MHC homolog and inhibits NK cells by interaction with an unknown ligand, presumably an NK cell inhibitory receptor. Host CTL responses are avoided by interference with MHC class I expression. The gp48 (*m06*) protein targets MHC class I for degradation in the lysosome, gp40 (*m152*) retains MHC class I in the endoplasmic reticulum-Golgi intermediate compartment (ERGIC), and gp34 (*m04*) binds to MHC class I and inhibits CTL function (and possibly NK cell function). The gene product from *m147.5* downregulates the co-stimulatory molecule B7.2 and functions with other unidentified genes to inhibit DC function and reduce T-cell priming and NK cell function. Note, NKG2D is also found on T cells as a co-stimulatory receptor, but the role of *m138*, *m145*, *m152*, and *m155*, if any, on T-cell response to MCMV has not been explored.

Cytokine responses are also crucial to the resolution of MCMV infection in mice. This is exemplified by the increased sensitivity of interferon-αβ receptor 1 chain (IFNAR1) and IFN-γR1 knockout mice to MCMV. IFNAR1/IFN-γR1 double-knockout mice are exquisitely sensitive to MCMV infection. This increased sensitivity is probably due to the combined effects of defective cellular innate and acquired immune responses due to loss of IFN-γ (type II IFN) signaling as well as the loss of innate intracellular responses resulting from a loss of type I IFN signaling. As a countermeasure to host type I and II IFNs, MCMV induces a state of type I and II IFN resistance by *M27*-mediated downregulation of STAT2. In addition, infection is enhanced by an MCMV-encoded chemokine and chemokine receptor. *M33* encodes a G protein-coupled receptor (GPCR) homolog that is constitutively active in an agonist-independent manner and promotes dissemination of the virus to the salivary gland, possibly by recruitment of macrophages. The MCMV-encoded (*m129/131*) chemokine MCK-2 serves a similar function in that it increases inflammation, and deletion of *m129/131* reduces viral titers in the salivary gland (**Table 3**).

## Rat Cytomegalovirus

CMVs have been identified in, and isolated from, several species of rat. The Maastricht strain of RCMV is the most widely studied and was isolated from the brown rat, *Rattus norvegicus*. Another strain of RCMV, the English strain, was isolated from *R. norvegicus*; however, its classification as a CMV has been questioned. Recently, more putative strains of RCMV have been isolated from *R. norvegicus*, *Rattus rattus* (black rat), and *Rattus argentiventer* (rice field rat). This article focuses on the Maastricht strain of RCMV.

The Maastricht strain of RCMV was isolated from a wild brown rat in 1982, and has been maintained *in vitro* and *in vivo* in inbred laboratory strains of *R. norvegicus*. The genome has been fully sequenced and consists of a unique region bounded by terminal repeats. Its genome is 230 138 bp in length and is collinear with that of MCMV, encoding at least 170 ORFs, approximately two-thirds of which share significant sequence homology with genes found in MCMV (**Figure 1**). One RCMV gene, *r127*, appears to be unique among CMVs, encoding a homolog of the parvovirus *rep* gene. As with MCMV, RCMV encodes several genes (e.g., the GPCRs *R33* and *R78*, the MHC class I homolog *r144*, and the CC chemokine homologs *r129* and *r131*) which are homologs of cellular genes and which have therefore presumably been appropriated by the virus during evolution.

As is the case with other CMVs, RCMV is able to modulate and subvert host immune responses. RCMV downregulates the level of MHC class I molecules on the surface of infected cells. However, unlike MCMV, this effect is short-lived and does not result in proteolytic degradation of MHC class I molecules, rather in their delayed exit from the endoplasmic reticulum (ER). The RCMV gene *r131* is a pro-inflammatory CC-chemokine homolog, and appears to act similarly to the MCMV gene *m129/131* in that deletion of these genes results in decreased inflammation at the site of inoculation. This *r131*-mediated inflammation may recruit susceptible target cells to the site of infection, allowing virus dissemination throughout the body. *R33* has been shown to promote smooth muscle cell migration; however, it may play other roles in viral disease.

## Pathogenesis

Following intraperitoneal inoculation of RCMV into rats, low levels of virus can be found in the visceral organs and bone marrow at 4 days post infection (p.i.). Virus can be detected in the salivary gland, in the striated duct cells, from 10 days p.i., with virus titers peaking at 28 days p.i., before gradually decreasing over several months. Interestingly, although at 6 months p.i. viral DNA can occasionally be detected in the visceral organs, infectious virus can only ever be isolated from the salivary gland at this time point. RCMV is capable of entering latency within the infected rat, although the major sites of latent infection are unknown. Virus can be reactivated from latency following immunological stress, such as immunosuppression or stimulation with allogeneic cells, resulting in virus replication.

As is the case with other CMVs, RCMV infection does not result in overt disease in immunocompetent hosts. However, in the immunocompromised animal, it may cause considerable morbidity and mortality. In immunocompetent rats, RCMV is capable of causing vascular injury, such as endothelial cell damage and leukocyte adherence to the aortic endothelium. RCMV infection may also lead to an influx of macrophages and lymphocytes and loosening of endothelial cells from the basement membrane. Inoculation of virus into immunocompromised animals, however, produces a different scale of pathology. Extensive cell necrosis results in organ damage, with multiple hemorrhages in lung, liver, spleen, and kidney. Virus also induces damage to the microvascular epithelium, vasculitis, and thrombotic occlusions.

The tendency of RCMV to induce vascular disease has made it a good model for studying the role of HCMV in these diseases in humans. Atherosclerosis is a chronic inflammatory disorder of large- and medium-sized arteries, and seroepidemiological and histopathological studies have implicated HCMV in the pathogenesis of this disease. RCMV induces vascular lesions with endothelial cell damage and leukocyte presence in the subendothelium, the presence of subendothelial foam cells, and morphological changes to the large blood vessels of infected rats. It has been suggested that RCMV induces smooth muscle cell migration, promotes leukocyte influx, and increases cellular expression of adhesion molecules, inflammatory cytokines, and chemokines, all components of the atherosclerotic process. Infection of rats with RCMV has also been shown to be a suitable model for arterial restenosis and transplant vascular sclerosis (chronic rejection). Both conditions have been linked to infection with HCMV in humans.

## Guinea Pig Cytomegalovirus

Classic viral inclusions were first seen in the salivary gland of guinea pigs in the 1920s. However, GPCMV was not isolated until 1957. Since this time, almost all research conducted on GPCMV has used the Hartley strain of the virus.

The genome of GPCMV, while not fully sequenced, is approximately 230 kbp in length and is collinear with those of other CMVs, sharing genes with HCMV, MCMV, and RCMV within the central two-thirds of its genome but having GPCMV-specific genes near the termini. Several genes have been fully sequenced, particularly those with significant homology to potential HCMV vaccine targets. Again, as with other CMVs, gene expression is temporally regulated with IE, E, and L kinetics, although expression in the latter categories has been less well studied.

Following inoculation with GPCMV, viremia occurs for approximately 10 days, with infectious virus being detectable, and disease evident, in lungs, spleen, liver, kidney, thymus, pancreas, and brain for about another 3 weeks. Virus may be detected in the salivary gland for up to 10 weeks p.i.

GPCMV is unique among rodent CMVs in its ability to cross the placenta and infect the fetus. Hence, it is widely used as a model of congenital infection with HCMV. The guinea pig placenta consists of a single trophoblast layer separating maternal and fetal circulation, and is histologically similar to the human placenta. Infection during early pregnancy leads to pup resorption, while infection late in pregnancy generally leads to pup mortality. Following infection mid-gestation, GPCMV can be detected in placental tissue, even in the face of maternal anti-GPCMV antibodies. However, even in this case, only a proportion of pups attached to infected placentas are infected with GPCMV. These infected pups exhibit CMV disease in the brain, visceral organs, and inner ear, analogous to congenital infection with HCMV. GPCMV has been extensively used in studies of vaccination against congenital CMV infection. Vaccines against both gB and GP83 have been shown to protect against congenital GPCMV infection, although protection against pup mortality was dependent on maternal antibody titer.

## Other Nonprimate Cytomegaloviruses

Swine cytomegalovirus is endemic in swine herds worldwide, and has also been shown to reactivate from pig-to-baboon xenotransplants. It causes rhinitis in young swine and is able to cross the placenta, resulting in generalized disease, runting, and fetal death. Initial sequence analysis of the DNA polymerase complex genes suggests that this virus is more closely related to HHV-6 than to CMVs. Tree shrew herpesvirus has been sequenced and has been found to resemble MCMV and other betaherpesviruses. A CMV has been isolated from deer mice (*Peromyscus maniculatas*) in North America, and has been characterized as a CMV based on physical and biological properties and genetic homology with several genes of other CMVs. CMVs have also been isolated from European and American ground squirrels, and designated as sciurid herpesvirus 1 and sciurid herpesvirus 2, respectively.

Agents resembling CMVs have also been described in hamsters, moles, voles, field mice, the Australian native rodent antechinus, cats, and dogs. These agents have not been characterized further and may not actually be CMVs.

*See also:* Cytomegaloviruses: Simian Cytomegaloviruses; Herpesviruses: General Features; Human Cytomegalovirus: General Features; Human Cytomegalovirus: Molecular Biology; Human Herpesviruses 6 and 7.

## Further Reading

Mocarski ES, Jr. (2004) Immune escape and exploitation strategies of cytomegaloviruses: Impact on and imitation of the major histocompatibility system. *Cellular Microbiology* 6: 707–717.

Rawlinson WD, Farrell HE, and Barrell BG (1996) Analysis of the complete DNA sequence of murine cytomegalovirus. *Journal of Virology* 70: 8833–8849.

Reddehase MJ (ed.) (2006) *Cytomegaloviruses Molecular Biology and Immunology.* Wymondham, UK: Caister Academic Press.

Schleiss MR (2002) Animal models of congenital cytomegalovirus infection: An overview of progress in the characterization of guinea pig cytomegalovirus (GPCMV). *Journal of Clinical Virology* 25: S37–S49.

Shellam GR, Redwood AJ, Smith LM, and Gorman S (2007) Mouse cytomegalovirus and other herpesviruses. In: Fox JG, Barthold SW, Davisson MT, Newcomer CE, Quimby FW, and Smith AL (eds.) *The Mouse in Biomedical Research, Vol. 2: Diseases,* 2nd edn., pp. 1–48. Amsterdam: Academic Press.

Vink C, Beuken E, and Bruggeman CA (2000) Complete DNA sequence of the rat cytomegalovirus genome. *Journal of Virology* 74: 7656–7665.

# Cytomegaloviruses: Simian Cytomegaloviruses

**D J Alcendor and G S Hayward,** Johns Hopkins School of Medicine, Baltimore, MD, USA

© 2008 Elsevier Ltd. All rights reserved.

## Glossary

**Paralogs** Genes that share homology through gene duplication and often have diverged functions or expression patterns.

**Stealth virus** An unproven chimeric virus that lacks or suppresses genes that trigger immune responses and is thereby able to go undetected by the immune system.

## Introduction

In addition to the well-studied cytomegaloviruses (CMVs) of human (HCMV; salivary gland virus) and mouse (MCMV), related agents with typical CMV-like characteristics have been described in rats (RCMV), guinea pigs (GpCMV), pigs (PCMV), elephants (EEHV, elephant endotheliotropic herpesvirus), Old and New World primates (collectively known as simian CMVs and represented by viruses such as RhCMV), and tree shrews (HVTupaia) (**Table 1**). GpCMV was originally recognized in 1920 and served as a model system for the biology and pathogenicity of HCMV disease for many years.

Most natural nonhuman primate populations studied harbor persistent or latent infections with host-specific simian CMVs. Isolates have been reported from almost all major primate groups, including gorilla, chimpanzee, bonobo, drill, baboon, rhesus and other macaques, African green monkey, spider monkey, owl monkey, capuchin, and marmoset. Infection is also common in captive breeding populations and, even when infection is usually inapparent in the absence of immunsuppression, virus is shed intermittently in urine and saliva. Indeed, simian CMVs are potential contaminants in primary cell cultures obtained from primate sources. Also, all simian CMVs can be adapted to grow in human fibroblasts. For research purposes, the main value of simian CMVs is for genome evolutionary comparisons and as animal models for HCMV disease, in which they are potentially superior to rodent CMVs. The genomes of primate CMVs differ among themselves and from HCMV to a surprisingly large degree, but nonetheless are more closely related to HCMV than are their equivalents in nonprimates.

## Classification and Evolution

CMVs are members of the family *Herpesviridae* and belong to subfamily *Betaherpesvirinae* (**Table 1**). Formally, this subfamily is divided into three genera, namely *Cytomegalovirus* (HCMV-like viruses from primates), *Muromegalovirus* (MCMV-like viruses from rodents), and *Roseolovirus* (HHV-6-like viruses, so far only from great apes). There is

**Table 1**  Primate and other viruses classified in or potentially belonging to the subfamily *Betaherpesvirinae*

| Genus | Host group | Host species | Common abbreviation | Formal name | RefSeq accession |
|---|---|---|---|---|---|
| *Cytomegalovirus* | Old World primate | Human | HCMV | Human herpesvirus 5 | NC_001347; NC_006273 |
| | | African green monkey | AgmCMV | Cercopithecine herpesvirus 5 | |
| | | Rhesus macaque | RhCMV | Cercopithecine herpesvirus 8 | NC_006150 |
| Possible members | | Chimpanzee | CzCMV | Pongine herpesvirus 4 | NC_003521 |
| | | Bonobo | BoCMV | | |
| | | Baboon | BaCMV | | |
| | | Vervet | SA6 | Cercopithecine herpesvirus 3 | |
| | | Vervet | SA15 | Cercopithecine herpesvirus 4 | |
| | New World primate | Marmoset | MaCMV | Callitrichine herpesvirus 2 | |
| | | Owl monkey | HVAotus type 1 | Aotine herpesvirus 1 | |
| | | Owl monkey | HVAotus type 3 | Aotine herpesvirus 3 | |
| | | Cebus | CeCMV | | |
| | | Capuchin | AL-5 | Cebine herpesvirus 1 | |
| | | Capuchin | AP-18 | Cebine herpesvirus 2 | |
| *Muromegalovirus* | Rodent | Mouse | MCMV | Murid herpesvirus 1 | NC_004065 |
| | | Rat | RCMV1 | Murid herpesvirus 2 | NC_002512 |
| Possible member | | Rat | RCMV2 | | |
| *Roseolovirus* | Old World Primate | Human | HHV-6A | Human herpesvirus 6 | NC_001664 |
| | | Human | HHV-6B | Human herpesvirus 6 | NC_000898 |
| | | Human | HHV-7 | Human herpesvirus 7 | NC_001716 |
| Possible member | | Chimpanzee | PaHV-6 | | |
| *Proboscivirus* (proposed) | Elephant | African elephant | EEHV | Elephantid herpesvirus 1 | |
| Unassigned | Various | Guinea pig | GpCMV | Caviid herpesvirus 2 | |
| | | European ground squirrel | SqCMV1 | Sciurid herpesvirus 1 | |
| | | American ground squirrel | SqCMV2 | Sciurid herpesvirus 2 | |
| | | Pig | PCMV | Suid herpesvirus 2 | |
| | | Tree shrew | HVTupaia | Tupaiid herpesvirus 2 | NC_002794 |

one genus pending, *Proboscivirus* (EEHV). Based on the relatively high G + C content of their genomes and their adaptability to grow in fibroblasts in cell culture, most traditionally recognized 'CMVs' are likely to be evolutionarily more similar to genera *Cytomegalovirus* or *Muromegalovirus* than to the (A + T)-rich genus *Roseolovirus*. The taxonomy of the *Betaherpesvirinae* will be shaped further with the continuing addition of members that are currently not fully classified, such as GpCMV, PCMV, and HVTupaia.

The genomes of nine betaherpesviruses are currently available, namely those of HCMV (strain AD169 in 1990, 229 354 bp; strain Merlin in 2004, 235 645 bp), CzCMV (in 2003, 241 087 bp), RhCMV (strain 68-1 in 2003, 221 459 bp; strain CMV 180.92 in 2006, 215 678 bp), MCMV (strain Smith, in 1996, 230 278 bp), RCMV (strain Maastricht in 2000, 229 896 bp), HHV6A (strain U1102 in 1995, 159 321 bp), HHV-6B (strain Z29 in 1999, 162 114 bp), HHV-7 (strain JI in, 1996, 144 861 bp; strain RK in, 1998, 153 080 bp), and HVTupaia (in 2001, 195 857 bp). The first complete betaherpesvirus genome sequenced was that of the AD169 isolate of HCMV. However, AD169 is a highly passaged laboratory strain that had suffered deletions and duplications during passaging in culture. Five more HCMV genomes have now been sequenced in the form of bacterial artificial chromosomes, and AD169 has been replaced by strains Merlin and FIX as the prototype HCMV genomes. Two strains of AgmCMV have now also been sequenced (GR2715 and Colburn), and work is in progress on EEHV as well as several other nonhuman primate CMV species.

All formally recognized and probable betaherpesviruses are listed in **Table 1**, with their common and official names and the RefSeq accession numbers of their complete sequences. Two very different muromegaloviruses (with somewhat different gene contents) are both known as RCMV, and are represented by the Maastricht (M) and English (E) strains as RCMV1 and RCMV2, respectively. CMV-like viruses have been associated with fatal hemorrhagic disease in young African and Asian elephants. Ongoing genetic analysis indicates that they are very distinct from members of the three recognized genera, and this has resulted in proposal of a new genus, *Proboscivirus*. An unusual herpesvirus isolate, referred to as 'stealth virus', was reported to have been recovered from a human patient with central nervous system (CNS) neuropathy, but genetic analysis has revealed that this virus is virtually identical to several characterized simian CMV isolates from African green and vervet monkeys (AgmCMV). Another AgmCMV isolate, 'Colburn', was also originally thought to have been of human origin, but it is unclear whether it is truly a human isolate rather than a contaminant from primary monkey cell cultures. Based on genome analyses, the bovine herpesvirus BHV-4 and two equine herpesviruses EHV-2 and EHV-5 were originally thought to be CMVs, but were reclassified into the genus *Rhadinovirus* in subfamily *Gammaherpesvirinae*.

Typically, betaherpesviruses have no more than 40 of their 110–165 genes in common with alpha- and gammaherpesviruses. Another 35 genes appear to be betaherpesvirus specific and are likely to be shared by all CMV and HHV-6-like viruses, whereas most of the remaining genes are unique to different virus genera or individual species. As judged by the presence of several large families of related genes in HCMV, AgmCMV, and RhCMV, it appears that the CMVs, which possess the largest of all mammalian herpesvirus genomes, underwent a rapid genomic expansion in the early stages of mammalian evolutionary radiation. In contrast, the roseoloviruses (HHV-6 and HHV-7) and HVTupaia can be considered to be mini-CMVs lacking most of the repeated gene families and the entire S segment of the genome. Unlike other herpesviruses, betaherpesviruses in general do not encode either thymidine kinase or the small subunit of ribonucleotide reductase, although EEHV has both of these. However, probably all betaherpesviruses encode a phosphotransferase (UL97) activity that is the target for the effective anti-CMV agent ganciclovir and its derivatives. Like all other known herpesviruses, primate CMVs encode the typical set of six core DNA replication proteins including DNA polymerase (POL), DNA primase (PRI), helicase (HEL), single-stranded DNA-binding protein (SSB), polymerase processivity factor (PPF), and primase accessory factor (PAF). In contrast to the CMVs and muromegaloviruses, the roseoloviruses and EEHV encode a homolog of the HSV UL9 origin-binding protein (OBP).

## Virion Structure

Primate CMV virions are structurally similar to those of other members of the *Herpesviridae*, and are essentially indistinguishable from those of HCMV in the electron microscope. However, many of the virion proteins of each species display characteristic size variations by polyacrylamide gel electrophoresis (PAGE) analysis. Large nuclear and cytoplasmic inclusion bodies detectable in lytically infected cells by light microscopy are a hallmark of all betaherpesviruses.

## Genome Structure

The genomes of all known CMV DNA molecules are c. 210–240 kbp in size, and most have a G + C content of around 58%, in contrast to the 145–166 kbp and 46% G + C content of HHV-6 and HHV-7. Structurally, two distinct CMV genome types can be discerned. The genomes of New World primate CMVs (HVAotus types 1 and 3), HCMV, and CzCMV represent one type, having internal inverted repeats flanking two unique regions (the L segment consisting of unique region $U_L$ flanked by inverted repeats $TR_L/IR_L$, and the S segment consisting of unique region $U_S$ flanked by inverted repeats $TR_S/IR_S$), which generate four isomeric arrangements of the segments, similar to the pattern observed in herpes simplex virus. On the other hand, AgmCMV, RhCMV, HVTupaia, and several rodent CMVs (MCMV, RCMV, and GpCMV) have noninverting DNA molecules. HHV-6 and HHV-7 lack the entire 40 kbp S segment of HCMV and the L segment is bounded by large direct terminal repeats of 5–10 kbp. MCMV has a similarly sized gene block in place of the S segment, but the region has very little organizational resemblance or similarity to the S segment of HCMV. Large blocks of genes at the left and right ends of the L segment of each genome are also unique to each of the three genera for which complete sequence data are available, including the segments mapping to the right of and adjacent to the major immediate-early (MIE) genes in nonhuman primate simian CMV genomes, which differ dramatically from each other as well as from the equivalent regions in HCMV.

DNA and amino acid sequence analysis indicates that species from different mammalian hosts are diverged in accordance with their host phylogeny, with different genes having diverged at different rates. For example, in a region that is highly conserved between HCMV and AgmCMV, the major single-stranded DNA-binding proteins (UL57) have 72% overall amino acid sequence identity, whereas a group of three adjacent betaherpesvirus-specific glycoproteins mapping to the left of the MIE region (UL118, UL120, and UL121) vary from as much as 35% identity overall to as little as 15% in one region. Similarly, the betaherpesvirus-specific IE2 (UL122) immediate-early (IE) regulatory proteins of HCMV and AgmCMV have 58% amino acid sequence identity, but only over the C-terminal half of the protein, and both are equally diverged from RhCMV in this region. Amino acid sequence differences between homologous CMV genes from even the most closely related host species, such as HCMV and CzCMV, are frequently at least 20–30%.

Even the DNA molecules of individual isolates of a single primate CMV species are often distinguishable on the basis of multiple restriction fragment length polymorphisms stemming from overall intraspecies nucleotide variations of up to 3–4%. In fact, different isolates or strains of HCMV and simian CMVs display a number of significant genomic differences. First, so-called laboratory strains of HCMV (such as AD169 and Towne), AgmCMV (Colburn), and RhCMV (68-1) have all undergone deletions and rearrangements. They have also accumulated frameshift or truncation mutations in a number of genes that appear to be critical for preserving natural tropism for endothelial and macrophage cell types, but are evidently selected against during adaptation to fibroblast cells. These changes particularly affect a set of three small genes (UL128, UL130, and UL131A). Even the less passaged HCMV (strain Toledo) has undergone a large inversion that affects UL128.

A second type of strain sequence difference has been documented in HCMV for many years within a subset of 'variable' genes, which include gB (UL55), gN (UL73), gO (UL74), exon 3 of UL37 (UL37ex3), UL9, RL12, RL13, UL144, UL146, and UL147. In each case, even when analyzed directly by polymerase chain reaction (PCR) sequencing of clinical samples, the proteins from different genomes can display up to 30–60% divergence at the amino acid level, with collections of isolates examined falling into clusters of 3, 5, 8, or even 15 subtypes, depending on the locus. The subtype patterns in different 'variable' gene loci are unlinked, and this is indicative of high levels of mixed infections, recombination, and chimerism in HCMV samples. The origin and biological significance of subtype patterns are not well understood, but are most likely related to founder and bottleneck effects during the recent evolutionary spread and migration of humans. Isolates of CMV from nonhuman primate host species also display similar patterns of variability.

The sequenced prototype betaherpesviruses contain several interesting, and in some cases genus-specific, genes that are related to cellular genes and appear to be used primarily for evasion of immune responses. For example, HCMV encodes four viral G protein-coupled receptors (vGPCRs; UL33, UL78, US27, and US28), at least one of which (US28) functions as a broad-spectrum chemokine receptor that promotes migration of smooth muscle cells, and three glycoproteins (US3, US10, and US11) that together function to destabilize or inhibit cellular HLA-mediated responses. HHV-6 and MCMV each encode two GPCR proteins. HCMV and MCMV also each encode diverged homologs of HLA-I (at different locations in their genomes), and HCMV also possesses two separate anti-apoptotic proteins known as vMIA (UL37) and vICA (UL36). Spliced genes in MCMV and RCMV encode two functional β-chemokines, whereas HCMV encodes two α-chemokines (UL146 and UL147). HHV-6 (but not HHV-7) encodes an REP protein apparently captured from adeno-associated virus, and both HHV-6 and HHV-7, as well as RCMV but not MCMV, encode different OX-2-related proteins. RhCMV and AgmCMV, but not HCMV or CzCMV, encode a highly spliced COX2 gene, and HCMV, RhCMV, and AgmCMV encode a spliced vIL-10 gene, whereas CzCMV does not.

The total genome contents of the two 'great ape' CMVs (HCMV and CzCMV) differ by at most a half dozen out of 165 clearly defined genes. However, differences between HCMV and the 'Old World primate' CMVs (RhCMV and AgmCMV) are numerous. One

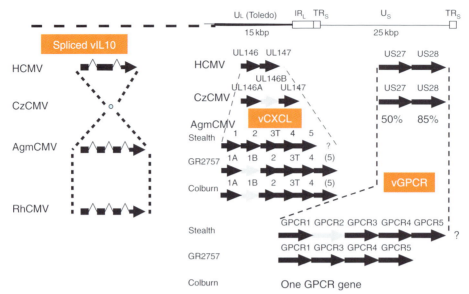

**Figure 1** Comparison of the organization of the spliced vIL10 genes, the UL146 vCXCL α-chemokine ligand gene cluster, and the US28 vGPCR chemokine gene cluster in CMV genomes. Amino acid sequence identity between HCMV and CzCMV is indicated for the US27 and US28 proteins. Stealth, GR2757, and Colburn are distinct isolates of AgmCMV. The position of the additional (but inverted) 15 kbp segment in HCMV strain Toledo $U_L$ compared to strain AD169 is indicated by the solid line. The spliced vIL10 gene is present in HCMV, AgmCMV, and RhCMV but absent from CzCMV. Another spliced gene (the COX2 gene) is present (at another location) in AgmCMV and RhCMV, but not in HCMV or CzCMV.

manifestation is the presence of either five or six tandemly repeated but highly diverged paralogs of the paired US27/US28 vGPCR genes of HCMV in multiple strains of RhCMV and AgmCMV, and the presence of either seven or eight tandemly repeated but highly diverged paralogs of the paired UL146/UL147 α-chemokine genes of HCMV in different strains of AgmCMV (**Figure 1**). Also, there are three, rather than two, of the α-chemokine genes in this cluster in CzCMV and at least three in RhCMV. Both the vGPCR and α-chemokine gene clusters appear to be undergoing highly dynamic evolution in Old World primate CMVs.

## Epidemiology and Physical Properties

Features of host range, virus propagation, virus transmission, tissue tropism, pathogenicity, histopathology, and immune responses of primate CMVs, together with the physical properties and assembly pathways of the virions, and a number of biochemical properties of their structural proteins, closely resemble those of HCMV and MCMV. In general, long-term persistent asymptomatic infection is nearly universal among natural primate populations in the wild, with primary infections occurring in infants and juveniles. Even captive colony-borne animals are rarely free of these viruses, which can be shed intermittently in saliva and urine throughout the host's lifespan. Like all other herpesviruses, primate CMVs have enveloped icosahedral capsids and replicate in the nucleus. Infection of humans with simian CMVs via prolonged contact with seropositive monkeys or apes, or from vaccines grown in unscreened primary cell cultures from primate sources, is plausible, but has not been documented unambiguously. Laboratory-adapted strains such as AgmCMV Colburn grow vigorously in human fibroblast cell culture (**Figure 2**) and have been known to outgrow HCMV in cell culture. However, primary isolates grow slowly in both human and simian fibroblasts.

## Replication Strategies

The lytic cycle pathway of gene expression for primate CMVs follows the typical herpesvirus cascade of IE mRNAs and proteins followed by activation of delayed-early (DE) then late (L) class genes, with synthesis of viral DNA. In cell culture, fully permissive host cell types for HCMV are restricted almost exclusively to diploid cells, including human fibroblasts, vascular endothelial cells (and the U373 astrocytoma cell line), smooth muscle cells, and differentiated macrophages. Fresh clinical HCMV isolates usually need to be adapted for efficient growth in fibroblasts by multiple rounds of passaging, which is accompanied by selection of inactivating point mutations (or deletions) in certain 'cell tropism' genes and a loss of ability to grow in endothelial cells or macrophages.

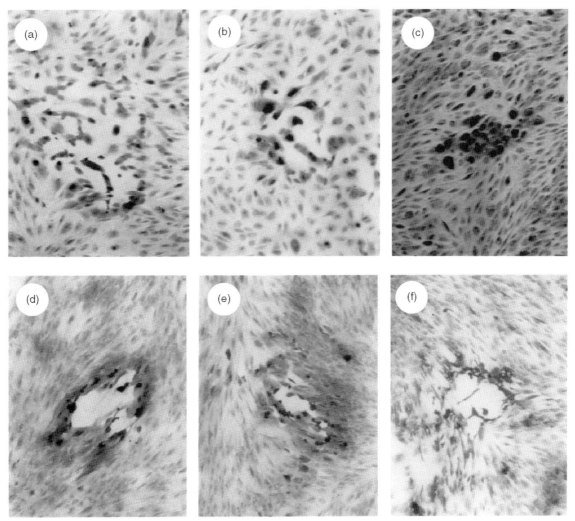

**Figure 2** Infection of human diploid fibroblasts (HF cells) with human and other primate CMVs. HF cells at passage level 15 were infected and examined daily for cytopathic effects. All infections produced characteristic cytomegalic cells in culture. (a) HCMV (Towne strain) plaque at 7 days after infection; (b) AgmCMV plaque at 8 days after infection; (c) BaCMV plaque at 11 days after infection; (d) CzCMV plaque at 14 days after infection; (e) RhCMV plaque at 14 days after infection; (f) CeCMV plaque at 8 days after infection.

Laboratory strains of HCMV replicate their DNA efficiently by 72 h in human fibroblasts, but express only the MIE proteins and fail to synthesize any viral DNA in rodent or monkey fibroblasts. AgmCMV, RhCMV, BaCMV, and CzCMV replicate their DNA and form plaques in both human and simian fibroblasts (**Figure 2**). In contrast, MCMV carries out viral DNA synthesis in human, monkey, and rodent fibroblasts. Most transformed human cell types are nonpermissive for HCMV replication, with the block occurring between DNA entry and synthesis of MIE mRNA. However, some differences between HCMV and AgmCMV have been observed at this stage. For example, HCMV fails to synthesize MIE proteins in infected human (NTera) or mouse (F9) teratocarcinoma stem cells, in human 293 or NBE cells, or in mouse L-cells. In contrast, AgmCMV produces MIE proteins in infected NTera, 293, and NBE cells, but not in F9 or L-cells. After differentiation with retinoic acid (RA), NTera stem cells can become permissive for HCMV MIE expression and infectious virus production. Similarly, F9 cells treated with RA induce AgmCMV, but not HCMV, MIE expression. These biological differences among the different species of Old World primate CMVs appear to coincide with structural differences in the organization of *cis*-acting transcriptional control elements and adjacent accessory domains located both upstream of and within the large first intron of the MIE genes. Although a major site *in vivo* for quiescent inactive infection for HCMV and MCMV is believed to involve monocytes, many other sites of inapparent noncytopathic infection occur, but whether any of these can be defined as the true site of reactivatable latent state infection has not yet been resolved.

## Control of Gene Expression

Similarly organized MIE transcription units, which encode regulatory proteins that trigger the lytic cycle, have been described for HCMV, CzCMV, AgmCMV, RhCMV, BaCMV, GpCMV, RCMV, MCMV, HHV-6, and HHV-7. In each case, these transcription units produce several multiply spliced mRNAs whose expression is controlled by powerful upstream *cis*-acting enhancer regions. The predominant viral mRNAs synthesized after infection of permissive cells in the presence of cycloheximide (an inhibitor of protein synthesis) are the MIE mRNAs, and these are also the only HCMV or AgmCMV mRNAs and proteins produced after infection of nonpermissive rodent fibroblasts. At least two types of phosphorylated nuclear regulatory proteins are encoded by the MIE transcription unit. These include the highly abundant, acidic IE1 (UL122) nuclear protein and the less abundant IE2 (UL123) DNA-binding transactivator/repressor protein. The IE2 DNA-binding transcription factor proteins are not only essential for stimulating transcription of downstream DE and L HCMV promoters, but they also specifically downregulate their own MIE promoters (negative autoregulation) and are probably also engaged in control of viral DNA replication, altering cell cycle function and blocking interferon and apoptotic responses. In particular, both the IE1 and IE2 proteins target to subnuclear domains known as PODs (protein oncogenic domains), which contain the PML proto-oncoprotein, and appear to modify these sites (also known as ND10 (nuclear domain 10)) for utilization by the viral genomes to initiate viral IE transcription and viral DNA synthesis.

In contrast to the rest of the viral genome, the IE1 and the IE2 coding regions of the MIE transcription units in HCMV, CzCMV, AgmCMV, MCMV, RCMV, HHV-6, and HHV-7 (and probably all other betaherpesviruses) are highly CpG-suppressed, which suggests that they are accessible to cellular methylation events at some stage of the viral life cycle during which all other viral genes are transcriptionally silent. The 490-amino-acid residue HCMV IE1 protein has only 15% amino acid sequence identity with AgmCMV or RhCMV IE1, and much less still is conserved between the primate and MCMV and RCMV versions, although all of the betaherpesvirus IE1 proteins (except for those of HHV-6 and HHV-7) have a highly acidic, Glu-rich C-terminus. In contrast, the 579-amino-acid residue IE2 proteins of HCMV, AgmCMV, RhCMV, MCMV, RCMV, and their much larger counterparts in HHV-6 and HHV-7, exhibit between 25% and 58% sequence identity over the C-terminal 270-residue conserved DNA-binding domain. As expected, the CzCMV IE1 and IE2 proteins are much more similar to the HCMV versions.

The upstream MIE promoter/enhancer regions, which serve to sense the intranuclear environment and control entry into or out of the lytic cycle, are large, complex, noncoding DNA sequence domains that consist of often multicopy, high-affinity binding sites for numerous constitutive or inducible cellular transcription factors. In HCMV, CzCMV, AgmCMV, RhCMV, BaCMV, GpCMV, and MCMV, these sites include response elements for cyclic AMP (CRE, PKA), phorbol esters (TPA, PKC), and RA, together with recognition motifs for SRF/ETS, CREB, SP-1, AP-1, and nuclear factor kappa B (NFκB). Even among the five Old World primate CMVs that have been examined in detail, the organization of these sites within the MIE enhancer regions differs significantly, and the number and pattern of the adjacent tandemly repeated 15 and 30 bp high-affinity NFI/YY1 motifs also differ greatly (**Figure 3**). Overall, the MIE control region in AgmCMV encompassing the BENT, NFI, ENH, and INTRON segments totals 2.3 kbp in size, much more than the approximately 1.1 kbp MIE control region in HCMV.

Other specific regulatory proteins common to HCMV and HHV-6, such as the UL36 and UL37 proteins, members of the US22 family, the UL82 and UL83 matrix phosphoproteins, and the UL84 replication-associated protein, are all conserved in primate CMVs. However, the second immediate early promoter (IES) and its novel, complex NFκB-containing enhancer is found in HCMV and CzCMV upstream from gene US3, and is not conserved in RhCMV or AgmCMV.

The CMV lytic cycle origin of DNA replication (Ori-Lyt) is located to the right of and adjacent to the single-stranded DNA-binding protein gene (UL57), near the center of the HCMV, CzCMV, AgmCMV, RhCMV, and MCMV genomes. It differs significantly from virus to virus in structural organization and apparently cannot be complemented by the protein replication machinery from another virus within this group. Although Ori-Lyt is located at the equivalent site in the genomes of HHV-6 and HHV-7 (and presumably also EEHV), it differs in structure and in its UL9-dependent mode of DNA replication from that in CMVs. It is still not known whether any betaherpesviruses utilize a latent cycle origin of DNA replication, such as that found among gammaherpesviruses.

## Future Perspectives

Persistent, inapparent infection with CMVs is probably virtually ubiquitous in most individuals of all mammalian species. However, the biological and pathological properties of primate CMVs have attracted the interest of herpesvirologists and clinicians interested in HCMV disease in acquired immune deficiency syndrome (AIDS) and organ transplant patients, not so much because of the serious morbidity or economic consequences of these infections in their own hosts, but more often as models for latency and pathogenesis or immunological responses in HCMV. MCMV, RCMV, GpCMV, RhCMV, and

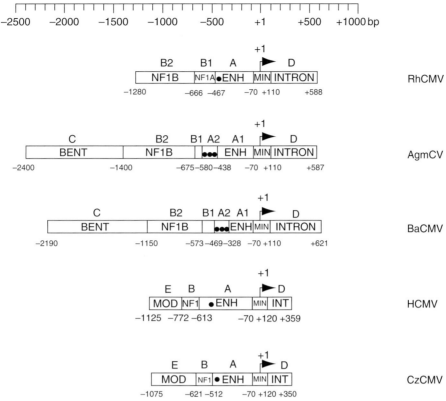

**Figure 3** Upstream control regions in the MIE genes of HCMV and other primate CMVs. The diagram illustrates the respective size and organization of domains containing specific transcription factor-binding motif patterns. These include clusters of nuclear factor 1 (NF1) and YY1 sites, the distal enhancer containing multiple CRE, SRE, and NFκB sites (ENH), the proximal promoter and TATATAA box region (MIN) that includes the start site of transcription at +1 (arrowed), and the 5' half of the large first intron (INTRON). Where appropriate, additional upstream AT-rich (BENT) and modulatory (MOD) regions are also included. Domain boundaries are designated as + or − nucleotide positions with reference to the start site of transcription and given in bp. The domains have also been designated A–E with subcategories 1 or 2 in the case of the NF1 and ENH block. Adapted from Chan YJ, Chiou CJ, Huang Q, and Hayward GS (1996) Synergistic interactions between overlapping binding sites for the serum response factor and ELK-1 proteins mediate both basel enhancement and phorbal ester responsiveness of primate cytomegalovirus major immediate-early promoters in monocyte and T-lymphocyte cell types. *Journal of Virology* 70: 8590–8605, with permission from American Society for Microbiology.

BaCMV have all been investigated as models for virus reactivation, and for control of acute and chronic disease by anti-CMV agents (such as ganciclovir, a nucleotide analog that specifically inhibits CMV DNA replication) in association with allografted organ transplants or immunosuppression. There is also interest in questions related to whether different CMVs can cross host species barriers with possible pathological consequences, which is of concern with BaCMV and PCMV for xenografted organ transplants. Primate CMVs also have comparative value in molecular genetics and biochemical analysis because of their similarities to and differences from HCMV or HHV-6 and HHV-7. Current basic scientific interest in genomic evolution (made possible by large-scale DNA sequencing), as well as the role of transcriptional gene regulation mechanisms in determining cell tropism and the switching between latent and lytic cycle virus–cell interaction pathways, and the many novel mechanisms used by betaherpesviruses for immune evasion, should lead to an expansion of these types of studies with nonhuman primate CMVs over the next several years.

*See also:* Cytomegaloviruses: Murine and Other Non-primate Cytomegaloviruses; Herpesviruses: General Features; Human Cytomegalovirus: General Features; Human Cytomegalovirus: Molecular Biology; Human Herpesviruses 6 and 7; Simian Alphaherpesviruses; Simian Gammaherpesviruses.

## Further Reading

Bahr U and Darai G (2001) Analysis and characterization of the complete genome of tupaia (tree shrew) herpesvirus. *Journal of Virology* 75: 4354–4370.

Bankier AT, Beck S, Bohni R, *et al.* (1991) The DNA sequence of the human cytomegalovirus genome. *DNA Sequence* 2: 1–12.

Chan YJ, Chiou CJ, Huang Q, and Hayward GS (1996) Synergistic interactions between overlapping binding sites for the serum response factor and ELK-1 proteins mediate both basal enhancement and phorbal ester responsiveness of primate cytomegalovirus major immediate-early promoters in monocyte and T-lymphocyte cell types. *Journal of Virology* 70: 8590–8605.

Chee MS, Bankier AT, Beck S, *et al.* (1990) Analysis of the protein-coding content of the sequence of human cytomegalovirus strain AD169. *Current Topics in Microbiology and Immunology* 154: 125–169.

Davison AJ, Dolan A, Akter P, *et al.* (2003) The human cytomegalovirus genome revisited: Comparison with the chimpanzee cytomegalovirus genome. *Journal of General Virology* 84: 17–28.

Dolan A, Cunningham C, Hector RD, *et al.* (2004) Genetic content of wild-type human cytomegalovirus. *Journal of General Virology* 85: 1301–1312.

Dominguez G, Dambaugh TR, Stamey FR, Dewhurst S, Inoue N, and Pellett PE (1999) Human herpesvirus 6B genome sequence: Coding content and comparison with human herpesvirus 6A. *Journal of Virology* 73: 8040–8052.

Gibson W (1983) Protein counterparts of human and simian cytomegaloviruses. *Virology* 128: 391–406.

Gompels UA, Nicholas J, Lawrence G, *et al.* (1995) The DNA sequence of human herpesvirus-6: Structure, coding content, and genome evolution. *Virology* 209: 29–51.

Hansen SG, Strelow LI, Franchi DC, Anders DG, and Wong SW (2003) Complete sequence and genomic analysis of rhesus cytomegalovirus. *Journal of Virology* 77: 6620–6636.

Kaplan AS (1973) *The Herpesviruses,* chs. 12 and 13. New York: Academic Press.

Megaw AG, Rapaport D, Avidor B, Frenkel N, and Davison AJ (1998) The DNA sequence of the RK strain of human herpesvirus 7. *Virology* 244: 119–132.

Nicholas J (1996) Determination and analysis of the complete nucleotide sequence of human herpesvirus 7. *Journal of Virology* 70: 5975–5989.

Rawlinson WD, Farrell HE, and Barrell BG (1996) Analysis of the complete DNA sequence of murine cytomegalovirus. *Journal of Virology* 70: 8833–8849.

Richman LK, Montali RJ, Garber RL, *et al.* (1999) Novel endotheliotropic herpesviruses fatal for Asian and African elephants. *Science* 283: 1171–1176.

Rivailler P, Kaur A, Johnson RP, and Wang F (2006) Genomic sequence of rhesus cytomegalovirus 180.92: Insights into the coding potential of rhesus cytomegalovirus. *Journal of Virology* 80: 4179–4182.

Staczek J (1990) Primate cytomegaloviruses. *Microbiological Reviews* 54: 247.

Vink C, Beuken E, and Bruggeman CA (2000) Complete DNA sequence of the rat cytomegalovirus genome. *Journal of Virology* 74: 7656–7665.

**Set**
978-0-12-373935-3

**Volume 1 of 5**
978-0-12-373936-0